RECEIVED

Y 19 2004

Library Serials

# REVIEWS in MINERALOGY and GEOCHEMISTRY

## VOLUME 56        2004

# EPIDOTES

**EDITORS:**

**Axel Liebscher**        *GeoForschungsZentrum Potsdam*
*Potsdam, Germany*

**Gerhard Franz**        *Technische Universität Berlin*
*Berlin, Germany*

**FRONT COVER:** Photomicrograph of an aggregate of zoned clinozoisite-epidote crystals with 0.40 to 0.56 cations $Fe^{3+}$ per formula unit from core to rim (crossed polarized light, sample no. 81-370, Min. Collection Techn. Univ. Berlin) from a hydrothermal epidote-quartz vein, Baja California. The picture shows the upper part of radially grown crystals, which are up to 10 mm long. At the tip the individual crystal are between ≈ 150 and ≈ 400 μm thick. Crystals are macroscopically pink colored due to a $Mn_2O_3$ content of 0.2–0.7 wt%. The outer rim of the crystals as well as crystals grown in an open crack in the lower part, are macroscopically green epidote with 0.85 to 1.04 $Fe^{3+}$ per formula unit, indicating a second growth stage. The crystals also show oscillatory zoning (see back cover and chapter 1 by Franz and Liebscher; photos by Irene Preuss; ZELMI at TUB).

*Series Editor:* **Jodi J. Rosso**

MINERALOGICAL SOCIETY of AMERICA
GEOCHEMICAL SOCIETY

Copyright 2004

MINERALOGICAL SOCIETY OF AMERICA

The appearance of the code at the bottom of the first page of each chapter in this volume indicates the copyright owner's consent that copies of the article can be made for personal use or internal use or for the personal use or internal use of specific clients, provided the original publication is cited. The consent is given on the condition, however, that the copier pay the stated per-copy fee through the Copyright Clearance Center, Inc. for copying beyond that permitted by Sections 107 or 108 of the U.S. Copyright Law. This consent does not extend to other types of copying for general distribution, for advertising or promotional purposes, for creating new collective works, or for resale. For permission to reprint entire articles in these cases and the like, consult the Administrator of the Mineralogical Society of America as to the royalty due to the Society.

# REVIEWS IN MINERALOGY
# AND GEOCHEMISTRY

( Formerly: REVIEWS IN MINERALOGY )

ISSN 1529-6466

## Volume 56

## *Epidotes*

Library
University of Texas
at San Antonio

ISBN 093995068-5

*Additional copies of this volume as well as others in
this series may be obtained at moderate cost from:*

THE MINERALOGICAL SOCIETY OF AMERICA
1015 EIGHTEENTH STREET, NW, SUITE 601
WASHINGTON, DC 20036 U.S.A.
WWW.MINSOCAM.ORG

# EPIDOTES

**56**     *Reviews in Mineralogy and Geochemistry*     **56**

## FROM THE SERIES EDITOR

The review chapters in this volume were the basis for a two-day short course on Epidote Minerals held prior (June 3–4, 2004) to the 2004 Goldschmidt Conference in Copenhagen, Denmark. The editors (and conveners of the short course) Axel Liebscher and Gerhard Franz and the other chapter authors/presenters have produced a comprehensive review volume that should be a standard used by scientists for many years to come.

Errata (if any) can be found at the MSA website *www.minsocam.org*.

Jodi J. Rosso, Series Editor
*West Richland, Washington*
March 2004

## PREFACE

Our understanding of rock forming geological processes and thereby of geodynamic processes depends largely on a sound basis of knowledge of minerals. Due to the application of new analytical techniques, the number of newly discovered minerals increases steadily, and what used to be a simple mineral may have turned into a complex group. A continuous update is necessary, and the "Reviews in Mineralogy and Geochemistry" series excellently fulfills this requirement. The epidote minerals have not yet been covered and we felt that this gap should be filled.

The epidote mineral group consists of important rock-forming minerals such as clinozoisite and epidote, geochemical important accessory minerals such as allanite, and minerals typical for rare bulk compositions such as hancockite. Zoisite, the orthorhombic polymorph of clinozoisite, is included here because of its strong structural and paragenetic similarity to the epidote minerals. Epidote minerals occur in a wide variety of rocks, from near-surface conditions up to high- and ultrahigh-pressure metamorphic rocks and as liquidus phases in magmatic systems. They can be regarded as the low-temperature and high-pressure equivalent of Ca-rich plagioclase, and thus are equally important as this feldspar for petrogenetic purposes. In addition, they belong to the most important $Fe^{3+}$ bearing minerals, and give important information about the oxygen fugacity and the oxidation state of a rock. Last but not least, they can incorporate geochemically relevant minor and trace elements such as Sr, Pb, REE, V, and Mn.

The epidote minerals are undoubtedly very important from a petrogenetic and geochemical point of view, and have received a lot of attention in the last years from several working groups in the field of experimental studies and spectroscopic work. As a result, the thermodynamic database of epidote minerals has been significantly enlarged during the last decade. Recent studies have revealed the importance of zoisite in subduction zone processes as a carrier of $H_2O$ and suggested zoisite to be the main $H_2O$ source in the pressure interval between about 2.0 and 3.0 GPa. Many studies have shown that an understanding of trace element geochemical processes in high-pressure rocks is impossible without understanding the geochemical influence of the epidote minerals. Recent advances in microanalytical techniques have also shown that epidote minerals record detailed information on their geological environment.

## Epidotes – Preface

W. A. Deer, R. A. Howie and J. Zussmann edited the last comprehensive review on this mineral group almost 20 years ago in 1986. In 1990, on the occasion of the 125th anniversary of the discovery of the famous Knappenwand locality in the Tauern/Austria, an epidote conference was held in Neukirchen/Austria organized by the Austrian Mineralogical Society by V. Höck and F. Koller. In 1999, there was a special symposium at the EUG 10 in Strasbourg, convened by R. Gieré and F. Oberli, entitled "Recent advances in studies of the epidote group" that highlighted the relevance of the epidote minerals for Earth science. However, there are many open questions in the community regarding the epidote minerals and there is a need for a new overview that brings together the recent knowledge on this interesting group of minerals. The present volume of the Reviews in Mineralogy and Geochemistry reviews the current state of knowledge on the epidote minerals with special emphasis on the advances that were made since the comprehensive review of Deer et al. (1986). We hope that it will serve to outline the open questions and direction of future research.

In the Introduction, we review the structure, optical data and crystal chemistry of this mineral group, all of which form the basis for understanding much of the following material in the volume. In addition, we provide some information on special topics, such as morphology and growth, deformation behavior, and gemology. Thermodynamic properties (Chapter 2, Gottschalk), the spectroscopy of the epidote minerals (Chapter 3, Liebscher) and a review of the experimental studies (Chapter 4, Poli and Schmidt) constitute the first section of chapters. These fields are closely related, and all three chapters show the significant progress over the last years, but that some of the critical questions such as the problem of miscibility and miscibility gaps are still not completely solved. This section concludes with a review of fluid inclusion studies (Chapter 5, Klemd), a topic that turned out to be of large interest for petrogenetic interpretation, and leads to the description of natural epidote occurrences in the second section of the book. These following chapters review the geological environments of the epidote minerals, from low temperature in geothermal fields (Chapter 6, Bird and Spieler), to common metamorphic rocks (Chapter 7, Grapes and Hoskin) and to high- and ultrahigh pressure (Chapter 8, Enami, Liou and Mattinson) and the magmatic regime (Chapter 9, Schmidt and Poli). Allanite (Chapter 10, Gieré and Sorensen) and piemontite (Chapter 11, Bonazzi and Menchetti), on which a large amount of information is now available, are reviewed in separate chapters. Finally trace element (Chapter 12, Frei, Liebscher, Franz and Dulski) and isotopic studies, both stable and radiogenic isotopes (Chapter 13, Morrison) are considered. We found it unavoidable that there is some overlap between individual chapters. This is an inherited problem in a mineral group such as the epidote minerals, which forms intensive solid solutions between the major components of rock forming minerals as well as with trace elements.

We thank the Mineralogical Society of America and the Geochemical Society for giving us the opportunity to edit this volume. We appreciate especially the editorial work of Jodi J. Rosso, the easy communication and her attitude towards deadlines. We also thank all the reviewers for the individual chapters: Irmgard Abs-Wurmbach, Michael Andrut, Dennis Bird, Masaki Enami, Anne Feenstra, Karl Thomas Fehr, Dirk Frei, Retro Gieré, Rodney Grapes, Matthias Gottschalk, Lutz Hecht, Wilhelm Heinrich, Jörg Hermann, Paul Hoskin, Dave Jenkins, Reiner Klemd, Monika Koch-Müller, Dominique Lattard, Friedrich Lucassen, Jean Morrison, Dave Jenkins, Michael Raith, Rolf Romer, Jane Selverstone, Holger Stünitz, Thomas Will, Antje Wittenberg, who have done an excellent job and helped us to complete the editorial work in time. Generous financial support from GeoForschungsZentrum Potsdam for both the editorial work and the short course is gratefully acknowledged, as well as the editorial assistance of Katrin Mai and Sarah Bernau.

***Axel Liebscher***
Potsdam, Germany

***Gerhard Franz***
Berlin, Germany

# TABLE OF CONTENTS

**1**   **Physical and Chemical Properties of the Epidote Minerals**
**–An Introduction–**

*Gerhard Franz & Axel Liebscher*

WITHDRAWN
UTSA Libraries

# 2        Thermodynamic Properties of Zoisite, Clinozoisite and Epidote

*Matthias Gottschalk*

# 3       **Spectroscopy of Epidote Minerals**

*Axel Liebscher*

# 4      Experimental Subsolidus Studies on Epidote Minerals

*Stefano Poli & Max W. Schmidt*

# 5   Fluid Inclusions in Epidote Minerals and Fluid Development in Epidote-Bearing Rocks

*Reiner Klemd*

# 6      Epidote in Geothermal Systems

*Dennis K. Bird & Abigail R. Spieler*

# 7 Epidote Group Minerals in Low–Medium Pressure Metamorphic Terranes

*Rodney M. Grapes & Paul W. O. Hoskin*

# 8    Epidote Minerals in High P/T Metamorphic Terranes: Subduction Zone and High- to Ultrahigh-Pressure Metamorphism

*M. Enami, J.G. Liou & C. G. Mattinson*

# 9            Magmatic Epidote

*Max W. Schmidt & Stefano Poli*

# 10    Allanite and Other REE-Rich Epidote-Group Minerals

*Reto Gieré & Sorena S. Sorensen*

# 11 Manganese in Monoclinic Members of the Epidote Group: Piemontite and Related Minerals

*Paola Bonazzi & Silvio Menchetti*

# 12     Trace Element Geochemistry of Epidote Minerals

*Dirk Frei, Axel Liebscher,*
*Gerhard Franz & Peter Dulski*

# 13 Stable and Radiogenic Isotope Systematics in Epidote Group Minerals

*Jean Morrison*

Reviews in Mineralogy & Geochemistry
Vol. 56, pp. 1-82, 2004
Copyright © Mineralogical Society of America

# Physical and Chemical Properties of the Epidote Minerals – An Introduction –

## Gerhard Franz

*Technische Universität Berlin*
*Fachgebiet Petrologie*
*D 10623 Berlin, Germany*
*gerhard.franz@tu-berlin.de*

## Axel Liebscher

*Department 4, Chemistry of the Earth*
*GeoForschungsZentrum Potsdam*
*Telegraphenberg*
*D 14473 Potsdam, Germany*
*alieb@gfz-potsdam.de*

## INTRODUCTION

Epidote minerals are known since the 18th century, but at that time the greenish to dark colored varieties were termed actinolite or schorl and not distinguished from the minerals to which these names apply today. Haüy defined the mineral species and introduced the name "epidote" in 1801, whereas Werner in 1805 used the term pistacite (quoted from Hintze 1897). Epidote is derived from greek *epidosis* = to increase, because the base of the rhombohedral prism has one side larger than the other and pistacite refers to its green color (all references for names after Lüschen 1979; Blackburn and Dennen 1997). Weinschenk (1896) proposed the name clinozoisite from its monoclinic symmetry and zoisite-like composition for those monoclinic members of the epidote family that are Fe poor, optically positive and have low refractive indices and birefringence.

Zoisite was probably confused with tremolite until the beginning of the 19th century. In 1804, Siegmund Zois, Baron von Edelstein 1747–1819, an Austrian sponsor of mineral collections, found and described a new mineral in a handspecimen from the Saualpe Mountains in Carinthia that was named zoisite by Werner.

Haüy (1822) interpreted zoisite as a variety of epidote and included it in his "epidote spezies." Weiss (1820) presented a theory of the epidote system and also discussed crystal morphological features of the epidote minerals (Weiss 1828). Rammelsberg (1856) studied the relationship between epidote and zoisite and presented a compilation of chemical analyses of zoisite. He already noticed that the relative concentrations of di-, tri- and tetravalent cations are identical in zoisite and epidote but that the Fe content in zoisite (about 2–3.5 wt% $Fe_2O_3$) is generally less than in epidote (about 9–12 wt% $Fe_2O_3$). Piemontite (see Bonazzi and Menchetti 2004) was probably first described in 1758 by Cronstedt as "röd Magnesia" and later termed "Manganèse rouge" by Chevalier Napione in 1790. Cordier (1803) first recognized the mineral as an "Épidote manganésifere" a name, which was adopted by Haüy (1822) who

1529-6466/04/0056-0001$10.00

included this mineral as a variety in his "Epidote spezies." Finally the name "piedmontite" (today transformed into piemontite) was proposed according to the type locality Piemont in Northern Italy. The first Cr rich epidote occurrence was described by Bleek (1907) who also introduced the term tawmawite according to the type locality Tawmaw in Burma. Brooke in 1828 (quoted by Hintze 1897) gave the name "thulite" to the pink colored Mn-minerals of the epidote group, named after Thule, the ancient name for Norway.

Allanite, the rare earth element epidote, was already described by Thomson (1810) and named after Thomas Allan, 1777–1833, a Scottish mineralogist, who noticed it in alkaline rocks of Greenland. The name was used for unaltered crystals with a tabular habit parallel to the orthopinacoid. Berzelius (1818, quoted in Hintze 1897) found a similar mineral, however altered and hydrated and with a prismatic habit, and named it "orthit," but this name was abandoned later (Hutton 1951). The rare earth element epidote minerals developed into a complex group (see Gieré and Sorensen 2004), and the newly discovered species are androsite-(La) (Bonazzi et al. 1996), named after its locality Andros Island/Greece; dissakisite-(Ce) (Grew et al. 1991), named from Greek *dissakis* twice over, alluding to the fact that this Mg analogue has been described twice; dollaseite-(Ce) (Peacor and Dunn 1988), named after W. A. Dollase, who carried out definitive research on the epidote minerals; khristovite-(Ce) (Pautov et al. 1993), named after the Russian geologist E. V. Khristov.

Among the rare epidote minerals, hancockite, with Pb and Sr (named after E. P. Hancock, a mineral collector from Franklin/USA) was also already known in the 19th century (Penfield and Warren 1899). All others, mukhinite, the $V^{3+}$ epidote, named after A. S. Mukhin, a geologist who worked in Western Siberia (Shepel and Karpenko 1969) and the Sr epidotes twedillite, named after S. M. Twedill, the first curator of the geological museum in Pretoria/RSA (Armbruster et al. 2002); strontiopiemontite (Bonazzi et al. 1990); and niigataite (Miyajima et al. 2003), named after its locality, were discovered late in the 20th century or recently.

The sum formula $Ca_2(Al,Fe^{3+})_3Si_3O_{13}H$ for epidote was presented as early as 1872 by Ludwig and was since then generally accepted. Finally Ito (1950) gave the structural interpretation of the sum formula as $Ca_2(Al,Fe)_3SiO_4Si_2O_7(O/OH)$. Since that time, our knowledge of occurrence, structure, and crystal chemistry of the epidote minerals increased considerably. X-ray diffraction studies with structure refinements established the structural relationships within and between the monoclinic and orthorhombic solid solution series, spectroscopic studies addressed and partly resolved the problems of site occupancies, and microanalytical techniques allowed to study chemical variations within the epidote minerals. What used to be rather simple has turned into a quite complex mineral group.

This chapter intends to serve as an introduction to the different mineralogical and geochemical aspects of epidote mineralogy addressed in the individual chapters. After nomenclatural aspects it reviews the structural, crystal chemical, and optical properties of the epidote minerals. Finally, it addresses some aspects of epidote mineralogy that are not covered by own chapters e.g., morphology, twinning, deformation behavior, and use as gemstones.

## NOMENCLATURE

Epidote group minerals are monoclinic sorosilicates with mixed $SiO_4$ tetrahedra and $Si_2O_7$ groups with the general formula $A1A2M1M2M3[O/OH/SiO_4/Si_2O_7]$, the M cations in octahedral and the A cations in larger coordination (group 9.BG 05 of Strunz and Nickel 2001). The orthorhombic mineral zoisite (group 9.BG 10 of Strunz and Nickel 2001), structurally very similar, is a polymorph of clinozoisite, and therefore also included here. The epidote minerals are structurally related to the pumpellyite-sursassite group minerals (Table 1). Besides their common silica building units, they all have the M position (filled predominantly with $M^{3+}$ but

**Table 1.** General formulae of epidote and structurally related minerals

| | | | | |
|---|---|---|---|---|
| epidote, zoisite | A1A2 | M1M2M3 | $[O/OH/$ | $SiO_4/Si_2O_7]$ |
| pumpellyite | A1A2 | M1M2M3 | $[(OH)_2/(O,OH)/$ | $SiO_4/Si_2O_6(O,OH)]$ |
| sursassite | A1A2 | M1M2M3 | $[(OH)_3/$ | $SiO_4/Si_2O_7]$ |
| ilvaite | A | M1M2M3 | $[O/(OH)_2/$ | $Si_2O_7]$ |
| gatelite-(Ce) | A1A(2,3,4) | M1M2a,bM3 | $[O/(OH)_2/$ | $(SiO_4)_3/Si_2O_7]$ |
| lawsonite | A | M1M2 | $[(OH)_2/$ | $Si_2O_7] \cdot H_2O$ |

also $M^{2+}$) in edge sharing chains. In ilvaite the octahedra form double chains, and in lawsonite the single octahedral chains have very similar M positions as in epidote. All minerals listed in Table 1 have (OH) bonded to the octahedra; lawsonite is the only mineral, which also has molecular water in the structure. In the newly described mineral gatelite-(Ce) (Bonazzi et al. 2003), modules of the epidote-type structure alternate with modules of another hydrous REE nesosilicate törnebohmite. These mineral groups are also chemically related because they can basically be regarded as hydrous Ca-Al-silicates, where Ca can be exchanged against REE and Sr, and Al against $Fe^{3+}$ and $Mn^{3+}$.

For a description of the nomenclature we use the concept of additive and exchange components. As the additive component we choose the composition $Ca_2Al_3[O/OH/SiO_4/Si_2O_7]$, or written as a sum formula $Ca_2Al_3Si_3O_{12}(OH)$, for both the monoclinic and orthorhombic forms, and we distinguish simple and combined exchange vectors (Table 2). There has been considerable confusion in the literature how to name the monoclinic end members and the different monoclinic solid solutions and how to express their respective compositions, especially with respect to the monoclinic Al-$Fe^{3+}$ solid solution series. Pistacite has been either used to describe Fe rich crystals with a composition near to $Ca_2Al_2Fe^{3+}Si_3O_{12}(OH)$ or to describe the hypothetical end member $Ca_2Fe^{3+}_3Si_3O_{12}(OH)$. Epidote has also been used for the composition $Ca_2Al_2Fe^{3+}Si_3O_{12}(OH)$. Some authors used epidote for the optically negative and clinozoisite for the optically positive solid solutions thus following Weinschenk (1896), while others reserved epidote for the whole mineral group and termed all monoclinic Al-$Fe^{3+}$ solid solutions as clinozoisite. Consequently, the composition of solid solutions was expressed as mol% or mol fraction of the $Ca_2Fe^{3+}_3Si_3O_{12}(OH)$ component, as $Fe^{3+}$ per formula unit, or as mol% or mol fraction of the $Ca_2Al_2Fe^{3+}Si_3O_{12}(OH)$ component. In case of $Cr^{3+}$ bearing monoclinic solid solutions the term tawmawite has been used for the end member composition $Ca_2Al_2Cr^{3+}Si_3O_{12}(OH)$ but was also applied generally to Cr rich solid solutions. For a discussion of the problems with the nomenclature of Mn bearing epidote minerals and of allanite see Bonazzi and Menchetti (2004) and Gieré and Sorensen (2004), respectively.

The confusion and problems with the nomenclature of epidote minerals are partly historically determined (different nomenclatural traditions in North America and Europe) and partly due to the structural and crystal chemical characteristics of the epidote minerals (see below). According to the standard guidelines for mineral nomenclature, end member names should be reserved to those members that have the respective element dominant at one crystallographic site. Applying this rule to epidote minerals with three octahedral and two A-sites (see below) will obviously lead to an uncomfortable number of possible end members, even if one considers only the major elements Ca, Sr, Pb, $REE^{3+}$, Al, $Fe^{3+}$, $Mn^{3+}$, $Cr^{3+}$, and V that constitutes most of the epidote minerals (see below). Another problem arises because some of these possible end members have not yet been found in nature and are thus only hypothetic and not minerals in the strict sense. To overcome at least partly the confusion with the nomenclature of epidote minerals, the CNMMN subcommittee of the International Mineralogical Association on the nomenclature of epidote minerals has set up some

**Table 2.** Exchange vectors and nomenclature for orthorhombic and monoclinic epidote minerals.

| EXCHANGE VECTORS | FORMULA | NAME |
|---|---|---|
| *Orthorhombic Pnma, additive component zoisite $Ca_2Al_3Si_3O_{12}(OH)$* | | |
| simple in A-position | | |
| $Ca_{-1}Sr$ | $CaSrAl_3Si_3O_{12}(OH)$ | no separate name |
| $Ca_{-1}Pb$ | $CaPbAl_3Si_3O_{12}(OH)$ | no separate name |
| simple in M-positions | | |
| $Al_{-1}Fe^{3+}$ | $Ca_2Al_2Fe^{3+}Si_3O_{12}(OH)$ | no separate name |
| $Al_{-1}Mn^{3+}$ | $Ca_2Al_2Mn^{3+}Si_3O_{12}(OH)$ | no separate name (var. "thulite") |
| $Al_{-1}V^{3+}$ | $Ca_2Al_2V^{3+}Si_3O_{12}(OH)$ | no separate name (var. "tanzanite") |
| $Al_{-1}Cr^{3+}$ | $Ca_2Al_2Cr^{3+}Si_3O_{12}(OH)$ | no separate name |
| combined in AM-positions | | |
| $Ca_{-1}Al_{-1}REE^{*)}(Mg,Fe^{2+})$ | $CaREEAl_2(Mg, Fe^{2+})Si_3O_{12}(OH)$ | no separate name |
| *Monoclinic $P2_1/m$, additive component clinozoisite $Ca_2Al_3Si_3O_{12}(OH)$* | | |
| simple in A-position | | |
| $Ca_{-1}Sr$ | $CaSrAl_3Si_3O_{12}(OH)$ | niigataite |
| $Ca_{-1}Pb$ | $CaPbAl_3Si_3O_{12}(OH)$ | hancockite |
| $Ca_{-1}Mn^{2+}$ | $CaMn^{2+}Al_3Si_3O_{12}(OH)$ | (manganoan clinozoisite) |
| simple in M-positions | | |
| $Al_{-1}Fe^{3+}$ | $Ca_2Al_2Fe^{3+}Si_3O_{12}(OH)$ | epidote |
| $Al_{-3}Fe^{3+}_3$ | $Ca_2Fe^{3+}_3Si_3O_{12}(OH)$ | ("pistazite", theor. endmember) |
| $Al_{-1}Mn^{3+}$ | $Ca_2Al_2Mn^{3+}Si_3O_{12}(OH)$ | piemontite |
| $Al_{-1}V^{3+}$ | $Ca_2Al_2V^{3+}Si_3O_{12}(OH)$ | mukhinite |
| $Al_{-1}Cr^{3+}$ | $Ca_2Al_2Cr^{3+}Si_3O_{12}(OH)$ | "tawmawite" |
| combined in AM-positions | | |
| $Ca_{-1}Al_{-1}SrMn^{3+}$ | $CaSrAl_2Mn^{3+}Si_3O_{12}(OH)$ | strontiopiemontite |
| $Ca_{-1}Al_{-2}SrMn^{3+}_2$ | $CaSrAlMn^{3+}_2Si_3O_{12}(OH)$ | twedillite |
| $Ca_{-2}Al_{-2}REEMn^{2+}Mn^{3+}Mn^{2+}$ | $Mn^{2+}REEAlMn^{3+}Mn^{2+}Si_3O_{12}(OH)$ | androsite |
| $Ca_{-1}Al_{-1}REEMg$ | $CaREEAl_2MgSi_3O_{12}(OH)$ | dissakisite |
| $Ca_{-1}Al_{-1}REEFe^{2+}$ | $CaREEAl_2Fe^{2+}Si_3O_{12}(OH)$ | allanite (syn.: "orthite") |
| $Ca_{-1}Al_{-2}REEFe^{3+}Fe^{2+}$ | $CaREEAlFe^3Fe^{2+}Si_3O_{12}(OH)$ | $Fe^{3+}$-analogue of allanite-(Ce) |
| combined in AMO-positions | | |
| $Ca_{-1}Al_{-1}O_{-1}REE Mg_2F$ | $CaREEAlMg_2Si_3O_{11}F(OH)$ | dollaseite |
| $Ca_{-2}Al_{-3}O_{-1}REE_2MgMn^{2+}F$ | $REE_2Mn_2MgSi_3O_{11}F(OH)$ | krishtovite |

\* includes all trivalent rare earth elements and Y; the corresponding monoclinic endmembers and minerals are then named, for example, allanite-(Ce) according to the dominantly present REE

preliminary guidelines for the nomenclature of epidote minerals (Armbruster, pers. com.). According to these guidelines the term epidote refers to the whole mineral group but also to the composition $Ca_2Al_2Fe^{3+}Si_3O_{12}(OH)$. The Al end member $Ca_2Al_3Si_3O_{12}(OH)$ is termed clinozoisite and tawmawite refers to the $Cr^{3+}$ end member $Ca_2Al_2Cr^{3+}Si_3O_{12}(OH)$. The term pistacite should generally be avoided. Consequently, the composition of binary or ternary solid solutions is given as mol fraction ($X$) of the respective end members:

$$X_{Taw} = \frac{Cr^{3+}}{Fe^{3+} + Al + Cr^{3+} - 2}$$

$$X_{Czo} = \frac{Al - 2}{Fe^{3+} + Al + Cr^{3+} - 2} \qquad X_{Ep} = \frac{Fe^{3+}}{Fe^{3+} + Al + Cr^{3+} - 2}$$

For the nomenclature of intermediate members the 50% rule applies (Fig. 1). Although these preliminary guidelines have some severe shortcomings, they are adopted in Table 2. These short comings are that the term epidote not only denotes the whole mineral group but also an end member as well as specific solid solutions, and that Al-M$^{3+}$ solid solutions with more than 1 M$^{3+}$ = Fe$^{3+}$ + Cr$^{3+}$ per formula unit (pfu) will have negative mole fractions of clinozoisite if one uses only the end members mentioned above. Unfortunately, the preliminary guidelines have been set up during the preparation of this volume and it was therefore impossible with respect to editorial deadlines to make all chapters consistent with the recommended nomenclature. The reader should therefore be aware that different chapters use different nomenclature.

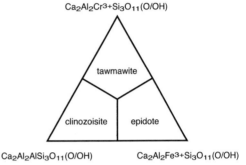

**Figure 1.** Nomenclature of monoclinic ternary Al-Fe-Cr epidote minerals. Analogue diagrams for other solid solutions (e.g., with V = mukhinite as an endmember) can be used.

Because the orthorhombic polymorph zoisite displays a much more restricted chemical variation and any substitution is restricted to small values far below mol fractions of $X = 0.5$ (see below), the nomenclature is much more consistent in the literature. The variety names "thulite" for Mn$^{3+}$ bearing and "tanzanite" for V bearing solid solutions should be avoided, although with regard to petrographic purposes these names are convenient and helpful as they refer to zoisite crystals with distinct optical properties (see below). Furthermore, the only epidote mineral that is a significant gem stone is the blue colored vanadium bearing zoisite, which in this context is generally termed tanzanite. We therefore recommend using the terms "thulite" and "tanzanite" only for petrographic or gemological purposes. The composition of orthorhombic solid solutions should be expressed as the mole fractions $X_{Fe}$, $X_{Mn}$ and $X_V$ of the respective theoretical end members, which are Ca$_2$Al$_2$Fe$^{3+}$Si$_3$O$_{12}$(OH), Ca$_2$Al$_2$Mn$^{3+}$Si$_3$O$_{12}$(OH), and Ca$_2$Al$_2$V$^{3+}$Si$_3$O$_{12}$(OH). Very often it is difficult to distinguish optically between orthorhombic and monoclinic, because of small grain size and inappropriate orientation of the crystals, and many authors have used "zoisite/clinozoisite." We recommend using instead "epidote minerals" which comprises both.

Recommended abbreviations for the common epidote minerals are:

Aln = allanite

Cz = clinozoisite

Ep = epidote

Pie = piemontite

Ps = pistacite (note, however, that it will not be a recommended name by the IMA)

Taw = tawmawite

Zo = zoisite

For the other only rarely occurring epidote minerals with exotic composition we recommend not to use abbreviations.

## CRYSTAL STRUCTURE

### The monoclinic epidote minerals

*Principal features and lattice constants.* The monoclinic structure was first determined by Ito (1950) and subsequently confirmed by Ito et al. (1954) and Belov and Rumanova (1953, 1954). Based on the similarities in diffraction pattern and by comparing the X-ray spectra intensities of Fe poor and Fe rich clinozoisite-epidote, Gottardi (1954) established the isomorphous relationship along the Al-Fe solid solution join. Ueda (1955) and later Rumanova and Nikolaeva (1960) refined the structure of allanite and found a structural topology similar to clinozoisite. In detailed studies, Dollase (1968, 1969, 1971) fully refined the structures of two crystals with $X_{Ep} = 0.03$ and $X_{Ep} = 0.81$, respectively, and those of hanckokite, allanite, and piemontite. Intermediate crystals ($X_{Ep} = 0.22, 0.40, 0.60, 0.81, 0.84$) were later refined by Gabe et al. (1973), Stergiou et al. (1987), Kvick et al. (1988), and Comodi and Zanazzi (1997). By means of single crystal X-ray diffraction with structure refinements, Carbonin and Molin (1980) systematically studied the geometrical and structural variations accompanying the Al-Fe$^{3+}$ substitution in natural crystals spanning the compositional range $X_{Ep} = 0.30$ to 0.86. Bonazzi and Menchetti (1995) extended this work by refining natural Al-Fe$^{3+}$ crystals of the compositional range $X_{Ep} = 0.24$ to 1.14 and natural REE and Fe$^{2+}$ bearing crystals to study the effects of the Fe$^{3+}$ and Fe$^{2+}$ substitution and the entry of REE. The only systematic structural study on synthetic Al-Fe$^{3+}$ crystals is from Giuli et al. (1999) on seven samples spanning the compositional range $X_{Ep} = 0.66$ to 1.09.

The monoclinic epidote minerals all have space group $P2_1/m$ (for a discussion of a possible symmetry reduction to $Pm$, $P\bar{1}$, or $P2_1$ as response to cation ordering see below). The monoclinic structure (Fig. 2) is built up by two types of endless octahedral chains that run parallel [010] and consist of three non-equivalent octahedra M1, M2, and M3. The M1 octahedra form endless, edge sharing single chains to which individual M3 octahedra are attached on alternate sides forming a zigzag pattern. The second type of octahedral chain is made up exclusively by edge sharing M2 octahedra. The octahedral chains are cross linked in [100] and [001] by isolated T3 tetrahedra and T1T2O$_7$ groups with two non-equivalent large A1 and A2 positions in between (Fig. 2). Depending on composition and study cited, the lattice parameters of natural and synthetic Al-Fe$^{3+}$ dominated solid solutions are $a = 8.861–8.922$ Å, $b = 5.577–5.663$ Å, $c = 10.140–10.200$ Å, $\beta = 115.31–115.93°$, and the cell volume is $V = 452.3–463.9$ Å$^3$ (Table 3), which correlate positively with Fe content, except for $\beta$ which shows a slightly negative correlation (Figs. 3, 4). Linear regression equations of the data are compiled in Table 3.

The correlation trends of individual structural parameters with increasing Fe content display changes in slope at $X_{Ep} = 0.6$ to 0.7 and suggest a change in the Al–Fe$^{3+}$ substitution mechanism from exclusive substitution on M3 for $X_{Ep} < 0.6$ to 0.7 to a combined substitution on M3 and M1 for $X_{Ep} > 0.6$ to 0.7 (see below). Such a change in the Al–Fe$^{3+}$ substitution mechanism should also appear in the correlation trends between lattice constants and Fe content. Unfortunately, due to the different techniques applied by the different authors and the problems inherent in studying natural samples (impurities, minor amounts of other elements, zoning etc.) the data scatter and mask this change (Fig. 3). To overcome this problem we plotted only the data for almost pure Al-Fe$^{3+}$ solid solutions (Fig. 4) from the systematic studies on Fe incorporation in clinozoisite-epidote from Carbonin and Molin (1980), Bonazzi and Menchetti (1995), and Giuli et al. (1999). Although the data still scatter, there appear subtle changes in the slopes of the correlation trends between lattice parameters $a$, $b$, $c$, and $\beta$ and Fe content in the compositional range $X_{Ep} = 0.5$ to 0.7. Linear regression equations of the data for $X_{Ep} < 0.6$ and $X_{Ep} > 0.6$ are compiled in Table 3.

The changes in slope are opposite for $a$ and $c$. Therefore, they largely cancel out and there appears no change in slope for $V$ within the quality of the data (Fig. 4e). The observed changes

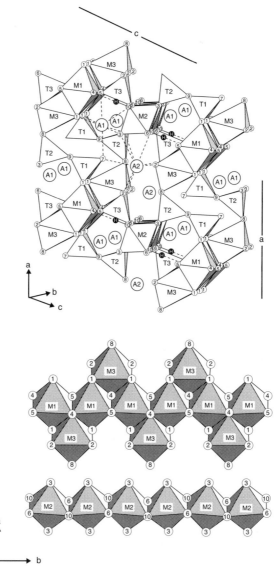

**Figure 2.** Coordination polyhedra of epidote (upper part) and notation of the M positions in the octahedral chains (lower part), with numbers for oxygen (adopted from Dollase 1968).

in slope of the lattice parameters at about $X_{Ep} = 0.6$ are in good accordance with the optical parameters that show discontinuities in this compositional range (Hörmann and Raith 1971; see below), with Mössbauer spectroscopic work that indicate Fe substitution exclusively on M3 for $X_{Ep} < 0.6$ and a combined substitution on M3 and M1 for $X_{Ep} > 0.6$ (Fehr and Heuss-Aßbichler 1997), and with the theoretical model for the intracrystalline Al–Fe$^{3+}$ distribution of Bird and Helgeson (1980) that indicates an increasingly important contribution of M1 to the overall Fe incorporation for Fe contents above $X_{Ep} = 0.6$ to 0.7 (see Gottschalk 2004).

**Table 3.** Linear regression equation for lattice constants as a function of composition of monoclinic and orthorhombic epidote minerals; data sources for the monoclinic epidote minerals see text.

### MONOCLINIC

|  | $0 < X_{Ep} < 1$ | $X_{Ep} < 0.6$ | $X_{Ep} > 0.6$ |
|---|---|---|---|
| $a$ [Å] | $(3.59 \times 10^{-2}) X_{Ep} + 8.868$ | $(4.75 \times 10^{-2}) X_{Ep} + 8.861$ | $(3.30 \times 10^{-2}) X_{Ep} + 8.869$ |
| $b$ [Å] | $(6.34 \times 10^{-2}) X_{Ep} + 5.583$ | $(7.83 \times 10^{-2}) X_{Ep} + 5.576$ | $(6.78 \times 10^{-2}) X_{Ep} + 5.580$ |
| $c$ [Å] | $(2.81 \times 10^{-2}) X_{Ep} + 10.143$ | $(2.61 \times 10^{-2}) X_{Ep} + 10.142$ | $(3.20 \times 10^{-2}) X_{Ep} + 10.137$ |
| $\beta$ [°] | $(-7.76 \times 10^{-2}) X_{Ep} + 115.48$ | $(-1.55 \times 10^{-1}) X_{Ep} + 115.51$ | $(-1.04 \times 10^{-1}) X_{Ep} + 115.49$ |
| $V$ [Å³] | $(8.70) X_{Ep} + 452.98$ | — | — |

### ORTHORHOMBIC

|  | Myer (1966) | Liebscher et al. (2002) | |
|---|---|---|---|
|  |  | zoisite I | zoisite II |
| $a$ [Å] | $(1.35 \times 10^{-1}) X_{Fe} + 16.201$ | $(-3.72 \times 10^{-2}) X_{Fe} + 16.1913$ | $(-8.26 \times 10^{-2}) X_{Fe} + 16.2061$ |
| $b$ [Å] | $(7.84 \times 10^{-2}) X_{Fe} + 5.550$ | $(6.43 \times 10^{-2}) X_{Fe} + 5.5488$ | $(8.14 \times 10^{-2}) X_{Fe} + 5.5486$ |
| $c$ [Å] | $(4.32 \times 10^{-2}) X_{Fe} + 10.035$ | $(3.43 \times 10^{-2}) X_{Fe} + 10.0320$ | $(1.18 \times 10^{-1}) X_{Fe} + 10.0263$ |
| $V$ [Å³] | $(24.2) X_{Fe} + 902.2$ | $(11.4) X_{Fe} + 901.3$ | $(19.3) X_{Fe} + 901.6$ |

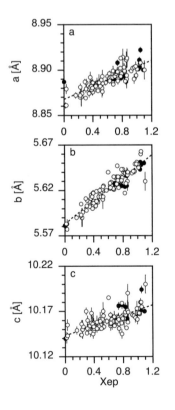

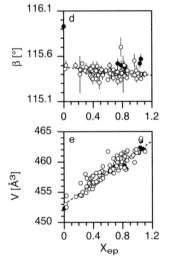

**Figure 3.** Lattice constants of monoclinic epidote, unselected data; open circles: natural; dots: synthetic; for data sources see Appendix A.

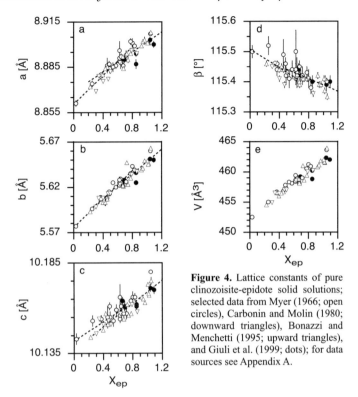

**Figure 4.** Lattice constants of pure clinozoisite-epidote solid solutions; selected data from Myer (1966; open circles), Carbonin and Molin (1980; downward triangles), Bonazzi and Menchetti (1995; upward triangles), and Giuli et al. (1999; dots); for data sources see Appendix A.

A detailed description and discussion of the structural changes accompanying the Al-$Fe^{3+}$ substitution is given below.

A comparable change in substitution mechanism is also seen in the lattice parameters of synthetic Al-$Mn^{3+}$ solid solutions (Anastasiou and Langer 1977, Langer et al. 2002; Fig. 5). Although some of the reported compositions of the synthetic Al-$Mn^{3+}$ solid solutions might be questionable because Anastasiou and Langer (1977) could not analyze the run products, but assumed synthesis on the bulk composition based on mass balance considerations, the data indicate a change in substitution mechanism at about 1 $Mn^{3+}$ pfu. The data also show that the substitution by $Mn^{3+}$ changes the lattice parameters differently than the substitution by $Fe^{3+}$ (Figs. 4, 5). For the same degree of Al–$M^{3+}$ substitution (with $M^{3+}$ = $Fe^{3+}$ or $Mn^{3+}$) *a* is generally shorter (it actually correlates negatively with Mn content for $Mn^{3+}$ < 1 cat. pfu) whereas *b*, *c*, and *V* are slightly larger in Al-$Mn^{3+}$ than in Al-$Fe^{3+}$ solid solutions. In contrast to Al-$Fe^{3+}$ solid solutions in which β correlates negatively with Fe content it correlates positively with Mn contents in Al-$Mn^{3+}$ solid solutions (Figs. 4d, 5d). The changes in slopes of the lattice parameters at about 1 $Mn^{3+}$ pfu indicate the change in substitution mechanism from incorporation of $Mn^{3+}$ on M3 to incorporation on M1 (for a detailed description of Mn bearing monoclinic epidote minerals see Bonazzi and Menchetti 2004).

To study the structural changes accompanying the Al–$Fe^{3+}$ substitution in monoclinic epidote minerals, we took the structure refinements of natural and synthetic samples for which precise chemical analyses are presented and which represent almost pure Ca-Al-$Fe^{3+}$ solid solutions (Dollase 1968, 1971; Gabe et al. 1973; Carbonin and Molin 1980; Stergiou et al. 1987; Kvick et al. 1988; Bonazzi and Menchetti 1995; Comodi and Zanazzi 1997;

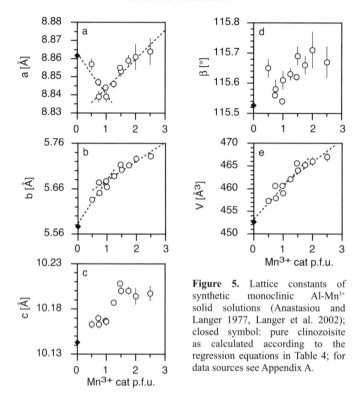

**Figure 5.** Lattice constants of synthetic monoclinic Al-Mn³⁺ solid solutions (Anastasiou and Langer 1977, Langer et al. 2002); closed symbol: pure clinozoisite as calculated according to the regression equations in Table 4; for data sources see Appendix A.

Giuli et al. 1999). We started from the refined fractional atom coordinates and recalculated the structural parameters. Below we briefly review the results for the individual polyhedra with special emphasis on the structural changes accompanying the Al–Fe³⁺ substitution.

*Octahedra.* The M1 octahedron lies on a symmetry centre and links the oxygen atoms O1, O4, and O5. The individual M1 octahedra share a common O4–O5 edge forming endless octahedral chains to which the individual M3 octahedra are attached. M1 and M3 share a common O1–O4 edge (Fig. 2). The M1 octahedron is fairly regular with a bond angle variance (Robinson 1971) $s^2 = 15$ to 20 (Fig. 6, Appendix B). Its mean bond length is about 1.90 to 1.94 Å and its volume 9.0 to 9.5 Å³ (Figs. 6a, b). Up to about $X_{Ep} = 0.6$ the M1 octahedron shows only minor structural changes with increasing Fe content: the mean bond length, the volume, and the distance between the two apical O1 atoms slightly increase (Fig. 6) as response to the expansion of the attached M3 (see below) due to increased Fe content on M3 (Bonazzi and Menchetti 1995). For $X_{Ep} > 0.6$ the structural changes are slightly more pronounced suggesting Fe³⁺ incorporation also on M1. With increasing Fe content, the bond angle variance $s^2$ generally decreases and M1 becomes more regular (Fig. 6d). The inclination of M1 to [001] calculated as the angle between O1–O1′ and [001] decreases only slightly with increasing Fe³⁺ content (Appendix B).

Like M1 the M2 octahedron lies on a symmetry centre. It links the oxygen atoms O3, O6, and O10. The M2 octahedra share a common O6–O10 edge and form endless, edge sharing single chains that are interconnected with the other structural units by the isolated T3O₄ tetrahedra, which are attached alternately on both sides of the M2 chains (Fig. 2). It is the smallest and most regular of all three octahedra and shows the smallest changes with

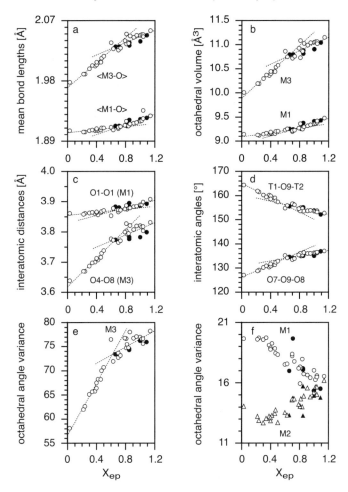

**Figure 6.** Bond length, volume and interatomic angles of monoclinic epidote minerals; open circles: natural; dots: synthetic; for data sources see Appendix B.

increasing Fe content. Its mean bond length is 1.88 to 1.89 Å and the volume 8.8 to 8.9 Å$^3$ (Appendix B). The bond angle variance $s^2$ is between 12 and 16 and indicates that M2 becomes more distorted with increasing Fe content (Fig. 6d). The inclination of M2 to [001] calculated as the angle between O3–O3′ and [001] decreases continuously with increasing $Fe^{3+}$ content indicating a rotation of M2 (Appendix B). The structural changes in M2 are due to the changes occurring in the other structural units especially in M1 and M3 and to the overall expansion of the structure with increasing Fe content (Bonazzi and Menchetti 1995). The structural data give no hints for $Fe^{3+}$ incorporation in M2, although Bonazzi and Menchetti (1995) found evidence for very small amounts of Fe in M2 for $X_{Ep} > 1.0$ and Giuli et al. (1999) found up to 0.08 $Fe^{3+}$ pfu in M2 in their synthetic samples. Incorporation of small amounts of $Fe^{3+}$ in M2 is also supported by Mössbauer spectroscopy (see Liebscher 2004).

The M3 octahedron is located on the mirror plane and coordinates the oxygen atoms O1, O2, O4, and O8. It is the largest and most distorted octahedron in the structure. Increasing

Fe content increases its mean bond length from 1.98 to 2.06 Å and enlarges its volume from 10.0 to 11.2 Å$^3$ (Figs. 6a,b; Appendix A). The distance between the two apical oxygen atoms O4 and O8 is considerable smaller than the comparable O1–O1 distance in M1, but increases significantly with increasing Fe content (Fig. 6c). Because M3 shares O4 with two M1 octahedra (Fig. 2), which therefore has a relatively fixed position, the increase in the O4–O8 distance primarily results from a significant shift of O8 perpendicular to the M1 octahedral chain. The general short O4–O8 distance points to a strong tetragonal distortion of M3. The strong distortion of M3 is also supported by its bond angle variance $s^2$, which is very large compared to M1 and M2 and increases from 58 to 79 with increasing Fe content (Fig. 6d). All structural changes of M3 are most pronounced for $X_{Ep} < 0.6$ and display a clear linear relationship with Fe content. For $X_{Ep} > 0.6$ the structural changes are less pronounced and the linear trend is partly masked by a considerable scatter of the data (Fig. 6). This different behavior for $X_{Ep} < 0.6$ and $X_{Ep} > 0.6$ is consistent with Fe$^{3+}$ substitution exclusively in M3 for $X_{Ep} < 0.6$ and a combined Fe$^{3+}$ substitution in M1 and M3 for $X_{Ep} > 0.6$. The scatter of the data for $X_{Ep} > 0.6$ most probably reflects different degrees of intracrystalline order-disorder at M1 and M3 due to different P and T of crystallization or metastable persistence of an ordered or disordered state (see below).

*Tetrahedra and Si$_2$O$_7$ group.* The tetrahedra are the most rigid units of the structure. Mean bond lengths and tetrahedral volumes are 1.626 (2) Å and 2.197 (9) Å$^3$ for T1, 1.614 (2) Å and 2.157 (8) Å$^3$ for T2, and 1.640 (2) Å and 2.248 (8) Å$^3$ for T3 (Appendix B). None of these parameters shows significant changes, which indicates no substantial substitution of Si by other cations, in accordance with the findings of Bonazzi and Menchetti (1995). The most evident and important feature is the rotation of T1 and T2 of the T1T2O$_7$ group relative to each other to compensate the structural changes in M3 accompanying the incorporation of Fe$^{3+}$. Fe substitution in M3 enlarges the M3 octahedron and results in a significant shift of O8 (see above). Because O8 is shared by M3 and T2, its shift leads to a rotation of T2. This rotation is mirrored by an increase of the O7–O9–O8 angle and a decrease of the T1–O9–T2 angle of the T1T2O$_7$ group (Fig. 6d). Like the other structural parameters the changes of both angles are most pronounced for $X_{Ep} < 0.6$, which gives further evidence that the substitution mechanisms change at about $X_{Ep} = 0.6$ from Fe incorporation purely in M3 to a combined incorporation in M1 and M3.

*A-positions.* In Al-Fe$^{3+}$ solid solutions A1 is mostly described as nine-fold coordinated whereas A2 is described as ten-fold coordinated. Only in an almost Fe free clinozoisite, Dollase (1968) described A1 as seven-fold and A2 as eight-fold coordinated. From a purely charge balance point of view A1 should best be described as seven-fold coordinated whereas A2 is generally under-bonded even in ten-fold coordination (Appendix B) according to charge balance calculations using the refined bond lengths and the bond valence parameters from Brese and O′Keefe (1991). In allanite A1 is nine-fold coordinated whereas A2 is most probably eleven-fold coordinated (Dollase 1971). The A positions show only minor changes with increasing Fe content. These changes depend mostly on structural changes occurring on other sites and on the overall expansion of the structure (Bonazzi and Menchetti 1995).

*Proton.* Based on crystal chemical considerations, Ito et al. (1954) proposed bonding of the proton to O10 of the M2 octahedron with a hydrogen bridge to O4 of the neighboring M1 octahedron (Fig. 2). This was later confirmed by X-ray diffraction with bond valence calculations (Dollase 1968; Gabe et al. 1973; Carbonin and Molin 1980), infrared spectroscopy (Hanisch and Zemann 1966; Langer and Raith 1974; see also Liebscher 2004), and neutron diffraction (Nozik et al. 1978; Kvick et al. 1988). With increasing Fe content the length of the hydrogen bridge as well as the O10–O4 distance increase (Fig. 7a; Appendix B). But independent from Fe content the length of the O10–H … O4 hydrogen bridge is about 0.02 Å longer than the O10–O4 distance. This is due to bending of the hydrogen bridge with

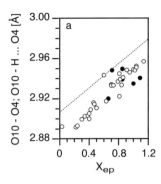 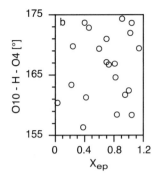

**Figure 7.** Interatomic distances and angles of the proton environment in monoclinic epidote; (a) O10 – O4 interatomic distance (symbols) and length of the O10 – H … O4 hydrogen bridge (dotted line). The hydrogen bridge is generally about 0.02 Å longer than the O10 – O4 distance due to its bent nature; (b) Bending of the hydrogen bridge with angles between 155 and 175°; open circles: natural; dots: synthetic; for data sources see Appendix B.

angles between 155 and 175° (Fig 7b) as response to the electrostatic repulsions from the M3 octahedron and the A2 position (Kvick et al. 1988). The lengthening of the hydrogen bridge with increasing Fe content was also found by Langer and Raith (1974) by means of infrared spectroscopy. These authors compared the length of the hydrogen bridge for different Fe contents, as calculated from the corresponding IR band, with published O10–O4 distances. Both values seem to converge with increasing Fe content and the authors therefore concluded, that the hydrogen bridge becomes more linear with increasing Fe content (Langer and Raith 1974). Contrary, the more recent diffraction data give no evidence for a change of the bending angle with increasing Fe content (Fig. 7b).

## Zoisite (the orthorhombic epidote minerals)

*Principal features and lattice constants.* The structure of zoisite was first determined by Fesenko et al. (1955, 1956) for an Fe-free zoisite and later refined and fully determined by Dollase (1968) for a zoisite with $X_{Fe}$ = 0.08–0.11. It is orthorhombic with space group Pnma. The structural features resemble those of the monoclinic epidote minerals, but zoisite has only one type of octahedral chain that consists of two non-equivalent octahedra M1,2 and M3. The M1,2 octahedra form endless, edge sharing single-chains parallel [010] to which individual M3 octahedra are attached exclusively on one side (Fig. 8a). As in the monoclinic forms, the octahedral chains are cross linked in [100] and [001] by isolated T3 tetrahedra and T1T2O$_7$ groups with two non equivalent large A1 and A2 positions in between (Fig. 8b). Depending on composition and study cited, the lattice constants of zoisite are $a$ = 16.15–16.23 Å, $b$ = 5.51–5.581 Å, $c$ = 10.0229–10.16 Å, and the cell volume ($V$) is 900.0–909 Å$^3$ (Appendix C). The data suggest a positive correlation between $b$, $V$, and Fe content whereas for $a$ and $c$ the data scatter and show no clear dependence on Fe content (Fig. 9). The data of Myer (1966) on five natural zoisite crystals with Fe contents ranging from 0.016 to 0.15 $X_{Fe}$ suggest a positive correlation between Fe content and lattice constants. Linear regression equations of his data are compiled in Table 3.

In contrast, in a study on five synthetic zoisite crystals with Fe contents from $X_{Fe}$ = 0 to 0.113, Liebscher et al. (2002) found discontinuities in refined lattice parameters at ~ 0.05 $X_{Fe}$ and attributed them to two orthorhombic modifications labeled zoisite I (< 0.05 $X_{Fe}$) and zoisite II (> 0.05 $X_{Fe}$). The correlation between lattice parameters and Fe content was derived for zoisite I and II (Fig. 9; dotted lines). Linear regression equations of the data are compiled in Table 3.

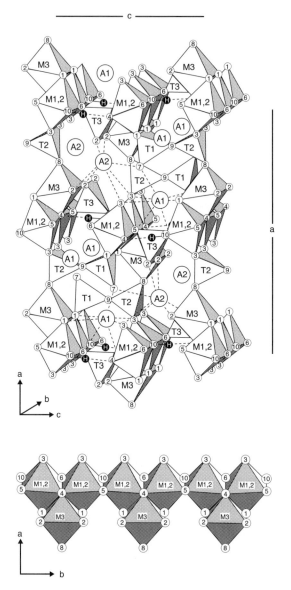

**Figure 8.** Coordination polyhedra of zoisite and notation of the M positions in the octahedral chains; adopted from Dollase (1968). Compare Figure 2.

Contrary to the data from Myer (1966) $a$ is negatively correlated with Fe content and only $b$, $c$, and $V$ show a positive correlation. At the transition from zoisite I to zoisite II $a$ and $V$ show positive, $b$ and $c$ negative offsets (Fig. 9). Independent on Fe content, zoisite II has the larger cell volume compared to zoisite I. The reason for the discrepancy between the results on synthetic zoisite from Liebscher et al. (2002) and those on natural zoisite is not clear but probably is due to small amounts of other elements like Mn, Mg, Ti, $Fe^{2+}$, and Sr in the natural

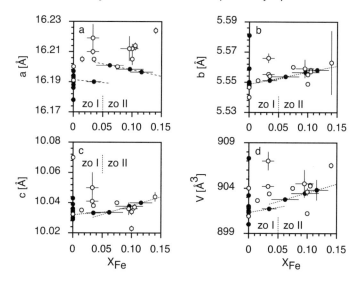

**Figure 9.** Lattice constants of natural (open circles) and synthetic (dots) zoisite; dotted lines are linear regressions of the data of Liebscher et al. (2002) for zoisite I and zoisite II (corresponding equations see Table 4); for data sources see Appendix C.

zoisite samples that mask the influence of increasing Fe content on the lattice constants. Below we give a brief description of the individual polyhedra with special emphasis on the structural changes accompanying the Al–Fe$^{3+}$ substitution.

*Octahedra.* Independent upon Fe content the M1,2 octahedron is relatively undistorted. Its minimum and maximum bond lengths deviate by less than 0.1 Å from the mean M1,2–O bond length of about 1.89 Å (Appendix D) and the interatomic angles are within ± 8° of the ideal octahedral values. Its edges show the typical feature of edge shearing octahedra. The mean M1,2 edge length is 2.67 Å, but the shared edges (O4–O6 and O5–O10 between M1,2 octahedra and O1–O4 between M1, 2 and M3 octahedra, Fig. 8) are only 2.47–2,60 Å whereas the edges parallel to the Al–Al vector (O4–O5 and O6–O10) are 2.78–2.79 Å (Appendix D). This length of the Al–Al vector is considerable larger than the mean edge length of the tetrahedra (~2.63 Å; Appendix D), which bridge between the O1 and O3, respectively, of the individual M1, 2 octahedra in [010]. To account for this misfit, the M1,2 octahedral chain is slightly undulated (Fig. 8). The structural features of M1,2 show only minor changes with increasing Fe content. Only O1–O6 and O1–O3 clearly correlate negatively with Fe content, displaying discontinuities at the transition from zoisite I to zoisite II (Fig. 10 b,c).

In contrast to M1,2 the M3 octahedron is strongly distorted. Its minimum and maximum bond lengths deviate up to 0.22 Å from the mean M3–O bond lengths of 1.96–1.97 Å and its interatomic angles deviate up to 18° from the ideal octahedral values (Appendix D). With 2.76 to 2.77 Å its mean edge length is about 0.1 Å larger than that of M1,2. The edges shared with M1,2 (O1–O4, Fig. 10d) are significantly shorter (2.48–2.56 Å) whereas O1–O8 which points away from M1,2 is significantly longer (up to 3.16 Å) than the mean edge length. As indicated by the larger mean bond lengths and mean edge lengths the volume of M3 (9.7 to 9.8 Å$^3$) is about 1 Å$^3$ larger than that of M1, 2 (8.8 to 9.0 Å$^3$). Almost all structural features of M3 show significant changes with increasing Fe content. Especially the O1–O1 distance decreases whereas the O4–O8 distance significantly increases with increasing Fe content and both show discontinuities at the transition from zoisite I to zoisite II (Fig. 10 c,d).

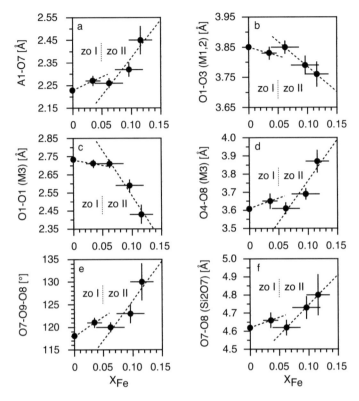

**Figure 10.** Interatomic distances and bond angles of synthetic zoisite (Liebscher et al. 2002); zo I and zo II refer to iron-poor and iron-rich zoisite in Figure 11; see Appendix D.

***Tetrahedra and Si$_2$O$_7$ group.*** The T3 tetrahedron and the T1T2O$_7$ group connect the octahedral chains in [100] and [001] (Fig. 8). T1 and T2 vary significantly and get more distorted with increasing Fe content whereas T3 shows only minor changes. The mean T1–O bond length increases from 1.64 to 1.73 Å mostly due to a significant increase of the T1–O1 bond length whereas the mean T2–O bond length decreases from 1.62 to 1.57 Å (Appendix D). The O–T–O angles in T1 and T2 partly deviate more than 13° from the ideal tetrahedral value of 109°. In T1 the mean O–O distance as well as the volume increase whereas in T2 they decrease. In T3 only O2–O2′ slightly increase with increasing Fe content; all other structural values are, within errors, approximately constant (Appendix D). The geometry of the T1T2O$_7$ group significantly changes with increasing Fe content, due to an opposite rotation of the T1 and T2 tetrahedra (Liebscher et al. 2002). The T1–O9–T2 and O1–O9–O3 angles slightly tighten whereas O7–O9–O8 opens significantly (Fig. 11). Consequently, the O7–O8 distance strongly increases (Appendix D).

***A-positions.*** Including all neighbors up to 2.85 Å, A1 and A2 are seven-fold coordinated independent on Fe content and are best described as trigonal prisms with a seventh and closer oxygen ligand lying outside of one of the equatorial prism faces (Dollase 1968; Appendix D). The mean A1–O bond length is generally shorter by about 0.1 Å than the mean A2–O bond length and both show almost no variation with Fe content. Only A1–O7 significantly increases with increasing Fe content (Appendix D) reflecting the rotation of T1 (see above).

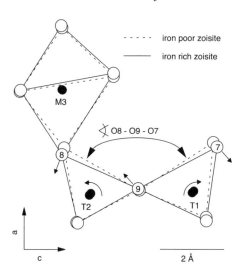

**Figure 11.** Rotation of tetrahedra T1 and T2 and of the M3 octahedron with increasing Fe content in zoisite.

**Proton.** The proton in zoisite is bonded to O10 (Fig. 8; Ito et al. 1954; Hanisch and Zemann 1966; Dollase 1968; Linke 1970; Gabe et al. 1973; Smith et al. 1987; Liebscher et al. 2002), which has the lowest bond strength (= 1.02 to 1.25 without hydrogen; Dollase 1968; Liebscher et al. 2002) of all oxygen atoms. The length of the hydrogen bond is 1.2 (2) Å (Dollase 1968). Based on the approximate position of the proton, the O10–O4 separation length and the fact that O4 is slightly under-bonded, Dollase (1968) concluded that the proton forms a hydrogen bridge with O4 of the neighboring octahedral chain. IR spectroscopic work proved this conclusion but also showed evidence for a second hydrogen bridge (Linke 1970; Langer and Raith 1974; Langer and Lattard 1980; Liebscher et al. 2002). Langer and Raith (1974) and Langer and Lattard (1980) speculated about an additional proton position in the structure with a corresponding second hydrogen bridge. However, the results of pressure dependent (Winkler et al. 1989) and temperature dependent (Liebscher et al. 2002) IR spectroscopy indicate that the proton is exclusively bonded to O10 but forms two hydrogen bridges with O4 and O2 of the neighboring octahedral chain. The hydrogen bridge to O2 must be bifurcated (for a review of spectroscopic work done on zoisite see Liebscher 2004). The O10–O4 hydrogen bridge has a length of 2.63 to 2.76 Å and slightly shortens with increasing Fe content (Liebscher et al. 2002; Appendix D).

## Monoclinic - orthorhombic structural relationship

Ito (1950) was the first to study the structural relationship between epidote and zoisite. He proposed a unit cell twinning model with $2a_{mcl}\sin\beta = a_{orth}$ that satisfactorily relates unit cell sizes and space group symmetries of the two structures (Fig. 12a). Later structural refinements showed that a simple *n*-glide unit cell twinning does not reproduce the relative atomic movements that occur during the monoclinic–orthorhombic transformation (Dollase 1968). Especially the M1,2 octahedron, that would be produced by this operation and which is formed by joining half the M1 and half the M2 octahedron of clinozoisite, would be distorted due to the different inclination of M1 and M2 to the twin plane (100) (see above). In the orthorhombic structure, however, the M1,2 octahedron is nearly regular (Dollase 1968). This regularity arises from shifts of O3, O6, and O10 that can be viewed as a rotation of half of the M1,2 octahedron. This rotation is responsible for the main difference between zoisite and the monoclinic structure, i.e. the larger A2–O10 bond length and therefore reduced coordination number of A2 in zoisite (sevenfold versus eightfold in clinozoisite) and the larger T1–O9–T2 angle in zoisite (173° versus 164° in clinozoisite) due to a rotation of the T2 tetrahedron (Dollase 1968). Therefore the simple model of unit cell-twinning (Ito 1950) is qualitatively useful for descriptive purposes but does not strictly hold in a quantitative manner (Dollase 1968).

Ray et al. (1986) discussed the polytypic relationship between the monoclinic and orthorhombic structure in terms of different stacking sequences of layers or modules of monoclinic unit cells in [100]. The authors, somewhat arbitrarily, choose the stacking module

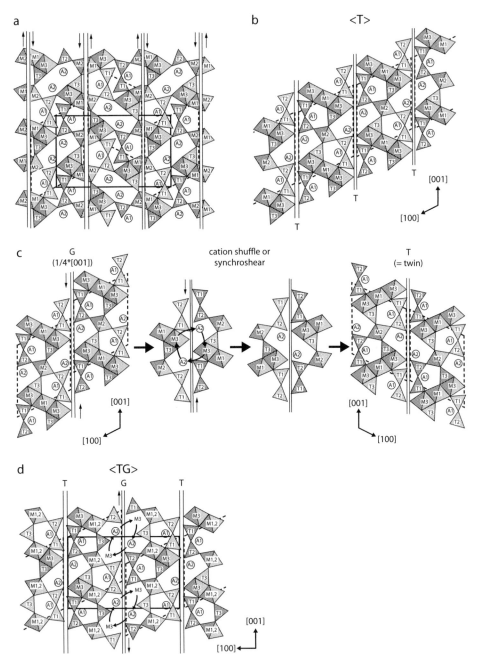

**Figure 12.** Relationship between clinozoisite and zoisite. (a) Model of Ito (1950) with a simple shearing. (b,c,d) Model of Ray et al. (1986); note the different position of the shearing planes. Monoclinic and orthorhombic unit cells are outlined (for detailed explanation see text).

in that way that its boundary coincides with a ($h$00) plane in the structure with minimal links between the structural framework (Fig. 12b). They show with high resolution TEM images that this plane corresponds well with the position of stacking faults in natural crystals. Each successive module might either be stacked by simple translation without any displacement between the modules (operator T = translation) or by translation with a glide component that introduces a shear displacement between the modules (operator G = gliding). Repeated stacking sequences can then be described by the respective operators in angular brackets, i.e. TTTT becomes <T>, GGGG becomes <G>, and TGTG becomes <TG> (Ray et al. 1986). A simple <T> stacking sequence results in the monoclinic structure (Fig. 12b). Introducing a glide component of 1/4 [001] between each module (<G> stacking sequence) results in the twin of <T> although a "cation shuffle" or "synchroshear" has to take place in order to reproduce a cation arrangement that resembles the original structure (Fig. 12c). If the glide operation together with the "cation shuffle" or "synchroshear" is only introduced every other stacking module a <TG> stacking sequence is produced which results in the orthorhombic structure (Fig. 12d; Ray et al. 1986).

The difference between the real orthorhombic structure and the theoretical structure produced by simple unit cell twinning also sheds some light on the monoclinic to orthorhombic transition as a function of composition. The main structural difference between the monoclinic and the orthorhombic structure, i.e. reduced coordination number of A2 and larger T1–O9–T2 angle in the latter, are due to the rotation of half of the M1,2 octahedron (see above). The data on the $Fe^{3+}$ incorporation in zoisite presented above show that increasing $Fe^{3+}$ content in zoisite enlarges M3, rotates T1 and T2, and decreases the T1–O9–T2 angle. The rotation of T2 also leads to a slight rotation of M1,2 in just the opposite sense to the rotation that occurs at the monoclinic to orthorhombic transition. Therefore, the effects of increasing $Fe^{3+}$ content on the orthorhombic structure are opposite to those of the monoclinic–orthorhombic transition and $Fe^{3+}$ stabilizes the monoclinic with respect to the orthorhombic structure. A structural stabilization of the monoclinic over the orthorhombic form with increasing $Fe^{3+}$ content, or generally speaking with increasing substitution of a larger cation for Al, is also supported by the inclination to (100) of the M1 and M2 octahedra of the monoclinic structure. With increasing $Fe^{3+}$ substitution for Al the inclination to (100) of the M2 octahedron continuously decreases from 66.5° to 61.7° whereas the inclination to (100) of the M1 octahedron only decreases from 76.9° to 75.7° (Appendix B). The structural misfit between M1 and M2 to produce a regular, orthorhombic M1, 2 octahedron therefore continuously increases and prohibits the monoclinic–orthorhombic transition.

### TEM investigations

Only very few TEM studies on epidote minerals have been carried out. Our own experience with ion thinning of zoisite, especially when the crystals are embedded in a soft matrix such as carbonate can be problematic, because it is often difficult to obtain a thin edge. Ray et al. (1986) and Heuss-Aßbichler (2000) mentioned that epidote is not stable under the electron beam and easily transforms into amorphous material, which may start along stacking faults. Allanite, in contrast, is quite stable and does not show electron-beam radiation damage (Janeczek and Eby 1993). TEM investigations turned out to be a very useful tool to study the degree of metamictization in allanite and the process of crystallization during heating experiments.

The best orientation to study planar defects on (100) is viewing down the [010] zone axis. Ray et al. (1986) showed lamellar twins on (100) to be common in euhedral clinozoisite vein crystals, and they propose that most of the twins occur as a result of stacking mistakes during growth. However, Müller (pers. comm.) did observe mechanical twinning in epidote (see below Fig. 25k). Stacking faults on (100) are observed sporadically and some of them are terminated by partial dislocations. Investigations on zoisite (from the Eclogite Zone, Tauern Window,

Austria; Ray et al. 1986) showed occasional isolated stacking faults in crystals, which do not show obvious signs of deformation in thin section. In crystals from deformed veins, stacking faults spaced at approximately 250 Å and terminated by partial dislocations are numerous. The stacking faults can produce the clinozoisite sequence TTT in zoisite, interpreted as the result of deformation. Another type of stacking faults on (100) was observed as lamellar intergrowths of iron rich clinozoisite in iron poor zoisite on a scale of 20–200 μm. In zoisite crystals from deformed high-pressure pegmatites, Franz and Smelik (1995) also observed lamellae of clinozoisite in areas of anomalously high Fe content (own unpublished data), which can be interpreted from the textures as a late stage transition from zoisite into clinozoisite. Heuss-Aßbichler (2000) investigated two epidote crystals with $X_{Ep} = 0.6$ and 0.39, respectively. Generally these crystals lack interesting features of realbau such as stacking faults. However, in *a-c* sections the crystal with $X_{Ep} = 0.6$ shows a slight bending of the planes parallel to the *c*-axis, whereas the planes parallel to *a* are perfectly straight. Oriented parallel to the [100] zone axis the crystal shows a mosaic microstructure caused by twisting of the octahedral chains, which results in stacking faults and dislocations. The other crystal with $X_{Ep} = 0.39$ is also rather perfect, only in one case modulations were observed, interpreted as the indication for a tweed structure, and additional reflections were interpreted as a superstructure.

## CRYSTAL CHEMISTRY

The crystal chemistry of the epidote minerals is especially interesting because they form solid solutions with end members, which are composed of the abundant components CaO-$Al_2O_3$-$Fe_2O_3$-$SiO_2$ and with end members whose components are present in a rock normally only at a subordinate level ($Mn_2O_3$) or at the trace element level (Sr, Pb, REE, Ba). We will restrict this review to the general features of the crystal chemistry of those elements that normally occur as major or minor constituents in epidote minerals. The trace element crystal chemistry of epidote minerals is presented by Frei et al. (2004), that of Mn bearing monoclinic epidote minerals by Bonazzi and Menchetti (2004), and that of allanite by Gieré and Sorensen (2004). The stable and radiogenic isotope geochemistry of epidote minerals is reviewed by Morrison (2004). For a detailed review of the compositional variation of the epidote minerals in different geological environments, the reader is referred to Bird and Spieler (2004), Grapes and Hoskin (2004), Enami et al. (2004), and Schmidt and Poli (2004). Compilations of analyses of epidote minerals can be found in Kepezhinskas and Khlestov (1971) and Deer et al. (1986).

### The monoclinic epidote minerals

*Compositional range.* Despite their comparable simple crystal structure, the monoclinic epidote minerals display a wide and diverse range of composition. The vast majority belongs to the Al-$Fe^{3+}$ solid solution series with $Fe^{3+}$ contents between about $X_{Ep} = 0.2$ and $X_{Ep} = 1.0$. Clinozoisite with $Fe^{3+}$ content below $X_{Ep} = 0.2$ is only rarely reported, examples include clinozoisite from Willsboro, New York, with about $X_{Ep} = 0.03$ (DeRudder and Beck 1964; this clinozoisite was used by Dollase 1968 for his structural refinement), and clinozoisite with about $X_{Ep} = 0.09$ from Alpe Arami, Western Alps (Ernst 1977), although the latter is questionable as Ernst (1977) did not mention how the monoclinic symmetry has been determined. More widespread are examples of epidote with $X_{Ep} > 1.0$, although these compositions are normally restricted to low-grade metamorphism and alteration or to hydrothermal systems. To the authors knowledge, the highest $Fe^{3+}$ content so far reported for epidote is $X_{Ep} = 1.48$ from a secondary epidote that occurs in granitoid rocks from New Zealand (Tullock 1979).

Natural binary Al-$Mn^{3+}$ solid solutions along the clinozoisite-piemontite join are rare as most samples also contain $Fe^{3+}$. Synthetic Al-$Mn^{3+}$ solid solutions suggest about 1.5 $Mn^{3+}$ pfu as the maximum $Mn^{3+}$ content in $Fe^{3+}$ free piemontite (Langer et al. 2002). In addition, they indicate an orthorhombic to monoclinic transition along the Al-$Mn^{3+}$ binary comparable to the Al-$Fe^{3+}$ binary but at considerable higher degrees of Al substitution; $Mn^{3+}$ content in zoisite is as high as 0.53 $Mn^{3+}$ pfu (Langer et al. 2002; Fig. 13). Natural ternary Al-$Fe^{3+}$-$Mn^{3+}$ epidote minerals mostly display Al contents between 1.5 and 2.5 Al pfu with a complete $Fe^{3+}$-$Mn^{3+}$ substitution (see Bonazzi and Menchetti 2004). The highest degree of Al substitution in ternary Al-$Fe^{3+}$-$Mn^{3+}$ epidote minerals (Al = 0.74 pfu, $Fe^{3+}$ = 0.81 pfu, and $Mn^{3+}$ = 1.42 pfu) is reported from the Sanbagawa belt, Japan (Enami and Banno 2001). This sample also contains considerable amounts of Sr, Pb, and Ba substituting for Ca.

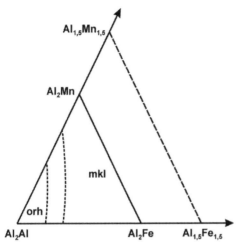

**Figure 13.** Schematic phase relations for monoclinic and orthorhombic epidote minerals in the ternary system Al-Fe-Mn; the majority of natural crystals lies within the composition $Al_2Al$-$Al_2Fe$-$Al_2Mn$ and near to the $Al_2Al$-$Al_2Fe$ and $Al_2Fe$-$Al_2Mn$ sides. The maximum amount of substitution for Al is in both cases near to 1.5, and solid solution on the join $Al_2Fe$-$Al_2Mn$ is continuous (see text). The position of the two-phase area along the $Al_2Al$-$Al_2Mn$ side was estimated from the data by Langer et al. (2002). Note that the position of the two-phase area is a projection from P and T, and varies strongly.

Epidote minerals containing $Cr^{3+}$ as major element have so far only been reported from few localities (e.g., from Burma by Bleek 1907; Finland by Eskola 1933 and Treolar 1987a, b; New Zealand and Australia by Ashley and Martin 1987; Grapes 1981; Cooper 1980; Challis et al. 1995; India by Devaraju et al. 1999; Spain by Sánchez-Viscaíno et al. 1995). The data suggest a complete solid solution series along the Al-$Cr^{3+}$ join towards tawmawite $Ca_2Al_2Cr^{3+}Si_3O_{12}(OH)$ but indicate a composition gap along the $Cr^{3+}$-$Fe^{3+}$ substitution (see Grapes and Hoskin 2004). The highest $V^{3+}$ content found so far is 11.29 wt% $V_2O_3$ (0.80 $V^{3+}$ pfu) in mukhinite from a marble associated with an iron deposit (Tashelginskoye deposit, Russia, Shepel and Karpenko 1969). The concentration of divalent cations $Fe^{2+}$, $Mn^{2+}$, and Mg on the octahedral sites is generally very low. Only in case of REE bearing epidote minerals it can reach significant values to charge balance the $REE^{3+}$ substitution on the normally divalent A sites. As the $REE^{3+}$ content in epidote minerals can be $\sum REE^{3+} > 1$ pfu, the amount of divalent cations can also be likewise high in REE bearing epidote minerals (see Gieré and Sorensen 2004). It is still an open question if there is complete miscibility along the join $Ca_{-1}M^{3+}_{-1}REE^{3+}_{+1}M^{2+}_{+1}$ or if there are compositional gaps (Gieré and Sorensen 2004).

Important homovalent substitutions on the divalent A sites include $Ca_{-1}Sr_{+1}$, $Ca_{-1}Pb_{+1}$, and $Ca_{-1}Ba_{+1}$. Sr bearing to Sr rich epidote minerals are reported from several localities (e.g., Grapes and Watanabe 1984; Brastad 1985; Mottana 1986; Akasaka et al. 1988; Harlow 1994; Perseil 1990; Enami and Banno 2001) and the highest Sr content so far reported is 16.3 wt% SrO from Itoigawa-Ohmi, Japan, which corresponds to 0.8 Sr pfu and thus almost represents niigataite endmember composition $CaSrAl_2(Al,Fe^{3+})Si_3O_{12}(OH)$ (Miyajima et al. 2003). The data suggest a complete miscibility of Ca and Sr in epidote minerals (see Grapes and Hoskin 2004). Epidote minerals that contain Pb as trace or minor constituent are widespread but Pb as major element is only reported from few localities (e.g., Penfield and Warren 1899; Dunn 1985; Neumann 1985; Holtstam and Langhof 1994; Jancev and Bermanec 1998; Enami and Banno 2001). Holtstam and Langhof (1994) described almost pure hancockite ($CaPbAl_2(Al,Fe^{3+})Si_3O_{12}(OH)$) with 32–33 wt% PbO (= 0.95 Pb pfu) from Jacobsberg, Sweden. In rocks from Nezilovo, Macedonia, Jancev and Bermanec (1998) found epidote with ~10 wt% PbO (= 0.23 Pb pfu) and hancockite with 26–27 wt% PbO (= 0.69–0.72 Pb pfu) reflecting different parageneses with different bulk chemistry. From the same locality also piemontite with Pb content up to 23 wt% PbO (~0.6 Pb pfu) is described (Bermanec et al. 1994; Jancev and Bermanec 1998). With its large cation radius Ba normally substitutes for Ca only in trace or minor amounts. The Ba content can only reach noticeable values with concomitant high degrees of ($Mn^{3+}$, $Fe^{3+}$) substitution for Al on M1 and M3, as incorporation of $Mn^{3+}$ and $Fe^{3+}$ on M1 and M3 expands the A2 site (Smyth and Bish 1988). Consequently, the highest Ba contents reported so far with 2.7–6.7 wt% BaO (= 0.1–0.25 Ba pfu) occur in Sr piemontite from the Sanbagawa belt, Japan, that also shows $Mn^{3+} + Fe^{3+} = 1.68$–2.23 pfu (Enami and Banno 2001).

***Site occupancy.*** The sites in which the different major and minor elements substitute in the monoclinic epidote structure are comparably well known. X-ray diffraction and spectroscopic data show that $Fe^{3+}$ strongly favors the largest and most distorted M3 site (e.g., Dollase 1971, 1973; Gabe et al. 1973; Kvick et al. 1988). At high total $Fe^{3+}$ content, it also substitutes on M1 (e.g., Dollase 1973), as increasing $Fe^{3+}$ on M3 enlarges the edge-sheared M1 octahedron (Bonazzi and Menchetti 1995). The ability of M1 to house $Fe^{3+}$ therefore correlates with the volume of M3 and Bonazzi and Menchetti (1995) claimed a critical volume for M3 of about 11 Å$^3$ for the incorporation of significant amounts of $Fe^{3+}$ in M1. This volume corresponds to a total $Fe^{3+}$ content of about 0.8 pfu (Bonazzi and Menchetti 1995) in good accordance with the data of Dollase (1973) that only show $Fe^{3+}$ in M1 for a total $Fe^{3+}$ content of > 0.8 pfu. Contrary, the structural data presented above suggest a substitution in M1 already at a total $Fe^{3+}$ content of about 0.6 pfu. In case of REE bearing epidote minerals in which divalent cations like $Fe^{2+}$ or Mg allow for a greater expansion of M3 (see below) significant amounts of $Fe^{3+}$ may be present in M1 at even lower total $Fe^{3+}$ content (Bonazzi and Menchetti 1995). As M2 is the smallest octahedron of the structure it is generally free of $Fe^{3+}$ (e.g., Dollase 1971, 1973; Gabe et al. 1973; Kvick et al. 1988). Only at very high degrees of substitution and/or high temperature it may contain small but considerable amounts of $Fe^{3+}$. Giuli et al. (1999) report up to 0.08 $Fe^{3+}$ pfu in M2 in synthetic Al-$Fe^{3+}$ epidote with $Fe^{3+}_{tot} = 1.08$ pfu, synthesized at 0.5 GPa/700°C, and found a positive correlation between $Fe^{3+}_{tot}$ and $Fe^{3+}$ on M2.

Structural refinements of piemontite by e.g., Dollase (1969), Catti et al. (1988, 1989), and Ferraris et al. (1989) show that $Mn^{3+}$ also has a strong preference for the M3 site and a less pronounced preference for M1 and that M2 is exclusively occupied by Al. Burns and Strens (1967) also interpreted their spectroscopic data on piemontite with a strong preference of $Mn^{3+}$ for M3 but contrary to the structural refinements postulated an only slightly less pronounced preference of $Mn^{3+}$ for M2 compared to M3 and only a very minor preference of $Mn^{3+}$ for M1. In contrast to $Fe^{3+}$ and $Mn^{3+}$, $Cr^{3+}$ has a strong preference for the M1 site and comparable but significantly minor preferences for M3 and M2 (Burns and Strens 1967). Octahedrally coordinated divalent cations as $Fe^{2+}$, $Mn^{2+}$, and Mg have a strong preference for the large M3 site although Mössbauer spectroscopy (Dollase 1971, 1973) and structural refinements (Bonazzi et al. 1992) indicate small amounts of $Fe^{2+}$ and Mg also on M1.

Stoichiometric considerations indicate that in some cases $Mn^{2+}$ must also substitute for Ca on the A sites. In these cases, it most probably substitutes in A1 (Smith and Albee 1967; Chopin 1978; Reinecke 1986; Bonazzi et al. 1992) in good accordance with zoisite that also indicate substitution of $Mn^{2+}$ for Ca in A1 (Ghose and Tsang 1971; see below). Due to their large cation size, the other divalent cations like Sr, Pb, and Ba will substitute for Ca in the larger A2 site. Like these the REE are expected to substitute in A2 although Cressey and Steel (1988) found evidence that the MREE and HREE also substitute in A1. For the smallest HREE these authors even speculated about incorporation in an octahedral site (Cressey and Steel 1988).

***Order-disorder phenomena.*** Studies on order-disorder phenomena in monoclinic epidote minerals are restricted to the Al-$Fe^{3+}$ solid solution series in which two different order-disorder processes must be distinguished, the non-convergent ordering of Al and $Fe^{3+}$ between the different crystallographic sites, which is not associated by a symmetry change, and their convergent ordering on M3 (Fehr and Heuss-Aßbichler 1997; Heuss-Aßbichler 2000) that leads to a splitting of M3 into two slightly different sites (labeled M3 and M3′ by the above authors) and a concomitant symmetry reduction.

As shown by the site occupancy data (see above) the non-convergent ordering of Al and $Fe^{3+}$ between the different octahedral sites only becomes important at total $Fe^{3+}$ contents higher than about $X_{Ep} = 0.6$ and is restricted to the M1 and M3 sites except for high temperature and high Fe content at which conditions small amounts of $Fe^{3+}$ may also enter the M2 site (Giuli et al. 1999). Based on the data available at that time, Bird and Helgeson (1980) derived a mixing model for the partitioning of Al and $Fe^{3+}$ between M1 and M3 and showed that the $Fe^{3+}$ content in M1 increases with both temperature and total Fe content (see Gottschalk 2004). Bird et al. (1988) and Patrier et al. (1991) found metastable disorder higher than that predicted from the theoretical model (Bird and Helgeson 1980) in natural hydrothermal Al-$Fe^{3+}$ solid solutions (see Bird and Spieler 2004). Contrary, Fehr and Heuss-Aßbichler (1997) by means of Mössbauer spectroscopy on natural heat-treated Al-$Fe^{3+}$ solid solutions (see Liebscher 2004) and Bonazzi and Menchetti (1995) by means of X-ray diffraction on natural Al-$Fe^{3+}$ solid solutions determined a significant lower disorder than theoretically predicted. In synthetic Al-$Fe^{3+}$ solid solutions Giuli et al. (1999) found good agreement between their data and the theoretical model for runs at 700°C but a considerable higher disorder than predicted in the runs at 600°C.

Gottschalk (2004) calculated the partition coefficient

$$K_D = \frac{^{M3}X_{Fe^{3+}} \times {}^{M1}X_{Al}}{^{M1}X_{Fe^{3+}} \times {}^{M3}X_{Al}}$$

for the reaction

$$^{M1}Fe^{3+} + {}^{M3}Al = {}^{M3}Fe^{3+} + {}^{M1}Al$$

based on the data of Dollase (1973), Bird and Helgeson (1980), Patrier et al. (1991), Fehr and Heuss-Aßbichler (1997), and Giuli et al. (1999). In a $\ln K_D$ vs. $1/T$ plot, the data are inconsistent (see Fig. 20 in Gottschalk 2004) leaving much room for speculation. Taking only the data from Dollase (1973), Fehr and Heuss-Aßbichler (1997), and the 700°C data of Giuli et al. (1999) the temperature dependence of $\ln K_D$ calculates to $\ln K_D = -8800/T + 5.4$ (T in K), which slightly differs from the data of Bird and Helgeson (1980). Assuming ideal behavior for $Fe^{3+}$ and Al on M3 and M1, respectively, and $\Delta V = 0$ and $\Delta c_p = 0$ for the above reaction, the temperature dependence of $\ln K_D$ results in $\Delta S = 45$ J/(mol K) and $\Delta H = 73$ kJ/mol for the intracrystalline $Fe^{3+}$–Al exchange. The curves for the calculated $Fe^{3+}$ content on M1 as a function of the total $Fe^{3+}$ content at different temperatures following the procedure described in Bird and Helgeson (1980) show (Fig. 14) that only for T > 600°C and/or $Fe^{3+}$ > 0.8 to 0.9 pfu M1 incorporates significant amounts of $Fe^{3+}$.

In addition to this non-convergent intracrystalline Al-$Fe^{3+}$ distribution, the data by Fehr and Heuss-Aßbichler (1997), Heuss-Aßbichler and Fehr (1997), and Heuss-Aßbichler (2000) indicate a convergent ordering process of Al and $Fe^{3+}$ exclusively on the M3 site. In a study on the intercrystalline Al-$Fe^{3+}$ distribution between epidote and grossularite-andradite Heuss-

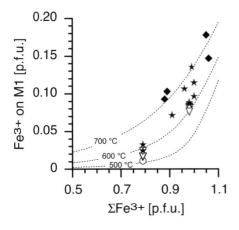

**Figure 14.** Theoretical partitioning curves of $Fe^{3+}$ on M1 in epidote in relation to total $Fe^{3+}$ content for different temperatures calculated with the temperature dependence of KD of the order-disorder reaction $Fe^{3+}(M1) + Al(M3) = Fe^{3+}(M3) + Al(M1)$ (see text); data points are from Dollase (1972); Fehr and Heuss-Aßbichler (1997); Giuli et al. (1999) for 500°C (open circles), 600°C (triangles), 650°C (asterisks), 700°C (diamonds).

Aßbichler and Fehr (1997) found two solvi along the clinozoisite–epidote solid solution series that are separated by an intermediate composition of about $X_{Ep} = 0.5$. In addition, the authors were only able to fit their measured Mössbauer spectra of Al-$Fe^{3+}$ solid solutions with $X_{Ep} = 0.39, 0.43, 0.48, 0.60,$ and $0.72$ by two doublets that both show the principal characteristics of $Fe^{3+}$ on M3 but display small differences in their quadrupol splitting (see Liebscher 2004). The authors interpreted this as due to two slightly different M3 sites, labeled M3 and M3′. These belong to two different monoclinic epidote phases, one with Al-$Fe^{3+}$ disorder on M3 (M3 doublet) and one with an ordered distribution of Al and $Fe^{3+}$ on M3 (M3′ doublet), the latter representing the intermediate composition of about $X_{Ep} = 0.5$ (Fehr and Heuss-Aßbichler 1997; Heuss-Aßbichler 2000). Depending on the geometric nature of the Al-$Fe^{3+}$ ordering in M3 the space group of the intermediate composition should reduce from $P2_1/m$ to either $Pm$ or $P\bar{1}$. In case of space group $P\bar{1}$ the lattice parameters $a$, $b$, and $c$ of the ordered phase will be twice that of the disordered phase (Heuss-Aßbichler 2000). However, up to now no structural refinement of epidote minerals showed space group $Pm$ or $P\bar{1}$.

**Zoisite (the orthorhombic epidote minerals)**

*Compositional range.* Zoisite displays a much more restricted deviation from the endmember composition $Ca_2Al_3Si_3O_{12}(OH)$ than the monoclinic epidote minerals (Franz and Selverstone 1992). Important octahedral substitutions include the incorporation of $Fe^{3+}$, $Mn^{3+}$, $Cr^{3+}$, and $V^{3+}$ for Al. The $Fe^{3+}$ content in zoisite normally does not exceed about $X_{Fe} = 0.15$. Higher $Fe^{3+}$ contents are reported, e.g., by Raith (1976) in zoisite from amphibolite to eclogite facies metabasites with $X_{Fe} = 0.17$, by Brunsmann et al. (2000) in zoisite from high-pressure segregations with $X_{Fe} = 0.18$, and by Cooper (1980) in chromian zoisite with $X_{Fe} = 0.20$. The highest $Fe^{3+}$ contents in zoisite so far reported are $X_{Fe} = 0.21$ (Brunsmann et al. 2000) and $X_{Fe} = 0.23$ (Vogel and Bahezre 1965) from high-pressure vein zoisites. However, it is not clear if these data are from pure zoisite, as the analysis from Brunsmann et al. (2000) was taken from a zoisite crystal which interference color displayed very fine lamellae suggesting fine, lamellar intergrowths of zoisite and clinozoisite, and the analysis from Vogel and Bahezre (1965) represents a wet chemical bulk analysis of a zoned zoisite crystal and might contain small undetected inclusions of other phases.

The $Mn^{3+}$ content of zoisite is normally very low and rarely exceeds 1 wt% $Mn_2O_3$ which corresponds to about 0.02 $Mn^{3+}$ pfu. The highest $Mn^{3+}$ contents in natural zoisite are reported from highly oxidized low-grade metamorphic rocks with 1.6 to 3.7 wt% $Mn_2O_3$ (0.09–0.22 $Mn^{3+}$ pfu; Reinecke 1986). Higher $Mn^{3+}$ contents are only known from synthetic zoisite samples. Anastasiou and Langer (1977) report 0.25 $Mn^{3+}$ pfu and Langer et al. (2002) found 0.50–0.53 $Mn^{3+}$ pfu in zoisite synthesized at 800°C/1.5 GPa and $f_{O_2}$ of the $Mn_2O_3/MnO_2$ oxygen buffer. As the latter coexists with piemontite it should represent the maximum $Mn^{3+}$ content in zoisite at these experimental conditions.

Like in the monoclinic epidote minerals the $Cr^{3+}$ content in zoisite rarely exceeds the trace element level. Game (1954) reports 0.33 wt% $Cr_2O_3$ in zoisite from Tanzania and Grapes (1981) 0.23 to 0.40 wt% $Cr_2O_3$ (0.014 to 0.024 $Cr^{3+}$ pfu) in zoisite from New Zealand. Own, unpublished data show up to 1.66 wt% $Cr_2O_3$ (0.10 $Cr^{3+}$ pfu) in zoisite from Tanganyika. The highest $Cr^{3+}$ content found in zoisite so far is 2.46 wt% $Cr_2O_3$ (0.15 $Cr^{3+}$ pfu) in a chromian zoisite in a fuchsite-margarite pseudomorph after kyanite from New Zealand (Cooper 1980).

Geochemical important homovalent substitutions for Ca in the A sites of zoisite include $Sr^{2+}$ and $Pb^{2+}$. As both A sites in zoisite have a size comparable to the monoclinic A1 site in which no $Sr^{2+}$ or $Pb^{2+}$ substitution occurs, zoisite normally contains $Sr^{2+}$ and $Pb^{2+}$ only at the trace or minor element level. In case of $Sr^{2+}$ this holds especially for normal rock compositions metamorphosed under *P–T* conditions where other potential $Sr^{2+}$ carriers such as calcic plagioclase are stable. However, in very $Sr^{2+}$ rich rock compositions and/or outside the stability fields of other potential $Sr^{2+}$ carriers, the $Sr^{2+}$ content in zoisite may reach several wt%. Brastad (1985) reports up to 7.4 wt% SrO (0.33 $Sr^{2+}$ pfu) in zoisite from metasomatized eclogite from Norway, which is the highest so far reported. Zoisite in eclogite from Su-Lu, China has up to 3.2 wt% SrO (Nagasaki and Enami 1998), zoisite in albitites from Guatemala and in zoisite-pegmatites from Spain has 1.2 to 1.4 wt% SrO (Maaskant 1985; Harlow 1994). Contrary to the monoclinic epidote minerals in which it can reach several weight percent the PbO content of zoisite, at least so far reported, is generally in the range of only a few hundred ppm (Frei et al. 2004).

***Site occupancy.*** Structural, electron paramagnetic resonance, and spectroscopic data show that $Fe^{3+}$, $Cr^{3+}$, and $Mn^{3+}$ substitute for Al exclusively in the M3 site (e.g., Hutton 1951; Ghose and Tsang 1971; Tsang and Ghose 1971; Schmetzer and Berdesinski 1978; Grapes 1981; Langer et al. 2002). Optical spectroscopy suggests that $V^{3+}$ may enter the M1,2 as well as the M3 site although it shows a preference for M3 (Tsang and Ghose 1971). As the A2 site has a slightly larger mean A–O distance and is more distorted than the A1 site (see above), any substitution of cations larger than Ca such as $Sr^{2+}$ and $Pb^{2+}$ will occur at the A2 site. In contrast, as indicated by electron paramagnetic resonance data, $Mn^{2+}$ is most likely located in A1 whereas $V^{2+}$ occupies both A sites with a preference for A1 (Ghose and Tsang 1971).

## Microanalysis

Minerals are commonly analyzed with the electron microprobe, and this is also recommended for epidote minerals. It is the only routine method, which allows resolving chemical zoning, but one should be aware that, for example, oscillatory zoning might be on a scale below the resolution of the electron microprobe. In addition to the microprobe, optical examination may give qualitative information about composition, because interference colors are very sensitive to small changes in Al-Fe substitution, especially well visible at low birefringence. It is not recommended, however, to use optical data to derive quantitative information about composition, because of the many possibilities of simple and coupled substitutions. Epidote minerals are stable under the electron beam at normal conditions of 10 to 20 kV, 15 to 20 nA, even with a fine focused beam.

We recommend to analyze epidote minerals with the following strategy: Distinguish optically between orthorhombic and monoclinic by a combination of observation of extinction angle and low birefringence; be aware that with small crystals this might be very difficult, and in many cases actually impossible, if the crystals are not oriented in the correct position. Oblique extinction is only visible in (010), whereas {010] always shows straight extinction.

Check your determination with the electron microprobe: If the $Fe_2O_3$ content is below $X_{Fe}$ 0.15 - 20, the mineral is probably zoisite. If your optical determination was "monoclinic" with $Fe_2O_3$ content significantly below $X_{Fe}$ 0.15, then you have one of the rare cases of very Fe-poor clinozoisite. Check with X-ray methods or transmission electron microscopy! If you have no exact optical and/or chemical characterization and in the absence of X-ray determination, characterize the minerals as "epidote minerals".

Analyze first for the common components $SiO_2$, $Al_2O_3$, $Fe_2O_3$, $Mn_2O_3$, $Cr_2O_3$, $TiO_2$, CaO, and MgO and calculate the formula on a basis of 12.5 or 25 oxygen pfu for common epidote minerals. Alternatively you can calculate on a basis of 8 cations pfu or on 3 Si pfu. All three methods yield very similar results. In a first step calculate all Fe and Mn as trivalent. Check the formula for internal consistency:

- The total should be 98 ± 1 wt% except for metamict allanite. If the sum is lower, there are possibly other less common elements present. It is best to use EDS parallel to WDS;

- If Ca is < 2 pfu check for Sr, REE, Pb; there might also be $Mn^{2+}$ instead of $Mn^{3+}$;

- If alkalis are present, the mineral might be strongly altered (except in some very rare cases of allanite);

- If there are REE present, check for Mg and recalculate the formula with charge balance for $Fe^{2+}$ by simultaneous normalization to 25 oxygen and 8 cations pfu; if in addition Mn is present, first assume all Mn as $Mn^{3+}$, then $Mn^{2+}$;

- If $\sum$A1,2 is still < 2.0, check calculations, if $Mn^{2+}$ is possible;

- If $\sum$M1,2,3 (including Ti) is < 3.0, check for V;

- If Si is below or above 3.0, the analysis is probably not reliable, except for metamict allanite;

- Anions: it is normally not necessary to analyze for Cl and F (for common epidote minerals below detection limit), only in the case of REE bearing minerals they must be checked;

- Selection of points to analyze: Optical zoning gives an important hint to chemical zoning. Also, the difference in Al-Fe as well as Al-Mn atomic weight is large enough to be seen quite well in back scattered electron images. However, Fe-Mn zoning is less visible. Be aware of the grain shape of the epidote minerals, with the highest growth rate in *b* (elongation of prismatic crystals), and be aware of possible sector and oscillatory zoning;

- Coexisting minerals: Grossularite-andradite, prehnite, pumpellyite and sodic amphibole also show Al-$Fe^{3+}$ substitution and the distribution of these elements can be used to check for equilibrium. In the case of Sr- rich epidote minerals check for distribution in coexisting pyroxene, plagioclase, carbonate, and apatite; for Cr-rich epidotes, check chlorite, mica, amphibole, and spinel.

## CHEMICAL ZONING

Like other neso- and soro-silicates with low diffusion rates chemical zoning is a common feature in minerals of the epidote group (see e.g., Grapes and Hoskins 2004; Enami et al. 2004). In thin section the zoning is sometimes obvious from the color intensity, but very typically shows up in the interference colors, because birefringence is a strong function of the chemical composition, especially the Fe-Al content. Small changes in Fe-Al, which are at the limit of the resolution of the electron microprobe, can be seen very well in crossed polarized light. Zoning is mostly growth zoning, less modified by diffusion, because the typical epidote occurrences are from rocks, which have not experienced temperatures above 700°C. Zoning is often modified by retrograde dissolution-precipitation. It reflects the changing *P-T* parameters (Grapes and Hoskins 2004), and is strongly influenced by the miscibility gaps in the solid solution series and by the zoisite-clinozoisite phase transition (Franz and Selverstone 1992). Very importantly it depends also on $f_{O_2}$, which can be controlled internally by the solid mineral phases with or without a fluid, or externally via the fluid phase. Zoning can also be controlled by the local presence of immobile cations such as $Cr^{3+}$. Sánchez-Viscaíno et al. (1995) described such a case, where the presence of relic chromite grains strongly influenced the local composition of epidote, which results in a very complex irregular zoning pattern. Cr content varies from below 4 wt% to > 10 wt% over a distance of 50 μm and the whole zoning pattern mimics the locally inherited composition. Similar irregular zoning patterns are frequently observed in low-temperature metamorphic rocks, where small individual grains of epidote minerals assembled into one single crystal. Gieré and Sorensen (2004) show an example of allanite-epidote core-rim zoning, where the outlines of the inner core zone shows such an irregular pattern.

Core-rim zoning is, however, the general case, and Grapes and Hoskins (2004) give some rules. In many cases, Fe rich epidote grows first at subgreenschist facies conditions and the Al content increases up to amphibolite facies conditions, but note that there are also many exceptions. In zoisite, cores are often rich in Fe, and the individual zones mimic the euhedral shape (e.g., Franz and Smelik 1995) such that in sections perpendicular to (010) it is of optically diagnostic value (Pichler and Schmitt-Riegraf 1993). Replacement of the Fe rich zones by Al rich zones parallel to (100) results in a lamellar appearance in sections parallel to the crystallographic *b* axis. Also the fan shaped growth of zoisite in segregations (Brunsmann et al. 2000) with several growth stages produces a lamellar appearance in sections parallel to (100) with alternating Fe rich and Fe poor lamellae.

Sector zoning was described in zoisite (Enami 1977, see also Enami et al. 2004 and back cover) in three sectors {100}, {010} and {001}, in which $X_{Fe}$ is highest in {100} ($\approx 0.09$) and lower in {010} and {001} (both $\approx 0.06$). Because [100] is the slowest growth direction is seems that Al is preferentially incorporated compared to Fe. In epidote, (Yoshizawa 1984; Banno and Yoshizawa 1993) the sectors are {100}, {110}, {001} and {101}, and the order of $X_{Ep}$ is {100} $\geq$ {110} $\geq$ {001} > {101}. Sector zoning was also observed in Cr-rich epidote from quartzitic schist (Treloar 1987a,b), in epidote from high-pressure segregations (own unpublished results), and in hydrothermal epidote from geothermal fields (Bird and Spieler 2004), and might in fact be a rather common phenomenon (Banno and Yoshizawa 1993).

Oscillatory zoning is rarely observed in common metamorphic epidote; an exception is oscillatory zoning of Cr-epidote superimposed on sector zoning (Treloar 1987a). In hydrothermal epidote it is not rare. Bird and Spieler (2004) mention several references, and we analyzed for this review a Mn-bearing clinozoisite from a hydrothermal quartz vein (Baja California; sample # 81/370, Fig. 15a-d; see also back cover). The overview (Fig. 15a) shows the outer part of radially grown clinozoisite ($X_{Ep} = 0.44$ to $\approx 0.50$ from core to rim) with an inner overgrowth of epidote ($X_{Ep} \approx 1.0$). This is separated by a discontinuity, which was created by dissolution, from an outer overgrowth of epidote with $X_{Ep} \approx 0.55$ to $0.85$. The oscillatory zoning is present in the Fe-rich outer part (Fig. 15b), where it is dominantly Al-Fe-zoning, but also in the clinozoisite, where it shows up already in plain polarized light as a herring-bone structure (Fig. 15c). A profile of 15 zones over a distance of 25 μm shows also Mn-Fe zoning (Fig. 15d), which is, however, not completely complimentary to each other. The size of the small zones of $\approx 1$–2 μm is below the resolution of the electron microprobe, but the profile shows that in general the $Mn_2O_3$ content varies between 0.2 and 0.3 wt%, whereas the

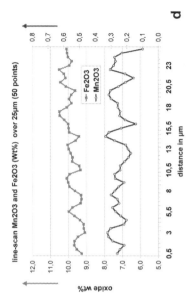

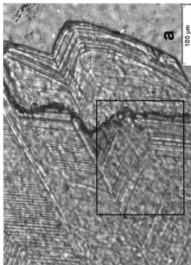

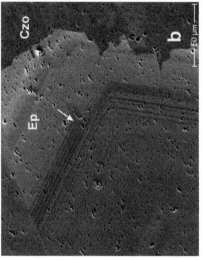

**Figure 15.** Oscillatory zoning in a Mn-bearing clinozoisite-epidote, from a hydrothermal quartz vein (sample 81-270, Baja California). (a) Outer part of a radially grown aggregate in quartz (qz) with a dissolution discontinuity (dark rim), outlined area is Figure (b); photomicrograph, plain polarized light. (b) Secondary electron image, showing the discontinuity, which separates epidote with $X_{Ep} \approx 1.0$ (from center to the left) from epidote with $X_{Ep} = 0.55$ to 0.85 (right part) and the oscillatory zoning in Fe-Al. Arrow points to an area where the width of five dark zones decreases from $\approx 12$ µm to $\approx 6$ µm. (c) Detail of an area in the crystal with $X_{Ep} \approx 0.50$ with oscillatory zoning. Line indicates line scan in Figure (d); arrows point to locations where the oscillatory zones show kinks. (d) Line scans for $Fe_2O_3$ and $Mn_2O_3$ (wt%).

differences in $Fe_2O_3$ content from zone to zone are $\approx 0.5$ wt% between 9 and 10.5 wt%. This sample also shows interesting growth features, such as a decrease in width of five zones (arrow in Fig. 15b) from 12 μm to 6 μm, and slight kinks in the zones (arrows in Fig. 15c). There are also parts of the crystal, which are free of oscillatory zones (upper left of Fig. 15c). These features indicate that the local environment significantly influences the growth mechanisms. It is beyond the purpose of the introduction to this volume to derive a conclusion about the growth conditions from this example, but we want to emphasize that there is a potential wealth of information hidden in such phenomena.

## OPTICAL PROPERTIES

### Monoclinic epidote minerals

Earliest studies on the optical properties of monoclinic epidote minerals are from Brewster (1819), Klein (1874), Forbes (1896), Weinschenk (1896), Zambonini (1903, 1920, quoted from Strens 1966), Andersen (1911), Becke (1914), Goldschlag (1917), and Orlov (1926). Compilations of the optical data are from Johnston (1948), Winchell and Winchell (1962), Tröger (1982), Deer et al. (1986), Strens (1966), and Kepezhinskas (1969). Thorough and systematic studies on the relationships between optical properties and composition of the Al-$Fe^{3+}$ solid solution series came from Myer (1966) and Hörmann and Raith (1971). The latter also compiled and re-evaluated the previously published optical data. Systematic studies on optical properties of Al-$Mn^{3+}$ solid solutions are restricted to a study on synthetic piemontite from Langer et al. (2002). Optical data for Al-$Fe^{3+}$-$Mn^{3+}$ solid solutions were published by Malmqvist (1929), Short (1933), Guild (1935), Otto (1935), Tsuboi (1936), Hutton (1938, 1940), Hirowatari (1956), Marmo et al. (1959), Nayak and Neuvonen (1963) and Ernst (1964), which were compiled, extended and systematically evaluated by Strens (1966). Below we give a brief review of the optical properties of the different monoclinic solid solutions. Based on the published data we extended and modified the well-known figures from Tröger (1982) in Figure 15.

*Al-$Fe^{3+}$ solid solutions.* All members of the Al-$Fe^{3+}$ solid solution series have $Y$ parallel to [010], $X$ approximately parallel to [001] and $Z$ slightly inclined to [100] (Fig. 16). With increasing $Fe^{3+}$ the indicatrix rotates around [010] and the extinction angle between $X$ and [001] decreases from $\sim +12°$ to $-5°$ and the angle between $Z$ and [100] increases from $\sim +13°$ to $+30°$ (Fig. 16). $Fe^{3+}$ poor solid solutions are almost colorless in thin section whereas with increasing $Fe^{3+}$ they become colored with the typical pistacite like yellowish to greenish colors. The pleochroism is weak to distinct and its scheme is $X$ = pale yellow, $Y$ = greenish yellow, and $Z$ = yellowish green (Burns and Strens 1967; Fig. 16). All three refractive indices increase with increasing $Fe^{3+}$. This increase is strongest in $n\gamma$ and $n\beta$ and smallest in $n\alpha$. Because of this difference, birefringence ($n\gamma$–$n\alpha$) significantly increases by one order of magnitude from about 0.005 in $Fe^{3+}$ poorest clinozoisite to about 0.05 in the most $Fe^{3+}$ rich samples (Fig. 17). $2V_z$ increases from about 40 to 60° in $Fe^{3+}$ poor clinozoisite to about 120° in $Fe^{3+}$ rich clinozoisite leading to a change in the optic sign at $X_{Ep}$ ~0.35 (Figs. 16, 17). All three refractive indices as well as the birefringence display changes in their correlation trends with $Fe^{3+}$ content at about $X_{Ep} = 0.6$ thus precluding linear correlation trends as proposed by Strens (1966) and Kepezhinskas (1969). These changes are in good accordance with the results from structural refinements (see above) and reflect the change in the Fe-Al substitution from exclusively in M3 to a combined one in M1 and M3. The dispersion of refractive indices and birefringence ($n\gamma$–$n\alpha$), ($n\gamma$–$n\beta$), and ($n\beta$–$n\alpha$) increases linearly with $Fe^{3+}$ content and is generally $v > r$. This leads to positive Ehringhaus numbers of $\leq +5$ and anomalous interference colors (Hörmann and Raith 1971). Contrary, Goldschlag (1917) determined the dispersion of ($n\gamma$–$n\alpha$) for two $Fe^{3+}$ rich epidotes with $X_{Ep} \geq 0.9$ as $v < r$. The dispersion of $2V_z$ is $v < r$ for $X_{Ep} < \sim0.35$ and $v > r$ for $X_{Ep} > \sim0.35$ (Hörmann and Raith 1971;

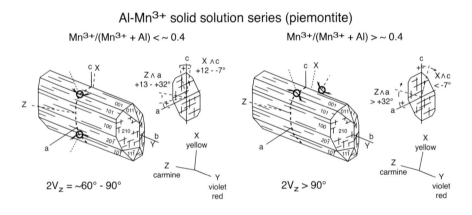

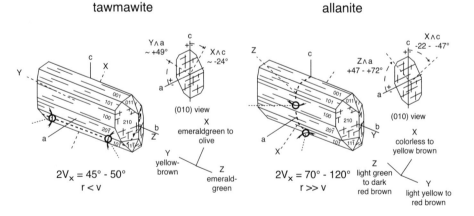

**Figure 16.** Optical orientation of monoclinic clinozoisite-epidote (upper part), epidote-piemontite (center), tawmawite and allanite (bottom); modified after Tröger 1982.

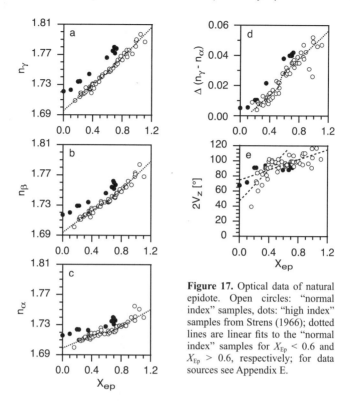

**Figure 17.** Optical data of natural epidote. Open circles: "normal index" samples, dots: "high index" samples from Strens (1966); dotted lines are linear fits to the "normal index" samples for $X_{Ep} < 0.6$ and $X_{Ep} > 0.6$, respectively; for data sources see Appendix E.

Fig. 16) and that of the extinction angle between $nx$ and [001] is generally $r > v$ (Klein 1874; Goldschlag 1917; Johnston 1948; Hörmann and Raith 1971).

***Al-Mn³⁺ solid solutions.*** Data on the optical properties of binary Al-$Mn^{3+}$ solid solutions are restricted to the studies on synthetic samples (Anastasiou and Langer 1977; Langer et al. 2002). Langer et al. (2002) stated that the optical data from Anastasiou and Langer (1977) are questionable and especially those for $n\beta$ are incorrect. We will therefore refer only to the data of Langer et al. (2002) who also provide electron microprobe analyses. As in the Al-$Fe^{3+}$ solid solution series $n\beta$ is generally parallel [010], $n\alpha$ approximately parallel to [001], and $n\gamma$ is slightly inclined to [100] (Fig. 16). With increasing $Mn^{3+}$ content the indicatrix rotates around [010] and the extinction angle between $X$ and [001] decreases from ~ +12° to < −7°and the angle between $Z$ and [100] increases from ~ +13° to > +32° (Fig. 16). Piemontites are generally strongly colored and show an impressive pleochroism in thin section with $X$ = yellow, $Y$ = violet red, and $Z$ = carmine (Fig. 16). All three refractive indices increase with increasing $Mn^{3+}$ content and show values comparable to those of Al-$Fe^{3+}$ solid solutions with comparable degrees of Al substitution (Fig. 18). As in Al-$Fe^{3+}$ solid solutions this increase is strongest in $n\gamma$ and $n\beta$ and smallest in $n\alpha$, and the birefringence significantly increases up to about 0.10 for $Mn^{3+}$ contents of 1.5 $Mn^{3+}$ pfu. $2V_z$ is < 90° for $Mn^{3+}/(Mn^{3+}+Al)$ < ~0.4 and > 90° for $Mn^{3+}/(Mn^{3+}+Al)$ > ~0.4 leading to optical positive and negative signs, respectively (Figs. 16, 18). The change in substitution mechanism, which is evident from the X-ray data (see above), is not apparent in the optical properties, because of the small number of data; for more details about piemontite see Bonazzi and Menchetti (2004).

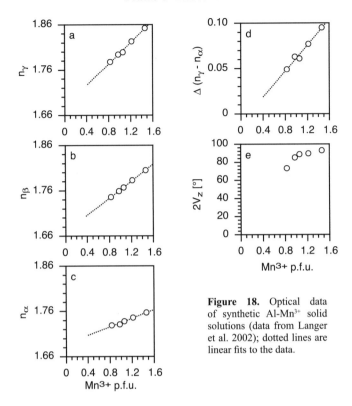

**Figure 18.** Optical data of synthetic Al-Mn³⁺ solid solutions (data from Langer et al. 2002); dotted lines are linear fits to the data.

***Al-Fe³⁺-Mn³⁺ solid solutions.*** Members of the ternary Al-Fe³⁺-Mn³⁺ solid solution series display optical properties that are mixtures of those of the respective binary solid solutions (Strens 1966) and are thus not suitable to derive optical properties of the binary solid solutions. Tröger (1982) for example published optical properties of piemontite based on the data from Marmo et al. (1959) for Al-Fe³⁺-Mn³⁺ solid solutions. These data indicate an optical negative sign for Mn³⁺-poor piemontite, which is obviously wrong (see Figs. 18, 19). Likewise, his extrapolation of the refractive indices to Mn³⁺-free composition yielded values, which are significantly higher than those derived for pure Al clinozoisite by extrapolation of the data for Al-Fe³⁺ solid solutions (Tröger 1982; see above). Published optical signs for Al-Fe³⁺-Mn³⁺ solid solutions (Fig. 19) as function of composition in a ternary plot define a field of optical positive sign on the Al-rich side of the ternary system, and this field extends to much higher degree of substitution for Mn³⁺-bearing solid solutions than for Fe³⁺-bearing solid solutions. Due to the lack of data for high Fe³⁺ + Mn³⁺ solid solutions it is not clear if the stippled lines in Figure 18 meet or if the field of optical positive sign extends to the Fe³⁺–Mn³⁺ binary. Like the other optical parameters both color and pleochroism of Al-Fe³⁺-Mn³⁺ solid solutions are mixtures of those of the two binaries. Their general pleochroic scheme in thin section can be described as $X$ = lemon or orange yellow, $Y$ = amethyst, violet, or pink, and $Z$ = bright red (Burns and Strens 1967; see also Bonazzi and Menchetti 2004).

***Cr³⁺-bearing solid solutions.*** Data on optical properties of Cr³⁺-bearing solid solutions (tawmawite) are sparse due to the rare occurrence of this mineral. According to Tröger (1982) refractive indices and birefringence increase with increasing Cr³⁺. They vary between $n\alpha$ = 1.695 and $n\gamma$ = 1.715 and birefringence is generally below 0.008, considerable smaller than in

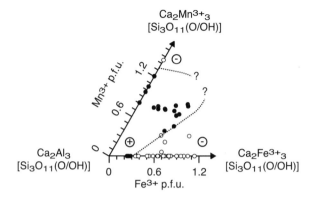

**Figure 19.** Optical character of monoclinic clinozoisite-epidote-piemontite solid solutions. Dots: optic positive, open circles: optic negative; dotted lines are only drawn somewhat arbitrarily for visualization; for data sources see Appendix E.

Al-$Fe^{3+}$-$Mn^{3+}$ solid solutions. The most significant difference is the position of the optic axial plane, which is not orientated perpendicular but parallel to [010] with $Z$ in [010] (Fig. 16). Assuming complete solid solution and thus continuous change of the optical properties along the Al–$Cr^{3+}$ binary, the position of the optic axial plane parallel to [010] with $Z$ in [010] implies that the [010] direction exhibits the strongest increase of its refractive indices with increasing $Cr^{3+}$. The orientation of the indicatrix, the optic axial angle, and the optic character will thus change in following manner along the Al–$Cr^{3+}$ binary (Fig. 16, upper left and lower left): At low $Cr^{3+}$ contents Al–$Cr^{3+}$ solid solutions have $2V_z < 90°$ and are optically positive with $Y$ in [010] comparable to $Fe^{3+}$ poor clinozoisite. With increasing $Cr^{3+}$ $n\beta$ in [010] significantly increases leading to $2V_z > 90°$ and an optical negative sign comparable to $Fe^{3+}$-rich clinozoisite (Fig. 20). Further increase of $Cr^{3+}$ increases $2V_z$ up to 180° at which composition $n\beta = n\gamma$ and

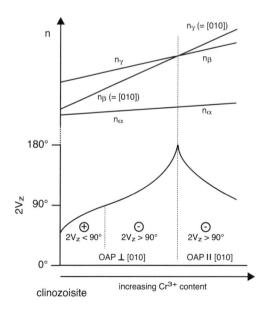

**Figure 20.** Change of n (upper part) and $2V_z$ (lower part) in clinozoisite-tawmawite solid solutions and the change in orientation of the optical axial plane; the diagram is not to scale due to the small number of data.

the Al–$Cr^{3+}$ solid solution is uniaxial negative. At still higher $Cr^{3+}$ contents $2V_z$ will decrease and the Al–$Cr^{3+}$ solid solution will be again biaxial negative with the optic axial plane parallel to [010] and $Z$ in [010]. Al–$Cr^{3+}$ solid solutions are generally pleochroic in thin section although the pleochroic schemes given in the literature significantly differ. Tröger (1982) describes two pleochroic schemes with $X = Z$ = emerald green and $Y$ = yellow and $X$ = olive, $Y$ = brown, and $Z$ = emerald green whereas Burns and Strens (1967) described $X = Y$ = emerald green and $Z$ = yellow. The reason for this discrepancy is probably a different composition of the samples and the corresponding change of $n\beta$ and $n\gamma$. Because Burns and Strens (1967) also give the exact crystallographic directions for the refractive indices, their data are adopted in Figure 15.

*Allanite and other REE-bearing epidote minerals.* There are many elements, which accompany REE in a coupled substitution for Ca and Al-$Fe^{3+}$ in epidote leading to a large chemical variety, different end members (Table 2) and very variable optical parameters. Among these elements are the radioactive elements Th and U, which lead to metamictization, and in addition to the variability caused by substitutions, different degrees of metamictization change the optical parameters. In completely amorphous metamict crystals refractive indices are minimal and double refraction is zero (see Gieré and Sorensen 2004).

## Zoisite (the orthorhombic epidote minerals)

Tschermak and Sipöcz (1882) were the first who systematically studied the optical properties of zoisite. Their work was extended by Weinschenk (1896) who studied two zoisite samples with different Fe content. He already concluded that the refractive indices as well as the birefringence of zoisite increase with increasing Fe content and noticed two different optical orientations in zoisite, one with the optic axial plane parallel (010) and dispersion $r > v$ and the other with the optic axial plane parallel (100) and dispersion $r < v$ (see below). These two different orientations were later studied by Termier (1898, 1900) who named β-zoisite the orientation with optic axial plane parallel (010), $Y$ (= $n\beta$) parallel [010] (this orientation was later termed "pseudozoisite" by Tröger 1982) according to the refractive index that lies parallel to the crystallographic b-axis (Fig. 20). The orientation with optic axial plane parallel (100), $X$ (= $n\alpha$) parallel [010] was named α-zoisite, and later termed zoisite by Tröger 1982. Orlov (1926) compiled the early optical data for zoisite and compared them with those of the monoclinic epidote minerals. Myer (1966), who studied the variation of the optical properties with composition on five natural zoisite samples spanning the compositional range $X_{Fe} = 0.015$ to 0.14, found that the optical properties change continuously from the β-orientation to the α-orientation with increasing $Fe^{3+}$-content and concluded that the two optical orientations of zoisite do not represent two polymorphs. To avoid confusion with the α–β-terminology normally used for polymorphs, he emphasized to use the terms zoisite for the β-orientation and ferrian zoisite for the α-orientation (Myer 1966; Fig. 21). A detailed study on the relationship between the optic axial angle and $Fe^{3+}$-content was performed by Maaskant (1985) on natural zoisite samples spanning the compositional range $X_{Fe} = 0.015$ to 0.14, however without publishing the refractive indices.

The published data indicate that all three refractive indices increase with increasing $Fe^{3+}$ whereas the birefringence slightly decreases (Fig. 22). The optic axial angle clearly shows the change in the optical orientation. With increasing $Fe^{3+}$ the optic axial angle decreases from about $2V_z = 55°$ for Fe-free zoisite to $2V_z = 0°$ for about $X_{Fe} = 0.09$. Further $Fe^{3+}$ substitution again increases the optic axial angle. At about $X_{Fe} = 0.165$ zoisite is optically neutral with an optic axial angle of 90°. At still higher $Fe^{3+}$-content zoisite becomes optically negative. Zoisite is normally colorless in thin section, has a strong dispersion, and displays pronounced anomalous interference colors.

Due to the low degrees of substitution by other elements the optical differences of the other orthorhombic varieties are only minor. Most noticeable is a change in color and pleochroism that accompanies the incorporation of small amounts of $Mn^{3+}$, $Cr^{3+}$, and V. $Mn^{3+}$ bearing zoisite

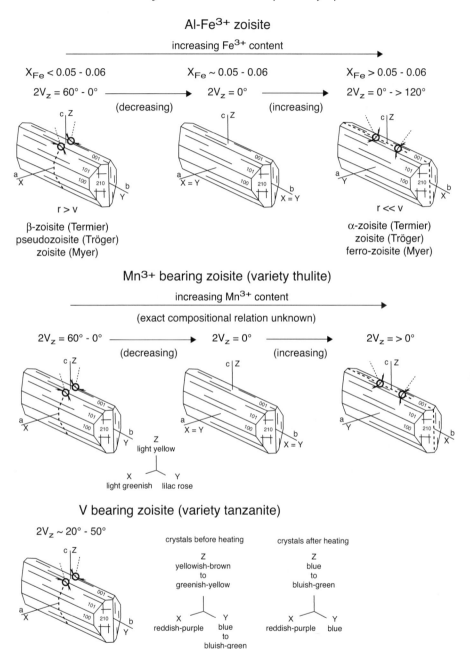

**Figure 21.** Optical orientation of zoisite and its $Mn^{3+}$ and V bearing varieties thulite and tanzanite. The terminology for Al-$Fe^{3+}$ solid solutions with the optic axial plane parallel (010) and (100), respectively, is from Termier (1898, 1900), Tröger (1982), and Myer (1966). The irreversible color change in V bearing zoisite occurs during heating above about 450°C (see text). The greenish colors in V bearing zoisite are probably due to small amounts of $Cr^{3+}$; modified after Tröger 1982.

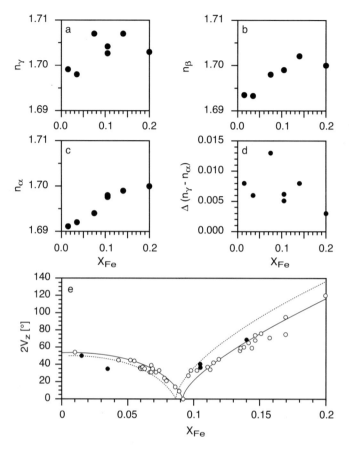

**Figure 22.** Optical data for zoisite; dots: data from Myer (1966), Seki (1959), and Kiseleva et al. (1974), open circles: data from Maaskant (1985) for 589 nm; solid and dotted lines in (e) are fits to the data from Maaskant (1985) for 589 nm and 470 nm (data not shown by symbols) respectively; see Appendix F.

("thulite") is violet to pink. It displays the two identical optic orientations as normal zoisite although the compositional dependence of these orientations is unknown. Its pleochroic scheme is $X$ = pale yellowish green, $Y$ = violet rose, and $Z$ = yellowish. $Cr^{3+}$-bearing zoisite has the typical emerald green colors (Weinschenk 1896; Game 1954). Its refractive indices do not differ from those of $Cr^{3+}$-free zoisite of comparable $Fe^{3+}$-content (Fig. 22). Small amounts of vanadium ("tanzanite") lead to blue colors (Bank et al. 1967; Fay and Nickel 1971; Tsang and Ghose 1971; Schmetzer and Bank 1979; see below under Gem Materials). Substitution of Sr for Ca in the A sites results in a lower optic axial angle (Maaskant 1985).

**Gem materials**

Though epidote frequently occurs in gem quality crystals, cut stones are generally unimportant on the market. This is due to the relatively low Moh's hardness of 6.5 and the fact that beautiful color varieties with the exceptions listed below are rare (Bauer 1968). The strong pleochroism of Fe rich crystals produces a rather unattractive olive-brownish to yellowish color (Fischer 1977). Crystals from the locality Knappenwand in Tyrol, Austria have been used for faceted stones, and (Fischer 1977) gives the angles of the appropriate orientation of the facets to obtain the best color. In the process of gem testing it was found that epidote is practically opaque to X-rays (Webster 1962).

The only variety, which is commercially important, is "tanzanite" (Anderson 1968, Hurlbut 1969), found in the Merelani area in NE Tanzania (Malisa 1987). Sapphire blue zoisite occurs in hydrothermally formed calc-silicate rocks, associated with graphitic gneiss with kyanite/sillimanite-quartz-anorthite, which were intruded by Pan-African granite pegmatites and quartz veins at ≈600 Ma. Conditions of metamorphism were in the upper amphibolite facies of ≈650°C and 0.8 GPa. The gneisses are rich in graphite and the trace elements V, U, Mo and W, and most silicates are enriched in $V_2O_3$ (Malisa 1987). Two color varieties exist, a brown and a blue one. Brown crystals can be converted into blue ones by heat treatment at 400 to 650°C, and most stones available on the market are heat-treated. The color change is caused by the disappearance of an absorption band in the blue part of the visible spectrum (Schmetzer and Bank 1979). Webster and Anderson (1983) note that heat treatment above 380°C must be carried out with care because the stones tend to disintegrate. Brown stones may also convert into colorless stones after heat treatment (Bank 1969). Tanzanite crystals also show bluish-green, yellow, pink and khaki colors. The pleochroism of two varieties is given in Table 4 (Schmetzer and Bank 1979). The interpretation of the cause of color was controversial; Ghose and Tsang (1971) interpreted it as $V^{2+}$, Schmetzer and Bank (1979) as $V^{3+}$ and in the bluish-green variety as a combination of $V^{3+} + Cr^{3+}$, and Faye and Nickel (1971) also as $V^{3+}$ in combination with $Ti^{4+}$. Zoisite crystals from this locality have been used frequently as starting material for experiments (e.g., Matthews 1985; Matthews and Goldsmith 1984; Perkins et al. 1980; Smelik et al. 2001) because they are free of inclusions and essentially pure zoisite.

Pink zoisite varieties from several localities worldwide ("thulite") are used as cabochons. Only rarely such crystals can be used as cut stones. Bank (1980a,b) described clear transparent zoisite and also clinozoisite from Norway, but this seems to be the exception.

Green zoisite from Longido Mts. (Kenya) is also important on the mineral market. The Cr-bearing zoisite forms the matrix for red corundum (ruby) crystals, and together with black amphibole, these rocks exhibit an impressive color contrast. Zoisite has $X_{Fe}$ = 0.04 to 0.12 and up to 0.10 Cr pfu. Clinozoisite is very rarely present, has $X_{Ep}$ = 0.48 to 0.55, and contains only up to 0.03 Cr pfu. The black amphibole is pargasite with 0.58 to 2.24 wt% $Cr_2O_3$. Spinel is a $Mg\text{-}Fe^{2+}\text{-}Cr^{3+}\text{-}Al$ solid solution with generally 0.08 to 0.44 Cr pfu, rarely up to 0.9 Cr pfu. Corundum (ruby) has also up to 0.035 Cr pfu. Anorthite, which occurs in thin layers as well as in reaction textures with amphibole, is almost pure (An 98-100), except in some cores (An 85-90) and in a late generation (~An 62). The textures indicate a (? retrograde) decomposition of corundum together with amphibole and the formation of zoisite and spinel (own unpublished results). For more information concerning the trace element content and distribution see Frei et al. (2004).

**Table 4.** Pleochroism of untreated and
heat treated "tanzanite" (Schmetzer and Bank 1979)

| color variety | X//a | Y//b | Z//c |
|---|---|---|---|
| brown (untreated) | reddish purple | blue | yellowish brown |
| blue (heat treated) | reddish purple | blue | blue |
| bluish green (untreated) | reddish purple | bluish green | greenish yellow |
| bluish green (heat treated) | reddish purple | blue | bluish yellow |

The mottled green-pink massive variety "unakite" of epidosite (from Unaka/North Carolina) is used as cabochon (Webster and Anderson 1983), and similar material is reported from Zimbabwe and Ireland.

## OCCURRENCE AND PHASE RELATIONS

Epidote minerals are stable over an extremely wide range of pressure and temperature. They are considered as typical metamorphic minerals, but are also known from igneous rocks, are very important as hydrothermal minerals and occur as detrital heavy minerals in sediments. It is not only the large range of pressure and temperature, but also the large range in composition of epidotes, which makes them important constituents in many rocks. The chemical variation concerns both the major components of crustal rocks $Al_2O_3$ and $Fe_2O_3$ and the minor and trace components $Mn_2O_3$, $Cr_2O_3$, SrO, PbO and REE. Therefore epidotes occur in rocks with a common composition as well as in rare rocks, which are enriched in the trace components such as ore deposits. However, a certain rock must not necessarily be enriched in the trace elements to stabilize epidote, it is often the preference of the epidote structure for these minor and trace elements which makes them their carrier in the accessory minerals.

Besides being major rock forming minerals epidotes are also very abundant in veins, segregations and cavities, which developed during various stages and in various shapes and dimension in metamorphic, igneous and hydrothermal rocks. This points to the fact that metasomatism and fluid-rock interaction can play an important role in the formation of epidote. Epidosite (Johannson 1937) is a rock, which consists mainly of epidote and quartz and its genesis is commonly explained as due to mass transport via a fluid phase. Epidotization—the formation of epidote from mainly feldspar—is observed in many igneous rocks during a low-temperature stage of hydrothermal overprint. They are very common in geothermal fields and in altered oceanic crust (Bird and Spieler 2004). The term "helsinkite" has also been applied to such rocks, and "saussuritization" for the process. Rosenbusch already noted in 1887 (Johannson 1937) that saussuritization is a metamorphic process, not a weathering phenomenon, and Johannson (1937) pointed to the important reaction albite + anorthite + water = zoisite + paragonite + quartz. However, this assemblage only forms at high pressure above approximately 0.8 GPa (Franz and Althaus 1977). In the presence of K, either as K-feldspar, as a K-feldspar component in ternary feldspars, or as $K^+$ in the fluid, muscovite-phengite instead or together with paragonite can form, $CO_2$ in the fluid can be fixed as carbonate, and also scapolite can be found, if $Cl^-$ is available. Often the retrograde overprint is accompanied by brittle deformation and results in the formation of epidote in cataclastic zones (forming a rock called "unakite").

In the sedimentary environment epidote minerals have not yet been reported as an authigenic low-temperature mineral, but may already form at the high-temperature part of diagenesis. In sediments they are also important as detrital minerals and were used for provenance studies (see the common textbooks on sedimentary petrology). They can make up to 90 vol% of the heavy mineral spectrum (v. Eynatten 2003). Epidote minerals are frequently attributed to source rocks of low to medium grade metamorphism, which must be considered carefully because of the extremely large stability field for the whole mineral group. Most recent studies show that a combination of chemical composition of the epidotes and their isotope ratios (e.g., Spiegel et al. 2002) will yield a better result, because they allow to distinguish between mantle derived rocks (roughly equivalent to a source with metabasites) and crustal derived rocks (roughly equivalent to metagranitoids). The most promising approach will be to use single grain analysis in heavy mineral concentrates by laser ablation coupled with ICP MS analysis.

The minimum pressure to stabilize epidote is not known. Epidote minerals were found in shallow geothermal fields at a few hundred meters depth, suggesting a pressure of much less than 100 MPa, and at a temperature below 200°C (Bird and Spieler 2004). On the other extreme, the maximum pressure at which zoisite is stable, is in the range of 7 GPa, far outside of what is possible even in extremely overthickened crust (Poli and Schmidt 2004). Common epidote–clinozoisite solid solutions break down in the range of 600 to 700°C, but they can be stabilized by incorporation of REE and thus allanite occurs in a variety of igneous rocks with high crystallization temperature, and has been reported from experimental studies at 1050°C at 3 to 4.5 GPa (Green and Pearson 1983; Hermann 2002). Thus from the petrogenetic point of view the epidote minerals are rather unspecific. On the other hand, they have a large potential for petrogenetic applications once their crystal chemical and thermodynamic properties are better understood. The major problem encountered by the petrologist is to distinguish between different stages of formation of epidote, especially on the retrograde path, but also on the prograde path through the peak stage of metamorphism, where complex zoning can be preserved (see also Grapes and Hoskins 2004, Enami et al. 2004). Here we give a general overview about the occurrences of epidote. For a detailed review of special members of the epidote minerals and for special localities, the reader is referred to the individual chapters of the volume, and also to Deer et al. (1986).

## Al-rich clinozosite and zoisite

In igneous rocks, Al-rich clinozosite and zoisite are very rare (for Fe-rich epidote see below). The only exception is zoisite in pegmatites from high-pressure areas, which are explained as melting products of eclogites and which crystallized at high pressure. In such an environment, the concentration of Fe in the melt is very low, and when the pressure is too high for plagioclase to crystallize, zoisite (with minor amounts of clinozoisite) is the liquidus phase.

In common regionally metamorphic rocks, Al-rich clinozosite and zoisite are found mainly in two types of environments, where Ca and Al are enriched (Fig. 23). The one is in meta-igneous rocks, where the enrichment occurs in the anorthite component of plagioclase. The other one is in metasediments, where Ca and Al are enriched in the sedimentary protolith (marl and impure limestone) in carbonate and clay minerals. A third type, much less abundant but of large petrogenetic importance, is the enrichment in metamorphic blasts of lawsonite. Characteristic assemblages of clinozoisite and zoisite are therefore for type a) as inclusions in mostly Na rich plagioclase, replacing the anorthite component, b) in calcsilicate rocks of all metamorphic grades, and c) in lawsonite pseudomorphs.

A typical occurrence for type a) is in metagabbro, especially in cumulate layers of such protoliths, and in fact clinozoisite and zoisite rich parts of metagabbro have often been interpreted as former cumulates. They are also found as part of

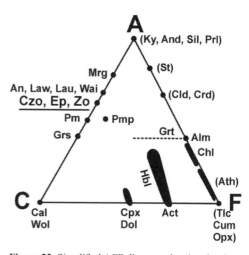

**Figure 23.** Simplified ACF diagram, showing the close chemical relationship of clinozoisite-epidote and zoisite with anorthite (component in plagioclase), lawsonite and Ca-zeolites. In metabasites at greenschist facies conditions, tie lines czo-chl-act-dol are stable, which are substituted at amphibolite facies conditions by plagioclase-amphibole. At eclogite facies conditions, where garnet can be rich in grossularite component, the common assemblage clinopyroxene-garnet limits the occurrence of epidote minerals. Only in compositions poor in mafic components epidote minerals are stable.

an inclusion assemblage in igneous porphyritic plagioclase crystals in metagranitoids. Often the distribution of the inclusion minerals mimics a former zoning of the plagioclase crystal. Because the anorthite component is not stable at high pressure due to the reaction anorthite + water = zoisite + kyanite + quartz, or albite + anorthite + water = zoisite + paragonite + quartz in many localities such occurrences of clinozoisite and zoisite are a good indication of a minimum pressure of approximately 0.7 to 1.0 GPa. Associated minerals are white mica or kyanite and quartz. Because anorthite is also not stable at a temperature below approximately 400°C due to the transformation of anorthite + water into epidote, and/or Ca-Al zeolites, margarite and sericitic white mica, the formation of inclusions in a plagioclase crystal can also be due to a retrograde metamorphism or to a hydrothermal overprint. Both processes, the decomposition of primary igneous plagioclase due to a higher pressure than in the protolith, as well as the decomposition due to a low temperature have been called saussuritization, and one should try to distinguish between them. Note however that it will be extremely difficult to distinguish between a retrograde low temperature metamorphism and a hydrothermal event, which can also be transitional to a weathering stage.

Occurrences of type b) in metasediments are found at all metamorphic grades, from low pressure subgreenschist facies rocks together with prehnite, pumpellyite up to the granulite facies where zoisite and clinozoisite occur together with spinel and corundum, in low pressure contact aureoles and also in the high and ultrahigh pressure rocks.

Occurrences of type c) are found mainly in metabasite. Lawsonite typically occurs in large blasts in former fine-grained basaltic rocks, thus creating a local chemical environment dominated by Ca and Al. If such a rock is heated and crosses the upper temperature stability limit of lawsonite it should decompose into an aggregate of zoisite, kyanite and quartz, but the association clinozoisite, white mica and quartz is much more common. Very often this heating of a rocks takes place without a penetrative deformation such that the shape relics of the lawsonite is preserved in clearly visible pseudomorphs, which have been described from many blueschist and eclogite terranes. Fe, Na and K are transported into the Ca-Al rich microdomain of the rock, and therefore kyanite is less frequently observed in the pseudomorphs, white mica is typically paragonite or a mixture of paragonite and phengite, and clinozoisite (also depending on the local position in the center or in the rim of the pseudomorph) is rich in Fe. Spear and Franz (1986) noted such pseudomorphs inside garnet, where they are completely protected from deformation, and argued that garnet must have been stable before lawsonite broke down, and the intersection of the garnet-in reaction with the lawsonite-out reaction gives a lower pressure limit for the prograde path, a very useful interpretation for the *P-T* path of a high pressure rock.

## Fe rich epidote

The formation of Fe rich epidotes requires in addition to the enrichment in Ca and Al enrichment of Fe and a relatively high $f_{O_2}$. They can therefore not be found in common igneous rocks of basaltic-andesitic composition, where $f_{O_2}$ is generally low, and where also plagioclase is stable instead of the clinozoisite component in epidote. Enrichment of Fe (relative to Mg) in an igneous rock with high Ca and Al contents is typical for granodiorite-granite, and epidote minerals are considered as primary igneous phases in such rocks (see Schmidt and Poli 2004). If, however, the epidote minerals are relatively rich in REE, this stabilizes the epidotes towards high temperature and also towards other compositions, where epidote minerals are normally absent, such as andesite (Gieré and Sorensen 2004). Epidote minerals are also not typical in metapelite with low Ca content, but may occur in metapsammite, especially in metagraywacke. Fe rich epidotes are also not typical for metamorphosed impure limestone and marl, because Mg generally dominates in these rocks.

In rocks with the appropriate composition and $f_{O_2}$ epidote may form at pressure-temperature conditions, where the anorthite component in plagioclase is not stable. The most common rock type is metabasite, i.e. a rock of basaltic-andesitic composition, either as a volcanic, subvolcanic, plutonic or pyroclastic igneous rock or volcanoclastic sediment, which is dominated by plagioclase and mafic minerals. Starting the discussion at surface conditions and going to high-grade metamorphism, in such rocks epidote can form even below very-low grade (<200°C, Bird and Spieler 2004) metamorphism. Associated Ca minerals are zeolites, grossularite and carbonate together with chlorite and quartz (see ACF diagram Fig. 23). Disequilibrium or only local equilibrium in small domains is the rule rather than the exception. In the lowermost greenschist facies the typical assemblage is epidote + chlorite + albite + quartz + carbonate. Epidote together with chlorite is a common constituent of pseudomorphs after mafic igneous minerals pyroxene and amphibole. The Al component for epidote is derived from breakdown of nearby plagioclase, and mass transfer on the grain scale is mainly from plagioclase to the mafic crystallization site. Note that in a coarse grained protolith it is of course possible to have both processes, the formation of Fe-rich epidote in mafic micro-domains as well as the formation of Al-rich clinozoisite or zoisite in Ca-Al rich micro-domains (plagioclase) present in the same thin section, as the product of the same mineral forming process. Only at sufficiently high temperature or strong deformation of the rock, equilibrium on a larger scale can be achieved.

The two dominant reactions, which produce more epidote in such a rock with heating, are chlorite + calcite = epidote + dolomite, and chlorite + dolomite = epidote + actinolite (Fig. 23). In the upper greenschist facies, transitional to the amphibolite facies the modal amount of epidote starts to decrease by continuous reactions of the type epidote + chlorite + quartz = Al-rich amphibole + anorthite (component in plagioclase) + $H_2O$ (Fig. 23). However this transition is very sensitive to total pressure (note that Fe rich epidote is a dense mineral with a density of up to 3.5 g/cm$^3$) and also to $f_{O_2}$. At high $P_{tot}$- $f_{O_2}$ epidote is favored, giving rise to the epidote-amphibolite facies (Spear 1993) between approximately 0.4 GPa/500°C and 1.1 GPa/650°C. For more details see Grapes and Hoskin (2004). Epidote minerals are common major rock-forming minerals in blueschist facies and in eclogite facies rocks (see Enami et al. 2004). The transition from greenschist to blueschist facies is given by the reaction actinolite + chlorite + albite = epidote + glaucophane + quartz, which can be written as a water-conserving reaction when the Tschermaks exchange vector $Al_2M_{-1}Si_{-1}$ is considered for amphibole and chlorite (Spear 1993). The transition from blueschist to eclogite facies is dominated by the lawsonite breakdown (with or without jadeite) to epidote minerals. At the *P-T* conditions of the eclogite facies, epidote minerals are stable, but their occurrence is chemically restricted because the common assemblage clinopyroxene + garnet (which can be rich in grossularite component, Fig. 23) covers a wide range of composition. Only in rocks that are relatively rich in Ca and Al epidote will be part of the assemblage (Spear 1993). At granulite facies, epidote is absent in common metabasite.

**Metasomatism**

Alternatively, the appropriate composition in the Ca-Al rich system as well as in the system rich in $Fe^{3+}$ can be achieved by mass transport. Starting from plagioclase as a common precursor mineral it can be transformed by the model reactions

3 anorthite + 1 $Ca^{2+}$ + 1 $CO_3^{2-}$ + $H_2O$ = 2 zoisite + 1 $CO_2$

3 anorthite + 1 $Ca^{2+}$ + 2 $Cl^-$ + 2 $H_2O$ = 2 zoisite + 2 HCl

3 anorthite + 1 $Ca^{2+}$ + 2 $F^-$ + 2 $H_2O$ = 2 zoisite + 2 HF

(see Klemd 2004). In these reactions, the Ca component can easily be transported as a $CaCO_3$ or as a $CaCl_2$ bearing solution. Fluorine is also a likely metasomatic agent, but will most probably result in the formation of fluorite. The equilibria indicate that in general the formation of epidosite requires Ca-metasomatism, with a low $CO_2$ and low HCl content in the fluid, i.e. with a low acidity (see Bird and Spieler 2004). Note that the reactions can also be used to describe the formation of zoisite and epidote inclusions inside plagioclase by mass transport on the thin section scale. The decarbonation reactions can be so important that $CO_2$ from this source was actually mined in the Salton Sea geothermal field from 1934 to 1954 (see Bird and Spieler 2004).

During contact metamorphism-metasomatism, especially in a situation where at the contact of an igneous body to limestone the typical skarn assemblages formed, model reactions for the Al-free hypothetical endmember are

$$2 \text{ calcite} + 1.5 \text{ hematite} + 3 \text{ quartz} + 0.5 \text{ H}_2\text{O} = 1 \text{ Ca}_2\text{Fe}_3\text{Si}_3\text{O}_{12}(\text{OH}) + 2 \text{ CO}_2$$

and

$$2 \text{ calcite} + 1 \text{ magnetite} + 3 \text{ quartz} + 0.5 \text{ H}_2\text{O} + 0.5 \text{ O}_2 = 1 \text{ Ca}_2\text{Fe}_3\text{Si}_3\text{O}_{12}(\text{OH}) + 2 \text{ CO}_2$$

In such a case the assemblage carbonate-oxide can be transformed into epidote component with a $SiO_2$ rich aqueous solution. Klemd (2004) gives examples for such reactions involving anorthite or grossularite, which represent the Al component in marl-limestone mixtures. $Fe^{2+}$ can also be present in the starting assemblage in carbonate or silicate, and many important metasomatic reactions are redox equilibria.

Alternatively, Fe may be the metasomatic agent as a late stage chloridic solution from an acid igneous rock

$$2 \text{ calcite} + 3 \text{ FeCl}_3 + 3 \text{ quartz} + 5 \text{ H}_2\text{O} = 1 \text{ Ca}_2\text{Fe}_3\text{Si}_3\text{O}_{12}(\text{OH}) + 9 \text{ HCl}$$

In summary, metasomatism of Ca, Si and/or Fe may be responsible for epidote formation.

The occurrence of epidote minerals in open cavities and druses, in veins and quartz-carbonate segregations of very variable shape and mineral association is different from metasomatism. In this case the fluid did not only transport specific components for a metasomatic solid-fluid reaction, but all components to precipitate the minerals. The very common occurrence of all types of epidote minerals, Fe-poor zoisite to Fe-rich epidote, in veins and segregations points to the fact that Al as a major component must be transported via a fluid phase and can not be considered as immobile. In summary, metasomatic reactions that form epidote minerals can be explained as mass transport of all major elements essential for epidote, Ca, Al, Fe and Si.

## Miscibility vs. miscibility gaps

One of the interesting and controversial discussed questions in research about the epidote minerals is that of immiscibility along the (monoclinic and orthorhombic) Al-$Fe^{3+}$ solid solution join. This subject is discussed in detail in Grapes and Hoskins (2004) for natural occurrences and in Poli and Schmidt (2004) from the experimental point of view. Gottschalk (2004) presents a calculated phase diagram based on newly derived thermodynamic data. The conclusions drawn from experimental studies and investigations on natural epidote minerals are not unequivocal. Experimental studies by Fehr and Heuss-Aßbichler (1997), Heuss-Aßbichler and Fehr (1997), Brunsmann (2000), and Brunsmann et al. (2002) point to two solvi along the Al-$Fe^{3+}$ solid solution join. Strens (1963, 1964), Hietanen (1974), and Raith (1976) provide good evidences for miscibility gaps of variable extend along the Al-$Fe^{3+}$ solid solution join from natural epidote parageneses (see also Grapes and Hoskin 2004). Contrary, Enami et al. (2004) and Bird and Spieler (2004) describe clinozoisite-epidote compositions from blueschist facies and geothermal clinozoisite-epidote occurrences, respectively, that lie within the proposed miscibility gaps and interpreted these as pointing to complete miscibility. However, it is generally accepted that a two-phase area ("two-phase loop") between clinozoisite and zoisite exists for the transition from the monoclinic to the orthorhombic polymorphs (Gottschalk 2004, his Fig. 21). At 0.5 GPa, the equilibrium temperature for the clinozoisite-zoisite transition in the Fe-free system is about 125°C. Above 730°C, there is continuous solid solution between disordered clinozoisite and disordered epidote; intermediate clinozoisite ($X_{Ep} = 0.5$) starts to order, and at ≈630°C two solvi start to open towards lower temperature. The situation can be compared with the jadeite-omphacite-diopside system or with the calcite-dolomite-magnesite system. The diagram, however, does not include possible complications like the potential intersection of the low Fe solvus with the two-phase area (see Poli and Schmidt 2004, their Fig. 5). The position of the solvi is quite sensitive to the chosen values of the Margules parameters, and it is possible that at 200 to 400°C the low Fe solvus intersects the two phase loop and changes the phase relations as shown schematically by Franz and Selverstone (1992).

The position of the proposed solvi is a strong function of the ordering in epidote. This may explain why in many low temperature environments, such as geothermal fields, a large range of compositions inside the theoretical solvi are found (see Bird and Spieler 2004). Metastable formation of disordered clinozoisite-epidote seems very likely to occur in a hydrothermal environment. Also small amounts of other components such as Sr, Mn, REE, Cr etc. will change the position of the solvi significantly, first by a change in the thermodynamic parameters, and also by their possible influence on the order-disorder behavior. This may explain the contradictory results from the natural occurrences. Not much information is available on the question of miscibility gaps in other systems, such as the Mn-bearing epidote minerals (Bonazzi and Menchetti 2004) and REE-bearing epidote minerals (Gieré and Sorensen 2004).

## MORPHOLOGY, TWINNING AND DEFORMATION

### Morphology, growth and dissolution

Freely grown epidote and zoisite crystals generally form elongated prisms, where the long axis is the *b*-axis, or platy crystals with the plates in the *b*-*c* plane (note that in old literature one might find for zoisite the orientation with the long axis = *c,* such that $a_{morph} = c_{X-ray}$; $b_{morph} = a_{X-ray}$; $c_{morph} = b_{X-ray}$; Game 1954). Therefore the relative growth velocities are $b>>c>a$. The fastest growth direction *b* is the direction of the octahedral chains, *a* is the direction in which the building units of octahedral chains are connected by the large A sites. An overview of the common euhedral crystal shapes of epidote, piemontite, allanite ("orthite") and zoisite is given by Goldschmidt (1916) who lists 372 crystal drawings of different forms of epidote and 18 of zoisite (Goldschmidt, 1923; Fig. 24). Common forms of the terminal faces of epidote are {210}, {011} and {111} (Fig. 25a). For the prism faces the common forms are {001}, {101}, {201}. Striation parallel to the long (= crystallographic *b*) axis due to subordinately developed {301} up to {801} is frequent. Striation of the prism faces is a rather regular alternation of two different {*h0l*} faces with mostly sharp edges (Fig. 25c). However, this is not always the case, and the edges may show systematic etch phenomena, which indicate that one of the faces is less stable (Figs. 25d,e,f). Scanning electron microscopy of single crystals of epidote (Fig. 25b) show that also terminal faces can be striated. Etching on the terminal faces is frequently observed (Figs. 25g,h) both in a regular (g) and an irregular (h) shape. Etching also occurs on the prism faces (Fig. 25i), where it can be confused with striation due to alternating faces.

Allan and Fawcett (1982) observed that faces in the [010] zone of synthetic zoisite (synthesized at 0.5 GPa, 500 to 700°C) are striated due to rapid growth, when the rate of spreading of surface layers by addition of material in steps is a slow process relative to the rate at which new layers nucleate. They also demonstrated in their study that zoisite is preferentially dissolved parallel to the *b* crystallographic direction, as expected from the Law of Bravais (i.e. the rate of dissolution in a particular direction is inversely proportional to

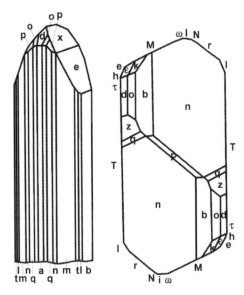

**Figure 24.** Crystal drawings for zoisite (*left*) and epidote (*right*); redrawn from Goldschmidt (1916, 1923).

the lattice dimension in that direction). This observation agrees with that shown above (Fig. 25i) that etching in natural crystals occurs parallel to *b*. In an experimental study using a natural zoisite (var. "tanzanite") at 650°C to 800°C/ 0.1 to 0.2 GPa, Matthews (1985) observed preferred dissolution on the terminal faces which produced saw-tooth shaped structures and in an advanced stage rod-like structures (his Fig. 6b,c,e). Etching (resorption) and subsequent growth of clinozoisite parallel to (100) was observed in pegmatitic zoisite from eclogites of the Münchberg area (Franz and Smelik 1995), which led to lamellar zoisite-clinozoisite intergrowths. Heuss-Aßbichler and Fehr (1997) interpreted a saw-tooth shape of synthetic epidote-clinozoisite prisms as a result of etching of the terminal faces and subsequent growth on these etch pits. Brandon et al. (1996) determined the epidote dissolution kinetics to constrain the rates of granitic magma transport.

Epidote grains as rock forming minerals often have a less distinct elongated form. Euhedral crystals are less common, but clinozoisite and zoisite, especially as small inclusions in plagioclase, often show well-developed platy crystals. Porphyroblast and poikiloblasts are common. Zoisite, epidote and piemontite may show radially grown prismatic crystals (Tröger 1982), especially when formed hydrothermally, zoisite from segregations often has a fan shape (Brunsmann et al. 2000), interpreted as a result of rapid growth. Synthetic zoisite (Allan and Fawcett 1982; Matthews and Goldsmith 1984; Liebscher et al. 2002) and epidote (Heuss-Aßbichler and Fehr 1997) varies from needle-shaped to prismatic and tabular.

Perfect cleavage in zoisite is parallel to (100), imperfect cleavage parallel to (001) (Tröger 1982; Deer et al. 1986) consistent with relative weak bonding at the A sites. In epidote cleavage parallel (100) is only weak, whereas cleavage parallel to (001) is perfect, due to the fact that the M3 at the octahedral chains alternate in two directions and therefore hinder a perfect cleavage in (100).

---

**Figure 25 (*on facing page*).** (a) SEM micrograph of the terminal faces of a single crystal of clinozoisite ($X_{Ep}$ = 0.28; analyses see Smelik et al. 2001) from Untersulzbachtal, Tyrol/Austria. (b) Enlarged part of a (lower arrow), showing rare striations on a terminal face. (c) SEM micrograph of a striated prism face of an epidote crystal ($X_{Ep}$ = 0.88; analyses see Smelik et al. 2001) from Knappenwand, Tyrol/Austria, with generally sharp edges between the two differently developed {*h0l*} faces; note however the irregular feature on the edge in the upper left. (d) SEM micrograph of a striated prism face of a clinozoisite-epidote crystal ($X_{Ep}$ = 0.48; analyses see Smelik et al. 2001) from Gilgat/Pakistan with etch phenomena. (e) Enlarged part of Figure d (lower center), showing the transition from an unstable (left part) to a stable {*h0l*} face. Note the transition from elongated etch pits in the right to arrow-like shapes and to crystallographically bound large steps. (f) Enlarged part of Figure d (upper right), showing the transition from an unstable (right) to a stable {*h0l*} face. Note the difference to (e) – preferred etching of the unstable face leaves step-like forms and elongated rods. *(continued on following page)*

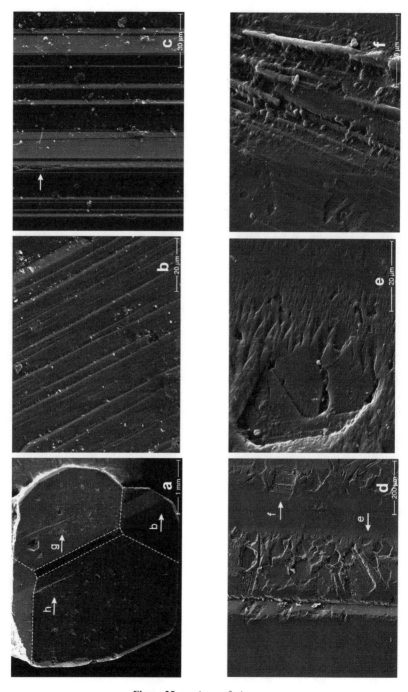

**Figure 25.** *caption on facing page*

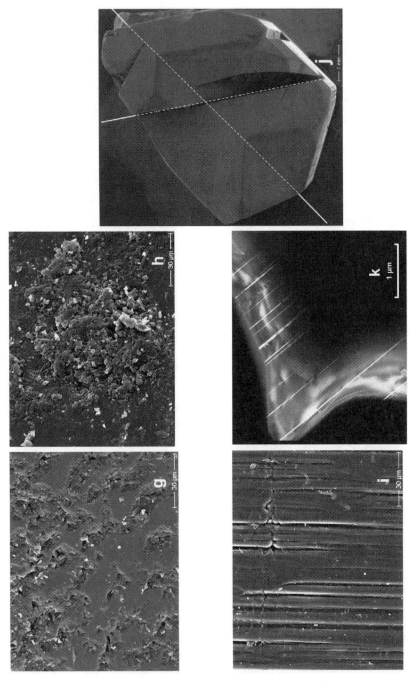

**Figure 25 continued**. *caption on facing page*

## Twinning

Both Tröger (1982) and Deer et al. (1986) list lamellar twinning in epidote as "not common", with the twin plane in (100), in allanite also rarely in (001), no twinning in zoisite. Own observations show that freely grown hydrothermal epidote crystals from certain localities such as the Knappenwand (Austria) are very often macroscopically twinned. Figure 25j shows the terminal faces of such a twin. The crystal also shows the phenomenon of parallel intergrowth, which is visible similarly as the twinning as incised terminal faces. In thin section parallel to the crystallographic *b* axis of such crystals the lamellar character of this twinning can be seen. Ray et al. (1986) observed in TEM investigations from a variety of epidote crystals with different Fe content also that growth twins are very common in freely grown euhedral crystals. As a rock-forming mineral, epidote may show simple and more rarely lamellar twinning in thin section, but it is not a good diagnostic feature. Twins are probably growth twins, because they occur in freely grown undeformed crystals.

TEM investigations (Müller, Darmstadt, pers. comm. 2003) on minerals from a strongly deformed eclogite body of the Frosnitztal/Tauern Window (Austria) indicate mechanical twinning with the twin lamellae parallel (100) and the twin law *m* parallel (100). Width of the twin lamellae varies from 40 nm down to 0.8 nm, which corresponds to a single elementary period. Distances are irregular between 30 nm and 0.5 μm (Fig. 25k).

## Deformation behavior

Little is known about the behavior of epidote minerals during deformation. Field observations show that it is relatively rigid. Epidote rich layers in a matrix of chlorite-amphibole-plagioclase show a strong competence contrast at deformation conditions of upper greenschist to lower amphibolite facies, and are boudinaged (Fig. 26a). In contrast, layers of epidote in an eclogite are harmonically folded together with pyroxene-garnet rich layers and quartz-rich layers at approximately 600°C/2 GPa (Fig. 26b). The fold shapes do not show great differences indicating a similar rheological behavior similar for epidote and pyroxene-garnet aggregates. Masuda et al. (1990, 1995) investigated the rheological behavior of piemontite, which is probably very similar to epidote, compared to quartz and albite. Piemontite prisms define the lineation and stretching direction of quartz-rich schist. The crystals are micro-boudinaged, i.e., cracked perpendicular to the elongation of the prisms [010] and torn apart, and the interstices are filled with quartz and albite. The deformation was of nearly pure shear type, no rotation of the piemontite crystals into the stretching direction was found, and the deformation temperature was above that of the brittle-plastic transition of albite. No ductile deformation was apparent at the edges of the sharp fracture planes of piemontite, consistent with the conclusion that epidote minerals are significantly more rigid than feldspar.

In thin section, apart from cracking mentioned above, potential development of subgrain boundaries, kinking and bending have been observed (Fig. 27). Monomineralic layers of epidote frequently show grains with low misorientation angles, possibly representing subgrain boundaries (Fig. 27a), oriented approximately parallel to the cleavage {100} and {001}.

---

**Figure 25 continued (*on facing page*).** (g) Enlarged part of Figure a (right large face) with regular etch pits. (h) Enlarged part of Figure a (left large face) with irregular etch pits. (i) Etch pits on prism face of an epidote crystal from Sustenpass/Switzerland; note the oblique bending at the end of the pits and their connection with crack-like features (center, upper part). (j) SEM micrograph of terminal faces of a twinned epidote from Knappenwand, Austria (88 mol% Ep) viewed almost perpendicular down the crystallographic *b* axis. Twin plane (100) runs from lower left to upper right (dashed line); the almost vertically running line separates parallel intergrowths of epidote crystals, marked by an open space. A similar parallel intergrowth is seen in the upper right (not marked by dashed line); composite of four individual pictures. (k) TEM micrograph of mechanical twinning of clinozoisite (sample F 146; see Müller et al, in press, for more information about the sample) from an eclogite body; the width of the twin lamellae are from 40 nm down to 0.8 nm, the width of one elementary period in (100); photo by courtesy W. Müller, Darmstadt.

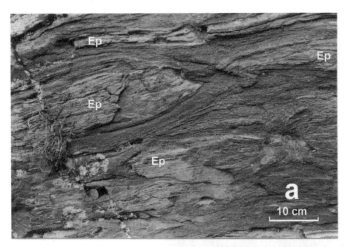

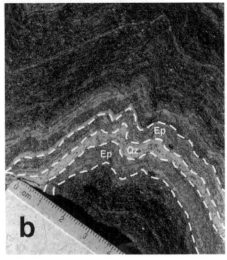

**Figure 26.** Deformation of epidote-rich rocks. (a) Boudinaged epidote rich layers in a greenschist; Upper Schieferhülle, Frosnitztal, Tauern Window/Austria. (b) Harmonic folding an in a layered eclogite; outlined are two layers of clinozoisite-epidote (Ep) separated by a quartz layer (Qz); upper and lower part is garnet-omphacite dominated rock. Eclogite Zone, Frosnitztal, Tauern Window/Austria).

However, the misorientation may also be due to microfracturing. Recrystallization has not been observed as a typical feature, but along cracks and subgrain boundaries, a few differently oriented small grains develop. Ray et al. (1986) also found polygonal subgrains with subhedral outlines dominated by the cleavage planes, as determined by TEM investigations. Dislocations can be observed, some of which are partials trailing stacking faults. These are inclined but their orientation is consistent with (100) planes. Partial dislocations form the subgrain walls. No fine scale multiple twins were found, but Müller (pers. comm. 2003; see Fig. 25k) observed mechanical twinning as a response to very strong deformation in epidote from a small eclogite body near to a shear zone in the Tauern Window.

Zoisite from sheared segregations in metabasite rocks (Lower Schieferhülle Tauern Window, Brunsmann et al. 2000) are rotated with their long axis into the foliation and, depending on the orientation, stretched in [010] or show bookshelf gliding on the (100) cleavage planes. These large crystals also show kinking with the kink axis [010] and the kink plane (001), and shearing parallel to (010). The same type of kinking was observed by

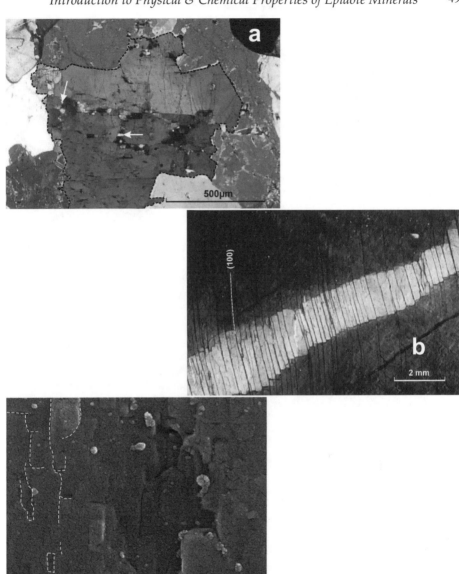

**Figure 27.** Deformation of epidote minerals on a microscopic scale. (a) Subgrain boundaries in epidote oriented approximately parallel to the cleavage; sample # 1427 Schriesheim, Odenwald/Germany, photomicrograph with crossed polarized light. (b) Kinkbands and book shelf gliding parallel to (100) with the kink axis [010] in zoisite; sample HU93-8 from Sterzing, Tauern/Italy; photomicrograph with crossed polarized light (c) SEM photograph of the concave upward bent cleavage plane (100) of zoisite (sample HU93-8), oriented perpendicular to the electron beam, with the axis of bending east-west, and the elongation (*b*) axis north-south. The plane shows a rough surface, structured by approximately rectangular domains in the size of 1 μm × 0.2 μm (outlined in the left part of the figure), which are slightly tilted against each other. *(continued on following page)*

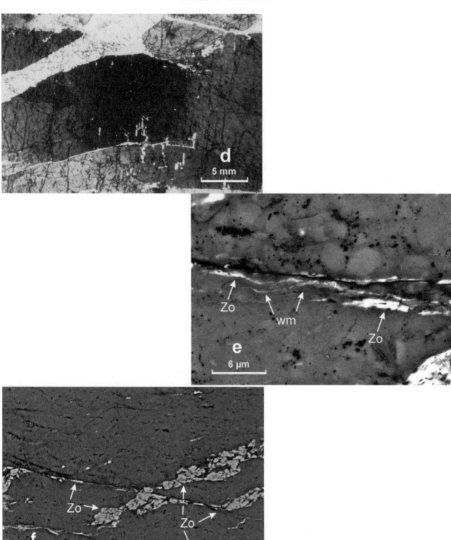

**Figure 27 continued.** (d) Bent zoisite crystal (grey to dark grey) with undulose extinction (white = plagioclase) from a deformed pegmatite (sample # 84-3, Weißenstein, Münchberg Massif, Bavaria/Germany). Deformation conditions were in the order of 1.0 GPa/500 to 600°C. Cracks were formed in a subsequent event of brittle deformation at lower greenschist to subgreenschist facies conditions. (e) BSE image of synthetic zoisite (white, Zo) with small amounts of white mica (thin white streaks, wm) in a matrix of reacted plagioclase (light grey An 54 = relict starting composition, dark grey An 35 = newly formed, experimentally produced at 1.5 GPa, 750°C; see Stünitz and Tullis 2001). The extremely small zoisite platelets are aligned in an east-west running shear band. (f) BSE image of synthetic zoisite (white, Zo) with small amounts of white mica (thin white streaks) in a matrix of reacted plagioclase (grey An 60 = starting composition, transformed into An 35), experimentally produced at 1.5 GPa, 750°C; see Stünitz and Tullis 2001). A WNW-ESE running shear band shears the larger zoisite crystals. Small grains of albitic plagioclase, white mica, and zoisite nucleate during deformation in the shear band (Figs. e and f courtesy of H. Stünitz).

Ray et al. (1986) in zoisite from the Eclogite Zone (Tauern Window). Large crystals of zoisite from another locality in the Tauern (Sterzing, sample HU 93-8) also show such kink bands (Fig. 27b) in thin section. The kinking produces glide on the cleavage plane (100) as a bookshelf gliding.

Large crystals of zoisite are also bent. Figure 27d shows such a crystal from a deformed pegmatite with continuous undulose extinction. The deformation conditions are estimated as epidote-amphibolite facies near 500 to 600°C/1.0 GPa. The zoisite crystal shown above, which shows kinking (Fig. 27b), also shows bending. When viewed with the scanning electron microscope onto the concave bent part of a cleavage plane (100) of these crystals (Fig. 27c), they show a surface, which is structured into small domains in the size of 1 μm × 0.2 μm. The domains are slightly tilted against each other. When viewed onto the convex face, they show a similar structure. In summary, the optical and electron microscopic investigations of bent and kinked crystals show that {100}, {010}, {001} planes play a major role in the deformation of epidote group minerals, either as slip planes or as cleavage planes for fracturing.

Stünitz and Tullis (2001) verified experimentally the observation from natural rocks that epidote minerals are relatively rigid. They performed deformation experiments of plagioclase, which reacts to zoisite (+ albitic plagioclase, kyanite, quartz and white mica) during deformation at 1.5 GPa and 750°C. The microstructures indicate that zoisite underwent preferred growth and some fracturing at low strain, whereas at high strain it develops shear bands, which separate long crystals (or crystal aggregates) of > 100 μm length in domains of < ≈ 30 μm. The shear bands themselves are only 1-3 μm wide and form an angle of ≈ 15 to 25° to the shear zone boundary. The deformation mechanism in the fine-grained aggregates of the shear bands is granular flow, a diffusion-assisted grain boundary sliding process, by which the rock deforms at low stress. This type of deformation mechanism operates in very fine-grained materials. The most effective way to produce such fine-grained aggregates is to recrystallize or neocrystallize the material. Plagioclase breakdown reactions producing epidote group minerals are classic examples for neocrystallization leading to very fine-grained material in shear zones (Stünitz 1993; Stünitz and Fitzgerald 1993). The main reason for the fine grain size is the rapid nucleation rate of these phases. Thus, epidote group minerals are important for affecting the deformation properties of plagioclase rocks, especially in the middle to upper crust. The plagioclase breakdown reactions involving epidote group minerals commonly cause a change in deformation mechanism to grain size sensitive mechanisms and in this way weaken the crust mechanically.

## ACKNOWLEDGMENTS

We thank J. Nissen, F. Galbert, I. Preuss at ZELMI at TU Berlin for SEM photographs, analyses and figures (Figs. 15, 25a-j, 27c); they were of invaluable help for our exploratory studies of surface and growth phenomena of epidotes. S. Herting-Agthe (Berlin) and G. Smelik (North Carolina) kindly provided specimens. V. Schenk (Kiel) and S. Herting-Agthe helped with literature research; J. Kruhl (München) and H. Stünitz (Basel) are thanked for comments on deformation, and M. Raith (Bonn) for comments on optical data. S. Bernau (Berlin) patiently helped in preparation of this manuscript (and others of this volume) and we thank R. Geffe (Berlin) for help in drawing and preparation of figures.

## REFERENCES

Akasaka M, Sakakibara M, Togari K (1988) Piemontite from the manganiferous hematite ore deposits in the Togoro belt, Hokkaido, Japan. Min Pet 38:105-116

Allen JM, Fawcett JJ (1982) Zoisite-anorthite-calcite stability relations in $H_2O$-$CO_2$ fluids at 500 bars - An experimental and SEM study. J Petrol 23:215-239

Anastasiou P, Langer K (1977) Synthesis and physical properties of piemontite $Ca_2Al_{3p}Mn_p^{3+}(Si_2O_7/SiO_4/O/OH)$. Contrib Mineral Petrol 60:225-245

Andersen O (1911) On epidote and other minerals from pegmatite veins in granulite at Notodden, Telemarken, Norway. Archiv Mathem Naturvidenskab B 31/15:313-362

Anderson BW (1968) Three items of interest to gemmologists. J Gemmol 11:1-6

Armbruster Z, Gnos E, Dixon R, Gutzmer J, Hejny C, Döbelin N, Medenbach O (2002) Manganvesuvianite and tweddilite, two new $Mn^{3+}$-silicate minerals from the Kalahari manganese fields, South Africa. Mineral Mag 66:137-150

Ashley PM, Martin JE (1987) Chromium-bearing minerals from a metamorphosed hydrothermal alteration zone in the Archean of Western Australia. N Jahrb Mineral Abh 157:81-111

Bank H (1969) Hellbraune bis farblos durchsichtige Zoisite aus Tansania. Z Dtsch Gemmol Ges 18:61-65

Bank H (1980a) Geschliffener durchsichtiger roter Klinozoisit aus Arendale/Norwegen. Z Dtsch Gemmol Ges 29:186-188

Bank H (1980b) Schleifwürdiger manganhaltiger durchscheinender roter Zoisit (Thulit) aus Norwegen. Z Dtsch Gemmol Ges 29:188-189.

Bank H, Beredesinski W, Nuber B (1967) Strontiumhältiger trichroitischer Zoisit von Edelsteinqualität. Z Dtsch Gemmol Ges 61:27-29

Banno S, Yoshizawa H (1993) Sector zoning of epidote in the Sanbagawa schists and the question of an epidote miscibility gap. Minal Mag 57:739-743

Bauer M (1968) Precious stones. Dover Publications, New York

Becke F (1914) Über den Zusammenhang der physikalischen, besonders der optischen Eigenschaften mit der chemischen Zusammensetzung der Silikate. *In*: Handbuch der Mineralchemie, 2, I Doelters (ed), Dresden

Belov NV, Rumanova JM (1953) The crystal structure of epidote $Ca_2Al_2FeSi_3O_{12}(OH)$. Dokl Akad Nauk SSSR 89:853-856 (Abstract in Struct Rep 17:567-568)

Belov NV, Rumanova JM (1954) The crystal structure of epidote. Trudy Inst Krist Akad Nauk SSSR 9:103-164 (Abstract in Struct Rep 18:544-545)

Bermanec V, Armbruster Th, Oberhänsli R, Zebec V (1994) Crystal chemistry of Pb- and REE-rich piemontite from Nezilovo, Macedonia. Schweiz Mineral Petrogr Mitt 74:321-328

Berzelius (1818) Afhandel i. Fys 5:32 (quoted in Hintze 1897)

Bird D, Spieler AR (2004) Epidote in geothermal systems. Rev Mineral Geochem 56:235-300

Bird DK, Cho M, Janik CJ, Liou JG, Caruso LJ (1988) Compositional, order/disorder, and stable isotope characteristics of Al-Fe epidote, State 2-14 Drill Hole, Salton Sea Geothermal System. J Geophys Res 93:13,135-13,144

Bird DK, Helgeson HC (1980) Chemical interaction of aqueous solutions with epidote-feldspar mineral assemblages in geologic systems. 1. Thermodynamic analysis of phase relations in the system CaO-FeO-$Fe_2O_3$-$Al_2O_3$-$SiO_2$-$H_2O$-$CO_2$. Am J Sci 280:907-941

Blackburn WH, Dennen WH (1997) Encyclopedia of mineral names. The Canadian Mineralogsit Spec Publ 1, Mineral Assoc Canada, Ontario, p 360

Bleek AWG (1907) Die Jadeitlagerstätten in Upper Burma. Z prak Geol:341-365

Bonazzi P, Bindi L, Parodi G (2003) Gatelite-(Ce), a new REE-bearing mineral from Trimouns, French Pyrenees; crystal structure and polysomatic relationships with epidote and törnebohmite-(Ce). Am Mineral 88:223-228

Bonazzi P, Garbarino C, Menchetti S (1992) Crystal chemistry of piemontites: REE-bearing piemontite from Monte Brugiana, Alpi Apuane, Italy. Eur J Mineral 4:23–34

Bonazzi P, Menchetti S (1995) Monoclinic members of the epidote group: effects of the Al $\Leftrightarrow$ $Fe^{3+}$ $\Leftrightarrow$ $Fe^{2+}$ substitution and of the entry of $REE^{3+}$. Mineral Petrol 53:133-153

Bonazzi P, Menchetti S (2004) Manganese in monoclinic members of the epidote group: piemontite and related minerals. Rev Mineral Geochem 56:495-552

Bonazzi P, Menchetti S, Palenzona A (1990) Strontiopiemontite, a new member of the epidote group, from Val Graveglia, Liguria, Italy. Eur J Mineral 2:519-523

Bonazzi P, Menchetti S, Reinecke T (1996) Solid solution between piedmontite and androsite-(La), a new mineral of the epidote group from Andros Island, Greece. Am Mineral 81:735-744

Brandon AD, Creaser RA, Chacko T (1996) Constraints on rates of granitic magma transport from epidote dissolution kinetics. Science 271:1845-1848

Brastad K (1985) Sr metasomatism and partition of Sr between the mineral phases of a meta-eclogite from Björkedalen, West Norway. Tschermaks Mineral Petrogr Mitt 34:87–103

Brese NE, O'Keeffe M (1991) Bond-valence parameters for solids. Acta Crystall B 47:192-197

Brewster (1819) On the absorption of polarized light by doubly refracting crystals. Phil Transact London 19

Brunsmann A (2000) Strukturelle, kristallchemische und phasenpetrologische Untersuchungen an synthetischen und natürlichen Zoisit und Klinozoisit Mischkristallen. Dissertation, Technical University of Berlin, Germany (http://edocs.tu-berlin.de/diss/2000/brunsmann_axel.htm)

Brunsmann A, Franz G, Erzinger J, Landwehr D (2000) Zoisite- and clinozoisite-segregations in metabasites (Tauern Window, Austria) as evidence for high-pressure fluid-rock interaction. J metam Geol 18:1–21

Brunsmann A, Franz G, Heinrich W (2002) Experimental determination of zoisite-clinozoisite phase equilibria in the system $CaO-Al_2O_3-Fe_2O_3-SiO_2-H_2O$. Contrib Mineral Petrol 143:115-130

Burns RG, Strens RGJ (1967) Structural interpretation of polarized absorption spectra of the Al-Fe-Mn-Cr epidotes. Min Mag 36:204-226

Carbonin S, Molin G (1980) Crystal-chemical considerations on eight metamorphic epidotes. N Jahrb Mineral Abh 139:205-215

Catti M, Ferraris G, Ivaldi G (1988) Thermal behavior of the crystal structure of strontian piemontite. Am Mineral 73:1370-1376

Catti M, Ferraris G, Ivaldi G (1989) On the crystal chemistry of strontian piemontite with some remarks on the Nomenclature of the epidote group. N Jahrb Mineral Monatsh 1989:357-366

Cech F, Vrána S, Povondra P (1972) A non-metamict allanite from Zambia. N Jahrb Mineral Abh 116:208-223

Challis A, Grapes R, Palmer K (1995) Chromian muscovite, uvarovite and zincian chromite: products of regional metasomatism in northwest Nelson, New Zealand. Can Mineral 33:1263-1284

Chatterjee ND, Johannes W, Leistner H (1984) The system $CaO-Al_2O_3-SiO_2-H_2O$: new phase equilibria data, some calculated phase relations, and their petrological applications. Contrib Mineral Petrol 88:1-13

Chopin C (1978) Les paragenèses réduites ou oxydées de concentrations manganésifères de "schistes lustrés" de Hate-Maurienne (Alps francaises). Bulletin Minéral 101:514-531

Comodi P, Zanazzi PF (1997) The pressure behavior of clinozoisite and zoisite: An X-ray diffraction study. Am Mineral 82:61-68

Cooper AF (1980) Retrograde alteration of chromian kyanite in metachert and amphibolite whiteschist from the Southern Alps, New Zealand, with implication for uplift on the Alpine fault. Contrib Mineral Petrol 75:153-164

Cordier L (1803) Analyse du Minéralconnu sus le nom de Mine de Manganese violet du Piédmont, faite au Labratoire de l'Ecole de Mines. J Mineral 13:135

Cressey G, Steel AT (1988) An EXAFS study on Gd, Er and Lu site location in the epidote structure. Phys Chem Mineral 15:304-312

Deer WA, Howie RA, Zussman J (1986) Epidote group. *In*: Deer WA, Howie RA, Zussman J Disilicates and ring silicates. Second edition. London, Longman Scientific and Technical, pp 2-179

DeRudder RD, Beck CW (1964) Clinozoisite from the Willsboro wollastonite deposit, New York. Special Paper, Geol Soc Am 76:42-43

Devaraju TC, Raith MM, Spiering B (1999) Mineralogy of the archean barite deposit of Ghattihosahalli, Karnataka, India. Can Mineral 37:603-617

Dollase WA (1968) Refinement and comparison of the structures of zoisite and clinozoisite. Am Mineral 53:1882-1898

Dollase WA (1969) Crystal structure and cation ordering of piemontite. Am Mineral 54:710-717

Dollase WA (1971) Refinement of the crystal structures of epidote, allanite and hancockite. Am Mineral 56:447-464

Dollase WA (1973) Mössbauer spectra and iron distribution in the epidote group minerals. Z Krist 138:41-63

Dunn PJ (1985) The lead silicates from Franklin, New Jersey: occurrence and composition. Mineral Mag 49:721-727

Enami M (1977) Sector zoning of zoisite from a metagabbro at Fujiwara, Sanbagawa metamorphic terrane in central Shikoku. J Geol Soc Japan 83:693-697

Enami M, Banno Y (2001) Partitioning of Sr between coexisting minerals of the hollandite- and piemontite-groups in a quartz-rich schist from the Sanbagawa metamorphic belt, Japan. Am Mineral 86:205-214

Enami M, Liou JG, Mattinson CG (2004) Epidote minerals in high P/T metamorphic terranes: Subduction zone and high- to ultrahigh-pressure metamorphism. Rev Mineral Geochem 56:347-398

Ernst WG (1964) Petrochemical study of coexisting minerals from low-grade schists, Eastern Shikoku, Japan. Geochim Cosmochim Acta 28:1631-1668

Ernst WG (1977) Mineralogic study of eclogitic rocks from Alpe Arami, Lepontine Alps, southern Switzerland. J Petrol 18:371-398

Eskola P (1933) On the chrome minerals of Outokumpu. Bulletin de la Commission Geologique de Finlande p 26-44

Faye GH, Nickel EH (1971) On the pleochroism of vanadium-bearing zoisite from Tanzania. Can Mineral 10: 812-821

Fehr KT, Heuss-Aßbichler S (1997) Intracrystalline equilibria and immiscibility gap along the join clinozoisite - epidote: An experimental and $^{57}$Fe Mössbauer study. N Jb Mineral Abh 172:43-67

Ferraris G, Ivaldi G, Fuess H, Gregson D (1989) Manganese/iron distribution in a strontian piemontite by neutron diffraction. Z Krist 187:145-151

Fesenko EG, Rumanova IM, Belov NV (1955) The crystal structure of zoisite. Dokl Akad Nauk SSSR 102: 275-278 (Abstract in Structural Report 19:464-465)

Fesenko EG, Rumanova IM, Belov NV (1956) Crystal structure of zoisite. Kristallografiya SSSR 1:132-151 (Abstract in Structural Report 20:396-398)

Fischer K (1977) Edelstein Epidot. Lapis 2:10-13

Fleischer M (1970) New Mineral Names. Am Mineral 55:317-323

Forbes EH (1896) Über den Epidot von Huttington, Mass., und über die optischen Eigenschaften des Epidots. Z Kryst 26:138-142

Franz G, Althaus E (1977) The stability relations of the paragenesis paragonite-zoisite-quartz. N Jb Mineral Abh 130:159-167

Franz G, Selverstone J (1992) An empirical phase diagram for the clinozoisite – zoisite transformation in the system $Ca_2Al_3Si_3O_{12}(OH) - Ca_2Al_2Fe^{3+}Si_3O_{12}(OH)$. Am Mineral 77:631–642

Franz G, Smelik, EA (1995) Zoisite-clinozoisite bearing pegmatites and their importance for decompressional melting in eclogites. Eur J Mineral 7:1421-1436

Frei D, Liebscher A, Franz G, Dulski P (2004) Trace element geochemistry of epidote minerals. Rev Mineral Geochem 56:553-605

Gabe EJ, Portheine FC, Whitlow SH (1973) A reinvestigation of the epidote structure: confirmation of the iron location. Am Mineral 58:218-223

Game PM (1954) Zoisite-amphibolite with corundum from Tanganyika. Mineral Mag 30:458-466

Ghose S, Tsang T (1971) Ordering of $V^{2+}$, $Mn^{2+}$, and $Fe^{3+}$ ions in zoisite, $Ca_2Al_3Si_3O_{12}(OH)$. Science 171: 374-376

Gieré R, Sorensen SS (2004) Allanite and other REE-rich epidote-group minerals. Rev Mineral Geochem 56: 431-494

Giuli G, Bonazzi P, Menchetti S (1999) Al-Fe disorder in synthetic epidotes: a single-crystal X-ray diffraction study. Am Mineral 84:933-936

Goldschlag M (1917) Über die optischen Eigenschaften der Epidote. Tscherm Mineral Petr Mitt NF 23-60

Goldschmidt V (1916) Atlas der Kristallformen, Tafeln. Vol. 3, Carl Winters Universitätsbuchhandlung, Heidelberg p 247

Goldschmidt V (1923) Atlas der Kristallformen, Tafeln. Vol. 9, Carl Winters Universitätsbuchhandlung, Heidelberg p 128

Gottardi G (1954) Dati ed osservazioni sulla struttura dell'epidoto. Period Mineral 23:245-250

Gottschalk M (2004) Thermodynamic properties of zoisite, clinozoisite and epidote. Rev Mineral Geochem. 56:83-124

Grapes R, Watanabe T (1984) Al – $Fe^{3+}$ and Ca – $Sr^{2+}$ epidotes in metagreywacke – quartzofeldspathic schist, Southern Alps, New Zealand. Am Mineral 69:490–498

Grapes RH (1981) Chromian epidote and zoisite in kyanite amphibolite, Southern Alps, New Zealand. Am Mineral 66:974-975

Grapes RH, Hoskin PWO (2004) Epidote group minerals in low–medium pressure metamorphic terranes. Rev Mineral Geochem 56:301-345

Green TH, Pearson NJ (1983) REE partitioning betweenb sphene, allanite and chevkinite and coexisting intermediate - felsic liquids at high P, T. Lithosphere dynamics and evolution of the continental crust, Abstract Vol, p. 157. Geological Soc Australia, Canberra

Grevel KD, Nowlan EU, Faßhauer DW, Burchard M (2000) *In situ* X-ray difraction investigation of lawsonite and zoisite at high pressures and temperatures. Am Mineral 85:206-216

Grew ES, Essene EJ, Peacor DR, Su S-C, Asami M (1991) Dissakisite-(Ce), a new member of the epidote group and the Mg analogue of allanite-(Ce), from Antartica. Am Mineral 76:1990-1997

Guild (1935) Piedmontite in Arizona. Am Mineral 20:679-692

Hanisch K, Zemann J (1966) Messung des Ultrarot-Pleochroismus von Mineralen. IV. Der Pleochroismus der OH-Streckfrequenz in Epidot. N Jb Mineral Monatshefte 1966:19-23

Harlow GE (1994) Jadeitites, albitites and related rocks from the Motagua Fault Zone, Guatemala. J Metam Geol 12:49-68

Haüy RJ (1822) Traité de Mineralogie. 2:575

Hermann J (2002) Allanite: thorium and light rare earth element carrier in subducted crust. Chem Geol 192: 289-306

Heuss-Aßbichler S (2000) Ein neues Ordnungsmodell für die Mischkristallreihe Klinozoisit-Epidot und das Granat-Epidot-Geothermometer. Habilitationthesis, Munich p105

Heuss-Aßbichler S, Fehr KT (1997) Intercrystalline exchange of Al and $Fe^{3+}$ between grossular – andradite and clinozoisite – epidote solid solutions. N Jahrb Mineral Abh 172:69–100

Hietanen (1974) Amphibole pairs, epidote minerals, chlorite and plagioclase in metamorphic rocks. Am Mineral 59:22-40

Hintze CH (1897) Handbuch der Mineralogie, Vol 2 Silicate und Titanate Verlag Veit & Comp. Leipzig

Hirowatari F (1956) Manganiferous epidote from the Kakinomoto Mine, Kochi Prefecture. J Min Soc Japan 2: 331-346 (quoted in Strens 1966)

Holdaway MJ (1972) Thermal stability of Al-Fe epidote as a function of $f_{O2}$ and Fe content. Contrib Mineral Petrol 37:307-340

Holland TJB, Redfern SAT, Pawley AR (1996) Volume behavior of hydrous minerals at high pressure and temperature: II. Compressibilities of lawsonite, zoisite, clinozoisite, and epidote. Am Mineral 81:341-348

Holtstam D, Langhof J (1994) Hancockite from Jacobsberg, Filipstad, Sweden: the second world occurrence. Mineral Mag 58:172-174

Hörmann PK, Raith M (1971) Optische Daten, Gitterkonstanten, Dichte und magnetische Susceptibilität von Al-Fe(III)-Epidoten. N Jahrb Mineral Abh 116:41-60

Hurlbut CS (1969) Gem zoisite from Tanzania. Am Mineral 54:702-709

Hutton CO (1938) On the nature of withamite from GlenCoe, Scotland. Mineral Mag 25:119-124

Hutton CO (1940) Metamorphism in the Lake Wakatipu region, western Otago, New Zealand. Dept Sci Ind Res, New Zealand, Geol Mem 5

Hutton CO (1951) Allanite from Yosemity National Park, Tuolumne Co, California. Am Mineral 36:233-248

Ito T (1950) X-ray studies on polymorphism. Tokyo, Maruzen Co., chapter 5

Ito T, Morimoto N, Sadanga R (1954) On the structure of Epidote. Acta Cryst 7:53-59

Jambor JL, Puziewicz J, Roberts AC (1995) New Mineral Names. Am Mineral 80:404-409.

Jancev S, Bermanec V (1998) Solid solution between epidote and hancockite from Nezilovo, Macedonia. Geol Croat 51/1:23-26

Janeczek J, Eby RK (1993) Annealing of radiation damage in allanite and gadolinite. Phys Chem Mineral 19: 343-356

Johannes W, Ziegenbein D (1980) Stabilität von Zoisit in $H_2O$-$CO_2$-Gasphasen. Fortschr Mineral 50:46-47

Johannson A (1937) A Descriptive Petrography of the Igneous Rocks. The University of Chicago Press, Chicago

Johnston RW (1948) Clinozoisite from Camaderry Mountain, Co. Wicklow. Min Mag 28:505-515

Kartashov PM, Ferraris G, Ivaldi G, Sokolova E, McCammon CA (2002) Ferriallanite-(Ce), $CaCeFe^{3+}AlFe^{2+}$ $(SiO_4)(Si_2O_7)O(OH)$, a new member of the epidote group: Description, X-ray and Mössbauer study. Can Mineral 40:1641-1648

Kepezhinskas KB (1969) Determination of the composition of minerals of the epidote group from their physical properties. Dokl Acad Sci, USSR, Earth Sci Sect 185:104-106

Kepezhinskas KB, Khlestov VV (1971) Statistical analysis of the epidote group minerals. Trudy Inst Geol Geifiz Akad Nauk, USSR. Nowosibirsk, p 310

Kiseleva JA, Topor ND, Andreyenko EO (1974) Thermodynamic parameters of minerals of the epidote group. Geochem Internat 11:389-398

Klein C (1874) Die optischen Eigenschaften des Sulzbacher Epidot. N Jahrb Mineral 1-21

Klemd R (2004) Fluid inclusions in epidote minerals and fluid development in epidote-bearing rocks. Rev Mineral Geochem. 56:197-234

Kvick Å, Pluth JJ, Richardson Jr. JW, Smith JV (1988) The ferric iron distribution and hydrogen bonding in epidote: a neutron diffraction study at 15 K. Acta Cryst B 44:351-355

Langer K, Lattard L (1980) Identification of a low-energy OH-valence vibration in zoisite. Am Mineral 65: 779-783

Langer K, Raith M (1974) Infrared spectra of Al-Fe(III)-epidotes and zoisites, $Ca_2(Al_{1-p}Fe^{3+}_p)Al_2O(OH)[Si_2O_7][SiO_4]$. Am Mineral 59:1249–1258

Langer K, Tillmanns E, Kersten M, Almen H, Arni RK (2002) The crystal chemistry of $Mn^{3+}$ in the clino- and orthozoisite structure types, $Ca_2M_3^{3+}[OH/O/SiO_4/Si_2O_7]$: A structural and spectroscopic study of some natural piemontites and "thulites" and their synthetic equivalents. Z Krist 217:563-580

Liebscher A (2004) Spectroscopy of epidote minerals. Rev Mineral Geochem. 56:125-170

Liebscher A, Gottschalk M, Franz G (2002) The substitution $Fe^{3+}$ - Al and the isosymmetric displacive phase transition in synthetic zoisite: A powder X-ray and infrared spectroscopy study. Am Mineral 87:909-921

Linke W (1970) Messung des Ultrarot-Pleochroismus von Mineralen. X. Der Pleochroismus der OH-Streckfrequenz in Zoisit. Tschermaks Mineral Petrogr Mitteil 14:61-63

Liou JG (1973) Synthesis and stability relations of epidote, $Ca_2Al_2FeSi_3O_{12}(OH)$. J Petrol 14:381-314

Lüschen H (1979) Die Namen der Steine. Ott Verlag, Thun (Switzerland), p 380

Maaskant P (1985) The iron content and the optic axial angle in zoisites from Galicia, NW Spain. Mineral Mag 49:97-100

Malisa E (1987) Geology of the tanzanite gemstone deposits in the Lelanterna area, NE Tanzania. Ann Acad Sci Fennicae A 111:146

Malmqvist D (1929) Studien innerhalb der Epidotgruppe mit besonderer Rücksicht auf die manganhältigen Glieder. Bull Geol Inst Uppsala 22:223-280

Marmo V, Neuvonen KJ, Ojanpera P (1959) The piemontite of Piedmont (Italy) Kajlidongri (India) and Marampa (Sierra Leone). Bull Comm géol Finlande 184:11-20 (quoted in Strens 1966)

Masuda T, Shibutani T, Kuriyama M, Igarashi T (1990) Development of microboudinage: an estimate of changing differential stress with increasing strain. Tectonophysics 178:379-387

Masuda T, Shibutani T, Yamaguchi H (1995) Comparative rheological behaviour of albite and quartz in siliceous schists revealed by the microboudinage of piedmontite. J Struc Geol 11:1523-1533

Matthews A (1985) Kinetics and mechanisms of the reaction of zoisite to anorthite under hydrothermal conditions: reaction phenomenology away from the equilibrium region. Contrib Mineral Petrol 89:110-121

Matthews A, Goldsmith JR (1984) The influence of metastability on reaction kinetics involving zoisite formation from anorthite at elevated pressures and temperatures. Am Mineral 69:848-857

Mingsheng P, Dien L (1987) Spectroscopy, genesis and process properties of partly metamict allanite. J Central-south Inst Mining Metallurgy 18:362-368

Miyajima H, Matsubara S, Miyawaki R, Hirokawa K (2003) Niigataite, $CaSrAl_3(Si_2O_7)(SiO_4)O(OH)$: Sr-analogue of clinozoisite, a new member of the epidote group from the Itoigawa-Ohmi district, Niigata Prefecture, central Japan. J Mineral Petrol Sci 98:118-129

Morrison J (2004) Stable and radiogenic isotope systematics in epidote group minerals. Rev Mineral Geochem 56:607-628

Mottana A (1986) Blueschist-facies metamorphism of manganiferous cherts: A review of the alpine occurrence. *In:* Blueschists and Eclogites. Evans BW, Brown EH (eds) Memoir Geol Soc Amer 164:267-300

Myer GH (1965) X-Ray determinative curve for epidote. Am J Sci 263:78-86

Myer GH (1966) New data on zoisite and epidote. Am J Sci 264:364-385

Nagasaki A, Enami M (1998) Sr-bearing zoisite and epidote in ultra-high pressure (UHP) metamorphic rocks from the Su-Lu province, eastern China: an important Sr reservoir under UHP conditions. Am Mineral 83:240–247

Nayak VK, Neuvonen KJ (1963). Some manganese minerals from India. Bull Comm géol Finlande 212:27-36 (quoted in Strens 1966)

Neumann H (1985) The norwegian minerals. Nor Geol Unders Skr 68, p 278

Newton RC (1965) The thermal stability of zoisite. J Geol 73:431-441

Nozik YK, Kanepit VN, Fykin LY, Makarov YS (1978) A neutron diffraction study of the structure of epidote. Geochem Int 15:66-69

Orlov A (1926) On the iron poor members of the zoisite-epidote group. Mem Soc Roy Bohème 19:1-42 (in Czech)

Otto H (1935) Tschermaks Min Petr Mitt 47:89 (quoted in Strens 1966)

Paesano A (1983) A $^{57}Fe$ Mössbauer study of epidote. Hyperfine Interactions 15/16:841-844

Patrier P, Beaufort D, Meunier A, Emery J-P, Petit S (1991) Determination of the nonequilibrium ordering state in epidote from the ancient geothermal field of Saint Martin: Application of Mössbauer spectroscopy. Am Mineral 76:602-610

Pautov LA, Khorov PV, Ignatenko KI, Sokolova EV, Nadezhina TN (1993) Khristovite (Ce) - (Ca,REE)-RE E(Mg,Fe)AlMnSi$_3O_{11}$(OH)(F,O): A new mineral in the epidote group. Zapsiki Vseross Mineral Obshch 122(3): 103-111 (in Russian)

Pawley AR, Chinnery NJ, Clark SM (1998) Volume measurements of zoisite at simultaneously elevated pressure and temperature. Am Mineral 83:1030-1036

Pawley AR, Redfern SAT, Holland TJB (1996) Volume behavior of hydrous minerals at high pressure and temperature: I. Thermal expansion of lawsonite, zoisite, clinozoisite, and diaspore. Am Mineral 81:335-340

Peacor DR, Dunn PJ (1988) Dollaseite-(Ce) (magnesium orthite redefined): Structure refinement and implications for F+M2+ substituitions in epidote group minerals. Am Mineral 73:838-842

Penfield SL, Warren CH (1899) Some new minerals from the zinc mines at Franklin, N. J. and note concerning the chemical composition of ganomalite. Amer Jour Sci 4th Ser 8:339-353

Perkins III D, Westrum EFJ, Essene EJ (1980) The thermodynamic properties and phase relations of some minerals in the system $CaO-Al_2O_3-SiO_2-H_2O$. Geoch Cosmochim Acta 44:61-84

Perseil EA (1990) Sur la présence du strontium dans les minérallisations manganésifères de Falotta et de Parsettens (Grisonsuisse) – Evolution des paragenèes. Schweiz Min Petrogr Mitt 70:315-320

Pichler S-R (1993) Gesteinsbildende Minerale im Dünnschliff. Enke Verlag, Stuttgart, p 233

Pistorius CWFT (1961) Synthesis and lattice constants of pure zoisite and clinozoisite. J Geol 69:604-609

Poli S, Schmidt MW (2004) Experimental subsolidus studies on epidote minerals. Rev Mineral Geochem 56: 171-195

Raith M (1976) The Al – Fe(III) epidote miscibility gap in a metamorphic profile through the Penninic Series of the Tauern Window. Contrib Mineral Petrol 57:99–117

Rammelsberg KF (1856) Über den Zoisit und seine Beziehung zum Epidot, so wie über die Zusammensetzung des letzteren. Monatsberichte der Königlich Preußischen Akademie der Wissenschaften zu Berlin 1856: 605-617

Ray NJ, Putnis A, Gillet P (1986) Polytypic relationship between clinozoisite and zoisite. Bull Mineral 109: 667-685

Reinecke T (1986) Crystal chemistry and reaction relations of piemontites and thulites from highly oxidized low grade metamorphic rocks at Vitali, Andros Island, Greece. Contrib Mineral Petrol 93:56-76

Robinson K, Gibbs GV, Ribbe PH (1971) Quadratic elongation: a quantitative measure of distortion in coordination polyhedra. Science 172:567-570

Rouse RC, Peacor DR (1993) The crystal structure of dissakisite-(Ce), the Mg analogue of allanite-(Ce). Can Mineral 31:153-157

Rumanova IM, Nikolaeva TV (1960) Crystal structure of orthite. Soviet Phys – Crystallography 4:789-795

Sánchez-Viscaíno VL, Franz G, Gomez-Pugnaire MT (1995) The behaviour of Cr during metamorphism of carbonate rocks from the Nevado-Filabride complex, Betic Cordilleras, Spain. Can Mineral 33:85-104

Schiffmann P, Liou JG (1983) Synthesis of Fe-pumpellyite and its stability relations with epidote. J Metamor Geol 1:91-101

Schmetzer K, Bank H (1979) Bluish-green zoisite from Merelani, Tanzania. J Gemmol 16:512-513.

Schmetzer K, Berdesinski W (1978) Das Absorptionsspektrum von $Cr^{3+}$ in Zoisit. Neu Jahrb Mineral Monatsh 1978:197-202

Schmidt MW, Poli S (1994) The stability of lawsonite and zoisite at high pressures: Experiments in CASH to 92 kbar and implications for the presence of hydrous phases in subducted lithosphere. Earth Planet Sci Lett 124:105-118

Schmidt MW, Poli S (2004) Magmatic epidote. Rev Mineral Geochem 56:399-430

Seki Y (1959) Relation between chemical composition and lattice constants of epidote. Am Mineral 44:720-730

Shepel AB, Karpenko MV (1969) Mukhinite, a new variety of epidote. Dokl Akad Nauk SSR 185:1342-1345 (in Russian)

Short AM (1933) A chemical study of piedmontite from Shadow Lake, madera County, California. Am Mineral 18:493-500

Smelik EA, Franz G, Navrotsky A (2001) A calorimetric study of zoisite and clinozoisite solid solutions. Am Mineral 86:80-91

Smith D, Albee AL (1967) Petrology of a piemontite-bearing gneiss, San Ggiorgonio Pass, California. Contrib Mineral Petrol 16:189-203

Smith G, Halenius U, Langer K (1982) Low temperature spectral studies of $Mn^{3+}$-bearing andalusite and epidote type minerals in the range 30,000-5,000 $cm^{-1}$. Phys Chem Mineral 8:136-142

Smith JV, Pluth JJ, Richardson Jr. JW, Kvick Å (1987) Neutron diffraction study of zoisite at 15 K and X-ray study at room temperature. Z Krist 179:305-321

Smyth JR, Bish DL (1988) Crystal structures and cation sites of the rock-forming minerals. Allen and Unwin, Boston, p 332

Spear FS (1993) Metamorphic Phase Equilibria and Pressure-Temperature-Time Paths. Min Soc Am, Washington DC

Spear FS, Franz G (1986) P-T-evolution of metasediments from the Eclogite Zone, south-central Tauern Window, Austria. Lithos 19:219-234

Spiegel C, Siebel W, Frisch W, Berner Z (2002) Nd and Sr isotopic ratios and trace element geochemistry of epidote from the Swiss Molasse Basin as provenance indicators: implications for the reconstruction of the exhumation history of the Central Alps. Chem Geol 189:231-250

Stergiou AC, Rentzeperis PJ, Sklavounos S (1987) Refinement of the crystal structure of a medium iron epidote. Z Krist 178:297-305

Storre B, Johannes W, Nitsch KH (1982) The stability of zoisite in $H_2O$-$CO_2$ mixtures. N Jahrb Mineral Monatsh 1982:395-406

Strens RGJ (1963) Some relationship between members of the epidote group. Nature 198:80-81

Strens RGJ (1964) Epidotes of the Borrowdale volcanic rocks of central Borrowdale. Mineral Mag 33:868-886

Strens RGJ (1966) Properties of the Al-Fe-Mn epidotes. Mineral Mag 25:928-944

Strunz H, Nickel EH (2001) Strunz mineralogical tables. E Schweizerbart'sche Verlagsbuchhandlung Stuttgart. p 870

Stünitz F (1993) Deformation of granitoids at low metamorphic grade. I: Reactions and grain size reduction. Tectonophysics 221:269-297

Stünitz F(1993) Transition from fracturing to viscous flow in a naturally deformed metagabbro. *In*: Defects and Processes in the Solid State: Geoscience Applications. JN Boland, JD Fitz Gerald (eds) Elsevier, Amsterdam, p 121-150

Stünitz T (2001) Weakening and strain localization produced by syn-deformational reaction of plagioclase. Int J Earth Sciences (Geol Rundsch) 90:136-148

Termier P (1898) Sur une variete de zoisite des schistes metamorphiques des Alpes et sur les properties optiques de la zoisite classique. Soc franc mineral Bull 21:148-170

Termier P (1900) Sur une association d´epidote et de zoisite et sur les rapports cristallographiques de ces especes minerales. Soc franc mineral Bull 21:148–170

Thomson T (1810) Experiments on allanite, a new mineral from Greenland. R Soc Edinburgh Trans 8:371-386

Treloar PJ (1987a) Chromian muscovites and epidotes from Outukumpu, Finland. Minal Mag 51:593-599

Treloar PJ (1987b) The Cr-minerals of Outukumpu - their chemistry and significance. J Petrol 28:867-886

Tröger WE (1982) Optische Bestimmung der gesteinsbildenden Minerale, Teil 1 Bestimmungstabellen. E Schweizerbarth'sche Verlagsbuchhandlung Stuttgart. p 188

Tsang T, Ghose S (1971) Ordering of transition metal ions in zoisite. EOS Transactions Am Geophys Union 52,4:380-381

Tschermak G, Sipöcz L (1882) Beitrag zur Kenntnis des Zoisits. Z Kryst Mineral 6:200-202

Tsuboi S (1936) Japan J Geol Geog 13:333 (quoted in Strens 1966)

Tullock AJ (1979) Secondary Ca-Al silicates as low-grade alteration products of granitoid biotite. Contrib Mineral Petrol 69:105-117

Ueda T (1955) The crystal structure of allanite, $OH(Ca,Ce)_2(Fe^{3+}Fe^{2+})Al_2OSi_2O_7SiO_4$. Mem Coll Sci Univ Kyotot B12:145-163

v Eynatten H (2003) Petrography and chemistry of sandstones from the Swiss Molasse Basin: an archive of the Oligocene to Miocene evolution of the Central Alps. Sedimentol 50:703-724

Vogel DE, Bahezre C (1965) The composition of partially zoned garnet and zoisite from Cabo Ortegal, N.W, Spain. N Jahrb Mineral Monatsh 1965:140-149

Webster R (1962) Gems: Their Sources, Descriptions and Identification. Butterworths, London

Webster R, Anderson BW (1983) Gems: Their Sources, Descriptions and Identification. Butterworths, London

Weinschenk E (1896) Über Epidot und Zoisit. Z Kryst Mineral 26:154-177

Weiss ChS (1820) Über die Theorie des Epidotsystemes. Abhandlungen der Königlich Preußischen Akademie der Wissenschaften zu Berlin 1818/1819 Phys:242-269

Weiss ChS (1828) Über die Verhältnisse in den Dimensionen der Krystallsysteme, und insbesondere des Quarzes, des Feldspathes, der Hornblende, des Augites und des Epidotes. Abh König Preuß Akad Wissensch Berlin 1825 Phys:163-200

Winchell AN, Winchell H (1962) Elements of optical mineralogy.- Part III: Determinative Tables. 2$^{nd}$ ed, Wiley and Sons, New York p 231

Winkler B, Langer K, Johannsen PG (1989) The influence of pressure on the OH valence vibration of zoisite. Physics and Chemistry of Minerals 16:668-671

Yoshizawa H (1984) Notes on petrography and rock-forming mineralogy; (16), Sector-zoned epidote from Sanbagawa Schist in central Shikoku, Japan. J Japan Assoc Mineral Petrol Econ Geol 79:101-110

Zambonini F (1903) Krystallographisches über den Epidot. Z Krist 37:70 (quoted in Strens 1966)

Zambonini F (1920) Sulla clinozoisite di Chiampernotto in Val d´Ala. Boll Com Geol Ital 47:65-99 (quoted in Strens 1966)

**Appendices found on the following pages.**

# APPENDIX A

## Lattice constants of natural and synthetic monoclinic epidote minerals

| Composition | Sample |
|---|---|
| **Al-Fe Epidotes** | |
| $Ca_2Al_3[Si_3O_{11}(O/OH)]$ | Synth. |
| $Ca_2(Al_{2.996}Fe^{3+}_{0.019})[Si_{2.989}O_{11}(O/OH)]$ | Nat. |
| $Ca_2(Al_{2.97}Fe^{3+}_{0.03})[Si_3O_{11}(O/OH)]$ | Nat. |
| $(Ca_{1.975}Fe^{2+}_{0.01}Mn^{2+}_{0.01})(Al_{2.79}Fe^{3+}_{0.16})[Si_{3.04}O_{11}(O/OH)]$ | Nat. |
| $Ca_2(Al_{2.78}Fe^{3+}_{0.22})[Si_3O_{11}(O/OH)]$ | Nat. |
| $Ca_{2.003}(Al_{2.809}Fe_{0.224}Mn_{0.009})[Si_{2.956}O_{11}(O/OH)]$ | Nat. |
| $(Ca_{1.975}Fe^{2+}_{0.04}Na_{0.005}K_{0.005}Mn^{2+}_{0.015}Mg_{0.005})(Al_{2.705}Fe^{3+}_{0.225}Ti_{0.003})[Si_{3.00}O_{11}(O/OH)]$ | Nat. |
| $Ca_{1.985}(Al_{2.73}Fe^{3+}_{0.235}Fe^{2+}_{0.02}Mn^{2+}_{0.04})[Si_{3.005}O_{11}(O/OH)]$ | Nat. |
| $(Ca_{2.015}Na_{0.015})(Al_{2.725}Fe^{3+}_{0.24}Fe^{2+}_{0.04}Mn^{2+}_{0.01})[Si_{2.95}Al_{0.05}O_{11}(O/OH)]$ | Nat. |
| $(Ca_{1.99}Fe^{2+}_{0.01}Na_{0.015}K_{0.02})(Al_{2.68}Fe^{3+}_{0.25}Mg_{0.015}Mn^{2+}_{0.02}Ti_{0.005})[Si_{3.00}P_{0.005}O_{11}(O/OH)]$ | Nat. |
| $(Ca_{1.925}Fe^{2+}_{0.04}Na_{0.005}K_{0.005}Mn^{2+}_{0.005}Mg_{0.035})(Al_{2.665}Fe^{3+}_{0.29}Ti_{0.025})[Si_{2.97}Al_{0.025}O_{11}(O/OH)]$ | Nat. |
| $Ca_2(Al_{2.7}Fe^{3+}_{0.3})[Si_3O_{11}(O/OH)]$ | Nat. |
| $(Ca_{1.985}Fe^{2+}_{0.01}Mn^{2+}_{0.005})(Al_{2.465}Fe^{3+}_{0.305}Ti_{0.015})[Si_{2.915}O_{11}(O/OH)]$ | Nat. |
| $(Ca_{1.965}Fe^{2+}_{0.045}Na_{0.015}K_{0.015}Mn^{2+}_{0.01}Mg_{0.035})(Al_{2.585}Fe^{3+}_{0.31}Ti_{0.055})[Si_{2.975}Al_{0.025}O_{11}(O/OH)]$ | Nat. |
| $Ca_2(Al_{2.66}Fe^{3+}_{0.34})[Si_3O_{11}(O/OH)]$ | Nat. |
| $(Ca_{1.982}Fe^{2+}_{0.005}Mn^{2+}_{0.007}Mg_{0.03})(Al_{2.644}Fe^{3+}_{0.348}Ti_{0.008})[Si_{3.002}O_{11}(O/OH)]$ | Nat. |
| $(Ca_{1.975}Fe^{2+}_{0.04}Na_{0.005}K_{0.005}Mn^{2+}_{0.005}Mg_{0.005})(Al_{2.56}Fe^{3+}_{0.36}Ti_{0.05})[Si_{2.96}Al_{0.04}O_{11}(O/OH)]$ | Nat. |
| $(Ca_2Fe^{2+}_{0.03}Na_{0.01}K_{0.02})(Al_{2.59}Fe^{3+}_{0.365}Mg_{0.01}Mn^{2+}_{0.005}Ti_{0.01})[Si_{2.995}O_{11}(O/OH)]$ | Nat. |
| $(Ca_{1.955}Fe^{2+}_{0.04}Na_{0.005}K_{0.005}Mn^{2+}_{0.005}Mg_{0.005})(Al_{2.595}Fe^{3+}_{0.365}Ti_{0.03})[Si_{2.98}Al_{0.02}O_{11}(O/OH)]$ | Nat. |
| $(Ca_{1.90}Fe^{2+}_{0.035}Na_{0.01}K_{0.005}Mn^{2+}_{0.01})(Al_{2.595}Fe^{3+}_{0.37}Mg_{0.045}Ti_{0.03})[Si_{2.97}Al_{0.025}P_{0.005}O_{11}(O/OH)]$ | Nat. |
| $(Ca_{1.955}Fe^{2+}_{0.035}Na_{0.005}K_{0.005}Mn^{2+}_{0.005})(Al_{2.595}Fe^{3+}_{0.38}Ti_{0.02})[Si_{2.995}O_{11}(O/OH)]$ | Nat. |
| $(Ca_{1.999}Fe^{2+}_{0.027}Mg_{0.018})(Al_{2.57}Fe^{3+}_{0.385})[Si_{2.955}Al_{0.045}O_{11}(O/OH)]$ | Nat. |
| $(Ca_{1.985}Fe^{2+}_{0.04}Na_{0.005}Mn^{2+}_{0.005})(Al_{2.53}Fe^{3+}_{0.39}Ti_{0.05})[Si_{2.985}Al_{0.015}O_{11}(O/OH)]$ | Nat. |
| $(Ca_{1.93}Fe^{2+}_{0.04}Na_{0.015}Mn^{2+}_{0.005})(Al_{2.93}Fe^{3+}_{0.395}Ti_{0.03})[Si_{3.04}O_{11}(O/OH)]$ | Nat. |
| $Ca_{1.99}(Al_{2.649}Fe_{0.396})[Si_{2.965}O_{11}(O/OH)]$ | Nat. |
| $Ca_2(Al_{2.60}Fe^{3+}_{0.40})[Si_3O_{11}(O/OH)]$ | Nat. |
| $Ca_2(Al_{2.6}Fe^{3+}_{0.4})[Si_3O_{11}(O/OH)]$ | Nat. |
| $(Ca_{2.00}Fe^{2+}_{0.015}Mn^{2+}_{0.005})(Al_{2.475}Fe^{3+}_{0.40}Ti_{0.01})[Si_{3.075}O_{11}(O/OH)]$ | Nat. |
| $Ca_{1.992}(Al_{2.624}Fe_{0.429}Mn_{0.007})[Si_{2.947}O_{11}(O/OH)]$ | Nat. |
| $(Ca_{1.955}Fe^{2+}_{0.01}Mn^{2+}_{0.005})(Al_{2.5}Fe^{3+}_{0.44}Ti_{0.01})[Si_{3.05}O_{11}(O/OH)]$ | Nat. |
| $(Ca_{1.925}Fe^{2+}_{0.045}Na_{0.005}K_{0.005}Mn^{2+}_{0.005})(Al_{2.48}Fe^{3+}_{0.44}Ti_{0.025})[Si_{3.05}O_{11}(O/OH)]$ | Nat. |
| $(Ca_{1.90}Fe^{2+}_{0.04}Na_{0.01}K_{0.005}Mn^{2+}_{0.005})(Al_{2.55}Fe^{3+}_{0.455}Mg_{0.055}Ti_{0.015})[Si_{2.91}Al_{0.08}P_{0.01}O_{11}(O/OH)]$ | Nat. |
| $(Ca_{1.978}Th_{0.004})(Al_{2.584}Fe_{0.458})[Si_{2.977}O_{11}(O/OH)]$ | Nat. |
| $Ca_2(Al_{2.53}Fe^{3+}_{0.47})[Si_3O_{11}(O/OH)]$ | Nat. |
| $(Ca_{1.93}Fe^{2+}_{0.04}Na_{0.01}K_{0.005}Mn^{2+}_{0.005})(Al_{2.475}Fe^{3+}_{0.485}Mg_{0.04}Ti_{0.02})[Si_{2.96}Al_{0.035}P_{0.005}O_{11}(O/OH)]$ | Nat. |
| $(Ca_{1.92}Fe^{2+}_{0.05}Na_{0.02}K_{0.01}Mn^{2+}_{0.005}Mg_{0.01})(Al_{2.465}Fe^{3+}_{0.485}Ti_{0.025})[Si_{3.015}O_{11}(O/OH)]$ | Nat. |
| $(Ca_{1.975}Fe^{2+}_{0.02}Na_{0.01}K_{0.025})(Al_{2.41}Fe^{3+}_{0.49}Mg_{0.01}Mn^{2+}_{0.01}Ti_{0.005})[Si_{3.045}P_{0.005}O_{11}(O/OH)]$ | Nat. |
| $Ca_2(Al_{2.5}Fe^{3+}_{0.5})[Si_3O_{11}(O/OH)]$ | Nat. |
| $(Ca_{1.95}Fe^{2+}_{0.02}Mn^{2+}_{0.01})(Al_{2.425}Fe^{3+}_{0.515})[Si_{3.05}O_{11}(O/OH)]$ | Nat. |
| $Ca_{1.855}(Al_{2.51}Fe^{3+}_{0.525}Fe^{2+}_{0.065}Mn^{2+}_{0.01}Mg_{0.04})[Si_{2.98}Al_{0.02}O_{11}(O/OH)]$ | Nat. |
| $(Ca_{1.981}Mn^{2+}_{0.018}Mg_{0.003})(Al_{2.461}Fe^{3+}_{0.53}Ti_{0.006})[Si_{2.997}Al_{0.003}O_{11}(O/OH)]$ | Nat. |
| $(Ca_{1.93}Fe^{2+}_{0.05}Na_{0.025}K_{0.02}Mn^{2+}_{0.015})(Al_{2.425}Fe^{3+}_{0.53}Ti_{0.025})[Si_{3.05}O_{11}(O/OH)]$ | Nat. |
| $(Ca_{1.905}Fe^{2+}_{0.04}Na_{0.01}K_{0.005}Mn^{2+}_{0.01})(Al_{2.40}Fe^{3+}_{0.535}Mg_{0.085}Ti_{0.02})[Si_{2.96}Al_{0.04}P_{0.01}O_{11}(O/OH)]$ | Nat. |
| $(Ca_{1.95}Fe^{2+}_{0.04}Sr_{0.015})(Al_{2.37}Fe^{3+}_{0.575}Mg_{0.015}Mn^{2+}_{0.005}Ti_{0.03}V_{0.005}Cr_{0.005})[Si_{3.005}O_{11}(O/OH)]$ | Nat. |
| $Ca_2(Al_{2.40}Fe^{3+}_{0.60})[Si_3O_{11}(O/OH)]$ | Nat. |
| $(Ca_{1.965}Fe^{2+}_{0.015}K_{0.02})(Al_{2.295}Fe^{3+}_{0.625}Mg_{0.015}Mn^{2+}_{0.005}Ti_{0.005})[Si_{3.045}P_{0.005}O_{11}(O/OH)]$ | Nat. |
| $(Ca_{1.99}Fe^{2+}_{0.02}Na_{0.015}K_{0.02})(Al_{2.225}Fe^{3+}_{0.625}Mg_{0.04}Mn^{2+}_{0.005}Ti_{0.005})[Si_{3.07}O_{11}(O/OH)]$ | Nat. |
| $(Ca_{1.98}Fe^{2+}_{0.025}Na_{0.01}K_{0.02})(Al_{2.265}Fe^{3+}_{0.625}Mg_{0.02}Ti_{0.005})[Si_{3.08}O_{11}(O/OH)]$ | Nat. |
| $Ca_2(Al_{2.37}Fe^{3+}_{0.63})[Si_3O_{11}(O/OH)]$ | Nat. |
| $(Ca_{1.96}Fe^{2+}_{0.03}Na_{0.01}K_{0.015})(Al_{2.30}Fe^{3+}_{0.64}Mg_{0.005}Mn^{2+}_{0.01}Ti_{0.01})[Si_{3.025}O_{11}(O/OH)]$ | Nat. |
| $(Ca_2Fe^{2+}_{0.02}Na_{0.01}K_{0.02})(Al_{2.295}Fe^{3+}_{0.645}Mg_{0.015}Mn^{2+}_{0.005}Ti_{0.005})[Si_{3.08}O_{11}(O/OH)]$ | Nat. |
| $(Ca_{1.992}Sr_{0.017})(Al_{2.413}Fe_{0.67})[Si_{2.908}O_{11}(O/OH)]$ | Nat. |
| $(Ca_{1.93}Fe^{2+}_{0.05}Na_{0.03}K_{0.005}Mn^{2+}_{0.015})(Al_{2.24}Fe^{3+}_{0.675}Mg_{0.065}Ti_{0.035})[Si_{2.93}Al_{0.07}O_{11}(O/OH)]$ | Nat. |
| $(Ca_{1.895}Fe^{2+}_{0.04}Na_{0.0025}K_{0.0025}Mn^{2+}_{0.005}Mg_{0.005})(Al_{2.17}Fe^{3+}_{0.675}Ti_{0.02})[Si_{2.96}Al_{0.04}O_{11}(O/OH)]$ | Nat. |
| $Ca_{1.967}(Al_{2.389}Fe_{0.678}Ti_{0.007})[Si_{2.960}O_{11}(O/OH)]$ | Nat. |
| $(Ca_{1.96}Fe^{2+}_{0.025}Mn^{2+}_{0.01})(Al_{2.185}Fe^{3+}_{0.70}Ti_{0.015})[Si_{3.07}O_{11}(O/OH)]$ | Nat. |
| $(Ca_{1.845}Fe^{2+}_{0.035}Na_{0.015}Mn^{2+}_{0.025})(Al_{2.245}Fe^{3+}_{0.715}Mg_{0.045}Ti_{0.01})[Si_{3.055}O_{11}(O/OH)]$ | Nat. |
| $(Ca_{1.945}Fe^{2+}_{0.02}Mn^{2+}_{0.005})(Al_{2.205}Fe^{3+}_{0.715}Ti_{0.01})[Si_{3.065}O_{11}(O/OH)]$ | Nat. |
| $(Ca_{1.945}Fe^{2+}_{0.02}Mn^{2+}_{0.005})(Al_{2.185}Fe^{3+}_{0.715}Ti_{0.03})[Si_{3.06}O_{11}(O/OH)]$ | Nat. |

| $a$ [Å] | $b$ [Å] | $c$ [Å] | $\beta$ [°] | $V$ [Å³] | Source |
|---|---|---|---|---|---|
| 8.887 (1) | 5.5810 (1) | 10.14 (2) | 115.93 (13) | 452.3 (25) | Pistorius 1961 |
| 8.861 (3) | 5.5830 (1) | 10.141 (6) | 115.46 (2) | 453.0 (1) | Pawley et al. 1996 |
| 8.879 (5) | 5.5830 (5) | 10.155 (6) | 115.50 (5) | 454.4 | Dollase 1969 |
| 8.872 (3) | 5.590 (2) | 10.14 (4) | 115.500 (35) | 454.25 (21) | Myer 1965 |
| 8.870 (1) | 5.5920 (1) | 10.144 (2) | 115.4 (2) | 454.5 (2) | Comodi and Zanazzi 1997 |
| 8.872 (1) | 5.5930 (1) | 10.144 (1) | 115.46 (1) | 454.5 (1) | Bonazzi and Menchetti 1995 |
| 8.881 (4) | 5.605 (2) | 10.156 (4) | 115.45 (3) | 456.49 (46) | Hörmann and Raith 1971 |
| 8.87 (1) | 5.59 (1) | 10.15 (1) | 115.45 (3) | 454.4 (1) | Seki 1959 |
| 8.87 (1) | 5.59 (1) | 10.15 (1) | 115.45 (3) | 454.4 (1) | Seki 1959 |
| 8.874 (2) | 5.596 (1) | 10.153 (3) | 115.52 (3) | 455.0 (12) | Myer 1966 |
| 8.887 (3) | 5.611 (1) | 10.170 (4) | 115.44 (3) | 457.98 (39) | Hörmann and Raith 1971 |
| 8.869 (1) | 5.598 (1) | 10.146 (1) | 115.450 (1) | 454.9 | Carbonin and Molin 1980 |
| 8.879 (4) | 5.603 (1) | 10.151 (5) | 115.477 (45) | 455.93 (23) | Myer 1965 |
| 8.886 (4) | 5.609 (2) | 10.160 (5) | 115.43 (3) | 457.27 (47) | Hörmann and Raith 1971 |
| 8.879 (1) | 5.608 (1) | 10.154 (1) | 115.46 (1) | 456.5 | Carbonin and Molin 1980 |
| 8.874 (4) | 5.602 (2) | 10.147 (5) | 115.45 (4) | 455.50 (51) | Holdaway 1972 |
| 8.881 (7) | 5.611 (3) | 10.157 (8) | 115.39 (5) | 457.25 (74) | Hörmann and Raith 1971 |
| 8.879 (2) | 5.609 (1) | 10.154 (2) | 115.44 (2) | 456.70 (11) | Myer 1966 |
| 8.892 (6) | 5.615 (2) | 10.168 (6) | 115.46 (5) | 457.34 (6) | Hörmann and Raith 1971 |
| 8.880 (1) | 5.607 (1) | 10.154 (2) | 115.46 (1) | 456.50 (8) | Myer 1966 |
| 8.882 (7) | 5.613 (3) | 10.156 (8) | 115.37 (5) | 457.44 (79) | Hörmann and Raith 1971 |
| 8.878 (4) | 5.600 (2) | 10.145 (5) | 115.44 (4) | 455.50 (51) | Holdaway 1972 |
| 8.881 (8) | 5.617 (3) | 10.157 (9) | 115.39 (6) | 457.70 (92) | Hörmann and Raith 1971 |
| 8.880 (8) | 5.617 (3) | 10.153 (9) | 115.35 (6) | 457.71 (87) | Hörmann and Raith 1971 |
| 8.880 (2) | 5.603 (1) | 10.148 (1) | 115.44 (1) | 456.0 (1) | Bonazzi and Menchetti 1995 |
| 8.8802 (10) | 5.6043 (8) | 10.1511 (13) | 115.455 (12) | 456.2 | Gabe et al. 1973 |
| 8.8756 (10) | 5.608 (1) | 10.151 (1) | 115.41 (1) | 456.4 | Carbonin and Molin 1980 |
| 8.886 (4) | 5.606 (2) | 10.155 (5) | 115.490 (47) | 456.61 (31) | Myer 1965 |
| 8.884 (1) | 5.603 (3) | 10.157 (1) | 115.45 (1) | 456.5 (2) | Bonazzi and Menchetti 1995 |
| 8.883 (4) | 5.608 (2) | 10.151 (4) | 115.438 (37) | 456.70 (34) | Myer 1965 |
| 8.882 (6) | 5.619 (2) | 10.158 (7) | 115.35 (4) | 458.17 (67) | Hörmann and Raith 1971 |
| 8.886 (5) | 5.608 (3) | 10.155 (6) | 115.49 (5) | 456.80 (29) | Myer 1966 |
| 8.885 (1) | 5.607 (2) | 10.151 (1) | 115.44 (1) | 456.7 (2) | Bonazzi and Menchetti 1995 |
| 8.877 (1) | 5.613 (1) | 10.153 (1) | 115.39 (1) | 457.0 | Carbonin and Molin 1980 |
| 8.887 (2) | 5.615 (1) | 10.161 (2) | 115.46 (2) | 457.80 (11) | Myer 1966 |
| 8.886 (9) | 5.620 (3) | 10.159 (10) | 115.31 (7) | 458.26 | Hörmann and Raith 1971 |
| 8.884 (3) | 5.613 (2) | 10.155 (3) | 115.42 (3) | 457.40 (17) | Myer 1966 |
| 8.882 (1) | 5.613 (1) | 10.151 (1) | 115.43 (1) | 457.0 | Carbonin and Molin 1980 |
| 8.889 (4) | 5.621 (2) | 10.161 (6) | 115.450 (43) | 458.44 (41) | Myer 1965 |
| 8.88 (1) | 5.61 (1) | 10.17 (1) | 115.42 (3) | 457.6 (1) | Seki 1959 |
| 8.876 (4) | 5.613 (2) | 10.160 (5) | 115.40 (4) | 457.30 (51) | Holdaway 1972 |
| 8.893 (7) | 5.624 (2) | 10.165 (8) | 115.39 (5) | 459.25 (82) | Hörmann and Raith 1971 |
| 8.885 (3) | 5.625 (1) | 10.153 (3) | 115.41 (3) | 458.30 (16) | Myer 1966 |
| 8.886 (2) | 5.620 (1) | 10.159 (3) | 115.44 (3) | 458.10 (15) | Myer 1966 |
| 8.893 (3) | 5.640 (1) | 10.185 (1) | 115.34 (2) | 461.70 (1) | Stergiou et al. 1987 |
| 8.889 (2) | 5.625 (1) | 10.154 (2) | 115.41 (2) | 458.60 (11) | Myer 1966 |
| 8.891 (2) | 5.622 (1) | 10.156 (2) | 115.44 (2) | 458.40 (11) | Myer 1966 |
| 8.896 (6) | 5.628 (3) | 10.166 (7) | 115.50 (7) | 459.4 (3) | Myer 1966 |
| 8.888 (1) | 5.623 (1) | 10.157 (1) | 115.41 (1) | 458.5 | Carbonin and Molin 1980 |
| 8.888 (4) | 5.629 (2) | 10.161 (5) | 115.43 (5) | 459.10 (25) | Myer 1966 |
| 8.892 (2) | 5.625 (1) | 10.157 (3) | 115.42 (2) | 458.80 (41) | Myer 1966 |
| 8.892 (1) | 5.622 (1) | 10.159 (1) | 115.40 (1) | 458.8 (1) | Bonazzi and Menchetti 1995 |
| 8.886 (2) | 5.621 (1) | 10.157 (2) | 115.45 (2) | 458.10 (11) | Myer 1966 |
| 8.899 (5) | 5.635 (2) | 10.164 (6) | 115.36 (4) | 460.54 (62) | Hörmann and Raith 1971 |
| 8.884 (1) | 5.621 (1) | 10.154 (1) | 115.39 (1) | 458.1 (1) | Bonazzi and Menchetti 1995 |
| 8.893 (2) | 5.634 (1) | 10.159 (2) | 115.415 (18) | 459.73 (17) | Myer 1965 |
| 8.893 (3) | 5.626 (2) | 10.157 (3) | 115.42 (3) | 459.00 (17) | Myer 1966 |
| 8.896 (2) | 5.627 (1) | 10.159 (3) | 115.408 (23) | 459.32 (20) | Myer 1965 |
| 8.898 (2) | 5.635 (1) | 10.164 (2) | 115.425 (23) | 460.22 (18) | Myer 1965 |

**Appendix A** continued.

| Composition | Sample |
|---|---|
| Al-Fe Epidotes (continued from previous page) | |
| $(Ca_{1.989}Pb_{0.017})(Al_{2.059}Fe_{0.724}Mn_{0.2})[Si_{3.011}O_{11}(O/OH)]$ | Nat. |
| $(Ca_{1.985}Fe^{2+}_{0.015})(Al_{2.12}Fe^{3+}_{0.73}Ti_{0.015})[Si_{3.095}O_{11}(O/OH)]$ | Nat. |
| $Ca_{1.99^-2.005}(Al_{2.22^-2.43}Fe^{3+}_{0.67^-0.80})[Si_{2.93^-2.985}O_{11}(O/OH)]$ | Synth. |
| $(Ca_{1.975}Fe^{2+}_{0.035}Na_{0.005}K_{0.005}Mn^{2+}_{0.015}Mg_{0.01})(Al_{2.24}Fe^{3+}_{0.735}Ti_{0.005})[Si_{2.98}Al_{0.02}O_{11}(O/OH)]$ | Nat. |
| $(Ca_{1.979}Fe^{2+}_{0.004}Mn^{2+}_{0.015}Mg_{0.003})(Al_{2.25}Fe^{3+}_{0.74}Ti_{0.01})[Si_{2.999}Al_{0.001}O_{11}(O/OH)]$ | Nat. |
| $Ca_{1.961}(Al_{2.34}Fe_{0.741})[Si_{2.958}O_{11}(O/OH)]$ | Nat. |
| $X_{Ep} \sim 0.75$ (composition determined from X-ray data) | Synth. |
| $(Ca_{2.025}Fe^{2+}_{0.005}K_{0.02})(Al_{2.14}Fe^{3+}_{0.775}Mg_{0.01}Mn^{2+}_{0.01}Ti_{0.005})[Si_{3.08}O_{11}(O/OH)]$ | Nat. |
| $Ca_{1.978}(Al_{2.247}Fe_{0.777}Mn_{0.012})[Si_{2.986}O_{11}(O/OH)]$ | Nat. |
| $(Ca_{1.99}Mn^{2+}_{0.007}Mg_{0.011})(Al_{2.192}Fe^{3+}_{0.779}Ti_{0.022})[Si_{2.995}Al_{0.005}O_{11}(O/OH)]$ | Nat. |
| $Ca_2(Al_{2.22}Fe^{3+}_{0.78})[Si_3O_{11}(O/OH)]$ ? (composition from Rietveld refinment) | Nat. |
| $Ca_{1.99^-2.025}(Al_{2.135^-2.32}Fe^{3+}_{0.71^-0.855})[Si_{2.985^-3.00}O_{11}(O/OH)]$ | Synth. |
| $Ca_2(Al_{2.21}Fe^{3+}_{0.79})[Si_3O_{11}(O/OH)]$ | Nat. |
| $(Ca_{1.985}Fe^{2+}_{0.005}Na_{0.01}K_{0.02})(Al_{2.125}Fe^{3+}_{0.80}Mn^{2+}_{0.01})[Si_{3.04}P_{0.005}O_{11}(O/OH)]$ | Nat. |
| $Ca_2(Al_{2.15}Fe^{3+}_{0.81}Ti_{0.02}Mn_{0.02})[Si_3O_{11}(O/OH)]$ | Nat. |
| $(Ca_{1.948}Fe^{2+}_{0.03}Mn^{2+}_{0.03}Mg_{0.003})(Al_{2.169}Fe^{3+}_{0.824}Ti_{0.007})[Si_{2.982}Al_{0.018}O_{11}(O/OH)]$ | Nat. |
| $Ca_{1.93^-2.00}(Al_{2.145^-2.34}Fe^{3+}_{0.78^-0.87})[Si_{2.945^-2.99}O_{11}(O/OH)]$ | Synth. |
| $(Ca_{1.935}Fe^{2+}_{0.055}Na_{0.055}K_{0.005})(Al_{2.075}Fe^{3+}_{0.83}Mg_{0.07}Mn^{2+}_{0.015}Ti_{0.015})[Si_3O_{11}(O/OH)]$ | Nat. |
| $Ca_2(Al_{2.16}Fe^{3+}_{0.84})[Si_3O_{11}(O/OH)]$ | Nat. |
| $(Ca_{1.985}Fe^{2+}_{0.035}Mn^{2+}_{0.015}Mg_{0.015})(Al_{2.12}Fe^{3+}_{0.855}Ti_{0.005})[Si_{2.955}Al_{0.045}O_{11}(O/OH)]$ | Nat. |
| $(Ca_{1.98}Na_{0.015})(Al_{1.985}Fe^{3+}_{0.86}Fe^{2+}_{0.08}Mn^{2+}_{0.02}Mg_{0.025})[Si_{3.06}O_{11}(O/OH)]$ | Nat. |
| $Ca_2(Al_{2.14}Fe^{3+}_{0.86})[Si_3O_{11}(O/OH)]$ | Nat. |
| $(Ca_{1.96}Fe^{2+}_{0.03}Na_{0.005}Mn^{2+}_{0.01}Mg_{0.025})(Al_{2.085}Fe^{3+}_{0.865}Ti_{0.025})[Si_{2.97}Al_{0.03}O_{11}(O/OH)]$ | Nat. |
| $(Ca_{1.935}Fe^{2+}_{0.05}Na_{0.005}K_{0.005}Mn^{2+}_{0.005}Mg_{0.01})(Al_{2.12}Fe^{3+}_{0.89}Ti_{0.005})[Si_{2.985}Al_{0.015}O_{11}(O/OH)]$ | Nat. |
| $Ca_{1.976}(Al_{2.13}Fe_{0.923})[Si_{2.971}O_{11}(O/OH)]$ | Nat. |
| $Ca_{1.968}(Al_{2.098}Fe_{0.942}Mn_{0.007})[Si_{2.985}O_{11}(O/OH)]$ | Nat. |
| $Ca_2(Al_{2.04}Fe^{3+}_{0.96})[Si_3O_{11}(O/OH)]$ ? (no complete analysis) | Nat. |
| $Ca_{1.996}(Al_{2.001}Fe^{3+}_{0.961}Fe^{2+}_{0.017}Mn^{2+}_{0.004}Cr^{3+}_{0.006}Ti_{0.002})[Si_{3.013}O_{11}(O/OH)]$ | Nat. |
| $Ca_2(Al_{2.053}Fe_{0.965}Ti_{0.012})[Si_{2.971}O_{11}(O/OH)]$ | Nat. |
| $(Ca_{1.995}Fe^{2+}_{0.01})(Al_{1.935}Fe^{3+}_{0.985})[Si_{3.055}O_{11}(O/OH)]$ | Nat. |
| $(Ca_{1.996}Mn^{2+}_{0.012}Mg_{0.012})(Al_{1.985}Fe^{3+}_{0.990}Ti_{0.004})[Si_{2.984}Al_{0.016}O_{11}(O/OH)]$ | Nat. |
| $Ca_2(Al_2Fe^{3+})[Si_3O_{11}(O/OH)]$ ? (no complete analysis) | Nat. |
| $Ca_{1.962}(Al_{2.029}Fe_{1.006}Mn_{0.016}Ti_{0.011})[Si_{2.976}O_{11}(O/OH)]$ | Nat. |
| $(Ca_{2.005}Sr_{0.036})(Al_{1.982}Fe_{1.015}Ti_{0.015})[Si_{2.948}O_{11}(O/OH)]$ | Nat. |
| $Ca_{1.96}(Al_{2.063}Fe_{1.022})[Si_{2.954}O_{11}(O/OH)]$ | Nat. |
| $Ca_{2.05^-2.06}(Al_{1.90^-2.04}Fe^{3+}_{0.955^-1.115})[Si_{2.96^-2.98}O_{11}(O/OH)]$ | Synth. |
| $(Ca_{1.99}Fe^{2+}_{0.03})(Al_{1.935}Fe^{3+}_{1.045})[Si_{2.945}Al_{0.055}O_{11}(O/OH)]$ | Nat. |
| $Ca_{2.02^-2.035}(Al_{1.93^-2.01}Fe^{3+}_{0.98^-1.12})[Si_{2.955^-2.99}O_{11}(O/OH)]$ | Synth. |
| $Ca_{1.959}(Al_{1.957}Fe_{1.105}Mn_{0.015})[Si_{2.963}O_{11}(O/OH)]$ | Nat. |
| $Ca_{2.03}(Al_{1.80}Fe^{3+}_{1.105}Fe^{2+}_{0.025}Mn^{2+}_{0.02})[Si_{3.03}O_{11}(O/OH)]$ | Nat. |
| Pb-bearing Epidote | |
| $(Ca_{1.17}Mn^{2+}_{0.17}Pb_{0.47}Sr_{0.21})(Al_{1.95}Fe_{0.88}Mn_{0.10}Mg_{0.07})[Si_{2.95}Al_{0.05}O_{11}(O/OH)]$ | Nat. |
| Sr-bearing Al-Fe-Mn Epidotes | |
| $(Ca_{1.84}Sr_{0.16})(Al_{1.87}Fe^{3+}_{0.35}Mn^{3+}_{0.78})[Si_3O_{11}(O/OH)]$ | Nat. |
| $(Ca_{1.80}Sr_{0.20})(Al_{1.91}Fe^{3+}_{0.33}Mn^{3+}_{0.82})[Si_3O_{11}(O/OH)]$ | Nat. |
| $(Ca_{1.82}Sr_{0.06})(Al_{1.86}Fe^{3+}_{0.35}Mn^{3+}_{0.82})[Si_{3.03}O_{11}(O/OH)]$ | Nat. |
| $(Ca_{1.88}Sr_{0.08})(Al_{1.88}Fe^{3+}_{0.13}Mn^{3+}_{0.98})[Si_{2.98}O_{11}(O/OH)]$ | Nat. |
| $(Ca_{1.74}Sr_{0.16})(Al_{1.83}Fe^{3+}_{0.06}Mn^{3+}_{1.15})[Si_{3.01}O_{11}(O/OH)]$ | Nat. |
| $(Ca_{1.62}Sr_{0.35})(Al_{1.57}Fe^{3+}_{0.31}Mn^{3+}_{1.12})[Si_{3.01}O_{11}(O/OH)]$ | Nat. |
| $(Ca_{1.05}Mn^{2+}_{0.22}Sr_{0.73})(Al_{1.80}M^{3+}_{1.20})[Si_3O_{11}(O/OH)]$; M = Fe$^+$Mn | Nat. |
| $(Ca_{1.38}Sr_{0.62})(Al_{1.74}M^{3+}_{1.26})[Si_3O_{11}(O/OH)]$; M = Fe$^+$Mn | Nat. |
| $(Ca_{1.87}Sr_{0.13})(Al_{1.97}Fe^{3+}_{0.309}Mn^{3+}_{0.721})[Si_3O_{11}(O/OH)]$ | Nat. |
| Al-Fe-Mn Epidotes | |
| $(Ca_{1.92}Mg_{0.02}Mn^{2+}_{0.05}Na_{0.01})(Al_{1.96}Fe^{3+}_{0.66}Mn^{3+}_{0.38})[Si_{2.97}Al_{0.03}O_{11}(O/OH)]$ | Nat. |
| $(Ca_{1.96}Mg_{0.03}Mn^{2+}_{0.01})(Al_{1.87}Fe^{3+}_{0.07}Mn^{3+}_{1.04}Ti_{0.02})[Si_{2.97}Al_{0.03}O_{11}(O/OH)]$ | Nat. |
| $(Ca_{1.92}Mg_{0.02}Na_{0.01})(Al_{1.81}Fe^{3+}_{0.28}Mn^{3+}_{0.88})[Si_{3.04}O_{11}(O/OH)]$ | Nat. |
| $(Ca_{1.92}Mg_{0.04})(Al_{1.75}Fe^{3+}_{0.14}Mn^{3+}_{1.08})[Si_{3.04}O_{11}(O/OH)]$ | Nat. |
| $(Ca_{1.93}Mg_{0.02})(Al_{1.70}Fe^{3+}_{0.45}Mn^{3+}_{0.82})[Si_{3.05}O_{11}(O/OH)]$ | Nat. |
| $(Ca_{1.98}Mg<_{0.01}Sr<_{0.01}Zn_{0.01})(Al_{2.43}Fe^{3+}_{0.43}Mn^{3+}_{0.14}Cu_{0.01})[Si_{2.99}O_{11}(O/OH)]$ | Nat. |
| $Ca_2(Al_{1.87}Fe^{3+}_{0.51}Mn^{3+}_{0.66})[Si_{2.97}O_{11}(O/OH)]$ | Nat. |
| $(Ca_{1.802}Mn^{2+}_{0.178}Mg_{0.025})(Al_{1.825}Fe^{3+}_{0.346}Mn^{3+}_{0.829})[Si_{2.992}Al_{0.008}O_{11}(O/OH)]$ | Nat. |

| $a$ [Å] | $b$ [Å] | $c$ [Å] | $\beta$ [°] | $V$ [Å$^3$] | Source |
|---------|---------|---------|-------------|-------------|--------|
| 8.894 (1) | 5.647 (1) | 10.162 (1) | 115.41 (1) | 461.0 (1) | Bonazzi and Menchetti 1995 |
| 8.898 (4) | 5.631 (2) | 10.163 (5) | 115.423 (43) | 459.94 (36) | Myer 1965 |
| 8.908 (3) | 5.628 (2) | 10.176 (4) | 115.52 (5) | 460.4 (3) | Liou 1973 |
| 8.898 (5) | 5.635 (2) | 10.161 (6) | 115.32 (4) | 460.53 (61) | Hörmann and Raith 1971 |
| 8.888 (4) | 5.630 (2) | 10.151 (5) | 115.34 (4) | 459.10 (51) | Holdaway 1972 |
| 8.890 (2) | 5.623 (1) | 10.156 (1) | 115.41 (1) | 458.6 (1) | Bonazzi and Menchetti 1995 |
| 8.900 (6) | 5.639 (3) | 10.165 (3) | 115.75 (7) | 459.5 (5) | Schiffman and Liou 1983 |
| 8.902 (2) | 5.635 (1) | 10.164 (2) | 115.43 (2) | 460.50 (11) | Myer 1966 |
| 8.894 (1) | 5.630 (1) | 10.157 (1) | 115.41 (1) | 459.4 (1) | Bonazzi and Menchetti 1995 |
| 8.893 (4) | 5.631 (2) | 10.145 (5) | 115.34 (4) | 459.10 (51) | Holdaway 1972 |
| 8.913 (1) | 5.643 (1) | 10.179 (1) | 115.7 (1) | 461.3 | Nozik et al. 1978 |
| 8.891 (3) | 5.625 (2) | 10.177 (4) | 115.50 (5) | 459.4 (3) | Liou 1973 |
| 8.896 (1) | 5.634 (1) | 10.162 (1) | 115.41 (1) | 460.0 | Carbonin and Molin 1980 |
| 8.901 (3) | 5.643 (1) | 10.166 (3) | 115.41 (3) | 461.20 (15) | Myer 1966 |
| 8.914 (9) | 5.640 (3) | 10.162 (9) | 115.4 (2) | 461.5 | Dollase 1971 |
| 8.903 (4) | 5.617 (2) | 10.169 (5) | 115.50 (4) | 459.00 (51) | Holdaway 1972 |
| 8.893 (3) | 5.624 (2) | 10.175 (4) | 115.48 (5) | 459.4 (3) | Liou 1973 |
| 8.899 (1) | 5.639 (1) | 10.166 (2) | 115.38 (2) | 460.90 (9) | Myer 1966 |
| 8.8877 (14) | 5.6275 (8) | 10.1517 (12) | 115.383 (14) | 458.7 | Gabe et al. 1973 |
| 8.911 (9) | 5.640 (5) | 10.171 (11) | 115.44 (7) | 461.6 (11) | Hörmann and Raith 1971 |
| 8.89 (1) | 5.63 (1) | 10.19 (1) | 115.40 (3) | 460.7 (1) | Seki 1959 |
| 8.894 (1) | 5.637 (1) | 10.158 (1) | 115.36 (1) | 460.2 | Carbonin and Molin 1980 |
| 8.913 (10) | 5.640 (4) | 10.169 (11) | 115.44 (6) | 461.85 (103) | Hörmann and Raith 1971 |
| 8.915 (7) | 5.650 (3) | 10.178 (9) | 115.40 (5) | 462.75 (81) | Hörmann and Raith 1971 |
| 8.899 (1) | 5.639 (1) | 10.166 (1) | 115.42 (1) | 460.8 (1) | Bonazzi and Menchetti 1995 |
| 8.902 (1) | 5.641 (1) | 10.165 (1) | 115.39 (1) | 461.1 (1) | Bonazzi and Menchetti 1995 |
| 8.96 (1) | 5.36 (1) | 10.3 (1) | 115.40 (2) | 446.8 | Paesano et al. 1983 |
| 8.890 (5) | 5.641 (3) | 10.164 (6) | 115.55 (7) | 459.9 (3) | Holland et al. 1996 |
| 8.903 (2) | 5.649 (1) | 10.163 (1) | 115.39 (1) | 461.8 (1) | Bonazzi and Menchetti 1995 |
| 8.904 (5) | 5.649 (3) | 10.173 (8) | 115.440 (57) | 462.04 (46) | Myer 1965 |
| 8.894 (4) | 5.651 (2) | 10.161 (5) | 115.35 (4) | 461.50 (51) | Holdaway 1972 |
| 8.96 (1) | 5.63 (1) | 10.3 (1) | 115.40 (2) | 469.4 | Ito et al. 1954 |
| 8.901 (1) | 5.649 (1) | 10.173 (1) | 115.41 (1) | 462.0 (1) | Bonazzi and Menchetti 1995 |
| 8.901 (1) | 5.645 (1) | 10.168 (1) | 115.37 (1) | 461.6 (1) | Bonazzi and Menchetti 1995 |
| 8.901 (1) | 5.646 (1) | 10.167 (1) | 115.41 (1) | 461.5 (1) | Bonazzi and Menchetti 1995 |
| 8.911 (3) | 5.643 (2) | 10.180 (4) | 115.52 (5) | 462.0 (3) | Liou 1973 |
| 8.907 (1) | 5.660 (1) | 10.180 (1) | 115.40 (1) | 463.60 (6) | Myer 1966 |
| 8.922 (3) | 5.648 (2) | 10.194 (4) | 115.57 (5) | 463.4 (3) | Liou 1973 |
| 8.908 (1) | 5.663 (1) | 10.175 (2) | 115.35 (1) | 463.9 (1) | Bonazzi and Menchetti 1995 |
| 8.9 (1) | 5.63 (1) | 10.2 (1) | 115.40 (3) | 461.7 (1) | Seki 1959 |
| 8.958 (20) | 5.665 (10) | 10.304 (20) | 114.4 (4) | 476.2 | Dollase 1971 |
| 8.884 (3) | 5.684 (1) | 10.202 (3) | 115.23 (2) | 466.0 | Ferraris et al. (1989) |
| 8.884 (2) | 5.684 (1) | 10.202 (3) | 115.23 (2) | 466.0 | Catti et al. 1988 |
| 8.879 (2) | 5.687 (1) | 10.187 (3) | 115.35 (2) | 464.9 | Catti et al. 1989 |
| 8.880 (3) | 5.6829 (9) | 10.187 (3) | 115.36 (2) | 464.5 | Catti et al. 1989 |
| 8.870 (4) | 5.699 (1) | 10.201 (4) | 115.30 (2) | 466.2 | Catti et al. 1989 |
| 8.897 (4) | 5.702 (2) | 10.232 (5) | 115.07 (4) | 470.2 | Catti et al. 1989 |
| 8.849 (2) | 5.671 (2) | 10.203 (2) | 114.63 (2) | 465.4 (2) | Bonazzi et al. 1990 |
| 8.870 (2) | 5.681 (1) | 10.209 (2) | 114.88 (2) | 466.7 (2) | Bonazzi et al. 1990 |
| 8.878 (10) | 5.692 (5) | 10.201 (10) | 115.4 (2) | 465.7 | Dollase 1969 |
| 8.85 (1) | 5.660 (6) | 10.18 (2) | 115.6 (1) | 459.9 (10) | Akasaka et al. 1988 |
| 8.881 (7) | 5.687 (5) | 10.180 (8) | 115.53 (3) | 464.0 (6) | Akasaka et al. 1988 |
| 8.864 (6) | 5.690 (3) | 10.189 (8) | 115.4 (1) | 464.2 (5) | Akasaka et al. 1988 |
| 8.877 (1) | 5.696 (1) | 10.198 (1) | 115.61 (1) | 465.0 (19) | Akasaka et al. 1988 |
| 8.881 (7) | 5.697 (6) | 10.194 (7) | 115.54 (3) | 465.4 (6) | Akasaka et al. 1988 |
| 8.8739 (11) | 5.6156 (8) | 10.1484 (13) | 115.49 (1) | 456.5 (1) | Langer et al. 2002 |
| 8.8756 (11) | 5.6734 (7) | 10.1686 (13) | 115.5 (1) | 462.2 (1) | Langer et al. 2002 |
| 8.847 (3) | 5.677 (2) | 10.159 (3) | 115.35 (7) | 461.1 (4) | Smith et al. 1982 |

**Appendix A** continued.

| Composition | Sample |
|---|---|

**Al-Fe-Mn Epidotes** (continued from previous page)

$(Fe_{0.61};Mn_{0.08})$
$(Fe_{0.61};Mn_{0.13})$
$(Fe_{0.39};Mn_{0.38})$
$(Fe_{0.21};Mn_{0.59})$
$(Fe_{0.54};Mn_{0.31})$
$(Fe_{0.71};Mn_{0.24})$
$(Fe_{0.30};Mn_{0.73})$
$(Fe_{0.64};Mn_{0.45})$
$(Fe_{0.40};Mn_{0.70})$
$(Fe_{0.41};Mn_{0.72})$
$(Fe_{0.41};Mn_{0.72})$
$(Fe_{0.62};Mn_{0.63})$
$(Fe_{0.68};Mn_{0.75})$
$(Fe_{0.66};Mn_{0.78})$

**Al-Mn Epidotes**

| Composition | Sample |
|---|---|
| $(Ca_{1.92}Mn^{2+}_{0.11})(Al_{2.26}Mn^{3+}_{0.73})[Si_{2.97}O_{11}(O/OH)]$ | Synth. |
| $Ca_{1.95}(Al_{2.05}Mn^{3+}_{0.975})[Si_{3.005}O_{11}(O/OH)]$ | Synth. |
| $Ca_{1.97}(Al_{1.57}Mn^{3+}_{1.465})[Si_{2.985}O_{11}(O/OH)]$ | Synth. |
| $Ca_2(Al_{2.5}Mn^{3+}_{0.5})[Si_3O_{11}(O/OH)]$ | Synth. |
| $Ca_2(Al_{2.25}Mn^{3+}_{0.75})[Si_3O_{11}(O/OH)]$ | Synth. |
| $Ca_2(Al_2Mn^{3+}_1)[Si_3O_{11}(O/OH)]$ | Synth. |
| $Ca_2(Al_{1.75}Mn^{3+}_{1.25})[Si_3O_{11}(O/OH)]$ | Synth. |
| $Ca_2(Al_{1.5}Mn^{3+}_{1.5})[Si_3O_{11}(O/OH)]$ | Synth. |
| $Ca_2(Al_{1.25}Mn^{3+}_{1.75})[Si_3O_{11}(O/OH)]$ | Synth. |
| $Ca_2(Al_1Mn^{3+}_{2.0})[Si_3O_{11}(O/OH)]$ | Synth. |
| $Ca_2(Al_{0.5}Mn^{3+}_{2.5})[Si_3O_{11}(O/OH)]$ | Synth. |

**REE bearing Epidotes**

| Composition | Sample |
|---|---|
| $(Ca_{1.20}Y_{0.02}La_{0.23}Ce_{0.48}Nd_{0.07})(Al_{1.65}Fe_{1.20}Ti_{0.06}Mn_{0.04}Mg_{0.04})[Si_3O_{11}(O/OH)]$ | Nat. |
| $(Ca_{1.08}Ce_{0.92})(Al_{0.66}Fe^{3+}_{1.24}Fe^{2+}_{0.93}Mn_{0.07}Ti_{0.14})[Si_{2.94}Al_{0.06}O_{11}(O/OH)]$ | Nat. |
| $(Ca_{1.186}REE_{0.7})(Al_{1.921}Fe_{0.953}Mg_{0.167}Mn_{0.016}Ti_{0.045})[Si_{3.012}O_{11}(O/OH/F_{0.038})]$ | Nat. |
| $(Ca_{1.271}REE_{0.7}Y_{0.021})(Al_{1.968}Fe_{0.944}Mg_{0.033}Ti_{0.051})[Si_{3.003}O_{11}(O/OH/F_{0.041})]$ | Nat. |
| $(Ca_{1.05}Ce_{0.57}La_{0.33}Nd_{0.07}Pr_{0.03})(Al_{1.91}Fe_{0.14}Mg_{0.93}Ti_{0.06})[Si_{2.94}O_{11}(O/OH_{0.94}/F_{0.06})]$ | Nat. |
| $(Ca_{1.05}Ce_{0.57}La_{0.33}Nd_{0.07}Pr_{0.03})(Al_{1.91}Fe_{0.14}Mg_{0.93}Ti_{0.06})[Si_{2.94}O_{11}(O/OH_{0.94}/F_{0.06})]$ | Nat. |
| $(Ca_{0.91}Ce_{0.45}La_{0.20}Nd_{0.20}Pr_{0.09}Sm_{0.08}Gd_{0.06})(Al_{0.97}Fe_{0.25}Mg_{1.81})[Si_3O_{10.99}(OH_{1.25}/F_{0.88})]$ | Nat. |
| $(Ca_{1.639}REE_{0.323}Sr_{0.007})(Al_{2.009}Fe_{0.833}Mg_{0.083}Mn_{0.021}Ti_{0.012})[Si_{2.982}O_{11}(O/OH)]$ | Nat. |
| $(Ca_{1.25}REE_{0.6}Th_{0.02})(Al_{2.036}Fe_{0.890}Mg_{0.137}Mn_{0.02}Ti_{0.025})[Si_{3.024}O_{11}(O/OH)]$ | Nat. |
| $(Ca_{1.47}Mn^{2+}_{0.22}REE_{0.28}Sr_{0.03})(Al_{1.69}M^{3+}_{1.03}M^{2+}_{0.20}Mg_{0.08})[Si_3O_{11}(O/OH)]; M = Fe^+Mn$ | Nat. |
| $(Ca_{1.71}Mn^{2+}_{0.20}REE_{0.05}Sr_{0.04})(Al_{1.77}M^{3+}_{1.18}M^{2+}_{0.04}Mg_{0.01})[Si_3O_{11}(O/OH)]; M = Fe^+Mn$ | Nat. |
| $(Ca_{0.96}Mn^{2+}_{0.33}REE_{0.50}Sr_{0.19}Th^{4+}_{0.02})(Al_{1.71}Fe^{3+}_{0.288}Mn^{3+}_{0.432}Fe^{2+}_{0.112}Mn^{2+}_{0.168}Mg_{0.16}Cu_{0.13})[Si_3O_{11}(O_{0.97}/F_{0.03}/OH)]$ | Nat. |
| $(Ca_{0.64}Mn^{2+}_{0.60}REE_{0.72}Sr_{0.04})(Al_1Fe^{3+}_{0.064}Mn^{3+}_{1.216}Fe^{2+}_{0.035}Mn^{2+}_{0.675}Cu_{0.01})[Si_3O_{11}(O_{0.97}/F_{0.03}/OH)]$ | Nat. |

| $a$ [Å] | $b$ [Å] | $c$ [Å] | $\beta$ [°] | $V$ [Å$^3$] | Source |
|---|---|---|---|---|---|
| | 5.632 | | | | Strens 1966 |
| | 5.632 | | | | Strens 1966 |
| | 5.650 | | | | Strens 1966 |
| | 5.679 | | | | Strens 1966 |
| 8.870 | 5.650 | 10.170 | 115.5 | 460.4 | Strens 1966; aus Ernst. 1964 |
| 8.87 | 5.650 | 10.150 | 115.4 | 459.5 | Strens 1966 |
| 8.885 | 5.687 | | | | Strens 1966 |
| 8.87 | 5.660 | 10.150 | 115.4 | 459.9 | Strens 1966; aus Marmo et al. 1959 |
| 8.89 | 5.670 | 10.220 | 115.6 | 464.8 | Strens 1966; aus Marmo et al. 1959 |
| 8.86 | 5.681 | 10.156 | 115.4 | 461.8 | Strens 1966 |
| 8.88 | 5.69 | 10.167 | 115.4 | 463.6 | Strens 1966 |
| 8.89 | 5.67 | 10.170 | 115.5 | 461.9 | Strens 1966; aus Marmo et al. 1959 |
| 8.88 | 5.66 | 10.160 | 115.5 | 460.9 | Strens 1966; aus Nayak & Neuvonen. 1964 |
| 8.885 | 5.69 | 10.160 | 115.4 | 464.0 | Strens 1966 |
| 8.847 (2) | 5.674 (1) | 10.170 (1) | 115.56 (1) | 460.6 (1) | Langer et al. 2002 |
| 8.844 (1) | 5.677 (1) | 10.167 (1) | 115.54 (1) | 460.6 (1) | Langer et al. 2002 |
| 8.855 (1) | 5.713 (1) | 10.208 (1) | 115.62 (1) | 465.6 (1) | Langer et al. 2002 |
| 8.857 (3) | 5.636 (3) | 10.163 (4) | 115.65 (3) | 457.3 (4) | Anastasiou and Langer 1977 |
| 8.839 (3) | 5.651 (3) | 10.163 (4) | 115.58 (3) | 457.9 (4) | Anastasiou and Langer 1977 |
| 8.839 (3) | 5.664 (2) | 10.166 (4) | 115.61 (3) | 459.0 (4) | Anastasiou and Langer 1977 |
| 8.846 (2) | 5.688 (2) | 10.187 (2) | 115.63 (2) | 462.1 (3) | Anastasiou and Langer 1977 |
| 8.853 (3) | 5.702 (3) | 10.200 (4) | 115.69 (3) | 464.0 (5) | Anastasiou and Langer 1977 |
| 8.859 (3) | 5.712 (3) | 10.200 (4) | 115.66 (3) | 465.2 (5) | Anastasiou and Langer 1977 |
| 8.861 (7) | 5.726 (7) | 10.194 (9) | 115.71 (6) | 466.0 (9) | Anastasiou and Langer 1977 |
| 8.864 (7) | 5.732 (6) | 10.197 (8) | 115.67 (5) | 467.0 (9) | Anastasiou and Langer 1977 |
| 8.927 (8) | 5.761 (6) | 10.150 (9) | 114.77 (5) | 474.0 | Dollase 1971 |
| 8.962 (2) | 5.836 (2) | 10.182 (2) | 115.02 (1) | 482.6 | Kartashov et al. 2002 |
| 8.902 (1) | 5.713 (1) | 10.127 (1) | 114.86 (1) | 467.3 (1) | Bonazzi and Menchetti 1995 |
| 8.883 (1) | 5.710 (1) | 10.092 (1) | 114.96 (1) | 464.1 (2) | Bonazzi and Menchetti 1995 |
| 8.916 (2) | 5.700 (8) | 10.140 (25) | 114.72 (14) | 468.1 (1) | Grew et al. 1991 |
| 8.905 (18) | 5.684 (1) | 10.113 (1) | 114.62 (2) | 465.3 | Rouse and Peacor (1993) |
| 8.934 (1) | 5.721 (7) | 10.176 (22) | 114.31 (12) | 474.0 | Peacor and Dunn 1988 |
| 8.891 (1) | 5.662 (1) | 10.130 (2) | 115.20 (1) | 461.4 (1) | Bonazzi and Menchetti 1995 |
| 8.902 (1) | 5.702 (1) | 10.132 (1) | 114.94 (1) | 466.3 (1) | Bonazzi and Menchetti 1995 |
| 8.881 (1) | 5.683 (1) | 10.150 (2) | 114.99 (2) | 464.3 (2) | Bonazzi et al. 1992 |
| 8.857 (1) | 5.671 (1) | 10.156 (1) | 115.29 (1) | 461.2 (1) | Bonazzi et al. 1992 |
| 8.890 (2) | 5.690 (1) | 10.135 (2) | 114.44 (2) | 466.7 (2) | Bonazzi et al. 1996 |
| 8.896 (1) | 5.706 (1) | 10.083 (1) | 113.88 (1) | 468.0 (1) | Bonazzi et al. 1996 |

## APPENDIX B

### (on following 5 pages)

Bond lengths [Å], polyhedra volumes [Å$^3$], selected interatomic distances [Å] and angles [°], and distortion parameters  in natural and synthetic monoclinic Al-Fe$^{3+}$ solid solutions. Calculated from the data of Dollase 1968, 1971 (D68, D71); Gabe et al. 1973 (G); Carbonin and Molin 1980 (C80); Stergiou et al. 1987 (S); Kvick et al. 1988 (K); Bonazzi and Menchetti 1995 (B&M); Comodi  and Zanazzi 1997 (C&Z); Giuli et al. 1999 (G99). The distortion parameters are calculated according to Robinson et al. (1971).

| Composition [$X_{Ep}$] | 0.03 | 0.22 | 0.24 | 0.30 | 0.34 | 0.38 | 0.40 | 0.40 |
|---|---|---|---|---|---|---|---|---|
| Sample | Nat. | Nat. | Nat. | Nat. | Nat. | Nat. | Nat. | Nat. |
| Source | D68 | C&Z | B&M | C80 | C80 | B&M | C80 | G |
| Sample-Nr. | | | CH | | | LP | | LEP |
| **A-Positions** | | | | | | | | |
| A1 - O1 (2×) | 2.490 | 2.485 | 2.485 | 2.4762 | 2.4758 | 2.478 | 2.4722 | 2.4780 |
| A1 - O3 (2×) | 2.369 | 2.355 | 2.354 | 2.3492 | 2.3497 | 2.353 | 2.3461 | 2.3449 |
| A1 - O5 | 2.522 | 2.526 | 2.526 | 2.5317 | 2.5389 | 2.535 | 2.5393 | 2.5343 |
| A1 - O6 | 2.745 | 2.764 | 2.766 | 2.7894 | 2.7998 | 2.786 | 2.8092 | 2.7893 |
| A1 - O7 | 2.282 | 2.277 | 2.282 | 2.2752 | 2.2836 | 2.286 | 2.2825 | 2.2841 |
| A1 - O9 (2×) | 2.9524 | 2.963 | 2.965 | 2.9694 | 2.9779 | 2.973 | 2.9797 | 2.9746 |
| Mean A1 - O | 2.575 | 2.575 | 2.576 | 2.576 | 2.581 | 2.580 | 2.581 | 2.5781 |
| A2 - O2 (2×) | 2.543 | 2.537 | 2.537 | 2.5369 | 2.5380 | 2.534 | 2.4661 | 2.5359 |
| A2 - O2´ (2×) | 2.819 | 2.814 | 2.816 | 2.8039 | 2.8070 | 2.813 | 2.7918 | 2.8097 |
| A2 - O3 (2×) | 2.531 | 2.552 | 2.555 | 2.5729 | 2.5807 | 2.572 | 2.6440 | 2.5754 |
| A2 - O7 | 2.267 | 2.263 | 2.262 | 2.2556 | 2.2539 | 2.260 | 2.3423 | 2.2616 |
| A2 - O8 (2×) | 3.0445 | 3.038 | 3.038 | 3.0324 | 3.0358 | 3.034 | 3.0569 | 3.0294 |
| A2 - O10 | 2.575 | 2.554 | 2.560 | 2.5511 | 2.5501 | 2.554 | 2.5373 | 2.5509 |
| Mean A2 - O | 2.672 | 2.670 | 2.671 | 2.670 | 2.673 | 2.672 | 2.680 | 2.6713 |
| **M-Positions** | | | | | | | | |
| M1 – O1 (2×) | 1.930 | 1.928 | 1.927 | 1.9311 | 1.9322 | 1.930 | 1.9329 | 1.9316 |
| M1 – O4 (2×) | 1.850 | 1.848 | 1.847 | 1.8451 | 1.8482 | 1.845 | 1.8471 | 1.8466 |
| M1 – O5 (2×) | 1.937 | 1.936 | 1.939 | 1.9413 | 1.9419 | 1.944 | 1.9463 | 1.9433 |
| Mean M1 – O | 1.906 | 1.904 | 1.904 | 1.906 | 1.907 | 1.906 | 1.909 | 1.9072 |
| Volume M1 | 9.146 | 9.130 | 9.12 | 9.146 | 9.169 | 9.15 | 9.191 | 9.168 |
| λ M1 | 1.0064 | 1.0058 | 1.0069 | 1.0066 | 1.0066 | 1.0069 | 1.0064 | 1.0064 |
| $\sigma^2$ M1 | 19.6 | 19.6 | 19.7 | 19.6 | 19.7 | 19.1 | 18.7 | 18.9 |
| Inclination to (100) | 76.9 | 76.8 | 76.8 | 76.5 | 76.5 | 76.5 | 76.4 | 76.5 |
| O1 – O1 | 3.8609 | 3.8579 | 3.8532 | 3.8622 | 3.8644 | 3.8592 | 3.8659 | 3.8633 |
| M2 – O3 (2×) | 1.859 | 1.854 | 1.854 | 1.8530 | 1.8552 | 1.854 | 1.8535 | 1.8582 |
| M2 – O6 (2×) | 1.923 | 1.926 | 1.927 | 1.9259 | 1.9268 | 1.928 | 1.9275 | 1.9262 |
| M2 – O10 (2×) | 1.852 | 1.859 | 1.858 | 1.8630 | 1.8659 | 1.861 | 1.8651 | 1.8642 |
| Mean M2 – O | 1.878 | 1.880 | 1.880 | 1.881 | 1.883 | 1.881 | 1.882 | 1.8829 |
| Volume M2 | 8.773 | 8.807 | 8.80 | 8.818 | 8.845 | 8.82 | 8.837 | 8.848 |
| λ M2 | 1.0047 | 1.0039 | 1.0045 | 1.0041 | 1.0042 | 1.0044 | 1.0042 | 1.0042 |
| $\sigma^2$ M2 | 14.0 | 13.2 | 12.9 | 12.7 | 13.0 | 13.2 | 12.8 | 13.1 |
| Inclination to (100) | 66.5 | 65.6 | 65.6 | 65.0 | 64.9 | 65.1 | 64.5 | 64.9 |
| M3 - O1 (2×) | 2.184 | 2.190 | 2.191 | 2.1994 | 2.2035 | 2.200 | 2.2085 | 2.2000 |
| M3 - O2 (2×) | 1.927 | 1.944 | 1.944 | 1.9485 | 1.9541 | 1.954 | 1.9574 | 1.9563 |
| M3 - O4 | 1.862 | 1.882 | 1.882 | 1.8906 | 1.8957 | 1.899 | 1.9004 | 1.9027 |
| M3 - O8 | 1.781 | 1.788 | 1.790 | 1.7993 | 1.8043 | 1.803 | 1.8138 | 1.8100 |
| Mean M3 - O | 1.978 | 1.990 | 1.990 | 1.998 | 2.003 | 2.002 | 2.008 | 2.0042 |
| Volume M3 | 10.009 | 10.193 | 10.19 | 10.295 | 10.369 | 10.36 | 10.443 | 10.396 |
| λ M3 | 1.0262 | 1.0260 | 1.0269 | 1.0273 | 1.0274 | 1.0270 | 1.0277 | 1.0271 |
| $\sigma^2$ M3 | 58.1 | 62.0 | 62.6 | 65.1 | 65.6 | 65.7 | 67.5 | 66.4 |
| O4 – O8 | 3.638 | 3.6694 | 3.6692 | 3.687 | 3.6976 | 3.700 | 3.7125 | 3.7109 |
| **T-Positions** | | | | | | | | |
| T1 - O1 (2×) | 1.652 | 1.650 | 1.651 | 1.6492 | 1.6527 | 1.652 | 1.6483 | 1.6523 |
| T1 - O7 | 1.566 | 1.564 | 1.564 | 1.5688 | 1.5676 | 1.565 | 1.5664 | 1.5619 |
| T1 - O9 | 1.628 | 1.634 | 1.630 | 1.6274 | 1.6335 | 1.633 | 1.6352 | 1.6327 |
| Mean T1 – O | 1.625 | 1.625 | 1.624 | 1.624 | 1.627 | 1.626 | 1.625 | 1.6248 |
| Volume T1 | 2.190 | 2.193 | 2.19 | 2.187 | 2.199 | 2.20 | 2.191 | 2.192 |
| T2 - O3 (2×) | 1.620 | 1.619 | 1.620 | 1.6195 | 1.6203 | 1.619 | 1.6197 | 1.6190 |
| T2 - O8 | 1.593 | 1.591 | 1.591 | 1.5887 | 1.5887 | 1.591 | 1.5913 | 1.5916 |
| T2 - O9 | 1.627 | 1.622 | 1.628 | 1.6260 | 1.6276 | 1.627 | 1.6244 | 1.6272 |
| Mean T2 – O | 1.615 | 1.613 | 1.615 | 1.613 | 1.614 | 1.614 | 1.614 | 1.6142 |
| Volume T2 | 2.157 | 2.151 | 2.16 | 2.152 | 2.156 | 2.16 | 2.154 | 2.155 |
| T3 - O2 (2×) | 1.629 | 1.624 | 1.624 | 1.6256 | 1.6280 | 1.627 | 1.6279 | 1.6253 |
| T3 - O5 | 1.661 | 1.667 | 1.664 | 1.6675 | 1.6686 | 1.663 | 1.6661 | 1.6669 |
| T3 - O6 | 1.657 | 1.646 | 1.643 | 1.6439 | 1.6451 | 1.643 | 1.6410 | 1.6479 |
| Mean T3 – O | 1.644 | 1.640 | 1.639 | 1.641 | 1.642 | 1.640 | 1.641 | 1.6414 |
| Volume T3 | 2.261 | 2.248 | 2.24 | 2.248 | 2.257 | 2.25 | 2.250 | 2.251 |
| **T1T2O₇ group** | | | | | | | | |
| O7 – O9 – O8 | 127.084 | 128.834 | 129.175 | 130.231 | 130.929 | 130.393 | 131.719 | 130.811 |
| T1 – O9 – T2 | 164.382 | 161.898 | 161.529 | 160.275 | 159.433 | 160.196 | 158.609 | 159.870 |
| **Proton environment** | | | | | | | | |
| H – O10 | 0.7689 | 0.6829 | 0.9611 | | | 0.8562 | | 0.8872 |
| O10 – O4 | 2.8922 | 2.8916 | 2.8930 | 2.8995 | 2.9024 | 2.9031 | 2.9052 | 2.8999 |
| O10 – H ... O4 | 2.9253 | 2.9135 | 2.9032 | | | 2.9546 | | 2.9037 |
| O10 – H – O4 | 160.405 | 163.377 | 169.772 | | | 156.358 | | 173.675 |

| Composition [$X_{Ep}$] | 0.42 | 0.46 | 0.47 | 0.50 | 0.60 | 0.63 | 0.66 | 0.70 |
|---|---|---|---|---|---|---|---|---|
| Sample | Nat. | Nat. | Nat. | Nat. | Nat. | Nat. | Synth. | Nat. |
| Source | B&M | B&M | C80 | C80 | S | C80 | G99 | B&M |
| Sample-Nr. | VO | DF | | | / | | CC11c | GI2 |
| **A-Positions** | | | | | | | | |
| A1 - O1 (2×) | 2.473 | 2.472 | 2.4680 | 2.4649 | 2.4705 | 2.4658 | 2.4610 | 2.464 |
| A1 - O3 (2×) | 2.353 | 2.346 | 2.3414 | 2.3388 | 2.3412 | 2.3336 | 2.3430 | 2.338 |
| A1 - O5 | 2.538 | 2.536 | 2.5401 | 2.5451 | 2.5538 | 2.5506 | 2.5591 | 2.549 |
| A1 - O6 | 2.794 | 2.798 | 2.8232 | 2.8243 | 2.8543 | 2.8416 | 2.8479 | 2.838 |
| A1 - O7 | 2.291 | 2.289 | 2.2840 | 2.2893 | 2.3152 | 2.2945 | 2.3079 | 2.295 |
| A1 - O9 (2×) | 2.979 | 2.977 | 2.9840 | 2.9851 | 3.0110 | 2.9938 | 2.9939 | 2.991 |
| Mean A1 - O | 2.581 | 2.579 | 2.582 | 2.582 | 2.597 | 2.586 | 2.590 | 2.585 |
| A2 - O2 (2×) | 2.535 | 2.535 | 2.5360 | 2.5341 | 2.5400 | 2.5366 | 2.5416 | 2.530 |
| A2 - O2´ (2×) | 2.811 | 2.807 | 2.7977 | 2.7978 | 2.7966 | 2.7891 | 2.7951 | 2.794 |
| A2 - O3 (2×) | 2.584 | 2.587 | 2.6030 | 2.6108 | 2.6466 | 2.6338 | 2.6291 | 2.624 |
| A2 - O7 | 2.259 | 2.259 | 2.2547 | 2.2511 | 2.2647 | 2.2437 | 2.2485 | 2.251 |
| A2 - O8 (2×) | 3.031 | 3.027 | 3.0268 | 3.0228 | 3.30331 | 3.0195 | 3.0190 | 3.020 |
| A2 - O10 | 2.554 | 2.548 | 2.5436 | 2.5405 | 2.5391 | 2.5398 | 2.5542 | 2.536 |
| Mean A2 - O | 2.673 | 2.672 | 2.673 | 2.672 | 2.684 | 2.674 | 2.677 | 2.672 |
| **M-Positions** | | | | | | | | |
| M1 – O1 (2×) | 1.928 | 1.932 | 1.9336 | 1.9352 | 1.9265 | 1.9387 | 1.9415 | 1.933 |
| M1 – O4 (2×) | 1.846 | 1.844 | 1.8460 | 1.8450 | 1.8442 | 1.8478 | 1.8453 | 1.842 |
| M1 – O5 (2×) | 1.945 | 1.945 | 1.9495 | 1.9500 | 1.9637 | 1.9350 | 1.9529 | 1.953 |
| Mean M1 – O | 1.906 | 1.907 | 1.910 | 1.910 | 1.911 | 1.907 | 1.913 | 1.909 |
| Volume M1 | 9.16 | 9.16 | 9.206 | 9.211 | 9.233 | 9.259 | 9.259 | 9.20 |
| λ M1 | 1.0061 | 1.0069 | 1.0064 | 1.0064 | 1.0064 | 0.9998 | 1.0063 | 1.0065 |
| σ² M1 | 19.3 | 18.8 | 18.4 | 18.3 | 17.5 | 17.5 | 17.0 | 18.2 |
| Inclination to (100) | 76.4 | 76.4 | 76.3 | 76.1 | 76.0 | 76.1 | 75.9 | 76.1 |
| O1 – O1 | 3.8560 | 3.8635 | 3.8672 | 3.8704 | 3.8531 | 3.8775 | 3.8830 | 3.8657 |
| M2 – O3 (2×) | 1.850 | 1.854 | 1.8550 | 1.8530 | 1.8596 | 1.8541 | 1.8492 | 1.852 |
| M2 – O6 (2×) | 1.929 | 1.927 | 1.9242 | 1.9266 | 1.9312 | 1.9275 | 1.9278 | 1.926 |
| M2 – O10 (2×) | 1.862 | 1.862 | 1.8654 | 1.8694 | 1.8733 | 1.8704 | 1.8693 | 1.865 |
| Mean M2 – O | 1.880 | 1.881 | 1.882 | 1.883 | 1.888 | 1.884 | 1.882 | 1.881 |
| Volume M2 | 8.81 | 8.82 | 8.827 | 8.851 | 8.919 | 8.863 | 8.836 | 8.82 |
| λ M2 | 1.0044 | 1.0044 | 1.0044 | 1.0041 | 1.0043 | 1.0043 | 1.0043 | 1.0044 |
| σ² M2 | 13.3 | 13.4 | 13.8 | 12.7 | 13.5 | 13.3 | 13.3 | 14.6 |
| Inclination to (100) | 64.9 | 64.7 | 64.1 | 63.9 | 63.0 | 63.4 | 63.7 | 63.6 |
| M3 - O1 (2×) | 2.206 | 2.206 | 2.2122 | 2.2151 | 2.2362 | 2.2205 | 2.2291 | 2.222 |
| M3 - O2 (2×) | 1.961 | 1.958 | 1.9648 | 1.9657 | 1.9847 | 1.9764 | 1.9777 | 1.978 |
| M3 - O4 | 1.900 | 1.905 | 1.9108 | 1.9120 | 1.9378 | 1.9264 | 1.9323 | 1.927 |
| M3 - O8 | 1.809 | 1.813 | 1.8199 | 1.8274 | 1.8404 | 1.8470 | 1.8423 | 1.840 |
| Mean M3 - O | 2.007 | 2.008 | 2.014 | 2.017 | 2.037 | 2.028 | 2.031 | 2.028 |
| Volume M3 | 10.43 | 10.44 | 10.534 | 10.575 | 10.864 | 10.749 | 10.796 | 10.74 |
| λ M3 | 1.0280 | 1.0278 | 1.0282 | 1.0283 | 1.0299 | 1.0279 | 1.0287 | 1.0286 |
| σ² M3 | 68.2 | 68.2 | 70.0 | 70.7 | 76.4 | 71.9 | 73.4 | 73.6 |
| O4 – O8 | 3.707 | 3.7158 | 3.7291 | 3.7380 | 3.7768 | 3.7724 | 3.7736 | 3.7658 |
| **T-Positions** | | | | | | | | |
| T1 - O1 (2×) | 1.654 | 1.652 | 1.6505 | 1.6497 | 1.6573 | 1.6469 | 1.6430 | 1.649 |
| T1 - O7 | 1.567 | 1.566 | 1.5681 | 1.5658 | 1.5509 | 1.5725 | 1.5697 | 1.566 |
| T1 - O9 | 1.631 | 1.632 | 1.6343 | 1.6377 | 1.6463 | 1.6319 | 1.6306 | 1.631 |
| Mean T1 – O | 1.627 | 1.626 | 1.626 | 1.626 | 1.628 | 1.625 | 1.622 | 1.624 |
| Volume T1 | 2.20 | 2.19 | 2.196 | 2.196 | 2.206 | 2.191 | 2.180 | 2.19 |
| T2 - O3 (2×) | 1.623 | 1.620 | 1.6187 | 1.6185 | 1.6125 | 1.6178 | 1.6240 | 1.617 |
| T2 - O8 | 1.592 | 1.593 | 1.5926 | 1.5906 | 1.5867 | 1.5868 | 1.5944 | 1.591 |
| T2 - O9 | 1.628 | 1.628 | 1.6254 | 1.6225 | 1.6440 | 1.6339 | 1.6277 | 1.633 |
| Mean T2 – O | 1.617 | 1.615 | 1.614 | 1.613 | 1.614 | 1.614 | 1.618 | 1.615 |
| Volume T2 | 2.16 | 2.16 | 2.155 | 2.149 | 2.155 | 2.156 | 2.170 | 2.16 |
| T3 - O2 (2×) | 1.624 | 1.626 | 1.6260 | 1.6262 | 1.6317 | 1.6250 | 1.6162 | 1.626 |
| T3 - O5 | 1.663 | 1.667 | 1.6664 | 1.6647 | 1.6600 | 1.6686 | 1.6688 | 1.666 |
| T3 - O6 | 1.641 | 1.645 | 1.6447 | 1.6413 | 1.6405 | 1.6409 | 1.6381 | 1.641 |
| Mean T3 – O | 1.638 | 1.641 | 1.641 | 1.640 | 1.641 | 1.640 | 1.635 | 1.640 |
| Volume T3 | 2.24 | 2.25 | 2.250 | 2.245 | 2.254 | 2.247 | 2.228 | 2.25 |
| **T1T2O₇ group** | | | | | | | | |
| O7 – O9 – O8 | 131.199 | 131.400 | 132.388 | 132.890 | 135.222 | 134.217 | 134.640 | 133.758 |
| T1 – O9 – T2 | 159.198 | 159.018 | 157.768 | 157.407 | 154.724 | 155.902 | 155.517 | 156.445 |
| **Proton environment** | | | | | | | | |
| H – O10 | 0.8486 | 0.8616 | | | 1.0547 | | | 0.9108 |
| O10 – O4 | 2.9090 | 2.9159 | 2.9144 | 2.9107 | 2.9432 | 2.9174 | 2.9199 | 2.9339 |
| O10 – H ... O4 | 2.9410 | 2.9206 | | | 2.9548 | | | 2.9418 |
| O10 – H – O4 | 161.296 | 172.832 | | | 169.369 | | | 170.995 |

| Composition [$X_{Ep}$] | 0.70 | 0.71 | 0.73 | 0.79 | 0.81 | 0.81 | 0.82 | 0.84 |
|---|---|---|---|---|---|---|---|---|
| Sample | Nat. | Synth. | Nat. | Nat. | Nat. | Nat. | Nat. | Nat. |
| Source | B&M | G99 | B&M | C80 | K | D71 | B&M | G |
| Sample-Nr. | PS | CC11b | GI1 | / | / | / | MRV | HEP |
| **A-Positions** | | | | | | | | |
| A1 - O1 (2×) | 2.464 | 2.4695 | 2.463 | 2.4592 | 2.4608 | 2.4647 | 2.458 | 2.4585 |
| A1 - O3 (2×) | 2.338 | 2.3343 | 2.336 | 2.3282 | 2.3272 | 2.3250 | 2.329 | 2.3227 |
| A1 - O5 | 2.551 | 2.5526 | 2.549 | 2.5594 | 2.5519 | 2.5622 | 2.556 | 2.5556 |
| A1 - O6 | 2.836 | 2.8512 | 2.838 | 2.8633 | 2.8590 | 2.8847 | 2.858 | 2.8612 |
| A1 - O7 | 2.295 | 2.3043 | 2.294 | 2.2967 | 2.2957 | 2.2932 | 2.296 | 2.2949 |
| A1 - O9 (2×) | 2.992 | 2.9933 | 2.994 | 3.0030 | 3.0019 | 3.0074 | 3.000 | 2.9999 |
| Mean A1 - O | 2.586 | 2.589 | 2.585 | 2.589 | 2.5874 | 2.593 | 2.587 | 2.5860 |
| A2 - O2 (2×) | 2.535 | 2.5435 | 2.531 | 2.5315 | 2.5193 | 2.5340 | 2.528 | 2.5270 |
| A2 - O2' (2×) | 2.794 | 2.7877 | 2.794 | 2.7819 | 2.807 | 2.7824 | 2.788 | 2.7842 |
| A2 - O3 (2×) | 2.627 | 2.6441 | 2.628 | 2.6557 | 2.6412 | 2.6817 | 2.648 | 2.6528 |
| A2 - O7 | 2.255 | 2.2455 | 2.253 | 2.2477 | 2.2508 | 2.2342 | 2.249 | 2.2480 |
| A2 - O8 (2×) | 3.022 | 3.0217 | 3.021 | 3.0164 | 3.0177 | 3.0140 | 3.017 | 3.0146 |
| A2 - O10 | 2.541 | 2.5242 | 2.536 | 2.5310 | 2.5206 | 2.5292 | 2.525 | 2.5307 |
| Mean A2 - O | 2.675 | 2.676 | 2.674 | 2.675 | 2.6729 | 2.679 | 2.674 | 2.6736 |
| **M-Positions** | | | | | | | | |
| M1 – O1 (2×) | 1.936 | 1.9400 | 1.936 | 1.9377 | 1.9407 | 1.9436 | 1.937 | 1.9393 |
| M1 – O4 (2×) | 1.844 | 1.8470 | 1.844 | 1.8444 | 1.8454 | 1.8460 | 1.844 | 1.8434 |
| M1 – O5 (2×) | 1.952 | 1.9521 | 1.955 | 1.9606 | 1.9566 | 1.9708 | 1.959 | 1.9559 |
| Mean M1 – O | 1.911 | 1.913 | 1.912 | 1.914 | 1.9142 | 1.920 | 1.913 | 1.9129 |
| Volume M1 | 9.22 | 9.248 | 9.23 | 9.274 | 9.274 | 9.364 | 9.26 | 9.252 |
| λ M1 | 1.0064 | 1.0069 | 1.0068 | 1.0063 | 1.0063 | 1.0061 | 1.0064 | 1.0065 |
| $\sigma^2$ M1 | 18.4 | 19.7 | 17.8 | 17.2 | 17.3 | 15.9 | 17.2 | 17.9 |
| Inclination to (100) | 76.1 | 76.1 | 76.0 | 75.8 | 76.1 | 75.8 | 75.8 | 75.9 |
| O1 – O1 | 3.8721 | 3.8800 | 3.8715 | 3.8753 | 3.8811 | 3.887 | 3.8747 | 3.8787 |
| M2 – O3 (2×) | 1.853 | 1.8523 | 1.854 | 1.8546 | 1.8533 | 1.8597 | 1.852 | 1.8539 |
| M2 – O6 (2×) | 1.926 | 1.9312 | 1.926 | 1.9270 | 1.9249 | 1.9303 | 1.927 | 1.9269 |
| M2 – O10 (2×) | 1.865 | 1.8657 | 1.866 | 1.8729 | 1.9726 | 1.8741 | 1.869 | 1.8696 |
| Mean M2 – O | 1.881 | 1.883 | 1.882 | 1.885 | 1.8836 | 1.888 | 1.883 | 1.8835 |
| Volume M2 | 8.82 | 8.844 | 8.83 | 8.872 | 8.855 | 8.917 | 8.84 | 8.853 |
| λ M2 | 1.0047 | 1.0048 | 1.0046 | 1.0045 | 1.0044 | 1.0045 | 1.0046 | 1.0045 |
| $\sigma^2$ M2 | 14.6 | 14.7 | 14.7 | 14.2 | 14.1 | 14.0 | 14.7 | 14.0 |
| Inclination to (100) | 63.6 | 63.2 | 63.5 | 62.8 | 63.0 | 62.3 | 62.9 | 62.7 |
| M3 - O1 (2×) | 2.218 | 2.2253 | 2.222 | 2.2309 | 2.2164 | 2.2398 | 2.227 | 2.2241 |
| M3 - O2 (2×) | 1.977 | 1.9835 | 1.980 | 1.9902 | 1.9871 | 1.9961 | 1.989 | 1.9855 |
| M3 - O4 | 1.926 | 1.9180 | 1.928 | 1.9402 | 1.9435 | 1.9475 | 1.937 | 1.9352 |
| M3 - O8 | 1.840 | 1.8516 | 1.843 | 1.8596 | 1.8601 | 1.8804 | 1.856 | 1.8600 |
| Mean M3 - O | 2.026 | 2.031 | 2.029 | 2.040 | 2.0351 | 2.050 | 2.038 | 2.0357 |
| Volume M3 | 10.72 | 10.794 | 10.76 | 10.927 | 10.731 | 11.085 | 10.89 | 10.863 |
| λ M3 | 1.0279 | 1.0285 | 1.0286 | 1.0291 | 1.0358 | 1.0287 | 1.0285 | 1.0283 |
| $\sigma^2$ M3 | 72.7 | 72.9 | 73.1 | 76.7 | 74.7 | 76.6 | 75.4 | 74.7 |
| O4 – O8 | 3.7648 | 3.7682 | 3.7706 | 3.7992 | 3.8031 | 3.8273 | 3.7927 | 3.7945 |
| **T-Positions** | | | | | | | | |
| T1 - O1 (2×) | 1.653 | 1.6468 | 1.651 | 1.6520 | 1.6536 | 1.6455 | 1.651 | 1.6499 |
| T1 - O7 | 1.567 | 1.5645 | 1.567 | 1.5676 | 1.5684 | 1.5855 | 1.567 | 1.5635 |
| T1 - O9 | 1.636 | 1.6336 | 1.634 | 1.6373 | 1.6409 | 1.6392 | 1.636 | 1.6343 |
| Mean T1 – O | 1.627 | 1.623 | 1.626 | 1.627 | 1.6291 | 1.629 | 1.626 | 1.6244 |
| Volume T1 | 2.20 | 2.186 | 2.19 | 2.202 | 2.209 | 2.211 | 2.20 | 2.190 |
| T2 - O3 (2×) | 1.619 | 1.6168 | 1.617 | 1.6175 | 1.6209 | 1.6117 | 1.619 | 1.6184 |
| T2 - O8 | 1.592 | 1.5758 | 1.589 | 1.5936 | 1.5924 | 1.5880 | 1.594 | 1.5876 |
| T2 - O9 | 1.630 | 1.6247 | 1.632 | 1.6309 | 1.6322 | 1.6282 | 1.632 | 1.6315 |
| Mean T2 – O | 1.615 | 1.609 | 1.614 | 1.615 | 1.6169 | 1.610 | 1.616 | 1.6140 |
| Volume T2 | 2.16 | 2.133 | 2.16 | 2.159 | 2.167 | 2.139 | 2.16 | 2.155 |
| T3 - O2 (2×) | 1.627 | 1.6160 | 1.626 | 1.6275 | 1.6283 | 1.6220 | 1.627 | 1.6274 |
| T3 - O5 | 1.667 | 1.6760 | 1.666 | 1.6653 | 1.6695 | 1.6616 | 1.665 | 1.6676 |
| T3 - O6 | 1.643 | 1.6354 | 1.643 | 1.6418 | 1.6408 | 1.6399 | 1.639 | 1.6383 |
| Mean T3 – O | 1.641 | 1.636 | 1.640 | 1.641 | 1.6417 | 1.636 | 1.640 | 1.6402 |
| Volume T3 | 2.25 | 2.231 | 2.25 | 2.250 | 2.255 | 2.232 | 2.25 | 2.249 |
| **T1T2O7 group** | | | | | | | | |
| O7 – O9 – O8 | 133.517 | 134.178 | 133.916 | 135.431 | 135.327 | 136.104 | 135.001 | 135.392 |
| T1 – O9 – T2 | 156.649 | 156.570 | 156.175 | 154.515 | 154.791 | 153.568 | 154.935 | 154.574 |
| **Proton environment** | | | | | | | | |
| H – O10 | 0.8525 | | 0.7491 | | 0.9753 | | 0.8479 | 0.9548 |
| O10 – O4 | 2.9329 | 2.9480 | 2.9331 | 2.9369 | 2.9222 | 2.9447 | 2.9391 | 2.9334 |
| O10 – H ... O4 | 2.9482 | | 2.9489 | | 2.9393 | | 2.9610 | 2.9790 |
| O10 – H – O4 | 167.145 | | 166.722 | | 166.872 | | 164.604 | 158.457 |

| Composition [$X_{Ep}$] | 0.85 | 0.85 | 0.86 | 0.91 | 0.95 | 1.00 | 1.0 | 1.02 |
|---|---|---|---|---|---|---|---|---|
| Sample | Synth. | Synth. | Nat. | Nat. | Nat. | Nat. | Synth. | Nat. |
| Source | G99 | G99 | C80 | B&M | B&M | B&M | G99 | B&M |
| Sample-Nr. | CC9e | 16b | | TRV | IG | MBN | 20c | CC |

| **A-Positions** | | | | | | | | |
|---|---|---|---|---|---|---|---|---|
| A1 - O1 (2×) | 2.4570 | 2.4547 | 2.4526 | 2.454 | 2.456 | 2.455 | 2.4564 | 2.454 |
| A1 - O3 (2×) | 2.3311 | 2.3248 | 2.3237 | 2.328 | 2.327 | 2.327 | 2.3265 | 2.327 |
| A1 - O5 | 2.5583 | 2.5555 | 2.5583 | 2.558 | 2.559 | 2.564 | 2.5664 | 2.560 |
| A1 - O6 | 2.8629 | 2.8758 | 2.8786 | 2.870 | 2.879 | 2.888 | 2.8730 | 2.882 |
| A1 - O7 | 2.3010 | 2.2900 | 2.2993 | 2.297 | 2.300 | 2.301 | 2.2876 | 2.298 |
| A1 - O9 (2×) | 3.0022 | 2.9960 | 3.0076 | 3.009 | 3.009 | 3.013 | 3.0108 | 3.012 |
| Mean A1 - O | 2.589 | 2.586 | 2.589 | 2.590 | 2.591 | 2.593 | 2.590 | 2.592 |
| A2 - O2 (2×) | 2.5424 | 2.5263 | 2.5292 | 2.530 | 2.529 | 2.531 | 2.5420 | 2.531 |
| A2 - O2′ (2×) | 2.7872 | 2.7860 | 2.7793 | 2.783 | 2.782 | 2.781 | 2.7782 | 2.779 |
| A2 - O3 (2×) | 2.6570 | 2.6544 | 2.6687 | 2.669 | 2.671 | 2.678 | 2.6777 | 2.677 |
| A2 - O7 | 2.2440 | 2.2418 | 2.2417 | 2.246 | 2.249 | 2.246 | 2.2479 | 2.245 |
| A2 - O8 (2×) | 3.0223 | 3.0147 | 3.0120 | 3.016 | 3.015 | 3.019 | 3.0202 | 3.016 |
| A2 - O10 | 2.5213 | 2.5210 | 2.5254 | 2.525 | 2.524 | 2.519 | 2.5138 | 2.524 |
| Mean A2 - O | 2.678 | 2.673 | 2.675 | 2.677 | 2.677 | 2.678 | 2.680 | 2.678 |

| **M-Positions** | | | | | | | | |
|---|---|---|---|---|---|---|---|---|
| M1 – O1 (2×) | 1.9478 | 1.9412 | 1.9395 | 1.942 | 1.942 | 1.945 | 1.9432 | 1.945 |
| M1 – O4 (2×) | 1.8493 | 1.8476 | 1.8456 | 1.847 | 1.847 | 1.848 | 1.8592 | 1.850 |
| M1 – O5 (2×) | 1.9677 | 1.9587 | 1.9625 | 1.963 | 1.966 | 1.967 | 1.9677 | 1.967 |
| Mean M1 – O | 1.922 | 1.916 | 1.916 | 1.917 | 1.918 | 1.920 | 1.923 | 1.921 |
| Volume M1 | 9.381 | 9.299 | 9.300 | 9.32 | 9.34 | 9.36 | 9.417 | 9.37 |
| λ M1 | 1.0064 | 1.0062 | 1.0062 | 1.0063 | 1.0059 | 1.0062 | 1.0055 | 1.0062 |
| $\sigma^2$ M1 | 17.2 | 17.1 | 16.7 | 16.6 | 16.4 | 17.0 | 15.3 | 16.2 |
| Inclination to (100) | 75.7 | 75.8 | 75.7 | 75.7 | 75.7 | 75.7 | 75.7 | 75.7 |
| O1 – O1 | 3.8955 | 3.8825 | 3.8789 | 3.8832 | 3.8846 | 3.8898 | 3.8863 | 3.8901 |
| M2 – O3 (2×) | 1.8548 | 1.8597 | 1.8532 | 1.853 | 1.853 | 1.852 | 1.8584 | 1.854 |
| M2 – O6 (2×) | 1.9271 | 1.9323 | 1.9241 | 1.926 | 1.929 | 1.930 | 1.9238 | 1.927 |
| M2 – O10 (2×) | 1.8671 | 1.8696 | 1.8719 | 1.869 | 1.873 | 1.874 | 1.8762 | 1.872 |
| Mean M2 – O | 1.883 | 1.887 | 1.883 | 1.883 | 1.885 | 1.885 | 1.886 | 1.884 |
| Volume M2 | 8.840 | 8.908 | 8.848 | 8.84 | 8.87 | 8.88 | 8.888 | 8.86 |
| λ M2 | 1.0050 | 1.0043 | 1.0044 | 1.0046 | 1.0048 | 1.0044 | 1.0046 | 1.0049 |
| $\sigma^2$ M2 | 15.7 | 13.3 | 14.9 | 15.6 | 14.8 | 15.2 | 14.9 | 15.6 |
| Inclination to (100) | 62.8 | 62.7 | 62.3 | 62.5 | 62.4 | 62.2 | 62.2 | 62.3 |
| M3 - O1 (2×) | 2.2367 | 2.2291 | 2.2346 | 2.236 | 2.236 | 2.236 | 2.2401 | 2.239 |
| M3 - O2 (2×) | 1.9861 | 1.9771 | 1.9947 | 1.994 | 1.997 | 1.997 | 1.9858 | 1.998 |
| M3 - O4 | 1.9240 | 1.9304 | 1.9454 | 1.944 | 1.947 | 1.949 | 1.9333 | 1.946 |
| M3 - O8 | 1.8532 | 1.8532 | 1.8759 | 1.864 | 1.872 | 1.872 | 1.8497 | 1.871 |
| Mean M3 - O | 2.037 | 2.033 | 2.047 | 2.045 | 2.048 | 2.048 | 2.039 | 2.049 |
| Volume M3 | 10.882 | 10.814 | 11.026 | 10.99 | 11.04 | 11.05 | 10.906 | 11.06 |
| λ M3 | 1.0292 | 1.0287 | 1.0290 | 1.0295 | 1.0291 | 1.0288 | 1.0297 | 1.0289 |
| $\sigma^2$ M3 | 74.3 | 74.5 | 78.0 | 76.7 | 77.1 | 77.0 | 76.2 | 76.6 |
| O4 – O8 | 3.7761 | 3.7828 | 3.8208 | 3.8074 | 3.8193 | 3.8205 | 3.7824 | 3.8160 |

| **T-Positions** | | | | | | | | |
|---|---|---|---|---|---|---|---|---|
| T1 - O1 (2×) | 1.6454 | 1.6516 | 1.6515 | 1.651 | 1.652 | 1.653 | 1.6538 | 1.651 |
| T1 - O7 | 1.5660 | 1.5710 | 1.5694 | 1.573 | 1.568 | 1.567 | 1.5705 | 1.569 |
| T1 - O9 | 1.6352 | 1.6304 | 1.6356 | 1.639 | 1.635 | 1.635 | 1.6412 | 1.637 |
| Mean T1 – O | 1.623 | 1.626 | 1.627 | 1.628 | 1.627 | 1.627 | 1.630 | 1.627 |
| Volume T1 | 2.185 | 2.198 | 2.201 | 2.21 | 2.20 | 2.20 | 2.212 | 2.20 |
| T2 - O3 (2×) | 1.6161 | 1.6131 | 1.6188 | 1.618 | 1.619 | 1.621 | 1.6142 | 1.617 |
| T2 - O8 | 1.5874 | 1.5900 | 1.5876 | 1.594 | 1.594 | 1.594 | 1.6013 | 1.594 |
| T2 - O9 | 1.6262 | 1.6247 | 1.6346 | 1.634 | 1.634 | 1.635 | 1.6268 | 1.634 |
| Mean T2 – O | 1.611 | 1.610 | 1.615 | 1.616 | 1.617 | 1.618 | 1.614 | 1.616 |
| Volume T2 | 2.145 | 2.140 | 2.160 | 2.16 | 2.16 | 2.17 | 2.156 | 2.16 |
| T3 - O2 (2×) | 1.6209 | 1.6315 | 1.6266 | 1.627 | 1.628 | 1.630 | 1.6281 | 2.626 |
| T3 - O5 | 1.6600 | 1.6681 | 1.6676 | 1.667 | 1.664 | 1.664 | 1.6635 | 1.666 |
| T3 - O6 | 1.6405 | 1.6293 | 1.6421 | 1.644 | 1.641 | 1.639 | 1.6528 | 1.644 |
| Mean T3 – O | 1.636 | 1.640 | 1.641 | 1.641 | 1.640 | 1.641 | 1.643 | 1.641 |
| Volume T3 | 2.229 | 2.251 | 2.252 | 2.25 | 2.25 | 2.25 | 2.261 | 2.25 |

| **T1T2O$_7$ group** | | | | | | | | |
|---|---|---|---|---|---|---|---|---|
| O7 – O9 – O8 | 135.320 | 135.034 | 136.409 | 136.008 | 136.059 | 136.274 | 135.066 | 136.292 |
| T1 – O9 – T2 | 154.778 | 155.131 | 153.521 | 153.646 | 153.793 | 153.522 | 153.720 | 153.474 |

| **Proton environment** | | | | | | | | |
|---|---|---|---|---|---|---|---|---|
| H – O10 | | | | 0.8536 | 0.8885 | 0.8658˙ | | 0.8781 |
| O10 – O4 | 2.9494 | 2.9390 | 2.9409 | 2.9483 | 2.9459 | 2.9488 | 2.9348 | 2.9480 |
| O10 – H … O4 | | | | 2.9512 | 2.9775 | 2.9776 | | 2.9542 |
| O10 – H – O4 | | | | 174.313 | 161.711 | 162.430 | | 171.965 |

| Composition [$X_{Ep}$] | 1.04 | 1.04 | 1.09 | 1.14 |
|---|---|---|---|---|
| Sample | Nat. | Nat. | Synth. | Nat. |
| Source | B&M | B&M | G99 | B&M |
| Sample-Nr. | NI | CZ | 12a | FC |
| **A-Positions** | | | | |
| A1 - O1 (2×) | 2.455 | 2.454 | 2.4463 | 2.455 |
| A1 - O3 (2×) | 2.328 | 2.326 | 2.3333 | 2.325 |
| A1 - O5 | 2.557 | 2.563 | 2.5569 | 2.560 |
| A1 - O6 | 2.884 | 2.892 | 2.8842 | 2.902 |
| A1 - O7 | 2.298 | 2.299 | 2.2950 | 2.299 |
| A1 - O9 (2×) | 3.011 | 3.015 | 3.0227 | 3.023 |
| Mean A1 - O | 2.592 | 2.594 | 2.593 | 2.596 |
| A2 - O2 (2×) | 2.534 | 2.533 | 2.5624 | 2.536 |
| A2 - O2′ (2×) | 2.777 | 2.774 | 2.9880 | 2.773 |
| A2 - O3 (2×) | 2.681 | 2.684 | 2.5030 | 2.700 |
| A2 - O7 | 2.251 | 2.244 | 2.2189 | 2.244 |
| A2 - O8 (2×) | 3.016 | 3.020 | 3.0498 | 3.019 |
| A2 - O10 | 2.526 | 2.518 | 2.3279 | 2.521 |
| Mean A2 - O | 2.679 | 2.678 | 2.675 | 2.682 |
| **M-Positions** | | | | |
| M1 – O1 (2×) | 1.949 | 1.950 | 1.9464 | 1.954 |
| M1 – O4 (2×) | 1.849 | 1.853 | 1.8508 | 1.855 |
| M1 – O5 (2×) | 1.968 | 1.970 | 1.9766 | 1.976 |
| Mean M1 – O | 1.922 | 1.924 | 1.925 | 1.928 |
| Volume M1 | 9.39 | 9.42 | 9.432 | 9.48 |
| λ M1 | 1.0062 | 1.0064 | 1.0059 | 1.0064 |
| $\sigma^2$ M1 | 16.6 | 16.5 | 15.5 | 16.5 |
| Inclination to (100) | 75.7 | 75.7 | 75.7 | 75.7 |
| O1 – O1 | 3.8980 | 3.8989 | 3.8927 | 3.9082 |
| M2 – O3 (2×) | 1.853 | 1.858 | 1.8592 | 1.855 |
| M2 – O6 (2×) | 1.927 | 1.932 | 1.9249 | 1.930 |
| M2 – O10 (2×) | 1.871 | 1.874 | 1.8788 | 1.876 |
| Mean M2 – O | 1.884 | 1.888 | 1.888 | 1.887 |
| Volume M2 | 8.85 | 8.92 | 8.910 | 8.90 |
| λ M2 | 1.0049 | 1.0043 | 1.0045 | 1.0047 |
| $\sigma^2$ M2 | 15.7 | 15.0 | 14.8 | 16.2 |
| Inclination to (100) | 62.2 | 62.1 | 62.1 | 61.7 |
| M3 - O1 (2×) | 2.239 | 2.242 | 2.2486 | 2.245 |
| M3 - O2 (2×) | 1.998 | 1.995 | 1.9943 | 2.003 |
| M3 - O4 | 1.945 | 1.943 | 1.9437 | 1.951 |
| M3 - O8 | 1.872 | 1.861 | 1.8560 | 1.881 |
| Mean M3 - O | 2.049 | 2.046 | 2.048 | 2.055 |
| Volume M3 | 11.06 | 11.02 | 11.044 | 11.15 |
| λ M3 | 1.0289 | 1.0295 | 1.0296 | 1.0295 |
| $\sigma^2$ M3 | 75.6 | 75.6 | 75.9 | 76.6 |
| O4 – O8 | 3.8175 | 3.8037 | 3.7988 | 3.8312 |
| **T-Positions** | | | | |
| T1 - O1 (2×) | 1.651 | 1.652 | 1.6512 | 1.654 |
| T1 - O7 | 1.564 | 1.566 | 1.5800 | 1.570 |
| T1 - O9 | 1.637 | 1.637 | 1.6461 | 1.638 |
| Mean T1 – O | 1.626 | 1.627 | 1.632 | 1.629 |
| Volume T1 | 2.19 | 2.20 | 2.221 | 2.21 |
| T2 - O3 (2×) | 1.616 | 1.617 | 1.6154 | 1.621 |
| T2 - O8 | 1.593 | 1.599 | 1.5811 | 1.597 |
| T2 - O9 | 1.633 | 1.636 | 1.6355 | 1.634 |
| Mean T2 – O | 1.615 | 1.617 | 1.612 | 1.618 |
| Volume T2 | 2.16 | 2.17 | 2.147 | 2.17 |
| T3 - O2 (2×) | 2.626 | 2.628 | 1.6249 | 1.630 |
| T3 - O5 | 1.667 | 1.669 | 1.6544 | 1.668 |
| T3 - O6 | 1.644 | 1.642 | 1.6528 | 1.644 |
| Mean T3 – O | 1.641 | 1.642 | 1.639 | 1.643 |
| Volume T3 | 2.25 | 2.26 | 2.245 | 2.26 |
| **T1T2O$_7$ group** | | | | |
| O7 – O9 – O8 | 136.244 | 136.571 | 136.896 | 137.054 |
| T1 – O9 – T2 | 153.631 | 153.151 | 152.112 | 152.693 |
| **Proton environment** | | | | |
| H – O10 | 0.7330 | 0.9465 | | 0.8622 |
| O10 – O4 | 2.9522 | 2.9509 | 2.9404 | 2.9569 |
| O10 – H … O4 | 2.9914 | 2.9549 | | 2.9673 |
| O10 – H – O4 | 158.394 | 173.633 | | 169.429 |

# APPENDIX C

Lattice constants of natural and synthetic zoisite at room temperature and pressure.

| Composition | Sample |
|---|---|
| $Ca_2Al_2Al[Si_3O_{11}(O/OH)]$ | Synth. |
| $Ca_2Al_2Al[Si_3O_{11}(O/OH)]$ | Synth |
| $Ca_2Al_2Al[Si_3O_{11}(O/OH)]$ | Synth. |
| $Ca_2Al_2(Al_{0.957-0.977}Fe^{3+}_{0.023-0.043})[Si_3O_{11}(O/OH)]$ | Nat. |
| $Ca_2Al_2(Al_{0.957-0.977}Fe^{3+}_{0.023-0.043})[Si_3O_{11}(O/OH)]$ | Nat. |
| $Ca_2Al_2Al[Si_3O_{11}(O/OH)]$ | Synth. |
| $Ca_2Al_2(Al_{0.953-0.977}Fe^{3+}_{0.023-0.047})[Si_3O_{11}(O/OH)]$ | Synth. |
| $Ca_2Al_2(Al_{0.915-0.961}Fe^{3+}_{0.039-0.085})[Si_3O_{11}(O/OH)]$ | Synth. |
| $Ca_2Al_2(Al_{0.881-0.927}Fe^{3+}_{0.073-0.119})[Si_3O_{11}(O/OH)]$ | Synth. |
| $Ca_2Al_2(Al_{0.865-0.903}Fe^{3+}_{0.097-0.135})[Si_3O_{11}(O/OH)]$ | Synth. |
| $(Ca_{2.025}Fe^{2+}_{0.01}Na_{0.005}K_{0.02})Al_2(Al_{0.94}Fe^{3+}_{0.015}Mg_{0.005}Ti_{0.005})[Si_{2.995}P_{0.005}O_{11}(O/OH)]$ | Nat. |
| $(Ca_{1.995}Fe^{2+}_{0.02}Na_{0.005}K_{0.02})Al_2(Al_{0.935}Fe^{3+}_{0.035}Mg_{0.005}Ti_{0.005})[Si_{2.955}P_{0.005}Al_{0.03}O_{11}(O/OH)]$ | Nat. |
| $(Ca_{2.07}Fe^{2+}_{0.015}Na_{0.005}K_{0.02})Al_2(Al_{0.815}Fe^{3+}_{0.105}Mg_{0.005}Ti_{0.005}Mn_{0.005})[Si_{2.985}P_{0.005}Al_{0.01}O_{11}(O/OH)]$ | Nat. |
| $(Ca_{2.005}Fe^{2+}_{0.015}Na_{0.005}K_{0.02})Al_2(Al_{0.85}Fe^{3+}_{0.105}Mg_{0.005}Ti_{0.01})[Si_{2.98}P_{0.01}Al_{0.01}O_{11}(O/OH)]$ | Nat. |
| $(Ca_{2.00}Fe^{2+}_{0.02}Na_{0.005}K_{0.02})Al_2(Al_{0.81}Fe^{3+}_{0.14}Mg_{0.005})[Si_{3.005}P_{0.005}O_{11}(O/OH)]$ | Nat. |
| $Ca_2Al_2Al[Si_3O_{11}(O/OH)]$ | Synth. |
| $Ca_2Al_2(Al_{0.9}Fe^{3+}_{0.1})[Si_3O_{11}(O/OH)]$ | Nat. |
| $Ca_2Al_2Al[Si_3O_{11}(O/OH)]$ | Synth. |
| $Ca_2Al_2Al[Si_3O_{11}(O/OH)]$ | Synth. |
| $Ca_2Al_2Al[Si_3O_{11}(O/OD)]$ | Synth. |
| $Ca_2Al_2Al[Si_3O_{11}(O/OH)]$ | Synth. |
| $Ca_2Al_2(Al_{0.89-0.92}Fe^{3+}_{0.08-0.11})[Si_3O_{11}(O/OH)]$ | Nat. |
| $Ca_2Al_2Al[Si_3O_{11}(O/OH)]$ | / |
| $Ca_2Al_2(Al_{0.925}Fe^{3+}_{0.075})[Si_3O_{11}(O/OH)]$ | Nat. |
| $Ca_2Al_2Al[Si_3O_{11}(O/OH)]$ | Synth. |
| $Ca_2Al_2Al[Si_3O_{11}(O/OH)]$ | Synth. |
| $Ca_{2.02}Al_2(Al_{0.92}Fe^{3+}_{0.075}Mn_{0.005})[Si_{2.89}Al_{0.11}O_{11}(O/OH)]$ | Nat. |
| $Ca_2Al_2Al[Si_3O_{11}(O/OH)]$ | Synth. |
| $(Ca_{1.97}Sr_{<0.01}Mg_{<0.01}Zn_{0.01})Al_2(Al_{0.86}Fe^{3+}_{0.10}Mn^{3+}_{0.03}Cu_{0.01})[Si_{3.02}O_{11}(O/OH)]$ | Nat. |
| $Ca_2Al_2(Al_{0.75}Mn^{3+}_{0.25})[Si_3O_{11}(O/OH)]$ | Synth. |
| $Ca_2Al_2(Al_{0.88}Fe^{3+}_{0.01}Mn^{3+}_{0.11})[Si_{2.99}O_{11}(O/OH)]$ | Nat. |
| $Ca_2Al_2(Al_{0.83}Fe^{3+}_{0.03}Mn^{3+}_{0.14})[Si_3O_{11}(O/OH)]$ | Nat. |
| $Ca_{1.99}Al_2(Al_{0.86}Mn^{3+}_{0.14})[Si_{2.99}O_{11}(O/OH)]$ | Nat. |
| $(Ca_{1.98}Sr_{0.02})Al_2(Al_{0.85}Fe^{3+}_{<0.01}Mn^{3+}_{0.14})[Si_{2.99}O_{11}(O/OH)]$ | Nat. |
| $Ca_2Al_2(Al_{0.82}Fe^{3+}_{<0.01}Mn^{3+}_{0.17})[Si_{3.01}O_{11}(O/OH)]$ | Nat. |
| $Ca_2Al_2(Al_{0.77}Fe^{3+}_{0.12}Mn^{3+}_{0.11})[Si_3O_{11}(O/OH)]$ | Nat. |
| $(Ca_{1.98}Sr_{<0.01}Mg_{0.01})Al_2(Al_{0.96}Fe^{3+}_{<0.01}V^{3+}_{0.01})[Si_3O_{11}(O/OH)]$ | Nat. |
| Tanzanite | Nat. |

| $a$ [Å] | $b$ [Å] | $c$ [Å] | $V$ [Å$^3$] | Source |
|---|---|---|---|---|
| 16.188 (6) | 5.550 (2) | 10.034 (4) | 901.5 (6) | Grevel et al. 2000 |
| 16.178 (6) | 5.547 (2) | 10.029 (5) | 900.0 (7) | Grevel et al. 2000 |
| 16.186 (7) | 5.548 (3) | 10.039 (6) | 901.5 (8) | Grevel et al. 2000 |
| 16.219 (9) | 5.566 (3) | 10.05 (1) | 907 (1) | Pawley et al. 1998 |
| 16.210 (2) | 5.5552 (8) | 10.041 (1) | 904.2 (2) | Pawley et al. 1996 |
| 16.1913 (4) | 5.5488 (1) | 10.0320 (3) | 901.30 (5) | Liebscher et al. 2002 |
| 16.1900 (6) | 5.5511 (2) | 10.0332 (3) | 901.70 (7) | Liebscher et al. 2002 |
| 16.2009 (5) | 5.5536 (2) | 10.0336 (3) | 902.76 (6) | Liebscher et al. 2002 |
| 16.1983 (8) | 5.5564 (2) | 10.0376 (5) | 903.43 (10) | Liebscher et al. 2002 |
| 16.1964 (11) | 5.5580 (3) | 10.0400 (6) | 903.8 (11) | Liebscher et al. 2002 |
| 16.205 (2) | 5.551 (1) | 10.035 (1) | 902.64 (18) | Myer 1966 |
| 16.205 (1) | 5.554 (1) | 10.038 (1) | 903.38 (16) | Myer 1966 |
| 16.213 (2) | 5.558 (1) | 10.037 (2) | 904.34 (23) | Myer 1966 |
| 16.214 (2) | 5.557 (1) | 10.037 (2) | 904.26 (22) | Myer 1966 |
| 16.224 (2) | 5.563 (21) | 10.044 (3) | 906.51 (27) | Myer 1966 |
| / | / | / | 901.86 (32) | Schmidt and Poli 1994 |
| 16.212 (3) | 5.555 (1) | 10.034 (2) | 903.6 (4) | Comodi and Zanazzi 1997 |
| 16.1903 (15) | 5.5487 (7) | 10.0337 (11) | 901.37 (13) | Chatterjee et al. 1984 |
| 16.193 (2) | 5.549 (1) | 10.036 (2) | 901.8 (1) | Langer and Lattard 1980 |
| 16.194 (2) | 5.550 (1) | 10.036 (2) | 902.0 (1) | Langer and Lattard 1980 |
| 16.15 (1) | 5.581 (5) | 10.06 (1) | 906.2 (24) | Pistorius 1961 |
| 16.212 (8) | 5.559 (6) | 10.036 (4) | 904.5 (15) | Dollase 1968 |
| 16.23 | 5.51 | 10.16 | 909 | Fesenko et al. 1955 |
| 16.20 | 5.54 | 10.07 | 904 | Newton 1965 |
| 16.188 (6) | 5.581 (3) | 10.043 (5) | 904.2 (6) | Storre et al. 1982 |
| 16.190 (14) | 5.559 (4) | 10.035 (10) | 903.2 (14) | Storre et al. 1982 |
| 16.20 | 5.56 | 10.04 | 904 | Seki 1959 |
| 16.188 (6) | 5.581 (3) | 10.043 (5) | 904.2 (6) | Johannes and Ziegenbein 1980 |
| 16.2051 (37) | 5.5488 (12) | 10.0229 (18) | 902.1 | Langer et al. 2002 |
| 16.177 (6) | 5.575 (2) | 10.045 (3) | 905.8 (4) | Anastasiou and Langer 1977 |
| 16.194 (3) | 5.559 (1) | 10.041 (2) | 903.9 (0.2) | Reinecke 1986 |
| 16.199 (2) | 5.565 (1) | 10.038 (2) | 904.8 (0.2) | Reinecke 1986 |
| 16.189 (2) | 5.562 (1) | 10.039 (1) | 904.0 (0.2) | Reinecke 1986 |
| 16.188 (4) | 5.562 (2) | 10.039 (3) | 903.9 (0.3) | Reinecke 1986 |
| 16.190 (4) | 5.563 (1) | 10.038 (2) | 904.1 (0.2) | Reinecke 1986 |
| 16.204 (2) | 5.560 (1) | 10.035 (1) | 904.1 (0.2) | Reinecke 1986 |
| 16.20 (1) | 5.55 (1) | 10.00 (1) | 899 (3) | Hurlbut 1969 |
| 16.1909 (15) | 5.5466 (5) | 10.0323 (6) | 900.9 (9) | Smith et al. 1987 |

## APPENDIX D

Bond lengths [Å], polyhedra volumes [Å³], selected interatomic distances [Å] and angles [°] in natural and synthetic Al-Fe³⁺ zoisite solid solutions. Calculated from the data of Dollase 1968 (D68); Brunsmann 2000 (B); Liebscher et al. 2002 (L).

| Composition [$X_{Fe}$] | 0.0 (0) | 0.035 (12) | 0.062 (23) | 0.096 (23) | 0.095 (15) | 0.116 (19) |
|---|---|---|---|---|---|---|
| Sample | synth. | synth. | synth. | synth. | nat. | synth. |
| Source | B, L | B, L | B, L | B, L | D68 | B, L |
| **A-Positions** | | | | | | |
| A1 - O1 (2×) | 2.475 (12) | 2.457 (16) | 2.468 (14) | 2.46 (2) | 2.51 (1) | 2.40 (3) |
| A1 - O3 (2×) | 2.415 (12) | 2.405 (18) | 2.409 (16) | 2.39 (2) | 2.42 (1) | 2.39 (4) |
| A1 - O5 | 2.598 (16) | 2.61 (2) | 2.62 (2) | 2.63 (3) | 2.59 (1) | 2.62 (6) |
| A1 - O6 | 2.565 (16) | 2.57 (2) | 2.57 (2) | 2.60 (3) | 2.55 (1) | 2.57 (5) |
| A1 - O7 | 2.229 (18) | 2.27 (2) | 2.26 (2) | 2.32 (3) | 2.25 (1) | 2.45 (6) |
| A1 - O9 * (2×) | 2.902 (5) | 2.906 (7) | 2.908 (6) | 2.915 (10) | 2.916 | 2.933 (18) |
| Mean A1 - O (without *) | 2.453 (14) | 2.453 (18) | 2.458 (17) | 2.46 (2) | 2.47 (1) | 2.46 (4) |
| A2 - O2 (2×) | 2.505 (12) | 2.504 (16) | 2.486 (16) | 2.45 (2) | 2.52 (1) | 2.50 (4) |
| A2 - O2' (2×) | 2.794 (12) | 2.824 (16) | 2.819 (14) | 2.83 (2) | 2.79 (1) | 2.84 (4) |
| A2 - O3 (2×) | 2.448 (12) | 2.437 (18) | 2.445 (16) | 2.47 (2) | 2.47 (1) | 2.49 (4) |
| A2 - O7 | 2.321 (16) | 2.33 (2) | 2.32 (2) | 2.31 (3) | 2.31 (1) | 2.26 (5) |
| A2 - O8 * (2×) | 3.015 (8) | 3.004 (10) | 3.018 (10) | 2.990 (14) | 3.007 | 2.96 (2) |
| A2 - O10 * | 3.013 (18) | 3.03 (3) | 3.00 (2) | 2.98 (4) | 3.01 (1) | 3.01 (6) |
| Mean A2 - O (without *) | 2.545 (13) | 2.551 (17) | 2.546 (16) | 2.54 (2) | 2.55 (1) | 2.56 (4) |
| **M-Positions** | | | | | | |
| M1,2 – O1 | 1.991 (12) | 1.979 (16) | 1.997 (14) | 1.96 (2) | 1.96 (1) | 1.93 (3) |
| M1,2 – O3 | 1.864 (12) | 1.850 (16) | 1.859 (14) | 1.83 (2) | 1.85 (1) | 1.84 (3) |
| M1,2 – O4 | 1.827 (12) | 1.835 (14) | 1.839 (14) | 1.85 (2) | 1.84 (1) | 1.84 (3) |
| M1,2 – O5 | 1.893 (12) | 1.917 (16) | 1.920 (14) | 1.90 (2) | 1.90 (1) | 1.88 (4) |
| M1,2 – O6 | 1.932 (12) | 1.900 (16) | 1.911 (14) | 1.92 (2) | 1.93 (1) | 1.91 (4) |
| M1,2 – O10 | 1.848 (10) | 1.871 (14) | 1.857 (12) | 1.86 (2) | 1.85 (1) | 1.90 (3) |
| Mean M1,2 – O | 1.893 (12) | 1.892 (15) | 1.897 (14) | 1.89 (2) | 1.89 (1) | 1.88 (3) |
| Volume | 9.0 (2) | 9.0 (2) | 9.0 (2) | 8.9 (3) | 8.9 | 8.8 (4) |
| M3 - O1 (2×) | 2.129 (12) | 2.142 (16) | 2.126 (16) | 2.10 (2) | 2.13 (1) | 2.07 (4) |
| M3 - O2 (2×) | 1.934 (12) | 1.938 (16) | 1.939 (14) | 1.94 (2) | 1.96 (1) | 1.90 (3) |
| M3 - O4 | 1.871 (18) | 1.90 (3) | 1.87 (2) | 1.87 (4) | 1.82 (1) | 1.92 (6) |
| M3 - O8 | 1.74 (2) | 1.75 (3) | 1.74 (2) | 1.82 (4) | 1.78 (1) | 1.95 (6) |
| Mean M3 - O | 1.956 (14) | 1.97 (2) | 1.957 (17) | 1.96 (3) | 1.97 (1) | 1.97 (4) |
| Volume | 9.7 (2) | 9.8 (2) | 9.7 (2) | 9.7 (3) | 9.9 | 9.8 (5) |
| O4 – O8 | 3.61 (3) | 3.65 (4) | 3.61 (3) | 3.69 (3) | 3.61 | 3.87 (6) |
| **T-Positions** | | | | | | |
| T1 - O1 (2×) | 1.659 (12) | 1.686 (16) | 1.672 (14) | 1.75 (2) | 1.66 (1) | 1.83 (4) |
| T1 - O7 | 1.607 (16) | 1.58 (2) | 1.57 (2) | 1.60 (3) | 1.59 (1) | 1.57 (5) |
| T1 - O9 | 1.634 (18) | 1.66 (3) | 1.68 (2) | 1.66 (4) | 1.65 (1) | 1.67 (6) |
| Mean T1 – O | 1.640 (15) | 1.65 (2) | 1.649 (17) | 1.69 (3) | 1.64 (1) | 1.73 (5) |
| Volume | 2.24 (5) | 2.30 (8) | 2.28 (8) | 2.46 (11) | 2.23 | 2.60 (17) |
| T2 - O3 (2×) | 1.614 (12) | 1.639 (18) | 1.630 (16) | 1.65 (2) | 1.62 (1) | 1.67 (4) |
| T2 - O8 | 1.63 (2) | 1.61 (3) | 1.61 (2) | 1.56 (4) | 1.58 (1) | 1.47 (6) |
| T2 - O9 | 1.620 (18) | 1.56 (2) | 1.56 (2) | 1.54 (3) | 1.62 (1) | 1.48 (5) |
| Mean T2 – O | 1.620 (16) | 1.61 (2) | 1.608 (18) | 1.60 (3) | 1.61 (1) | 1.57 (5) |
| Volume | 2.17 (5) | 2.14 (7) | 2.12 (7) | 2.09 (10) | 2.14 | 1.97 (14) |
| T3 - O2 (2×) | 1.632 (10) | 1.620 (14) | 1.637 (14) | 1.651 (18) | 1.62 (1) | 1.64 (3) |
| T3 - O5 | 1.635 (18) | 1.62 (2) | 1.61 (2) | 1.65 (3) | 1.65 (1) | 1.70 (6) |
| T3 - O6 | 1.660 (18) | 1.67 (2) | 1.68 (2) | 1.62 (3) | 1.67 (1) | 1.65 (5) |
| Mean T3 – O | 1.640 (12) | 1.633 (17) | 1.641 (17) | 1.64 (2) | 1.64 (1) | 1.66 (4) |
| Volume | 2.23 (5) | 2.21 (7) | 2.24 (8) | 2.26 (10) | 2.23 | 2.31 (15) |
| **T1T2O₇ Group** | | | | | | |
| O7 – O8 | 4.62 (3) | 4.66 (4) | 4.62 (4) | 4.73 (6) | 4.61 | 4.80 (11) |
| T1 - O9 - T2 | 174.5 (12) | 173.1 (16) | 173.2 (14) | 172 (2) | 172.6 (8) | 170 (4) |
| O7 – O9 – O8 | 118 (1) | 121 (1) | 120 (1) | 123 (2) | 120 | 130 (4) |
| **Proton environment** | | | | | | |
| H – O10 | / | / | / | / | 1.2 (2) | / |
| O10 – H ... O4 | 2.74 (3) | 2.69 (4) | 2.72 (4) | 2.70 (6) | 2.76 (2) | 2.63 (11) |

# APPENDIX E

Optical properties of natural and synthetic monoclinic epidote minerals.

| Composition | Sample |
|---|---|
| ***Al-Fe Epidote*** | |
| $Fe^{3+} = 0$ | |
| $Fe^{3+} = 0.10$ | |
| $Ca_2(Al_{2.92}Fe^{3+}_{0.115})[Si_{2.97}O_{11}(O/OH)]$ | Nat. |
| $(Ca_{1.975}Fe^{2+}_{0.01}Mn_{0.01})(Al_{2.79}Fe^{3+}_{0.16})[Si_{3.04}O_{11}(O/OH)]$ | Nat. |
| $Ca_{1.99}(Al_{2.865}Fe^{3+}_{0.185}Mg_{0.01}Mn_{0.005})[Si_{2.945}O_{11}(O/OH)]$ | Nat. |
| $Fe^{3+} = 0.20$ | |
| $(Ca_{1.975}Fe^{2+}_{0.04}Na_{0.005}K_{0.005}Mn^{2+}_{0.015}Mg_{0.005})(Al_{2.705}Fe^{3+}_{0.225}Ti_{0.003})[Si_{3.00}O_{11}(O/OH)]$ | Nat. |
| $Ca_{1.98}(Al_{2.35}Fe^{3+}_{0.225})[Si_{2.945}O_{11}(O/OH)]$ | Nat. |
| $Ca_{1.985}(Al_{2.73}Fe^{3+}_{0.235}Fe^{2+}_{0.02}Mn^{2+}_{0.04})[Si_{3.005}O_{11}(O/OH)]$ | Nat. |
| $(Ca_{2.015}Na_{0.015})(Al_{2.725}Fe^{3+}_{0.24}Fe^{2+}_{0.04}Mn^{2+}_{0.01})[Si_{2.95}Al_{0.05}O_{11}(O/OH)]$ | Nat. |
| $(Ca_{1.99}Fe^{2+}_{0.01}Na_{0.015}K_{0.02})(Al_{2.68}Fe^{3+}_{0.25}Mg_{0.015}Mn^{2+}_{0.02}Ti_{0.005})[Si_{3.00}P_{0.005}O_{11}(O/OH)]$ | Nat. |
| $(Ca_{1.923}K_{0.023})(Al_{2.633}Fe^{3+}_{0.276}Ti_{0.004}Fe^{2+}_{0.038}Mg_{0.071})[Si_{3.028}O_{11}(O/OH)]$ | Nat. |
| $(Ca_{1.925}Fe^{2+}_{0.04}Na_{0.005}K_{0.005}Mn^{2+}_{0.005}Mg_{0.035})(Al_{2.665}Fe^{3+}_{0.29}Ti_{0.025})[Si_{2.995}Al_{0.005}O_{11}(O/OH)]$ | Nat. |
| $(Ca_{1.985}Fe^{2+}_{0.02}Mn^{2+}_{0.005})(Al_{2.465}Fe^{3+}_{0.305}Ti_{0.015})[Si_{2.915}O_{11}(O/OH)]$ | Nat. |
| $(Ca_{1.965}Fe^{2+}_{0.045}Na_{0.015}K_{0.015}Mn^{2+}_{0.01}Mg_{0.035})(Al_{2.585}Fe^{3+}_{0.31}Ti_{0.055})[Si_{2.975}Al_{0.025}O_{11}(O/OH)]$ | Nat. |
| $(Ca_{1.982}Fe^{2+}_{0.005}Mn^{2+}_{0.007}Mg_{0.03})(Al_{2.644}Fe^{3+}_{0.348}Ti_{0.008})[Si_{3.002}O_{11}(O/OH)]$ | Nat. |
| $Fe^{3+} = 0.35$ | |
| $(Ca_{1.975}Fe^{2+}_{0.04}Na_{0.005}K_{0.005}Mn^{2+}_{0.005}Mg_{0.005})(Al_{2.56}Fe^{3+}_{0.36}Ti_{0.05})[Si_{2.96}Al_{0.04}O_{11}(O/OH)]$ | Nat. |
| $(Ca_2Fe^{2+}_{0.03}Na_{0.01}K_{0.02})(Al_{2.59}Fe^{3+}_{0.365}Mg_{0.01}Mn^{2+}_{0.005}Ti_{0.01})[Si_{2.995}O_{11}(O/OH)]$ | Nat. |
| $(Ca_{1.955}Fe^{2+}_{0.04}Na_{0.005}K_{0.005}Mn^{2+}_{0.005}Mg_{0.005})(Al_{2.595}Fe^{3+}_{0.365}Ti_{0.03})[Si_{2.98}Al_{0.02}O_{11}(O/OH)]$ | Nat. |
| $Ca_{1.98}(Al_{2.695}Fe^{3+}_{0.38}Mn_{0.015})[Si_{2.945}O_{11}(O/OH)]$ | Nat. |
| $(Ca_{1.90}Fe^{2+}_{0.035}Na_{0.01}K_{0.005}Mn^{2+}_{0.01})(Al_{2.595}Fe^{3+}_{0.37}Mg_{0.045}Ti_{0.03})[Si_{2.97}Al_{0.025}P_{0.005}O_{11}(O/OH)]$ | Nat. |
| $(Ca_{1.823}Sr_{0.08}Na_{0.01})(Al_{2.641}Fe^{3+}_{0.38}Ti_{0.011}Fe^{2+}_{0.051}Mg_{0.03})[Si_3O_{11}(O/OH)]$ | Nat. |
| $(Ca_{1.955}Fe^{2+}_{0.035}Na_{0.005}K_{0.005}Mn^{2+}_{0.005})(Al_{2.595}Fe^{3+}_{0.38}Ti_{0.02})[Si_{2.995}Al_{0.005}O_{11}(O/OH)]$ | Nat. |
| $(Ca_{1.999}Fe^{2+}_{0.027}Mg_{0.018})(Al_{2.57}Fe^{3+}_{0.385})[Si_{2.955}Al_{0.045}O_{11}(O/OH)]$ | Nat. |
| $(Ca_{1.985}Fe^{2+}_{0.04}Na_{0.005}K_{0.005}Mn^{2+}_{0.005})(Al_{2.53}Fe^{3+}_{0.39}Ti_{0.05})[Si_{2.985}Al_{0.015}O_{11}(O/OH)]$ | Nat. |
| $(Ca_{1.95}Fe^{2+}_{0.04}Na_{0.015}Mn^{2+}_{0.005})(Al_{2.93}Fe^{3+}_{0.395}Ti_{0.03})[Si_{3.04}O_{11}(O/OH)]$ | Nat. |
| $(Ca_{2.00}Fe^{2+}_{0.015}Mn^{2+}_{0.005})(Al_{2.475}Fe^{3+}_{0.40}Ti_{0.01})[Si_{3.075}O_{11}(O/OH)]$ | Nat. |
| $Ca_{1.965}(Al_{2.605}Fe^{3+}_{0.415})[Si_{3.015}O_{11}(O/OH)]$ | Nat. |
| $(Ca_{1.925}Fe^{2+}_{0.045}Na_{0.005}K_{0.005}Mn^{2+}_{0.005})(Al_{2.48}Fe^{3+}_{0.44}Ti_{0.025})[Si_{3.05}O_{11}(O/OH)]$ | Nat. |
| $(Ca_{1.955}Fe^{2+}_{0.01}Mn^{2+}_{0.005})(Al_{2.5}Fe^{3+}_{0.44}Ti_{0.01})[Si_{3.05}O_{11}(O/OH)]$ | Nat. |
| $(Ca_{1.90}Fe^{2+}_{0.04}Na_{0.01}K_{0.005}Mn^{2+}_{0.005})(Al_{2.55}Fe^{3+}_{0.455}Mg_{0.055}Ti_{0.015})[Si_{2.91}Al_{0.08}P_{0.01}O_{11}(O/OH)]$ | Nat. |
| $Ca_{2.04}(Al_{2.525}Fe^{3+}_{0.455}Mg_{0.035}Ti_{0.005}Mn_{0.005})[Si_{2.935}O_{11}(O/OH)]$ | Nat. |
| $(Ca_{1.93}Fe^{2+}_{0.04}Na_{0.01}K_{0.005}Mn^{2+}_{0.005})(Al_{2.475}Fe^{3+}_{0.485}Mg_{0.04}Ti_{0.02})[Si_{2.96}Al_{0.035}P_{0.005}O_{11}(O/OH)]$ | Nat. |
| $(Ca_{1.92}Fe^{2+}_{0.05}Na_{0.02}K_{0.01}Mn^{2+}_{0.005}Mg_{0.01})(Al_{2.465}Fe^{3+}_{0.485}Ti_{0.025})[Si_{3.015}O_{11}(O/OH)]$ | Nat. |
| $(Ca_{1.975}Fe^{2+}_{0.02}Na_{0.01}K_{0.025})(Al_{2.41}Fe^{3+}_{0.49}Mg_{0.01}Mn^{2+}_{0.01}Ti_{0.005})[Si_{3.045}P_{0.005}O_{11}(O/OH)]$ | Nat. |
| $(Ca_{1.95}Fe^{2+}_{0.02}Mn^{2+}_{0.01})(Al_{2.425}Fe^{3+}_{0.515})[Si_{3.05}O_{11}(O/OH)]$ | Nat. |
| $Ca_{1.855}(Al_{2.51}Fe^{3+}_{0.525}Fe^{2+}_{0.065}Mn^{2+}_{0.01}Mg_{0.04})[Si_{2.98}Al_{0.02}O_{11}(O/OH)]$ | Nat. |
| $(Ca_{1.981}Mn^{2+}_{0.018}Mg_{0.003})(Al_{2.461}Fe^{3+}_{0.53}Ti_{0.006})[Si_{2.997}Al_{0.003}O_{11}(O/OH)]$ | Nat. |
| $(Ca_{1.93}Fe^{2+}_{0.05}Na_{0.025}K_{0.02}Mn^{2+}_{0.015})(Al_{2.425}Fe^{3+}_{0.53}Ti_{0.025})[Si_{3.005}O_{11}(O/OH)]$ | Nat. |
| $(Ca_{1.905}Fe^{2+}_{0.045}Na_{0.01}K_{0.005}Mn^{2+}_{0.01})(Al_{2.40}Fe^{3+}_{0.535}Mg_{0.085}Ti_{0.02})[Si_{2.96}Al_{0.04}P_{0.01}O_{11}(O/OH)]$ | Nat. |
| $Fe^{3+} = 0.54$ | |
| $(Ca_{1.95}Fe^{2+}_{0.04}Sr_{0.01})(Al_{2.37}Fe^{3+}_{0.575}Mg_{0.015}Mn^{2+}_{0.005}Ti_{0.03}V_{0.005}Cr_{0.005})[Si_{3.005}O_{11}(O/OH)]$ | Nat. |
| $Fe^{3+} = 0.59$ | |
| $(Ca_{1.965}Fe^{2+}_{0.015}K_{0.02})(Al_{2.295}Fe^{3+}_{0.625}Mg_{0.015}Mn^{2+}_{0.005}Ti_{0.005})[Si_{3.045}P_{0.005}O_{11}(O/OH)]$ | Nat. |
| $(Ca_{1.99}Fe^{2+}_{0.02}Na_{0.015}K_{0.02})(Al_{2.225}Fe^{3+}_{0.625}Mg_{0.04}Mn^{2+}_{0.005}Ti_{0.005})[Si_{3.07}O_{11}(O/OH)]$ | Nat. |
| $(Ca_{1.98}Fe^{2+}_{0.025}Na_{0.01}K_{0.02})(Al_{2.265}Fe^{3+}_{0.625}Mg_{0.02}Ti_{0.005})[Si_{3.08}O_{11}(O/OH)]$ | Nat. |
| $(Ca_{1.96}Fe^{2+}_{0.03}Na_{0.01}K_{0.013})(Al_{2.30}Fe^{3+}_{0.64}Mg_{0.005}Mn^{2+}_{0.01}Ti_{0.01})[Si_{3.025}O_{11}(O/OH)]$ | Nat. |
| $(Ca_2Fe^{2+}_{0.02}Na_{0.01}K_{0.02})(Al_{2.295}Fe^{3+}_{0.645}Mg_{0.015}Mn^{2+}_{0.005}Ti_{0.005})[Si_{3.08}O_{11}(O/OH)]$ | Nat. |
| $Fe^{3+} = 0.67$ | |
| $(Ca_{1.93}Fe^{2+}_{0.05}Na_{0.03}K_{0.005}Mn^{2+}_{0.015})(Al_{2.24}Fe^{3+}_{0.675}Mg_{0.065}Ti_{0.035})[Si_{2.93}Al_{0.07}O_{11}(O/OH)]$ | Nat. |
| $(Ca_{1.895}Fe^{2+}_{0.04}Na_{0.0025}K_{0.0025}Mn^{2+}_{0.015}Mg_{0.005})(Al_{2.17}Fe^{3+}_{0.675}Ti_{0.02})[Si_{2.96}Al_{0.04}O_{11}(O/OH)]$ | Nat. |
| $Fe^{3+} = 0.69$ | |
| $(Ca_{1.96}Fe^{2+}_{0.025}Mn^{2+}_{0.01})(Al_{2.185}Fe^{3+}_{0.70}Ti_{0.015})[Si_{3.07}O_{11}(O/OH)]$ | Nat. |
| $Fe^{3+} = 0.70$ | |
| $Fe^{3+} = 0.70$ | |
| $(Ca_{1.845}Fe^{2+}_{0.035}Na_{0.015}Mn^{2+}_{0.025})(Al_{2.245}Fe^{3+}_{0.715}Mg_{0.045}Ti_{0.01})[Si_{3.055}O_{11}(O/OH)]$ | Nat. |
| $(Ca_{1.945}Fe^{2+}_{0.02}Mn^{2+}_{0.005})(Al_{2.205}Fe^{3+}_{0.715}Ti_{0.01})[Si_{3.065}O_{11}(O/OH)]$ | Nat. |

| $n_\alpha$ | $n_\beta$ | $n_\gamma$ | $n_\gamma - n_\alpha$ | $2V_z$ | Source |
|---|---|---|---|---|---|
| 1.716 | 1.717 | 1.721 | 0.0052 | 66.2 | Strens 1966 |
| 1.718 | 1.720 | 1.723 | 0.0056 | 81.7 | Strens 1966 |
| 1.7176 | 1.7195 | 1.7232 | 0.0056 | 81 | Johnston 1948 |
| 1.700 | 1.701 | 1.709 | 0.009 | 40 | Myer 1965 |
| 1.706 | 1.709 | 1.712 | 0.006 | | Johnston 1948 |
| 1.724 | 1.729 | 1.734 | 0.0105 | 89.3 | Strens 1966 |
| 1.710 | 1.711 | 1.714 | 0.0040 | 76 | Hörmann and Raith 1971 |
| 1.7238 | 1.7291 | 1.7343 | 0.0105 | 89 | Johnston 1948 |
| 1.710 | 1.713 | 1.719 | 0.009 | 65 | Seki 1959 |
| 1.710 | 1.714 | 1.719 | 0.009 | 76 (2) | Seki 1959 |
| 1.709 | 1.711 | 1.717 | 0.008 | 60 | Myer 1966 |
| 1.710 | | 1.718 | 0.008 | 87.5 (15) | Kiseleva et al. 1974 |
| 1.712 | 1.715 | 1.719 | 0.0074 | 86 | Hörmann and Raith 1971 |
| 1.713 | 1.717 | 1.726 | 0.013 | 70 | Myer 1965 |
| 1.714 | 1.717 | 1.721 | 0.0070 | 87 | Hörmann and Raith 1971 |
| 1.712 | 1.718 | 1.724 | 0.012 | 90 | Holdaway 1972 |
| 1.723 | 1.734 | 1.744 | 0.0216 | 93.4 | Strens 1966 |
| 1.717 | 1.722 | 1.727 | 0.0100 | 92 | Hörmann and Raith 1971 |
| 1.713 | 1.718 | 1.725 | 0.012 | 76 | Myer 1966 |
| 1.715 | 1.719 | 1.724 | 0.0088 | 93 | Hörmann and Raith 1971 |
| 1.714 | 1.716 | 1.724 | 0.01 | 89 | Johnston 1948 |
| | 1.726 | | | | Myer 1966 |
| 1.711 | 1.714 | 1.719 | 0.008 | 86 (1) | Kiseleva et al. 1974 |
| 1.717 | 1.722 | 1.727 | 0.0104 | 94 | Hörmann and Raith 1971 |
| 1.713 | 1.720 | 1.727 | 0.014 | 90 | Holdaway 1972 |
| 1.716 | 1.721 | 1.726 | 0.0102 | 94 | Hörmann and Raith 1971 |
| 1.719 | 1.726 | 1.731 | 0.0123 | 95 | Hörmann and Raith 1971 |
| 1.715 | 1.724 | 1.730 | 0.015 | 102 | Myer 1965 |
| 1.714 | 1.7196 | 1.725 | 0.011 | 90 | Johnston 1948 |
| 1.719 | 1.726 | 1.731 | 0.0122 | 96 | Hörmann and Raith 1971 |
| 1.716 | 1.726 | 1.733 | 0.017 | 100 | Myer 1965 |
| | 1.728 | | | | Myer 1966 |
| 1.712 | 1.723 | 1.735 | 0.023 | | Johnston 1948 |
| | 1.729 | | | | Myer 1966 |
| 1.720 | 1.728 | 1.734 | 0.0147 | 98 | Hörmann and Raith 1971 |
| 1.716 | 1.729 | 1.739 | 0.023 | 98 | Myer 1966 |
| 1.718 | 1.728 | 1.740 | 0.022 | 85 | Myer 1965 |
| 1.726 | 1.735 | 1.741 | 0.015 | < 130 | Seki 1959 |
| 1.721 | 1.732 | 1.743 | 0.022 | 90 | Holdaway 1972 |
| 1.723 | 1.733 | 1.741 | 0.0185 | 99 | Hörmann and Raith 1971 |
| | 1.737 | | | | Myer 1966 |
| 1.753 | 1.781 | 1.812 | 0.059 | 90 | Strens 1966 |
| 1.718 | 1.734 | 1.746 | 0.028 | 100 | Myer 1966 |
| 1.728 | 1.746 | 1.766 | 0.038 | 80 | Strens 1966 |
| 1.720 | 1.734 | 1.747 | 0.027 | 93 | Myer 1966 |
| 1.718 | 1.735 | 1.750 | 0.032 | 95 | Myer 1966 |
| 1.722 | 1.738 | 1.751 | 0.029 | 98 | Myer 1966 |
| 1.720 | 1.735 | 1.748 | 0.028 | 95 | Myer 1966 |
| 1.720 | 1.736 | 1.750 | 0.03 | 95 | Myer 1966 |
| 1.735 | 1.754 | 1.775 | 0.040 | 75 | Strens 1966 |
| | 1.741 | | | | Myer 1966 |
| 1.723 | 1.742 | 1.752 | 0.0280 | 105 | Hörmann and Raith 1971 |
| 1.740 | 1.762 | 1.780 | 0.040 | | Strens 1966 |
| 1.721 | 1.741 | 1.756 | 0.035 | 100 | Myer 1965 |
| 1.731 | 1.752 | 1.772 | 0.041 | 85 | Strens 1966 |
| 1.739 | 1.760 | 1.779 | 0.040 | 80 | Strens 1966 |
| 1.721 | 1.740 | 1.756 | 0.035 | 96 | Myer 1966 |
| 1.721 | 1.741 | 1.756 | 0.035 | 100 | Myer 1965 |

**APPENDIX E.** continued

| Composition | Sample |
|---|---|
| ***Al-Fe Epidote (continued)*** | |
| $(Ca_{1.945}Fe^{2+}_{0.02}Mn^{2+}_{0.005})(Al_{2.185}Fe^{3+}_{0.715}Ti_{0.03})[Si_{3.06}O_{11}(O/OH)]$ | Nat. |
| $Fe^{3+} = 0.72$ | |
| $(Ca_{1.975}Fe^{2+}_{0.035}Na_{0.005}K_{0.005}Mn^{2+}_{0.015}Mg_{0.01})(Al_{2.24}Fe^{3+}_{0.735}Ti_{0.005})[Si_{2.98}Al_{0.02}O_{11}(O/OH)]$ | Nat. |
| $(Ca_{1.979}Fe^{2+}_{0.004}Mn^{2+}_{0.015}Mg_{0.003})(Al_{2.25}Fe^{3+}_{0.74}Ti_{0.01})[Si_{2.999}Al_{0.001}O_{11}(O/OH)]$ | Nat. |
| $Ca_{1.99-2.005}(Al_{2.22-2.43}Fe^{3+}_{0.67-0.80})[Si_{2.93-2.985}O_{11}(O/OH)]$ | Synth. |
| $(Ca_{2.025}Fe^{2+}_{0.005}K_{0.02})(Al_{2.14}Fe^{3+}_{0.775}Mg_{0.01}Mn^{2+}_{0.01}Ti_{0.005})[Si_{3.08}O_{11}(O/OH)]$ | Nat. |
| $(Ca_{1.99}Mn^{2+}_{0.007}Mg_{0.011})(Al_{2.192}Fe^{3+}_{0.779}Ti_{0.022})[Si_{2.995}Al_{0.005}O_{11}(O/OH)]$ | Nat. |
| $Ca_{1.99-2.025}(Al_{2.135-2.32}Fe^{3+}_{0.71-0.855})[Si_{2.985-3.00}O_{11}(O/OH)]$ | Synth. |
| $(Ca_{1.985}Fe^{2+}_{0.005}Na_{0.01}K_{0.02})(Al_{2.125}Fe^{3+}_{0.80}Mn^{2+}_{0.01})[Si_{3.04}P_{0.005}O_{11}(O/OH)]$ | Nat. |
| $(Ca_{1.948}Fe^{2+}_{0.019}Mn^{2+}_{0.03}Mg_{0.003})(Al_{2.169}Fe^{3+}_{0.824}Ti_{0.007})[Si_{2.982}Al_{0.018}O_{11}(O/OH)]$ | Nat. |
| $(Ca_{1.935}Fe^{2+}_{0.055}Na_{0.055}K_{0.005})(Al_{2.075}Fe^{3+}_{0.83}Mg_{0.07}Mn^{2+}_{0.015}Ti_{0.01})[Si_3O_{11}(O/OH)]$ | Nat. |
| $Ca_{1.93-2.00}(Al_{2.145-2.34}Fe^{3+}_{0.78-0.87})[Si_{2.945-2.99}O_{11}(O/OH)]$ | Synth. |
| $(Ca_{1.985}Fe^{2+}_{0.035}Mn^{2+}_{0.015})(Al_{2.12}Fe^{3+}_{0.855}Ti_{0.005})[Si_{2.955}Al_{0.045}O_{11}(O/OH)]$ | Nat. |
| $(Ca_{1.98}Na_{0.015})(Al_{1.985}Fe^{3+}_{0.86}Fe^{2+}_{0.08}Mn^{2+}_{0.02}Mg_{0.025})[Si_{3.06}O_{11}(O/OH)]$ | Nat. |
| $(Ca_{1.96}Fe^{2+}_{0.03}Na_{0.005}K_{0.005}Mn^{2+}_{0.01}Mg_{0.025})(Al_{2.085}Fe^{3+}_{0.865}Ti_{0.025})[Si_{2.97}Al_{0.03}O_{11}(O/OH)]$ | Nat. |
| $(Ca_{1.935}Fe^{2+}_{0.05}Na_{0.005}K_{0.005}Mn^{2+}_{0.005}Mg_{0.01})(Al_{2.12}Fe^{3+}_{0.89}Ti_{0.005})[Si_{2.985}Al_{0.015}O_{11}(O/OH)]$ | Nat. |
| $Ca_{1.805}(Al_{2.183}Fe^{3+}_{0.936}Ti_{0.012}Fe^{2+}_{0.023}Mg_{0.018})[Si_{3.017}O_{11}(O/OH)]$ | Nat. |
| $Ca_{1.944}(Al_{2.092}Fe_{0.981}Mn_{0.009})[Si_{2.969}O_{11}(O/OH)]$ | Nat. |
| $(Ca_{1.995}Fe^{2+}_{0.01})(Al_{1.935}Fe^{3+}_{0.985})[Si_{3.055}O_{11}(O/OH)]$ | Nat. |
| $Ca_{2.05-2.06}(Al_{1.90-2.04}Fe^{3+}_{0.955-1.115})[Si_{2.96-2.98}O_{11}(O/OH)]$ | Synth. |
| $(Ca_{1.996}Mn^{2+}_{0.012}Mg_{0.012})(Al_{1.985}Fe^{3+}_{0.990}Ti_{0.004})[Si_{2.984}Al_{0.016}O_{11}(O/OH)]$ | Nat. |
| $(Ca_{1.99}Fe^{2+}_{0.03})(Al_{1.935}Fe^{3+}_{1.045})[Si_{2.945}Al_{0.055}O_{11}(O/OH)]$ | Nat. |
| $Ca_{2.02-2.035}(Al_{1.93-2.01}Fe^{3+}_{0.98-1.12})[Si_{2.955-2.99}O_{11}(O/OH)]$ | Synth. |
| $Ca_{2.03}(Al_{1.80}Fe^{3+}_{1.105}Fe^{2+}_{0.025}Mn^{2+}_{0.02})[Si_{3.03}O_{11}(O/OH)]$ | Nat. |

| | |
|---|---|
| ***Al-Fe-Mn Epidote*** | |
| $(Fe_{0.45};Mn_{0.17})$ | |
| $(Fe_{0.67};Mn_{0.06})$ | |
| $(Fe_{0.64};Mn_{0.21})$ | |
| $(Fe_{0.67};Mn_{0.19})$ | |
| $(Fe_{0.55};Mn_{0.38})$ | |
| $(Fe_{0.23};Mn_{0.73})$ | |
| $(Fe_{0.30};Mn_{0.71})$ | |
| $(Fe_{0.28};Mn_{0.75})$ | |
| $(Fe_{0.40};Mn_{0.70})$ | |
| $(Fe_{0.64};Mn_{0.45})$ | |
| $(Fe_{0.39};Mn_{0.71})$ | |
| $(Fe_{0.59};Mn_{0.61})$ | |
| $(Fe_{0.53};Mn_{0.68})$ | |
| $(Fe_{0.92};Mn_{0.30})$ | |
| $(Fe_{0.62};Mn_{0.63})$ | |
| $(Fe_{0.51};Mn_{0.76})$ | |
| $(Fe_{0.68};Mn_{0.75})$ | |
| $(Fe_{0.66};Mn_{0.78})$ | |

| | |
|---|---|
| ***Al-Mn Epidote*** | |
| $Ca_{1.92}Mn^{2+}_{0.11}(Al_{2.26}Fe_{0.981}Mn^{3+}_{0.73})[Si_{2.79}O_{11}(O/OH)]$ | Synth. |
| $Ca_{1.98}(Al_{2.01}Mn^{3+}_{1.00})[Si_{3.00}O_{11}(O/OH)]$ | Synth. |
| $Ca_{1.98}(Al_{1.93}Mn^{3+}_{1.11})[Si_{2.98}O_{11}(O/OH)]$ | Synth. |
| $Ca_{1.96}(Al_{1.82}Mn^{3+}_{1.23})[Si_{2.99}O_{11}(O/OH)]$ | Synth. |
| $Ca_{1.97}(Al_{1.59}Mn^{3+}_{1.45})[Si_{2.98}O_{11}(O/OH)]$ | Synth. |

| | |
|---|---|
| ***REE Epidote*** | |
| $(Ca_{1.08}Ce_{0.92})(Al_{0.66}Fe^{3+}_{1.24}Fe^{2+}_{0.93}Mn_{0.07}Ti_{0.14})[Si_{2.94}Al_{0.06}O_{11}(O/OH)]$ | Nat. |
| $(Ca_{1.05}Ce_{0.57}La_{0.33}Nd_{0.07}Pr_{0.03})(Al_{1.91}Fe_{0.14}Mg_{0.93}Ti_{0.06})[Si_{2.94}O_{11}(O/OH_{0.94}/F_{0.06})]$ | Nat. |

| $n_\alpha$ | $n_\beta$ | $n_\gamma$ | $n_\gamma - n_\alpha$ | $2V_z$ | Source |
|---|---|---|---|---|---|
| 1.721 | 1.741 | 1.757 | 0.036 | 98 | Myer 1965 |
| 1.736 | 1.757 | 1.778 | 0.042 | 85 | Strens 1966 |
| 1.726 | 1.746 | 1.758 | 0.0320 | 107 | Hörmann and Raith 1971 |
| 1.728 | 1.746 | 1.763 | 0.035 | 93 | Holdaway 1972 |
| 1.730 | | 1.775 | 0.035 | | Liou 1973 |
| 1.729 | 1.749 | 1.764 | 0.035 | 100 | Myer 1966 |
| 1.729 | 1.751 | 1.768 | 0.039 | 99 | Holdaway 1972 |
| 1.738 | | 1.770 | 0.032 | | Liou 1973 |
| 1.728 | 1.748 | 1.763 | 0.035 | 100 | Myer 1966 |
| 1.730 | 1.753 | 1.772 | 0.042 | 97 | Holdaway 1972 |
| | 1.763 | | | | Myer 1966 |
| 1.735 | | 1.778 | 0.043 | | Liou 1973 |
| 1.731 | 1.756 | 1.770 | 0.0390 | 106 | Hörmann and Raith 1971 |
| 1.729 | 1.754 | 1.776 | 0.047 | 106.5 (5) | Seki 1959 |
| 1.729 | 1.754 | 1.767 | 0.0377 | 108 | Hörmann and Raith 1971 |
| 1.734 | 1.762 | 1.776 | 0.0423 | 109 | Hörmann and Raith 1971 |
| 1.740 | 1.757 | 1.771 | 0.031 | 111.5 (15) | Kiseleva et al. 1974 |
| 1.755 | 1.768 | 1.781 | 0.026 | 107 (1) | Kiseleva et al. 1974 |
| 1.736 | 1.773 | 1.788 | 0.052 | 115 | Myer 1965 |
| 1.748 | | 1.785 | 0.037 | | Liou 1973 |
| 1.741 | 1.771 | 1.790 | 0.049 | 105 | Holdaway 1972 |
| 1.751 | 1.784 | 1.797 | 0.046 | 116 | Myer 1966 |
| 1.745 | | 1.786 | 0.041 | | Liou 1973 |
| 1.740 | 1.768 | 1.787 | 0.047 | 106 | Seki 1959 |
| | 1.730 | | 0.028 | 73 | Strens 1966 |
| 1.733 | 1.750 | 1.762 | 0.029 | 102 | Strens 1966 |
| 1.730 | 1.746 | 1.765 | 0.035 | 98 | Strens 1966 |
| | 1.749 | | | 104 | Strens 1966 |
| 1.740 | 1.752 | 1.779 | 0.039 | 87 | Strens 1966 |
| 1.732 | 1.750 | 1.778 | 0.046 | | Strens 1966 |
| 1.746 | 1.764 | 1.806 | 0.0604 | 68.2 | Strens 1966 |
| 1.739 | 1.765 | 1.799 | 0.0600 | 69.5 | Strens 1966 |
| 1.742 | 1.767 | 1.805 | 0.063 | 81 | Strens 1966 |
| 1.751 | 1.781 | 1.812 | 0.061 | 86 | Strens 1966 |
| 1.746 | 1.776 | 1.793 | 0.047 | 87 | Strens 1966 |
| 1.754 | 1.772 | 1.795 | 0.041 | 85 | Strens 1966 |
| 1.751 | 1.783 | 1.833 | 0.082 | 79.5 | Strens 1966 |
| 1.743 | 1.779 | 1.808 | 0.065 | 97 | Strens 1966 |
| 1.756 | 1.783 | 1.823 | 0.067 | 72 | Strens 1966 |
| 1.750 | 1.782 | 1.832 | 0.0823 | 80.6 | Strens 1966 |
| 1.768 | 1.803 | 1.843 | 0.075 | 88 | Strens 1966 |
| 1.772 | 1.813 | 1.860 | 0.088 | 88 | Strens 1966 |
| 1.729 (7) | 1.746 (7) | 1.778 (8) | 0.049 | | Langer et al. 2002 |
| 1.731 (7) | 1.759 (9) | 1.794 (8) | 0.063 | | Langer et al. 2002 |
| 1.738 (5) | 1.767 (5) | 1.799 (9) | 0.061 | | Langer et al. 2002 |
| 1.746 (5) | 1.783 (5) | 1.823 (11) | 0.077 | | Langer et al. 2002 |
| 1.757 (5) | 1.805 (5) | 1.852 (5) | 0.095 | | Langer et al. 2002 |
| 1.825 (2) | 1.855 (5) | 1.880 (5) | 0.055 | 97 | Kartashov et al. 2002 |
| 1.735 (3) | 1.741 (3) | 1.758 (3) | 0.023 | 64.2 (3) | Grew et al. 1991 |

# APPENDIX F

Optical properties of natural zoisite.

---

**Composition**

---

$(Ca_{2.025}Fe^{2+}_{0.01}Na_{0.005}K_{0.02})Al_2(Al_{0.94}Fe^{3+}_{0.015}Mg_{0.005}Ti_{0.005})[Si_{2.995}P_{0.005}O_{11}(O/OH)]$

$(Ca_{1.995}Fe^{2+}_{0.02}Na_{0.005}K_{0.02})Al_2(Al_{0.935}Fe^{3+}_{0.035}Mg_{0.005}Ti_{0.005})[Si_{2.955}P_{0.005}Al_{0.03}O_{11}(O/OH)]$

$(Ca_{2.07}Fe^{2+}_{0.015}Na_{0.005}K_{0.02})Al_2(Al_{0.815}Fe^{3+}_{0.105}Mg_{0.005}Ti_{0.005}Mn^{3+}_{0.005})[Si_{2.985}P_{0.005}Al_{0.01}O_{11}(O/OH)]$

$(Ca_{2.005}Fe^{2+}_{0.015}Na_{0.005}K_{0.02})Al_2(Al_{0.85}Fe^{3+}_{0.105}Mg_{0.005}Ti_{0.01})[Si_{2.98}P_{0.01}Al_{0.01}O_{11}(O/OH)]$

$(Ca_{2.00}Fe^{2+}_{0.02}Na_{0.005}K_{0.02})Al_2(Al_{0.81}Fe^{3+}_{0.14}Mg_{0.005})[Si_{3.005}P_{0.005}O_{11}(O/OH)]$

$Ca_{2.02}Al_2(Al_{0.93}Fe^{3+}_{0.075}Mn^{3+}_{0.005})[Si_{2.89}Al_{0.11}O_{11}(O/OH)]$

$(Ca_{1.791}K_{0.011}Na_{0.018})(Al_{2.455}Fe^{3+}_{0.20}Ti_{0.011}Mg_{0.374})[Si_{3.129}O_{11}(O/OH)]$

$Ca_{2.07}Al_2(Al_{1.04}Fe^{3+}_{0.1}Cr^{3+}_{0.02})[Si_{3.02}O_{11}(O/OH)]$

$(Ca_{1.987}Mg_{0.008}Sr_{0.002})Al_2(Al_{0.985}Fe^{3+}_{0.002}V^{3+}_{0.012})[Si_3O_{11}(O/OH)]$

"tanzanite" (pure zoisite, no analysis given)

---

| $n_\alpha$ | $n_\beta$ | $n_\gamma$ | $n_\gamma - n_\alpha$ | $2V_z$ | Source |
|---|---|---|---|---|---|
| 1.6911 | 1.6935 | 1.6991 | 0.0080 | 50 (1) | Myer 1966 |
| 1.6920 | 1.6933 | 1.6980 | 0.0060 | 35 (1) | Myer 1966 |
| 1.6976 | 1.6990 | 1.7027 | 0.0051 | 40 (1) | Myer 1966 |
| 1.6980 | 1.6990 | 1.7042 | 0.0062 | 37 (1) | Myer 1966 |
| 1.6990 | 1.7021 | 1.7070 | 0.0080 | 69 (2) | Myer 1966 |
| 1.694 | 1.698 | 1.707 | 0.013 | 44 (5) | Seki 1959 |
| 1.700 | 1.700 | 1.703 | 0.003 | 55 (2) | Kiseleva et al.1974 |
| 1.700 | 1.700 | 1.705 | 0.005 | 25 (8) | Game 1954 |
| 1.6925 | 1.6943 | 1.7015 | 0.009 | 53 | Hurlbut 1969 |
| 1.6917 | 1.6927 | 1.7005 | 0.0088 | n.d. | Anderson 1968 |

Reviews in Mineralogy & Geochemistry
Vol. 56, pp. 83-124, 2004
Copyright © Mineralogical Society of America

**2**

# Thermodynamic Properties of Zoisite, Clinozoisite and Epidote

## Matthias Gottschalk

*Section of Experimental Geochemistry and Mineral Physics*
*GeoForschungsZentrum*
*Telegrafenberg*
*14473 Potsdam, Germany*

## INTRODUCTION

The natural occurrence of epidote minerals is widespread over a large variety of geological settings. Thus epidote minerals are part of numerous phase equilibria, which need to be evaluated to understand the geological processes in general. There are two principal approaches to evaluate phase equilibria in the deep earth. The first uses direct experimental investigations, while the second is by thermodynamic calculations and modeling. Performing and evaluating experiments is often a tedious procedure and by far not all systems can be studied at the required physical and chemical conditions, considering all of the possible variables. Therefore experimental investigations are in many instances only case studies in simplified systems, but the thermodynamic framework provides a powerful tool to perform calculations of complex phase equilibria, if the required parameters are available. However, these two approaches are not necessarily independent, because experimental results are often used to evaluate and to calibrate physical-chemical parameters for such calculations.

Many physical-chemical textbooks treat the principles of thermodynamics, and in addition some texts (e.g., Anderson and Crerar 1993; Nordstrom and Munoz 1994) introduce its application to the geological sciences. Therefore only the fundamental equations and their relationship to the required parameters are treated here briefly.

The evaluation of phase equilibria and/or stable phase assemblages involves the calculation of the apparent chemical potential $\mu_{i(P,T)}$ for each component $i$ present at the $P$ and $T$ of choice according to

$$\mu_{i(P,T)} = \Delta_f h^\circ_{i(P_r,T_r)} - T s^\circ_{i(P_r,T_r)} + \int_{T_r}^{T} c^\circ_{Pi(T)} dT - T \int_{T_r}^{T} \frac{c^\circ_{Pi(T)}}{T} dT + \int_{P_r}^{P} v^\circ_{i(P,T)} dP + RT \ln a_i \quad (1)$$

The calculation involves the following molar standard state properties (note small letters designate molar quantities): absolute enthalpy of formation from the elements $\Delta_f h_i^\circ$ and third law entropy $s_i^\circ$ at reference conditions $P_r$ and $T_r$ (i.e., 0.1 MPa, 298.15 K), the heat capacity $c_P{}^\circ_i$ at constant pressure as function of temperature, and the volume $v_i^\circ$ as a function of temperature and pressure for each pure phase component $i$. If the phase involved is a mixture of phase components (i.e., solid solution, fluid, melt) an additional term involving the activity $a$ has to be accounted for, which considers any entropy and enthalpy change due to mixing. The used symbols and their appropriate units are listed in Table 1.

A complete set of parameters is required to calculate $\mu_i$ at elevated temperatures and

1529-6466/04/0056-0002$05.00

**Table 1.** Used abbreviations, symbols, indexes and units.

| | |
|---|---|
| $i$ | phase component $i$ |
| $T$ | temperature (K, °C) |
| $P$ | pressure (MPa, GPa) |
| $G$ | Gibbs free energy (J, kJ) |
| $H$ | enthalpy (J, kJ) |
| $S$ | entropy (J/K) |
| $V$ | volume (cm³, J/MPa) |
| $R$ | gas constant (J/K/mol) |
| $T_r, P_r$ | reference temperature and pressure (298.15 K, 0.1 MPa) |
| $y°$ | molar standard state property (pure phase component) |
| $\Delta_f h$ | enthalpy of formation from the elements (kJ/mol) |
| $\mu$ | chemical potential (J/mol, kJ/mol) |
| $h$ | molar enthalpy (J/Mol, kJ/mol) |
| $s$ | molar entropy (J/K/mol) |
| $c_P$ | molar heat capacity (J/K/mol) |
| $v$ | molar volume (cm³, J/MPa) |
| $\alpha$ | thermal expansion coefficient (1/K) |
| $\beta$ | compressibility (1/MPa, 1/GPa) |
| $K$ | bulk modulus (MPa, GPa) |
| $n$ | number of moles (mol) |
| $X$ | mole fraction |
| $a$ | activity |
| $f$ | fugacity |
| $\gamma$ | activity coefficient |
| $K_{red}$ | reduced equilibrium constant |
| $w_{BW}$ | Bragg-Williams exchange enthalpy (kJ/mol) |
| $w_{RS}$ | regular solution exchange enthalpy (kJ/mol) |
| $\zeta$ | Bragg-Williams order parameter |

pressures. Because of the large experimental effort and the limited availability of pure or nearly pure phases as samples, which are required for their experimental determination, such values and their functional dependences are not always available. In many cases these are only measured for rock-forming phases. Therefore the following review is restricted only to the endmember compositions zoisite, clinozoisite, and epidote*. For other epidote minerals (e.g., piemontite, allanite, hancockite, dollaseite), which are usually not rock-forming phases, only molar volumes have been measured at reference conditions while other required parameters are not available. Furthermore volumes of these phases are not those of endmembers, but often only such of complex solid solutions which cannot be necessarily extrapolated to endmember compositions.

For the thermodynamic treatment of zoisite, clinozoisite, and epidote one has to bear in mind the following crystallographic features and peculiarities of these minerals. While zoisite is orthorhombic, clinozoisite and epidote have identical monoclinic structures. A continuous solid solution series exist between clinozoisite and epidote. Starting from clinozoisite, $Fe^{3+}$ is substituted for $Al^{3+}$ mainly at the M3-site and to a smaller extend at the M1-site. The distribution of $Fe^{3+}$ at the M1- and M3-sites is controlled by intracrystalline nonconvergent ordering. Experimental indications also exist, that at least for intermediate compositions

---

* throughout this chapter the formula $Ca_2Al_2Fe^{3+}[SiO_4/Si_2O_7/O/OH]$ is used to describe the composition of the endmember epidote

within the clinozoisite-epidote solid solution series, convergent ordering operates by splitting of the M3-sites into two energetically distinct M3 an M3′ sites. A limited amount of $Fe^{3+}$-substitution at the M3-site is also observed in orthorhombic zoisite forming a solid solution series between zoisite and a hypothetical endmember "orthoepidote." As a consequence of these crystallographic constrains the pure and totally order phase components listed in Table 2 are chosen and define the applied standard state structures and compositions.

In the following mainly parameters for the phase components zoisite, clinozoisite, epidote, and orthoepidote are considered whereas those for the other hypothetical endmembers in Table 2 are mentioned when appropriate equilibria are discussed and only parameters relative to those of zoisite, clinozoisite, epidote, and orthoepidote are given.

**Table 2.** Totally ordered pure phase components used as standard state.

| Phase Component | Occupancies of the M1,M2 and M3 sites |
|---|---|
| *orthorhombic* | |
| zoisite | $Ca_2Al_2{}^{M1,2}Al^{M3}[SiO_4/Si_2O_7/O/OH]$ |
| orthoepidote | $Ca_2Al_2{}^{M1,2}Fe^{M3}[SiO_4/Si_2O_7/O/OH]$ |
| *monoclinic* | |
| clinozoisite | $Ca_2Al^{M1}Al^{M2}Al^{M3}[SiO_4/Si_2O_7/O/OH]$ |
| epidote | $Ca_2Al^{M1}Al^{M2}Fe^{M3}[SiO_4/Si_2O_7/O/OH]$ |
| M1-epidote | $Ca_2Fe^{M1}Al^{M2}Al^{M3}[SiO_4/Si_2O_7/O/OH$ |
| split M3-M3′ clinozoisite | $Ca_2Al^{M1}Al^{M2}Al_{1/2}{}^{M3}Al_{1/2}{}^{M3'}[SiO_4/Si_2O_7/O/OH]$ |
| split M3-M3′ epidote | $Ca_2Al^{M1}Al^{M2}Fe_{1/2}{}^{M3}Fe_{1/2}{}^{M3'}[SiO_4/Si_2O_7/O/OH]$ |

## VOLUME PROPERTIES

### Volume at reference conditions

From those parameters required (Eqn. 1) to calculate the chemical potential at elevated $T$ and $P$ the most accessible value is that of the molar volume at ambient conditions, which is usually measured by X-ray diffraction. The required quantity of a sample can be as low as 1 mg or even less and precise lattice parameters can still be determined. Interpretation of the measurements requires, however, precise characterization of the chemical composition and the structural state of the sample. Any chemical deviation from the composition of the ideal endmember due to cation substitutions or cation order-disorder will shift the observed volume away from its true value. This will be especially problematic for natural samples because the chemistry of such samples is usually complex. Therefore determinations of molar volumes from synthetic samples should be used preferentially. In synthetic samples the chemical variability is drastically reduced and the determined volumes should reflect those of the pure endmember, if it can be synthesized. Difficulties may arise in certain samples, however, due to order-disorder and crystal imperfections.

***Zoisite.*** For pure zoisite and solid solutions involving the $Fe^{3+}$ and $V^{3+}$ substitution for $Al^{3+}$ at the M3 site, determinations of molar volumes are available for both synthetic and natural samples, which are plotted in Figure 1a. The measured range covers compositions between of $X_{zo}$ 1.0 and 0.8 (i.e., $X_{zo} = n_{zo} / (n_{zo} + n_{oep})$).

Volumes for pure synthetic zoisite are available from Pistorius (1961), Langer and Lattard (1980), Storre et al. (1982), Chatterjee et al. (1984), Heinrich and Althaus (1988), Schmidt

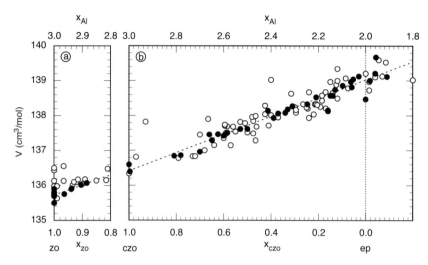

**Figure 1.** Molar volume as a function of composition at reference conditions for a) zoisite-orthoepidote solid solutions, open circles: Seki (1959), Pistorius (1961), Newton (1965, 1966), Dollase (1968), Kiseleva et al. (1974), Storre et al. (1982), Heinrich and Althaus (1988), Smith et al. (1987), Shannon and Rossman (1992), Pawley et al. (1996), Comodi and Zanazzi (1997); filled circles: Langer and Lattard (1980), Chatterjee et al. (1984), Schmidt and Poli (1994), Grevel et al. (2000), Liebscher et al. (2002) and for b) clinozoisite-epidote solid solutions, open circles: Ito (1947), Belov and Rumanova (1953, 1954), Fesenko et al. (1955), Seki (1959), Pistorius (1961), Myer (1965, 1966), Dollase (1968, 1971), Hörmann and Raith (1971), Holdaway (1972), Gabe et al. (1973), Liou (1973), Kiseleva et al. (1974), Nozik et al. (1978), Chatterjee et al. (1984), Stergiou et al. (1987); filled circles: Carbonin and Molin (1980), Bonazzi and Menchetti (1995), Holland et al. (1996), Pawley et al. (1996, 1998), Comodi and Zanazzi (1997), Giuli et al. (1999). For discussion see text.

and Poli (1994), Grevel et al. (2000), and Liebscher et al. (2002). The published volumes are in the rather large range between 135.50–136.43 cm$^3$. If the high volume by Pistorius (1961) of 136.43 cm$^3$ is neglected, which is nearly identical to the one of clinozoisite in the same article, the observed range is reduced to 135.50–135.90 cm$^3$. The volumes given by Langer and Lattard (1980), Chatterjee et al. (1984), Schmidt and Poli (1994), Grevel et al. (2000), Liebscher et al. (2002) yield a consistent and reliable value of 135.72 ± 0.07 cm$^3$. For conventional lattice parameter evaluation methods for powders the observed scatter is mostly within the error range of the measurements if $2\sigma$ errors are considered. For evaluations using the Rietveld structure refinement method the errors are typically smaller. Using Rietveld refinements, Liebscher et al. (2002) determined a value of 135.694 ± 0.008 cm$^3$ that is in the range of the average value given above.

According to Shannon (1976) the ionic radii for $Al^{3+}$, $Fe^{3+}$, and $V^{3+}$ in octahedral coordination are 0.535, 0.78 and 0.64 Å, respectively. Therefore any substitution of $Fe^{3+}$ and $V^{3+}$ for $Al^{3+}$ increases the volume of zoisite. Reported volumes for nearly pure natural zoisite ($X_{zo} > 0.99$) vary between 135.63 and 135.98 cm$^3$ for minor $V^{3+}$ substitution (i.e., variety tanzanite, Smith et al. 1987; Shannon and Rossman 1992) and is given as 135.90 cm$^3$ by Perkins et al. (1980) for minor $Fe^{3+}$ substitution. Further incorporation of $Fe^{3+}$ increases the volume significantly and values between 135.90 and 136.50 cm$^3$ are observed (Seki 1959; Newton 1965; Myer 1966; Dollase 1968; Kiseleva et al. 1974; Pawley et al. 1996; Comodi and Zanazzi 1997).

Even for the rather simple phase zoisite, the reported molar volumes show a respectable scatter that is especially large for the pure endmember (Fig. 1a). Furthermore Liebscher et al. (2002) and Liebscher and Gottschalk (2004) reported an isosymmetric displacive phase transition at $X_{zo}$ of approximately 0.95 at ambient conditions with a volume change of 0.1 cm$^3$. Therefore the linear extrapolation to the volume of pure zoisite using volumes of iron bearing phases is hampered. Nevertheless based on the most reliably determinations a value of $135.72 \pm 0.05$ cm$^3$ is recommended (see also Table 3). This value agrees with the values of 135.88 and 135.75 cm$^3$ which are used by Berman (1988) and Holland and Powell (1998) for their consistent databases, respectively. Hemingway et al. (1982) and Robie and Hemingway (1995) list a value of 136.52 cm$^3$ which was also adopted by Gottschalk (1997). According to the present evaluation of available volumes this value is definitely too large. For the evaluation of phase equilibria at metamorphic conditions in the crust involving a fluid, the difference between the recommended and the latter value is of minor importance. For example, by using the value of 136.52 cm$^3$ at a pressure of 500 MPa the enthalpy of zoisite will be overestimated by approximately 0.4 kJ/mol, but at higher pressures, relevant for e.g., processes in subducting plates, a value in the range of 135.72 cm$^3$ must be used.

***Clinozoisite and epidote.*** A large body of molar volumes is available for clinozoisite-epidote solid solutions (Fig. 1b). As a first approximation the volumes seem to be a linear function of composition, but the available data scatter considerably. The error of the volumes is reported to be in the range of 0.03 to 0.35 cm$^3$ but is usually well below 0.15 cm$^3$. Therefore it seems unlikely that the observed scatter is due to high errors of the volumes but stems from insufficient precision of the compositional characterization of the phases. Considerable analytical efforts are required to determine high precision chemical composition with the electron microprobe. Furthermore many of the volumes were published prior to the introduction of the electron microprobe. It is especially obvious that results from older determinations show a larger scatter than newer ones. If only the more recent experimental results by Carbonin and Molin (1980), Bonazzi and Menchetti (1995), Holland et al. (1996), Pawley et al. (1996, 1998), Comodi and Zanazzi (1997), and Giuli et al. (1999) are considered (filled symbols in Fig. 1b) the volumes show much less scatter. Additional factors might influence the volume of the clinozoisite-epidote solid solutions. With increasing iron content, $Fe^{3+}$ is incorporated at the M3 and also to a certain but much lesser degree at the M1 site (e.g., Bird and Helgeson 1980; Bonazzi and Menchetti 1995; Fehr and Heuss-Aßbichler 1997; Giuli et al. 1999). Minor amounts of $Fe^{3+}$ were also observed at the M2-site by Giuli et al. (1999). All this makes it also possible to incorporate more than one $Fe^{3+}$ per formula unit into epidote. Besides nonconvergent order-disorder schemes, convergent ordering at the M3-site has been proposed (e.g., Hörmann and Raith 1971; Fehr and Heuss-Aßbichler 1997; Brunsmann et al. 2002; Franz and Liebscher 2004) both of which might have a volume effect.

Because pure clinozoisite is metastable with respect to zoisite (see below) at conditions usually used for synthesis and seems to be rare in nature, only a few volumes have been reported for the pure endmember. The reported values are 136.41, 136.59 and 136.39 cm$^3$ by Pistorius (1961), Chatterjee et al. (1984), and Pawley et al. (1998), respectively.

For epidotes with a composition close to one $Fe^{3+}$ per formula unit Seki (1959), Holdaway (1972), Liou (1973), Bonazzi and Menchetti (1995), and Giuli et al. (1999) report volumes of 138.73, 138.97, 139.10, 139.00 and 139.20 cm$^3$, respectively. In principle the volume of ordered epidote at its standard state is of thermodynamic interest, i.e., $Fe^{3+}$ only at the M3 site. The measured volumes may reflect some degree of disorder, however. For epidote it is unclear how much the volume changes by disordering of $Fe^{3+}$ at the M3 and M1 site, but it can be assumed that the effect will be minor. Using the linear compositional trend which is especially pronounced for the most recently determined volumes (Carbonin and Molin 1980; Bonazzi and Menchetti 1995; Holland et al. 1996; Pawley et al. 1996, 1998; Comodi and Zanazzi 1997; Giuli et al. 1999) molar volumes of $136.40 \pm 0.07$ for clinozoisite and $139.01 \pm 0.07$ for ordered epidote are derived.

**Table 3.** Molar volumes, coefficients of thermal expansion and compressibilities.

| source of experimental results | zoisite | clinozoisite | epidote |
|---|---|---|---|
| *molar volume* $v^o_{i(P_r,T_r)}$ *(J/MPa/mol)* | | | |
| | 135.72(7) | 136.40(14) | 139.01(14) |
| *constant thermal expansion coefficient (Eqn. 6)* $\alpha^o_{0,i} \times 10^{-6}$ *(1/K), (* $\alpha^o_{1,i} = 0$ *)* | | | |
| Pawley et al. (1996)[a] | 37.1(3) | 29.5(4) | |
| *constant compressibility (Eqn. 11)* $\beta^o_{i(P_r,T_r)} \times 10^6$ *(1/MPa)* | | | |
| Comodi and Zanazzi (1997)[a] | 9.09(20) | 7.80(7) | |
| Pawley et al. (1998)[a] | 6.67(25) | | |
| Holland et al. (1996)[a] | 3.38(4) | 5.96(11) | 5.80(10) |
| Grevel et al. (2000)[a] | 7.55(12) | | |
| Winkler et al. (2001) | 8.47(14)[b] | 7.35(21)[b] | |
| | 7.34(16)[c] | 6.48(19)[d] | |
| *constant bulk modulus (Eqn. 11)* $K^o_{i(P_r,T_r)}$ *(GPa)] (e.g.* $1/\beta^o_{i(P_r,T_r)}$ *)* | | | |
| Comodi and Zanazzi (1997)[a] | 110(3) | 128(1) | |
| Pawley et al. (1998)[a] | 150(5) | | |
| Holland et al. (1996)[a] | 296(3) | 168(3) | 172(3) |
| Grevel et al. (2000)[a] | 132(2) | | |
| Winkler et al. (2001) | 118(2)[b] | 136(4)[b] | |
| | 136(3)[c] | 154(4)[d] | |
| *bulk modulus (Eqn. 16)* $K^o_{i(0,T_r)}$ *(GPa), Murnaghan EoS,* $K^{o'}_{i(0,T_r)} = 4$ | | | |
| Comodi and Zanazzi (1997)[a] | 104(2) | 120(1) | |
| Pawley et al. (1998)[a] | 141(6) | | |
| Holland et al. (1996)[a] | 275(5) | 155(3) | 164(3) |
| Grevel et al. (2000)[a] | 123(2) | | |
| | 127(3)[c] | 142(4)[d] | |
| *bulk modulus (Eqn. 20)* $K^o_{i(0,T_r)}$ *(GPa), 2. ord. Birch-Murnaghan EoS,* $K^{o'}_{i(0,T_r)} = 4$ | | | |
| Comodi and Zanazzi (1997)[a] | 104(2) | 120(1) | |
| Pawley et al. (1998)[a] | 138(6) | | |
| Holland et al. (1996)[a] | 267(8) | 155(3) | 163(3) |
| Grevel et al. (2000)[a] | 123(2) | | |
| | 125(3)[c] | 141(4)[d] | |
| *bulk modulus* $K^o_{i(0,T_r)}$ *(GPa),* $\left(\partial K^o_{i(0,T)}/\partial T\right)_{P=0}$ *(GPa/K), Murnaghan EoS,* $K^{o'}_{i(0,T_r)} = 4$ | | | |
| | 129(2), −0.029(4)[f] | | |
| *bulk modulus* $K^o_{i(0,T_r)}$ *(GPa),* $\left(\partial K^o_{i(0,T)}/\partial T\right)_{P=0}$ *(GPa/K), 2nd ord. B.-M. EoS,* $K^{o'}_{i(0,T_r)} = 4$ | | | |
| | 127(2), −0.026(4)[f] | | |

[a] fitted using experimental results from the source cited
[b] theoretical value based on first principles calculations
fitted using combined experimental results from:
[c] Comodi and Zanazzi (1997), Pawley et al. (1998) and Grevel et al. (2000) at $T=T_r$
[d] Comodi and Zanazzi (1997) and Pawley et al. (1998) at $T=T_r$
[f] results from [c] at all $T$ using respective value of $\alpha_{0,i}^o$ listed above

**Volume at elevated temperature and pressure**

The calculation of the chemical potential of a phase component at elevated pressure and temperature by Equation (1) requires the knowledge of its volume as function of pressure and temperature. The determination of this functional dependence is approached experimentally in three ways, which differ also in the effort to derive the respective parameters. The first is the determination of the temperature dependence of volume at reference pressure, its coefficient of thermal expansion $\alpha$. The second is the measurement of the pressure dependence at reference temperature, its compressibility $\beta$ or bulk modulus $K$. And third, the evaluation of volume at both elevated temperature and pressure provides information about the required temperature dependence of $\beta$ or $K$.

**Coefficient of thermal expansion**

The coefficient of thermal expansion of a component $\alpha_i^o$ at $P$ and $T$ is defined as

$$\alpha_i^o = \frac{1}{V_i^o}\left(\frac{\partial V_i^o}{\partial T}\right)_P \tag{2}$$

Using the molar volume instead of the total volume $v_i^o$

$$v_i^o = \frac{V_i^o}{n_i} \tag{3}$$

integration of Equation (2) leads to Equation (4)

$$v_{i(P_r,T)}^o = v_{i(P_r,T_r)}^o\, e^{\int_{T_r}^{T} \alpha_i^o dT} \tag{4}$$

which is used to calculate the molar volume at elevated temperature. The functional temperature dependence of $\alpha_i^o$ is usually expressed in the form

$$\alpha_{i(P_r,T)}^o = \alpha_{0,i}^o + \frac{\alpha_{1,i}^o}{\sqrt{T}} \tag{5}$$

with the constants $\alpha_{0,i}^o$ and $\alpha_{1,i}^o$. Using this functional dependence Equation (4) becomes:

$$v_{i(P_r,T)}^o = v_{i(P_r,T_r)}^o\, e^{\alpha_{0,i}^o(T-T_r)+2\alpha_{1,i}^o(\sqrt{T}-\sqrt{T_r})} \tag{6}$$

The value of $\alpha_i^o$ is usually low and in the range of $10^{-5}$–$10^{-6}$ K$^{-1}$, therefore the following approximation is often used:

$$v_{i(P_r,T)}^o = v_{i(P_r,T_r)}^o\left[1+\alpha_{0,i}^o(T-T_r)+2\alpha_{1,i}^o(\sqrt{T}-\sqrt{T_r})\right] \tag{7}$$

Molar volumes at elevated temperature are typically determined by in situ X-ray diffraction measurements. It is important to note that minerals of the epidote group due to their incorporation of the OH-group, are not stable at reference pressure and high temperature and tend to decompose at these conditions as reaction kinetics become enhanced. This might limit the time available for measurement.

***Zoisite.*** The temperature dependence of the volume of zoisite has only been determined experimentally by Pawley et al. (1996) in the range of 298 to 1023 K (Fig. 2). Using Equation (7) and setting $\alpha_{1,zo}^o$ to 0, Pawley et al. (1996) derived a value $38.6 \pm 0.5 \times 10^{-6}$ K$^{-1}$ for $\alpha_{0,zo}^o$. If the same experimental results are reevaluated using Equation (6) instead and also setting $\alpha_{1,zo}^o$ to 0, a value of $37.1 \pm 0.3 \times 10^{-6}$ K$^{-1}$ is derived (see Table 3). Considering $\alpha_{1,zo}^o$ as an

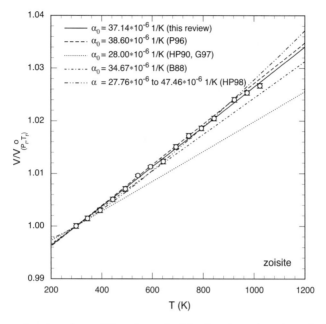

**Figure 2.** Temperature dependence of the volume of zoisite. Experimental results are by Pawley et al. (1996). The curves drawn are based on values for $\alpha_0^\circ$ derived here and used by Pawley et al. (1996) P96, Holland and Powell (1990) (HP90), Gottschalk (1997) G97, Berman (1988) B88, Holland and Powell (1998) HP98.

additional parameter yields an error for $\alpha^\circ_{1,zo}$, which is about 400% of its value and does not improve the representation of the experimental data. The derived parameters would then be $36.1 \pm 3.9 \times 10^{-6}$ K$^{-1}$ for $\alpha^\circ_{0,zo}$ and $23 \pm 92 \ 10^{-6}$ K$^{-1/2}$ for $\alpha^\circ_{1,zo}$.

It has been stated by Pawley et al. (1996) that many phases appear to follow the relationship between $\alpha^\circ_{0,i}$ and $\alpha^\circ_{1,i}$

$$\alpha^\circ_{1,i} = -10\,\alpha^\circ_{0,i} \tag{8}$$

Using Equation (8) in conjunction with Equation (7) a value of $65.9 \pm 0.9 \times 10^{-6}$ K$^{-1}$ for $\alpha^\circ_{0,zo}$ is derived by fitting these equations to the original experimental data. As a consequence $\alpha^\circ_{zo}$ varies between 27.8 and 47.5 $\times 10^{-6}$ K$^{-1}$ at 298 and 1298 K, respectively. At infinite temperatures $\alpha^\circ_{zo}$ reaches the limit of $65.9 \times 10^{-6}$ K$^{-1}$ for which $\alpha^\circ_{zo}$ is nearly doubled if compared to reference conditions. This leads to a noticeable increase of the temperature dependence of the respective volumes, which is not supported by the experimental results. Therefore this form seems to have no advantage and Equation (8) is at least questionable for epidote group minerals. However, Holland and Powell (1998) applied this approach using a $\alpha^\circ_{0,zo}$ value of $67 \times 10^{-6}$ K$^{-1}$ and a respective $\alpha^\circ_{1,zo}$ value according to Equation (8).

Before the experimental results by Pawley et al. (1996) became available for zoisite, Berman (1988) used in his data set an $\alpha^\circ_{0,zo}$-value of $34.67 \times 10^{-6}$ K$^{-1}$, and Holland and Powell (1990) estimated its value to be $28.0 \times 10^{-6}$ K$^{-1}$, which was also adopted by Gottschalk (1997). While the last value is close to that of clinozoisite (see below) it is definitely too low if the experimental results Pawley et al. (1996) are considered. According to the arguments given above a value of $\alpha^\circ_{0,zo}$ of $37.1 \pm 0.3 \times 10^{-6}$ K$^{-1}$ ($\alpha^\circ_{1,zo} = 0$) is recommended for zoisite.

***Clinozoisite and epidote.*** As for zoisite the coefficient of thermal expansion for clinozoisite has been determined only by Pawley et al. (1996) between 298 and 1173 K (Fig. 3). Using Equation (7) they derived a value for $\alpha^\circ_{0,czo}$ of $29.4 \pm 0.5 \times 10^{-6}$ K$^{-1}$. A reevaluation of their data using Equation (6) yields a value of $29.5 \pm 0.4 \times 10^{-6}$ K$^{-1}$. In both cases $\alpha^\circ_{1,czo}$ has been set to 0. Thus the value of $\alpha^\circ_{0,czo}$ for clinozoisite is about 20% lower than for zoisite. If $\alpha^\circ_{1,czo}$ is included as an additional parameter then values of $36.6 \pm 6.9 \times 10^{-6}$ K$^{-1}$ and $49 \pm 167 \times 10^{-6}$ K$^{-1/2}$ are derived for $\alpha^\circ_{0,czo}$ and $\alpha^\circ_{1,czo}$, respectively. In conjunction with the evaluation of the experimental results for zoisite the large error in $\alpha^\circ_{1,czo}$ shows that the precision of the experimental data is insufficient to extract the temperature dependence of $\alpha^\circ_{0,czo}$.

Fitting the experimental results to Equations (7) and (8) yields a value of $50.9 \pm 0.9 \times 10^{-6}$ K$^{-1}$ for $\alpha^\circ_{0,czo}$ which is in contrast to the value of $46 \times 10^{-6}$ K$^{-1}$ applied by Holland and Powell (1998) using the same assumptions. The latter value yields volumes that are systematically lower than those derived experimentally. Using the value of $50.9 \times 10^{-6}$ K$^{-1}$ derived here, as for zoisite, $\alpha^\circ_{czo}$ would vary over a rather large range of values between 21.4 and $36.74 \times 10^{-6}$ K$^{-1}$ for temperatures of 298 and 1298 K, respectively.

In contrast to zoisite, values for $\alpha^\circ_{0,czo}$ were quite well predicted by the databases of Holland and Powell (1990) and Gottschalk (1997), who used an estimated value of $27.8 \times 10^{-6}$ K$^{-1}$. In contrast Berman (1988) chose the same value as that used for zoisite ($34.67 \times 10^{-6}$ K$^{-1}$), which yields volumes at elevated temperatures which are definitely too high. As for zoisite the best choice for $\alpha^\circ_{0,czo}$ is that derived from Equation (6) applied to the available experimental results, which yields the above mentioned value for $\alpha^\circ_{0,czo}$ of $29.5 \pm 0.4 \times 10^{-6}$ K$^{-1}$ ($\alpha^\circ_{1,czo} = 0$).

No experimental data are available for epidote. The only estimated $\alpha^\circ_{0,ep}$ value is that by Holland and Powell (1998) of $50.5 \times 10^{-6}$ K$^{-1}$ for the use in Equations (7) and (8). This value

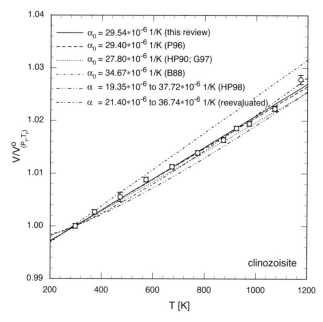

**Figure 3.** Temperature dependence of the volume of clinozoisite. Experimental results are by Pawley et al. (1996). The curves drawn are based on values for $\alpha_0^\circ$ derived here and used by Pawley et al. (1996) P96, Holland and Powell (1990) HP90, Gottschalk (1997) G97, Berman (1988) B88, Holland and Powell (1998) HP98 as well the reevaluation of the their approach.

is very close to the one derived for clinozoisite using the same equations and indicates that the compositional effect is minor. Because of the lack of experimental results it seems best to use the same $\alpha°_{0,ep}$ value as for clinozoisite, namely $29.5 \pm 0.4 \times 10^{-6}$ K$^{-1}$, and to set $\alpha°_{1,ep}$ to 0.

## Compressibility

The compressibility $\beta_i°$ is defined as

$$\beta_i° = -\frac{1}{V_i°}\left(\frac{\partial V_i°}{\partial P}\right)_T \tag{9}$$

and is linked to bulk modulus $K_i°$ by the expression $K_i° = 1/\beta_i°$ or

$$K_i° = -V_i°\left(\frac{\partial P}{\partial V_i°}\right)_T \tag{10}$$

The bulk modulus $K_i°$ is used preferentially in conjunction with crystalline substances and is therefore used here exclusively.

***Constant $K_i°$.*** If $K_i°$ is not a function of pressure, Equation (10) can be integrated to yield the following expression for molar volume $v_i$ as a function of pressure at any temperature $T$:

$$v_{i(P,T)}° = v_{i(P_r,T)}° e^{-\frac{(P-P_r)}{K_{i(P_r,T)}°}} \tag{11}$$

In this case combining Equations (6) and (11) results in the complete expression for $v_i°$ as a function of pressure and temperature:

$$v_{i(P,T)}° = v_{i(P_r,T_r)}° e^{\alpha_{0,i}°(T-T_r)+2\alpha_{1,i}°(\sqrt{T}-\sqrt{T_r})-\left((P-P_r)/K_{i(P_r,T)}°\right)} \tag{12}$$

The ultimate goal is the calculation of the Gibbs free energy, e.g., the chemical potential (Eqn. 1), of a phase at elevated temperatures and pressures. To do this the integral

$$\mu_{i(P,T)}° - \mu_{i(P_r,T)}° = \int_{P_r}^{P} v_i° \, dP \tag{13}$$

has to be solved and the result to be inserted into Equation (1). If $K_i°$ is not a function of pressure the analytical expression for Equation (13) is:

$$\int_{P_r}^{P} v_i° \, dP = K_{i(P_r,T)}° v_{i(P_r,T_r)}° e^{\alpha_{0,i}°(T-T_r)+2\alpha_{1,i}°(\sqrt{T}-\sqrt{T_r})}\left(1-e^{-\frac{(P-P_r)}{K_{i(P_r,T)}°}}\right) \tag{14}$$

***Linear pressure dependence of $K_i°$.*** If $K_i°$ is a linear function of pressure

$$K_{i(P,T)}° = K_{i(0,T)}° + K_{i(T)}^{o'} P \tag{15}$$

the integral of Equation (10) takes the following form:

$$v_{i(P,T)}° = v_{i(P_r,T)}° \left(\frac{K_{i(0,T)}° + K_{i(T)}^{o'}P}{K_{i(0,T)}° + K_{i(T)}^{o'}P_r}\right)^{-\frac{1}{K_{i(T)}^{o'}}} \tag{16}$$

Often the reference pressure $P_r$ is set to 0 MPa because one deals usually with high pressures and the low reference pressure of 0.1 MPa for $P_r$ can be neglected. This yields the typical form of the Murnaghan equation (Murnaghan 1937):

$$v^o_{i(P,T)} = v^o_{i(P_r,T)} \left( 1 + \frac{K^{o'}_{i(T)}P}{K^o_{i(0,T)}} \right)^{-\frac{1}{K^{o'}_{i(T)}}} \tag{17}$$

Integration of Equation (17) yields:

$$\int_{P_r}^{P} v^o_i \, dP = v^o_{i(P_r,T)} \frac{\left( K^o_{i(0,T)} + K'_{i(T)}P_r \right)}{1 - K^{o'}_{i(T)}} \left( 1 - \left( \frac{K^o_{i(0,T)} + K^{o'}_{i(T)}P}{K^o_{i(0,T)} + K^{o'}_{i(T)}P_r} \right)^{\frac{K^{o'}_{i(T)}-1}{K^{o'}_{i(T)}}} \right) \tag{18}$$

To get the complete analytical expressions as a function of $P$ and $T$, Equation (6) has to be inserted into Equations (16), (17) and (18).

**Birch-Murnaghan equation.** Another approach for an equation of state for a solid is the use of the Birch-Murnaghan equation. The Birch-Murnaghan equation (Birch 1947) is based upon the assumption that the strain energy of the crystal lattice can be expressed as a Taylor series in the finite strain $f_E$ (see also Angel 2000):

$$f_E = \left( \left( \frac{v^o_{i(P_r,T)}}{v^o_{i(P,T)}} \right)^{\frac{2}{3}} - 1 \right) / 2 \tag{19}$$

The expansion to the third order in strain yields

$$P = 3K^o_{i(0,T)} f_E (1 + 2f_E)^{\frac{5}{2}} \left( 1 + \frac{3}{2} \left( K^{o'}_{i(0,T)} - 4 \right) f_E \right) \tag{20}$$

where $K^{o'}_i$ is the pressure derivative of $K^o_i$ at the pressure of 0 MPa, that is

$$K^{o'}_{i(0,T)} = \left( \frac{\partial K^o_i}{\partial P} \right)_T \tag{21}$$

Truncation of the expansion (Eqn. 20) to the second order implies a $K^{o'}_i$ value of 4, which is often used in the case that the database is not sufficiently dense for the extraction of this value. Because of the formulation of Equation (20) the value of 4 seems to have at least some physical meaning and is therefore also often used in the Murnaghan equation (Eqn. 17).

Equation (20) cannot be solved directly for $v_i^o$, therefore for the use in Equation (1) the volume integral (Eqn. 13) must be either integrated numerically or evaluated by using the expression which involves the Helmholtz free energy $A$:

$$G = A + PV \tag{22}$$

Equation (20) can now be integrated directly over $V$ and not $P$:

$$\mu^o_{i(P,T)} = \mu^o_{i(P_r,T)} - P_r v^o_{i(P_r,T)} - \int_{v^o_{i(P_r,T)}}^{v^o_{i(P,T)}} P \, dv + P v^o_{i(P,T)} \tag{23}$$

The analytical expression for the integral in Equation (23) is:

$$
\int_{v^o_{i(P_r,T)}}^{v^o_{i(P,T)}} Pdv = \frac{3}{8} K^o_{i(0,T)} \left( v^o_{i(P,T)} - v^o_{i(P_r,T)} \right) \left( \left( \frac{v^o_{i(P_r,T)}}{v^o_{i(P,T)}} \right)^3 - \left( \frac{v^o_{i(P,T)}}{v^o_{i(P,T)}} \right)^{\frac{7}{3}} \right)
$$

$$
\times \left( \left( 3K^{o'}_{i(0,T)} - 12 \right) + \left( 16 - 3K^{o'}_{i(0,T)} \right) \left( \frac{v^o_{i(P_r,T)}}{v^o_{i(P,T)}} \right)^{-\frac{2}{3}} \right)
$$

(24)

Besides $v_i^o$ at $P_r$ and $T$ which is calculated using Equation (6), Equation (24) requires the knowledge of $v_i^o$ at $P$ and $T$, which is calculated by solving Equation (20) iteratively once for the required elevated conditions.

***Temperature dependence of $K_i^o$ and $K_i^{o\prime}$.*** Both $K_i^o$ and $K_i^{o\prime}$ can be functions of temperature. The simplest approach would be a linear relationship:

$$
K^o_{i(0,T)} = K^o_{i(0,T_r)} + (T - T_r) \left( \frac{\partial K^o_{i(0,T)}}{\partial T} \right)_{P=0}
$$

(25)

and

$$
K^{o'}_{i(0,T)} = K^{o'}_{i(0,T_r)} + (T - T_r) \left( \frac{\partial K^{o'}_{i(0,T)}}{\partial T} \right)_{P=0}
$$

(26)

If values for the temperature derivatives in Equations (25) and (26) are available the respective values of $K_i^o$ and $K_i^{o\prime}$ at elevated temperatures can be inserted into Equations (14), (18) or (24).

***Zoisite.*** The dependence of the molar volume of zoisite on pressure at reference temperature has been determined experimentally by Holland et al. (1996), Comodi and Zanazzi (1997), Pawley et al. (1998) and Grevel et al. (2000) for pressures up to 14 GPa (Fig. 4). In addition, theoretical values of $K^o_{zo}$ and $K^o_{czo}$ based on first-principles calculations are available (Winkler et al. 2001, Table 3). Besides the direct experimental results the cited publications provide extracted parameters for $K^o_{zo}$ and/or $K^{o\prime}_{zo}$. As the authors used various procedures and different models to compare their results, the respective parameters have all been redetermined here using the Equations (11), (16) and (20) for the different models and the respective experimental results. The parameters were obtained with the *NonLinearFit* package of *Mathematica®* and are presented with the respective errors in Table 3.

Whereas the results by Comodi and Zanazzi (1997), Pawley et al. (1998) and Grevel et al. (2000) are consistent within the error of the measurements, those by Holland et al. (1996) differ significantly. If $K^o_{zo}$ is assumed to be constant (Eqn. 11) the derived values range from 110 to 150 GPa except for the data of Holland et al. (1996), which lead to a value of 296 GPa. This is surprising given that Holland et al. (1996) used the same powdered sample as that of Pawley et al. (1998). It has been argued by Pawley et al. (1998) that there might have been difficulties with the pressure distribution within the experimental assembly used by Holland et al. (1996). The $K^o_{zo}$ values based on the results by Comodi and Zanazzi (1997), Pawley et al. (1998), and Grevel et al. (2000) differ by approximately 35%. An average value of 136 ± 3 GPa using these latter data has been calculated which is very close to the value found by Grevel et al. (2000).

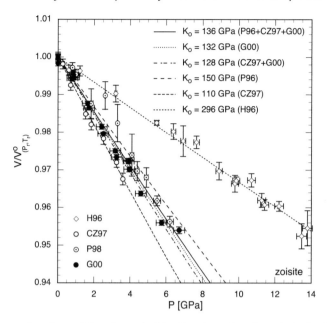

**Figure 4.** Pressure dependence of the volume of zoisite. Experimental results are by Holland et al. (1996) H96, Comodi and Zanazzi (1997) CZ97, Pawley et al. (1998) P98 and Grevel et al. (2000) G00. Curves are calculated using the assumption of constant $K°_{zo}$ (Eqn. 11).

Application of the Murnaghan equation (Eqn. 16) with consideration of $K°'_{zo}$ as an adjustable parameter, produces extracted $K°'_{zo}$ values that vary between −19 to +9 for the different sets of experimental results. The same applies to the Birch-Murnaghan equation (Eqn. 20) for which $K°'_{zo}$ values between −12 and +7 were derived. This implies that the precision of the experimental results is insufficient for the extraction of $K°'_{zo}$. Therefore $K°'_{zo}$ has been set to 4 as suggested by Equation (20) for both the Murnaghan and the Birch-Murnaghan equation (see Table 3). The extracted values for $K°_{zo}$ using the more or less consistent results by Comodi and Zanazzi (1997), Pawley et al. (1998), and Grevel et al. (2000) are then 127 ± 3 and 125 ± 3 GPa for these two equations, respectively (Table 3).

Volume measurements at elevated pressures as well as temperatures are provided by Pawley et al. (1998) and Grevel et al. (2000). These allow the determination of the temperature derivative of $K°_{zo}$ (Eqn. 25). Using all experimental results except those of Holland et al. (1996) and a $\alpha°_{0,zo}$ of 37.1 × 10⁻⁶ K⁻¹ (see above), for $(\partial K°_{zo}/\partial T)$ and $K°_{zo}$ values of −0.029 GPa/K and 129 GPa were extracted for the Murnaghan equation, and −0.026 GPa/K and 127 GPa for the Birch-Murnaghan equation, respectively.

Figure 5 illustrates the residual deviations from the experimental results for the following 3 models: constant $K°_{zo}$, the Murnaghan, and the Birch-Murnaghan equation. For the latter two a $K°'_{zo}$ value of 4 and the respective fitted values for $(\partial K°_{zo}/\partial T)$ mentioned above were used. All experimental results except those of Holland et al. (1996) were considered. All results are reproduced within 1%, but the distribution of the residuals are much more in favor of the Murnaghan and the Birch-Murnaghan equations for which the distributions are more of a gaussian shape than for the case of constant $K°_{zo}$. For these two equations most results are within the range of 0.2%. The results using the Murnaghan and the Birch-Murnaghan equations are indistinguishable, however.

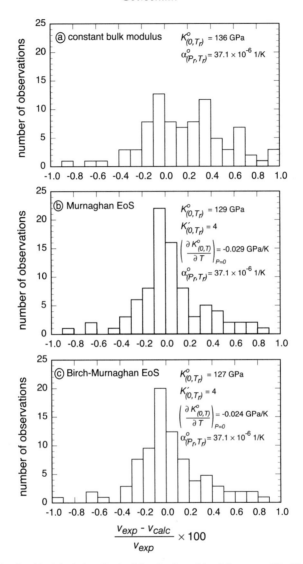

**Figure 5.** Calculated residual deviations for the following 3 models of a) constant $K^\circ{}_{zo}$, b) the Murnaghan, and c) the Birch-Murnaghan equation for zoisite from experimental results by Comodi and Zanazzi (1997), Pawley et al. (1998), and Grevel et al. (2000).

Estimations of $K^\circ{}_{zo}$ prior to the availability of the experimental results are shown in Figure 6. Berman (1988) used a quadratic polynomial to describe the pressure dependence of the volume. If this is applied to pressures above 3 GPa the calculated volumes bend noticeably upward, which is not supported by most experimental results except by those of Holland et al. (1996). The estimated $K^\circ{}_{zo}$ value used by Holland and Powell (1990) leads to volumes which are only slightly higher than the experimental results (Comodi and Zanazzi 1997; Pawley et al. 1998; Grevel et al. 2000), whereas these are well reproduced by the formalisms used by Gottschalk (1997) and Holland and Powell (1998).

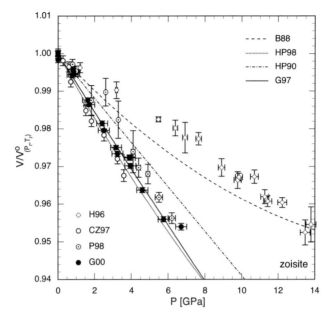

**Figure 6.** Comparison of the pressure dependence of the volume of zoisite as used by the internally consistent databases by Berman (1988) B88, Holland and Powell (1990) HP90, Gottschalk (1997) G97 and Holland and Powell (1998) HP98 with available experimental results by Holland et al. (1996) H96, Comodi and Zanazzi (1997) CZ97, Pawley et al. (1998) P98 and Grevel et al. (2000) G00.

***Clinozoisite and epidote.*** Experimental results (Fig. 7) for the pressure dependence of the volume of clinozoisite are available from Comodi and Zanazzi (1997) and Holland et al. (1996), while that for epidote has only measured by Holland et al. (1996). Comodi and Zanazzi (1997) and Holland et al. (1996) both report that the monoclinic structure seems to be stiffer than the orthorhombic structure. This is also supported by first-principles calculations by Winkler et al. (2001) (Table 3). Assuming a constant $K_i^\circ$ the experimental results by Comodi and Zanazzi (1997) yield a $K^\circ_{czo}$ value of 128 ± 1 vs. 110 ± 3 GPa for $K^\circ_{zo}$, whereas these values are 168 ± 3 vs. 150 ± 5 GPa if the results by Holland et al. (1996) and Pawley et al. (1998) are used. As for zoisite this is a difference of approximately 30% between these two datasets for clinozoisite. If all results are considered together a value for $K^\circ_{czo}$ of 154 ± 4 GPa is calculated.

According to Holland et al. (1996) there is no significant difference in $K_i^\circ$ between clinozoisite and epidote (168 ± 3 vs. 172 ± 3 GPa).

As for zoisite the extraction of $K^{\circ\,\prime}_{czo}$ from the available data is unreasonable because of the large uncertainties involved. Therefore a value of 4 is again assumed for clinozoisite and for epidote. No measurements at elevated pressures and temperatures are available for clinozoisite or epidote, therefore $(\partial K^\circ_{czo}/\partial T)$ and $(\partial K^\circ_{ep}/\partial T)$ cannot be evaluated. As a first approximation it can be assumed that these values are at least similar to that of zoisite, -0.029 and −0.026 GPa/K for the Murnaghan and the Birch-Murnaghan equations, respectively.

Values of $K^\circ_{czo}$ in internally consistent databases by Holland and Powell (1990) and Gottschalk (1997), which were estimated prior to the work of Comodi and Zanazzi (1997) predicted volumes at high pressure excellently, whereas the approach by Berman (1988) based on a quadratic polynomial overestimates $K^\circ_{czo}$ considerably (Fig. 8). Holland and Powell

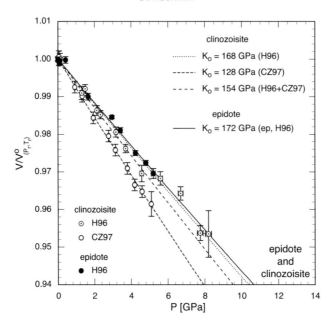

**Figure 7.** Pressure dependence of the volume of clinozoisite and epidote. Experimental results are by Holland et al. (1996) H96 and Comodi and Zanazzi (1997) CZ97. Curves are calculated using the assumption of constant $K°_{czo}$ and $K°_{ep}$ (Eqn. 11).

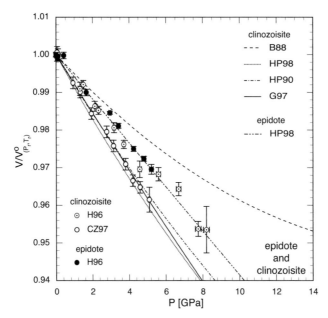

**Figure 8.** Comparison of the pressure dependence of the volume of clinozoisite and epidote as used by the internally consistent databases by Berman (1988) B88, Holland and Powell (1990) HP90, Gottschalk (1997) G97 and Holland and Powell (1998) HP98 with available experimental results by Holland et al. (1996) H96 and Comodi and Zanazzi (1997) CZ97.

(1998) also uses a value of $K^\circ_{czo}$, which is consistent with the experimental results of Comodi and Zanazzi (1997).

**General comments.** Considerable differences exist between experimentally determined volumes as a function of pressure by different authors. It is clear that at least one series of measurements, that of Holland et al. (1996) for zoisite, is inconsistent with other studies. Determination of the real *in situ* hydrostatic pressure during experimental determination of the lattice parameters seems to be the largest source of error. Even within a single set of results the precision of the measurements is insufficient to derive values for $K_i^{\circ\prime}$.

## THERMAL PROPERTIES

### Heat capacity

The calculation of enthalpy and entropy of a phase component at elevated temperature and $P_r$ using the integrals in Equation (1) requires the knowledge of the heat capacity $c_{P\,i}^\circ$ at reference conditions and its temperature dependence from $T_r$ to the respective temperature $T$. The temperature dependence of $c_{P\,i}^\circ$ is usually expressed in some polynomial form like

$$c_{P_r,i}^\circ = \sum_j c_{i,j} T^{m_j} \qquad (27)$$

In many cases $c_{P\,i}^\circ$ is given in the form proposed by Haas and Fisher (1976):

$$c_{P_r,i}^\circ = a + bT + cT^2 + \frac{d}{T^{1/2}} + \frac{f}{T^2} \qquad (28)$$

The extrapolation of $c_{P\,i}^\circ$ presented by Equation (28) to temperatures above the experimental range is impossible, or at least, very problematic (Berman and Brown 1985) because, according to this equation the $c_{P\,i}^\circ$ becomes infinite at infinite temperature. Extrapolation, however, is often required because $c_{P\,i}^\circ$ is measured only over a limited temperature range. A mathematically better behaved but not perfect polynomial of the following form

$$c_{P_r,i}^\circ = a + \frac{d}{T^{1/2}} + \frac{e}{T} + \frac{f}{T^2} + \frac{g}{T^3} \qquad (29)$$

was proposed by Berman and Brown (1985), except for the parameter $e$, which they set to 0. Equation (29) mimics the high temperature limit predicted by lattice vibrational theory according to the law of Dulong and Petit at elevated temperatures,

$$c_{P_r,i}^\circ = 3 n_{pfu} R + \alpha_{i,(P_r,T)}^{\circ 2} K_{i,(P_r,T)}^\circ V_{i,(P_r,T)}^\circ T \qquad (30)$$

in which $n_{pfu}$ represents the number of atoms per formula unit.

**Zoisite.** A large number of different $c_{P\,zo}^\circ$ equations are available for zoisite (Table 4). All of them are based either on the calorimetric results by Kiseleva and Topor (1973a,b) and Kiseleva et al. (1974) (Fig. 9) and/or by Perkins et al. (1980) (Fig. 10). Drop calorimetry in the range of 298-1100 K was used by Kiseleva and Topor (1973a,b) and Kiseleva et al. (1974), whereas Perkins et al. (1980) applied differential scanning calorimetry (DSC) from ambient temperatures up to 729 K. With the drop calorimetry method enthalpy differences between $T$ and $T_r$ are directly measured and $c_{P\,zo}^\circ$ is the first derivative with respect to temperature. Therefore extraction of $c_{P\,zo}^\circ$ requires the evaluation of the slope to the experimental results. The raw data are not absolutely smooth (Fig. 9), which makes it difficult to find a strong monotonous derivative. This is required, however, because $c_{P\,zo}^\circ$ increases monotonously with temperature. In addition it is also ambiguous to define the derivative values at the temperature

**Table 4.** Heat capacities as a function of temperature

| phase | $c^\circ_p$-function (J/K/mol) | $T$-range | $c^\circ_p$ (298 K) | $c^\circ_p$ (730 K) | $c^\circ_p$ (2000 K) |
|---|---|---|---|---|---|
| *zoisite* | | | | | |
| Kiseleva et al. (1974) | $497.478 + 3.67774\times10^{-2}\,T - 1.31796\times10^{7}\,T^{-2}$ | 298–1100 | 360.18 | 499.59 | 567.74 |
| Perkins et al. (1980) | $413.881 + 1.5213\times10^{-1}\,T - 1.00751\times10^{7}\,T^{-2}$ | 298–750 | 345.90 | 506.03 | 715.62 |
| Helgeson et al. (1978) | $443.998 + 1.05495\times10^{-1}\,T - 1.13575\times10^{7}\,T^{-2}$ | 298–700 | 347.69 | 499.70 | 652.15 |
| Haas et al. (1981) | $834.622 - 3.96894\times10^{-2}\,T - 8.14875\times10^{3}\,T^{-0.5}$ | 200–1250 | 350.86 | 504.05 | 573.03 |
| Berman and Brown (1985) | $751.37 - 6.5857\times10^{3}\,T^{-0.5} - 1.9308 + 5.143\times10^{7}\,T^{-3}$ | 298–2000* | 350.19 | 504.13 | 603.63 |
| Berman (1988) | $749.17 - 6.5093\times10^{3}\,T^{-0.5} - 2.3805\times10^{6}\,T^{-2} + 1.2486\times10^{8}\,T^{-3}$ | 298–2000* | 350.12 | 504.10 | 603.04 |
| Robie and Hemingway (1995) | $1134.0 - 4.523\times10^{-1}\,T - 1.1570\times10^{4}\,T^{-0.5} + 2.391\times10^{-4}\,T^{2}$ | 298–750 | 350.34 | 503.01 | 927.09 |
| Holland and Powell (1998) | $595.7 + 6.2297\times10^{-2}\,T - 3.3947\times10^{3}\,T^{-0.5} - 5.9213\times10^{6}\,T^{-2}$ | 298–2000* | 351.06 | 504.42 | 642.91 |
| Chatterjee et al. (1998) | $789.825 - 7.58567\times10^{3}\,T^{-0.5}$ | 298–729 | 350.51 | 509.06 | 620.20 |
| Poli and Schmidt (1998) | $718.11 - 5.1848\times10^{3}\,T^{-0.5} - 1.3593\times10^{7}\,T^{-2} + 2.3077\times10^{9}\,T^{-3}$ | 298–2000* | 352.00 | 506.64 | 599.07 |
| † | $733.034 - 5.97952\times10^{3}\,T^{-0.5} - 5.07966\times10^{6}\,T^{-2} + 5.52314\times10^{8}\,T^{-3}$ | 200–2000* | 350.43 | 503.61 | 598.13 |
| † | $-2.699\times10^{-1}\,T + 2.1848\times10^{-2}\,T^{-2} - 1.0071\times10^{-4}\,T^{3} + 1.4937\times10^{-7}\,T^{4}$ | 0–240 | — | — | — |
| *clinozoisite* | | | | | |
| Kiseleva et al. (1974) | $502.833 + 3.22586\times10^{-2}\,T - 1.3849\times10^{7}\,T^{-2}$ | 298–1100 | 356.66 | 500.39 | 563.89 |
| Holland and Powell (1998) | $567.0 + 1.8063\times10^{-2}\,T - 2.603\times10^{3}\,T^{-0.5} - 7.034\times10^{6}\,T^{-2}$ | 298–2000* | 342.51 | 470.65 | 543.16 |
| *epidote* | | | | | |
| Kiseleva et al. (1974) | $476.056 + 6.1463\times10^{-2}\,T - 1.21001$ | 298–1100 | 358.26 | 498.22 | 595.96 |
| Helgeson et al. (1978) | $492.13 + 5.36221\times10^{-2}\,T - 1.33319\times10^{7}\,T^{-2}$ | 298–1100 | 358.14 | 506.26 | 596.04 |
| Berman and Brown (1985) | $643.85 - 3.087\times10^{3}\,T^{-0.5} - 1.40917\times10^{7}\,T^{-2} + 1.54457\times10^{9}\,T^{-3}$ | 298–2000* | 364.82 | 507.12 | 571.49 |
| Holland and Powell (1998) | $544.6 + 2.4781\times10^{-2}\,T - 1.1921\times10^{3}\,T^{-0.5} - 1.123\times10^{7}\,T^{-2}$ | 298–2000* | 356.62 | 497.50 | 564.70 |
| ‡ | $728.954 - 5.56532\times10^{3}\,T^{-0.5} - 5.93736\times10^{6}\,T^{-2} + 6.10589\times10^{8}\,T^{-3}$ | 200–2000* | 362.89 | 513.40 | 603.10 |
| ‡ | $-1.0074 - 0.17488\,T + 2.5412\times10^{-2}\,T^{2} - 1.5343\times10^{-4}\,T^{3} + 3.9018\times10^{-7}\,T^{4} - 3.657\times10^{-10}\,T^{5}$ | 40–300 | 361.00 | — | — |
| † | $2.4105\,T - 0.29447\,T^{2} + 1.5139\times10^{-3}\,T^{3} - 3.6018\times10^{-4}\,T^{4} + 4.201\times10^{-6}\,T^{5} - 1.9317\times10^{-8}\,T^{6}$ | 0–50 | — | — | — |

* max $T$ might be higher, † refitted using original experimental data (see text), ‡ approximated using $c^\circ_p$ functions of corundum and hematite

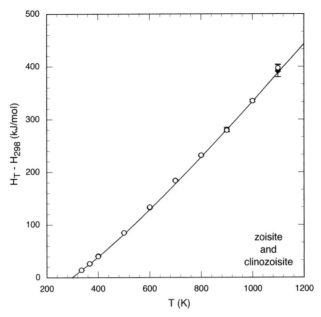

**Figure 9.** Experimental results from drop calorimetry measurements $(H^{\circ}_T - H^{\circ}_{298})$ of zoisite and clinozoisite from Kiseleva and Topor (1973a,b) and Kiseleva et al. (1974). The solid curve is calculated by integrating the derived $c_P{}^{\circ}_{zo}$ equation from Table 4.

boundaries of the measurements. In contrast using differential scanning calorimetry $c_P{}^{\circ}_{zo}$ is measured directly and errors are usually lower than for the drop calorimetry method. It is important to note, however, that the measurements by Perkins et al. (1980) range only up to 729 K compared to the measurements by Kiseleva and Topor (1973a,b) and Kiseleva et al. (1974), which go up to 1100 K. It is obvious (Fig. 10) that the $c_P{}^{\circ}_{zo}$ values using the polynomial form derived by Kiseleva et al. (1974) are substantially higher than the values observed by Perkins et al. (1980) in the range between 300 and 600 K. Most equations in Table 4 were derived using only the DSC measurements of Perkins et al. (1980). A severe deficiency is, however, the low maximum temperature achieved by Perkins et al. (1980). The extrapolation of the equations by Helgeson et al. (1978), Perkins et al. (1980), Haas et al. (1981), Robie and Hemingway (1995), and Chatterjee et al. (1998) fail to satisfy the high temperature limit predicted by Equation (30) and are strictly applicable only below 800 to 900 K. In contrast the equations by Berman and Brown (1985), Berman (1988), and Poli and Schmidt (1998) approach the theoretical high temperature limit and can be used with some confidence at least up to 2000 K. The high temperature limit of $c_P{}^{\circ}_{zo}$ in conjunction with the available experimental results of Kiseleva and Topor (1973a,b), Kiseleva et al. (1974), and Perkins et al. (1980) has been used to derive an expression in the form of Equation (29). In the range of 298 to 2000 K this equation (Table 4) is practically identical to the expression by Poli and Schmidt (1998), but is in addition also valid between 200 and 298 K. The integral of the derived equation is also plotted in Figure 9 and shows that the calorimetric results of Kiseleva and Topor (1973a,b) and of Kiseleva et al. (1974) are also well reproduced. Therefore the results of Kiseleva and Topor (1973a,b) and Kiseleva et al. (1974) are within error compatible with those of Perkins et al. (1980).

*Clinozoisite and epidote.* For clinozoisite only drop calorimetric measurements are available by Kiseleva et al. (1974) (Fig. 9). Within error they are indistinguishable from those of zoisite. In the absence of other experimental determinations it is therefore common practice

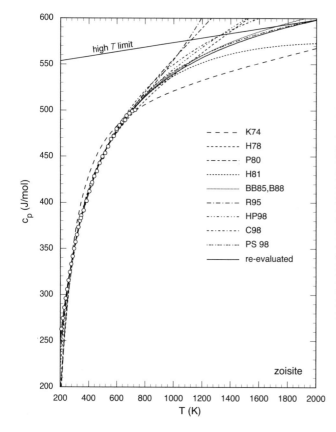

**Figure 10.** Experimental results from differential scanning calorimetry heat capacity measurements of zoisite from Perkins et al. (1980). Curves are evaluated expressions by Kiseleva et al. (1974) K74, Helgeson et al. (1978) H78, Perkins et al. (1980) P80, Haas et al. (1981) H81, Berman and Brown (1985) BB85, Berman (1988) B88, Robie and Hemingway (1995) R95, Holland and Powell (1998) HP98, Chatterjee et al. (1998) C98, Poli and Schmidt (1998) PS98 and the derived expression from Table 4.

to use the same heat capacity function for clinozoisite and zoisite. Only Holland and Powell (1998) used a slightly different expression for clinozoisite.

For epidote, as for clinozoisite, there are only a few experimental $c_P^\circ{}_{ep}$ values available. These are again drop calorimetry measurements by Kiseleva et al. (1974) and some low temperature DSC measurements ($T < 300$ K) by Gurevich et al. (1990) (Figs. 11, 12). The available functional expressions for $c_P^\circ{}_{ep}$ (Kiseleva et al. 1974; Helgeson et al. 1978; Berman and Brown 1985; Holland and Powell 1998) are consistent with the few calorimetric results by Gurevich et al. (1990) above 298 K. At temperatures above 600 K, however, the extrapolated values for $c_P^\circ{}_{ep}$ differ significantly.

As a first approximation the heat capacity of a phase can be estimated by using some sort of components for which the $c_P^\circ{}_i$ can be added or subtracted to obtain the propper chemical composition. Starting from $c_P^\circ{}_{zo}$ of zoisite, subtracting $c_P^\circ{}_{cor}$ of corundum and adding $c_P^\circ{}_{hem}$ of hematite in the form of

$$c_{P_r\,ep}^\circ = c_{P_r\,zo}^\circ - \frac{1}{2}c_{P_r\,cor}^\circ + \frac{1}{2}c_{P_r\,hem}^\circ \qquad (31)$$

the heat capacity of epidote $c_P^\circ{}_{ep}$ can be approximated. The equation for $c_P^\circ{}_{ep}$ in Table 4 has been obtained by using the function for zoisite derived here and the expressions for corundum and $\alpha$-hematite by Berman and Brown 1985. The so derived expression for $c_P^\circ{}_{ep}$ reproduces

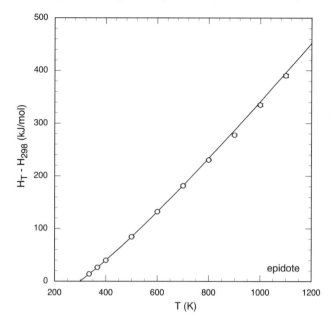

**Figure 11.** Experimental results from drop calorimetry measurements ($H°_T$ - $H°_{298}$) of epidote from Kiseleva et al. (1974). The solid curve is calculated by integrating the approximated $c_P°_{ep}$ equation from Table 4.

the calorimetrical results by Gurevich et al. (1990) in the range of 200 to 303 K (Fig. 12) as well as the results by Kiseleva et al. (1974) up to 1100 K (Fig. 11). This expression represents $c_P°_{ep}$ for ordered epidote, i.e., $Fe^{3+}$ only at the M3 site. For partially disordered epidotes additional energy contributions have to be considered.

**Third law entropy**

The third law of thermodynamics states that for a totally ordered solid phase at 0 K the entropy is 0 J/K. Therefore the absolute (third law) entropy at $T_r$ can be determined, if no phase transition occurs below $T_r$, by measuring $c_P°_i$ at low temperatures starting from close to 0 K up to $T_r$.

$$s°_{i(P_r,T_r)} = \int_0^{T_r} \frac{c_{Pi}°}{T} dT \tag{32}$$

***Zoisite.*** For zoisite low temperature $c_P°_{zo}$ values have been determined by Perkins et al. (1980) from 5 to 347 K (Fig. 13) using adiabatic calorimetry and DSC. By integration Perkins et al. (1980) got a value for $s°_{zo}$ of 295.85 ± 2.2 J/K/mol. These experimental results have been re-evaluated here. Using the derived expression for $c_P°_{zo}$ from 0 to 220 K (Table 4) in conjunction with the high temperature equation for $c_P°_{zo}$ a value for $s°_{zo}$ of 294.9 ± 2.5 J/K/mol is obtained at reference conditions. This value is within error identical to the one given by Perkins et al. (1980).

***Clinozoisite and epidote.*** There are no direct $s°_{czo}$ measurements available for clinozoisite. Therefore a value for $s°_{czo}$ must be derived by other means using results from phase equilibria (see below). For epidote Gurevich et al. (1990) measured $c_P°_{ep}$ between 6 and 303 K using

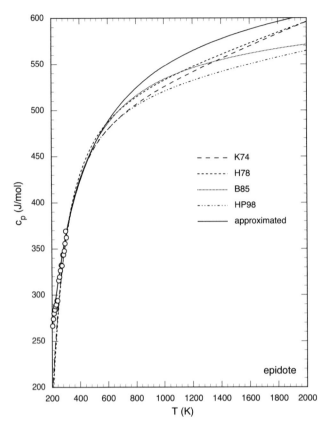

**Figure 12.** Experimental results from calorimetry measurements of epidote from Gurevich et al. (1990). Curves are evaluated expressions by Kiseleva et al. (1974) K74, Helgeson et al. (1978) H98, Berman and Brown (1985) B85, Holland and Powell (1998) HP98 and the approximated expression from Table 4.

adiabatic calorimetry and DSC (Fig. 14). By integrating these results they report a value of $328.9 \pm 3.2$ J/K/mol for $s°_{ep}$ at reference conditions. As for zoisite the calorimetric results by Gurevich et al. (1990) have been re-evaluated in this review and the resulting equations (Table 4) integrated to yield a substantially higher value for $s°_{ep}$ of $332.8 \pm 3.9$ J/K/mol at reference conditions.

## Enthalpy of formation

Evaluations of enthalpies of formation from the elements $\Delta_f h°_i$ are available through solution calorimetry measurements. One must bear in mind, however, that this method of extracting $\Delta_f h°_i$ values involves chemical cycles and relies on the knowledge of the solution enthalpies of the substances involved in these cycles. Therefore the determination of $\Delta_f h°_i$ by solution calorimetry are not really direct measurements. This is important for the error assessment.

***Zoisite.*** Solution calorimetry measurements for zoisite are available from Kiseleva and Ogorodova (1986, 1987) and Smelik et al. (2001). Kiseleva and Ogorodova (1986, 1987) report a $\Delta_f h°_{zo}$ value of $-6891.7 \pm 5.8$ kJ/mol whereas Smelik et al. (2001) gives a measured $\Delta_f h°_{zo}$ value of $-6879.9 \pm 6.9$ kJ/mol and an extrapolated value based on additional measurements of solid solutions of $-6878.5 \pm 6.8$ kJ/mol.

***Clinozoisite and epidote.*** Kiseleva and Ogorodova (1987) and Smelik et al. (2001) also provide solution calorimetry determinations for clinozoisite and epidote. Because pure

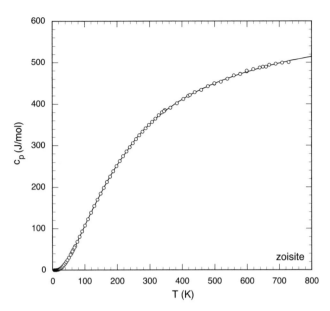

**Figure 13.** Experimental results from low temperature calorimetry measurements of zoisite by Perkins et al. (1980). The solid curve is calculated using the fitted low temperature equation and the reevaluated $c_P^\circ{}_{zo}$ equations from Table 4.

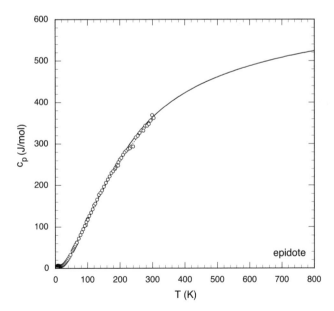

**Figure 14.** Experimental results from low temperature calorimetry measurements of epidote by Gurevich et al. (1990). The solid curve is calculated using the fitted low temperature equation and the reevaluated $c_P^\circ{}_{ep}$ equations from Table 4.

clinozoisite specimens are rare, measurements for clinozoisite were conducted with iron bearing phases. By correcting the results to pure clinozoisite, Kiseleva and Ogorodova (1987) and Smelik et al. (2001) obtained −6888.7 ± 11.1 kJ/mol and −6882.5 ± 6.7 kJ/mol, respectively. For epidote with a near endmember composition Kiseleva and Ogorodova (1987) and Smelik et al. (2001) measured $\Delta_f h°_{czo}$ values of −6462.6 ± 8.1 kJ/mol and −6461.9 ± 6.9 kJ/mol, respectively.

### Third law entropy and enthalpy of formation from databases

Internally consistent datasets (Helgeson et al. 1978; Halbach and Chatterjee 1984; Berman 1988; Holland and Powell 1985, 1990, 1998; Gottschalk 1997; Chatterjee et al. 1998), which use results from experimentally determined phase equilibria in conjunction with available calorimetric measurements are another source for $s_i°$ and $\Delta_f h_i°$ values. The extracted $s_i°$ and $\Delta_f h_i°$ values for zoisite, clinozoisite and epidote (Table 5) in these datasets depend ultimately on the thermodynamic properties used for all other phase components involved in the considered phase equilibria, i.e.. differences in these datasets in the properties of anorthite translate directly in differences into those of zoisite.

*Zoisite.* Because the experimental determination of $s°_{zo}$ by Perkins et al. (1980) is the only one available, this value is highly weighted and appears, within the limits of its reported experimental error, in all the datasets. The extracted values range between 295.88 and 297.85 J/K/mol and there seems to be a consensus about the real value of $s°_{zo}$. This does not hold for $\Delta_f h°_{zo}$, however. The extracted values in the different datasets range between −6879.044 and −6901.082 kJ/mol. While this difference of about 3‰ seems to be small, it is far too large to be acceptable for the calculation of phase equilibria. The value proposed by Helgeson et

**Table 5.** $s_i°$ and $\Delta_f h_i°$ from internally consistent datasets

| | $v°_{i(P_r, T_r)}$ | $\Delta_f h°_{i(P_r, T_r)}$ | $s°_{i(P_r, T_r)}$ |
|---|---|---|---|
| | [cm³/mol] | [kJ/mol] | [J/K/mol] |
| *zoisite* | | | |
| Helgeson et al. (1978) | 135.9 | −6879.044 | 295.98 |
| Haas et al. (1981)[a,b] | 136.52 | −6891.117 | 295.88[c] |
| Halbach and Chatterjee (1984) | 135.70 | −6901.082 | 296.45 |
| Berman (1988) | 135.88 | −6889.488 | 297.58 |
| Robie and Hemingway (1995)[d] | 136.5 | −6901.100 | 295.85 |
| Gottschalk (1997) | 136.52 | −6891.340 | 297.38 |
| Holland and Powell (1998) | 135.75 | −6898.570 | 297.00 |
| Chatterjee et al. (1998) | 135.70[d] | −6897.551 | 296.09 |
| *clinozoisite* | | | |
| Helgeson et al. (1978) | 136.2 | −6879.421 | 295.56 |
| Berman (1988) | 136.73 | −6894.968 | 287.08 |
| Gottschalk (1997) | 136.12 | −6896.478 | 286.66 |
| Holland and Powell (1998) | 136.30 | −6898.110 | 301.00 |
| Chatterjee et al. (1998) | 135.59[d] | −6903.986 | 283.71 |
| *epidote* | | | |
| Helgeson et al. (1978) | 139.2 | −6461.900 | 314.97 |
| Robie and Hemingway (1995)[b] | 138.1 | — | 328.90 |
| Holland and Powell (1998) | 139.10 | −6463.200 | 328.00 |

[a] see also Hemingway et al. (1982)
[b] datasets by Haas et al. (1981) and Robie and Hemingway (1995) are not internally consistent
[c] Perkins et al. (1980)
[d] Chatterjee et al. (1984)

al. (1978) of $-6879.044$ kJ/mol is close to the value determined by Smelik et al. (2001) of $-6878.5 \pm 6.8$ kJ/mol, whereas the values proposed by Berman (1988) and Gottschalk (1997) of $-6889.488$ kJ/mol and $-6891.340$ kJ/mol, respectively, are close to the reported value by Kiseleva and Ogorodova (1986, 1987) of $-6891.7$ kJ/mol. Halbach and Chatterjee (1984), Chatterjee et al. (1998), and Holland and Powell (1998) determined even higher values of $-6901.082$ kJ/mol, $-6897.551$ kJ/mol and $-6898.570$ kJ/mol, respectively, which are not within error of the results from Kiseleva and Ogorodova (1986, 1987). From these results it is therefore unclear what the real value of $\Delta_f h°_{zo}$ is.

***Clinozoisite.*** Whereas extensive experimental phase equilibria involving zoisite exist, experimental studies with clinozoisite are rare. For example Gottschalk (1997) considered 17 different equilibria with zoisite but only 3 with clinozoisite. Additionally no direct experimental determination of $s°_{czo}$ is available. Consequently values of $s°_{czo}$ in internally consistent datasets range from 283.71 J/K/mol to 301.00 J/K/mol. As for zoisite $\Delta_f h°_{czo}$ values range over a large interval from $-6879.421$ kJ/mol to $-6903.986$ kJ/mol. Except for the dataset of Holland and Powell (1998), which gives the highest $s°_{czo}$ value of 301.00 J/K/mol, all other datasets report that $\Delta_f h°_{czo}$ is more negative than $\Delta_f h°_{zo}$. Furthermore all but the $\Delta_f h°_{czo}$ value of Helgeson et al. (1978) are significantly more negative than those determined calorimetrically by Kiseleva and Ogorodova (1987) and Smelik et al. (2001). As for zoisite it remains therefore unclear what the values of $\Delta_f h°_{czo}$ and $s°_{czo}$ really are (see below for further discussion).

***Epidote.*** Only the two internally consistent datasets by Helgeson et al. (1978) and Holland and Powell (1998) present data for epidote. The value of $s°_{ep}$ of 328.00 J/K/mol by Holland and Powell (1998) is the one determined by Gurevich et al. (1990) whereas the value used by Helgeson et al. (1978) of 314.97 JK/mol is considerably lower. Surprisingly the reported $\Delta_f h°_{ep}$ values of $-6461.9$ kJ/mol (Helgeson et al. 1978) and $-6463.2$ kJ/mol (Holland and Powell 1998) are basically identical to the calorimetric values of $-6462.6$ kJ/mol and $-6461.9$ kJ/mol provided by Kiseleva and Ogorodova (1987) and Smelik et al. (2001), respectively. The value by Smelik et al. (2001) is a compositional extrapolated value to epidote endmember composition, however.

## PHASE EQUILIBRIA

Experimental results on phase equilibria involving epidote minerals are thoroughly discussed in this volume by Poli and Schmidt (2004). In the following only the reactions involving both zoisite and/or clinozoisite are considered. Because absolute values for $s°_{czo}$, $\Delta_f h°_{zo}$ and $\Delta_f h°_{czo}$ seem to be debatable (see above), experimental results for reactions involving both zoisite and clinozoisite provide further information and relative constraints on these values.

### Pure clinozoisite and zoisite

Jenkins et al. (1983, 1985) conducted experiments to confine the equilibria of the reactions

$$1 \text{ kyanite} + 1 \text{ quartz} + 2 \text{ zoisite} \leftrightarrow 4 \text{ anorthite} + 1 H_2O \tag{33}$$

$$1 \text{ kyanite} + 1 \text{ quartz} + 2 \text{ clinozoisite} \leftrightarrow 4 \text{ anorthite} + 1 H_2O \tag{34}$$

in the region between 590 to 845 MPa and 500 to 600°C using synthetic zoisite and clinozoisite. The observed difference in the location of the equilibria in $P$-$T$ space can be directly related to differences in $s_i°$ and $\Delta_f h_i°$ for zoisite and clinozoisite. These differences can be obtained by plotting the available experimental results on a $1/T$ vs. $\ln K_{red}$ diagram. The respective $\ln K_{red}$ values are calculated (see e.g., Gottschalk 1997) according to

$$\ln K_{red} = \sum_i n_i \left( \int_{T_r}^{T} c_{P_i}^o \, dT - T \int_{T_r}^{T} \frac{c_{P_i}^o}{T} \, dT + \int_{P_r}^{P} v_i^o \, dP + RT \ln a_i \right) \qquad (35)$$

The last two terms in the sum of Equation (35) are only used for solids, for fluid phase components they have to be replaced by

$$\int_{P_r}^{P} v_i^o \, dP + RT \ln a_i \rightarrow RT \ln \frac{f_i}{P_r} \qquad (36)$$

In a $1/T$ vs. $\ln K_{red}$ diagram a straight line should separate points representing forward and back reaction. This straight line obeys the following relationship

$$\ln K_{red} = -\frac{\Delta_R H^o}{RT} + \frac{\Delta_R S^o}{R} \qquad (37)$$

The available experimental results for reactions (33) and (34) (Jenkins et al. 1983, 1985; Best and Graham 1978; Johannes and Ziegenbein 1980; Goldsmith 1981; Chatterjee et al. 1984) are plotted in Figure 15. For the calculation of the $\ln K_{red}$ values (Eqns. 35 and 36) for zoisite and clinozoisite the data from Table 3 and 4 and the data from Gottschalk (1997) for all other phase components were used. Because direct measurements of $c_P^o{}_{czo}$ are not available the $c_P^o$ expression for zoisite was applied.

The solid line in Figure 15 separates as required the open and filled square symbols representing the forward reaction and reverse reaction direction for the equilibrium with zoisite (Eqn. 33). Only the experimental results by Best and Graham (1978) are not consistent with all others and are well below the lines defined by Equation (37). Clinozoisite reacts at lower temperatures than zoisite (open and filled circles) and is therefore metastable at the conditions of the experiments. Due to the lack of calorimetric measurements and paucity of results for phase equilibria for clinozoisite, values of $s^o{}_{czo}$ and $\Delta_f h^o{}_{czo}$ are less well defined. Three basically different options are available to draw lines which separate the experimental results for the clinozoisite equilibrium (Eqn. 34), leading to different choices for $s^o{}_{czo}$ and $\Delta_f h^o{}_{czo}$. Because the use of $s^o{}_{zo}$ and $\Delta_f h^o{}_{zo}$ values from the dataset by Gottschalk (1997) is somewhat arbitrary and $s^o{}_{czo}$ and $\Delta_f h^o{}_{czo}$ would therefore be biased, only entropy and enthalpy values for clinozoisite relative to zoisite ($\Delta_f h^o{}_{zo} - \Delta_f h^o{}_{czo}$, $s^o{}_{zo} - s^o{}_{czo}$) are discussed below. While there is an infinite number of options the following three differ mainly in the value for $s^o{}_{zo} - s^o{}_{czo}$, which is chosen to be larger than, equal to, and smaller than 0 J/K, respectively.

The first option (labeled 1 in Fig. 15) uses the observation by Jenkins et al. (1985) that at pressures of 1.3 to 1.6 GPa the equilibrium of the reaction

$$\text{clinozoisite} \leftrightarrow \text{zoisite} \qquad (38)$$

is at low temperature ($T < 350°C$). Using the experimental results and this constraint one solution of $\Delta_f h^o{}_{zo} - \Delta_f h^o{}_{czo}$ and $s^o{}_{zo} - s^o{}_{czo}$ values ($\Delta_R H^o$ and $\Delta_R S^o$ for reaction 38) would be 5.138 kJ and 10.225 J/K, respectively. As a result for reaction (38) the equilibrium temperature at $P_r$ would be 229°C. This is, however in contrast to experimental results by Holdaway (1972) and Holland (1984), which indicate equilibrium temperatures of $635 \pm 75°C$ at 300 MPa and $500 \pm 25°C$ at 1.5 GPa, respectively. Because the change of volume $\Delta_R V^o$ for reaction (44) at $P_r$ is negative and remains negative at higher pressures and $\Delta_R S^o$ is positive, for option 1 the Clapeyron slope of this reaction would be negative. Also for option 1 the $s^o{}_{zo} - s^o{}_{czo}$ value is 0.5 J/K smaller than the one used by Gottschalk (1997).

The second option would be to choose a parallel to the line of the zoisite equilibrium. As a result $\Delta_f h^o{}_{zo} - \Delta_f h^o{}_{czo}$ and $s^o{}_{zo} - s^o{}_{czo}$ would be 0 kJ and 4.5 J/K, respectively. Consequently the

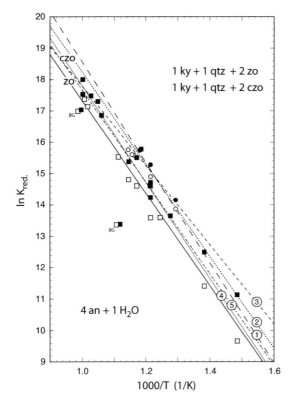

Figure 15. $1/T$ vs. ln $K_{red}$ plot for reactions (33) and (34) involving zoisite (squares) and clinozoisite (circles). The solid line separates as required the open and filled symbols representing the forward reaction and reverse reaction direction for the equilibrium with zoisite (Eqn. 33). Experimental results for the reaction with zoisite are from Best and Graham (1978), Johannes and Ziegenbein (1980), Goldsmith (1981), Jenkins et al. (1983), Jenkins et al. (1985), Chatterjee et al. (1984) and with clinozoisite from Jenkins et al. (1983, 1985). Lines for different choices for $\Delta_f h^\circ_{zo}-\Delta_f h^\circ_{czo}$ and $s^\circ_{zo}-s^\circ_{czo}$ so that: 1. $T_{eq}$ for czo↔zo is below 350°C; 2. line is parallel to the zo equilibrium ($\Delta_R H=0$), i.e. czo always metastable; 3. czo is stable above 1000°C; 4. as published by Brunsmann et al. (2002); 5. as derived here (for more details see text).

equilibrium lines for the reactions (33) and (34) never cross and due to the volume behavior as a function of $P$ and $T$ of both phases, pure clinozoisite would always be metastable with respect to zoisite at $P_r$ and also at higher pressure.

The third option would be to pick a flatter slope than that for the zoisite reaction (33) leading to −8.8 kJ and −6.9 J/K for $\Delta_f h^\circ_{zo}-\Delta_f h^\circ_{czo}$ and $s^\circ_{zo}-s^\circ_{czo}$, respectively. In this case at $P_r$ clinozoisite would be stable with respect to zoisite above 1000°C and, because of the negative value of $s^\circ_{zo}-s^\circ_{czo}$, the Clapeyron slope of reaction (38) would be positive. There is however no field evidence to support this option (Jenkins et al 1985; Brunsmann et al. 2002). The lines labeled 4 and 5 in Figure 15 are discussed below.

### Clinozoisite and zoisite solid solutions

Zoisite and clinozoisite are both able to incorporate $Fe^{3+}$ at their M3 sites and form solid solutions. Observations of natural samples (e.g., Enami and Banno 1980; Franz and Selverstone 1992), experimental studies (Prunier and Hewitt 1985; Brunsmann et al. 2002) as well as thermodynamic considerations have established that a two phase field, i.e., a phase loop, exists. For coexisting zoisite/clinozoisite pairs clinozoisite is always the $Fe^{3+}$-richer phase.

The phase loop has been investigated experimentally by Prunier and Hewitt (1985) and Brunsmann et al. (2002) (Fig. 16). The compositions at the phase loop are calculated by simultaneous evaluation of equilibrium (38) and

$$\text{epidote} \leftrightarrow \text{orthoepidote} \tag{39}$$

In reaction (39) orthoepidote designates a hypothetical phase component of epidote

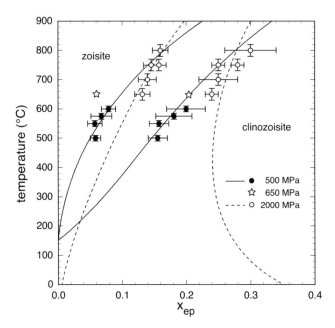

**Figure 16.** Experimental results for compositions of coexisting synthetic zoisite and clinozoisite at 500 MPa and 2000 MPa by Brunsmann et al. (2002) (circles) and at 650 MPa by Prunier and Hewitt (1985) (stars). Calculated equilibria using derived values of $\Delta_R H°$ and $\Delta_R S°$ (see Fig. 19) for reactions defined by Equations (40) and (41).

endmember composition with the orthorhombic structure of zoisite. Simultaneous evaluation of equilibria (38) and (39) requires assumptions about the mixing behavior in both phases, zoisite and clinozoisite solid solutions.

In general the calculation of an equilibrium involves the change in Gibbs free energy of a reaction $\Delta_R G$:

$$\Delta_R G = \sum_i n_i \mu_i = 0 \tag{40}$$

which requires the chemical potentials of each phase component involved:

$$\mu_i = \mu_i^o + RT \ln a_i \tag{41}$$

For reactions (38) and (39) this leads to the following expressions which must be solved simultaneously for the calculation of the phase compositions along the phase loop:

$$\Delta\mu_R^{(38)} + RT \ln \frac{a_{zo}}{a_{czo}} = 0 \tag{42}$$

$$\Delta\mu_R^{(39)} + RT \ln \frac{a_{oep}}{a_{ep}} = 0 \tag{43}$$

For epidote minerals and mixing only at the M3 site the activity can be expressed by

$$a_i = X_i \gamma_i \tag{44}$$

which requires assumptions about the mixing properties, i.e., activity coefficients $\gamma_i$, of phase components in zoisite and clinozoisite solid solutions.

## MIXING PROPERTIES

### Excess enthalpies for mixing at M3

One possibility to determine activity coefficients $\gamma_i$ is from solution calorimetry. This requires the measurement of $\Delta_f h_i^\circ$ values of the endmembers and a set of solid solutions to obtain the excess enthalpy $h_{excess}$ of mixing as a function of composition:

$$h_{excess}^{mix} = \Delta_f h_{ss} - \left( X_{ep} \Delta_f h_{ep}^\circ + X_{czo} \Delta_f h_{czo}^\circ \right) \tag{45}$$

Differentiation with respect to $n_i$ yields $\gamma_i$:

$$\left( \frac{\partial \left( n h_{excess}^{mix} \right)}{\partial n_i} \right)_{P,T,n_j \neq n_i} = RT \ln \gamma_i \tag{46}$$

A set of experimentally determined excess enthalpies in the system zoisite-clinozoisite is available from Smelik et al. (2001) and one value is given by Kiseleva and Ogorodova (1986, 1987). The calorimetric results for $\Delta_f h_{ss}$ and the derived $h_{excess}$ values are plotted in Figures 17 and 18.

For the system clinozoisite-epidote Smelik et al. (2001) modeled their $h_{excess}$-values using an asymmetric Margules equation

$$h_{excess}^{mix} = X_{czo} X_{ep} \left( w_{czo\text{-}ep} X_{ep} + w_{ep\text{-}czo} X_{czo} \right) \tag{47}$$

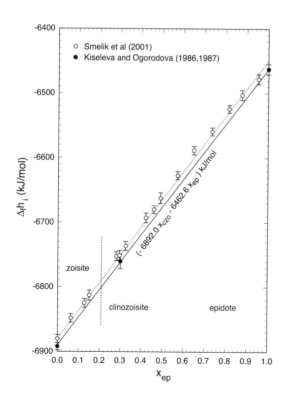

**Figure 17.** Experimentally derived values of $\Delta_f h_{zo}$, $\Delta_f h_{czo}$ and $\Delta_f h_{ep}$ for zoisite and clinozoisite-epidote solid solutions by Kiseleva and Ogorodova (1986, 1987) (filled circles) and Smelik et al. (2001) (open circles). Solid line reflects enthalpies of a mechanical mixture using the values for clinozoisite by Gottschalk (1997) and for epidote by Kiseleva and Ogorodova (1987) as endmember values. Parallel dotted line is defined by a systematic offset of 15 kJ.

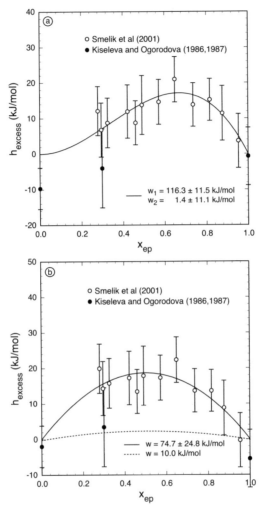

**Figure 18.** Excess enthalpies using a) the asymmetric Margules equation by Smelik et al. (2001) with −6882.5 kJ/mol and −6461.9 kJ/mol for $\Delta_f h°_{czo}$ and $\Delta_f h°_{ep}$ b) a regular solution model with −6894.7 ± 8.2 kJ/mol and −6457.0 ± 3.5 kJ/mol $\Delta_f h°_{czo}$ and $\Delta_f h°_{ep}$.

and extracted values of −6882.5 ± 6.9 kJ/mol and −6461.9 ± 6.8 kJ/mol for $\Delta_f h°_{czo}$ and $\Delta_f h°_{ep}$, respectively, and 116.3 ± 11.5 kJ/mol and 1.4 ± 11.1 kJ/mol for $w_{czo\text{-}ep}$ and $w_{ep\text{-}czo}$, respectively (Fig. 18a). Using these parameters Smelik et al. (2001) calculated a critical temperature for the corresponding miscibility gap of 6279°C. This unreasonable high critical temperature means that for example at 700°C the maximum epidote content $X_{ep}$ in clinozoisite would be 0.095 and no clinozoisite dissolves in epidote ($X_{czo} < 10^{-6}$). Considering also the zoisite-clinozoisite phase loop no clinozoisite-epidote solid solutions should therefore exist. This is obviously not in agreement with natural and especially experimental observations like those by Prunier and Hewitt (1985) and Brunsmann et al. (2002). Smelik et al. (2001) recognized that this calculated critical temperature is completely unreasonable in the view of field data and that an asymmetric Margules equation might not be appropriate to model clinozoisite solid solutions.

The observed asymmetry of $h_{excess}$ by Smelik et al. (2001) is strongly influenced by their selection of $\Delta_f h°_{ep}$ and especially $\Delta_f h°_{czo}$ which seems to be too low (see also Table 5). A simpler model would be that of a regular solution

$$h_{excess}^{mix} = X_{czo} X_{ep} w_{RS} \qquad (48)$$

Fitting the experimental results of Smelik et al. (2001) to Equations (45) and (48) and using $\Delta_f h°_{czo}$ and $\Delta_f h°_{ep}$ as additional parameters yields values of $-6894.7 \pm 8.2$ kJ/mol and $-6457.0 \pm 3.5$ kJ/mol for $\Delta_f h°_{czo}$ and $\Delta_f h°_{ep}$, respectively, and of $74.7 \pm 24.8$ kJ/mol for $w_{RS}$. While the derived values for $\Delta_f h°_{czo}$ and $\Delta_f h°_{ep}$ are reasonable and in agreement with many internally consistent datasets the high $w_{RS}$ value leads again to a very high critical temperature of 4218°C. Using this model at 700°C the maximum epidote content in clinozoisite would be only 0.1‰ $Fe^{3+}$ per formula unit. Again this is not in agreement with natural and experimental observations.

Closer examination of the experimental results by Smelik et al. (2001) and comparison with results by Kiseleva and Ogorodova (1986, 1987) as well as with values for endmembers from internally consistent datasets (Table 5) indicates that these results are systematically more positive by a value of about 10 to 15 kJ/mol (Fig. 17). Consideration of this possible offset would lead to a corrected value of $w_{RS}$ in the range of 10 to 17 kJ/mol, which would be an adequate value for the system clinozoisite-epidote (see below).

Solid solutions in the system zoisite-orthoepidote are stable only over a limited compositional range, observed orthoepidote contents ($X_{oep}$) are below 0.2. In addition the hypothetical endmember orthoepidote, i.e., epidote in orthorhombic structure, is not a stable phase. Therefore a very limited compositional range is accessible to $\Delta_f h_{ss}$ measurements. Few $\Delta_f h_{ss}$ determinations were made by Smelik et al. (2001) between $X_{oep}$ values of 0 and 0.152. Because measurements for $\Delta_f h°_{oep}$ are totally lacking it seems to be impossible to evaluate $h_{excess}$ and therefore $\gamma_{ep}$ and $\gamma_{oep}$ in the system zoisite-orthoepidote. One might speculate from the denser structure of zoisite and the larger ionic radius of $Fe^{3+}$ that the $h_{excess}$ values for zoisite-orthoepidote solid solutions should be higher than for clinozoisite-epidote solid solutions.

**Ideal mixing at M3**

While the results regarding $h_{excess}$ in the system zoisite-clinozoisite are at least controversial they are totally lacking for the system epidote-orthoepidote. Therefore as a first approximation in the treatment of reactions (38) and (39) one can assume ideal mixing, i.e., $h_{excess}$ is set to 0 in both systems, and as a consequence the activity coefficients for all components are 1. Additional parameters for the fictive endmember orthoepidote are estimated as follows. The iron-free phases zoisite has a smaller volume than clinozoisite. As a consequence a molar volume of 138.52 J/MPa is chosen for orthoepidote, which is 0.49 J/MPa smaller than for epidote. The values for $\alpha°_{oep}$ and $\beta°_{oep}$ are set identical to those of zoisite and the $c_P°_{oep}$ expression is the same as for epidote. As for the evaluation of reactions (33) and (34) the $c_P°_{zo}$ equation is the same as $c_P°_{czo}$.

The calculated $K_{red}$ values for reactions (38) and (39) using the compositions of coexisting pairs of zoisite and clinozoisite by Brunsmann et al. (2002) and assuming ideal mixing are shown in Figure 19. If the coexisting phases are in equilibrium, then a linear fit through the respective data points should yield the values for $\Delta_R H°$ and $\Delta_R S°$ ($\Delta_f h°_{zo}-\Delta_f h°_{czo}$, $s°_{zo}-s°_{czo}$ and $\Delta_f h°_{oep}-\Delta_f h°_{ep}$, $s°_{oep}-s°_{ep}$, respectively) for these reactions. For reaction (38) values for $\Delta_R H°$ and $\Delta_R S°$ of $1.3 \pm 0.6$ kJ and $2.5 \pm 0.5$ J/K, respectively, and for reaction (39) values for $\Delta_R H°$ and $\Delta_R S°$ of $13.1 \pm 1.4$ kJ and $8.0 \pm 2.5$ J/K, respectively, are derived. Using a similar approach Brunsmann et al. (2002) determined for reaction (38) values of $2.8 \pm 1.3$ kJ and 4.5 $\pm 1.4$ J/K for $\Delta_R H°$ and $\Delta_R S$, respectively, which are twice the values extracted here. The only difference between the approach taken here and that of Brunsmann et al. (2002) is the selected

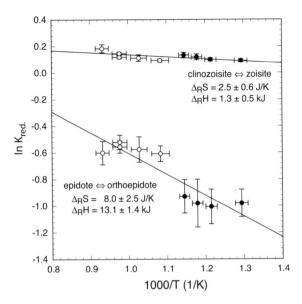

**Figure 19.** $1/T$ vs. ln $K_{red}$ plot for the equilibria (38) and (39). The data are experimental results by Brunsmann et al. (2002) for 500 MPa (filled circles) and 2000 MPa (open circles).

set of values for $v°_{ep}$, $v°_{oep}$, $\alpha°_{ep}$, $\alpha°_{oep}$, $\beta°_{ep}$, and $\beta°_{oep}$. The calculation of $K_{red}$ is, however, very sensitive to the choice of these values.

The derived values for $\Delta_f h°_{zo} - \Delta_f h°_{czo}$ and $s°_{zo} - s°_{czo}$ should be identical from both the $Fe^{3+}$-free and the $Fe^{3+}$-bearing system. This is, however not the case. The lines for reaction (34) involving clinozoisite (labeled as lines 4 and 5 in Fig. 15) are calculated using the values $\Delta_f h°_{zo} - \Delta_f h°_{czo}$ and $s°_{zo} - s°_{czo}$ using reaction (38) from above (lines 4 and 5 in Figure 15). They plot significantly closer to the solid line of the zoisite equilibrium (Eqn. 33) and are therefore no longer consistent with the results for clinozoisite reported by Jenkins et al. (1983, 1985). Lines 4 and 5 in Figure 15 would predict equilibrium temperatures for reaction (34) which are 20°C higher than observed experimentally. Under the assumptions made above, no other choice of $v°_{ep}$, $v°_{oep}$, $\alpha°_{ep}$, $\alpha°_{oep}$, $\beta°_{ep}$, and $c_P°_{czo}$ can make these two sets of experimental results compatible. The reason for this might be twofold. The assumption of ideal mixing might not be appropriate and/or the equilibrium temperatures as determined by Jenkins et al. (1983, 1985) are too low by 20°C. An explanation could be their method of equilibrium determination. They used relative X-ray diffraction intensities to determine the direction of reaction. The nucleation of margarite, which was observed by Jenkins et al. (1985), should occur at the expense of clinozoisite and lower its X-ray intensities. This could lead to an underestimation of the observed equilibrium temperatures.

The values of $\Delta_R H°$ and $\Delta_R S$ derived here are used to calculate the zoisite-clinozoisite phase loop at 500 and 2000 MPa (Fig. 16). Overall the experimental results are well reproduced. But while the loop at 500 MPa looks as expected, the two phase field at 2000 MPa opens to lower temperatures. There is nothing wrong about this, because due to the negative Clapeyron slope the equilibrium temperature of the reaction (38) is already below 0 K at 2000 MPa.

### General comments on results for $\Delta_f h°_{zo} - \Delta_f h°_{czo}$ and $s°_{zo} - s°_{czo}$

Several sets of information exist for the determination of values for $\Delta_f h°_{zo} - \Delta_f h°_{czo}$ and $s°_{zo} - s°_{czo}$: the experimental results for reactions (33) and (34) by Jenkins et al. (1983, 1985) and the compositions of coexisting zoisite and clinozoisite solid solutions by Brunsmann et al. (2002). In addition Smelik et al. (2001) present results on excess enthalpies and activity coefficients for these solid solutions. All three sets of information are presently not completely internally consistent and additional data are needed. The determined activities (Smelik et al. 2001) are not usable because they preclude the known existence of a zoisite-clinozoisite solid solution series. The results by Jenkins et al. (1983, 1985) and by Brunsmann et al. (2002) show that clinozoisite is the stable phase at reference conditions. Using ideal mixing for clinozoisite-zoisite and epidote-orthoepidote solid solutions the results by Jenkins et al. (1983, 1985) and Brunsmann et al. (2002) are inconsistent. As a consequence the actual values for $\Delta_f h°_{zo} - \Delta_f h°_{czo}$ and $s°_{zo} - s°_{czo}$ can not be determined until the activity coefficients for the 4 involved phase components zoisite, orthoepidote, clinozoisite and epidote become available. The absolute enthalpy and entropy differences between clinozoisite and zoisite remain therefore, in a relative narrow range, however, uncertain.

## ORDER-DISORDER

Two different types of intracrystalline order-disorder mechanisms are operating in clinozoisite-epidote solid solutions (see also Franz and Liebscher 2004, Liebscher 2004). One is the nonconvergent ordering of $Al^{3+}$ and $Fe^{3+}$ at the M1 and M3 sites (Dollase 1973; Raith 1976; Bird and Helgeson 1980; Patrier et al. 1991; Bonazzi and Menchetti 1995; Fehr and Heuss-Aßbichler 1997; Bird et al. 1988; Giuli et al. 1999). The other (Heuss-Aßbichler and Fehr 1997; Fehr and Heuss-Aßbichler 1997) is convergent ordering of $Al^{3+}$ and $Fe^{3+}$ at the M3 and M3′ sites which are created by splitting the M3 site. As a result of the convergent ordering miscibility gaps become stable which are observed in natural samples (Strens 1965; Hietanen 1974; Raith 1976; Schreyer and Abraham 1978) and experimentally (Heuss-Aßbichler and Fehr 1997; Brunsmann et al. 2002).

### Nonconvergent ordering at the M3-M1 sites

The following reaction describes the nonconvergent ordering of $Al^{3+}$ and $Fe^{3+}$ at the M1 and M3 sites

$$Al_{M1}Fe^{3+}_{M3} \leftrightarrow Fe^{3+}_{M1}Al_{M3} \tag{49}$$

for which the equilibrium is defined by:

$$0 = \Delta\mu_R^o + RT \ln \frac{a^{M1}_{Fe^{3+}} a^{M3}_{Al}}{a^{M1}_{Al} a^{M3}_{Fe^{3+}}} \tag{50}$$

As already discussed the activity-composition relations for mixing of $Al^{3+}$ and $Fe^{3+}$ at the M3 site determined by Smelik et al. (2001) seems not to be applicable, and no such information exists for $Al^{3+}$ and $Fe^{3+}$ mixing at the M1 site. Therefore, as a first approximation the respective activity coefficients are set to 1 and Equation (50) becomes:

$$0 = \Delta\mu_R^o + RT \ln \frac{X^{M1}_{Fe^{3+}} X^{M3}_{Al}}{X^{M1}_{Al} X^{M3}_{Fe^{3+}}} \tag{51}$$

If it is further assumed that $\Delta V°_R$ and $\Delta c_P°_R$ are both 0, then the values for $K_{red}$ can be readily calculated. These calculated $K_{red}$ values from experimentally determined intracrystalline $Al^{3+}$ and $Fe^{3+}$ distributions between M1 and M3 (Dollase 1973; Fehr and Heuss-Aßbichler 1997; Giuli et al. 1999) as well as determinations from natural samples for which the formation temperatures are known (Bird et al. 1988; Partier et al. 1991) are plotted into an $1/T$ vs. $\ln K_{red}$ diagram (Fig. 20). If the samples were in equilibrium and the assumptions made above are

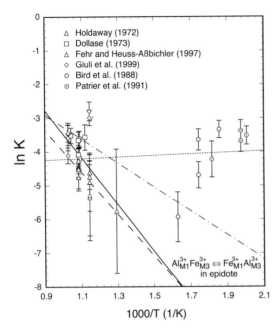

**Figure 20.** $1/T$ vs. ln $K_{red}$ plot for the non convergent intracrystalline order-disorder exchange reactions (49) involving $Al^{3+}$ and $Fe^{3+}$ at the M1 and M3 sites. The data are experimental results (Holdaway 1972; Dollase 1973; Fehr and Heuss-Aßbichler 1997; Giuli et al. 1999) as well as determinations from natural samples for which the formation temperatures are known (Bird et al. 1988; Partier et al. 1991). The solid line represents $\Delta_R H$ and $\Delta_R S$ values of $57.5 \pm 18.9$ kJ and $28.6 \pm 20.9$ J/K, the dashed line $51.3 \pm 12.0$ kJ and $18.1 \pm 13.7$ J/K, and the dotted line $-1.7 \pm 3.5$ kJ and $-36.9 \pm 3.5$ kJ, respectively (for discussion see text).

correct the data should plot on a single straight line, which is not the case. Preferential use of the experimental results alone would yield a line with a negative slope whereas the results from natural samples would yield a positive slope. The respective value of $\Delta\mu°_R$, i.e., its $\Delta H°_R$ and $\Delta S°_R$ values, can be determined by a linear fit. Using all experimental results and weighting each results with the reciprocal value of its error, which is assumed to be $\pm 0.02\ X_{Fe}$, yields values for $\Delta H_R°$ and $\Delta S_R°$ of $57.5 \pm 18.9$ kJ and $28.6 \pm 20.9$ J/K, respectively. Using only the results by Fehr and Heuss-Aßbichler (1997) values of $51.3 \pm 12.0$ kJ and $18.1 \pm 13.7$ J/K are calculated. Both fits lead to a strong temperature dependence of $Fe^{3+}$ at the M1 site. Below 400°C an epidote with an overall composition of $X_{ep}$ equal to 1 would be practically totally ordered, at 700°C about 13% of all $Fe^{3+}$ would be at the M1 site. A fit using all results yields $\Delta H_R°$ and $\Delta S_R°$ values of $-1.7 \pm 3.5$ kJ and $-36.9 \pm 3.5$ kJ, respectively. In this case for epidote ($X_{ep} = 1$) the $Fe^{3+}$-content at the M1 site would be practically independent of temperature and always about 10%. In comparison Bird and Helgeson derived for reaction (49) values of 29.2 kJ and $-2.1$ J/K for $\Delta H_R°$ and $\Delta S_R°$, respectively.

The evaluation of the available results shows that those from experimentally temperature treated and those from natural samples are not consistent and the actual distribution of the $Fe^{3+}$ between the M1 and M3 sites remains still unclear. It might be, however, possible that cation distribution in natural epidotes, which formed hydrothermally at relatively low temperatures, is not in equilibrium.

**Convergent ordering at the M3-M3′ sites**

The convergent ordering of $Al^{3+}$ and $Fe^{3+}$ at the M3 and M3′ sites can be expressed by the reaction

$$\left(Al_{(1-X_{ep})/2}Fe^{3+}_{X_{ep}/2}\right)_{M3}\left(Al_{(1-X_{ep})/2}Fe^{3+}_{X_{ep}/2}\right)_{M3'} \leftrightarrow \left(Al_{1-X_{ep}}\right)_{M3}\left(Fe^{3+}_{X_{ep}}\right)_{M3'} \qquad (52)$$

which describes the process in which a random distribution of $Al^{3+}$ and $Fe^{3+}$ goes into an ordered state. The occurrence of such ordering along the clinozoisite-epidote solid solution series is indicated by the observation of two miscibility gaps in natural and experimental samples (Strens 1965; Hietanen 1974; Raith 1976; Schreyer and Abraham 1978; Heuss-Aßbichler and Fehr 1997; Fehr and Heuss-Aßbichler 1997; Brunsmann et al. 2002). The miscibility gaps are separated by an intermediate phase with a $X_{ep}$ of around 0.5 (Dollase 1973; Langer and Raith 1974; Bonazzi and Menchetti 1995; Fehr and Heuss-Aßbichler 1997) which is additionally supported by observations from Mössbauer spectroscopy, IR-spectroscopy, and X-ray diffraction.

As a first approximation convergent ordering can be approached by using a Bragg-Williams model (e.g., Moelwyn-Hughes 1965; Navrotsky and Loucks 1977; Gottschalk and Metz 1992). In such a model the enthalpy change associated with ordering is calculated as

$$h_{BW}^{\zeta} - h_{BW}^{\zeta=0} = -w_{BW}\zeta^2 X_{ep}^2 \tag{53}$$

The parameter $\zeta$ in this equation defines the degree of equilibrium long range order, ranging from 1 for total order (Al at M3 and $Fe^{3+}$ only at M3') to 0 for total disorder (equal distribution of Al and $Fe^{3+}$ at M3 and M3').

The energy parameter $w_{BW}$ is directly related to the critical temperature, i.e., the temperature at which total disorder is reached, for a phase of intermediate composition ($X_{ep} = 0.5$):

$$w_{BW} = 2RT_{crit} \tag{54}$$

The degree of order $\zeta$ at any $T$ and $X_{ep}$ can be calculated by solving the following equation:

$$\frac{(1+\zeta)\left[1-(1-\zeta)X_{ep}\right]}{(1-\zeta)\left[1-(1+\zeta)X_{ep}\right]} = \exp\left(\frac{4x_{ep}w_{BW}\zeta}{RT}\right) \tag{55}$$

Knowing the degree of order $\zeta$ the entropy of mixing can be calculated from the expression:

$$\begin{aligned} s_{mix} = -\frac{R}{2}\Big\{ &(1+\zeta)X_{ep}\ln\left[(1+\zeta)X_{ep}\right]+\left[1-(1+\zeta)X_{ep}\right]\ln\left[1-(1+\zeta)X_{ep}\right] \\ &+(1-\zeta)X_{ep}\ln\left[(1-\zeta)X_{ep}\right]+\left[1-(1-\zeta)X_{ep}\right]\ln\left[1-(1-\zeta)X_{ep}\right]\Big\} \end{aligned} \tag{56}$$

At a miscibility gap due to ordering at the M3 and M3′ sites the following two reactions operate for the two coexisting phases *I* and *II*:

$$czo^{inI} \leftrightarrow czoi_{nII} \tag{57}$$

$$ep^{inI} \leftrightarrow ep^{inII} \tag{58}$$

The phase equilibrium can then be calculated by solving the following two equations simultaneously:

$$\mu_{czo}^{I} = \mu_{czo}^{II} \tag{59}$$

$$\mu_{ep}^{I} = \mu_{ep}^{II} \tag{60}$$

In addition to the change of energy due to ordering, an order-independent term on the basis of a regular solution (Eqn. 48) can be considered leading to the model of the generalized point approximation (e.g., Capobianco et al. 1987; Burton and Davidson 1988). In this case the activity coefficients are calculated from the expression

$$RT\ln\gamma_{czo} = X_{ep}^2 w_{RS} \tag{61}$$

and

$$RT \ln \gamma_{ep} = X_{czo}^2 w_{RS} \tag{62}$$

Using the sum of Equations (48), (53) and (56) for the Gibbs free energy of mixing $g_{mix}$ in addition to the energy of mechanical mixing $g°_{mech}$, leads, after differentiation, to:

$$\left( \frac{\partial \left( n \left( g_{mech}^o + g_{mix} \right) \right)}{\partial n_i} \right)_{P,T,n_j \neq n_i} = \mu_i \tag{63}$$

which gives the chemical potentials $\mu_{ep}$ and $\mu_{czo}$. For compositions of $X_{ep} < 0.5$ the following two equations are valid:

$$\mu_{ep} = \mu_{ep}^o - \zeta^2 w_{BW} \left( 2 - X_{ep} \right) X_{ep} + w_{RS} \left( 1 - X_{ep} \right)^2$$
$$+ \frac{RT}{2} \left\{ (1-\zeta) \ln \left[ (1-\zeta) X_{ep} \right] + (1+\zeta) \ln \left[ (1+\zeta) X_{ep} \right] \right. \tag{64}$$
$$\left. - \zeta \ln \left[ 1 - (1+\zeta) X_{ep} \right] + \zeta \ln \left[ 1 - (1-\zeta) X_{ep} \right] \right\}$$

$$\mu_{czo} = \mu_{czo}^o + \zeta^2 w_{BW} \left( 1 - X_{czo} \right)^2 + w_{RS} \left( 1 - X_{czo} \right)^2$$
$$+ \frac{RT}{2} \left\{ \ln \left[ X_{czo} - (1 - X_{czo}) \zeta \right] + \ln \left[ X_{czo} + (1 - X_{czo}) \zeta \right] \right\} \tag{65}$$

Whereas due to symmetry considerations for compositions $X_{ep} > 0.5$ the equations

$$\mu_{ep} = \mu_{ep}^o + \zeta^2 w_{BW} \left( 1 - X_{ep} \right)^2 + w_{RS} \left( 1 - X_{ep} \right)^2$$
$$+ \frac{RT}{2} \left\{ \ln \left[ X_{ep} - (1 - X_{ep}) \zeta \right] + \ln \left[ X_{ep} + (1 - X_{ep}) \zeta \right] \right\} \tag{66}$$

$$\mu_{czo} = \mu_{czo}^o - \zeta^2 w_{BW} \left( 2 - X_{czo} \right) X_{czo} + w_{RS} \left( 1 - X_{czo} \right)^2$$
$$+ \frac{RT}{2} \left\{ (1-\zeta) \ln \left[ (1-\zeta) X_{czo} \right] + (1+\zeta) \ln \left[ (1+\zeta) X_{czo} \right] \right. \tag{67}$$
$$\left. - \zeta \ln \left[ 1 - (1+\zeta) X_{czo} \right] + \zeta \ln \left[ 1 - (1-\zeta) X_{czo} \right] \right\}$$

have to be used.

The result of the application of the generalized point approximation model is shown in Figure 21 for which a critical temperature of 730°C and a value for $w_{RS}$ of 10 kJ were used. The selection of these values is somewhat arbitrary and is actually not supported by experimental measurements. They are chosen for simplicity in such a way that in Figure 21 the clinozoisite-zoisite phase loop and the miscibility gaps do not intersect. For a totally disordered clinozoisite solid solution of intermediate composition, however, these parameters lead to a $h_{excess}$ value of 2.5 kJ which is consistent with the results of Kiseleva and Ogorodova (1987) (Fig. 18). The calculated solvi reproduce the experimental results quite well at 300 and 500 MPa by Heuss-Aßbichler and Fehr (1997) and Brunsmann et al. (2002), respectively. No excess volumes are available because of the large scatter in the volumes of clinozoisite-epidote solid solutions (Fig. 1). Therefore the pressure dependence of the solvi is not accessible and at 300 and 500 MPa the same locations are calculated.

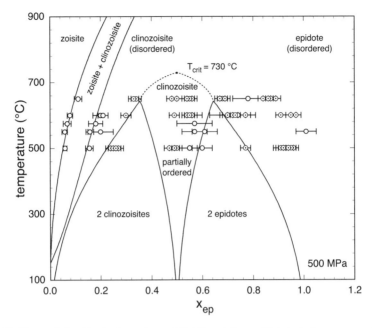

**Figure 21.** Phase relationships in the system zoisite/clinozoisite – epidote including convergent order-disorder at the split M3-M3' positions. For the calculation of the miscibility gaps in the system clinozoisite-epidote a $W$ value of 10 kJ and for the Bragg-Williams parameter a critical temperature for the intermediate phase of 730°C was used. Zoisite/clinozoisite equilibria are the same as in Figure 16. Experimental results are from Heuss-Aßbichler and Fehr (1997) (circles with dots) and Brunsmann et al. (2002) (circles).

## CONCLUSIONS

As discussed above thermodynamic properties for the phase components zoisite, clinozoisite, and epidote are available. At a first glance these results might indicate a well-examined system. However, for many properties discrepancies exist between different sources and types of data and some important pieces of information are still missing and a lot of questions remain.

Molar volume measurements, especially those of solid solutions, suffer from insufficient compositional characterization of the samples and unknown structural states of order. Therefore excess volumes of mixing are not available. Coefficients of thermal expansion for epidote are also missing. Despite considerable effort different studies on the variation of volume as a function of pressure show substantial differences. It is impossible to decide which data set is more reliable. Even within a given data set, errors are too large to derive values of the pressure and temperature derivatives of the bulk modulus. Consequently it is an academic question if either a Murnaghan or a Birch-Murnaghan equation is better to use for epidote group minerals.

Only the high-temperature heat capacities of zoisite were well measured. Heat capacities of all other epidote group phases rely on estimations. Calorimetric determinations of the third law entropy and enthalpy of formation for clinozoisite are not available. The two bodies of experimental results on phase equilibria, which would be sufficient to determine their values relative to those of zoisite, are not consistent. The only existing data on excess enthalpies for clinozoisite-zoisite solid solutions predicts phase equilibria that are contrary to natural and experimental observations. In addition, in contrast to M1-M3 disorder in clinozoisite-epidote solid solutions, M3-M3′ disorder phenomena are still poorly understood and experimental results with well-characterized samples are not yet available.

Silicate structures are complicated and often hard to deal with experimentally. In addition, the closer one looks the more is discovered like the isostructural phase transition in zoisite (Liebscher et al. 2002; Liebscher and Gottschalk 2004) or the M3-M3′ disorder (Heuss-Aßbichler and Fehr 1997; Brunsmann et al. 2002), which might or might not be of thermodynamic importance. There are still ample possibilities to do research on thermodynamic properties of epidotes minerals.

Listed in Tables 6 and 7 is a set of recommended thermodynamic parameters, which

**Table 6.** Recommended values for zoisite and orthoepidote

| *zoisite* | |
|---|---|
| $v^o_{zo(P_r,T_r)}$ | 135.72(7) J/MPa/mol[a] |
| $\alpha^o_{0,zo}$ | 37.1(3) 1/K[b] |
| $\alpha^o_{1,zo}$ | 0.0 1/K$^{1/2}$ |
| $K^o_{zo(0,T_r)}$ | 129(2) GPa[c,d] |
| $K^{o'}_{zo(0,T_r)}$ | 4[c,d] |
| $\left(\partial K^o_{zo(0,T)}/\partial T\right)_{P=0}$ | −0.029(4) GPa/K[c,d] |
| $c^o_{P\,zo(P_r,T)}$ | 733.034 −5.97952×10$^3$ $T^{-0.5}$ −5.07966×10$^6$ $T^{-2}$ + 5.52314×10$^8$ $T^{-3}$ J/K/mol[e] |
| $s^o_{zo(P_r,T_r)}$ | 297.38 J/K/mol[f] |
| $\Delta_f h^o_{zo(P_r,T_r)}$ | −6891.34 kJ/mol[g] |
| *orthoepidote* | |
| $v^o_{oep(P_r,T_r)}$ | 138.52 J/MPa/mol[h] |
| $\alpha^o_{0,oep}$ | 37.1(3) 1/K[i] |
| $\alpha^o_{1,oep}$ | 0.0 1/K$^{1/2}$ |
| $K^o_{oep(0,T_r)}$ | 129(2) GPa[c,i] |
| $K^{o'}_{oep(0,T_r)}$ | 4[c,i] |
| $\left(\partial K^o_{oep(0,T)}/\partial T\right)_{P=0}$ | −0.029(4) GPa/K[i] |
| $c^o_{P\,oep(P_r,T)}$ | 728.954 − 5.56532×10$^3$ $T^{-0.5}$ − 5.93736×10$^6$ $T^{-2}$ + 6.10589×10$^8$ $T^{-3}$ J/K/mol[j] |
| $s^o_{oep(P_r,T_r)}$ | 340.8 J/K/mol[k] |
| $\Delta_f h^o_{oep(P_r,T_r)}$ | −6449.5 kJ/mol[l] |

[a] estimated from Figure 1
[b] reevaluated using data by Pawley et al. (1996)
[c] reevaluated Murnaghan equation
[d] reevaluated using data by Comodi and Zanazzi (1997), Pawley et al. (1998) and Grevel et al. (2000)
[e] reevaluated using data by Perkins et al. (1980) (temperature range 200-2000K)
[f] adapted from Gottschalk (1997), within error in agreement with value by Perkins et al. (1980)
[g] adapted from Gottschalk (1997), in agreement with Kiseleva and Ogorodova (1987)
[h] estimated
[i] estimated, same value as for zoisite
[j] estimated, same value as for epidote
[k] 8.0 J/K/mol lower than for zoisite, from reaction (39) (see Fig. 19)
[l] 13.1 kJ/mol more negative than for zoisite, from reaction (39) (see Fig. 19)

is a summary of the results in this article. However, the specific choice of a value might be somewhat subjective and reflects the opinion of the author.

## ACKNOWLEDGMENTS

I would like to thank Thomas Will for his review. The manuscript benefited from comments by Dennis K. Bird. I am most grateful, however, for the thorough review and useful advice provided by David M. Jenkins, which improved the manuscript considerably. I also want to thank the editors Gerhard Franz and Axel Liebscher for their efforts and their patience.

**Table 7**. Recommended values for clinozoisite and epidote

| *clinozoisite* | |
|---|---|
| $v^o_{czo(P_r,T_r)}$ | 136.40(14) J/MPa/mol[a] |
| $\alpha^o_{0,czo}$ | 29.5(4) 1/K[b] |
| $\alpha^o_{1,czo}$ | 0.0 1/K$^{1/2}$ |
| $K^o_{czo(0,T_r)}$ | 142(4) GPa[c,d] |
| $K^{o'}_{czo(0,T_r)}$ | 4[c,d] |
| $\left(\partial K^o_{czo(0,T)} / \partial T\right)_{P=0}$ | −0.029(4) GPa/K[e] |
| $c^o_{P\,czo(P_r,T)}$ | $733.034 - 5.97952{\times}10^3\,T^{-0.5} - 5.07966{\times}10^6\,T^{-2} + 5.52314{\times}10^8\,T^{-3}$ J/K/mol[f] |
| $s^o_{czo(P_r,T_r)}$ | 294.88 J/K/mol[g] |
| $\Delta_f h^o_{czo(P_r,T_r)}$ | −6892.04 kJ/mol[h] |
| *epidote* | |
| $v^o_{ep(P_r,T_r)}$ | 139.01(14) J/MPa/mol[a] |
| $\alpha^o_{0,ep}$ | 29.5(4)1/K[i] |
| $\alpha^o_{1,ep}$ | 0.0 1/K$^{1/2}$ |
| $K^o_{ep(0,T_r)}$ | 164(3) GPa[c,j] |
| $K^{o'}_{ep(0,T_r)}$ | 4[c,j] |
| $\left(\partial K^o_{ep(0,T)} / \partial T\right)_{P=0}$ | −0.029(4) GPa/K[e] |
| $c^o_{P\,ep(P_r,T)}$ | $728.954 - 5.56532{\times}10^3\,T^{-0.5} - 5.93736{\times}10^6\,T^{-2} + 6.10589{\times}10^8\,T^{-3}$ J/K/mol[k] |
| $s^o_{ep(P_r,T_r)}$ | 332.8 ± 3.9 J/K/mol[l] |
| $\Delta_f h^o_{ep(P_r,T_r)}$ | −6462.6 ± 8.1 kJ/mol[m] |

[a] estimated from Figure 1
[b] reevaluated using data by Pawley et al. (1996)
[c] reevaluated Murnaghan equation
[d] reevaluated using data by Comodi and Zanazzi (1997) and Pawley et al. (1998)
[e] estimated, same value as for zoisite
[f] reevaluated using data by Perkins et al. (1980) (temperature range 200-2000K)
[g] 2.5 J/K/mol lower than for zoisite from reaction (38) (see Figure 19)
[h] 1.3 kJ/mol more negative than for zoisite from reaction (38) (see Fig. 19)
[i] estimated, same value as for clinozoisite
[j] reevaluated using data by Holland et al. (1996)
[k] approximated using $c^o_P$ functions of corundum and hematite
[l] reevaluated using data by Gurevich et al. (1990)
[m] calorimetric value by Kiseleva and Ogorodova (1987)

# REFERENCES

Anderson GM, Crerar DA (1993) Thermodynamics in geochemistry. Oxford University Press, New York

Angel RJ (2000) Equations of state. Rev Mineral Geochem 41:35-59

Belov NV, Rumanova, IM (1953) The crystal structure of epidote $Ca_2Al_2FeSi_2O_{12}(OH)$. Dokl Akad Nauk SSSR 89:853-856

Belov NV, Rumanova IM (1954) The crystal structure of epidote. Trudy Instituta Kristallografii 9:103-163

Berman RG (1988) Internally-consistent thermodynamic data for minerals in the system $Na_2O-K_2O-CaO-MgO-FeO-Fe_2O_3-Al_2O_3-SiO_2-TiO_2-H_2O-CO_2$. J Petrol 29:445-522

Berman RG, Brown TH (1985) Heat capacity of minerals in the system $Na_2O-K_2O-CaO-MgO-FeO-Fe_2O_3-Al_2O_3-SiO_2-TiO_2-H_2O-CO_2$: representation, estimation and high temperature extrapolation. Contrib Mineral Petrol 89:168-183

Best NF, Graham CM (1978) Redetermination of the reaction 2 zoisite + quartz + kyanite ↔ 4 anorthite + $H_2O$. Prog Exp Petrol 11:153-155

Birch F (1947) Finite strain of cubic crystals. Phys Rev 71:809-824

Bird DK, Cho M, Janik CJ, Liou JG, Caruso, LJ (1988) Compositional, order-disorder, and stable isotope characteristics of Al-Fe epidote, State 2-14 drill hole, Salton Sea geothermal system. J Geophys Res 93: 13135-13144

Bird DK, Helgeson HC (1980) Chemical interaction of aqueous solutions with epidote-feldspar mineral assemblages in geologic systems. 1. Thermodynamic analysis of phase relations in the system $CaO-FeO-Fe_2O_3-Al_2O_3-SiO_2-H_2O-CO_2$. Am J Sci 280:907-941

Bonazzi P, Menchetti S (1995) Monoclinic members of the epidote group: effects of the Al ↔ $Fe^{3+}$ substitution and the entry of $REE^{3+}$. Mineral Petrol 53:133-153

Brunsmann A, Franz G, Heinrich W (2002) Experimental investigation of zoisite-clinozoisite phase equilibria in the system $CaO-Fe_2O_3-Al_2O_3-SiO_2-H_2O$. Contrib Mineral Petrol 143:115-130

Burton BP, Davidson PM (1988) Multicritical phase relations in minerals. *In*: Structural and Magnetic Phase Transitions in Minerals, Vol. 7. Ghose S, Coey JMD, Salje E (eds) Springer-Verlag, New York, p 60-90

Capobianco C, Burton BP, Davidson PM, Navrotsky A (1987) Structural and calorimetric studies of order-disorder in $CdMg(CO_3)_2$. J Solid State Chem 71:214-223

Carbonin S, Molin G (1980) Crystal-chemical considerations on eight metamorphic epidotes. N Jb Mineral Abh 139:205-215

Chatterjee ND, Johannes W, Leistner H (1984) The system $CaO-Al_2O_3-SiO_2-H_2O$: new phase equilibria data, some calculated phase relations, and their petrological applications. Contrib Mineral Petrol 88:1-13

Chatterjee ND, Krüger R, Haller G, Olbricht W (1998) The Bayesian approach to an internally consistent thermodynamic database: theory, database, and generation of phase diagrams. Contrib Mineral Petrol 133:149-168

Comodi P, Zanazzi PF (1997) The pressure behavior of clinozoisite and zoisite: an X-ray diffraction study. Am Mineral 82:61-68

Dollase WA (1968) Refinement and comparison of the structures of zoisite and clinozoisite. Am Mineral 53: 1882-1898

Dollase WA (1971) Refinement of the crystal structures of epidote, allanite and hancockite. Am Mineral 56: 447-464

Dollase WA (1973) Mössbauer spectra and iron distribution in the epidote-group minerals. Z Krist 138:41-63

Enami M, Banno S (1980) Zoisite-clinozoisite relations in low- to medium grade high-pressure metamorphic rocks and their implications. Min Mag 43:1005-1013

Fehr KT, Heuss-Aßbichler S (1997) Intracrystalline equilibria and immiscibility along the join clinozoisite-epidote. An experimental and $^{57}Fe$ Mossbauer study. N Jb Mineral Abh 172:43-67

Fesenko EG, Rumanova IM, Belov NV (1955) The crystal structure of zoisite. Dokl Akad Nauk SSSR 102: 275-278

Franz G, Liebscher A (2004) Physical and chemical properties of the epidote minerals–an introduction. Rev Mineral Geochem 56:1-82

Franz G, Selverstone J (1992) An empirical phase diagram for the clinozoisite-zoisite transformation in the system $Ca_2Al_3Si_3O_{12}(OH)-Ca_2Al_2Fe^{3+}Si_3O_{12}(OH)$. Am Mineral 77:631-642

Gabe EJ, Porteine JC, Whitlow SH (1973) A reinvestigation of the epidote structure: confirmation of the iron location. Am Mineral 58:218-223

Giuli G, Bonazzi P, Menchetti S (1999) Al-Fe disorder in synthetic epidotes: a single-crystal diffraction study. Am Mineral 84:933-936

Goldsmith JR (1981) The join $CaAl_2Si_2O_8-H_2O$ (anorthite-water) at elevated pressures and temperatures. Am Mineral 66:1183-1188

Gottschalk M (1997) Internally consistent thermodynamic data set for rock forming minerals in the system $SiO_2$-$TiO_2$-$Al_2O_3$-$Fe_2O_3$-CaO-MgO-FeO-$K_2O$-$Na_2O$-$H_2O$-$CO_2$. Eur Journ Mineral 9:175-223

Gottschalk M, Metz P (1992) The system calcite-dolomite: a model to calculate the Gibbs free energy of mixing on the basis of existing experimental data. N Jb Mineral Abh 164:29-55

Grevel KD, Nowland EU, Fasshauer DW, Burchard M (2000) *In situ* X-ray investigation of lawsonite and zoisite at high pressures and temperatures. Am Mineral 85:206-216

Gurevich VM, Semenov YV, Sidorov YI, Gorbunov VE, Gavrichev KS, Zhdanov VM, Turdakin VA, Khodakovskii IL (1990) Low temperature heat capacity of epidote $Ca_2FeAl_2Si_3O_{12}(OH)$. Geochem Int 27:111-114

Haas JL, Fisher JR (1976) Simultaneous evaluation and correlation of thermodynamic data. Am J Sci 276: 525-545

Haas JL, Robinson GR, Hemingway BS (1981) Thermodynamic tabulations for selected phases in the system calcium oxide-aluminum oxide-silicon dioxide-water at 101.325 kPa (1 atm) between 273.15 and 1800 K. J Phys Chem Ref Data 10:575-669

Halbach H, Chatterjee ND (1984) An internally consistent set of thermodynamic data for twenty-one CaO-$Al_2O_3$-$SiO_2$-$H_2O$ phases by linear programming. Contrib Mineral Petrol 88:14-23

Heinrich W, Althaus E (1988) Experimental determination of the reactions 4 Lawsonite + 1 Albite ↔ 1 Paragonite + 2 Zoisite + 2 Quartz + 6 $H_2O$ and 4 Lawsonite + 1 Jadeite ↔ 1 Paragonite + 2 Zoisite + 1 Quartz + 6 $H_2O$. N Jb Mineral Mh 1988:516-528

Helgeson HC, Delany JM, Nesbitt HW, Bird DK (1978) Summary and critique of the thermodynamic properties of rock-forming minerals. Am J Sci 278A:229

Hemingway BS, Haas JL, Robinson GR (1982) Thermodynamical properties of selected minerals in the system $Al_2O_3$-CaO-$SiO_2$-$H_2O$ at 298.15 K and 1 bar ($10^5$ pascal) pressure and at higher temperatures. US Geol Surv Bull 1544

Heuss-Aßbichler S, Fehr KT (1997) Intracrystalline exchange of Al and $Fe^{3+}$ between grossular-andradite and clinozoisite-epidote solid solutions. N Jb Mineral Abh 172:69-70

Hietanen A (1974) Amphibole pairs, epidote minerals, chlorite and plagioclase in metamorphic rocks, northern Sierra Nevada, California. Am Mineral 69:22-40

Holdaway MJ (1972) Thermal stability of Al-Fe epidote as a function of $f_{O2}$ and Fe content. Contrib Mineral Petrol 37:307-340

Holland TJB (1984) Stability relations of clino- and orthozoisite. Prog Rep NERC, 6:185-186

Holland TJB, Powell R (1985) An internally consistent thermodynamic dataset with uncertainties and correlations: 2. Data and results. J Metamorphic Geol 3:343-370

Holland TJB, Powell R (1990) An enlarged and updated internally consistent thermodynamic dataset with uncertainties and correlations: the system $K_2O$-$Na_2O$-CaO-MgO-MnO-FeO-$Fe_2O_3$-$Al_2O_3$-$TiO_2$-$SiO_2$-C-H-$O_2$. J Metamorphic Geol 8:89-124

Holland TJB, Powell R (1998) An internally consistent thermodynamic data set for phases of petrological interest. J Metamorphic Geol 16:309-343

Holland TJB, Redfern SAT, Pawley AR (1996) Volume behavior of hydrous minerals at high pressure and temperature: II. Compressibilities of lawsonite, zoisite, clinozoisite, and epidote. Am Mineral 81:341-348

Hörmann P-K, Raith M (1971) Optische Daten, Gitterkonstanten, Dichte und magnetische Suszeptibilität von Al-Fe(III)-Epidoten. N Jb Mineral Abh 116:41-60

Ito T (1947) The structure of epidote ($HCa_2(Al,Fe)Al2Si_2O_{13}$). Am Mineral 32:309-321

Jenkins DM, Newton RC, Goldsmith JR (1983) Fe-free clinozoisite stability relative to zoisite. Nature 304: 622-623

Jenkins DM, Newton RC, Goldsmith JR (1985) Relative stability of Fe-free zoisite and clinozoisite. J Geol 93:663-672

Johannes W, Ziegenbein D (1980) Stabilität von Zoisit in $H_2O$-$CO_2$-Gasphasen. Fortschr Miner 58:61-63

Kiseleva IA, Ogorodova, LP (1987) Calorimetric data on the thermodynamics of epidote, clinozoisite, and zoisite. Geochem Int 24:91-98

Kiseleva IA, Ogorodova LP (1986) Thermochemical determination of the enthalpy of formation of zoisite. Geochem Int 23:43-49

Kiseleva IA, Topor ND (1973a) On the thermodynamic properties of zoisite. Geochem Int 10:1169

Kiseleva IA, Topor ND (1973b) Thermodynamic properties of zoisite. Geokhimiya:1547-1555

Kiseleva IA, Topor ND, Andreyenko ED (1974) Thermodynamic parameters of minerals of the epidote group. Geochem Int 11:389-398

Langer K, Lattard D (1980) Identification of a low-energy OH-valence vibration in zoisite. Am Mineral 65: 779-783

Langer K, Raith M (1974) Infrared spectra of Al-Fe(III)-epidotes and zoisites, $Ca_2(Al_{1-p}Fe^{3+}_p)Al_2O(OH)(Si_2O_7)$ ($SiO_4$). Am Mineral 59:1249-1258

Liebscher A (2004) Spectroscopy of epidote minerals. Rev Mineral Geochem. 56:125-170

Liebscher A, Gottschalk M (2004) The *T-X* dependence of the isosymmetric displacive phase transition in synthetic $Fe^{3+}$-Al zoisite: a temperature-dependent infrared spectroscopy study. Am Mineral 89:31-38

Liebscher A, Gottschalk M, Franz G (2002) The substitution $Fe^{3+}$-Al and the isosymmetric dosplacive phase transition in synthetic zoisite: a powder X-ray and infrared spectroscopy study. Am Mineral 87:909-921

Liou JG (1973) Synthesis and stability relations of epidote, $Ca_2Al_2FeSi_3O_{12}(OH)$. J Petrol 14:381-413

Moelwyn-Hughes EA (1965) Physical Chemistry. 2nd rev. Edition, Pergamon Press, Oxford

Murnaghan FD (1937) Finite deformations of an elastic solid. Am J Math 49:235-260

Myer GH (1965) X-ray determinative curve for epidote. Am J Sci 263:78-86

Myer GH (1966) New data on zoisite and epidote. Am J Sci 264:364-385

Navrotsky A, Loucks D (1977) Calculation of subsolidus phase relations in carbonates and pyroxenes. Phys Chem Minerals 1:109-127

Newton RC (1965) The thermal stability of zoisite. J Geol 73:431-440

Nordstrom DK, Munoz JL (1994) Geochemical thermodynamics. Blackwell Scientific Publications, Oxford

Nozik YZ, Kanepit VN, Fykin LY, Makarov YS (1978) A neutron-diffraction study of the structure of epidote. Geochem Int 15:66−69

Patrier P, Beaufort D, Meunier A, Eymery J-P, Petit S (1991) Determination of the nonequilibrium ordering state in epidote from the ancient geothermal field of Saint Martin: Application of Mössbauer spectroscopy. Am Mineral 76:602−610

Pawley AR, Chinnery NJ, Clark SM (1998) Volume measurements of zoisite at simultaneously elevated pressure and temperature. Am Mineral 83:1030-1036

Pawley AR, Redfern SAT, Holland TJB (1996) Volume behavior of hydrous minerals at high pressure and temperature: I. Thermal expansion of lawsonite, zoisite, clinozoisite, and diaspore. Am Mineral 81:335-340

Perkins D, Westrum EF, Essene EJ (1980) The thermodynamic properties and phase relations of some minerals in the system $CaO-Al_2O_3-SiO_2-H_2O$. Geochim Cosmochim Acta 44:61-84

Pistorius CWFT (1961) Synthesis and lattice constants of pure zoisite and clinozoisite. J Geol 69:604−609

Poli S, Schmidt MW (1998) The high-pressure stability of zoisite and phase relationships of zoisite-bearing assemblages. Contrib Mineral Petrol 130:162-175

Poli S, Schmidt MW (2004) Experimental subsolidus studies on epidote minerals. Rev Mineral Geochem 56: 171-196

Prunier AR, Hewitt DA (1985) Experimental observations on coexisting zoisite-clinozoisite. Am Mineral 70: 375-378

Raith M (1976) The Al-Fe(III) epidote miscibility gap in a metamorphic profile through the Penninic series of the Tauern window. Contrib Mineral Petrol 57:99-117

Robie RA, Hemingway BS (1995) Thermodynamic properties of minerals and related substances at 298.15 K and 1 bar ($10^5$ pascals) pressure and at higher temperatures. US Geol Surv Bull 2131

Schmidt MW, Poli S (1994) The stability of lawsonite and zoisite at high pressures: experiments in CASH to 92 kbar and implications for the presence of hydrous phases in subducted lithosphere. Earth Planet Sci Lett 124:105-118

Schreyer W, Abraham K (1978) Prehnite/chlorite and actinolite/epidote mineral assemblages in the metamorphic igneous rocks of IA Helle and halles, Venn-Stavelot-Massif, Belgium. Ann Soc Geol Belg 101:227-241

Seki Y (1959) Relation between chemical composition and lattice constants of epidote. Am Mineral 44:720-730

Shannon RD (1976) Revised effective ionic radii and systematic studies of interatomic distances in halides and chalcogenides. Acta Crystallogr A32:751-767

Shannon RD, Rossman GR (1992) Dielectric constants of apatite, epidote, vesuvianite, and zoisite, and the oxide additivity rule. Phys Chem Minerals 19:157-165

Smelik EA, Franz G, Navrotsky A (2001) A calorimetric study of zoisite and clinozoisite solid solutions. Am Mineral 86:80-91

Smith JV, Pluth JJ, Richardson JW (1987) Neutron diffraction study of zoisite at 15K and x-ray study at room temperature. Z Krist 179:305-321

Stergiou AC, Rentzeperis PJ, Sklavounos S (1987) Refinement of the crystal structure of a medium iron epidote. Z Krist 178:297-305

Storre B, Johannes W, Nitsch K-H (1982) The stability of zoisite in $H_2O-CO_2$ mixtures. N Jb Mineral Mh 1982: 395-406

Strens RGJ (1965) Some relationships between members of the epidote group. Nature 198:80-81

Winkler B, Milman V, Nobes RH (2001) A theoretical investigation of the relative stabilities of Fe-free clinozoisite and orthozoisite. Phys Chem Minerals 28:471-474

Reviews in Mineralogy & Geochemistry
Vol. 56, pp. 125-170, 2004
Copyright © Mineralogical Society of America

# 3

# Spectroscopy of Epidote Minerals

## Axel Liebscher

*Department 4, Chemistry of the Earth*
*GeoForschungsZentrum Potsdam*
*Telegraphenberg*
*D 14473 Potsdam, Germany*
*alieb@gfz-potsdam.de*

## INTRODUCTION

Numerous spectroscopic techniques have been applied to the epidote minerals to characterize their structure and crystal chemistry. The substitution of transition metal ions Mn, Cr, and V, besides Fe, in the different crystallographic sites of epidote minerals and with different valence states has been studied by optical absorption spectroscopy. These studies mainly focused on the determination of (i) the site preferences of the different transition metal ions within the epidote minerals (e.g., Burns and Strens 1967; Tsang and Ghose 1971), (ii) the physical and structural characteristics of these sites as a function of composition, temperature and/or pressure (e.g., Taran and Langer 2000; Langer et al. 2002), (iii) their crystal field stabilization energy (e.g., Burns and Strens 1967; Langer et al. 2002), and (iv) the cause of the color and pleochroism in some epidote minerals (e.g., Faye and Nickel 1971). Major topics of infrared spectroscopic studies have been the proton environment and its changes with composition, temperature, and pressure (e.g., Langer and Raith 1974; Winkler et al. 1989; Della Ventura et al. 1996; Liebscher et al. 2002) and the phase transition within the orthorhombic solid solution series (e.g., Liebscher and Gottschalk 2004). Mössbauer spectroscopy has been used (i) to resolve the valence state of Fe in the different epidote minerals and its site location (e.g., Dollase 1973; Kartashov et al. 2002) and (ii) to study the intracrystalline Al-Fe partitioning between the different octahedral sites and the kinetic of this ordering process (e.g., Patrier et al. 1991; Fehr and Heuss-Aßbichler 1997).

This chapter reviews the different spectroscopic studies and techniques applied to epidote minerals with emphasis given to the crystal chemical results. An in-depth presentation and discussion of the different spectroscopic techniques and their theoretical framework is beyond the scope of this chapter. Additional information about spectroscopic studies of piemontite and allanite can be found in Bonazzi and Menchetti (2004) and Gieré and Sorensen (2004), respectively.

## ULTRAVIOLET, VISIBLE AND NEAR-INFRARED SPECTROSCOPY

Most spectroscopic studies in the ultraviolet (UV), visible (VIS), and near-infrared (NIR) spectral range focused on the monoclinic epidote minerals. After the first spectral measurements of an Al-$Fe^{3+}$ solid solution of unspecified composition reported by Grum-Grzhimailo et al. (1963), White and Keester (1966) published the 300 K spectrum below 25,000 cm$^{-1}$ of a natural epidote with 17 wt% $Fe_2O_3$ and Marfunin et al. (1967) the spectra of epidote of unspecified composition. The first systematic spectroscopic study between

25,000 and 4,540 cm$^{-1}$ stems from Burns and Strens (1967) who recorded the polarized absorption spectra of several natural Al-Fe-Mn-Cr epidotes of variable composition. Wood and Strens (1972) then used the piemontite spectra of Burns and Strens (1967) to test their method for calculating the crystal field splitting in distorted coordination polyhedra. Abu-Eid (1974) recorded the powder spectra of a natural piemontite at 10$^{-4}$ and 19.7 GPa and Langer et al. (1976) and Langer and Abu-Eid (1977) reported the first polarized room temperature absorption spectra of synthetic piemontite single crystals. The first temperature dependent polarized absorption spectra between 23,000 and 7,000 cm$^{-1}$ were published by Parkin and Burns (1980) at 293 to 573 K for an epidote with 15.8 wt% Fe$_2$O$_3$. Smith et al. (1982) recorded the polarized absorption spectra at room and liquid nitrogen (100 ± 10 K) temperature of a natural piemontite. Kersten et al. (1988) and Langer et al. (2002) performed further detailed spectroscopic studies on natural and synthetic piemontite samples. Taran and Langer (2000) recorded polarized absorption spectra, slightly extending the temperature range studied by Parkin and Burns (1980) to 600 K, and were the first to publish pressure dependent absorption spectra up to 9.6 GPa for an epidote with $X_{Ep} \sim 0.78$ [$X_{Ep} = Fe^{3+}/(\Sigma M^{3+} - 2)$].

So far, spectroscopic studies of the orthorhombic epidote minerals in the UV, VIS, and NIR spectral range are rare. Faye and Nickel (1971) studied the room temperature (polarized and unpolarized) absorption spectra of a tan (untreated and heat treated) and a blue colored natural V bearing zoisite. Schmetzer and Berdesinski (1978) recorded the absorption and remission spectra of natural Al-Fe$^{3+}$ and of natural Cr$^{3+}$ and V$^{3+}$ bearing zoisite and Schmetzer and Bank (1979) studied the effect of heat treatment on the pleochroism of V$^{3+}$ and V$^{3+}$-Cr$^{3+}$ bearing zoisite. The polarized room temperature absorption spectra of natural and synthetic Mn$^{3+}$ bearing zoisite were then reported by Petrusenko et al. (1992) and Langer et al. (2002).

The different studies cited display the recorded spectra either in the form of absorbance, linear absorption coefficient $\alpha$, or molar extinction coefficient $\varepsilon$ as a function of the wavelength or the wavenumber. To allow for easier comparison between the different studies, the wavelength values are generally converted to the corresponding wavenumbers throughout this review. Because the wavenumber is the reciprocal of the wavelength, the reader should be aware that the converted wavenumber scale of those spectra displayed in the figures that were originally presented as a function of wavelength is not linear. The other values, i.e., absorbance, linear absorption coefficient $\alpha$, or molar extinction coefficient $\varepsilon$ are not converted into a common unit as not all studies provide the necessary values (i.e., sample thickness, concentration). Likewise, the different studies use different terminologies to describe the respective polarization schemes. To be consistent, the terms X-, Y-, and Z-spectrum or X-, Y-, and Z-polarization are used throughout and designate polarization schemes in which the electrical vector **E** is parallel to one of the principal vibration directions **X**, **Y**, or **Z** of the indicatrix.

## Optical absorption spectra of the monoclinic epidote minerals

*Spectra of Al-Fe$^{3+}$ solid solutions*. Investigations of Al-Fe$^{3+}$ solid solutions are restricted to the studies of Grum-Grzhimailo et al. (1963), White and Keester (1966), Burns and Strens (1967), Parkin and Burns (1980), Petrusenko et al. (1992), and Taran and Langer (2000). The reported spectra closely resemble each other. Although in the case of polarized spectra each polarization is distinctive, the spectra are generally characterized by weak bands in the VIS and NIR region and bands in the violet that are situated on the shoulder of the UV absorption edge. Typical polarized spectra of Fe$^{3+}$ in epidote are shown in Figure 1a,b together with pressure dependent spectra (Fig. 1c) recorded with unpolarized light. All three polarizations are characterized by two bands at ~22,000 (band d) and ~21,200 (band c) cm$^{-1}$, with intensities in the order Z > Y > X (Fig. 1a,b). In addition, the X-spectrum contains a weak inflection at ~17,900 cm$^{-1}$ (band b) and a weak band at ~12,000 cm$^{-1}$ (band a), the Y-spectrum two pronounced inflections at ~17,900 cm$^{-1}$ (band b) and 11,100 cm$^{-1}$ (band a), and the Z-spectrum

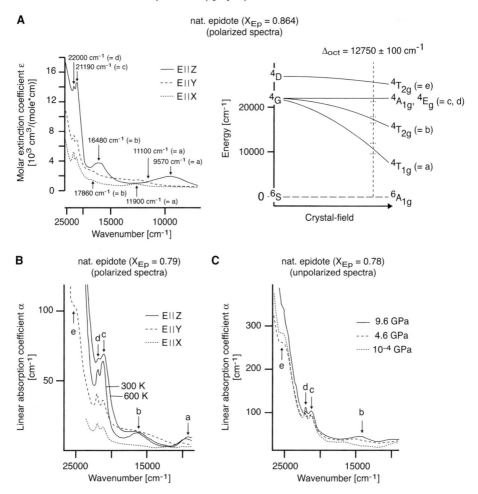

**Figure 1.** (A) Polarized room-temperature absorption spectra of a natural epidote with $X_{Ep} = 0.864$. The spectra show four bands a to d that arise from spin forbidden *dd* transitions. The simplified energy level diagram shows only those spin forbidden *dd* transitions that are relevant for the observed absorption bands. (B) Polarized absorption spectra of a natural epidote with $X_{Ep} = 0.79$ at 300 and 600 K. For clarity only the Z-spectrum is shown for 600 K. In addition to the bands a to d, the Y-spectrum at 300 K displays a shoulder on the UV absorption edge (band e). (C) Unpolarized absorption spectra of a natural epidote with $X_{Ep} = 0.78$ at $10^{-4}$, 4.6, and 9.6 GPa. The position of the bands c, d, and e is independent on pressure whereas the band b shows a notable pressure shift to lower wavenumber. Redrawn and modified after Burns and Strens (1967) and Taran and Langer (2000).

two well-defined bands at $\sim$16,500 cm$^{-1}$ (band b) and 9,600 cm$^{-1}$ (band a) (Fig. 1a,b). In the Y-spectrum a fifth band at $\sim$25,100 cm$^{-1}$ (band e) occurs as a prominent shoulder on the absorption edge (Fig. 1b,c; Taran and Langer 2000).

The Fe$^{3+}$ ion has a d$^5$ electronic configuration and thus only spin forbidden *dd* transitions occur. The observed bands below 25,000 cm$^{-1}$ in the spectra of Fe$^{3+}$ in epidote are generally assigned to the spin forbidden *dd* transitions $^6A_{1g} \rightarrow {}^4T_{1g}$ ($^4G$) (band a), $^6A_{1g} \rightarrow {}^4T_{2g}$ ($^4G$) (band b), $^6A_{1g} \rightarrow {}^4E_g$ ($^4G$) (band c), and $^6A_{1g} \rightarrow {}^4A_{1g}$ ($^4G$) (band d) (Fig. 1a; Burns and Strens 1967;

Taran and Langer 2000). The band e at about 25,100 cm$^{-1}$ observed in the Y-spectrum by Taran and Langer (2000; Fig. 1b,c) may arise from the spin forbidden *dd* transition $^6A_{1g} \rightarrow {}^4T_{2g}$ ($^4D$) (see Fig. 1a; Taran and Langer 2000). The absorption edge is due to $O^{2-} \rightarrow Fe^{3+}$ charge-transfer transitions (Petrusenko et al. 1992; Taran and Langer 2000). From the position of the observed bands shown in Figure 1a, Burns and Strens (1967) calculated the crystal field splitting parameter $\Delta_{oct}$ to 12,750 ± 100 cm$^{-1}$ for the investigated epidote with $X_{Ep} = 0.864$.

Bruns and Strens (1967) studied the compositional dependence of the spectral properties of $Fe^{3+}$ in monoclinic Al-$Fe^{3+}$ solid solutions within the range $X_{Ep} = 0.155$ to 0.915. At low Fe contents the weak inflections in the Y- and X-spectra become inconspicuous and only the weak to well-defined bands in the X- and Z-spectra are still apparent. With increasing Fe content the absorption maxima of bands a and b in the Z-spectra move to lower wavenumber whereas the absorption maximum of band a in the X-spectra moves to higher wavenumber. The two bands c and d are not resolved in the spectra of low Fe epidote but appear as only one band at ~22,200 cm$^{-1}$, which is asymmetric on the low wavenumber side. With increasing Fe content the bands c and d become resolved and their absorption maxima are shifted to lower wavenumber. Over the compositional range $X_{Ep} = 0.155$ to 0.915 $\Delta_{oct}$ varies from 13,200 ± 100 to 12,700 ± 100 cm$^{-1}$ (Burns and Strens 1967).

The influence of temperature on the spectral properties of $Fe^{3+}$ in epidote has been studied by Parkin and Burns (1980) up to 573 K and by Taran and Langer (2000) up to 600 K. With increasing temperature, the intensity of the absorption edge strongly increases (Fig. 1b; Taran and Langer 2000). The data of Taran and Langer (2000) indicate that (i) the absorption maxima of bands a and b shift with about 0.85 cm$^{-1}$ K$^{-1}$ upon heating, (ii) the half-width of both bands broadens with increasing temperature although with a different temperature dependence (band a: 0.03 cm$^{-1}$ K$^{-1}$, band b: 0.74 cm$^{-1}$ K$^{-1}$), and (iii) the integral intensities of both bands slightly shift with temperature in an opposite sense (band a: $-1\times10^{-5}$ K$^{-1}$, band b: $8\times10^{-4}$ K$^{-1}$) (Fig. 1b). These results are partly in contradiction to the findings of Parkin and Burns (1980) for a temperature increase from 293 to 473 K, who found (i) a shift of the absorption maximum of band a (Z-spectrum) to lower wavenumber of ~700 cm$^{-1}$ (i.e., $-3.9$ cm$^{-1}$ K$^{-1}$), (ii) a constant position for the absorption maximum of band b (Z-spectrum), (iii) constant half-widths of bands a and b, and (iv) a slight intensity increase for band a. The spectra recorded by Taran and Langer (2000) further show that band c shifts with $-0.4$ cm$^{-1}$ K$^{-1}$ to lower wavenumber on heating whereas the position of band d remains practically unchanged resulting in a slight increase of the splitting between bands c and d (Fig. 1b). The half-width of the bands c and d broadens with 0.6 cm$^{-1}$ K$^{-1}$ and 0.8 cm$^{-1}$ K$^{-1}$, respectively, and their integral intensities decrease with $-2.2\times10^{-4}$ K$^{-1}$ and $-0.9\times10^{-4}$ K$^{-1}$, respectively, with raising temperature (Taran and Langer 2000). These findings for bands c and d are in agreement with the results of Parkin and Burns (1980) although the latter also found a slight shift of band d to lower wavenumber upon heating.

Taran and Langer (2000) recorded the unpolarized absorption spectra of an epidote with $X_{Ep}$ ~0.78 at 10$^{-4}$, 4.6, and 9.6 GPa (Fig. 1c). In these spectra the bands b, c, d, and e appear as well defined absorption maxima. In the spectrum taken at 10$^{-4}$ GPa the positions of these bands closely resemble those of the corresponding bands in the polarized spectra (compare Fig. 1b). With increasing pressure, only the absorption maximum of band b significantly shifts with $-242$ cm$^{-1}$ GPa$^{-1}$ to lower wavenumber and its half-width decreases with $-32.3$ cm$^{-1}$ GPa$^{-1}$ (Fig. 1c; Taran and Langer 2000). Neither band c nor band d show any change in position or half-width with increasing pressure (Fig. 1c; Taran and Langer 2000). The integral intensities of all bands remain almost constant (Taran and Langer 2000). Up to about 5 GPa the intensity of the absorption edge slightly decreases followed by a strong increase upon further pressure increase. At 9.6 GPa the shoulder that it caused by band e and that is clearly seen at lower pressure is almost hidden by the absorption edge (Fig. 1c).

Based on the observed temperature dependence of the absorption bands, Taran and Langer (2000) concluded that the M3 octahedron retains its distortion or gets only slightly more distorted with increasing temperature. From the temperature shift of band a they calculated the local thermal expansion coefficient $\alpha_{loc}$ for the M3 octahedron to about $0.9 \times 10^{-5}$ $K^{-1}$. The lack of a pressure shift of the position of bands c and d confirms that the covalent bonding of $Fe^{3+}$–O in the M3 octahedron does not change significantly with pressure (Taran and Langer 2000). From the pronounced pressure shift of band b Taran and Langer (2000) obtained a compression modulus $K_{poly}$ for the $Fe^{3+}$ bearing M3 octahedron of about 75 GPa, which implies a comparable high compressibility of M3 within the epidote structure.

***Spectra of natural and synthetic piemontite***. After Burns and Strens (1967) had recorded the first polarized VIS and NIR absorption spectra of six natural piemontite samples, Abu-Eid (1974) published the powder spectra of a natural piemontite of unspecified composition at $10^{-4}$ and 19.7 GPa. Later, Langer et al. (1976) recorded the first polarized absorption spectra of a synthetic piemontite. Smith et al. (1982) then studied the polarized absorption spectra of a natural piemontite at room and liquid nitrogen temperature within the spectral range 30,000 to 10,000 $cm^{-1}$ and Taran et al. (1984) those of natural Mn clinozoisites. Pressure dependent polarized spectra of a natural piemontite up to 7.7 GPa are reported by Langer (1990). Most recently, Langer et al. (2002) performed a detailed spectroscopic study in the spectral range 35,000 to 5,000 $cm^{-1}$ on several natural and synthetic piemontite samples.

All reported spectra of $Mn^{3+}$ in natural as well as synthetic piemontite closely resemble each other in their principal spectral properties (Figs 2a, 3). They are dominated by a slightly polarized UV absorption edge. Within the VIS and NIR spectral range, the spectra are characterized by three intense and strongly polarized absorption bands $\nu_I$, $\nu_{II}$, and $\nu_{III}$ (Figs 2, 3, 5). Band $\nu_I$ is centered between 13,000 and 12,000 $cm^{-1}$ and is strongly polarized with intensities in the order Y-spectrum >> Z- and X-spectrum. Band $\nu_{II}$ has its maximum absorption between 19,000 and 18,000 $cm^{-1}$ and can be identified in all three polarizations

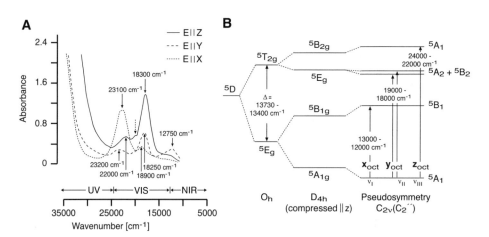

**Figure 2.** (A) Polarized room-temperature absorption spectra of synthetic piemontite between 35,000 and 10,000 $cm^{-1}$. Solid arrows denote bands $\nu_I$, $\nu_{II}$, and $\nu_{III}$ that can be assigned to spin allowed *dd* transitions of $Mn^{3+}$. The stippled arrow at ~20,000 $cm^{-1}$ denotes a band that probably arises from a spin forbidden *dd* transition of $Mn^{3+}$. (B) Energy level diagram for the proposed pseudosymmetry $C_{2v}(C_2'')$ showing the spin allowed *dd* transitions responsible for the bands $\nu_I$, $\nu_{II}$, and $\nu_{III}$. $x_{oct}$, $y_{oct}$, and $z_{oct}$ denote the polarization of the bands $\nu_I$, $\nu_{II}$, and $\nu_{III}$ with respect to the internal axes of the M3 octahedron. For the relation of the internal axes to the principal axes of the indicatrix see text. Redrawn and modified after Langer et al. (1976) and Langer et al. (2002).

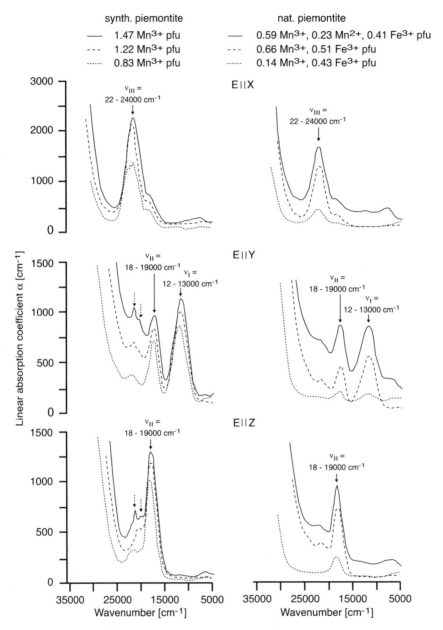

**Figure 3.** Polarized room temperature absorption spectra of synthetic (left) and natural (right) piemontite. Solid arrows denote the bands $\nu_I$, $\nu_{II}$, and $\nu_{III}$ that arise from the spin allowed *dd* transitions of $Mn^{3+}$ in M3. Stippled arrows denote bands that either arise from spin forbidden *dd* transitions of $Mn^{3+}$ in M3 or from spin allowed *dd* transitions of $Mn^{3+}$ in M1. Redrawn and modified after Langer et al. (2002).

with intensities Z-spectrum $\gg$ Y-spectrum $>$ X-spectrum. The third band $v_{III}$ appears between 24,000 and 22,000 cm$^{-1}$ and has strong polarization with intensities in the order X-spectrum $\gg$ Y- and Z-spectrum. The strong polarization of the main bands $v_I$, $v_{II}$, and $v_{III}$ accounts for the trichroic scheme in piemontite with $\mathbf{X}$ = yellow, $\mathbf{Y}$ = violet red, and $\mathbf{Z}$ = carmine (Langer et al. 2002). In addition to these main bands, the Y- and Z-spectra of some synthetic and natural piemontite samples display weak but distinct bands or shoulders in the spectral range 22,000 to 20,000 cm$^{-1}$ in (Figs 2,3; Langer et al. 1976; Langer and Abu-Eid 1977; Smith et al. 1982; Petrusenko et al. 1992; Langer et al. 2002).

The strong and almost complete polarization of the bands $v_I$, $v_{II}$, and $v_{III}$ cannot be explained on the basis of the site symmetry $C_s$ of the M3 octahedron (Langer et al. 2002) and the authors postulate the pseudosymmetry $C_{2v}(C_2'')$ for M3. The derived energy level diagram for the pseudosymmetry $C_{2v}(C_2'')$ is well in accord with the number of the observed bands and their polarizations (Fig. 2b) as the orientation of the internal octahedral axes $\mathbf{x}_{oct}$, $\mathbf{y}_{oct}$, and $\mathbf{z}_{oct}$ shown in Figure 2b relative to the optical indicatrix axes is $\mathbf{x}_{oct}$ = $\mathbf{Y}$, $\mathbf{y}_{oct}$ ~ $\mathbf{Z}$, and $\mathbf{z}_{oct}$ ~ $\mathbf{X}$ (Langer et al. 2002). Based on the energy level diagram for the pseudosymmetry $C_{2v}(C_2'')$, Langer et al. (2002) assigned the bands $v_I$, $v_{II}$, and $v_{III}$ to the spin allowed $dd$ transitions $^5A_1$ $\rightarrow$ $^5B_1$ ($^5D$) (band $v_I$), $^5A_1$ $\rightarrow$ $^5A_2$ + $^5B_2$ ($^5D$) (band $v_{II}$), and $^5A_1$ $\rightarrow$ $^5A_1$ ($^5D$) (band $v_{III}$) (Fig. 2b). The weak but distinct bands between ~22,000 and 20,000 cm$^{-1}$ can be assigned to either spin allowed $dd$ transitions of Mn$^{3+}$ in the M1 octahedron or to spin forbidden $dd$ transitions of Mn$^{3+}$ in M3 (Langer et al. 2002). As M1 is centrosymmetric, the intensity of the bands that arise from the spin allowed $dd$ transitions of Mn$^{3+}$ in M1 will be significantly lower (at least one order of magnitude) than the intensity of the bands that arise from the spin allowed $dd$ transitions of Mn$^{3+}$ in M3 and will be partly or fully superimposed by the latter (Burns and Strens 1967; Langer et al. 2002).

The influence of composition on the spectral properties of Mn$^{3+}$ in piemontite has been studied by Burns and Strens (1967) and Langer et al. (2002) for natural piemontite and by Langer et al. (2002) also for synthetic piemontite. With increasing Mn$^{3+}$ content the UV absorption edge shifts to lower wavenumber (Figs 3, 4b). This shift is most pronounced in the Y-spectra (Fig. 4b). Within the VIS and NIR spectral range the intensity and thus the linear absorption coefficient $\alpha$ of the bands $v_I$, $v_{II}$, and $v_{III}$ generally increase with increasing Mn$^{3+}$ substitution (Fig. 3, 4a) and the absorption maxima of the bands $v_I$, $v_{II}$, and $v_{III}$ slightly shift to lower wavenumber (Fig. 4a). The crystal field splitting $\Delta_{oct}$ (calculated as $\Delta_{oct}$ = $v_{III}$ – $2(v_{III} - v_{II})/3 - v_I/2$; Langer et al. 2002) decreases with increasing Mn$^{3+}$ substitution (Burns and Strens 1967; Langer et al. 2002). Consequently, the octahedral crystal field stabilization energy (CFSE) of Mn$^{3+}$ in M3 (calculated as CFSE$_{Mn3+(M3)}$ = $6\Delta_{oct}/10 + v_I/2$; Langer et al. 2002) decreases slightly on increasing Mn$^{3+}$ substitution in M3 by about 200 cm$^{-1}$ or 2.4 kJ/g-atom per 0.1 Mn$^{3+}$ (Fig. 4c; Langer et al. 2002).

Smith et al. (1982) recorded the polarized absorption spectra of a natural piemontite with 0.829 Mn$^{3+}$ and 0.346 Fe$^{3+}$ pfu (per formula unit) at room and liquid nitrogen temperature (Fig. 5). The spectra display the typical features of Mn$^{3+}$ in M3 as outlined above: In the spectra recorded at liquid nitrogen temperature the bands $v_I$, $v_{II}$, and $v_{III}$ appear at 12,300 cm$^{-1}$ ($v_I$), 18,600 to 17,600 cm$^{-1}$ ($v_{II}$), and 22,000 to 21,600 cm$^{-1}$ ($v_{III}$) with their characteristic intensity ratios and polarizations. Beside these main bands, Smith et al. (1982) further observed a band at 15,600 cm$^{-1}$ in the X- and Z-spectra, a fine structure of $v_{III}$ in the Y- and Z-spectra displaying shoulders at about 22,500 and 20,600 cm$^{-1}$, and two weak bands at 24,600 and 23,700 cm$^{-1}$ in the liquid nitrogen Z-spectrum (Fig. 5). By comparing the recorded spectra with those reported by Burns and Strens (1967) for Fe$^{3+}$ in epidote and by Langer et al. (1976) for synthetic piemontite, Smith et al. (1982) tentatively assigned these additional bands to the following spin forbidden $dd$ transitions in Mn$^{3+}$ and Fe$^{3+}$: (i) bands at 24,600 and 23,700 cm$^{-1}$: $^6A_{1g}$ $\rightarrow$ $^4T_{2g}$ ($^4D$) in Fe$^{3+}$, (ii) band at 22,500 cm$^{-1}$: $^6A_{1g}$ $\rightarrow$ $^4A_{1g}$, $^4E_g$ ($^4G$) in Fe$^{3+}$, band at ~20,600 cm$^{-1}$: $^5A_{1g}$ $\rightarrow$ $^3T_{1g}$ ($^3H$) in Mn$^{3+}$, and (iv) band at 15,600 cm$^{-1}$: $^6A_{1g}$ $\rightarrow$ $^4T_{2g}$ ($^4G$) in Fe$^{3+}$.

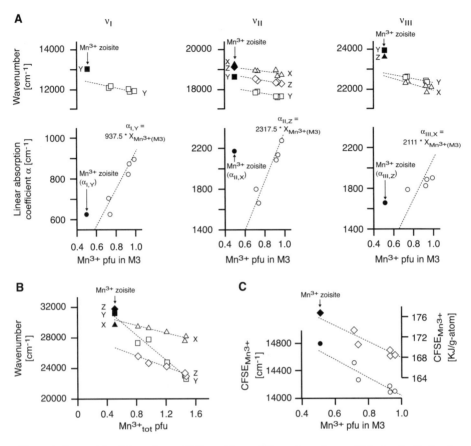

**Figure 4.** Compositional dependence of (A) position and linear absorption coefficient $\alpha$ for the bands $\nu_I$, $\nu_{II}$, and $\nu_{III}$, (B) the position of the UV absorption edge, and (C) the crystal field stabilization energy $\text{CFSE}_{Mn3+(M3)}$ in synthetic piemontite (open symbols) and $Mn^{3+}$ zoisite (closed symbols). X, Y, and Z denote the polarization. The linear absorption coefficient $\alpha$ is calculated for the polarization with maximum intensity for the respective band. The position of the absorption edge is defined as the wavenumber at which the linear absorption coefficient $\alpha$ is 1,200 $cm^{-1}$ above the background. Note the different x-axis values: $Mn^{3+}$ pfu in M3 in (A) and (C) and $Mn^{3+}_{tot}$ pfu in (B). Stippled lines are linear fits to the piemontite data. Redrawn and modified after Langer et al. (2002).

The influence of pressure on the spectral properties of piemontite has so far only be studied for two natural piemontite samples by Abu-Eid (1974) and Langer (1990). Abu-Eid (1974) reports the powder spectra of a natural piemontite of unspecified composition at $10^{-4}$ and 19.7 GPa. On raising pressure the absorption maxima of the bands $\nu_I$ (12,000 → 12,250 $cm^{-1}$), $\nu_{II}$ (18,170 → 20,200 $cm^{-1}$), and $\nu_{III}$ (22,000 → 23,250 $cm^{-1}$) shift to higher wavenumber and the crystal field splitting $\Delta_{oct}$ increases from 13,450 to 15,092 $cm^{-1}$ (Abu-Eid 1974). Assuming a linear behavior, these values correspond to pressure shifts of ~13 $cm^{-1}$/GPa ($\nu_I$), ~103 $cm^{-1}$/GPa ($\nu_{II}$), ~63 $cm^{-1}$/GPa ($\nu_{III}$), and ~83 $cm^{-1}$/GPa ($\Delta_{oct}$). From the pressure dependence of $\Delta_{oct}$, Abu-Eid (1974) calculated a compression modulus $K_{poly}$ for the $Mn^{3+}$ bearing M3 octahedron of about 280 GPa. This compression modulus is rather high compared to the bulk modulus of piemontite, which Abu-Eid (1974) reports to 107 GPa, and the author therefore concludes, "the octahedral site is rather incompressible suggesting that the Ca-polyhedra take up most

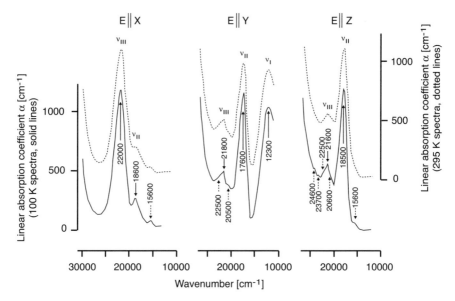

**Figure 5.** Polarized room temperature (stippled) and liquid-nitrogen (solid) spectra of natural piemontite. Solid arrows denote bands $\nu_I$, $\nu_{II}$, and $\nu_{III}$ arising from spin allowed $dd$ transitions of $Mn^{3+}$ in M3 whereas stippled arrows denote bands arising from spin forbidden $dd$ transitions of $Mn^{3+}$ (band at ~20,600 cm$^{-1}$) and $Fe^{3+}$ (bands at ~24,600, 23,700, 22,500, and 15,600 cm$^{-1}$) in M3. Redrawn and modified after Smith et al. (1982).

of the change in volume." This high compression modulus and the conclusion of Abu-Eid (1974) are in sharp contrast to the results obtained by Taran and Langer (2000) for $Fe^{3+}$ in M3 (see above). Unfortunately, with the data at hand this discrepancy cannot be resolved. Langer (1990) reports the Y-spectra of a natural piemontite of unspecified composition at $10^{-4}$, 1.7, 3.7, and 7.7 GPa. Due to the experimental setup only the positions of the UV absorption edge and the bands $\nu_{II}$ and $\nu_{III}$ were recorded. With increasing pressure the UV absorption edge shifts to lower wavenumber whereas the absorption maxima of the bands $\nu_{II}$ (17,600 cm$^{-1}$ at $10^{-4}$ GPa → 19,000 cm$^{-1}$ at 7.7 GPa) and $\nu_{III}$ (21,800 cm$^{-1}$ at $10^{-4}$ GPa → 22,500 cm$^{-1}$ at 7.7 GPa) shift to higher wavenumber, consistent with the results of Abu-Eid (1974). The different pressure shift of the bands $\nu_{II}$ (=182 cm$^{-1}$/GPa) and $\nu_{III}$ (= 91 cm$^{-1}$/GPa) leads to a decrease of the energy difference between $\nu_{III}$ and $\nu_{II}$ with increasing pressure. The observed pressure shifts of the UV absorption edge and the bands $\nu_{II}$ and $\nu_{III}$ reflect the decrease of the mean O–$Mn^{3+}$ distance with pressure whereas the decrease in the energy difference between $\nu_{III}$ and $\nu_{II}$ indicates a decreasing distortion of M3 (Langer 1990). Because the experimental setup did not allow measuring the absorption maximum of the band $\nu_I$ as a function of pressure, Langer (1990) was not able to calculate the pressure dependence of the crystal field splitting $\Delta_{oct}$ for $Mn^{3+}$ in M3.

***Spectra of tawmawite.*** Burns and Strens (1967) recorded the polarized absorption spectra of a natural tawmawite with 0.51 $Cr^{3+}$ pfu. Each polarization displays two moderately intense bands $\nu_I$ and $\nu_{II}$ situated in the vicinity of ~16,000 cm$^{-1}$ ($\nu_I$) and 24,000 ($\nu_{II}$) (Fig. 6). The absorption maxima of these bands occur at slightly different wavenumbers in the different polarizations. The molar extinction coefficients $\varepsilon$ of the observed bands are small ($\varepsilon$ ~ 30 l mole$^{-1}$ cm$^{-1}$) compared to that of the bands arising from $Mn^{3+}$ ($\varepsilon$ ~ 100 to 140 l mole$^{-1}$ cm$^{-1}$; Burns and Strens 1967). From these small molar extinction coefficients and the fact

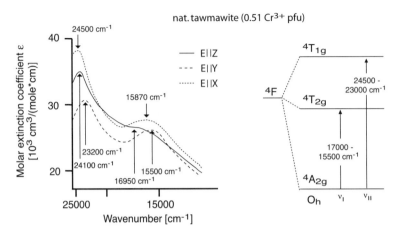

**Figure 6.** Polarized room temperature spectra of a natural $Cr^{3+}$ epidote ("tawmawite") and the corresponding energy level diagram. Redrawn and modified after Burns and Strens (1967).

that the absorption maxima of the bands occur at not widely separated wavenumbers in the different polarizations, Burns and Strens (1967) concluded that the $Cr^{3+}$ ion substitutes in the centrosymmetric M1 site with $D_{2h}$ site symmetry.

Burns and Strens (1967) assigned the observed bands to the spin allowed *dd* transitions $^4A_{2g} \rightarrow {}^4T_{2g}$ $(^4F)$ (band $\nu_I$) and $^4A_{2g} \rightarrow {}^4T_{1g}$ $(^4F)$ (band $\nu_{II}$) (Fig. 6). From the position of $\nu_I$, Burns and Strens (1967) calculated a crystal field splitting $\Delta_{oct}$ for $Cr^{3+}$ in M1 of 16,225 to 16,410 $cm^{-1}$ and a $CFSE_{Cr3+(M1)}$ of 19,470 to 19,692 $cm^{-1}$ or 233 to 236 kJ/g-atom. However, the combination of the derived $\Delta_{oct}$ with the position of $\nu_{II}$ results in a Racah parameter B that is higher than that of the free $Cr^{3+}$ ion and therefore physically senseless. The interpretation of the spectra by Burns and Strens (1967) is thus at least questionable.

## Optical absorption spectra of the orthorhombic epidote minerals

***Spectra of natural and synthetic Mn bearing zoisite ("thulite").*** Polarized room temperature absorption spectra of natural and synthetic $Mn^{3+}$ bearing zoisite (variety "thulite") are reported by Petrusenko et al. (1992) in the spectral range 25,000 to 5,000 $cm^{-1}$ and by Langer et al. (2002) in the spectral range 35,000 to 5,000 $cm^{-1}$. The spectra of $Mn^{3+}$ in zoisite closely resemble those of $Mn^{3+}$ in piemontite and like these they are characterized by a slightly polarized absorption edge in the UV and three intense and strongly polarized bands in the spectral range 30,000 to 10,000 $cm^{-1}$ labeled $\nu_I$, $\nu_{II}$, and $\nu_{III}$ (Fig. 7a,b). Like in piemontite, the strong polarization of $\nu_I$, $\nu_{II}$, and $\nu_{III}$ accounts for the trichroic scheme of $Mn^{3+}$ zoisite with **X** = light greenish, **Y** = lilac rose, and **Z** = light yellow (Langer et al. 2002). The close similarity

**Figure 7 (*on facing page*).** (A) Polarized room-temperature absorption spectra of natural $Mn^{3+}$ zoisite ("thulite"). Like in piemontite, the bands $\nu_I$, $\nu_{II}$, and $\nu_{III}$ arise from spin allowed *dd* transitions. (B) Polarized room-temperature absorption spectra of synthetic $Mn^{3+}$ zoisite ("thulite"). Beside the bands $\nu_I$, $\nu_{II}$, and $\nu_{III}$ (solid arrows) the stippled arrow denotes a band that arises from a spin forbidden *dd* transition of $Mn^{3+}$ in M3 (compare Fig. 2). (C) Polarized absorption spectra of untreated and heat treated natural $V^{3+}$ zoisite ("tanzanite"). On heating the pronounced band at ~22,500 $cm^{-1}$ in the Z-spectrum disappears. (D) Unpolarized absorption spectrum of an untreated natural $V^{3+}$ zoisite ("tanzanite"). The spectrum shows an additional band at ~27,000 $cm^{-1}$ and like the heat-treated specimen in (C) lacks the pronounced band at ~22,500 $cm^{-1}$. (E) Unpolarized absorption spectra of natural $Cr^{3+}$ zoisite. The two envelopes centered at ~21,800 $cm^{-1}$ and ~15,000 $cm^{-1}$ are due to spin allowed *dd* transitions of $Cr^{3+}$. The stippled arrows denote bands that arise probably from spin forbidden *dd* transitions of $Fe^{3+}$. Redrawn and modified after Faye and Nickel (1971), Schmetzer and Berdesinski (1978), and Langer et al. (2002).

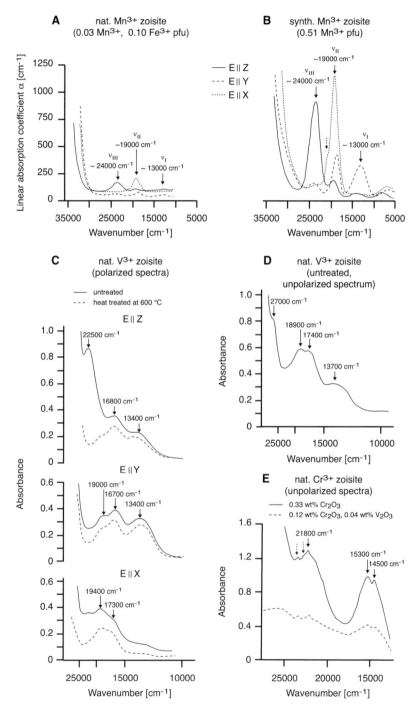

**Figure 7.** *caption on facing page*

between the spectra of Mn in zoisite and piemontite proves the trivalent state of Mn in zoisite (Kersten et al. 1988; Langer et al. 2002). Due to the different orientation of the indicatrix axes **X** and **Z** with respect to the M3 octahedron in $Mn^{3+}$ zoisite compared to piemontite (see Figs. 16 and 21 of Franz and Liebscher 2004), the polarization of the bands $v_I$ (Y-spectrum >> X- and Z-spectrum), $v_{II}$ (X-spectrum >> Y-spectrum > Z-spectrum), and $v_{III}$ (Z-spectrum >> X- and Y-spectrum) in $Mn^{3+}$ zoisite differ from that in piemontite but their relative intensities are identical to those found in piemontite (compare Fig. 7a,b with Figs 2a, 3). However, the positions of the absorption edge and the absorption maxima of the bands $v_I$, $v_{II}$, and $v_{III}$ are at considerable higher wavenumbers than expected from an extrapolation of the piemontite data with respect to $Mn^{3+}$ in M3 (Fig. 4a,b). In $Mn^{3+}$ zoisite $v_I$ occurs at ~13,000 $cm^{-1}$, $v_{II}$ at ~19,000 $cm^{-1}$, and $v_{III}$ at ~24,000 $cm^{-1}$ (Fig. 7a,b; Petrusenko et al. 1992; Langer et al. 2002). These higher wavenumbers in $Mn^{3+}$ zoisite are due to the notably smaller mean $Mn^{3+}$–O distance in $Mn^{3+}$ zoisite compared to piemontite (Langer et al. 2002). Comparing the spectra of synthetic $Mn^{3+}$ zoisite (Fig. 7b) with those of synthetic piemontite (Fig. 3) it is obvious that also the linear absorption coefficient $\alpha$ is distinctly higher in synthetic $Mn^{3+}$ zoisite than expected from an extrapolation of the piemontite data with respect to $Mn^{3+}$ in M3 (Fig. 4a). Langer et al. (2002) ascribe this difference to the different distortion of the M3 octahedron in $Mn^{3+}$ zoisite compared to piemontite. In both structure types the M3 octahedron has site symmetry $m$, but the mean quadratic elongation $\lambda_{oct}$ and the bond angle variance $\sigma_{oct}^2$ (Robinson et al. 1971) significantly differ. Based on their derived structural data, Langer et al. (2002) calculated $\lambda_{oct} = 1.0211$ and $\sigma_{oct}^2 = 29.1$ for $Mn^{3+}$ zoisite and $\lambda_{oct} = 1.0349$ and $\sigma_{oct}^2 = 83.7$ for piemontite. From the positions of the absorption maxima of the bands $v_I$, $v_{II}$, and $v_{III}$, Langer et al. (2002) calculated a crystal field splitting $\Delta_{oct}$ for $Mn^{3+}$ in M3 in the studied synthetic $Mn^{3+}$ zoisite of 13,780 $cm^{-1}$ corresponding to a CFSE of 14,790 $cm^{-1}$ or 176.9 kJ/g-atom (Fig. 4c). These values are slightly higher than those expected from the extrapolation of the piemontite data (Fig. 4c).

Like the Z-spectrum in synthetic piemontite (Fig. 2a), the X-spectrum of synthetic $Mn^{3+}$ zoisite displays a weak but distinct shoulder at ~20,000 $cm^{-1}$ on the high wavenumber side of the $v_{II}$ band (Fig. 7b). As $Mn^{3+}$ is confined exclusively to M3 in zoisite, this band must arise from a spin forbidden *dd* transition in $Mn^{3+}$ in M3 (Langer et al. 2002). This spin forbidden *dd* transition is also found in the spectra of a natural $Mn^{3+}$ zoisite reported by Petrusenko et al. (1992). These latter spectra also show an additional weak but distinct band at ~22,000 $cm^{-1}$. Like in the natural piemontite studied by Smith et al. (1982) this band may be assigned to the spin forbidden $^6A_{1g} \rightarrow {}^4A_{1g}, {}^4E_g \, (^4G)$ transition in $Fe^{3+}$.

***Spectra of V bearing zoisite ("tanzanite").*** Optical absorption studies of V bearing zoisite (variety "tanzanite") aimed primarily at elucidating the valence state of the V ion in octahedral coordination and the peculiar color change observed in V zoisite upon heating. Polarized and unpolarized spectra of natural V bearing zoisite specimens are reported by Tsang and Ghose (1971) and Faye and Nickel (1971). The latter studied a tan and a blue colored (in unpolarized white light) specimen. The tan colored specimen contains 0.24 to 0.30% V and ~0.04% Ti, lacks Fe and Mn, and is trichroic in polarized light (before heat treatment), whereas the blue colored specimen contains ~0.19% V, lacks Ti, Fe, and Mn, and is dichroic in polarized light. In case of the tan colored specimen, Faye and Nickel (1971) recorded polarized absorption spectra between 25,000 and 10,000 $cm^{-1}$ before and after heat treatment at 600°C. The polarized spectra of the untreated tan colored specimen are characterized by two broad multi-component envelopes centered at ~13,400 $cm^{-1}$ (Y- and Z-spectrum) and ~18,000 $cm^{-1}$ (X-, Y-, and Z-spectrum), and a strongly polarized band at ~22,500 $cm^{-1}$ (Z-spectrum) (Fig. 7c). These absorption characteristics in the VIS region account for the trichroic scheme observed in the untreated tan colored specimen with **X** = red-violet, **Y** = blue, and **Z** = yellow-green. The X- and Y-spectrum of the heat-treated specimen closely resemble those of the untreated specimen whereas the Z-spectrum of the heat-treated specimen lacks the pronounced band at ~22,500 $cm^{-1}$ (Fig. 7c). As the Z-spectrum of the heat-treated specimen is essentially identical to the Y-spectrum, this accounts for the change in the pleochroic scheme from trichroic in the untreated specimen to dichroic in the heat-treated specimen with $X$ = red-violet, $Y = Z$ = blue (Faye and Nickel 1971). The unpolarized spectra of both the tan colored (not shown in Fig. 7c) and blue colored (Fig. 7d) specimens extend into the UV region and display a fifth band at $\sim$27,000 cm$^{-1}$ (Faye and Nickel 1971). Beside this band at $\sim$27,000 cm$^{-1}$, the unpolarized spectrum of the blue colored specimen shows two broad multi-component envelopes at $\sim$13,700 and 18,000 cm$^{-1}$ in good agreement with the polarized spectra of the heat-treated specimen (Fig. 7c,d).

Except the band at $\sim$22,500 cm$^{-1}$, Faye and Nickel (1971) attribute the observed bands to spin allowed $dd$ transitions in V$^{3+}$ and assign the broad multi component envelopes centered at $\sim$13,500 and $\sim$18,000 cm$^{-1}$ to the $^3T_{1g} \rightarrow {}^3T_{2g}$ ($^3F$) and $^3T_{1g} \rightarrow {}^3T_{1g}$ ($^3P$) transitions and the band at 27,000 cm$^{-1}$ to the $^3T_{1g} \rightarrow {}^3A_{2g}$ ($^3F$) transition. The band at $\sim$22,500 cm$^{-1}$ is the only one that is affected by heat treatment and it is therefore unlikely that it comprises a part of the $dd$ spectrum of V$^{3+}$. As V and Ti are the only transition metal ions present in the tan colored specimen, Faye and Nickel (1971) assign this band to the $^2T_{2g} \rightarrow {}^2E_g$ transition in either V$^{4+}$ or Ti$^{3+}$, however favor Ti$^{3+}$ because of its charge and because the band at $\sim$22,500 cm$^{-1}$ is only found in the Ti bearing specimen. A Ti$^{3+} \rightarrow$ Ti$^{4+}$ charge transfer that is proposed responsible for the pleochroism of Ti bearing andalusite by Faye and Harris (1969) can be ruled out in case of zoisite as the polarization of the band at $\sim$22,500 cm$^{-1}$ can neither be related to the M1,2–M3 nor the M1,2–M1,2 vector between adjacent octahedral sites. The disappearance of the band at $\sim$22,500 cm$^{-1}$ on heating suggests the oxidation of Ti$^{3+}$ to Ti$^{4+}$ or of V$^{4+}$ to V$^{5+}$. As both Ti$^{4+}$ and V$^{5+}$ have a d$^0$ configuration, they do not give rise to any $dd$ absorption spectrum. However, because the band at $\sim$22,500 cm$^{-1}$ disappears on heating in air as well as in hydrogen the proposed oxidation of Ti$^{3+}$ or V$^{4+}$ would occur in an oxidizing as well as in a reducing environment and Faye and Nickel (1971) state that "it is difficult to rationalize a mechanism that accounts for this proposal." Based on their data, Faye and Nickel (1971) conclude that V$^{3+}$ is strongly ordered in the M3 site whereas Ti$^{3+}$ is probably ordered in M1,2. Contrary, Tsang and Ghose (1971) conclude that V$^{3+}$ is approximately randomly distributed between M1,2 and M3.

***Spectra of Cr$^{3+}$ bearing zoisite***. Studies of Cr$^{3+}$ in zoisite are restricted to the remission and absorption spectra of natural Cr$^{3+}$ bearing zoisite recorded by Schmetzer and Berdesinski (1978). The unpolarized absorption spectra of two polycrystalline samples are characterized by two broad bands centered at $\sim$15,100 cm$^{-1}$ ($v_I$) and $\sim$21,800 cm$^{-1}$ ($v_{II}$) (Fig. 7e) resembling the principal features of the tawmawite spectra (see Fig. 6). Additional weak bands or shoulders can be identified on the high wavenumber side of $v_{II}$ at $\sim$23,400 and 22,100 cm$^{-1}$ whereas $v_I$ is clearly split into two components at 15,300 and 14,500 cm$^{-1}$ (Fig. 7e). Schmetzer and Berdesinski (1978) assigned the bands $v_I$ and $v_{II}$ to the spin allowed $^4A_{2g} \rightarrow {}^4T_{2g}$ and $^4A_{2g} \rightarrow {}^4T_{1g}$ transitions in Cr$^{3+}$. The weak features at $\sim$23,400 and 22,100 cm$^{-1}$ are most probably due to spin forbidden $dd$ transitions in Fe$^{3+}$ whereas the fine structure of the band $v_I$ is either due to a splitting of the $^4A_{2g} \rightarrow {}^4T_{2g}$ transition or to an overlap with the spin forbidden $^4A_{2g} \rightarrow {}^2T_{1g}, {}^2E_g$ transition in Cr$^{3+}$ (Schmetzer and Berdesinski 1978). From the center of the band $v_I$ Schmetzer and Berdisinski (1978) calculated a crystal field splitting $\Delta_{oct}$ of 15,100 cm$^{-1}$ for Cr$^{3+}$ in zoisite, which is lower than that derived for Cr$^{3+}$ in tawmawite ($\sim$16,300 cm$^{-1}$, see above; Burns and Strens 1967). The low $\Delta_{oct}$ suggests a strong preference of Cr$^{3+}$ for the M3 compared to the M1,2 site in zoisite as the smaller mean O–M distance in M1,2 would result in a higher $\Delta_{oct}$ (Schmetzer and Berdisinski 1978).

## INFRARED SPECTROSCOPY

Moenke (1962) published a first infrared (IR) spectrum of an epidote. Polarized single crystal IR spectroscopy focused on the determination of the proton location within the monoclinic and orthorhombic crystal structures (Hanisch and Zemann 1966; Linke 1970).

The first systematic study on the spectral properties in the spectral range 5,000 to 250 cm$^{-1}$ of the monoclinic and orthorhombic Al-Fe$^{3+}$ solid solution series as function of the Fe$^{3+}$ content stems from Langer and Raith (1974). The spectrum of a deuterated zoisite was presented by Langer and Lattard (1980), and Winkler et al. (1989) studied the influence of pressure on the OH environment in zoisite. Janeczek and Sachanbinski (1989) report the IR spectrum of an Mn bearing clinozoisite. Later, Gavorkyan (1990) and Petrusenko et al. (1992) studied a suite of natural epidote minerals including zoisite, clinozoisite, Mn bearing clinozoisite, epidote, and allanite by IR spectroscopy and published the first IR spectrum of a deuterated clinozoisite. The structural changes of the OH environment that accompany the incorporation of Sr in the A sites and of Mn$^{3+}$ in the M3 site in the monoclinic solid solution series has then been studied by Della Ventura et al. (1996). The systematic work of Langer and Raith (1974) was extended by Heuss-Aßbichler (2000) for the monoclinic series and by Brunsmann (2000) and Liebscher et al. (2002) for the orthorhombic series. Bradbury and Williams (2003) used pressure dependent infrared spectroscopy to study the response of the monoclinic structure to changes in pressure with special emphasis given to the bonding environment of the hydrogen. Recently, Liebscher and Gottschalk (2004) studied the temperature evolution of the IR spectra of zoisite. IR spectra of allanite were recorded by Cech et al. (1972), Mingshen and Daoyuan (1986), and Mingshen and Dien (1987). Cech et al. (1972) compared the spectrum of allanite with that of an epidote whereas Mingshen and Daoyuan (1986) and Mingshen and Dien (1987) studied the evolution of the allanite IR spectrum with the degree of metamictization.

## IR spectroscopy of monoclinic epidote minerals

All published IR spectra of the monoclinic epidote minerals are consistent with respect to their general features. They are characterized by the main OH stretching vibration between about 3,450 and 3,300 cm$^{-1}$ (depending on composition) and the lattice vibrations below about 1,200 cm$^{-1}$. Langer and Raith (1974) identified a total of 34 bands in the studied spectral range of 5,000 to 200 cm$^{-1}$ and numbered them consecutively from high to low wavenumber. This numbering is also adopted here. Band 1 corresponds to the main OH stretching vibration whereas the bands 2 to 34 correspond to vibrations below 1,200 cm$^{-1}$. It should be noted that Le Cleac'h et al. (1988) observed the lowest IR active mode in clinozoisite at 80 cm$^{-1}$.

*Lattice vibrational range.* Typical room temperature powder (KBr pellets) IR spectra below 1,200 cm$^{-1}$ of natural and synthetic Al-Fe$^{3+}$ solid solutions with Fe$^{3+}$ contents between $X_{Ep} = 0.00$ and $X_{Ep} = 0.98$ from Langer and Raith (1974) and Heuss-Aßbichler (2000) are shown in Figure 8. They are characterized by strong absorption (Fig. 8a) and low transmission (Fig. 8b), respectively, between about 1,200 and 820 cm$^{-1}$ (bands 2 to 12) and below about 700 cm$^{-1}$ (bands 14 to 34; Fig. 8). Between about 820 and 700 cm$^{-1}$ there is only one very weak band (band 13) present in the monoclinic epidote minerals independent on Fe content. This is in contrast to the orthorhombic forms, which show two sharp bands of medium intensity in this region (see below). In the 1,200 to 820 cm$^{-1}$ range the bands 5, 7, and 9 occur as single marked bands throughout the entire compositional range whereas bands 6, 8, 10, 11, and 12 only show up as weak to very weak shoulders of the main bands (Fig. 8). Bands 2, 3, and 4 partly strongly overlap and display noticeable changes with increasing Fe content. Band 2 can be traced throughout the entire compositional range whereas the bands 3 and 4 significantly change their relative intensities. From these two bands only band 3 can be resolved as single band in Fe free clinozoisite, band 4 is either absent or its intensity is to low to be resolved (Fig. 8b, lowermost spectrum). With increasing Fe content the intensity of band 4 increases whereas that of band 3 decreases. Consequently, in medium Fe clinozoisite–epidote both bands strongly overlap, display a comparable intensity, and occur as shoulders of band 2 (spectra with $X_{Ep} = 0.39$, 0.44, and 0.48 in Fig. 8). In Fe rich epidote the band 4 shows up as a single band whereas band 3 cannot be resolved with confidence due to the strong overlap of bands 2 and 4 (spectra with $X_{Ep} > 0.72$ in Fig. 8) and might even be absent in the most Fe rich samples.

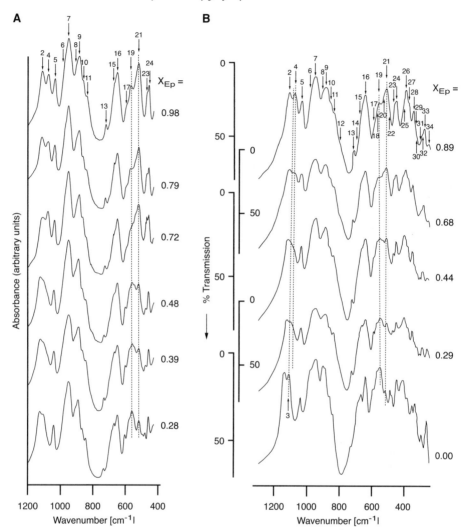

**Figure 8.** Typical IR spectra of monoclinic Al-Fe$^{3+}$ solid solutions with different iron contents from (A) Heuss-Aßbichler (2000) and (B) Langer and Raith (1974). Note that the spectra in (A) are displayed in the form of absorbance whereas those in (B) are displayed in the form of transmission. To allow for easier visual comparison between (A) and (B) the spectra in (B) are shown in an unconventional orientation with the transmission increasing downwards. Dotted lines correspond to IR bands that significantly change with increasing iron content and that are discussed in detail in the text. Redrawn and modified after Langer and Raith (1974) and Heuss-Aßbichler (2000).

Below 700 cm$^{-1}$ the spectra of monoclinic Al-Fe$^{3+}$ solid solutions remarkably change with increasing Fe content and only the bands 13, 16, 26, and 28 can be confidentially resolved as single bands throughout the entire compositional range (Fig. 8). Like the bands 3 and 4 some of the bands below 700 cm$^{-1}$ significantly change their relative intensities. Band 21 occurs only as a shoulder of the much more intense bands 19 and 20 at low Fe contents but is one of the most prominent bands below 700 cm$^{-1}$ in Fe rich epidote. In these Fe rich epidotes bands

19 and 20 only show up as shoulders of band 21 (Fig. 8). Band 22 is a single sharp band of medium intensity in Fe free clinozoisite (lowermost spectra of Fig. 8b) but occurs as a weak shoulder of band 21 at high Fe content. Likewise, band 23 is a weak shoulder of band 22 in Fe free clinozoisite, a single sharp band of low intensity in medium Fe clinozoisite–epidote, and occurs as a shoulder of band 24 in Fe rich epidote (Fig. 8). Band 29 can be resolved as a single band in Fe free clinozoisite with an intensity slightly higher than that of band 28 (lowermost spectra of Fig. 8b). With increasing Fe content the intensity of band 29 decreases relative to that of band 28 and already at an Fe content of $X_{Ep} = 0.29$ band 29 occurs only as a shoulder of band 28. The bands 32, 33, and 34 display a similar change in their relative intensities. In Fe free clinozoisite band 34 is very sharp and pronounced compared to the low intensity maximum which is formed either by band 32 or 33 (Langer and Raith 1974 were not able to unequivocally assign this low intensity maximum to one of these two bands). With increasing Fe content the intensity of band 34 significantly decreases relative to the intensity of bands 32 and 33. In Fe rich epidote the bands 32 and 33 can be resolved as individual bands of medium intensity whereas band 34 only occurs as a band of low intensity (uppermost spectra in Fig. 8b).

Those bands that can be confidentially resolved as single bands throughout the entire compositional range (bands 5, 7, 9, 13, 16, 26, and 28) either shift to lower wavenumber with increasing Fe content (bands 5, 9, 13, 26, and 28) or display a constant position over the entire compositional range (bands 7 and 16; Fig. 9a; see also Fig. 4 of Langer and Raith 1974). The shift to lower wavenumber of the bands 5, 9, 13, 26, and 28 can well be used to estimate the Fe content of monoclinic Al-Fe$^{3+}$ solid solutions from IR spectra. Linear regressions relating band position to Fe content show two groups of bands with different slopes (Fig. 9a; see also Table 5 of Langer and Raith 1974). The bands 5 and 9 (Fig. 9a) and 28 ($\nu = 364.2 - 13.2X_{Ep}$; Langer and Raith 1974) display negative slopes of $-12.5$ to $-13.2$ whereas the bands 13 (Fig. 9a) and 26 ($\nu = 419.4 - 27.2X_{Ep}$; Langer and Raith 1974) have steeper negative slopes of $-23.8$ to $-27.2$. According to Langer and Raith (1974) the determination of the Fe content using these linear correlations is accurate to about $\pm 0.04\ X_{Ep}$ for bands 13 and 26 and about $\pm 0.08\ X_{Ep}$ for bands 5, 9, and 28.

Bradbury and Williams (2003) studied the room temperature evolution of the IR spectrum of a clinozoisite with $X_{Ep} = 0.46$ with pressure up to 36 GPa. Within the 1,250 to 550 cm$^{-1}$ spectral region they tracked the evolution of the bands 2, 5, 7, 9, 10, 11, 15, and 16 with pressure (only bands 2, 5, 7, 9, 15, and 16 are shown in Fig. 9b). All bands shift to higher wavenumber with compression in accordance with the anticipated response of vibrations to simple bond compaction. This shift was found to be continuous up to 16 GPa and fully reversible on decompression, indicating that no phase change occurred up to this pressure (Bradbury and Williams 2003). Except band 15 which shows a notably small pressure shift of only 0.6 cm$^{-1}$/GPa, the rate at which the bands shift upon compression ranges from 3.1 cm$^{-1}$/GPa (bands 7, 9, and 11) to 4.5 cm$^{-1}$/GPa (bands 2 and 16) and is in general accord with the pressure shift of stretching vibrations in other ortho- and sorosilicates (Bradbury and Williams 2003). From the band positions ($\omega_0$) at ambient pressure, the shift of the bands with pressure ($d\omega/dP$), and the bulk moduli ($K_0$) for clinozoisite from Holland et al. (1996; $K_0 = 154$ GPa) and Comodi and Zanazzi (1997; $K_0 = 127$ GPa) Bradbury and Williams (2003) calculated the Grüneisen parameter $\gamma = (K_0/\omega_0)(d\omega/dP)$ for each band. The small pressure shift of band 15 corresponds to a small Grüneisen parameter of only 0.12 (for $K_0 = 127$ GPa) and 0.14 (for $K_0 = 154$ GPa). Depending on the bulk modulus chosen, the other bands have Grüneisen parameters between 0.41 (band 7, calculated for $K_0 = 127$ GPa) and 1.04 (band 16, calculated for $K_0 = 154$ GPa). The average mode Grüneisen parameter is 0.48 for $K_0 = 127$ GPa and 0.58 for $K_0 = 154$ GPa. For comparison, Bradbury and Williams (2003) calculated a bulk thermodynamic Grüneisen parameter from $\gamma = \alpha K_0/\rho C_P$ with the data for $\alpha$ and $C_P$ from Pawley et al. (1996) and Kiseleva and Ogorodnikova (1987), respectively. With $\gamma = 0.75$ for $K_0 = 127$ GPa and $\gamma = 0.91$ for $K_0 = 154$ GPa, the bulk thermodynamic Grüneisen parameter is slightly higher than the average

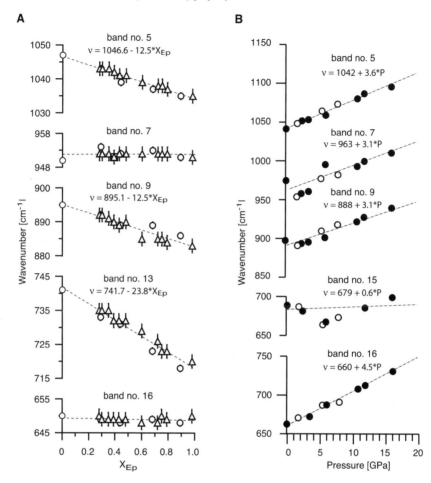

**Figure 9.** Compositional (A) and pressure (B) shifts of some IR bands in the lattice vibrational range of monoclinic Al-Fe$^{3+}$ solid solutions. (A) The bands either shift to lower wavenumber with increasing iron content or display a constant position over the entire compositional range. The equations that relate the band positions to the iron content represent linear fits to the data points. Data sources for (A) are: O = Langer and Raith (1974), △ = Heuss-Aßbichler (2000). (B) Except for band 15, which shows a comparable small pressure shift, the bands shift to higher wavenumber with pressure with a slope of about 3.1 to 4.5 cm$^{-1}$/GPa. The equations are linear fits to the data. In (B), ● represent data that were determined during compression, O are data that were determined during decompression (redrawn and modified after Bradbury and Williams 2003).

mode Grüneisen parameter derived from the mid infrared modes. This indicates that bands that were not observed by Bradbury and Williams (2003) either at lower frequencies, in the Raman spectrum, or off-Brillouin zone center modes must have larger average Grüneisen parameters than the observed ones (Bradbury and Williams 2003).

**Band assignment.** Up to now, no rigorous band assignment has been performed for the lattice vibrational range of the spectra of monoclinic epidote minerals. Langer and Raith (1974) assigned the bands 2, 3, 5, 7, 9, 11, and 16 to Si–O valence vibrations of the single SiO$_4$ tetrahedron or the Si$_2$O$_7$ group. Bradbury and Williams (2003) adopted these assignments except for band 16 and assigned additionally band 10 to a Si–O valence vibration. Contrary to Langer and Raith (1974), Bradbury and Williams (2003) assigned band 16 to either an Al–O stretching or a Si–O bending vibration. The spectral evolution of the bands 3 and 4 (see above) sheds additional light on the interpretation of these two bands. It is reasonable to assume that not only band 3 but also band 4 is due to a Si–O valence vibration. As band 4 is missing in the spectrum of Fe free clinozoisite and its intensity increases with increasing Fe content whereas that of band 3 decreases, both bands should arise from a Si–O bond that is influenced by the substitution of $Fe^{3+}$ in a neighboring octahedron. The most likely candidate is the Si2–O8–M3 bridge with bands 3 and 4 representing the Si2–O8 valence vibration for Al and $Fe^{3+}$ in M3, respectively. This interpretation was already discussed by Heuss-Aßbichler (2000).

By analogy with some layer silicates, band 8 was interpreted as due to an OH bending vibration by Langer and Raith (1974). Because of its small pressure shift, Bradbury and Williams (2003) assigned band 15 to an OH bending vibration. Gavorkyan (1990) and Petrusenko et al. (1992) observed additional bands at 780 and 665 $cm^{-1}$ in synthetic fully deuterated clinozoisite that are missing in the spectra of undeuterated clinozoisite and therefore arise from OD bending vibrations. Taking a $v_{OH}/v_{OD}$ factor of 1.34 to account for the isotope effect, Petrusenko et al. (1992) calculated the corresponding OH bending vibrations in the undeuterated form to occur at about 1045 and 890 $cm^{-1}$. These values closely correspond to the positions of the bands 5 and either band 8 or 9. The interpretation of band 5 as arising from an OH bending vibration is in contradiction to the interpretation of this band as due to a Si–O valence vibration by Langer and Raith (1974) and Bradbury and Williams (2003). In the spectrum of the fully deuterated clinozoisite (Gavorkyan 1990; Petrusenko et al. 1992) band 5 still appears but with a significantly smaller intensity as in the spectra of the undeuterated forms. Band 5 therefore most probably originates from two strongly overlapping bands that are due to an OH bending and a Si–O valence vibration of almost identical energy. The occurrence of an OH bending vibration at about 890 $cm^{-1}$ agrees well with the interpretation of Langer and Raith (1974), although it is not clear from the data, which of the two bands 8 and 9 actually correspond to the second OH bending vibration. The interpretation of band 15 as due to an OH bending vibration by Bradbury and Williams (2003) could not be verified by the work of Gavorkyan (1990) and Petrusenko et al. (1992).

As the spectra of monoclinic and orthorhombic epidote minerals are completely different between 820 to 700 $cm^{-1}$ and 540 to 320 $cm^{-1}$ (see below), bands in these spectral regions most probably originate from M–O vibrations within the different octahedra of the two structure types (Langer and Raith 1974). Only band 34 of the monoclinic epidote minerals has an equivalent in the orthorhombic structure (band 48, see below) and was thus assigned to a Ca–O vibration by Langer and Raith (1974).

***OH stretching vibrational region.*** In the monoclinic Al-$Fe^{3+}$ solid solutions the OH stretching vibration occurs between 3,330 and 3,380 $cm^{-1}$ (Figs 10a,b and 11a; Langer and Raith 1974; Della Ventura et al. 1996; Heuss-Aßbichler 2000) and is characterized by the single sharp band 1 (nomenclature of Langer and Raith 1974). Hanisch and Zemann (1966) measured the pleochroic scheme of the OH stretching vibration in epidote by means of polarized single crystal IR spectroscopy and observed the maximum absorption parallel [001] and minor absorption parallel [100] and [010] with the absorption parallel [100] being slightly larger than parallel [010]. They described the three dimensional absorption figure of the OH stretching vibration as a lemniscate that rotates around [001] and confirmed the bonding of the proton to O10 with a hydrogen bridge to O4 as already proposed by earlier structure refinement studies (e.g., Ito et al. 1954; Belov and Rumanova 1954). The data from Langer and Raith (1974) and Heuss-Aßbichler (2000) show that the OH stretching vibration of Al-$Fe^{3+}$ solid solutions correlates linearly with the Fe content according to $v = 3{,}327.1 + 49.4\,X_{Ep}$ (Fig. 11a). The determination of the Fe content in Al-$Fe^{3+}$ solid solutions with this band is accurate to about $\pm\,0.04\,X_{Ep}$ (Langer and Raith 1974).

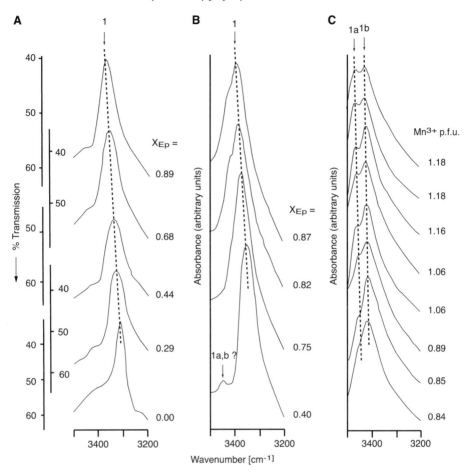

**Figure 10.** Typical IR spectra of monoclinic Al-Fe$^{3+}$ solid solutions (A), (B) and of piemontites (C) in the region of the OH stretching vibration. Note that the spectra are displayed in the form of transmission in (A) and in the form of absorbance in (B) and (C). Like in Figure 8, the spectra in (A) are shown in an unconventional orientation with the transmission increasing downwards. In the monoclinic Al-Fe$^{3+}$ solid solutions the OH stretching vibration is characterized by a single sharp maximum between about 3380 and 3330 cm$^{-1}$. The spectra of piemontite (C) show a splitting of the main OH stretching vibration into two components 1a and 1b and a significant shift to higher wavenumber (about 3460 to 3410 cm$^{-1}$) compared to the Al-Fe$^{3+}$ solid solutions. Redrawn and modified after Langer and Raith (1974) and Della Ventura et al. (1996).

In natural Sr bearing piemontite samples with 0.836 to 1.197 Mn$^{3+}$ pfu, 0.056 to 0.379 Fe$^{3+}$ pfu, and 0.09 to 0.291 Sr pfu, Della Ventura et al. (1996) found a shift of the OH stretching vibration to significantly higher wavenumbers compared to the Al-Fe$^{3+}$ solid solution series as well as a splitting of the main band into two components. Such a splitting was also observed by Langer et al. (1976) in synthetic piemontite with more than about 1.0 Mn$^{3+}$ pfu. To be in line with the numbering of Langer and Raith (1974) these two components are labeled as bands 1a and 1b (Fig. 10c). Independent on composition the band separation between bands 1a and 1b is 35 to 40 cm$^{-1}$ (Fig. 11a). Band 1b is the main component and appears as a sharp maximum in all spectra. It occurs between 3,412 and 3,423 cm$^{-1}$. Band 1a is the higher energy component

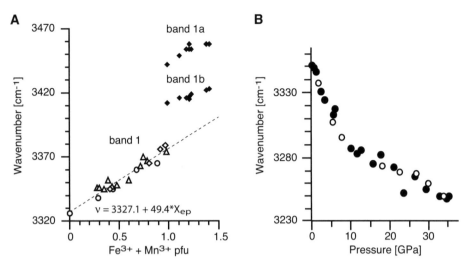

**Figure 11.** Compositional (A) and pressure (B) shifts of the OH stretching vibrations of monoclinic epidote minerals. (A) Band 1 of the monoclinic Al-Fe$^{3+}$ solid solutions (open symbols) shifts linearly to higher wavenumber with increasing iron content. The stippled line represents a linear fit to the data. In piemontite (filled symbols) band 1 is splitted into the two bands 1a and 1b that also shift to higher wavenumber with increasing iron content although with a slightly smaller compositional dependence. The band separation between bands 1a and 1b is about 35 to 40 cm$^{-1}$ in all samples. Circles are from Langer and Raith (1974), triangles from Heuss-Aßbichler (2000), and diamonds from Della Ventura et al. (1996). (B) Band 1 of monoclinic Al-Fe$^{3+}$ solid solutions generally shifts with pressure but shows a break in slope at about 15 GPa. Below 15 GPa the shift is $-5.1$ cm$^{-1}$/GPa, above 15 GPa the average shift is only $-1.5$ cm$^{-1}$/GPa. Filled circles represent data that were determined during compression, open circles data that were determined during decompression (data are for a clinozoisite with $X_{Ep} = 0.46$; redrawn after Bradbury and Williams 2003).

and occurs between 3,442 and 3,458 cm$^{-1}$ (Figs 10c, 11a). With increasing Mn$^{3+}$ content its intensity increases relative to that of band 1b and consequently it occurs as a shoulder at the high energy side of band 1b at low Mn$^{3+}$ content (Mn$^{3+}$ + Fe$^{3+}$ = 0.99 to 1.25 pfu; lower four spectra in Fig. 10c) but forms a medium to sharp maximum at higher Mn$^{3+}$ + Fe$^{3+}$ content (Mn$^{3+}$ + Fe$^{3+}$ = 1.21 to 1.41 pfu; upper four spectra in Fig. 10c). Both bands display a positive linear shift with the Mn$^{3+}$ + Fe$^{3+}$ content of almost identical slope (Figs 10c, 11a).

Assuming only Al in the M2 site, Della Ventura et al. (1996) interpreted the general shift of the OH stretching vibration to higher wavenumber in Sr bearing piemontite as being due to the substitution of Sr for Ca in the A2 site. According to the structural data of Dollase (1969), Catti et al. (1989), Ferraris et al. (1989), and Bonazzi et al. (1990) the incorporation of Sr has a major effect on the expansion of the A2 polyhedron and especially the A2–O10 distance increases linearly with the Sr content at least up to 0.5 Sr pfu in A2 (see Fig. 6 of Della Ventura et al. 1996). The increasing A2–O10 distance leads to a decrease of the incident bond strength at O10 and an increase of the O10–H bond strength and consequently shifts the OH stretching vibration to higher wavenumber (Della Ventura et al. 1996). This interpretation is in accord with the results of Perseil (1987) who found a shift of the OH stretching vibration to higher wavenumber with increasing Sr content in piemontite samples with an identical Mn$^{3+}$ + Fe$^{3+}$ content. The observed splitting of the OH stretching vibration is due to the Jahn-Teller effect that the Mn$^{3+}$ cation exerts on the M3 octahedron (Della Ventura et al. 1996). The substitution of Mn$^{3+}$ for Al in M3 flattens the M3 octahedron along the O4–O8 direction and shortens the M3–O4 bond length compared to the Al-Fe$^{3+}$ solid solution series (see Fig. 5 of Della Ventura et al. 1996). As a consequence the length of the H...O4 hydrogen bridge increases and the stretching vibration is shifted to a higher wavenumber. Della Ventura et al. (1996) therefore assigned band 1a to the configuration $^{[M2]}Al_2$–O10–H...O4–$^{[M1]}(Al,Fe^{3+})_2$$^{[M3]}Mn^{3+}$. The band separation of 35 to 40 cm$^{-1}$ between the bands 1a and 1b corresponds to a difference in the length of the underlying hydrogen bridge of about 0.02 to 0.03 Å (Bellamy and Owen 1969). This is consistent with the structural data for piemontite that indicate an about 0.04 Å shorter M3–O4 bond length compared to Al-Fe$^{3+}$ solid solutions of comparable substitutional degree (see Fig. 5 of Della Ventura et al. 1996). Based on the spectral properties of the Al-Fe$^{3+}$ solid solutions and piemontite, respectively, Della Ventura et al. (1996) speculated about two different types of behavior of the OH stretching vibration in monoclinic epidote minerals: a continuous shift of a single band in Al-Fe$^{3+}$ solid solutions as a function of the Fe content (their so-called "one-mode behavior") and a splitting of the OH stretching vibration into two separate bands in piemontite as a function of the (Al,Fe$^{3+}$)–Mn$^{3+}$ distribution (their so-called "two-mode behavior"). However, the authors did not exclude a second explanation, i.e., that the OH stretching vibration is also split into two separate bands in the Al-Fe$^{3+}$ solid solutions but that the band separation due to the different octahedral configurations is too small to be resolved in their samples and by their methods, respectively. By analogy with the results of IR studies on the OH stretching vibration in orthorhombic Al-Fe$^{3+}$ solid solutions that show two bands with a small band separation due to the different octahedral configurations (see below), this second explanation seems to be much more likely.

At room temperature the OH stretching vibration shifts to lower wavenumber with increasing pressure whereas the peak markedly broadens (Fig. 11b; Bradbury and Williams 2003 for a clinozoisite with $X_{Ep}$ = 0.46). The shift is not linear but shows a break in slope at about 15 GPa: below about 15 GPa it is −5.1 cm$^{-1}$/GPa whereas the average shift above 15 GPa is only −1.5 cm$^{-1}$/GPa. For $P$ < 15 GPa the corresponding Grüneisen parameter is $\gamma$ = −0.19 and −0.23 for a bulk modulus of $K_0$ = 127 and 154 GPa, respectively (for the calculation of $\gamma$ see above). The shift of the stretching vibration as well as the broadening of the peak are significantly less than those observed by Winkler et al. (1989) for the OH stretching vibration in zoisite (see below). This difference probably reflects the weaker, longer, and more bent hydrogen bond in clinozoisite compared to zoisite, which may be therefore less sensitive to pressure (Bradbury and Williams 2003). But compared to other hydrous phases that have comparable ambient pressure hydrogen bond strengths (e.g., lawsonite, chondrodite, and hydrated β-phase; Williams 1992; Cynn and Hofmeister 1994; Scott and Williams 1999) and compared to its own Si–O stretching vibrations, the broadening of the OH stretching vibration in clinozoisite is markedly larger (Bradbury and Williams 2003).

To explain the shift and the broadening of the OH stretching vibration in clinozoisite with pressure, Bradbury and Williams (2003) modeled its hydrogen bond potential (Fig. 12). At ambient pressure this model describes a double potential well in which the hydrogen atom is preferentially bound to O10 (Fig. 12). With increasing pressure the minimum of the deeper well moves away from O10 and suggests an increasing O10–H bond length with a concomitant decrease in energy, which is in accord with the observed shift of band 1 to lower wavenumber with pressure. Calculating the first excited state of the hydrogen atom as 3/2 the energy of the OH stretching vibration above the well energy minimum (Herzberg 1950; Bradbury and Williams 2003), it is also evident from Figure 12 that the potential well significantly broadens with pressure. This broadening increases the amount of thermal disorder of the hydrogen atom along the axis of the hydrogen bond and is responsible for the broadening of the OH stretch peak in the IR spectra (Bradbury and Williams 2003). Based on the modeled hydrogen bond potential the hydrogen bond in clinozoisite might become symmetric at pressures above 50 GPa.

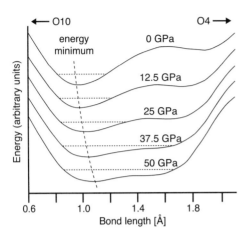

**Figure 12.** Modeled hydrogen potential for the O10–H...O4 hydrogen bond in monoclinic Al-Fe$^{3+}$ solid solution for different pressures. At low pressure the potential is a double well with one pronounced energy minimum and the proton is preferentially bound to O10. With increasing pressure this minimum moves away from O10 and suggests an increasing O10–H bond length (stippled line). Calculating the first excited state of the proton as 3/2 the energy of the OH stretching vibration above the well energy minimum it is evident that the potential well also significantly broadens with pressure. Redrawn and modified after Bradbury and Williams (2003).

### IR spectroscopy of zoisite

Moenke (1962) published a zoisite IR spectrum between 4,000 to 400 cm$^{-1}$. Later, Linke (1970) studied the proton environment in zoisite by means of polarized single crystal IR spectroscopy and Langer and Raith (1974) investigated systematically the powder IR spectral properties of zoisite and their dependence on composition. They identified a total of 36 bands in the spectral range 5,000 to 200 cm$^{-1}$. In detailed temperature dependent ($T = -170$ to 250°C) IR spectroscopic studies in the range 5,000 to 350 cm$^{-1}$, Brunsmann (2000), Liebscher et al. (2002), and Liebscher and Gottschalk (2004) were able to identify 12 additional bands resulting in a total of 48 bands in the spectral range 5,000 to 200 cm$^{-1}$ (Fig. 13). To be in line with the numbering of the bands in the monoclinic epidote minerals and with the numbering introduced by Langer and Raith (1974), these bands will be consecutively numbered from high to low wavenumber. Like the IR spectra of the monoclinic epidote minerals all published IR spectra of different varieties of zoisite resemble each other in their principal features. The IR spectrum is characterized by the OH stretching vibration at 3,300 to 3,000 cm$^{-1}$, a low intensity band at 2,300 to 2,100 cm$^{-1}$ which is most probably related to an OH bending vibration (see below), and the lattice vibrations below 1,200 cm$^{-1}$.

*Lattice vibrational range.* Typical powder IR spectra in the lattice vibrational range below 1,200 cm$^{-1}$ from Langer and Raith (1974) and Liebscher and Gottschalk (2004) are shown in Figure 13. The spectra from Liebscher and Gottschalk (2004) were recorded at −150°C (Fig. 13a) whereas those from Langer and Raith (1974) were recorded at room temperature conditions (Fig. 13b). All spectra are characterized by strong absorption and low transmission, respectively, between about 1,180 and 1,080 cm$^{-1}$ (bands 11 to 13), 1,000 and 800 cm$^{-1}$ (bands 14 to 23), 800 to 720 cm$^{-1}$ (bands 24 and 25), and below 720 cm$^{-1}$ (bands 26 to 48). In the low temperature spectra most of the bands are single sharp maxima of medium to high intensity and only the bands 11, 14, 15, 17, 19, 22, 23, 30, 35, 36, 37, and 42 occur as weak shoulders of the main bands (Fig. 13a). With increasing temperature the bands broaden and strongly overlap and only the main bands are still observable at high temperature (Liebscher and Gottschalk 2004). Already in the room temperature spectra of Langer and Raith (1974) only the bands 24, 25, 26, 27, 31, 38, 40, 41, 46, 47, and 48 are visible as single sharp maxima. Due to this strong overlap, Langer and Raith (1974) were unable to resolve bands 12 and 13 as single bands and erroneously interpreted them to represent only one band (6) and claimed this feature as a main difference to the spectra of monoclinic epidote minerals, in which they were able to recognize two separate maxima in this region (bands 2 and 3 or 4 in Fig. 8). The best suited and most easily recognizable criteria to distinguish between the spectra of monoclinic and orthorhombic

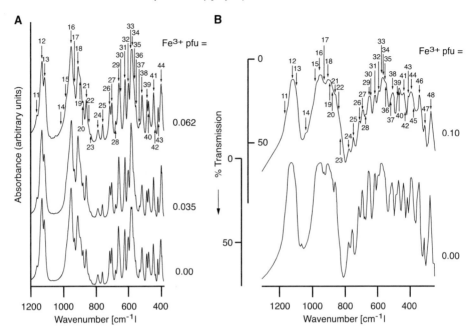

**Figure 13.** Typical IR spectra of orthorhombic Al-Fe³⁺ solid solutions with different iron contents at (A) −150°C (Liebscher and Gottschalk 2004) and (B) room-temperature (Langer and Raith 1974). Note that the spectra in (A) are displayed in the form of absorbance whereas those in (B) are displayed in the form of transmission. To allow for easier visual comparison between (A) and (B) the spectra in (B) are shown in an unconventional orientation with the transmission increasing downwards. Redrawn and modified after Langer and Raith (1974) and Liebscher and Gottschalk (2004).

epidote minerals are, however, the bands 5 in the spectra of monoclinic epidote minerals, which is missing in the zoisite spectra, and 24 and 25 of the zoisite spectra, which are missing in the spectra of the monoclinic epidote minerals (compare Figs 8 and 13).

Due to the only small compositional variation in zoisite, the positions of the bands show no interpretable shift with the Fe content but rather scatter. Liebscher and Gottschalk (2004) and Liebscher (unpublished data) studied the temperature evolution from −150 to 250°C of the IR spectra of synthetic zoisite below 1200 cm⁻¹ (Fig. 14). From those bands they were able to trace over the almost entire temperature range the bands 21 and 26 shift to higher wavenumbers with temperature whereas bands 16, 18, 27, 29, and 41 shift to lower wavenumbers (Fig. 14). Band 27 displays the highest temperature shift. The shift of the bands 29 and 41 is linear over the entire temperature interval whereas that of bands 16, 18, 21, 26, and 27 displays a change in slope at about 50°C (Fig. 14), which is related to the zoisite I–zoisite II phase transition (see below).

***Band assignment.*** As is the case for the monoclinic epidote minerals no rigorous band assignment has yet been performed for the lattice vibrational range of the IR spectra of zoisite. Langer and Raith (1974) assigned the bands 12, 13, 14, 16, 18, 21, 26, and 27 to Si–O valence vibrations of the $SiO_4$ tetrahedron and/or the $Si_2O_7$ group. With respect to the $Si_2O_7$ group, Liebscher and Gottschalk (2004) calculated the theoretical positions of the $\nu_{as}$Si–O–Si and $\nu_s$Si–O–Si vibrations (see Farmer 1974 for terminology) using the dependence of both vibrations on the Si–O–Si bond angle θ from Lazarev (1972) and Farmer (1974) and the values for θ from Liebscher et al. (2002). The $\nu_{as}$Si–O–Si and $\nu_s$Si–O–Si should occur at about

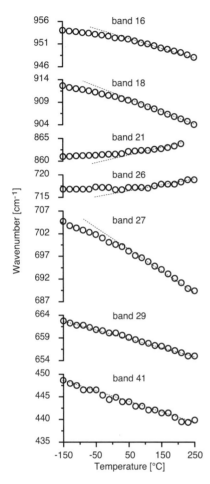

**Figure 14.** Temperature shift of some IR bands in the lattice vibrational range of synthetic iron free zoisite between −150 and 250°C. Bands that can be tracked over the entire temperature range either shift to higher (bands 21 and 26) or to lower (bands 16, 18, 27, 29, and 41) wavenumber with temperature. The temperature shift of the bands 29 and 41 is linear over the entire temperature interval whereas that of bands 16, 18, 21, 26, and 27 displays a change in slope at about 50°C which relates to the zoisite I–zoisite II phase transition. Dotted lines only visualize these changes in slope. Data are from Liebscher and Gottschalk (2004) and Liebscher (unpublished data).

1,115 and 880 cm$^{-1}$, respectively (Liebscher and Gottschalk 2004). Therefore either band 12 or 13 can be assigned to $\nu_{as}$Si–O–Si and either band 19 or 20 to $\nu_s$Si–O–Si.

By analogy with results from some layer silicates, band 17 at about 930 cm$^{-1}$ was tentatively assigned to an OH bending vibration by Langer and Raith (1974). In a later IR spectroscopy study on a synthetic deuterated zoisite sample Langer and Lattard (1980) abounded this tentative assignment. In the spectrum of the deuterated form they found three additional bands at 815, 730, and 705 cm$^{-1}$ that arise from OD bending vibrations. Applying a $\nu_{OH}/\nu_{OD}$ factor of 1.34 as a first approximation to account for the isotope effect, the corresponding OH bending vibrations in undeuterated zoisite should occur at 1,090, 975, and 945 cm$^{-1}$ (Langer and Lattard 1980). The proposed band at 1,090 cm$^{-1}$ is most probably obscured by the high intensities of the bands 12 and 13 in the zoisite spectra (see Fig. 13). But consistent with a band at about 1,090 cm$^{-1}$ in the spectrum of undeuterated zoisite, Langer and Lattard (1980) observed a greater width of the bands 12 and 13 in the spectrum of the OH than in the spectrum of the OD form. The proposed OH bending vibration at 975 cm$^{-1}$ coincides with band 15 in the spectra of undeuterated zoisite. Unfortunately, the authors were not able to identify any spectral feature in the spectrum of undeuterated zoisite that could account for the proposed third OH bending vibration at 945 cm$^{-1}$ (Langer and Lattard 1980). As band 27 of the spectrum of undeuterated zoisite is missing in the spectrum of the deuterated form, Langer and Lattard (1980) speculated that this band is due to a fourth OH bending vibration. This interpretation is consistent with two different hydrogen bridges in the zoisite structure (Liebscher et al. 2002). Two hydrogen bridges would result in four theoretical OH bending vibrations in the lattice vibrational range (Langer and Lattard 1980).

Based on the differences between the IR spectra of zoisite and the monoclinic epidote minerals, Langer and Raith (1974) concluded that the zoisite IR bands between 820 to 700 cm$^{-1}$ and 540 to 320 cm$^{-1}$ mainly originate from M-O vibrations within the two different octahedra M1,2 and M3.

***OH stretching vibrational region.*** The OH stretching vibration in zoisite occurs between

3,150 and 3,100 $cm^{-1}$, at roughly 200 $cm^{-1}$ lower wavenumbers than in the monoclinic Al-$Fe^{3+}$ solid solutions. By means of polarized single crystal IR spectroscopy, Linke (1970) measured the pleochroic scheme of the OH stretching vibration in two zoisite samples that had the optic axial plane parallel (100) ($\alpha$ orientation) and (010) ($\beta$ orientation), respectively (for a discussion of the $\alpha$ and $\beta$ orientation see Franz and Liebscher 2004). In both samples he observed a strong pleochroism in thin sections that were cut parallel (100) and (010) with the maximum absorption parallel [001]. Thin sections that were cut parallel (001) did not show any pleochroism (Linke 1970). From his results Linke (1970) concluded that the three dimensional absorption figure of the OH stretching vibration in zoisite, like that in the monoclinic epidote minerals (see above), is an elongated lemniscate that rotates around [001] and that the OH dipole is oriented parallel [001]. This orientation is consistent with an O10–H...O4 hydrogen bridge and confirms the previous findings of Dollase (1968). Linke (1970) also noticed that the three dimensional absorption figure of the OH stretching vibration is identical in the $\alpha$ and $\beta$ orientation, respectively, and therefore cannot be responsible for the different orientation of the optic axial plane in zoisite.

In their IR spectroscopic study on natural and synthetic zoisite samples of different composition, Langer and Raith (1974) fitted the OH stretching region of the spectra by three bands (bands 2, 5, and 7 in Fig. 15b). They assigned band 5 to the OH stretching vibration and interpreted the bands 2 and 7 as resulting from a coupling of band 5 with lattice vibrations at about 100 $cm^{-1}$. Contrary, in their study on synthetic Fe free and Fe bearing zoisite Liebscher et al. (2002) fitted the OH stretching region in synthetic Fe free zoisite with four bands (bands 2, 3, 5, and 7; lowermost spectrum in Fig. 15a) and that in Fe bearing zoisite with seven bands (bands 1 to 7; upper two spectra in Fig. 15a). In accordance with Langer and Raith (1974) they assigned band 5 to the OH stretching vibration of the configuration $^{[M1,2]}Al_2$–O10–H...O4–$^{[M1,2]}Al_2$$^{[M3]}Al$. As band 4 is only present in Fe bearing zoisite and the relative intensities of the bands 4 and 5 correlate well with the amounts of $Fe^{3+}$ (band 4) and Al (band 5) in the samples as determined by electron microprobe, Liebscher et al. (2002) assigned band 4 to the OH stretching vibration of the configuration $^{[M1,2]}Al_2$–O10–H...O4–$^{[M1,2]}Al_2$$^{[M3]}Fe^{3+}$. In line with the interpretation of Langer and Raith (1974), the bands 1, 3, 6, and 7 result from a coupling of bands 4 and 5 with lattice vibrations at about 100 $cm^{-1}$. Because in pure Al-$Fe^{3+}$ zoisite only the above mentioned configurations for the O10–H...O4 hydrogen bridge can occur, band 2 which is present in all spectra can neither be due to this hydrogen bridge nor to the incorporation of $Fe^{3+}$. By comparing the compositional shift of band 2 with the corresponding X-ray diffraction data, Liebscher et al. (2002) concluded that band 2 arises from the stretching vibration of a second hydrogen bridge of the configuration $^{[M1,2]}Al_2$–O10–H...O2–$^{[M3]}(Al,Fe^{3+})$. This second hydrogen bridge O10–H...O2 was already discussed although not proven by Smith et al. (1987) based on their neutron and X-ray diffraction study. As O10 and the proton both lie on a mirror plane whereas O2 has a general position, the O10–H...O2 hydrogen bridge should be bifurcated between the two symmetrically arranged O2 atoms (Liebscher et al. 2002).

Due to the only small compositional variations within the zoisite solid solutions, the compositional shift of the main OH band 5 is not as clear as in the monoclinic members. Nevertheless, the data indicate a slightly positive correlation between the Fe content and the position of band 5 (Fig. 16). The influence of pressure and temperature on the main OH stretching vibration (band 5) has been studied by Winkler et al. (1989) and Liebscher (unpublished data), respectively. The data of Winkler et al. (1989) for the pressure dependence at room temperature up to 11.6 GPa show a significant linear shift of band 5 to lower wavenumber with increasing pressure (Fig. 17a). With $-33.8$ $cm^{-1}$/GPa this shift is markedly more pronounced than in clinozoisite (see above). Taking the bulk modulus for zoisite from Comodi and Zanazzi (1997) of $K_0 = 114$ GPa, the corresponding Grüneisen parameter for band 5 is $\gamma = -1.22$, about six times smaller than in clinozoisite (Bradbury and Williams 2003; see above). Like in clinozoisite not only a shift of band 5 with pressure is observed in

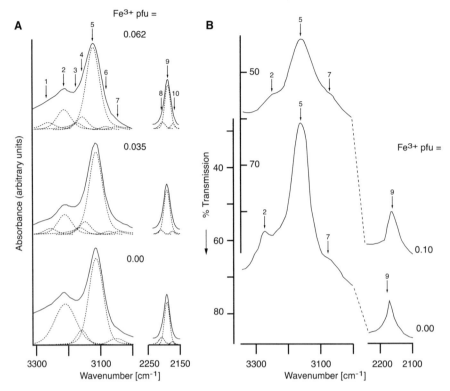

**Figure 15.** Typical IR spectra of orthorhombic Al-$Fe^{3+}$ solid solutions in the region of the OH stretching vibration and between 2250 and 2100 cm$^{-1}$ at (A) $-150°C$ (Liebscher et al. 2002) and (B) room-temperature (Langer and Raith 1974). Note that the spectra are displayed in the form of absorbance in (A) and in the form of transmission in (B). The spectra in (B) are shown with the transmission increasing downwards and are to scale. In the orthorhombic Al-$Fe^{3+}$ solid solutions the OH stretching vibration is characterized by a maximum between about 3200 and 3100 cm$^{-1}$ (band 5; configuration [M1,2]Al$_2$-O10-H…O4-[M1,2]Al$_2$[M3]Al). Band 2 arises from a second hydrogen bridge O10–H…O2. Band 4 is only present in iron bearing zoisite and arises from the configuration [M1,2]Al$_2$-O10-H…O4-[M1,2]Al$_2$[M3]$Fe^{3+}$. Redrawn and modified after Langer and Raith (1974) and Liebscher et al. (2002).

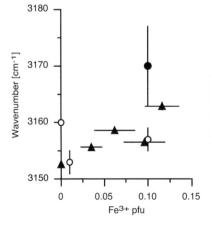

**Figure 16.** Compositional shift of the zoisite IR band 5. The data scatter but indicate a positive compositional shift of band 5. Data sources are as follows: O = Langer and Raith (1974), ● = Winkler et al. (1989), and ▲ = Liebscher et al. (2002).

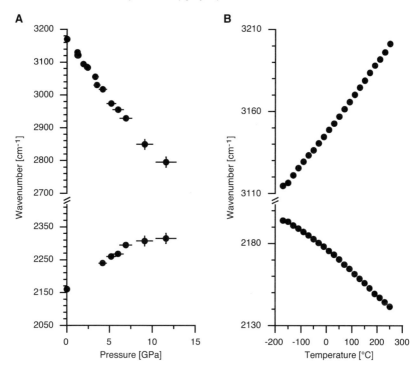

**Figure 17.** Pressure (A) and temperature (B) shifts of the zoisite IR bands 5 and 9. Band 5 shifts linearly to lower wavenumber with pressure and to higher wavenumber with temperature whereas band 9 displays the opposite behavior.

zoisite, but also a significant broadening of band 5 (Winkler et al. 1989). The half width of band 5 increases by about 27 cm$^{-1}$/GPa, about three-times the value observed in clinozoisite (Bradbury and Williams 2003). Based on their model derived for clinozoisite, Bradbury and Williams (2003) speculated that the dramatic shift and broadening of the zoisite OH stretching vibration may be related to the initially shorter hydrogen bond in zoisite and a correspondingly closer initial approach of the two potential minima (see Fig. 12). The temperature shift to higher wavenumber of band 5 is linear with 0.21 cm$^{-1}$/°C over the entire investigated temperature range from −170 to 250°C (Fig. 17b; Liebscher unpublished data).

***Band 9 at 2,300 to 2,100 cm$^{-1}$.*** The band 9 between 2,300 and 2,100 cm$^{-1}$ is a unique feature of the IR spectrum of zoisite compared to the spectra of the monoclinic epidote minerals. Compared to the main OH stretching vibration (band 5) the intensity of band 9 is very low (Fig. 15b; note that the spectra in Fig. 15a are not to scale). Contrary to band 5, which shifts to higher wavenumber with increasing Fe content, band 9 shifts to lower wavenumber (Liebscher et al. 2002). Langer and Raith (1974) assigned band 9 to a second OH stretching vibration that arises from a second hydrogen bridge in the zoisite structure. Band 9 has a notable low energy for an OH stretching vibration, and Langer and Raith (1974) calculated an O–O distance for the proposed underlying second hydrogen bridge of only 2.56 Å. Such a short O–O distance exists only between oxygen atoms that form edges of the coordination polyhedra and Langer and Raith (1974) therefore concluded, "it is uncertain which oxygen atoms are involved in the short bridge." The spectrum of a deuterated zoisite recorded by Langer and Lattard (1980) shows a shift of band 9 to a lower wavenumber by a factor of 1.35

and thus proved band 9 to arise from some proton related vibration. The interpretation of band 9 as an OH stretching vibration by Langer and Raith (1974) was later questioned by Winkler et al. (1989) based on the pressure shift of band 9 and polarized single crystal IR data. Contrary to band 5 that shifts to lower wavenumber with increasing pressure, band 9 shifts to higher wavenumber on compression (Fig. 17a) consistent with an OH bending vibration and Winkler et al. (1989) therefore assigned band 9 to the first overtone of the in-plane bending vibration of the O10–H...O4 hydrogen bridge. Nevertheless, as the first overtone of the OH bending vibration should occur at a lower energy of about 2,000 to 1,950 cm$^{-1}$ (Langer and Raith 1974), the interpretation of Winkler et al. (1989) could not explain the comparable high energy of band 9. Consistent with the interpretation of band 9 as a bending vibration, Liebscher et al. (2002) and Liebscher (unpublished data) found a shift to lower wavenumber of band 9 with temperature (Fig. 17b). As there is also evidence for a second hydrogen bridge O10–H...O2 that must be bifurcated (band 2; see above), band 9 is interpreted as the first overtone of the in-plane bending vibration of the O10–H...O2 hydrogen bridge, and the bifurcated nature of this hydrogen bridge well explains the comparable high energy of band 9 (Liebscher et al. 2002).

*Phase transition in zoisite.* Brunsmann (2000), Liebscher et al. (2002), and Liebscher and Gottschalk (2004) used IR spectroscopy to study the isosymmetric displacive phase transition in zoisite (see Franz and Liebscher 2004). In good accordance with the results from X-ray diffraction the IR spectra of the studied synthetic zoisite samples show discontinuities in their compositional shift at about 0.05 Fe$^{3+}$ pfu. To study the temperature dependence of the phase transition, Liebscher and Gottschalk (2004) performed a temperature dependent IR spectroscopic study between −150 and 170°C. Whereas some of the IR bands show a linear shift with temperature over the entire temperature interval, others show a smaller linear positional shift with temperature at low than at high temperature (see Fig. 14). This change in slope has been interpreted as the transition from zoisite I to zoisite II. To account for strongly overlapping bands especially at higher temperature that may place some doubt on the exact determination of the band positions, the authors analyzed the spectra in the spectral range 1,050 to 820 cm$^{-1}$ (bands 14 to 23) by means of autocorrelation (Salje et al. 2000). Like the band positions, the autocorrelation analyses show a different low and high temperature evolution of the spectra in all samples (Fig. 18a). This different temperature evolution is best displayed by

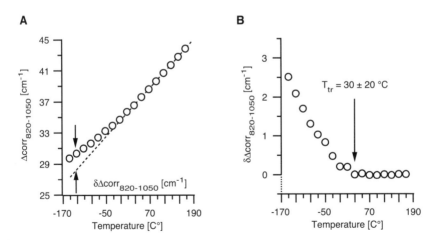

**Figure 18.** (A) Autocorrelation analysis of the spectral range 1050 to 820 cm$^{-1}$ and (B) the corresponding excess line width parameter $\delta\Delta_{corr}$ for an iron free zoisite. The data show a different low and high temperature evolution of the spectrum. The change from the low to the high temperature evolution marks the transition from zoisite I to zoisite II (redrawn after Liebscher and Gottschalk 2004).

the excess line width parameter $\delta\Delta_{corr}$ (Salje et al. 2000) that clearly shows the transition from zoisite I to zoisite II (Fig. 18b). The data of Liebscher and Gottschalk (2004) indicate that the transition temperature $T_{tr}$ correlates negatively with the Fe content and a tentative linear fit to their data yields $T_{tr} = 57 - 1223\ X_{Fe}$ [°C].

## MÖSSBAUER SPECTROSCOPY

So far, studies on epidote minerals are restricted to the monoclinic members and to $^{57}$Fe-Mössbauer spectroscopy. De Coster et al. (1963) published the first Mössbauer spectrum of an epidote of unspecified composition and origin. Later, Marfunin et al. (1967) reported the 80 and 300 K spectra of an epidote of unspecified composition and Bancroft et al. (1967) studied the spectra of two epidotes with $X_{Ep} = 0.84$ and 0.87, respectively, and one piemontite with 0.63 $Mn^{3+}$ and 0.33 $Fe^{3+}$ pfu. Dollase (1971) published the first Mössbauer spectrum of allanite and later studied systematically the spectra of and the Fe distribution in 33 samples of natural and synthetic epidote minerals including Al-$Fe^{3+}$ solid solutions with $X_{Ep} = 0.36$ to 1.00, piemontites, allanites, and oxyallanites (Dollase 1973). Temperature dependent spectra down to 2 K were recorded by Pollak and Bruyneel (1975) and Paesano et al. (1983). Whereas these early studies focused their work primarily on the determination of the valence state of Fe and the Fe site assignment, Bird et al. (1988) and Patrier et al. (1991) used $^{57}$Fe-Mössbauer spectroscopy to study the equilibrium or nonequilibrium compositional order/disorder in natural epidote samples as a function of the physicochemical conditions during epidote formation. Fehr and Heuss-Aßbichler (1997) and Heuss-Aßbichler (2000) later extended these studies and determined the intracrystalline ordering of Al and $Fe^{3+}$ in Al-$Fe^{3+}$ solid solutions by means of $^{57}$Fe-Mössbauer spectroscopy on natural heat treated samples with $X_{Ep} = 0.39$ to 0.98. A theoretical treatment of the Mössbauer spectra of Fe rich Al-$Fe^{3+}$ solid solutions by molecular orbital calculations was then presented by Grodzicki et al. (2001). The results of the different studies in terms of the fitted Mössbauer parameters are summarized in Table 1.

### Mössbauer spectra of the different epidote minerals

***Monoclinic Al-Fe$^{3+}$ solid solutions.*** In the monoclinic Al-$Fe^{3+}$ solid solutions Fe is almost exclusively in the ferric state and shows a very strong affinity for the M3 site compared to the M1 and M2 sites (see Franz and Liebscher 2004). Their Mössbauer spectra are therefore in most cases simple (Dollase 1973). All spectra of the monoclinic Al-$Fe^{3+}$ solid solutions are characterized by one very prominent doublet (Fig. 19a,b) and the reported Mössbauer parameters for this doublet are consistent between the different studies. Its isomer shift (IS) relative to α-Fe ranges from 0.31 to 0.44 mm/sec (note that all isomer shift data presented in this review were recalculated relative to α-Fe to allow for comparison) and the quadrupole splitting (QS) is 1.89 to 2.096 mm/sec (Fig. 20a). This isomer shift is characteristic for $Fe^{3+}$ in octahedral coordination whereas the quadrupole splitting is one of the largest observed for a high-spin $Fe^{3+}$ compound and points to a strongly distorted site (Bancroft et al. 1967; the high-spin electronic state of $Fe^{3+}$ in epidote group minerals was deduced by Burns and Strens 1967 based on magnetic susceptibility measurements). Consequently, this doublet has been assigned to $Fe^{3+}$ in M3 by all studies (Fig. 19a,b; for a splitting of this doublet into two components M3 and M3′ in Al-$Fe^{3+}$ solid solutions with $X_{Ep} < 0.7$ see below).

Although the earliest studies fitted the recorded epidote spectra exclusively with one doublet for $Fe^{3+}$ in M3 (De Coster et al. 1963; Bancroft et al. 1967; Marfunin et al. 1967), later and more detailed studies clearly show doublets of minor intensity in addition to that for $Fe^{3+}$ in M3 in most of the studied spectra of monoclinic Al-$Fe^{3+}$ solid solutions. It should be noted that of the early studies mentioned at least the epidote spectrum reported by Marfunin et al. (1967) suggests a second doublet that was not fitted by the authors. The Mössbauer parameters for these additional minor doublets determine three distinct groups (Fig. 20a). In addition to

**Table 1.** Isomer shift IS [mm/sec] relative to $\alpha$-iron, quadrupole splitting QS [mm/sec], and full width at half maximum FWHM [mm/sec] for the different doublets fitted to the spectra of epidote minerals

| | Al-Fe$^{3+}$ solid solutions | Piemontite | Allanite | Ferriallanite | Oxyallanite |
|---|---|---|---|---|---|
| ***Fe$^{3+}$ in M1*** | | | | | |
| IS | 0.22–0.36[1, 3, 7] | | 0.29–0.29[2] | 0.36[6] | 0.34–0.36[1] |
| QS | 1.46–1.67[1, 3, 7] | | 1.01–1.36[2, 4, 11] | 1.78[6] | 1.84–1.91[1] |
| FWHM | 0.23–0.51[1, 3, 7] | | 0.40–0.74[2, 4] | | 0.41–0.45[1] |
| ***Fe$^{3+}$ in M2 (as recommended in this review, see text)*** | | | | | |
| IS | 0.31–0.37[5, 8] | 0.25–0.33[1] | | 0.334[6] | |
| QS | 0.80–1.0[5, 8] | 0.83–1.13[1] | | 0.85[6] | |
| FWHM | 0.40–0.50[5, 8] | 0.40–0.52[1] | | | |
| ***Fe$^{3+}$ in M3*** | | | | | |
| IS | 0.31–0.44 [1, 3, 5, 7, 8, 9, 10, 12] | 0.34–0.37 [1, 10] | 0.37–0.55 [1, 2, 11] | | 0.35–0.36 [1] |
| QS | 1.89–2.10 [1, 3, 5, 7, 8, 9, 10, 12] | 2.0–2.02 [1, 10] | 1.61–2.00 [1, 2, 4, 11] | | 2.26–2.31 [1] |
| FWHM | 0.25–0.49 [1, 3, 5, 7, 8, 9, 10, 12] | 0.40–0.52 [1, 10] | 0.30–0.58 [1, 2, 4, 11] | | 0.41–0.45 [1] |
| ***Fe$^{3+}$ in M3´*** | | | | | |
| IS | 0.24–0.36[7] | | | | |
| QS | 1.99–2.32[7] | | | | |
| FWHM | 0.22–0.35[7] | | | | |
| ***Fe$^{3+}$ in unspecified site*** | | | | | |
| IS | | | 0.35–0.39[1] | 1.07[6] | |
| QS | | | 1.87–1.97[1] | 1.62[6] | |
| FWHM | | | 0.37–0.47[1] | | |
| ***Fe$^{2+}$ in M3*** | | | | | |
| IS | 1.09–1.40[7] | | 1.07–1.08[1, 2, 11] | 1.07[6] | |
| QS | 1.78–2.42[7] | | 1.63–1.83[1, 2, 4, 11] | 1.62[6] | |
| FWHM | 0.26–0.73[7] | | 0.35–0.58[1, 2, 4, 11] | | |
| ***Fe$^{2+}$ in M1*** | | | | | |
| IS | | | 1.24–1.24[2, 11] | | |
| QS | | | 1.93–2.07[2, 4, 11] | | |
| FWHM | | | 0.40–0.58[2, 4, 11] | | |
| ***Fe$^{2+}$ in unspecified site*** | | | | | |
| IS | 1.07–1.30[1, 8] | | | | 1.09–1.14[1] |
| QS | 1.8–2.28[1, 8] | | 1.87–2.14[1] | | 1.62–1.72[1] |
| FWHM | 0.32–0.44[1, 8] | | 0.35–0.42[1] | | 0.34–0.60[1] |

*Note.* Small numbers in parentheses refer to following sources: [1] Dollase (1973); [2] Dollase (1971); [3] Bird et al. (1988); [4] Mingsheng and Daoyuan (1986) and Mingsheng and Dien (1987); [5] Artioli et al. (1995); [6] Kartashov et al. (2002); [7] Fehr and Heuss-Aßbichler (1997) and Heuss-Aßbichler (2000); [8] Patrier et al. (1991); [9] De Coster et al. (1963); [10] Bancroft et al. (1967); [11] Lipka et al. (1995); [12] Marfunin et al. (1967)

the doublet for $Fe^{3+}$ in M3, Dollase (1973), Bird et al. (1988), and Fehr and Heuss-Aßbichler (1997) fitted some of their spectra with a second doublet that has IS = 0.217 to 0.355 mm/sec and QS = 1.455 to 1.666 mm/sec (Fig. 20a). The range in the isomer shift of this second doublet is comparable to the main doublet for $Fe^{3+}$ in M3 but its quadrupole splitting is smaller and suggests a slightly less distorted site (Bancroft et al. 1967). The authors therefore assigned this doublet to $Fe^{3+}$ in M1 (Fig. 19a). Contrary to the aforementioned authors, Patrier et al. (1991) and Artioli et al. (1995) evaluated most of their spectra with a second doublet that has a comparable isomer shift (0.31 to 0.37 mm/sec) but a notable smaller QS of only 0.8 to 1.0 mm/sec (Figs19b, 20a). Notwithstanding this significant difference in QS, Patrier et al. (1991) and Artioli et al. (1995) also assigned the second doublet to $Fe^{3+}$ in M1. A comparison with the spectrum of a ferriallanite (Fig. 19e, Kartashov et al. 2002; see below) indicates that this assignment by Patrier et al. (1991) and Artioli et al. (1995) might be incorrect. One of the doublets in the spectrum of the ferriallanite has IS = 0.334 mm/sec and QS = 0.85 mm/sec. These values are identical to those found by Patrier et al. (1991) and Artioli et al. (1995) for the second, low intensity doublet. As the doublet in the ferriallanite spectrum must be assigned to $Fe^{3+}$ in M2 (Kartashov et al. 2002; see below) the second doublet of Patrier et al. (1991) and Artioli et al. (1995) probably also arises from $Fe^{3+}$ in M2. This potentially incorrect assignment of the second doublet to $Fe^{3+}$ in M1 by Patrier et al. (1991) would also explain the erroneously high disorder these authors claimed for the intracrystalline partitioning of Al and $Fe^{3+}$ between M3 and M1 as the authors actually studied the partitioning of Al and $Fe^{3+}$ between M3 and M2.

Beside the doublets that can be assigned to $Fe^{3+}$ based on the isomer shift, Dollase (1973), Patrier et al. (1991), and Fehr and Heuss-Aßbichler (1997) found evidence in some of the recorded spectra for small amounts of $Fe^{2+}$ in the studied samples (Fig. 19a,b). These spectra show a very low intensity doublet with IS = 1.065 to 1.295 mm/sec and QS = 1.8 to 2.28 mm/sec (Fig. 20a). This isomer shift is characteristic for $Fe^{2+}$ in octahedral coordination (Bancroft et al. 1967). Unfortunately, due to the low intensity of this doublet in the studied spectra, neither Dollase (1973) nor Patrier et al. (1991) were able to assign it unequivocally to $Fe^{2+}$ in a specific site. Dollase (1973) even discussed the possibility that this doublet may arise from some undetected Fe bearing impurity phases. In contrast, Fehr and Heuss-Aßbichler (1997) assigned this doublet to $Fe^{2+}$ in the largest and most distorted octahedron M3.

Each of the three doublets assigned to $Fe^{3+}$ has very homogeneous Mössbauer parameters. The range in the isomer shift is identical for all three doublets and shows no dependence on composition. Likewise, the quadrupole splitting although of a characteristic value for each doublet is independent on composition (Fig. 21). This indicates that changes in the distortion of the individual octahedra with increasing Fe content are too small to be resolved. It is also evident from Figure 21 that the doublet for $Fe^{3+}$ in M3 appears throughout the entire compositional range and that it is the only doublet present in the spectra of solid solutions with $X_{Ep} < 0.7$. This is consistent with the structural and optical data for the monoclinic Al-$Fe^{3+}$ solid solution series that also indicate exclusive incorporation of $Fe^{3+}$ in M3 for Fe contents below about $X_{Ep} = 0.7$ (see Franz and Liebscher 2004). The other two doublets for $Fe^{3+}$ in M1 and, potentially, M2 are only found in the spectra of Al-$Fe^{3+}$ solid solutions with $X_{Ep} > 0.7$. For these high Fe contents, the structural data as well as detailed crystal chemical studies (e.g., Giuli et al. 1999) also provide good evidence for incorporation of appreciable amounts of Fe in M1 and, under specific conditions, of small but notable amounts of Fe in M2 (see Franz and Liebscher 2004). In this context it should be noted that none of the Mössbauer studies has yet resolved both minor doublets for $Fe^{3+}$ in M1 and M2 in one and the same spectrum although the data of Giuli et al. (1999) strongly suggest that Fe substitutes simultaneously in M1 and M2. This inconsistency might be partly due to the low intensity of the two doublets in question that makes their fitting very difficult but most probably reflects the pre-assumption of most studies that the M2 position is entirely filled with Al, which let the authors not to search for a third doublet.

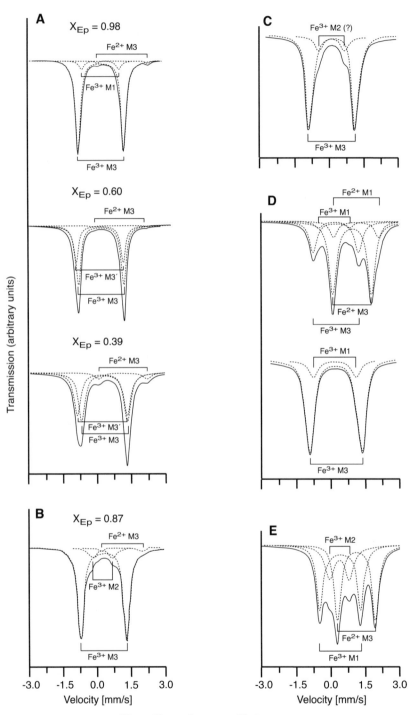

**Figure 19.** *caption at top of facing page*

**Figure 19 (*on facing page*).** Typical room-temperature Mössbauer spectra of (A) and (B) Al-Fe³⁺ solid solutions, (C) piemontite, (D) allanite and oxyallanite, and (E) ferriallanite. For the assignment of the two doublets to $Fe^{3+}$ in M3 and M3′ in the two lower spectra of (A) see text. The low intensity $Fe^{3+}$ doublets in (B) and (C) are assigned to M2 by comparison with the ferriallanite spectrum (E). In the original papers, these doublets are assigned to $Fe^{3+}$ in M1. The two spectra in (D) represent one and the same allanite sample before (upper spectrum) and after (lower spectrum) heat treatment and complete oxidation. Redrawn and modified after (A) Fehr and Heuss-Aßbichler (1997) and Heuss-Aßbichler (2000), (B) Patrier et al. (1991), (C) and (D) Dollase (1973), and (E) Kartashov et al. (2002).

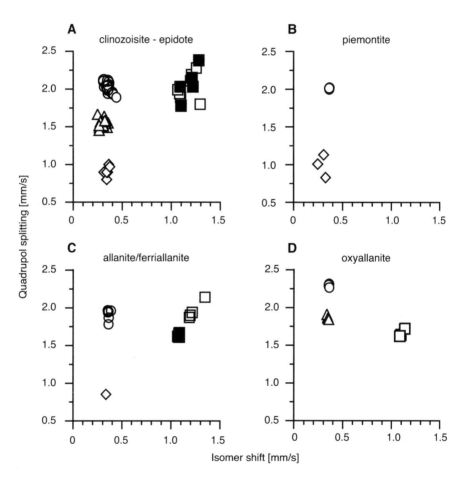

**Figure 20.** Quadrupole splitting and isomer shift (recalculated relative to α-iron) of the different doublets fitted to the Mössbauer spectra of (A) Al-Fe³⁺ solid solution, (B) piemontite, (C) allanite and ferriallanite, and (D) oxyalllanite. O = $Fe^{3+}$ in M3, △ = $Fe^{3+}$ in M1, ◇ = $Fe^{3+}$ in M2 (as recommended in this review, the original papers assigned these doublets also to $Fe^{3+}$ in M1, see text), ■ = $Fe^{2+}$ in M3, □ = $Fe^{2+}$ in unspecified site. Data from Bancroft et al. (1967), Dollase (1971, 1973), Bird et al. (1988), Patrier et al. (1991), Artioli et al. (1995), Fehr and Heuss-Aßbichler (1997), Heuss-Aßbichler (2000), and Kartashov et al. (2002).

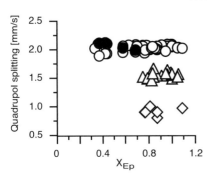

**Figure 21.** Quadrupole splitting of $Fe^{3+}$ in the different octahedral sites in Al-$Fe^{3+}$ solid solutions as a function of composition. O,● = $Fe^{3+}$ in M3 (● refer to $Fe^{3+}$ in M3´, see text), △ = $Fe^{3+}$ in M1, ◇ = $Fe^{3+}$ in M2 (as recommended in this review, the original papers assigned these doublets also to $Fe^{3+}$ in M1, see text). Data from Bancroft et al. (1967), Dollase (1973), Bird et al. (1988), Patrier et al. (1991), Artioli et al. (1995), Fehr and Heuss-Aßbichler (1997), and Heuss-Aßbichler (2000).

Temperature dependent Mössbauer spectra are recorded by Pollak and Bruyneel (1975) for an epidote of unspecified composition between 300 and 4.2 K and by Paesano et al. (1983) for an epidote with $X_{Ep} = 0.96$ between 300 and 2 K (Fig. 22a). Down to 100 K the spectra recorded by Pollak and Bruyneel (1975) show no significant changes. All spectra could be fitted with one doublet with Mössbauer parameters typical for $Fe^{3+}$ in M3 (Fig. 22b). Contrary, the spectrum recorded at 4.2 K display three superposed symmetric doublets that were tentatively assigned to $Fe^{3+}$ in M3, M2, and M1 by Pollak and Bruyneel (1975). Whereas the doublet assigned to $Fe^{3+}$ in M3 shows the typical Mössbauer parameters, the two doublets assigned to $Fe^{3+}$ in M2 and M1, respectively, are characterized by a notable small quadrupole splitting and isomer shift (Fig. 22c). Unfortunately, Pollak and Bruyneel (1975) did not find a conclusive explanation for the appearance of the two additional doublets in the 4.2 K spectrum. Paesano et al. (1983) recorded Mössbauer spectra down to 2 K with special emphasis given to the low temperature region. Down to 77 K the spectra show a well-defined and symmetric doublet that can be attributed to $Fe^{3+}$ in M3 based on the Mössbauer parameters (Fig. 22b; Paesano et al. 1983). Below 10 K the character of the spectra change, the doublet for $Fe^{3+}$ in M3 becomes increasingly asymmetric and below 4 K a well-defined magnetic hyperfine structure appears (Fig. 22a). But throughout the entire investigated temperature range, Paesano et al. (1983) did not find any evidence for $Fe^{3+}$ in another site than M3 (Fig. 22b,c). From the results of their fits to the low temperature spectra, they concluded that the magnetic hyperfine structure observed below 4 K results from "a freezing of the paramagnetic moment on $Fe^{3+}$." As the spectra recorded at 3 and 2 K show a constant hyperfine field, no magnetic transition has occurred and they speculated that the additional structures observed by Pollak and Bruyneel (1975) in the 4.2 K spectrum arise from a paramagnetic hyperfine structure instead of $Fe^{3+}$ in the M1 and M2 sites.

Grodzicki et al. (2001) carried out cluster molecular orbital calculations in local spin density approximation to understand the observed quadrupole splittings of $Fe^{3+}$ in M3 and M1 and of $Fe^{2+}$ in M3. To obtain Mössbauer parameters that are consistent with the experimental results, the chosen clusters have to be rather large. Beside the first coordination shell, they must at least include the coordination spheres of the oxygen atoms bound to Fe. Based on clusters fulfilling these requirements, the authors calculated quadrupole splittings of $-1.89$ to $-2.09$ mm/sec for $Fe^{3+}$ in M3 (for Al-$Fe^{3+}$ solid solutions with $X_{Ep} = 0.60$ to 0.89 and an allanite), $-1.30$ mm/sec for $Fe^{3+}$ in M1 (for an Al-$Fe^{3+}$ solid solution with $X_{Ep} = 0.76$) and $+2.03$ mm/sec for $Fe^{2+}$ in M3. These values quantitatively agree with the experimentally determined ones (see Table MS 1). The large quadrupole splitting calculated and observed for $Fe^{3+}$ in M3 is due to the strong tetragonal compression of this site whereas the smaller quadrupole splitting of $Fe^{3+}$ in M1 reflects the smaller tetragonal compression of M1 (Grodzicki et al. 2001). The strong tetragonal compression of M3 is also responsible for the observed and calculated quadrupole splitting of $Fe^{2+}$ in M3 (~2.0 mm/sec), which is considerable smaller than the 2.60 to 3.60 mm/sec range commonly observed for the quadrupole splitting of $Fe^{2+}$ in minerals (Grodzicki et al. 2001).

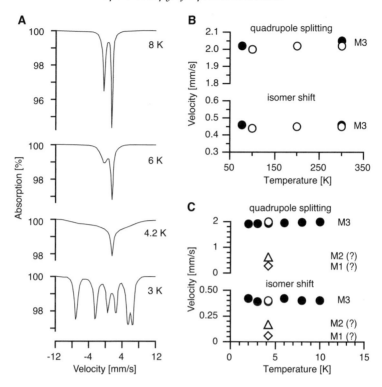

**Figure 22.** Evolution of the Mössbauer spectra of Al-Fe³⁺ solid solutions with decreasing temperature. (A) Low-temperature Mössbauer spectra of an epidote with $X_{Ep}$ = 0.96 (redrawn and modified after Paesano et al. 1983). (B) and (C) Temperature dependence of quadrupole splitting and isomer shift (relative to α-iron) of Fe³⁺ in Al-Fe³⁺ solid solutions between 300 and 50 K (B) and between 10 and 2 K (C). Data from Pollak and Bruyneel (1975; open symbols) and Paesano et al. (1983; filled symbols). Pollak and Bruyneel (1975) fitted their 4.2 K spectrum with three doublets and assigned them to Fe³⁺ in M3, M2 (diamond), and M1 (triangle). The Mössbauer parameter for Fe³⁺ in M3 (filled and open circles) are constant over the entire temperature range.

***Piemontite.*** Like in the monoclinic Al-Fe³⁺ solid solutions, Fe is predominantly ferric in the piemontite solid solutions. Only in REE bearing piemontite samples, Fe²⁺ might occur in appreciable amounts. The Mössbauer spectra of piemontite solid solutions therefore closely resemble those of the Al-Fe³⁺ solid solutions (Fig. 19a–c). Bancroft et al. (1967) studied a piemontite with 0.332 Fe³⁺ pfu and 0.625 Mn³⁺ pfu and evaluated it with only one doublet with IS = 0.34 mm/sec and QS = 2.02 mm/sec. Later, Dollase (1973) analyzed the spectra of three piemontite samples with 0.12 to 0.33 Fe³⁺ pfu and 0.72 to 0.86 Mn³⁺ pfu. All three spectra are consistent and in contrast to the results of Bancroft et al. (1967) had to be evaluated by two doublets (Fig. 19c). The main doublet has an isomer shift of 0.362 to 0.365 mm/sec and a quadrupole splitting of 2.0 to 2.02 mm/sec. The second doublet has a much smaller intensity (Fig. 19c), a slightly smaller isomer shift (0.245 to 0.325 mm/sec) and a significantly smaller quadrupole splitting (0.83 to 1.13 mm/sec; Fig. 20b). By analogy with the results from the Al-Fe³⁺ solid solution series, the main doublet can be assigned to Fe³⁺ in M3 whereas the second doublet may arise from Fe³⁺ in M2 (Fig. 19c).

***Allanite, oxyallanite, and ferriallanite.*** In contrast to the Al-Fe³⁺ solid solutions and to piemontite, in which the overwhelming majority of the Fe is normally ferric, it is generally present as both ferric and ferrous in allanite with $Fe^{2+}$ being the major component (see Gieré and Sorensen 2004). The ferric and ferrous Fe might be distributed over the different octahedral sites and the Mössbauer spectra of allanite are therefore more complicated and often only poorly resolved and ambiguous (Dollase 1973). Dollase (1971) published the first Mössbauer spectrum of an allanite with $Fe_{tot} = 1.2$ pfu and also provided a first preliminary interpretation of its spectral properties. Later, he studied the Mössbauer spectra of this and three more allanite samples in greater detail and, based on the changes of the spectral properties, analyzed the effects of oxidation and reduction on allanite (Dollase 1973). The Mössbauer spectrum of a partly metamict allanite was presented by Mingsheng and Daoyuan (1986) and Mingsheng and Dien (1987). More recently, Lipka et al. (1995) performed a Mössbauer spectroscopic study on three allanite samples and Kartashov et al. (2002) reported the Mössbauer spectrum of a ferriallanite.

The spectra of allanite are typically evaluated with three or four doublets (Fig. 19d upper spectrum). Based on their isomer shifts these doublets can be confidentially assigned to either ferrous or ferric Fe (Fig. 20c). The main $Fe^{2+}$ doublet that normally displays the highest intensity of all doublets has very consistent Mössbauer parameters throughout the different studies. Its IS ranges from 1.068 to 1.083 mm/sec with a QS of 1.660 to 1.666 mm/sec (Fig. 20c). This doublet is generally assigned to $Fe^{2+}$ in M3. If present, the second $Fe^{2+}$ doublet has a much lower intensity and is therefore only poorly constrained with respect to its peak location. Consequently, its Mössbauer parameters show a greater scatter and range from 1.185 to 1.345 mm/sec (IS) and 1.87 to 2.14 mm/sec (QS) (Fig. 20c). Dollase (1971), Mingsheng and Daoyuan (1986), Mingsheng and Dien (1987), and Lipka et al. (1995) assigned this second doublet to $Fe^{2+}$ in M1 whereas Dollase (1973) claimed, "no unequivocal assignment seems possible with the data at hand." But as the Mössbauer spectra of the oxidized allanite samples (Dollase 1973; see below) exclusively show $Fe^{3+}$ in M3 and M1 and because it can be assumed that no significant Fe diffusion occurred during the oxidization procedure it is reasonable to assume that $Fe^{2+}$ in the unoxidized allanite samples is also restricted to M3 and M1.

Beside the doublets for ferrous Fe, the Mössbauer spectra of allanite typically show one or two doublets that can be assigned to ferric Fe. Dollase (1971), Mingsheng and Daoyuan (1986), Mingsheng and Dien (1987), and Lipka et al. (1995) evaluated their spectra with two doublets for $Fe^{3+}$ and assigned them to M3 and M1, respectively (Fig. 19d upper spectrum). The corresponding Mössbauer parameters are 0.37 to 0.55 mm/sec (IS) and 1.61 to 1.97 mm/sec (QS) for $Fe^{3+}$ in M3 and 0.22 to 0.29 mm/sec (IS) and 1.33 to 1.36 mm/sec (QS) for $Fe^{3+}$ in M1. Contrary, Dollase (1973) failed in fitting two ferric doublets to the studied allanite spectra and reports only a single ferric doublet without specific site assignment. The parameters of this doublet are IS = 0.351 to 0.385 mm/sec and QS = 1.78 to 1.965 mm/sec. But as this quadrupole splitting is substantially smaller than that attributed to $Fe^{3+}$ in M3 in the Al-$Fe^{3+}$ solid solutions, there is good indication that the single ferric doublet actually represents two separate but strongly overlapping doublets (Dollase 1973).

To study the effect of oxidation on allanite, Dollase (1973) heat-treated two of the studied allanite samples in air at 370 to 700°C for 46 to 163 hours and analyzed the evolution of the ferric/ferrous ratios by means of Mössbauer spectroscopy (for a more general and thorough discussion of oxyallanite see Gieré and Sorensen 2004). Up to $T = 400°C$ he did not observe any change in the ferrous/ferric ratio (Fig. 23). For $T > 400°C$ the intensities of the doublets assigned to $Fe^{3+}$ continuously increase with increasing temperature whereas those assigned to $Fe^{2+}$ decrease. In one sample this results in an increase of the calculated $X_{Fe3+} = Fe^{3+}/Fe_{tot}$ from 0.32 in the untreated sample to 0.88 in the sample that was heat-treated at 680°C and in almost complete oxidation at 700°C (Fig. 23). The corresponding Mössbauer spectrum of the fully oxidized allanite exhibits only two doublets that can be confidentially assigned to $Fe^{3+}$ in M3 and M1 (Fig. 19d lower spectrum). As none of the spectra of the partly to fully oxidized allanite

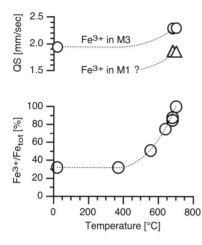

**Figure 23.** Oxidation of $Fe^{2+}$ to $Fe^{3+}$ in natural allanite as determined by Mössbauer spectroscopy. Up to 400°C no oxidation is observed. Further heating results in a significant increase of $Fe^{3+}$ and at about 700°C complete oxidation is achieved. In the fully oxidized sample the quadrupole splitting of $Fe^{3+}$ in M3 is notable larger than in the unoxidized sample. The same is probably valid for $Fe^{3+}$ in M1. The stippled lines in the upper part of the figure are drawn to parallel the trend in the $Fe^{3+}/Fe_{tot}$ ratio. Data from Dollase (1973).

samples shows evidence for $Fe^{3+}$ in a site other than M3 or M1, Dollase (1973) concluded that $Fe^{2+}$ in the unoxidized allanite samples is also restricted to M1 and M3. Longer heating at the same temperature did not result in any further oxidation and thus the degree of oxidation is a function of temperature only. Heating an oxidized allanite in a reducing $H_2$ atmosphere at 680°C for 6.5 hours results in a Mössbauer spectrum identical to that of the untreated sample and indicates that the oxidation process is fully reversible (Dollase 1973). The Mössbauer parameters of the $Fe^{3+}$ doublets in the oxyallanite samples are very consistent (Fig. 20d). Their isomer shifts of 0.353 to 0.36 mm/sec for $Fe^{3+}$ in M3 and 0.335 to 0.355 mm/sec for $Fe^{3+}$ in M1 are comparable to the corresponding ones in the $Al$-$Fe^{3+}$ solid solutions but their quadrupole splittings of 2.264 to 2.311 mm/sec for $Fe^{3+}$ in M3 and 1.84 to 1.91 mm/sec for $Fe^{3+}$ in M1 are significantly larger than in the $Al$-$Fe^{3+}$ solid solutions (Figs 20, 23). Unfortunately, Dollase (1973) only provided the Mössbauer parameters for the almost fully oxidized samples ($X_{Fe3+} > 0.88$) but not for the partly oxidized ones. The evolution of the Mössbauer parameters with increasing heating/oxidation can therefore only be assumed (Fig. 23).

Kartashov et al. (2002) reported the Mössbauer spectrum of a ferriallanite with 1.24 $Fe^{3+}$ and 0.93 $Fe^{2+}$ pfu. The spectrum shows three well-resolved doublets (Fig. 19e). Based on their isomer shifts, one of the doublets can be assigned to $Fe^{2+}$ whereas the other two must be assigned to $Fe^{3+}$. Due to its high Fe content, the Fe must be distributed over all three octahedral sites. The authors therefore assigned the $Fe^{2+}$ doublet (IS = 1.068 mm/sec, OS = 1.62 mm/sec) to $Fe^{2+}$ in M3 and the two $Fe^{3+}$ doublets to $Fe^{3+}$ in M1 and M2, respectively. The Mössbauer parameters of the $Fe^{3+}$ doublets (IS = 0.334 and 0.359 mm/sec and OS = 0.85 and 1.78 mm/sec for $Fe^{3+}$ in M2 and M1, respectively) closely resemble those of the two different low intensity $Fe^{3+}$ doublets in the spectra of the $Al$-$Fe^{3+}$ solid solutions (see above). It is reasonable to assume that these low intensity $Fe^{3+}$ doublets in the $Al$-$Fe^{3+}$ solid solutions must therefore be also assigned to M2 and M1, respectively.

**Intracrystalline $Al$-$Fe^{3+}$ partitioning within the $Al$-$Fe^{3+}$ solid solution series**

Dollase (1973), Bird et al. (1988), Patrier et al. (1991), and Fehr and Heuss-Aßbichler (1997) used Mössbauer spectroscopy to study the intracrystalline $Al$-$Fe^{3+}$ partitioning in $Al$-$Fe^{3+}$ solid solutions. In good accordance with the structural data, the Mössbauer spectroscopic studies have shown that Fe partitions between the different octahedral sites only for $X_{Ep} > 0.7$ but is confined exclusively to M3 for $X_{Ep} < 0.7$. Consequently, for $X_{Ep} > 0.7$ there will be a non-convergent ordering process of Al and $Fe^{3+}$ between the different octahedral sites whereas for $X_{Ep} < 0.7$ a convergent ordering process occurs (Heuss-Aßbichler 2000).

As discussed above, the Mössbauer parameters provide evidence that Patrier et al. (1991) might have interpreted their recorded spectra incorrectly and these data are therefore neglected in the following discussion. Furthermore, only the Mössbauer spectroscopic data of those samples will be considered, for which $P$ and $T$ of equilibration are known. A detailed discussion and thermodynamic treatment of the non-convergent ordering of Al and $Fe^{3+}$ between the different octahedral sites is given in the chapter on the thermodynamic properties of epidote minerals by Gottschalk (2004; his Figure 20). Dollase (1973) and Fehr and Heuss-Aßbichler (1997) studied the Al–$Fe^{3+}$ partitioning between the M3 and M1 sites in natural, heat treated epidote samples with $X_{Ep}$ = 0.79 to 1.0. Calculating the concentrations of $Fe^{3+}$ in the two sites from the relative areas of the doublets, the data indicate that the amount of $Fe^{3+}$ in M1 increases with increasing total $Fe^{3+}$ content as well as with temperature. For high $X_{Ep}$ values, the relative amount of $Fe^{3+}$ in M1 may reach up to 10 %. Bird et al. (1988) studied natural epidote samples from the Salton Sea geothermal system (see also Bird and Spieler 2004). Based on the probable downhole temperatures, Bird et al. (1988) inferred equilibration temperatures of 260 to 400°C for the studied samples. Some of the estimated $Fe^{3+}$ contents in M1 are notable higher than predicted by the data of Dollase (1973) and Fehr and Heuss-Aßbichler (1997) for these temperatures. Bird et al. (1988) interpreted this higher disorder as a metastable persistence of a former equilibrium stage. Likewise, some of the synthetic samples of Liou (1973) that were studied by Dollase (1973) also show a higher disorder than expected for the synthesis conditions. This probably reflects a metastable higher disorder inherited during the fast growth of the synthetic crystals (Fehr and Heuss-Aßbichler 1997).

Fehr and Heuss-Aßbichler (1997) studied the kinetics of the non-convergent ordering process of Al and $Fe^{3+}$ between M3 and M1. They heat treated two epidote samples with $X_{Ep}$ = 0.79 and 0.98 at 500, 600, and 650°C/0.3 GPa for 1 to 22 days under controlled oxygen fugacities of the hematite/magnetite buffer and determined the $Fe^{3+}$ partitioning between M3 and M1 by means of Mössbauer spectroscopy. The data indicate that already after 5 days equilibrium is achieved. The Mössbauer parameters for $Fe^{3+}$ in M3 and M1, respectively, did not change during the course of the heating experiments and the authors concluded that no change in the bonding character and the local environments of M1 and M3 occurred (Fehr and Heuss-Aßbichler 1997).

The spectra of the Al-$Fe^{3+}$ solid solutions with $X_{Ep}$ = 0.39, 0.43, 0.46, 0.60, and 0.70 studied by Fehr and Heuss-Aßbichler (1997) and Heuss-Aßbichler (2000) are remarkably asymmetric and display an intense peak at high velocity and a broad low velocity peak (Fig. 19a lower two spectra). To account for this asymmetry the authors evaluated the spectra with two only slightly different doublets for ferric Fe besides a very low intensity doublet for $Fe^{2+}$. The Mössbauer parameters for the two $Fe^{3+}$ doublets are very similar. Their isomer shift is 0.37 to 0.44 and 0.24 to 0.36 mm/sec and their quadrupole splitting is 1.89 to 1.97 and 1.99 to 2.32 mm/sec. In accordance with the general high quadrupole splitting and because $Fe^{3+}$ is exclusively confined to M3 at $X_{Ep}$ < 0.70 (see above), Fehr and Heuss-Aßbichler (1997) and Heuss-Aßbichler (2000) assigned both doublets to a M3 site (labeled M3 and M3′). As the quadrupole splitting of M3′ is slightly higher than that of M3, it must be slightly more distorted. They concluded that in the analyzed samples two epidote phases with slightly different local environments of their respective M3 sites coexist and that at least in the compositional range $X_{Ep}$ = 0.39 to 0.70 immiscibility occurs along the clinozoisite–epidote solid solutions series. This interpretation is consistent with the results of Dollase (1973) who also found asymmetric doublets for $Fe^{3+}$ in M3 in the samples with $X_{Ep}$ = 0.36 and 0.42, but ascribed this unequal broadening of the peaks to relaxation broadening due in part to the long average Fe-Fe distances in the low-Fe clinozoisites.

Because the samples are homogeneous on the microscale according to microprobe analyses and X-ray diffraction data, the immiscibility must form domains in the nanoscale range (Fehr and Heuss-Aßbichler 1997). Unfortunately, the authors were not able to determine the composition of at least one of the two coexisting phases. By comparison with their results on the intercrystalline exchange of Al–$Fe^{3+}$ between the grossular–andradite and clinozoisite–epidote solid solution series (Heuss-Aßbichler and Fehr 1997) they proposed two solvi in the compositional range $X_{Ep} = 0.3$ to 0.75 along the clinozoisite–epidote solid solution series that are separated by a small stability field of an intermediate Al-$Fe^{3+}$ solid solution of about $X_{Ep} = 0.5$.

To test their interpretation with two solvi, Fehr and Heuss-Aßbichler (1997) and Heuss-Aßbichler (2000) heat-treated some of the samples. With increasing temperature, the intensities of the M3 and M3′ doublets changed in accordance with the proposed closure of the solvi and indicate stable phase relations. Stable phase relations were also proven by reversal experiments at different temperatures for one of the studied samples. Although the exact positions of the solvi were not determined, their data strongly suggest that immiscibility occurs along the Al-$Fe^{3+}$ solid solution series (for a structural and thermodynamic treatment of this immiscibility see Franz and Liebscher 2004 and Gottschalk 2004, respectively).

## OTHER SPECTROSCOPIC METHODS

Beside the aforementioned spectroscopic methods that were applied to the different epidote minerals by numerous and detailed studies, there are other spectroscopic methods that were only applied by few and singular studies. Therefore these spectroscopic methods will only be shortly reviewed below.

### X-ray absorption spectroscopy

X-ray absorption spectroscopy has been applied to Al and $Fe^{3+}$ in natural monoclinic Al-$Fe^{3+}$ solid solutions and to the REE Gd, Er, and Lu in REE rich epidote. Waychunas et al. (1983, 1986) published the first $K$-edge XANES (X-ray absorption near edge structure) and $K$-edge EXAFS (extended X-ray absorption fine structure) powder spectra of $Fe^{3+}$ in epidote of unspecified composition. Later, Waychunas and Brown (1990) recorded polarized single crystal $Fe^{3+}$ $K$-edge XANES and EXAFS spectra of epidote of unspecified composition and Artioli et al. (1995) published the $Fe^{3+}$ $K$-edge XANES powder spectrum of an epidote with $X_{Ep} \sim 1.1$. X-ray absorption spectroscopy of Al in epidote minerals is restricted to the Al $K$-edge XANES spectra of two epidotes with $X_{Ep} \sim 0.54$ and $\sim 0.79$ recorded by Li et al. (1995). The $Fe^{3+}$ $K$-edge XANES powder spectra recorded by Waychunas et al. (1983) and Artioli et al. (1995) reveal three and four main features, respectively. They are characterized by a pre-edge feature at uniform $\sim 7,113$ eV and one or two shoulders at $\sim 7,122$ eV and $\sim 7,122$ and 7,128 eV, respectively. The edge-crest occurs at $\sim 7,129$ and $\sim 7,133$ eV. The polarized single crystal $Fe^{3+}$ $K$-edge XANES spectra of epidote recorded by Waychunas and Brown (1990) with E parallel [010] and E normal to (100) display the pre-edge feature at 7,112.5 eV (Fig. 24) consistent with the powder spectra. Two shoulders at 7,121.0 and 7,124.0 eV appear only in the E parallel [010] spectrum whereas both spectra show a first large feature at 7,126.0 and 7,126.5 eV, respectively, and a second one at $\sim 7,132$ eV (Fig. 24). From the polarized EXAFS spectra Waychunas and Brown (1990) calculated $Fe^{3+}$–O distances of $Fe^{3+}$–O2 = 1.952 Å and $Fe^{3+}$–O1 = 2.225 Å (E parallel [010] spectrum) and mean $Fe^{3+}$–(O4, O8) = 1.959 Å and $Fe^{3+}$–O1 = 2.236 Å (E normal (100) spectrum). The unpolarized EXAFS spectrum of Waychunas et al. (1986) gives $Fe^{3+}$–O1 = 2.253 Å and mean $Fe^{3+}$–(O2, O4, O8) = 1.954 Å. These results are in fairly good agreement with the XRD data (see Franz and Liebscher 2004).

The Al $K$-edge XANES spectra recorded by Li et al. (1995) yield uniform edge–crest positions of 1,568.0 eV for $X_{Ep} \sim 0.54$ and 1,567.8 eV for $X_{Ep} \sim 0.79$. Unfortunately, the authors did not publish the spectra nor provide details on further spectral properties.

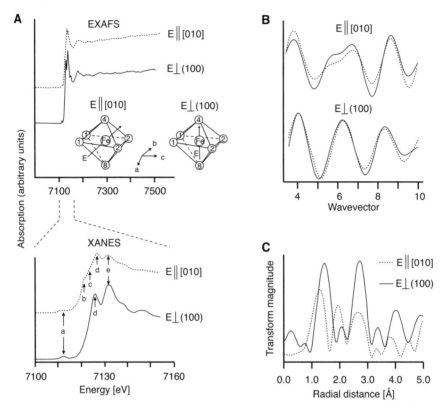

**Figure 24.** (A) Polarized single crystal XANES and EXAFS spectra of epidote. The polarizations E parallel [010] and E normal to (100) are synonymous with probing the $Fe^{3+}$–O1, O2 and $Fe^{3+}$–O4, O8 bonds of the M3 octahedron, respectively (see polyhedra sketches). (B) Epidote polarized EXAFS fitted filtered spectra. Upper part for E parallel [010] and lower part for E normal to (100). Solid lines represent the observed EXAFS, stippled lines the model calculation fit based on two separate Fe–O distances. (C) Polarized EXAFS structure functions of the EXAFS spectra shown in (A). The phase shift functions are not considered and therefore the peaks do not coincide with actual absorber-backscatterer distances. The peak at about 2.0 Å is non-physical and is produced by the large amplitude beat in the EXAFS due to the $Fe^{3+}$–O1 and $Fe^{3+}$–O2 distances. Redrawn and modified after Waychunas and Brown (1990).

Cressey and Steel (1988) recorded $L_{III}$ edge EXAFS spectra of Gd, Er, and Lu in synthetic epidote with composition $CaLa_{0.9}X_{0.1}Al_2MgSi_3O_{12}(OH)$ (X = Gd or Er or Lu). Based on detailed multiple shell analysis of EXAFS pair distribution functions and comparing them with those calculated for natural epidote they assigned Gd to the A2 site, Er to A1, and Lu to an octahedral environment, probably M3.

**Electron paramagnetic resonance spectroscopy**

Electron paramagnetic resonance spectroscopy has been applied to zoisite by Ghose and Tsang (1971), Hutton et al. (1971), and Tsang and Ghose (1971). Ghose and Tsang (1971) and Tsang and Ghose (1971) studied natural V and Mn bearing Al-$Fe^{3+}$ zoisite. Based on the recorded electron paramagnetic resonance spectra they concluded that V is both di- and trivalent and Mn divalent in the studied samples. They assigned $Mn^{2+}$ and $Fe^{3+}$ exclusively to the A1 and M3 sites, respectively. In case of $V^{2+}$ they found evidence that it occupies two different crystallographic sites and assigned it to the A1 and A2 sites, respectively, with a preference for A1. The electron paramagnetic resonance spectroscopic results of Hutton et al. (1971) on natural untreated and heat-treated zoisite are at variance with those of Ghose and Tsang (1971) and Tsang and Ghose (1971). Hutton et al. (1971) interpreted the line that was attributed to $Fe^{3+}$ by Ghose and Tsang (1971) and Tsang and Ghose (1971) as due to $Cr^{3+}$. Additionally, Hutton et al. (1971) preclude the existence of $V^{2+}$ but assigned the observed V lines to $VO^{2+}$ and $V^{4+}$ located in M1,2 and M3.

## Raman spectroscopy

Le Cleac'h et al. (1988) studied the Raman spectra of zoisite and clinozoisite between 40 and 4,000 cm$^{-1}$. They observed the lowest optic mode at 115 cm$^{-1}$ in zoisite and at 87 cm$^{-1}$ in clinozoisite. Wang et al. (1994) report the Raman spectrum of an epidote and Huang (1999) the Raman spectrum of a zoisite between 150 and 1,500 cm$^{-1}$ and between 2,800 and 3,800 cm$^{-1}$. Because Wang et al. (1994) did not provide further details, the band positions can only be estimated. The epidote spectrum is characterized by intense bands at ~400 to 420, ~550, ~910, and ~1080 cm$^{-1}$ and a lack of bands between ~600 and 820 cm$^{-1}$ (Fig. 25). The zoisite spectrum of Huang (1999) displays 19 bands in the range 150 to 1,500 cm$^{-1}$. Of these bands the most intense one (band 9) occurs at 490 cm$^{-1}$ and is assigned to a Si–O bending mode (Huang 1999). Compared to epidote, the zoisite spectrum is further characterized by band 14 at ~680 cm$^{-1}$, band 15 at ~870 cm$^{-1}$, and the bands 18 and 19 at ~1,070 and 1,091 cm$^{-1}$ (Fig. 25). The OH stretching mode in zoisite appears at 3,151 cm$^{-1}$ (Huang 1999).

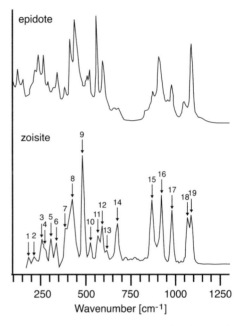

**Figure 25.** Raman spectra of natural epidote and zoisite of unspecified composition. Redrawn and modified from Wang et al. (1994) and Huang (1999).

## Nuclear magnetic resonance spectroscopy

Nuclear magnetic resonance (NMR) spectroscopy on epidote minerals is restricted to the studies by Brinkmann et al. (1969) and Alemany et al. (2000) on zoisite. Brinkmann et al. (1969) studied the nuclear magnetic resonance of $^{27}Al$ and $^{1}H$ in zoisite. For Al in M1,2 and M3 they found quadrupole coupling constants ($e^2qQ/h$) and asymmetry parameters ($\eta$) of $e^2qQ/h = 8.05 \pm 0.12$ MHz and $\eta = 0.46 \pm 0.01$ for M1,2 and $e^2qQ/h = 18.50 \pm 0.05$ MHz and $\eta = 0.160 \pm 0.005$ for M3. The nuclear magnetic resonance of $^{1}H$ confirms the proton position as deduced from X-ray diffraction data (Brinkmann et al. 1969). Alemany et al. (2000) studied the nuclear magnetic resonance of $^{27}Al$ in zoisite by single-puls magic-angle spinning (MAS), selective Hahn echo MAS, and 3QMAS NMR spectroscopy and determined the isotropic chemical shift ($\delta_{CS}$), $e^2qQ/h$, and $\eta$ for Al in M1,2 and M3. With $e^2qQ/h = 7.9 \pm 0.1$ MHz and $\eta = 0.51 \pm 0.05$ for M1,2 and $e^2qQ/h = 18.4 \pm 0.1$ MHz and $\eta = 0.16 \pm 0.01$ for M3 their data agree well with those of Brinkmann et al. (1969). The isotropic chemical shift is $\delta_{CS} = 10.7 \pm 0.2$ ppm for M1,2 and $\delta_{CS} = 8.0 \pm 0.2$ ppm for M3. From their data Alemany et al. (2000) conclude that the signal of the distorted M3 site can be observed in MAS NMR spectra that were obtained with fields of at least 11.7 T and spinning speeds over 30 kHz. The signal intensity of the distorted M3 site is increased significantly in MQMAS NMR experiments using amplitude-modulated pulses to generate double frequency sweeps.

## CONCLUDING REMARKS

The numerous spectroscopic studies applied to the epidote minerals have established a profound knowledge of the crystal chemical characteristics of these minerals. Optical absorption and Mössbauer spectroscopy elucidate the valence states and site preferences of the transition metal ions in epidote minerals. Infrared spectroscopy has been proven a powerful tool to study structural changes within the different solid solution series and to distinguish between orthorhombic and monoclinic epidote minerals as well as to determine the respective compositions. However, there are still some open questions. Within the well-established framework, promising and also challenging topics for future spectroscopic work include:

(i)   Only few studies recorded pressure dependent optical absorption spectra and their results are inconsistent. More studies on this topic will help to solve this inconsistency and also to derive precise compression moduli for individual octahedral.

(ii)  X-ray data indicate that $Fe^{3+}$ also substitute in M2 in monoclinic Al-$Fe^{3+}$ solid solutions. Mössbauer studies have so far assigned all $Fe^{3+}$ doublets in monoclinic Al-$Fe^{3+}$ solid solutions to either M3 or M1. As recommended in this review, one kind of doublet may be assigned to M2. Future Mössbauer studies should try to verify or falsify this recommendation. Best suited for this purpose are synthetic monoclinic Al-$Fe^{3+}$ solid solutions for which precise X-ray data indicate $Fe^{3+}$ in M3, M1, and also M2.

(iii) The intracrystalline ordering of $Fe^{3+}$ in monoclinic Al-$Fe^{3+}$ solid solutions can be studied by Mössbauer spectroscopy. Unfortunately, the exact positions and closure temperatures of the proposed solvi are still not known. To unravel the question of immiscibility along the clinozoisite–epidote solid solution join is probably one of the most important and promising topics for future studies.

(iv)  Natural zoisite and clinozoisite samples often show very fine lamellae that are too fine to be resolved by electron microprobe. The different interference colors of these lamellae suggest them to represent coexisting zoisite and clinozoisite. Infrared spectroscopic studies on such specimens would help to distinguish between zoisite and clinozoisite and to determine the composition of both.

(v)   Mössbauer spectra of zoisite are completely lacking. As the iron content in natural zoisite is generally low, Mössbauer spectra of zoisite should be recorded on synthetic zoisite doped with $^{57}Fe$.

## ACKNOWLEDGMENTS

This review benefited from thorough and critical readings by M. Koch-Müller and M. Andrut. Fruitful discussions during the preparation of the manuscript with M. Koch-Müller, I. Abs-Wurmbach, and M. Wilke are gratefully acknowledged. Prof. Jan Stanek from Kraków kindly provided some literature. Throughout the preparation of the manuscript, K. Mai and S. Bernau helped with literature search, figure drawing, and proof reading.

# REFERENCES

Abu-Eid (1974) Absorption spectra of transition metal-bearing minerals at high pressures. *In*: The Physics and Chemistry of Minerals and Rocks. Strens RGJ (ed) Wiley, New York, p 641-675

Alemany LB, Callender RL, Barron AR, Steuernagel S, Iuga D, Kentgens APM (2000) Single-pulse MAS, selective Hahn echo MAS, and 3QMAS NMR Studies of the mineral zoisite at 400, 500, 600, and 800 MHz. Exploring the limits of Al NMR detectability. J Phys Chem B 104:11612-11616

Artioli G, Quartieri S, Deriu A (1995) Spectroscopic data on coexisting prehnite-pumpellyite and epidote-pumpellyite. Can Mineral 33:67-75

Bancroft GM, Maddock AG, Burns RG (1967) Applications of the Mössbauer effect to silicate mineralogy. I. Iron silicates of known crystal structure. Geochim Cosmochim Acta 31:2219-2246

Bellamy LJ, Owen AJ (1969) A simple relationship between the infrared stretching frequencies and the hydrogen bond distance in crystals. Spectrochim Acta 25A:329-333

Belov NV, Rumanova JM (1954) The crystal structure of epidote. Trudy Inst Krist Akad Nauk SSSR 9:103-164 (Abstract in Struct Rep 18:544-545)

Bird D, Spieler AR (2004) Epidote in geothermal systems. Rev Mineral Geochem 56:235-300

Bird DK, Cho M, Janik CJ, Liou JG, Caruso LJ (1988) Compositional, order/disorder, and stable isotope characteristics of Al-Fe epidote, state 2-14 drill hole, Salton Sea Geothermal System. J Geophys Res 93: 13135-13144

Bonazzi P, Menchetti S (2004) Manganese in monoclinic members of the epidote group: piemontite and related minerals. Rev Mineral Geochem 56:495-552

Bonazzi P, Menchetti S, Palenzona A (1990) Strontiopiemontite, a new member of the epidote group, from Val Graveglia, Liguria, Italy. Eur J Mineral 4:519-523

Bradbury SE, Williams Q (2003) Contrasting bonding behavior of two hydroxyl-bearing metamorphic minerals under pressure: Clinozoisite and topaz. Am Mineral 88:1460-1470

Brinkmann D, Staehli JL, Ghose S (1969) Nuclear magnetic resonance of $^{27}$Al and $^{1}$H in zoisite, $Ca_2Al_3Si_3O_{12}(OH)$. J Chem Phys 51:5128-5133

Brunsmann A (2000) Strukturelle, kristallchemische und phasenpetrologische Untersuchungen an synthetischen und natürlichen Zoisit und Klinozoisit Mischkristallen. Dissertation, Technical University of Berlin, Germany (*http://edocs.tu-berlin.de/diss/2000/brunsmann_axel.htm*)

Burns RG, Strens RGJ (1967) Structural interpretation of polarized absorption spectra of the Al-Fe-Mn-Cr epidotes. Min Mag 36:204-226

Catti M, Ferraris G, Ivaldi G (1989) On the crystal chemistry of strontian piemontite with some remarks on the nomenclature of the epidote group. N Jahrb Mineral Monatsh 1989:357-366

Cech F, Vrána S, Povondra P (1972) A non-metamict allanite from Zambia. N Jahrb Miner Abh 116:208-223

Comodi P, Zanazzi PF (1997) The pressure behavior of clinozoisite and zoisite: An X-ray diffraction study. Am Mineral 82:61-68

Cressey G, Steel AT (1988) An EXAFS study of Gd, Er, and Lu site location in the epidote structure. Phys Chem Mineral 15:304-312

Cynn H, Hofmeister A (1994) High-pressure IR spectra of lattice modes and OH vibrations in Fe-bearing wadsleyite. J Geophys Research B 99:717-727

De Coster M, Pollak H, Amelinckx S (1963) A study of Mössbauer absorption in iron silicates. Physica Stat Solidi 3:283-288

Della Ventura G, Mottana A, Parodi GC, Griffin WL (1996) FTIR spectroscopy in the OH-stretching region of monoclinic epidotes from Praborna (St. Marcel, Aosta Valley, Italy). Eur J Mineral 8:655-665

Dollase WA (1968) Refinement and comparison of the structures of zoisite and clinozoisite. Am Mineral 53: 1882-1898

Dollase WA (1969) Crystal structure and cation ordering of piemontite. Am Mineral 54:710-717

Dollase WA (1971) Refinement of the crystal structures of epidote, allanite and hancockite. Am Mineral 56: 447-464

Dollase WA (1973) Mössbauer spectra and iron distribution in the epidote group minerals. Z Kristallogr 138: 41-63

Farmer VC (1974) Orthosilicates, pyrosilicates, and other finite-chain silicates. *In:* The Infrared Spectra of Minerals 4. Monograph of the Mineralogical Society. Farmer VC (ed) Mineralogical Society, London, p 285-303

Faye GH, Harris DC (1969) On the origin of colour and pleochroism in andalusite from Brazil. Can Mineral 10:47-56

Faye GH, Nickel EH (1971) On the pleochroism of vanadium-bearing zoisite from Tanzania. Can Mineral 10: 812-821

Fehr KT, Heuss-Aßbichler S (1997) Intracrystalline equilibria and immiscibility along the join clinozoisite–epidote: An experimental and $^{57}$Fe Mössbauer study. N Jahrb Mineral Abh 172:43-76

Ferraris G, Ivaldi G, Fuess H, Gregson D (1989) Manganese/iron distribution in a strontian piemontite by neutron diffraction. Z Krist 187:145-151

Franz G, Liebscher A (2004) Physical and chemical properties of the epidote minerals–an introduction. Rev Mineral Geochem 56:1-82

Gavorkyan SV (1990) IR spectra of epidote-group minerals. Mineralogiceskij zurnal 12:63-66 (in Russian)

Ghose S, Tsang T (1971) Ordering of $V^{2+}$, $Mn^{2+}$, and $Fe^{3+}$-ions in zoisite, $Ca_2Al_3Si_3O_{12}$ (OH). Science 171:374-376

Gieré R, Sorensen SS (2004) Allanite and other REE-rich epidote-group minerals. Rev Mineral Geochem 56:431-494

Giuli G, Bonazzi P, Menchetti S (1999) Al-Fe disorder in synthetic epidotes: a single-crystal X-ray diffraction study. Am Mineral 84: 933-936

Gottschalk M (2004) Thermodynamic properties of zoisite, clinozoisite and epidote. Rev Mineral Geochem. 56:83-124

Grodzicki M, Heuss-Aßbichler S, Amthauer G (2001) Mössbauer investigations and molecular orbital calculations on epidote. Phys Chem Mineral 28:675-681

Grum-Grzhimailo SV, Brilliantov NA, Sviridov DT, Sviridova RK, Sukhanova ON (1963) Absorption spectra of crystals containing $Fe^{3+}$ at temperatures down to 1.7 K. Opt Spect 14:118-120 (in Russian)

Hanisch K, Zemann J (1966) Messung des Ultrarot-Pleochroismus von Mineralen. IV. Der Pleochroismus der OH-Streckfrequenz in Epidot. N Jahrb Miner Monatsh:19-23

Herzberg G (1950) Molecular Spectra and Molecular Structure. Van Nostrand, New York

Heuss-Aßbichler S (2000) Ein neues Ordnungsmodell für die Mischkristallreihe Klinozoisit-Epidot und das Granat-Epidot-Geothermometer. Habilitationthesis, München, 105 pp

Heuss-Aßbichler S, Fehr KT (1997) Intercrystalline exchange of Al and $Fe^{3+}$ between grossular-andradite and clinozoisite-epidote solid solutions. N Jahrb Mineral Abh 172:69-100

Holland TJB, Redfern SAT, Pawley AR (1996) Volume behavior of hydrous minerals at high pressure and temperature: II. Compressibilities of lawsonite, zoisite, clinozoisite, and epidote. Am Mineral 81:341-348

Huang E (1999) Raman spectroscopic study of 15 gem minerals. J Geol Soc China 42:301-318

Hutton DR, Troup GJ, Stewart GA (1971) Paramagnetic ions in zoisite. Science 174:1259

Ito T, Morimoto N, Sadanga R (1954) On the structure of Epidote. Acta Crystallographica 7:53-59

Janeczek J, Sachanbinski M (1989) Chemistry and zoning of thulite from the Wiry magnesite deposit, Poland. N Jahrb Miner Monatsh:325-333

Kartashov PM, Ferraris G, Ivaldi G, Sokolova E, McCammon CA (2002) Ferriallanite-(Ce), $CaCeFe^{3+}AlFe^{3+}$ $(SiO_4)(Si_2O_7)O(OH)$, a new member of the epidote group: Description, X-ray and Mössbauer study. Can Mineral 40:1641-1648

Kersten M, Langer K, Almen H, Tillmanns E (1988) The polarized single crystal spectra and structures of synthetic thulite and piemontites, $Ca_2Al_{3-p}Mn^{3+}{}_p[O/OH/SiO_4/Si_2O_7]$, with $0.5 \leq p \leq 1.6$. Z Kristall 185:111

Kiseleva IA, Ogorodnikova LP (1987) Calorimetric data on the thermodynamics of epidote, clinozoisite, and zoisite. Geochem Int 24:91-98

Langer K (1990) High pressure spectroscopy. *In*: Absorption Spectroscopy in Mineralogy. Mottana A, Burragato F (eds) Elsevier, Amsterdam, p 228-284

Langer K, Abu-Eid RM (1977) Measurements of the polarized absorption spectra of synthetic transition metal-bearing silicate microcrystals in the spectral range 44,000-4,000 cm$^{-1}$. Phys Chem Mineral 1:273-299

Langer K, Lattard D (1980) Identification of a low-energy OH-valence vibration in zoisite. Am Mineral 65:779-783

Langer K, Raith M (1974) Infrared spectra of Al-Fe(III)-epidotes and zoisites, $Ca_2(Al_{1-p}Fe^{3+}{}_p)Al_2O(OH)[Si_2O_7]$ $[SiO_4]$. Am Mineral 59:1249-1258

Langer K, Abu-Eid RM, Anastasiou P (1976) Absorptionsspektren synthetischer Piemontite in den Bereichen 43000-11000 cm$^{-1}$ (232,6-909,1 nm) und 4000-250 cm$^{-1}$ (2,5-40 μm). Z Kristall 144:434-436

Langer K, Tillmanns E, Kersten M, Almen H, Arni RK (2002) The crystal chemistry of $Mn^{3+}$ in the clino- and orthozoisite structure types, $Ca_2M^{3+}{}_3[OH/O/SiO_4/Si_2O_7]$: A structural and spectroscopic study of some natural piemontites and "thulites" and their synthetic equivalents. Z Kristall 217:563-580

Lazarev AN (1972) Vibrational Spectra and Structure of Silicates. Plenum, New York.

Le Cleac'h A, Gillet P, Putnis A (1988) IR, Raman spectra and thermodynamic properties of zoisite and clinozoisite. Terra Cognita 8:70

Li D, Bancroft GM, Fleet ME, Feng XH, Pan Y (1995) Al K-edge XANES spectra of aluminosilicate minerals. Am Mineral 80:432-440

Liebscher A, Gottschalk M (2004) The *T-X* dependence of the isosymmetric displacive phase transition in synthetic $Fe^{3+}$-Al zoisite: A temperature-dependent infrared spectroscopy study. Am Mineral 89:31-38

Liebscher A, Gottschalk M, Franz G (2002) The substitution $Fe^{3+}$-Al and the isosymmetric displacive phase transition in synthetic zoisite: A powder X-ray and infrared spectroscopy study. Am Mineral 87:909-921

Linke W (1970) Messung des Ultrarot-Pleochroismus von Mineralen. X. Der Pleochroismus der OH-Streckfrequenz in Zoisit. Tschermaks Mineral Petrogr Mitt 14:61-63

Liou JG (1973) Synthesis and stability relations of epidote, $Ca_2Al_2FeSi_3O_{12}(OH)$. J Petrol 14:381-314

Lipka J, Petrik I, Tóth I, Gajdosová M (1995) Mössbauer study of allanite. Acta Phys Slovaca 45:61-66

Marfunin AS, Mineyeva RM, Mkrtchyan AR, Nyussik YM, Fedorov VY (1967) Optical and Mössbauer spectroscopy of iron in rock-forming silicates. Seriya Geologicheskaya 10:86-102 (in Russian)

Mingsheng P, Daoyuan R (1986) A spectroscopic study on the allanite from a granite body in northeastern Guangdong. J Central-South Inst Mining Metall 4:1-9 (in Chinese)

Mingsheng P, Dien L (1987) Spectroscopy, genesis, and process properties of partly metamict allanite. J Central-South Inst Mining Metall 18:362-368

Moenke H (1962) Mineral-Spektren. Akademie Verlag Berlin

Paesano A, Kunrath JI, Vasquez A (1983) A $^{57}Fe$ Mössbauer study of epidote. Hyperfine Interactions 15/16: 841-844

Parkin KM, Burns RG (1980) High temperature crystal field spectra of transition metal-bearing minerals: Relevance to remote-sensed spectra of planetary surfaces. Proc 11th Lunar Planet Sci Conf 1:731-755

Patrier P, Beaufort D, Meunier A, Eymery JP, Petit S (1991) Determination of the nonequilibrium ordering state in epidote from the ancient geothermal field of Saint Martin: Application of Mössbauer spectroscopy. Am Mineral 76:602-610

Pawley AR, Redfern SAT, Holland TJB (1996) Volume behavior of hydrous minerals at high pressure and temperature: I. Thermal expansion of lawsonite, zoisite, clinozoisite, and diaspore. Am Mineral 81:335-340

Perseil EA (1987) Particularités des piémontites de Saint-Marcel-Praborna (Italie); spectres IR. Actes du 112. Congrès National des Sociétés Savantes. Edition du CTHS, Paris.

Petrusenko S, Taran MN, Platonov AN Gavorkyan SV (1992) Optical and infrared spectoscopic studies of epidote group minerals from the Rhodope region. Rev Bulgarian Geol Soc 53:1-9 (in Russian)

Pollack H, Bruyneel W (1975) On the iron distribution between sites in epidote. *In*: Hrynkiewicz AZ, Sawicki JA (eds) Proceedings of the international Conference on Mössbauer Spectroscopy, Cracow, Poland, 259-260

Robinson K, Gibbs GV, Ribbe PH (1971) Quadratic elongation: A quantitative measure of distortion in coordination polyhedra. Science 172:567-570

Salje EKH, Carpenter MA, Malcherek T, Boffa Ballaran T (2000) Autocorrelation analysis of infrared spectra from minerals. Eur J Mineral 12:503-519

Schmetzer K, Bank H (1979) Bluish-green zoisite from Merelani, Tanzania. J Gemm 16:512-513

Schmetzer K, Berdesinski W (1978) Das Absorptionsspektrum von $Cr^{3+}$ in Zoisit. N Jb Mineral Monatsh: 197-202

Scott H, Williams Q (1999) An infrared spectroscopic study of lawsonite to 20 GPa. Phys Chem Mineral 26: 437-445

Smith G, Hålenius U, Langer K (1982) Low temperature spectral studies of $Mn^{3+}$-bearing andalusite and epidote type minerals in the range 30,000-5,000 $cm^{-1}$. Phys Chem Mineral 8:136-142

Smith JV, Pluth JJ, Richardson Jr JW, Kvick A (1987) Neutron diffraction study of zoisite at 15 K and X-ray study at room temperature. Z Kristall 179:305-321

Taran MN, Langer K (2000) Electronic absorption spectra of $Fe^{3+}$ in andradite and epidote at different temperatures and pressures. Eur J Mineral 12:7-15

Taran MN, Platonov AN, Petrusenko SI, Khomenko VM, Belichenko VP (1984) Optical absorption spectra of $Mn^{3+}$ ion in natural clinozoisites. Geochem Mineral Petrol 19:43-51 (in Russian)

Tsang T, Ghose S (1971) Electron paramagnetic resonance of $V^{2+}$, $Mn^{2+}$, $Fe^{3+}$ and optical spectra of $V^{3+}$ in blue zoisite, $Ca_2Al_3Si_3O_{12}(OH)$. J Chem Phys 4,3:856-862

Wang A, Han J, Guo L, Yu J, Zeng P (1994) Database of standard Raman spectra of minerals and related inorganic crystals. Applied Spectroscopy 48:959-968

Waychunas GA, Brown Jr GE (1990) Polarized X-ray absorption spectroscopy of metal ions in minerals. Phys Chem Mineral 17:420-430

Waychunas GA, Apted MJ. Brown Jr GE (1983) X-ray K-edge absorption spectra of Fe minerals and model compounds: Near-edge structure. Phys Chem Mineral 10:1-9

Waychunas GA, Brown Jr GE, Apted MJ (1986) X-ray K-edge absorption spectra of Fe minerals and model compounds: II. EXAFS. Phys Chem Mineral 13:31-47

White WB, Keester KL (1966) Optical absorption spectra of iron in the rock-forming silicates. Am Mineral 51:774-791

Williams Q (1992) A vibrational spectroscopic study of hydrogen in high-pressure mineral assemblages. *In*: High-Pressure Research: Application to Earth and Planetary Sciences. Syono Y, Manghnani MH (eds) Am Geopyhs Union, Washington D.C., p 289-296

Winkler B, Langer K, Johannsen PG (1989) The influence of pressure on the OH-valence vibration of zoisite. Phys Chem Mineral 16:668-671

Wood BJ, Strens RGJ (1972) Calculation of crystal field splittings in distorted coordination polyhedra: spectra and thermodynamic properties of minerals. Mineral Mag 38:909-917

Reviews in Mineralogy & Geochemistry
Vol. 56, pp. 171-195, 2004
Copyright © Mineralogical Society of America

**4**

# Experimental Subsolidus Studies on Epidote Minerals

## Stefano Poli

*Dipartimento di Scienze della Terra*
*Università degli Studi di Milano*
*Via Botticelli 23*
*20133 Milano, Italy*

## Max W. Schmidt

*Institute for Mineralogy and Petrology*
*ETH*
*8092 Zürich, Switzerland*

## INTRODUCTION

Despite the fact that epidote group minerals are very typical for metamorphism at very low pressure, e.g., in geothermal fields (Bird and Spieler 2004), the first successful synthesis of zoisite and epidote$_{ss}$ was reported by Coes (1955) in a paper in the Journal of American Ceramic Society entitled "High pressure minerals." Synthesis conditions were 1 GPa at 800°C; zoisite was obtained from a mixture of kaolin, $SiO_2$, CaO, and $CaCl_2$, whereas epidote was formed by adding $FeCl_2 \cdot H_2O$ to the previous mixture. Once experimental facilities enabled pressures exceeding a few hundred MPa, zoisite and epidote minerals were easily obtained from a variety of starting materials, made of oxides, gels and glasses.

Historically, early experimental studies on epidote focused on the formation at low pressure conditions, and then ventured into the simple system $CaO$-$Al_2O_3$-$SiO_2$-$H_2O$ at conditions attainable by piston cylinder equipment (Newton and Kennedy 1963; Boettcher 1970) in which zoisite was found to have an extremely large temperature stability. Then, the role of $Fe^{3+}$ was investigated systematically at pressures typical for the middle and lower continental crust (Holdaway 1972; Liou 1973). Epidote minerals in bulk compositions directly applicable to natural rocks were not investigated experimentally until the early 70's (Liou et al. 1974; Apted and Liou 1983). Subsequent studies in the context of the very popular hydrous phase stabilities at subduction conditions in the 90's extended the experimentally determined stability of epidote$_{ss}$ in natural compositions to 3.5 GPa. With the relatively easy access to multi-anvil machines, the pressure stability of zoisite was defined (Poli and Schmidt 1998).

The increasing number of experimental studies on epidote minerals reveals that the members of this group of ubiquitous rock forming minerals have huge stability fields, which extend to 7 GPa in pressure and to more than 1200°C in temperature. However, most of the experimental studies were performed at subsolidus conditions, thus melting relations are still partially unknown.

In this chapter we focus on experimental studies in the subsolidus performed both on simplified model systems and on natural (or close to natural) rock compositions where epidote$_{ss}$ is described by the components (clino)zoisite and epidote. Experimental investigations

1529-6466/04/0056-0004$05.00

clarifying the role of epidote in magmatic systems are discussed in Schmidt and Poli (2004). Experimental studies on Mn bearing epidote minerals are presented by Bonazzi and Menchetti (2004), those on REE bearing allanite by Gieré and Sorensen (2004).

Unless differently stated the term epidote$_{ss}$ employed in this chapter is intended to describe epidote mineral compositions along the pseudobinary join (clino)zoisite-epidote.

## THE MODEL SYSTEM CaO-Al$_2$O$_3$-SiO$_2$-H$_2$O (CASH)
## AND PHASE RELATIONS INVOLVING ZOISITE

This chemical system includes a variety of major rock forming minerals (e.g., kyanite/ andalusite/sillimanite, grossular, anorthite, wollastonite, margarite, prehnite, lawsonite; Fig. 1). It thus has received particular attention in petrology, for modeling both low temperature transformations including Ca-poor bulk compositions (e.g., metapelites) and high temperature processes where Ca-Al-bearing hydrous silicates strongly affect melting of crustal rocks. Furthermore, the four component CASH system provides the basis for reaction bundles in the system CMASH (CaO-MgO-Al$_2$O$_3$-SiO$_2$-H$_2$O) where pyroxene, amphibole and garnet solid solutions are involved in additional five component reactions. This latter chemical system constitutes the most simple system sufficient for understanding the basic phase relations of both mafic and intermediate rocks.

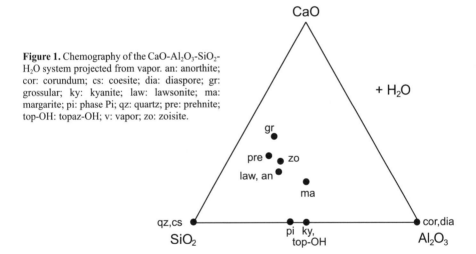

**Figure 1.** Chemography of the CaO-Al$_2$O$_3$-SiO$_2$-H$_2$O system projected from vapor. an: anorthite; cor: corundum; cs: coesite; dia: diaspore; gr: grossular; ky: kyanite; law: lawsonite; ma: margarite; pi: phase Pi; qz: quartz; pre: prehnite; top-OH: topaz-OH; v: vapor; zo: zoisite.

### Zoisite – clinozoisite

Since the first attempts to synthesize epidote minerals in Fe-free systems, zoisite was found to be largely dominant as a product phase whereas nucleation of clinozoisite was observed randomly distributed in a wide *P-T* range by Pistorius et al. (1962) from 1.6 GPa, 500°C to 3.0 GPa, 700°C, starting from anorthite glass, scolecite + water, and lawsonite + coesite, by Matthews and Goldsmith (1984) at 450 and 600°C and 1 to 2 GPa, by Newton and Kennedy (1963) at 4.0 GPa, 450°C from laumontite + water, and by Schmidt and Poli (1994) at 2.3 GPa, 500°C from anorthite + 2 wt% water. Jenkins et al. (1983, 1985) conclude that the equilibrium transition clinozoisite-zoisite situates below 200°C at pressures up to 1 GPa and clinozoisite is a metastable phase in Fe-free systems at most geologically relevant conditions.

Experiments performed at pressures up to 4.5 GPa at 700°C and to 7 GPa at 1000°C, starting from zoisite-bearing crystalline mixtures yield zoisite (Poli and Schmidt 1998) and give no indication for an enlarged stability field of clinozoisite at higher pressures. Thus, we maintain that zoisite is the stable Fe-free epidote mineral at experimentally accessible and geologically relevant conditions. Further and more precise, though indirect, indications about the relative stabilities of clinozoisite and zoisite have been gained from experiments in the system CaO-$Fe_2O_3$-$Al_2O_3$-$SiO_2$-$H_2O$-$O_2$ (CFASHO) confirming a very limited stability of pure CASH-clinozoisite. This will be discussed below.

### Building the CASH phase diagram on the basis of experimental results

Given the amount of experiments performed in the model system CASH in the last forty years, this section is intended to offer a synthesis of information available on zoisite-bearing assemblages and to construct a phase diagram (Fig. 2) based on a critical overview (obviously somewhat biased by personal evaluation) of experimental results. First, it should be noted that

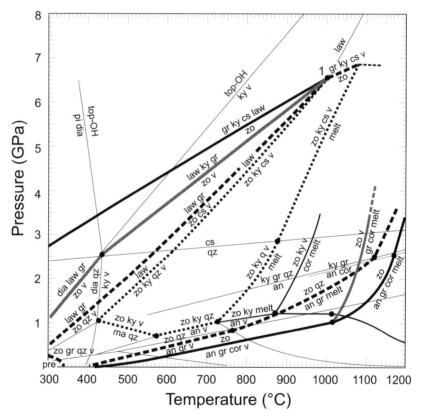

**Figure 2.** Experimentally determined reactions in the CaO-$Al_2O_3$-$SiO_2$-$H_2O$ system. The maximum stability fields of zoisite (heavy black lines), zoisite + vapor (gray lines), zoisite + quartz/coesite (dashed lines) and zoisite + kyanite + quartz/coesite + vapor (dotted lines) are outlined. Note the positive slope and therefore the "solidification" character with temperature of reaction zoisite + kyanite + melt = anorthite + vapor at low pressure and the "hydration with temperature" of reaction grossular + kyanite + coesite + $H_2O$ = zoisite at high pressure. For details of melting relations compare Figure 2 in Schmidt and Poli (2004). Mainly based on Boettcher (1970), Goldsmith (1981), Chatterjee et al. (1984), Schmidt and Poli (1994), Poli and Schmidt (1998), Wunder et al. (1993).

inconsistencies between experimental works in CASH are minor and can be largely ascribed to developments in experimental techniques (e.g., differences in pressure determination by Boettcher 1970 and Goldsmith 1982, see below) or to intrinsic experimental uncertainties. Such uncertainties are in the order of 5°C and 30 MPa in piston cylinder experiments and 20°C and 200 MPa in multianvil experiments at *P-T* conditions within the zoisite stability field. The advent of the so-called "internally consistent" thermodynamic databases has led to some underestimation of the importance of the original experimental results and the corresponding phase diagram computed using most popular datasets (e.g., Holland and Powell 1998, update 2002) largely differ both in the high pressure and in the high temperature regions (see also Gottschalk 2004 for a discussion on thermodynamic data). It is not the purpose of this chapter to discuss thermodynamic properties of phases in CASH and we will therefore focus on phase relations deduced on the basis of experimental results and Schreinemakers' analysis.

Because of a stability field which ranges from less than 300 to >1200°C in temperature and from 0.1 to 7 GPa in pressure, zoisite is omnipresent in phase relationships in CASH. The fundamental layout of this model system was determined and discussed by Boettcher (1970) and Chatterjee et al. (1984) in two comprehensive experimental studies mainly devoted to the high and low temperature region, respectively. High pressure phase relationships mainly rely on the data by Schmidt and Poli (1994) and Poli and Schmidt (1998).

Figure 2 emphasizes the reaction curves limiting the stability fields of zoisite, of zoisite + quartz/coesite, of zoisite + vapor (i.e., an aqueous fluid), and of zoisite + kyanite + quartz/coesite + vapor, which are most relevant to common zoisite-bearing rocks. It is worth noting, that the breakdown of zoisite both at high pressure below ca. 1000°C, as well as at high-temperature conditions above ca. 1 GPa occurs via fluid-absent ($H_2O$-conserving) reactions. This poses some experimental problems due to sluggish kinetics at fluid absent conditions, coupled to the well documented persistence and/or metastable growth of zoisite/clinozoisite (Matthews and Goldsmith 1984; Poli and Schmidt 1997, 1998). Such an extended metastable persistence is not only of interest for the interpretation of experiments but might also have some geological consequences, especially in subducting slabs at low temperature.

The relatively low pressure, central portion of the phase diagram (Fig. 2) is strongly constrained by the location of the upper pressure stability limit of anorthite + vapor through the equilibrium (the high temperature assemblage is on the right side of all equations which follow, unless differently stated)

$$2 \text{ zoisite} + 1 \text{ kyanite} + 1 \text{ quartz} = 4 \text{ anorthite} + 1H_2O \tag{1}$$

as determined by Goldsmith (1981), Chatterjee et al. (1984), and Jenkins (1985). Goldsmith's study also significantly modifies pressure estimates given by Boettcher (1970) for eutectic melting of zoisite + kyanite + quartz + vapor and therefore places a fundamental constraint on all melting reactions in CASH. Discrepancies between the studies of Chatterjee et al. (1984), Matthews and Goldsmith (1984) and Storre and Nitsch (1974) are found for the stability of margarite + quartz, nevertheless, the substantial agreement between most experimental work performed in this system is remarkable (Pistorius and Kennedy 1960; Pistorius 1961; Pistorius et al. 1962; Newton and Kennedy 1963; Newton 1966; Nitsch 1968; Strens 1968; Nitsch and Winkler 1965; Best and Graham 1978; Nitsch et al. 1981; Skrok et al. 1994; and references previously cited).

### High pressure stability of zoisite

The fluid absent reaction in $H_2O$ undersaturated systems

$$3 \text{ lawsonite} + 8 \text{ kyanite} + 7 \text{ grossular} + 1 \text{ coesite} = 12 \text{ zoisite} \tag{2}$$

defines the upper pressure stability of zoisite (Poli and Schmidt 1998) and thus the upper pressure limit of zoisite + lawsonite bearing assemblages (Figs. 2 and 3). For sake of

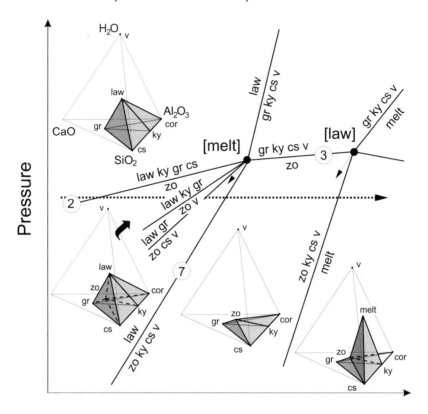

**Figure 3.** Arrangement of univariant curves and divariant parageneses around two invariant points of the CASH system (after Poli and Schmidt 1997). Invariant point [melt] corresponds to invariant point 1 (Fig. 2 at 6.7 GPa/1000°C). The shaded surfaces within the CaO-Al₂O₃-SiO₂-H₂O tetrahedra represent fluid-saturated parageneses. The broken arrow indicates a P-*T* trajectory which causes a change in the hydrous phases present at fluid-undersaturated conditions, the passage from fluid-undersaturated to fluid-saturated conditions when zoisite is the only hydrous phase stable, and again back to fluid-undersaturated conditions with melting (for bulk compositions located in the quadrangle zo-law-cs-ky). Small arrows departing from invariant point illustrate the shift when grossular is diluted in garnet by addition of FeO and/or MgO.

simplicity, the stability field of zoisite + H₂O-fluid (Fig. 2) does not overlap the field of topaz-OH, but this has still to be experimentally demonstrated.

It is worth remembering, that H₂O-undersaturation and fluid-absent conditions are experimentally obtained by employing a bulk composition well within the composition space that is H₂O-undersaturated and fluid-absent *at the pressure and temperature conditions of interest*. In this context, the physical character of the starting material (e.g., with water or with hydrates) is irrelevant as long as it is suited to reach equilibrium, as only the bulk composition of the starting material is related to the actual chemical potential of H₂O at pressure and temperature. The same holds for all chemical components and phases added in a starting material, e.g., for silica: periclase and quartz may well be present in the starting material, but the assemblage olivine + periclase is far away from silica-saturation. The importance of studying $H_2O$-conservative reactions in retrieving thermodynamic properties of hydrates stems from their independence on the equation of state of fluids adopted, and secondly from the possibility to avoid the large uncertainties in defining the activity of $H_2O$ in complex high-pressure fluids. Such fluids are solute-rich (e.g., 5 to 20 wt% $SiO_2$ at 1 to 5 GPa, 700 to 900°C in a binary $H_2O$-$SiO_2$-system, Manning 1994), and the content of solute and hence the activity of $H_2O$ in the fluid vary strongly with pressure and temperature. The effect of reduced $H_2O$-activity in the fluid is generally unaccounted for when thermodynamic data are retrieved from high pressure experiments. It is thus the fluid-absent Reaction (2), which poses the most reliable constrains for deriving thermodynamic data of zoisite at high pressures (and temperatures).

Towards higher temperatures, the reaction

$$4 \text{ grossular} + 5 \text{ kyanite} + 1 \text{ coesite} + 3H_2O = 6 \text{ zoisite} \qquad (3)$$

located at ca. 6.7 GPa delimits the zoisite pressure stability between the lawsonite breakdown and the solidus (Schmidt and Poli 1994). This reaction was proposed as a geobarometer for kyanite saturated eclogites by Chopin et al. (1991) and Okay (1995). It is worth noting that Reaction (3) is a hydration reaction with increasing temperature. Zoisite shows a similar anomalous behavior at much lower pressures (ca. 1 GPa), where it participates in an almost unique example of "solidification" reaction with increasing temperature. Substantially consistent experimental results (neglecting discrepancies in absolute pressure determinations) by Boettcher (1970) and Goldsmith (1982) indicate that the reaction

$$\text{zoisite} + \text{kyanite} + \text{melt} = \text{anorthite} + H_2O \qquad (4)$$

has a positive slope with anorthite + vapor on the high temperature side (Fig. 2). Whether such unusual features of these zoisite-bearing reactions might have geological consequences in more complex systems is still unknown.

## High temperature stability of zoisite

The low pressure, high temperature stability of zoisite (Newton 1966; Boettcher 1970; Chatterjee et al. 1984; Fig. 2) is delimited by

$$6 \text{ zoisite} = 6 \text{ anorthite} + 2 \text{ grossular} + 1 \text{ corundum} + 3H_2O \qquad (5)$$

which transforms at 1020°C, 0.7 GPa into the melting reaction

$$\text{zoisite} = \text{anorthite} + \text{grossular} + \text{corundum} + \text{melt} \qquad (6)$$

Between 2 and 7 GPa, the upper thermal stability limit of zoisite is not yet defined by experiments. It reaches at least 1200°C (Boettcher 1970), which is among the highest temperatures for hydrous rock forming minerals in oceanic or continental crust. The melting reactions and melt compositions involved in this pressure range are discussed in Schmidt and Poli (2004).

## Zoisite vs. lawsonite and fluid-saturation vs. $H_2O$-undersaturation

In this review, we use the term "saturation" or "undersaturation" with respect to a phase (e.g., fluid) or a chemical component (e.g., $H_2O$). With respect to a phase, the practical significance is then equivalent to "present" or "absent," with respect to a chemical component "saturation" is achieved, when the chemical potential of the component is at maximum (i.e. equivalent to the chemical potential in the *pure* one-component phase). In a strict sense, $H_2O$-saturation cannot be achieved in silicate systems at any condition, as there is always some dissolved species present in the fluid. However, at low pressure and solubility, saturation in an aqueous fluid is almost equivalent to $H_2O$-saturation, whereas, at high pressure, the activity and chemical potential of $H_2O$ in an aqueous fluid is significantly reduced.

At high pressure and with increasing temperature, zoisite is a product of the lawsonite breakdown (Fig. 2)

$$4 \text{ lawsonite} = 2 \text{ zoisite} + \text{kyanite} + \text{quartz/coesite} + 7H_2O \qquad (7)$$

At fluid-saturated conditions, the coexistence of zoisite and lawsonite in CASH is restricted to a limited *P-T* range and a very small compositional field. However, a comparably larger pressure range exists (Fig. 3), where zoisite and lawsonite may be in equilibrium at $H_2O$-undersaturated conditions between Reactions (2) and (7). In more complex systems, the reaction from a lawsonite to a zoisite bearing paragenesis could include either Na-clinopyroxene and paragonite (Heinrich and Althaus 1988) or Na-amphibole (Pognante 1991; Poli and Schmidt 1995—their Rxns. 9 and 10). Assemblages of zoisite + garnet + clinopyroxene after lawsonite are often regarded as indication for a clockwise prograde *P-T* path, however might as well be interpreted in terms of changes in $H_2O$-activity or $H_2O$-availability.

This is illustrated by the example in Figure 3. At pressures below Reaction (2), three possible divariant, silica-saturated assemblages with different degree of $H_2O$-undersaturation occur: grossular + zoisite + kyanite + coesite, lawsonite + zoisite + grossular + coesite, and lawsonite + zoisite + kyanite + coesite. These assemblages are all fluid-absent. In a closed system, a fluid-absent bulk composition in the quadrangle lawsonite-zoisite-kyanite-coesite transforms with increasing temperature into a fluid-present zoisite + kyanite + coesite + fluid assemblage. With further increasing temperature, a fluid-absent assemblage zoisite + kyanite + coesite + melt may develop again. The exact composition of melt (Fig. 3) is unknown, as chemographic constraints on melt compositions at high pressure are essentially absent (see also Schmidt and Poli 2004). Nevertheless, these examples show that zoisite and lawsonite should not be regarded as mutually exclusive as often reported for blueschist facies rocks.

## THE JOIN (CLINO-)ZOISITE-EPIDOTE AND THE ROLE OF OXYGEN FUGACITY IN CaO-FeO-Al$_2$O$_3$-SiO$_2$-H$_2$O-O$_2$ (CFASHO)

As largely recognized since the earliest experimental studies on epidote minerals (Ehlers 1953), iron plays a fundamental role in phase relationships of epidote-bearing rocks. The chemography (Fig. 4) shows the extent of solid solution in epidote, prehnite, and garnet in the system CaO-Al$_2$O$_3$-Fe$_2$O$_3$ projected from SiO$_2$ and H$_2$O. Because epidote group minerals host iron in the fully oxidized state (see Franz and Liebscher 2004), high oxygen fugacities (equal to or above the hematite-magnetite oxygen buffer) are expected to maximize epidote stability. In this section we first discuss complexities present along the join (clino-)zoisite-epidote and then review the dependence of epidote$_{ss}$ on oxygen fugacity.

### The binary zoisite-epidote Ca$_2$Al$_3$Si$_3$O$_{12}$(OH)-Ca$_2$Fe$^{3+}$Al$_2$Si$_3$O$_{12}$(OH)

The extent of one and two phase fields along the zoisite-epidote pseudo-binary join have been mainly proposed on the basis of field studies (e.g., Raith 1976; Enami and Banno 1980; Franz and Selverstone 1992; see also Grapes and Hoskin 2004) as experiments specifically devoted to the solution of this problem are scarce (Heuss-Aßbichler and Fehr 1997; Brunsmann et al. 2002). Brunsmann et al. (2002), using equilibrium overgrowths on zoisite and clinozoisite seeds, attempted to reconcile experimental data and natural evidences in the schematic *T-X* diagram (Fig. 5). It is worth noting, that Brunsmann (2000) demonstrated the existence of two zoisite modifications, zoisite I and zoisite II (see Franz and Liebscher 2004 for discussion). Because structural differences between the two forms are minor, for our purpose it is sufficient to generically use the term zoisite.

***The transformation zoisite to clinozoisite.*** Brunsmann et al. (2002) found that orthorhombic zoisite extends to no more than 0.11 Fe$^{3+}$ per formula unit (pfu) at 0.5 GPa,

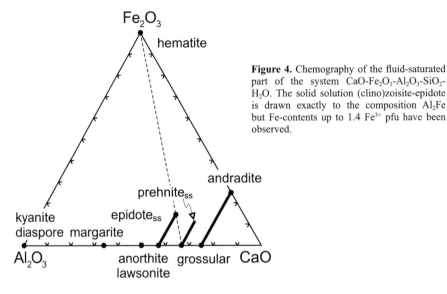

Figure 4. Chemography of the fluid-saturated part of the system $CaO$-$Fe_2O_3$-$Al_2O_3$-$SiO_2$-$H_2O$. The solid solution (clino)zoisite-epidote is drawn exactly to the composition $Al_2Fe$ but Fe-contents up to 1.4 $Fe^{3+}$ pfu have been observed.

Figure 5. Schematic pseudobinary (clino)zoisite-epidote projected from quartz and vapor, including solvi and high temperature breakdown reactions at 0.3-0.5 GPa, based on experimental data from Brunsmann et al. (2002) and Heuss-Assbichler and Fehr (1997). The one phase and two phase fields to 600-650°C are experimentally constrained, but the shape of the two phase loop between zoisite and epidote above 600-650°C is uncertain. Pure zoisite (+quartz) breaks down to grossular + anorthite + $H_2O$ at 610-620°C (Rxn. 22). In the Fe-bearing system, the reaction products are $gar_{ss}$+an+qtz+hem+$H_2O$ (Rxn. 10) and the temperature of reaction depends on the epidote composition. czoI: clinozisite I; czoII: clinozoisite II; gar: garnet; hm: hematite.

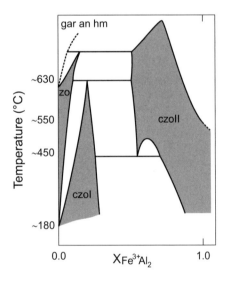

650°C and ca. 0.16 $Fe^{3+}$ pfu at 2 GPa, 800°C. The pressure-temperature dependence of the Al = $Fe^{3+}$ exchange in coexisting zoisite and clinozoisite (clinozoisite I in Brunsmann et al. 2002) was empirically described by the equations (with $P$ in GPa and $T$ in°C):

$$X_{Fe^{3+}Al_2}^{\text{zoisite}} = 1.9 \times 10^{-4}T + 3.1 \times 10^{-2}P - 5.63 \times 10^{-2}$$

$$X_{Fe^{3+}Al_2}^{\text{clinozoisite}} = (4.6 \times 10^{-4} - 4 \times 10^{-5}P)T + 3.82 \times 10^{-2}P - 8.76 \times 10^{-2}$$

Extrapolation of these limbs to $X_{Fe^{3+}Al_2}$ =0 (i.e. $X_{epi}$ = 0, where $X_{epi}$ = $Fe^{3+}$/($Fe^{3+}$+$Al_2$) see Fig. 5), i.e. in the Fe-free system, offers an estimate of the transformation temperature from clinozoisite to zoisite, although direct experimental constraints in the system CASH are still lacking. Data from Brunsmann et al. (2002) support previous determinations by Jenkins et al. (1983) at pressures to 0.8 GPa and by Franz and Selverstone (1992): the transformation clinozoisite-zoisite is expected to occur at ca. $180 \pm 50°C$ at 0.5 GPa and at 0 to $100°C$ at 2 GPa, confirming that zoisite is the stable polymorph in the CASH system at all geologically relevant conditions.

***The high temperature breakdown of zoisite and clinozoisite.*** Currently available data (Holdaway 1972; Brunsmann et al. 2002) support preferential incorporation of $Fe^{3+}$ in epidote$_{ss}$ in the Fe-poor region of the pseudobinary join (Fig. 5). As a result the equilibrium which describe the two phase loop between zoisite and garnet is

Fe-poor zoisite + quartz = Fe-rich zoisite + anorthite + garnet + $H_2O$ (8)

Thermal breakdown of zoisite is described by the reaction (at $T < 650°C$):

Fe-zoisite = clinozoisite II + garnet + anorthite + $H_2O$ (9)

The maximum stability limit of epidote at $T = 700°C$ is given by the reaction

clinozoisite II = garnet + anorthite + hematite + $H_2O$ (10)

***Extent of immiscibility along the clinozoisite-epidote join.*** This is more controversial. Heuss-Aßbichler and Fehr (1997) proposed two solvi between $X_{epi} = 0.24$ and $X_{epi} = 0.76$ at 0.3 GPa separated by a narrow single-phase region centered at ca. $X_{epi} = 0.52$. Their experimental data covers the temperature range from 500 to 650°C, and Heuss-Aßbichler and Fehr (1997) suggest a possible extension of immiscibility up to ca. 750°C. However, data by Brunsmann et al. (2002) at 0.5 GPa do not support immiscibility in the range $X_{epi} = 0.52$ to 0.76 at 600°C and more experimental data are required to resolve this discrepancy.

## The dependence of epidote-stability on oxygen fugacity

Breakdown of epidote minerals of the join (clino)zoisite-epidote at high temperature is constrained by $Fe^{3+}$-partitioning between garnet and epidote$_{ss}$. Complexities and controversies in the description of epidote$_{ss}$ combine to an extreme sensitivity of the epidote composition to conditions in oxygen fugacity (Figs. 6 and 7). Fyfe (1960), Merrin (1962), and Nitsch and Winkler (1965) investigated the temperature stability of epidote at unbuffered conditions, whereas Holdaway (1972), Liou (1973) and Liou et al. (1983) explored the stability of Fe-rich epidotes over a large range of oxygen fugacities spreading from those of the cuprite-tenorite equilibrium (extremely oxidizing conditions, 7-8 log units above hematite-magnetite buffer) down to those of the iron-magnetite.

The uppermost temperature stability of clinozoisite-epidote at oxygen fugacities buffered to hematite-magnetite is in discussion. Both Brunsmann et al. (2002) and Heuss-Aßbichler and Fehr (1997) found a decrease in $Fe^{3+}$ pfu in epidote when approaching the thermal breakdown reaction (maximum Fe-contents are $X_{epi} = 0.78$ at 650°C, 0.5 GPa, Brunsmann et al. 2002; and $X_{epi} = 0.88$ at 650°C, 0.3 GPa, Heuss-Aßbichler and Fehr 1997). Contrasting, the systematic survey of Liou (1973) indicated a stability of clinozoisite with $X_{epi} = 1.0$ up to 680°C at 0.3 GPa and to 750°C at 0.5 GPa. At present, this contradiction cannot be resolved and most experimental data on the CFASHO system are limited to pressures below 0.7 GPa, clearly indicating the necessity of further experimentation.

Despite the general agreement that epidote becomes more aluminous as oxygen fugacity decreases, there is still substantial uncertainty about the quantitative relationship between epidote composition and oxygen fugacity. For example, Holdaway (1972) proposed that the ultimate composition of epidote at nickel-bunsenite buffer (NNO) is $X_{epi} = 0.24$ at 642°C, 0.3 GPa, whereas Liou (1973) found $X_{epi} = 0.75$ at 600°C, 0.3 GPa.

Liou et al. (1983) report synthesis of epidote at oxygen fugacity buffered to iron-

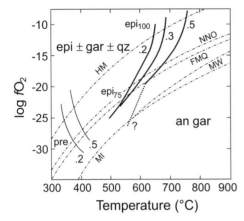

**Figure 6.** Upper thermal stability limits of epidote ± garnet ± quartz (bold lines) and prehnite + hematite/magnetite + epidote + garnet (thin lines) in the temperature- $f_{O_2}$ space based on experimental data from Liou (1973) and Liou et al. (1983). Labels on stability limits indicate pressure in GPa and approximate composition of epidote in mol% $Fe^{3+}Al_2$. The dotted line with a question mark represents the interpolation of epidote stability at 0.5 GPa on the basis of synthesis experiments at MI buffer (Liou et al. 1983). Dashed-dotted lines give various $f_{O_2}$ buffers: HM = hematite-magnetite; NNO = nickel-bunsenite; FMQ = fayalite-magnetite+quartz; MW = magnetite-wuestite; MI = magnetite-iron.

**Figure 7.** Upper thermal stability limits of epidote ± garnet ± quartz (bold lines) and prehnite + hematite/magnetite + epidote + garnet (thin lines) in the pressure-temperature space based on experimental data from Liou (1973) and Liou et al. (1983) (compare to Fig. 6). Labels give $f_{O_2}$ for the epidote-out reactions. HM = hematite-magnetite; NNO = nickel-bunsenite; FMQ = fayalite-magnetite+quartz.

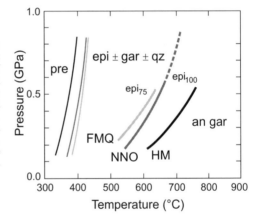

magnetite at 500 to 550°C, 0.5 GPa, whereas at 575°C, 0.5 GPa the assemblage anorthite + garnet is found (Fig. 6, Liou 1973). This thermal stability of epidote at low oxygen fugacity is somewhat incoherent with Liou's data at higher oxygen fugacity, probably pointing towards the unfavorable reaction kinetics at low temperature and the technical difficulty to control low oxygen fugacity. Furthermore, the stoichiometry of the thermal epidote breakdown reactions are matter of debate, as these strongly depend on epidote composition which in turn is a function of oxygen fugacity. Figure 7 is based on the data of Liou (1973), however, extrapolation of epidote breakdown curves to higher pressures, where epidote is a major rock forming mineral in a variety of bulk compositions, are speculative. This, and in particular the intersection of the $H_2O$-saturated solidus in simple but Fe-bearing systems, represents a substantial lack of knowledge.

### $Fe^{3+}$ partitioning between epidote and garnet solid solutions

The complexities observed in the $P$-$T$-$f_{O_2}$ evolution of epidote$_{ss}$ are also related to the complexity in the partitioning of $Fe^{3+}$ between epidote and garnet (Fig. 8). At $f_{O_2}$ buffered to hematite-magnetite, Heuss-Aßbichler and Fehr (1997) demonstrated the inversion of the temperature dependence of $Fe^{3+}$ partitioning when moving from the intermediate to the Fe-rich portion of the clinozoisite-epidote solid solution (see also Fig. 7 in Heuss-Aßbichler and Fehr

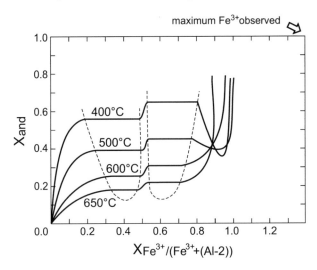

**Figure 8.** Partitioning of Al and $Fe^{3+}$ between the grossular-andradite (grandite) solid solution and the clinozoisite-epidote solid solution at $f_{O_2}$ fixed by the HM buffer. Note that $X_{and}$ corresponds to $Fe^{3+}/(Fe^{3+}+Al^{total})$ in garnet solid solution but $X_{epi}^2$ corresponds to $Fe^{3+}/(Fe^{3+}+Al_2)$ in epidote solid solution. Thus, partitioning curves do not have to end in the upper right corner. The right hand limit of the diagram was (arbitrarily) located at the highest $Fe^{3+}$ content measured in epidote s.l. and does not represent an absolute limit within compositional space. The two intermediate solvi in the epidote join are indicated by stippled lines. Note that the study from Heuss-Aßbichler and Fehr (1997) did not investigate the compositional space at $X_{epi}<0.23$, where Brunsmann et al. (2002) identified an additional solvus at lower $Fe^{3+}$ contents. (Partitioning curves redrawn and modified after Heuss-Aßbichler and Fehr 1997).

1997). At relatively low Fe contents ($0.24 < Fe^{3+} < 0.34$ pfu) a temperature decrease favors the incorporation of $Fe^{3+}$ into the garnet structure. On the contrary, at epidote compositions approaching $X_{epi}=1$, a decrease in temperature promotes the incorporation of $Fe^{3+}$ in epidote. Brunsmann et al. (2002) and Holdaway (1972) also found preferential $Fe^{3+}$ partitioning into zoisite at very low-Fe bulk contents, pointing towards a fairly complex behavior of $Fe^{3+}$ in the CFASHO system. Unfortunately, comparison of mixing models describing $Fe^{3+}$ partitioning between garnet and epidote$_{ss}$ is somewhat complicated by different terminologies used to express end-members in epidote$_{ss}$ (Perchuk and Aranovich 1979; Bird and Helgeson 1980; Heuss-Aßbichler and Fehr 1997; Brunsmann et al. 2002).

Such very intricate relationships and poor knowledge of partitioning at lower oxygen fugacities (e.g., NNO or FMQ) more relevant for most geological environments, hinders the practical usage of geothermobarometers based on epidote$_{ss}$. The sensitivity of epidote and garnet solid solutions to redox conditions can be described by a variety of reactions, which on principle represent widely applicable oxygen barometers (e.g., Donohue and Essene 2000):

$$2 \text{ epidote} = 2 \text{ garnet } (Ca_2FeAl_2Si_3O_{12}) + H_2O + 0.5O_2 \tag{11}$$

$$8 \text{ epidote} = 6 \text{ anorthite} + 2 \text{ andradite} + 0.67 \text{ magnetite} + 0.5H_2O + 0.25O_2 \tag{12}$$

$$12 \text{ epidote} + 16 \text{ kyanite} + 8 \text{ quartz} = 4 \text{ almandine} + 24 \text{ anorthite} + 6H_2O + 3O_2 \tag{13}$$

Nevertheless, simplified models of the highly non-ideal epidote solid solutions should be regarded with skepticism. A summary of phase relationships in CFASHO at 0.5 GPa is shown in Figure 9.

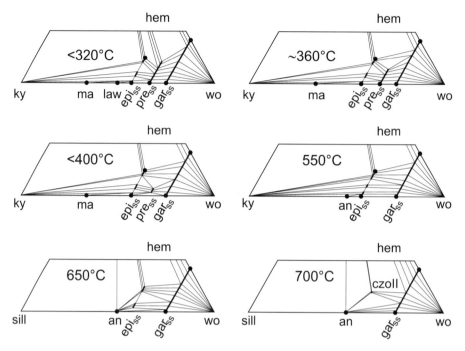

**Figure 9.** Compositional fields and chemography for epidote-bearing assemblages at 0.5 GPa projected from quartz and vapor into the lower portion of the CaO-Al$_2$O$_3$-Fe$_2$O$_3$ ternary (Fig. 4). Note the occurrence and variation of various solvi in the epidote solid solution, the disappearance of prehnite around 400°C, and the appearance of anorthite at $T > 490$ to 500°C. The composition of the epidote at its highest temperature stability of ca. 700°C is $X_{Fe^{3+}Al_2} < 0.75$ (redrawn and modified after Brunsmann et al. 2002 and Liou et al. 1983).

# THE SYSTEMS NCASH, CMASH, NCFMASHO
# AND PHASE RELATIONSHIPS IN MAFIC ROCKS

## Extension of CASH to more complex systems

The application of phase relationships in simple systems to natural compositions can be illustrated by stepwise addition of chemical components. Reaction (2) and (3) (Fig. 3) involve zoisite, lawsonite, kyanite, silica, grossular, and vapor. Because epidote and garnet are the only two minerals which show extensive solid solution in more complex, Fe-bearing systems, element partitioning between these two phases controls the displacement of these reactions and therefore affects the actual pressure stability of zoisite in natural rock compositions.

The addition of FeO and/or MgO causes dilution of grossular by almandine and pyrope whereas these elements will not be incorporated into zoisite in any significant amounts (Franz and Liebscher 2004). Reaction (2) and (3) shift to lower pressure and temperature, which results in a significantly reduced stability of zoisite in natural rock compositions (Fig. 3). However at highly oxidizing conditions (hematite-magnetite) Fe$^{3+}$ stabilizes zoisite at the expense of garnet and/or melt and therefore Reactions (2) and (7) will shift to lower temperature and Reaction (3) to higher pressure (Brunsmann et al. 2002) with the effect of enlarging the zoisite stability field. As a net result, highly oxidizing conditions might increase the stability field of zoisite compared to the model system CASH.

The addition of $Na_2O$ and MgO to the model systems CASH and CFASHO (NCFMASHO) permits a more extended chemical description of pyroxene, amphibole, plagioclase, and garnet, providing a basis for the discussion of phase relationships in a variety of mafic and intermediate compositions. Experimental data on natural mafic compositions are usually modeled in the system NCFMASH as other constituents are only of minor relevance (e.g., MnO) or they strongly partition in a single phase (e.g., $K_2O$ in biotite or phengite) and do not significantly alter phase relations. The experimental data on epidote composition in such complex systems are extremely scarce and almost randomly scattered over a large *P-T-X* space. Because of analytical difficulties in complex mineral assemblages having often more than six phases present, characterization of epidote minerals is usually poor and a comparison to the data obtained in CFASHO is almost impossible. Additionally, the principal difficulties in determining the oxidation state of iron in coexisting synthetic garnet, pyroxene, and amphibole has often led experimentalists to consider epidote as a phase of fixed composition.

Epidote participates in most reactions defining the boundaries of metamorphic facies, but not in reactions delimiting the granulite facies which is beyond the temperature stability of zoisite in common rock compositions. A brief discussion of experimentally derived constraints is given in the following sections.

### The formation of epidote and the greenschist facies

The appearance of epidote in mafic bulk compositions at low temperature and pressure (Fig. 10) is mainly related to the breakdown of the join pumpellyite + chlorite resulting in coexistence of epidote + actinolite through the reaction

$$\text{pumpellyite + chlorite + quartz = epidote + actinolite + } H_2O \tag{14}$$

which takes place at 350°C below 0.7 GPa (Nitsch 1971). This reaction is responsible for the appearance of the so called "COMMON" assemblage, typical of the greenschist facies, i.e., actinolite + epidote + albite + chlorite + quartz (Laird 1980). Experimental determinations of the reactions

$$\text{Mg-Al pumpellyite = zoisite + grossular + chlorite + quartz + vapor} \tag{15}$$

$$\text{Mg-Al pumpellyite = clinozoisite + grossular + chlorite + quartz + vapor} \tag{15b}$$

by Hinrichsen and Schürmann (1969) and Schiffman and Liou (1980) in CMASH, reveal a sensitivity to the starting material used, Reaction (15) being located approximately 50°C lower than Reactions (15b) and (14).

Schiffman and Liou (1983) suggested that the formation of epidote from pumpellyite in iron-bearing systems is a redox reaction where $Fe^{3+}$ is preferentially partitioned into epidote. Experimental data by Schiffman and Liou (1980, 1983) support an extension of the stability of epidote + vapor down to temperatures as low as 250 to 270°C at 0.5 GPa. Moody et al. (1983) synthesized epidote in a fluid-saturated basaltic composition only at $f_{O_2}$ of the hematite-magnetite buffer above ca. 350°C through the reactions

$$\text{wairakite + hematite = epidote + quartz + fluid} \tag{16}$$

and

$$\text{pumpellyite + quartz + hematite = epidote + chlorite + fluid} \tag{17}$$

Experiments at 0.2 and 0.4 GPa (Moody et al. 1983) at $f_{O_2}$ below HM did not yield epidote in the temperature range from 300 to 550°C, though their results are in conflict with epidote-bearing assemblages reported by Liou et al. (1974) at 0.2 GPa in the temperature range 450 to 550°C. However, these latter experiments were not reversed, Liou et al. (1974) observed epidote only when a natural greenschist was used as a starting material, but could not obtain epidote growth from a hornblende + plagioclase mixture (natural amphibolite).

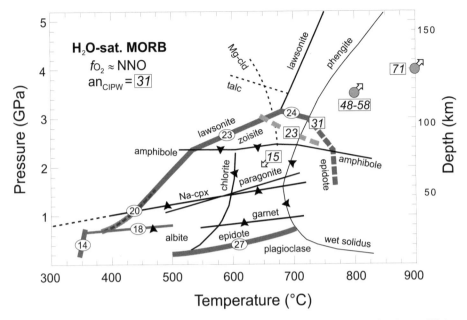

**Figure 10.** Phase stabilities in hydrated MORB compositions with emphasis on epidote-bearing equilibria. Thick grey lines delimit the stability fields of different epidote-bearing assemblages, as defined in Poli and Schmidt (1995, 2002) and Schmidt and Poli (1998). Arrows on reactions indicate the direction of increasing amounts of epidote. Bold numbers indicate normative anorthite content in weight percent from studies that synthesised epidote (circles; an = 48-58: Thompson and Ellis 1994; an = 71: Wittenberg et al. 2003) or defined an epidote-out reaction (an = 31: Poli and Schmidt 1995; Schmidt and Poli 1998; an = 23: Forneris and Holloway 2003, broken bold line). Experiments performed on a Mg-rich andesite, with an = 15 (Poli and Schmidt (1995)), did not yield epidote, pointing to a low pressure stability of epidote in such composition. Note the systematic increase in normative anorthite content with increasing pressure. Numbers in circles correspond to reaction numbering in the text.

Because of the intrinsic limits of such very low temperature experiments and the strong dependence on oxygen fugacity and the starting material employed, it will probably remain impossible to unequivocally define a unique temperature for an "epidote-in" in metamorphosed basalts and gabbros.

### Greenschist to blueschist transition

The transformation from greenschist to blueschist facies rocks was experimentally investigated by Maruyama et al. (1986). This study employed mixtures of crystalline materials to demonstrate growth of glaucophane/Mg-riebeckite instead of tremolite on amphibole seeds present in the starting material. The reactions

$$6 \text{ tremolite} + 9 \text{ chlorite} + 50 \text{ albite} =$$
$$6 \text{ clinozoisite} + 25 \text{ glaucophane} + 7 \text{ quartz} + 14H_2O \qquad (18)$$

and

$$4 \text{ tremolite} + 10 \text{ albite} + 7 \text{ hematite} + 7H_2O =$$
$$4 \text{ epidote} + 5 \text{ Mg-riebeckite} + \text{chlorite} + 7 \text{ quartz} \qquad (19)$$

might represent upper (18) and lower (19) pressure bounds for the transition of greenschists to epidote-blueschists (stoichiometries are after Maruyama et al. 1986, but strongly depend on individual mineral compositions; the high pressure blueschist assemblage is on the right side of equations 18 and 19). Reaction (18), which can be approximated in NCMASH, was bracketed to 0.7 GPa, 300°C and 0.8 GPa, 450°C (Maruyama et al. 1986), indicating a very gentle *P-T* slope. Note that, similar to Reaction (3) in CASH, Reaction (18) is a hydration reaction with increasing temperature. Experimental results on Reaction (19) suggest that strongly oxidizing conditions (HM buffer) again favor epidote and amphibole formation, Mg-riebeckite growth was detected at pressures as low as 0.47 GPa at 300°C.

Though Reactions (18) and (19) have fluid and chlorite on opposite sides of the equations, epidote is unequivocally a reaction product in both of them and thus is produced at the greenschist to blueschist facies transformation, its abundance is expected to be maximized in blueschists.

### Subdivision of the blueschist facies

The blueschist facies is commonly divided into a lawsonite- and an epidote-subfacies. The related reaction was experimentally studied in the model system NCASH and involves albite or jadeite at lower and higher pressure, respectively (Heinrich and Althaus 1988):

$$4 \text{ lawsonite} + 1 \text{ albite} = 1 \text{ paragonite} + 2 \text{ zoisite} + 2 \text{ quartz} + 6H_2O \tag{20}$$

$$4 \text{ lawsonite} + 1 \text{ jadeite} = 1 \text{ paragonite} + 2 \text{ zoisite} + 1 \text{ quartz} + 6H_2O \tag{20b}$$

The corresponding reaction curves are situated at about 420°C at 0.9 GPa, and about 470°C at 1.5 GPa. Franz and Althaus (1977) experimentally investigated the stability of the assemblage paragonite + zoisite + quartz and found that Reaction (1) modifies to

$$\text{zoisite} + \text{paragonite} + \text{quartz} = \text{plagioclase} + \text{vapor} \tag{21}$$

which defines, combined with experiments by Heinrich and Althaus (1988) a low pressure limit for zoisite + paragonite + quartz at ≈ 0.7 to 0.8 GPa, 400°C.

Because most mafic to intermediate rock compositions have albite and/or jadeite in excess to lawsonite, the actual transformation from lawsonite blueschist/eclogite to epidote blueschist/eclogite is controlled by Reaction (20) which is at slightly lower temperature (10 to 40°C) than the model Reaction (7) in CASH. Extrapolation of the experimental data performed by Poli and Schmidt (1995) between 2.2 and 2.6 GPa and temperatures as low as 550°C suggests that the transformation from lawsonite to epidote in glaucophane-bearing mafic rocks (Fig. 10) occurs very close to the boundary defined by extrapolation of Reactions (20) at the amphibole breakdown pressure.

### Epidote-eclogites and the pressure stability limit of epidote minerals

The generalized reaction:

$$\text{glaucophane} + \text{epidote} = \text{omphacite} + \text{garnet} + \text{paragonite} \tag{22}$$

governs the transition from blueschist to eclogite facies (Ridley 1984). This reaction occurs around 450 to 550°C in blueschist-eclogite terranes but there has been no attempt to experimentally determine the complex multivariant loops generated by the coexistence of epidote, amphibole, pyroxene and garnet solid solution in a *P-T* range where reaction rates are still very sluggish. The high pressure breakdown of epidote in $H_2O$-saturated systems can be efficiently modeled by the reactions

$$13 \text{ lawsonite} + 3 \text{ diopside} = 8 \text{ zoisite} + 1 \text{ pyrope} + 5 \text{ quartz/coesite} + 22 \ H_2O \tag{23}$$

and, above 680°C at 3 GPa beyond the lawsonite stability,

$$15 \text{ diopside} + 12 \text{ zoisite} = 13 \text{ grossular} + 5 \text{ pyrope} + 12 \text{ quartz/coesite} + 6 \ H_2O \tag{24}$$

In CMASH, these reactions are univariant but become multivariant over a large range of temperature in Fe- and Na-bearing systems where the bulk composition controls the actual disappearance of epidote minerals. Normative anorthite contents may be taken as representative of the potential amount of (clino)zoisite and epidote that could be formed, and it is thus expected that differentiated plagioclase rich gabbros (e.g., hydrated troctolite) with high normative anorthite content show epidote over a much larger pressure range than hydrated tholeiitic basalt. Currently available data on complex synthetic systems or natural compositions support this conclusion: Forneris and Holloway (2003) found epidote to break down at ca. 2.7 GPa, 700°C in a bulk composition containing 23 wt% (CIPW-) normative anorthite. Poli and Schmidt (1995) determined the breakdown of epidote in average MORB with an$^{CIPW}$ = 31 wt% to ca. 3.1 GPa, 700°, whereas Thompson and Ellis (1994) and Wittenberg et al. (2002) were able to synthesize epidote-bearing assemblages at 3.5 GPa, 800°C and 4 GPa, 900 to 1000°C in synthetic mafic bulk compositions having an$^{CIPW}$ = 48 to 58 wt% and an$^{CIPW}$ = 71 wt%, respectively, though both of these latter studies did not extend experiments to pressures beyond the stability of epidote coexisting with garnet + clinopyroxene. In this context, it is completely coherent that high pressure experiments on a peculiar high Mg-andesite composition with an$^{CIPW}$ = 15 wt%, did not yield epidote at conditions ranging from 2.2 to 3.6 GPa, 600°C to 725°C (Poli and Schmidt 1995), suggesting an upper epidote stability limit in this composition below 2.2 GPa.

The composition of epidote crystallizing at its uppermost stability in the different experimental studies is quite variable: Poli and Schmidt (1995) found $X_{Fe3+Al2}$ = 0.15, Pawley and Holloway (1993) $X_{Fe3+Al2}$ = 0.70, and Forneris and Holloway (2003) $X_{Fe3+Al2}$ = 0.63. A possible explanation for such discrepancies are differences in oxygen fugacity (which in turn should shift epidote delimiting reactions considerably) and/or in the starting material used, which in part contained seeds of epidote or zoisite.

In the last decade there has been an increasing interest in epidote properties and stability at high pressure, high temperature conditions because of the ability of epidote-group minerals to store significant amounts of trace elements (Frei et al. 2004) and $H_2O$, thus providing a source for metasomatism. All of the experimental studies to date (other than Liu et al. 1996, who unexpectedly could not find any epidote at 0.8 to 3 GPa, 600 to 950°C) unequivocally indicate that epidote is the ultimate fluid source in mafic compositions (neglecting small amounts of phengite which are related to the small average bulk $K_2O$-contents) in an oceanic crust subducting at intermediate temperatures, i.e. at temperatures above the lawsonite stability but below the solidus. As extensively discussed in Schmidt and Poli (2003) and Poli and Schmidt (2002), most high pressure breakdown reactions of hydrous minerals in mafic compositions are continuous reactions where crystalline phases, e.g., epidote, disappear progressively without generating a fluid pulse, but rather producing the metasomatizing fluid over a large depth range as a function of the width of such continuous reactions. In the case of epidote, variations in oxygen fugacities at the top of the slab might strongly control its stability and further complicate the dehydration signal.

### The amphibolite facies

Below ≈ 1 GPa, and temperature conditions from 500 to 700°C, the NCASH-system model reaction

plagioclase$_1$ (An-poor) + zoisite + kyanite + quartz = plagioclase$_2$ (An-rich) + $H_2O$   (25)

represents the transition from epidote-amphibolite to amphibolite facies (Goldsmith 1982). Goldsmith observed that up to ca. 1 GPa, $H_2O$-saturated conditions, plagioclase with anorthite contents beyond the peristerite gap (An$_{40}$ to An$_{100}$) breaks down along the same $P$-$T$ curve as pure anorthite (Rxn. 1 in CASH). However, in amphibolite garnet is much more frequent than kyanite, and the appropriate reaction zoisite + quartz = anorthite + grossular + $H_2O$ (26)

might show a similar behavior of plagioclase in $Al_2SiO_5$-undersaturated systems. In fact, the position of the epidote-out reaction

epidote + albite + hornblende$_1$ + quartz = oligoclase + hornblende$_2$ +$H_2O$ (27)

after Apted and Liou (1983, $f_{O_2}$ = NNO, natural basaltic bulk composition) closely approaches Reaction (26) in CASH. This could be explained by free energy changes of plagioclase having an overwhelming role for the epidote breakdown at high temperature. In Reaction (27) hornblende$_2$ is enriched in tschermakite with respect to hornblende$_1$.

The wide stability field over which epidote and other Ca-bearing phases such as amphibole, pyroxene, and garnet coexist in mafic and intermediate rocks, makes epidote a suitable buffering phase for a variety of potential geothermobarometers. Reactions (3) and (24) are implicitly used as geobarometers on the basis of available thermodynamic data (e.g., Chopin et al. 1991; Okay 1995; Poli and Schmidt 1997, 1998). The following further equilibria were experimentally investigated in the model system CMASH by Cho and Ernst (1991), Hoschek (1995), and Quirion and Jenkins (1998), respectively. These equilibria buffer the tschermak component in amphibole:

2 zoisite + 3 tremolite + 7 kyanite + 1$H_2O$ = 5 tschermakite + quartz (28)

2 zoisite + 3 tremolite + 7 corundum + 1$H_2O$ = 5 tschermakite (29)

6 zoisite + 10 tremolite = 9 hornblende + 14 diopside + 7 quartz + 4$H_2O$ (30)

Because most experimental results (see also Najorka et al. 2002) and natural occurrences show an increase in octahedral aluminum in amphibole with increasing temperature, Reactions (28) and (29) imply hydration with temperature, adding further evidence for a somewhat counterintuitive behavior of some epidote-involving reactions (see also Rxns. 3, 4, and 19). Tschermak component in amphibole, though strongly dependent on the buffering assemblage, was also shown to decrease with pressure for Reactions (28), (29), and (30). This results in a production of epidote (and fluid) with pressure as found for Reactions (18) and (19).

## Maximizing epidote/zoisite abundance

Reaction (18) and (19) describing the greenschist to blueschist transition result in a progressive increase of epidote abundance with pressure (Fig. 11). It is only when amphibole is consumed via the reactions

barroisite + epidote = omphacite + garnet + quartz + $H_2O$ (31)

amphibole + epidote = chloritoid + omphacite + quartz + $H_2O$ (32)

(Poli 1993; Schmidt 1993), that epidote decreases in abundance. Equilibrium with garnet and clinopyroxene as described by Reaction (24) is responsible for consumption of epidote until complete exhaustion and increases grossular component in garnet with pressure. The transition from garnet-amphibolite to eclogite via the reaction

garnet + plagioclase + $H_2O$ = amphibole + omphacite + epidote (33)

(products are on the high pressure side, Poli 1993) is responsible for epidote production at the transition to the eclogite facies. An analogous reaction in CMASH occurring in bulk compositions typical for metarodingite is

4 grossular + 13 anorthite + 1 spinel + 6$H_2O$ = 12 zoisite + 1 diopside (34)

(products are on the high pressure side) which was experimentally investigated by Cheng and Greenwood (1989). As the transition from blueschist to eclogite occurs via an epidote-

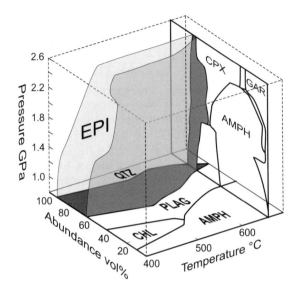

**Figure 11.** Temperature-pressure-abundance diagram for hydrated MORB compositions. The epidote volume is emphasized by grey shading. The highest abundance of epidote (30-35 wt%) occurs in the epidote-blueschist facies. Data from Poli (1993) and Apted and Liou (1983).

consuming Reaction (22), the highest modal abundance of epidote occurs in the epidote-blueschist facies (Fig. 11). At temperatures above the epidote-blueschist facies (650°C), epidote reaches its highest volumetric importance within the amphibole-eclogite facies in mafic rocks.

## EPIDOTE AND MICAS IN PELITIC AND FELSIC COMPOSITIONS

Experimental studies on epidote$_{ss}$ in potassium-rich systems are extremely scarce. This is mostly a result of the common use of a simplified metapelite system, which neglects the CaO-component. In quartz + aluminosilicate saturated metapelite, a variety of model reactions as previously discussed will define the reactions between anorthite[plagioclase], grossular[garnet], and (clino)zoisite[epidote]. In fact, an extended field of epidote$_{ss}$ in metapelite and metagranitoid is confirmed by natural occurrences (see Grapes and Hoskin 2004), however, due to relatively low CaO-contents, epidote in metapelite is only a minor constituent.

Johannes (1980) investigated melting and subsolidus equilibria involving zoisite in the system $K_2O-CaO-Al_2O_3-SiO_2-H_2O$ (KCASH). Apart from the significant lowering of melting temperature with respect to CASH, muscovite saturation causes an increase of the necessary pressure to form epidote by ca. 0.2 GPa. The Reactions (26) and (1), defining the minimum pressure or maximum temperature of the assemblages zoisite + quartz and zoisite + quartz + kyanite in CASH (Fig. 2), become

$$\text{zoisite} + \text{quartz} + \text{muscovite} = \text{anorthite} + \text{orthoclase} + H_2O \tag{35}$$

This latter reaction should also delimit the zoisite stability in metapelite, which generally has excess muscovite at the relevant conditions. However, extensive solid solution in zoisite, muscovite, and plagioclase will shift also this reaction in *P-T* space.

In KCMASH, Hoschek (1995) determined the reaction

$$\text{muscovite} + \text{talc} + \text{tremolite} = \text{phlogopite} + \text{zoisite} + \text{quartz} + H_2O \tag{36}$$

to 700°C, 1.7 GPa, and Hermann (2002) delimited the stability of zoisite + talc and zoisite

+ amphibole in the presence of a phengitic mica to ≈ 680°C at 3.6 GPa and ≈ 800°C at 2.2 GPa (Fig. 12). Although these reactions delimit important assemblages in the KCMASH model system, they are limited to relatively restricted bulk compositions, and because of the extensive solid solutions involved, are not directly applicable to natural rock compositions.

An experimental study on an intermediate tonalitic composition with an$^{CIPW}$ = 26 wt% containing about 20 wt% micas resulted in an epidote stability field almost identical to that of average hydrated MORB composition (Schmidt 1993; Poli and Schmidt 1995). In these studies, biotite occurs up to 1.3 GPa, and at higher pressures phengite is present, however, the micas do not appear to significantly influence epidote stability.

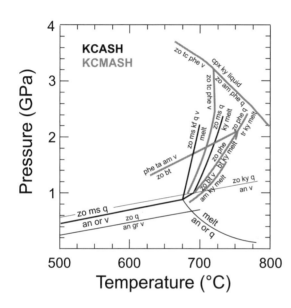

**Figure 12.** Pressure-temperature diagram of zoisite-bearing reactions in the KCASH (black lines) and KCMASH (grey lines) systems. After Johannes (1980), Schliestedt and Johannes (1984), Hoschek (1990), Hermann (2002).

## EPIDOTE, CARBONATES AND C-O-H FLUIDS

Epidote minerals are also ubiquitous in calcsilicate rocks. The upper temperature stability of zoisite in carbonate-bearing assemblages is mainly modeled by the reaction

$$2\ zoisite + 1CO_2 = anorthite + calcite + 1H_2O \qquad (37)$$

which has been experimentally investigated by Storre and Nitsch (1972), Johannes and Ziegenbein (1980), Storre et al. (1982), and Allen and Fawcett (1982). Despite differences in thermodynamic modeling of Reaction (37) (Fig. 13), the available experiments consistently indicate an enlargement of the stability field of zoisite in a $T$-$X_{CO_2}$ section with increasing pressure. At 0.2 GPa the maximum $CO_2$ content of the fluid in equilibrium with zoisite is in the order of 0.05 mole fraction at ca. 400°C, whereas at 0.5 GPa it moves to $X_{CO_2}$ of 0.2 at 450 to 500°C.

Data about equilibria in $CO_2$-bearing systems at higher pressure are essentially absent. To our knowledge, the only experimental study available is on epidote-amphibolite and epidote-amphibole-eclogite in the presence of a C-O-H fluid (Molina and Poli 2000). Epidote was found to be much more abundant in a Mg-enriched basaltic composition than in a MORB composition and absent in a Fe-rich basalt throughout a pressure-temperature range of 1.2 to 2.0 GPa and 665 to 730°C. Though this study was not focused on epidote phase relationships,

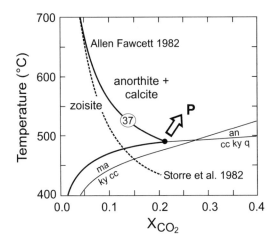

**Figure 13.** Temperature $- X_{CO_2}$ section delimiting the zoisite field at 0.5 GPa. The presence of zoisite in calcsilicates indicates moderate $CO_2$-contents in the fluid. The bold arrow indicates the shift with pressure of the invariant point indicating the maximum in $X_{CO_2}$ of a fluid coexisting with zoisite. The subhorizontal line delimits the lower temperature stability of anorthite+fluid in this system.

it should be noted that, consistently with other theoretical and experimental studies, Molina and Poli (2000) suggest a strong tendency for carbon to fractionate into the solids (i.e. into calcite up to $\approx 1$ GPa, and then into dolomite and magnesite with increasing pressure). As a result the fluid is progressively enriched in $H_2O$ and the stability of hydrous phases, including epidote, should not differ significantly from $H_2O$-saturated systems, beyond 2 GPa, although experimental investigations are still lacking. Nevertheless the complex interplay of C-O-H speciation, oxygen fugacity, and the availability of volatile components (i.e. mass-balance constraints) can be responsible for variable scenarios. Common coexistence of hydrates, carbonates and graphite/diamond in ultra-high pressure rocks suggest very complex phase relationships.

### Zoisite morphology in experiments on meta-calcsilicates

Allen and Fawcett (1982) provided a detailed description of morphological features of zoisite during growth and dissolution processes in their experiments (Fig. 14). One merit of this study is to show how morphological features can be used for the interpretation of reaction progress in sluggish reactions at low temperature (as is the case for many of the zoisite-bearing equilibria). Allen and Fawcett (1982) found that the morphology of zoisite crystals was strongly dependent on the experimental setup. When a rhomb of calcite was placed in fine-grained anorthite mixture, zoisite develops in plates, 10-40 mm in size, only slightly elongated parallel to the *b*-axis and less than 3 mm thick (Fig. 14A). But when a single crystal of anorthite was placed in a fine-grained powder of calcite, nucleation and growth of needles, strongly elongated along the *b*-axis and flattened parallel to (100) was observed (Fig. 14D). Faces in the [010] zone were striated forming steps parallel to the *b*-axis. One set of striae on (001) is parallel to the *b*-axis and a second set on (100) is parallel to *c* (Fig. 14D). Stepped forms result when the rate of spreading of surface layers by addition of material in the steps is a slow process relative to the rate at which new layers nucleate (Allen and Fawcett 1982). A completely different striated appearance forms when zoisite is dissolved (Fig. 14E) as a result of preferential etching parallel to *b*-axis.

As a conclusion of this study, dissolution features of zoisite are easily recognizable, also when comparative powder diffraction would not yield an unequivocal direction of reaction progress. In contrast, growth features of zoisite are expressed in a multitude of different morphologies dependant on starting materials and the experimental setup adopted.

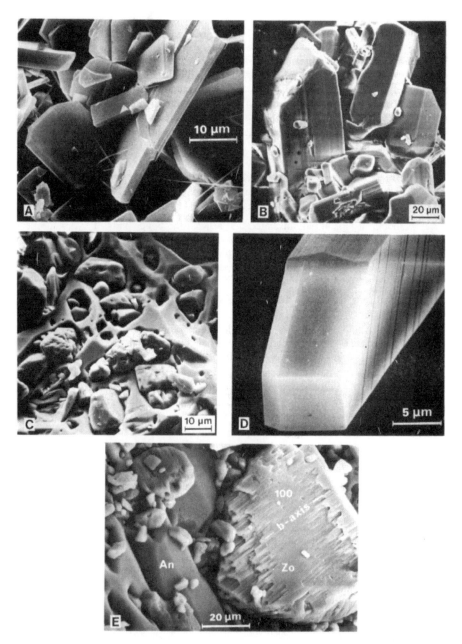

**Figure 14.** Morphology and growth/dissolution features of zoisite. (A) Zoisite plates spontaneously nucleated from calcite and anorthite. (B) Overgrowth on small zoisite seeds. Striae are parallel to the *b*-axis. (C) Large overgrown zoisite seed with inclusions of partially dissolved anorthite and calcite. (D) Opposing sets of striae or steps on zoisite growing elongated parallel to the *b*-axis. (E) Zoisite with edge striae parallel to the b-axis lying on and being included in growing anorthite. [used by permission of Oxford University Press from Allen and Fawcett (1982), *Journal of Petrology*, Vol. 23, Plate 2, p. 227]

## OUTLOOK – OPEN TOPICS FOR FUTURE WORK

Experimental results constrain a wide field of $P$-$T$-$f_{O_2}$-$a_{H_2O}$ conditions for the occurrence of epidote$_{ss}$ in nature. In fact, (clino)zoisite and epidote are ubiquitous in metamorphic rocks of the oceanic and continental crust, and only the high temperatures of the upper amphibolite or granulite facies are sufficient to avoid epidote$_{ss}$. Within the epidote$_{ss}$ field, only very particular bulk compositions (e.g., eclogite lying exactly on the join garnet-omphacite) may not have epidote at particular pressure-temperature conditions. Nevertheless, there are significant gaps in our understanding of equilibria involving epidote and/or zoisite:

- The simple system CFASHO is not yet investigated at high pressure. Experiments on reactions involving the breakdown of epidote$_{ss}$ are necessary at varying oxygen fugacities, as well as on the binary (clino)zoisite-epidote at high pressure. At higher pressures, the upper thermal stability limit of epidote-group minerals is shifted to higher temperatures, and thus experiments on these minerals would be facilitated by the improved reaction kinetics.

- Although it is generally accepted that at least the upper parts of hydrated oceanic crust is somewhat oxidized (Parkinson and Arculus 1999), experimental studies on this subject were either unbuffered or buffered to oxygen fugacities of NNO or FMQ equilibria. In the $P$-$T$ region where epidote becomes critical for volatile transport, a possible significant extension of the epidote-eclogite field at high oxygen fugacities remains unknown.

- Continuous reactions involving the common assemblage epidote + garnet + clinopyroxene ± kyanite are suitable geobarometers but are not yet correctly modeled when using the common internally consistent thermodynamic databases, thus underlining the necessity of additional experiments.

- The partitioning of $Fe^{3+}$ between epidote and other minerals in Mg-bearing complex systems needs much more detailed work carefully controlling equilibration and redox conditions. Solving the tremendous task of unraveling $Fe^{3+}$ distribution among epidote and coexisting phases is a prerequisite for reliable calculations in iron-bearing systems at variable oxygen fugacities.

## ACKNOWLEDGMENTS

We would like to thank Axel Liebscher, Gerhard Franz, Dominique Lattard and Matthias Gottschalk for their reviews that have significantly contributed to improving the manuscript. Support for this work was provided in part by the MIUR-COFIN2003 program.

## REFERENCES

Allen JM, Fawcett JJ (1982) Zoisite-anorthite-calcite stability relations in $H_2O$-$CO_2$ fluids at 500 bars - An experimental and SEM study. J Petrol 23:215-239

Apted MJ, Liou JG (1983) Phase relations among greenschist, epidote-amphibolite, and amphibolite in a basalt system. Am J Sci 283A:328-354

Best NF, Graham CM (1978) Redetermination of the reaction 2 zoisite + quartz + kyanite = 4 anorthite + $H_2O$. Progr Expt Petr NERC ser D 11:153-154

Bird DK, Helgeson HC (1980) Chemical interaction of aqueous-solutions with epidote-feldspar mineral assemblages in geologic systems. 1. Thermodynamic analysis of phase-relations in the system CaO-FeO-$Fe_2O_3$-$Al_2O_3$-$SiO_2$-$H_2O$-$CO_2$. Am J Sci 280:907-941

Bird D, Spieler AR (2004) Epidote in geothermal systems. Rev Mineral Geochem 56:235-300

Boettcher AL (1970) The system CaO-$Al_2O_3$-$SiO_2$-$H_2O$ at high pressures and temperatures. J Petrol 11:337-379

Bonazzi P, Menchetti S (2004) Manganese in monoclinic members of the epidote group: piemontite and related minerals. Rev Mineral Geochem 56:495-552

Brunsmann A (2000) Strukturelle, kristallchemische und phasenpetrologische Untersuchungen an synthetischen und naturlichen Zoisit- un Klinozoisit-Mischkristallen. PhD Thesis, Technical University of Berlin. (*http://edocs.tu-berlin.de/diss/2000/brunsmann_axel.htm*)

Brunsmann A, Franz G, Heinrich W (2002) Experimental investigation of zoisite-clinozoisite phase equilibria in the system $CaO-Fe_2O_3-Al_2O_3-SiO_2-H_2O$. Contrib Mineral Petrol 143:115-130

Chatterjee ND, Johannes W, Leistner H (1984) The system $CaO-Al_2O_3-SiO_2-H_2O$: new phase equilibria data, some calculated phase relations, and their petrological applications. Contrib Mineral Petrol 88:1-13

Cheng W, Greenwood HJ (1989) The stability of the assemblage zoisite + diopside. Can Mineral 27:657-662

Cho MS, Ernst WG (1991) An experimental-determination of calcic amphibole solid-solution along the join tremolite-tschermakite. Am Mineral 76:985-1001

Chopin C, Henry C, Michard A (1991) Geology and petrology of the coesite-bearing terrain, Dora Maira massif, Western Alps. Eur J Mineral 3:263-291

Coes L (1955) High-pressure minerals. J Am Ceram Soc 38:298

Donohue CL, Essene EJ (2000) An oxygen barometer with the assemblage garnet-epidote. Earth Planet Sci Lett 181:459-472

Ehlers EG (1953) An investigation of the stability relations of the Al-Fe members of the epidote group. J Geol 61:231-251

Enami M, Banno S (1980) Zoisite – clinozoisite relations in low- to medium-grade high-pressure metamorphic rocks and their implications. Mineral Mag 43:1005-1013

Forneris JF, Holloway JR (2003) Phase equilibria in subducting basaltic crust: implications for $H_2O$ release from the slab. Earth Planet Sci Lett 6716:1-15

Franz G, Althaus E (1977) The stability relations of the paragenesis paragonite-zoisite-quartz. N Jahrb Mineral Abh 130:159-167

Franz G, Liebscher A (2004) Physical and chemical properties of the epidote minerals–an introduction. Rev Mineral Geochem 56:1-82

Franz G, Selverstone J (1992) An empirical phase diagram for the clinozoisite-zoisite transformation in the system $Ca_2 Al_3 Si_3 O_{12} (OH)-Ca_2 Al_2 Fe^{3+} Si_3 O_{12} (OH)$ Am Mineral 77:631-642

Frei D, Liebscher A, Franz G, Dulski P (2004) Trace element geochemistry of epidote minerals. Rev Mineral Geochem 56:553-606

Fyfe WS (1960) Stability of epidote minerals. Nature 187:497-497

Gieré R, Sorensen SS (2004) Allanite and other REE-rich epidote-group minerals. Rev Mineral Geochem 56:431-494

Goldsmith JR (1981) The join $CaAl_2Si_2O_8-H_2O$ (anorthite-water) at elevated pressures and temperatures. Am Mineral 66:1183-1188

Goldsmith JR (1982) Plagioclase stability at elevated temperatures and water pressures. Am Mineral 67:653-675

Gottschalk M (2004) Thermodynamic properties of zoisite, clinozoisite and epidote. Rev Mineral Geochem. 56:83-124

Grapes RH, Hoskin PWO (2004) Epidote group minerals in low–medium pressure metamorphic terranes. Rev Mineral Geochem 56:301-346

Heinrich W, Althaus E (1988) Experimental-determination of the reactions 4 lawsonite + 1 albite = 1 paragonite + 2 zoisite + 2 quartz + 6H$_2$O and 4 lawsonite + a jadeite = 1 paragonite + 2 zoisite + 1 quartz + 6H$_2$O. N Jahrb Mineral Monatsh (11):516-528

Hermann J (2002) Experimental constraints on phase relations in subducted continental crust. Contrib Mineral Petrol 143:219-235

Heuss-Aßbichler S, Fehr KT (1997) Intercrystalline exchange of Al and $Fe^{3+}$ between grossular-andradite and clinozoisite-epidote solid solutions. N Jahrb Mineral Abh 172:69-100

Holdaway MJ (1972) Thermal-stability of Al-Fe epidote as a function of $f_{O_2}$ and Fe content. Contrib Mineral Petrol 37(4):307-340

Holland TJB, Powell R (1998) An internally consistent thermodynamic data set for phases of petrological interest. J Metamorph Geol 16:309-343

Hoschek G (1995) Stability relations and Al content of tremolite and talc in CMASH assemblages with kyanite plus zoisite plus quartz plus $H_2O$. Eur J Mineral 7:353-362

Jenkins DM, Newton RC, Goldsmith JR (1983) Fe-free clinozoisite stability relative to zoisite. Nature 304:622-623

Jenkins DM, Newton RC, Goldsmith JR (1985) Relative stability of Fe-free zoisite and clinozoisite. J Geol 93:663-672

Johannes (1980) Melting and subsolidus reactions in the system $K_2O$-CaO-$Al_2O_3$-$SiO_2$-$H_2O$. Contrib Mineral Petrol 74:29-34

Johannes W, Ziegenbein D (1980) Stabilität von Zoisit in H2O-CO2-Gasphasen. Fortschr Mineral 58:61-63

Laird J (1980) Phase equilibria in mafic schists from Vermont. J Petrol 21:1-37

Liou JG (1973) Synthesis and stability relations of epidote, $Ca_2Al_3FeSi_3O_{12}(OH)$. J Petrol 14:381-413

Liou JG, Kuniyoshi S, Ito K (1974) Experimental studies of the phase relations between greenschist and amphibolite in a basaltic system. Am J Sci 274: 613-632

Liou JG, Lan CY, Supp J, Ernst WG (1977) Taiwan ophiolite: its occurrence, petrology, metamorphism and tectonic setting. Rept Mining Res Service Organization (Industr Techn Res Inst Taipei, Formosa) 1:212

Liou JG, Kim HS, Maruyama S (1983) Prehnite-epidote equilibria and their petrologic applications. J Petrol 24:321-342

Liu J, Bohlen SR, Ernst WG (1996) Stability of hydrous phases in subducting oceanic crust. Earth Planet Sci Lett 143:161-171

Manning CE (1994) The solubility of quartz in $H_2O$ in the lower crust and upper mantle. Geochim Cosmochim Acta 58:4831-4839

Maruyama S, Cho M, Liou JG (1986) Experimental investigations of blueschist-greenschist transition equilibria: pressure dependence of $Al_2O_3$ contents in sodic amphiboles – A new geobarometer. Geol Soc Am Mem 164:1-16

Matthews A, Goldsmith JR (1984) The influence of metastability on reaction kinetics involving zoisite formation from anorthite at elevated pressures and temperatures. Am Mineral 69:848-857

Merrin S (1962) Experimental studies of iron-aluminum epidote group. J Geophy Res 67:3580

Molina JF, Poli S (2000) Carbonate stability and fluid composition in subducted oceanic crust: an experimental study on $H_2O$-$CO_2$-bearing basalts. Earth Planet Sci Lett 176:295-310

Moody JB, Meyer D, Jenkins JE (1983) Experimental characterization of the greenschist/amphibolite boundary in mafic systems. Am J Sci 283:48-92

Najorka J, Gottschalk M, Heinrich W (2002) Composition of synthetic tremolite-tschermakite solid solutions in amphibole+anorthite- and amphibole+zoisite-bearing assemblages. Am Mineral 87:462-477

Newton RC (1966) Some calc-silicate equilibrium relations. Am J Sci 264:204-222

Newton RC, Kennedy GC (1963) Some equilibrium reactions in the join $CaAl_2Si_2O_8$-$H_2O$. J Geophys Res 68: 2967-2983

Nitsch KH (1968) Die Stabilität von Lawsonit. Naturwissenschaften 55:388

Nitsch KH (1971) Stabilitätsbeziehungen von Prehnit- und Pumpellyit-haltigen Paragenesen. Contrib Mineral Petrol 30:240-260

Nitsch KH, Winkler HGF (1965) Bildungsbedingungen von Epidot und Orthozoisit. Beitr Mineral Petrol 11: 470-486

Nitsch KH, Storre B, Töpfer U. (1981) Experimentelle Bestimmung der Gleichewichtsdaten der Reaktion 1 Margarit + 1 Quarz = 1 Anorthit + 1 Andalusit/Disthen + 1 $H_2O$. Fortschr Min 59:139-140

Okay AI (1995) Paragonite eclogites from Dabie Shan, China: re-equilibration during exhumation? J Metamorph Geol 13:449-460

Parkinson IJ, Arculus RJ (1999) The redox state of subduction zones: insights from arc-peridotites. Chem Geol 160:409-423

Pawley AR, Holloway JR (1993) Water sources for subduction zone volcanism - new experimental constraints. Science 260:664-667

Perchuk LL, Aranovich LY (1979) Thermodynamics of minerals of variable composition: andradite-grossularite and pistacite-clinozoisite solid solution. Phys Chem Mineral 5:1-14

Pistorius CWFT (1961) Synthesis and lattice constants of pure zoisite and clinozoisite. J Geol 69:604-609

Pistorius CWFT, Kennedy GC (1960) Stability relations of grossularite and hydrogrossularite at high temperatures and pressures. Am J Sci 258:247-257

Pistorius CWFT, Kennedy GC, Sourirajan S (1962) Some relations between the phases anorthite, zoisite and lawsonite at high temperatures and pressures. Am J Sci 260:44-56

Pognante U (1991) Petrological constraints on the eclogite- and blueschist- facies metamorphism and *P-T-t* paths in the Western Alps. J Metamorph Geol 9:5-17

Poli S (1993) The amphibolite-eclogite transformation - an experimental study on basalt. Am J Sci 293:1061-1107

Poli S, Schmidt MW (1995) $H_2O$ transport and release in subduction zones - experimental constraints on basaltic and andesitic systems. J Geophys Res-Sol EA 100 (B11):22299-22314

Poli S, Schmidt MW (1997) The high-pressure stability of hydrous phases in orogenic belts: An experimental approach on eclogite-forming processes. Tectonophysics 273:169-184

Poli S, Schmidt MW (1998) The high-pressure stability of zoisite and phase relationships of zoisite-bearing assemblages. Contrib Mineral Petrol 130:162-175

Poli S, Schmidt MW (2002) Petrology of subducted slabs. Annual Rev Earth Planet Sci 30:207-235

Quirion DM, Jenkins DM (1998) Dehydration and partial melting of tremolitic amphibole coexisting with zoisite, quartz, anorthite, diopside, and water in the system $H_2O$-CaO-MgO-$Al_2O_3$-$SiO_2$. Contrib Mineral Petrol 130:379-389

Raith M (1976) The Al-Fe(III) epidote miscibility gap in a metamorphic profile through the Penninic series of the Tauern window, Austria. Contrib Mineral Petrol 57:99-117

Ridley J (1984) Evidence of a temperature-dependent 'blueschist' to 'eclogite' transformation in high-pressure metamorphism of metabasic rocks. J Petrol 25:852-870

Schiffman P, Liou JG (1980) Synthesis and stability relations of Mg-Al pumpellyite, $Ca_4Al_5MgSi_6O_{21}(OH)_7$. J Petrol 21:441-474

Schiffman P, Liou JG (1983) Synthesis of Fe-pumpellyite and its stability relations with epidote. J Metamorph Geol 1:91-101

Schliestedt M, Johannes W (1984) Melting and subsolidus reactions in the system $K_2O$-CaO-$Al_2O_3$-$SiO_2$-$H_2O$: corrections and additional experimental data. Contrib Mineral Petrol 88:403-405

Schmidt MW (1993) Phase-relations and compositions in tonalite as a function of pressure - an experimental-study at 650°C. Am J Sci 293:1011-1060

Schmidt MW, Poli S (1994) The stability of lawsonite and zoisite at high-pressures - experiments in cash to 92 kbar and implications for the presence of hydrous phases in subducted lithosphere. Earth Planet Sci Lett 124:105-118

Schmidt MW, Poli S (1998) Experimentally based water budgets for dehydrating slabs and consequences for arc magma generation. Earth Planet Sci Lett 163:361-379

Schmidt MW, Poli S (2003) Generation of Mobile Components During Oceanic Crust Subduction. *In*: Holland HD Turekian KK (Executive eds.) Treatise on Geochemistry. Vol. 3: The Crust (Rudnick RL ed.), ch 21. Elsevier Science, The Netherlands

Schmidt MW, Poli S (2004) Magmatic epidote. Rev Mineral Geochem 56:399-430

Skrok V, Grevel KD, Schreyer W (1994) Die Stabilität von Lawsonit, $CaAl_2[Si_2O_7](OH)_2 \cdot H_2O$, bei Drücken bis zu 50 kbar. Beihefte Eur J Mineral 6:270

Storre B, Nitsch KH (1972) Die Reaktion 2 Zoisit + 1 $CO_2$ <==> 3 Anorthit + 1 Calcit + 1 $H_2O$. Contrib Mineral Petrol 35:1-10

Storre B, Nitsch KH (1974) Zur Stabilität von Margarit im System CaO-$Al_2O_3$-$SiO_2$-$H_2O$. Contrib Mineral Petrol 43:1-24

Storre B, Johannes W, Nitsch KH (1982) The stability of zoisite in $H_2O$-$CO_2$ mixtures. N Jahrb Mineral Monatsh (9):395-406

Strens RGJ (1968) Reconnaissance of the prehenite stability field. Am Min 36:864-867

Thompson AB, Ellis DJ (1994) CaO+MgO+$Al_2O_3$+$SiO_2$+$H_2O$ to 35 kb - amphibole, talc, and zoisite dehydration and melting reactions in the silica-excess part of the system and their possible significance in subduction zones, amphibolite melting, and magma fractionation. Am J Sci 294:1229-1289

Wittenberg A, Schussler J, Koepke J (2002) Kinetic studies on the trace element distribution of Ca-Al-rich silicates. Geochim Cosmochim Acta 66 (15A):A841-A841 Suppl

Wunder B, Medenbach O, Krause W, Schreyer W (1993) Synthesis, properties and stability of $Al_3Si_2O_7(OH)_3$ (phase Pi), a hydrous high-pressure phase in the system $Al_2O_3$-$SiO_2$-$H_2O$ (ASH). Eur J Mineral 5:637-649

Reviews in Mineralogy & Geochemistry
Vol. 56, pp. 197-234, 2004
Copyright © Mineralogical Society of America

# 5

# Fluid Inclusions in Epidote Minerals
# and Fluid Development
# in Epidote-Bearing Rocks

**Reiner Klemd**

*Institut für Mineralogie*
*Universität Würzburg*
*Am Hubland*
*97074 Würzburg, Germany*
*reiner.klemd@mail.uni-wuerzburg.de*

## INTRODUCTION

The widespread occurrence of epidote minerals in metamorphic and igneous rocks as well as in many ore deposit types makes it a promising candidate for fluid inclusion studies. Apart from high- to very high-temperature and low- to intermediate-pressure conditions, epidote minerals are stable over a wide range of pressure and temperature in the continental and oceanic crust (e.g., Poli and Schmidt 1998). Yet fluid inclusion studies on epidote minerals are surprisingly scarce, even in fluid-saturated environments like certain vein-type deposits or hydrothermal-volcanic vugs and druses. For example, epidote minerals are not mentioned in the subject index of Roedder's (1984) outstanding summary and review of fluid inclusion studies and occurrences, which lists more than sixty different host minerals for fluid inclusions. Nonetheless, more recent studies showed fluid inclusions in epidote minerals to be the only direct witness of the physiochemical and compositional fluid evolution during certain geodynamic processes mainly found in fossil geothermal systems, ore deposits and high-pressure to ultra-high pressure rocks.

The aim of this review is to outline and summarize some aspects and interpretations of geodynamic processes, which are based on temperature ($T$), pressure ($P$), molar volume ($V$) and composition ($X$) data from fluid inclusions in epidote minerals as well as associated host minerals from various geological environments.

The review starts with a chapter on some typical mixed volatile solid-fluid equilibria involving epidote minerals, which are relevant for the here discussed environments. This is followed by a short introduction into the basic concepts of fluid inclusion research and the role of epidote minerals. The next section covers fluid inclusion studies on epidote minerals from active and fossil geothermal systems as well as low-grade metamorphic rocks and constraints on the $P$-$T$-$X$ properties of the fluids present in these systems. This is followed by a short introduction on skarn deposits and the role of ore-forming fluids during the prograde and retrograde skarn evolution, with the focus on those skarn deposits, in which epidote minerals were investigated for fluid inclusions. In addition fluid inclusion studies on epidote minerals are described from other ore deposits and plutonic rocks. The next part deals with high pressure (HP) and ultra-high pressure (UHP) metamorphic rocks, which frequently preserve prograde and retrograde epidote minerals containing fluid inclusions. The presence and composition of a free fluid phase during HP and UHP metamorphism is discussed in terms of the consequences for mineral stabilities, $P$-$T$ conditions, fluid-phase equilibria and fluid flow mechanisms.

1529-6466/04/0056-0005$05.00

## MIXED VOLATILE REACTIONS INVOLVING EPIDOTE MINERALS

In addition to pressure and temperature, epidote mineral stability also depends on the $Al/Fe^{3+}$ ratio, the oxygen fugacity, the bulk and fluid composition and the solution pH. Here the emphasis is mainly placed on some typical mixed volatile solid-fluid equilibria involving epidote minerals, which are relevant for the subsequent chapters, so that the composition and development of the fluid phase in equilibrium with the epidote-bearing mineral parageneses can be discussed and compared with the composition of the fluid inclusions in epidote minerals.

### Hydrothermal and low-grade conditions

Epidote occurs in active geothermal systems at temperatures <200°C as well as under low-grade conditions at temperatures below 360°C (Deer et al. 1986; Bird and Spieler 2004). Epidotization is usually interpreted to be a result of mass transport via a fluid phase in such low temperature environments. Epidote is therefore often found to have precipitated in open fractures and cavities, but also to have replaced the immediate country rocks. It is typically associated with laumontite, heulandite, wairakite, prehnite, pumpellyite, illite, anorthite, chlorite, calcite, quartz and hematite. Some -under the above mentioned conditions- important general reactions in the $CaO-FeO-Fe_2O_3-Al_2O_3-H_2O$-system occur during the transition to greenschist-facies conditions (summarized in Deer et al. 1986):

pumpellyite = epidote + $H_2O$                                                             (1)

pumpellyite = epidote + actinolite + $H_2O$                                                (2)

pumpellyite + chlorite + quartz = epidote + actinolite + $H_2O$                            (3)

prehnite = epidote + grossularite + $H_2O$                                                  (4)

prehnite + hematite = epidote + $H_2O$                                                      (5)

laumontite + hematite = epidote + quartz + $Al_2O_3$ + $H_2O$                               (6)

wairakite + hematite = epidote + quartz + $H_2O$                                            (7)

Metamorphism frequently proceeds directly into the greenschist-facies without the development of lower grade minerals such as prehnite, pumpellyite, Fe-poor epidote or laumontite. The reason for this being that even small amounts of $CO_2$ in the fluid phase suppress the formation of these lower grade facies minerals (e.g., Thompson 1971; Bird and Helgeson 1981; Yardley 1989). However in the presence of quartz and calcite, epidote solid solutions of intermediate composition form over a wide range of $X_{CO_2}$ values at low-pressure conditions as was shown by Bird and Helgeson (1981). For example, the mineral assemblage epidote-K-feldspar-muscovite-calcite-quartz is common in the Salton Sea geothermal system. Bird and Helgeson (1981) demonstrated that the epidote of this mineral assemblage displays an increase in Fe with increasing $X_{CO_2}$ under constant pressure and temperature conditions. The following reaction (Arnason et al. 1993) represents the equilibrium conditions of the mineral assemblage:

muscovite + calcite + quartz = $epidote_{ss}$ + K-feldspar + $CO_2$ + $H_2O$                (8)

### Greenschist-facies conditions

Epidote minerals are characteristically associated with chlorite, quartz, albite, actinolite, muscovite, biotite, garnet, corundum and calcite. Typical mixed volatile reactions (summarized in Tracy and Frost 1991; Labotka et al. 1988) in the $CaO-Al_2O_3-CO_2-H_2O$ (CASCH)-system are:

2 zoisite + 1$CO_2$ = 3 anorthite + 1 calcite + 1$H_2O$                                    (9)

4 zoisite + 1 quartz = 1 grossularite + 5 anorthite + 2H$_2$O     (10)

6 zoisite = 2 grossularite + 6 anorthite + 1 corundum + 3H$_2$O     (11)

2 zoisite + 2 calcite = 1 corundum + 2 grossularite + 2CO$_2$ + 1H$_2$O     (12)

2 zoisite + 2 quartz + corundum = 4 anorthite + 1H$_2$O     (13)

2 zoisite + 4CO$_2$ = 4 calcite + 6 quartz + 3 corundum + 1H$_2$O     (14)

2 zoisite + 5 calcite + 3 quartz = 3 grossularite + 1H$_2$O + 5CO$_2$     (15)

and in the CaO-(FeO,MgO)-Al$_2$O$_3$-CO$_2$-H$_2$O (C(FM)ASCH)-system:

5 vesuvianite + 1 clinozoisite + 4 quartz = 29 grossularite + 10 diopside + 23H$_2$O     (16)

3 chlorite + 10 calcite + 21 quartz = 3 actinolite + 2 clinozoisite + 10CO$_2$ + 1H$_2$O     (17)

The fluid phase in all of these reactions has a relatively low $X_{CO_2}$. This is demonstrated by a $T$-$X_{CO_2}$ projection in the CASCH end-member system at a constant pressure of 2 kbar and with an excess fluid phase (Fig. 1). The $T$-$X_{CO_2}$ projection was calculated for $0 < X_{CO_2} < 0.3$ and with temperatures of between 300 and 650°C using the internally-consistent thermodynamic dataset of Holland and Powell (1990) with the endmembers zoisite, grossularite, quartz, wollastonite, corundum, anorthite, calcite and an H$_2$O-CO$_2$ fluid (computer program THERMOCALC v.3.1; Powell and Holland 1988). The zoisite-bearing assemblages, which are stable only at $X_{CO_2} < 0.1$ between 280 and 580°C, are delimited by reactions (9), (11) and (14).

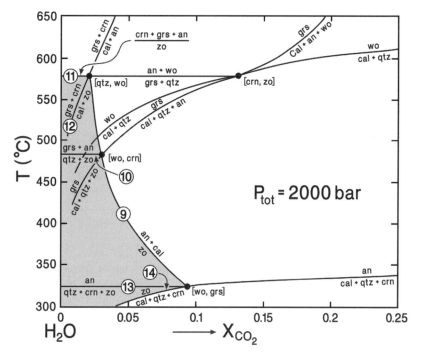

**Figure 1.** *T*-$X_{CO2}$ projection calculated at 2000 bar in the system CASCH and with an excess fluid phase. Pure endmember compositions were used. Shaded field outlines zoisite stability. Numbers refer to reactions listed in the text. Abbreviations follow Kretz (1983).

A similar result was obtained by Will et al. (1990), who calculated a $T$-$X_{CO_2}$ petrogenetic grid for the system CFMASCH with quartz and fluid in excess at a fixed pressure of 2 kbar and with temperatures of between 380 and 510°C involving the minerals amphibole, chlorite, anorthite, clinozoisite, dolomite, chloritoid, garnet, margarite, andalusite and calcite. In this diagram the stability field of Fe-free clinozoisite-bearing assemblages was restricted to $X_{CO_2} < 0.15$, which is also shown by a pseudosection for a specific bulk composition (Will et al. 1990). Even when adding Na (mainly to plagioclase) and $Fe^{3+}$ (mainly to epidote) to the system the divariant field of the clinozoisite-epidote-bearing assemblage is restricted to a very narrow field and a low $X_{CO_2}$. However, temperature and $X_{CO_2}$ values increase with pressure.

## High- and ultra-high pressure metamorphic conditions

Experimental data indicate that orthorhombic zoisite is the Fe-free epidote mineral at *experimentally and geologically relevant conditions* (Poli and Schmidt 1998). Nevertheless, in natural rocks clinozoisite as well as epidote are believed to occur at low to intermediate pressure (Poli and Schmidt 1998). Furthermore, besides zoisite clinozoisite and/or epidote are frequently reported to occur in eclogite-facies rocks (e.g., Holland 1979; Giaramita and Sorensen 1994; Klemd et al. 1994; Scambelluri et al. 1998; Fu et al. 2001, 2003). Several (mixed) volatile reactions involving epidote minerals occur under high- to very-high pressure conditions. The transition of very-low grade rocks to blueschist-facies conditions is displayed by the following general reaction, which involves zoisite (Spear 1993):

$$\text{prehnite} + \text{calcite} = \text{lawsonite} + \text{zoisite} + H_2O + CO_2. \tag{18}$$

The transition of greenschist- to blueschist-facies conditions is marked by the NCFMASH divariant reaction (Maruyama et al. 1986; Will et al. 1998):

$$\text{actinolite} + \text{chlorite} + \text{albite} = \text{glaucophane} + \text{zoisite (or epidote)} + \text{quartz} + H_2O \tag{19}$$

Blueschist facies can be subdivided into a lawsonite- and epidote-bearing subfacies, which is displayed by the following NCFASCH reactions (Franz and Althaus 1977; Winkler 1979; Heinrich and Althaus 1988):

$$4 \text{ lawsonite} + 1 \text{ albite} = 1 \text{ paragonite} + 2 \text{ zoisite} + 2 \text{ quartz} + 6H_2O \tag{20}$$

$$4 \text{ lawsonite} + 1 \text{ jadeite} = 1 \text{ paragonite} + 2 \text{ zoisite} + 1 \text{ quartz} + 6H_2O \tag{21}$$

$$3 \text{ lawsonite} + 1 \text{ calcite} = 2 \text{ zoisite} + 1CO_2 + 5H_2O \tag{22}$$

$$5 \text{ lawsonite} + 1 \text{ albite} = 1 \text{ margarite} + 2 \text{ zoisite} + 2 \text{ quartz} + 8H_2O \tag{23}$$

At high pressure and with increasing temperatures the following important reactions, which are relevant for epidote minerals, occur in the NCMASH system (Gao and Klemd 2001; Poli and Schmidt 2004):

$$6 \text{ zoisite} + 4 \text{ pyrope} + 7 \text{ quartz} = 13 \text{ kyanite} + 12 \text{ diopside} + 3H_2O \tag{24}$$

$$13 \text{ lawsonite} + 3 \text{ diopside} = 8 \text{ zoisite} + 1 \text{ pyrope} + 5 \text{ quartz/coesite} + 22H_2O \tag{25}$$

$$15 \text{ diopside} + 12 \text{ zoisite} = 13 \text{ grossularite} + 5 \text{ pyrope} + 12 \text{ quartz/coesite} + 6H_2O \tag{26}$$

$$13 \text{ glaucophane} + 5 \text{ zoisite} =$$
$$26 \text{ jadeite} + 12 \text{ diopside} + 9 \text{ pyrope} + 19 \text{ quartz} + 15H_2O \tag{27}$$

$$12 \text{ lawsonite} + 1 \text{ glaucophane} =$$
$$2 \text{ paragonite} + 1 \text{ pyrope} + 6 \text{ zoisite} + 5 \text{ quartz} + 20H_2O \tag{28}$$

All of these reactions are believed to occur under very low $X_{CO_2}$ conditions (e.g., Franz and Spear 1983; Klemd et al. 1994; Gao and Klemd 2001; Boundy et al. 2002). This is supported by experimental investigations by Poli and Schmidt (1998) who came to the conclusion that at low temperature (<800°C)-high pressure conditions lawsonite or zoisite coexist with an $H_2O$-rich fluid phase. Boundy et al. (2002) suggested three reasons for the low $X_{CO_2}$ of the fluid phase present during eclogite-facies metamorphism: 1) Lack of progress of decarbonation reactions during subduction with depths less than 100 km; 2) Dehydration reactions dominate during subduction; and 3) $CO_2$ fractionation into solid phases during eclogite-facies conditions.

However, several fluid inclusion studies on eclogite-facies rocks (see below) indicate the presence of considerable amounts of dissolved salts in primary aqueous fluid inclusions in high-pressure minerals, which implies that at least some eclogite-facies fluids will have reduced water-activities and increased $SiO_2$ solubilities (e.g., Shmulovich and Graham 1999).

### Low-pressure contact metamorphic-metasomatic fluids in skarn deposits

Typically, epidote minerals are associated with contact metamorphic/metasomatic open-system processes, as is the case of skarn deposits. Due to advecting hydrothermal solutions, which are at least partly derived from nearby magmatic intrusions, a compositional gradient is established in addition to the thermal gradient. Typical minerals of many skarn deposits are—besides the epidote minerals—diopside-hedenbergite-rich pyroxene, grossularite-andradite-rich garnet, actinolite, wollastonite, vesuvianite, anorthite, phlogopite, calcite, dolomite, magnetite and other Ca-Al-Fe-Mn-Mg silicates (see below). In addition to the reactions above, in skarns and skarn deposits epidote minerals are the result of several other mixed volatile reactions, as for example in the CFMASCH-system (Labotka et al. 1988):

5 vesuvianite + 1 clinozoisite + 4 quartz = 29 grossularite + 10 diopside + $23H_2O$    (16)

3 vesuvianite + $14CO_2$ =
$\qquad$ 9 grossularite + 6 diopside + 5 clinozoisite + 14 calcite + $11H_2O$    (29)

3 vesuvianite + $87CO_2$ =11 clinozoisite + 85 calcite + 9 quartz + 6 diopside + $8H_2O$    (30)

These phase equilibria indicate very low $X_{CO_2}$, as is derived from $T$-$X_{CO_2}$ diagrams (Labotka et al. 1988). The authors suggested that almost $CO_2$-free aqueous fluids were in equilibrium with calcareous argillites in the Big Horse Limestone, west-central Utah, which had locally undergone contact metamorphism.

However, even more characteristic for skarn deposits is the involvement of $Fe^{3+}$-bearing phases in mixed volatile reactions, which occur almost in all types of skarn deposits and are generally related to the ore-forming process (Einaudi et al. 1981; see below). Some characteristic epidote-involving reactions in the CaO-$Fe_2O_3$-$Al_2O_3$-$SiO_2$-$CO_2$-$H_2O$ (CF*ASCH) system are:

2 epidote + 1 grossularite + $1CO_2$ = 1 calcite + 3 anorthite + 1 andradite + $1H_2O$    (31)

2 epidote + 1 calcite + 1 quartz = 2 anorthite + 1 andradite + $1CO_2$ + $1H_2O$    (32)

2 epidote + 1 calcite + 1 hedenbergite =
$\qquad$ 2 grossularite + 2 quartz + 1 magnetite + $1CO_2$ + $1H_2O$    (33)

2 epidote + 5 calcite + 3 quartz = 2 grossularite + 1 andradite + $5CO_2$ + $1H_2O$    (34)

4 epidote + 1 quartz + 1 grossularite = 5 anorthite + 2 andradite + $2H_2O$    (35)

8 epidote + 1 hedenbergite =
$\qquad$ 1 quartz + 8 anorthite + 1 magnetite + 3 andradite + $4H_2O$    (36)

2 epidote + 3 calcite + 1 hedenbergite + $3CO_2$ =
$\qquad$ 3 calcite + 2 anorthite + 1 magnetite + 4 quartz + $1H_2O$    (37)

These oxygen-conserving mixed volatile reactions are displayed on a $T$-$X_{CO_2}$ projection for the

CF*ASCH end-member system at a constant pressure of 1 kbar (Fig. 2). The $T$-$X_{CO_2}$ projection was calculated for $0 < X_{CO_2} < 0.45$ and temperatures between 300 and 650°C using the internally-consistent thermodynamic dataset of Holland and Powell (1990) and the computer program THERMOCALC v.3.1 (Powell and Holland 1988). The diagram is calculated for zoisite, grossularite, quartz, wollastonite, andradite, hedenbergite, anorthite, magnetite, calcite and quartz and the fluid phase as excess phases. The diagram is highly schematic seeing that solid solutions between $Fe^{3+}$ and Al in epidote and garnet were neglected, but it still serves to show the principles of the phase relations. Epidote is clearly stable up to much higher $X_{CO_2}$ values as in the CASCH-system.

Metasomatic reactions of epidote can be formulated as exchange reactions with a fluid of the type

$$3 \text{ anorthite} + 1CaCl_2 + 1H_2O = 2 \text{ zoisite} + 2HCl \tag{38}$$

$$3 \text{ anorthite} + 1CaF_2 + 1H_2O = 2 \text{ zoisite} + 2HF \tag{39}$$

$$3 \text{ anorthite} + 1Ca^{2+} + 1CO_3^{2-} + 1H_2O = 2 \text{ zoisite} + 2CO_2 \tag{40}$$

On the one hand these reactions show that epidote becomes less stable with increasing acidity, on the other hand that $CaCl_2$ can be an important salt in fluids, which are responsible for the formation of epidote minerals. This is supported by the fluid inclusion studies below.

In summary it can be stated that Fe-poor epidote is a good indicator for low $X_{CO_2}$ values, at both low- and high-pressure conditions. However, Poli and Schmidt (1998) also suggested that zoisite can coexist with a $CO_2$-rich fluid phase at very high pressures of 40 kbar and temperatures above 800°C. Furthermore, epidote solid solutions of intermediate composition may be stable over a wide range of $CO_2$ concentrations in the fluid phase of hydrothermal, low-pressure environments as proposed by Bird and Helgeson (1981). In addition, under

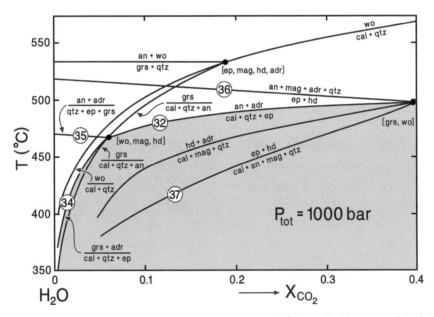

**Figure 2.** $T$-$X_{CO_2}$ projection calculated at 1000 bar in the system CF*SCH and with quartz and the fluid phase in excess. Pure endmember compositions were used. Shaded field outlines epidote-calcite-quartz stability. Numbers refer to reactions listed in the text. Abbreviations follow Kretz (1983).

highly oxidizing conditions, as is the case in skarns, epidote shows an enlarged stability range with respect to $X_{CO_2}$ (Fig. 2). These results contradict the findings of Taylor and Liou (1978) and Liou (1993) who after evaluating the stability of epidote in the CaO-Al$_2$O$_3$-SiO$_2$-FeO$_x$-H$_2$O-CO$_2$-system at 2 kbar concluded that epidote together with quartz can not be stable at $X_{CO_2}$ < 0.2. These contrasting conclusions are the consequence of the use of different thermodynamic datasets, different endmembers as well as different chemical systems. However, fluid inclusions found in epidote minerals in low-grade rocks are usually H$_2$O-salt rich and CO$_2$ poor (see below). The dissolved salts frequently result in a considerable decrease in the H$_2$O-activity. Therefore the low $X_{CO_2}$ value—as displayed by the epidote minerals' stability in the $T$-$X_{CO_2}$ projection—of the above discussed CASCH-example do not necessarily indicate that pure H$_2$O fluids were in equilibrium with the epidote minerals.

## FLUID INCLUSION STUDIES:
## INTRODUCTION OF SOME BASIC CONCEPTS

Fluid inclusions are minute samples of fluids trapped in cavities generally smaller than 100 μm in diameter. These fluids were trapped at a certain event in the geological history of the host rock. Such inclusions may therefore provide information on the $P$-$V$-$T$-$X$ properties of the fluid at the time of trapping. All of these variables are significant for application in a variety of geological and mineralogical fields, such as the study of ore deposits, igneous and metamorphic petrology, structural geology, sedimentology and gemology, to name just a few (for a more comprehensive list see Roedder 1984). In order to determine $P$-$V$-$T$-$X$ variables during trapping conditions microthermometric studies of fluid inclusions have to be undertaken. However, before conducting microthermometric measurements with a heating-cooling stage a detailed petrographic description and classification of the fluid inclusion occurrences is imperative. Some fluid inclusions—usually H$_2$O-salt and/or CO$_2$—are trapped during crystal growth and are referred to as primary or pseudosecondary (for textural criteria see Roedder 1984). Secondary fluid inclusions form by sealing and healing of fractures subsequent to the growth of the host mineral. They occur in minerals and rocks that have been exposed to tectonic and/or thermal stresses. Temperatures of phase transitions (microthermometry) in fluid inclusions are determined with heating-freezing stages. If the composition of an inclusion fluid is known, then its density can be calculated from the homogenization temperature ($T_h$) of the different phases into one fluid phase. Thereafter an isochore, which is defined as line of constant volume and hence constant density for a homogenous fluid, can be plotted on a $P$-$T$ diagram. Homogenization temperatures are the entrapment temperatures of fluids, only if the hydrostatic pressure did not exceed their equilibrium vapor pressure. In this case a "pressure correction" is necessary to obtain true entrapment temperatures. After freezing the fluid inclusions are heated to obtain the initial and final melting temperature, both of which provide information on the composition of the inclusion fluid. In the case of a H$_2$O-salt solution, the final ice melting temperature ($T_m$) corresponds to the freezing point depression of pure H$_2$O, while the initial melting temperature may correspond to the eutectic temperature for the respective salt system. The salinity of aqueous inclusions is inferred from the temperature of final ice melting and given as weight % NaCl equivalent (wt. % NaCl eq.). A detailed description of microthermometrical techniques and basic information concerning the application of the microthermometric data are summarized in several publications (Hollister and Crawford 1981; Roedder 1984; Shepherd et al. 1986; Goldstein and Reynolds 1994; Andersen et al. 2001). Two requirements have to be fulfilled before unambiguous $P$-$T$-$X$ conditions can be determined from fluid inclusion studies: Firstly, the fluid inclusions must have remained a chemically closed system throughout the geological history of the host mineral, and secondly, the molar volume of the fluid inclusion must have remained constant throughout the geological history of the host mineral. Both assumptions have been shown to be violated for certain soft, easily-cleaved host minerals such as halite, barite, fluorite and carbonates when exposed to mechanical or thermal stress (e.g., Shepherd et al. 1986). Nonetheless, the widespread occurrence of quartz allows fluid inclusion studies in various geological fields to be conducted such as ore petrology, igneous and metamorphic petrology, structural geology and sedimentology. This is due to the fact that fluid inclusions are most commonly observed in quartz, although fluid inclusions have also been observed in other minerals such as apatite, halite, carbonate, topaz, barite, fluorite, olivine and less commonly in garnet, pyroxene, amphibole, kyanite, feldspar and epidote minerals.

Most fluid inclusions in epidote minerals are texturally primary or pseudosecondary. They are either aligned with their long dimension parallel to the b-axis of the epidote host or occur as large isolated single inclusions (Fig. 3). However secondary fluid inclusions occurring along healed microfractures, which crosscut grain boundaries, are also observed in epidote minerals. The shape of all inclusions ranges from irregular to tabular and negative crystal shapes are also common. Most fluid inclusions display a large range in size from <3 to 100 μm, while some of the isolated larger inclusions may reach up to 300 μm in size. All reported fluid inclusions in epidote minerals are aqueous and consist of a liquid and a vapor bubble. In addition one or more solid phases such as rutile, apatite, quartz, zircon and calcite were found in these inclusions. These solids are usually believed—according to the criteria established by Roedder (1984)—to have been accidentally trapped. However, especially in metamorphic rocks some of these solids may have formed from "back reactions" of the trapped inclusion fluid with the epidote mineral host (cf. Heinrich and Gottschalk 1995). Consequently, quartz seems to be a perfect host mineral since it does not react with the entrapped fluid to form daughter minerals by "back reactions". In addition it almost always contains fluid inclusions and has a widespread stability. Frequently a mismatch between pressure and compositional estimates— from mineral geothermobarometry and fluid-phase equilibria with fluid inclusion data (mainly from quartz)—is encountered in amphibolite-facies to high-grade metamorphic rocks (e.g., Crawford and Hollister 1986; Touret 1992). This is the result of decrepitation (fluid inclusions loose fluid along fractures in the host mineral during implosion or explosion), leakage (form of compositional and physical re-equilibration in which parts of the inclusion fluid is lost) or stretching (non-elastic volume change which is accommodated by plastic deformation of the host mineral) due to pressure differences between internal pressure (within inclusions) and confining pressure e.g., during exhumation of the host rock (e.g., Crawford and Hollister 1986; Bodnar et al. 1989; Sterner et al. 1995; Touret 1992; Küster and Stöckhert 1997).

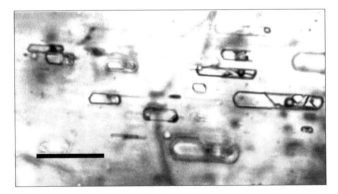

**Figure 3.** Photomicrograph show-ing primary high-salinity fluid inclusions parallel to the b-axis of epidote in the Sujiahe eclogites from the NW Dabie Shan HP-belt in China (Fu 2002). The inclusion contains— besides a liquid as well as a vapor-bubble—one or more unidentified solid phase, which did not dissolve during heating (for further explanation see text). Scale bar = 20 μm.

In contrast to other high-pressure minerals such as garnet and omphacite, epidote for example was suspected to be susceptible to H-diffusion (leakage) during eclogite-facies metamorphism (Giaramita and Sorensen 1994), which was however not verified in various fluid inclusion studies on epidote-group minerals from HP-UHP rocks (see discussion on HP and UHP metamorphism below). Besides primary fluid inclusions in syn-metamorphic vein minerals, fluid inclusions in metamorphic rocks are usually secondary and mainly occur along healed fractures in matrix quartz. Syn-metamorphic quartz veins represent fractures— ranging in size from microcracks to veins—along which fluid flow was accommodated and channelized (e.g., Walther and Orville 1982; Yardley 1986). The fluid in secondary inclusions along healed fractures in the matrix quartz and the quartz veins may have been derived by prograde devolatilization reactions of the immediate adjacent rock and, thus, would have been in equilibrium with the source rock at this *P-T* condition. These inclusions may therefore preserve the same information as texturally primary inclusions formed during the growth of syn-metamorphic vein minerals, which formed at sites of hydraulic fracturing. Nonetheless, it is also possible that fluids were trapped far away from the source rock and postdate the devolatilization of the host rock (e.g., Crawford and Hollister 1986; Yardley 1986). This is supported by the solubility of $SiO_2$ in $H_2O$-salt solutions, which decreases with decreasing temperatures, and thus allows the formation of quartz veins as the fluid approaches cooler rocks (e.g., Yardley 1986; Newton and Manning 2000). In addition to these uncertainties and the above described post-entrapment modifications in high-grade metamorphic rocks, quartz is also highly susceptible to recrystallization by grain boundary migration under low-grade conditions (e.g., Johnson and Hollister 1995). This recrystallization favors entrapment of $CO_2$ inclusions, whereas microfracturing favors entrapment of aqueous inclusions. $CO_2$ can therefore only be removed by hydrofracturing, while $H_2O$ ± salt due to its strong polarity wets the quartz surface and remains outside the advancing crystal front during recrystallization. Due to the low dihedral angle of $H_2O$ ± salt (<60°) it can migrate along quartz-quartz grain boundaries, whereas $CO_2$ with high dihedral angles (>60°) will stay behind (e.g., Holness 1993). Therefore, pure $CO_2$ inclusions in quartz in low- to high-grade metamorphic rocks as well as in deformed and heated hydrothermal/metamorphic quartz veins are most likely to represent the residue of a deformation and recrystallization process rather than a primary hydrothermal or metamorphic fluid (e.g., Johnson and Hollister 1995; Klemd 1998). As a result fluid inclusion investigations on metamorphic rocks should preferentially be conducted on matrix and vein minerals, which are less susceptible to post-trapping modifications such as garnet, pyroxene, kyanite and epidote minerals, or on quartz inclusions in these minerals, in order to obtain more accurate *P-T-X* information for peak metamorphic fluids. However, daughter minerals in fluid inclusions of these minerals may result from "back reactions" of the host mineral with the entrapped fluid, which may have altered the density and composition of the remaining fluid (cf. Heinrich and Gottschalk 1995; Kleinefeld and Bakker 2002).

One of the most significant achievements of fluid inclusion studies is that it has shed new light on the major role of fluid immiscibility, for example in the formation of ore deposits and oil fields, in the fractionation of siliceous melts and during metamorphism. Hollister (1981) summarized three types of fluid immiscibility based on the composition of the fluid phase:

I. The first type is boiling of predominantly aqueous fluids, which results in simultaneous entrapment of liquid-water and/water-rich vapor. In order to prove that boiling had occurred, both types of inclusion have to homogenize at the same temperature. Fluid inclusion investigations on epidote have revealed that boiling had occurred in active and fossil geothermal systems (e.g., De Vivo et al. 1989; Kelley et al. 1993) as well as in skarn deposits (e.g., Bukharev et al. 1982).

II. The presence of NaCl and/or other electrolytes in the $H_2O$-$CO_2$ system on the one hand enlarges the solvus between $CO_2$- and $H_2O$-rich fluids (Fig. 4) and on the other

*Klemd*

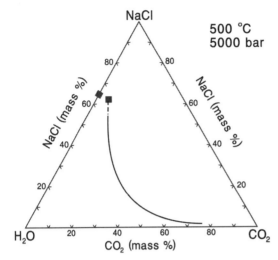

**500 °C**
**5000 bar**

**Figure 4.** Isobaric isothermal phase diagram of the $H_2O$-$CO_2$-NaCl system at 5000 bar and 500°C. The thick solid line shows the interpreted immiscibility boundary between the one-fluid field and the two-fluid field for the $H_2O$-$CO_2$-salt system. Increasing NaCl content raises the immiscibility boundary considerably. Filled squares = halite saturation under experimental conditions (modified after Shmulovich and Graham 2003).

hand increases the non-ideality of $H_2O$-$CO_2$ fluid mixing. Thus mineral assemblages that are stable at low temperature in the presence of a pure $H_2O$-$CO_2$ fluid may also be stable at higher temperature in the presence of an $H_2O$-$CO_2$-salt fluid. Furthermore fluid-phase equilibria will fail to reveal the correct fluid composition at immiscibility conditions if the additional presence of electrolytes in the $H_2O$-$CO_2$ system are not recognized (Bowers and Helgeson 1983). A similar effect exists in the $H_2O$-$CH_4$ system, where the addition of salts dramatically increases the immiscibility field of $H_2O$- and $CH_4$-rich fluids towards higher *P-T* conditions (Goldstein and Reynolds 1994). Fluid inclusions found in epidote minerals in low-grade rocks are usually $H_2O$-salt rich and $CO_2$ poor (e.g., Stalder and Rykart 1980), which is in accordance with fluid-phase equilibria under these *P-T* conditions (see discussion above). Yet in some low-grade metamorphic rocks epidote seems to occur in textural equilibrium with other minerals, which contain $H_2O$-$CO_2$-rich fluids with a $CO_2$-content ranging from 60 to 90 mol % (Kreulen 1980, 1989). This may indicate epidote growth during or after loss of $CO_2$ due to unmixing (i.e., the mole fraction of $H_2O$ is increased in the fluid phase)

III. The third type concerns the separation of a low-density fluid from a silicate melt. The solubility of an aqueous fluid is higher in silicate melts when compared to the solubility of a carbonic fluid. Accordingly a $CO_2$-rich fluid phase can separate early from a crystallizing melt. Hollister (1981) suggests that this is a possible explanation for the common occurrence of nearly pure $CO_2$ occurring with glass (quenched silicate melt) inclusions in phenocrysts and xenocrysts. However, another possibility is that $CO_2$ in glass inclusions are formed by the trapping of external $CO_2$ through melt migrating along grain boundaries (Witt-Eickschen et al. 2003). This type of immiscibility as well as melt inclusions have not yet been described for inclusions in epidote.

## GEOTHERMAL SYSTEMS

### Introduction

Many geothermal systems are associated with active or recently active volcanism and plutonism, which act as heat sources to enhance fluid circulation and, thereby, mass transport in the continental and oceanic crust (e.g., Norton 1977, 1987; Cathles 1997). Irreversible reactions between hydrothermal fluid and immediate host rocks produce mineral alteration in order to reach equilibrium between fluid and rock (e.g., Helgeson 1979; Norton 1987). The presence or absence of certain minerals in the alteration assemblage depends on factors such as $CO_2$ and $O_2$ fugacities, solution pH as well as bulk rock composition (e.g., Henley and Ellis 1983; Hedenquist and Henley 1985; Barnes 1997). Fluid inclusions are often trapped during the growth of the alteration minerals, and are therefore able to give direct evidence on the physicochemical properties of the hydrothermal fluid responsible for the alteration. In active hydrothermal systems fluid inclusion data allow a comparison with directly measured temperature and fluid composition in order to assess short time scale variations (De Vivo et al. 1989; Hedenquist et al. 1992; Lecuyer et al. 1999). Fluid inclusion data from minerals of fossil geothermal systems provide direct evidence concerning the composition and the evolution of the hydrothermal fluids (Kelley et al.1993; Barnes 1997).

This review is restricted to studies involving fluid inclusions in epidote minerals from several well-known geothermal systems, including those associated with active magmatic hydrothermal systems as well as submarine hydrothermal systems. For more details on geothermal systems along with extensive reference lists see Barnes (1997) and Bird and Spieler (2004).

## ACTIVE MAGMATIC HYDROTHERMAL SYSTEMS

World-wide numerous active geothermal systems are formed by penetrating hydrothermal fluids, which are driven by the heat of associated volcanic or plutonic activity. The hydrothermal solutions are usually mixtures of meteoric and magmatic fluids as revealed by stable isotope investigations. All described occurrences document the physicochemical properties and composition of inclusion fluids, which represent pristine samples of the hydrothermal fluids and, if possible, its evolution in time and space.

### Los Azufres geothermal fields, Mexico

The Los Azufres geothermal field is a well-described geothermal system in the Trans-Mexican volcanic belt and has been the subject of several geothermal fluid and fluid inclusion studies (Cathelineau and Nieva 1986; Cathelineau et al. 1989, Izquierdo et al. 1997; Torres-Alvarado 2002) conducted on rock samples from different wells and depths. The up to 3 km thick stratigraphy of the geothermal field consists of andesitic lavas, which are overlain by rhyolitic and dacitic lavas and a volcano-sedimentary sequence (Cathelineau and Nieva 1986; Cathelineau et al. 1989). The authigenic minerals quartz, calcite, epidote and anhydrite have primary inclusions with low salinity (1–7 wt. % NaCl eq.) aqueous fluids. Homogenization temperatures are in close agreement with downhole temperatures, with some local evidence for cooling (30–40°C) since peak thermal conditions (Cathelineau and Nieva 1986; Cathelineau et al. 1989). A mineralogical sequence from the surface to the deepest levels of four temperature zones was established by a combination of fluid inclusion data and mineral assemblages (Izquierdo et al. 1997):

alunite-amorphous silica-gypsum-native sulfur (80–150°C),

kaolinite-smectite-clinoptilolite-laumontite (100–180°C),

illite-chlorite-calcite-pyrite-wairakite-anhydrite (190–250°C), and

chlorite-mica-quartz-epidote-hematite-diopside-prehnite-adularia (220–320°C)

Epidote is part of the high-temperature mineral assemblage and occurs in vesicles and veins and appears to have formed together with chlorite, quartz, prehnite, and hematite. Reaction (5) represents the equilibrium conditions of this mineral assemblage.

## Chipilapa geothermal field, El Salvador

The Chipilapa geothermal volcanic field is located in western El Salvador and resulted from the subduction of the Cocos plate under the Caribbean plate (Molner and Sykes 1969). Geothermal activity is displayed by fumaroles and solfatares (Bril et al. 1996; Patrier et al. 1996). Mineral assemblage and fluid inclusion studies were undertaken on eight drill cores, which revealed the following vertical zoning of alteration (Papapanagiotou 1994; Bril et al.1996): an early propylitic alteration (epidote-chlorite-quartz-prehnite-adularia) followed by a clay-phyllic alteration (illite/smectite-chlorite/saponite-calcite), which is prominent in the upper part of the drill core. Fluid inclusions were investigated in early and late quartz, calcite and epidote. All three minerals contain two-phase primary and secondary aqueous inclusions, while quartz and calcite additionally contain one-phase vapor inclusions. Homogenization into the liquid ranges from 196 to 300°C for the two-phase inclusions in all of the minerals, while final melting temperatures range between −1.5 and −0.4°C. Average $T_h$ values are 260°C for quartz, 230°C for epidote and 220°C for calcite. Bril et al. (1996) interpreted the fluid inclusion data to display successive crystallization events of quartz, epidote and calcite all of which had precipitated from a low-salinity/low-temperature hydrothermal fluid. Quartz deposition was interpreted to have started at about 270°C, whereas calcite crystallized at about 230°C. Epidote precipitated before calcite at temperatures just above 230°C. Authigenic quartz and epidote commonly precipitated from the low-salinity/low-temperature hydrothermal fluids as vug-fillings with increasing grain size towards greater depth (Fig. 5). Prehnite and adularia are associated with quartz and epidote at the greatest depth (2556 m), while wairakite additionally occurs at shallower depths (above 1400m). The homogenization temperature in the primary and secondary fluid inclusions was found to be higher than the temperature measured directly in the drill holes. This was considered as evidence for the cooling of the geothermal field. The presence of the one-phase vapor-rich secondary inclusions was considered to record local boiling events. The reconstruction of the geothermal evolution in the Chipilapa area was achieved by combining the alteration and fluid inclusion data (Bril et al. 1996; Patrier et al. 1996): Initially the propyllitic alteration was formed by a conductive thermal gradient, which was followed by the influx of meteoric water, thus causing local boiling (Fig. 5). During the last stage a new geothermal gradient was formed and clay minerals were converted to illite/chlorite (Fig. 5).

## Dixie Valley Geothermal system, Nevada

The Dixie Valley geothermal system in west-central Nevada is a fault-related geothermal system (e.g., Parry et al. 1991). The fluids from wells have maximum temperatures of 265–275°C and were formerly believed to be of meteoric origin only (Nimz et al. 1999; Blackwell et al. 2000). The source of the heat for the geothermal system was thought to be due to the circulation of the hydrothermal fluids through an area with an anomalously high geothermal gradient (Blackwell et al. 2000). Fluid inclusion microthermometry and gas analysis by quadropole mass spectrometry were used in order to gain information on the *P-T* condition and composition of the hydrothermal fluid phase (Lutz et al. 2002). The relationship between $CO_2/CH_4$, $N_2/Ar$, and $H_2S$ in fluid inclusions from epidote-bearing fault-gauges, and hematite- and actinolite-bearing veins indicates a mixing of meteoric and magmatic fluids, which was further supported by $N_2/Ar$ ratios of up to 300. Consequently, the hydrothermal fluids of the Dixie Valley geothermal system must have been at least partially related to Miocene volcanism (Lutz et al. 2002).

## Phlegrean geothermal field, Italy

The geothermal field of Phlegrean is located west of Naples in southern Italy within Pliocene graben structures of the Campanian Plain of the western Apennines. This geothermal system formed subsequently to the opening of the Tyrrhenian Basin (e.g., De Vivo et al. 1989).

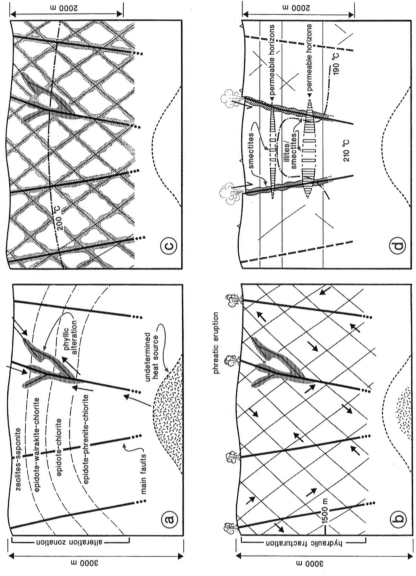

**Figure 5.** The evolution of the Chipilapa geothermal system as determined by the alteration mineralogy and fluid inclusion data. (a) The first stage involves a propylitic alteration and quartz crystallization. (b) The second stage shows an influx of meteoric water. (c) The third stage displays a self-sealing of the fracture-network by carbonate and clays and the formation of a new geothermal gradient. (d) The last stage indicates active fluid circulation controlled by fault systems, along which smectite precipitated independently of the temperature (modified after Bril et al. 1996).

In order to conduct a geothermal exploration program, several shallow and deep (~3 km) wells were drilled in areas displaying a high thermal gradient (Belkin and De Vivo 1987; De Vivo et al. 1989). The wells penetrated hydrothermally altered volcanic rocks, such as trachytic lavas, volcaniclastics and sedimentary rocks in shallow depths. Below about 900 m the effect of thermometamorphism was encountered. Five studied drill cores contain four main alteration zones as a function of decreasing depth: 1) Argillic zone, 2) Illite-chlorite zone, 3) Ca-Al-silicate zone, and 4) Thermometamorphic zone. Hydrothermal minerals usually occur as open fracture- and vug-filling. Fluid inclusion and Sr isotope studies were undertaken on the drill core in order to constrain the evolution of the fluid composition and the geothermal field (Barbieri et al. 1986; Belkin et al. 1986; Belkin and De Vivo 1987; De Vivo et al. 1989). Primary fluid inclusions were observed in hydrothermal quartz, K-feldspar, calcite and epidote. The fluid inclusions are aqueous and occur as two-phase (liquid- and/or vapor-rich) and multiphase inclusions (liquid + vapor + solids). The contemporaneous occurrences of liquid- and vapor-rich fluids were found to represent boiling conditions. Salinity determined by ice and halite dissolution temperatures range from 0.1 to 49 wt. % NaCl eq. The initial melting temperatures range from $-52$ to $-22°C$ indicating the presence of $CaCl_2$, in addition to NaCl and KCl. Reactions (38) and (40) have already shown that $CaCl_2$ can be an important salt in fluids from which epidote minerals are formed. This salinity range corresponds to the composition of the encountered fluid from the wells (Carella and Guglielminetti 1983). In general homogenization temperatures increase with depth and are consistent with those in the wells. It should be mentioned that temperatures derived from one well are lower than those from the fluid inclusions. This was found to be related to the cooling of the geothermal system. The $^{87}Sr/^{86}Sr$ study on carbonate and leached residue suggested an interaction with seawater, which was probably introduced along aquifers at depth.

## FOSSIL SUBMARINE HYDROTHERMAL SYSTEMS

Circulation of seawater through mid-ocean ridges is believed to account for at least 25% of the heat flux from the interior of the earth (e.g., Thompson 1983; Scott 1997). Large-scale circulation of heated seawater in the vicinity of ocean or back-arc ridges results in wide spread alteration of the oceanic crust. In order to extend our knowledge on the relationship between hydrothermal and magmatic systems at oceanic ridges microthermometric investigations were undertaken on fluid inclusions in alteration minerals such as epidote and anhydrite. Such investigations reveal the composition of the hydrothermal fluid and give information on the evolution of the magmatic-hydrothermal fluid. Fluid expelled from hydrothermal vents on the ocean floor exhibits a salinity ranging from 0.1 to 200% of the seawater value (e.g., von Damm et al. 1985, 1997). The composition of the fluid rising to the surface is mainly determined by fluid/rock interaction in the deep high-temperature reaction zones of the geothermal system (Bischoff and Rosenhauer 1989; Seyfried et al. 1991; Seyfried and Ding 1995). The process, which is responsible for the large spread of salinity in the hydrothermal fluids emanating from the vents is still a matter of discussion. Phase separation and/or fluid mixing may play an important role in fixing the fluid chemistry (Bischoff and Rosenbauer 1984, 1989; Butterfield et al. 1990; Nehlig 1991; Kelley and Robinson 1990; Kelley et al.1993; Lecuyer et al. 1999; Lüders et al. 2002).

### Ophiolites

Ophiolites are tectonically obducted mafic and ultramafic rocks, which are understood to represent relics of oceanic or back arc basin crust (Coleman 1977). They form at crustal rifts, which are associated with extensive magmatism and related geothermal systems. Accordingly, hydrothermal alteration due to fluid circulation has been observed in ophiolites world-wide (e.g., Gillis and Bangerjee 2000). The hydrothermally altered rocks—including epidosites—

were the subject of numerous fluid inclusion studies in order to contribute to the current discussion on the evolution of submarine hydrothermal systems.

Nehlig (1991) conducted an extensive fluid inclusion study on quartz, epidote, plagioclase, amphibole, anhydrite and sphalerite from hydrothermally altered plutonic and volcanic rocks of the Semail (Oman) and Trinity (California) ophiolites. The plagiogranites of the ophiolite sequence were almost completely altered to epidosites, which were formed due to metamorphic equilibrium reactions such as reaction (9), or to metasomatic exchange reactions such as reactions (38) and (40), which require the introduction of $CaCl_2$. The salinity of primary aqueous fluid inclusions in all minerals in the plagiogranites show a wide range from 0.3 to 52 wt. % NaCl eq. However, more than 60% of the mean are within the range of seawater. Most minerals have inclusions with a salinity somewhat higher than seawater, for example that of primary fluid inclusions in secondary epidote, quartz and sphalerite ranges between 3.9 and 10.3 wt. % NaCl eq., while some secondary inclusions in quartz have salinity of up to 52 wt. % NaCl eq. The very high salinity is restricted to rocks occurring in the transition zone between the magma chamber and the sheeted dyke complex. Homogenization of the low-salinity inclusions occurs between 213 and 452°C into the liquid or the vapor phase. High-salinity fluid inclusions with halite always homogenize into the liquid phase by halite dissolution between 333 and 446°C ($T_h$ (V $\rightarrow$ L) < $Tm_{halite}$). One measured initial melting temperature occurred at temperatures <26.7°C indicating the presence of $CaCl_2$, apart from NaCl in the inclusion fluid. This supports the theory that the hydrothermal fluids, which were responsible for the epidotisation of the plagiogranites, were enriched in $CaCl_2$. Nehlig (1991) interpreted the fluid inclusion data to reflect phase separation in hydrothermal or magmatic fluids at a temperature in excess of 500°C within the transition zone of the hydrothermal system and the magma chamber.

Further fluid inclusion studies on plutonic and diabase samples from the Troodos ophiolite (Cyprus) were conducted by Kelley and Robinson (1990) and Kelley et al. (1992). These authors observed high-salinity (30–61 wt. % NaCl eq.) fluid inclusions in quartz and epidote from plagiogranites and associated epidosites. Halite dissolution always occurred into the liquid phase at a temperature between 400 to 500°C. Low- salinity (2–7 wt. % NaCl eq.) liquid- and vapor-dominated aqueous inclusions occur in secondary fluid inclusion trails, which often crosscut high-salinity inclusion trails. The former have $T_h$ values from 200 to 400°C. Initial melting temperatures range from –76 to –61°C indicating the possible presence of $CaCl_2$, $FeCl_2$ and $MgCl_2$, in addition to NaCl and KCl in the inclusion fluid. However, Kelley et al. (1992) found the low initial melting temperatures to be related to the presence of mainly $CaCl_2$, in addition to NaCl and KCl. They suggested that this fluid represented the hydrothermal fluid, which was responsible for the extreme epidote alteration of the plagiogranites. Kelley and Robinson (1990)—in accordance with Nehlig (1991)—interpreted the high-salinity inclusions to have resulted by phase separation of seawater or an exsolved magmatic aqueous fluid phase at >500°C. The low-salinity fluid inclusions however were considered to have their origin in seawater circulating along microfractures in the upper crust. Hydration reactions with primary magmatic minerals and/or mixing with the already phase-separated fluids resulted in the salinity enrichment of these seawater-derived fluids. Although phase separation of brine and vapor seemed to have been a well established model to explain the occurrence of the high- and low-salinity fluids (see above), Kelley et al. (1992) preferred the model of direct exsolution of the high-salinity fluids from a magmatic source. They suggested that these brines were trapped as high-salinity inclusions while migrating along microfractures in the deep-seated, high temperature portion of the hydrothermal system. The later low-salinity fluids were interpreted as seawater-derived fluids trapped at a temperature of >200–400°C.

## Mid-Atlantic Ridge (MARK area)

Kelley et al. (1993) undertook a fluid inclusion study on hydrothermally altered gabbros, quartz-breccias and basalts from the western intersection of the Mid-Atlantic Ridge and the Kane Fracture Zone in order to shed some light on the thermal and compositional evolution of hydrothermal fluids in oceanic-rift magma-hydrothermal systems. Multiple generations of primary and secondary fluid inclusions occur in plagioclase, clinopyroxene, amphibole, epidote and apatite. Apatite in plutonic rocks contains the most complex fluid inclusion inventory (Kelly and Delaney 1985, 1986; Kelley et al. 1993). Type 1 are primary vapor- and liquid-dominated, halite ± pyrite-bearing inclusions. Total homogenization occurs at a temperature of >700°C, while halite dissolves between 352 and 408°C indicating a salinity of 41–47 wt. % NaCl eq. Some vapor-dominated, halite-bearing inclusions contain $CO_2$ in variable amounts as indicated by their freezing behavior and confirmed by Raman spectroscopy. Associated with the brine inclusions are Type 2 low-salinity (1–2 wt. % NaCl eq.) vapor-dominated $H_2O$-$CO_2$ inclusions, which homogenize between 364 and 420°C, and Type 3 are liquid-dominated, low-salinity, secondary inclusions which homogenize between 263 and 309°C and have a salinity of 0.4–7.6 wt. % NaCl eq. Epidote, which mainly formed at the expense of plagioclase (see reactions (9) or (38)) or occurs in veins, contains primary Type 2 and Type 3 inclusions, both of which homogenize at about 400°C and have a salinity of between 4.4 and 5.9 wt. % NaCl eq. Secondary Type 2 and 3 inclusions mainly occur in plagioclase and may contain up to four solids. Quartz from the quartz-breccias also contains secondary to primary Type 3 inclusions with somewhat lower $T_h$ values from 187 to 343°C and a salinity from 3.5–6.5 wt. % NaCl eq., when compared with the fluid inclusion data from the plutonic rocks.

As a consequence of these fluid inclusion data, Kelley et al. (1993) suggested the following scenario (Fig. 6): During solidification and cooling of the magma a $CO_2$-rich vapor (Type 1, vapor-rich inclusions) was initially exsolved. This was followed by $H_2O$-$CO_2$ vapor-rich fluids (Type 2 inclusions) and formation of a cogenetic $H_2O$-NaCl ± $CO_2$ ± Fe-rich immiscible brine (Type 1, liquid-rich inclusions). The exsolution of an early vapor-rich fluid phase may have initiated fracturing close to the magma chamber at a temperature >700°C. The transition from magmatic to hydrothermal conditions was marked by initial penetration of low-salinity seawater-like fluids (Type 3 inclusions) at a temperature >400°C. The fluid flow came to an end at a temperature of 180–340°C represented by Type 3 inclusions in plagioclase and quartz.

## Discussion

All fluid inclusion studies were conducted on oceanic or back arc basin crust, which formed in the vicinity of crustal rifting due to extensive magmatism. The magma chamber acted as a heat source in order to keep up the fluid flow in the geothermal system. Consequently all microthermometric studies from the different localities revealed inclusion fluids with low and high salinity. The high-salinity fluid inclusions contain $CaCl_2$ besides NaCl and KCl and are thought to be responsible for the extreme epidotization of the plagiogranites. These fluids— whether or not derived by phase separation and/or fluid unmixing—were always considered to reflect the participation of a magmatic component in the seawater-dominated hydrothermal systems. This is in accordance with many other fluid inclusion studies from other localities with similar geotectonic settings (e.g., Vanko 1988; Vanko et al. 1992; Lecuyer et al. 1999) as well as with "segregation of brine and vapor" processes in natural vent emanations (von Damm et al. 1997). Nevertheless, in this context it is interesting to note that the combined use of Cl/Br ratios and $\delta^{37}Cl$ values of inclusion fluid in sphalerite from fracture fillings in felsic igneous rocks with massive sulfide mineralization from the Jade field (Central Okinawa Trough) and the North Fidji basin exclude any significant magmatic component (Lüders et al. 2002). These authors consequently interpreted the salinity variations in the fluid inclusions of sphalerite to be the result of phase separation of seawater only.

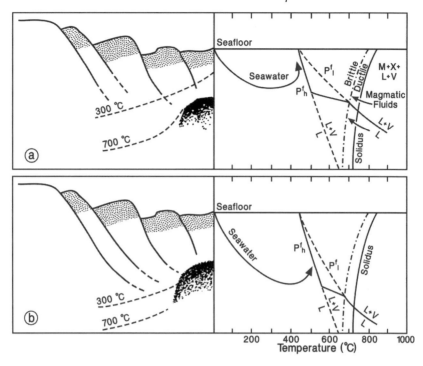

**Figure 6.** Model of the fluid evolution in the MARK-geothermal system. (a) Crossection with the 300 and 700°C isotherms adjacent to an axial magma chamber. The right side of the sketch shows the two-phase curves for NaCl-bearing fluids under hydrostatic ($p_h^f$) and lithostatic ($p_l^f$) conditions. Also shown is the solidus. The intersection of the solidus and the two-phase curves determines the field of the magmatic fluids coexisting with melt (M) and crystals (X). Fracturing near the solidifying magma chamber allowed the migration of exsolved brine and vapor phases at temperatures >700°C. (b) Brittle fracturing occurred during cooling at 400–550°C and allowed the penetration of seawater-derived hydrothermal fluids (modified after Kelley et al. 1993).

## LOW-GRADE METAMORPHIC ROCKS

Kreulen (1980) undertook fluid inclusion studies in a succession of metamorphic rocks in the metamorphic core complex on Naxos (Greece). He reported the presence of mainly $CO_2$-rich inclusions of varying density in quartz segregations and matrix quartz from metamorphic rocks. About 70% of all fluid inclusions have an optically estimated $CO_2$-content of between 60 to 70 mol. %, which is independent of metamorphic grade and lithology. $CO_2$-rich inclusions were found besides in quartz in various other minerals such as kyanite, andalusite, corundum, feldspar and dolomite. However, Kreulen (1980, 1989) also observed pure aqueous inclusions in epidote from associated lower-grade rocks and associated lenses of vesuvianite-grossularite-diopside, which is intimately intergrown with quartz containing mainly primary $CO_2$-rich inclusions and only few aqueous inclusions. This indicates that epidote and quartz did not grow during different stages of metamorphism, and accordingly should have grown in contact with the same fluid phase, unless $H_2O$ and $CO_2$ were immiscible under the specific low-grade metamorphic conditions (Kreulen 1980). Yet the reactions (15), (16), (29) and (30), which represent equilibrium conditions for the above described mineral assemblage, indicate in accordance with the fluid inclusions found in epidote a low $X_{CO_2}$ content for the fluid phase.

# SKARN DEPOSITS

## Introduction

Fluid inclusion studies are a major tool for the investigation of all types of ore deposits, in order to determine the composition and the $P$-$V$-$T$-$X$ properties of the ore-forming fluids. As a result of numerous fluid inclusion studies the fluid environment for certain ore deposits has been tightly constrained, and thus fluid inclusion data are used as exploration tools for these deposits (e.g., Klemd et al. 1993; Roedder 1979; Roedder and Bodnar 1997). Despite the large number of fluid inclusion studies only a few were conducted on epidote, although it is widely distributed in many ore deposits. An exception is epidote in skarn deposits, on which this brief review focuses. In the following I will describe some of the main characteristics of skarn deposits, however, for a more detailed discussion see the reviews of Einaudi et al. (1981) and Meinert (1992). The terms skarns and skarn deposits—i.e., those skarns, which have an economic ore mineralization—should be used in a purely descriptive manner, since both types are the product of regional and contact metamorphism as well as metasomatism involving metamorphic, magmatic, meteoric, and/or marine fluids (Einaudi et al. 1981; Meinert 1992). Skarns are characterized by coarse-grained Ca-Al-Fe-Mn-Mg silicates such as andraditic garnet, clinopyroxene, wollastonite, epidote and amphibole, which mainly formed by the replacement of a carbonate-rich lithology. However, skarns also may form in shale, sandstone, granite, basalt and komatiite. The terms exoskarn and endoskarn refer to a sedimentary or igneous origin, respectively. Einaudi et al. (1981) classified skarn deposits in terms of their dominant economic metals. Seven major skarn types are distinguished: Fe-, Au-, W-, Cu-, Zn-, Mo-, and Sn skarn deposits. Almost all occur in tectonic settings with igneous activity, such as magmatic arcs above subduction zones, and often along with porphyry copper deposits at convergent plate boundaries (Fig. 7a,b,c). However, Sn and some Zn skarns may be related to rift settings and

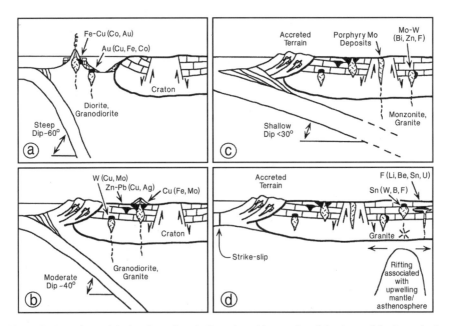

**Figure 7.** Tectonic models for skarn deposit formation: (a) oceanic subduction and back arc basin development; (b) continental subduction; (c) transitional low-angle subduction; (d) rifting (modified after Meinert 1992).

anorogenic magmatism (Fig. 7d). Thus skarn deposit formation is a process associated with the intrusion of a plutonic body, which is accompanied by regional to contact metamorphism and metasomatism of the immediate country rocks. Metamorphism will be most affective at deeper levels (1-3 kbar, corresponding to 3 to 10 km depth), while retrograde alteration will affect skarns at shallower depths (0.3 to 1 kbar, corresponding to 1 to 3 km depth). This dynamic process often involves an early isochemical metamorphism followed by later metasomatism, which are overprinted by a subsequent retrograde stage. A complex metasomatic fluid evolution results from the early high-temperature metamorphic stage towards the late low-temperature stages. The formation of skarn deposits involves several continuous stages, initially exceeding 700°C, during which the skarn and initial ore mineralization formed, followed by subsequent phases of cooling, which are accompanied by a retrograde stage and continued ore deposition. Depending on the geotectonic setting and the rock's composition these stages may correspond to the early potassic and later phyllic/argillic alteration.

Formation of epidote minerals can occur at the early high-temperature metasomatic stage and/or the later lower temperature retrograde stage, both of which may be associated with ore mineralization (e.g., Einaudi et al. 1981; Brown et al. 1985). Einaudi et al. (1981) suggested that epidote minerals can form at medium to high temperatures only at low $X_{CO_2}$ values. This conclusion was based on a $T$-$X_{CO_2}$ diagram, which was constructed for the CASCH system at 2 kbar. However under highly oxidizing conditions it is more appropriate (see above) to use the CF*ASCH-system when constructing a $T$-$X_{CO_2}$ projection (Fig. 2), which displays some of the most characteristic epidote-involving equilibrium reactions (32)-(37) in skarns. In contrast to the results of Einaudi et al. (1981) the mineral-fluid phase reactions in Figure 2 clearly demonstrate that epidote is stable for a wide range of $CO_2$ concentrations in the aqueous fluid phase. However epidote in skarn deposits may also be formed by a metasomatic exchange with the infiltrating fluid phase as is shown by the simplified reaction (38), (39) and (40), which usually requires the introduction of Ca and $Fe^{3+}$ by a magmatic/hydrothermal fluid phase.

Fluid inclusion studies on skarn deposits do not only provide information on the *P-V-T-X* conditions of the ore-forming fluid phase, but also on the change of these parameters from prograde to retrograde conditions as well as on the metasomatic mass transfer of elements. Summaries of the fluid inclusion literature on skarn deposits are given in Kwak (1986) and Meinert (1992).

## Iron skarn deposits

Iron skarn deposits are mined exclusively for Fe, although they do have minor concentrations of other metals such as Au, Co and Cu. Magnetite is the dominant mineral in these skarn deposits, whereas silicate mineral assemblages frequently play a minor role. Calcic-Fe skarn deposits, which are formed by the replacement of limestone, display two generations of minerals: Early prograde, high-temperature minerals such as andraditic garnet, clinopyroxene, magnetite and epidote and later retrograde low-temperature minerals such as amphibole, chlorite, calcite, quartz, magnetite, epidote and ilvaite. The equilibrium relations of the early magnetite- and epidote-involving prograde mineral assemblage are represented by the reactions (33), (36) and (37). In contrast magnesian-Fe skarns, which have formed by the replacement of dolomitic country rocks, preferentially contain—besides garnet and magnetite—early forsterite, diopside, epidote and periclase. Several fluid inclusion studies were conducted on Fe skarn deposits (for summaries see Kwak 1986; Meinert 1992). $T_h$ of primary aqueous fluid inclusions of early garnet and clinopyroxene ranges from 370 to >700°C and 300 to 690°C, respectively. Homogenization always occurred into the liquid phase. Dissolution temperature of halite and sylvite daughter minerals indicate a salinity of up to 50 wt. % NaCl eq. Crosscutting quartz veins contain epidote, in which the aqueous fluid inclusions display a somewhat lower $T_h$ of about 250°C with a salinity <25 wt. % NaCl eq. (Meinert 1992). However, fluid inclusion data from skarn deposits vary somewhat from deposit to deposit. Beuline (1976) reported high $T_h$ values for the Tagarsk Fe skarn deposit in Russia, between 520–610°C for primary aqueous fluid inclusions in andraditic garnet and clinopyroxene, while primary aqueous inclusions in epidote showed $T_h$ between 140 and 330°C. Even higher $T_h$ of 350 to 380°C for primary aqueous inclusion in epidote were reported by Vorontsov et al. (1978) and Ayshford et al. (1997) from the Korshunvskoe (Russia) and Tallaway (Tasmania) Fe skarn deposits, respectively. Primary aqueous fluid inclusions in pyroxene and garnet from the Blagodat (Ural) Fe skarn deposit revealed $T_h$ values from 350 to 780°C, while primary aqueous fluid inclusions in epidote showed lower $T_h$ values, between 480 and 580°C into the gas and liquid phase, indicating boiling (Bukharev et al. 1982). Li et al. (1989) and Feng and Chang (1996) observed primary aqueous fluid inclusions in prograde diopside and garnet with $T_h$ of 540–660°C and a salinity between 25 and 65 wt. % NaCl eq. as determined by the dissolution of halite in the Fe-skarn deposits in the Shanxi province of NE-China. Epidote, which had precipitated with the magnetite ore, has an average $T_h$ of 422°C and an average salinity of 41 wt. % NaCl eq. Besides halite and sylvite, Ca- and Fe-rich daughter minerals were observed in the high-salinity inclusions, thus indicating the introduction of these elements by the hydrothermal fluids.

It can be concluded that magnetite, the major ore mineral in the Fe skarn deposits, precipitated early during the high-temperature metasomatic stage as well as during an intermediate temperature stage before the onset of retrograde alteration. Fluid inclusion data derived from epidote clearly support this evolution of the ore-forming fluid, by means of decreasing salinity during decreasing temperature. Furthermore, indications for boiling (homogenization in the gas and liquid phase at the same temperature) were observed in some of the high-temperature fluids.

## Gold skarn deposits

Gold skarn deposits occur in a large variety of tectonic settings and range in age between Archean and Phanerozoic (e.g., reviews by Meinert 1992, 1998). Four different types of Au skarn deposits have been distinguished:

1. reduced Au skarn deposits, which are associated with reduced (ilmenite-bearing, $Fe^{3+}/Fe^{2+} < 0.75$) dioritic to granodioritic plutons and dike/sill complexes;

2. oxidized Au skarn deposits (high garnet/pyroxene ratios; Fe-poor garnet and pyroxene; low total sulfides);

3. magnesian-Au skarn deposits replacing dolomitic country rocks;

4. metamorphic Au skarn deposits associated with shallow plutonic rocks, which intruded the sedimentary country rocks.

All of these deposits have common characteristics—besides economic Au concentrations—such as distal and early biotite ± potassium feldspar hornfels, garnet-pyroxene alteration and a Au association with various Bi and Te minerals. Epidote usually is a late stage low-temperature product.

Fluid inclusion studies in reduced and oxidized Au skarn deposits (Hickey 1990; Ettlinger et al. 1992; Brooks 1994, all of which are summarized in Meinert, 1992) revealed the presence of high-temperature brines in prograde garnet and pyroxene with $T_h$ of between 210 to 730°C and a salinity of 5 to 40 wt. % NaCl eq. As postulated for several Fe skarn deposits some of the high temperature fluids provided evidence for boiling (see Vargunina and Andrusenko 1983). The fluid inclusions in retrograde epidote have lower $T_h$, of between 255 and 450°C (305–450°C pressure-corrected, Hickey 1990), and a salinity of up to 28 wt. % NaCl eq. Similar $T_h$-values between 380 and 390°C were measured for fluid inclusions in retrograde epidote from a Au skarn deposit from NE-Russia (Vargunina and Andrusenko 1983). Gold deposition from aqueous saline fluids was usually accompanied by sulfide mineralization and retrograde hydrous minerals such as amphibole, chlorite and epidote at a temperature of >300°C, as evidenced by the fluid inclusion studies above and several fluid-phase and mineral equilibria (summarized in Meinert 1998). Meinert (1992, 1998) interpreted these features as evidence for the transport of Au in chloride complexes at a temperature >300°C and at a high salinity.

## Tungsten skarn deposits

Tungsten skarn deposits are found worldwide in Precambrian to Triassic limestone and are usually associated with calc-alkaline intrusives in orogenic belts (Fig. 7b,c). For details see the recent summary and review by Newberry (1998). W skarn deposits are almost always associated with coarse-grained granodioritic to quartz monzonitic stocks and batholiths. Newberry and Einaudi (1981) differentiated between reduced and oxidized W skarn deposits. Reduced W skarn deposits have early andradite garnet and pyroxene, which are associated with Mo-rich scheelite. Later retrograde mineral assemblages include $Mn$-$Fe^{2+}$-rich garnet, biotite, hornblende, epidote, sulfides and low-Mo scheelite. In oxidized W skarn deposits andraditic garnet dominates over clinopyroxene and the main retrograde phases are epidote and amphibole, both of which are associated with the main ore mineralization. Fluid inclusion studies of both deposit types are summarized in Kwak (1986) and Meinert (1992). Garnet and clinopyroxene in W skarns from three deposits (MacMillan, Yukon; King Island, Tasmania and Salau, Pyrenees) contain primary aqueous inclusions with a wide range of $T_h$ values and salinity, from 290 to 800°C (with evidence of boiling) and 23 to 65 wt. % NaCl eq., respectively. Late retrograde minerals such as epidote, quartz and amphibole have lower $T_h$ values and salinity, between 250 and 470°C and 2–28 wt. % NaCl eq., respectively. Besides NaCl and KCl, $CaCl_2$- and $NaAlCO_3(OH)_2$-minerals were observed in the high-salinity inclusions in garnet and clinopyroxene, while the low-salinity fluids are relatively $CaCl_2$-poor as is shown by the fluid inclusions in epidote, amphibole and quartz. Other observed solids in the high-salinity fluid inclusions are unidentified Zn, W and Cu minerals representing the metal characteristics of the hydrothermal, ore-forming fluid (Kwak 1986). Primary aqueous inclusions in scheelite show $T_h$ values between 270 and 600°C and a salinity between 15 and 61 wt. % NaCl eq. (see Kwak 1986; Meinert 1992). A relative high $T_h$ of 448°C was observed for aqueous fluid inclusions in epidote from the Sangdong W skarn deposit in Korea (Kwak 1986). Epidote and scheelite in the Yaguki W skarn deposit in Japan have primary fluid inclusions with almost identical $T_h$ values from 220 to 330°C and 230 to 330°C, respectively (Muramatsu and Nambu 1982). Similar $T_h$ values—from 213 to 364°C—and low salinity were obtained from primary inclusions in epidote and clinozoisite from the Copina W skarn deposit in Argentina (Ocanto et al. 2001). Consequently, the scheelite precipitation must have occurred over a large temperature range, from the waning prograde stage until the late retrograde alteration.

## Zinc and tin skarn deposits

Most Zn skarn deposits are associated with continental subduction, while related rocks range from diorite to a high-silica granitic composition. Sn skarn deposits are predominantly related to high-silica granites, which were derived by partial melting of continental crust associated with rifting (Fig. 7d). A brief summary and review of both types of deposits are given by Einaudi et al. (1981) and by Kwak (1986, 1987), respectively. In contrast to most other skarn deposits Zn skarn deposits usually have low fluid inclusion $T_h$ values (<600°C), even for the prograde mineral assemblage, which is consistent with boiling and a relatively shallow (0.2–1 kbar) and distal geotectonic setting (Kwak 1986; Meinert 1992). Fluid inclusion studies on Zn skarn deposits from four different mining districts Verladenia and Naica, Mexico (Megaw et al. 1998; Haynes and Kesler 1988); Groundhog Mine, New Mexico (Meinert 1987), and Kamioka Mine, Japan (Takeno et al. 1999) revealed $T_h$ values from 290 to 586°C for primary inclusions in garnet and clinopyroxene with a salinity of 2–23 wt. % NaCl eq. $T_h$ values of primary aqueous inclusions in sphalerite range from between 230 and 420°C, while $T_h$ values of primary and secondary inclusions in epidote from the Kamioka Mine range from 240 to 300°C. Similar $T_h$ values of between 260 and 362°C and an average salinity of 9.5 wt. % NaCl eq. were displayed by primary inclusions in epidote from the Orphid Mine Zn skarn deposit in Utah (Wilson and Parry 1986).

Fluid inclusion investigations on quartz and fluorite from five Sn Skarn deposits (summarized in Kwak 1986, and Meinert 1992) displayed $T_h$ values (into the liquid phase only) ranging from 100 to 492°C and salinity from 6 to 40 wt. % NaCl eq. Garnet and pyroxene however accommodated fluid inclusions, which homogenized into the gas and liquid phase at a temperature of 300 to >600°C. Primary fluid inclusions in epidote of the JC mine (Yukon Territory, Canada) revealed a $T_h$ range of 243 to 420°C and salinity between 0.27 and 34 wt. % NaCl eq., while cassiterite showed somewhat higher temperature (Layne et al. 1987; Layne and Spooner 1991). Daughter minerals in the high-salinity inclusions are mainly NaCl, KCl, $CaF_2$, thus reflecting the importance of metasomatic exchange reactions such as (38) and (39).

**Discussion**

Fluid inclusion studies in ore deposits (as well as in metamorphic rocks) usually focus on secondary inclusions in quartz. Due to the common post-trapping modifications of fluid inclusions in quartz and quartz recrystallization, the interpretation of fluid inclusion data is often problematic and ambiguous (see Introduction). In contrast, skarn deposits usually have primary fluid inclusions in prograde skarn minerals such as andraditic garnet and clinopyroxene as well as in later retrograde minerals such as epidote and amphibole. This offers a unique opportunity to study the *P-T-X* evolution of the skarn-forming fluid, which is displayed by high salinity at high temperature and declining salinity during decreasing temperature. Boiling is usually evident during the early phase of skarn formation but rare during later retrograde processes. Furthermore, systematic quantitative variations of the salt composition in single fluid inclusions from prograde and retrograde host minerals of different skarn deposits (Haynes and Kesler 1988) combined with the results of $\delta^{18}O$ and $\delta D$ values of later retrograde minerals such as epidote (Layne et al. 1991) indicate mixing of multiple fluids of magmatic and meteoric origin.

## OTHER ORE DEPOSITS

The geology and hydrothermal alteration of the porphyry copper deposit at Ann Mason (Yerrinton, Nevada) have been described in detail by Dilles and Einaudi (1992) and Dilles et al. (1992): The Ann Mason porphyry copper deposit contains 495 mt of 0.4 wt. % Cu and ca. 0.01 wt. % Mo, which mainly occurs within or near the contact of the Jurassic Yerrington batholith, a highly differentiated granite. At least three different hydrothermal fluids caused three alteration zones in the plutonic rocks: 1) an early potassic alteration caused by fluids of magmatic origin; 2) a contemporaneous sodic-calcic (propylitic) alteration and epidote veins, mainly caused by formation brines, and 3) a late sodic to sericitic-chloritic alteration probably as a result of seawater interaction. Primary liquid-rich aqueous inclusions (L + V + halite) in quartz from the potassic alteration zone (1) were homogenized by halite dissolution from 150 to 550°C, thus indicating a salinity of between 32 and 62 wt. % NaCl eq. This inclusion fluid is linked to the copper deposition and believed to represent relics of magmatic fluids, as is supported by stable isotope studies of K-feldspar and biotite. Epidote and quartz from the alteration zone (2) contain primary aqueous inclusions (L + V + halite), which also homogenized by halite dissolution between 170 and 340°C, thereby indicating a salinity of 31 to 41 wt. % NaCl eq. The inclusion fluids of both minerals were interpreted to be relics of elevated formation water, which circulated convectively through the granitoid, and by this means leaching elements such as K, Fe, Cu and S. Lower temperature ($T_h$ = 100–250°C) aqueous fluid inclusions in quartz, which exhibit a salinity from 2–5 wt. % NaCl eq., are believed to be of seawater origin and to have caused the late sodic and sericite-chlorite alteration (3).

Hagemann et al. (1996) investigated the Wiluna lode-gold deposit in the Archean Norseman-Wiluna greenstone belt in western Australia. Using a combined microthermometric and laser Raman and ion chromatographic study on quartz and epidote from unmineralized pillow lavas, they mainly observed aqueous fluids, which were interpreted to represent unchanged relics of evolved seawater.

Gem quality zoisite ("tanzanite" = vanadiferous zoisite) from Merelani (Tanzania) has primary aqueous fluid inclusions with up to three unidentified daughter minerals (Malisa et al. 1986). Homogenization into the liquid phase occurred from 37 to 51°C, however, one vapor-rich inclusion homogenized into the vapor phase at 242°C. The daughter minerals only began dissolving at 350°C.

## PLUTONIC ROCKS

The Hercynian calc-alkaline plutonic complex of Charroux-Civray (NW-Massif Central, France) displays several stages of hydrothermal alteration (Freiberger et al. 2001a,b). The primary igneous mineralogy and an early postmagmatic Ca-Al silicate alteration were overprinted by a multiphase illite-chlorite-phengite-carbonate alteration. Cooling of the calc-alkaline pluton started at a solidus temperature of about 650°C at about 4 kbar, as was indicated by the isochores of the primary aqueous fluid inclusions in subsolidus epidote (Freiberger et al. 2001a,b). These authors further observed decompressional cooling of the pluton to 2–3 kbar at 200–280°C, with a subsequent greisenization by an $H_2O$-$CO_2$-rich fluid of varying density at a temperature of between 400 to 450°C, which was indicated by primary and secondary fluid inclusions mainly in quartz. This high temperature was related to the late intrusion of one or several leucogranitic bodies, that also caused the low temperature illite-chlorite-phengite-carbonate alteration (Freiberger et al. 2001a).

Aqueous fluid inclusions in epidote of the granite-pegmatites from the Strzegom Massif and the Karkonosie Mts. in Poland indicate a late crystallization temperature between 360 to 120°C (Karwowski and Wlodyka 1981; Lendowski 1983). Kuznetzova and Gostyaeva (1982) reported very low $T_h$ values of about 120°C for two-phase aqueous inclusions in epidote from albitized granites in the Ukranian Shield.

## HIGH- AND ULTRA-HIGH PRESSURE (HP AND UHP) METAMORPHISM

### Introduction

One of the problems when investigating high- and ultra-high pressure metamorphism is the effect of the possible presence of a free fluid phase. Syn-metamorphic quartz veins with coarse grained eclogite-facies minerals such as omphacite, kyanite, garnet and epidote were taken as evidence for the presence of such a free aqueous fluid phase in equilibrium with eclogite-facies conditions (e.g., Essene and Fyfe 1967; Okrusch et al. 1978; Holland 1979; Klemd 1989). Several fluid inclusion studies on syn-metamorphic high- to ultrahigh pressure vein minerals and matrix minerals of the surrounding host rocks revealed a large variety of fluid compositions. Reported fluid compositions include aqueous low-salinity fluids (e.g., Giaramita and Sorensen 1994; Vallis and Scambelluri 1996; El-Shazly and Sisson 1999; Franz et al. 2001; Gao and Klemd 2001), high-salinity aqueous fluids and/or $CO_2$-$N_2$- or $N_2$-rich fluids (e.g., Andersen et al. 1989; Selverstone et al. 1992; Philippot 1993; Klemd et al. 1992, 1995; Scambelluri et al. 1998; Xiao et al. 2000; Fu et al. 2001, 2002, 2003). The presence of a fluid as well as the exact composition is of primary importance when considering *P-T* estimates and quantitative phase diagrams such as *P-T* pseudosections (phase diagrams that are constructed for the specific bulk composition of a rock). Together with *P* and *T* the composition of the fluid phase is important when assessing its capability to transport and move certain major and trace elements during eclogite-facies metamorphism. The reason for this being that element transport and fluid-rock interaction during blueschist- to eclogite-facies conditions play a key-role in understanding the fluid storage and recycling during subduction and continent-continent collision (e.g., Peacock 1993; Brunsmann et al. 2000, Scambelluri and Philippot 2001). Several models based on theoretical considerations suggest that large amounts of $H_2O$-rich fluids, which were released by the dehydration of subducted lithospheric serpentinized mantle and/or oceanic crust—for instance during the transition from blueschist to eclogite—act as element carriers from the slab to the overlying mantle wedge and are thus responsible for the trace element signatures of island arc magmas (Fig. 8; e.g., Peacock 1993; Schmidt and Poli 1998; Mibe et al. 1999, 2003; Draper et al. 1999; Rüpke et al. 2002; John and Schenk 2003). The latter theoretical considerations concerning the dehydration of the oceanic crust are supported by recent findings in the Tianshan high-pressure belt (NW-China). Large eclogite-facies veins in blueschist were found to be the product of hydrofracturing induced by devolatilisation of minerals such as glaucophane, paragonite and epidote during blueschist-eclogite transition (see reactions (24) to (28)) and thus, represent former fluid pathways within a paleosubduction zone (Gao and Klemd 2001). However, significant differences do not only exist as far as the fluid composition is concerned but also with respect to the kind of fluid flow under eclogite-facies conditions. A first model suggests the influx of relatively large quantities of aqueous fluids in an open system on a kilometer-scale (e.g., Bebout and Barton 1993; Giaramita and Sorensen 1994; Nelson 1995), whereas a second model—based mainly on fluid inclusion, stable isotope and connectivity studies—demonstrates that fluid flow under eclogite-facies condition was limited and restricted to a millimeter to centimeter scale (e.g.,

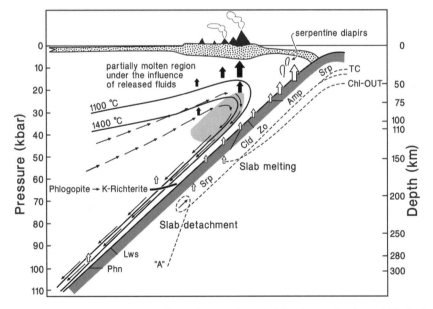

**Figure 8.** Schematic diagram showing the subduction of oceanic lithosphere and regions of dehydration along with partial melting in the mantle wedge as well as the associated island arc volcanism (modified after Schmidt and Poli 1998).

Philippot and Selverstone 1991; Selverstone et al. 1992; Klemd et al. 1992, 1995; Philippot 1993; Xiao et al. 2000; Mibe et al. 2003).

Some of the established results and discrepancies, which are based on fluid inclusion studies in eclogite-facies rocks, may be ambiguous given that they were conducted on fluid inclusions in quartz which are prone to post-entrapment modification during post-peak metamorphic and recrystallization processes (e.g., Bakker and Jansen 1991; Sterner et al. 1995; Johnson and Hollister 1995; Küster and Stöckhert 1997). However, even fluid inclusion studies on eclogite-facies minerals such as garnet, omphacite, kyanite and epidote, which are less susceptible to recrystallization and thus to re-equilibration and leakage, revealed fluid compositional differences and density modifications of fluid inclusions, as is shown by a mismatch between pressure estimates based on conventional geothermobarometry of silicate minerals and fluid inclusions (e.g., Philippot and Selverstone 1991; Klemd et al. 1992, 1995; Giaramita and Sorensen 1994; Selverstone et al. 1992; Scambelluri et al. 1998; Svensen et al. 1999; Xiao et al. 2000; Franz et al. 2001; Gao and Klemd 2001; Fu et al. 2001, 2002, 2003; Schmid et al. 2003).

Although the density of texturally primary eclogite-facies fluid inclusions in HP and UHP minerals (Fig. 3) may have re-equilibrated during exhumation, the peak metamorphic fluid composition can be preserved. This is indicated by the constant composition of different-sized fluid inclusions in individual as well as in different minerals such as garnet, omphacite, kyanite and epidote from single rock samples, furthermore by the preserved fluid heterogeneity between samples, which have undergone an identical *P-T* evolution but come from different localities or even from the same outcrop (see studies below). Moreover, theoretical considerations and fluid-phase equilibria based on the actual eclogite-facies mineral assemblage (Holland 1979) and model calculations in the $H_2O$-$CO_2$-NaCl-system (Selverstone et al. 1992) with the aim to establish the composition of the fluid phase during eclogite-facies metamorphism, show that these results correspond with fluid inclusion composition from eclogite-facies vein minerals. Nevertheless, Philippot and Selverstone (1991) pointed out the large uncertainty in pressure determination from inclusions with complex daughter minerals, some of which may have formed by "back reactions," thus changing the original density and the composition of the fluid (e.g., Heinrich and Gottschalk 1995; Svensen et al. 1999). The following section provides an overview on fluid inclusions in epidote in eclogite-facies rocks and can hopefully shed some light on the above described observations and discrepancies.

**Tauern Window, Austria**

In one of the early fluid inclusion studies on HP-metamorphic rocks Luckscheiter and Morteani (1980) investigated fluid inclusions in Alpine epidote-apatite-bearing quartz veins in amphibolites and amphibolite-facies gneisses (Zentralgneis area), in order to compare them with fluid inclusions in matrix quartz of eclogites from the lower and upper Schieferhülle of the Tauern Window in Austria. Epidote and apatite contained texturally primary aqueous inclusions with a relatively low salinity of 3 to 9 wt. % NaCl eq. Homogenization always occurred into the liquid phase, between 240 and 220°C. Secondary to pseudosecondary fluid inclusions in associated quartz display the same composition. In contrast, fluid inclusions in quartz of syn-metamorphic high-pressure veins contain secondary to pseudosecondary $CO_2$-$H_2O$-rich inclusions with a highly variable $CO_2$-content of between 20 and 80 vol. % and a salinity of 5 to 21 vol. % NaCl eq. for the fluid phase. The $CO_2$-density, which was derived exclusively from the $CO_2$ homogenization temperature, ranges from between 0.16 and 1.15 g/cm³. The presence of syn-metamorphic fluids with such a high $CO_2$-and NaCl-content stands in direct contrast to fluid-phase equilibria considerations and a fluid inclusion study by Holland (1979), who suggested that the Tauern Window eclogites with the peak-mineral assemblage paragonite-glaucophane-epidote-talc-magnesite-dolomite in association with garnet, omphacite, kyanite and quartz necessitates the coexistence of a free aqueous fluid phase during eclogite-facies metamorphism. Furthermore, quartz-omphacite-kyanite high pressure veins and the presence of aqueous inclusions ($X_{CO_2} < 0.1$; $X_{NaCl} \leq 0.02$) in omphacite and epidote suggest that crystallization took place under conditions comprising a high water activity at ca. 19.5 kbar and ca. 620°C. The high water activity during peak metamorphic conditions was supported by fluid-phase equilibria calculations in eclogite-facies siliceous dolomites from this area (Franz and Spear 1983). Yet Selverstone et al. (1992) calculated fluid activity ratios with strongly varying $H_2O$ activities at peak metamorphic conditions for millimeter- to centimeter-scale layers in banded mafic eclogites from the Austrian Tauern Window, whereas $CO_2$ activities were almost constant between the same layers (Fig. 9).

The banding is displayed by alternations of omphacite-, garnet-, clinozoisite-, zoisite-, dolomite-, and phengite-rich assemblages. Their model concerning the $H_2O$-$CO_2$-NaCl-system is consistent with the presence of different saturated saline brines, carbonic fluids, or immiscible water-rich and $CO_2$- ± $N_2$-rich fluid phases. Fluid inclusion investigations in quartz, omphacite, kyanite, apatite, epidote and magnesite (see summary in Selverstone et al. 1992) revealed the presence of highly saline brine inclusions (salinity of up to 39 wt. % NaCl eq.) with up to six daughter minerals, and thus supported their model calculations. Low initial melting temperature indicated—besides NaCl and KCl—the presence of $CaCl_2$ and $MgCl_2$. $X_{CO2}$ varied from between 0–0.18 in the brines and between 0.2–1 in the $CO_2$ ± $N_2$ fluid. The density of all inclusions was modified due to re-crystallization and partial decrepitation during unroofing (Selverstone et al. 1992).

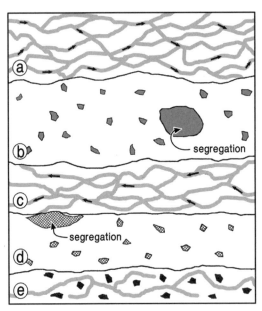

**Figure 9.** Schematic diagram of a model for the fluid distribution in banded eclogites. Layer (a) and (c) have an interconnected fluid flow (fluid phase has a dihedral angle of <60° in the polycrystalline eclogite layers); arrows indicate direction of fluid flow. Layers (b) and (d) show isolated fluid pockets (fluid has a dihedral angle of >60° in the polycrystalline eclogite layers). Layer (e) contains immiscible fluids, one with a low and one with a high dihedral angle in the polycrystalline eclogite layer (modified after Selverstone et al. 1992).

## Münchberg Gneiss Complex, Germany

The Münchberg Gneiss Complex, which contains eclogites and eclogite-facies rocks in the its upper parts, is believed to be an inverted nappe in the Central European Variscides. The eclogite-facies conditions are estimated at pressures of >24 kbar and temperatures of between 600 and 700°C (Klemd 1989; Klemd et al. 1991, 1994; O'Brien 1993). The eclogite-facies stage was followed by an amphibolite-facies overprint of between 10 to 14 kbar at the same temperature range, thereby suggesting an almost isothermal exhumation to higher crustal levels (Franz et al. 1986; Klemd 1989; Klemd et al. 1994; O'Brien 1993). Pegmatitic segregations from the contact zone to garnet-amphibolites (retrograded eclogites) have been the subject of a detailed field and laboratory study by Franz and Smelik (1995). These authors interpreted the pegmatites, which have a leuco-tonalitic composition, to be the product of partial melting of the surrounding eclogites during the isothermal decompression in the presence of a free $H_2O$-rich fluid phase. In order to prove the formation of a water-saturated melt a mass balance calculation for $H_2O$ production was undertaken by Franz and Smelik (1995). They estimated that a partial melting of about 2.5 vol. % could generate enough melt to produce the volume of the pegmatites observed in the Münchberg Gneiss Complex. The matrix of the pegmatites mainly consist of graphic intergrowth of quartz and albite, which are associated with zoisite and minor clinozoisite, phengite, Ca-amphibole and biotite. The occurrence of such water-saturated partial melts within subduction zones is of considerable importance with regards to element transport in this environment (see above). In order to verify the presence of a former water-saturated melt, a fluid inclusion study on zoisite, which contains abundant fluid inclusions, was conducted (Klemd unpub. data). The size of the fluid inclusions range from <3 to 280 μm. Most fluid inclusions are primary and are normally two-phase (liquid and vapor) aqueous inclusions, which are aligned parallel to the b-axis of the zoisite or occur as large isolated single inclusions (Fig. 10). However secondary fluid inclusions occurring along healed microfactures were also observed. Some primary and secondary fluid inclusions contain—apart from liquid and vapor—up to two daughter minerals, which did not dissolve during heating (Fig. 10). The liquid/solid ratios are strongly variable, whereas the liquid/vapor ratio of most of the inclusions is relatively constant between 0.75–0.90. According to the criteria of Roedder (1984) this suggests the accidental trapping of the daughter minerals. Primary and secondary fluid inclusions behaved identical during the microthermometric study. Final melting temperatures range between –4.0 and –0.1°C (103 measurements), thereby indicating a low salinity of <6 wt. % NaCl. The apparent initial melting was observed between –28 and –10°C, implying a NaCl and KCl dominated aqueous fluid. Yet rare clathrate melting at between 10 and 20°C indicates the presence of minor amounts of other gases such as $CO_2$ and/or $CH_4$. $T_h$ values range from 155 to 280°C and thus, densities of <1 g/cm$^3$. The presence of low-salinity and -density aqueous fluid inclusions is typical for late-stage fluids in pegmatites (e.g., London 1986). The presence of such fluid inclusions furthermore provide unambiguous evidence for the exsolution of a low-density fluid from the water-saturated

**Figure 10.** Photomicrograph showing a large isolated primary three-phase fluid inclusion (inclusion diameter = 25 μm) in a pegmatitic segregation from the contact zone to garnet-amphibolites (retrograded eclogites) from the Münchberg Gneiss Complex in Germany. The inclusion contains—besides a liquid and a vapor-bubble—one unidentified solid phase, which did not dissolve during heating (for further explanation see text).

leuco-tonalitic melt during zoisite crystallization and consequently confirm the hypothesis of Franz and Smelik (1995).

## Mt. Emilius, Italy

Eclogite-facies metabasites of the Mt. Emilius continental unit (western Italian Alps) were the subject of a more recent study by Scambelluri et al. (1998). The Alpine eclogite-facies metamorphism occurred at a temperature between 450 and 550°C and pressure of 11 to 13 kbar as was derived from conventional geothermobarometry (Dal Piaz et al. 1983). Scambelluri et al. (1998) investigated two rock types, namely eclogites with the mineral assemblage omphacite-garnet-glaucophane-epidote-phengite/paragonite and eclogitized granulites with the mineral assemblage omphacite-garnet-epidote-amphibole-chlorite. Epidote was considered to have formed prograde by reaction (21) and to have been stable under peak eclogite-facies conditions (Dal Piaz et al. 1983). The high-pressure foliation of both rock types is crosscut by eclogite-facies garnet-omphacite-epidote veins, which were all the subject of fluid inclusion studies. In some cases Scambelluri et al. (1998) also investigated fluid inclusions in garnet and omphacite from the eclogitized granulite domains surrounding the high-pressure veins. All vein minerals contain texturally primary to pseudosecondary high-salinity inclusions with up to 50 wt. % NaCl eq. No compositional difference or varying density was found for fluid inclusions in different minerals of the same vein, thereby excluding a possible H-diffusion exclusively in epidote (cf. Giaramita and Sorensen 1994). The high-salinity aqueous fluids were either two-phase (liquid + vapor) or multiphase (liquid + vapor + salt + quartz). Initial melting temperature between −50 to −34°C was interpreted to indicate the presence of $NaCl\text{-}CaCl_2\text{-}MgCl_2$ as the main chloride components. In the eclogitized granulites primary inclusions in garnet and omphacite however displayed much lower salinity between 9 and 23 wt. % NaCl eq. Estimated inclusion density varies between 0.8 and 1.1 $g/cm^3$. The isochores for all inclusions pass below peak metamorphic conditions. The highest-density isochore falls at least 5 kbar short of the estimated *P-T* conditions for the eclogite-facies metamorphism. This indicates that the fluid density must have been modified (see above Selverstone et al. 1992) during the post-peak metamorphic evolution. However, textural evidence indicates that the inclusion fluid must have been trapped under eclogite-facies conditions (Scambelluri et al. 1998). Despite these density modifications Scambelluri et al. (1998) excluded a compositional change of the inclusion fluid during post-peak metamorphic conditions. They considered the distinct fluid salinity difference in omphacite, garnet and epidote of the high-pressure veins as well as in omphacite and garnet of the eclogitized granulite to be the result of progressive hydration during the eclogitization of the granulites. Thus the increasing salinity resulted from extensive fluid-rock interaction during eclogitization of the granulites, i.e., the water gets incorporated into the hydrous eclogite-facies minerals, thereby causing an increasing salinity in the residual inclusion fluid (Scambelluri et al. 1998).

## Dabie-Sulu terrane, China

The first fluid inclusion study on coesite-bearing UHP-eclogites from the Central Dabie Shan involving several eclogite-facies minerals such as kyanite, omphacite and clinozoisite revealed the presence of $NaCl \pm CaCl_2 \pm MgCl_2$-rich brines without significant $CO_2$-contents in texturally primary and secondary aqueous as well as $CH_4\text{-}N_2$-rich inclusions (Cong and Touret 1993). Since then several fluid inclusion studies, which were often combined with oxygen-isotope studies, have been carried out on the HP and UHP rocks of the Dabie Shan and Sulu terranes (e.g., You et al. 1996; Xiao et al. 2000, 2001; Franz et al. 2001; Fu et al. 2001, 2002, 2003; Schmid et al. 2003).

The Dabie-Sulu terranes constitute the eastern part of a Triassic suture between the Sino-Korean and Yangtse cratons. The eastern part of this metamorphic terrane was displaced more than 500 km by the NNE-SSW trending sinistral strike-slip Tan-Lu fault (Xu et al. 1992).

Recent chronological studies on the HP and UHP rocks give ages of between 240 and 220 Ma for the peak metamorphic event (e.g., Hacker et al. 2000). The UHP unit mainly consists of gneisses, marble, jadeite quartzite, minor eclogite and garnet clinopyroxenite layers and lenses as well as associated ultramafic rocks. The extremely low-$\delta^{18}$O whole-rock signature in some of the HP and UHP eclogites and garnet clinopyroxenites from the Dabie-Sulu terrains—which is interpreted to be due to interaction of the meteoric fluids with the pre-metamorphic protolith of the eclogites and garnet clinopyroxenites before subduction (for discussion and extensive reference list see Zheng et al. 2003)—encouraged Fu et al. (2001, 2002, 2003) to undertake fluid inclusion studies on these rocks. Their aim was to search for possible relics of such a pre-metamorphic fluid in eclogite-facies minerals such as garnet, omphacite and epidote. The UHP eclogites have the following peak-metamorphic mineral assemblage: garnet-omphacite-coesite/quartz-phengite±rutile±epidote±kyanite±amphibole-apatite-ilmenite-magnetite. Some garnet porphyroblasts contain epidote, dolomite, kyanite, K-feldspar and apatite, which are interpreted to have been trapped under prograde (pre-peak) metamorphic conditions (see reactions (17) to (21)). Therefore the fluid inclusions, which were found in epidote and kyanite, may contain relics of a very early fluid phase (Fu et al. 2003). The peak metamorphic assemblage of the HP-eclogites is garnet-omphacite-epidote-quartz-rutile±amphibole-apatite while the garnet clinopyroxenites reveal the assemblage garnet-diopside-calcite/dolomite-ilmenite/magnetite-apatite. The HP and UHP eclogites and garnet clinopyroxenites display a wide range of whole-rock $\delta^{18}$O values, from –10 to 11‰, whereby no differences were reported either for the HP- or the UHP rocks. Fu et al. (2001, 2003) established two generations of primary fluid inclusions, early pre-peak metamorphic low-salinity inclusions (<14 wt. % NaCl eq.; $T_h$ = 95–340°C), which occur in quartz inclusions in epidote and in epidote cores, and syn-peak metamorphic, texturally primary inclusions in garnet, omphacite and epidote (Fig. 3), which mainly revealed aqueous inclusions with a wide variety of salinity (>14 wt. % NaCl eq. to halite saturation) and homogenization temperature (110 to 390°C) as well as $N_2 \pm CO_2 \pm CH_4$-rich inclusions. No difference in composition or density was reported for individual samples concerning the inclusion fluids among the investigated peak-metamorphic minerals. In order to assign primary inclusion fluids to certain rock types, Fu et al. (2002, 2003) reported that high-$\delta^{18}$O HP/UHP eclogites and clinopyroxenites have primary fluid inclusions with high-salinity brines. Low initial melting temperature <−35.5°C indicate the presence of divalent ions such as $Ca^{2+}$, $Mg^{2+}$ and $Fe^{2+}$ besides Na. Furthermore $N_2$-rich inclusions are associated with the high-salinity brines. In contrast low-$\delta^{18}$O HP/UHP eclogites have low-salinity inclusions with up to two accidentally trapped solid inclusions in quartz inclusions in epidote and in cores of epidote porphyroblasts. These low-salinity inclusions reveal high initial melting temperature >−20°C, thereby indicating a NaCl and KCl dominant system. Consequently the authors assumed a correlation between $\delta^{18}$O values and fluid salinity. They concluded that peak metamorphic fluids in high-$\delta^{18}$O HP/UHP eclogites would be $CaCl_2$-$MgCl_2$-dominated high-salinity brines, while NaCl-dominated low-salinity brines would occur in low-$\delta^{18}$O HP/UHP rocks. Due to the extremely low-$\delta^{18}$O values of some of these rocks Fu et al. (2003) suggested that the low-salinity fluids in epidote could be remnants of meteoric water, which had interacted with the protoliths of the high-pressure rocks prior to plate subduction, and which were subsequently modified during peak metamorphic conditions. Nonetheless I want to point out that the correlation between fluid salinity and oxygen-isotope composition is ambiguous, since some low-$\delta^{18}$O eclogites from the Central Dabie Shan have syn-peak omphacite and epidote, which contain texturally primary inclusions with high-salinity brines (Fu et al. 2002). It should furthermore be noted that all primary inclusion fluids underwent a density modification, since none of the calculated fluid isochores correlate with peak metamorphic conditions. Furthermore, some of the "accidentally trapped" solids in epidote (as described above) may have been formed by back reactions with the inclusion fluid (cf. Heinrich and Gottschalk 1995).

**Discussion**

As shown above, the presence of a discrete and free fluid phase during HP and UHP metamorphism was substantiated by the finding of syn-peak metamorphic low- to high-salinity aqueous fluid inclusions and or gaseous $CO_2$-$N_2$-rich inclusions in eclogite-facies minerals in veins or in the matrix of these rocks. This is in accordance with several other fluid inclusion studies on HP and UHP rocks (e.g., Philippot and Selverstone 1991; Philippot 1993; Giaramita and Sorensen 1994; El-Shazly and Sisson 1999; Franz et al. 2001; Gao and Klemd 2001). Consequently the composition of the eclogite-facies fluid strongly varies between the different eclogite-facies terranes, as is indicated by all the above-mentioned studies. For example, even within single terranes such as the Dabie-Sulu in eastern China, where the timing of peak metamorphism is tightly constrained, fluid inclusion investigation revealed strong fluid activity gradients, which apparently persisted during high-pressure metamorphism. This is in support of other fluid inclusion studies on HP/UHP rocks world-wide (e.g., Philippot and Selverstone 1991; Selverstone et al. 1992; Klemd et al. 1992, 1995; Xiao et al. 2000; Schmid et al. 2003). For example, eclogite-facies mineral assemblages in textural equilibrium with epidote minerals always contain $H_2O$-salt inclusions, while those without epidote minerals may contain $CO_2$ (Selverstone et al. 1992). Such fluid gradients suggest local fluid production and/or internal buffering as well as limited fluid flow on a relatively small scale during eclogite-facies metamorphism in subduction zones. This is in agreement with the preservation of small scale isotopic heterogeneities in eclogites as observed by Scambelluri and Philippot (2001), who proposed that fluids mostly remain entrapped in the subducting rocks within subduction zones and are released into the overlying mantle only at depths greater than those revealed by most exposed eclogites (>120 km). The isolation of aqueous and $CO_2$-rich fluids in pockets in eclogites would—due to their large dihedral angle (>65°)—prevent any large-scale flow along grain boundaries from the slab to the overlying mantle wedge until a depth of about 150 km (e.g., Severstone et al. 1992; Mibe et al. 2003). Therefore the transport of aqueous fluid and thereby trace elements from the subducting oceanic crust into the overlying mantle wedge, which is necessary for the formation of island arc magmas, can be achieved mainly by hydrofracturing, since the influence of shear deformation on aqueous fluid connectivity is rather limited at geological strain rates (Davies 1999). This is supported by the findings of Gao and Klemd (2001). They observed large eclogite veins, which were generated from hydrofracturing related to the dehydration of the surrounding blueschist host (cf. Kirby et al. 1996), thereby implying the possibility of massive fluid transfer in the subduction zone at about 60 km depth.

However, limited fluid flow as well as the local fluid production and/or buffering will prevent petrologists from assuming a distinct fluid composition for eclogite-facies metamorphism in order to calculate *P-T* conditions or quantitative phase diagrams, which demand the input of a certain $H_2O$ and/or $CO_2$ activities from fluid-phase equilibria. Furthermore differences in fluid regimes between eclogite-facies rocks cannot exclusively be related to different tectonic settings, and thus geodynamic settings—when derived from fluid inclusion studies only—will remain ambiguous (e.g., Andersen et al. 1993; Klemd and Bröcker 1999).

## CONCLUDING REMARKS

Fluid inclusion investigations on epidote minerals in combination with other petrological studies such as stable and radiogenic isotope or geothermobarometrical investigations provide important information on geodynamic processes for a range of hydrothermal and metamorphic processes which are related to the formation of epidote mineral-bearing assemblages. Furthermore fluid inclusions in epidote minerals provide direct evidence on the composition of the fluid phase involved in the formation of these minerals and thus on fluid-rock interaction processes. Consequently, microthermometrical and analytical studies of fluid inclusions in epidote minerals should be conducted in regional metamorphic rocks, in skarn and other hydrothermal ore deposits and in metasomatically influenced rocks such as epidosites in order to gain information on fluid related epidote mineral-forming processes. Such data will provide a wealth of information for example on fluid immiscibility and composition during the formation of these minerals in these environments. In addition laboratory-simulated experiments should be conducted in order to quantify possible re-equilibration and/or the role of volatile diffusion behavior of fluid inclusions in epidote minerals during internal over- and underpressure conditions.

## ACKNOWLEDGMENTS

This manuscript was considerably improved by the constructive reviews of D. K. Bird (Stanford University) and W. Heinrich (GFZ Potsdam). In addition the author is grateful to G. Franz (TU Berlin), who patiently provided helpful comments and continuous editorial support. Furthermore it also benefitted from comments and/or reviews of T. John (Universität Kiel), M. Klemd (Würzburg), S.E.Wille, T.M. Will, A. Zeh (all Universität Würzburg) on various aspects. K.-P. Kelber (Universität Würzburg) is thanked for preparing the diagrams and B. Fu (James Cook University) for providing Figure 3.

## REFERENCES

Andersen T, Burke EAJ, Austrheim H (1989) Nitrogen-bearing, aqueous fluid inclusions in some eclogites from the Western Gneiss Region of the Norwegian Caledonides. Contrib Mineral Petrol 103:153-156

Andersen T, Burke EAJ, Austrheim H, Elvevold S (1993) $N_2$ and $CO_2$ in deep crustal fluids: evidence from the Caledonides of Norway. Chem Geol 103:153-165

Andersen T, Frezzotti ML, Burke EAJ (2001) Fluid inclusions: phase relationships-methods-applications. Lithos 55:1-320

Arnason JG, Bird DK, Liou JG (1993) Variables controlling epidote composition in hydrothermal and low pressure regional metamorphic rocks. Abh Geol BA 49:17-25

Ayshford S, Offler R, Seccombe PK (1997) Geology and origin of the Tallawang magnetite skarn, Gulgong, NSW. Geol Soc Australia-Abstracts 44:5

Bakker RJ, Jansen JBH (1991) Experimental post-entrapment water loss from synthetic $CO_2$-$H_2O$ inclusions in natural quartz. Geochim Cosmochim Acta 55:2215-2230

Barnes HL (1997) Geochemistry of Hydrothermal Ore Deposits. John Wiley & Sons, New York

Barbieri M, Belkin HE, Chelini W, de Vivo B, Lattanzi P, Lima A, Tolomeo L (1986) Fluid inclusions and Sr isotopes from Mofete 1 and San Vito 1 geothermal wells, Campania, Italy. 5th Int symp water-Rock Interaction, Reykjavik, Iceland, p 86

Bebout GE, Barton MD (1993) Metasomatism during subduction: products and possible paths in the Catalina Schist, California. Chem Geol 108:61-92

Belkin HE, de Vivo B (1987) The Phlegrean fields (Italy) water-dominated geothermal system: Fluid inclusions and Sr isotopes. Am Current Res on Fluid Incl, Socorro, Program with Abstract (unpaginated)

Belkin, HE, Chelini W, de Vivo B, Lattanzi P (1986) Fluid inclusions in hydrothermal minerals from Mufti 2, Mufti 5 and San Vita 3 geothermal wells, Phlegrean Fields, Campania, Italy. Symposium 5 Int Volcanol Congress, Auckland p 7-12

Beuline MV (1976) Conditions of formation of the Tagarsk iron-ore deposit (central near-Angarsk) according to inclusions in minerals. COFFI-abstracts 9:16

Bird DK, Helgeson HC (1981) Chemical interaction of aqueous solutions with epidote-feldspar mineral assemblages in geologic systems. II. Equilibrium constraints in metamamorphic/geothermal processes. Am J Sci 281:576-614

Bird D, Spieler AR (2004) Epidote in geothermal systems. Rev Mineral Geochem 56:235-300

Bischoff JL, Rosenhauer RJ (1984) The critical point and two-phase boundary of seawater, 200-500°C. Earth Planet Sci Lett 68:172-180

Bischoff JL, Rosenhauer RJ (1989) Salinity variations in submarine hydrothermal systems by layered double-diffusive convection. J Geol 97:613-623

Blackwell DD, Golan B, Benoit D (2000) Temperatures in the Dixie valley, Nevada geothermal system. Geother Res Coun Trans 24:223-228

Bodnar RJ, Binns PR, Hall DL (1989) Synthetic fluid inclusions. VI. Quantitative evaluation of the description behaviour of fluid inclusions in quartz at one atmosphere confining pressure. J Metamorph Geol 7:229-242

Boundy TM, Donohue CL, Essene EJ, Mezger K, Austrheim H (2002) Discovery of eclogite facies carbonate rocks from the Lindas Nappe, Caledonides, Western Norway. J Metamorph Geol 20:649 667

Bowers TS, Helgeson HD (1983) Calculation of the thermodynamic and geochemical consequences of nonideal mixing in the system $H_2O$-$CO_2$-NaCl on phase relations in geologic systems: Metamorphic equilibria at high pressures and temperatures. Am Mineral 68:1059-1075

Bril H, Papapanagiotou P, Patrier P, Lenain JF, Beaufort D (1996) Fluid-rock interaction in the geothermal field of Chipilapa (El Salvador): contribution of fluid inclusion data. Eur J Mineral 8:515-531

Brooks JW (1994) Petrology and geochemistry of the McCoy gold skarn, Lander County, Nevada. Unpublished Ph.D. thesis, Washington State University, Pullman, Washington, 607 p

Brown PE, Bowman JR, Kelly WC (1985) Petrologic and stable isotope constraints on the source and evolution of skarn-forming fluids at Pine Creek, California. Econ Geol 80:72-95

Brunsmann A, Franz G, Erzinger J, Landwehr D (2000) Zoisite- and clinozoisite-seggregations in metabasites (Tauern window, Austria) as evidence for high-pressure fluid-rock interaction. J Metamorph Geol 18:1-21

Bukharev VP, Gostyaeva NM, Naumenko VV, Shemyakina TI (1982) Temperature conditions of formation of ore and metasomatite minerals of the skarn-magnetite deposit Blagodat Mt. (Middle Urals). Geo Rudoobrazovaniye 9:29-36

Butterfield DA, Massoth RE, McDuff RE, Lupton JE, Lilley MD (1990) Geochemistry of hydrothermal fluids from axial seamount hydrothermal emissions study vent field, Juan de Fuca Ridge. J Geophys Res 95: 12895-12921

Carella R, Guglielminetti M (1983) Multiple reservois in the Mofete fields, Naples. 9th Workshop on Geotherm Res Eng, Univ Stanford 12pp

Cathelineau M, Nieva D (1986) Geothermometry of hydrothermal alteration in the Los Azufres geothermal system: Significance of fluid inclusion data. Inter Sympos Water-Rock Interaction 5:104-107

Cathelineau M, Izqierdo G, Nieva D (1989) Thermobarometry of hydrothermal alteration in the Los Azufres geothermal system (Michoacan, Mexico): Significance of fluid inclusion data. Chem Geol 76:229-238

Cathles NM, III (1997) Thermal aspects of ore formation. *In:* Geochemistry of Hydrothermal Ore Deposits. Barnes HL (ed) John Wiley & Sons, New York, p 191-227

Coleman RG (1977) Ophiolites. Springer-Verlag, New York

Cong Y, Touret JLR (1993) Fluid inclusions in eclogites from Dabie Mountains, eastern China. Terra Abstracts 1:475

Crawford ML, Hollister (1986) Metamorphic fluids: the evidence from fluid inclusions. *In*: Fluid-rock Interaction During Metamorphism. Walther JV, Wood BJ (eds) Springer-Verlag, New York, p 1-35

Dal Piaz GV, Gosso G, Lombardo B (1983) Metamorphic evolution of the Mt. Emilius klippe, Dent Blanche nappe, western Alps. Am J Sci 283:438-458

Davies JH (1999) The role of hydraulic fractures and intermediate-depth earth-quakes in generating subduction-zone magmatism. Nature 398:142-145

Dilles JH, Einaudi, MT (1992) Wall-rock alteration and hydrothermal flow paths about the Ann-Mason porphyry copper deposit, Nevada—A 6- km vertical reconstruction. Econ Geol 87:1963-2001

Dilles JH, Solomon GC, Taylor HP Jr, Einaudi, MT (1992) Oxygen and hydrogen isotope characteristics of hydrothermal alteration at the Ann-Mason porphyry copper deposit, Yerrington, Nevada. Econ Geol 87: 44-63

Deer WA, Howie RA, Zussman J (1986) Disilicates and ring silicates. Longman Scientific & Technical, Harlow

De Vivo B, Belkin, HE, Barbieri M, Chelini W, Lattanzi P, Lima A, Tolomeo L (1989) The Camp Flegrei (Italy) geothermal system: a fluid inclusion study of the Mofete and San Vito fields. J Volcanol Geotherm Res 36:303-326

Draper DS, Brandon AD, Becker H (1999) Interactions between slab and sub-arc mantle: dehydration, melting and element transport in subduction zones. Chem Geol 160:251-253

El-Shazly AK, Sisson VB (1999) Retrograde evolution of eclogite facies rocks from NE Omen: evidence from fluid inclusions and petrological data. Chem Geol 154:193-223

Einaudi MT, Meinert LD, Newberry RJ (1981) Skarn deposits. Econ Geol, 75th Anniv. Vol., p 317-391

Essene EJ, Fyfe WS (1967) Omphacite in Californian metamorphic rocks. Contrib Mineral Petrol 15:1-23

Ettlinger AD, Meinert LD, Ray GE (1992) Gold skarn mineralization and fluid evolution in the Nickel Plate Deposit, Hedley, District, British Columbia. Econ Geol 87:1541-1565

Feng Z, Chang Z (1996) Mineralization and rock alteration associated with igneous intrusions in the southern eastern Shanxi province, China. 30th Inter Geol Cong Abstracts 2:620

Franz G, Althaus E (1977) The stability relations of the paragenesis paragonite-zoisite-quartz. N Jahrb Mineral Abh 130:159-167

Franz G, Spear FS (1983) High pressure metamorphism of siliceous dolomites from the Central Tauern Window, Austria. Am J Sci 283:396-413

Franz G, Smelik EA (1995) Zoisite-clinozoisite bearing pegmatites and their importance for decompressional melting in eclogites. Eur J Mineral 7:1421-1436

Franz G, Thomas S, Smith DC (1986) High-pressure phengite decomposition in the Weissenstein eclogite, Münchberg Gneiss Massif, Germany. Contrib Mineral Petrol. 92:71-85

Franz L, Romer RL, Klemd R, Schmid R, Oberhänsli R, Wagner T, Shuwen D (2001) Eclogite-facies quartz veins within metabasites of the Dabie Shan (eastern China): P-T-t-d-x conditions and fluid flow during exhumation of high-pressure rocks. Contrib Mineral Petrol 141:322-346

Freiberger R, Boiron MC, Cathelineau M, Cuney M (2001a) Late Hercynian fluid circulation in the Charroux-Civray plutonic complex, NW Massif Central, France. *In*: Water-Rock Interaction WR 1-10. Cidu R (ed) Swets & Zeitlinger Publications, p. 705-708

Freiberger R, Boiron MC, Cathelineau M, Cuney M, Buschaert S (2001b) Retrograde P-T evolution and high temperature-low pressure fluid circulation in relation to late Hercynian intrusions: a mineralogical and fluid inclusion study of the Charroux-Civray plutonic complex (north-western Massif Central, France). Geofluids 4:241-262

Fu B (2002) Fluid regime during high- and ultrahigh-pressure metamorphism in the Dabie-Sulu terranes, eastern China. PhD-thesis. Vrije Universiteit Amsterdam, 144 p

Fu B, Touret JLR, Zheng YF (2001) Fluid inclusions in coesite-bearing eclogites and jadeite quartzite at Shuanhe, Dabie Shan (China). J Metamorph Geol 19:529-545

Fu B, Zheng YF, Touret JLR (2002) Petrological, isotopic and fluid inclusion studies of eclogites from Sujiahe, NW Dabie Shan (China). Chem Geol 187:107-128

Fu B, Touret JLR, Zheng YF (2003) Remnants of premetamorphic fluid and oxygen isotopic signatures in eclogites and garnet clinopyroxenite from the Dabie Sulu terrane, eastern China. J Metamorph Geol 21: 561-578

Gao J, Klemd R (2001) Primary fluids entrapped at blueschist to eclogite transition: evidence from the Tianshan meta-subduction complex in northwestern China. Contrib Mineral Petrol 142:1-14

Goldstein RH, Reynold TJ (1994) Systematics of fluid inclusions in diagenetic minerals. SEPM Short Course 31, p 199

Giaramita MJ, Sorensen SS (1994) Primary fluids in low-temperature eclogites: evidence from two subduction complexes (Dominican Republic and California, USA). Contrib Mineral Petrol 117:279-292

Gillis KM, Bangerjee NR (2000) Hydrothermal alteration patterns in supra-subduction zone ophiolites. Geol Soc Am Special Paper 349:283-297

Hacker BR, Ratschbacher L, Webb LE, McWilliams M, Ireland T, Dong S, Calvert A, Wenk HR (2000) Exhumation of the ultrahigh-pressure continental crust in east-central China: Late Triassic-Early Jurassic extension. J Geophys Res 105:13339-13364

Hagemann SG, Bray C, Brown PE, Spooner ETC (1996) Combined gas and ion chromatography of fluid inclusions and sulfides from the Archean Epizonal Wiluna lode-gold deposit, western Australia. PACROFI VI, Wisconsin-Madison, p 69-70

Haynes FM, Kesler SE (1988) Compositions and sources of mineralizing fluid for chimney and manto limestone-replacement ores in Mexico. Econ Geol 83:1985-1992

Hedenquist JW, Henley RW (1985) Hydrothermal eruptions in the Waiotapu geothermal system, New Zealand: Their origin, associated breccias, and relation to precious metal mineralization. Econ Geol 80:1640-1668

Hedenquist JW, Reyes, AG, Simmons SF, Taguchi S (1992) The thermal and geochemical structure of geothermal and epithermal systems. Eur J Mineral 4:989-1015

Helgeson HC (1979) Mass transfer among minerals and hydrothermal solutions. *In*: Geochemistry of Hydrothermal Ore Deposits. Barnes HL (ed) John Wiley & Sons, New York, p 568-610

Heinrich W, Althaus E (1988) Experimental-determination of the rections 4 lawsonite + 1 albite = 1 paragonite + 2 zoisite + 1 quartz + 6 $H_2O$ and 4 lawsonite + 1 jadeite = 1 paragonite + 2 zoisite + 1 quartz + 6 $H_2O$. Neues Jahrb Mineral Mh 11:516-528

Heinrich W, Gottschalk M (1995) Metamorphic reactions between fluid inclusions and mineral hosts. I. Progress of the reaction calcite + quartz = wollastonite + $CO_2$ in natural wollastonite-hosted fluid inclusions. Contrib Mineral Petrol 122:51-61

Henley RW, Ellis AL (1983) Geothermal systems ancient and modern: A geochemical review. Earth Planet Sci Rev Lett 19:1-50

Hickey RJ (1990) The geology of the Buckhorn mountain gold skarn, Okanagon County, Washington. Unpublished M.S. thesis, Washington State University, Pullman, Washington, 171 p

Holland TJB (1979) High water activities in the generation of high-pressure kyanite eclogites of the Tauern window, Austria. J Geol 87:1-27

Holland TJB, Powell R (1990) An internally consistent thermodynamic data set with uncertainties and correlations: the system $K_2O-Na_2O-CaO-MgO-MnO-FeO-Fe_2O_3-Al_2O_3-TiO_2-SiO_2-C-H_2-O_2$. J Metamorph Geol 8:89-124

Hollister LS (1981) Information intrinsically available from fluid inclusions. In: Hollister LS, Crawford ML (eds) Fluid inclusions: Applications to petrology. Mineral Ass Canada, Calgary, Short Course Handbook 6:1-12

Hollister LS, Crawford ML (1981) Fluid inclusions: Applications to petrology. Mineral Ass Canada, Calgary, Short Course Handbook 6, 304 p

Holness MB (1993) Temperature and pressure dependence of quartz-aqueous dihedral angles: The control of absorbed $H_2O$ on the permeability of quartzites. Earth Planet Sci Lett 117:363-377

Izquierdo G, Arellano VM, Cathelineau M (1997) *P-T-X* conditions of formation of zeolites, clays and calc-silicates in the Los Azufres geothermal field. ECROFI XIV-abstracts, Nancy, France, p151-152

John T, Schenk V (2003) Partial eclogitisation of gabbroic rocks in a late Precambrian subduction zone (Zambia): prograde metamorphism triggered by fluid infiltration. Contrib Mineral Petrol 146:174-191

Johnson EL, Hollister LS (1995) Syndeformational fluid trapping in quartz: determining the pressure-temperature conditions of deformation from fluid inclusions and the formation of pure $CO_2$ fluid inclusions during grain boundary migration. J Metamorph Geol 13:239-249

Karwowski L, Wlodyka R (1981) Conditions of formation of drusy minerals at Michatowice (Karkonosie Mts.): Postmagmatic processes in plutonic and volcanic rocks. COFFI-abstracts 14:99

Kelley DF, Delaney JR (1985) High temperature, high salinity aqueous fluids from the Kane fracture zone, Mid-Atlantic Ridge. Geol Soc Am Abstracts with Programs 171:626

Kelley DF, Delaney JR (1986) Fluid inclusion evidence for multiple fracturing events in gabbros from the Mid-Atlantic Ridge 23°N. EOS 67:1283

Kelley DF, Robinson PT (1990) Development of a brine-dominated hydrothermal system at temperatures of 400-500°C in the upper level plutonic sequence, Troodos ophiolite, Cyprus. Geochim Cosmochim Acta 54:653-661

Kelley DF, Robinson PT, Malpas JG (1992) Processes of brine generation and circulation in the oceanic crust: Fluid inclusion evidence from the Troodos Ophiolite, Cyprus. J Geophys Res 97:9307-9322

Kelley DF, Gillis KM, Thompson JG (1993) Fluid evolution in submarine magma-hydrothermal systems at the Mid-Atlantic Ridge. J Geophys Res 98:19579-19596

Kirby S, Engdahl ER, Denlinger R (1996) Intermediate-depth intraslab earthquakes and arc volcanism as physical expressions of crustal and uppermost mantle metamorphism in subducting slabs. Geophy Mono 96:195-214

Kleinefeld B, Bakker RJ (2002) Fluid inclusions as microchemical systems: evidence and modeling of fluid-host interactions in plagioclase. J Metamorph Geol 20:845-859

Klemd R (1989) P-T evolution and fluid inclusion characteristics of retrograded eclogites, Münchberg Gneiss Complex, Germany. Contrib Mineral Petrol 102:221-229

Klemd R (1998) Comment on the paper by Schmidt Mumm et al.: High $CO_2$ content of fluid inclusions in gold mineralisations in the Ashanti belt, Ghana: a new category of ore forming fluids? Mineral Deposita 33: 317-319.

Klemd R, Matthes S, Okrusch M (1991) High-pressure relics in meta-sediments intercalated with the Weissenstein eclogite, Münchberg gneiss complex, Bavaria. Contrib Mineral Petrol 107:328-342

Klemd R, van den Kerkhof AM, Horn EE (1992) High-density $CO_2-N_2$ inclusions in eclogite-facies metasediments of the Münchberg Gneiss complex, SE Germany. Contrib Mineral Petrol 111:409-441

Klemd R, Hirdes W, Olesch M, Oberthür T (1993) Fluid inclusions in quartz-pebbles of the gold-bearing Tarkwaian conglomerates of Ghana as guides to their provenance area. Mineral Deposita 28:334-343

Klemd R, Matthes S, Schüssler U (1994) Reaction textures and fluid behaviour in very high-pressure calc-silicate rocks of the Münchberg gneiss complex, Bavaria, Germany. J Metamorph Geol 12:735-745

Klemd R, Bröcker M, Schramm J (1995) Characterisation of amphibolite-facies fluids of Variscan eclogites from the Orlica-Snieznik dome (Sudetes, SW Poland). Chem Geol 119:101-113

Klemd R, Bröcker M (1999) Fluid influence on mineral reactions in ultrahigh-pressure granulites: a case study in the Snieznik Mts. (West Sudetes, Poland). Contrib Mineral Petrol 136:358-373

Kretz R (1983) Symbols for rock-forming minerals. Am Mineral 68:277-279

Kreulen R (1980) $CO_2$-rich fluids during regional metamorphism on Naxos (Greece): carbon isotopes and fluid inclusions. Am J Sci 280:745-771

Kreulen R (1989) High integrated fluid rock ratios during metamorphism at Naxos: reply. Contrib Mineral Petrol 103:127-129

Küster M, Stöckhert B (1997) Density changes of fluid inclusions in high-pressure low-temperature metamorphic rocks from Crete: A thermobarometric approach based on the creep strength of the host mineral. Lithos 41:151-167

Kuznetsov SV, Gostyaeva NM (1982) Studies of gas-liquid inclusions in metasomatic albitites of Precambrian age. Geokh Rudoobrazonaniye 10:27-35

Kwak TAP (1986) Fluid inclusions in skarns (carbonate replacement deposits). J Metamorph Geol 4:363-384

Kwak TAP (1987) W-Sn skarn deposits and related metamorphic skarns and granitoids: Developments in Economic Geology, Vol 24. Elsevier, Amsterdam

Labotka TC, Nabelek PI, Papike JJ (1988) Fluid infiltration through the Big Horse Limestone Member in the Notch Peak contact-metamorphic aureole, Utah. Am Mineral 73:1302-1324

Layne GD, Spooner ETC (1991) The JC tin skarn deposit, southern Yukon Territory: I. Geology, paragenesis, and fluid inclusion microthermometry. Econ Geol 86:29-47

Layne GD, Spooner ETC, Longstaffe FJ (1987) Mineralogical, fluid inclusion and stable isotope studies of the JC tin skarn, Yukon. Geol Soc Am Abstracts with Programs 19:742

Layne GD, Longstaffe FJ, Spooner ETC (1991) The JC tin skarn deposit, southern Yukon Territory: II. A carbon, oxygen, hydrogen, and sulfur stable isotope study. Econ Geol 86:48-65

Lendowski W (1983) Physico-chemical condition of crystallisation of the low- and moderate-temperature mineral parageneses in the Strzegom massif. Archiw Mineral 39:53-66

Lecuyer C, Dubois M, Margnac C, Gruau G, Fouquet Y, Ramboz C (1999) Phase separation and fluid mixing in subseafloor back arc hydrothermal systems: A microthermometric and oxygen isotope study of fluid inclusions in the barite-sulfide chimneys of the Lau basin. J Geophys Res 104:17911-17927

Li N, Feng Z, Yu F (1989) Genesis of the Beiluoxia skarn deposit in Shanxi province: Fluid inclusion evidence. Mineral Deposits 8:43-54

Liou JG (1993) Stabilities of natural epidotes. Abh Geol B-A 49:7-16

London D (1986) Formation of tourmaline-rich gem pockets in miarolitic pegmatites. Am Mineral 71:396-405

Luckscheiter B, Morteani G (1980) Microthermometrical and chemical studies of fluid inclusions from Alpine veins from the penninic rocks of the central and western Tauern Window, Austria. Lithos 13:61-77

Lüders V, Banks DA, Halbach P (2002) Extreme Cl/Br and $\delta^{37}Cl$ isotope fractionation in fluids of modern submarine hydrothermal systems. Mineral Deposita 37:765-771

Lutz SJ, Moore JN, Blamey NJF, Norman DI (2002) Fluid inclusion gas chemistry of the Dixie Valley (NV) geothermal system. Proc 27th Worksh Geotherm Reserv Eng, Stanford Univ, SGP-TR-171

Malisa E, Kinnunen K, Koljonen T (1986) Notes on fluid inclusions of vanadiferous zoisite (tanzanite) and green grossular in Merelani area, northern Tanzania. Bull Geol Soc Finland 58:53-58

Maruyama S, Cho M, Liou JG (1986) Experimental investigations of blueschist-greenschist transition equilibria: pressure dependence of $Al_2O_3$ contents in sodic amphiboles – A new geobarometer. Geol Soc Am Mem 164:1-16

Megaw PKM, Ruiz J, Titley SR (1988) High-temperature, carbonate-hosted Ag-Pb-Zn (Cu) deposits of northern Mexico. Econ Geol 83:1856-1885

Meinert LD (1987) Skarn zonation and fluid evolution in the Groundhog Mine, Central Mining District, New Mexico. Econ Geol 82:523-545

Meinert LD (1992) Skarns and skarn deposits. Geoscience Canada 19:145-162

Meinert LD (1998) A review of skarns that contain gold. *In:* Mineralized Intrusion-Related Skarn Systems. Lentz DR (ed) Min Assoc Can Short Course 26:359-414

Mibe K, Fujii T, Yasuda A (1999) Control of the location of the volcanic front in island arcs by aqueous fluid connectivity in the mantle wedge. Nature, 401:259-262

Mibe K, Yoshino T, Ono S, Yasuda A, Fujii T (2003) Connectivity of aqueous fluid in eclogite and its implications for fluid migration in the Earth's interior. J Geophys Res 108:2295-3006

Molnar P, Sykes LR (1969) Tectonics of the Caribbean and Middle America regions from focal mechanism and seismicity. Geol Soc Am Bull 80:1639-1684

Muramatsu Y, Nambu M (1982) Fluid inclusion study on the contact metamorphic tungsten ore deposits of the Yaguki mine, Kushima prefecture, Japan. Mining Geol 32:107-116

Nehlig P (1991) Salinity of oceanic hydrothermal fluids: a fluid inclusion study. Earth Planet Sci Lett 102:310-325

Nelson BK (1995) Fluid flow in subduction zones: evidence from Nd- and Sr-isotope variations in metabasalts of the Franciscan complex, California. Contrib Mineral Petrol 119:247-262

Newberry RJ (1998) W- and Sn-skarn deposits: a 1998 Status Report. *In:* Mineralized Intrusion-Related Skarn Systems. Lentz DR (ed) Min Assoc Can Short Course 26:289-335

Newberry RJ, Einaudi MT (1981) Tectonic and geochemical setting of tungsten skarn mineralization in the Cordillera. Arizona Geol Soc Digest 14:99-112

Newton RC, Manning CE (2000) Quartz solubility in $H_2O$-NaCl and $H_2O$-$CO_2$ solutions at deep crust-upper mantle pressures and temperatures: 2-15 kbar and 500-900°C. Geochim Cosmochim Acta 64:2993-3005

Nimz G, Janik C, Goff F, Dunlap C, Huebner M, Counce D, Johnson SD (1999) Regional hydrology of the Dixie Valley geothermal field, Nevada: preliminary interpretation of chemical and isotope data. Geotherm Res Coun Trans 23:333-338

Norton DL (1977) Fluid circulation in the Earth's crust. Am Geophys Union, Mono 20:693-704

Norton DL (1987) Advective metasomatism. In Helgeson HC (ed) NATO ASI Series. Series C: Mathematical and Physical Sciences, 218, 123-132. D. Reidel Publishing Company, Dordrecht, Boston

Ocanto CA, Gomez GM, Lira R (2001) Microthermometric data on epidote-clinozoisite from metasomatic-hydrothermal environments, Oriental Pampean Ranges, Argentina. ECROFI XVI-abstracts, 347-348, Porto

O'Brien PJ (1993) Partially retrograded eclogites of the Münchberg Massif, Germany: records of a multi-stage Variscan uplift history in the Bohemian Massif. J Metamorph Geol 11:241-260

Okrusch M, Seidel E, Davies EN (1978) The assemblage jadeite-quartz in the glaucophane rocks of Sifnos. N Jahrb Mineral Abh 132:284-308

Parry WT, Hedderly-Smith D, Bruhn R L (1991) Fluid inclusions and hydrothermal alteration on the Dixie Valley fault, Nevada. J Geophys Res 96:19733-19748

Patrier P, Papapanagiotou P, Beauford D, Traineau H, Bril H, Rojas J (1996) Past and present thermal regime of the active geothermal field of Chipilapa (Salvador): contribution of the <0.2 μm clay fraction. J Volc Geoth Res 72:101-107

Papapanagiotou P (1994) Evolution des mineraux argileux en relation avec la dynamic des champs gepthermiques haute enthalpie: l'exemple du champ de Chipilapa (Salvador). PhD-thesis, Univ de Poitiers, 189 p

Peacock SM (1993) The importance of blueschist → eclogite dehydration reactions in subducting oceanic crust. Geol Soc Am Bull 105:684-694

Philippot P (1993) Fluid-melt-rock interaction in mafic eclogites and coesite-bearing metasediments: constraints on volatile recycling during subduction. Chem Geol 108:93-112

Philippot P, Selverstone J (1991) Trace-element-rich brines in eclogitic veins; implications for fluid composition and transport during subduction. Contrib Mineral Petrol 106:417-430

Poli S, Schmidt MW (1998) The high-pressure stability of zoisite and phase relationships of zoisite-bearing assemblages. Contrib Mineral Petrol 130:162-175

Poli S, Schmidt MW (2004) Experimental subsolidus studies on epidote minerals. Rev Mineral Geochem 56: 171-195

Powell R, Holland TJB (1988) An internally consistent thermodynamic data set with uncertainties and correlations: 3. Applications to geobarometry, worked examples and a computer program. J Metamorph Geol 6:173-204

Roedder E (ed) (1984) Fluid inclusions. Reviews in Mineralogy, Vol 12. Mineral Soc Amer, Washington DC

Roedder E (1979) Fluid inclusions as samples of ore fluids. *In*: Geochemistry of Hydrothermal Ore Deposits. Barnes HL (ed) John Wiley & Sons, New York, p 684-737

Roedder E, Bodnar RJ (1997) Fluid inclusion studies of hydrothermal ore deposits. *In*: Geochemistry of Hydrothermal Ore Deposits. Barnes HL (ed) John Wiley & Sons, New York, p 657-689

Rüpke LH, Morgan JP, Hort M, Connolly JAD (2002) Are the regional variations in Central American arc lavas due to differing basaltic versus peridotitic slab sources of fluids. Geology 30:1035-1038

Scambelluri M, Philippot P (2001) Deep fluids in subduction zones. Lithos 55:213-227

Scambelluri M, Pennachioni G, Phillipot P (1998) Salt-rich aqueous fluids formed during eclogitization of metabasites in the Alpine continental crust (Austroalpine Mt. Emilius unit, Italian Western Alps). Lithos 43:151-167

Schmid R, Klemd R, Franz L, Oberhänsli R, Dong S (2003) UHP metamorphism and associated fluid evolution: a case study in the Bixiling area (Dabie Shan, China). Bh. z. Eur J Mineral 15:175

Schmidt MW, Poli S (1998) Experimentally based water budgets for dehydrating slabs and consequences for arc magma generation. Earth Planet Sci Lett 163:361-379

Scott SD (1997) Submarine hydrothermal systems and deposits. *In*: Geochemistry of Hydrothermal Ore Deposits. Barnes HL (ed) John Wiley & Sons, New York, p 797-875

Selverstone J, Franz G, Thomas S, Getty S (1992) Fluid variability in 2 GPa eclogites as an indicator of fluid behaviour during subduction. Contrib Mineral Petrol 112:341-357

Seyfried Jr WE, Ding K (1995) Phase equilibria in subseafloor hydrothermal systems: A review of the role of redox, temperature, pH and dissolved Cl on the chemistry of hot spring fluids at mid-ocean ridges. *In:* Seafloor Hydrothermal Systems: Physical, Chemical, Biologic and Geological Interactions. Humphris SE, Zierenberg RA, Mullineaux LS, Thompson JRE (eds) Geophys Mono 91:248-273. Am Geophys Union.

Seyfried Jr WE, Ding K, Berndt ME (1991) Phase equilibria constraints on the chemistry of hot spring fluids at mid-ocean ridges. Geochim Cosmochim Acta 55:3559-3580

Shepherd TJ, Rankin AH, Alderton DHM (1986) A Practical Guide to Fluid Inclusion Studies. Blackie, Glasgow-London

Shmulovich KI, Graham CM (1999) An experimental study of phase equilibria in the system $H_2O$-$CO_2$-NaCl at 800°C and 9 kbar. Contrib Mineral Petrol 136:247-257

Shmulovich KI, Graham CM (2003) An experimental study of phase equilibria in the systems $H_2O$-$CO_2$-$CaCl_2$ and $H_2O$-$CO_2$-NaCl at high pressures and temperatures (500−800°C, 0.5−0.9 Gpa): geological and geophysical applications. Contrib Mineral Petrol online: 10.1007/s00410-003-0507-5

Spear FS (1993) Metamorphic Phase Equilibria and Pressure-Temperature-Time Paths. Mineral Soc Am, Washington

Stalder HA, Rykart R (1980) Negative (crystal) forms of epidote in quartz crystals. Schweizer Strahler 5: 320-327

Sterner SM, Hall DL, Keppler H (1995) Compositional re-equilibration of fluid inclusions in quartz. Contrib Mineral Petrol 119:1-15

Svensen H, Jamtveit B, Yardley B, Engvik AK, Austrheim H, Broman C (1999) Lead and bromine enrichment in eclogite-facies fluids: extreme fractionation during lower-crustal hydration. Geology 27:467-470

Takeno N, Sawaki T, Murakami H, Miyake K (1999) Fluid inclusion study of skarns in the Maruyama Deposit, the Kamioka Mine, Central Japan. Res Geol 49:233-242

Taylor BE, Liou JG (1978) The low-temperature stability of andradite in COH fluids. Am Mineral 63:378-393

Thompson AB (1971) $P_{CO2}$ in low grade metamorphism: zeolite, carbonate, clay mineral, prehnite relations in the system $CaO$-$Al_2O_3$-$CO_2$-$H_2O$. Contrib Mineral Petrol 33:145-161

Thompson G (1983) Basalt-seawater-interaction. *In:* Hydrothermal Processes at Seafloor Spreading Centers. Rona PA, Bostrom K, Laubier L, Smith KLJr (eds), Plenum, New York, p 225-278

Torres-Alvarado IS (2002) Chemical equilibrium in hydrothermal systems: the case of Los Azufres geothermal field. Int Geol Rev 44:639-652

Tracy RJ, Frost BR (1991) Phase equilibria and thermobarometry of calcareous, ultramafic and mafic rocks, and iron formations. Rev Mineral 26:207-289

Vargunina NP, Andrusenko NI (1983) Mineralogical-geochemical peculiarities of a polygenic gold-silver deposit. Dokl Akad Nauk SSR 269:419-423

Vallis F, Scambelluri M (1996) Redistribution of high-pressure fluids during retrograde metamorphism of eclogite-facies rocks (Voltri Massif, Italian West Alps). Lithos 39:81-92

Vanko DA (1988) Temperature, pressure, and composition of hydrothermal fluids, with their bearing on the magnitude of tectonic uplift at mid-ocean ridges, inferred from fluid inclusions in oceanic layer 3 rocks. J Geophys Res 93:4595-4611

Vanko DA, Griffith JD, Erickson CL (1992) Calcium-rich brines and other hydrothermal fluids in fluid inclusions from plutonic rocks, Oceanographer Transform, Mid-Atlantic Ridge. Geochim Cosmochim Acta 56:35-47

Von Damm KL, Edmond JM, Grant B, Measures, Walden B, Weiss RF (1985) Chemistry of submarine hydrothermal solutions at 21° North, East Pacific Rise. Geochim Cosmochim Acta 49:2197-2220

Von Damm KL, Buttermore LG, Oosting SE, Bray AM, Fornari DJ, Lilley MD, Shanks WD (1997) Direct observation of the evolution of a seafloor "black smoker" from vapor to brine. Earth Planet Sci Lett 149: 1001-111

Vorontsov AY, Pukhnarevich MM, Afonina GG, Makagon VM, Smirnov VN, Zav'yanova LL (1978) Hydrothermal feldspars from the Korshunovskoe iron ore deposit. Dok Akad Nauk SSR 241:1171 1174

Walther JV, Orville PM (1982) Volatile production and transport in regional metamorphism. Contrib Mineral Petrol 79:252-257

Will TM, Powell R, Holland TJB, Guiraud M (1990) Calculated greenschist facies mineral equilibria in the system -$CaO$-$FeO$-$MgO$-$Al_2O_3$-$TiO_2$-$SiO_2$-$CO_2$-$H_2O$. Contrib Mineral Petrol 104:353-368

Will TM, Okrusch M, Schmädicke E, Chen G (1998) Phase relations in greenschist-blueschist-amphibolite-eclogite facies: Calculated mineral equilibria in the system $Na_2O$-$CaO$-$FeO$-$MgO$-$Al_2O_3$-$SiO_2$-$H_2O$, with applications to the PT evolution of metamorphic rocks from Samos, Greece. Contrib Mineral Petrol 132: 85-102

Wilson PN, Parry WT (1986) Petrologic and fluid inclusion studies of the Ophir Hill mine Pb-Zn-(Ag) skarn deposit, Ophir district, Tooele County, Utah. Geol Soc Am Abstracts with Programs 18:423

Winkler HGF (1979) Petrogenesis of Metamorphic Rocks. 5th edition, Springer Verlag, New York, Heidelberg, Berlin

Witt-Eickschen G, Klemd R, Seck HA (2003) Density contrast of fluid inclusions associated with melt (glass) from two distinct suites of mantle peridotites from the West Eifel, Germany: Implications for melt origin. Eur J Mineral 15:95-103

Xiao YL, Hoefs J, van den Kerkhof AM, Fiebig J, Zheng Y (2000) Fluid history of UHP metamorphism in Dabie Shan, China: a fluid inclusion and oxygen isotope study on the coesite-bearing eclogite from Bixiling. Contrib Mineral Petrol 139:1-16

Xiao YL, Hoefs J, van den Kerkhof AM, Li S (2001) Geochemical constraints of the eclogite and granulite facies metamorphism as recognized in the Raobazhi complex from North Dabie Shan, China. J Metamorph Geol 19:3-19

Xu S, Okay AI, Ji S, Sengör AMC, Su W, Liu Y, Jiang L (1992) Diamond from the Dabie Shan metamorphic rocks and its implication for tectonic setting. Science 256:80-82

Yardley BWD (1986) Fluid migration and veining in the Connemara Schists, Ireland. *In:* Advances in Physical Geochemistry 5. Walther JV, Wood BJ (eds) Springer, New York, p 109-131

Yardley BWD (1989) An introduction to metamorphic petrology. Longman Earth Sciences Harlow, 248 p

You Z, Han Y, Yang W, Zhang Z, Wie B, Liu R (1996) The high-pressure and ultra-high-pressure metamorphic belt in the East Quinling and Dabie Mountains, China. China University of Geoscience Press, Wuhan, 150 pp

Zheng YF, Fu B, Gong B, Li L (2003) Stable isotope geochemistry of ultrahigh pressure metamorphic rocks from the Dabie-Sulu orogen in China: implications for geodynamics and fluid regime. Earth Sci Rev 62: 105-161

Reviews in Mineralogy & Geochemistry
Vol. 56, pp. 235-300, 2004
Copyright © Mineralogical Society of America

**6**

# Epidote in Geothermal Systems

## Dennis K. Bird and Abigail R. Spieler

*Department of Geological and Environmental Sciences*
*Stanford University*
*Stanford, California 94305, U.S.A.*

## INTRODUCTION

Early in the 20th Century epidote was readily recognized as a common rock-forming mineral of metamorphic and hydrothermal processes (Becke 1903; Grubenmann 1904; Van Hise 1904; Goldschmidt 1911; Eskola 1915). Its distribution is widespread in the Earth's crust, including metamorphic environments of pumpellyite-prehnite, greenschist, epidote-amphibolite, and blueschist facies (Seki 1972; Liou 1993). In lower-pressure hydrothermal environments epidote is a common mineral in skarns, in propylitic altered volcanic rocks and in late-stage veins related to silicic intrusions (Lindgren 1933; Coats 1940; Nakovnik 1963). Within obducted segments of oceanic crust (ophiolites) and in large igneous provinces epidote is found in veins and replacement bodies (epidosites, cf. Dana 1875) associated with intrusion of dolerite dikes and gabbros (Coleman 1977). It was not until the 1960's that epidote was first discovered in drill hole samples from active geothermal systems (Naboko and Piip 1961; Sigvaldason 1963; White et al. 1963; Steiner 1966; Keith et al. 1968; Marinelli 1969). Geothermal drill holes provided the first samples of epidote-altered rocks and coexisting hydrothermal fluids at measured temperatures and pressures (White and Sigvaldason 1963; Naboko 1964). Formation of epidote in such low-pressure geologic environments was initially questioned (Rusinov 1966), due in part to geologic observations (Korzhinskiy 1963) and to the sluggish nature of epidote synthesis at low pressures and temperatures (Fyfe et al. 1958; Coombs et al. 1959; Merrin 1960; Fyfe 1960). Epidote is now recognized as a key index mineral related to temperature, permeability, and fluid composition in geothermal systems worldwide (Browne 1978; Giggenbach 1981; Henley and Ellis 1983; Bird et al. 1984; Reyes 1990; Absar 1991; Reed 1994; Muramatsu and Doi 2000).

In general, hydrothermal epidote exhibits a wide range in octahedral substitution of $Al^{3+}$ for $Fe^{3+}$, typically with notable oscillatory or irregular zoning. Complex paragenetic relations and coexisting phase relations are not uncommon. Partitioning of Al and $Fe^{3+}$ between coexisting epidote and garnet or prehnite often reflect metastable compositional relations. These mineralogic characteristics, presented in the review below, are a natural consequence of the evolution of magma-hydrothermal processes and the sensitive dependence of epidote composition to temporal and spatial variations in fluid flux, permeability, temperature, redox conditions, $CO_2$ concentrations, pH, and the aqueous speciation of $Al^{3+}$ and $Fe^{3+}$ complexes.

We begin our review of epidote in hydrothermal systems with a brief summary of magma-hydrothermal processes and epidote paragenesis. Here we discuss the consequences of extrema in transport and the thermodynamic properties of hydrothermal solutions on intensive and extensive thermodynamic variables that control compositional and phase relations of epidote. This is followed by a review of epidote parageneses in active geothermal systems organized by global tectonic setting: first, geothermal systems within convergent plate boundaries of the

1529-6466/04/0056-0006$10.00

Circum-Pacific Margin and in the Mediterranean Region; second, geothermal systems in rifting environments, including Iceland, Ethiopia, and the Salton Trough of California and Mexico. A brief review of epidote parageneses in fossil geothermal systems is considered next, restricted to ophiolites and to the Early Tertiary North Atlantic igneous province. They represent some of the largest geothermal systems on Earth, related to oceanic spreading centers and mantle plumes, for which there are no drill holes into zones of active epidote formation (excluding Iceland). Epidote in fossil geothermal systems associated with silicic intrusions, environments represented in our analysis of active geothermal systems of convergent plate margins, are not presented here. We include in this review evidence of epidote parageneses, phase relations, spatial and temporal distributions, compositions of epidote and geothermal fluids, and related theoretical phase diagrams as per published observations and interpretations. Our objective is to present epidote parageneses in terms of active magma-hydrothermal processes characteristic of subduction and rifting environments in the Earth's crust.

This review is not conclusive; it is intended as an introduction to the extreme dependence of epidote parageneses, phase relations and elemental mass transfer on the transitory properties of hydrothermal systems related to high-level igneous intrusions. Below, we adopt mineralogic terminology of the papers being reviewed. For example, K-rich dioctahedral layer silicates are variably referred to as illite, sericite, K-mica, muscovite or white mica by various authors. We have adopted the notation of $X_{ps}$ to denote the mol fraction of the pistacite ($Ca_2Fe_3Si_3O_{12}(OH)$) component in epidote.

## MAGMA-HYDROTHERMAL PROCESSES AND EPIDOTE PARAGENESIS

Epidote paragenesis in the geothermal environment is a function of the physical and chemical processes that characterize magma-hydrothermal systems. Intrusion of magmas into upper levels of the Earth's crust generates potentials for irreversible transport of energy and mass (Norton 1977, 1984; Cathles 1977, 1997). Heat transfer, fracture formation, fluid flow, and fluid-rock reaction are not independent processes in these systems; they are all related to one another through a series of intricate couplings (feedback loops) as described by Norton (1987, 1988, 1990) and Bredehoeft and Norton (1990). Potentials for irreversible reaction between hydrothermal solutions and their mineralogic environment are developed as fluids flow along variable temperature-pressure paths and encounter rocks of differing composition (Norton 1979). Extrema in the thermodynamic and transport properties of water in the near critical region of $H_2O$ has an important influence on advective transport rates of thermal energy and chemical components, as well as on the thermodynamic properties of aqueous electrolyte solutions, and thus, water-rock reaction (Helgeson 1981; Norton 1984; Johnson and Norton 1991). Numerical experiments of heat and mass transfer in magma-hydrothermal systems reported by Norton and Dutrow (2001) demonstrate temporal oscillatory and chaotic variations in the transport and thermodynamic properties of aqueous solutions in the near critical region of water, and the resonant effect of these variations on chemical reactions (i.e., oscillatory zoning in minerals) and fracture propagation. Coupled with a geologic history of multiple intrusions and related brittle deformation, and with boiling and mixing of fluids from different sources (i.e., magmatic water, meteoric water, seawater, metamorphic water; Giggenbach and Steward 1982; Henley and Ellis 1983; Hedenquest et al. 1992), it is evident that the evolution of hydrothermal solutions and the paragenesis of secondary minerals is complex.

Epidote is reported in active geothermal systems at temperatures <200°C, but is a common and abundant phase above 230–260°C in volcanic rocks, and at slightly higher temperatures in calcareous sediments (Seki 1972; Browne 1978; Bird et al. 1984; see below). Under hydrothermal conditions it is stable over a range of temperature and pressure that includes the critical point for $H_2O$ and the critical point for most natural electrolyte solutions (Liou 1993). Epidote is formed by precipitation in veins and cavities, and by replacement of silicates, carbonates and Fe-oxides. In some cases its paragenesis is simple, in others there is evidence of multiple stages of epidote formation and dissolution. The crystal size of epidote generally increases with increasing temperature (Patrier et al. 1990; Beaufort et al. 1992; see review below), and metastable substitutional disorder of octahedral Al and $Fe^{+3}$ has been documented in geothermal epidotes formed at <300°C (Bird et al. 1988a; Patrier et al. 1991). Hydrothermal epidote exhibits a wide range in $Al^{3+}$ and $Fe^{3+}$ substitution, commonly with complex oscillatory or irregular zoning on a scale of 10's of microns. Compositional variation in both hydrothermal and metamorphic epidote has been attributed to a variety of intensive and extensive thermodynamic variables including temperature, fugacities of $O_2$ and $CO_2$, and bulk rock and fluid composition (Brown 1967; Holdaway 1972; Liou 1973; Raith 1976; Cavarretta et al. 1980a; Giggenbach 1981; Shikazono 1984; Arnason and Bird 1992; Arnason et al. 1993).

Arnason et al. (1993) present a thermodynamic analysis of variables affecting epidote composition in hydrothermal and low-pressure regional metamorphic rocks. They conclude that temperature, along with $O_2$ and $CO_2$ fugacities are the most important intensive variables controlling local variations in epidote composition. However, cause and effect relationships are not readily apparent in many cases. For example, the temperature dependence of epidote composition associated with dehydration, decarbonation, and/or hydrolysis reactions is largely determined by the mineral assemblage and the standard molar enthalpy of reaction. The reaction of K-feldspar, muscovite and calcite to form epidote, biotite and quartz, a common assemblage related to the biotite isograd in the Salton Sea geothermal system (see below), has a positive enthalpy of reaction. Under conditions of constant fugacity of $CO_2$ and biotite composition, the $Al^{3+}$ content of epidote will increase with increasing temperature, and near the critical point of water, the enthalpy of this reaction becomes a very large positive number, increasing the sensitivity of epidote composition to only minor changes in temperature (see enthalpy-temperature diagram of Fig. 1 in Arnason et al. 1993). In contrast, the reaction describing the assemblage epidote, hematite, calcite, and quartz, a vein assemblage in the Salton Sea geothermal system, has a negative enthalpy of reaction, thus the $Al^{3+}$ content of epidote decreases with increasing temperature for this assemblage. Similar relations are observed for the effect of $CO_2$ fugacity on the composition of epidote, in that mineral assemblage and reaction stoichiometry determines if the $Al^{3+}$ content of epidote will increase or decrease with increasing fugacity of $CO_2$. The $Fe^{3+}$ content of epidote is also a function of oxygen fugacity. At oxygen fugacities near the hematite-magnetite buffer, Fe-rich epidote is stable, and epidote composition under these conditions is relatively insensitive to changes in oxygen fugacity. However, under reducing conditions, approaching the quartz-fayalite-magnetite equilibrium, the composition of Al-rich epidote is a sensitive function of changes in the fugacity of oxygen (Bird and Helgeson 1981).

The composition of epidote is also a function of bulk rock and fluid composition (Shikazono 1984; Arnason et al. 1993). The dependence of epidote composition on the $Fe_2O_3$ content of the rock, and on aqueous concentrations of $Fe^{3+}$ and $Al^{3+}$ are intuitively obvious. Other relations are not qualitatively apparent, such as solution pH and the aqueous speciation of $Fe^{3+}$ and $Al^{3+}$. Thermodynamic relations and irreversible mass transfer models illustrating the dependence of epidote composition on pH and fluid composition are presented by Arnason and Bird (1992) and Arnason et al. (1993). In dilute, near-neutral pH hydrothermal solutions, where aqueous $Al(OH)_4^-$ and $Fe(OH)_3^0$ are predominant species of $Fe^{3+}$ and $Al^{3+}$, the Fe-content of epidote will increase with increasing pH and decrease with decreasing pH. Figure 1 illustrates these relations for two irreversible mass transfer reaction models summarized by the arrows on the activity-activity phase diagram at 300°C and 86 bar (reaction paths computed using EQ6, Wolery 1992; see Fig. 1 caption for details). The arrow marked by the number 4 in the figure denotes the theoretical reaction path and composition of epidote formed by

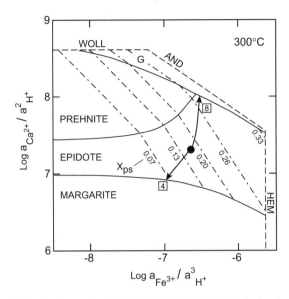

**Figure 1.** Phase relations in the system $CaO-Al_2O_3-Fe_2O_3-SiO_2-H_2O-HCl$ showing the stabilities of minerals as a function of cation to hydrogen ion activity ratios in the coexisting fluid at 300°C and 86 bar in the presence of quartz and a fluid with unit activity of water. Fields labeled epidote, prehnite and garnet (G) represent the stability of $Al^{3+}-Fe^{3+}$ solid solutions for these minerals, the dot-dashed lines corresponding to isopleths of constant $Fe^{3+}$ content, which are labeled in the epidote field as values of $X_{ps}$. Dashed lines labeled WOL (wollastonite), AND (andradite), and HEM (hematite) denote values of cation to hydrogen ion activity ratios in the fluid required for saturation with these phases. Arrows represent changes in epidote composition predicted by theoretical irreversible reaction path model of anorthite reaction with aqueous solution (labeled 4 in the diagram) and of calcite reaction with aqueous solution (labeled 8 in the diagram). Diagram computed from equations and data reported by Helgeson et al. (1978), Bird and Helgeson (1980), Rose and Bird (1987), and Johnson et al. (1992). Reaction paths computed using EQ6 computer package (Wolery 1992). The numbers 4 and 8 refer to initial conditions for reactions given in Table 1 of Arnason and Bird (1992). Modified after Figure 2 of Arnason and Bird (1992).

irreversible hydrolysis of anorthite in a solution initially in equilibrium with epidote ($X_{ps}$ = 0.15, solid symbol). Epidote formed by this reaction is zoned from $X_{ps}$ = 0.15 (corresponding to the core) to $X_{ps}$ = 0.07 (corresponding to the rim). Zoning is associated with a decrease in pH of only several tenths (pH decreases during anorthite hydrolysis because $OH^-$ is consumed to form Al-hydroxide species). The arrow marked 8 in Figure 1 represents a model for irreversible reaction of calcite with a dilute aqueous solution initially in equilibrium with epidote ($X_{ps}$ = 0.15). During this reaction pH increases (by several tenths) due to calcite dissolution, and the equilibrium epidote composition changes from $X_{ps}$ = 0.15 to 0.27 (arrow 8). These irreversible reaction path models serve to illustrate the dependence of epidote composition on pH and aqueous speciation of $Fe^{3+}$ and $Al^{3+}$, and may offer partial explanations for the common observation that geothermal epidote replacing plagioclase in volcanic rocks are zoned with Al-rich rims, and epidote formed in calcareous sediments are zoned with Fe-rich rims (see review below). Nevertheless, Arnason et al. (1993) conclude that the sensitivity of epidote composition to fluid chemistry, and the multivariance of many epidote-bearing assemblages, makes chemically zoned epidote unreliable indicators of the chemical and physical evolution of hydrothermal systems.

Drill holes in active geothermal systems provide samples of altered rock and geothermal fluids at measured temperatures and pressures. However, as noted by Zen (1974), observed mineral alteration may not have formed under contemporary conditions. This is evident for a number of geothermal systems considered in the following sections, where epidote is found below 200°C, or to occur with acid or $CO_2$-rich hydrothermal solutions that are not in equilibrium with epidote. In addition, Arnórsson (1995) notes that the number of hydrothermal minerals reported in a single drill hole sample is commonly greater than permitted by consideration of the phase rule, indicating that secondary mineral assemblages may form over a range of temperatures and fluid compositions, and that metastable phase relations are not uncommon. In contrast, vein-filling epidote typically occurs as multivariant assemblages of only a few phases. Nevertheless, in a number of geothermal systems equilibrium between major components of hydrothermal solutions and epidote-bearing assemblages has been demonstrated (Giggenbach 1981; Bird and Norton 1981; Arnórsson et al. 1983; see review below). Although solution-mineral reactions during the dynamic conditions of a magma-hydrothermal system are irreversible processes, local equilibrium is temporally approached for epidote-bearing assemblages in a number of active geothermal systems.

## CIRCUM-PACIFIC MARGIN GEOTHERMAL SYSTEMS

Epidote has been reported in the Pacific Margin region associated with high-level intrusions related to late Cenozoic andesite to rhyolite volcanic centers and calderas near convergent plate boundaries. Below we limit our review to selected geothermal areas in Japan, Philippines, New Zealand, and Central America. In general, the hydrology of these geothermal systems is controlled by regional volcanic and sedimentary stratigraphy, and locally by dikes and faults. Hydrothermal solutions typically represent varying mixtures of magmatic and meteoric fluids that have locally experienced boiling, phase separation, and remixing. The case studies summarized below illustrate the effects of multiple intrusions, and the mixing and boiling of hydrothermal solutions on epidote parageneses.

### Japan

Of the numerous active geothermal systems in Japan (Sumi and Takashima 1976), we restrict our review to two geothermal areas in the Honshu region, where epidote compositions and parageneses are well documented.

***Onikobe geothermal system.*** The late Pliocene to middle Pleistocene Onikobe caldera is located on the eastern edge of the Green Tuff Basin in north-central Honshu. Miocene green tuffs together with Plio-Pleistocene andesitic and dacitic extrusive deposits overlie a basement complex of Paleozoic pelitic schists and Cretaceous granodiorites (Yamada 1972). The most recent magmatic activity is a 0.35 Ma dacitic lava dome intruded into the southern part of the caldera (Yamada et al. 1978). Geothermal mineralogy and fluid chemistry are presented for 11 drill holes by Seki et al. (1983), together with downhole temperature measurements. The maximum temperature of about 275°C is observed at a depth of 1250 m. Temperature gradient reversals are common in the upper 500 m of the drill holes.

Composition of geothermal fluids from springs and drill holes is documented by Ozawa and Nagashima (1975), Yamada (1976), Nakamura et al. (1977), Ozawa et al. (1980), and Seki et al (1983). In general, fluids produced from deep drill holes are described as dilute Na-Cl and Na-Ca-Cl solutions that have evolved through variable mixing and boiling of meteoric and magmatic fluids. According to analyses summarized by Seki et al. (1983) the stoichiometric ionic strength of the near neutral pH geothermal fluids is <0.1 and for acid solution it is <0.4. Sulfur isotopes suggest a component of fossil seawater from Miocene basement volcanogenic sediments (Seki et al. 1983). Compositions of geothermal fluids indicate the release of magmatic gas rich in HCl from depth, its mixing with meteoric water to form acidic, high-Cl thermal water and the neutralization of this water by interaction with volcanic rocks. Acid thermal fluids have been produced from depths of 1100–1300 m; however, changes in the composition of production fluids in shallow wells from near neutral pH (5–7) to acid solutions (pH 3–4) over a period of several months suggest of an active magmatic source of HCl-rich fluids. Recent drill holes are producing fluids at depths >1000 m with pH's of 8.3–8.5, suggesting that the flux of acid magmatic fluids is localized and sporadic.

The general depth and temperature zoning of geothermal minerals in the Onikobe geothermal area are summarized in Figure 2. The epidote isograde is observed at depths of approximately 100-150 m in six drill holes (P-5, -7, -8 ,-10, and wells 123 and 124) where present-day temperatures range from 110–210°C. In drill holes GO-7, -8, -10, and -11, epidote first occurs at depths between 315 and 500 m where measured temperatures range from 175–200°C. As noted by Seki et al. (1983) and Liou et al. (1985), these temperatures are significantly lower than expected for the formation of epidote based on experimental and geologic observations (Seki 1972; Liou 1993). This is consistent with the depth distribution of geothermal minerals reported within the epidote zone in these drill holes (Seki et al. 1983). In four of the six wells where epidote is first found at depths of 100–150 m, low-temperature zeolite assemblages with mordenite and yugawaralite are reported, as is pyrophyllite at greater depth (250 m in Fig. 2), a phase indicative of low-pH solutions. In the wells were epidote is reported at depths greater than 315–500 m, three of the drill holes have pyrophyllite within the epidote zone. Two of these wells produced fluids with pH's <3, one at a temperature of 240°C.

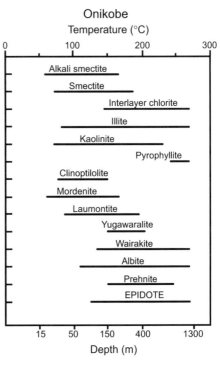

**Figure 2.** Secondary minerals as a function of depth and temperature in the Onikobe geothermal system. Modified after Figure 36 of Seki et al. (1983).

Throughout most of the epidote zone calcite and gypsum are reported (Seki et al. 1983). These observations all indicate that present-day temperature and pH conditions are lower than expected for the formation of epidote.

Compositions and parageneses of epidote reported by Liou et al. (1985) and Seki et al. (1983) are summarized below. Epidote occurs as tabular crystals or as spongy crystal aggregates replacing plagioclase and volcanic matrix or as fan-shaped, coarse grained aggregates. Characteristic yellow to light-yellow pleochroism and very high birefringence is noted. Assemblages include epidote + wairakite + chlorite + albite + calcite + quartz + pyrite; epidote + wairakite + chlorite + albite + prehnite + quartz + pyrite; and epidote + laumontite + calcite + chlorite + pyrite ± prehnite. Epidote associated with wairakite + calcite + albite + chlorite + quartz + pyrite is zoned and ranges from $X_{ps}$ = 0.19–0.37 with most analyses around $X_{ps}$ = 0.29–0.33; however, the authors do not characterize the zoning. Changes in epidote composition with increasing depth and temperature are not systematic, but reflect decreasing iron content with depth. Analysis of phase relations, compositions and parageneses using the projection procedure proposed by Seki et al. (1983) for the ternary Ca-2Al-2Fe$^{3+}$ for the system CaO-Fe$_2$O$_3$-Al$_2$O$_3$-(FeO + MgO)-SiO$_2$-H$_2$O-CO$_2$-S leads to the following conclusions:

(1) iron-rich epidote and prehnite are restricted to low temperatures, and become more Al-rich with increasing temperature, (2) Al end-member epidote and prehnite are not stable within the physical and chemical conditions of this geothermal system, (3) for coexisting epidote and prehnite, epidote always contains a higher concentration of iron, and (4) coexisting epidote and prehnite in the assemblage epidote + prehnite + wairakite contain less iron than in the assemblage epidote + prehnite + calcite.

***Sumikaw-Ohnuma geothermal system.*** The Hachimantai volcanic region (also referred to as the Sengan geothermal area) is one of the largest geothermal areas in Japan consisting of a number of active volcanoes and three caldera structures in the Nasu volcanic belt of northeastern Japan. The area consists of Miocene volcanoclastic and sedimentary rocks, welded tuffs (2–1 Ma) and andesitic lavas and pyroclastic rocks (1–0 Ma; Tamanyu and Suto 1978; Suto 1987). Below we summarize the occurrence of epidote in two of the hydrothermal areas in this region, the Sumikaw-Ohnuma and Kakkonda geothermal systems.

The Sumikaw-Ohnuma geothermal system is located on the northern flank of the active volcano of Mt Yakeyama and is hosted in andesitic and dacitic volcanic and minor sedimentary rocks of Miocene and Quaternary age which fill a north-south graben (Inoue and Ueda 1965; Sato et al. 1981). Tonolitic to granodioritic intrusions (6–7 Ma, NEDO 1990) that are encountered at depths of 1000–2000 m are significantly older than granitic intrusions found elsewhere in the Hachimantai area (see Kakkonda below). Subsurface temperatures are >300°C at 2000 m. The geothermal reservoir is characterized by a local two-phase zone below a caprock of lacustrine sediments, and a deeper liquid dominated zone where temperature-depth profiles follow the boiling point curve, with local temperature inversions (Kubota 1985). In general, fluid inclusion homogenization temperatures of hydrothermal minerals are similar to measured downhole temperatures (Ueda et al. 1991; Sakai et al. 1993; de Vivo and Sasada 1992). However, temperature inversions in drill holes in the northern portion of the geothermal system suggest the incursion of cooler fluids (Kubota 1985; Sakai et al. 1993). Fluids are of variable composition, but are characterized as Na-Cl fluids with <3000 mg/l total dissolved solids, pH's between 5.8 and 8.5 (at room temperature), and fugacities of $CO_2$ between $10^{-0.11}$ and $10^{-1.13}$ bars (Sakai et al. 1986; Ueda et al. 1991). Fluids are of meteoric origin with only a minor magmatic component (Matsubaya et al. 1983; Ueda et al. 1991).

The first occurrence of epidote within geothermal systems of Japan was reported by Shimazu and Yajima (1973) from the Sumikaw-Ohnuma geothermal system. They report epidote in drill holes east of the Sumikaw River penetrating a sequence of welded dacitic tuffs, andesitic lavas and pyroclastics. Epidote occurs in veins and as a replacement mineral in the volcanics, in association with sericite, chlorite, and calcite, (±) laumontite and wairakite (one occurrence) but the paragenetic relations are not reported. The first appearance of epidote is at depths ranging from 634–1050 m in four drill holes. The epidote isograd in the drill holes studied closely parallels the measured 200°C isotherm (Fig. 3); however, pressure-corrected homogenization temperatures of two-phase fluid inclusions in wairakite from the epidote zone range from 220–245°C. These temperatures are up to 50°C greater than the current downhole temperatures, suggesting that this portion of the geothermal system has experienced cooling since formation of wairakite and presumably epidote. Two epidote analyses are reported. Both are of unzoned vein epidotes (yellow-green idoblastic crystals) in association with prehnite and chlorite, with $X_{ps} = 0.27$ and 0.28.

Inoue et al. (1999) report geothermal mineral occurrences and parageneses in a portion of the Sumikaw-Ohnuma geothermal system to the west of the area studied by Shimazu and Yajima (1973). Epidote, prehnite and actinolite are usually associated and have a complex spatial distribution (Inoue et al. 1999). The first appearance of epidote closely parallels the 230°C isotherm (Bamba et al. 1987) except in the northern portion of the geothermal system, where there appears to be incursion of colder waters (Kubota 1985; Sakai et al.

Sumikaw-Ohnuma

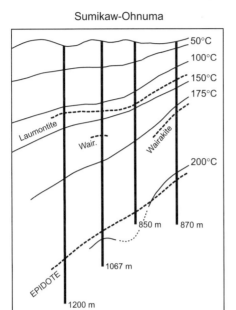

**Figure 3.** Schematic cross section of a portion of the Sumikaw-Ohnuma geothermal system, showing isotherms and isograds for laumontite, wairakite and epidote. Thick vertical lines represent drill holes, with depth marked in meters. Modified from Figure 4 of Shimazu and Yajima (1973).

1993). Epidote is commonly associated with chlorite and illite and, to a lesser degree, with calcite and anhydrite. K-feldspar is locally distributed within the epidote zone, but albite is widespread (Inoue et al. 1999). Epidote ($X_{ps}$ = 0.05–0.39, with an average value of 0.26) is found in veins, disseminated in the matrix and replacing plagioclase phenocrysts. There are no obvious compositional trends with respect to temperature or paragenesis, except that vein epidote exhibits a wider range in composition than does epidote replacing plagioclase or the volcanic matrix. Zoning is variable, with both homogeneous grains and those with complex compositional zoning patterns; however, vein epidote is typically zoned with Fe-rich rims. Thermodynamic analysis of mineralogic phase relations and measured geothermal fluid compositions by Inoue et al. (1999) suggest that activities of the aqueous species $Na^+$, $K^+$, $Ca^{2+}$ and $HCO_3^-$, as well as the solution pH, are buffered by assemblages of K-feldspar, illite, calcite and calc-silicates including epidote, prehnite and/or wairakite at temperatures between about 250 and 275°C. They suggest that the abundance of epidote and prehnite veins and the local formation of K-feldspar are a response to an increase in the pH of ascending fluids that are undergoing boiling and loss of $CO_2$.

***Kakkonda geothermal system.*** This geothermal system is centered above the Quaternary Kakkonda granitoid intrusion and is located about 15 km southeast of the Sumikawa-Ohuma geothermal system. The stock is a composite intrusion of tonalite, granodiorite, quartz diorite and granite, several tens of km in area that is encountered in drill holes at depths between 1.5 and 3 km (Doi et al. 1991). It intrudes pre-Tertiary and Tertiary volcanic and sedimentary rocks and older intrusions. K-Ar age dates of hornblende, biotite and K-feldspar from the intrusion range from 0.24–0.01 Ma (Doi et al. 1998). Well WD-1 encountered the intrusion at 2.8 km depth and was drilled 0.87 km into the granite, where temperatures exceeded 500°C and hypersaline magmatic fluids were sampled (Fig. 4; Ikeuchi et al. 1998; Kasai et al. 1998). The Kakkondo granitoid is considered to be the heat source for the extensive biotite-grade contact aureole and for the present-day hydrothermal activity (Doi et al. 1991, Muraoka et al. 1998). The geologic cross section (Fig. 4) shows present day isotherms relative to the geometry of the Kakkondo granitoid.

## Kakkonda

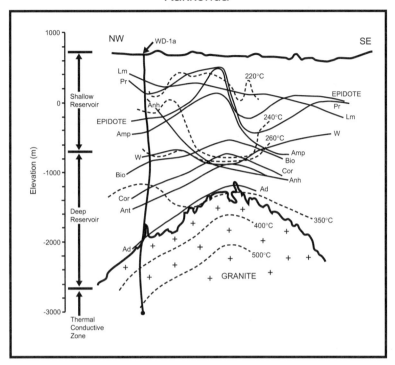

**Figure 4.** Generalized cross section of a portion of the Kakkonda geothermal system illustrating isotherms (dashed lines), distribution of contact metamorphic and hydrothermal minerals (solid lines), location of drill hole WD-1a, and the location of the shallow and deep geothermal reservoirs relative to the Kakkonda granitoid intrusion. Isograds are shown for prehnite (Pr), epidote (EPIDOTE), amphibole (Amp), biotite (Bio), cordierite (Cor), anthophyllite (Ant), and andalusite (Ad). Lines marked Lm, Anh, and W, denote the deepest occurrence of laumontite, anhydrite, and wairakite, respectively. Modified from Figure 9 of Muraoka et al. (1998).

The geothermal reservoir is highly fractured and separated into two parts based on fracture distributions, temperature, and fluid chemistry (Fig. 4; Kato and Doi 1993; Kajiwara et al. 1993; Nakamura and Sumi 1981; Yanagiya et al. 1996). The higher-permeability shallow reservoir, <1500 m depth at 230–260°C, has dilute Na-Cl fluids (<2000 mg/l total dissolved solids) and pH's between 8.8 and 9.0. The deep reservoir, >1500 m at 300–350°C, has lower permeability and hosts acidic geothermal fluids (pH 3.7–4.5) and Na-Cl solutions with total dissolved solids of approximately 3000 mg/l. Present-day reservoir fluids are primarily of meteoric origin (Yanagiya et al. 1996). Below the geothermal convection system, which includes fracture systems in the pre-Tertiary and Tertiary formations and the upper margin of the Kakkonda granitoid, is a zone of conductive heat transfer where recovered fluids are of magmatic origin, with total salinities of about 40 wt% NaCl equivalent (Kasai et al. 1998). Fluid inclusion analyses in hydrothermal and magmatic minerals have been reported by Komatsu and Muramatsu (1994), Sasaki et al. (1995), Muramatsu and Komatsu (1996), Sawaki et al. (1999), Muramatsu and Doi (2000), and Muramatsu et al. (2000). In general, these studies have identified an early stage of magmatic fluids with salinities ranging from 24–75 wt% that were expelled from the Kakkonda granite, and they all conclude that the temperature, $CO_2$ content and salinity of geothermal fluids circulating above the granite decreased with time as a consequence of mixing of the hypersaline fluids with dilute meteoric waters and boiling of the ascending fluids.

Hydrothermal mineral parageneses in the Kakkonda magma-hydrothermal system are considered to result, first, from extensive contact metamorphism related to the intrusion of the granitoid, and later, by fracture-controlled hydrothermal fluid circulation (Doi et al. 1998; Sasaki et al. 1998; Takeno et al. 2001). Contact metamorphism is identified by whole-rock alteration that forms mineral isograds closely paralleling the intrusive contact. Isograds extending away from the deepest-drilled portions of the intrusive complex include andalusite, orthopyroxene, cummingtonite, anthophyllite, cordierite and biotite. They are subparallel to the intrusive contact of the granite, but crosscut present-day isotherms (Fig. 4). It appears from alteration types and mineral associations (Kato and Doi 1993, Doi et al. 1998) that trace amounts of epidote occur throughout these contact alteration zones, but phase relations, parageneses and compositions are not reported. According to Figure 8 of Doi et al. (1998), modally abundant epidote is restricted to the upper, shallow geothermal reservoir within the temperature range of approximately 220–260°C. The epidote isograd is shown in Figure 4, where it can be compared to isograds for prehnite and amphibole, and the deeper limits for the occurrence of laumontite, anhydrite, and wairakite (Muraoka et al. 1998). Note that the shallowest occurrence of epidote is centered above the granite intrusion, and that the epidote isograd crosscuts present-day geotherms. Comparison of mineral distributions for well 13 (Kato and Doi 1993) and well 4 (Muramatsu and Doi 2000) with the general alteration mineralogy summarized by Doi et al. (1998) and Sasaki et al. (1998) suggests that the appearance of epidote corresponds to a modal decrease in calcite; that epidote overlaps with the occurrence of prehnite, but extends to higher temperature; that actinolite forms at slightly higher temperature than the first appearance of epidote, and that illite (muscovite) and chlorite are found throughout the epidote zone. Laumontite is not present in the epidote zone, and wairakite is most abundant above the epidote zone but does occur with prehnite in epidote-altered rocks.

During the initial opening of a shallow production hole (well 4, 1448 m deep), located to the southeast of well WD-1a in Figure 4, large amounts of chips of hydrothermal minerals were ejected with steam and water (Muramatsu and Doi 2000). The ejectae were derived from a high-permeability fracture zone near the bottom of the well and consist of euhedral quartz, epidote ($X_{ps}$ = 0.22–0.25), prehnite ($X_{Fe}$ = 0) and wairakite with trace amounts of pyrite, rutile, titanite and apatite. Two types of ejectae were observed, one consisting of euhedral quartz, prehnite and wairakite without epidote, and the other of euhedral quartz, chlorite, prismatic green epidote and prehnite without wairakite. The first mineral to form in the latter ejecta was epidote, followed by prehnite. The calc-silicate assemblages in the ejectae are similar to mineral assemblages forming a scale inside well casing at the Ngatamarike geothermal system in New Zealand (Browne et al. 1989), and to epidote-rich (epidosite) well ejectae produced from the high-permeability production zone of the Salton Sea geothermal system in California (well State 2-14; Elders and Sass 1988). Based on thermodynamic analysis of phase relations, present-day geothermal fluids and analyses of fluid inclusions (Muramatsu et al. 2000), Muramatsu and Doi (2000) conclude that the present-day geothermal reservoir fluids are close to equilibrium with prehnite, quartz and calcite (Fig. 5, triangles), but that earlier geothermal fluids represented by fluid inclusion analyses (Fig. 5 solid circles) were in equilibrium with epidote or prehnite and wairakite as well as quartz and calcite. They conclude that this paragenesis results from cooling and degassing (lower $CO_2$ fugacities) of the reservoir after emplacement of the Kakkonda granitoid.

Shikazono (1984) summarizes the compositional variation of epidote in active and fossil geothermal systems, reporting on eight localities in Japan not reviewed above. In general it is concluded that the iron content of epidote roughly correlates with the bulk rock $Fe_2O_3$ content

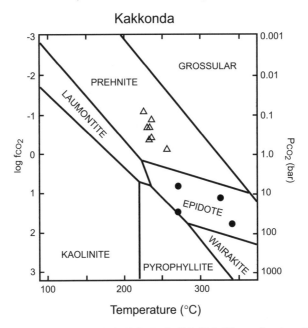

**Figure 5.** Phase relations in the system $CaO-Al_2O_3-Fe_2O_3-SiO_2-H_2O-CO_2$ as a function of temperature and the fugacity of $CO_2$. All minerals are stoichiometric with the exception of epidote ($X_{ps} = 0.22$). Symbols represent fluids from the Kakkonda geothermal system. Open triangles represent measured temperatures and fluid compositions of Well-4 and other shallow drill holes, and the solid symbols denote temperature and gas analytical data of fluid inclusions within anhydrite reported by Muramatsu et al. (2000). Modified from Figure 7 of Muramatsu and Doi (2000). Solid solution model for epidote from Bird and Helgeson (1980), thermodynamic data for minerals are not cited.

of unaltered rock, that epidote associated with hematite has the highest iron content, and that epidote in assemblages of prehnite, pyrite and pyrrhotite has the lowest iron content, reflecting the dependence of $Fe^{3+}$ substitution in epidote on the redox conditions of the hydrothermal system.

**Philippines**

There are over 30 explored high-temperature geothermal systems in the Philippines. Most are located within a region of plate convergence related to subduction zones east and west of the archipelago (Sussman et al. 1993). Geothermal systems are associated with Pleistocene to Recent andesitic volcanic centers or dacite-andesite domes (Reyes 1990), and some are sites of historic eruptions (1991 at Pinatubo; 1988 at Canlaon; 1860 at Cagua; 1939 at Biliran; 1895 at Mahagnao). Drill holes typically transect >2000 m Pliocene to Recent andesitic and dacitic flows and hyaloclastites and minor basalt, locally overlying Miocene to Late Pliocene carbonaceous sedimentary rocks, that have experienced repeated intrusions of dikes, stocks, and diorite or quartz monzodiorite plutons (Leach et al. 1983; Reyes and Giggenbach 1992; Reyes 1990, 1995b; Reyes et al. 1993). Temperatures encountered by drilling are ≈340°C (2400 m) at Tongonan, and >410°C (1770 m) at Mt. Cagua (Reyes 1995b). Comparisons of mineral parageneses, measured temperatures and fluid compositions, and fluid inclusion experiments (Leach et al. 1983; Reyes 1989, 1990, 1995a, 1995b; Reyes et al. 1995) provide evidence of repeated stages of heating and cooling that is largely due to multiple high-level intrusion of dikes and plutons.

Hydrothermal fluids are typically of meteoric origin, but have been modified by local mixing with magmatic fluids ("andesitic water"), boiling, phase separation and remixing with colder meteoric waters (Giggenbach 1992; Reyes and Giggenbach 1992; Reyes et al. 1993; Alvis-Isido et al. 1993; Gerardo et al. 1993; Ruaya et al. 1992; Ramos-Candelaria et al. 1995). Based on analyses of surface and well discharges, fluid inclusion experiments, and isotopic composition of gas and water, a model has been constructed for the distribution of fluids in the Alto Peak geothermal system that includes a central core ($\approx$1 km wide and 3 km in height) of high gas vapor (1.1–5.6 molal $CO_2$) that is surrounded by intermediate salinity water (7000 mg/kg Cl, near-neutral pH, 250–350°C), which in turn is surrounded by bicarbonate waters (Reyes et al. 1993). Evidence for magmatic fluids, characterized by high $CO_2$ and Cl contents, low pH, enriched $^{18}O$ and D, and $^3He/^4He$ ratios >4–7, are encountered at the Biliran, Mt. Pinatubo, Mt. Cagua, Alto Peak, Mahagnao, Palinpinon and Bacon-Manito geothermal systems, and fluid inclusion data suggest such fluids were temporally present in other Philippine geothermal systems (Reyes and Giggenbach 1992; Giggenbach 1992; Ramos-Candelaria et al. 1995). Chloride contents of the hydrothermal fluids vary between 3000 and 10,000 mg/kg (Reyes 1995a), with much higher concentrations observed in fluid inclusions, that are considered to represent either magmatic fluids, or fluids that have experienced boiling and phase separation (Leach et al. 1983; Hedenquist et al. 1992; Reyes et al. 1993; Reyes 1995b; Reyes et al. 1995).

Reyes (1990) identifies approximately one hundred secondary minerals in Philippine geothermal systems, grouped by their parageneses including weathering, diagenesis, hydrothermal metasomatism (<340°C; including alteration by near neutral-pH and acid solutions), magmatic-hydrothermal alteration (340–520°) and contact metamorphism (Reyes et al. 1993, 1995; Reyes 1995b, 1998). The near-neutral pH alteration is the most pervasive type at depth and contains locally abundant epidote group minerals including four texturally distinct types of epidote, together with minor clinozoisite and zoisite (Reyes 1990). Mineral isograds for the near-neutral pH alteration generally parallel present-day geotherms, whereas acid alteration is structurally controlled by faults and other high permeability zones (Leach et al. 1983; Reyes 1990; Reyes et al. 1993; Ramos-Candelaria et al. 1995). The neutral-pH alteration has been divided into three zones (Reyes et al. 1993); smectite, transitional, and illite; and into subzones based on isograds of epidote, amphibole, biotite and pyroxene as illustrated in the cross section of the Alto Peak geothermal system (Fig. 6). The temperature distribution of hydrothermal minerals associated with near-neutral pH alteration (Fig. 7) is based on measured well temperatures and fluid inclusion experiments. It appears that minor chemical mass transfer is associated with the near-neutral pH type of alteration (see Fig. 11 of Reyes 1990).

It should be noted that epidote, clinozoisite and allanite are reported to occur in zones of magmatic-hydrothermal alteration and contact metamorphism (see Fig. 7 of Reyes et al. 1993). In the Mt. Cagua geothermal system, magmatic-hydrothermal alteration (340–520°C) is associated with biotite, muscovite, illite, quartz, orthoclase, oligoclase, albite, epidote, clinozoisite, hornblende and actinolite, as well as assemblages of REE-rich minerals including metamict gadolinite and well-crystallized allanite, associated with danburite, tourmaline, fluorite, topaz, fluoroapatite, quartz and anhydrite (Reyes 1995b). In well AP-2D (−850 to −100 m elevation sea level) at the Alta Peak geothermal system, argillaceous and calcareous sediments intercalated with the volcanics are altered to the following assemblages, all of which contain epidote: grandite garnet, pyroxene, orthoclase, anhydrite, wollastonite, amphibole and quartz; grandite garnet, wollastonite, vesuvianite and orthoclase; biotite, anhydrite, amphibole, and pyroxene, and; quartz, biotite and anhydrite. These assemblages are considered by Reyes et al. (1993) to reflect local interaction of hydrothermal and magmatic fluids related to dike emplacement.

## Alto Peak

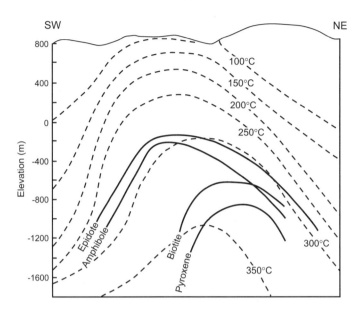

**Figure 6.** Schematic cross section of a portion of the Alto Peak geothermal system showing isotherms and isograds for epidote, amphibole, biotite and pyroxene. Modified after Figure 6 of Reyes et al. (1993).

## Philippine Geothermal Systems

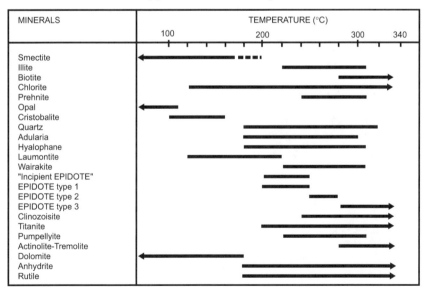

**Figure 7.** The temperature distribution of common geothermal minerals associated with near-neutral pH hydrothermal solutions in Philippine geothermal systems. Modified from data provided in Figure 6 of Reyes (1990), Figure 7 of Reyes et al. (1993), and Figures 3 and 29 of Reyes (1998).

In regions of near-neutral pH alteration, Reyes (1990, 1998) identifies four kinds of epidote based on crystallinity, mode and paragenesis, they are referred to as "incipient epidote" and epidote types 1, 2 and 3. Incipient epidote develops from titanomagnetite or the volcanic rock matrix between 180 and 220°C, is poorly crystalline, "sphene-like" in appearance and brownish in plane-polarized light. Type 1 epidote (240–260°C) is <5 μm in size, but well-crystalline, replacing plagioclase, ferromagnesian minerals and the volcanic matrix. Type 2 is similar to Type 1 but has a modal abundance of 5–10%. The highest temperature epidote (>280°C; type 3) is an abundant rock-forming mineral, occurring as a pervasive replacement and as euhedral vein fillings. Retrograde alteration of epidote is recorded by etched crystal surfaces and staining with goethite.

***Tiwi geothermal system.*** An excellent example of multiple generations of vein epidote is provided by the Tiwi geothermal system (Hoagland and Bodell 1991; Gambill and Beraquit 1993), where drill core was recovered to a depth of 2438 m (well Matalibong-25), encountering a maximum temperature of 275°C at 1829 m (Moore et al. 2000). Core samples provide the opportunity to examine mineral and vein parageneses; relationships that are difficult to evaluate when investigating drill cuttings from production wells. Below a depth of 908 m the volcanic stratigraphy is altered to epidote (becoming abundant at depths >1900 m), carbonate, chlorite and sericite. Moore et al. (2000) identified ten episodes of fracture mineralization based on cross cutting relations and vein mineral parageneses. Alteration is grouped into six stages: (1) deposition of clays and chalcedony; (2) sericite veins; (3) alternating cycles of epidote + adularia + quartz + pyrite veins with veins of calcite and/or anhydrite (± barite); (4) sericite veins; (5) wairakite + epidote veins followed by actinolite then calcite; and (6) sericite veins. Vein mineral parageneses, fluid inclusion experiments and gas compositions, $^{40}Ar/^{39}Ar$ dating of adularia, and thermal models based on adularia age spectra and kinetic data were used to predict the thermal and fluid evolution of the geothermal system penetrated by the Matalilbong-25 drill hole. The main phase of hydrothermal alteration (Stages 3, 4 and 5 above) began about 0.314–0.279 Ma (Stage 3), and epidote bearing veins formed at about 330°C by upwelling and boiling of deep hydrothermal fluids that contain magmatic gases. The intervening calcite and anhydrite veins formed between 333°C to ≈270°C by marginal recharge of the system with fluids containing components of meteoric or crustal origin. Stage 5 formed between 0.2 and 0.220 Ma at approximately 300°C. The system cooled from 0.190±0.020 Ma to about 250°C, stabilizing at approximately 235°C until about 0.050–0.010 Ma (Stage 6), to the present-day geothermal system. This model provides an excellent example of the dynamics of magma-hydrothermal systems as determined by repeated intrusion, brittle deformation, fracture sealing and the mixing and boiling of various fluid types. It clearly shows the short comings and potential pit falls in unraveling the physicochemical conditions of epidote formation in geothermal systems and thus of the geothermal systems themselves when only investigating drill cuttings from production wells.

## New Zealand

There are more than 20 geothermal areas in the Taupo Volcanic Zone (TVZ) of the North Island of New Zealand. The magmatic, hydrothermal, and tectonic activity of the TVZ is related to westward subduction of the Pacific Plate beneath a thin continental crust of the Indo-Australian Plate (Stern 1987; Gamble et al. 1993). The zone is composed of andesitic-dacitic volcanoes, and a complex tectonic depression hosting silicic caldera structures and related deposits (Cole 1981, 1984). Andesitic volcanism began at about 2 Ma and voluminous rhyolitic volcanism from about 1.6 Ma (Wilson et al. 1995). Continental basement marginal to the TVZ consists of greywacke and argillite of late Paleozoic to late Mesozoic age. Nearly 98 vol% of the volcanic rocks are rhyolitic lavas, pyroclastics and sediments associated with calderas, with approximately 2 vol% andesite, 0.1 vol% dacite and <0.1 vol% basalt (Cole 1981; Browne et al. 1992 and references cited therein). Many of the geothermal systems are associated with silicic calderas that developed between 0.35 and 0.15 Ma (Hedenquist 1986; Wood 1995). In the Ngatamariki geothermal area a diorite pluton dated at 0.55 ka was encountered at a depth of 2460 m (Browne et al. 1992; Arehart et al. 2002). Temperatures of geothermal reservoirs are typically 240–280°C, however more than 300°C have been encountered at the Waiotapu and Rotokawa geothermal systems (Hedenquist 1986; Krupp and Seward 1987; Hedenquist and Browne 1989; Reyes 1995a).

High-temperature (>250°C) geothermal fluids are characterized as dilute alkaline near-neutral chloride solutions, with variable $CO_2$ concentration. They are primarily of meteoric origin, having a pH between 5.5 and 6.5, and a salinity between 0.1 and 0.2 wt% NaCl, which are less than in andesite-hosted geothermal systems of the Philippines where salinity is about 1–2 wt% NaCl (Hedenquist 1986). Giggenbach (1995) identified two types of geothermal fluids in the TVZ based on variations in the chemical and isotopic compositions; near the eastern margin of the TVZ hydrothermal solutions have high gas contents and trace element and isotopic signatures suggesting volatiles were derived from andesitic magmas, but the western portion of the TVZ geothermal systems has low gas content with characteristics of mantle-derived volatiles. Below we review the occurrence of epidote in five geothermal systems in the TVZ.

***Wairakei geothermal system.*** One of the earliest discoveries of epidote in active geothermal systems was by Steiner (1966) in the Wairakei geothermal system. Steiner (1953; 1968; 1977) identifies three zones of hypogene alteration: the montmorillonite zone (Ca-montmorillonite, ptilolite, and laumontite), an intermediate zone (wairakite, albite, calcite, mixed-layer illite-montmorillonite, and chlorite), and a K-feldspar zone (K-feldspar associated with intense silicification near fault zones). Epidote is a common alteration mineral in the intermediate zone, associated with trace amounts of clinozoisite, at temperatures >235°C (Steiner 1977). It is found as granules, prismatic crystals and radiating aggregates, and occurs in five paragenetically distinct associations: (1) replacing andesine phenocrysts with albite, wairakite, calcite, micaceous clay and minor K-feldspar, (2) replacing hypersthene and hornblende with chlorite, micaceous clay, apatite, and pyrite, (3) replacing the volcanic groundmass with chloritic and micaceous clay and quartz, (4) replacing magnetite, and (5) filling vesicles and veins. Steiner (1977) reports the density, optical, X-ray properties of geothermal epidote. He also describes progressive development of incipient epidote about opaque granules and magnetite in the volcanic groundmass, beginning with a beige "leucoxene"-like alteration product that becomes increasingly translucent, birefringent, then pistachio-green with increasing crystallinity. This transformation is similar to that described by Reyes (1990, 1998) for the formation of "incipient" and Type 1 epidote in Philippine geothermal systems (see Fig. 7). Reyes and Giggenbach (1999) propose that the western margin of the Wariakei geothermal field (Poihipi Sector) has cooled by approximately 90°C based on comparison of present-day downhole temperatures with mineralogic and fluid inclusion temperatures (240–290°C) associated with an early assemblage of epidote, clinozoisite, illite, chlorite, albite, adularia, calcite and titanite. Ellis (1969) suggests that the abundance of epidote and wairakite in the Wairakei geothermal system is due in part to the low concentration of $CO_2$ (about 0.01 m) that favors their formation relative to calcite (see Broadland-Ohaaki geothermal system below).

***Waiotapu geothermal system.*** Epidote is associated with alteration related to the "deep" near-neutral chloride fluids, and exhibits mineral associations similar to those described above for Wairakei (Hedenquist and Browne 1989). Epidote is first found at temperatures that are now about 200–220°C, but fluid inclusion evidence suggests that this portion of the geothermal system has experienced cooling. In well Wt4, epidote is abundant at depths >300 m and about 250°C, replacing andesine and filling fractures together with wairakite. At depths between 604 an 665 m in this well, epidote crystals are zoned with $X_{ps}$ ranging from 0.15 at the base of crystals to 0.32 at the tip. Thermodynamic analysis of fluid-mineral equilibria suggests that the geothermal production fluids are now in local equilibrium with white mica (hydromuscovite), consistent with the late-stage overprinting of alteration assemblages by this mineral. Hedenquist and Browne (1989) propose that this paragenesis is a consequence of mixing of a boiling high-temperature chloride solution with steam heated ground waters. Marginal influx of these lower-pH, $CO_2$-rich hybrid solutions results in the observed paragenesis of late-stage white mica (see Fig. 8 below).

*Rotokakwa geothermal system.* Drill holes in the Rotokawa geothermal system transect rhyolites and andesites allowing comparison of epidote parageneses in these two rock types. Epidote and clinozoisite are more common in the andesites according to Krupp and Seward (1987). These authors report alteration assemblages of phenocrysts and of the volcanic groundmass in rhyolitic and andesitic rocks. Andesine in the andesites is replaced by epidote, adularia, albite, calcite, and illite. The same assemblage is found in the replacement of andesine in rhyolites with the addition of wairakite and anhydrite. Mafic minerals including hypersthene and biotite are replaced by epidote, chlorite, anatase, and titanite in rhyolites, and in the andesites this assemblage also includes calcite and illite. The groundmass in both rock types is altered to epidote, chlorite and quartz, but albite, adularia, and illite are included in the rhyolite assemblage, whereas clinozoisite, titanite and pyrite are also found in the andesite assemblage. Browne (1989) provides a summary of different alteration assemblages in rhyolitic and andesitic rocks in the New Zealand geothermal systems that is largely in accord with the observations of Krupp and Seward (1987).

*Broadlands-Ohaaki geothermal system.* Hydrothermal alteration of the Broadlands-Ohaaki geothermal system has been extensively studied. The hydrothermal minerals in the "deep" central upflow zone (>250°C and >600 m) consist primarily of quartz, K-feldspar, albite, chlorite, calcite and pyrite (Browne and Ellis 1970, Lonker et al. 1990; Hedenquist 1990; Simmons and Browne 2000). Epidote is rare due to the high concentrations of $CO_2$ (Browne and Ellis 1970), ranging from 0.3–0.75 mol/kg (Hedenquist and Stewart 1985; Hedenquist 1990; Simmons and Christenson 1994). However, there are several notable features of epidote paragenesis during the evolution of this geothermal system. Absar (1991) reports the occurrence of epidote in 24 of the 44 drill holes. It forms as a replacement of plagioclase and groundmass of the rhyolitic rocks, locally making up to 5% of the mineral mode, and occurring with albite, illite, calcite, chlorite, wairakite and quartz (Simmons and Browne 2000). In veins and vugs it occurs with calcite (in some cases calcite is deposited directly on epidote crystals, Browne and Ellis 1970; Simmons and Browne 2000) or with a paragenesis of illite, then epidote + K-feldspar, followed by chlorite, or illite then epidote + K-feldspar + illite (Lonker et al. 1990). Lonker et al. (1990) notes that there are no systematic variations of epidote composition within single crystals or with depth in the geothermal system, however, they do note an increase in the Mn content of epidote near the high-permeability fracture zones.

The depth of the epidote isograd is recorded in 24 drill holes in the Broadlands-Ohaaki geothermal system by Absar (1991). Assuming that epidote forms "generally" at temperatures >250°C and closer to 270°C (Browne 1978) Absar compares depth of the epidote isograd to the depth of present-day 270°C isotherm to evaluate regions of cooling within the geothermal system. In a similar kind of analysis Hedenquist (1990) compares primary fluid inclusion homogenization temperatures (mostly in hydrothermal quartz) with present-day measured downhole temperatures and demonstrates that the geothermal system has cooled 10–30°C in the south and about 10–20°C in the north. Although the results of Absar's (1991) and Hedenquist's (1990) studies differ in detail, they both demonstrate evidence of marginal cooling of the Broadlands-Ohaaki geothermal system, and illustrate the utility of epidote as an indicator of paleo-temperatures in active geothermal systems as noted by Reyes and Giggenbach (1999) for Wariakei. The marginal cooling indicated by these studies is considered to be due to the influx of $CO_2$-rich steam heated groundwaters formed by the boiling of upwelling near-neutral chloride hydrothermal solutions, and the descent and local mixing of the steam-heated solutions with the deep chloride waters (Hedenquist 1990; Lonker et al. 1990; Simmons and Christenson 1994; Simmons and Browne 2000). The present-day alkaline, near-neutral pH, chloride solutions are in equilibrium with K-mica and/or K-feldspar (Fig. 8). The high-$CO_2$ content precludes present-day formation of epidote.

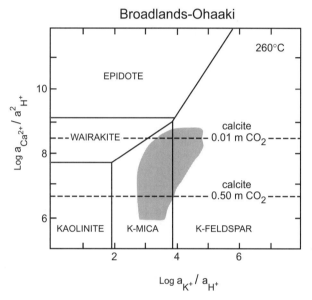

**Figure 8.** Phase relations among minerals and aqueous solutions in the system $K_2O$-$CaO$-$Al_2O_3$-$Fe_2O_3$-$SiO_2$-$H_2O$-$CO_2$ at 260°C, 47 bars, and quartz saturation as a function of cation to hydrogen ion activity ratios in the fluid. Dashed lines denote calcite saturation in terms of the molality of $CO_2$, and the shaded area represents the range in composition of alkaline, near-neutral pH, chloride solutions in the Broadlands-Ohaaki geothermal system. Modified after Figure 14 of Simmons and Christenson (1994). Source of thermodynamic data not cited.

*Ngatamariki geothermal system.* Browne et al. (1989) report an interesting example of epidote forming as scale inside drill casing and slotted liners of a well (NM2) in the Ngatamariki geothermal system. The drill hole is 2404 m deep, penetrating rhyolitic ignimbrite and minor andesites. It is located approximately 1 km from drill hole NM4, which encountered a young diorite intrusion surrounded by hornfels (Browne et al. 1992; Arehart et al. 2002). Well NM2 was cased to 888 m, slotted liners emplaced below this, and cuttings were not returned during drilling below a depth of 1350 m. Upon completion, the well was shut-in for 472 days before discharge, which produced several kilograms of ejectae consisting of pale-gray, spongy, euhedral, crystalline aggregates of wairakite, prehnite, and epidote, with minor quartz and pyrite. Approximately 3% of the ejectae (0.5–2 cm in diameter) have a smooth "cast" shape indicating that they formed within the slotted liner and in the drill casing. Browne et al. (1989) conclude that the ejectae were probably derived from a well depth corresponding to the permeable zone between 1580 and 1600 m. Cores from this interval are ignimbrite breccia. The breccia matrix is altered to quartz, albite, chlorite, illite, and pyrite, and numerous wairakite + quartz veins are found in the core. Andesine phenocrysts are replaced by adularia followed by illite, or by wairakite and epidote. Cavities formed from pumice clasts are partly filled with calc-silicates exhibiting a paragenesis of wairakite, wairakite + epidote, then epidote + prehnite. In the crystalline ejecta, wairakite was the first mineral to form followed by epidote and prehnite. The authors estimate the minimum growth rates for the minerals based on the time the well was shut-in, an approximate heating rate of the well, the temperature of the first occurrence of these minerals in New Zealand geothermal systems, and the size of the crystals. The growth rate for epidote parallel to the (100) face is 0.2 μm/day, for prehnite parallel to (110) the rate is 0.2–0.5 μm/day, and for wariakite 190 μm/day. They also provide a rough estimate of the minimum amount of hydrothermal fluid required to produce 5 kilograms of calc-silicates in the well liner based on the Al content of New Zealand geothermal fluids (alkaline near-neutral chloride solutions with 0.34 ppm Al; Ellis and Mahon 1977), and the mass of Al in the calc-silicates; a minimum of 3000 $m^3$ of solution passing through the slotted casing liner at 6.4 $m^3$/day.

**Central America**

*Miravalles geothermal system, Costa Rica.* This geothermal system is located on the south-west flank of the Miravalles Volcano, one of many strato-volcanoes forming Guanacaste Cordillera in north-western Costa Rica. Volcanic activity is related to subduction along the Middle American Trench of the Cocos Plate underneath the Caribbean Plate (Bowin 1976). The Nicoya Ophiolite Complex and overlying Cretaceous to Miocene marine sediments separate the Middle American Trench from the Guanacaste volcanic arc that consists of Tertiary andesite, including pyroclastics and ignimbrite, and Quaternary strato-volcanoes (Instituto Costariccense de Electricidad, unpublished reports cited by Milodowski et al. 1989). The active geothermal area, with some 50 $km^2$ of surface thermal activity, is confined to the 600 ka Caldera de Guayabo and related graben structures located to the southwest of the <50 ka Miravalles strato-volcano. Nine wells were drilled into the geothermal system as of 1989. Chemical and isotopic analyses of fluids from five wells (Giggenbach and Soto 1992), and analyses of steam and liquid discharge for a similar number of wells (Milodowski et al. 1989) are reported. Detailed well-temperature logs are not reported, but it appears that production fluids from the epidote-zone are in the vicinity of about 250°C or greater. Fluids produced from the deep wells are near-neutral pH, Na-Cl solutions, with approximately 7000 mg/kg of total dissolved solids and between 1000 and 2500 mg/kg of total $CO_2$. In general, the variety of hydrothermal fluids produced from wells and hot-springs are similar to those reported from geothermal systems in Panama (Bath and Williamson 1983) and in the Philippines (Giggenbach 1992; Reyes and Giggenbach 1992; Reyes et al. 1993), derived primarily from meteoric waters, mixing with magmatic fluids and experiencing boiling and remixing with steam-heated fluids. One well (PGM-2) has produced acid (pH 2.8) Na-Cl-$SO_4$ solutions at depths between 1600 and 2000 m, suggesting an active input of magmatic fluids in this system (Milodowski et al. 1989; Giggenbach and Soto 1992).

Milodowski et al. (1989) provide a detailed analysis of epidote parageneses, phase relations, and compositions in the deep wells of the Miravalles Geothermal System. Rochelle et al (1989) and Yardley et al. (1991) provide additional information on the vein mineralogy. The review presented below is based on their observations and analyses. Alteration has been divided into seven zones, not unlike alteration types described above for geothermal systems in Japan, Philippines and New Zealand. With increasing depth, and presumably temperature, the zones are referred to as (1) acid-silicic, (2) argillic, (3) chlorite-sericite-calcite, (4) chlorite-sericite-laumontite-epidote, (5) chlorite-sercite-wairakite-epidote, (5A) quartz-sulfide-anhydrite, (6) prehnite-epidote, (7) garnet-epidote-magnetite. Zone 5A corresponds to a region of deep acid alteration associated with the production of Na-Cl-$SO_4$ solutions (pH 2.8) at >1600 m in well PGM-2.

Alteration of the volcanic rocks and volcano-clastic sediments are similar in Zones 4 and 5, with the exception of the index Ca-zeolite phases, laumontite or wairakite. These zones are best developed in the northern portion of the geothermal field. Associated minerals include quartz, K-feldspar, albite, titanite, calcite, pyrite, anhydrite and local hematite. Trace minerals are chalcopyrite, galena, sphalerite, chalcocite, Cu-telluride, Ag, Au-telluride, native gold, anatase, barite, baritocelestite, xenotime, dolomite, bastnaesite, and Ba-plagioclase. In the lavas, vesicles are filled first with chlorite + titanite + quartz, followed by quartz + sericite, and finally epidote that locally replaces chlorite and titanite, with calcite being the final phase (zoned Mn-rich to Mn-poor). Plagioclase is dissolved and replaced, forming secondary porosity and subsequent complex mineral parageneses; first involving albitization along microcracks, dissolution of the plagioclase and formation of K-feldspar ± quartz ± sericite, followed by finely zoned epidote. Epidote typically has Fe-rich brecciated cores that are enclosed by Al-rich epidote, and finally a rim of Fe-rich epidote (oscillatory zoning is on the scale of about 1μm). Wairakite or laumontite with calcite are followed by anhydrite as the final replacement of plagioclase. The authors also describe hydrothermal mineral parageneses of the volcanic rock matrix involving chlorite, titanite, sericite, K-feldspar, epidote, and pyrite, and late-stage calcite veins.

Rock alteration associated with Zone 6 is distinguished from the chlorite + sericite + epidote + wairakite alteration described above (Zone 5) by the abundance of epidote + prehnite in veins and cavities. There is a inverse relationship of epidote and prehnite modes in the altered rocks, epidote predominates in the wall rock, and prehnite in veins and cavities. Vein paragenesis is complex. First, euhedral quartz + K-feldspar with extensive K-metasomatism of the wall rock, followed by prismatic or fibrous epidote, local replacement of K-feldspar, and massive infilling of the vein with prehnite accompanied by epidote + prehnite replacement of wall rock mineral assemblages, with epidote extending further away from the vein than prehnite. Further fracturing, with a second generation of quartz + K-feldspar, then prehnite replacing K-feldspar, locally followed by the sequence of quartz, then prehnite + quartz, and, finally anhydrite or calcite. Vein epidote and prehnite exhibit notable, fine scale zoning, oscillatory on a scale of about 1 μm. Such zoning is not observed in metasomatic wall rock assemblages marginal to the veins.

Epidote compositions range from $X_{ps}$ = 0.18–0.32, and prehnite from $X_{Fe}$ = 0.04–0.24. Zoning is described as "sympathetic" in both minerals and, in general, there is an Fe-rich core, overgrown by a more Al-rich phase, and finally Fe-rich rim. LREE-enrichment is observed in the early epidote replacing the matrix and in vesicle fillings. It is noted that the earlier Fe-rich zones are often "brecciated and corroded", being replaced by the later more Al-rich phase. Coexisting compositions of epidote and prehnite in vesicule-filling, metasomatic wall rock alteration, and veins are shown in Figure 9 for well PGM-11 at a depth of 974 m. Epidote and prehnite crystals exhibit both macro-scale and micro-scale zoning with Fe-rich cores and rims, and a mid-zone of the Al-rich phase. Compositions of coexisting epidote and prehnite solid solutions are compared with equilibrium isotherms (Fig. 9) for the exchange reaction

$$Ca_2FeAl_2Si_3O_{12}(OH) + Ca_2Al(AlSi_3O_{10})(OH)_2 \leftrightarrow Ca_2Al_3Si_3O_{12}(OH) + Ca_2Fe(AlSi_3O_{10})(OH)_2$$

   *Epidote*     *Prehnite*      *Clinozoisite*    *Fe-Prehnite*

as predicted by equations and data presented by Helgeson et al. (1978), Bird and Helgeson (1980, 1981), and Rose and Bird (1987). As noted by Milodowski et al. (1989), compositions of the paragenetically distinct types of epidote and prehnite (vesicule fill, metasomatic, and vein fill) each plot on or near specific isotherms, suggesting that the observed compositional zoning is developed under near isothermal conditions. Milodowski et al. (1989), Rochelle et al. (1989) and Yardley et al. (1991) conclude that the complex macro-scale and micro-scale zoning in epidote and prehnite is a consequence of rapid changes in fluid composition under near isothermal conditions, in accord with theoretical predictions presented by Arnason and Bird (1992) and Arnason et al. (1993). Temporal fluxations in the pH or $CO_2$ of the fluid phase appear to be controlling variables in the complex zoning of coexisting epidote and prehnite

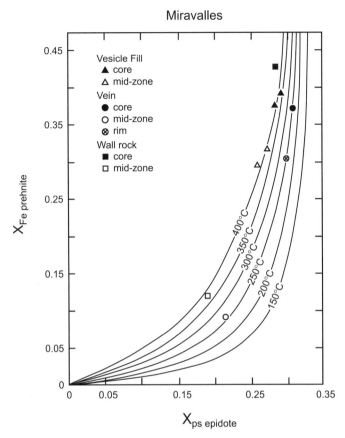

**Figure 9.** Compositions of coexisting epidote and prehnite from drill hole PGM 11 at 974 m in the Miravalles geothermal system. Isotherms computed from equations and data reported by Helgeson et al. (1978), Bird and Helgeson (1980) and Rose and Bird (1987). Modified from Figure 19 of Milodowski et al. (1989).

(Milodowski et al. 1989; Rochelle et al. 1989; Arnason et al.1993). It is important to note, with regard to the data shown in Figure 9, that non-equilibrium partitioning of Al and Fe between coexisting epidote and prehnite is common in many hydrothermal systems (see Fig. 6 of Rose and Bird 1987).

The highest grade of alteration was found in the deepest core of well PGM-11 (1453 m) that has extensive epidote replacement of the volcanic host rock accompanied by andradite-grossular, magnetite, and rare plagioclase replacing earlier chlorite, titanite, albite, and quartz alteration. Epidote replaces chlorite and igneous Ti-magnetites, together with grandite garnet and euhedral hydrothermal Ti-magnetite. Garnets are euhedral, ranging in size from 10–50 μm and range in composition form 60–65 mol% andradite, with a slightly Fe-enriched core. Epidotes are extensively zoned on both a macro- and micro-scale, ranging in composition from $X_{ps}$ = 0.18–0.48, typically with Fe-rich cores. In some cases low-Fe epidote replaces euhedral grandite garnets. Small prismatic crystals of bytownite (77 mole percent anorthite) are associated with the grandite garnet and magnetite assemblages.

***Los Azufres geothermal system, Mexico.*** This geothermal system is one of many thermal areas located within the Trans-Mexican Volcanic belt in central Mexico. At Los Azufres volcanic activity began about 18 Ma with eruption of up to 3 km³ of andesitic lavas and related pyroclastic deposits with minor basalts, overlying Late Mesozoic to Oligocene metamorphic and sedimentary rocks (Dobson and Mahood 1985; Ferrari et al. 1991). Following eruption of silicic volcanism (between 1.0 and 0.15 Ma) formed a broad caldera structure and related extrusives and domes of rhyolite, rhyodacite and dacites (Dobson and Mahood 1985; Ferrari et al. 1991). High-angle normal faults appear to have a controlling effect on ascending geothermal fluids (Cathelineau et al. 1985; Torres-Alvarado 2002), illustrated by two thermal domes in the north-south cross section (Fig. 10A; Cathelineau et al. 1985). Geothermal fluids are near neutral, Na-Cl rich, primarily of meteoric origin, with total chloride contents between 2000–4000 mg/kg (Torres-Alvarado 2002). Carbon dioxide represents >90 wt% of the noncondensable gases (Santoyo et al. 1991). The northern portion of the geothermal system produces a mixture of vapor and liquids at 300–320°C, and in the southern portion of the field, vapor dominates over the liquid phase and temperatures are typically 260–280°C (Torres-Alvarado 2002). Fluid inclusion analyses from calcite, quartz and epidote indicate temperatures close to present-day drill hole measurements, with evidence for local cooling (30–40°C), and the presence of an early saline (2–7 wt % NaCl) hydrothermal solution at depth (Cathelineau and Nieva 1986; Cathelineau et al. 1989).

Bulk rock and mineralogic alteration of andesite in the Los Azufres geothermal system (Cathelineau and Nieva 1985; Cathelineau et al. 1985, 1991; Izquierdo et al. 1995; Torres-

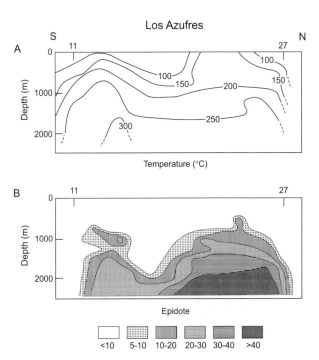

**Figure 10.** Present-day isotherms (A) and the distribution of epidote (B) in a north-south cross section of a portion of the Los Azufres geothermal system. Numbers assigned to the patterns represent an arbitrary scale for the relative modal abundance of epidote as denoted by Cathelineau et al. (1985). Surface locations of drill holes 11 and 27 are shown for geographic reference. Modified from Figure 5 of Cathelineau et al. (1985).

Alvarado 1998, 2002) allow identification of three zones (Fig. 11A). In general the alteration is similar to many other geothermal systems in the circum-Pacific margin discussed above. The distribution and relative modal abundance of epidote (Fig. 10B) is illustrated in a cross section, where it can be seen that epidote first appears between 210 and 250°C (Cathelineau et al. 1985). Torres-Alvarado (2002) notes that epidote is more abundant in the northern portion of the field (see Fig. 10B) because of higher temperatures (not apparent in the cross section of Fig. 10A), and the lower vapor content in the fluids relative to the southern portion of the geothermal field. The first appearance of epidote occurs as anhedral fine-grained aggregates, and at higher temperatures, epidote forms idomorphic, tabular, radiating and fibrous masses in vesicules and veins. Epidote forms together with chlorite and hematite in pseudomorphs

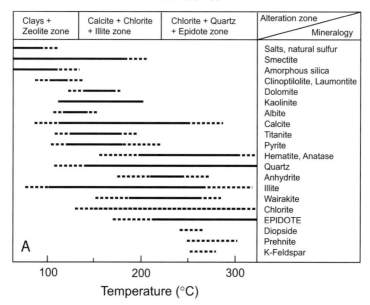

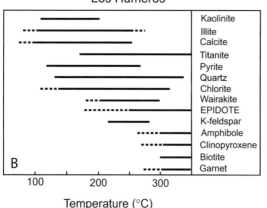

**Figure 11.** (A) Present-day temperature distribution of hydrothermal minerals and mineral zones in the Los Azufres geothermal system. Modified after Figure 2 of Cathelineau et al. (1985). (B) Fluid inclusion homogenization temperature distribution of hydrothermal minerals in the Los Humeros geothermal system. Modified after Figure 5 of Martinez-Serrano (2002).

after igneous pyroxene and amphiboles, and with sericite, calcite, albite, adularia, and chlorite after igneous feldspar (Torres-Alvarado 2002). Epidote with $X_{ps}$ up to 0.40 have been reported (Torres-Alvarado 2002). Hydrothermal pyroxenes and amphiboles have a localized distribution in the epidote zone (Cathelineau et al. 1985), and andradite-rich garnet is found only in the highest-temperature portions of the system (Torres-Alvarado 2002). Thermodynamic analysis of solution-mineral equilibria suggest that the present-day geothermal fluids are in equilibrium with mineral assemblages of the epidote zone (Torres-Alvarado 1998; 2002), in accord with stable isotope equilibrium between these fluids and calcite (Torres-Alvarado et al. 1995).

***Los Humeros geothermal system, Mexico.*** This geothermal system is located within a resurgent caldera complex in the eastern portion of the Plio-Pleistocene Trans-Mexican Volcanic Belt. The caldera complex formed about 500,000 years BP and volcanic activity continued until about 20,000 years BP (Ferriz and Mahood 1984). Drill holes up to 2.2 km deep have penetrated andesites, dacites, rhyolitic tuffs, rhyolites and minor basalts that have experienced varying degrees of hydrothermal alteration at temperatures <350°C. Regional basement of the volcanic sequence is composed of Paleozoic metamorphic and intrusive rocks, overlain by Cretaceous limestones. The geothermal reservoir is centered about fault-controlled grabens and horsts within the caldera complex, and geophysical data suggests the presence of young intrusions at depths of about 0.5 km (Arredondo 1987).

The geothermal system is characterized by complex temporal variations in mineral parageneses and fluid chemistry. These variations are recorded in several stages of hydrothermal alteration observed in drill cores and cuttings; in fluid inclusion homogenization temperatures and alteration mineralogy that indicate regions of subsurface cooling; and in fluid chemistry of production fluids that are variable throughout the field and have changed during the period of geothermal exploitation (Prol-Ledesma and Browne 1989; Martinez-Serrano et al. 1996; Prol-Ledesma 1998; Martinez-Serrano 2002). Geothermal fluids are dilute (Na and Cl <500 ppm, and Cl varies from 10–500 ppm), contain excess steam, and high concentrations of $CO_2$ (0.05–1.7 m). The deep production fluids are chloride solutions, but the shallow production wells discharge mixtures of chloride and bicarbonate waters indicative of boiling and mixing of fluids within the hydrothermal system. These fluids are largely meteoric waters that have reacted with the basement rocks before entering the hydrothermal systems, and volatile species (S and $CO_2$) have both magmatic and sedimentary components (Martinez-Serrano et al. 1996; Prol-Ledesma 1998; Martinez-Serrano 2002). In general, the above studies conclude that water-rock equilibrium has not been attained in this geothermal system.

Hydrothermal alteration is described in detail by Prol-Ledesma and Browne (1989), Martinez-Serrano et al. (1996), Prol-Ledesma (1998), Martinez-Serrano and Dubois (1998) and Martinez-Serrano (2002). Martinez-Serrano (2002) identifies three alteration zones with increasing depth: 1. Argillic zone characterized by kaolinite, montmorillonite, chlorite and zeolites at depths typically <400 m; 2. Propylitic zone containing epidote, amphibole, chlorite, calcite montmorillonite-illite, sulfides and iron-oxides at depths ranging from about 400–500 to 1800 m; and 3. Skarn zone identified by garnet, wollastonite and pyroxene in the basement limestones at depths >1800 m. The propylitic zone is further divided (Prol-Ledesma 1998) into a low-permeability upper zone of epidote, chlorite, montmorillonite-illite and hematite-pyrite (400–500 to 1000 m), and a high-permeability lower zone of epidote, wairakite, chlorite, amphibole, pyroxene, illite, and pyrrhotite-pyrite (about 1200–1800 m). Mineralogic modes of epidote in the lower zone locally exceed 15% (Martinez-Serrano 2002). Several stages of hydrothermal alteration have been identified in the above cited studies; early alteration minerals include formation of epidote, amphibole, pyroxene, and chlorite decreasing the overall permeability, followed by fracturing and later formation of zeolites, montmorillonite-illite, wairakite. In the skarn zone early garnet is replaced by wollastonite and pyroxene. In addition to paragenetic relations, local cooling is also indicated by the shallow occurrence of epidote and garnet at measured down-hole temperatures (120° and 150°C, respectively) that are well below their common formation temperatures in geothermal environments (Prol-Ledesma 1998). Figure 11B illustrates the distribution of common hydrothermal minerals as a function of fluid inclusion homogenization temperatures (primarily in quartz and calcite). Dashed lines in Figure 11B denote temperatures where specific minerals (i.e., epidote, amphibole, clinopyroxene, garnet) are considered by Martinez-Serrano (2002) to be "outside" there usual temperature range in geothermal environments; presumably quartz and/or calcite in these samples are paragenetically later phases.

Epidote occurs in veins and cavities, and as replacement of plagioclase and the volcanic groundmass. At depths between 1000 and about 1400 m epidote forms as fine-grained radiating acicular or anhedral crystals, but at depths >1500 m it occurs as prismatic crystals up to 1 mm in size (Martinez-Serrano 2002). More than 180 electron microprobe analyses of epidote between depths of 1100 and 2440 m where conducted by Martinez-Serrano (2002), compositions range from $X_{ps}$ = 0.13–0.34, and there is no obvious compositional gap within this range of Al-Fe substitution. Based on homogeneous optical properties of the analyzed epidote crystals, it is concluded that variation in $X_{ps}$ of individual crystals is <0.02, however, significant compositional ranges are reported between epidotes grains within single samples. Epidote compositions do not correlate with temperature, permeability, or to the varying volcanic lithology. However, it is noted that in the Calapso region of the geothermal system $X_{ps}$ varies from 0.18–0.33 where fluid inclusion homogenization temperatures are around 250°C, and in the Xalapazco region $X_{ps}$ ranges from 0.13–0.26 and fluid inclusion temperatures are >270°C.

## MEDITERRANEAN REGION GEOTHERMAL SYSTEMS: LARDERELLO

Of the many geothermal areas within the Mediterranean Region we focus our review on the Larderello geothermal system in Tuscany, Italy. This long active geothermal area is located in the Northern Apennines, formed by continental collision between the Corsica-Sardinia and Adria microplates. Subsequent extension and crustal thinning in the Late Miocene produced anatectic melts and the emplacement of granitic batholiths (Gianelli and Puxeddu 1994; Franceschini 1995; Boccaletti et al. 1997). Geophysical evidence suggests a partially molten batholith is still present below the Larderello geothermal area (Foley et al. 1992), and drill holes (to depths of 4.5 km) have encountered temperatures in excess of 400°C (Gianelli and Ruggieri 2002). Stratigraphy is varied and complex, consisting of (from top to bottom) Neogene sediments, allochthonous ophiolite and flysch sequences of Jurassic-Eocene age, the Tuscan Nappe of Upper Triassic to Oligo-Miocene siliciclastic, carbonate and evaporitic sequences, a complex of tectonic slices that include portions of the Tuscan Nappe and the underlying Hercynian metamorphic basement complex. The latter consists of phyllite, micaschist, amphibolite, gneiss, and marble (Gianelli et al. 1978; Bagnoli et al. 1979; Pandeli et al. 1994). Granites, pegmatites, and aplites intrude the metamorphic basement, as evident from several deep drill holes (Del Moro et al. 1982; Villa and Puxeddu 1994; Gianelli and Laurenzi 2001). The granite intrusions are considered to be of a regional extent and of several km in thickness (Gianelli et al. 1997a; Manzella et al. 1995, 1998). Granite intrusions and related contact metamorphic minerals range in age from 3.8–1.0 Ma (Del Moro et al. 1982; Villa and Puxeddu 1994; Villa et al. 1997; 2002). The 3–4 Ma longevity of the geothermal area is considered a consequence of continuous magma input from a volatile-rich upper mantle (Gianelli and Puxeddu 1994; Gianelli and Laurenzi 2001).

Contact metamorphism related to the granite emplacement overprints earlier regional metamorphism related to the Alpine orogeny and to Hercynian or pre-Hercynian events (Batini et al. 1985). Contact metamorphic minerals formed at depths between 1.5 and 4.5 km and include varying assemblages of biotite, andalusite, cordierite, tourmaline, corundum, forsterite, wollastonite, periclase, clinopyroxene, hornblende and feldspar (Cavarretta et al. 1983, 1986; Cavarretta and Puxeddu 1990; Franceschini 1995, 1998; Gianellli and Ruggieri 2002). Fluid inclusion experiments and stable isotope analyses of silicates suggest that contact metamorphism occurred between 425 and 670°C, under lithostatic pressures, in the presence of several generations of fluids including, magmatic Li-Na-Cl-rich brines (up to 50–60 wt% total dissolved solids) and metamorphic $H_2O$-$CO_2$-$CH_4$-$N_2$-rich vapors and liquids (Valori et al. 1992a; Iacumin et al. 1992; Cathelineau et al. 1994, 1995; Petrucci et al. 1994; Ruggieri and Gianelli 1995).

The present-day geothermal system has an upper steam production zone, predominantly within lithologies of the Tuscan Nappe, and locally significant fluid production zones in the deeper metamorphic basement (Batini et al. 1985; Ruggieri et al. 1999; Gianelli and Ruggieri 2002). There have been extensive studies of hydrothermal alteration at Larderello (Marinelli 1969; Cavarretta et al. 1980a, 1980b, 1982, 1983; D'Amore et al. 1983; D'Amore and Gianelli 1984; Bertini et al. 1985; Ruggieri et al. 1999). Gianelli and Ruggieri (2002) categorize hydrothermal alteration in the deeper geothermal reservoir of the metamorphic basement into two basic types: sericite, chlorite, and quartz alteration of contact metamorphic cordierite-bearing assemblages, and a pervasive propylitic alteration consisting of varying proportions of vein epidote, chlorite, quartz, calcite, K-feldspar, titanite, actinolite, anhydrite, albite, and pyrite. The latter is considered by Cavarretta et al. (1982) to have formed from near neutral-pH solutions with partial pressures of $CO_2$ between 0.2 and 8 MPa, in equilibrium with oxygen and sulfur fugacities at 200–400°C of the present-day reservoir (D'Amore and Gianelli 1984). However, it should be noted that fluid inclusion analyses indicate a diverse nature of hydrothermal solutions during the evolution of this system, including moderate to low salinity $H_2O$-NaCl-$CO_2$ liquids, high salinity $H_2O$-NaCl-$CaCl_2$ fluids, high-salinity $H_2O$-NaCl solutions and low-density $H_2O$-$CO_2$ vapors produced by boiling, and nearly pure $H_2O$ condensates (Belkin et al. 1985; Valori et al. 1992a; Gianelli et al. 1997b; Ruggieri et al. 1999; Ruggieri and Gianelli 1999). These fluids are considered to be largely of meteoric origin, and are modified by water-rock interaction (with evaporates, carbonates and silicates), boiling and mixing, including components of magmatic fluids and perhaps metamorphic fluids (Petrucci et al. 1993; D'Amore and Bolognesi 1994).

The geothermal system evolved to the present condition where the shallow reservoir produces fluids that are 20–60 wt% steam (D'Amore and Celati 1983), conditions interpreted by Bertini et al. (1985) as integrated values of steam from major fractures, and liquid from micro-fractures where the hydrothermal mineralization occurs (cf. Truesdell and White 1973). Marinelli (1969), Cavarretta et al. (1980b) and Ruggieri et al. (1999) suggest that the observed hydrothermal alteration reflects an older liquid dominated or two phase (boiling) geothermal system. In general, fluid inclusion analyses and the retrograde alteration of contact mineral assemblages suggest an overall cooling of the geothermal system.

The distribution of hydrothermal minerals (Fig. 12) is a function of temperature. Cavarretta et al. (1982) identify three groups of hydrothermal assemblages. Calcite, anhydrite, chlorite, pyrite and quartz are characteristic of the shallowest zone (Group A, 150–250°C). Group B includes the same minerals, in addition to K-feldspar, K-bearing mica, hematite, titanite, wairakite, and datolite (200–300°C). Group C is characterized by epidote, chlorite, titanite, together with local prehnite, datolite, clinopyroxene, actinolite, albite, and minor calcite, K-bearing mica, and anhydrite (250–350°C). Prehnite is paragenetically later than epidote (Ruggieri et al. 1999). The depth to the epidote isograd is near 1000 m (Fig. 13), between 230 and 290°C (Bertini et al. 1985). Isograds for K-feldspar are at shallower depth, for actinolite and albite below the epidote isograd. The chlorite-out isograd (Ruggieri et al. 1999) is near 2.2–2.5 km. Typically the minimum temperature for the epidote isograd is 270°C (Fig. 13; Petrucci et al. 1994), and closely parallel to the upper contact of the metamorphic

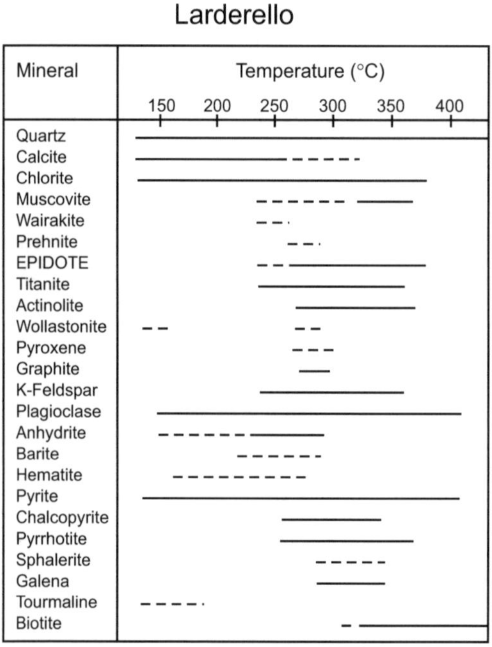

Larderello
Mineral
Temperature (°C)
150 200 250 300 350 400
Quartz
Calcite
Chlorite
Muscovite
Wairakite
Prehnite
EPIDOTE
Titanite
Actinolite
Wollastonite
Pyroxene
Graphite
K-Feldspar
Plagioclase
Anhydrite
Barite
Hematite
Pyrite
Chalcopyrite
Pyrrhotite
Sphalerite
Galena
Tourmaline
Biotite

**Figure 12.** Temperature distribution of hydrothermal minerals in the Larderello geothermal system. Modified from Figure 3 of Cavarretta et al. (1982) and Figure 2 of Petrucci et al. (1994).

Larderello
S
N
SP2
C2B
S22
Ser
Depth (m)
0
1000
2000
3000
m
Kfs
250°C
EPIDOTE
EPIDOTE
Act
Act
300°C
Ab
350°C
400°C
Chl-X
Chl-X
Chl-X
350°C
0 1 2 km

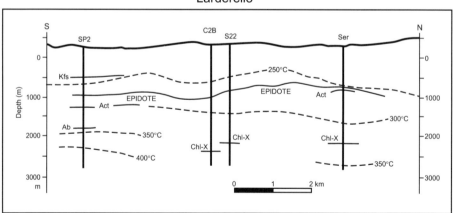

**Figure 13.** North-south cross section of a portion of the Larderello geothermal system showing isotherms (stippled lines) and isograds (solid lines) for K-feldspar (Kfs), epidote (EPIDOTE), amphibole (Act) and albite (Ab; after Bertini et al. 1985) and the deepest occurrence of chlorite (Chl-X; after Ruggieri et al. 1999). Thick solid vertical lines represent drill holes (Sp2, San Pompeo 2; C2B, Capannoli 2B; S22, Sasso 22; Ser, Serrazano). Modified after Figure 2 of Ruggieri et al. (1999).

basement (phyllite and quartzite). Local andradite garnet-, wollastonite- and clinopyroxene-assemblages are considered to be relic phases of a previous higher temperature hydrothermal stage (Cavarretta et al. 1982; 1983).

Cavarretta et al. (1982) report that epidote is very abundant in the Group C assemblages, occurring as xenoblastic to idioblastic yellow-green crystals up to 0.8 mm long, with weak pleochroism, and prismatic or acicular habits. Radiating fan-shaped aggregates are common. Epidosite type replacement of the host rock, up to several centimeters thick, is reported in two wells (Sasso 22, 1600 m; San Pompeo, 2219 m). Epidote is zoned, commonly with iron-rich rims (Cavarretta et al. 1980a; 1982). With the exception of two samples (Profundo, 1503 m and Sasso 22, 1984 m) the compositions of the cores of epidote crystals span a narrow range of $X_{ps} = 0.18–0.22$, however the rims of these crystals are more variable ranging from $X_{ps} = 0.23–0.37$. No obvious trend in epidote composition is apparent with depth or temperature.

Thermodynamic analysis of phase relations (Cavarretta et al. 1982; D'Amore and Gianelli 1984; Gianelli and Calore 1996) suggest that observed epidote assemblages in the Group C alteration are likely mineralogic buffers for fugacities of oxygen and sulfur, and for the partial pressure of carbon dioxide in the geothermal reservoir. Temperature of measured gas compositions of production fluids and predicted fugacities for equilibrium among epidote, chlorite, quartz, actinolite and pyrite (Fig. 14; D'Amore and Gianelli 1984) are strongly correlated. Despite the uncertainties in the thermodynamic data of Fe-chlorite, and solid solution mixing approximations for chlorite and actinolite as discussed by the authors, the close correlation between measured and predicted oxygen and sulfur fugacities is remarkable. In a similar kind of analysis, Cavarretta et al. (1982) and Gianelli and Calore (1996) show that the near constant partial pressures of $CO_2$ observed in the geothermal production wells over time is likely due to decarbonation reactions in the metamorphic basement that are buffered by epidote assemblages including: 1. epidote, calcite, quartz and prehnite; and 2. muscovite, calcite quartz, epidote and K-feldspar. A close correlation between observed partial pressures of $CO_2$ and values calculated from mineral equilibria is obtained for epidote with $X_{ps} = 0.275$ and 0.2 in the first assemblage, and for epidote ($X_{ps} = 0.275$) and of muscovite between $X_{muscovite} = 0.3–1.0$ in the second assemblage.

Epidote, not in the Group C assemblages discussed above, is reported by Cavarretta et al. (1983) and Cavarretta and Puxeddu (1990) from the San Pompeo 2 well. At depths between 2200 and 2900 m, where present-day temperatures are >400°C, quartzites and phyllites of the metamorphic basement complex contain extensive anastomosing veins of biotite, tourmaline, calcic plagioclase, epidote and K-feldspar. Several generations of mineralization are recognized; first, tourmaline, biotite and minor apatite; second, actinolite, tourmaline, biotite, plagioclase, and minor K-feldspar, quartz, epidote (locally resorbed textures), titanite and apatite; and third, fibrous actinolite, K-feldspar, epidote, tourmaline, biotite, titanite and ilmenite. In samples from 2389 m there are two types of plagioclase in the assemblages with $An_1$ and $An_{14}$, but in the sample from 2580 m there is one, with $An_{45}$. This alteration sequence records early contact metamorphic and later hydrothermal history of the system, involving varying components of magmatic, metamorphic, and meteoric waters over a temperature range of 425–670°C, temperatures that are 100–200°C greater than the highest measured downhole temperatures (Cavarretta et al. 1983; Cavarretta and Puxeddu 1990; Cathelineau et al. 1994).

## GEOTHERMAL SYSTEMS IN RIFTING ENVIRONMENTS

### Iceland

There are more than twenty high-temperature (>200°C at <1 km depth) geothermal areas in Iceland; all are located within Iceland's active volcanic zone, a landward extension

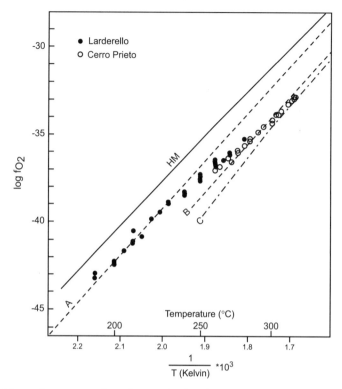

**Figure 14.** Measured temperature dependence of the fugacity of oxygen in the Larderello (Italy, solid symbols) and Cerro Prieto (Mexico, open symbols) geothermal systems reported by D'Amore and Gianelli (1984). The dashes lines labeled A and B represent calculated equilibrium of epidote ($X_{ps}$ = 0.275 and 0.20, respectively), chlorite ($X_{Fe^{2+}} = X_{Mg^{2+}} = 0.5$), quartz and actinolite (using averaged compositions). The dot-dashed line labeled C represents calculated equilibrium for the assemblage epidote ($X_{ps}$ = 0.125), chlorite ($X_{Fe^{2+}} = X_{Mg^{2+}} = 0.5$), and Ca-clinopyroxene ($X_{Fe^{2+}} = X_{Mg^{2+}} = 0.5$). HM denotes the hematite-magnetite buffer. Modified from Figure 1 of D'Amore and Gianelli (1984). Thermodynamic data are from Helgeson et al. (1978) and Bird and Helgeson (1980), site mixing models were employed for chlorite, amphibole and pyroxene solid solutions, and the thermodynamic data for chlorite end members were modified from values reported by Helgeson et al. (1978), see D'Amore and Gianelli (1984) for details.

of the Mid-Atlantic Ridge in Iceland that has been modified by interaction with the Iceland mantle plume (Bödvarsson 1961; Palmason et al. 1979; Fridleifsson 1979, 1991; Tryggvason et al. 1983; Steinthórsson et al. 1987; Arnórsson 1995; Gudmundsson 2000). Most of the high-temperature systems host water of meteoric origin, but near coastal regions geothermal fluids are modified seawater or mixed seawater and meteoric water (Björnsson et al. 1972; Arnórsson 1978, 1985; 1995; Darling and Ármannsson 1989; Sveinbjörnsdóttir et al. 1986; 1995). The highest permeability zones in the geothermal systems are associated with near vertical dikes and faults (Ármannsson et al. 1987; Arnórsson 1995). We restrict our review of epidote parageneses to four geothermal systems located within the active volcanic zone, each characterized by distinct hydrothermal fluids: 1. Nesjavellir, a meteoric water hydrothermal system with temperatures up to >380°C; 2. Krafla, a meteoric water dominated system that has experienced recent dike intrusions and influxes of magmatic gases, locally producing reservoir fluids with pH's as low as 2 and temperatures up to ≈340°C; 3. Reykjanes, a seawater-dominated geothermal system with temperatures up to ≈320°C; and 4. Svartsengi, a system with mixed seawater-meteoric water hydrothermal fluids with temperatures up to 240°C. Finally, we mention the occurrence of epidote in two fossil geothermal systems in eastern Iceland, the Geitafell central volcano and basaltic crust penetrated by the Iceland Research Drilling Project.

*Nesjavellir geothermal system.* This geothermal area located on the northern flanks of the Hengill central volcano, but within the fissure swarm associated with the volcanic complex. The field is underlain by a sequence of 2000 a to <1 Ma old hyaloclastites and interbedded olivine tholeiitic or tholeiitic lavas (Saemundsson 1967; Hardardóttir 1983; Franzson et al. 1986; Fridleifsson 1991). Dikes and sills make up <5 vol% of the upper 600 m, and increase in abundance downward to >50 vol% below 1500 m. Permeability of the system ranges from 1–50 millidarcies (Bödvarsson et al. 1990). The highest subsurface temperature recorded in Iceland is from well 11 in Nesjavellir (>380°C). The temperature distribution and geologic structures within the geothermal field suggest that the heat source lies beneath the well-field where ascending boiling water forms an elongate two phase reservoir over the heat source that is encroached upon by invading cold ground water. Geothermal fluids are of meteoric origin, and characterized by total dissolved solids of <2000 ppm (Arnórsson 1995). Commonly silica is the most abundant dissolved solid, sodium the most abundant cation, and chloride and sulfate the major anions. Dissolved gases ($CO_2$, $H_2S$, and $H_2$) may be present in concentrations exceeding those of any dissolved solid, especially in the hottest waters (Arnórsson 1995). The pH of reservoir fluids ranges from 6.7–7.25, values typical of dilute (<500 ppm total chloride) meteoric-hydrothermal solutions in Iceland (Fig. 15A; Stefánsson and Arnórsson 2002).

Based on mineral parageneses in vesicules and veins, together with age determinations of the Hengill central volcano and intrusive rocks, Franzson (2000) proposes a time sequence for

## Iceland Geothermal Systems

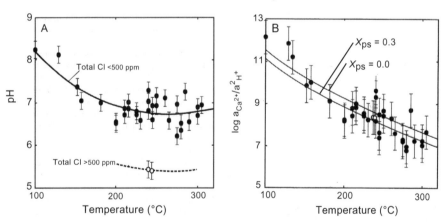

**Figure 15.** Temperature dependence of (A) pH and (B) the activity ratio of $Ca^{2+}$ to $(H^+)^2$ in hydrothermal fluids from Iceland geothermal systems (Stefánsson and Arnórsson 2002). Solution pH's and activity ratio of $Ca^{2+}$ to $(H^+)^2$ are computed from analyzed fluids and aquifer temperatures using the WATCH program (Arnórsson et al. 1982; uncertainties correspond to ± 0.2 pH units). Solid symbols denote dilute geothermal fluids (Total Cl < 500 ppm, including fluids from Nesjavellir and Krafla geothermal systems), and the open symbols denote saline geothermal fluids (Total Cl > 500 ppm) from Svartsengi and Reykjanes geothermal systems. The lines in diagram A denote pH-temperature functions proposed by Stefánsson and Arnórsson (2002). Lines in diagram B are computed for equilibrium among prehnite, epidote and quartz for values of $X_{ps}$ of 0.0 and 0.3. See Stefánsson and Arnórsson (2002) for summary of thermodynamic properties employed in the calculations. Modified after Figures 1 and 2 of Stefánsson and Arnórsson (2002).

the evolution of the Nesjavellir geothermal system. The high-temperature geothermal system developed less than 100,000 years ago with dioritic intrusions. Progressive heating along NE-SW trending faults and eruptive fissures led to maximum temperatures about 5000 years ago. Much of the system is cooling, as reflected with late-stage calcite deposition, except for the southern portion of the system which is undergoing heating.

Hydrothermal mineralogy at Nesjavellir (Franzson et al. 1986; Franzon 1988, 1998, 2000; Fridleifsson 1991; Hreggvidsdóttir 1987; Steingrímsson et al. 1986a, 1986b, 1990; Schiffman and Fridleifsson 1991; Larsson et al. 2002) display systematic depth zonation ranging from zeolite to amphibolite facies metamorphism. The degree of basalt alteration increases near high permeability structures such as brecciated-vesiculated flow tops and fracture zones. The most common hydrothermal minerals and their range in temperature can be summarized as follows (Arnórsson 1995): calcite and pyrite occur over a wide temperature range; chalcedony <200°C, at higher temperatures quartz is formed; below about 100°C zeolites including chabazite, thompsonite, analcime, scolecite, stilbite, heulandite, clinoptilolite, and mordenite form systematic zones with depth, but laumontite (>120°C) and wairakite (>200°C) occur at higher temperatures; at shallow depths smectite-type clay minerals are abundant, at about 200°C they are replaced by regularly or randomly interstratified smectite-chlorite; chlorite forms at 230–240°C; prehnite appears at >200°C, epidote at slightly higher temperature (but becomes abundant at >360°C); salite, actinolite and garnet at >280°C; actinoliltic-hornblende and hornblende at >350°C. Albite is the most common hydrothermal feldspar, but andesine and oligoclase also occur in the deeper portions of the system. Titanite, pyrrhotite, chalcopyrite, sphalerite, and apatite are accessory minerals. Figure 16 illustrates the depth dependence of secondary minerals in well NJ-15.

Hreggvidsdóttir (1987) identifies eight distinct secondary mineral assemblages in drill hole NJ-11 at depths between 1000 and 2265 m and temperatures between 260° and 380°C, five of which contain epidote, and two of the epidote assemblages include hornblende. With increasing temperature and depth the observed assemblages are: 1. Chlorite, titanite, albite, oligoclase, andesine, salite, Fe-Ti oxide; 2. Chlorite, titanite, epidote, albite, Fe-Ti oxide; 3. Chlorite, titanite, epidote, prehnite, albite, K-feldspar ($Or_{90-95}$), actinolite, Fe-Ti-oxide; 4. Chlorite, titanite, epidote, albite, actinolite, actinolitic hornblende, Fe-Ti oxide; 5. Chlorite, titanite, epidote, oligoclase, andesine, K-feldspar ($Or_{79-80}$), actinolite, actinolitic hornblende, hornblende; 6. Titanite, oligoclase, andesine, actinolite, actinolitic hornblende, hornblende, salite; 7. Quartz, talc, Fe-Ti oxide; and 8. Quartz, epidote, albite, oligoclase, andesine, K-feldspar ($Or_{90-93}$), actinolite, actinolitic hornblende, hornblende, Fe-Ti oxide. All these assemblages occur in intrusive or extrusive basaltic rocks with the exception of assemblage 8 which is found in felsic intrusions. Epidote occurs as both replacement of plagioclase and matrix material and as an abundant vein fill phase together with prehnite, chlorite, quartz, zeolites, calcite, and amphibole. Hreggvidsdóttir (1987) reports epidote compositions from $X_{ps} = 0.10$–$0.41$. There is no apparent trend in epidote composition with temperature in these samples. Calcic amphiboles with $^{IV}Al$ contents up to 1.2 moles per formula unit (based on 23 oxygens) occur with epidote in assemblages 5 and 8. As shown by Hreggvidsdóttir (1987) these amphiboles are complex intergrowths of actinolite and hornblende lamellae (<10 μm in width). Franzson (2000) notes the abundance of epidote together with chlorite, wollastonite, garnet, quartz, amphibole and sulfides in the Nesjavellir geothermal system. The andradite-rich garnet is considered to be associated with contact metamorphism related to dike emplacement (Fridleifsson 1991).

***Krafla geothermal system.*** This geothermal system is located within the caldera of the Krafla central volcano in Northern Iceland. The caldera formed about $10^5$ years ago and volcanic activity in the Krafla region is ongoing (Saemundsson 1974, 1991). Since 1975 magma has intruded into chambers 3–8 km below the geothermal field, and the most recent

# Nesjavellir

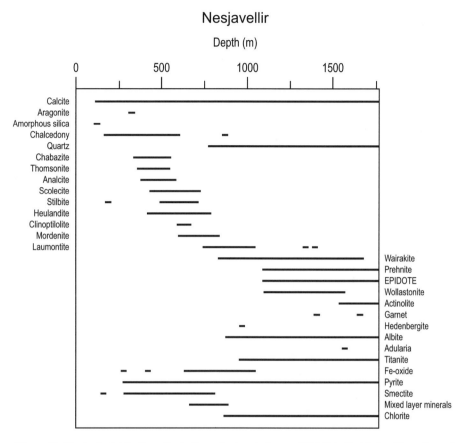

**Figure 16.** The depth distribution of hydrothermal minerals from well NJ-15 in the Nesjavellir geothermal system. Modified after Figure 13 of Arnórsson (1995).

subaerial eruption was in 1984 (Björnsson et al. 1977; Einarsson 1978; Ármannsson et al. 1987; Arnórsson 1995). Permeability in the Krafla system is low (2 millidarcies: Bödvarsson et al. 1984), possibly due to the abundance of igneous intrusions at depth. Basalt and dolerite intrusions dominate the stratigraphy below 1200–1300 m depth and gabbro occurs below 1800 m in some areas. Minor silicic intrusions are associated with the gabbro and there are "rhyolite ridges" near the eastern and western parts of the caldera rim (Ármannsson et al. 1987). The temperature distribution is complex, and in the Leirbotnar area there are two reservoirs separated by an aquaclude, with the cooler upper reservoir fed by the lower boiling reservoir, but also recharged by colder groundwater flowing in from the north (Stefánsson 1981; Ármannsson et al. 1987).

There are >34 wells at Krafla with maximum temperatures of approximately 350°C. Geothermal fluids are dilute (TDS <1500 pm) and of meteoric origin (Sveinbjörnsdóttir et al. 1986; Arnórsson 1995; Gudmundsson and Arnórsson 2002). Although most of the deep hydrothermal solutions are near-neutral pH, several of the wells have discharged fluids contaminated by magmatic vapors, containing anomalous concentrations of $CO_2$, HCl and $SO_2$, resulting in solution pH's as low as 2 (Ármannsson et al. 1982, 1987, 1989; Arnórsson 1995). Kristmannsdóttir (1979) and Ármannsson et al. (1987) have defined five secondary mineral zones, similar to those described above for Nesjavellir geothermal systems: smectite-zeolite, near the top of the system; mixed-layer chlorite-smectite, from 100–150 m; chlorite, from 150–250 m; chlorite-epidote, from 250–450 m; and epidote-amphibolite, below 450 m depth. Pyrite and pyrrhotite are abundant in zones of boiling (Steinthórsson and Sveinbjörnsdóttir 1981); calcite has overprinted alteration patterns in the Leirbotnar area.

Epidote is found as a trace mineral at temperatures as low as 200°C, but is an abundant phase, up to 15–20% of the mineral mode as replacement of plagioclase and in veins, at temperatures >260°C (Kristmannsdóttir 1975, 1979). Sveinbjörnsdóttir (1992) reports that epidote forms xenoblastic crystals up to 0.75 mm, yellow in color and pleochroic from yellow to colorless, and as aggregates of columnar form with crystals up to 150 μm in length. Compositions range from $X_{ps}$ = 0.18–0.32 and there are no obvious trends with depth and temperature. It is noted that there is a correlation among the Al content of epidote and that of $^{IV}$Al in coexisting chlorite in the Krafla and Reykjanes geothermal systems.

The concentrations of $CO_2$, $H_2S$ and $H_2$ in geothermal reservoir fluids in Iceland (including Krafla) appear to be controlled by equilibria with mineral buffers that include epidote. Arnórsson and Gunnlaugsson (1985) and Stefánsson and Arnórsson (2002) demonstrate, by comparison of gas analyses with thermodynamic analysis of mineral-gas equilibria, that $CO_2$ concentrations at temperatures >230°C are in equilibrium with the assemblage epidote + prehnite + calcite + quartz, and for dilute meteoric hydrothermal solutions (<500 ppm chloride, as in Krafla and Nesjavellir), $H_2$ and $H_2S$ concentrations are buffered by the assemblage epidote + prehnite + pyrite + pyrrhotite. Ármannsson et al. (1982) suggests that modally abundant epidote, formed by reaction of basalts with near-neutral pH meteoric hydrothermal solutions, provides a neutralizing buffer for the acid-$SO_2$-rich magmatic emanations that are periodically released from the high-level magma chamber (3–5 km deep) at Krafla. Based on analyses of gases from drill hole production fluids it appears that magmatic $SO_2$ is removed from the ascending fluids by formation of pyrite and pyrrhotite during reaction with the hydrothermally altered basalts. Ármannsson et al. (1982) proposed a reaction scheme involving epidote hydrolysis and reduction of ferric iron, leading to the formation of secondary iron sulfides and $H_2S$, consistent with the temporal variations of gases produced from drill holes.

***Reykjanes and Svartsengi geothermal systems.*** These geothermal systems are located on the southwest tip of the Reykjanes peninsula, where the active volcanic zone of Iceland merges with the Mid-Atlantic. Dominant structures in the area are northeast-southwest-trending faults and shallow grabens, although north-south and northwest-southeast-trending faults and fractures become more common with depth (Franzson 1990). The stratigraphy consists mainly of hyaloclastites, breccias, tuffaceous sediments and basaltic flows (including pillow lavas; Tómasson and Kristmannsdóttir 1972; Franzson 1983). Ages of extrusive rocks range from the Stampar fissure eruption of 1226 BC to approximately 0.1–0.5 Ma (Franzson 1983; Franzson et al. 2002). Igneous intrusions are common below 2500 m at Reykjanes (Kristmannsdóttir 1983), but are abundant below about 800 m at Svartsengi (Franzson 1983). At Reykjanes, temperature follows the boiling point curve from 400 m to 1000 m depth, but is approximately constant at greater depths (maximum temperature about 320°C at 2 km), consistent with good vertical permeability and the scarcity of shallow intrusions (Arnórsson 1995). Subsurface temperatures at Svartsengi are nearly uniform at 230–240°C below 600 m depth, indicative of high structural permeability (Björnsson and Steingrímsson 1991), however fluid inclusion measurements on hydrothermal minerals suggests that the system is locally cooling (Franzson 1990).

The source of geothermal fluids in the Reykjanes system is largely seawater, and at Svartsengi a mixture of about two-thirds seawater and one-third meteoric water; both fluids have been modified chemically by reactions with the basaltic host rocks (Olafsson and Riley 1978; Arnórsson 1978; Arnórsson et al. 1983; Sveinbjörnsdóttir et al. 1986). Compared to seawater the geothermal fluids at Reykjanes have 90% less F, lower pH, less Mg, $SO_4$ and Na, more Ca, Fe, Si, and K and slightly less Cl. The Svartsengi geothermal fluids are enriched in K, Ca, B, Fe, Si, $CO_2$ and $H_2S$, and depleted in Na, Mg, Al, F and $SO_4$ relative to the ideal mixture of seawater and meteoric water (Ragnarsdóttir et al. 1984). Reservoir fluids from both geothermal systems have more calcium and a lower pH (generally between 5 and 6) than the dilute meteoric water geothermal systems at Nesjavellir and Krafla (see Fig. 15A; Arnórsson 1978; Arnórsson et al. 1978, 1983). There is mineralogical and isotopic evidence for repeated incursions of seawater, mixing of seawater with the geothermal brine or cold meteoric water, and seawater-basalt reaction (Olafsson and Riley 1978; Sveinbjörnsdóttir et al. 1986; Lonker et al. 1993).

Hydrothermal alteration at Reykjanes and Svartsengi are similar in many respects. With increasing depth and temperature both exhibit mineral zones of smectite-zeolite, mixed layer clay, and chlorite-epidote (Tómasson and Kristmannsdóttir 1972; Franzson 1983; Franzson et al. 2002). The top of the chlorite-epidote zone occurs around 600–700 m depth at ≈260°C in Reykjanes, and 230–240°C in Svartsengi; epidote is abundant below these depths with modes up to 15% in Reykjanes (Tómasson and Kristmannsdóttir 1972) and 19% in Svartsengi (Ragnarsdóttir et al. 1984). Epidote is most abundant in high-permeability zones. Secondary vein and replacement minerals in the epidote zone of both geothermal systems include chlorite, garnet, albite, K-feldspar, calcite, titanite, hematite, and sulfides including pyrite, chalcopyrite, and rare sphalerite (Lonker et al. 1993). Alteration at Reykjanes is distinguished from that at Svartsengi by wollastonite and prehnite, and a zone of epidote-amphibole that occurs at >1100 m depth and >285°C (Tómasson and Kristmannsdóttir 1972; Lonker et al. 1993; Franzson 1983; Franzson et al. 2002). Mineral parageneses in the Reykjanes system (Fig. 17) are interpreted by Franzson et al. (2002) to indicate that the system is young and progressively heating, presumably due to the lack of retrograde mineralization. In contrast, mineral parageneses at Svartengi (Franzson 1983; Lonker et al. 1993) indicate several generations of epidote mineralization followed by varying stages of calcite, chlorite and zeolite formation, indicative of local cooling or fluid mixing events.

Epidote ($X_{ps}$ = 0.19–0.44) typically occurs as yellowish green, slightly pleochroic, xenoblastic crystals (<0.8 mm) and as radiating crystal aggregates (25–250 μm) replacing plagioclase and filling vesicules and veins (Tómasson and Kristmannsdóttir 1972; Sveinbjörnsdóttir 1992; Ragnarsdóttir et al. 1984; Lonker et al. 1993). There are no obvious

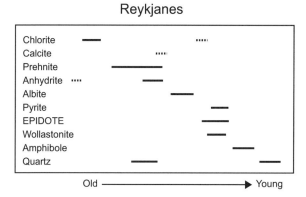

**Figure 17.** Paragenetic sequence of hydrothermal mineral formation at depths between 600 and 2000 m in well RN-10 of the Reykjanes geothermal system. The dotted lines denote probable paragenetic relations. Modified after Figure 4 of Franzson et al. (2002).

compositional trends with depth or temperature. Epidote replacing plagioclase is often zoned with Al-rich rims, and epidote with hematite is zoned with Fe-rich rims (Lonker et al. 1993). Strontium concentrations in epidote from Reykjanes and the Nesjavellir (Fig. 18) range from 100–1500 ppm, values that are up to an order of magnitude larger than in unaltered basalt and several orders of magnitude greater than in the fluid from Reykjanes. Textural and paragenetic relations suggest a complex history of mineral formation, dissolution and local phase disequilibrium in Reykjanes and Svartsengi geothermal systems. Albite replacing plagioclase is commonly corroded and partially replaced by epidote (Franzson et al. 2002; Lonker et al. 1993), a replacement probably related to an increase in temperature (Rose et al. 1992; Rose and Bird 1994). In addition, epidote (as well as calcite) crystals display irregular and embayed contacts with chlorite that mantles epidote, indicating a stage of epidote dissolution. In many cases chlorite, epidote and calcite parageneses are ambiguous, apparently reflecting multiple generations of secondary mineralization and replacement. Based on thermodynamic calculations and geologic observations of basalt-seawater interaction (Reed 1982, 1983; Mottl 1983), Lonker et al. (1993) suggest that dissolution of epidote and calcite followed by formation of chlorite is due to local seawater incursion into the geothermal system and seawater-basalt interaction at high water to rock ratios. Lonker et al. (1993) also demonstrates that most compositions of coexisting epidote and garnet do not exhibit equilibrium partitioning of $Fe^{3+}$ and $Al^{3+}$ as represented by equations and data presented by Bird and Helgeson (1980).

Although there is convincing petrologic evidence of complex mineral parageneses and local mineral disequilibrium in Reykjanes and Svartsengi geothermal systems, thermodynamic analysis of phase relations among geothermal solutions and epidote-bearing mineral assemblages indicate a close approach to equilibrium. Ragnarsdóttir et al. (1984) demonstrate local equilibrium between the major components of the geothermal fluid at Svartsengi and secondary phases epidote, chlorite, albite, quartz, and calcite. Calculations presented by Reed and Spycher (1984) predict that the present-day reservoir fluids at Reykjanes (drill hole RN8) are saturated with respect to epidote, chlorite, albite, K-feldspar, illite and pyrite. Analysis

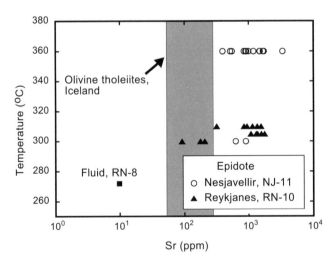

**Figure 18.** Strontium concentration in epidote from Nesjavellir (well NJ-11) and Reykjanes (well RN-10) geothermal systems (data from SHRIMP-RG, Stanford University, present study), and geothermal fluid (well RN-8, Reykjanes; Elderfield and Greaves 1981) as a function of measured drill hole temperature at the sampling depth. Range of Sr content in Iceland olivine tholeiite is shown for comparison (Hemond et al. 1993).

of coexisting minerals and solution-mineral equilibrium presented for the Reykjanes and Svartsengi systems (Lonker et al. 1993) indicate that: 1. Epidote and Na-rich feldspar are in equilibrium with the present-day fluids, but garnet and prehnite are metastable, garnet requiring higher temperatures than the present-day to obtain equilibrium; and 2. Predicted initial fluid discharge concentrations of $H_2S$ and $H_2$ are close to equilibrium with the observed assemblage epidote + chlorite + pyrite and epidote + chlorite + hematite.

Finally, we note that calculations presented by Arnórsson (1999) demonstrate that the deep geothermal reservoir fluids of Iceland geothermal systems (including Nesjavillir, Krafla, Reykjanes and Svartsengi) are saturated, or close to saturation with respect to epidote ($X_{ps}$ = 0.3). In addition, values of the activity ratios of $Ca^{2+}$ to $(H^+)^2$, computed from aqueous speciation of geothermal reservoir fluids in Iceland closely correspond to values required for equilibrium of epidote ($X_{ps}$ = 0–0.3), prehnite, quartz, and solution (Fig. 15B; Stefánsson and Arnórsson 2002).

***Geitafell Central Volcano.*** Located in southeastern Iceland, this tholeiitic central volcano formed within the Iceland rift zone approximately 5–6 Ma, first by uplift during emplacement of gabbros, then caldera formation, and subsequent uplift during regional flexuring (Fridleifsson 1983). The area was deeply incised by glaciation during the past 3 Ma, exposing the fossil geothermal reservoir and the magmatic heat source. Hydrothermal alteration related to the intrusive and structural history of this system is described by Fridleifsson (1983, 1984) and Fridleifsson and Björnsson (1986). The authors identify twelve intrusive phases, including several episodes of gabbros in the central portion of the complex, plus a variety of doleritic, and lesser felsic, dikes and sills. Spatially and temporally related to these intrusions are five mineral zones (four of which include epidote: chlorite + epidote zone, epidote zone, garnet + epidote zone , and actinolite + epidote zone), and seven generations of vein assemblages (four of which contain epidote). Alteration sequence and mineral parageneses described are similar to the active geothermal systems in Iceland, however, exposures within this uplifted and eroded fossil geothermal system provide excellent examples of the relationship of multiple intrusions to vein formation and hydrothermal mineralization that is not evident in drill cores or cuttings from active geothermal systems.

***Iceland Deep Drilling Project.*** In 1978 a 1920 m deep research drill hole was completed in Eastern Iceland (Fridleifsson et al. 1982). The drill hole is located within the dike swarm that extends northwards from the Breiddalur central volcano, and is located just east of the Thingmuli central volcano. Stratigraphy consists of subaerial basaltic lava flows, volcanoclastics, and dikes (ages 8.9–10.9 Ma, Albertsson et al. 1982) that have experienced hydrothermal alteration by meteoric-water at temperatures as high as 300°C (at the base of the drill hole) due to the fossil geothermal system of the Thingmuli central volcano, and later contact metamorphism related to the Breiddalur dike swarm (Mehegan et al. 1982; Schmincke et al. 1982; Viereck et al. 1982). Alteration minerals and parageneses are similar to active geothermal systems in Iceland described above (Kristmannsdóttir 1982). Epidote first occurs at about 780 m depth in volcanoclastic rocks and 900 m depth in basalt flows; epidote becomes abundant at depths greater than 1200–1400 m, completely replacing the brecciated flow tops of the lavas near the bottom of the drill hole (yellowish-greenish crystals up to 7 mm in length together with quartz and chlorite; Mehegan et al. 1982; Viereck et al. 1982). At depths greater than 900 m, epidote + chlorite + quartz assemblages replace earlier low temperature alteration of clay minerals and zeolites, and are subsequently overprinted by late-stage calcite + laumontite + anhydrite formed during the cooling of the geothermal system (Mehegan et al. 1982; Viereck et al. 1982).

Epidote ranges in composition between $X_{ps}$ = 0.15–0.50, there is no obvious compositional trend with depth, and individual crystals are zoned with higher Fe and Ti cores, and Al-, Sr-, and Mn-rich rims (Exley 1982; Viereck et al. 1982). Zoning is classified as continuous, discontinuous (with two or more growth stages), and sector. Exley (1982) interprets variations of Fe, Al, Sr and Mn in the sector zones in terms of crystal structure and growth rates, suggesting that observed compositional trends supports the "protosite" model of Nakamura (1973). Mn content in epidote ranges form <0.02–0.62 wt% (Exley 1982), a range that overlaps with the Mn compositions of epidotes reported by Dickin et al. (1980) for low-Mn epidotes in altered mafic rocks and the high-Mn epidotes in altered felsic rocks from the Isle of Skye. Sr content in epidote ranges from below detection (0.016 wt %) to 0.90 wt%, with an average of 0.23 wt% (for 447 analyses; Exley 1982). Sr in epidote shows no obvious trend with depth, but Exley (1982) observed variations with respect to epidote parageneses in high permeability zones (veins and brecciated flow tops) relative to massive central portions of dikes and flows, suggesting that Sr uptake in epidote is a complex function of permeability, the abundance of epidote, and water to rock ratios. At 1180 m depth allanite occurs with epidote in the replacement of plagioclase phenocrysts, here the central portion of a radiating fan-shape prismatic crystal aggregate is allanite, while the outer portion is epidote (Viereck et al. 1982).

## Ethiopia

*Aluto-Langano geothermal system.* This geothermal system is located within the Quaternary Alto central volcanic complex in the Main Ethiopian Rift southeast of the Afar triple junction in East Africa (Di Paola 1972). Drill holes have encountered silicic units of the Alto central volcanic complex (0.155–0.002 Ma), overlying rift-lake basin sediments, Pliocene basalt (1.6 Ma) and a silicic unit termed the "Tertiary ignimbrite" (2.3 Ma; Gebregzabher 1986; Electroconsult 1986). Temperatures up to 335°C have been recorded at about 2 km depth. In cross section the present day thermal structure is defined by a narrow (>1 km wide) fault controlled upflow zone where temperatures range from 300–335°C. Marginal to the upflow zone, isotherms spread out laterally into the surrounding Pliocene basalt aquifer where maximum temperatures are 150–270°C (Teklemariam et al. 1996). Geothermal fluids produced from drill holes are dilute (0.1–0.3 wt% NaCl equivalent) alkali-bicarbonate-chloride solutions, with near-neutral to slightly alkaline pH, and partial pressures of $CO_2$ between 0.6 and 1.3 MPa in the central upflow zone and values up to 5.8 MPa in lateral outflow region (Mekuria et al. 1987; Gizaw 1989).

Fluid inclusion and mineralogic investigations of drill hole samples (Gebregzabher 1986; Valori et al. 1992b; Gianelli and Teklemariam 1993; Teklemariam et al. 1996) indicate a complex evolution of fluid chemistry and temperature in this geothermal system. In the central up flow region, below a zone of Ca-zeolites (mordenite, heulandite and laumontite), the predominate alteration phases are epidote, calcite, quartz and chlorite. Biotite and actinolite occur with epidote assemblages at >300°C in altered ophitic alkaline basalt, as does garnet. Epidote is found replacing plagioclase in basalts and as vein fillings at 250–332°C. Vein and cavity filling assemblages of epidote, hematite, quartz, chlorite, and calcite are common. Epidote, calcite, and quartz filled veins commonly have extensive metasomatic replacement of the adjacent basalts by epidote. Prehnite occurs in both vein and cavity filling assemblages, but is rare. Plagioclase is partially replaced by albite and epidote, and mafic minerals are replaced by hematite, titanite, epidote and quartz. Similar kinds of assemblages occur in the basalt aquifer marginal to the central up flow region (the out flow zone), where epidote is found in the temperature range of 90°–220°C. Here epidote and garnet are partially replaced by calcite and clay minerals, and calcite veins cross cut veins with epidote-bearing assemblages.

Thermodynamic analysis of epidote stability in the Aluto-Langano geothermal system (Valori et al. 1992b; Gianelli and Teklemariam 1993; Teklemariam et al. 1996) demonstrate that epidote and garnet assemblages are in equilibrium with present-day fluids of the deep up flow zone, but in the lateral out flow zones the fluids are in equilibrium with calcite and clays, consistent with paragenetic observations (Fig. 19). However, fluid inclusion experiments reported by these authors indicate that the geothermal fluids and temperature have evolved with time. Specifically, there has been slight heating in the central up flow zone, cooling of up to 171°C in the lateral out flow zones, a decrease in salinity (early fluids have 0.8–2.3 wt% NaCl equivalent), and an increase in $CO_2$ content with time. These changes are interpreted by Teklemariam et al. (1996) to be a consequence of an influx of early magmatic fluids and later cooling and dilution by solutions of meteoric origin.

**Tendaho geothermal system.** This geothermal system is located in a landward extension of the Red Sea rift system in the Afar region of Ethiopia (Abbate et al. 1995; Gresta et al. 1997). Details of water-rock interaction and geothermal fluid characteristics for four geothermal drill holes are given by Gianelli et al. (1998) and D'Amore et al. (1997). Drill holes transect a sequence of Quaternary lacustrine and alluvial sediments (4–9 ka; Semmel 1971) and Pliocene to Early Pleistocene basalt flows of the "Afar Stratoid Series" (1–4 Ma; Abbate et al. 1995). Recent volcanic activity of the rhyolitic Kurub volcano occurred between 4 and 10 ka. Subsurface temperatures between 221 and 270°C are encountered at depths between 500 and 2000 m depth. Production fluids are dilute near-neutral Na-Cl meteoric hydrothermal solutions

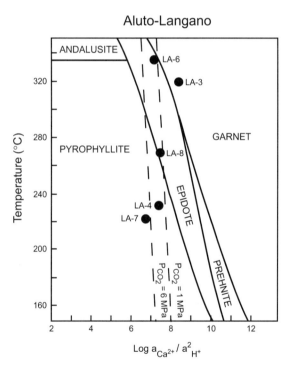

**Figure 19.** Phase relations in the system $CaO$-$Al_2O_3$-$Fe_2O_3$-$SiO_2$-$H_2O$-$CO_2$ as a function of temperature and the activity ratio of $Ca^{2+}$ to $(H^+)^2$ in aqueous solutions. Dashed lines represent calcite saturation for the specified partial pressures of $CO_2$. Solid symbols represent fluid compositions and temperatures from drill holes LA-3 and LA-6 in the central up flow zone, and drill holes LA-4, LA-7 and LA-8 within the lateral outflow zone of the Aluto-Langano geothermal field. Phase relations depict the stability of epidote ($X_{ps} = 0.3$) and iron-rich garnet (composition not specified). Modified from Figure 8 of Teklemariam et al. (1996). Phase relations are computed using thermodynamic data and the program SUPCRT92 reported by Johnson et al. (1992). Activity-composition relations for epidote, prehnite and garnet are from Bird and Helgeson (1980).

with total dissolved solids between 1620 and 2070 ppm, and isotopic compositions of C and He suggesting a magmatic origin of gas components (D'Amore et al. 1997).

Gianelli et al. (1998) identify calcite, wairakite, laumontite, garnet, epidote, clinopyroxene, amphibole, quartz and prehnite as common hydrothermal minerals in the geothermal reservoir. Epidote and garnet are iron-rich, with $X_{ps}$ of 0.27–0.36, and $X_{Fe}$ of 0.62–0.90, respectively. The mole fraction of diopside in clinopyroxene range from 0.27–0.70, and the mole fraction of $Fe^{3+}$ in prehnite is about 0.12. Quartz occurs with assemblages of calcite, wairakite, laumontite, epidote and prehnite. In some cases quartz forms before calcite. In the wells with the highest present-day temperatures (TD1, TD2 and TD4, Fig. 20) calcite, zeolites (wairakite or laumontite) and quartz form before epidote. Here calcite is found as relics within zeolites and textures indicate several stages of dissolution and precipitation of calcite during zeolite formation. Fluid inclusion analyses of quartz, calcite, and laumontite indicate a salinity between 0.15 and 0.20 wt% NaCl equivalent, similar to the present-day geothermal fluids. In addition, fluid inclusion analyses provide details of temperature evolution associated with

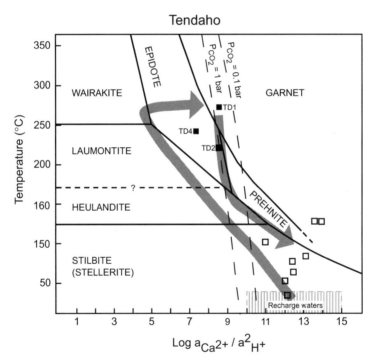

**Figure 20.** Phase relations in the system $CaO-Al_2O_3-Fe_2O_3-SiO_2-H_2O-CO_2$ as a function of temperature and the activity ratio of $Ca^{2+}$ to $(H^+)^2$ in aqueous solutions. Dashed lines represent calcite saturation for the specified partial pressures of $CO_2$. Solid squares marked TD1, TD2 and TD4 represent fluid compositions and temperatures in drill holes in the central upflow zone of the Tendaho geothermal system. Open squares denote fluids from hot springs and shallow temperature gradient drill holes, and the patterned area represents recharge waters to the geothermal system. Arrows denote hypothetical temperature-fluid compositions paths of recharge and discharge waters in the geothermal system. Phase relations depict the stability of epidote ($X_{ps}$ = 0.33) and iron-rich garnet (composition not specified). Modified from Figure 12 of Gianelli et al. (1998). Phase relations are computed using thermodynamic data and the program SUPCRT92 reported by Johnson et al. (1992), with the exception of heulandite and stilbite (data from Berman 1988). Activity-composition relations for epidote, prehnite and garnet are from Bird and Helgeson (1980).

observed mineral parageneses of the geothermal system. In drill hole TD1 at 1594 m, quartz and laumontite formed at 210–230°C, heating to about 270°C resulted in the replacement of laumontite and formation of wairakite, epidote and prehnite. At a depth of 2015 m in the same drill hole, early quartz and calcite formed at 180–240°C, followed by epidote, prehnite and calcite as temperatures increased to about 270°C. In drill hole TD2, calcite and wairakite formed at 210–240°C, followed by zeolite breakdown with formation of epidote, prehnite, clinopyroxene, garnet and amphibole as temperatures increased. In drill hole TD3, which appears to be marginal to the present-day upflow zone of the geothermal system, epidote, calcite, quartz, garnet, clinopyroxene and amphibole formed at temperatures <290°C, followed by cooling (up to 110°C) leading to the formation of laumontite. Comparative analysis of bulk rock compositions of altered and unaltered basalts, using the isocon method of Grant (1986), demonstrate that alteration forming epidote-bearing assemblages is accompanied by an increase in the components Ca, Fe, Mg and Al (due to calc-silicate formation) and a loss of Na, K, Si and Ti (due to glass and plagioclase dissolution).

A model of the thermal and chemical evolution of the geothermal system was proposed by Gianelli et al. (1998; Fig. 20). The diagram shows the stability of calc-silicate minerals as a function of temperature and activity ratio of $Ca^{2+}$ to $(H^+)^2$, as well as fluids produced from geothermal wells (TD1, TD2 and TD4), recharge waters from the local highlands, and fluids from shallow gradient holes and hot springs. Reaction of the recharge waters with sediments and basalt during heating produces zeolite zones of stilbite, heulandite, laumontite and wairakite, boiling with loss of $CO_2$, and an increase in solution pH moves the fluid into the epidote and epidote + garnet stability fields (arrow in Fig. 20). Fluid decompression and cooling occurs along the epidote + prehnite and later zeolite + prehnite phase boundaries producing fluids with the compositions observed in the shallow gradient wells and hot springs (solid squares in Fig. 20).

### Salton Trough, California and Baja California

Located at the northern end of the Gulf of California, the Salton Trough is a continental rift basin filled mostly with terrigeneous sediments of the Colorado River delta that are up to 6 km in thickness (Merriam and Brandy 1965; Muffler and Doe 1968 Fuis et al. 1982). Structurally the Gulf of California area is a zone of transition between the divergent oceanic plate boundary of the East Pacific Rise to the south and the continental transform plate boundary of the San Andreas fault to the north. It is characterized by short spreading ridges offset by en echelon right-lateral transform faults (Atwater 1970; Larson et al. 1972; Elders et al. 1972; Moore 1973). Of the many geothermal systems in the Salton Trough area we restrict our review of epidote parageneses to two well-studied geothermal systems; Cerro Prieto (Mexico) and Salton Sea (California). These two geothermal systems are characterized by: 1. geothermal reservoirs within terrigeneous sedimentary basins (Elders et al. 1972); 2. Maximum subsurface temperatures encountered by drill holes of >350°C at depths of approximately 2 km; 3. Recent volcanic domes extruded onto Quaternary sediments (rhyolite at the Salton Sea with xenoliths of basalt and gabbro, Robinson et al. 1976; dacite at Cerro Prieto, Reed 1984; Herzig 1990); and 3. Drill holes at Cerro Prieto and the Salton Sea, as well as in other geothermal systems of the Salton Trough, have intersected dikes and sills of diabase, microgabbro, basalt, andesite, dacite and rhyolite, all interpreted to have been derived from partial melting of mantle peridotite that has undergone fractionation and varying degrees of contamination by continental crust (Robinson et al 1976; Browne and Elders 1976; Keskinen and Sternfeld 1982; Herzig and Elders 1988; Herzig 1990). The two systems differ in terms of local hydrology and the amount of evaporates in sediments hosting the hydrothermal systems.

***Cerro Prieto geothermal system, Mexico.*** The upper 2–4 km are hosted by Pliocene to middle Pleistocene alluvial, deltaic, estuarine and shallow-marine sediments (divided into two units based largely on the degree of induration due to hydrothermal metamorphism) that overlie a granodioritic crystalline basement (Puente and De La Pena 1979; Vonder Haar and Howard 1981; Halfman et al. 1982). The thermal structure of the geothermal field in a N-S cross section (Fig. 21A), shows a symmetric central thermal dome with temperatures of 350°C at depths of about 2 km, and complex temperature inversions on the margins of the geothermal system. In contrast, a southwest to northeast profile through the geothermal field reveals an asymmetric distribution of isotherms, with a pronounced shallow thermal plume extending to the west (see for example, Fig. 4 of Elders et al. 1984). Consideration of the temperature distribution of the geothermal reservoir, together with stable isotope analyses of carbonate minerals and fluids, vitrinite reflectance geothermometry of kerogens, Na-K-Ca geothermometry of hydrothermal fluids, fission track annealing analyses of detrital apatite, and numerical analysis of heat and mass transfer, the following features of the geothermal system have been inferred: 1. Deep geothermal fluids originate in the eastern portion of the field, flowing westward and rising along faults, and mixing with cooler fluids in the western portion of the field; and 2. The geothermal system is young (<50 ka), heating occurred pencontemporaneously across the central portion of the geothermal system, and there is local cooling of 50–100°C in the northern portion of the system (Mercado 1976; Olson 1979; Truesdell et al. 1979; Manon et al. 1979; Elders et al. 1980; Barker and Elders 1981; Sanford and Elders 1981; Halfman et al. 1982; Williams and Elders 1984; Elders et al. 1984; Izquierdo et al. 2001).

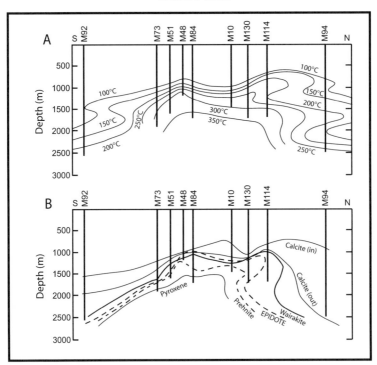

**Figure 21.** North-south cross section of a portion of the Cerro Prieto geothermal system showing the distribution of (A) isotherms, and (B) selected hydrothermal minerals. Isograds are shown for wairakite, epidote, prehnite and pyroxene. The lines marked Calcite (in) and Calcite (out) denote the area of abundant calcite mineralization. Modified after Figure 13 of Bird et al. (1984).

Production of the geothermal reservoir at Cerro Prieto, which began in 1973, has resulted in expanding zones of boiling in the aquifer and incursion of cooler dilute groundwater, precluding accurate analysis of the original geothermal fluid chemistry (Truesdell et al. 1981; Nehring and D'Amore 1984). Fluid analyses reported by Reed (1975) and Fausto et al. (1979) suggest that the geothermal reservoir fluids are Na-K-Cl brines with approximately 1.0 to1.5 wt% total dissolved solids and pH in the range of 5.2–5.5. Carbon dioxide is the dominant gas in the deep geothermal reservoir ($P_{CO2}$ of 5–10 bars, Nehring and D'Amore 1984). Based on stable isotope analyses, Cl content, and Cl/Br ratios, Truesdell et al. (1981, 1984) suggests that the geothermal brines originated from a mixture of partially evaporated Colorado River water and a saline brine of seawater origin that has been modified by interaction with the sediments of the Colorado River delta within the hydrothermal system. Hydrocarbons in the geothermal fluid are derived from metamorphism of organic matter in the sediments (Nehring and Fausto 1979; Des Marais et al. 1981), C and S are likely also derived by hydrothermal metamorphism of the sediments (Truesdell et al. 1979), but He originates from magmatic (mantle) sources (Welhan et al. 1979). Oxygen fugacities within the geothermal reservoir are considered to be buffered by $CO_2$ and carbon in the form of lignite in the sediments (Nehring and D'Amore 1984).

The Colorado River sediments within the geothermal reservoir consist of continentally derived sandstone, siltstone, and mudstones. Detrital mineralogy is composed primarily of quartz, and feldspar with subordinate amounts of calcite and dolomite and clay minerals illite, montmorillonite and kaolinite, as well as lithic fragments (chert and volcanics; Muffler and Doe 1968; Van de Kamp 1973). Systematic changes in the mineralogy and texture of the sediments are caused by interaction with the geothermal brine in the hydrothermal system (Reed 1975; Hoagland and Elders 1978). Based on petrologic analyses of cores and cuttings from approximately fifty drill holes, Elders et al. (1979, 1981; 1984) and Schiffman et al. (1984) have defined four alteration zones, and several subzones, with increasing temperature (Fig. 22). The mineral zones are with increasing depth and temperature: 1. Montmorillonite–

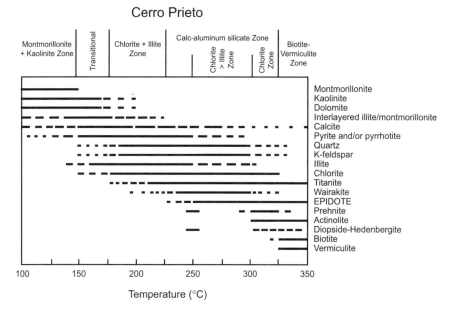

**Figure 22.** The distribution of hydrothermal mineral zones and individual minerals as a function of temperature in the Cerro Prieto geothermal system. Modified after Elders et al. (1984).

kaolinite zone (<150°–180°C); 2. Illite-chlorite zone (150°–180°C to 230°–250°C); 3. Calc-aluminum silicate zone (230°–250°C to >350°C); and 4. an overlapping biotite zone (315°–325°C to >350°C) where biotite is the predominant layer silicate which coexists with epidote, actinolite and clinopyroxene. Schiffman et al. (1985) divides the calc-aluminum silicate zone (Figs. 21B, 22) into subzones characterized by the index minerals wairakite, epidote, prehnite and clinopyroxene, and notes that the overall alteration sequence is characterized by temperature-telescoped dehydration and decarbonation reactions spanning the clay-carbonate, zeolite, greenschist and amphibolite facies of low-pressure metamorphism. The distribution of calcite and isograds for wairakite, epidote, prehnite and clinopyroxene are shown in the cross section (Fig. 21B), in general the isograds closely parallel the present-day isotherms. Garnet is rare in this geothermal system, with only one occurrence of grandite garnet (mole fraction of grossular $\approx$0.5) being reported together with biotite, actinolite, epidote and titanite in a sandstone matrix at 1581 m, 325°C, in well E2 (Schiffman et al. 1985).

Epidote is the most abundant calc-silicate in the Cerro Prieto geothermal system, locally making up to >20% of the mineral mode, and occurring within veins and as interstitial cement within the sediments at >230°C (Schiffman et al. 1985). At the lowest temperature occurrence of epidote, it forms fine-grained acicular crystals within carbonate pore-filling cement and is intergrown with fine-grained granular titanite. It appears to nucleate on the surface of detrital quartz grains, or on fine-grained oxide inclusions in the carbonate cement. Above 300°C, epidote is coarse grained (<0.5 mm maximum dimension) prismatic, with a subidioblastic texture of densely intergrown masses replacing geothermal carbonates and wairakite. Epidote is rarely porphyroblastic, and never poikiloblastic relative to the detrital minerals.

Epidotes exhibit a wide range in $Al^{3+}$-$Fe^{3+}$ substitution. Over 400 epidote analyses on a compositional frequency histogram show a normal distribution for $X_{ps}$ between 0.11 and 0.31, with no evidence of a compositional gap (Bird et al. 1984). In individual drill cuttings epidote compositions typically vary by less than 5 mole percent of the pistacite component, however analyses of epidotes in drill cutting recovered from the same depth will have a much larger range of composition. It appears that compositional homogeneity is spatially restricted to a scale of centimeters or less. Compositional variations in individual crystals is complex, with continuous, oscillatory and sector zoning. Vein epidote coexisting with K-feldspar, prehnite and quartz (1500 m depth, well E2) is zoned with Fe-rich cores and Al-rich rims, whereas epidote (with prehnite) replacing carbonate cement in the sandstone matrix is zoned with Fe-rich cores, Al-rich mantles, and Fe-rich outer rims (Arnason and Bird 1992). The average composition of epidote exhibits subtle, but systematic, changes with increasing depth, temperature, and coexisting mineralogy. The average composition of epidote coexisting with calcite and wairakite is typically $X_{ps} < 0.2$, and remains constant or decreases slightly with increasing depth to the prehnite isograd. The $Fe^{3+}$ content of epidote coexisting with prehnite increases with depth and temperature to values slightly greater than $X_{ps} = 0.2$. The most $Fe^{3+}$-rich epidotes are found at the biotite isograd ($X_{ps} \approx 0.24$) and with increasing temperature and depth the average value of $X_{ps}$ decreases.

Schiffman et al. (1985) noted that the system-wide correlation between present-day isotherms and calc-silicate isograds (Fig. 21), together with the inferred young age of the geothermal system, suggests that solution-mineral equilibrium has been attained on a regional scale. The authors note that partitioning of $Fe^{2+}$ an $Mg^{2+}$ between coexisting biotite and actinolite in the Cerro Prieto geothermal system are consistent with observations from many calc-silicate metamorphic rocks of greenschist and amphibolite facies. In addition, partitioning of $Al^{3+}$ and $Fe^{3+}$ between coexisting epidote and prehnite are close to theoretical and experimental equilibrium values (Liou et al. 1983; Bird et al. 1984; Rose and Bird 1987), as is partitioning of the same elements between epidote and grandite garnet (Schiffman et al. 1985). Calculated activity ratios of aqueous $Ca^{2+}$ to $Mg^{2+}$ required for equilibrium among epidote, actinolite and clinopyroxene are consistent with measured ratios of these cations in the geothermal fluids between 310° and 340°C, but similar ratios required for equilibrium among epidote, prehnite and Ca-clinopyroxene at 340°C are not (Bird et al. 1984; Schiffman et al. 1984).

D'Amore and Gianelli (1984) propose that the fugacities of $O_2$ and $H_2S$ in the geothermal reservoir at Cerro Prieto are buffered by epidote-bearing mineral assemblages. There is a close correlation between the measured fugacity of oxygen and calculated values for the assemblage epidote ($X_{ps}$ = 0.20) chlorite, actinolite and pyrite below 300°C (line B in Fig. 14), and for the assemblage epidote ($X_{ps}$ = 0.125), actinolite, Ca-clinopyroxene and pyrite at temperatures >300°C (line C in Fig. 14). Similar relations between $O_2$ fugacity and temperature for the assemblage epidote and clinopyroxene are reported by Bird et al. (1984), who noted a linear relation between log $f_{CO2}$ and the composition of epidote represented by $X_{ps}$.

***Salton Sea geothermal system, California.*** The Salton Sea geothermal system (SSGS) is located near the southeast shore of the Salton Sea in the Imperial Valley of California. Five Quaternary rhyolite domes (containing xenoliths of basalt and gabbro), and diabase dikes that have been sampled during drilling, provide evidence of a heat source at depth, apparently due to magmatic activity related to the landward extension of the Gulf of California rift system (Robinson et al. 1976; Herzig and Elders 1988). The SSGS is hosted by a >6000-meter-thick section of Miocene-Pleistocene deltaic sediments from the Colorado River, and lies within an isolated sedimentary basin, where the intermittent formation and evaporation of Lake Cahuilla in the Pleistocene formed nonmarine evaporite deposits (Merriam and Bandy 1965; Muffler and Doe 1968; Van De Kamp 1973; Babcock 1974; Fuis et al. 1982; McKibben et al. 1988b). The dissolution of evaporites probably contributed to the concentrated brine of the geothermal field (see below; McKibben et al. 1988b). Drill holes to depths of 3.2 km have identified (in the central portion of the geothermal field) an upper sequence of unconsolidated lacustrine clay-silt-gravel-evaporite (0.4 km thick), and a lower sequence of consolidated deltaic-lacustrine sandstone-siltstone-shale-evaporite, at depths >1 km the sediments are recrystallized to dense hornfels that are highly fractured (McDowell and Elders 1980, 1983; Herzig et al. 1988). Sandstone is predominately quartz-feldspar-lithic wackes, with calcite cement and clay matrix. Between 1934 and 1954, the SSGS was mined for $CO_2$ produced by decarbonation reactions forming chlorite and epidote at the expense of calcite and clays in the sediments (Muffler and White 1969). Beginning in the late fifties the SSGS was exploited as a source of geothermal energy, leading to the discovery of a reservoir of hypersaline, base metal-enriched brines, with up to 25 wt% total dissolved solids (White et al. 1963; Craig 1966; Skinner et al. 1967; Helgeson 1967, 1968).

The SSGS is distinguished from other sediment-hosted geothermal systems in the Salton Trough by: 1. High temperatures (365°C at 2.2 km depth); 2. High salinity geothermal brines (20–25 wt% total dissolved solid, mostly NaCl); 3. High metal content of the brine, notably Fe ($\approx$1500 ppm), Mn ($\approx$1000 ppm), Zn ($\approx$500 ppm), Li ($\approx$200 ppm), and Pb (100 ppm); and, 4. Active ore mineralization (Helgeson 1967, 1968; Skinner et al. 1967; McKibben and Elders 1985; McKibben et al. 1987, 1988a; Thompson and Fournier 1988; Charles et al. 1988; Elders et al. 1992). In the central part of the geothermal field, the 300°C isotherm lies within 1000 m of the surface, and it declines below 2000 m on the edges of the field, about 4 km away (Helgeson 1968). Fluid pressure in the system is approximately hydrostatic and reaches 275 bars at 3000 m (Helgeson 1968). The geothermal system consists of a hypersaline metal-rich brine (20–25 wt% total dissolved solids) overlain by lower salinity solutions (<10–13 wt% total dissolved solids and containing less that 100 ppm each of Fe, Mn and Zn; McKibben et al. 1987; 1988a; Williams and McKibben 1989). The interface between the two brines closely follows the 250°C isotherm, and is thus domal in shape. Geothermal brines are meteoric water of the Colorado River that have undergone variable near surface evaporation and interaction with sediments in the geothermal system (Williams and McKibben 1989). The brines acquired their high salinity primarily by dissolution of nonmarine evaporites (McKibben et al. 1987, 1988b; Williams and McKibben 1989), and the dissolved metals were largely derived by reaction with the detrital sediments (Doe et al. 1966; McKibben and Elders 1985; Shearer et al. 1988). Sedimentary anhydrite appears to be the main source of sulfur in the brine (McKibben and Eldridge 1989). Williams and McKibben (1989) propose that the geothermal system evolved first from non-marine basinal brines in a closed sedimentary basin, heating by a subvolcanic intrusions led to dissolution of evaporites in the section, and diapiric rise of the hypersaline geothermal brine to within a few hundreds of meters below the surface, with most of the ore-mineral genesis produced in zones of mixing near the brine interface (McKibben et al. 1987, 1988a).

Drill cuttings and core recovered from the River Ranch 1, Magmamax 2; Elmore 1, Sportsman, I.I.D. No. 1, and State 2-14 drill holes are the basis of a number of studies of hydrothermal metamorphic phase relations, mineralogy and ore genesis in the SSGS (e.g. Skinner et al. 1967; Keith et al. 1968; Muffler and White 1969; McDowell and Elders 1980, 1983; Bird and Helgeson 1981; Bird and Norton 1981; McKibben and Elders 1985; McDowell and Paces 1985; McDowell 1986; Caruso et al. 1988; Cho et al. 1988; McKibben et al. 1987; 1988a, 1988b; Shearer et al. 1988; Bird et al. 1988a; Charles et al. 1988; Enami et al. 1992, 1993). The delineation of four major metamorphic mineral zones (dolomite-ankerite zone at <190°C, chlorite-calcite zone between 190° and 325°C, biotite zone between 325° and 360°C, and garnet zone at >360°C) was first made by McDowell and Elders (1980) through study of drill cuttings from Well Elmore 1. The recovery of drill core from more than 3 km depth in well State 2-14 allowed Cho et al. (1988) and Shearer et al. (1988) to elaborate on the phase relations and geochemistry of these zones in sandstone and shale. In sandstones, the dominant mineral assemblage in the chlorite-calcite zone of the greenschist facies (190–325°C) in the SSGS is quartz + epidote + chlorite + calcite + K-feldspar + albite ± K-mica, with pyrite, titanite, and hematite and anhydrite as accessory phases. Abundant epidote first appears at approximately 300°C and is produced by decarbonation reactions such as 30 muscovite + 16 calcite + 18 quartz ⇔30 K-feldspar + 3 chlorite + 8 epidote + 14 $H_2O$ + 16 $CO_2$ (Cho et al. 1988). The biotite zone of the greenschist facies (325–350°C) is characterized by the presence of biotite + quartz + K-feldspar + albite + epidote + pyrite + actinolite + titanite, with minor muscovite, chlorite, anhydrite, and sphalerite. McDowell and Elders (1980) also observed a decrease in interlayer vacancies in phyllosilicates in the Elmore 1 well, as the pyrophyllite component of illite decreases and the K content of biotite increases with increasing metamorphic grade. Biotite-forming reactions can also produce epidote and $CO_2$ while consuming chlorite, e.g., 6 muscovite + 3 chlorite + 8 calcite ↔ 6 biotite + 4 epidote + 10 $H_2O$ + 8 $CO_2$ (Cho et al. 1988). K-feldspar modally decreases near the biotite isograd. At higher metamorphic grade (≥350°C), Cho et al. (1988) identified a clinopyroxene zone, where the presence of oligoclase and actinolitic hornblende, together with epidote, indicates the transition zone between greenschist and epidote-amphibolite facies metamorphism (Maruyama et al. 1983).

Caruso et al. (1988) examined epidote mineral assemblages in veins of drill core between 906 and 2955 m depth from the State 2-14 and found that phase relations as a function of depth in veins are, in general, similar to the phase relations observed in sandstone (Cho et al. 1988). Vein minerals include epidote, pyrite, calcite, quartz, K-feldspar, anhydrite, chlorite, hematite, sulfides, actinolite, titanite, trace allanite, and zircon. Paragenetic relations indicate that K-feldspar (when present) is the first vein mineral to form followed by epidote. The last minerals to form in the veins include pyrite, hematite, calcite quartz, sulfides and anhydrite. In epidote-bearing veins, calcite is restricted to less than 2000 m depth. K-feldspar occurs between 1700 and 2745 m, consistent with the production of biotite by the reaction of muscovite, quartz and calcite in the chlorite zone and the disappearance of K-feldspar in sandstone in the clinopyroxene zone below 3000 m. Vein anhydrite occurs in a restricted range (2195–2745 m) in the lower chlorite-calcite and upper biotite zones; anhydrite-bearing assemblages are characteristic of sandstones in the biotite zone (Cho et al. 1988). Actinolite first appears at

2890 m. Titanite occurs in anhedral masses replacing epidote; euhedral apatite is associated with vein epidote. Hydrothermal zircon occurs as small inclusions (5–15 μm) in vein epidote from the clinopyroxene zone but rare, euhedral hydrothermal zircon up to 50 μm in diameter, is also present in vein epidote (Charles et al. 1988; Spieler 2003).

During the December 1985 flow test of the ≈1867 m deep fluid production zone in the State 2-14 drill hole large masses (on the order of kilograms) of rounded rock samples were ejected from the well (Elders and Sass 1988). The rock samples are largely epidosite, composed of open-space filling, porous (up to 20% porosity) aggregates of acicular epidote (individual crystals up to 1.3 mm in length) together with varying amounts of quartz, K-feldspar ($Or_{90}Ab_{10}$), chlorite, pyrite, hematite, and traces of titanite and allanite (Charles et al. 1988; Bird et al. 1988a). Epidotes range in composition between $X_{ps}$ = 0.22–0.38, they are zoned with Al-rich cores and Fe-rich rims, and the most Fe-rich epidotes occur with hematite (Bird et al. 1988a). Drill core was not obtained from the epidosite zone, but cores above and below are indurated gray to black mudstone containing epidote, pyrite, K-feldspar, chlorite, quartz, and traces of apatite, titanite, zircon, and albite, and the modal abundance of epidote increases as the flow zone is approached from both above and below (Charles et al. 1988). Charles et al. (1988) notes that the epidosite ejectae consist primarily of two phase assemblages, epidote + pyrite and epidote + hematite, and that epidote, pyrite and hematite do not coexist in the same sample. The formation of epidosite in high permeability production zones noted here are similar to epidosite discussed earlier in the Ngatamariki geothermal system in New Zealand (Browne et al. 1989) and in Kokkonda geothermal system in Japan (Muramatsu and Doi 2000). Fluid analyses from the 1867 m flow zone are reported by Charles et al. (1988) and Thompson and Fournier (1988) for major elements and most metals, and by Michard (1989) for Sr and REE.

Epidote is the most abundant calc-silicate in the sediments of the SSGS, with modes locally exceeding 70% (Shearer et al. 1988). Compositions of epidote in the matrix of sandstone and in veins (Fig. 23) from the State 2-14 drill hole ranges from $X_{ps}$ = 0.11–0.42. Epidote is generally zoned with Al-rich cores and $Fe^{+3}$-rich rims, although oscillatory zoning is also common, and reverse zoning (Al-rich rims) has been noted in some epidotes from the sedimentary matrix (Caruso et al. 1988; Shearer et al. 1988; Cho et al. 1988; Arnason and Bird 1992). The most iron-rich epidotes in metasandstone occurs at about 900 m, and with increasing depth in the chlorite + calcite zone the average Fe content of epidote decreases with minima at about 1220 m and 2480 m, near the biotite isograd $X_{ps}$ in epidote increases from 0.15–0.29 at 2479 m to 0.29–0.35 at 2650 m. Vein epidote displays a range of $X_{ps}$ from 0.21–0.41, and is usually more iron rich than epidote in the sandstone and shale matrix (Fig. 23; Caruso et al. 1988; Shearer et al. 1988). With increasing depth the minimum value of $X_{ps}$ decreases from 0.33 at 906 m to 0.21 at 3179 m.

Bird et al. (1988a) report the octahedral distribution of $Fe^{3+}$ in the M(1) and M(3) sites of four epidotes from the State 2-14 drill hole samples based on $^{57}Fe$ Mössbauer spectroscopy. Ferric iron in the M(1) site ranges from 7.5–11.4% of the total iron. The most ordered epidote occurs in veins at 2618 m in the biotite zone, which corresponds to an equilibrium state of substitutional order/disorder of approximately 390°C (based on equations and data reported by Bird and Helgeson 1980), a value close to the downhole temperature of ≈340°C. In contrast, epidotes from veins at 1420 m (≈265°C), and from epidosite ejectae from the fluid production zone at 1867 m (≈300°C), both in the chlorite-calcite mineral zone, are more disordered, corresponding to calculated equilibrium states of order/disorder exceeding 450°C. The data suggest that epidote in the chlorite-calcite zone is in a metastable state of substitutional order/disorder with respect to the distribution of octahedral $Fe^{3+}$ and $Al^{3+}$. Patrier et al. (1991) also note metastable disordering in epidote of the fossil geothermal system of Saint Martin at <300°C (also see Liebscher 2004).

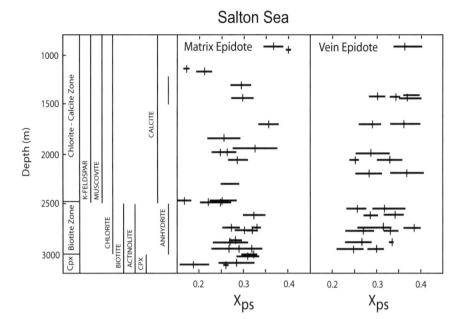

**Figure 23.** Depth distribution of hydrothermal K-feldspar, muscovite, calcite, anhydrite, chlorite, biotite, actinolite and clinopyroxene (CPX) within the chlorite-calcite, biotite and clinopyroxene (Cpx) mineral zones (Cho et al. 1988), and the composition of epidote in the matrix of metasandstones and in veins from drill hole State 2-14 in the Salton Sea geothermal system. Symbols denote the mean (vertical bar) and one standard deviation (horizontal bar) of epidote compositions analyzed in individual samples (Cho et al. 1988; Caruso et al. 1988; Bird et al. 1988a). Modified after Figure 1 of Bird et al. (1988a).

Trace elements, including Sr and the REE, have been analyzed in epidote from samples of the State 2-14 drill hole in the Salton Sea geothermal system by the authors (see Spieler 2003). Concentrations of Sr in epidote, sediments and geothermal fluids as a function of depth is shown in Figure 24. Concentrations of Sr in the sediments and mineral zones is generally <300 ppm at depths <2000 m, and increase with increasing depth as the biotite isograd is approached (Shearer et al. 1988). Strontium in the geothermal brine ($\approx$450 ppm) is similar to the maximum concentrations in the sediments. In contrast, epidote exhibits a wide range of Sr content (Fig. 24), ranging from values similar to bulk rock and fluid compositions to values of $\approx$6000 ppm. The epidosite at 1867 m fluid production zone has about 1800 ppm Sr. Individual crystals of epidote exhibit a wide range in Sr concentration, and no systematic relationships were observed with depth, mineral zones, or Fe content. Epidote also exhibits a wide range in REE concentrations (Fig. 25). The abundance of La ranges from 0.4–763 ppm and Yb ranges from 0.1–15.2 ppm, and La and Yb appear to vary independently of each other. Extreme variations in concentration and ratios of heavy to light REE is found within individual crystals, and no obvious correlations were apparent among REE distributions and temperature, mineral zones or $X_{ps}$ of epidote. Concentrations of REE in epidotes are three to six orders of magnitude larger than in the geothermal brine from the 1867 m flow zone in the State 2-14 drill hole reported by Michard (1989). For detailed discussion of the trace element geochemistry of epidote minerals see Frei et al. (2004).

Oxygen and hydrogen isotope analyses of four epidotes from the State 2-14 drill hole are reported by Bird et al. (1988a). Vein epidotes from 1867 m (epidosite), 2227 m and 2618 m

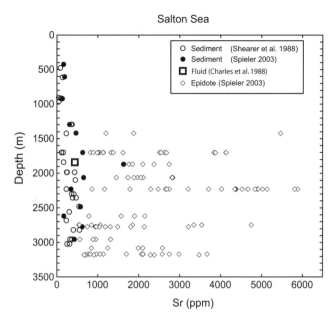

**Figure 24.** Concentrations of Sr in epidote, sediments, and geothermal fluids from the State 2-14 drill hole in the Salton Sea geothermal system as a function of depth. Data for epidotes are measured on the SHRIMP-RG, Stanford University, present study and Spieler (2003).

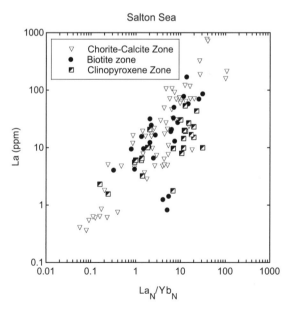

**Figure 25.** Concentrations of La (ppm) as a function of the chrondite normalized ratio of La to Yb in epidotes from the chlorite-calcite, biotite and clinopyroxene zones of the State 2-14 drill hole in the Salton Sea geothermal system. Analyses from SHRIMP-RG, Stanford University, present study and Spieler (2003). Concentrations of REE in chrondritic meteorites from Anders and Grevesse (1989).

have $\delta D$ of $-90$ to $-91‰$ (relative to SMOW) and $\delta^{18}O$ between 2.0 and 2.7‰. Epidote from 1420 has a $\delta D$ of $-96‰$. When compared to the stable isotope composition of the geothermal brine at 1867 m (C. J. Janik and A. H. Truesdell, personal communication 1987) the oxygen isotope fractionation between epidote and the geothermal brine is about zero at 300°C. Hydrogen isotope fractionation (represented by $1000 \ln \alpha_{\text{epidote-solution}}$) is equal to $-19‰$, a value comparable to experimental and geothermal hydrogen isotope fractionations between epidote and concentrated aqueous electrolyte solutions reported by Graham (1981) and Graham and Sheppard (1980). For detailed discussion of stable isotope systematics in epidote group minerals see Morrison (2004).

Thermodynamic analysis of phase relations among epidote-bearing mineral assemblages and the geothermal brines in the Salton Sea geothermal system are reported by Bird and Helgeson (1981), Bird and Norton (1981), McKibben and Elders (1985), Caruso et al. (1988), and Charles et al. (1988). In general, there is a consensus that the geothermal brines are in equilibrium with epidote-bearing assemblages at temperatures around 300°C, where comparative analyses between theoretical calculations based on compositions of coexisting minerals and compositions of production fluids are possible. Calculated fugacities of $CO_2$ consistent with the chlorite-calcite zone mineral assemblage epidote, K-mica, K-feldspar, calcite and quartz using mineral compositions from the Elmore 1 well samples (Bird and Helgeson 1981; Bird and Norton 1981) are similar to values reported for production fluids in the IID 2 well (Helgeson 1968) at 300°C. Similarly, $CO_2$ fugacities computed for the vein assemblage epidote ($X_{ps} = 0.33$), calcite, hematite and quartz (Caruso et al. 1988) are in agreement with fugacities calculated from analyses of production fluids reported by Janik et al. (1987) for the State 2-14 drill hole at 1867 m. Fugacities of $CO_2$ in the chlorite-calcite zone of the geothermal system are variable (2.5–14 bars at 300°C), and are predicted to increase with increasing temperature as the biotite zone is approached (Bird and Norton 1981). At 300°C, fugacities of oxygen compatible with epidote ($X_{ps} = 0.33$), pyrite, and hematite assemblages are approximately $10^{-30}$ bar, and fluid pH between $\approx 5.4$ and 5.6 (McKibben and Elders 1985, Caruso et al. 1988; Charles et al. 1988; see Charles et al. 1988 for details of the pH calculations). Charles et al. (1988) predict that the geothermal brine produced from the 1867 m zone of the State 2-14 drill hole is in equilibrium with the epidote + hematite assemblage of epidosite in this high permeability zone, but that the fluids are not in equilibrium with the epidote + pyrite assemblages occurring above and below the production zone. McKibben and Elders (1985) suggest that the redox state of the present-day geothermal brine is controlled by reactions between hematite, pyrite and iron-bearing silicates (chlorite and epidote).

## FOSSIL HYDROTHERMAL SYSTEMS IN RIFTING ENVIRONMENTS

We conclude our review of epidote in hydrothermal systems by briefly noting the occurrence of epidote in fossil hydrothermal systems associated with ophiolites and the early Tertiary North Atlantic igneous province (NAIP). Ophiolites and the NAIP both represent regions of crustal rifting related to mafic magmatism that formed extensive geothermal systems. Erosion following tectonic emplacement of ophiolites, and uplift of continental margins in the North Atlantic region, have exposed deep portions of these once active geothermal systems, thus allowing observations of relationships between multiple intrusions and extensional tectonics on the evolution of hydrothermal fluid flow and water-rock reactions. In general, the types of hydrothermal alteration in these systems is similar to the alteration of basaltic rocks described above for active geothermal systems in Iceland. With increasing depth or temperature, alteration of basaltic rocks in ophiolites and in the NAIP generally follows a sequence of zeolite to greenschist to amphibolite facies of hydrothermal metamorphism (Almond 1964; Bird et al. 1985; Manning and Bird 1995; Alt and Teagle 2000; Gillis and Banerjee 2000).

## Ophiolites

Recent reviews of hydrothermal alteration in ophiolites, and comparison with alteration in drill hole samples of mid-ocean ridge hydrothermal systems, are presented by Harper (1999), Alt (1999), Alt and Teagle (2000), and Gillis and Banerjee (2000). Of interest here is the occurrence of localized Ca-metasomatism associated with the formation of epidosites. Epidosite forms by replacement of mafic rocks (dikes and lavas) or plagiogranite. They are characterized by modally abundant granoblastic epidote with lesser quantities of quartz, chlorite, titanite and magnetite. As noted by Gillis and Banerjee (2000) epidosite occurs in ophiolites worldwide and have been described in Tethyan ophiolites at Troodos (Richardson et al. 1987; Schiffman et al. 1987; Schiffman and Smith 1988; Bettison-Varga et al. 1992, 1995) and Semail (Nehlig and Juteau 1988), the Caledonian ophiolite at Karmoy (Scott 1991; Pedersen and Malpas 1984), and ophiolites of the circum-Pacific margin including Josephine (Harper et al. 1988; Harper 1995), Del Puerto (Evarts and Schiffman 1983), and Tonga (Banerjee et al. 2000).

Epidosites in sheeted dike swarms may form zones hundreds of meters wide, commonly elongate and parallel to individual dikes. They are typically concentrated in the lower portion of the dike complex (Richardson et al. 1987; Schiffman et al. 1987; Schiffman and Smith 1988; Nehlig et al. 1994). Based on fluid inclusion experiments, and elemental and isotope analyses, epidosites are considered to have formed in the temperature range of approximately 310–440°C by metasomatic processes that added Ca and Sr, and removed Na, Mg, K, Cu, Zn and S from the host rock, presumably by Mg-depleted, Ca-enriched, hydrothermal solutions with $\delta^{18}O$, $Sr^{87}/Sr^{86}$ and $\delta^{34}S$ values shifted toward seawater values, thus indicating very high water to rock ratios along zones of concentrated fluid upflow (Richardson et al. 1987; Harper et al. 1988; Schiffman and Smith 1988; Seyfried et al. 1988; Bickle and Teagle 1992; Alt 1994; Nehlig et al. 1994; Bickle et al. 1998). Epidosites may represent deep fluid flow zones responsible for the formation of seafloor massive sulfide deposits (Haymon et al. 1989; Nehlig et al. 1994). Paleomagnetic, textural and structural relationships suggest that epidosites form shortly after dike emplacement (Varga et al. 1999). However, paragenetic relations demonstrate that albite + chlorite alteration precedes formation of epidosites at some localities (Bettison-Varga et al. 1995; Harper 1999), a paragenesis noted above in the Reykjanes geothermal system (Fig. 17). Problematic features of epidosite formation include the evolution of porosity and permeability that concentrated fluid flow, and the chemical driving force for the extensive metasomatism. Channeling of fluids along dike parallel faults and fractures, generation of porosity by metasomatic reactions, magmatic degassing during dike solidification, albitization preceding Ca-metasomatism, and fluid flow along a positive temperature gradient have been proposed as possible explanations for the localized extreme Ca-metasomatism (Harper et al. 1988; Harper 1995; Rose et al. 1992; Teagle 1993; Bettison-Varga et al. 1992, 1995; Alt 1999; Gillis and Banerjee 2000).

### North Atlantic igneous province

Continental rifting between Greenland and Eurasia in the early Tertiary was accompanied by extensive regional extrusion of flood basalt associated with partial melting of the mantle and initiation of the Iceland mantle plume (White and McKenzie 1989; 1995; Smallwood and White 2002). Intrusions of gabbroic plutons and mafic dike swarms, as well as minor but significant silicic and alkaline plutons, into the volcanic pile generated enormous meteoric-water hydrothermal systems (Forester and Taylor 1977a,b; Taylor and Forester 1979).

Hydrothermal alteration of basaltic rocks intruded by gabbroic plutons of the Cuillin (Scotland) and the Skaergaard (Greenland) intrusions consists of contact metamorphic pyroxene hornfels, grading outward away from the intrusion to actinolite + albite + epidote + chlorite + quartz assemblages, followed at further distance by epidote + chlorite + quartz

± prehnite ± albite ± K-feldspar, then calcite + K-feldspar + albite + quartz ± prehnite, and finally a zone of regional zeolite alteration (Almond 1964; Bird et al. 1985, Manning and Bird 1991, 1995; Manning et al. 1993; Neuhoff et al. 1997, 2000). Waning of the geothermal systems due to solidification, cooling and fracturing of gabbro intrusions resulted in retrograde alteration of early high-temperature hydrothermal hornblende + pyroxene veins within the fractured gabbros to assemblages including epidote + chlorite ± amphibole ± prehnite ± albite, and the formation of epidosites in gabbroic pegmatites (Norton et al. 1984; Bird et al. 1986, 1988b; Rose and Bird 1987). Dickin and Jones (1983) document the extent of major and trace element mobility associated with the formation of a 6 mm wide epidote + calcite vein in an albite + calcite + chlorite + epidote altered basic sill on the Isle of Skye. In addition, Dickin et al. (1980) provide evidence of variable Sr isotope mobility associated with vein epidote alteration of the Coire Uaigneich granophyre, also on the Isle of Skye. Analysis of REE mobility related to formation of hydrothermal allanites in altered Tertiary granite and skarn on the Isle of Skye is presented by Exley (1980) and Smith et al. (2002). Extreme Ca-metasomatism within sheeted mafic dike swarms of east Greenland has produced nearly complete dike replacement by prehnite together with lesser amounts of epidote, amphibole, salite, titanite, calcite and chlorite, a prehnite-rich analog of epidosites found in ophiolites (Rose and Bird 1994). Albitization of the dikes preceded Ca-metasomatism.

## CONCLUDING REMARKS

Alteration in geothermal systems is characterized by systematic mineralogic zoning with increasing temperature (Browne 1978; Henley and Ellis 1983; Bird et al. 1984; Reed 1994). In general, there is a progressive transformation from shallow clay-carbonate alteration to a prograde series of zones characterized by index minerals such as zeolite(s), chlorite, prehnite, epidote, amphibole (actinolite to hornblende), biotite, pyroxene, and garnet (Figs. 3, 4, 5, 7, 11, 12, 16, 21, 22, 23). Analysis of divariant mineral assemblages led Schiffman et al. (1984) to describe geothermal mineral zoning as low-pressure/temperature metamorphic series characterized by "temperature-telescoped" dehydration and decarbonation reactions. This analogy between geothermal alteration zones and metamorphic facies should be applied with caution. The physical and chemical environment of geothermal systems is determined by the dynamics of magma-hydrothermal processes, and is highly variable over short time periods. Superposition of prograde and retrograde mineral assemblages in geothermal systems is a consequence of repeated episodes of high-level igneous intrusions, brittle deformation, metasomatism and fracture sealing, and mixing and boiling of fluids from diverse sources. Mineral formation is commonly controlled by mass transfer and kinetic processes; the former largely determined by the thermodynamic and transport properties of water in the near supercritical region. Although systematic mineral zones are documented in geothermal systems worldwide, details of mineral paragenesis are complex, reflecting the dynamic nature of these systems.

Epidote is an abundant and common mineral formed in geothermal systems. Its paragenesis and composition reflect the dynamic physical and chemical environment created by intrusion and cooling of magmas into the upper portions of the Earth's crust. Spatial distribution and modal abundance of epidote has proven useful as a geothermometer, and for identifying regions of local cooling and zones of high permeability in active geothermal systems. The composition of geothermal epidote is highly variable, and metastable phase relations are not uncommon. The extreme sensitivity of epidote composition to intensive and extensive thermodynamic variables, and the multi-variance of many epidote-bearing assemblages, suggests that interpretation of the chemical and physical evolution of magma-hydrothermal systems based on chemically zoned epidotes must be considered with caution.

## ACKNOWLEDGMENTS

We thank Gehard Franz, Lutz Hecht, Reiner Klemd, Axel Liebscher and Phil Neuhoff for thoughtful and constructive reviews of this manuscript, and Jodi Rosso for her editorial expertise.

## REFERENCES

Abbate E, Passerini P, Zan L (1995) Strike-slip faults in a rift area: a transect in the Afar Triangle, East Africa. Tectonophys 241:67-97

Absar A (1991) Hydrothermal epidote - an indicator of temperature and fluid composition. J Geol Soc India 38:625-628

Albertsson KJ, Hooker PJ, Miller JA (1982) A brief K-Ar study of the IRDP borehole, Reydarfjordur, Eastern Iceland. J Geophy Res 87:6566-6590

Almond DC (1964) Metamorphism of Tertiary lavas in Strathaird, Skye. Trans Royal Soc Edinburgh 65:413-434

Alt JC (1994) A sulfur isotopic profile through the Troodos ophiolite, Cyrpus: Primary composition and the effects of seawater hydrothermal alteration. Geochim Cosmochim Acta 58:1825-1840

Alt JC (1999) Hydrothermal alteration and mineralization of oceanic crust: Mineralogy, geochemistry, and processes. *In*: Volcanic-associated massive sulfide deposits: Processes and examples in modern and ancient settings. Barrie CT, Hannington MD (eds) Rev Econ Geol 8:133-155

Alt JC, Teagle DAH (2000) Hydrothermal alteration and fluid fluxes in ophiolites and oceanic crust. *In*: Ophiolites and Oceanic Crust: New Insights from Field Studies and Ocean Drilling Program. Dilek Y, Moores EM, Elthon D, Nicolas A (eds) Geol Soc Am Spec Paper #349 p. 273-282

Alvis-Isidro R, Solana R, D'Amore F, Nuti S, Gonfiantini R (1993) Hydrology of the Greater Tongonan geothermal system, Philippines as deduced from geochemical and isotopic data. Geothermics 22:435-449

Anders E, Grevesse N (1989) Abundances of elements: Meteoritic and solar. Geochim Cosmochim Acta 53:197-214

Arehart GB, Christenson BW, Wood CP, Foland KA, Browne PRL (2002) Timing of volcanic, plutonic and geothermal activity at Ngatamariki, New Zealand. J Volcan Geothermal Res 116:201-214

Ármannsson H, Benjamínsson J, Jeffrey WAJ (1989) Gas changes in the Krafla geothermal system, Iceland. Chem Geol 76:175-196

Ármannsson H, Gíslason G, Hauksson T (1982) Magmatic gases in well fluids aid the mapping of flow pattern in a geothermal system. Geochim Cosmochim Acta 46:167-177

Ármannsson H, Gudmundsson Á, Steingrímsson B (1987) Exploration and development of the Krafla geothermal area. Jokull 37:13-30

Arnason JG, Bird DK (1992) Formation of zoned epidote in hydrothermal systems. Internat Symp Water-Rock Interaction 7:1473-1476

Arnason JG, Bird DK, Liou JG (1993) Variables controlling epidote composition in hydrothermal and low-pressure regional metamorphic rocks. Abhand Geol Bund 49:17-25

Arnórsson S (1978) Major element chemistry of the geothermal sea-water at Reykjanes and Svartsengi, Iceland. Mineral Mag 42:209-220

Arnórsson S (1985) The use of mixing models and chemical geothermometers for estimating underground temperatures in geothermal systems. J Volcan Geothermal Res 23:299-235

Arnórsson S (1995) Geothermal systems in Iceland; structure and conceptual models; I, High-temperature areas. Geothermics 24:561-602

Arnórsson S (1999) Progressive water-rock interaction and mineral-solution equilibria in groundwater systems. *In*: Geochemistry of the Earth's Surface. Ármannsson H (ed) Balkema, Rotterdam, p 471-474

Arnórsson S, Groenvold K, Sigurdsson S (1978) Aquifer chemistry of four high-temperature geothermal systems in Iceland. Geochim Cosmochim Acta 42:523-536

Arnórsson S, Gunnlaugsson E (1985) New gas geothermometers for geothermal exploration; calibration and application. Geochim Cosmochim Acta 49:1307-1325

Arnórsson S, Gunnlaugsson E, Svavarsson H (1983) The chemistry of geothermal waters in Iceland; II, Mineral equilibria and independent variables controlling water compositions. Geochim Cosmochim Acta 47:547-566

Arnórsson S, Sigurdsson S, Svavarsson H (1982) The chemistry of geothermal waters in Iceland; I, Calculation of aqueous speciation from 0 degrees to 370 degrees C. Geochim Cosmochim Acta 46:1513-1532

*Bird & Spieler*

Atwater T (1970) Implications of plate tectonics for the Cenozoic tectonic evolution of western North America. Bull Geol Soc Am 81:3513-2536

Arredondo A (1987) Estudio gravimetrico a detalle para el desarrollo del campo geotermico de la caldera de Los Humeros. Geotermia 3:53-63

Babcock EA (1974) Geology of the northeast margin of the Salton Trough, Salton Sea, California. Bull Geol Soc Am 85:321-332

Bagnoli G, Gianelli G, Puxeddu M, Rau M, Squarci P, Tongiorgi M (1979) A tentative stratigraphic reconstruction of the Tuscan Paleozoic Basement. Mem Soc Geol Italiana 20:99-116

Bamba M, Hatakeyama K, Katoh H, Kizawa Y, Ozeki T (1987) Thermal history at North Hachimantai-Yakeyama area from alteration and fluid inclusion thermometry (in Japanese). Annu Mtg Jap Geothermal Soc C5

Banerjee NR, Gillis KM, Muehlenbachs K (2000) Discovery of epidosites in a modern oceanic setting, the Tonga forearc. Geology 28:151-154

Barker CE, Elders WA (1981) Vitrinite reflectance geothermometry and apparent heating duration in the Cerro Prieto geothermal field. Geothermics 10:207-223

Bath AH, Williamson KH (1983) Isotopic and chemical evidence for water sources and mixing in the Cerro Pando geothermal area, Republic of Panama. Geothermics 12:117-184

Batini F, Bertini G, Gianelli G, Pandeli E, Puxeddu M, Villa IM (1985) Deep structure, age and evolution of the Larderello-Travale geothermal field. Trans Geothermal Resource Counc 9:253-259

Beaufort D, Patrier P, Meunier A, Ottaviani MM (1992) Chemical variations in assemblages including epidote and/or chlorite in the fossil hydrothermal system of Saint Martin (Lesser Antilles). J Volcan Geothermal Res 51:95-114

Becke F (1903) Über Mineralbestand und Struktur der kristallinischen Schiefer. Akad Wiss (Vienna) Denkschr, Math-Natv Kl 75:1-53

Belkin HE, de Vivo B, Gianelli G, Lattanzi P (1985) Fluid inclusions in minerals from the geothermal fields of Tuscany, Italy. Geothermics 14:59-72

Berman RG (1988) Internally consistent thermodynamic data for minerals in the system $Na_2O-K_2O-CaO-MgO-FeO-Fe_2O_3-Al_2O_3-SiO_2-TiO_2-H_2O-CO_2$. J Petrol 29:445-522

Bertini G, Gianelli G, Pandeli E, Puxeddu M (1985) Distribution of hydrothermal minerals in Larderello-Travale and Mt. Amiata geothermal fields, Italy. Trans Geothermal Resource Counc 9:261-266

Bettison-Varga L, Schiffman P, Janecky DR (1995) Fluid-rock interaction in the hydrothermal upflow zone of the Solea Graben, Troodos Ophiolite, Cyprus. *In*: Low Grade Metamorphism of Mafic Rocks. Schiffman P, Day H (eds) Geol Soc Am Spec Paper # 296, p 81-100

Bettison-Varga L, Varga RJ, Schiffman P (1992) Relation between ore-forming hydrothermal systems and extensional deformation in the Solea Graben spreading center, Troodos Ophiolite, Cyprus. Geology 20: 987-990

Bickle MJ, Teagle DAH (1992) Strontium alteration in the Troodos ophiolite: Implications for fluid fluxes and geochemical transport in mid-ocean riddge hydrothermal systems. Earth Planet Sci Lett 113:219-237

Bickle MJ, Teagle DAH, Beynon J, Chapman HJ (1998) The structure and controls on fluid-rock interactions in ocean ridge hydrothermal systems: constraints from the Troodos ophiolite. *In*: Modern Ocean Floor Processes and the Geological Record. Vol. 148: Mills RA, Harrison K (eds) Geol Soc, London, p 127-152

Bird DK, Cho M, Janik CJ, Liou JG, Caruso LJ (1988a) Compositional, order-disorder, and stable isotope characteristics of Al-Fe epidote, State 2-14 drill hole, Salton Sea geothermal system. J Geophy Res 93: 13135-13144

Bird DK, Helgeson HC (1980) Chemical interaction of aqueous solutions with epidote-feldspar mineral assemblages in geologic systems: Thermodynamic analysis of phase relations in the system $CaO-FeO-Fe_2O_3-Al_2O_3-SiO_2-H_2O-CO_2$. Am J Sci 280:907-941

Bird DK, Helgeson HC (1981) Chemical interaction of aqueous solutions with epidote-feldspar mineral assemblages in geologic systems: II. Equilibrium constraints in metamorphic/geothermal processes. Am J Sci 281:576-614

Bird DK, Manning CE, Rose NM (1988b) Hydrothermal alteration of Tertiary layered gabbros, East Greenland. Am J Sci 288:405-457

Bird DK, Norton DL (1981) Theoretical prediction of phase relations among aqueous solutions and minerals: Salton Sea geothermal system. Geochim Cosmochim Acta 45:1479-1493

Bird DK, Rogers RD, Manning CE (1986) Mineralized fracture systems of the Skaergaard intrusion, East Greenland. Medd Grønland 16:1-68

Bird DK, Rosing MT, Manning CE, Rose NM (1985) Geologic field studies of the Miki Fjord area, East Greenland. Bull Geol Soc Denmark 34:219-236

Bird DK, Schiffman P, Elders WA, Williams AE, McDowell SD (1984) Calc-silicate mineralization in active geothermal systems. Econ Geol 79:671-695

Björnsson A, Saemundsson K, Einarsson P, Tryggvason E, Gronvöld K (1977) Current rifting episode in north Iceland. Nature 266:318-323

Björnsson G, Steingrímsson B (1991) Temperature and pressure in the Svartsengi geothermal system. Initial conditions and changes due to production (in Icelandic with English summary). Iceland National Energy Authority, Report, OS-91016/JHD-04, Reykjavik, Iceland, p 69

Björnsson S, Arnórsson S, Tómasson J (1972) Economic evaluation of the Reykjanes thermal brine area, Iceland. Bull Am Ass Petroleum Geol 56:2380-2391

Boccaletti M, Gianelli G, Sani F (1997) Tectonic regime, granite emplacement and crustal structure in the inner zone of the Northern Apennines (Tuscany, Italy): A new hypothesis. Tectonophys 270:127-143

Bödvarsson G (1961) Physical characteristics of natural heat resources in Iceland. Jokull 11:29-38

Bödvarsson G, Björnsson S, Gunnarsson Á, Gunnlaugsson E, Sigurdsson Ó, Stefánsson V, Steingrímsson B (1990) The Nesjavellir geothermal field, Iceland, Part 1. Field characteristcs and development of a three dimensional numerical model. Geothermal Sci Tech 2:189-228

Bödvarsson GS, Benson SM, Sigurdsson Ó, Stefánsson V, Elíasson ET (1984) The Krafla geothermal field, Iceland 1. Analysis of well test data. Water Resour Res 20:1515-1530

Bowin C (1976) Caribbean gravity field and plate tectonics. Geol Soc Am Spec Paper 169:1-79

Bredehoeft JD, Norton DL (1990) Mass and energy transport in a deforming Earth's crust. *In:* The Role of Fluids in Crustal Processes. Bredehoeft JD, Norton DL (eds) Natl Acad Press, Washington, p 27-41

Brown EH (1967) The greenschist facies in part of Eastern Otago, New Zealand. Contrib Mineral Petrol 14: 259-292

Browne PRL (1978) Hydrothermal alteration in active geothermal systems. Annu Rev Earth Planet Sci 6: 229-250

Browne PRL (1989) Contrasting alteration styles of andesitic and rhyolitic rocks in geothermal fields of the Taupo Volcanic Zone, New Zealand. Proc 11th New Zealand Geothermal Workshop 11:111-116

Browne PRL, Courtney SF, Wood CP (1989) Formation rates of calc-silicate minerals deposited inside drillhole casing, Nagatamariki geothermal field, New Zealand. Am Mineral 74:759-763

Browne PRL, Elders WA (1976) Hydrothermal alteration of diabase, Heber geothermal field, Imperial Valley, California. Geol Soc Am, Abt w Prog 8:793

Browne PRL, Ellis AJ (1970) The Ohaki-Broadlands hydrothermal area, New Zealand; mineralogy and related geochemistry. Am J Sci 269:97-131

Browne PRL, Graham IJ, Parker RJ, Wood CP (1992) Subsurface andesite lavas and plutonic rocks in the Rotokawa and Ngatamariki geothermal systems, Taupo Volcanic Zone, New Zealand. J Volcan Geothermal Res 51:199-215

Caruso LJ, Bird DK, Cho M, Liou JG (1988) Epidote-bearing veins in the State 2-14 drill hole; implications for hydrothermal fluid composition. J Geophy Res 93:13123-13133

Cathelineau M, Izquierdo G, Nieva D (1989) Thermobarometry of hydrothermal alteration in the Los Azufres geothermal system (Michoacan, Mexico): Significance of fluid-inclusion data. Chem Geol 76:229-238

Cathelineau M, Izquierdo G, Vazquez GR, Guevara M (1991) Deep geothermal wells in the Los Azufres (Mexico) caldera: Volcanic basement stratigraphy based on major-element analysis. J Volcan Geothermal Res 47:149-159

Cathelineau M, Marignac C, Boiron M, Yardley B, Gianelli G, Puxeddu M (1995) Use of fluid inclusions for the discrimination of multi-source components and P-T-X reconstruction in geothermal systems: Application to Larderello. World Geothermal Congress 2:1093-1097

Cathelineau M, Marignac C, Boiron M-C, Gianelli G, Puxeddu M (1994) Evidence for Li-rich brines and early magmatic fluid-rock interaction in the Larderello geothermal system. Geochim Cosmochim Acta 58:1083-1099

Cathelineau M, Nieva D (1985) A chlorite solid solution geothermeter: The Los Azufres (Mexico) geothermal system. Contrib Mineral Petrol 91:235-244

Cathelineau M, Nieva D (1986) Geothermometry of hydrothermal alteration in the Los Azufres geothermal system: Significance of fluid inclusion data. Internat Symp Water-Rock Interaction 5:104-107

Cathelineau M, Oliver R, Nieva D, Garfias A (1985) Mineralogy and distribution of hydrothermal mineral zones in the Los Azufres (Mexico) geothermal field. Geothermics 14:49-57

Cathles III LM (1977) An analysis of the cooling intrusives by ground-water convection which includes boiling. Econ Geol 72:804-826

Cathles III LM (1997) Thermal aspects of ore formation. *In*: Geochemistry of Hydrothermal Ore Deposits (3rd edition). Barnes HL (ed) John Wiley & Sons, New York, p 191-227

Cavarretta G, Gianelli G, Puxeddu M (1980a) Hydrothermal metamorphism in the Larderello geothermal field. Geothermics 9:297-314

Cavarretta G, Gianelli G, Puxeddu M (1980b) New data on hydrothermal metamorphism in the Larderello-Travale geothermal field. Internat Symp Water-Rock Interaction 3:112-114

Cavarretta G, Gianelli G, Puxeddu M (1982) Formation of authigenic minerals and their use as indicators of the physicochemical parameters of the fluid in the Larderello-Travale geothermal field. Econ Geol 77: 1071-1084

Cavarretta G, Gianelli G, Puxeddu M, Seki Y (1983) Hydrothermal and contact metamorphism in the Larderello geothermal field (Italy): A new contribution from San Pompeo 2 deep well. Internat Symp Water-Rock Interaction 4:82-86

Cavarretta G, Puxeddu M (1990) Schorl-dravite-ferridravite tourmalines deposited by hydrothermal magmatic fluids during early evolution of the Larderello geothermal field, Italy. Econ Geol 85:1236-1251

Cavarretta G, Puxeddu M, Hitchon B (1986) Tourmalines in hydrothermal mineral assemblages from Larderello geothermal field (Italy) Internat Symp Water-Rock Interaction 5:108-111

Charles RW, Janecky DR, Goff F, McKibben MA (1988) Chemographic and thermodynamic analysis of the paragenesis of the major phases in the vicinity of the 6120-foot (1866 m) flow zone, California State Well 2-14. J Geophy Res 93:13145-13157

Cho M, Liou JG, Bird DK (1988) Prograde phase relations in the State 2-14 Well metasandstones, Salton Sea geothermal field, California. J Geophy Res 93:13081-13103

Coats RR (1940) Propylitization and related types of alteration and the Comstock Lode. Econ Geol 35:1-16

Cole JW (1981) Genesis of lavas of the Taupo Volcanic Zone, North Island, New Zealand. J Volcan Geothermal Res 10:317-337

Cole JW (1984) Andesites of the Tongariro volcanic center, North Island, New Zealand. J Volcan Geothermal Res 3:121-153

Coleman RG (1977) Ophiolites. Springer-Verlag, New York

Coombs DS, Ellis AJ, Fyfe WS, Taylor AM (1959) The zeolite facies, with comments on the interpretation of hydrothermal synthesis. Geochim Cosmochim Acta 17:53-107

Craig H (1966) isotopic composition and origin of the Red Sea and Salton Sea geothermal brines. Science 154: 1544-1548

D'Amore F, Bolognesi L (1994) Isotopic evidence for a magmatic contribution to fluids of the geothermal systems of Larderello, Italy, and The Geysers, California. Geothermics 23:21-32

D'Amore F, Celati R (1983) Methodology for calculating steam quality in geothermal reservoirs. Geothermics 12:129-140

D'Amore F, Gianelli G (1984) Mineral assemblages and oxygen and sulphur fugacities in natural water-rock interaction processes. Geochim Cosmochim Acta 48:847-857

D'Amore F, Gianelli G, Seki Y (1983) Oxygen and sulphur fugacity buffering in geothermal systems. Internat Symp Water-Rock Interaction 4:103-107

D'Amore F, Giusti D, Berhanu G (1997) Geochemical assessment of the northern Tendaho rift, Ethiopia. Workshop on Geothermal Reservoir Engineering 22, Stanford University, California, p 435-445

Dana JD (1875) System of Mineralogy. John Wiley and Sons, New York

Darling G, Ármannsson H (1989) Stable isotopic aspects of fluid flow in the Krafla, Namafjall and Theistareykir geothermal systems of northeastern Iceland. Chem Geol 76:197-213

de Vivo B, Sasada M (1992) Fluid inclusion from deep borehole SN-7D, Sumikawa geothermal field, Sengan area, North Honshu, Japan. J Geothermal Resource Soc Japan 14:101-114

Del Moro A, Puxeddu M, Radicati di Brozolo F, Villa IM (1982) Rb-Sr and K-Ar ages of minerals at temperatures of 300-400°C from deep wells in the Larderello geothermal field (Italy). Contrib Mineral Petrol 81:340-349

Des Marais DJ, Donchin JH, Nehring NL, Truesdell AH (1981) Molecular carbon isotopic evidence for the origin of geothermal hydrocarbons. Nature 22:826-828

Di Paola GM (1972) The Ethiopian Rift Valley (between 7°00' and 8°40' lat. North). Bull Volcan 36:571-559

Dickin AP, Exley RA, Smith BM (1980) Isotopic measurement of Sr and O exchange between meteoric-hydrothermal fluid and the Coire Uaigneich Granophyre, Isle of Skye, N. W. Scotland. Earth Planet Sci Lett 51:58-70

Dickin AP, Jones NW (1983) Relative elemental mobility during hydrothermal alteration of a basic sill, Isle of Skye, N.W. Scotland. Contrib Mineral Petrol 82:147-153

Dobson PF, Mahood GA (1985) Volcanic stratigraphy of the Los Azufres geothermal area, Mexico. J Volcan Geothermal Res 25:273-287

Doe BR, Hedge CE, White DE (1966) Preliminary investigation of the source of lead and strontium in deep geothermal brines underlying the Salton Sea geothermal area. Econ Geol 61:462-483

Doi N, Kato O, Ikeuchi K, Komatsu R, Miyazaki S, Akaku K, Uchida T (1998) Genesis of the plutonic-hydrothermal system around Quaternary granite in the Kakkonda geothermal system, Japan. Geothermics 27:663-690

Doi N, Kato O, Muramatsu Y (1991) On the neo-granite pluton and the deep geothermal reservoir at the Kakkonda geothermal field, Iwate Prefecture, Japan. Proc Mini Geothermal Symp, Geol Surv Japan, Tsukuba, p. 9

Einarsson P (1978) S-wave shadows in the Krafla caldera in NE-Iceland, evidence for magma chamber in the crust. Bull Volcan 43:1-9

Elderfield H, Greaves MJ (1981) Strontium isotope geochemistry of Icelandic geothermal systems and implications for sea water chemistry. Geochim Cosmochim Acta 45:2201-2212

Elders WA, Bird DK, Williams AE, Schiffman P (1984) Hydrothermal flow regime and magmatic heat source of the Cerro Prieto geothermal system, Baja California, Mexico. Geothermics 13:27-47

Elders WA, Hoagland JR, McDowell SD, Cobo JM (1979) Hydrothermal mineral zones in the geothermal reservoir of Cerro Prieto. Geothermics 8:201-209

Elders WA, Hoagland JR, Williams AE (1981) Distribution of hydrothermal mineral zones in the Cerro Prieto geothermal field of Baja California, Mexico. Geothermics 10:245-253

Elders WA, Hoagland JR, Williams AE, Berge CW (1980) Hydrothermal alteration as an indicator of temperature and flow regime in the Cerro Prieto geothermal field of Baja California. Trans Geothermal Resource Counc 4:121-124

Elders WA, McKibben MA, Williams AE (1992) The Salton Sea hydrothermal system, California, USA; a review. Internat Symp Water-Rock Interaction 7:1283-1288

Elders WA, Rex RW, Meidav T, Robison PT, Biehler S (1972) Crustal spreading in southern California. Science 178:15-24

Elders WA, Sass JH (1988) The Salton Sea scientific drilling project. J Geophy Res 93:12,953-12,968

Electroconsult ELC (1986) Exploitation of Langano-Aluto Geothermal Resources. Milan, Italy

Ellis AJ (1969) Present-day hydrothermal systems and mineral deposition. 9th Commonwealth Mining Metallurgy Congress, Inst Mining & Metallurgy, London, p 1-30

Ellis AJ, Mahon WAJ (1977) Chemistry and Geothermal Systems. Academic Press, New York

Enami M, Liou JG, Bird DK (1992) Cl-bearing amphibole in the Salton Sea geothermal system, California. Canad Mineral 30:1077-1092

Enami M, Suzuki K, Liou JG, Bird DK (1993) Al-Fe$^{3+}$ and F-OH substitutions in titanite and constraints on their P-T dependence. Eur J Min 5:219-231

Eskola P (1915) On the relation between the chemical and mineralogical composition in the metamorphic rocks of the Orijarvi region. Bull Comm Geol Finlande 44:109-145

Evarts P, Schiffman P (1983) Submarine hydrothermal metamorphism of the Del Puerto ophiolite, California. Am J Sci 283:289-340

Exley RA (1980) Microprobe studies on REE-rich accessory minerals; implications for Skye granite petrogenesis and REE mobility in hydrothermal systems. Earth Planet Sci Lett 48:97-110

Exley RA (1982) Electron microprobe studies of Iceland Research Drilling Project high-temperature hydrothermal mineral geochemistry. J Geophy Res 87:6547-6558

Fausto JJ, Sanchez A, Jimenez ME, Esquer I, Ulloa F (1979) Hydrothermal geochemistry of the Cerro Prieto geothermal field. Proc 2nd Symp Cerro Prieto Geothermal Field, Baja California, Mexico, p. 199-223, Mexcali, Baja California, Mexico

Ferrari L, Garduno VH, Pasquare G, Tibaldi A (1991) Geology of Los Azufres caldera, Mexico, and its relationships with regional tectonics. J Volcan Geothermal Res 47:129-148

Ferriz H, Mahood GA (1984) Eruption rates and compositional trends at Los Humeros volcanic center. J Geophy Res 89:8511-8524

Foley JE, Toksoz MN, Batini F (1992) Inversion of teleseismic travel time residuals for velocity structure in the Larderello geothermal field, Italy. Geophys Res Lett 19:5-8

Forester RW, Taylor H, P., Jr. (1977a) $^{18}$O-depleted rocks from the Tertiary complex of the Isle of Mull, Scotland. Earth Planet Sci Lett 32:11-17

Forester RW, Taylor H, P., Jr. (1977b) $^{18}$O/$^{16}$O, D/H, and $^{13}$C/$^{12}$C studies of the Teritary igneous complex of Skye, Scotland. Am J Sci 277:136-177

Frei D, Liebscher A, Franz G, Dulski P (2004) Trace element geochemistry of epidote minerals. Rev Mineral Geochem 56:553-605

Franceschini F (1995) The Larderello plutono-metamorphic core complex: Petrographic data. World Geothermal Congress 2:667-672

Franceschini F (1998) Evidence of an extensive Pliocene-Quaternary contact metamorphism in southern Tuscany. *In*: Memorie della Societa Geologica Italiana. Vol 52. Pialli G, Barchi M, Minelli G (eds) Societa Geologica Italiana, Rome, p 479-492

Franzson H (1983) The Svartsengi high-temperature field, Iceland. Subsurface geology and alteration. Trans Geothermal Resource Council 7:141-145

Franzson H (1988) Nesjavellir, borehole geology and permeability in the reservoir (in Icelandic). Icelandic National Energy Authority, Report OS-88046/JHD-09, Reykjavik, Iceland, p 58

Franzson H (1990) Svartsengi. Geological model of a high temperature system and its surroundings (in Icelandic with English summary). Iceland National Energy Authority, Report, OS90050/JHD-08, Reykjavik, Iceland, p 41

Franzson H (1998) Reservoir geology of the Nesjavellir high-temperature field in SW-Iceland. Annu PNOC-EDC Geothermal Conf Manila, Philippines 19:13-20

Franzson H (2000) Hydrothermal evolution of the Nesjavellir high-temperature system, Iceland. Proc World Geothermal Congress, Kyushu-Tohoku, Japan, p 2075-2080

Franzson H, Gudmundsson Á, Fridleifsson GÓ, Tómasson J (1986) Nesjavellir high-T field, SW-Iceland, Reservoir geology. Internat Symp Water-Rock Interaction 5:210-213

Franzson H, Thordarson S, Björnsson G, Gudlaugsson ST, Richter B, Fridleifsson GÓ, Thorhallsson S (2002) Reykjanes high-temperature field, SW-Iceland. Geology and hydrothermal alteration of well RN-10. Workshop on Geothermal Reservoir Engineering 27. Stanford University, Stanford, California p 233-240

Fridleifsson GÓ (1983) Mineralogical evolution of a hydrothermal system. Trans Geothermal Resource Counc 7:147-152

Fridleifsson GÓ (1984) Mineralogical evolution of a hydrothermal system, II. Heat sources - fluid interactions. Trans Geothermal Resource Counc 8:119-123

Fridleifsson GÓ (1991) Hydrothermal systems and associated alteration in Iceland. Geol Survey Japan Report 277:83-90

Fridleifsson GÓ, Björnsson S (1986) Geothermal activity in the Geitafell central volcano. Internat Symp Water-Rock Interaction 5:214-217

Fridleifsson IB (1979) Geothermal activity in Iceland. Jokull 29:47-56

Fridleifsson IB, Gibson IL, Hall JM, Johnson HP, Christensen NI, Schmincke H, Schonharting G (1982) The Iceland Research Drilling Project. J Geophy Res 87:6359-6361

Fuis GS, Mooney WD, Healy JH, McMechan GA, Lutter WJ (1982) Crustal structure of the Imperial Valley, The Imperial Valley, California earthquake of October 15, 1979. US Geol Surv Prof Paper 1254:25-49

Fyfe WS (1960) Stability of epidote minerals. Nature 187:497-498

Fyfe WS, Turner FJ, Verhoogen J (1958) Metamorphic reactions and metamorphic facies. Geol Soc Am Mem 73:1-259

Gambill DT, Beraquit DB (1993) Development history of the Tiwi geothermal field, Philippines. Geothermics 22:403-416

Gamble JA, Smith IEA, McCulloch MT, Graham IJ, Kokelaar BP (1993) The geochemistry and petrogenesis of basalts from the Taupo Volcanic Zone and Kermadec Island Arc, SW Pacific. J Volcan Geothermal Res 54:265-290

Gebregzabher Z (1986) Hydrothermal alteration minerals in the Aluto Langano geothermal wells, Ethiopia. Geothermics 15:735-740

Gerardo JY, Nuti S, D'Amore F, Seastres JS, Gonfiantini R (1993) Isotopic evidences for magmatic and meteoric water recharge and the processes affecting reservoir fluids in the Palinpinon geothermal system, Philippines. Geothermics 22:521-533

Gianelli G, Calore C (1996) Models for the origin of carbon dioxide in the Larderello geothermal field. Bull Soc Geol Italiana 115:75-84

Gianelli G, Laurenzi M (2001) Age and cooling rate of the geothermal system at Larderello. Trans Geothermal Resource Council 25:731-735

Gianelli G, Manzella A, Puxeddu M (1997a) Crustal models of the geothermal areas of southern Tuscany (Italy). Tectonophys 281:221-239

Gianelli G, Mekuria N, Battaglia S, Chersicla A, Garofalo P, Ruggieri G, Manganelli M, Gebregziabher Z (1998) Water-rock interaction and hydrothermal mineral equilibria in the Tendaho geothermal system. J Volcan Geothermal Res 86:253-276

Gianelli G, Puxeddu M (1994) Geological comparison between Larderello and the Geysers geothermal fields. Memorie della Societa Geologica Italiana 48:715-717

Gianelli G, Puxeddu M, Squarci P (1978) Structural setting of the Larderello-Travale geothermal region. Memorie della Societa Geologica Italiana 19:469-476

Gianelli G, Ruggieri G (2002) Evidence of a contact metamorphic aureole with high-temperature metasomatism in the deepest part of the active geothermal field of Larderello, Italy. Geothermics 31:443-474

Gianelli G, Ruggieri G, Mussi M (1997b) Isotopic and fluid inclusion study of hydrothermal and metamorphic carbonates in the Larderello geothermal field and surrounding areas, Italy. Geothermics 26:393-417

Gianelli G, Teklemariam M (1993) Water-rock interaction processes in the Aluto-Langano geothermal field (Ethiopia). J Volcan Geothermal Res 56:429-445

Giggenbach WF (1981) Geothermal mineral equilibria. Geochim Cosmochim Acta 45:393-410

Giggenbach WF (1992) Isotopic shifts in waters from geothermal and volcanic systems along convergent plate boundaries and their origin. Earth Planet Sci Lett 113:495-510

Giggenbach WF (1995) Variations in the chemical and isotopic composition of fluids discharged from the Taupo Volcanic Zone, New Zealand. J Volcan Geothermal Res 68:89-116

Giggenbach WF, Soto RC (1992) Isotopic and chemical composition of water and steam discharges from volcanic-hydrothermal systems of the Guanacaste geothermal province, Costa Rica. Appl Geochem 7: 309-332

Giggenbach WF, Stewart MK (1982) Processes controlling the isotopic composition of steam and water discharges from steam vents and steam-heated pools in geothermal areas. Geothermics 11:71-80

Gillis KM, Banerjee NR (2000) Hydrothermal alteration patterns in supra-subduction zone ophiolites. Geol Soc Am Special Paper 349:283-297

Gizaw B (1989) Geochemical investigation of the Alto-Langano geothermal field, Ethiopian Rift Valley. M Phil Thesis. University of Leeds, Leeds, United Kingdom

Goldschmidt VM (1911) Die Gesetze der Mineralassociation vom Standpunkt der Phasenregel. Zeits Anorgan Chem 71:313-322

Graham CM (1981) Experimental hydrogen isotope studies, III, Diffusion of hydrogen in hydrous minerals, and stable isotope exchange in metamorphic rocks. Contrib Mineral Petrol 76:216-228

Graham CM, Sheppard SMF (1980) Experimental hydrogen isotope studies, II, Fractionation in the systems epidote-NaCl-$H_2O$, epidote-$CaCl_2$-$H_2O$ and epidote-seawater, and th hydrogen isotope compositions of natural epidotes. Earth Planet Sci Lett 49:237-251

Grant JA (1986) The isocon diagram - A simple solution to the Gresen's equation for metasomatic alteration. Econ Geol 81:1976-1982

Gresta S, Patane D, Daniel A, Zan L, Carletta A, Befekadu O (1997) Seismological evidence of active faulting in the Tendaho Rift (Afar Triangle, Ethiopia). Pure Appl Geophy 149:357-374

Grubenmann U (1904) Die kristallinen Schiefer. Bornträger, Berlin

Gudmundsson A (2000) Dynamics of volcanic systems in Iceland: Example of tectonism and volcanism at juxtaposed hot spot and mid-ocean ridge systems. Annu Rev Earth Planet Sci 28:107-140

Gudmundsson A, Arnórsson S (2002) Geochemical monitoring of the Krafla and Namafjall geothermal areas, N-Iceland. Geothermics 31:195-243

Halfman SE, Lippmann MJ, Zelwer R, Howard JH, Lacy RG (1982) Identification of fluid flow paths in the Cerro Prieto geothermal field. Trans Geothermal Resource Counc 6:265-268

Hardardottir V (1983) The petrology of the Hengill volcanic system, southern Iceland. MSc Thesis. McGill University, Montreal, Quebec, Canada

Harper GD (1995) Pumpellyosite and prehnitite associated with epidosite in the Josephine ophiolite - Ca metasomatism during upwelling of hydrothermal fluids at a spreading axis. Geol Soc Am Special Paper 296:101-122

Harper GD (1999) Structural styles of hydrothermal discharge in ophiolite/sea-floor systems. Rev Econ Geol 8:53-73

Harper GD, Bowman JR, Kuhns R (1988) Field, chemical, and isotopic aspects of submarine hydrothermal metamorphism of the Josephine ophiolite, Klamath Mountains, California-Oregon. J Geophy Res 93: 4625-4657

Haymon RM, Koski RA, Adams MJ (1989) Hydrothermal discharge zones beneath massive sulfide deposits mapped in the Oman ophiolite. Geology 17:531-535

Hedenquist JW (1986) Geothermal systems of the Taupo Volcanic Zone: Their characteristics and relation to volcanism and mineralization. *In*: Late Cenozoic Volcanism In New Zealand. Smith IEM (ed) Royal Soc New Zealand Bull 23:134-168

Hedenquist JW (1990) The thermal and geochemical structure of the Broadlands-Ohaaki geothermal system, New Zealand. Geothermics 19:151-185

Hedenquist JW, Browne PRL (1989) The evolution of the Waiotapu geothermal system, New Zealand, based on the chemical and isotopic composition of its fluids, minerals and rocks. Geochim Cosmochim Acta 53:2235-2257

Hedenquist JW, Reyes AG, Simmons SF, Taguchi S (1992) The thermal and geochemical structure of geothermal and epithermal systems: A framework for interpreting fluid inclusion data. Eur J Min 4: 989-1015

Hedenquist JW, Stewart MK (1985) Natural $CO_2$-rich steam-heated waters at Broadlands, New Zealand: Their chemistry, distribution and corrosive nature. Trans Geothermal Resource Counc 9:245-250

Helgeson HC (1967) Solution chemistry and metamorphism. *In*: Researches in Geochemistry. Ableson PH (ed) John Wiley & Sons, New York, p 362-404

Helgeson HC (1968) Geologic and thermodynamic characteristics of the Salton Sea geothermal system. Am J Sci 266:129-166

Helgeson HC (1981) Prediction of the thermodynamic properties of electrolytes at high pressures and temperatures. In: Physics and Chemistry of the Earth: Chemistry and Geochemistry of Solutions at High Temperatures and Pressures 13-14. Rickard DT, Wickman FE (eds) Pergamon Press, New York, p 133-178

Helgeson HC, M. DJ, Nesbitt HW, Bird DK (1978) Summary and critique of the thermodynamic properties of rock-forming minerals. Am J Sci 278-A:1-229

Hemond C, Arndt NT, Lichtenstein U, Hofmann AW (1993) The heterogeneous Iceland plume: Nd-Sr-O isotopes and trace element constraints. J Geophy Res 98:15,833-15,850

Henley RW, Ellis AJ (1983) Geothermal systems ancient and modern: A geochemical review. Earth Planet Sci Lett 19:1-50

Herzig CT (1990) Geochemistry of igneous rocks from the Cerro Prieto geothermal field, northern Baja California, Mexico. J Volcan Geothermal Res 42:261-270

Herzig CT, Elders WA (1988) Nature and significance of igneous rocks cored in the State 2-14 research borehole; Salton Sea Scientific Drilling Project, California. J Geophy Res 93:13,069-13,080

Herzig CT, Mehegan JM, Stelting CE (1988) Lithostratigraphy of the State 2-14 borehole: Salton Sea Scientific Drilling Project. J Geophy Res 93:12969-12980

Hoagland JR, Bodell JM (1991) The Tiwi geothermal reservoir: Geologic characteristics and response to production. Petromineralogy 20:28-35

Hoagland JR, Elders WA (1978) Hydrothermal mineralogy and isotopic geochemistry in the Cerro Prieto geothermal field, Mexico; I, Hydrothermal mineral zonation. Trans Geothermal Resource Counc 2:283-286

Holdaway MJ (1972) Thermal stability of Al-Fe epidote as a function of $f_{O_2}$ and Fe content. Contrib Mineral Petrol 37:307-340

Hreggvidsdóttir H (1987) The greenschist to amphibolite facies transition in the Nesjavellir hydrothermal system. MSc Thesis, Stanford University, Stanford, California

Iacumin P, Petrucci E, Gianelli G, Puxeddu M (1992) An oxygen isotope study of silicates at Larderello, Italy. Internat Symp Water-Rock Interaction 7:935-937

Ikeuchi K, Doi N, Sakagawa Y, Kamenosono H, Uchida T (1998) High-temperature measurements in well WD-1a and the thermal structure of the Kakkonda geothermal system, Japan. Geothermics 27:591-607

Inoue T, Ueda R (1965) On the Hanawa fault, Akita, Japan (in Japanese with English abstract). J Mining Coll Akita Univ Ser A III:15-29

Inoue T, Utada M, Shimazu M (1999) Mineral-fluid interactions in the Sumikawa geothermal system, northeastern Japan. In: The Japanese island Arc -- Its Hydrothermal and Igneous Activities. Shikazono N, Shimizu M, Inoue T, Utada M, (eds) Resource Geology Special Issue 20, Soc Resourc Geol, Japan, p 79-98

Izquierdo G, Cathelineau M, Garcia A (1995) Clay minerals, fluid inclusions and stabilized temperature estimates in two wells from Los Azufres geothermal field, Mexico. World Geothermal Congress 2:1083-1086

Izquierdo G, Portugal E, Aragon A, Torres I, Alvarez J (2001) Hyrothermal mineralogy, isotopy and geochemistry in the area of Cerro Prieto IV Baja California Norte, Mexico. Trans Geothermal Resource Council 25:353-356

Janik CJ, Shigeno H, Cheatam T, Truesdell AH (1987) Gas geothermometers applied to separated stam from the December 1985 flow test of the SSSDP well. EOS Trans Am Geophy Union 68:440

Johnson JW, Norton D (1991) Critical phenomena in hydrothermal systems; state, thermodynamic, electrostatic, and transport properties of $H_2O$ in the critical region. Am J Sci 291:541-648

Johnson JW, Oelkers EH, Helgeson HC (1992) Supcrt92: A software package for calculating the standard molal thermodynamic properties of minerals, gases, aqueous species, and reactions from 1 bar to 5000 bar and 0 degrees Celsius to 1000 degrees Celsius. Comput Geosci 18:899-947

Kajiwara R, Hanano M, Ikeushi K, Sakagawa Y (1993) Permeability structure at the Kakkonda geothermal field, Iwate Prefecture, Japan (in Japanese). Abst Annu Mtg Geothermal Res Soc Japan B 30

Kasai K, Sakagawa Y, Komatsu R, Kato O, Sasaki M, Akaku K, Uchida T (1998) The origin of hypersaline liquid in the Quaternary Kakkonda granite, sampled from WD-1a Kakkonda geothermal system, Japan. Geothermics 27:631-645

Kato O, Doi N (1993) Neo-granite pluton and later hydrothermal alteration at the Kakkonda geothermal field, Japan. New Zealand Geothermal Workshop 15:155-161

Keith TC, Muffler LJP, Creamer M (1968) Hydrothermal epidote formed in the Salton Sea geothermal system, California. Am Mineral 53:1635-1644

Keskinen M, Sternfeld J (1982) Hydrothermal alteration and tectonic setting of intrusive rocks from East Brawley, Imperial Valley: An application of petrology to geothermal reservoir analysis. Workshop on Geothermal Reservoir Engineering 8. Stanford University, California p 39-44

Komatsu R, Muramatsu Y (1994) Fluid inclusion study of the deep reservoir at the Kakkonda geothermal field, Japan. New Zealand Geothermal Workshop 16:91-96

Korzhinskiy DS (1963) Effect of depth on metamorphism of volcanic formations (in Russian). Trudy Lab Vulkan AN SSSR 19:5-11

Kristmannsdóttir H (1975) Hydrothermal alteration of basaltic rocks in Icelandic geothermal areas. 2nd United Nations Symposium on Development and Use of Geothermal Resources 2. United Nations, San Francisco, California, p 441-445

Kristmannsdóttir H (1979) Alteration of basaltic rock by hydrothermal activity at 100-300°C. *In*: International Clay Conference 1978. Mortland MM, Farmer VC (eds) Elsevier, Amsterdam,p 359-367

Kristmannsdóttir H (1982) Alteration in the IRDP drill hole compared with other drill holes in Iceland. J Geophy Res 87:6525-6531

Kristmannsdóttir H (1983) Chemical evidence from Icelandic geothermal systems as compared to submarine geothermal systems. *In*: Hydrothermal Processes at Seafloor Centers. Rona P, Bostrom K, Laubier L, Smith KL (eds) Plenum, New York, p 291-320

Krupp RE, Seward TM (1987) The Rotokawa geothermal system, New Zealand: an active epithermal gold-depositing environment. Econ Geol 82:110901121

Kubota Y (1985) Conceptual model of the northern Hachimantai-Yakeyama geothermal area (in Japanese with English abstract). J Geothermal Resource Soc Japan 7:231-245

Larson PA, Mudie JD, Larson RL (1972) Magnetic anomalies an fracture zone trends in the Gulf of California. Bull Geol Soc Am 83:3361-3368

Larsson D, Gronvöld K, Oskarsson N, Gunnlaugsson E (2002) hydrothermal alteration of plagioclase and growth of secondary feldspar in the Hengill volcanic center, SW-Iceland. J Volcan Geothermal Res 114: 275-290

Leach TM, Wood CP, Reyes AG (1983) Geology and hydrothermal alteration of the Tongonan geothermal field, Leyte, Republic of the Philippines. Internat Symp Water-Rock Interaction 4:275-278

Liebscher A (2004) Spectroscopy of epidote minerals. Rev Mineral Geochem. 56:125-170

Lindgren W (1933) Mineral deposits. McGraw-Hill Book Co., New York

Liou JG (1973) Synthesis and stability relations of epidote $Ca_2Al_2Fe\ Si_3O_{12}(OH)$. J Petrol 14:381-413

Liou JG (1993) Stabilities of natural epidotes. Abhand Geol Bund 49:7-16

Liou JG, Kim HS, Maruyama S (1983) Prehnite-epidote equilibria and their petrologic applications. J Petrol 24:321-342

Liou JG, Seki Y, Gillemette R, Sakai H (1985) Compositions and parageneses of secondary minerals in the Onikobe geothermal system, Japan. Chem Geol 49:1-20

Lonker SW, Fitzgerald JD, Hedenquist JW, Walshe JL (1990) Mineral-Fluid Interactions in the Broadlands-Ohaaki Geothermal System, New-Zealand. Am J Sci 290:995-1068

Lonker SW, Franzson H, Kristmannsdóttir H (1993) Mineral-fluid interactions in the Reykjanes and Svartsengi geothermal systems, Iceland. Am J Sci 293:605-670

Manning CE, Bird DK (1991) Porosity evolution and fluid flow in the basalts of the Skaergaard magma-hydrothermal system, East Greenland. Am J Sci 291:201-257

Manning CE, Bird DK (1995) Porosity, permeability and basalt metamorphism. *In*: Low-Grade Metamorphism of Mafic Rocks. Schiffman P, Day H (eds) Geol Soc Am Special Paper #296, Boulder, CO, p 123-140

Manning CE, Ingebritsen SE, Bird DK (1993) Missing mineral zones in contact metamorphosed basalt. Am J Sci 293:894-938

Manon MA, Sanchez AA, Fausto LJJ, Jimenez SME, Jacobo RA, Esquer PI (1979) Preliminary geochemical model of the Cerro Prieto geothermal field. Geothermics 8:211-222

Manzella A, Gianelli G, Puxeddu M (1995) Possible models of the deepest part of the Larderello geothermal field. World Geothermal Congress 2:1279-1282

Manzella A, Ruggieri G, Gianelli G, Puxeddu M (1998) Plutonic-geothermal systems of southern Tuscany: A review of the crustal models. Memorie della Societa Geologica Italiana 52:283-294

Marinelli G (1969) Some geological data on the geothermal areas of Tuscany. Bull Volcan 33:319-333

Martinez-Serrano RG (2002) Chemical variations in hydrothermal minerals of the Los Humeros geothermal system, Mexico. Geothermics 31:579-612

Martinez-Serrano RG, Dubois M (1998) Chemical variations in chlorite at the Los Humeros geothermal system, Mexico. Clays Clay Min 46:615-628

Martinez-Serrano RG, Jacquier B, Arnold M (1996) The $\delta^{34}S$ composition of sulfates and sulfides at the Los Humeros geothermal system, Mexico and their applications to physicochemical fluid evolution. J Volcan Geothermal Res 73:99-118

Maruyama S, Suzuki K, Liou JG (1983) Greenschist-amphibolite transition equilibria at low pressures. J Petrol 24:583-604

Matsubaya O, Etchu H, Momuro S (1983) Isotopic study of hot springs in Akita Prefecture. Report Res Underground Resource, Akita Univ 48:11-24

McDowell SD (1986) Composition and structural state of coexisting feldspars, Salton Sea geothermal field. Mineral Mag 50:75-84

McDowell SD, Elders WA (1980) Authigenic layer silicate minerals in borehole Elmore 1, Salton Sea geothermal field, California, USA. Contrib Mineral Petrol 74:293-310

McDowell SD, Elders WA (1983) Allogenic layer silicate minerals in borehole Elmore 1, Salton Sea geothermal field, California. Am Mineral 68:1146-1159

McDowell SD, Paces JB (1985) Carbonate alteration minerals in the Salton Sea geothermal system, California. Mineral Mag 49:469-479

McKibben MA, Andes JP, Jr., Williams AE (1988a) Active ore formation at a brine interface in metamorphosed deltaic lacustrine sediments; the Salton Sea geothermal system, California. Econ Geol 83:511-523

McKibben MA, Elders WA (1985) Fe-Zn-Cu-Pb mineralization in the Salton Sea geothermal system, Imperial Valley, California. Econ Geol 80:539-559

McKibben MA, Eldridge CS (1989) Sulfur isotopic variations among minerals and aqueous species in the Salton Sea geothermal system; a SHRIMP ion microprobe and conventional study of active ore genesis in a sediment-hosted environment. Am J Sci 289:661-707

McKibben MA, Williams AE, Elders WA, Eldridge CS, Hanor JS, Kharaka YK, Land LS (1987) Saline brines and metallogenesis in a modern sediment-filled rift; the Salton Sea geothermal system, California, U.S.A. Appl Geochem 2:563-578

McKibben MA, Williams AE, Okubo S, Norman DI (1988b) Metamorphosed Plio-Pleistocene evaporites and the origins of hypersaline brines in the Salton Sea geothermal system, California; fluid inclusion evidence. Geochim Cosmochim Acta 52:1047-1056

Mehegan JM, Robinson PT, Delaney JR (1982) Secondary mineralization and hydrothermal alteration in the Reydarfjordur drill core, Eastern Iceland. J Geophy Res 87:6511-6524

Mekuria N, Gizaw B, Teklu A, Gizaw T (1987) Geochemistry of Aluto-Langano geothermal field. Ethiopian Inst Geol Surv, Internal Report 55 p

Mercado S (1976) Movement of geothermal fluids and temperature in the Cerro Prieto geothermal field, Baja California, Mexico. Proc 2nd U N Symp Development and Use of Geothermal Resources 1:487-492

Merriam R, Bandy OL (1965) Source of upper Cenozoic sediments in Colorado delta region. J Sed Petrol 4: 911-916

Merrin S (1960) Synthesis of epidote and its apparent P-T stability curve. Bull Geol Soc Am 71:1229

Michard A (1989) Rare earth element systematics in hydrothermal fluids. Geochim Cosmochim Acta 53:745-750

Milodowski AE, Savage D, Bath AH, Fortey NJ, Nancarrow PHA, Shepherd TJ (1989) Hydrothermal mineralogy in geothermal assessment: studies of Miravalles field, Costa Rica and experimental simulations of hydrothermal alteration. Brit Geol Surv, Tech Rep WE/89/63

Moore DG (1973) Plate edge deformation and crustal growth of Gulf of California structural province. Bull Geol Soc Am 84:1883-1906

Moore JN, Powell TS, Heizler MT, Norman DI (2000) Mineralization and hydrothermal history of the Tiwi geothermal system, Philippines. Econ Geol 95:1001-1023

Morrison J (2004) Stable and radiogenic isotope systematics in epidote group minerals. Rev Mineral Geochem 56:607-628

Mottl MJ (1983) Metabasalts, axial hot springs, and the structure of hydrothermal systems at mid-ocean ridges. Bull Geol Soc Am 94:161-180

Muffler LJP, Doe BR (1968) Composition and mean age of detritus of the Colorado River delta in the Salton Trough, southeastern California. J Sed Petrol 38:384-399

Muffler LJP, White DE (1969) Active metamorphism of Upper Cenozoic sediments in the Salton Sea geothermal field and the Salton Trough, Southeastern California. Bull Geol Soc Am 80:157-182

Muramatsu Y, Doi N (2000) Prehnite as an indicator of productive fractures in the shallow reservoir, Kakkonda geothermal system, northeast Japan. J Mineral Petrol Sci 95:32-42

Muramatsu Y, Komatsu R (1996) Fluid inclusions in the Kakkonda shallow geothermal reservoir, Iwate Prefecture, northeastern Japan: a fluid inclusion study (in Japanese with English abstract). J Petrol Min Econ Geol 91:145-161

Muramatsu Y, Komatsu R, Sawaki T, Sasaki M, Yanagiya S (2000) Geochemical study of fluid inclusions in anhydrite from the Kakkonda geothermal system, northeast Japan. Geochem J 34:175-193

Muraoka H, Uchida T, Sasada M, Masahiko Y, Akaku K, Sasaki M, Yasukawa K, Miyazaki S, Doi N, Saito S, Sato K, Tanaka S (1998) Deep geothermal resources survey program: Igneous, metamorphic and hydrothermal processes in a well encountering 500°C at 3729 m depth, Kakkonda, Japan. Geothermics 27:507-534

Naboko SI (1964) Hydrothermal metamorphism of rocks in the volcanic provinces (in Russian). Izd SO An SSSR:129-135

Naboko SI, Piip BI (1961) Recent metamorphism of volcanic rocks in the region of the Pauzhetsk hot springs, Kamchatka (in Russian). Akad Nauk SSSR, Trudy Lab Vulkanol 19:99-114

Nakamura H, Sumi K (1981) Exploration and development at Takinoue, Japan. *In*: Geothermal Systems: Principles and Case Histories. Rybach L, Muffler LJP (eds) John Wiley and Sons Ltd, New York, p 247-272

Nakamura H, Sumi K, Ozawa T (1977) Characteristics of geothermal resources from geological and geochemical viewpoints in Japan. Japan Geothermal Energy Ass 14:3-19

Nakamura Y (1973) Origin of sector zoning of igneous clinopyroxene. Am Mineral 58:986-990

Nakovnik NI (1963) Vertical zonation of products of postmagmatic metasomatism, and the place in it of secondary quartz and prophylites (in Russian). Zap Vses Mineralog Obshch 92:394-409

NEDO (1990) Report for developing evaluation methods of geothermal reservoir, IV Sumikawa (in Japanese). New Energy Development Organization, Japan, p 261-496

Nehlig P, Juteau T (1988) Deep crustal seawater penetraton and circulation at ocean ridges: Evidence from the Oman ophiolite. Marine Geol 84:209-228

Nehlig P, Juteau T, Bendel V, Cotten J (1994) The root zones of oceanic hydrothermal systems: Constraints from the Samail ophiolite (Oman). J Geophy Res 99:4703-4713

Nehring NL, D'Amore F (1984) Gas chemistry and thermometry of the Cerro Prieto, Mexico, geothermal field. Geothermics 13:75-89

Nehring NL, Fausto L JJ (1979) Gases in steam from Cerro Prieto geothermal wells with a discussion of steam/ gas ratio measurements. Geothermics 8:253-255

Neuhoff PS, Fridrikksson T, Bird DK (2000) Zeolite parageneses in the North Atlantic igneous province: Implications for geotectonic and groundwater quality in basaltic crust. Int Geol Rev 42:15-44

Neuhoff PS, Watt WS, Bird DK, Pedersen AK (1997) Timing and structural relations of regional zeolite zones in basalts of the East Greenland continental margin. Geology 25:803-806

Norton D (1977) Fluid circulation in the Earth's crust. Am Geophs Union, Monograph 20:693-704

Norton D (1979) Transport phenomena in hydrothermal systems: the redistribution of chemical components around cooling magmas. Bull Mineral 102:471-486

Norton DL (1984) Theory of hydrothermal systems. Annu Rev Earth Planet Sci 12:155-177

Norton DL (1987) Advective metasomatism. *In*: NATO ASI Series. Series C: Mathematical and Physical Sciences, Vol. 218. Helgeson HC (ed) D. Reidel Publishing Company, Dordrecht-Boston, p 123-132

Norton DL (1988) Metasomatism and permeability. Am J Sci 288:604-618

Norton DL (1990) Pore fluid pressure near magma chambers. *In:* The Role of Fluids in Crustal Processes. Bredehoeft JD, Norton DL (eds) Natl Acad Press, Washington, p 42-49

Norton DL, Dutrow BL (2001) Complex behavior of magma-hydrothermal processes; role of supercritical fluid. Geochim Cosmochim Acta 65:4009-4017

Norton DL, Taylor Jr. HP, Bird DK (1984) The geometry and high-temperature brittle deformation of the Skaergaard intrusion. J Geophy Res 89:10178-10192

Olafsson J, Riley JP (1978) Geochemical studies on the thermal brine from Reykjanes (Iceland). Chem Geol 21:219-237

Olson ER (1979) Oxygen and carbon isotope studies of calcite from the Cerro Prieto geothermal field. Geothermics 8:245-251

Ozawa T, Nagashima S (1975) Geochemical studies for geothermal activity on Onikobe basin. Japan Geothermal Energy Ass 12:35-38

Ozawa T, Nagashima S, Iwasaki I (1980) Geochemical studies for geothermal activity in Onikobe basin. Bull Volcan 43:207-223

Palmason G, Arnórsson S, Fridleifsson IB, Kristmannsdóttir H, Saemundsson K, Stefánsson V, Steingrímsson B, Tómasson J, Kristjansson L (1979) The Iceland crust; evidence from drillhole data on structure and processes. *In*: Maurice Ewing Series, no.2. Talwani M, Harrison CG, Hayes DE (eds) Am Geophy Union, Washington, p 43-65

Pandeli E, Gianelli G, Puxeddu M, Elter FM (1994) The Paleozoic basement of the Northern Apennines; stratigraphy, tectono-metamorphic evolution and Alpine hydrothermal processes. *In*: Memorie Della Societa Geologica Italiana. Vol. 48. Bortolotti V, Chiari M (eds) Societa Geologica Italiana, Rome, p 627-654

Patrier P, Beaufort D, Meunier A, Eymery JP, Petit S (1991) Determination of nonequilibrium ordering state in epidote from the ancient geothermal field of Saint Martin: Application of Mössbauer Spectroscopy. Am Min 76:602-610

Patrier P, Beaufort D, Touchard G, Fouillac AM (1990) Crystal size of epidotes: A potential exploitable geothermometer in geothermal fields? Geology 18:1126-1129

Pedersen RB, Malpas J (1984) The origin of oceanic plagiogranites from the Karmoy ophiolite, Western Norway. Contrib Mineral Petrol 88:36-52

Petrucci E, Gianelli G, Puxeddu M, Iacumin P (1994) An oxygen isotope study of silicates in the Larderello geothermal field, Italy. Geothermics 23:327-337

Petrucci E, Sheppard SMF, Turi B (1993) Water/rock interaction in the Larderello geothermal field (southern Tuscany, Italy); an $^{18}O/^{16}O$ and D/H isotope study. J Volcan Geothermal Res 59:145-160

Prol-Ledesma RM, Browne PRL (1989) Hydrothermal alteration and fluid inclusion geothermometry of Los Humeros geothermal field, Mexico. Geothermics 18:677-690

Prol-Ledesma RM (1998) Pre- and post-exploitation variations in hydrothermal activity in Los Humeros geothermal field, Mexico. J Volcan Geothermal Res 83:313-333

Puente C I, de la Pena L A (1979) Geology of the Cerro Prieto geothermal field. Geothermics 8:155-175

Ragnarsdóttir KV, Walther JV, Arnórsson S (1984) Description and interpretation of the composition of fluid and alteration mineralogy in the geothermal system, at Svartsengi, Iceland. Geochim Cosmochim Acta 48:1535-1553

Raith M (1976) The Al-Fe(III) epidote miscibility gap in a metamorphic profile through the Penninic Series of the Tauern Window, Austria. Contrib Mineral Petrol 57:99-117

Ramos-Candelaria M, Sznchez DR, Salonga ND (1995) Magmatic contributions to Philippine hydrothermal systems. World Geothermal Congress 2:1337-1341

Reed MH (1982) Calculation of multicomponent chemical equilibria and reaction processes in systems involving minerals, gases, and an aqueous phase. Geochim Cosmochim Acta 46:513-528

Reed MH (1983) Seawater-basalt reaction and the origin of greenstones and related ore deposits. Econ Geol 78:466-485

Reed MH (1994) Hydrothermal alteration in active continental hydrothermal systems. Geol Ass Canada, Short Course Notes 11:315-337

Reed MH, Spycher N (1984) Calculation of pH and mineral equilibria in hydrothermal waters with application to geothermometry and studies of boiling and dilution. Geochim Cosmochim Acta 48:1479-1492

Reed MJ (1975) Geology and hydrothermal metamorphism in the Cerro Prieto geothermal field, Mexico. Proc 2nd U N Symp Development and Use of Geothermal Resources 1:539-547

Reed MJ (1984) Relationship between volcanism an hydrothermal activity at Cerro Prieto, Mexico. Trans Geothermal Resource Counc 8:217-221

Reyes AG (1989) Cooling in Philippine geothermal systems. Internat Symp Water-Rock Interaction 6:573-576

Reyes AG (1990) Petrology of Philippine geothermal systems and the application of alteration mineralogy to their assessment. J Volcan Geothermal Res 43:279-309

Reyes AG (1995a) Geothermal systems in New Zealand and the Philippines; why are they so different? Australasian Institute Mining Metallurgy 9/95:485-490

Reyes AG (1995b) Interaction of fluids and rocks at the magmatic-hydrothermal interface of the Mt Cagua geothermal system, Northeastern Luzon Island, the Philippines. Internat Symp Water-Rock Interaction 7:537-541

Reyes AG (1998) Petrology and mineral alteration in hydrothermal systems: From diagenesis to volcanic catastrophes. The United Nations University, Geothermal Training Programme, Reykjavik, Iceland

Reyes AG, Giggenbach WF (1992) Petrology and fluid chemistry of magmatic-hydrothermal systems in the Philippines. Internat Symp Water-Rock Interaction 7:1341-1344

Reyes AG, Giggenbach WF (1999) Condensate-formation and cold water incursion in the Piohipi Sector, Wairakei Geothermal System. New Zealand Geothermal Workshop 21:29-36

Reyes AG, Giggenbach WF, Saleras JR, Salonga ND, Vergara MC (1993) Petrology and geochemistry of Alto Peak, a vapor-cored hydrothermal system, Leyte Province, Philippines. Geothermics 22:479-519

Reyes AG, Zaide-Delfin MC, Bueza EL (1995) Petrological identification of multiple heat sources in the Bacon-Manito geothermal system, the Philippines. World Geothermal Congress 2:713-717

Richardson CJ, Cann JR, Richards HG, Cowan JG (1987) metal-depleted root zones of the Troodos ore-forming hydrothermal system, Cyprus. Earth Planet Sci Lett 84:243-253

Robinson PT, Elders WA, Muffler LJP (1976) Quaternary volcanism in the Salton Sea geothermal field, Imperial Valley, California. Bull Geol Soc Am 87:347-360

Rochelle CA, Milodowski AE, Savage D, Corella M (1989) Secondary mineral growth in fractures in the Miravalles geothermal system, Costa Rica. Geothermics 18:279-286

Rose NM, Bird DK (1987) Prehnite-epidote phase relations in the Nordre Aputiteq and Kruuse Fjord layered gabbros, East Greenland. J Petrol 28:1193-1218

Rose NM, Bird DK (1994) Hydrothermally altered dolerite dykes in East Greenland: implications for Ca-metasomatism of basaltic protoliths. Contrib Mineral Petrol 116:420-432

Rose NM, Bird DK, Liou JG (1992) Experimental investigation of mass transfer - albite, CaAl-silicates, and aqueous solutions. Am J Sci 292:21-57

Ruaya JA, Ramos MN, Gonfiantini R (1992) Assessment of magmatic components of the fluids at Mt. Pinatubo volcanic geothermal system, Philippines from chemical and isotopic data. Geol Survey Japan 279:68-79

Ruggieri G, Cathelineau M, Boiron M-C, Marignac C (1999) Boiling and fluid mixing in the chlorite zone of the Larderello geothermal system. Chem Geol 154:237-256

Ruggieri G, Gianelli G (1995) Fluid inclusion data from the Carboli 1 well, Larderello geothermal field, Italy. World Geothermal Congress 2:1088-1091

Ruggieri G, Gianelli G (1999) Multi-stage fluid circulation in a hydraulic fracture breccia of the Larderello geothermal field (Italy). J Volcan Geothermal Res 90:241-261

Rusinov VL (1966) Findings of prehnite and clastic nature of epidote in regions of hydrothermal metamorphism today. Int Geol Rev 8:731-738

Saemundsson K (1967) Vulkanismus und Tektonik des Hengill-Gebietes inn Suedwest Island. Acta Naturalia Islandica 7:1-105

Saemundsson K (1974) Evolution of the axial rifting zone in northern Iceland and the Tjornes fracture zone. Bull Geol Soc Am 85:495-504

Saemundsson K (1991) The geology of the Krafla system (in Icelandic). *In*: The Natural History of Lake Myvatn. Gardarsson A, Eirnarson P (eds) The Icelandic Natural History Society, Reykjavik, Iceland, p 24-95

Sakai H, Kubota Y, Hatakeyama K (1986) Geothermal exploration at Sumikawa, north Hachimantai, Akita (in Japanese with English abstract). Chinetsu 23:281-302

Sakai S, Matsunsaga E, Kubota Y (1993) Geothermal energy development in the Sumikawa field, northeast Japan (in Japanese with English abstract). Japan Resource Geol 43:409-425

Sanford SJ, Elders WA (1981) Dating thermal events at Cerro Prieto using fission track annealing. Lawrence Berkeley National Laboratory, Report LBL-11967, Berkeley, California, p 114-119

Santoyo E, Verma SP, Nieva D, Portugal E (1991) Variability in the gas phase composition of fluids discharged from Los Azufres geothermal field, Mexico. J Volcan Geothermal Res 47:161-181

Sasaki M, Fujimoto K, Sawaki T, Tsukimura K, Muraoka H, Sasada M, Ohtani T, Yagi M, Kurosawa M, Doi N, Kato O, Kasai K, Komatsu R, Muramatsu Y (1998) Characterization of a magmatic/meteoric transition zone at the Kakkonda geothermal system, northeast Japan. Internat Symp Water-Rock Interaction 9: 483-486

Sasaki M, Sasada M, Fujimoto K, Muramatsu Y, Komatsu R, Sawaki T (1995) History of post-intrusive hydrothermal system indicated by fluid inclusions occuring in the young granitic rocks at the Kakkonda and Nyuto geothermal systems, northern Honshu, Japan (in Japanese with English abstract). Resource Geol 45:303-312

Sato K, Ando S, Iae T, Takanohashi M, Saito S, Chiba Y, Doi N, Iwata T (1981) Geology of the Tamagawa welded tuffs in the Hachimantai volcanic area, northeast Japan (in Japanese with English abstract). J Geol Soc Japan 87:267-275

Sawaki T, Sasaki M, Komatsu R, Muramatsu Y, Sasada M (1999) Gas compositions of fluid inclusions from the shallow geothermal reservoir in the Kakkonda geothermal system, northeast Japan (in Japanese with English abstract). J Geothermal Resource Soc Japan 21:127-141

Schiffman P, Bird DK, Elders WA (1985) Hydrothermal mineralogy of calcareous sandstones from the Colorado River delta in the Cerro Prieto geothermal system, Baja California, Mexico. Min Mag 49: 435-449

Schiffman P, Elders WA, Williams AE, McDowell SD, Bird DK (1984) Active metasomatism in the Cerro Prieto geothermal system, Baja California, Mexico; a telescoped low-pressure, low-temperature metamorphic facies series. Geology 12:12-15

Schiffman P, Fridleifsson GO (1991) The smectite-chlorite transition in drillhole NJ-15, Nesjavellir geothermal field, Iceland: XRD, BSE, and electron microprobe investigatioins. J Meta Petrol 9:679-696

Schiffman P, Smith BM (1988) Petrology and oxygen isotope geochemistry of a fossil seawater hydrothermal system within the Solea Graben, northern Troodos Ophiolite, Cyprus. J Geophy Res 93:4612-4624

Schiffman P, Smith BM, Varga RJ, Moores EM (1987) Geometry, conditions, and timing of off-axis hydrothermal metamorphism and ore-deposition in the Solea graben. Nature 325:423-425

Schmincke H, Viereck LG, Griffin BJ, Pritchard RG (1982) Volcanoclastic rocks of the Reydarfjordur drill hole, Eastern Iceland, 1, Primary features. J Geophy Res 87:6437-6458

Scott JL (1991) Mineralization in the Karmoy ophiolite, southwest Norway., p. 221. MSc Thesis, St. John's memorial University of Newfoundland

Seki Y (1972) Lower-grade stability limit of epidote in light of natural occurrences. J Geol Soc Japan 78: 405-413

Seki Y, Liou JG, Gillemette R, Sakai H, Oki Y, Hirano T, Onuki H (1983) Investigation of Geothermal Systems in Japan I. Onikobe Geothermal Area. Hydroscience and Geotechnology Laboratory, Saitama University

Semmel A (1971) Zur jungquartären Klima - und Releifentwicklung in der Danakilwuste (Athiopen) und ihren westlichen Randgebeiten. Erdkunke 25:199-209

Seyfried WE, Jr., Berndt ME, Seewald JS (1988) hydrothermal alteration processes at mid-ocean ridges: Constraints for diabase alteration experiments, hot-springs, and composition of the oceanic crust. Canadian Min 26:787-804

Shearer CK, Papike JJ, Simon SB, Davis BL (1988) Mineral reactions in altered sediments from the California State 2-14 well: Variations in the modal mineralogy, mineral chemistry and bulk composition of the Salton Sea Scientific Drilling Project core. J Geophy Res 93:13104-13122

Shikazono N (1984) Compositional variations in epidote from geothermal areas. Geochem J 18:181-187

Shimazu M, Yajima J (1973) Epidote and wairakite in drill cores at the Hachimantai geothermal area, northeastern Japan. J Japan Ass Mineral Petrol Econ Geol 68:363-371

Sigvaldason GE (1963) Epidote and related minerals in two deep geothermal drill holes, Reykjavik and Hveragerdi, Iceland. US Geol Surv Prof Paper 450-E:77-79

Simmons SF, Browne PRL (2000) Hydrothermal minerals and precious metals in the Broadlands-Ohaaki geothermal system: Implications for understanding low-sulfidation epithermal environments. Econ Geol 95:971-999

Simmons SF, Christenson BW (1994) Origins of calcite in a boiling geothermal system. Am J Sci 294:361-400

Skinner BJ, White DE, Rose HJ, Jr., Mays RE (1967) Sulfides associated with the Salton Sea geothermal brine. Econ Geol 62:316-330

Smallwood JR, White RS (2002) Ridge-plume interaction in the North Altantic and its influence on continental breakup and seafloor spreading. In: The North Atlantic Igneous Province: Stratigraphy, Tectonic, Volcanic and Magmatic Processes. Jolley DW, Bell BR (eds) Geol Soc Special Pub, London, p 15-37

Smith MP, Henderson P, Jeffries T (2002) The formation and alteration of allanite in skarn from the Beinn an Dubhaich Granite aureole, Skye. Eur J Min 14:471-486

Spieler AR (2003) Rare earth elements in hydrothermal systems: Rare earth element speciation in hydrothermal fluids and the origin of rare earth element zoning in epidote. MSc Thesis, Department of Geological and Environmental Sciences, Stanford University, Stanford, California

Stefánsson A, Arnórsson S (2002) Gas pressures and redox reactions in geothermal fluids in Iceland. Chem Geol 190:251-271

Stefánsson V (1981) The Krafla geothermal field, Northeast Iceland. In: Geothermal Systems and Case Histories. Rybach L, Muffler LJP (eds) John Wiley, New York, p 273-294

Steiner A (1953) Hydrothermal rock alteration at Wariakei, New Zealand. Econ Geol 48:1-13

Steiner A (1966) On the occurrence of hydrothermal epidote at Wariakei, New Zealand (in Russian). Akad Nak SSSR Izv Ser Geol 1:167

Steiner A (1968) Clay minerals in hydrothermally altered rocks at Wairakei, New Zealand. Clays and Clay Min 16:193-213

Steiner A (1977) The Wairakei geothermal area, North Island, New Zealand. New Zealand Geol Survey Bull 60:1-136

Steingrímsson B, Fridleifsson GÓ, Sverrisdóttir G, Tulinius H, Sigurdsson Ó, Gunnlaugsson E (1986a) Nesjavellir, well NJ-15, Drilling, investigations and production characteristics (in Icelandic). National Energy Authority, Report OS-86029/JHD-09, Reykjavik, Iceland

Steingrímsson B, Gudmundsson Á, Franzson H, Gunnlaugsson E (1990) Evidence of a supercritical fluid at depth in the Nesjavellir field. Proc Workshop on Geothermal Reservoir Engineering 15, p. 81-88, Stanford, California

Steingrímsson B, Gudmundsson A, Sigurdsson O, Gunnlaugsson E (1986b) Nesjavellir - well NJ-ll (in Icelandic). p. 60. National Energy Authority, Report OS86025/JHD-05, Reykjavik, Iceland

Steinthórsson S, Oskarsson N, Arnórsson S, Gunnlaugsson E (1987) Metasomatism in Iceland; hydrothermal alteration and remelting of oceanic crust. In: NATO ASI Series C: Math Phys Sci Helgeson HC (ed) D. Reidel, Dordrecht-Boston, p 355-387

Steinthórsson S, Sveinbjörnsdóttir AE (1981) Opaque minerals in geothermal wells No. 7, Krafla, Northern Iceland. J Volcan Geothermal Res 10:245-261

Stern TA (1987) Asymetric back-arc spreading, heating flux and structure associated with the Central Volcanic Region of New Zealand. Earth Planet Sci Lett 85:265-276

Sumi K, Takashima I (1976) Absolute ages of the hydrothermal alteration halos and associated volcanic rocks in some Japanese geothermal fields. Proc 2nd U. N. Symposium on Development and Use of Geothermal Resources, United Nations, San Francisco, p 625-634

Sussman D, Javellina SP, Benavidez PJ (1993) Geothermal energy development in the Philippines: An overview. Geothermics 22:357-367

Suto S (1987) Large scale felsic pyroclastic flow deposits in the Sengan geothermal area, northeast Japan. Tamagawa and Old Tamagawa Welded Tuffs (in Japanese with English abstract). Report Geol Survey Japan 266:43-76

Sveinbjörnsdóttir ÁE (1992) Composition of geothermal minerals from saline and dilute fluids-Krafla and Reykjanes, Iceland. Lithos 27:301-315

Sveinbjörnsdóttir ÁE, Coleman ML, Yardley BWD (1986) Origin and history of hydrothermal fluids of the Reykjanes and Krafla geothermal fields, Iceland: A stable isotope study. Contrib Mineral Petrol 94:99-109

Sveinbjörnsdóttir ÁE, Johnsen S, Arnórsson S (1995) The use of stable isotopes of oxygen and hydrogen in geothermal studies in Iceland. World Geothermal Congress 2:1043-1048

Takeno N, Muraoka H, Sawaki T, Sasaki M (2001) Thermodynamic framework of the contact metamorphism around the Kakkonda granite in a active geothermal field, northeast Japan. Internat Symp Water-Rock Interaction 10:765-768

Tamanyu S, Suto S (1978) Stratigraphy and geochronology of Tamagawa Welded Tuff in the western part of Hachimantai, Akita Prefecture (in Japanese with English abstract). Bull Geol Surv Japan 29:159-173

Taylor H, P., Jr., Forester RW (1979) An oxygen isotope study of the Skaergaard intrusion and its country rocks: a description of a 55-m.y. old fossil hydrothermal system. J Petrol 20:355-419

Teagle DAH (1993) The formation of epidosite in the Phterykhiudhi Potamos Gorge, Troodos ophiolite, Cyprus. EOS Trans Am Geophy Union 74:666

Teklemariam M, Battaglia S, Gianelli G, Ruggieri G (1996) Hydrothermal alteration in the Aluto-Langano geothermal field, Ethiopia. Geothermics 25:679-702

Thompson JM, Fournier RO (1988) Chemistry and geothrmometry of brine produced from the Salton Sea Scientific Drill Hole, Imperial Valley, California. J Geophy Res 93:12,165-12,173

Tómasson J, Kristmannsdóttir H (1972) High temperature alteration minerals and thermal brines, Reykjanes, Iceland. Contrib Mineral Petrol 36:123-134

Torres-Alvarado IS (1998) Chemical stability of the hydrothermal silicates at the Los Azufres geothermal field, Mexico. Internat Symp Water-Rock Interaction 9:697-700

Torres-Alvarado IS (2002) Chemical equilibrium in hydrothermal systems: The case of Los Azufres geothermal field, Mexico. Int Geol Rev 44:639-652

Torres-Alvarado IS, Satir M, Fortier S, Metz P (1995) An oxygen-isotope study of hydrothermally altered rocks at the Los Azufres geothermal field, Mexico. World Geothermal Congress 2:1049-1052

Truesdell AH, Nehring NL, Thompson JM, Janik CJ, Coplen TB (1984) A review of progress in understanding the fluid geochemistry of the Cerro Prieto geothermal system. Geothermics 13:65-74

Truesdell AH, Rye RO, Pearson FJ, Jr., Olson ER, Nehring NL, Whelan JF, Huebner MA, Coplen TB (1979) Preliminary isotopic studies of fluids from the Cerro Prieto geothermal field. Geothermics 8:223-229

Truesdell AH, Thompson JM, Coplen TB, Nehring NL, Janik CJ (1981) The origin of the Cerro Prieto geothermal brine. Geothermics 10:225-238

Truesdell AH, White DE (1973) Production of superheated steam from vapor-dominated geothermal reservoirs. Geothermics 2:154-173

Tryggvason K, Husebye ES, Stefánsson R (1983) Seismic image of the hypothesized Icelandic hot spot. Tectonophys 100:97-118

Ueda R, Kubota Y, Katoh H, Hatakeyama K, Matsubaya O (1991) Geochemical characteristics of the Sumikawa geothermal system, northeast Japan. Geochem J 25:223-244

Valori A, Cathelineau M, Marignac C (1992a) Early fluid migration in a deep part of the Larderello geothermal field; a fluid inclusion study of the granite sill from well Monteverdi 7. J Volcan Geothermal Res 51: 115-131

Valori A, Teklemariam M, Gianelli G (1992b) Evidence of temperature increase of $CO_2$-bearing fluids from Aluto-Langano geothermal field (Ethiopia); a fluid inclusions study of deep wells LA-3 and LA-6. Eur J Min 4:907-919

Van de Kamp PC (1973) Holocene continental sedimentation in the Salton Basin, California: A reconnaissance. Bull Geol Soc Am 84:827-848

Van Hise CR (1904) A treatise on metamorphism. U S Geol Surv Monogr 47

Varga RJ, Gee JS, Bettison-Varga L, Anderson RS, Johnson CL (1999) Early establishment of seafloor hydrothermal systems during structural extension; paleomagnetic evidence from the Troodos Ophiolite, Cyprus. Earth Planet Sci Lett 171:221-235

Viereck LG, Griffin BJ, Schmincke H, Pritchard RG (1982) Volcanoclastic rocks of the Reydarfjordur drill hole, Eastern Iceland, 2, Alteration. J Geophy Res 87:6459-6476

Villa IM, Puxeddu M (1994) Geochronology of the Larderello geothermal field: new data and the "closure temperature" issue. Contrib Mineral Petrol 115:415-426

Villa IM, Ruggieri G, Puxeddu M (1997) Petrological and geochronological descrimination of two white mica generations in a granite cored from Larderello-Travale geothermal field (Italy). Eur J Min 9:563-568

Villa IM, Ruggieri G, Puxeddu M (2002) Geochronology of magmatic and hydrothermal micas from the Larderello geothermal field, Italy. Internat Symp Water-Rock Interaction 10:1589-1592

Vonder Haar S, Howard JH (1981) Intersecting faults and sandstone stratigraphy at the Cerro Prieto geothermal field. Geothermics 10:145-167

Welhan JA, Poreda R, Lupton JE, Craig H (1979) Gas chemistry and helium isotopes at Cerro Prieto. Geothermics 8:241-244

White DE, Anderson ET, Grubbs DK (1963) Geothermal brine well-mile-deep drill hole may tap ore-bearing magmatic water and rocks undergoing metamorphism. Science 139:919-922

White DE, Sigvaldason GE (1963) Epidote in hot-spring systems, and depth of formation of propylitic epidote in epithermal ore deposits. US Geol Survey Prof Paper 450-E:80-84

White RS, McKenzie D (1989) Magmatism at rift zones: The generation of volcanic continental margins and flood basalts. J Geophy Res 94:7685-7729

White RS, McKenzie D (1995) Mantle plumes and flood basalts. J Geophy Res 100:17543-17585

Williams AE, Elders WA (1984) Stable isotope systematics of oxygen and carbon in rocks and minerals from the Cerro Prieto geothermal anomaly, Baja California, Mexico. Geothermics 13:49-63

Williams AE, McKibben MA (1989) A brine interface in the Salton Sea geothermal system, California; fluid geochemical and isotopic characteristics. Geochim Cosmochim Acta 53:1905-1920

Wilson CJN, Houghton BF, McWilliams MO, Lanphere MA, Weaver SD, Briggs RM (1995) Volcanic and structural evolution of Taupo Volcanic Zone, New Zealand: A review. J Volcan Geothermal Res 68:1-28

Wolery TJ (1992) EQ6, A computer program for reaction path modeling of aqueous geochemical systems: Theoretical manual, user's guide, and related documentation. Lawrence Livermore Laboratory UCRL-110662:1-246

Wood CP (1995) Calderas and geothermal systems in the Taupo Volcanic Zone, New Zealand. Proc World Geothermal Congress 2:1331-1336

Yamada E (1972) Study on the stratigraphy of Onikobe area, Miyagi Prefecture. Japan Geol Survey Bull 23: 217-231

Yamada E (1976) Geological development of the Onikobe caldera and its hydrothermal systems. Proc 2nd U N Symp Development and Use of Geothermal Resources. 1, United Nations, San Francisco. p 665-672

Yamada E, Okada H, Nishimura S, Taniguchi M, Natori H (1978) Hydrothermal alteration of Katayama and Narugo geothermal areas, Tamatsukuri-gun, Miyagi Prefecture. Report Geol Survey Japan 259:341-375

Yanagiya S, Kasai K, Brown KL, Giggenbach WF (1996) Chemical characteristics of deep geothermal fluid in the Kakkonda geothermal system, Iwate Prefecture, Japan (in Japanese with English abstract). J Japan Geothermal Energy Ass 33:1-18

Yardley B, Rochelle CA, Barnicoat AC, Lloyd GE (1991) Oscillatory zoning in metamorphic minerals: an indicator of infiltration metasomatism. Mineral Mag 55:357-365

Zen E (1974) Burial metamorphism. Canadian Min 12:445-455

Reviews in Mineralogy & Geochemistry
Vol. 56, pp. 301-345, 2004
Copyright © Mineralogical Society of America

# 7

# Epidote Group Minerals in Low–Medium Pressure Metamorphic Terranes

## Rodney H. Grapes and Paul W. O. Hoskin

*Institut für Mineralogie, Petrologie und Geochemie*
*Albert-Ludwigs-Universität Freiburg, Albertstrasse 23 b*
*D-79104 Freiburg, Germany*
*rodney.grapes@minpet.uni-freiburg.de*
*paul.hoskin@minpet.uni-freiburg.de*

## INTRODUCTION

Epidote group minerals are common in metamorphosed mafic to intermediate igneous rocks, quartzofeldspathic sediments and calc-alumina silicate (marl) rocks of higher grade zeolite to medium grade amphibolite facies of low–medium pressure contact and regional metamorphic terranes (i.e., pressure and temperature conditions below the calcite-to-aragonite transition). Within any one rock, epidote composition in terms of $Fe^{3+}/(Fe^{3+} + Al)$ can be variable, but in general, is limited by whole-rock composition, such that epidote group minerals in metabasite lithologies are more Fe-rich than those in marls that tend to be more Al-rich and typically include zoisite. Because of their wide range of *P-T* stability, epidote group minerals of variable composition may form in a single rock during several stages of metamorphic re-equilibration. Slow rates of intra-crystalline $Fe^{3+}$-Al exchange, especially at low temperatures, preserve complex zonation patterns in individual grains that can serve as a "tape-recorder" providing evidence for continuous or discontinuous prograde and retrograde reactions and the *P-T*-fluid-redox conditions of metamorphism. Thus, relic lower grade epidote (typically Fe-rich) often form cores over which new (typically less Fe-rich) higher grade epidote rims form. In such a case, the boundary between the two generations of epidote may be sharp or gradational depending on the temperature history of the rock and $Fe^{3+}$-Al diffusivity. Often compositional differences are blurred across the boundary. In addition to zoning is the spread of individual epidote grain compositions within a rock (even on a thin-section scale). This is related to variation in the composition of coexisting phases or reactants (e.g., quartz; Ca-Al silicates such as plagioclase, margarite, lawsonite; mafic silicates such as chlorite, pumpellyite, amphiboles; other Ca-$Fe^{3+}$ silicates such as prehnite, andraditic garnet; carbonates; Fe-oxides; relict volcanic glass), which serve as compositional micro-domains in which epidote may form. Thus, where epidote is formed from the albitization of plagioclase (e.g., in sub-greenschist facies rocks), it tends to be more aluminous (clinozoisite) and may be zoned to more Fe-rich rims, or if it forms after An-rich plagioclase in amphibolite grade rocks it is typically clinozoisite or zoisite. This affects the correct interpretation of the extent of miscibility gaps in both the orthorhombic-monoclinic and within the monoclinic members of the epidote series. The effect of sector zoning may also be a complication, as suggested by Banno and Yoshizawa (1993).

Two other important variables that control epidote group mineral composition, referred to above, are fluid composition ($H_2O$ or $CH_4$-rich verses $CO_2$-rich) and the local oxidation

1529-6466/04/0056-0007$05.00

state, as epidote $Fe^{3+}$-content is strongly dependent on $f_{O_2}$ (Holdaway 1972; Liou 1973). Again, consideration of these variables is relevant both to whole-rock compositions and sub-domains in rocks. For example, in sub-greenschist facies metabasites, epidote with different compositions may occur in amygdules with or without calcite, and in the rock matrix where it may replace pumpellyite or has replaced volcanic glass and grown together with pumpellyite, chlorite, etc. In the two sub-domains (amygdules and matrix), $X_{CO_2}$, $X_{H_2O}$ and $f_{O_2}$ can significantly differ.

These interrelated controls, as well as other kinetic factors, determine the composition of single growth-generation epidote at any particular pressure and temperature. Nevertheless, general trends of epidote compositional changes in a number of metamorphic terranes have established that in metabasaltic and quartzofeldspathic rocks epidote tends to become more aluminous with increasing grade (Miyashiro and Seki 1958; Apted and Liou 1983). On the other hand, zoisite and clinozoisite, for example in meta-marl and Ca-rich metapelite, become more Fe-rich with increasing metamorphic grade.

This chapter reviews the changes in composition, the textures, epidote-producing and consuming reactions, conditions of oxidation, fluid composition, and *P-T* conditions during zoisite and clinozoisite-epidote formation in various rock types from selected low to medium pressure regional and contact metamorphic terranes. The review is intended to be broad but not comprehensive; it is focused on occurrences where epidote composition and textual data for a range of related rocks within a single terrane are available. Additional data for a range of epidote group mineral occurrences can be found in the summary of Deer et al. (1997). Data on compositional varieties such as Mn- (excluding piemontite; see Bonazzi and Menchetti 2004), Sr- and Cr-bearing members of the epidote group are included. Evidence for composition gaps in the zoisite-clinozoisite and clinozoisite-epidote series for natural occurrences is discussed in relation to experimental verification of miscibility gaps in the epidote group series. Finally, critical mineral associations and reactions relevant to epidote group mineral paragenesis in metabasaltic and Al-marl rocks are summarized.

### Nomenclature, mineral abbreviations

In this chapter, epidote group mineral compositions are expressed as $100(Fe^{3+}/(Fe^{3+} + Al))$ and denoted as $X_{Fe}$ throughout the text. Mineral abbreviations used in the text and figures are listed in Table 1.

## EPIDOTE GROUP MINERALS IN DIFFERENT LITHOLOGIES

### Metabasite

***Karmutsen Metabasite, Vancouver Island, Canada.*** In the low grade Karmutsen metabasite, Vancouver Island, epidote is one of the most common Ca-Al silicates and occurs in amygdules, veins and as fine-grained aggregates in the rock matrix. Cho et al. (1986) show that in low-variance (calcite-free, > 3-phase assemblages with excess chlorite + quartz) amygdule assemblages of epidote, pumpellyite, chlorite, quartz ± laumontite ± prehnite, there is a systematic change in epidote compositions together with an increase in modal abundance with increasing metamorphic grade. In zeolite facies metabasites, epidote $X_{Fe}$ content decreases toward the laumontite-out isograd, whereas upgrade of the isograd epidote becomes more Fe-rich (Fig. 1A). The metamorphic field gradient over which this change occurs is estimated to be ~160–250°C and 1.1–1.4 kbar (Cho et al. 1986), the high temperature gradient of ~80–90°C/km evidently relates to intrusion of the Jurassic Coast Range batholith. Greenschist facies assemblages ($T > {\sim}350°C$) are developed within 3.6 km of the batholith contact and contain unzoned epidote of $X_{Fe_{25-26}}$ (Kuniyoshi and Liou 1976a,b).

**Table 1.** Mineral abbreviations and compositions.

| | | |
|---|---|---|
| Ab | albite | $NaAlSi_3O_8$ |
| AS | *aluminum silicate* | $Al_2SiO_5$ |
| Act | actinolite | $Ca_2(Fe,Mg)_5Si_8O_{22}(OH)_2$ |
| And | andalusite | $Al_2SiO_5$ |
| An | anorthite | $CaAl_2Si_2O_8$ |
| Bt | biotite | $K_2(Fe,Mg,Al)_6(Si,Al)_6O_{20}(OH)_4$ |
| Cc | calcite | $CaCO_3$ |
| Ch | chlorite | $(Fe,Mg)_{6-x}Al_x(Si_{4-x}Al_x)O_{10}(OH)_8$ |
| Cz | clinozoisite | $Ca_2Al_3Si_3O_{12}(OH)$ |
| Do | dolomite | $CaMg(CO_3)_2$ |
| Ep | epidote | $Ca_2(Al,Fe^{3+})_3Si_3O_{12}(OH)$ |
| Grs | grossular | $Ca_3Al_2Si_3O_{12}$ |
| Hb | hornblende | $NaCa_2(Mg,Fe,Al)_5Si_6Al_2O_{22}(OH)_2$ |
| Jd | jadeite | $NaAlSi_2O_6$ |
| Ka | kaolinite | $Al_4Si_4O_{10}(OH)_8$ |
| Kf | K-feldspar | $KAlSi_3O_8$ |
| Ky | kyanite | $Al_2SiO_5$ |
| Lm | laumontite | $CaAl_2Si_4O_{12}(H_2O)_4$ |
| Lw | lawsonite | $CaAl_2Si_2O_7(OH)_2(H_2O)$ |
| Mrg | margarite | $Ca_2Al_4Si_4Al_4O_{20}(OH)_4$ |
| Ms | muscovite | $K_2Al_4Si_6O_{20}(OH)_4$ |
| Pg | paragonite | $Na_2Al_4Si_6Al_2O_{20}(OH)_4$ |
| Ph | phengite | $K_2(Mg,Al)_4(Si,Al)_6O_{20}(OH)_4$ |
| Pr | prehnite | $Ca_2(Al,Fe)_2Si_3O_{10}(OH)_2$ |
| Pp | pumpellyite | $Ca_4(Al,Fe)_5MgSi_6O_{21}(OH)_7$ |
| Prl | pyrophyllite | $Al_2Si_4O_{10}(OH)_2$ |
| Q | quartz | $SiO_2$ |
| Sil | sillimanite | $Al_2SiO_5$ |
| Stp | stilpnomelane | $(K,Na,Ca)_{0.6}(Mg,Fe^{2+},Fe^{3+})_6Si_8Al(O,OH)_{27}(H_4O)_{2-4}$ |
| Tm | tremolite | $Ca_2Mg_5Si_8O_{22}(OH)_2$ |
| Tn | titanite | $CaTiSiO_4(O,OH,F)$ |
| V | *vapor* | $H_2O$ |
| Zo | zoisite | $Ca_2Al_3Si_3O_{12}(OH)$ |

In other areas of the Karmutsen metabasite, epidote composition has been determined by Terabayashi (1988) and Starkey and Frost (1990). In pumpellyite-prehnite facies metabasite matrix and amygdules occurring within 3.5 km of the granite contact, Terabayashi (1988) shows that epidote $X_{Fe}$ decreases with increasing grade (Fig. 1B). This is the opposite trend to that determined by Cho et al. (1986), and is attributed to lower $f_{O_2}$ conditions related to lower $X_{CO_2}$ in the fluid phase. Al-enrichment also continues upgrade through prehnite-actinolite facies rocks, following a shift to more Fe-rich compositions across the actinolite-in isograd (Fig. 1B). In coarse-grained contact metamorphic amygdule assemblages of highest grade zeolite to lowest prehnite-pumpellyite facies rocks (~260°C and 1.5 kbar) studied by Starkey and Frost (1990), epidote is typically strongly and irregularly zoned with rim or near rim compositions of $X_{Fe_{10-29}}$. In lowest grade (± zeolite) calcite-bearing assemblages compositions range from $X_{Fe_{24-29}}$. Those in slightly higher grade (zeolite-absent) rocks are less Fe-rich at $X_{Fe_{10-22}}$.

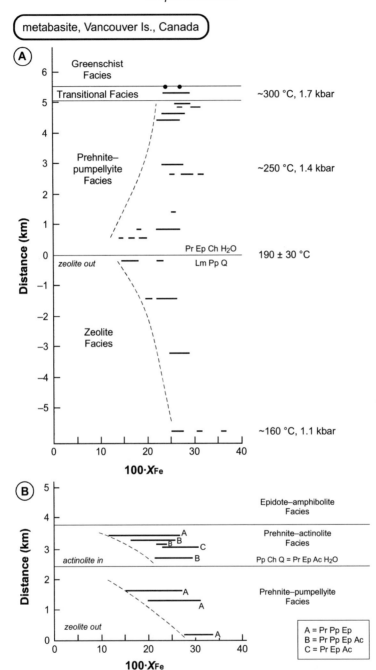

**Figure 1.** Plots of epidote $X_{Fe}$ with respect to map distance (km) from the zeolite-out isograd, Karmutsen metabasite, Vancouver Island, BC, Canada. (**A**) Data from Cho et al. (1986), Cho and Liou (1987), Kuniyoshi and Liou (1976a,b). The two filled circles in the Transitional facies are epidote compositions from Kuniyoshi and Liou (1976a). (**B**) Data from Terabayashi (1988). For mineral abbreviations see Table 1.

***Hamersley Basin, Australia.*** In the Hamersley Basin, Western Australia, Smith et al. (1982) document changes in epidote grain-size, texture and composition in basaltic and andesitic flows, and pyroclastics, with increasing grade from prehnite-pumpellyite, through pumpellyite-actinolite to greenschist facies (~210–430°C and 0.5–2.6 kbar; thermal gradient of ~40°C/km). In prehnite-pumpellyite facies rocks, epidote first appears as isolated 100–200 μm pale-yellow anhedra in the groundmass, and at higher grade it becomes more abundant forming yellow-colored domains together with pumpellyite, quartz and titanite. In pumpellyite-actinolite facies rocks grain-size is typically 200–250 μm and in greenschist facies lavas the epidote is essentially colorless so that epidote-rich domains have a grey to light-grey color resembling irregular quartzite patches.

The change in epidote $X_{Fe}$ with metamorphic grade, depth of burial and temperature, and in terms of Al-$Fe^{3+}$-Mg, is shown in Figure 2. Prehnite-pumpellyite and pumpellyite-actinolite facies epidote core-rim composition can range over 0.2 $Fe^{3+}$ and Al apfu, whereas greenschist facies epidote has a restricted range of <0.1 $Fe^{3+}$ and Al apfu. Overall, there is a gradual shift

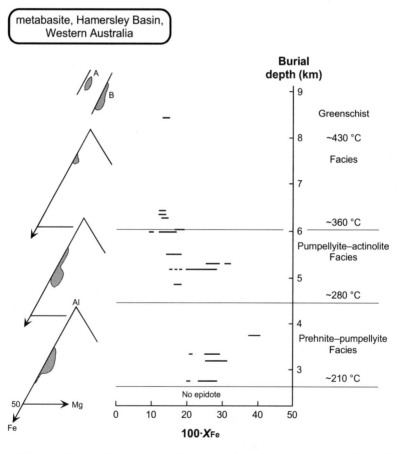

**Figure 2.** Plot of epidote $X_{Fe}$ in mafic–intermediate metabasites with respect to depth of burial (km) and in terms of isofacial Al-$Fe^{3+}$-Mg variation, Hamersley Basin, WA, Australia (data from Smith et al. 1982). Shaded area A = variation in a single grain; shaded area B = maximum variation within a single thin-section.

to less Fe-rich compositions with increasing grade. Some epidote from the middle grade part of the prehnite-pumpellyite facies zone are notably more ferric with $X_{Fe\sim40}$ (Fig. 2). Although not specifically commented on by Smith et al. (1982), such compositions probably occur in the conspicuous yellow-colored epidote-rich domains at this grade and formed under conditions of high $f_{O_2}$.

A number of epidote grains are observed to contain isolated twin lamellae (5 μm-wide) and in others with faint, very fine checker-board type textures that may represent exsolution. Unfortunately, any compositional differences between "twins" and host could not be resolved, and from the grains analyzed there is no evidence for the presence of a composition gap in the clinozoisite-epidote series.

***Northern California, U.S.A.*** Prehnite-pumpellyite, prehnite-actinolite and greenschist facies grade epidotes were analyzed by Springer et al. (1992) in a metabasite succession of the Smartville Complex, northern Sierra Nevada, California, USA, that formed at between ~250–350°C and 2.5 ± 0.5 kbar. Epidote occurs as xenoblastic grains within plagioclase, as idioblastic to xenoblastic grains with chlorite, and in amygdules and veins. Compositional ranges (Fig. 3) demonstrate a change from $X_{Fe_{18-32}}$ (prehnite-pumpellyite facies), through $X_{Fe_{8-18}}$ (prehnite-actinolite facies), to $X_{Fe_{13-32}}$ (greenschist facies). A gap between $X_{Fe_{21}}$ and $X_{Fe_{14}}$ is evident in greenschist facies epidote (upper panel of Fig. 3). The decrease in epidote $X_{Fe}$ from prehnite-pumpellyite to prehnite-actinolite facies is similar to that recorded by Terabayashi (1988) in the Karmutsen volcanics (Fig. 1B) and by Smith et al. (1982) in metabasites of the Hamersley Basin (Fig. 2).

Optically determined epidote compositions (accurate within ± 5 mol% $X_{Fe}$) are reported from regional metamorphic greenschist to epidote-amphibolite facies metabasite, Klamath Mountains, California, USA, by Holdaway (1965). In greenschist (with albite), a gap between

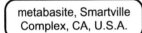

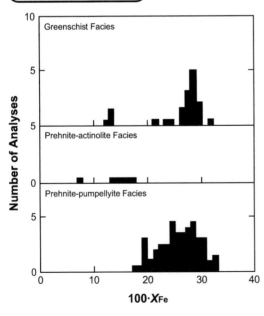

**Figure 3.** Frequency plots of epidote $X_{Fe}$ in metabasite of the Smartville Complex, northern Sierra Nevada, CA, USA (data from Springer et al. 1992). Prehnite-pumpellyite facies epidotes occur with and without amphibole and in one sample with andradite-grossular garnet in amygdules.

$X_{Fe_{16}}$ and $X_{Fe_{29}}$ is evident. In low grade epidote-amphibolite metabasite (first appearance of oligoclase with albite), epidote shows a continuous range from $X_{Fe_{21}}$ to $X_{Fe_{34}}$ and in higher grade rocks of the epidote-amphibolite facies the ranges are $X_{Fe_{18-24}}$ and $X_{Fe_{11-14}}$ (Fig. 4).

Hietanen (1974) determined the composition of epidote, clinozoisite and zoisite in greenschist (~340°C) and upper amphibolite (~590°C) facies metavolcanic and plutonic rocks of the northern Sierra Nevada, California, USA. Composition with respect to distance from the Granite Basin Pluton contact (Fig. 5) highlights the presence of a gap between epidote-clinozoisite and between clinozoisite-zoisite. Epidote in greenschist facies metadacite and meta-andesite show gaps between $X_{Fe_{12}}$ and $X_{Fe_{22}}$ and between $X_{Fe_{11}}$ and $X_{Fe_{25}}$, respectively. In these rocks, epidote ± clinozoisite occur as inclusions in albite, in aggregates with amphiboles and chlorite or with quartz, albite ± chlorite, and as pseudomorphs of pyroxene and plagioclase. In the high grade metagabbro, ~500 m from the granite contact, clinozoisite ($X_{Fe_{4-9}}$), epidote ($X_{Fe_{23}}$) and Fe-poor-zoisite ($X_{Fe_{0.2-0.4}}$) are present. Clinozoisite and epidote occur as individual grains or in clusters, with stringers, lamellae and domains of epidote within clinozoisite, and surrounded by small grains of zoisite (Fig. 5) that also occurs in later-formed veinlets and pods. Abundant dust-like inclusions in the zoisite may be Fe-oxide that exsolved when zoisite formed from clinozoisite. Zoisite is absent in another metagabbro, mineralogically and texturally similar to the former, ~640 m from the contact. Up to 1.3 km from the igneous contact the metamorphic grade is upper greenschist facies as the rocks contain only albite.

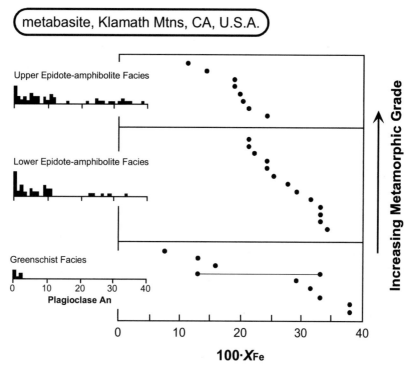

**Figure 4.** Epidote $X_{Fe}$ with respect to increasing metamorphic grade in metabasite, Klamath Mountains, CA, USA (after Holdaway 1965). Compositions are estimated optically from $2V_\alpha$ or β-refractive index. The thin horizontal line connects coexisting epidote and clinozoisite in the greenschist facies. Frequency plots of coexisting plagioclase compositions indicate metamorphic grade ranges from greenschist facies to lower and higher grade parts of the epidote-amphibolite facies.

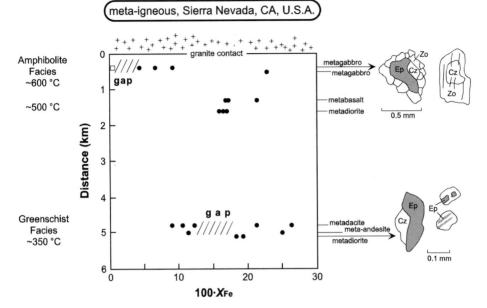

**Figure 5.** Epidote, clinozoisite and zoisite $X_{Fe}$ in metavolcanic and metaplutonic rocks with respect to map distance (km) from a granite pluton contact, northern Sierra Nevada, CA, USA (data from Hietanen 1974). Textural relations between epidote-clinozoisite and epidote-clinozoisite-zoisite in two samples and estimated temperatures of metamorphism are shown. Cross-hatched areas in the plot delineate composition gaps between epidote-clinozoisite (at ~4.9 km) and clinozoisite-zoisite (at ~0.3 km). Zoisite compositions of $X_{Fe_{0.2-0.4}}$ are indicated by the open square symbol.

Within 1 km of the contact, however, the plagioclase is anorthite-bytownite and the high temperature stable amphibole is hornblende. This indicates a very steep thermal gradient of perhaps >100°C/km within 1 km of the pluton resulting in the formation of zoisite and with Ca-rich plagioclase ($An_{72-93}$) in metagabbro near the contact possibly formed from the breakdown of zoisite.

*North and South Island, New Zealand.* Allocthonous metabasites in prehnite-pumpellyite facies Torlesse greywacke terrane of the Wellington area, North Island, are characterized by epidote associated with matrix chlorite, pumpellyite and titanite. The epidote has cores of $X_{Fe_{20-21}}$ and rims of $X_{Fe_{23-24}}$ and also contains ~0.86 and 3.05 wt% SrO, respectively (Roser and Grapes 1990). In the schistose equivalent of the Torlesse terrane rocks in South Island (the Otago and Haast schists), epidote in lowest grade (chlorite zone) greenschist facies metavolcanic horizons of the Otago Schist has more Fe-rich cores of $\sim X_{Fe_{27}}$ surrounded by $X_{Fe_{13-14}}$ rims (Bishop 1972), whereas higher grade greenschist facies chlorite and biotite zone metabasite epidote has compositions between $X_{Fe_{21}}$ and $X_{Fe_{28}}$ (Brown 1967) (Fig. 6). In the Haast Schist, Cooper (1972) records that zoning to Fe-poor rims is common in highest grade chlorite zone epidote, but with increasing grade the grains are more homogeneous and lack the Fe-rich cores typical of chlorite zone epidote. Also at higher grades, epidote is a common constituent occurring as distinctive yellowish-green laminations and as segregations of coarse prismatic crystals up to 1 cm long. Evidence of epidote breakdown to form plagioclase comes from the common occurrence of thin haloes of oligoclase around epidote inclusions in albite porphyroblasts in garnet zone metabasites. The disappearance of albite at the onset of amphibolite facies grade is marked by randomly oriented vermicular bodies of quartz within

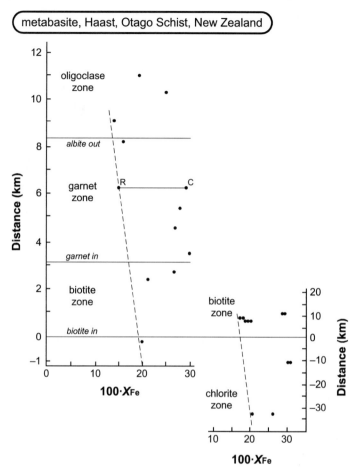

**Figure 6.** Metabasite epidote $X_{Fe}$ plotted against map distance (km) from the biotite-in isograd, eastern Otago Schist (right side) and Haast Schist (left side), South Island, New Zealand (data from Brown 1967 and Cooper 1972, respectively). Note the distance scale difference between the two plots. The core and rim analyses of a single crystal are denoted C and R, respectively. Dashed lines delineate minimum $X_{Fe}$ content with increasing metamorphic grade.

oligoclase that can be explained by the reaction:

$$2 \text{ albite} + 2 \text{ clinozoisite} + 2H^+ = 4 \text{ anorthite} + 4 \text{ quartz} + 2Na^+ + 2H_2O \tag{1}$$

The composition of epidote in biotite zone (upper greenschist facies), garnet zone (transitional to amphibolite facies) and oligoclase zone (amphibolite facies) metabasites is a function of distance upgrade of the biotite isograd (Fig. 6). There is an overall decrease of $X_{Fe}$ (from about $X_{Fe_{20}}$ to $X_{Fe_{16}}$) with increasing grade, as in chlorite to biotite zone metabasite epidote analyzed by Brown (1967). In the higher grade metabasites, the compositional change is not regular because of a positive relation between maximum epidote $X_{Fe}$ and bulk rock oxidation ratio (see section on rock oxidation influence below).

**_Yap Islands, western Pacific Ocean._** Epidote composition in metabasaltic basement rocks transitional from low $T$-$P$ (400–450°C and ~3 kbar) greenschist to amphibolite facies of Yap Islands, western Pacific, is recorded by Maruyama et al. (1983). Frequency plots (Fig. 7) indicate overall decreasing $X_{Fe}$ with increasing grade from greenschist facies (with chlorite, actinolite, albite, quartz, sulfides), through transitional facies (epidote-amphibolite; with chlorite, actinolite, hornblende, albite, oligoclase, quartz ± magnetite), to amphibolite facies (with chlorite, hornblende, oligoclase, quartz, magnetite). Zoning is from $Fe^{3+}$-rich cores to less Fe-rich rims. No grains with more than $X_{Fe_{30}}$ are recorded in greenschist metabasites although a relic composition of $\sim X_{Fe_{35}}$ persists in amphibolite facies rocks that also contain the most Fe-poor epidote ($X_{Fe_{19}}$).

***Tauern Window, Austria.*** Compositional variation of epidote from a prograde sequence of metabasaltic rocks (greenschist, garnet-amphibolite, eclogite) in the Eastern Alps, Austria, is detailed by Raith (1976). With increasing grade and decreasing oxidation conditions, zoned crystals become less Fe-rich and core-rim/outermost (retrograde) rim relations define a distinct asymmetric composition gap that narrows from $X_{Fe_{18-24}}$ to $X_{Fe_{18-21}}$ with increasing grade throughout greenschist facies rocks. The gap closes in garnet amphibolite rocks with a critical composition of $X_{Fe_{19}}$ (Figs. 8A and 8B). The *T-P* conditions of the gap vary from ~400–550°C and 5–6 kbar. β-zoisite ($\sim X_{Fe_2}$) occurs in

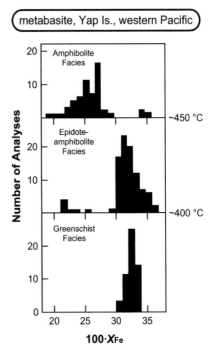

**Figure 7.** Frequency plots of epidote $X_{Fe}$ in greenschist, epidote-amphibolite and amphibolite facies metabasites, Yap Islands, western Pacific (after Maruyama et al. 1983).

greenschist, garnet-amphibolite and eclogite rocks and in the latter is joined by α-zoisite ($X_{Fe_{5-7}}$) as thin lamellae in β-zoisite and as individual grains (Fig. 8A).

## Quartzofeldspathic rocks

***Metagreywacke and schist, New Zealand.*** Unzoned matrix epidote in low grade Torlesse terrane prehnite-pumpellyite facies greywacke sandstone and black argillite, southern North Island, New Zealand, typically range $X_{Fe_{16-21}}$ with most grains characterized by narrow Sr-rich (1–3 wt% SrO) replacement rims (Roser and Grapes 1990; Grapes, unpublished data). In interlayered red argillite (hematite-bearing) and in greenish-grey argillites, homogeneous

---

**Figure 8** (*on facing page*). **(A)** Prograde epidote group mineral compositions plotted against map distance (km) in a metabasite sequence, together with schematic sketches of zoning profiles in single grains, Tauern Window, Eastern Alps, Austria (data from Raith 1976). Symbols: epidote (horizontal bars and dashes), β-zoisite (filled circles) and α-zoisite (open circles). Zoisite nomenclature after Myer (1966) with β-zoisite ($X_{Fe_{0-5}}$; $n\gamma\|c$, $n\beta\|b$, $n\alpha\|a$) and α-zoisite ($X_{Fe_{5-7}}$; $n\gamma\|c$, $n\beta\|a$, $n\alpha\|b$). Thin sub-vertical dashed and solid lines mark the extent of composition gaps in the clinozoisite-epidote series and between β-zoisite and α-zoisite. The left-most line through β-zoisite compositions indicates a trend of slightly increasing $Fe^{3+}$ with increasing metamorphic grade. **(B)** Retrograde epidote $X_{Fe}$ plotted against map distance (km) in a metabasite sequence, together with schematic sketches of zoning profiles in single grains, Tauern Window, Eastern Alps, Austria (data from Raith 1976). Here, epidote $X_{Fe}$ is plotted as small filled circles and not horizontal bars and dashes as in Fig. 8A. Dashed lines mark the extent of a composition gap. Solid lines indicate relative Al and $Fe^{3+}$ enrichment of Fe- and Al-rich compositions respectively, resulting in a narrowing of the epidote compositional range with increasing grade.

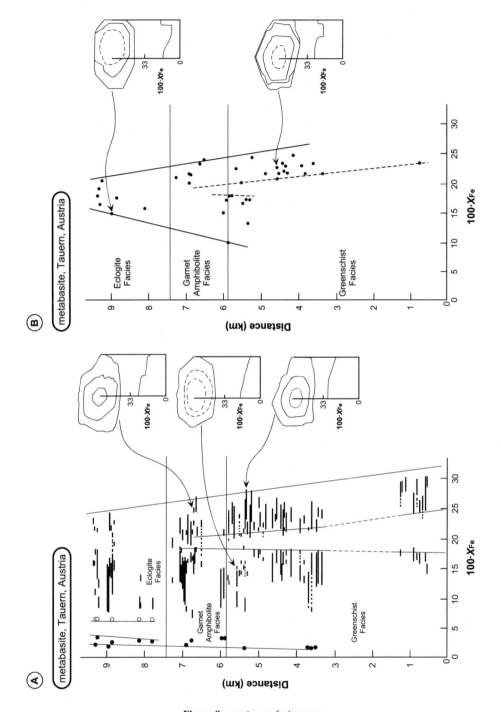

**Figure 8.** *caption on facing page*

epidote is more Fe-rich ($X_{Fe_{19-29}}$) and lacks Sr-rich rims (Roser and Grapes 1990). Several generations of epidote are present (Fig. 9), including one involving Sr-replacement, in higher grade prehnite-pumpellyite facies and lower grade pumpellyite-actinolite facies greywacke of central North Island, New Zealand (Grapes et al. 2001). First generation epidote is Fe-rich ($X_{Fe_{34-38}}$). Second generation epidote ($X_{Fe_{24-28}}$) occurs as clusters up to 100 μm in diameter. A third generation forms homogeneous subhedral–anhedral grains of $X_{Fe_{18-22}}$ and in volcanogenic greywacke it occurs as overgrowths on earlier generations. In non-volcanogenic greywacke, fourth generation and the most Fe-poor epidote ($X_{Fe_{12-15}}$) occur as small (<10 μm) grains in the matrix and as overgrowths on earlier epidote; this composition represents growth at the highest temperature of metamorphism. Epidote of all generations show incipient Sr-replacement along margins and cracks with SrO up to 5.6 wt%.

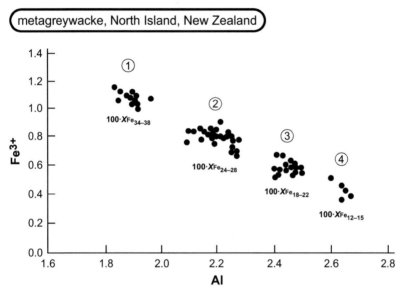

**Figure 9.** $Fe^{3+}$-Al plot of epidote from prehnite-pumpellyite facies metagreywacke, central North Island, New Zealand (data from Grapes et al. 2001). Various generations of epidote (with $100X_{Fe}$ range indicated) discussed in the text are labeled 1 through 4 with increasing temperature.

In South Island, New Zealand, epidote composition in schistose equivalents of North Island occurrences described above, has been determined by Bishop (1972) and Kawachi (1975) and compared with North Island examples (Fig. 10). In north Otago (South Island), authigenic epidote ($\sim X_{Fe_{20}}$) in high grade prehnite-pumpellyite facies and schistose pumpellyite-actinolite facies greywacke form tiny grains or spongy aggregates often associated with pumpellyite, and overgrowths on detrital epidote (Bishop 1972). Epidote in weakly schistose prehnite-pumpellyite–pumpellyite-actinolite facies rocks of the Upper Wakatipu (South Island) area described by Kawachi (1975) have a maximum at $\sim X_{Fe_{29}}$ and show no systematic difference between quartzofeldspathic and intercalated mafic lithologies. Nevertheless, in all occurrences, there is a marked shift to lower $X_{Fe}$ in the high grade part of the pumpellyite-actinolite facies (effectively the pumpellyite-actinolite to greenschist facies transition; Fig. 10). The later formed, highest temperature epidote rims in non-volcanogenic metasediment are less Fe-rich than those in volcanogenic metasediment reflecting the influence of different bulk rock compositions.

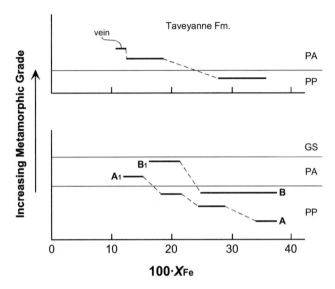

**Figure 10.** Schematic plot showing changes in epidote $X_{Fe}$ for two or more generations versus metamorphic grade (PP = prehnite-pumpellyite facies; PA = pumpellyite-actinolite facies; GS = greenschist facies). The letter A represents epidote $X_{Fe}$ in central North Island, New Zealand, metagreywacke and $A_1$ represents the latest generation epidote formed under pumpellyite-actinolite facies conditions (data from Grapes et al. 2001). The letter B represents epidote $X_{Fe}$ in Upper Wakatipu area, South Island, New Zealand, volcanogenic metasediments and mafic volcanics, with undistinguished earlier generations at prehnite-pumpellyite facies conditions and latest (rim) epidotes formed under the higher grade part of pumpellyite-actinolite facies conditions indicated by $B_1$ (data from Kawachi 1975). The upper part of the diagram shows epidotes from the Taveyanne Formation volcanogenic sandstone, Switzerland (data from Coombs et al. 1976). In this meta-sandstone, authigenic epidote ($100X_{Fe_{16-19}}$) overgrows spongy crystals and aggregates of early Fe-rich epidote ($100X_{Fe_{28-37}}$) some of which may be detrital. Vein epidote is the most Al-rich with $100X_{Fe_{11-13}}$. Metamorphic *T-P* conditions of 300–360°C and ~2.5–5 kbar are postulated by Schmidt et al. (1997).

Brown (1967) shows a slight enrichment in the maximum $X_{Fe}$ of epidote with increasing grade in greenschist facies (chlorite and biotite zone) quartzofeldspathic schist in east Otago (Fig. 11A), that also reflects weak zoning to more Fe-rich rims. In chlorite zone rocks a small number of relic Fe-rich epidote grains ($\sim X_{Fe_{30}}$) form cores to rims of $X_{Fe_{15-18}}$ and the rim composition is also measured for independent unzoned grains. These more Fe-poor epidotes formed under low grade greenschist facies conditions with a frequency maximum at $X_{Fe_{19}}$ (Fig. 11B). Similarly, Bishop (1972) records two generations of epidote as grains with cores of $X_{Fe_{20-21}}$ in sharp contact with rims of $X_{Fe_{13-14}}$ in lowest grade greenschist facies metagreywacke, although in some grains the core-rim zonation is gradual.

A detailed study of epidote paragenesis in greenschist facies chlorite and biotite zone metagreywacke of the Alpine Schist, Southern Alps of New Zealand, where mineral zones are steeply dipping, has been made by Grapes and Watanabe (1984). Several growth generations are recognized. The *T-P* conditions of metamorphism are between ~260–370°C and ~4.5–7 kbar (Grapes 1995). Upgrade of the pumpellyite-out isograd, in chlorite zone rocks, there is a small epidote composition gap that closes in the lowest grade part of the biotite-albite zone

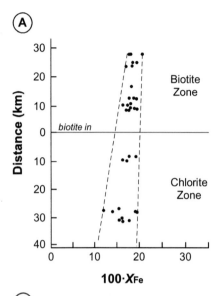

quartzofeldspathic schist, Otago, New Zealand

**Figure 11. (A)** Epidote $X_{Fe}$ in greenschist facies (chlorite and biotite zones) quartzofeldspathic schist, eastern Otago, South Island, New Zealand, plotted against map distance (km) from the biotite-in isograd. **(B)** Frequency plot of the data in (A) (chlorite and biotite zone quartzofeldspathic rocks, $100X_{Fe} \sim$ < 20) and more $Fe^{3+}$-rich epidote from quartzose schist and metabasite rocks (see Fig. 6) (data from Brown 1967).

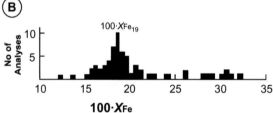

and the range of epidote $X_{Fe}$ within any one rock narrows with increasing grade (Fig. 12). The characteristics of distinct epidote generations are:

(i) Earliest generation epidote ($X_{Fe_{25-29}}$) occur as individual grains (chlorite zone), cores (biotite-albite zone), and as overgrowths on detrital epidote and allanite. Many grains are cracked as a result of cataclasis that accompanies the transition from metagreywacke to schist. They may represent relics of prehnite-pumpellyite or pumpellyite-actinolite facies metamorphism (cf. Brown 1967; Kawachi 1975).

(ii) First generation epidote is overgrown and partially replaced by second generation epidote ($X_{Fe_{20-25}}$) that also forms individual grains weakly zoned to less Fe-rich rims.

(iii) Third generation epidote is Sr-rich as a result of replacement along margins and cracks of pre-existing epidote and does not form new grains. The Sr enrichment comprises a distinct "time" marker.

(iv) Fourth generation epidote mantles earlier grains and Sr-replacement rims. Composition ranges $X_{Fe_{6-15}}$ (chlorite zone), through $X_{Fe_{9-20}}$ and $X_{Fe_{17-20}}$ in lower and higher grade parts of the biotite-albite zone, respectively. Individual grains show continuous zoning from Fe-rich cores to lower Fe rims whereas overgrowths on

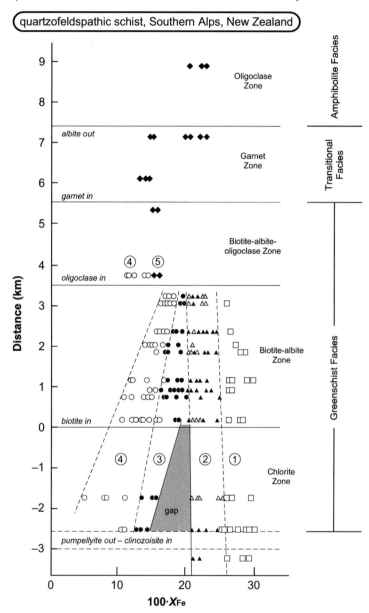

**Figure 12.** Epidote $X_{Fe}$ in quartzofeldspathic schist, central Southern Alps, South Island, New Zealand, plotted against structural distance (km) between steeply dipping isograds from the biotite-in isograd (data from Grapes and Watanabe 1984, 1994). Epidote generations are numbered 1 through 5 with increasing temperature. Symbols: open squares = first generation; filled and open triangles = second and third (Sr-rich) generations, respectively; filled circles = fourth generation rims on earlier generations; open circles = fourth generation weakly zoned to homogeneous individual grains; filled diamonds = fifth generation (see text for explanation). In the greenschist facies chlorite and biotite-albite zones, solid lines define a composition gap between second–third generation and fourth generation epidote, while dashed lines define compositional limits for various epidote generations.

earlier epidote tend to be homogeneous. In the lowest grade part of the biotite-albite-oligoclase zone, fourth generation epidote is notably more aluminous with cores of $X_{Fe_{10}}$ grading to rims of $X_{Fe_{13}}$, the opposite trend of lower grade epidote.

Fourth generation Fe-poor epidote first appears in the vicinity of the pumpellyite-out isograd and is accompanied by a decrease in modal chlorite and an increased amount of actinolite suggesting the following reaction, resulting in a greenschist facies assemblage:

$$\text{pumpellyite} + \text{chlorite} + \text{quartz} = \text{actinolite} + \text{clinozoisite} + H_2O \qquad (2)$$

(v) Above the oligoclase-in isograd which more or less coincides with the development of a true schistosity characterized by >2 mm-thick quartz–plagioclase and mica-rich lamellae, fourth generation epidote is surrounded by discontinuous rims of $X_{Fe_{14-16}}$ (fifth generation). The rims typically extend the grains in the direction of the schistosity indicating they grew during its development. Fe-enrichment of fifth generation epidote coincides with the appearance of small amounts of oligoclase ($An_{20-24}$) and an abrupt decrease in modal epidote, suggesting the reaction:

$$\begin{aligned}\text{epidote} + \text{muscovite} + \text{chlorite} + \text{quartz} \\ = \text{anorthite} + \text{Fe-richer epidote} + \text{biotite} + H_2O \qquad (3)\end{aligned}$$

At higher grades (transitional facies garnet zone and amphibolite facies oligoclase zone), homogeneous or weakly zoned (to more Fe-rich rims) epidote remains a minor constituent in some rocks. It becomes more Fe-rich ($\sim X_{Fe_{9-15}}$ to $X_{Fe_{20-23}}$; Fig. 12) in schist almost on the albite-out isograd (equal to the closure of the peristerite gap) above which associated plagioclase becomes progressively more An-rich.

***Hohe Tauern area, Austria.*** Epidote compositions from a sequence of greenschist to amphibolite facies psammite, pelite, paragneiss, orthogneiss, metatuff and amphibolite in the Hohe Tauern area, Tyrol, Austria (Fig. 13) were determined by Hörmann and Raith (1973). Epidote $X_{Fe}$ decreases in amphibolite facies rocks above the albite-out isograd consistent with decreasing whole-rock oxidation ratios. Associated plagioclase generally remains within the range of $An_{15-28}$, but unlike the New Zealand Alpine Schist, epidote does not become more Fe-rich and the Hohe Tauern amphibolite grade rocks contain titanite ± rutile and calcite rather than ilmenite. Epidote typically has Fe-rich cores with respect to rims, but in amphibolite grade paragneiss and orthogneiss, reverse zoned grains also occur.

## Calc-silicate rocks

***Northwest Scotland.*** Thin (cm-scale) calc-silicate pods, lenses or bands within the Moinian amphibolite facies rocks of northwest Scotland contain zoisite and clinozoisite. The association of both phases with oligoclase, andesine and bytownite-anorthite with prograde metamorphism illustrates their relative stabilities. The composition, texture and habit are described by Kennedy (1949) and Tanner (1976) and are summarized here:

(i) Weakly zoned needle-like zoisite ($X_{Fe_{3.4-4.3}}$) is stable in lower grade (oligoclase-zoisite-calcite zone) rocks, occurring within oligoclase or, more commonly, along plagioclase–plagioclase grain-boundaries. Associated clinozoisite ($X_{Fe_{14-17}}$) forms equant crystals within a fine-grained quartz-feldspar matrix.

(ii) In higher grade (andesine-zoisite-biotite zone) rocks, the zoisite forms stumpy prisms in both plagioclase and matrix together with minor clinozoisite that replaces some zoisite. The zoisite contains rare, minute quartz inclusions.

(iii) In andesine-zoisite-hornblende zone rocks, zoisite with quartz inclusions, forms stumpy prismatic to rounded anhedral grains within plagioclase. Coexisting clinozoisite forms irregular, rounded strongly zoned granules. It may replace zoisite.

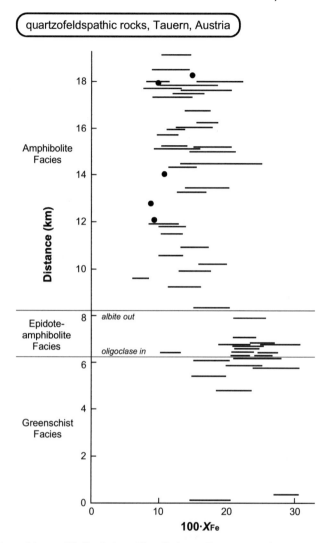

**Figure 13.** Greenschist–amphibolite facies epidote $X_{Fe}$ in a sediment-magmatite sequence plotted against map distance (km), Hohe Tauern area, Tyrol, Austria (data from Hörmann and Raith 1973). Symbols: thick lines = compositional ranges; filled circles = single analyses.

(iv) At the highest grade (calcic plagioclase–hornblende ± zoisite zone), zoisite either occurs as large spongy aggregates that are rimmed by and altered along cracks to clinozoisite, or is generally absent with clinozoisite occurring in cracks within plagioclase and intergrown with amphibole and garnet. The disappearance of zoisite may be related to Reaction (1).

***Central-south Maine, U.S.A.*** Fe-zoisite without clinozoisite in argillaceous calcareous layers of a regional metamorphic sequence (370–550°C and 2.5–3.5 kbar), central-south Maine, USA, is described by Ferry (1976a,b) who mapped a zoisite-in isograd within high grade staurolite–cordierite–andalusite zone rocks (Fig. 14). Zoisite (ranging $X_{Fe_{7-10}}$ with

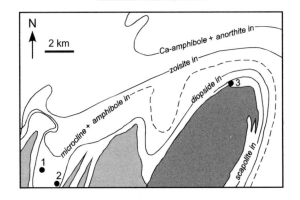

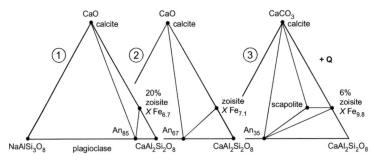

**Figure 14.** Map showing distribution of isograds around biotite granite (darker shading) and quartz monzonite (lighter shading), Waterville-Vassalboro area, south-central Maine, USA, together with chemographic relations (projections through $SiO_2$, $H_2O$, $CO_2$) between zoisite, plagioclase, calcite and scapolite coexisting with quartz from map localities 1, 2 and 3 representing differences in $T$, $f_{CO_2}$, $f_{H_2O}$ (after Ferry 1976a) (see text). Textural evidence indicates that scapolite is produced by the Reaction: 3 anorthite + calcite = scapolite and does not involve zoisite. Compositions in the figure cited as $X_{Fe}$ are $100X_{Fe}$.

increasing grade) first appears as overgrowths on calcic plagioclase and at plagioclase–calcite grain-boundaries implying the reaction:

$$3 \text{ anorthite} + \text{calcite} + H_2O = 2 \text{ zoisite} + CO_2 \tag{4}$$

Zoisite zone rocks contain the assemblage zoisite, andesine/anorthite, calcic amphibole, quartz, calcite, ± biotite, ± microcline, and zoisite persists into higher grade diopside zone metacarbonates so that Reaction (4) could be expanded in order to better represent the Fe-zoisite composition (Ferry 1983):

0.07 Fe-comp. in amphibole + 0.59 Fe-comp. in calcite + 1.31 anorthite + 0.01HCl + 0.59$H_2O$
$$= 1.00 \text{ zoisite} + 0.002 \text{ titanite} + 0.10 \text{ quartz} + 0.01\text{NaCl} + 0.59CO_2 + 0.16H_2 \tag{5}$$

Zoisite with the highest Fe-content of $X_{Fe_{10}}$ occurs in a scapolite zone that is restricted to a narrow area adjacent to biotite-granodiorite (Fig. 14), implying temperatures of at least 565°C. This supports the contention that zoisite becomes more $Fe^{3+}$-rich with increasing temperature (Enami and Banno 1980).

Temperatures for the zoisite zone rocks are between ~500–520°C with $X_{CO_2}$ in the fluid phase between ~0.09 and 0.12, and for diopside zone rocks the values are >520°C and $X_{CO_2}$ between 0.06 and 0.13 (Ferry 1983). Unlike the Moinian calc-silicates (Tanner 1976), zoisite in the central-south Maine sequence forms at a higher temperature (~60°C higher) than calcic plagioclase ($An_{70}$) (± calcic amphibole) (Fig. 14). This can be related to differences in the $X_{CO_2}/X_{H_2O}$ ratio of the metamorphic fluids of the two terranes. In the case of the Moinian rocks, prograde metamorphism occurred at nearly constant or possibly increasing $X_{CO_2}$, while prograde conditions for the Maine rocks occurred with the introduction of $H_2O$ according to Reactions (4) and (5) and the metacarbonates were in equilibrium with $H_2O$-rich fluids at all times during metamorphism.

***Lepontine Alps, Switzerland and Italy.*** Frank (1983) records assemblages comprising clinozoisite only, zoisite–clinozoisite and zoisite only in progressively metamorphosed amphibolite facies calcareous schists of the western Lepontine Alps, Switzerland and Italy. There is an overall trend of Fe-enrichment of zoisite and clinozoisite with increasing grade (Fig. 15) although core–rim analyses of single grains indicate lower Fe in rims suggesting

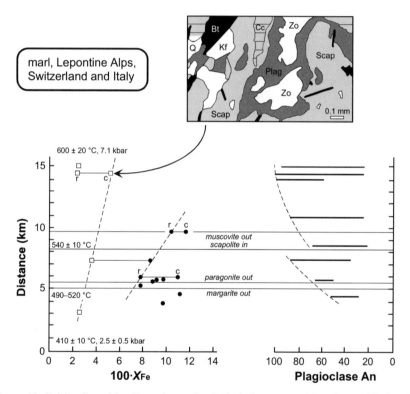

**Figure 15.** Zoisite-clinozoisite $X_{Fe}$ and associated plagioclase compositions in amphibolite facies calcareous schist, western Lepontine Alps, Switzerland and Italy, plotted against map distance (km) (data from Frank 1979, 1983). Symbols: open squares = zoisite; filled circles = clinozoisite; thick lines = range of plagioclase compositions. Thin horizontal tie-lines join core (c) and rim (r) compositions in individual zoisite and clinozoisite grains and, in one case, coexisting zoisite–clinozoisite. Dashed lines in the zoisite-clinozoisite plot delineate a possible composition gap. Curved dashed lines in the plagioclase plot indicate the maximum An composition of plagioclase in a single rock. Approximate positions of isograds are also shown (labeled thin horizontal lines). The textural sketch of a thin-section photomicrograph shows zoisite (Zo) rimmed by calcic plagioclase (Plag) separating it from calcite (Cc) and scapolite (Scap); Bt = biotite; Kf = K-feldspar; Q = quartz.

decreasing temperature. At the highest grade (above the muscovite-out isograd) only zoisite (as anhedral grains possibly indicating instability) is present where it is typically rimmed by anorthitic plagioclase separating the zoisite from matrix calcite (Fig. 15) and is explained by reversal of Reaction (4), similar perhaps to that in the Moinian rocks. Scapolite is also present and is for the most part separated from zoisite by plagioclase which can be strongly zoned from $An_{92}$ near zoisite to $An_{61}$ with scapolite. In addition to Reaction (4), the textural relations (Fig. 15) suggest that plagioclase with calcite, and possibly zoisite, have been involved in scapolite-producing reactions:

$$3 \text{ anorthite} + \text{calcite} = \text{scapolite} \tag{6}$$

$$2 \text{ zoisite} + CO_2 = \text{scapolite} + H_2O \tag{7}$$

Reaction (6) is also indicated by a reverse in the trend of increasing An-content of plagioclase across the scapolite-in isograd and Reaction (7) may also have contributed to a decrease in the Al-content of clinozoisite (Fig. 15).

Just downgrade of the scapolite-in isograd one sample contains both zoisite and clinozoisite (Fig. 15). Compositions upgrade and downgrade of this sample are suggestive of an up-temperature diverging asymmetric composition gap between zoisite and clinozoisite, similar to that delineated by Franz and Selverstone (1992) and demonstrating the Fe-enrichment of zoisite and clinozoisite with increasing metamorphic grade.

***Hohe Tauern, Austria.*** High grade marls in the western and central parts of the Hohe Tauern area contain a number of zoisite/clinozoisite/epidote-bearing assemblages ($X_{Fe}$ values not reported) that allow determination of a zoisite-in isograd according to the reactions (Höck and Hoschek 1980; Hoschek 1980a):

$$\text{muscovite} + \text{chlorite} + \text{calcite} + \text{quartz} = \text{biotite} + \text{zoisite} + H_2O + CO_2 \tag{8}$$

$$\text{muscovite} + \text{dolomite} + \text{quartz} + H_2O = \text{biotite} + \text{zoisite} + \text{calcite} + CO_2 \tag{9}$$

In the western area, the epidote mineral-bearing assemblages (+ quartz) are:

dolomite–calcite–zoisite/epidote–chlorite–plagioclase
calcite–zoisite
calcite–biotite–zoisite/epidote

and in the central part:

calcite–zoisite
calcite–chlorite–paragonite–clinozoisite–plagioclase
calcite–dolomite–chlorite–zoisite/clinozoisite
calcite–dolomite–chlorite–paragonite–margarite–chloritoid–zoisite
calcite–dolomite–chlorite–zoisite– ± paragonite–garnet
calcite–chlorite–zoisite/clinozoisite–garnet

Associated siliceous dolomite may contain the assemblage:

calcite–dolomite–quartz–tremolite–zoisite–chlorite

At the same pressure (5 kbar), the inferred $T$- $X_{CO_2}$ conditions for biotite-zoisite-bearing Fe-poor marls of the western Hohe Tauern lie in the range 480–560°C and 0.03–0.1 $X_{CO_2}$ (Höck and Hoschek 1980). Differences between the two sets of assemblages, such as the scarcity of biotite and the relative abundance of margarite, paragonite and/or grossular in the central Hohe Tauern area, reflect lower temperatures coupled with relatively higher bulk-rock Al, Na, Ca and lower K contents.

## Mn, Cr, Sr AND Mg ABUNDANCES OF LOW–MEDIUM PRESSURE METAMORPHIC EPIDOTE GROUP MINERALS

### Manganoan epidote

***Occurrences.*** Manganoan zoisite-clinozoisite-epidote with abundances between 0.05 and 4.0 wt%, reported as MnO or $Mn_2O_3$, are typically found in veins, mineral cleavage planes, pegmatite phases and cavities, as well as rock matrix in greenschist–amphibolite facies rocks (e.g., Brown 1967; Smith and Albee 1967; Grapes and Hashimoto 1978; Abrecht 1981 and references therein; Keskinen 1981; Kawachi et al. 1983; Pouliot et al. 1984; Reinecke 1986; Janeczek and Sachanbinski 1989; Grapes 1996; Liebscher and Franz, unpublished data). The occurrences indicate both a hydrothermal origin that in most cases appears to be related to granite intrusion, and Mn-rich precursor sediments, e.g., chert, carbonate, pelite.

An unusual occurrence of possibly retrograde (epidote-amphibolite facies) zoned manganoan clinozoisite is described by Heinrich (1964) from Ruby Mountains, Montana, USA. The crystals have pink cores and colorless rims and occur as porphyroblastic radial aggregates replacing plagioclase in banded dolomitic marble and associated lenses of corundum gneiss intercalated within amphibolite grade biotite schist. A wet chemical analysis gave 0.45 wt% $Mn_2O_3$, 1.83 wt% $Fe_2O_3$ and 0.08 wt% FeO. Unzoned manganoan zoisite with 1.6–3.7 wt% $Mn_2O_3$ coexists with Al-piemontite in retrograde greenschist facies (400–500°C and 5–6 kbar) pelitic and calcareous schists at Andros Island, Greece (Reinecke 1986). Mn-zoisite–Al-piemontite pairs show a composition gap that unlike that between clinozoisite and ferrian manganoan zoisite, varies in position and extent in $Fe^{3+}$-$Mn^{3+}$ space depending on $f_{O_2}$ and/or $X_{H_2O}$ (Reinecke 1986).

Epidote ($X_{Fe_{22-32}}$) in greenschist facies (chlorite zone) Mn-rich metacherts, Otago schist, South Island of New Zealand, is characterized by MnO ranging 0.89–2.15 wt%, with $Mn^{2+}$ probably substituting for Ca as they lack the pink hue characteristic of manganoan zoisite/ clinozoisite or piemontite (Brown 1967; Kawachi et al. 1983). The grains are weakly zoned with lower Mn abundances in rims than in cores. The manganoan epidote occurs with spessartine garnet and one example described by Kawachi et al. (1983) contains a core of piemontite. The *T-P* conditions for the manganoan epidote, spessartine, albite, quartz, white mica, chlorite, titanite, tourmaline, rutile ± hematite ± calcite assemblage are approximately 400°C and 6.4 kbar, similar to greenschist facies conditions at Andros Island, Greece.

Coexisting zoisite and clinozoisite-epidote with variable manganese (as $Mn_2O_3$) contents of between 0.22–4.04 wt%, 0.35–3.41 wt%, and 0.06–0.53 wt% occur in three amphibolite grade calcsilicate rocks respectively, Norway (Liebscher and Franz, unpublished data). The compositional variation is related to single grain core–rim and patchy zoning.

In terms of Al-$Fe^{3+}$-Mn cation proportions manganoan epidote group minerals show a maximum Mn-content of $\sim X_{Mn_{10}}$ (Fig. 16; see also Bonazzi and Menchetti 2004).

***Color and $f_{O_2}$ control.*** The typical pale pink color of manganoan zoisite/clinozoisite and epidote has been ascribed to $Mn^{3+}$ or $Mn^{2+}$ with no influence by $Fe^{3+}$. Manganese is widely believed to be present as $Mn^{3+}$ (Tillmanns et al. 1984) in manganoan zoisite-clinozoisite (>1.5 wt%) with very low iron content (<0.6 wt%). Burns and Strens (1967) suggest site preferences of $Mn^{3+}$: $M3 \geq M2 >> M1$ (see Liebscher 2004). In oxidized piemontite-bearing metasediments at Andros Island, Greece, manganoan zoisite shows a good negative correlation between (Al$Fe^{3+}$) and $Mn^{3+}$ (as total Mn) with *M*-site occupancies of 3.001 ± 0.014 apfu on the basis of 12.5 oxygens (Reinecke 1986). On the other hand, in a low Mn-zoisite (<0.5 wt%) with up to 1.85 wt% $Fe_2O_3$, there is a negative correlation between Ca and Mn (as $Mn^{2+}$; Abrecht 1981). In other manganoan zoisite and clinozoisite with intermediate Mn values and low Fe that occur in less oxidized (ilmenite or magnetite or bearing) assemblages lacking piemontite,

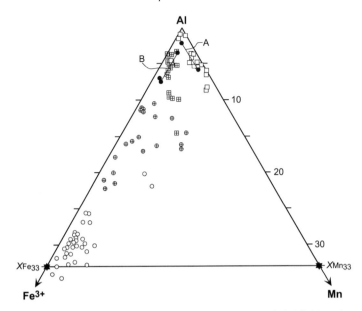

**Figure 16.** Compositions of manganoan epidote group minerals in terms of Al-$Fe^{3+}$-Mn cation proportions. Symbols: □ = zoisite; ● = clinozoisite, and those with tie lines connect pink and colorless lamellae compositions in single clinozoisite grains. The two different trends probably reflect different oxidation conditions (see text) (A, data from Grapes (1996), variable Mn, constant $Fe^{3+}$; B, data from Janeczek and Sachanbinski (1989), variable $Fe^{3+}$, constant Mn). ○ = epidote. Sources of the above data: Neumann and Svinndal (1955), Heinrich (1964), Smith and Albee (1967), Grapes and Hashimoto (1978), Abrecht (1981), Keskinen (1981), Kawachi et al. (1983), Pouliot et al. (1984), Reinecke (1986), Janeczek and Sachanbinski (1989) and Grapes (1996). Crossed squares and crossed circles are zoisite and clinozoisite-epidote compositions, respectively, in three amphibolite grade calc-silicate samples from Norway (Telemark, Leksvika and an unspecified locality; mineral collection, Technische Universität Berlin) (unpublished data of Franz and Leibscher).

both $Mn^{3+}$ and $Mn^{2+}$ appear to be present. For example, the data of Janeczek and Sachanbinski (1989) for zoned pink and colorless clinozoisite show that although Mn contents of colorless and pink zones are essentially the same (0.63 and 0.60 wt% MnO (total Mn), respectively), Fe contents are different (1.10 and 3.17 wt% $Fe_2O_3$, respectively). From stoichiometry and charge balance, colorless zones appear to be characterized by a $^{A2}Mn^{2+} = {}^{A2}Ca^{2+}$ substitution and the pink zones by a $^{M3}Mn^{3+} = {}^{M3}Fe^{3+}$ substitution, indicating that Mn is present with both valencies with the $Mn^{3+}/Mn^{2+}$ ratio varying from zone to zone and $Mn^{3+}$ predominant in the pink zones. Janeczek and Sachanbinski (1989) consider that higher $Fe^{3+}$ in the pink zones, compared with the colorless zones, may reflect broad, system-wide oxidation changes during clinozoisite growth. Another clinozoisite, occurring in blocks of manganiferous calc-silicate in serpentinite (Grapes 1996), is characterized by patchy zoning and alternating pink and colorless lamellae (Fig. 17), the $Fe_2O_3$ content is the same (0.48 wt%) and the MnO content is different at 2.25 (pink) and 0.45 wt% (colorless). In this case, stoichiometry and charge balance imply that a significant amount of the Mn, as $Mn^{3+}$, replaces Al rather than $Fe^{3+}$ in the pink zones and virtually all Mn, as $Mn^{2+}$, replaces Ca in the colorless zones. These examples underscore the effect of $f_{O_2}$ in determining Mn valency and $Mn^{3+}/Mn^{2+}$ ratios in these minerals.

   Piemontite-bearing metasediments typically lack manganoan epidote, clinozoisite and zoisite (the Andros Island association described above is an exception) reflecting the occurrence of Mn as $Mn^{3+}$ at high $f_{O_2}$ (Dollase 1969; Keskinen and Liou 1979). In such

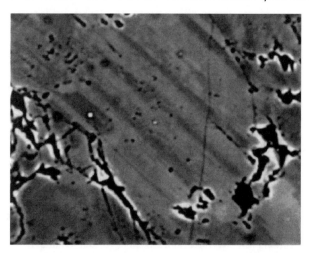

**Figure 17.** Backscattered electron image of clinozoisite with Mn-rich (lighter grey) and Mn-poor (darker grey) lamellae in a manganese-rich calc-silicate inclusion in serpentinite, Cobb Ultramafic Complex, northwest Nelson, New Zealand. The Mn-rich (lighter grey) lamellae are pink coloured in transmitted light while the Mn-poor (darker grey) lamellae are colorless. Lamellae compositions are plotted in Figure 16. Width of image is 60 μm.

conditions, $Fe^{3+}$ partitions into an oxide (hematite) or Fe-silicate phase (e.g., braunite). Other greenschist to upper amphibolite facies grade occurrences of manganoan epidote include those in which oxidized piemontite-bearing layers, lamellae or lenses are intercalated with less oxidized manganoan epidote-bearing quartz-rich layers or mafic rocks (Smith and Albee 1967, Grapes and Hashimoto 1978; Keskinen 1981). Of these occurrences, the highest MnO abundance (as total Mn) recorded is 3.52 wt% in a yellowish-green epidote (Grapes and Hashimoto 1978). Recalculation on the basis of 12.5 oxygens with all Mn as $Mn^{2+}$ (0.232 apfu) results in an excess of 0.073 cations in the $A$-site indicating that some Mn may also be present as $Mn^{3+}$. The host rock to this epidote is less oxidized than associated piemontite-bearing whole-rock samples. Keskinen (1981) notes that epidote in mafic rocks and chert associated with piemontite-bearing tuffs, is yellow-green, weakly pleochroic, sometimes showing a slight pink tinge (in the mafic rocks), and in cherts has pink–greenish intermediate pleochroism that is attributed to the presence of small amounts of $Mn^{3+}$.

**Chromian epidote**

Chromian epidote group minerals occur in low–medium pressure amphibolite grade rocks that are closely associated with ultramafics from which the Cr is derived. Epidote with 6.79 wt% $Cr_2O_3$ was described by Eskola (1933) in quartzite from Otukumpu, Finland, and with 11.16 wt% $Cr_2O_3$ from a jadeite-bearing rock from Tawmaw, Myanmar. Later investigation of the Otukumpu occurrence by Treloar (1987a,b) describes the epidote as pale-yellow to pale-green euhedral crystals up to 0.5 mm, but sometimes as large as 2 mm, exhibiting both sector and oscillatory zoning and containing up to 15.4 wt% $Cr_2O_3$. This composition is close to the theoretical Cr end-member (tawmawite, $X_{Cr_{33}}$) of the clinozoisite-epidote series. The *T-P* conditions of amphibolite facies regional metamorphism of the Outukumpu rocks are estimated at $600 \pm 50°C$ and $3.5 \pm 1$ kbar (Treloar et al. 1981). At Tawmaw, epidote occurs with zoisite, Na-amphibole, chlorite and chloritoid in a contact zone between jadeite-albite rock and peridotite (Bleeck 1907). At Karnataka, India, yellowish-green Cr-epidote is found with uvarovitic garnet in barite layers within amphibolite grade quartzite in the upper part of an ophiolite sequence (Devaraju et al. 1999). In one occurrence, $Cr_2O_3$ contents range 9.5–11.9

wt% and in another 0.9–4.7 wt%. Similar variation is present in epidote from thin Cr-rich layers in metacarbonates of the Nevado-Filábride Complex, Betic Cordilleras, southern Spain, where $Cr_2O_3$ can vary from 0.51–5.61 wt% in a single sample with differences of up to nearly 3 wt% $Cr_2O_3$ in individual grains (Sánchez-Vizcaíno et al. 1995). Such differences invariably occur in grains that contain inclusions of chromian zincian spinel, with Cr in epidote increasing towards the chromite, although in other examples chromite can also be intergrown with low Cr-epidote.

Coexisting Cr-bearing zoisite and epidote occurrences in kyanite-amphibolite and associated calc-silicate bands and metachert, Southern Alps, New Zealand, are described by Cooper (1980) and Grapes (1981). In kyanite-amphibolite, zoisite and epidote occur as individual, 30–90 μm-long homogeneous grains with chromite and green hornblende in bytownite. Zoisite contains 0.23–0.40 wt% $Cr_2O_3$ and epidote 3.72–7.12 wt% $Cr_2O_3$ (Grapes 1981). Two varieties of zoisite are associated in this rock with margarite alteration of kyanite (Cooper 1980). One, a yellow ($v > r$) β-zoisite ($X_{Fe_{6-7}}$) contains 1.35–3.20 wt% $Cr_2O_3$, and the other, a colorless ($r > v$) α-zoisite ($X_{Fe_{1-3}}$) has 0.16–0.81 wt% $Cr_2O_3$. In the calc-silicate, zoisite ($X_{Fe_{0.3}}$) is included in bytownite-anorthite and has 0.20 wt% $Cr_2O_3$. Cooper (1980) suggests maximum *T-P* conditions of 650–700°C and ~10 kbar for the kyanite–zoisite assemblage in the calc-silicate bands. A further occurrence of chromium-bearing zoisite (a green variety with 1.75 wt% $Fe_2O_3$ and 0.33 wt% $Cr_2O_3$) is described by Game (1954) in a corundum-bearing high grade amphibolite from northern Tanganyika, Tanzania.

A plot of chromian zoisite and clinozoisite-epidote in terms of Al-Cr-$Fe^{3+}$ cation proportions (Fig. 18) indicates substitution of $Al^{3+}$ by $Cr^{3+}$ at constant $Fe^{3+}$ up to $X_{Cr_{33}}$. Burns and Strens (1967) suggest site preference of $Cr^{3+}$ as $M1 \gg M3 = M2$ in a "tawmawite" with

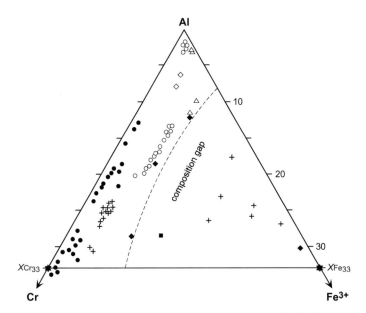

**Figure 18.** Compositions of chromian epidote group minerals in terms of Al-$Fe^{3+}$-$Cr^{3+}$ cation proportions. Symbols: ● = Outukumpu, Finland (Treloar 1987b); ■ = Western Australia (Ashley and Martyn 1987); ○ = Southern Alps, New Zealand (Grapes 1981); △ = Southern Alps, New Zealand (Cooper 1980); ◇ = northwest Nelson, New Zealand (Challis et al. 1995); + = Karnataka, India (Devaraju et al. 1999); ◆ = Betic Cordilleras, Spain (Sánchez-Vizcaíno et al. 1995). The dashed line indicates the maximum $Fe^{3+}$ content of Cr-bearing zoisite-clinozoisite.

0.51 wt% $Cr^{3+}$ (see also Sánchez-Vizcaíno et al. 1995). The existing data suggest a composition gap for compositions $>X_{Al_{73}}$ in the Al-Cr-Fe series as indicated in Figure 18, although it is unclear whether the gap is due to crystal chemistry, bulk-rock composition or reactant mineral (e.g. chromite) control.

**Strontian epidote**

The most Sr-rich epidote ($X_{Fe_{12-20}}$) with 13.9–16.3 wt% SrO (corresponding to the niigataite end-member) and clinozoisite ($X_{Fe_{0.8-2.4}}$) with 2.2–6.5 wt% SrO occur in druses with chlorite and diaspore in a prehnite-rich tectonic fragment within serpentinite at Niigata, Japan (H. Miyajima, pers. com., March 2003; Miyajima et al. 2003). Harlow (1994) reports Sr-bearing epidote ($X_{Fe_{24}}$) containing 10.27 wt% SrO and zoisite with 0.5–2.51 wt% SrO in metabasite and albite-mica blocks, respectively, that occur in jadeitite from Guatemala.

In the Alpine Schist of New Zealand, Sr-epidote ($X_{Fe_{20-29}}$ and $X_{Fe_{11-13}}$) with up to 8.4 wt% SrO (40 mol% of the niigataite end-member) occurs as a replacement along cracks in and around the margins of previously formed epidote (Grapes and Watanabe 1984; Fig. 19A). Single epidote grains exhibit a range of $^{A2}Sr^{2+} = ^{A2}Ca^{2+}$ replacement that can vary by as much as 4 wt% SrO. Sr content variability of epidote with metamorphic grade shows a marked increase across the pumpellyite-out isograd (Fig. 19B), with high Sr abundances in greenschist facies chlorite and biotite-albite zone rocks, that decrease to < 2.3 wt% SrO in epidote where oligoclase first appears (biotite-albite-oligoclase zone). The source of the Sr-bearing fluid responsible for Sr-replacement in epidote is believed to be related to albitization and cataclastic reduction of detrital protolith plagioclase which have up to 0.25 wt% SrO. The breakdown of detrital plagioclase begins in prehnite-pumpellyite facies metagreywacke and is significant in weakly schistose pumpellyite-actinolite facies rocks. In biotite-albite-oligoclase zone schist,

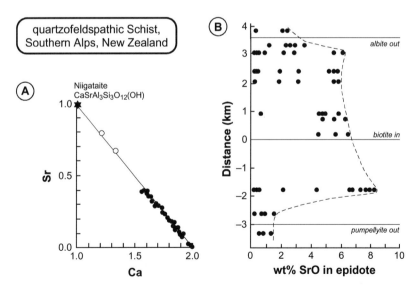

**Figure 19. (A)** Sr replacement of Ca in epidote $A2$ sites (apfu based on 12.5 oxygens). Symbols: solid circles = epidote from Alpine Schist, Southern Alps, New Zealand; open circles = Sr-rich epidote compositions from prehnite-rich tectonic fragments in serpentinite, Niigata, Japan (H. Miyajima, pers. com., March 2003). **(B)** Plot of wt% SrO in epidote in metagreywacke with distance (structural thickness between isograds) from the biotite-in isograd, central Southern Alps, New Zealand (data from Grapes and Watanabe 1984). See text for discussion.

Sr is again preferentially partitioned into newly formed oligoclase (~0.10–0.15 wt% SrO) when coexisting with epidote, and with 0.3–0.4 wt% SrO in the absence of epidote.

## Mg abundance

For the vast majority of analyzed epidote group minerals, Mg replacement of Ca is insignificant and MgO contents are typically much less than 0.10 wt%. However, epidote in prehnite-pumpellyite grade metabasites can have higher MgO contents. Examples include epidote with 0.95–1.22 wt% MgO in groundmass and in chlorite-filled cavities in metabasite from Western Australia (Smith et al. 1982) and epidote associated with prehnite-pumpellyite and prehnite-actinolite in the Tal y Fan metabasite of north Wales which has 0.03–0.39 wt% MgO (Bevins and Merriman 1988). Cho and Liou (1987) measured 0.2–1.2 wt% MgO in amoeboidal epidote filling amygdules or forming radiating sheaf-like aggregates in multiphase amygdules, noting that it commonly preserves the habit of pumpellyite from which the MgO is presumably derived. Epidote and pumpellyite are optically similar, and in view of the fact that epidote may form from pumpellyite (see Table 2) it is possible that elevated MgO in some epidote analyses is due to the accidental analysis of unobserved pumpellyite inclusions or intergrowths.

**Table 2.** Univariant epidote-producing and consuming reactions in the model metabasaltic system (CANMS-H) during low–medium grade metamorphism.*

| A. Epidote-producing reactions | B. Epidote-consuming reactions |
|---|---|
| 1. Lm + Pr = Cz + Q + V | 7. Cz + Ch + Tr + Q = Hb + V |
| 2. Lm + Pp = Cz + Ch + Q + V | 8. Cz + Ch + Ab + Q = Og + Tr + V |
| 3. Pp + Q = Cz + Pr + Ch + V | 9. Cz + Ch + Ab + Q = Og + Hb + V |
| 4. Pr + Ch + Q = Pp + Tr + V | |
| 5. Pr + Ch + Q = Cz + Tr + V | |
| 6. Pp + Ch + Q = Cz + Tr + V | |

* Mineral abbreviations and compositions are listed in Table 1; Og is oligoclase.

## ROCK OXIDATION INFLUENCE ON
## EPIDOTE GROUP MINERAL COMPOSITION

As well as the temperature and pressure of metamorphism, bulk-rock $f_{O_2}$ and major-element composition exert control on the $X_{Fe}$ composition of epidote group minerals (Holdaway 1972; Liou 1973; see also Poli and Schmidt 2004). The importance of $f_{O_2}$ on epidote composition was first noted by Strens (1965) at the Borrowdale volcanics, England, where quartz–epidote veins with hematite ± magnetite contain epidote of $X_{Fe_{22-29}}$ only and graphite-bearing veins have only clinozoisite ($X_{Fe_{11}}$). Throughout most of the Tal y Fan metabasite of north Wales, which has been metamorphosed to pumpellyite-actinolite facies (~310°C and 1.9 kbar), epidote ranges $X_{Fe_{16-20}}$ although in $Fe_2O_3$-rich horizons near the lower and top margins of the metabasite (the latter being a ferrodolerite lense) epidote is notably Fe-enriched at $X_{Fe_{26}}$ and $X_{Fe_{30}}$, respectively (Fig. 20). There is strong correlation between epidote $X_{Fe}$ and whole-rock Fe-content (Bevins and Merriman 1988). In greenschist–amphibolite facies metabasites of the Haast Schist, South Island, New Zealand, Cooper (1972) noted the relation between average epidote $X_{Fe}$ composition and the identity of coexisting oxide-sulfide phases. Epidote compositions > $X_{Fe_{27}}$ are associated with ilmenite + magnetite and compositions of $X_{Fe_{14-16}}$ coexist with ilmenite + pyrrhotite (Fig. 21A). An analogous relation between epidote $X_{Fe}$ content, whole-rock oxidation and mineral assemblage was noted by Beddoe-Stephens

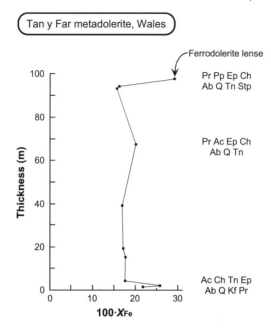

**Figure 20.** Variation of epidote $X_{Fe}$ in the metamorphosed Tan y Far dolerite intrusion, north Wales (data from Bevins and Merriman 1988). Epidote is typically unzoned and shows Fe-enrichment in two $Fe_2O_3$-rich horizons. Mineral assemblages in these two horizons and for the central part of the intrusion are listed (see Table 1 for abbreviations). Fe-rich epidote in the ferrodolerite lense coexists with stilpnomelane (Stp).

(1981) for the metamorphosed Rossland volcanic rocks or British Columbia, Canada (Fig. 21A). In the metabasaltic rocks of the Tauern Window, Austria, Raith (1976) records more Fe-rich epidote in oxidized hematite + magnetite-bearing assemblages than in reduced sulfide-bearing rocks. In both cases, the epidote becomes more Al-rich with increasing metamorphic grade: with hematite + magnetite, epidote composition ranges $X_{Fe22-30}$ at greenschist facies, $X_{Fe8-23}$ at garnet-amphibolite facies and $X_{Fe15-19}$ at eclogite facies; with magnetite ± sulfides, epidote composition ranges $X_{Fe14-27}$ at greenschist facies, $X_{Fe8-23}$ at garnet-amphibolite facies and $X_{Fe8-18}$ at eclogite facies.

Zoning profiles across epidote from the Tauern Window also vary according to oxidation conditions and metamorphic grade (Raith 1976). In strongly oxidized greenschist assemblages with hematite, the epidote $Fe^{3+}$ content decreases progressively from core ($X_{Fe31}$) to rim ($X_{Fe22}$) (lower zoning profile in Fig. 8A). In moderately oxidized garnet amphibolite assemblages with magnetite + quartz, zoning is discontinuous (upper zoning profile in Fig. 8A). In strongly reduced assemblages without Fe-oxides, epidote is relatively Fe-poor (middle zoning profile in Fig. 8A) and shows a continuous $Fe^{3+}$ decrease from core ($X_{Fe18}$) to rim ($X_{Fe8}$). Latest generation (retrograde) overgrowths having a sharp yet corroded contact with zoned cores (Fig. 8B) are characterized by more Fe-rich compositions in both oxidized ($X_{Fe25}$; lower zoning profile in Fig. 8B) and reduced assemblages ($X_{Fe20}$; upper zoning profile in Fig. 8B). In the same rock, these compositional changes reflect local and possibly system-wide changes in the redox potential of the metamorphic fluid phase during prograde and retrograde conditions.

In the schistose metagreywackes of the Alpine Schist, New Zealand, rock oxidation ratio decreases with increasing grade and the range of epidote $X_{Fe}$ of four epidote generations coexisting with albite in any one rock narrows (Fig. 21B; Grapes and Watanabe 1984). In

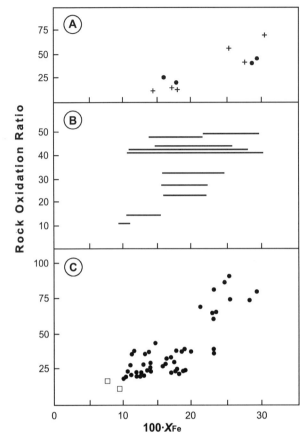

**Figure 21.** Epidote $X_{Fe}$ plotted against rock oxidation ratio for a variety of occurrences worldwide. The oxidation ratio is calculated as: mol. $2Fe_2O_3 \times 100/(2Fe_2O_3 + FeO)$. In the three plots, metamorphic grade increases in the direction of lower rock oxidation ratio. (**A**) Epidote-amphibolite to amphibolite facies metabasite, Haast Schist, South Island, New Zealand (crosses; Cooper 1972); sub-greenschist to lower amphibolite facies Rossland volcanic rocks, southern British Columbia, Canada (filled circles; Beddoe-Stephens 1981). (**B**) Greenschist to amphibolite facies quartzofeldspathic schist, Alpine Schist, South Island, New Zealand (Grapes and Watanabe 1984). (**C**) Greenschist facies quartzofeldspathic and metavolcanic rocks, Hohe Tauern area, Austria (Hörmann and Raith 1973); open squares = zoisite-bearing samples.

uppermost greenschist facies rocks where oligoclase + ilmenite first appear, rock oxidation ratios are <20, and there is a shift to less Fe-rich epidote compositions as in the Hohe Tauern epidote (described below), although individual grains are zoned to more Fe-rich rims (Figs. 12 and 13).

A positive relation between epidote composition and whole-rock oxidation ratio is documented by Hörmann and Raith (1973) for greenschist–amphibolite grade rocks in the Hohe Tauern area, Austria (Fig. 21C). Epidote coexisting with oligoclase in rocks with oxidation ratios below 50 have lower Fe abundances ($\sim X_{Fe_{10-20}}$) than those coexisting with albite ($\sim X_{Fe_{20-30}}$) in rocks with higher oxidation ratios (>50). Whole-rock samples with the lowest oxidation ratio contain zoisite. Hörmann and Raith (1973) suggest that the relation between epidote composition and host rock oxidation ratio can be explained by the redox reactions:

$$6 \text{ epidote} + 6 \text{ muscovite} + 6 \text{ quartz} + n\text{Ab-plagioclase} + 6 \text{ hematite} =$$
$$(12\text{An} + n\text{Ab})\text{-plagioclase} + \text{biotite} + 3H_2O + 4.5O_2 \qquad (10)$$

$$6 \text{ epidote} + 6 \text{ muscovite} + 6 \text{ quartz} + n\text{Ab-plagioclase} =$$
$$(12\text{An} + n\text{Ab})\text{-plagioclase} + 2 \text{ biotite} + 4 \text{ K-feldspar} + 7H_2O + 1.5O_2 \qquad (11)$$

## COMPOSITION (MISCIBILITY) GAPS BETWEEN
## EPIDOTE GROUP MINERALS

### Clinozoisite-epidote

Prior to experimental verification, zoning textures and analyzed compositions provided evidence for a composition gap in the clinozoisite-epidote series. While such supposed evidence may in fact represent the metastable persistence of earlier epidote or sluggish growth kinetics during prograde metamorphism (Bird and Helgeson 1981), perhaps the best textural evidence for unmixing is that given by Heitanen (1974) for epidote in a greenschist facies metavolcanic rock (Fig. 5). However, the textures in Figure 5 are not unequivocal and could also be interpreted as co-genetic growth of two phases or as reaction textures.

Coexisting clinozoisite ($X_{Fe_{10-12}}$) and epidote ($X_{Fe_{22-28}}$) in hydrothermally-altered andesitic lavas and tuffs of the Borrowdale Volcanic Series and in veins adjacent to a granite intrusion in the English Lake District, led to the proposal of a composition gap in the monoclinic epidote series (Strens 1963, 1964). In these occurrences, when coexisting in epidotized lavas and veins, epidote is rimmed by clinozoisite with sharp contacts yet essentially the same compositional ranges. The range of epidote $X_{Fe}$ in these rocks is shown in Figure 22. From the associated low grade greenschist facies assemblage of quartz, albite, chlorite ± white mica ± K-feldspar, calcite and hematite, formation temperatures of between 250 and 350°C at 1 kbar or less are probable (Strens 1964; see also Patrier et al. 1991). With the absence of compositions in the range of $X_{Fe_{13-21}}$ and two widely separated frequency maxima at ~$X_{Fe_{11}}$ and $X_{Fe_{27}}$ in greenschist facies metamorphic rocks, it appears that the miscibility gap exists and is centered about $X_{Fe_{19}}$ (Fig. 22).

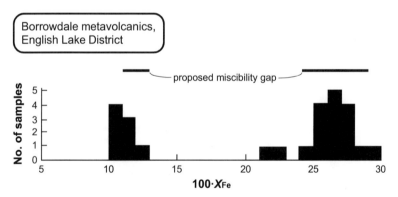

**Figure 22.** Frequency of epidote $X_{Fe}$ in metamorphosed andesitic lavas and tuffs of the Borrowdale Volcanic Series, English Lake District (data from Tables III and IV of Strens 1964). The width of the proposed miscibility gap is optically determined from coexisting epidote-clinozoisite pairs (Strens 1963, 1964).

On crystal chemical grounds, a solvus in the clinozoisite-epidote series is expected because of Fe-Al order-disorder relations within the structure. This occurs because two of the three octahedral sites, M1 and M3, can be occupied by both $Al^{3+}$ and $Fe^{3+}$, which can be represented by the relation: $^{M3}Fe^{3+} + {}^{M1}Al^{3+} = {}^{M1}Fe^{3+} + {}^{M3}Al^{3+}$. Structural strain is relaxed when separation (ordering) takes place into two components rich in Al and Fe respectively, and where large and small ions do not occupy adjacent sites (Strens 1965). The existence of composition gaps in the clinozoisite-epidote series is confirmed experimentally at 3 kbar under HM-buffer conditions above 500°C (Fehr and Heuss-Aßbichler 1997, Heuss-Aßbichler and Fehr 1997). Gaps between $X_{Fe_{10}}$ and $X_{Fe_{17}}$ and between $X_{Fe_{17}}$ and $X_{Fe_{25}}$ were delineated, with critical (metastable) solvus temperatures of ~740°C or higher at $X_{Fe_{14}}$ and $X_{Fe_{19}}$, respectively. At temperatures below 500°C the extrapolated solvus extends between $X_{Fe_{8-27}}$ which is similar to the gap suggested by Strens (1963, 1965) for natural samples formed at lower temperature.

In Figure 23, clinozoisite-epidote composition gaps from low–medium pressure greenschist facies metabasite rock occurrences worldwide are plotted with corresponding estimates of formation $T$ and $P$. These are compared to the composition gap at 5 kbar, HM-buffer conditions, as deduced by Strens (1965) and the solvii determined by Fehr and Heuss-Aßbichler (1997) and Heuss-Aßbichler and Fehr (1997). There is no consistency between the width of the estimated composition gaps and that suggested by Strens (1965) which extends

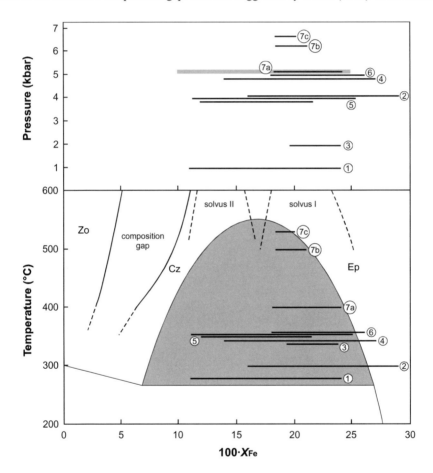

**Figure 23.** Composition gaps, represented as thick horizontal lines, in the clinozoisite-epidote series from greenschist facies metabasite occurrences worldwide plotted with corresponding estimates of average pressure (upper plot) and temperature (lower plot). In the lower plot, the grey-shaded area is the clinozoisite(Cz)-epidote(Ep) solvus at 5 kbar, HM-buffer conditions, as deduced by Strens (1965); solvi I and II are from Fehr and Heuss-Aßbichler (1997); the zoisite(Zo)-clinozoisite(Cz) composition gap is from Franz and Selverstone (1992). In the upper diagram, the grey-shaded line represents the width of the epidote miscibility gap at 5 kbar and 400°C as derived from the lower diagram. See text for discussion. Numbered data sources: 1. Strens 1964; 2. Holdaway 1965; M. J. Holdaway, pers. com., June 2003; 3. Schreyer and Abraham 1978; 4. Bishop 1972; 5. Hietanen 1974; 6. Katagas and Panagos 1979; 7a,b,c. Raith 1976.

to lower $X_{Fe}$. The data do indicate, however, that the gap generally narrows with increasing metamorphic grade as delineated by Raith (1976) (Fig. 8A). While the variability shown in the extent of estimated composition gaps in rocks of similar composition and grade (Fig. 23) could be due to factors such as the presence of metastable relics, sluggish kinetics at greenschist facies temperatures, or the effect of minor substituents such as Mn, Cr and the REE, the gaps are nevertheless considered to reflect an underlying composition gap(s) in the clinozoisite-epidote series. When the metabasite composition gaps are plotted in $T$-$P$-$X_{Fe}$ space, the surface of the solvus "envelope" is a distorted asymmetric hump, the accessible range of $X_{Fe}$ diminishing with increasing $T$ and $P$ (Fig. 24). Thus, the epidote composition in rocks formed at different $T$ and $P$ would not delineate a fixed-width composition gap. Also, the higher $T$ and $P$ data of Raith (1976) only reflect solvus I of Fehr and Heuss-Aßbichler (1997) (Fig. 23) and suggest a maximum closure temperature of that solvus at ~550°C which is considerably lower than 740°C as experimentally determined (Fehr and Heuss-Aßbichler 1997). Data presented here, and elsewhere (e.g., Liou et al. 1983), indicate that no composition gap exists for clinozoisite-epidote in prehnite-pumpellyite and pumpellyite-actinolite facies rocks. This could indicate metastable growth of highly disordered epidote at low temperature.

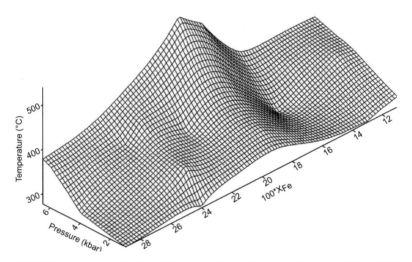

**Figure 24.** $T$-$P$-$X_{Fe}$ plot showing the form of the clinozoisite-epidote solvus from about 300 to 550°C and 1 to 6.5 kbar. The solvus is defined by data from natural occurrences as plotted in Figure 23.

## Zoisite-clinozoisite

In low–medium pressure rocks, zoisite and clinozoisite-epidote have been recorded occurring together in individual rocks or in separate but interlayered rocks (Myer 1966; Ackermand and Raase 1973; Tanner 1976; Coombs et al. 1977; Frey 1978; Frank 1983; Franz and Selverstone 1992). The occurrences recorded by Tanner (1976) and Frank (1983) have been described above. Other occurrences are summarized here:

(i) Myer (1966) describes Fe-zoisite ($X_{Fe_{3.5}}$) in amphibolite interlayered with calc-silicate from Trollheimem, Norway, that also contains clinozoisite of $X_{Fe_{12.5}}$ with both varieties coexisting in some calc-silicates as slender prisms (zoisite) and stubby crystals (epidote) together with quartz, diopside ± oligoclase ($An_{25-29}$) ± calcite. The zoisite and epidote are abundant (22 and 25 modal%, respectively) where plagioclase is absent suggesting operation of Reaction (4).

(ii) Zoisite in contact with clinozoisite occurs in epidote-amphibolite to low grade amphibolite facies (~550°C) calc-silicate and biotite schists from Hohe Tauern, Austria (Ackermand and Raase 1973). Edges of contacting grains in two samples have compositions of $X_{Fe_{4-9}}$ and $X_{Fe_{4-11}}$. Also from Hohe Tauern, a tourmaline-epidote segregation from a granitoid contact with metasediment (metamorphosed at ~550°C and 2 kbar) contains weakly zoned zoisite ($X_{Fe_{3-4}}$) and clinozoisite ($X_{Fe_{11-12}}$). Elongate porphyroblasts (several mm in length) of zoisite ($X_{Fe_4}$) are associated with clinozoisite ($X_{Fe_9}$) in marl from the northern part of the Lukmanier area, Swiss Alps, with *T-P* conditions of metamorphism estimated at 500–550°C and 5 kbar (Frey 1978).

(iii) Zoisite ($X_{Fe_{1-5}}$) coexisting with epidote ($X_{Fe_{11-18}}$) and locally in contact with more Fe-rich epidote of $X_{Fe_{26-28}}$, occurs within altered plagioclase phenocrysts of mafic dykes at Upper Eglington Valley, New Zealand (Coombs et al. 1977). The coexisting zoisite-epidote may represent products of deuteric alteration of plagioclase during cooling at temperatures equivalent to upper greenschist facies conditions as indicated by the presence of actinolitic hornblende. The dykes have subsequently been metamorphosed to pumpellyite-actinolite facies grade with the development of matrix quartz, albite, chlorite, pumpellyite, actinolite, epidote ($X_{Fe_{22-34}}$), and grandite.

(iv) Several coexisting (same fabric generation) zoisite-clinozoisite occurrences have been described by Franz and Selverstone (1992). In amphibolite from an ophiolite complex in northwestern Sudan (metamorphosed at 400 ± 50°C and <5 kbar), zoisite and clinozoisite occur in the matrix ($X_{Fe_{3-5}}$ and $X_{Fe_9}$) and a deformed vein ($X_{Fe_{2-3}}$ and $X_{Fe_{7-8}}$). Also from northern Sudan, zoned zoisite ($X_{Fe_{1-4}}$) with Fe-rich rims occurs with clinozoisite ($X_{Fe_{12-14}}$) in a calc-silicate nodule within amphibolite (metamorphosed at 650°C and 5 kbar). Although both phases are in contact, the zoisite is inferred to have formed during earlier granulite facies metamorphism.

(v) A low grade calcite-phengite rock from a metabasite/pelite contact (metamorphosed at <350°C and low pressure) contains needles of zoned zoisite ($X_{Fe_{0.3-4}}$) mostly associated with calcite, and larger clinozoisite ($X_{Fe_{2-15}}$) which is mainly associated with phengite (Franz and Selverstone 1992). Prograde core–rim zoning typically shows Fe-enrichment at the rim. In a clinozoisite grain from the contact, the rim in contact with calcite is $X_{Fe_{3-5}}$ and where it contacts phengite it is $X_{Fe_{10}}$ (see Fig. 8 of Franz and Selverstone 1992).

From prograde zoisite-clinozoisite in calcareous rocks from the Lepontine Alps (Frank 1983), a funnel-like composition gap in the zoisite-clinozoisite series is observed that widens with increasing *T* and *P* (Fig. 15). The shape of the gap is the same as "form B" of Enami and Banno (1980) and with that determined by Franz and Selverstone (1992) for zoisite-clinozoisite formed under various *T-P* conditions in a variety of rock types (eclogite, amphibolite, calc-silicate nodules, tourmaline-epidote segregations). The <5 kbar zoisite-clinozoisite composition gap (the so-called "two-phase loop") constructed from data given by Franz and Selverstone (1992) (Fig. 25), is consistent with compositional data from other sources (Ackermann and Raase 1973; Frey 1978; Frank 1983) where temperatures have been estimated (for experimental determination and thermodynamical calibration of the zoisite-clinozoisite two-phase loop see Brunsmann et al. 2002; Gottschalk 2004; Poli and Schmidt 2004). These data and the constructed two-phase loop permit temperatures to be inferred for examples where *T* estimates are not available (Fig. 25). As examples, the Norwegian sample described by Myer (1966) yields a temperature of ~650°C, considerably lower than the 750 ± 75°C estimate by Holdaway (1972) for the same rock; the data of Tanner (1976) from the Moinian calc-silicates yield 650–700°C; and zoisite-epidote of inferred deuteric origin in the mafic dykes analyzed by Coombs et al. (1977) yield 550–600°C.

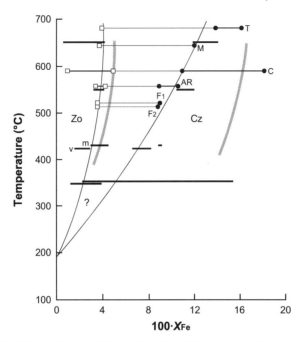

**Figure 25.** $T$-$X_{Fe}$ diagram for coexisting zoisite-clinozoisite in low–medium pressure metamorphic rocks. Symbols: open squares = zoisite; filled circles = clinozoisite. The form of the zoisite-clinozoisite miscibility gap is determined on the basis of analytical data from three samples described by Franz and Selverstone (1992) and are considered to have been metamorphosed at <5 kbar (thick horizontal lines; see text). The letters m and v refer to matrix and vein compositions, respectively (see text). Composition gaps from previous studies where temperatures were estimated are plotted as $X_{Fe}$ ranges connecting filled circles; thin horizontal lines show the extent of composition gaps; AR = Ackermand and Raase (1973); $F_1$ = Frey (1978); $F_2$ = Frank (1983). The estimated positions of composition gaps by other workers are denoted: T = Tanner (1976); M = Myer (1966); C = Coombs et al. (1977). The grey-shaded lines show a composition gap shift to higher $X_{Fe}$ and a wider gap at higher pressure (>10 kbar) (after Franz and Selverstone 1992).

## EPIDOTE GROUP MINERAL FORMING AND CONSUMING REACTIONS IN CANMS-H (METABASITE) AND KNCAS-HC, KCMAS-HC (MARL) SYSTEMS

The stability of epidote-zoisite in low–medium pressure rocks can be illustrated using three model systems, a mafic (metabasite) system, $CaO$-$Al_2O_3$-$Na_2O$-$MgO$-$SiO_2$-$H_2O$ (CANMS-H) and two systems for calcareous pelitic (marl) rocks, $K_2O$-$Na_2O$-$CaO$-$Al_2O_3$-$SiO_2$-$H_2O$-$CO_2$ (KNCAS-HC) and $K_2O$-$CaO$-$MgO$-$Al_2O_3$-$SiO_2$-$H_2O$-$CO_2$ (KCMAS-HC).

### Metabasite

The CANMS-H system is shown in Figure 26 assuming excess quartz, albite and chlorite. Addition of FeO and especially $Fe_2O_3$, shifts univariant lines and invariant points as shown in the lower diagram of Figure 26. Epidote is formed and consumed by several reactions in this system (Table 2) and participates in several key facies transitions:

- zeolite facies to prehnite-pumpellyite facies (Rxn. 2, Fig. 26):

    laumontite + pumpellyite = epidote + chlorite + quartz + $H_2O$          (12)

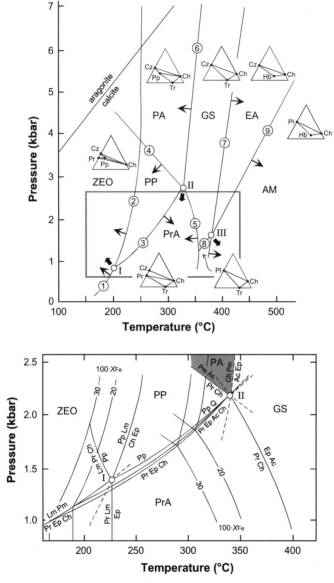

**Figure 26.** *Upper diagram*: *P-T* petrogenetic grid for various low–medium pressure metamorphic facies and their assemblages (+ quartz, albite and chlorite) in the CANMS-H model metabasite system (after Liou et al. 1985). Circled numbers refer to the epidote producing and consuming reactions listed in Table 2. Mineral abbreviations are given in Table 1. Arrows indicate the displacement directions of reaction curves (thin arrows) and invariant points (thick arrows) caused by introduction of Fe$_2$O$_3$ into the model system. *Lower diagram*: *T-P*$_{fluid}$ diagram showing continuous reactions around invariant points I and II in the upper diagram and displacement of these points by addition of Fe$_2$O$_3$ (the area of the plot is indicated in the upper diagram by the grey-line rectangle). Isopleths of epidote $X_{Fe}$ for the continuous reactions are also plotted (dashed lines indicate metastable extensions of these reactions) (after Liou et al. 1985). Facies abbreviations: ZEO = zeolite; PP = prehnite-pumpellyite; PA = pumpellyite-actinolite; PrA = prehnite-actinolite; GS = greenschist; EA = epidote-amphibolite; AM = amphibolite. Note that the PA field is shaded for clarity.

- prehnite-actinolite facies to greenschist facies at low pressure (Rxn. 5, Fig. 26):

$$\text{prehnite} + \text{chlorite} + \text{quartz} = \text{epidote} + \text{actinolite} + H_2O \tag{13}$$

- pumpellyite-actinolite facies to greenschist facies at medium pressure (Rxn. 6, Fig. 26):

$$\text{pumpellyite} + \text{chlorite} + \text{quartz} = \text{epidote} + \text{actinolite} + H_2O \tag{14}$$

- greenschist facies to epidote-amphibolite facies (Rxn. 7, Fig. 26):

$$\text{epidote} + \text{chlorite} + \text{actinolite} + \text{quartz} = \text{hornblende} + H_2O \tag{15}$$

- epidote-amphibolite facies to amphibolite facies (Rxn. 9, Fig. 26):

$$\text{epidote} + \text{chlorite} + \text{albite} + \text{quartz} = \text{oligoclase} + \text{hornblende} + H_2O \tag{16}$$

The lower diagram of Figure 26 predicts that epidote $X_{Fe}$ will decrease in buffered zeolite, prehnite-actinolite and prehnite-pumpellyite facies metabasite assemblages with increasing metamorphic grade (as seen for the Karmutsen metabasite, Figs. 1A (zeolite facies only) and 1B, and metabasite of the Hamersley Basin, Fig. 2). Although epidote can persist metastably in low grade assemblages, compositional trends will broadly vary as a function of thermal gradient, bulk-rock and fluid composition, and coexisting mineral assemblages. In the Karmutsen metabasite, minimum epidote $X_{Fe}$ increases with grade in prehnite-pumpellyite facies rocks (Cho et al. 1986) (Fig. 1A), and delineates a positive prehnite-pumpellyite-epidote amygdule assemblage $T$-$X_{Fe}$ loop above ~190°C (zeolite-out isograd).

## Marl

Metamorphism of Al-rich marl in the KNCAS-HC system involves the phases albite, anorthite, zoisite-clinozoisite, grossular, K-feldspar, margarite, paragonite, muscovite and pyrophyllite (with excess quartz + calcite). A computed petrogenetic grid shown in Figure 27 uses pure mineral compositions except for zoisite-clinozoisite that has an activity of 0.64. This has the effect of enlarging the zoisite-clinozoisite stability field bounded by Reaction (4) to higher $X_{CO_2}$ and adding the reaction:

$$2 \text{ clinozoisite} + 3 \text{ quartz} + 5 \text{ calcite} = 3 \text{ grossular} + H_2O + 5 \text{ } CO_2 \qquad \text{(Rxn. 5, Fig. 27)}$$

The grid is used to illustrate a hypothetical prograde path in the KNCAS-HC system. The path, with a low initial fluid $X_{CO_2}$ of 0.02, proceeds along points labeled **a** through **k** (Fig. 27) and is similar to a path determined for marls in the Lepontine Alps (Frank, 1983). Mineral reactions, fluid composition, and mol% changes in clinozoisite, plagioclase, margarite, paragonite and pyrophyllite at each step along the path are as follows:

a–b  an initial marl assemblage of Prl-Ms-Pg-Q-Cc is heated at constant $X_{CO_2}$ of 0.02.

b–c  margarite forms according to Reaction 1 of Table 3 with $X_{CO_2}$ increasing to 0.025 between 359–364°C.

c–d  remaining pyrophyllite is consumed at point **c** and heating continues at constant $X_{CO_2}$ = 0.025. The Mrg-Ms-Pg-Q-Cc assemblage represents lower greenschist facies.

d–e  clinozoisite appears between 425–436°C according to Reaction 2 of Table 3 and $X_{CO_2}$ increases to 0.034.

e–f  at point **e** the remaining margarite is consumed and the resultant Ms-Pg-Cz-Q-Cc assemblage represents medium grade greenschist facies.

f–g  between **points f** and **g**, andesine ($An_{33-36}$) forms at 450–458°C according to Reaction 3 of Table 3. Fluid $X_{CO_2}$ increases (to 0.045) as does mol% clinozoisite.

g–h  at point **g** the remaining paragonite is consumed and the Ms-Pl-Cz-Q-Cc assemblage at 458–507°C represents high grade greenschist facies. Fluid $XCO_2$ remains at 0.045

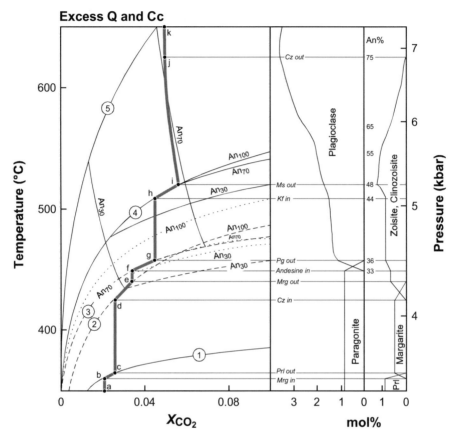

**Figure 27.** *T-P-X* diagram for part of the model Al-marl KNCAS-HC system with excess quartz and calcite up to $X_{CO_2} = 0.10$ (after Bucher and Frey 2002). Polybaric divariant equilibria for plagioclase compositions (labeled $\text{Ån}_{30-100}$) are shown as solid lines for muscovite-involving reactions, dotted lines for paragonite-involving reactions, and dashed lines for margarite-involving reactions. A prograde path, labeled **a–k**, is described in the text. Circled numbers refer to reactions listed in Table 3. Isograd positions are indicated. On the right-side of the diagram, the mol% of clinozoisite and plagioclase are shown together with changes in plagioclase An-content with increasing grade.

**Table 3.** Dehydration-decarbonation reactions in the model Al-rich marl system (KNCAS-HC).*

1. $2Prl + Cc = Mrg + 6Q + H_2O + CO_2$

2. $3Mrg + 5Cc + 6Q = 4Cz + H_2O + CO_2$

3. $Pg + 2Q + Cc = Ab + An + H_2O + CO_2$

4. $3Ms + 6Q + 4Cc = 3Kf + 2Cz + 2H_2O + 4CO_2$

5. $2Cz + 3Q + 5Cc = 3Grs + H_2O + 5CO_2$

* Mineral abbreviations and compositions are listed in Table 1.

and there is no change in mol% clinozoisite although mol% plagioclase slightly increases.

h–i  K-feldspar plus additional clinozoisite and plagioclase form by Reaction 4 of Table 3 at 507–520°C, associated with fluid $XCO_2$ increase to 0.058.

i–j  at point **i** all muscovite is consumed and the Kf-Pl-Cz-Q-Cc assemblage is characteristic of lower–middle amphibolite facies. Mol% clinozoisite decreases with gradually decreasing fluid $X_{CO_2}$ concomitant with increasing mol% plagioclase and increasing An-content from $An_{48}$ to $An_{75}$.

j–k  point **k** marks the upper stability of clinozoisite at ~625°C resulting in a Kf-Pl-Q-Cc upper amphibolite facies assemblage.

This path is not necessarily typical for the metamorphism of marls as each case will depend on the initial fluid $X_{CO_2}$ and the nature of the system as closed or open. Nonetheless, this hypothetical example demonstrates that with prograde metamorphism of Al-rich marl compositions with very low fluid $X_{CO_2}$ the increasing mol% abundance of zoisite-clinozoisite corresponds with small increases in fluid $X_{CO_2}$ and that this occurs at the expense of mol% increase and An-enrichment of associated plagioclase.

Zoisite stability (together with biotite + quartz) in the KCMAS-HC system at 5 kbar with $F/(F + OH)_{biotite} = 0.5$ and $Fe/(Fe + Mg)_{biotite} = 0.5$ was determined by Hoschek (1980a,b) for marls of the western Hohe Tauern area, Austria (Fig. 28). The stability field of zoisite is defined by Reaction (4) and:

$$3 \text{ margarite} + 5 \text{ calcite} + 6 \text{ quartz} = 4 \text{ zoisite} + H_2O + 5 \text{ } CO_2 \tag{17}$$

extending to a maximum fluid $X_{CO_2}$ of 0.30 at ~490°C. A $T$-$X_{CO_2}$ area between 440 and 565°C at $X_{CO_2} = 0.00$ and 500 and 535°C at $X_{CO_2} = 0.20$ is determined for the occurrence of biotite–zoisite–quartz ± calcite assemblages in the western Hohe Tauern area (shaded area of Fig. 28). Addition of F to biotite (with $X_{Fe_{Bt}} = 0.0$) results in an expansion of the biotite–zoisite–quartz ± calcite stability field to both higher and lower temperatures (dashed curves in Fig. 28).

## RETROGRADE ALTERATION OF ECLOGITE AND BLUESCHIST FACIES ASSEMBLAGES

Medium–low pressure retrograde alteration of eclogitic calc-silicate and metabasite and blueschist metabasite rocks results either in the breakdown of zoisite to form margarite and other phases, or in the formation of clinozoisite-epidote from lawsonite. For kyanite marble from the Eclogite Zone of the south-central Tauern Window, Austria, Spear and Franz (1986) describe the breakdown of zoisite as a result of nearly isothermal decompression according to the reactions:

$$\text{zoisite} + \text{dolomite} + \text{quartz} + H_2O = \text{chlorite} + \text{margarite} + CO_2 \tag{18}$$

$$\text{zoisite} + \text{dolomite} + \text{quartz} + H_2O = \text{chlorite} + \text{calcite} + CO_2 \tag{19}$$

$$\text{zoisite} + \text{dolomite} + \text{NaCl/KCl} + H_2O =$$
$$\text{chlorite} + \text{paragonite} + \text{muscovite} + CaCl_2 + CO_2 \tag{20}$$

implying *T-P* conditions of 460–530°C and 5–7 kbar (TW1; Fig. 29). A subsequent breakdown of zoisite to kaolinite:

$$\text{zoisite} + HCl + H_2O = \text{kaolinite} + CaCl_2 \tag{21}$$

together with sudoite (in kyanite), occurred during a final phase of retrogression estimated at 250–310°C and pressure less than 3 kbar (TW2; Fig. 29).

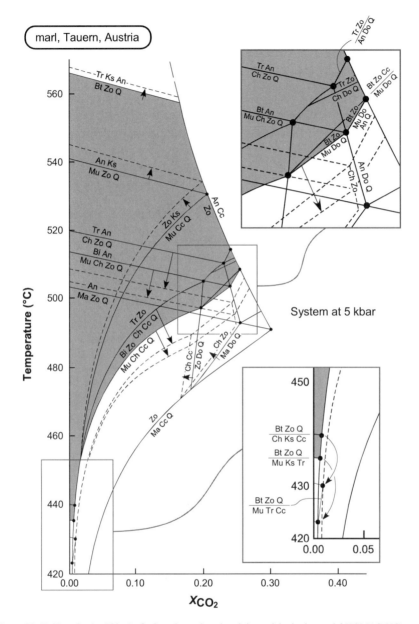

**Figure 28.** *T-* $X_{CO_2}$ plot (at 5 kbar) of mineral reactions involving zoisite in the model KCMAS-HC system for marl from the western Hohe Tauern, Austria (data from Hoschek 1980a,b). Equilibria are calculated with excess quartz and vapor, and with end-member mineral compositions (abbreviations are given in Table 1). Solid lines represent Fe-Mg substitution in biotite corresponding to Fe/(Fe + Mg)$_{biotite}$ = 0.5 with no F-OH substitution. Dashed lines represent F-OH substitution in biotite corresponding to F/(F + OH)$_{biotite}$ = 0.5. Arrows indicate the displacement of reactions that involve F-bearing biotite. Small filled circles are reaction invariant points. The stability field of biotite + zoisite + quartz is shaded.

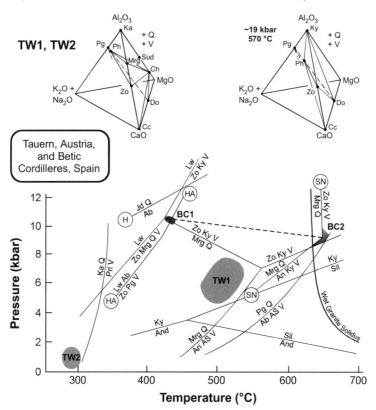

**Figure 29.** Inferred *P-T* stability fields of zoisite-breakdown assemblages in eclogitic kyanite marble, south-central Tauern Window, Austria (TW1, TW2; Spear and Franz 1986) and epidote-bearing pseudomorphs after lawsonite in amphibolitised eclogite, Betic Cordilleras, Spain (BC1, BC2; Gomez-Pugnaire et al. 1985). The dashed line connecting BC1 and BC2 indicates that the assemblage can exist at two different *P-T* conditions (see text). The mineral reaction curves are from Chatterjee et al. (1984) except for those labeled HA (Heinrich and Althaus 1980), SN (Storre and Nitsch 1974) and H (Holland 1980). Phase relations of primary high pressure (right) and secondary lower pressure (left) mineral assemblages in kyanite-zoisite marble (Tauern Window) within the system $(K_2O,Na_2O)$-CaO-MgO-$Al_2O_3$ (projection from quartz and fixed $H_2O/CO_2$ ratio) are also shown. Note that there are different compositions for primary and secondary phengite in these diagrams. Mineral abbreviations are given in Table 1.

Amphibolitized eclogites from the Betic Cordilleras, Spain, contain epidote, kyanite, margarite, paragonite and quartz pseudomorphs inferred to have formed after lawsonite (Gomez-Pugnaire et al. 1985). One type of pseudomorph is rhombohedral to prismatic with a core of white mica, margarite, paragonite, kyanite and a corona of epidote ($X_{Fe_{14-26}}$). Another pseudomorph comprises rectangular areas of epidote, quartz and paragonite that contain relic kyanite rimmed by margarite. The former pseudomorph assemblage suggests operation of these three reactions:

$$4 \text{ lawsonite} = 2 \text{ zoisite} + \text{kyanite} + \text{quartz} + 7H_2O \tag{22}$$

$$5 \text{ lawsonite} = 2 \text{ zoisite} + \text{margarite} + 2 \text{ quartz} + 8H_2O \tag{23}$$

$$2 \text{ zoisite} + 5 \text{ kyanite} + 3H_2O = 4 \text{ margarite} + 3 \text{ quartz} \tag{24}$$

representing a univariant assemblage (Fig. 29), although this becomes bivariant with addition of $Fe^{3+}$. Subsequent reaction of some of the margarite (with Na and Si in a fluid phase) to form paragonite could have released Ca and Al to produce the observed monominerallic epidote coronas which explains the absence of quartz which is predicted as a product from the above reactions. The latter pseudomorph assemblage can be explained by the reaction:

$$4 \text{ lawsonite} + \text{albite} = \text{paragonite} + 2 \text{ zoisite} + 2 \text{ quartz} + 6H_2O \tag{25}$$

Relevant experimental data for the above reactions, summarized in Figure 29, show that in the Fe-free system both pseudomorph assemblages could have formed at minimum *T-P* conditions of 420–430°C and ~10 kbar (field labeled BC1 in Fig. 29). However, if paragonite-margarite solid-solution is accounted for (in this case up to 50 mol%), the pseudomorphs could represent a significantly higher temperature of ~650°C (the lawsonite-out reactions being overstepped by nearly 250°C) at ~9 kbar (field labeled BC2 in Fig. 29). This estimate is consistent with a minimum *T-P* estimate of 615°C and 6.5 kbar from associated metapelites and with incipient anatexis (Gomez-Pugnaire 1979).

Another example of epidote-bearing pseudomorphs after lawsonite is reported by Selverstone and Spear (1985) from amphibolitized metabasite in the southwest of the Tauern Window, Austria. The pseudomorphs occur as rectangular, diamond and triangular-shaped aggregates of plagioclase (albite and oligoclase), epidote ($X_{Fe_{10-12}}$ and $X_{Fe_{19-21}}$), chlorite, minor quartz and rutile (see also Fry 1973 and Selverstone et al. 1984). Epidote compositions and sharp core-rim boundaries support a composition gap in the clinozoisite-epidote series. The likely pseudomorphic reaction is:

$$100 \text{ lawsonite} + 32Na_{0.3}(Ca_{1.7}Na_{0.3})(Mg_{3.9}Fe^{2+}{}_{0.4}Al_{0.7})(Al_{1.4}Si_{6.6})O_{22}(OH)_2 \text{ (hornblende)} =$$
$$20(0.2An_5 + 0.8An_{25}) + 76(0.5Ep + 0.5Cz) + 11 \text{ chlorite} + 17 \text{ quartz} + H_2O \tag{26}$$

Data for the greenschist–amphibolite facies transition deduced by Maruyama et al. (1983) together with the stability field of lawsonite implies that the pseudomorphs formed at a maximum of ~500°C and ~5 kbar at mid-greenschist facies conditions due to uplift and heating of an early moderate-pressure blueschist assemblage (lawsonite–albite–chlorite subfacies) (Fig. 30).

## CONCLUSIONS AND OUTLOOK

The epidote group minerals are common constituents of low–medium grade metamorphosed mafic–intermediate igneous rocks, quartzofeldspathic rocks and marls. They exhibit a wide range of *T-P* stability and occurrence. In low grade rocks, epidote is common either as an amygdule, vein or matrix phase. Although a number of workers have reported no relation between epidote composition and whole-rock composition particularly for low-grade rocks, a consideration of epidote-forming and consuming reactions in metabasite (Fig. 26) and marl rocks (Figs. 27 and 28) predicts that bulk-rock and fluid composition will control epidote group mineral composition and stability. This is observed, for example, in the prehnite-pumpellyite to prehnite-actinolite facies Karmutsen metabasites of Vancouver Island (Fig. 1) and the prehnite-pumpellyite to greenschist facies metabasites of the Hamersley Basin, Australia (Fig. 2). In both localities, epidote $X_{Fe}$ generally decreases as the Al content increases with increasing metamorphic grade, illustrating at least in part, bulk-rock compositional control. Generally, epidote in metabasite lithologies has more Fe-rich compositions than those in marl which tend to be more Al-rich and include zoisite.

Epidote group mineral compositional variation is dominated by Fe and Al and to a lesser extent by Mn (Fig. 16), Cr (Fig. 18), Sr (Fig. 19) and Mg. Within single crystals, compositional variation may be observed as oscillatory or sector zoning. Such zoning is often observed in

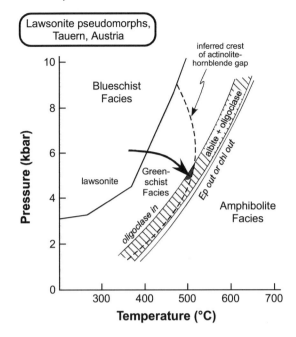

**Figure 30.** Inferred *P-T* path (thick arrow) followed by amphibolitized lawsonite-bearing metabasite resulting in higher temperature-lower pressure replacement of lawsonite by clinozoisite-epidote, albite-oligoclase and chlorite, southwestern Tauern Window (Selverstone and Spear 1985). Lawsonite field from Chatterjee (1976). Phase relations of the greenschist–amphibolite transition in metabasite as related to the stabilities of albite-oligoclase, actinolite-hornblende, epidote and chlorite are from Maruyama et al. (1983).

epidote from low–medium pressure metamorphic rocks and may preserve evidence of net-transfer reactions or fluid-flow during metamorphic petrogenesis. Unfortunately, many studies neither describe zoning textures fully nor provide electron backscattered electron images and thus overlook a potentially useful record of genetic processes. Moreover, inadequate descriptions accompanying electron microprobe analyses diminish the usefulness of microprobe data; descriptions should include the textural relations and co-existing mineralogy of analyzed epidote group minerals. Such petrographic information is essential, for example, for accurate identification of composition gaps in the zoisite-clinozoisite and clinozoisite-epidote series.

Composition gaps between members of the epidote group have been noted for occurrences worldwide and have been experimentally investigated and verified. A comparison of the width of composition gaps in the zoisite-clinozoisite and clinozoisite-epidote series from natural occurrences and experiment (Figs. 23 and 25) shows that an accurate and general delineation of a compositional gap (or gaps) is hindered perhaps by metastable persistence and sluggish reaction kinetics. However, in a broad sense, compositional gaps between clinozoisite and epidote do indicate that the gap narrows with increasing metamorphic grade (Fig. 24). In the zoisite-clinozoisite series, the composition gap is funnel-like and asymmetrically widens with increasing *T* and *P* (Figs. 15 and 25).

The ability to accurately determine major and trace-elements at a μm-scale by techniques such as electron microprobe and laser-ablation ICP-MS will certainly allow for a greater understanding of epidote group mineral occurrences. An important area of future research are the abundances and normalized patterns of rare-earth elements (REE) in epidote group minerals. It is probable that the REE are also oscillatory and sector zoned as are Ca, Al and Fe, such that experimental determination of REE diffusion rates should allow for estimation of the time required to blur primary zoning at a particular temperature. The occurrence of sector zoning due to inter-sectoral REE partitioning may be temperature sensitive and could be calibrated as a thermometer.

*Grapes & Hoskin*

## ACKNOWLEDGMENTS

Anne Feenstra is thanked for a detailed and constructive review. Axel Liebscher, Gerhard Franz and Jodi Rosso are gratefully acknowledged for their editorial efforts and oversight.

## REFERENCES

Abrecht J (1981) Pink zoisite from the Aar Massif, Switzerland. Mineral Mag 44:45-49
Ackermand D, Raase P (1973) Coexisting zoisite and clinozoisite in biotite schist from Hohe Tauern, Austria. Contrib Mineral Petrol 42:333-341
Apted MJ, Liou JG (1983) Phase relations amongst greenschist, epidote-amphibolite, and amphibolite in a basaltic system. Am J Sci 283-A:328-354
Ashley PM, Martyn JE (1987) Chromium-bearing minerals from a metamorphosed hydrothermal alteration zone in the Archean of Western Australia. Neues Jahr Mineral Abh 157:81-111
Banno S, Yoshizawa H (1993) Sector-zoning of epidote in the Sambagawa schists and the question of an epidote miscibility gap. Mineral Mag 57:739-743
Beddoe-Stephens B (1981) Metamorphism of the Rossland volcanic rocks, southern British Columbia. Can Mineral 19:631-641
Bevins RE, Merriman RJ (1988) Compositional controls on coexisting prehnite-actinolite and prehnite-pumpellyite faces assemblages in the Tal y Fan metabasite intrusion, North Wales: implications for Caledonian metamorphic field gradients. J Meta Geol 6:17-39
Bird DK, Helgeson HC (1981) Chemical interaction of aqueous solutions with epidote-feldspar mineral assemblages in geologic systems. II. Equilibrium constraints in metamorphic-geothermal processes. Am J Sci 281:576-614
Bishop DG (1972) Progressive metamorphism of prehnite-pumpellyite to greenschist facies in the Dansey Pass area, Otago, New Zealand. Bull Geol Soc Am 83:3177-3198
Bleeck AWG (1907) Die Jadeitlagerstätten in Upper Burma. Zeit Prakt Geol 15:341-365
Bonazzi P, Menchetti S (2004) Manganese in monoclinic members of the epidote group: piemontite and related minerals. Rev Mineral Geochem 56:495-552
Brown EH (1967) The greenschist facies in part of eastern Otago, New Zealand. Contrib Mineral Petrol 14:259-292
Brunsmann A, Franz G, Heinrich W (2002) Experimental investigation of zoisite-clinozoisite phase equilibria in the system $CaO-Fe_2O_3-Al_2O_3-SiO_2-H_2O$. Contrib Mineral Petrol 143:115-130
Bucher K, Frey M (2002) Petrogenesis of Metamorphic Rocks. 7th edn Springer-Verlag, Heidelberg
Burns RG, Strens RGJ (1967) Structural interpretation of polarized absorption spectra on the Al-Fe-Mn-Cr epidotes. Mineral Mag 36:204-226
Challis GA, Grapes RH, Palmer K (1995) Fuchsite, uvarovite, and zincian chromite: products of regional metasomatism in northwest Nelson, New Zealand. Can Mineral 33:1263-1284
Chatterjee ND (1976) Margarite stability and compatibility relations in the system $CaO-Al_2O_3-SiO_2-H_2O$ as a pressure-temperature indicator. Am Mineral 61:699-709
Chatterjee ND, Johannes W, Leistner H (1984) The system $CaO-Al_2O_3-SiO_2-H_2O$: new phase equilibria data, some calculated phase relations, and their petrological applications. Contrib Mineral Petrol 88:1-13
Cho M, Liou JG, Maruyama S (1986) Transition from the zeolite to prehnite-pumpellyite facies in the Karmutsen metabasites, Vancouver Island, British Columbia. J Petrol 27:467-494
Cho M, Liou JG (1987) Prehnite-pumpellyite to greenschist facies transition in the Karmutsen metabasites, Vancouver Island, B.C. J Petrol 28:417-443
Coombs DS, Nakamura Y, Vuagnat M (1976) Pumpellyite-actinolite facies schists of the Taveyanne Formation near Loèche, Valais, Switzerland. J Petrol 17:440-471
Coombs DS, Kawachi Y, Houghton BF, Hyden GM, Pringle IJ (1977) Andradite and andradite-grossular solid solutions in very low-grade regionally metamorphosed rocks in southern New Zealand. Contrib Mineral Petrol 63:229-246
Cooper AF (1972) Progressive metamorphism of metabasic rocks from the Haast Schist Group of southern New Zealand. J Petrol 13:457-492
Cooper AF (1980) Retrograde alteration of chromian kyanite in metachert and amphibolite whiteschist from the Southern Alps, New Zealand, with implications for uplift on the Alpine Fault. Contrib Mineral Petrol 75:153-164
Deer, WA, Howie RA, Zussman J (1997) Disilicates and Ring Silicates. Vol. 1B, 2nd end. The Geological Society, London

Devaraju TC, Raith MM, Spiering B (1999) Mineralogy of the Archean barite deposit of Ghattihosahalli, Karnataka, India. Can Mineral 37:603-617

Dollase WA (1969) Crystal structure and cation ordering of piemontite. Am Mineral 54:710-717

Enami M, Banno S (1980) Zoisite-clinozoisite relations in low- to medium-grade high-pressure metamorphic rocks and their implications. Mineral Mag 43:1005-1013

Eskola P (1933) On the chrome minerals of Outukumpu. Comm Geo Finland Bull 103:26-44

Fehr KT, Heuss-Aßbichler S (1997) Intercrystalline equilibria and immiscibility along the join clinozoisite-epidote: an experimental and $^{57}$Fe Mössbauer study. Neues Jahr Mineral Abh 172:43-67

Ferry JM (1976a) Metamorphism of calcareous sediments in the Warerville-Vassalboro area, south-central Maine: mineral reactions and graphical analysis. Am J Sci 276:841-882

Ferry JM (1976b) $P$, $T$, $f_{CO_2}$, and $f_{H_2O}$ during metamorphism of calcareous sediments in the Waterville-Vassalboro area, South-central Maine. Contrib Mineral Petrol 57:119-143

Ferry JM (1983) Regional metamorphism of the Vassalboro Formation, south-central Maine, USA: a case study of the role of fluid in metamorphic petrogenesis. J Geol Soc London 140:551-576

Frank E (1979) Metamorphose mesozoischer Gesteine im Querprofil Brig-Verampo: mineralogisch-petrographische und isotopengeologische Untersuchungen. PhD dissertation, University of Bern, Switzerland

Frank E (1983) Alpine metamorphism of calcareous rocks along a cross-section in the Central Alps: occurrence and breakdown of muscovite, margarite and paragonite. Schweiz Mineral Petrogr Mitt 63:37-93

Franz G, Selverstone J (1992) An empirical phase diagram for the clinozoisite-zoisite transformation in the system $Ca_2Al_3Si_3O_{12}(OH)$–$Ca_2Al_2Fe^{3+}Si_3O_{12}(OH)$. Am Mineral 77:631-642

Frey M (1978) Progressive low-grade metamorphism of a Black Shale formation, central Swiss Alps, with special reference to pyrophyllite and margarite-bearing assemblages. J Petrol 19:95-135

Fry N (1973) Lawsonite pseudomorphed in Tauern greenschist. Min Mag 39:121-122

Game PM (1954) Zoisite-amphibolite with corundum from Tanganyika. Mineral Mag 30:458-466

Gomez-Pugnaire MT (1979) Some considerations on the highest temperature reached in the out-cropping rocks of the Nevado-Filábride Complex in the Sierra de Baza area during Alpine metamorphism. Neues Jahr Mineral Abh 135:75-87

Gomez-Pugnaire MT, Visona D, Franz G (1985) Kyanite, margarite and paragonite in pseudomorphs in amphibolitized eclogites from the Beltic Cordilleras, Spain. Chem Geol 50:129-141

Gottschalk M (2004) Thermodynamic properties of zoisite, clinozoisite and epidote. Rev Mineral Geochem. 56:83-124

Grapes RH (1981) Chromian epidote and zoisite in kyanite amphibolite, Southern Alps, New Zealand. Am Mineral 66:974-975

Grapes RH (1995) Uplift and exhumation of Alpine Schist, Southern Alps, New Zealand. NZ J Geol Geophys 38:525-533

Grapes RH (1996) Occurrences of unrecorded minerals and new localities of known minerals in New Zealand (3). Mineral Soc NZ News 5:2-5

Grapes RH, Hashimoto S (1978) Manganiferous schists and their origin, Hidaka Mountains, Hokkaido, Japan. Contrib Mineral Petrol 68:23-35

Grapes RH, Roser B, Kifle K (2001) Composition of monocrystalline detrital and authigenic minerals, metamorphic grade, and provenance of Torlesse and Waipapa greywacke, central North Island, New Zealand. Inter Geol Rev 43:139-175

Grapes RH, Watanabe T (1984) Al-Fe$^{3+}$ and Ca-Sr$^{2+}$ epidotes in metagreywacke-quartzofeldspathic schist, Southern Alps, New Zealand. Am Mineral 69:490-498

Grapes RH, Watanabe T (1994) Mineral composition variation in Alpine Schist, Southern Alps, New Zealand: implications for recrystallisation and exhumation. Island Arc 3:163-181

Harlow GE (1994) Jadeitites, albitites and related rocks from the Motagua fault zone, Guatemala. J Meta Geol 12:49-68

Heinrich EW (1964) Thulite from Camp Creek, Ruby Mountains, Montana. Am Mineral 49:430-435

Heinrich W, Althaus E (1980) Die obere Stabilitätgrenze von Lawsonit plus Albit bzw. Jadeit. Fortschr Mineral 58 (Beih. 1):49-50

Hietanen A (1974) Amphibole pairs, epidote minerals, chlorite and plagioclase in metamorphic rocks, Northern Sierra Nevada, California. Am Mineral 59:22-40

Heuss-Aßbichler S, Fehr KT (1997) Intercrystalline exchange of Al-Fe$^{3+}$ between grossular-andradite and clinozoisite-epidote solid solutions. Neues Jahr Mineral Abh 172:69-100

Höck V, Hoschek G (1980) Metamorphism of Mesozoic calcareous metasediments in the Hohe Tauern, Austria. Mitt Öster Geol Ges 71/72:99-118

Holdaway MJ (1965) Basic regional metamorphic rocks in part of the Klamath Mountains, Northern California. Am Mineral 50:953-977

Holdaway MJ (1972) Thermal stability of Al-Fe epidote as a function of $f_{O_2}$ and Fe content. Contrib Mineral Petrol 37:307-340

Holland TJB (1980) The reaction albite = jadeite + quartz determined experimentally in the range 600-1200°C. Am Mineral 65:129-134

Hörmann P-K, Raith M (1973) Bildungsbedingungen von Al-Fe(III)-Epidoten. Contrib Mineral Petrol 38: 307-320

Hoschek G (1980a) Phase relations of a simplified marly rock system with application to the Western Hohe Tauern (Austria). Contrib Mineral Petrol 73:53-68

Hoschek G (1980b) The effect of Fe–Mg substitution on phase relations in marly rocks of the western Hohe Tauern (Austria). Contrib Mineral Petrol 75:123-128

Janeczek J, Sachanbinski M (1989) Chemistry and zoning of thulite from the Wiry magnesite deposit, Poland. Neues Jahr Mineral Mh H7:325-333

Katagas C, Panagos AG (1979) Pumpellyite-actinolite and greenschist facies metamorphism Lesvos Island (Greece). Tscher Mineral Petrol Mitt 26:235-254

Kawachi Y (1975) Pumpellyite-actinolite and contiguous facies metamorphism in part of Upper Wakatipu district, South Island, New Zealand. NZ J Geol Geophys 18:401-441

Kawachi Y, Grapes RH, Coombs DS, Dowse M (1983) Mineralogy and petrology of a piemontite-bearing schist, western Otago, New Zealand. J Meta Geol 1:353-372

Kennedy WQ (1949) Zones of progressive regional metamorphism in the Moine Schists of the Western Highlands of Scotland. Geol Mag 86:43-56

Keskinen M (1981) Petrochemical investigation of the Shadow Lake piemontite zone, eastern Sierra Nevada, California. Am J Sci 281:896-921

Keskinen M, Liou JG (1979) Synthesis and stability relations of Mn-Al piemontite, $Ca_2MnAl_2Si_3O_{12}(OH)$. Am Mineral 64:317-328

Kuniyoshi S, Liou JG (1976a) Contact metamorphism of the Karmutsen volcanics, Vancouver Island, British Columbia. J Petrol 17:73-99

Kuniyoshi S, Liou JG (1976b) Burial metamorphism of the Karmutsen volcanic rocks, northeastern Vancouver Island, British Columbia. Am J Sci 276:1096-1119

Liebscher A (2004) Spectroscopy of epidote minerals. Rev Mineral Geochem. 56:125-170

Liou JG (1973) Synthesis and stability relations of epidote, $Ca_2Al_2FeSi_3OH_{12}(OH)$. J Petrol 14:381-413

Liou JG, Kim HS, Maruyama S (1983) Prehnite-epidote equilibria and their petrologic applications. J Petrol 24:321-342

Liou JG, Maruyama S, Cho M (1985) Phase equilibria and mineral paragenesis of metabasites in low-grade metamorphism. Mineral Mag 49:321-333

Maruyama S, Suzuki K, Liou JG (1983) Greenschist-amphibolite transition equilibria at low pressures. J Petrol 24:583-604

Miyajima H, Matsubara S, Miyawaki R, Hirokawa K (2003) Niigataite, $CaSrAl_3(Si_2O_7)(SiO_4)O(OH)$: Sr-analogue of clinozoisite, a new member of the epidote group from Itoigawa–Ohmi district, Niigata Prefecture, central Japan. J Mineral Petrol Sci 98:118-129

Miyashiro A, Seki Y (1958) Enlargement of the composition field of epidote and piemontite with rising temperature. Am J Sci 256:423-430

Myer GH (1966) New data on zoisite and epidote. Am J Sci 264:354-385

Neumann H, Svinndal S (1955) The cyprin-thulite deposit at Øvstebø, near Kleppan in Sauland, Telemark, Norway. Norsk Geol Tidsskr 34:139-156

Patrier P, Beaufort D, Meunier A, Elmery J-P, Petit S (1991) Determination of the nonequilibrium ordering sate in epidote from the ancient geothermal field of Saint Martin: application of Mössbauer spectroscopy. Am Mineral 76:602-610

Poli S, Schmidt MW (2004) Experimental subsolidus studies on epidote minerals. Rev Mineral Geochem 56: 171-195

Pouliot G, Trudel P, Valiquette G, Samson P (1984) Armenite-thulite-albite veins at Rémigny, Quebec: the second occurrence of armenite. Can Mineral 22:453-464

Raith M (1976) The Al-Fe(III) epidote miscibility gap in a metamorphic profile through the Penninic series of the Tauern Window, Austria. Contrib Mineral Petrol 57:99-117

Reinecke T (1986) Crystal chemistry and reaction relations of piemontites and thulites from highly oxidized low grade metamorphic rocks at Vitali, Andros Island, Greece. Contrib Mineral Petrol 93:56-76

Roser BP, Grapes RH (1990) Whole-rock and mineral analyses of volcanic, pelagic and turbidite lithologies from Red Rocks, Wellington. Geol Board Studies Pub 6, Victoria University of Wellington, New Zealand

Sánchez-Viscaíno VL, Franz G, Gómez-Pugnaire MT (1995) The behaviour of Cr during metamorphism of carbonate rocks from the Nevado-Filabride Complex, Betic Cordilleras, Spain. Can Mineral 33:85-104

Schmidt D, Schmidt S-Th, Mullis J, Mählmann RF, Frey M (1997) Very low grade metamorphism of the Taveyanne formation, Switzerland. Contrib Mineral Petrol 129:385-403

Schreyer W, Abraham K (1978) Prehnite/chlorite and actinolite/epidote bearing mineral assemblages in the metamorphic igneous rocks of La Helle and Challes, Vann-Stavelot-Massif, Belgium. Ann Soc Géol Belgique 101:277-241

Selverstone J, Spear F, Franz G, Morteani G (1984) High-pressure metamorphism in the SW Tauern Window, Austria: P-T paths from hornblende-kyanite-staurolite schists. J Petrol 25:501-531

Selverstone J, Spear FS (1985) Metamorphic P-T paths from pelitic schists and greenstones from the southwest Tauern Window, Eastern Alps. J Meta Geol 3:439-465

Smith D, Albee AL (1967) Petrology of piemontite-bearing gneiss, San Gorgonio Pass, California. Contrib Mineral Petrol 16:189-203

Smith D, Parks TC (1982) Burial metamorphism in the Hamersley Basin, Western Australia. J Petrol 23:75-102

Spear FS, Franz G (1986) P-T evolution of metasediments from the Eclogite Zone, south-central Tauern Window, Austria. Lithos 19:219-234

Springer RK, Day HW, Beiersdorfer E (1992) Prehnite-pumpellyite to greenschist facies transition, Smartville Complex, near Auburn, California. J Meta Geol 10:147-170

Starkey RJ Jr, Frost BR (1990) Low-grade metamorphism of the Karmutsen volcanics, Vancouver Island, British Columbia. J Petrol 31:167-195

Storre B, Nitsch KH (1974) Zur Stabilität von Margarit in System $CaO-Al_2O_3-SiO_2-H_2O$. Contrib Mineral Petrol 43:1-24

Strens RCJ (1963) Some relationships between members of the epidote group. Nature 198:80-81

Strens RCJ (1964) Epidotes of the Borrowdale volcanic rocks of central Borrowdale. Mineral Mag 33:868-886

Strens RCJ (1965) Stability and relations of the Al-Fe-epidotes. Mineral Mag 35:464-475

Tanner PWG (1976) Progressive regional metamorphism of thin calcareous bands from Moinian rocks of N.W. Scotland. J Petrol 17:100-134

Terabayashi M (1988) Actinolite-forming reaction at low pressure and the role of $Fe^{2+}$-Mg substitution. Contrib Mineral Petrol 100:268-280

Tillmanns E, Langer K, Arni R, Abraham K (1984) Crystal structure refinements of coexisting thulite and piemontite, $Ca_2Al_{3-p}M^{3+}_p$ $[OH/O/SiO_4/Si_2O_7]$ ($M^{3+} = Mn^{3+} + Fe^{3+}$). Acta Crystal A 40:C258

Treloar PJ, Koistinen TJ, Bowes DR (1981) Metamorphic development of cordierite-amphibole rocks and mica schists in the vicinity of the Outukumpu ore deposit, Finland. Trans Roy Soc Edinburgh Earth Sci 72:201–215

Treloar PJ (1987a) The Cr-minerals of Outukumpu—their chemistry and significance. J Petrol 28:867–886

Treloar PJ (1987b) Chromian muscovites and epidotes from Outukumpu, Finland. Mineral Mag 51:593–599

Reviews in Mineralogy & Geochemistry
Vol. 56, pp. 347-398, 2004
Copyright © Mineralogical Society of America

# 8

# Epidote Minerals in High P/T Metamorphic Terranes: Subduction Zone and High- to Ultrahigh-Pressure Metamorphism

## M. Enami

*Department of Earth and Planetary Sciences*
*Nagoya University*
*Nagoya 464-8602, Japan*

## J. G. Liou and C. G. Mattinson

*Department of Geological and Environmental Sciences*
*Stanford University*
*Stanford, California 94305, U.S.A.*

## INTRODUCTION

Epidote minerals—the monoclinic epidote group minerals together with the orthorhombic polymorph zoisite—are important Ca-Al-silicates in many metabasites, metapelites and metacherts that are characterized by high $P/T$ ratios. Such high $P/T$ ratios are typical for subduction zones and the high-pressure (HP) and ultrahigh-pressure (UHP) metamorphism during continent-continent collisions (e.g., Liou 1973, 1993). All of these $P$-$T$ conditions can be described by geothermal gradients between 5 and 20°C/km, that therefore provide a rough framework for the $P$-$T$ conditions covered by this review (Fig. 1). Depending on the actual thermal structure of a subduction zone, the subducting plate will encounter subgreenschist, greenschist, blueschist, epidote-amphibolite, amphibolite, HP granulite, and/or eclogite facies conditions during its travel down into the mantle (Fig. 1). The $P$-$T$ regime of the eclogite facies can further be subdivided into amphibole eclogite, epidote eclogite, lawsonite eclogite, and dry eclogite facies (Fig. 1). HP metamorphism refers to metamorphic pressure in excess of ~1.0 GPa and includes parts of the blueschist, epidote-amphibolite, and HP granulite facies as well as the eclogite facies (Fig. 1). UHP refers to the metamorphism of crustal rocks (both continental and oceanic) at $P$ high enough to crystallize the index minerals coesite and/or diamond. HP and UHP metamorphism are separated conveniently by the quartz-coesite equilibrium which implies a minimum $P > 2.7$ GPa at $T > 600$°C for UHP metamorphism (Fig. 1). The equilibrium boundary for the graphite-diamond transition can be used to further subdivide the UHP region into diamond-grade and coesite-grade. The stability of coesite and other UHP minerals in a metamorphic regime requires abnormally low temperatures at depths greater than 100 km. Such environments can be attained only by the subduction of cold oceanic crust-capped lithosphere ± pelagic sediments or of continental crust.

Epidote minerals occur as characteristic phases in the greenschist, blueschist and epidote-amphibolite facies and in the lawsonite and epidote eclogite facies. They have also been reported from coesite-bearing eclogites (e.g., Liou 1993). Occurrence of polycrystalline clinozoisite aggregates, interpreted as pseudomorphs after former lawsonite porphyroblasts, are identified in HP rocks from several localities (e.g., Krogh 1982; Droop 1985; Gomez-Pugnaire

1529-6466/04/0056-0008$10.00

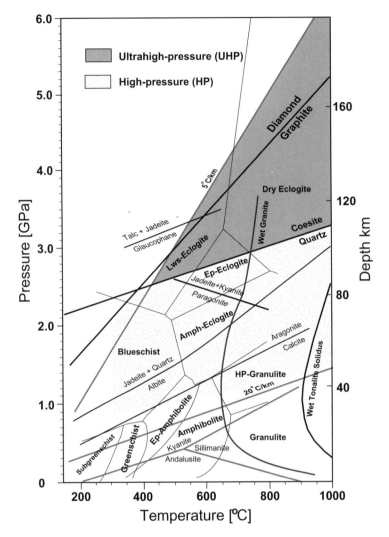

**Figure 1.** *P-T* regimes of UHP and HP metamorphism. The blank area below ≈ 1 GPa is the field of medium to low *P* metamorphism. Geotherms of 5°C/km and 20°C/km are indicated (modified from Fig. 1 of Liou and Zhang 2002).

et al. 1985) and were recently described in Dabie quartz eclogites with peak *P-T* estimated at 2.4 GPa and 700°C (e.g., Castelli et al. 1998). At appropriate bulk rock composition and $f_{O_2}$ and $X_{CO_2}$ conditions, epidote minerals might even coexist with microdiamond (+ coesite) as in the diamond-grade UHP rocks of the Kokchetav Massif where metamorphic microdiamond was first documented (Sobolev and Shatsky 1990).

The composition of epidote minerals is highly variable and mainly controlled by temperature, pressure, oxygen fugacity ($f_{O_2}$), and bulk rock composition. In metabasites they tend to be more Fe-rich than in metapelites, which commonly contain graphite. Metacherts, which generally recrystallize under much higher $f_{O_2}$ than other rock-types, typically contain epidote or piemontite. The compositional range of epidote minerals in terms of their $X_{Fe}$ [$Fe^{3+}$/ ($Al + Cr + Fe^{3+} + Mn^{3+}$)] value enlarges towards the low $X_{Fe}$-side with increasing metamorphic grade. Enlargement of the compositional field combined with the two-phase loop between zoisite and clinozoisite leads to the common occurrence of zoisite in HP metamorphic rocks, especially in eclogites.

Another compositional variation of epidote minerals important for the rocks reviewed in this chapter is the element proportions in their largest A(2)-site. High concentrations of Sr and REE in epidote minerals reported from HP and UHP terranes strongly suggest that (1) zoisite and epidote are the most important Sr and REE reservoirs at HP and UHP conditions where they are a major Ca-Al-silicate, and (2) their stability strongly controls the recycling of Sr and REE in HP-UHP environments (e.g., Sorensen and Grossman 1989; Nagasaki and Enami 1998).

**Major themes to be addressed**

This chapter focuses mainly on modes of occurrences of zoisite, clinozoisite-epidote and piemontite in various rock types from selected subduction zones and HP to UHP metamorphic terranes. The amount of information on these rocks has increased dramatically during the last 20 years and it is virtually impossible to completely review all. We therefore restrict this review mainly to those localities where we have our own experience, and add some information from other localities. Epidote minerals in metamorphic rocks that followed classical barrowian- or Buchan-type metamorphism are reviewed by Grapes and Hoskin (2004). Nevertheless, as the *P-T* conditions encountered during subduction zone and Barrovian-type metamorphism are not thus distinct in the *P-T* range of the greenschist, blueschist, to epidote-amphibolite facies transition, some overlaps between Grapes and Hoskin (2004) and this chapter were found to be unavoidable. As subduction zone metamorphism represents also a potential prograde path of many HP and UHP rocks we will first review some selected and well studied examples of subduction zone metamorphism and then prograde into the field of the peculiar HP and UHP metamorphic rocks.

Specifically, this chapter deals with the following major subjects: (1) Prograde sequences from lawsonite to epidote zone blueschists and the corresponding epidote-forming reactions are described from the Franciscan, New Caledonian and Sanbagawa metamorphic rocks; these examples are used to illustrate the role of epidote minerals in typical subduction zone metamorphic successions. (2) Phase relations of zoisite-bearing assemblages in calcareous metasediments are briefly reviewed for eclogite facies rocks in the European Alps. (3) The control of bulk rock composition on paragenesis and compositions of UHP phases is exemplified by various zoisite-bearing UHP eclogites from Dabie (central China) with an AFM diagram. (4) The common break down of plagioclase in HP and UHP rocks in the presence of $H_2O$ to form the assemblage zoisite + kyanite + sodic clinopyroxene + quartz is illustrated in some HP to UHP coronitic metagranites and metagabbros. (5) The common but selective occurrence of epidote minerals in Sulu-Dabie eclogite, quartzite and gneiss may be related to the oxidation of supracrustal rocks by a fossil hydrothermal system. (6) The pseudomorphic replacements of epidote/zoisite + amphibole + biotite + sodic plagioclase after garnet and of zoisite + albite after kyanite as indicators for a metasomatic and retrograde overprint during exhumation are described using kyanite-phengite-coesite eclogites from Dabie. (7) The coexistence of lawsonite and epidote in some HP metabasites is related to bulk composition, $f_{O_2}$ and metastable persistence. (8) Composition of epidote minerals from buffered assemblages is sensitive to *P*, *T* and $f_{O_2}$, and its use as a geobarometer or geothermometer is discussed. (9) Epidote minerals are a key to discuss fluid-rock interaction and partial melting of metamorphic rocks at high-*P* conditions. (10) Unusually high Sr and REE contents in epidote minerals including allanite suggest that they are the main reservoirs of these elements in subduction zones and HP to UHP environments. (11) Sector-zoning of zoisite and epidote and their petrological and mineralogical significance is also briefly discussed.

Mineral abbreviations employed in this chapter are after Kretz (1983) and Miyashiro (1994) other than Amph (amphibole), Bar (barroisite), Cro (crossite), Coe (coesite), Coe-ps (quartz pseudomorph after coesite), Dia (diamond), Fgln (ferroglaucophane), Mrb (magnesioriebeckite), Sr-Pie (strontiopiemontite) and Wnc (winchite).

## EPIDOTE MINERALS IN A TYPICAL BLUESCHIST FACIES METAMORPHISM: THE FRANCISCAN COMPLEX/CALIFORNIA

Major lithologies of the Franciscan Complex of the California Coast Range, the classic example of a subduction complex, are metaclastic rocks plus minor metacherts and metabasites. These rocks were subjected to Cretaceous blueschist-facies metamorphism and contain combinations of pumpellyite, lawsonite, epidote, phengite, albite, chlorite, sodic amphibole, quartz and titanite. Lawsonite and minor epidote and pumpellyite are the dominant Ca-Al-silicates. Epidote is restricted to the metabasites and has not been recognized as a neoblastic phase in metagraywackes (e.g., Ernst 1965; Ernst et al. 1970; Jayko et al. 1986). Within the *P-T* conditions of the blueschist facies, characteristic metabasite assemblages contain lawsonite at lower temperature and epidote at higher temperature together with aragonite and jadeitic clinopyroxene + quartz in addition to sodic amphibole. Franciscan epidote-bearing rocks include greenstone, greenschist, blueschist, high-grade blueschist "knockers" and eclogites. Many eclogite blocks have experienced a counter-clockwise *P-T* path starting from epidote-amphibolite through eclogite- to blueschist-facies re-equilibration (Moore 1984; Krogh et al. 1994). Epidote also occurs in both pre- and post-eclogite facies stages.

The following discussion focuses on paragenesis and compositions of epidote in blueschists and eclogites from Ward Creek of the Central belt as the most representative Franciscan epidote-bearing metabasites occur in this area. This area has been intensively studied since aragonite was first identified here (Coleman and Lee 1963; Liou and Maruyama 1987; Maruyama and Liou 1987, 1988; Oh and Liou 1990; Oh et al. 1991; Shibakusa and Maekawa 1997; Banno et al. 2000).

### Prograde zoning of Franciscan metabasite in Ward Creek of the Cazadero region

The Jurassic-Cretaceous blueschist sequence of Cazadero lies within the Central Franciscan belt of the northern California Coast Range. Coleman and Lee (1963) classified the metabasites into 4 types: (I) unmetamorphosed, (II) metamorphosed and non-foliated, (III) metamorphosed and foliated, and (IV) coarsely crystalline tectonic blocks. Type II and III metabasites are the most abundant and are continuously exposed for more than 2 km in Ward Creek. Within the metabasites three prograde metamorphic zones (lawsonite, pumpellyite, and epidote zone) and their mineral paragenesis were mapped (Fig. 2; Maruyama and Liou 1988; Oh et al. 1991). All metabasites contain minor quartz, white mica, albite, chlorite, titanite, and aragonite. Both the lawsonite and pumpellyite zones are equivalent to the Type II non-foliated metabasite of Coleman and Lee (1963). The epidote zone includes Type III blueschists, *in situ* eclogite and Type IV tectonic blocks. In some Type III epidote-zone blueschists garnet occurs together with omphacite, epidote, glaucophane and other epidote-zone minerals, indicating the onset of transition to the eclogite facies (Oh and Liou 1990).

Epidote zone metabasites are characterized by the assemblages epidote + clinopyroxene + two amphiboles + chlorite, lawsonite + pumpellyite + actinolite + chlorite and epidote + pumpellyite + two amphiboles + chlorite depending on bulk rock $Fe_2O_3$ content. With increasing grade, winchite appears, Fe-free lawsonite is stable with epidote, pumpellyite disappears, and omphacite contains a very low acmite component. The common assemblages are epidote + winchite + lawsonite and lawsonite + omphacite + winchite. Epidote becomes Al-rich where lawsonite is no longer stable. Hence epidote + glaucophane + omphacite + garnet is characteristic and was named high-epidote zone eclogite (Oh et al. 1991).

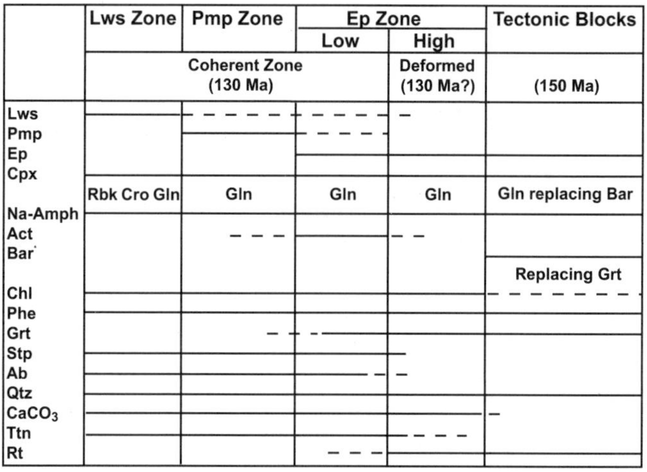

| | Lws Zone | Pmp Zone | Ep Zone Low | Ep Zone High | Tectonic Blocks |
|---|---|---|---|---|---|
| | | Coherent Zone (130 Ma) | | Deformed (130 Ma?) | (150 Ma) |
| Lws | | – – – – – | – – – – – | – | |
| Pmp | | | – – – – | | |
| Ep | | | | | |
| Cpx | | | | | |
| Na-Amph | Rbk Cro Gln | Gln | Gln | Gln | Gln replacing Bar |
| Act | | | | | |
| Bar | | – – – | | – – | |
| Chl | | | | | Replacing Grt |
| Phe | | | | | – – – – – – |
| Grt | | | – – | – | |
| Stp | | | | – | |
| Ab | | | | – | – |
| Qtz | | | | | |
| CaCO₃ | | | | | – |
| Ttn | | | | | – – – |
| Rt | | | – – – | | |

**Figure 2.** Mineral paragenesis in Ward Creek metabasites/Franciscan, California (modified from Table 2 of Oh et al. 1991). Three metamorphic zones increasing from left to right occur in a continuous exposure in Ward Creek (Maruyama and Liou 1988) and tectonic blocks are enclosed within serpentinite or in epidote-zone metabasites.

Serpentinite mélange of the Franciscan Complex contains Type IV tectonic blocks including eclogite, high-grade blueschist, and garnet amphibolite; these rocks are different from the epidote-zone metabasites mentioned above. Epidote is abundant in these rocks ranging from 15 to 29 vol% (mineral paragenesis in Fig. 2, compiled from Tables 3 and 4 of Oh et al. 1991). These rocks are older (150 Ma) than type III metabasites (130 Ma) (Lee et al. 1964; Coleman and Lanphere 1971).

Paragenesis and composition of major phases in Ward Creek blueschist and eclogitic rocks and *P-T* estimates are discussed in detail using a modified epidote-projection ACF diagram (Liou and Maruyama 1987; Maruyama and Liou 1987, 1988; Oh et al. 1991). The lawsonite zone appears to be stable at $T < 200°C$ in a pressure range of 0.4–0.65 GPa; the pumpellyite zone between 200-290°C and the epidote zone above 290°C at 0.65 to 0.9 GPa (Maruyama and Liou 1987). The *P-T* estimate for *in situ* eclogitic schist is $290°C < T < 350°C$, $0.8 < P < 0.9$ GPa, whereas that of type IV eclogite is $500°C < T < 540°C$, $P > 1.0–1.15$ GPa.

**Epidote texture and composition**

Maruyama and Liou (1987) describe two distinct modes of epidote occurrence in Type III epidote-zone metabasite. The first type is fine-grained epidote aggregates that replace primary augite phenocrysts. Individual epidote grains have an irregular habit and a dusty appearance. Some grains contain abundant inclusions of titanite, clinopyroxene and minor pumpellyite. The textural relations, the inverse modal proportion of epidote and clinopyroxene, and the sudden decrease in the modal abundance of both clinopyroxene + pumpellyite at the onset of the epidote zone suggest that the granular epidote have formed by a reaction such as clinopyroxene + pumpellyite = epidote + amphibole. This epidote type lacks the pleochroic colors for common epidote due to small grain size and constitutes less than 5 vol%; it is not surprising that Coleman and Lee (1962) did not identify epidote in Type III metabasite. The second type of epidote occurs as rims around xenoblastic, coarse-grained pumpellyite and lawsonite. This subidioblastic epidote is free from inclusions. Although lawsonite is abundant and coarse-grained in some epidote-zone metabasites, textural relations suggest that lawsonite is not in equilibrium with epidote. A third type of epidote can be described for the

Type III epidote-zone *in situ* eclogites. This is coarse-grained, idioblastic, and exhibits the characteristic interference colors.

Representative analyses of epidote from blueschists and eclogites from this area are listed in Table 1; their $X_{Fe}$ content is plotted against the metamorphic grade in Figure 3. Epidote contains low total Mn (0.05–0.6 wt% MnO) and $TiO_2$ and negligible amounts of $Na_2O$, MgO and $Cr_2O_3$. Epidote in blueschists varies substantially in composition, but has consistently higher $X_{Fe}$ in the core than in the rim suggesting progressive growth with increasing temperature. However, the overall trend with increasing grade does not show such a relationship (Fig. 3). The epidote core has $X_{Fe} = 0.18$ to 0.22 whereas the epidote rim has $X_{Fe} = 0.13$ to 0.17. Epidote of *in situ* eclogites has the same core to rim relation as that in the associated blueschists. However, it has generally higher $X_{Fe}$ (0.24 to 0.29) and MnO contents (0.37 to 0.67 wt%; Fig. 3) suggesting higher $Fe^{3+}$ and Mn in the whole rock, which is consistent with the occurrence of crossite and Mn-rich garnet (Oh et al. 1991). Contrary to epidote from the Type III epidote-zone rocks, epidote from Type IV tectonic blocks has the inverse core to rim relation with higher $Fe^{3+}$ content in the rim than in the core. This can be interpreted as retrograde re-equilibration at lower *T* during later blueschist facies overprint.

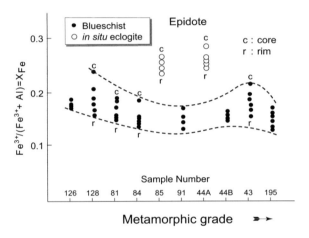

**Figure 3.** Compositional ranges of epidote from blueschist, and *in-situ* eclogite from Ward Creek. Both core and rim compositions are shown and they are arranged according to the inferred metamorphic grade for the blueschist (modified from Fig. 10 of Maruyama and Liou 1988).

## EPIDOTE MINERALS IN LAWSONITE BLUESCHIST-ECLOGITE FACIES METAMORPHISM: PAM-OUÉGOA-POUÉBO REGION/NEW CALEDONIA

Palaeogene high *P/T* metamorphic rocks are exposed in an elongate anticlinal range of northwestern New Caledonia (e.g., Brothers 1974; Black 1977). Based on the mineral paragenesis of metapelites, Yokoyama et al. (1986) grouped them into five zones of increasing metamorphic grade: low-grade lawsonite, Mn-garnet, lawsonite-epidote transition, epidote, and omphacite zones. Clarke et al. (1997) divided the metamorphic rocks into two terranes: (1) the structurally higher Diahot terrane of blueschist facies metasediments and metavolcanics; and (2) the structurally lower Pouébo terrane of eclogites and their hydration product glaucophanite and rare quartzite. The Diahot terrane is subdivided into a lower-grade ferroglaucophane-lawsonite zone and higher-grade albite-epidote-omphacite zone. Schists of the ferroglaucophane-lawsonite zone have mineral assemblages similar to the

**Table 1.** Representative analyses of epidote minerals from the Franciscan complex.

| Locality | Ward Creek | | | | | | | Diablo Range | | Oregon | |
|---|---|---|---|---|---|---|---|---|---|---|---|
| Rock Type | IEc | EBl | EBl | EBl | HEEc | GAm | VIEc | EBl | EBl | HBl | Hs |
| Sample | W-85 | W-91 | W-91 | W-43 | W-06 | W-32 | WT-73 | GL1 | GL1 | Ore11 | Ore8 |
| Mineral | Ep | Ep (c) | Ep (r) | Ep | Ep | Ep | Ep | Ep (c) | Ep (r) | Ep (r) | Ep (r) |
| $SiO_2$ | 37.55 | 38.70 | 38.99 | 38.06 | 36.63 | 38.53 | 38.09 | 37.2 | 38.2 | 38.11 | 37.77 |
| $TiO_2$ | 0.11 | 0.11 | 0.12 | 0.02 | 0.00 | 0.15 | 0.19 | 0.1 | 0.1 | 0.08 | 0.07 |
| $Al_2O_3$ | 24.40 | 26.69 | 28.13 | 27.23 | 24.86 | 26.17 | 26.28 | 22.5 | 25.8 | 27.77 | 25.93 |
| $Cr_2O_3$ | 0.07 | 0.07 | 0.00 | 0.12 | 0.00 | 0.00 | 0.00 | n.d. | n.d. | n.d. | n.d. |
| $Fe_2O_3$* | 11.26 | 8.39 | 6.71 | 8.07 | 12.71 | 8.82 | 7.41 | 15.5 | 11.4 | 8.80 | 11.20 |
| MnO | 0.51 | 0.30 | 0.08 | 0.10 | 0.44 | 0.04 | 0.10 | 0.3 | 0.3 | 0.12 | 0.18 |
| MgO | 0.04 | 0.17 | 0.00 | 0.13 | 0.00 | 0.07 | 0.08 | n.d. | 0.1 | 0.01 | 0.04 |
| CaO | 22.16 | 22.89 | 23.21 | 22.04 | 22.86 | 23.22 | 23.24 | 22.8 | 22.9 | 23.76 | 23.59 |
| $Na_2O$ | n.d. | 0.14 | n.d. | 0.00 | n.d. | n.d. | n.d. | n.d. | 0.2 | n.d. | n.d. |
| $K_2O$ | n.d. | 0.00 | n.d. | 0.00 | n.d. | n.d. | n.d. | n.d. | n.d. | n.d. | n.d. |
| Total | 96.10 | 97.46 | 97.24 | 95.77 | 97.50 | 97.00 | 95.39 | 98.40 | 99.00 | 98.65 | 98.78 |
| $X_{Fe}$** | 0.23 | 0.17 | 0.13 | 0.16 | 0.25 | 0.18 | 0.15 | 0.31 | 0.22 | 0.17 | 0.22 |
| Ref. | O91 | M88 | M88 | M88 | O91 | O91 | O91 | M84 | M84 | M89 | M89 |

* Total Fe as $Fe_2O_3$, $X_{Fe}$ = $Fe^{3+}/(Al + Cr + Fe^{3+} + Mn^{3+})$.

Abbreviations of rock-types: IEc, *in-situ* eclogite; EBl, epidote-blueschist; HEEc, HE-eclogite; GAm, garnet-amphibolite; VIEc, type VI-eclogite; Hs, hornblende schist.

Other abbreviations: c, crystal core; r, crystal rim; n.d., not determined.

References: O91, Oh et al. (1991); M88, Maruyama and Liou (1988); M84, Moore (1984); M89, Moore and Blake (1989).

Mn-garnet zone of Yokoyama et al. (1986) and their peak $P/T$ conditions are estimated at 0.7–0.9 GPa/340–460°C (Clarke et al. 1997). Schists of the albite-epidote-omphacite zone are significantly coarser grained than those in the ferroglaucophane-lawsonite zone and are equivalent in mineral paragenesis to schists of the lawsonite-epidote transition, epidote, and the lower-grade part of the omphacite zone of Yokoyama et al. (1986). Their estimated peak $P/T$ conditions are 1.1–1.4 GPa/530–610°C. The eclogite paragenesis of the Pouébo terrane roughly correspond to those of the omphacite zone of Yokoyama et al. (1986) with estimated peak $P/T$ conditions of 2.1–2.7 GPa/~600°C (Clarke et al. 1997). Clarke et al. (1997) recognized three rock-types of eclogite facies having bulk rock compositions consistent with protoliths of basalt (Type I), gabbro cumulate (Type II), and chert (Type III), respectively. The following review is mainly based upon data of Yokoyama et al. (1986) and Clarke et al. (1997) unless otherwise noted.

### Epidote-bearing mineral assemblages

Epidote minerals are common in metabasites and metapelites from the albite-epidote-omphacite zone of the Diahot terrane and in eclogites of the Pouébo terrane, but are absent in meta-acidite. As epidote minerals are unstable in the lower-grade ferroglaucophane-lawsonite zone, Black (1977) proposed the following epidote-producing reactions between the ferroglaucophane-lawsonite and albite-epidote-omphacite zones:

Lws + Fe-rich Chl + Cro = Ep + Alm + Mg-rich Chl + Gln              for metabasite,

Lws + Fe-rich Chl + Fgln = Ep + Alm + Mg-rich Chl + Gln              and/or

Lws + Fe-rich Chl + Ab = Ep + Mg-rich Chl + Pg              for metapelite.

Zoisite occurs in about 15% of the samples from the omphacite zone (Yokoyama et al. 1986) and usually coexists with Al-rich epidote. Some epidote grains are clearly retrograde products that form aggregates with chlorite and albite replacing glaucophane and paragonite. The stable epidote-bearing mineral assemblages in the different rock types as function of metamorphic grade are:

*Metabasite.* In the lower-grade part of the albite-epidote-omphacite zone, epidote coexists with lawsonite, omphacite, glaucophane, actinolite, chlorite, and titanite and ± phengite, albite, and/or quartz. With increasing grade lawsonite and actinolite phase out and Fe-rich garnet and rutile join the above assemblage. The eclogites of the Pouébo terrane consists of epidote with omphacite, glaucophane, barroisitic hornblende, Fe-rich garnet, paragonite, and rutile and ± titanite, phengite and/or quartz.

*Metapelite.* In the lower-grade part of the albite-epidote-omphacite zone, epidote coexists with lawsonite, glaucophane, Mn-rich garnet, chlorite, phengite, paragonite, titanite, rutile, quartz, graphite, and albite. With increasing grade lawsonite and Mn-rich garnet disappear and Fe-rich garnet becomes stable. In the eclogite facies metapelites of the Pouébo terrane zoisite and omphacite occur as major constituent phases, and chlorite is absent in most samples.

*Meta-acidite.* Epidote minerals are generally absent in meta-acidite. Black et al. (1988) discussed the relationship between bulk rock chemistry and stability of clinozoisite, omphacite, and jadeite in metamorphosed siliceous sediments at eclogite facies conditions in the ½Al-Na-Ca ternary (Fig. 4). This ternary shows that the stability of these minerals is almost exclusively a function of the $X_{Na}^{whole\,rock}$ [= Na/(Na + Ca)] of the host rock. Clinozoisite is restricted to $X_{Na}^{whole\,rock} < 0.27$, omphacite occurs in rocks with $X_{Na}^{whole\,rock} = 0.27$-0.55, and jadeite crystallizes in rocks with $X_{Na}^{whole\,rock} > 0.55$ (Fig. 4). These data indicate that epidote and zoisite become less stable with increasing $X_{Na}^{whole\,rock}$ of the host rock. As the meta-acidites from New Caledonia have $X_{Na}^{whole\,rock} > 0.55$ jadeitic pyroxene and/or its pseudomorph (albite) occur as Al-rich silicates instead of epidote and/or zoisite.

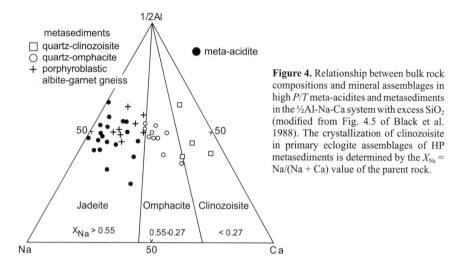

**Figure 4.** Relationship between bulk rock compositions and mineral assemblages in high $P/T$ meta-acidites and metasediments in the ½Al-Na-Ca system with excess $SiO_2$ (modified from Fig. 4.5 of Black et al. 1988). The crystallization of clinozoisite in primary eclogite assemblages of HP metasediments is determined by the $X_{Na} = Na/(Na + Ca)$ value of the parent rock.

### Compositional range of epidote minerals

Zoisite and clinozoisite-epidote in metabasite and metapelite have been described by Yokoyama et al. (1986) and Ghent et al. (1987). The overall range in $X_{Fe}$ of clinozoisite-epidote is 0.09 to 0.30. Zoisite displays a much smaller scatter and has generally $X_{Fe} < 0.06$. Epidote minerals in eclogitic rocks contain less $Fe^{3+}$ than those in common metabasite; coexisting zoisite and Al-rich epidote usually occur in the Type II eclogite (gabbroic cumulate protolith) of Clarke et al. (1997). The epidote minerals in eclogite and associated metabasite contain minor amounts of $TiO_2$ (0.07 to 0.22 wt%), MnO (< 0.03 to 0.32 wt%) and MgO (< 0.05 to 0.13 wt%) (Ghent et al. 1987). $X_{Fe}$ between whole rock and epidote shows a strong positive correlation (Potel et al. 2002).

Mineralogical and chemical characteristics of the New Caledonia high $P/T$ metamorphic rocks are very similar to those reported from Sanbagawa metamorphic rocks (see below): (1) the compositional range of zoisite and clinozoisite-epidote series, (2) the common occurrence of coexisting zoisite and clinozoisite in higher-grade zone, and (3) greater abundance of zoisite in metagabbro than metabasalt.

## EPIDOTE MINERALS IN GREENSCHIST/BLUESCHIST-EPIDOTE AMPHIBOLITE-ECLOGITE FACIES METAMORPHISM: THE SANBAGAWA BELT/JAPAN

The Sanbagawa high $P/T$ metamorphic belt extends throughout the Outer Zone (the Pacific Ocean side) of Southwest Japan for a length of roughly 800 km (Miyashiro 1994). In central Shikoku the Sanbagawa belt is divided into chlorite, garnet, albite-biotite and oligoclase-biotite zones based on the matrix mineral paragenesis of pelitic schists in ascending order of metamorphic grade (Enami 1983; Higashino 1990). Peak $P$-$T$ conditions of the lower-grade part of the chlorite zone correspond to that of the high-$P$ pumpellyite-actinolite facies. High-grade chlorite and low-grade garnet zones approximate to the high-$P$ greenschist and/or low-$P$ blueschist facies, and high-grade garnet zone and albite- and oligoclase-biotite zones are equivalent to the epidote-amphibolite facies. Peak $P/T$ conditions of the chlorite, garnet, albite-biotite and oligoclase-biotite zones are estimated at 0.5–0.7 GPa/~360°C, 0.7–1.0 GPa/

430-500°C, 0.8–1.1 GPa/470–570°C, and 0.9–1.1 GPa/580–630°C, respectively (Enami et al. 1994; Wallis et al. 2000). Within the region of epidote-amphibolite facies metamorphism of central Shikoku there are numerous ultramafic and mafic masses. The mafic rocks have protoliths of gabbro and basalt (Kunugiza et al. 1986; Takasu 1989). The mafic-ultramafic complex in the Besshi region in central Shikoku underwent extensive equilibration under epidote-amphibolite facies conditions; however, the rocks also locally preserve evidence of earlier eclogite facies metamorphism (c. 1.5–3.8 GPa/600-800°C; e.g., Takasu 1989; Wallis and Aoya 2000; Aoya 2001; Enami et al. 2004).

### Composition of host rocks and epidote minerals

Epidote minerals are common and important phases throughout the different lithologies of the Sanbagawa belt. They are confirmed in almost all examined pelitic samples from the different zones although they represent only a minor phase in the individual samples (Goto et al. 2002). This extensive presence of epidote minerals is in marked contrast to the rare occurrence of epidote minerals in low to medium-grade pelitic schists of other regional metamorphic terranes (Deer et al. 1986). It probably reflects the *P-T* conditions at which epidote fulfills the role as main Ca-Al-silicate instead of calcic plagioclase. This interpretation is consistent with the occurrence of Ca-poor sodic plagioclase throughout the Sanbagawa belt.

Epidote minerals commonly show concentric zoning with decreasing or increasing $X_{Fe}$ value from the core to margin, which are normally considered to represent prograde and retrograde growths, respectively (e.g., Otsuki and Banno 1990). Some grains, however, clearly show sector-zoning (see below). The following discussion is mainly based on data reported by Enami (1978), Otsuki (1980), Higashino et al. (1981, 1984) and Otsuki and Banno (1990).

***Pelitic schists.*** The pelitic schists are mainly composed of quartz, sodic plagioclase, phengite, epidote, and calcite; chlorite is the main ferromagnesian mineral of the chlorite zone. In the chlorite zone, most epidote grains are too fine-grained to be positively identified using polarizing microscope (usually < 30 μm: Goto et al. 2002) and must be confirmed by SEM observation. At higher-grade, pelitic schists of the garnet zone are characterized by the coexistence of chlorite and garnet. In the albite- and oligoclase-biotite zones, these minerals are joined by biotite and hornblende/pargasite. Lawsonite appears sporadically in the lower-grade part of the chlorite zone. Zoisite coexists with Al-poor epidote in the oligoclase-biotite zone.

Zoisite has limited compositional range of $X_{Fe}$ = 0.02–0.05 and CaO = 22.8–24.8 wt% (Fig. 5). Epidote has variable compositions ranging from epidote sensu strictu, through rare earth element (REE)-bearing epidote (Sakai et al. 1984) to allanite with CaO = 9.6–24.4 wt%. The $X_{Fe}$ values of epidote are usually 0.10-0.20 except for epidote from chlorite zone ($X_{Fe}$ = 0.2–0.35) and REE-bearing epidote and allanite (Fig. 5). The low $X_{Fe}$ of epidote in garnet and higher-grade zones is probably due to low-$f_{O_2}$ conditions inferred from the common presence of fully ordered graphite in pelitic schist (Itaya 1981, Tagiri 1985). The $X_{Fe}$ (up to 0.40; Fig. 5), MnO (up to 1.2 wt%) and MgO (up to 0.8 wt%) of REE-epidote and allanite tend to increase with decreasing CaO. This trend suggests that some amounts of iron and manganese of REE-epidote and allanite are divalent related to the coupled substitution $REE_{+1}(Fe^{2+},Mn^{2+},Mg)_{+1}Ca_{-1}Fe^{3+}_{-1}$ (Deer et al. 1986). Most REE-epidotes occur as overgrowths on allanite, and are in turn sometimes rimmed by REE-poor epidote (Fig. 6a). REE-epidote shows grayish blue abnormal interference color and commonly forms a pleochroic halo in host biotite and chlorite, so that it is easily identified under the polarizing microscope. Average grain size of the REE-epidote systematically increases from 30 μm in the chlorite zone, through 55 μm in the garnet zone, to 65 μm in the albite-biotite zone (Sakai et al. 1984). Total REE and $ThO_2$ contents of most REE-epidotes are 8–15 wt% and up to 1.0 wt%, respectively. Chondrite-normalized REE-patterns for the REE-epidote are similar to those of coexisting allanite (Fig. 6b). In the Sanbagawa belt, allanite occurs

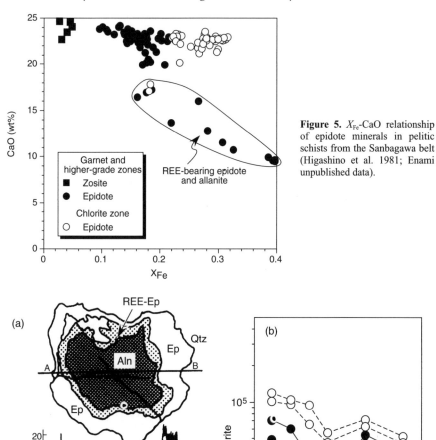

**Figure 5.** $X_{Fe}$-CaO relationship of epidote minerals in pelitic schists from the Sanbagawa belt (Higashino et al. 1981; Enami unpublished data).

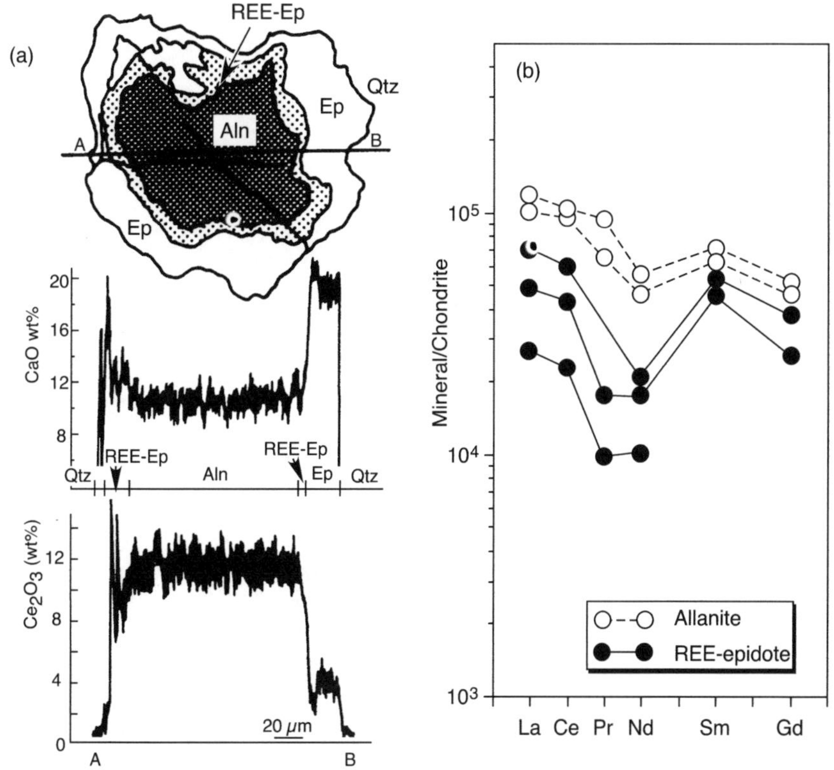

**Figure 6.** (a) Sketch of a epidote-allanite crystal with line scan for CaO and $Ce_2O_3$ and (b) Leedey chondrite (Masuda et al. 1973)-normalized REE patterns of coexisting REE-epidote and allanite in pelitic schists from the Sanbagawa belt (modified from Figs. 7 and 9 of Sakai et al. 1984).

only in metasediments and is probably detrital, and a major source of REE and Th for the crystallization of REE-epidote during prograde metamorphism.

   ***Siliceous schist.*** Most siliceous schists of the Sanbagawa are metamorphosed impure cherts, and have been commonly referred to as quartz schists or quartzitic schists. Silicate minerals are similar to those of high-$f_{O_2}$ mafic schists except for rare occurrences of sodic pyroxene ($X_{Jd} = 0.02$–$0.35$ and $X_{Acm} = 0.19$–$0.84$) and Mn-rich garnet (up to 38 wt% MnO).

   Most siliceous schists equilibrated under unusually high-$f_{O_2}$ conditions that are inferred from common occurrences of braunite, ardennite and other highly oxidized phases (e.g., Enami 1986). The epidote minerals within the siliceous schists can be classified into three groups based on the atomic proportions in the octahedral M-sites (Fig. 7). One group is Mn- and $Fe^{3+}$-poor epidote with $X_{Fe} = 0.11$–$0.21$ and $X_{Mn} < 0.01$, that occurs in siliceous schists intercalated with pelitic schists. This group probably formed under low-$f_{O_2}$ conditions. Another group most common in the siliceous schists, has compositions lying between $X_{Fe} = 0.33$ and $X_{Mn} = 0.33$ showing complete occupation of the M(3)-site with $Fe^{3+}$ and $Mn^{3+}$. A third group is unusually poor in Al suggesting $Fe^{3+}$ and $Mn^{3+}$ incorporation into M(1) as well as M(3) (Enami and Banno 2001). This group characteristically contains significant amounts of Sr and Ba in A(2) and can be termed strontiopiemontite as will be described later (Table 2).

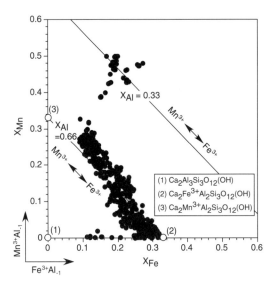

**Figure 7.** Compositional variations of epidote and piemontite of siliceous schists in the $X_{Al}$-$X_{Fe}$-$X_{Mn}$ system (Enami and Banno 2001; Enami unpublished data).

   ***Mafic schist.*** The typical mineral assemblage of mafic schist throughout the Sanbagawa belt is amphibole, chlorite, epidote, phengite, sodic plagioclase, and quartz. Amphibole compositions systematically change with increasing metamorphic grade from actinolite → winchite → crossite → barroisite → hornblende/pargasite under high-$f_{O_2}$ conditions in which hematite is stable, and actinolite → hornblende/pargasite under low-$f_{O_2}$ conditions. Pumpellyite occurs only in low-$f_{O_2}$ mafic schists such as a pyrrhotite-bearing sample from the lower-grade part of the chlorite zone. Garnet becomes stable only in the albite- and oligoclase-biotite zones.

   Epidote minerals in mafic schist are clinozoisite-epidote solid solutions (hereafter simply denoted as epidote) and have CaO > 20.4 wt% (> 1.78 pfu), MnO < 1.0 wt% and MgO < 0.9 wt%. The $X_{Fe}$ value shows a wide range of 0.1 to 0.36 that is closely related to the composition

**Table 2.** Representative analyses of epidote minerals in the Sanbagawa metamorphic rocks.

| Locality | Kotsu | Besshi | | Asemi | | Besshi | | | Tomisato | | |
|---|---|---|---|---|---|---|---|---|---|---|---|
| Sample | 0504 | TOCP-01 | | AS582 | M1 | GE1501b | | HU0311 | | GSJM31217 | |
| Mineral | Ep | Zo | REE-Ep | Ep | Pie | Ep | Zo | Ep | Sr-Pie | Sr-Pie | Sr-Pie |
| $SiO_2$ | 36.7 | 39.5 | 35.60 | 36.5 | 37.1 | 38.1 | 39.4 | 38.3 | 31.1 | 30.0 | 31.1 |
| $TiO_2$ | 0.11 | 0.10 | n.d. | n.d. | 0.00 | 0.00 | 0.08 | 0.15 | 0.48 | 0.27 | 0.53 |
| $Al_2O_3$ | 21.6 | 32.5 | 24.55 | 20.9 | 18.8 | 25.5 | 31.4 | 28.6 | 8.92 | 6.25 | 8.95 |
| $Cr_2O_3$ | 0.00 | n.d. | n.d. | n.d. | n.d. | 0.03 | 0.09 | 0.03 | 0.02 | n.d. | n.d. |
| $Fe_2O_3$* | 15.5 | 1.41 | 8.87 | 15.1 | 5.70 | 10.7 | 2.73 | 6.26 | 9.10 | 10.7 | 9.19 |
| MnO | 0.18 | 0.02 | n.d. | 0.70 | 14.9** | 0.06 | 0.00 | 0.06 | 17.5** | 18.7** | 17.4** |
| MgO | 0.00 | 0.02 | n.d. | n.d. | 0.05 | 0.22 | 0.04 | 0.03 | 0.02 | 0.01 | 0.01 |
| PbO | n.d. | n.d. | n.d. | n.d. | n.d. | n.d. | n.d. | n.d. | 5.44 | 7.55 | 5.61 |
| SrO | 2.24 | n.d. | n.d. | 2.83 | 0.13 | 1.11 | 0.14 | 0.14 | 8.59 | 9.98 | 8.68 |
| BaO | n.d. | n.d. | n.d. | n.d. | n.d. | n.d. | n.d. | n.d. | 6.44 | 3.66 | 6.71 |
| CaO | 21.0 | 24.7 | 16.05 | 21.0 | 22.0 | 22.4 | 24.0 | 23.0 | 10.2 | 10.1 | 10.3 |
| $Na_2O$ | 0.01 | 0.00 | n.d. | n.d. | 0.00 | n.d. | 0.00 | 0.00 | 0.03 | 0.01 | 0.00 |
| $K_2O$ | 0.00 | 0.00 | n.d. | n.d. | 0.00 | n.d. | 0.00 | 0.00 | 0.00 | 0.00 | 0.00 |
| $REE_2O_3$ | n.d. | n.d. | 14.89† | n.d. | n.d. | n.d. | n.d. | n.d. | n.d. | n.d. | n.d. |
| Total | 97.34 | 98.25 | 99.96 | 97.03 | 98.68 | 98.12 | 97.88 | 96.57 | 97.84 | 97.23 | 98.48 |
| $X_{Fe}$§ | 0.31 | 0.03 | 0.19 | 0.31 | 0.11 | 0.21 | 0.05 | 0.12 | 0.22 | 0.27 | 0.23 |
| Ref. | Eu | Eu | S84 | Eu | Eu | Eu | Eu | Eu | EB01 | EB01 | EB01 |

* Total Fe as $Fe_2O_3$,

** Total Mn as $Mn_2O_3$, § $X_{Fe} = Fe^{3+}/(Al + Cr + Fe^{3+} + Mn^{3+})$.

† Includes $La_2O_3$ (3.21 wt%), $Ce_2O_3$ (7.05 wt%), $Nd_2O_3$ (1.55 wt%), $Sm_2O_3$ (1.70 wt%) and $Gd_2O_3$ (1.38 wt%).

Abbreviations for rock types are: Bs, mafic schist; Ps, pelitic schist; Ss, siliceous schist; KEc, kyanite-eclogite; EAm, epidote-amphibolite. Other abbreviation: n.d., not determined.

References are: Eu, Enami (unpublished data); S84, Sakai et al. (1984); EB01, Enami and Banno (2001).

of associated opaque minerals. Epidote in pyrrhotite-bearing mafic schist mostly has $X_{Fe}$ value of 0.10–0.22 whereas epidote that coexists with hematite typically has $X_{Fe}$ value of 0.23–0.36 (Fig. 8). Stability of these two opaque minerals is strongly controlled by $f_{O_2}$; the occurrence of pyrrhotite indicates distinctly lower $f_{O_2}$ than hematite-bearing assemblages (e.g., Itaya et al. 1985). Thus, the close relationship between the composition of epidote and coexisting opaque minerals clearly suggests that $X_{Fe}$ is strongly controlled by $f_{O_2}$ during metamorphism (see also Poli and Schmidt 2004, Grapes and Hoskin 2004). A more detailed inspection of the relationships between epidote composition and coexisting mineral paragenesis in terms of the Al-Ca-Fe$^{3+}$ system with chlorite, albite, quartz, and $H_2O$-dominant fluid (Fig. 9; Nakajima 1982) indicates that under high-$P$ pumpellyite-actinolite facies conditions the maximum and minimum $X_{Fe}$ of epidote in mafic schist are defined by hematite + chlorite + sodic amphibole and pumpellyite + chlorite + actinolite, respectively.

***Epidote-amphibolite and eclogite.*** Protoliths of these rock types are mainly mafic plutonic and volcanic rocks; some occur as layers within the metasediments (Banno et al. 1976a, b; Kunugiza et al. 1986; Takasu 1989). These rocks underwent extensive recrystallization under epidote-amphibolite facies conditions, and their typical mineral assemblages are similar to those of the surrounding mafic schists except for the common occurrence of garnet in the epidote-amphibolites. However, they also locally preserve evidence of a prior stage of eclogite facies metamorphism (e.g., Takasu 1989; Wallis and Aoya 2000). Most epidote minerals in epidote-amphibolite and eclogite (hereafter denoted as epidote-amphibolite) are clinozoisite-epidote with $X_{Fe}$ = 0.07–0.27 (Fig. 10), CaO > 20.8 wt% (> 1.85 pfu), MnO < 0.7 wt% and MgO < 0.9 wt%. Zoisite occurs in an epidote-amphibolite derived from an inferred gabbroic protolith. Zoisite-rich layers (6 m in maximum thickness) intercalated with thin hornblende-rich layers (usually less than 1 m thick) occur in several epidote-amphibolite bodies (Banno et al. 1976a, b). The zoisite-rich rocks also contain subordinate amounts of kyanite, paragonite, phengite and quartz (Enami 1980), and are considered to have originally been cumulate anorthosite layers (Banno et al. 1976b; Yokoyama 1976). Zoisite has a limited compositional range of $X_{Fe}$ = 0.02–0.05, CaO = 23.5–24.5 wt% (1.93–2.02 pfu), MnO < 0.08 wt% and MgO < 0.05 wt% (Fig. 10).

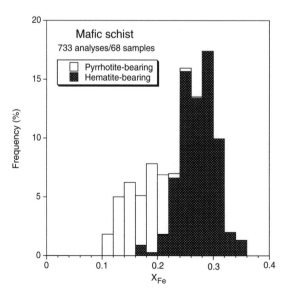

**Figure 8.** Frequency distribution of $X_{Fe}$ of epidote in hematite- and pyrrhotite-bearing mafic schists from the Sanbagawa belt (Higashino et al. 1981; Higashino et al. 1984; Enami unpublished data).

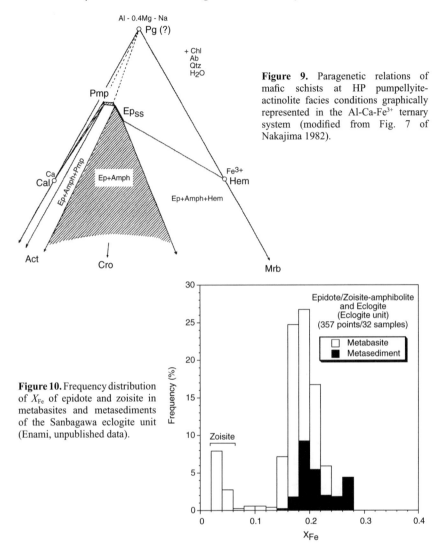

**Figure 9.** Paragenetic relations of mafic schists at HP pumpellyite-actinolite facies conditions graphically represented in the Al-Ca-Fe³⁺ ternary system (modified from Fig. 7 of Nakajima 1982).

**Figure 10.** Frequency distribution of $X_{Fe}$ of epidote and zoisite in metabasites and metasediments of the Sanbagawa eclogite unit (Enami, unpublished data).

## Compositional change of epidote minerals with metamorphic grade

Epidote composition varies with temperature by (1) decreasing $X_{Fe}$ and (2) enlargement of the compositional range towards lower $X_{Fe}$ with increasing metamorphic grade. Coexisting epidote and pumpellyite in the mafic schists of the Sanbagawa belt were first documented by Nakajima et al. (1977) and Nakajima (1982). $X_{Fe}$ of epidote coexisting with pumpellyite, actinolite, chlorite, albite, and quartz systematically decreases with increasing metamorphic grade (Fig. 11) and is controlled by the pumpellyite-consuming reaction (Fig. 12)

41 Pmp + 2 Chl + 47 Qtz = 71 Cz + 11 Act + 109 $H_2O$

A similar compositional change occurs in pelitic schists from the chlorite to garnet zone (Fig. 13). In this case, epidote is the only Ca-Al-silicate and the compositional shift cannot

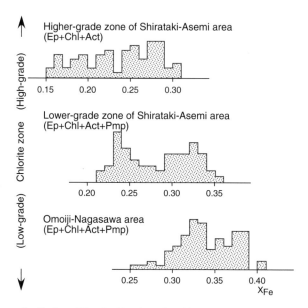

**Figure 11.** Frequency distribution of $X_{Fe}$ of epidote in mafic schists of the HP pumpellyite-actinolite facies and lower-grade areas of the Sanbagawa belt (modified from Fig. 8 of Nakajima 1982).

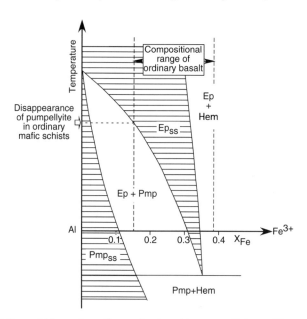

**Figure 12.** Al-Fe$^{3+}$ pseudobinary phase diagram for the epidote-pumpellyite equilibrium (modified from Fig. 6 of Nakajima 1982).

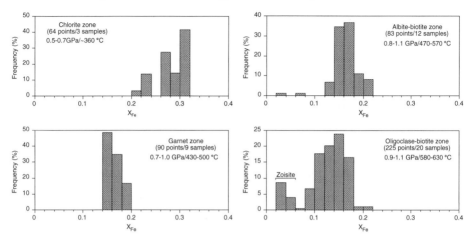

**Figure 13.** Frequency distribution of $X_{Fe}$ of epidote in pelitic schists from the Sanbagawa belt (Higashino et al. 1981; Enami unpublished data).

be explained by a net transfer reaction. Poorly crystalline carbonaceous matter or disordered-graphite is common in pelitic schists of the chlorite zone instead of fully ordered graphite in garnet and higher-grade schist (Itaya 1981; Tagiri 1985). They might not have completely buffered $f_{O_2}$ to reduced conditions, which may explain the high $X_{Fe}$ of epidote in chlorite zone pelitic schists. The breakdown of detrital Ca-plagioclase in low-grade pelitic schists could possibly explain the decrease in epidote $X_{Fe}$. Goto et al. (2002) emphasized that calcite is common in the Sanbagawa pelitic schists and thus its occurrence has a significant role in mineral reactions. They proposed possible dehydration-decarbonation reactions to produce biotite and calcic amphibole, respectively, which are common in pelitic schists of the albite-biotite and oligoclase-biotite zones, as follows:

$$3\ Ms + 2\ Chl + 16\ Cal + 24\ Qtz = 3\ Bt + 8\ Cz + 12\ H_2O + 16\ CO_2, \qquad \text{and}$$

$$4\ Chl + 9\ Ab + 22\ Cal + 13\ Qtz = 9\ Prg + 2\ Cz + 22\ H_2O + 22\ CO_2.$$

These reactions form clinozoisite as a high-$T$ product, and can explain (1) increasing modal abundance of epidote with increasing metamorphic grade (Fig. 3 of Goto et al. 2002) and (2) support a general consensus that the zoning of epidote with decreasing $X_{Fe}$ towards the margin is a prograde feature.

Enlargement of the epidote compositional field with increasing metamorphic grade has been documented by Miyashiro and Seki (1958) with optic axial angle measurements. They concluded that (1) $X_{Fe}$ of epidote in pelitic and psammitic schists tends to be lower than that in mafic schist, and (2) the compositional range enlarges towards lower $X_{Fe}$ with increasing metamorphic grade. These findings are confirmed by electron microprobe analyses of epidote (Fig. 13). The enlargement of the epidote compositional field and the presence of a two-phase loop between orthorhombic and monoclinic epidote minerals (Enami and Banno 1980) probably cause the appearance of zoisite in the albite-biotite and oligoclase-biotite zones (Figs. 10 and 13).

Studies on coexisting zoisite and epidote mainly in the Sanbagawa epidote-amphibolites and basic schists (Enami and Banno 1980) show that (1) clinozoisite contains distinctly higher $X_{Fe}$ than coexisting zoisite suggesting the presence of a two-phase loop between zoisite and clinozoisite, (2) the compositional range of this two-phase loop shifts towards $Fe^{3+}$-rich compositions with increasing metamorphic grade (Fig. 14), (3) zoisite solid solution, in relation to epidote solid solution, represents the higher $T$ phase, and thus (4) the frequency of appearance of zoisite should increase with increasing metamorphic grade (Fig. 14; Table 2). Franz and Selverstone (1992) reported similar zoisite-clinozoisite phase relations and suggested that the clinozoisite-side limb of the loop shifts towards $Fe^{3+}$-rich compositions with increasing metamorphic pressure.

Single epidote grains commonly display zonal discontinuities that have been attributed to a compositional gap in the clinozoisite-epidote series with a crest composition at $X_{Fe} = 0.17$ to $0.20$ (e.g., Strens 1965; Raith 1976). The compositional range of Sanbagawa epidotes, however, extends across the proposed compositional gap implying complete miscibility along the clinozoisite-epidote join under blueschist and high-$P$ pumpellyite-actinolite facies conditions (cf. Grapes and Hoskins 2004).

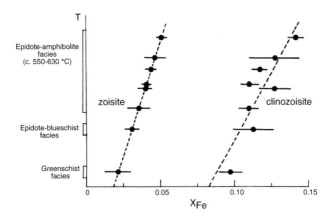

**Figure 14.** Relationship between $X_{Fe}$ of coexisting zoisite and clinozoisite and metamorphic grade (modified from Fig. 5 of Enami and Banno 1980).

## EPIDOTE MINERALS IN CALCAREOUS MEATASEDIMENTS OF THE ECLOGITE FACIES: EUROPEAN ALPS

Metacarbonates and calcareous metasediments are minor rock-types in HP metamorphic terranes that are closely related to the oceanic subduction system. The eclogite facies metasediments including impure marbles, siliceous dolomites, calc-mica schists and calc-schists are, however, widely distributed in the European Alps; these are rare lithological sequence in a HP terrane (e.g., Droop et al. 1990). Petrologic and mineralogical characteristics of the zoisite-bearing calcareous HP metasediments have been well documented from the Eclogite Zone of the Tauern Window, eastern Alps (Franz and Spear 1983; Spear and Franz 1986) and the Adula nappe, central Alps (Heinrich 1982). Many of the eclogite facies rocks are of early Alpine age of the Cretaceous. The eclogite facies event was followed by decompression and cooling to blueschist facies conditions.

Metamorphic conditions of the Eclogite Zone rocks in the Tauern Window during the high-$P$ event have been estimated at 1.9–0.2 GPa/590–630°C (e.g., Hoscheck 2001). Typical eclogite facies assemblages include zoisite + dolomite + kyanite + quartz in kyanite marbles, zoisite + phengitic muscovite + dolomite + calcite + quartz in zoisite marbles (Fig. 15; Spear and Franz 1986), zoisite + diopside + tremolite + dolomite + calcite + quartz in siliceous dolomites (Franz and Spear 1983), and zoisite + dolomite + calcite + phengitic muscovite +

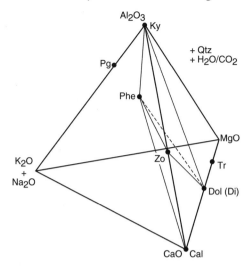

**Figure 15.** Phase relations of primary mineral assemblages in kyanite-zoisite marble from the Eclogite Zone, Tauern Window in the system $Al_2O_3$-CaO-MgO-($Na_2O$, $K_2O$). Projection was made from quartz and a fluid phase of fixed $H_2O/CO_2$ (modified from Fig. 9 of Spear and Franz 1986).

garnet + quartz in calc-mica schists (Hoscheck 2001). Zoisite ($X_{Fe}$ = 0.02–0.05) forms laths up to several mm in length, and is partly intergrown with garnet in the calc-mica schists. Zoisite in the zoisite marbles decomposed into paragonite, secondary muscovite, chlorite and kaolinite according to the following reactions:

$$Zo + Dol + Qtz + H_2O = Chl + Cal + CO_2,$$

$$Zo + Dol + NaCl/KCl + H_2O = Chl + Pg + Ms + CaCl_2 + CO_2,$$ and

$$Zo + HCl + H_2O = Kln + CaCl_2$$

(Spear and Franz 1986). In the calc-mica schists, clinozoisite/epidote occur as minor phases in the matrix ($X_{Fe}$ = 0.15–0.17) and as inclusions in the garnet ($X_{Fe}$ = 0.15–0.17).

Pressure and temperature conditions of the Adula nappe systematically increase from northern Vals area (1.0–1.3 GPa/450–550°C) through Confin-Trescolmen area (1.2–2.2 GPa/450–550°C) to southern Gagnone-Arami-Duria area (1.5–3.5 GPa/600–900°C) (Heinrich 1982, Droop et al. 1990). The zoisite-bearing calc-schists occur in the central and northern areas, and include assemblages of zoisite + dolomite + garnet + omphacite + kyanite + quartz, zoisite + dolomite + garnet + omphacite + quartz + hornblende + phengite + paragonite, and zoisite + calcite + garnet + phengite + quartz (Heinrich 1982).

## PARAGENESIS OF EPIDOTE MINERALS IN HP TO UHP ROCKS

The generalized *P-T* path of HP to UHP rocks shows at least four discrete stages of metamorphic crystallization: (I) the prograde stage during subduction from (sub)greenschist to blueschist, epidote-amphibolite, and even eclogite facies conditions, (II) the peak stage from blueschist or epidote-amphibolite to various eclogite facies conditions, even up to depths otherwise typical for the Earth's mantle, (III) a first retrograde epidote-amphibolite or amphibolite facies overprint during exhumation to crustal depths of < 35 km, and (IV) a second retrograde greenschist facies overprint during final uplift. Such a four-stage metamorphic evolution is exemplified by kyanite-bearing and kyanite-absent eclogites of the western Dabie (Fig. 16; Eide and Liou 2000). The prograde assemblage (stage I) of these eclogites comprises amphibole, epidote ($X_{Fe}$ = 0.21–0.28), phengite, quartz, and rutile, all occurring as inclusions

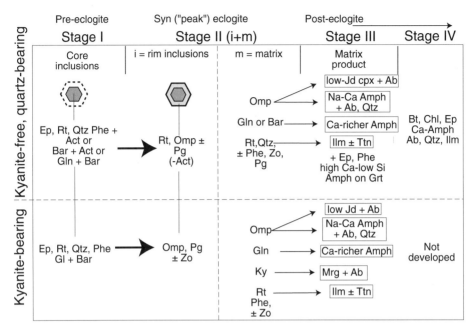

**Figure 16.** Four-stage metamorphic evolution of kyanite-bearing and kyanite-absent eclogites from the Dabie terrane showing the occurrence of epidote/zoisite in 4 different stages of HP and UHP metamorphism (see text for explanation). The two eclogite types in the rows are respectively UHP kyanite-bearing eclogite and HP quartz-bearing kyanite-free eclogite from western Dabie (modified from Fig. 4 of Eide and Liou 2000).

in garnet. The eclogite facies peak assemblage (stage II) consists of garnet, omphacite, rutile, amphibolite, coesite/quartz, and/or kyanite and ± phengite, zoisite ($X_{Fe}$ = 0.03–0.07), and paragonite. The stage II assemblage can be further subdivided into IIi and IIm, distinguished on textural grounds by eclogite facies inclusions in garnet rims (IIi) and by matrix minerals (IIm). In stage III, titanite (± ilmenite) replaces rutile and epidote ($X_{Fe}$ = 0.21 ± 0.05) and Ca-amphibole replace garnet. Stage IV of upper greenschist facies conditions includes epidote ($X_{Fe}$ = 0.31–0.33), albite, quartz, and titanite or biotite + chlorite pseudomorphs after garnet. This example underlines the importance of epidote minerals for the study of HP to UHP rocks. Due to their wide *P-T* stability field and occurrence in a variety of bulk rock compositions, epidote minerals could form during each of the four stages normally encountered by these rocks. They typically occur from small inclusions (< 50 μm) to large porphyroblasts (up to 2 mm) in the pre-, syn- and post-peak metamorphic assemblages of coesite-bearing eclogites and associated rocks. Because of the well known sluggish reaction kinetics of epidote minerals, it is likely that once they have formed they might persist metastably without re-equilibration, such that their chemical zoning in major and trace elements records the complex *P-T* history of these rocks.

### The prograde evolution of HP to UHP rocks

During the prograde stage, the protoliths of UHP rocks may have passed through (sub) greenschist, greenschist, blueschist, and/or epidote-amphibolite facies and developed the typical successions of mineral paragenesis described above for subduction zone metamorphism. All protoliths of UHP rocks, however, pass and probably recrystallize under quartz-eclogite facies conditions before entering the UHP domain. In mafic rocks the prograde quartz-eclogite facies recrystallization results in a combination of garnet, katophoritic amphibole, paragonite, omphacite, rutile, and epidote minerals. The prograde phases of HP to UHP rocks typically occur as inclusions in garnet porphyroblasts but also in omphacite, and/or kyanite. These inclusions are abundant in quartz eclogite but are comparatively rare in coesite eclogite. Enami et al. (1993) and Zhang et al. (1995a) used such prograde inclusions in garnet together with the garnet core compositions to constrain the prograde *P-T* path of the Donghai UHP eclogite through epidote-amphibolite and quartz eclogite facies conditions. One very prominent texture of the prograde evolution of UHP rocks is found in coronitic meta-granitoids and metagabbros in which primary magmatic minerals, especially plagioclase, are pseudomorphically replaced by typical HP mineral paragenesis including zoisite (Fig. 17). Well-preserved igneous textures and pseudomorphs after original magmatic phases are described in the UHP meta-granitoids of the Brossaco-Isasca unit of the Dora-Maira Massif (western Alps; Bruno et al. 2001) and the Sulu terrane of eastern China (Hirajima et al. 1993). In these rocks, magmatic plagioclase is replaced by zoisite, jadeite, quartz, K-feldspar, and kyanite and various coronitic reactions developed between biotite and adjacent minerals. Zoisite occurs as needle-like crystals < 10 μm long, jadeite as stumpy crystals from 5 to 20 μm across and kyanite as prismatic crystals of variable size included in amoeboid patches of K-feldspar ($Or_{90}Ab_{10}$). Zoisite and kyanite are pure, sodic pyroxene is a solid solution of jadeite, Ca-Tschermak and Ca-Eskola. A pseudomorphic reaction is

$$4\ Pl + x\ H_2O = 2x\ Zo + x\ Ky + (x+4y)\ Qtz + 4y\ Jd + (4 - 4x - 4y)\ Kfs$$

where $x$ and $y$ are the molar proportions of anorthite and albite, respectively in a plagioclase of composition $(Ca_xNa_yK_{1-x-y})(Al_{1+x}Si_{1-x})Si_2O_8$ (Bruno et al. 2001). Zhang and Liou (1997a) describe the preserved prograde gradational sequence from incipiently metamorphosed gabbro (referred to as coronitic metagabbro) over partly to strongly recrystallized gabbro (referred to as transitional metagabbro) to completely recrystallized coesite eclogite within a single boulder (about 30 m in diameter) from Sulu. This boulder was found adjacent to the above-mentioned

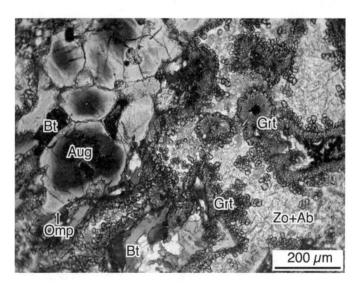

**Figure 17.** Photomicrograph (open nicol) of a meta-granitoid from Yangkou of the Sulu UHP terrane (Enami unpublished). Igneous augite with tiny inclusion of Fe-oxides is partly rimmed by very thin omphacite. Coronitic garnet is developed between igneous biotite and plagioclase that is pseudomorphed by fine aggregates of zoisite and albite. Biotite is partly replaced by aggregates of Ti-rich phengite (cf. Hirajima et al. 1993).

meta-granitoid described by Hirajima et al. (1993). The coronitic metagabbro preserved relict gabbroic texture and minerals. Primary plagioclase broke down to zoisite, albite, kyanite, phengite, and ± amphibole. In the transitional metagabbro, augite and orthopyroxene are totally replaced by omphacite, and the lower-$P$ assemblage zoisite, albite, kyanite, phengite, and garnet coexists with domains of omphacite, kyanite, and phengite in pseudomorphs after plagioclase. The coesite eclogite contains the typical UHP assemblage omphacite, kyanite, garnet, phengite, coesite/quartz, and rutile but still preserve a faint gabbroic texture. Estimated $P$-$T$ conditions for the different prograde stages recorded in this unique rock are 540 ± 50°C at ~1.3 GPa for the coronal stage, 600-800°C at ≥ 1.5–2.5 GPa for the transitional stage and 800-850°C at >3.0 GPa for the UHP peak metamorphism.

A nearly monomineralic epidosite was suggested to be the parent rock for an UHP grossular-rich garnetite boudin that contains ~80 vol% garnet ($Grs_{63}Alm_{25}Prp_7And_4$; 0.10–0.18 wt% $Na_2O$) and 15 vol% hercynite + anorthite symplectite after kyanite in migmatitic gneiss of the Gföhl unit in the easternmost Moldanubian Zone, the Bohemian Massif (Vrána and Gryda 2003). This suggestion was based on nearly identical major composition of epidote ($X_{Fe}$ = 0.23–0.48) to grossular-rich garnetite and supported by about 4 times higher whole-rock REE abundances than those in Moldanubian paragneisses. Moreover, epidosite occurs in some 'stratiform' garnet-pyroxene Ca-Fe-Mg skarns in the Moldanubian Zone. In the process of epidotization, epidote would scavenge REE from fluids or surrounding rocks, due to the isomorphism in the series epidote-allanite. Although the index mineral coesite was not found, the elevated Na content in the garnet provides independent evidence of the UHP history of the rock. An approximate $P$-$T$ estimate yields pressure ≈ 4 GPa, $T$ of 970–1100°C. A reaction based on the paragenesis and composition of garnetite is

$$0.37 \text{ Ttn} + 6 \text{ Ep } (X_{Fe} = 0.25) + \text{Mag} + 1.25SiO_2 =$$
$$4.25 \text{ Grs} + 2.50 \text{ Alm} + 0.75 \text{ Rt} + 3H_2O + 1.37O_2$$

**The peak metamorphism of HP to UHP rocks**

Only those epidote minerals that contain inclusions of coesite and/or coesite pseudomorphs can be unequivocally considered stable at UHP conditions. Nevertheless, previous descriptions often also list zoisite or clinozoisite in the peak UHP assemblage based on petrographic observation and experimental studies. Zoisite (with $X_{Fe}$ below 0.05) is a common phase in HP and UHP rocks as it is known to be stable up to almost 7 GPa and ≈ 1000°C (Poli and Schmidt 2004). It coexists with garnet and omphacite in many zoisite-kyanite eclogites and even in coesite-bearing eclogites. Yao et al. (2000) describe coesite inclusions in zoisite from different mineral assemblages (in estimated vol%) within a banded eclogite from northern Sulu:

(a) normal eclogite (30–60% Grt, 20–50% Omp, 2–10% Qtz, 0–5% Ky, 0–5% Zo, 1–2% Rt, and <2% carbonate)

(b) garnet-rich eclogite (60–90% Grt, 5–20% Qtz, 5–15% Omp, 0–2% Zo, 0–5% Ky, 0–5% Phe, and 2–5% Rt)

(c) zoisite- and kyanite-rich eclogite (20–70% Zo, 20–35% Ky, >10% Qtz, > 10% Phe, with minor Grt, Omp, and Rt)

(d) leucocratic thin layer (> 50% Qtz, 5–10% Ky, 10–15% Zo, 5–10% Phe, 20–30% Pl, with minor Grt, Cpx, Amph, and Rt)

This example illustrates the stable coexistence of zoisite with coesite and kyanite. This assemblage is common to most HP and UHP terranes and results from the breakdown of plagioclase (i.e., in its simplest form, An + $H_2O$ = Zo + Ky + Coe/Qtz; Newton and Kennedy 1963).

Compared to zoisite, epidote (typically with $X_{Fe}$ > 0.14) is less common in eclogite facies rocks. The maximum *P*-limit of epidote is not well defined. However, Zhang (1992) first reported inclusions of pseudomorphs of fine-grained quartz aggregates after coesite in a euhedral epidote crystal (Figs. 18A and B) with $X_{Fe} = 0.23$ from Sulu and suggested that epidote was stable together with coesite, garnet, and omphacite at about 850°C and 3.2 GPa. Inclusions of coesite pseudomorphs in epidote were then described by Hirajima et al. (1992) in nyböite eclogite in Donghai of the southern Sulu belt and recently in Himalayan coesite-bearing eclogite by Massonne and O'Brien (2003). These findings established the stability of epidote at UHP conditions. Since the first description of Zhang (1992), other localities with comparable inclusions of coesite in epidote minerals have been recognized.

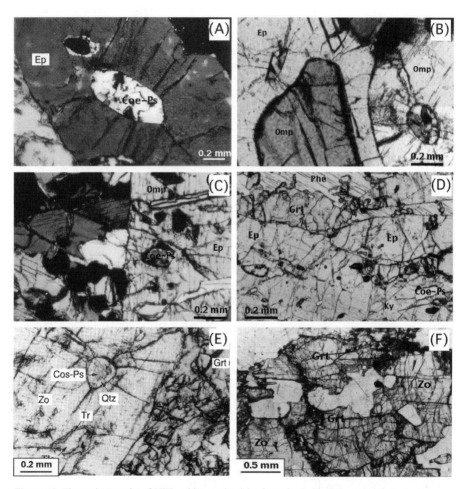

**Figure 18.** Photomicrographs of UHP epidote and zoisite from the Dabie-Sulu and Kokchetav terranes. (A) Porphyroblastic epidote with inclusions of coesite-pseudomorphs in eclogites from Sulu (after Zhang et al. 1995a), (B) Coesite-pseudomorphs in epidote and omphacite from Qinglongshan eclogite (A and B crossed polarized light; Zhang R. unpublished), (C) Porphyroblastic epidote with inclusions of omphacite, garnet, kyanite and coesite-pseudomorph cross-cut the foliation defined by alignment of kyanite, omphacite and talc in Qinglongshan eclogite (C to F plain polarized light; after Zhang et al. 1995a), (D) Coesite-pseudomorph in epidote from foliated Qinglongshan eclogite (Zhang R. unpublished), (E) Coesite inclusions in zoisite of Bixiling talc-kyanite eclogite (after Zhang et al. 1995b), (F) Coarse-grained zoisite in equilibrium with diamond-bearing garnet from Kokchetav gneiss (Zhang R. unpublished).

***Epidote and zoisite with inclusions of coesite or coesite pseudomorphs in Dabie-Sulu.***
Inclusions of coesite or coesite pseudomorphs in epidote and zoisite have been described
from several Dabie-Sulu eclogites (Zhang 1992; Hirajima et al. 1992; Zhang and Liou 1994;
Zhang et al. 1995a; Rolfo et al. 2000; Yao et al. 2000). Most of these eclogites contain kyanite
± talc in addition to the common assemblage of garnet, omphacite, rutile, and coesite/quartz.
Inclusions of coesite are exclusively described in zoisite whereas epidote contains only
inclusions of coesite pseudomorphs.

The Bixiling metamorphic complex is the largest UHP body in Dabie. It consists of
different types of eclogites that contain thin layers of garnet-bearing cumulate ultramafic
rocks. Differences in the bulk composition result in a distinct layering of some of the eclogitic
rocks defined by different mineral assemblages and extent of retrograde recrystallization.
These layered eclogites range in thickness from about 2 to 10 m and grade from (a) beige-
colored kyanite-rich eclogite, (b) kyanite-bearing eclogite that contains minor zoisite and talc
through (c) dark brown barroisite-bearing eclogite to (d) rutile-rich brown eclogite. In the type
(b) kyanite-bearing eclogite zoisite belongs to the peak stage assemblage whereas epidote is
a retrograde phase in both the barroisite-bearing and rutile-rich Fe-Ti eclogites. The coherent
layering of the eclogitic rocks suggests a cumulate origin with anorthositic plagioclase to
olivine-rich gabbro cumulate as the protolith. Eclogite facies metamorphism of these layered
gabbros occurred at 610–700°C and $P > 2.7$ GPa, whereas amphibolite-facies retrograde
metamorphism is characterized by symplectite of plagioclase and hornblende after omphacite
and replacement of tremolite after talc at $P < 0.6$–1.5 GPa and $T < 600$°C.

Beside the layered eclogites, a zoisite-rich talc-bearing coesite eclogite occurs in Bixiling.
This eclogite is composed of zoisite (20–50 vol%), garnet, omphacite, coesite/quartz, talc,
± kyanite, and minor rutile and phengite and low in Fe and Ti contents. Zoisite ($X_{Fe} \sim 0.03$)
forms prismatic crystals up to 5 mm long and contains inclusions of coesite and coesite
pseudomorphs with well-developed radial fractures (Fig. 18E).

The variation in paragenesis and composition of minerals as a function of bulk
composition for the zoisite-bearing eclogite assemblages of Bixiling can be best discussed in
the Al$_2$O$_3$-MgO-FeO-CaO tetrahedron projected from H$_2$O, SiO$_2$, jadeite, and rutile (Zhang et
al. 1995b). To delineate the Fe/Mg ratio for garnet and omphacite for eclogite assemblages,
the composition of these phases is projected from zoisite onto the AFM triangle (Fig. 19). In
this projection, the position of garnet depends on the amount of grossular component and the
Mg/(Mg + Fe) ratio. Eclogitic garnet with high grossular content has low (Al$_2$O$_3$ − 0.75CaO)/
[(Al$_2$O$_3$ − 0.75CaO) + MgO + FeO] and a high Mg/(Mg + Fe) ratio. Garnet of zoisite-talc-rich
eclogites contains a very high grossular component (35 to 40%) compared to garnet from the
other eclogites (18 to 26 %). Omphacite plots at negative co-ordinates, the extent depending
on the amount of the jadeite component.

Several features are apparent from Fig. 19: (1) eclogitic assemblages show a systematic
distribution of tie-lies among coexisting phases; (2) tie lines for eclogitic garnet - clinopyroxene
systematically shift from Fe-Ti rich to the Mg-Al rich samples and are nearly parallel to one
another. Garnet in the Fe-Ti rich layers is higher in the almandine component, and coexisting
clinopyroxene contains higher Fe than counterparts in the Mg-Al rich layers. Garnet
coexisting with talc and zoisite has a higher grossular content than garnet with kyanite + talc.
The systematic disposition of these tie-lines suggests that they are equilibrium assemblages.
Mafic rocks must have either high Mg content or high $f_{O_2}$ in order to stabilize talc in the
coesite stability field (see below). In the Donghai area of southern Sulu, several different HP to
UHP eclogites and associated gneiss and quartzite contain trace to moderate amounts of either
epidote or zoisite (e.g., Hirajima et al. 1992; Zhang et al. 1995a; Zhang et al. 2003):

- In a banded nyböite-bearing eclogite epidote occurs together with variable amounts
  of garnet, clinopyroxene, kyanite, phengite, Al-rich titanite, rutile, apatite, and quartz

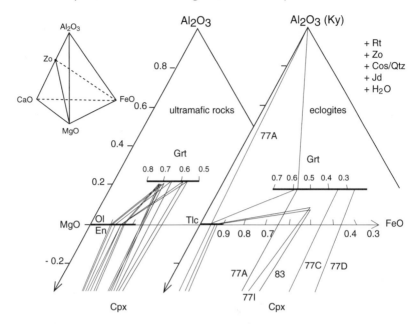

**Figure 19.** AFM diagram showing paragenesis and Fe-Mg partitioning between garnet and clinopyroxene from mafic-ultramafic layers of various compositions at nearly consistent *P-T* conditions. The AFM diagram is a projection from zoisite of the phase relations in the ACFM diagram. Sample numbers refer to eclogites of different compositions; some contain talc and kyanite as UHP assemblage in addition to garnet + omphacite + rutile + coesite. Garnets from both mafic-ultramafic layers are also grossular rich. For details see text (modified from Fig. 8 of Zhang et al. 1995b).

pseudomorphs after coesite. This eclogite was formed at about 740 ± 60°C and *P* > 2.8 GPa. Epidote commonly has allanitic cores with several wt% of REE. In an epidote porphyroblast-rich band inclusions of polycrystalline quartz aggregates after coesite are found in epidote with $X_{Fe}$ = 0.24 (Figs. 18A–D).

- Epidote eclogites display the representative assemblage of garnet, omphacite, kyanite, coesite/quartz, rutile, epidote, ± phengite and barroisite and are characterized by >5 vol% porphyroblastic epidote. They are foliated and exhibit two stages of equilibration within the coesite stability field. The earlier stage consists of medium-grained garnet, omphacite, kyanite, epidote, and rutile that define the rock foliation. The second stage resulted in the formation of coarse-grained epidote ± barroisite porphyroblasts that cut the foliation and contain inclusions of earlier minerals. Both medium-grained eclogitic minerals and porphyroblastic epidote contain inclusions of quartz pseudomorphs after coesite with well-developed radial fractures. A quartz-rich, massive variety of the epidote eclogites contains abundant inclusions of pre-eclogite stage minerals such as paragonite, amphibole, albite, and rare aggregates of paragonite, barroisite, magnesite, and zoisite in garnet.

- Mattinson et al. (in review) differentiated several stages of epidote from epidote-rich talc-kyanite-phengite eclogites from Qinglongshan. Epidote I occurs as inclusion in garnet and later epidote; large epidote poikiloblast II, which enclose garnet, omphacite, epidote I, quartz (including coesite pseudomorphs); epidote III with a vermicular intergrowth of optically continuous quartz but lacking coesite pseudomorphs; and retrograde rims in extensively amphibolitized samples (Ep IV). Some of the epidote grains are aligned with the foliation defined by omphacite and phengite, but most poikiloblasts cross cut it. Analyses of the various epidote types (Table 3, Fig. 20) show that poikiloblasts of Ep II and Ep III in fresh eclogites contain $X_{Fe} = 0.15$ to $0.18$; the composition is similar in different samples, though one epidote lath which contains talc and phengite inclusions has an Fe-poorer medial zone suggesting incomplete replacement of low-Fe epidote by high-Fe epidote. Prograde Ep I and retrograde Ep IV grains contain up to $X_{Fe} = 0.26$. Skeletal epidote in gneiss and quartzite is zoned with $X_{Fe} = 0.13$ cores surrounded by thin rims of $X_{Fe} = 0.31$ and up to 1 wt% MnO.

- Zoisite-rich kyanite eclogite contains >5 vol% kyanite and locally up to 30 vol% zoisite instead of epidote. Some foliated eclogites are defined by elongated coarse-grained zoisite and omphacite, and fine-grained kyanite, garnet and rutile; coarse-grained zoisite porphyroblasts have grown across the foliation. Inclusions of quartz pseudomorphs after coesite occur in omphacite. Multi-mineral inclusions of garnet, hornblende, zoisite and Mg-chlorite occur in coarse-grained garnet.

- Talc-bearing eclogite consists of abundant kyanite and zoisite/clinozoisite or epidote, and contains polycrystalline quartz inclusions within eclogitic minerals. In Qinglongshan, it contains porphyroblastic euhedral epidote (2–10 mm) in a foliated fine-grained matrix of garnet, omphacite, kyanite, epidote (0.5–3 mm), quartz, and rutile. Porphyroblastic epidote-zoisite (c. 5 vol%) developed within and across the foliation contains numerous inclusions of eclogite facies minerals and polycrystalline quartz aggregates after coesite.

Kyanite quartzite occurs as thin layers (type I) with eclogite or as a country rock (type II) of eclogite pods. Massive kyanite quartzite of type II consists of 70 vol% quartz, 12 vol% epidote, 13 vol% kyanite + omphacite, 3 vol% garnet, 2 vol% phengite, and minor rutile. Epidote aggregates are locally abundant and contain coesite pseudomorphs as inclusions. Foliated kyanite quartzite of type I contain similar assemblages but lacks omphacite. Kyanite occurs either as euhedral coarse-grained discrete crystals or as fine-grained inclusions in phengite and epidote. Phengite and kyanite are mantled by a thin (0.02–0.05 mm) corona of biotite, and white mica (inner) + albite (outer), respectively. Inclusions of polycrystalline quartz aggregates occur in kyanite and epidote.

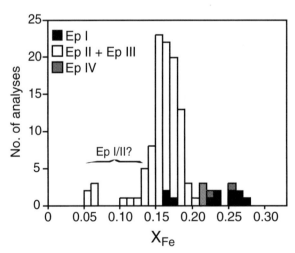

**Figure 20.** Frequency distribution of $X_{Fe}$ of epidote in eclogites from Qinglongshan, Su-lu terrane Most epidote poikiloblasts (Ep II + Ep III) have $X_{Fe} = 0.15$ to $0.18$; The Fe-poor medial region of a grain (Ep I) with prograde inclusions contains the lowest analyzed $X_{Fe}$; the Fe-richer core and rim of this grain are similar to the poikiloblast compositions. Fe-rich ($X_{Fe} = 0.20–0.26$) analyses include both prograde (Ep I) and late retrograde (Ep IV) grains (Mattinson et al. in review).

**Table 3.** Representative analyses of epidote minerals from Dabie, Sulu and Kokchetav terranes.

| Terrane | Dabie | | | | | | Sulu | | | | |
|---|---|---|---|---|---|---|---|---|---|---|---|
| Locality | Bixiling | | Shuanghe | | | | Qinglongshan | | | | |
| Rock Type | ZEc | ZEc | CEc | EEc | GBG | OG | TEc | Ec | Ec | Ec | KQz |
| Sample | 77l | 77l | CD29 | CD36 | 92H22 | SH6-1 | SL-25F | QL1C-r | QL3-i | QL32-r | SL25 |
| Mineral | Zo (r) | Zo (c) | Ep | Ep | Ep | Ep | Ep | Ep | Ep | Ep | Ep |
| $SiO_2$ | 39.38 | 39.31 | 39.97 | 38.68 | 38.62 | 38.83 | 37.86 | 37.64 | 37.56 | 38.43 | 37.55 |
| $TiO_2$ | 0.04 | 0.04 | 0.08 | 0.09 | 0.09 | 0.11 | 0.30 | 0.14 | n.d. | n.d. | 0.13 |
| $Al_2O_3$ | 31.82 | 32.35 | 27.11 | 25.57 | 25.83 | 27.93 | 24.50 | 23.79 | 24.31 | 26.77 | 25.18 |
| $Cr_2O_3$ | 0.03 | 0.05 | 0.34 | 0.03 | 0.02 | 0.03 | n.d. | n.d. | n.d. | n.d. | 0.00 |
| $Fe_2O_3$* | 1.79 | 1.33 | 7.34 | 10.13 | 10.02 | 6.68 | 13.35 | 13.80 | 13.12 | 9.46 | 11.96 |
| $MnO$ | 0.02 | 0.00 | 0.07 | 0.23 | 0.17 | 0.11 | 0.01 | 0.16 | 0.16 | 0.07 | 0.04 |
| $MgO$ | 0.02 | 0.01 | 0.16 | 0.03 | 0.04 | 0.00 | 0.37 | 0.10 | 0.20 | 0.22 | 0.21 |
| $SrO$ | n.d. | n.d. | n.d. | n.d | n.d. | n.d. | n.d. | n.d. | n.d. | n.d. | n.d. |
| $CaO$ | 24.29 | 23.55 | 22.31 | 23.96 | 23.01 | 24.41 | 22.43 | 22.94 | 22.04 | 23.17 | 21.90 |
| $Na_2O$ | 0.00 | 0.00 | 0.15 | 0.00 | 0.02 | 0.01 | 0.02 | n.d. | n.d. | n.d. | 0.00 |
| $K_2O$ | 0.00 | 0.00 | 0.09 | 0.01 | 0.01 | 0.00 | 0.01 | n.d. | n.d. | n.d. | 0.00 |
| Total | 97.39 | 96.64 | 97.62 | 98.73 | 97.83 | 98.11 | 98.85 | 98.57 | 97.39 | 98.12 | 96.97 |
| $X_{Fe}$** | 0.03 | 0.03 | 0.15 | 0.20 | 0.20 | 0.13 | 0.26 | 0.27 | 0.26 | 0.18 | 0.23 |
| Ref. | Z95b | Z95b | C95 | C95 | C95 | C95 | Z95a | M | M | M | Z95a |

**Table 3.** *continued on following page*

**Table 3.** *continued from previous page*

| Terrane | Caihu | | Sulu | | | | | | Kokchetav | | |
|---|---|---|---|---|---|---|---|---|---|---|---|
| Locality | | | JC | CZ | MZ | HT | YK | WH | | | |
| Rock Type | Ec | Ec | SEc | ZEc | ZEc | KQz | MGa | CSr | GZG | Ec | Ws |
| Sample | 91CH08b | | 88J5 | 88C2 | 87D2 | HZ9 | YK3 | 91-6E | K8-41 | K26 | K21 |
| Mineral | Zo | Ep | Ep | Zo | Zo | Zo | Zo* | Zo | Zo | Ep (i) | Zo (i) |
| $SiO_2$ | 38.4 | 37.6 | 38.39 | 39.52 | 39.73 | 38.45 | 39.94 | 39.40 | 39.57 | 37.54 | 39.59 |
| $TiO_2$ | 0.11 | 0.06 | n.d. | n.d. | n.d. | 0.08 | 0.03 | 0.00 | 0.01 | 0.08 | 0.03 |
| $Al_2O_3$ | 31.1 | 28.4 | 24.66 | 32.12 | 32.29 | 31.58 | 32.54 | 32.72 | 33.13 | 26.44 | 33.48 |
| $Cr_2O_3$ | 0.20 | 0.04 | n.d. | n.d. | n.d. | n.d. | 0.02 | 0.04 | 0.01 | 0.05 | 0.42 |
| $Fe_2O_3$* | 2.14 | 5.09 | 13.56 | 1.41 | 2.04 | 1.89 | 1.62 | 0.22 | 1.03 | 8.57 | 1.19 |
| $MnO$ | 0.03 | 0.04 | n.d. | n.d. | n.d. | n.d. | 0.01 | 0.01 | 0.07 | n.d | 0.07 |
| $MgO$ | 0.00 | 0.14 | 0.26 | 0.44 | 0.00 | 0.04 | 0.04 | 0.02 | 0.15 | 0.04 | 0.03 |
| $SrO$ | 1.20 | 2.66 | n.d. | n.d. | n.d. | n.d. | n.d. | n.d. | n.d. | n.d. | n.d. |
| $CaO$ | 23.1 | 22.0 | 22.29 | 23.85 | 24.09 | 23.38 | 23.61 | 24.31 | 24.45 | 23.25 | 24.44 |
| $Na_2O$ | n.d. | n.d. | n.d. | 0.00 | 0.00 | n.d. | 0.50 | 0.00 | 0.01 | 0.03 | 0.02 |
| $K_2O$ | n.d. | n.d. | n.d. | n.d. | n.d. | n.d. | 0.04 | 0.02 | 0.00 | 0.11 | 0.00 |
| Total | 96.50† | 96.03 | 99.16 | 97.34 | 98.15 | 95.42 | 98.35 | 96.74 | 98.43 | 96.11 | 99.27 |
| $X_{Fe}$** | 0.04 | 0.10 | 0.26 | 0.03 | 0.04 | 0.04 | 0.03 | 0.00 | 0.02 | 0.17 | 0.02 |
| Ref. | NE98 | NE98 | Z95a | Z95a | Z95a | Z95a | ZL97 | Z95c | Z97 | Z97 | Z97 |

* Total Fe as $Fe_2O_3$    ** $X_{Fe}$ = $Fe^{3+}/(Al + Cr + Fe^{3+} + Mn^{3+})$    n.d. = not determined

† Includes $Ce_2O_3$ (0.14 wt%) and $Nd_2O_3$ (0.08 wt%)

Abbreviations for rock-types: ZEc, zoisite eclogite; CEc, carbonate eclogite; EEc, epidote eclogite; CBG, garnet-biotite gneiss; OG, orthogneiss, TEc, Talc eclogite; Ec, eclogite; KQtz, kyanite quartzite; SEc, Al-titanite eclogite; MGa, metagabbro; CSr, calc-silicate rock; GZG, Grt-Zo gneiss; Ws, whiteschist.

Abbreviations for localities: JC, Jianchang; CZ, Chizhuang; MZ, Mengzhuang; HT, Hetang; YK, Yangko; WH, Weihai.

References: Z95b, Zhang et al. (1995b); C95, Cong et al. (1995); Z95a, Zhang et al. (1995a); M, Mattinson et al. (in press); NE98, Nagasaki and Enami (1998); ZL97, Zhang and Liou (1997); Z95c, Zhang et al. (1995c); Z97, Zhang et al. (1997).

Microprobe analyses show no significant difference in the $X_{Fe}$ value between epidote inclusions ($X_{Fe}$ = 0.19–0.24) and porphyroblastic epidote ($X_{Fe}$ = 0.18–0.25). A few porphyroblastic epidotes display compositional zoning with a slight decrease from cores of $X_{Fe}$ = 0.25 to rims of $X_{Fe}$ = 0.22. Epidote from kyanite quartzite has an $X_{Fe}$ value of 0.23. Only a few eclogitic zoisites have been analyzed; their $X_{Fe}$ values range from 0.03 to 0.04. Zoisite in kyanite quartzite has $X_{Fe}$ value of 0.04. Clinozoisite of garnet-quartz-jadeite rocks ranges from $X_{Fe}$ = 0.10–0.14.

***Epidote minerals in Kokchetav diamond-bearing gneiss.*** Microdiamond inclusions are mainly restricted to garnet, clinopyroxene, and zircon from felsic gneiss, marble and garnet-clinopyroxene rocks. Zhang et al. (1997, 2002) reports garnet-zoisite gneiss with quartz, plagioclase, potassium feldspars (50 vol%), garnet (25 vol%), zoisite (5 vol%), and tourmaline (5 vol%) in addition to minor biotite, chlorite, opaque, apatite, and zircon (Table 3). Garnet contains abundant inclusions of diamond; idioblastic and porphyroblastic zoisite crystals are in sharp contact with porphyroblastic diamond-bearing garnet (Fig. 18F) suggesting that zoisite is stable with garnet in the stability field of diamond. The associated eclogites exhibit the common assemblage garnet, omphacite, rutile, coesite/quartz, ±zoisite and kyanite. Inclusions of quartz pseudomorphs after coesite were identified in 2 to 3 mm large garnet. Zoisite-rich eclogite displays abundant S-shaped inclusion trails of minute titanite crystals in the core and zoisite ± quartz in the rim of garnet.

Diamond-bearing clinozoisite gneiss at the western extremity of the Kokchetav massif (Korsakov et al. 2002) contains mainly retrograde phases including quartz (10–60 vol%), biotite (10–40 vol%), garnet (5–30 vol%), clinozoisite (5–20 vol%), clinopyroxene (0–15 vol%) and K-feldspar (5–10 vol%) with minor kyanite, amphibole, chlorite, calcite, tourmaline, and zircon, rutile, titanite, apatite, graphite as accessory minerals. Diamond, coesite, and other UHP minerals formed at 950–1000°C and >4.0 GPa are preserved only as inclusions in garnet, kyanite, and zircon. Bulk rock composition suggests that they are metamorphosed Ca-rich clays. Clinozoisite ($X_{Fe}$ < 0.02) occurs either as single elongated crystals or forms as a quartz-clinozoisite symplectite after grossular; clinozoisite shows an oblique extinction in thin sections, with an angle $X \wedge c$ less than 12°. In spite of abundant clinozoisite in the rock, only one sample contains this phase as inclusion in garnet. Both types of clinozoisite are not in direct contact with diamond-bearing garnet and kyanite suggesting clinozoisite and the quartz-clinozoisite symplectite formed in retrograde re-equilibration during rapid exhumation estimated to be less than 0.1 Ma. *P-T* estimates for garnet, clinopyroxene, clinozoisite, kyanite, biotite, potassium feldspar, quartz and for garnet, clinopyroxene, clinozoisite, biotite, potassium feldspar, quartz assemblages are 950°C at 2.0 GPa and 800°C at 0.7 GPa, respectively. Korsakov et al. (2002) suggests that clinozoisite either crystallized directly from melt or transformed from grossular according to a reaction, Cz + Qtz = Grs + melt, within the quartz stability field.

***Epidote minerals in UHP felsic rocks.*** Compagnoni and Rolfo (1999) summarize some characteristics of UHP metapelite, gneiss and other unusual rocks from several UHP terranes. They pointed out that epidote minerals are generally stable in these lithologies at UHP conditions, including piemontite, which grows in Mn-rich compositions. Despite the felsic rocks described above for both Dabie-Sulu and Kokchetav, meta-sediments of the western Italian Alps contain clinozoisite-epidote ± sodic amphibole in addition to garnet and phengite as UHP stage minerals. Some oxidized manganiferous quartz-schists and quartzites have piemontite ($X_{Mn}$ = 0.15–0.21 and $X_{Fe}$ = 0.18–0.21) together with hematite, coesite, garnet, braunite, phengite, and rutile.

Pelagic metasediments and MORB-type metabasalts of the former Tethyan oceanic crust at Cignana, Zermatt-Saas Zone experienced UHP metamorphism at 615 ± 15°C and 2.8 ± 1.0 GPa (Reinecke 1991, 1998). The metamorphism resulted in the formation of coesite-glaucophane-eclogite in the basaltic layer and of garnet-dolomite-aragonite-lawsonite-coesite-phengite-bearing calc-schist and garnet-phengite-coesite-schist with variable amounts of epidote, talc, dolomite, sodic pyroxene, and sodic amphibole in the overlying metasediments. In contrast to the well-preserved UHP metamorphic record of the coesite-glaucophane eclogite, the HP/UHP assemblages of the metasediments have been largely obliterated during exhumation. Relics from which the metamorphic evolution of the rocks can be retrieved are restricted to rigid low-diffusion minerals like garnet, tourmaline, and apatite (Reinecke 1998). Assemblages with dolomite, aragonite, lawsonite, phengite, apatite, tourmaline, talc and epidote/piemontite occurring in the calc schists and quartz-schists indicate that the metasedimentary rocks contained significant amounts of $H_2O$ and $CO_2$ during the UHP stage. These assemblages may persist to more than 4 GPa; these hydrous phases may act as major carriers of fluids into mantle (e.g., Schmidt and Poli 1998).

***Factors controlling the common occurrence of UHP epidote minerals***. Paragenesis and compositions of epidote minerals in UHP rocks are controlled by $P$, $T$, $f_{O_2}$, and bulk rock composition. The common occurrence of UHP epidote in Dabie-Sulu is apparently related to the oxidized protoliths revealed from mineral assemblages. Such protoliths were obtained by hydrothermal alteration revealed from exciting findings of extraordinarily low $\delta^{18}O$ and $\delta D$ values of UHP minerals, including epidote-zoisite (see Morrison 2004). Reported $\delta^{18}O$ values for eclogites worldwide range from +1.5 to +12.0‰ (see summary by Yui et al. 1995). However, analyses of $\delta^{18}O$ values for mineral separates from eclogite, gneiss, and quartzite from Dabie-Sulu yielded world-record negative $\delta^{18}O$ values of −2.1 to −11.1‰ for epidote, −4 to −11‰ for garnet, −2.2 to −11.2‰ for omphacite, −1.2 to −10.7‰ for phengite, 0.8 to −7.7‰ for quartz and −5.3 to −14.8 ‰ for rutile (Yui et al. 1995; Zheng et al. 1996; Baker et al. 1997; Rumble and Yui 1998; Zheng et al. 1999). Compiling the available data, Zheng et al. (2003) show that UHP eclogites and associated gneisses have $\delta^{18}O$ values ranging from −11‰ to +22‰. These data cover the whole range of the major water reservoirs such as seawater (0 ‰), meteoric water ($< -2$ ‰, depending on latitude), magmatic water (+5 to +7‰), and marls (+25 to +35 ‰). Most of the unaltered samples preserve oxygen isotope equilibrium fractionations only on a mm-scale, yielding reasonable $T$ estimates of 650–750°C responsible for UHP metamorphism. Some UHP rocks preserve pre-metamorphic differences in $\delta^{18}O$ at the outcrop scale, suggesting that closely spaced rocks have not achieved meter-scale isotopic re-equilibration during the entire metamorphic process (Rumble and Yui 1998). Together with low $\delta D$ values (from −101 to −127 ‰), these characteristics are interpreted as follows. (1) The protoliths were once at or near the Earth's surface and were subjected to alteration in a geothermal area charged with meteoric water from a Neoproterozoic snowball Earth (Yui et al. 1995; Zheng et al. 1996; 1999, 2003; Rumble et al. 2002). (2) The persistence of pre-metamorphic differences in oxygen and hydrogen isotopic composition between different rocks and the preservation of high-$T$ isotope fractionation between UHP minerals indicate a lack of flowing fluid to mediate isotope exchange between rocks during UHP and retrograde metamorphism (e.g., Getty and Selverstone 1994; Rumble and Yui 1998; Philippot and Rumble 2000). Thus, the environment for subduction of old supracrustal rocks constitutes an example of a fluid-deficient region of metamorphism (Yoder 1955). Other recent studies indicate that O-isotope disequilibrium between UHP minerals may have been caused by retrograde hydration (Yui et al. 1997; Zheng et al. 1999) and the difference in isotopic composition of eclogite protoliths in various areas was not only controlled by different degrees of water-rock interaction before metamorphism, but also by channelized flow of fluids during retrograde metamorphism due to rapid exhumation (Zheng et al. 1999, 2003).

Magmatic-hydrothermal alteration of the protoliths of Dabie-Sulu UHP rocks has been proposed (e.g., Rumble et al. 2002; Zheng et al. 2003). One example is rift-related magmatism along the northern margin of the Yangtze craton resulting in melting of glacier ice and triggering meteoric-hydrothermal circulation during Snowball Earth time. Channelized fluid flow caused a variable extent of water-rock interaction and different degrees of $^{18}O$-depletion around the Neoproterozoic magmatic system (e.g., see Bird and Spieler 2004 for fossil and active hydrothermal systems for epidote formation). In addition to lower $\delta^{18}O$ for some protoliths, such alterations caused oxidation of the protoliths including quartzite, granite, paragneiss, and associated basaltic rocks. These oxidized rocks are the sites for the formation of UHP epidote during Triassic subduction. Common occurrences of UHP epidote in both eclogites and epidote ± acmite in gneiss at the type locality of the $\delta^{18}O$ anomaly in the Qinglongshan area mentioned above and piemontite schist in the Mulanshan area of the southern Hongan block (e.g., Eide and Liou 2000) supports such a model. This process excludes the common occurrence of microdiamond in the Dabie-Sulu terrane in spite of UHP metamorphism within the diamond *P-T* stability field.

The effect of $f_{O_2}$ on the stability of epidote is illustrated in Figure 21 for the epidote-bearing eclogites from Qinglongshan (Mattinson et al. in review). During eclogite-facies metamorphism, $f_{O_2}$ is estimated using the reaction

$$12 \text{ Ep } (X_{Fe} = 0.33) = 8 \text{ Grs} + 4 \text{ Alm} + 3 \text{ O}_2 + 6 \text{ H}_2\text{O}$$

calculated using the THERMOCALC v2.5 program and data set of Holland and Powell (1990). Mineral activities were calculated from rim compositions of peak-stage garnet and epidote using the a-X program of Holland and Powell (1998); unit $H_2O$ activity was assumed. The results indicate consistent, oxidized conditions for eclogite samples, approximately 2.5 log $f_{O_2}$ units above the hematite-magnetite (HM) buffer over the calculated temperature range (Fig. 21a). This is consistent with the Mg-rich composition of early-retrograde amphibole. Conditions were too oxidizing for the persistence of graphite or the formation of diamond under peak conditions, although *P-T* estimates for several samples lie within the diamond stability field. The graphite-$CO_2$ buffer (at $X_{CO_2}$ = 1) is approximately one log $f_{O_2}$ unit below the HM buffer (Fig. 21a), and for $X_{CO_2}$ appropriate for epidote stability (<0.04 at $P$ = 3 GPa: see also Boundy et al. 2002), the graphite-$CO_2$ curve is approximately 1.5log $f_{O_2}$ units lower.

The presence of Ti-hematite and the absence of magnetite and graphite in gneiss and quartzite samples are consistent with the $f_{O_2}$ calculated for the eclogites. Rim compositions of

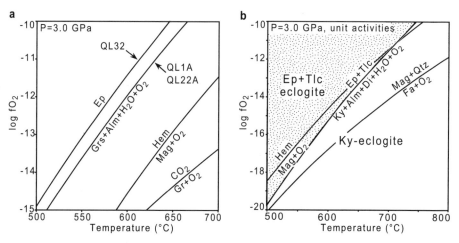

**Figure 21.** Log $f_{O_2}$-*T* diagrams calculated using THERMOCALC v. 2.5 program and data set of Holland and Powell (1990): (a) $f_{O_2}$ estimates for Qinglongshan epidote-kyanite-talc eclogite samples (sample numbers indicated), calculated from mineral rim compositions. Diamond is stable below 515°C if $f_{O_2}$ is sufficiently low; (b), $f_{O_2}$-*T* diagram calculated for unit activities. Ep + Tlc eclogites (stippled field) are the low temperature, high $f_{O_2}$ equivalents of common kyanite eclogites.

garnet and epidote in gneiss indicate slightly more oxidized conditions (if calculated at 3 GPa) than eclogitic samples; compositions of Ti-hematite and garnet included within it indicate $f_{O_2}$ conditions identical to that of eclogitic garnet-epidote pairs. The garnet-epidote reaction shifts by approximately 0.005log $f_{O_2}$ unit/GPa relative to the HM buffer over the pressure range 0.1 to 2.0 GPa (Donohue and Essene 2000). The reaction

$$8 \text{ Tlc} + 12 \text{ Ep} (X_{Fe} = 0.33) = 24 \text{ Di} + 4 \text{ Alm} + 8 \text{ Ky} + 3 \text{ O}_2 + 14 \text{ H}_2\text{O}$$

(calculated for unit activities) shows that epidote-talc eclogites are low-temperature oxidized equivalents of the more common kyanite eclogites (Fig. 21b). Garnet-epidote estimates of $f_{O_2}$ in eclogite-facies carbonate rocks (1.8–2.0 GPa/580–690°C) from the Norwegian UHP terrane indicate oxidizing conditions 0.3 to 1.0 log $f_{O_2}$ unit above the HM buffer (Donohue and Essene 2000, Boundy et al. 2002). This suggests that calculated $f_{O_2}$ for the Qinglongshan samples are not unreasonable.

Interaction with oxidized, meteoric water in a hydrothermal system such as that hypothesized by Rumble et al. (2000) may be responsible for oxidation and hydration, as well as $\delta^{18}$O depletion, of the Qinglongshan epidote-talc eclogite protoliths. If this is correct, combined oxygen isotope analysis and $f_{O_2}$ calculation should reveal that the $\delta^{18}$O of relatively reduced eclogites is higher than that of oxidized, epidote-talc eclogites. Heterogeneity of hydrothermal alteration combined with lack of extensive fluid flow during UHP metamorphism can explain the presence of epidote-free eclogites surrounded by epidote-talc eclogites, as well as the preservation of anomalously low, pre-metamorphic isotopic values in this region.

### Retrograde stages (stages III and IV)

Epidote minerals are common products of retrograde amphibolite to greenschist facies overprint during exhumation. Depending on the bulk composition of the host rock, retrograde epidote minerals in veins or in the matrix vary in composition. Some gneisses in Dabie are transformed from quartz-rich eclogite through retrograde hydration, metasomatism and crystallization of secondary minerals to hornblende-biotite-epidote gneiss (Zhang et al. 2003). The layered rock series consists of various intercalated eclogite, paragneiss, granitic gneiss, jadeite quartzite and marble with or without eclogite nodules (Cong et al. 1995). The eclogite contains abundant coesite relics in garnet and omphacite; all lithologies including epidote-bearing gneisses were subjected to *in situ* UHP metamorphism and multistage retrograde re-equilibration (Carswell et al. 2000).

Zhang et al. (2003) investigated a series of variably retrograded eclogites and gneissic rocks from a single outcrop of about 20 thin layers of mafic and felsic rocks with various compositions and extent of retrogression and its nearby localities. These eclogites are extensively retrograded; most omphacite grains are completely replaced by amphibole and plagioclase. Several layers of epidote-bearing gneissic rocks concordant with eclogite layers consist mainly of quartz and aggregates (or domains) of fine-grained amphibole, plagioclase, epidote, biotite, titanite, and ± calcite. Based on mineral constituents and shapes of domains, they are interpreted as pseudomorphs after garnet, omphacite, kyanite and rutile. Gneiss with coarse-grained quartz and pseudomorphs after garnet and omphacite exhibit strong deformation. Some plagioclase-epidote-amphibolite-biotite pseudomorphs after garnet contain skeletal to irregular garnet relics. Garnet domains with garnet relics are surrounded by amphibole and plagioclase or by amphibole, plagioclase, epidote, and biotite. Based on analyzed compositions and texture, such retrogressive reaction of garnet and omphacite may be written

$$\text{Grt} + \text{Omp} + \text{H}_2\text{O} = \text{Pl} + \text{Amph} \pm \text{Cpx}$$

Another possible transformation involves the reaction of almandine and grossular components of garnet with $H_2O$ to produce epidote ± amphibole, whereas the excess Fe and

Mg react with K of the fluid or phengitic mica to form biotite. Subsequent hydration reaction for the replacement of garnet by epidote and biotite can be written as

Amph + Grt = Ep + Bt                                                                                      or

Amph + Grt = Chl + Pl + Bt

The replacement of kyanite by zoisite, muscovite/biotite, and sodic plagioclase can be written as

$6\ Ky + 8\ Omp + K_2O + 2\ SiO_2 + 3\ H_2O = 2\ Zo + 2\ Ms + 4\ Ab + 4\ MgO$                      or

$4\ Ky + 8\ Grt + 8\ SiO_2 + 2\ K_2O + Na_2O + 7\ H_2O = 6\ Zo + 4\ Bt + 2\ Ab.$

Epidote biotite gneiss is characterized by pronounced foliation, and near equigranular texture. It is composed of biotite flakes (15 vol%) (+ minor muscovite), quartz, plagioclase, microcline, apatite, and coarse-grained epidote (up to 1 mm) pseudomorphic after coarse-grained garnet. Ghost habits of garnet and omphacite domains become obscure; the size of individual crystals including plagioclase, epidote and biotite in these domains are larger than those in the extensively retrograded eclogites. Only rare garnet domains could be recognized as trace garnet relics are embayed by aggregates of plagioclase + epidote + hornblende + biotite. Such aggregates in gneissic rocks represent products of retrograde reaction in a closed system with introduction of $H_2O$. However, for pseudomorphs with high biotite contents, additional potassium metasomatism must be evoked.

## EPIDOTE MINERALS AS AN INDICATOR OF
## FLUID-ROCK INTERACTION AND PARTIAL MELTING

Zoisite and clinozoisite occur in segregations and pegmatitic veins in HP-UHP metamorphic rocks. They obviously point to high fluid activity at these conditions, and are a key to discuss high-*P* fluid-rock interaction and partial melting process of metamorphic rocks during exhumation stage. Zoisite and clinozoisite segregations are widely distributed in HP-UHP metamorphic rocks of the Tauern Window, and have been well documented (e.g., Selverstone et al. 1992; Brunsmann et al. 2000).

In the Eclogite Zone of the Tauern Window, abundant small (cm-10 cm scale) segregations with variable mineral assemblages, including zoisite ($X_{Fe}$ = 0.03–0.06) + clinozoisite ($X_{Fe}$ = 0.10–0.22) + rutile, are observed in banded and massive eclogites (Selverstone et al. 1992). These are localized fractures and are not linked to any through-going fractures. The formation conditions of the segregates are roughly estimated as 2 GPa and 600°C. Zoisite and clinozoisite occur as euhedral blasts up to 1 mm in length and smaller subhedral to euhedral grains radiating inwards from the wall of the segregation, respectively. Zoisite frequently shows sector-zoning and patchy extinction. Close similarities in mineralogy and mineral chemistry between the segregations and their host eclogites, and the absence of metasomatic halos surrounding the segregations imply that the zoisite and clinozoisite segregations crystallized from a metamorphic fluid phase that was in equilibrium with the host rocks. Selverstone et al. (1992) interpreted the segregations to be fractures produced by local solution and precipitation of material either during hydrofracturing associated with devolatilization in the host eclogites, or in response to ponding of non-wetting fluids in certain of the layers. Thus the segregations might be an evidence for at least instantaneous conditions of $P_{fluid} > 2$ GPa.

High-*P* zoisite- and clinozoisite-bearing segregations (mm-cm scale) are common in garnet- and albite-bearing amphibolites of the Lower Schieferhülle of the Tauern Window (Brunsmann et al. 2000), which equilibrium conditions are slightly lower (0.6–1.2 GPa/400–550°C) than the Eclogite Zone. Formation history of these segregations is more complicated than in the Eclogite Zone. The zoisite segregations (primary assemblage: zoisite with $X_{Fe}$ = 0.03–0.06 + quartz + calcite) formed during an early to pre-Hercynian high-*P* event (> 0.6 GPa/500–550°C) by hydrofracturing as a result of protolith dehydration. Most contacts between the zoisite segregations and host rock are sharp, but some are slightly diffuse over a few mm in width. The clinozoisite segregations (primary assemblage: clinozoisite with $X_{Fe}$ = 0.10–0.19 + quartz + omphacite + titanite + chlorite + calcite) are commonly crosscutting the fabrics of the host amphibolite. They formed during the Eoalpine high-*P* event at 0.9–1.2 GPa/400–500°C.

Zoisite- and clinozoisite-bearing pegmatites in eclogites are well documented from the Münchberg Massif (Franz and Smelik 1995). The peak metamorphic conditions of the host eclogites are estimated at 2.5 GPa/600–700°C (e.g., O'Brien 1993). Zoisite occurs as large (up to 10 cm in length) euhedral crystals in a matrix of quartz and plagioclase with abundant graphic intergrowths. It is zoned with core ($X_{Fe}$ = 0.06) and inner rim ($X_{Fe}$ = 0.04)-outer rim ($X_{Fe}$ = 0.03), which boundary is very sharp and mimics the outline of the entire crystal (Fig. 22). Most clinozoisite crystals ($X_{Fe}$ = 0.10–0.13) occur as parallel intergrowths within the zoisite crystal with very sharp interface, and are commonly associated with albite and occasionally with calcic amphibole. Small (on the order of 100 μm) crystals of clinozoisite have grown at the rims of the zoisite crystal, especially where they are corroded. The graphic texture of the zoisite- and clinozoisite-bearing pegmatites, their large grain size, and their contact relationships with the country rocks all indicate igneous origin for the pegmatites, showing strong evidence for partial melting caused by isothermal decompression during exhumation (Fig. 22).

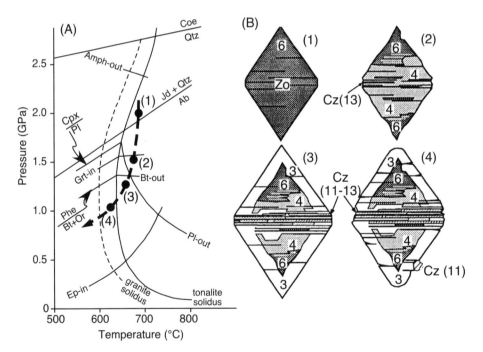

**Figure 22.** (A) *P-T* path of eclogites from the Münchberg Massif during exhumation leading to partial melting and pegmatite formation, and (B) progression of zoisite growth during four stages of pegmatite formation (modified from Fig. 4 of Franz and Smelik 1995). Numbers indicate $X_{Fe} \times 100$.

## SOME PECULIAR CRYSTALCHEMICAL
## ASPECTS OF HP TO UHP EPIDOTE MINERALS

### Compositional variations within the A(2)-site

Monoclinic epidote minerals have two distinct sites commonly occupied by Ca: the seven- to ninefold-coordinated A(1) and eight- to tenfold-coordinated A(2) sites. The A(2) site is slightly larger than the A(1) site, and its size increases with incorporation of $Fe^{3+}$ and $Mn^{3+}$ in octahedral M(1) and M(3)-sites. The A(2)-site can readily incorporate Sr (1.36 Å) and larger cations as indicated by the occurrence of niigataite, a Sr analog of clinozoisite in high *P/T* metamorphic rocks (Miyajima et al. 2003). In the Sanbagawa schists described above, considerable amounts of Sr and other divalent elements are detected in epidote minerals (Fig. 23, Table 3). The maximum Sr content of Al-$Fe^{3+}$ solid solutions ($Fe^{3+} > Mn^{3+}$) is 0.24 pfu (5.0 wt% SrO), whereas the Ba content is usually below the detection limit of the microprobe analysis (0.08 wt% for $2\sigma$ level). In normal piemontite ($Fe^{3+} + Mn^{3+} \geq 1$ and $Mn^{3+} > Fe^{3+}$) Sr ranges from 0.00 to 0.36 pfu (0.0–7.3 wt% SrO). Even higher Sr contents are found in strontiopiemontite. Strontium, Ba, and Pb contents are 0.43–0.71 pfu (up to 13.1 wt% SrO),

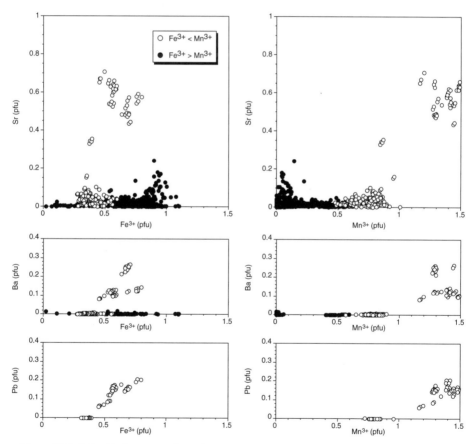

**Figure 23.** Relationship between ($Fe^{3+}$, $Mn^{3+}$) and (Sr, Ba, Pb) of epidote minerals from the Sanbagawa belt (Enami and Banno 2001; Enami unpublished data). Pb contents of epidote with $Fe^{3+} > Mn^{3+}$ are not analyzed.

0.08–0.26 pfu (up to 7.0 wt% BaO), and 0.06–0.20 pfu (up to 7.7 wt% PbO), respectively. The $X_{Al}$ value of strontiopiemontite is $0.36 \pm 0.05$, and the $X_{Fe}$ and $X_{Mn}$ values are 0.15–0.27 and 0.36–0.48, respectively. This implies as ideal formula $(Sr,Ba,Pb)Ca(Mn^{3+},Fe^{3+})_2AlSi_3O_{12}(OH)$. The Ba and Pb substitutions in the strontiopiemontite are described by the hypothetical end-members for epidote minerals "Ba-piemontite" $[BaCa(Mn^{3+}, Fe^{3+}, Al)_2AlSi_3O_{12}(OH)]$ and "Pb-piemontite $[PbCa(Mn^{3+}, Fe^{3+}, Al)_2AlSi_3O_{12}(OH)]$" (Fig. 24, Table 2; Enami and Banno 2001).

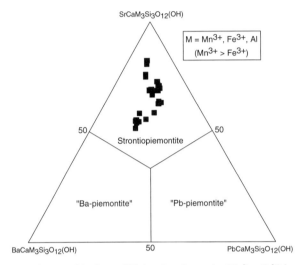

**Figure 24.** Compositional range of Sr, Ba, and Pb-bearing piemontite ($Mn^{3+} > Fe^{3+}$) in quartz schist from the Sanbagawa belt (modified from Fig. 7 of Enami and Banno 2001).

## Sector-zoning of epidote minerals

Sector-zoned zoisite is described in a metagabbro at Fujiwara, central Shikoku, from the Sanbagawa belt (Enami 1977). The metagabbro occurs near the boundary between the garnet and albite-biotite zone and its metamorphic conditions are equivalent to lower-grade epidote-amphibolite facies. The typical mineral assemblage coexisting with zoisite is diopside, hornblende, chlorite, albite, and quartz. It is difficult to find this sector structure using polarizing microscopy in normal thin sections. However, in slightly thicker thin sections, the sector structure is easily observed from the differences of interference colors among the sectors (Fig. 25). Zoisite is composed of three square pyramids of {100}, {010} and {001} named after their growing surfaces. Each sector is composed of a core and $Fe^{3+}$-richer mantle. In the core of each sector, $X_{Fe}$ decreases towards the outer margin. The $X_{Fe}$ of sectors increases in the order of {001} < {010} < {100}.

Sector-zoning is also observed in epidote of the Sanbagawa mafic schists from the chlorite zone (Fig. 26; Yoshizawa 1984). The epidote is fine-grained (less than 100 μm in length), and composed of sector-zoned cores and non-sector-zoned rims. The sector structure is best seen using polarizing microscopic observation of polished thin sections. $X_{Fe}$ decreases in the order {100}, {110}, {001} and {10$\bar{1}$} sectors from the core, and non-sector-zoned rim areas.

It is likely that compositional sector-zoning of epidote minerals is easily overlooked in regular thin sections and it may be more common in epidote of low-grade zone in HP metamorphic rocks than currently realized (Banno and Yoshizawa 1993). Compositions of epidote especially in low-grade metamorphic rocks are typically quite variable both in single

**Figure 25.** Photomicrograph (crossed polarized light) of a sector-zoned zoisite in (010) section from the Fujiwara metagabbro in the Sanbagawa belt reported by Enami (1977). The sector-zoned zoisite is composed of three types of rectangular pyramid. Symbol {*hkl*} indicates basal plane of the pyramid.

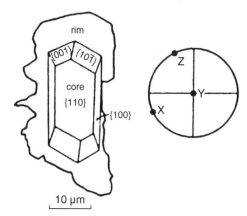

**Figure 26.** Sketch of sector-zoned epidote in (010) section from mafic schist in the Sanbagawa belt (modified from Fig. 3 of Yoshizawa 1984). The sector-zoned core is mantled by non-sector-zoned rim. Optical orientation of the section is shown in stereographic projection.

grains and within a single thin section (e.g., Fig. 11), and this may be in part due to sector-zoning. Thus, in Sanbagawa rocks sector-zoned epidote indicates disequilibrium growth, and care needs to be exercised in establishing equilibrium compositions.

## LAWSONITE-EPIDOTE RELATIONS

The *P-T* region of the blueschist facies is subdivided into lawsonite and epidote blueschist facies and this subdivision is qualitatively correlated with increasing grade. Lawsonite-bearing blueschist forms at a temperature below 375°C and epidote bearing, lawsonite-free blueschist at a higher temperature of up to 500°C (Evans 1990).

Textural replacement of lawsonite by later epidote or lawsonite growth from epidote have been documented in many major blueschist terranes, for example, in the Swiss, Tauern Window, Austrian-Italian Alps by Selverstone and Spear (1985), in Alaska by Thurston (1985), and in the Armorican Massif, France by Ballevre et al. (2003). Similarly, retrograde replacement of epidote by lawsonite in high-grade tectonic blocks is also common in the Franciscan Complex as described above. These relationships suggest two episodes of metamorphic re-equilibration where epidote and lawsonite are not in equilibrium. On the other hand, because of the slight compositional difference between these two phases, minor differences in $f_{O_2}$ during blueschist facies metamorphism, and the small domain size of equilibrium in low-$T$ metamorphic rocks, both lawsonite and epidote can be in equilibrium in blueschist (e.g., Brown 1977; see also discussion of this topic in Poli and Schmidt 2004).

Ca-Al-silicates in high $P/T$ metamorphic terranes include lawsonite, pumpellyite, epidote minerals and prehnite. Their occurrence has been used as an index mineral to map progressive metamorphic zones for metabasite, e.g., Franciscan, Sanbagawa, New Zealand, and New Caledonia (e.g., Maruyama et al. 1996). For the Franciscan metabasites in Ward Creek as described above, there is a progressive sequence from lawsonite through pumpellyite to an epidote zone. Progressive order of pumpellyite through lawsonite to epidote also occurs in New Caledonia, pumpellyite to epidote zone in Sanbagawa, and prehnite through pumpellyite to epidote with local lawsonite in New Zealand.

In blueschist-facies metamorphism, lawsonite occurs in low-grade blueschist, whereas epidote is common in high-grade blueschist, epidote-amphibolite and eclogite (Evans 1990). The relationship between zoisite and lawsonite has been experimentally determined (e.g., Schmidt and Poli 1994; Okamoto and Maruyama 1999) and thermodynamically calculated. The occurrence of epidote vs. lawsonite in Franciscan blueschist-facies rocks was investigated by Brown and Ghent (1983) and Maruyama and Liou (1987). Lawsonite occurs in nearly all lithologies in Pacheco Pass (e.g., Ernst et al. 1970) which display a high $P/T$ gradient ($< 12°C/$km) whereas epidote is ubiquitous in Shuksan blueschist (Washington) that has a relatively low $P/T$ gradient ($> 12°C/km$). An intermediate $P/T$ gradient of the South Fork Mountain Schist ($\sim 0.7$ GPa $/250–300°C$) of the northern California Coast Range has produced a variety of assemblages including epidote + lawsonite + pumpellyite together with other blueschist-facies minerals. Metaclastics contain the dominant assemblage quartz, albite, chlorite, phengite, and lawsonite with sporadic occurrence of aragonite and epidote. Mafic schists at Black Butte of northern California have the assemblage quartz, albite, chlorite, pumpellyite, sodic amphibole, and titanite. Actinolite is uncommon and does not coexist with epidote or sodic amphibole. Lawsonite occurs in some mafic schists but is not associated with sodic amphibole. Mafic schists from the Ball Rock area of northern California have similar assemblages with the notable exception that (1) lawsonite is common and coexists with sodic amphibole, and (2) epidote is less common. This control of the bulk composition on the appearance of lawsonite, pumpellyite, and/or epidote in blueschist-facies assemblages is apparent from the Ca-Al-Fe$^{3+}$ diagram (Fig. 27; Brown 1986; Maruyama and Liou 1987, 1988).

In addition to the effect of bulk rock composition, differences in $f_{O_2}$ and $X_{CO_2}$ also control the occurrence of lawsonite vs. epidote in different domains of a single outcrop. Chlorite and calcite at high $X_{CO_2}$ replace both lawsonite and epidote. A simple reaction such as

Lws + Fe-oxides = Ep + Qtz + Ky + H$_2$O

was used to depict the lawsonite-epidote transition at low $X_{CO_2}$. This is shown schematically for a hypothetically metamorphosed pillow-lava overlain by pelagic sediment (Fig. 28; Liou 1993). The assemblage lawsonite, magnetite, and glaucophane (+ chlorite, albite, and quartz) occurs in pillow cores, epidote, crossite, and magnetite in pillow rims and hematite, epidote, acmite, and riebeckite (+ quartz) in overlying metapelagic sediments. The variation in $f_{O_2}$ during blueschist-facies metamorphism may have been inherited from differences in the Fe$^{3+}$/Fe$^{2+}$ ratio of the protoliths, which in turn may be controlled by differences in the degree of water/rock interaction and effective water/rock ratio. Variation in the mineral assemblage

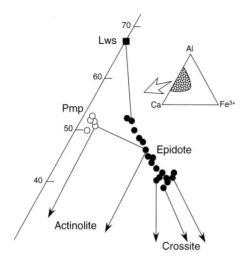

**Figure 27.** Phase relations of epidote (filled circles), lawsonite (filled square), and pumpellyite (open circles) in a Ca-Al-Fe$^{3+}$ diagram projected from quartz, albite, chlorite, and H$_2$O, from the Shuksan Suite, Washington (modified from Fig. 7 of Brown 1986).

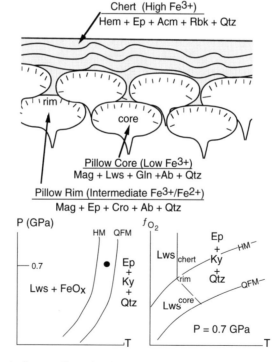

**Figure 28.** Schematic diagrams illustrating the effect of different $f_{O_2}$ in local domains on the occurrence of lawsonite assemblages in a metamorphosed pillow core, epidote assemblages in the pillow rim and in overlying chert at constant P and T (modified from Fig. 7 of Liou 1993)

at constant $P$ of about 0.7 GPa and 300°C in a single outcrop is due to difference in $f_{O_2}$ (Fig. 28). Under highly oxidizing conditions such as in metapelagic sediments including hematite-bearing chert, the hematite, epidote, acmite assemblage appears in the lawsonite $P$-$T$ stability field at $T < 300$°C.

## PARAGENESIS OF EPIDOTE-CLINOZOISITE IN THE BLUESCHIST-GREENSCHIST TRANSITION

The $P$-$T$ field of the blueschist facies is bounded by 4 to 5 other facies including the eclogite, epidote-amphibolite, greenschist, pumpellyite-actinolite facies and probably the prehnite-pumpellyite facies. Petrogenetic grids for these facies have been established (e.g., Oh and Liou 1990; Spear 1993). The blueschist facies occupies a large $P$-$T$ space at $T < 500$°C and $P > 0.5$ GPa. All blueschist-facies boundary reactions contain both epidote and glaucophane on the blueschist-facies side. Thus, the blueschist-facies $P$-$T$ field expands towards higher-$T$ and lower-$P$ in $Fe^{3+}$-rich rocks as epidote and glaucophane tend to preferentially incorporate $Fe^{3+}$ compared to other silicates that define facies boundary reactions. Boundary reactions depend on what oxide phases develop; if magnetite or hematite is included, the $P$-$T$ field of the blueschist facies expands (e.g., Banno and Sakai 1989).

Occurrences of two amphiboles (actinolite + sodic amphibole) together with epidote, chlorite, albite, titanite, and quartz in high $P/T$ rocks are common. Depending on the textural relationship between the amphiboles, two stages of metamorphic re-equilibration have been documented with blueschist-facies conditions followed by those of the greenschist facies, such as in the Swiss Alps (e.g., Ernst 1973; Dal Piaz and Lombardo 1986), Sanbagawa (e.g., Ernst et al. 1970; Otsuki and Banno 1990), and Catalina Island (Sorensen 1986), or vice versa as in Anglesey, U.K. (Gibbons and Gyopari 1986). However, both blueschist and greenschist facies assemblages may be stable in a transitional $P$-$T$ field, and coexisting sodic amphibole and actinolite define a miscibility gap in calcic amphibole (Liou and Maruyama 1987). Depending on bulk rock $Fe^{3+}/Al$ ratio, blueschist, greenschist and transitional facies assemblages may be interlayered on a cm-scale and may form during a single metamorphic episode (Brown et al. 1982; Dungan et al. 1983; Oberhänsli 1986; Barrientos and Selverstone 1993). The textural relationship of the two amphiboles from the Ward Creek epidote-zone metabasites exemplifies their coexistence.

We use the blueschist-greenschist transition equilibria to illustrate the usage of epidote as a possible geobarometer described below. The boundary reaction

Cz + Gln + Qtz + $H_2O$ = Act + Chl + Ab

was first proposed by Miyashiro and Banno (1958). The reaction is trivariant for natural metabasite; hence greenschist and blueschist assemblages could be interlayered as mentioned above. For a model basaltic system illustrated by the $CaO$-$Al_2O_3$-$MgO$ diagram, the transition between blueschist and greenschist assemblages is defined by the univariant reaction

$$Cz + Gln + Qtz + H_2O = Tr + Chl + Ab \qquad (1)$$

If $Fe^{3+}$ is introduced into the system, the reaction

$$Ep + Mrb + Chl + Qtz = Tr + Ab + Hem + H_2O \qquad (2)$$

The reaction (2) limits the low-$P$ stability of sodic amphibole in metabasite. At $P$ lower than that for this discontinuous reaction, metabasites do not contain magnesioriebeckite due to the stable coexistence of hematite and tremolite.

The blueschist-greenschist facies transition equilibria are bounded by the above two reactions in the model-$Fe_2O_3$ system (Fig. 29). Their $P$-$T$ locations were estimated by Brown (1974,

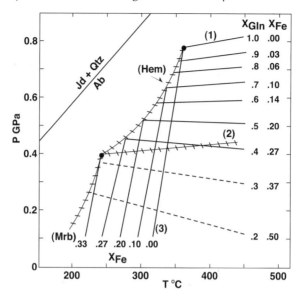

**Figure 29.** *P-T* diagram showing calculated compositions of coexisting epidote ($X_{Fe}$) and sodic amphibole ($X_{Gln}$) for the blueschist and greenschist transition assemblage of reaction (1) 6 Cz + 25 Gln + 7 Qtz + 14 $H_2O$ = 6 Tr + 9 Chl + 50 Ab. Lines with hatch marks designated as (Hem), (Mrb) and (2) 4 Ep + 5 Mrb + Chl + 7 Qtz = 4 Tr + 10 Ab + 7 Hem + 7 $H_2O$ are discontinuous reactions for the $Fe_2O_3$-saturated system. Lines designated as (3) are $X_{Fe}$ isopleths of epidote for the pumpellyite-actinolite and greenschist transition assemblage (modified from Fig. 10 of Maruyama et al. 1986).

1977) based on natural paragenesis and were experimentally determined by Maruyama et al. (1986). Both reactions possess similar *P-T* slopes but reaction (1) occurs at much higher pressure than reaction (2). Therefore, a greenschist-blueschist transitional assemblage of sodic amphibole, actinolite, and epidote (+ albite, chlorite, and quartz) occurs in the *P-T* region between these two reactions. Because compositions and relative proportions of these phases for a given bulk composition change systematically along the continuous reaction (1) as a function of *P* and *T*, blue amphibole and epidote can be used to estimate the pressure.

## POTENTIAL OF EPIDOTE MINERALS FOR THERMOBAROMETRY

The compositions of sodic amphibole and epidote are a function of pressure (Fig. 30, at 300°C). At $f_{O_2}$ conditions above the HM buffer, sodic amphibole may vary along the join glaucophane – magnesioriebeckite, tremolite may have a very limited $Fe^{2+}$ substitution, and the stable iron oxide is hematite. $P_3$ (Fig. 30) refers to the equilibrium pressure at 300°C for the reaction (1) in the model system. With gradual introduction of $Fe_2O_3$ into the model system, this reaction is displaced continuously toward lower pressure, and both epidote and sodic amphibole become more $Fe^{3+}$-rich. In the $Fe^{3+}$-saturated system, the discontinuous reaction (2) occurs at about 0.4 GPa. As long as $f_{O_2}$ is maintained above that defined by HM, the discontinuous reaction remains at fixed pressure isothermally and has fixed mineral compositions for a given bulk composition, and in a $P$-$X_{Fe}$ diagram the discontinuous reaction appears as a horizontal line (Fig. 30). If compositional variations of sodic amphibole of the reaction assemblage can be calibrated with pressure, the composition of clinozoisite-epidote could be used as a geobarometer (see below).

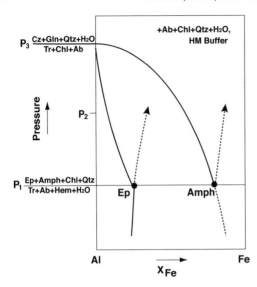

**Figure 30.** P-$X_{Fe}$ plot at 300°C showing changes in compositions of sodic amphibole and epidote for buffered assemblages in the blueschist-greenschist transition: (1) the solid lines are for Ep + Tr + sodic Amph, and (2) the dashed lines for Ep + Hem + sodic Amph. $P_1$ and $P_3$ are, respectively, the equilibrium pressure for the discontinuous and continuous reactions for the facies transition (modified from Fig. 3.11 of Liou et al. 1987).

Phase assemblages with schematic tie-lines and compositions of sodic amphibole at $P_1$, $P_2$, and $P_3$ at 300°C are illustrated in ternary diagrams projected from chlorite (Maruyama et al. 1986; Fig. 31). Complete solid solution is assumed for the glaucophane – magnesioriebeckite and clinozoisite-epidote ($X_{Fe} = 0.33$) joins. These three diagrams illustrate the paragenetic and compositional variations of the blueschist-greenschist reaction assemblage as a function of not only pressure described above but also of bulk composition. For example, basaltic rocks have compositions between those of epidote and tremolite whereas ironstones may be very oxidized, containing abundant $Fe_2O_3$. At $P = P_3$ where the reaction (1) in the model system occurs (Fig. 31), basaltic rocks may contain the typical blueschist assemblage epidote and glaucophane (+ albite, chlorite, and quartz), whereas ironstones contain epidote, magnesioriebeckite, and hematite. At intermediate pressure (e.g., $P = P_2$), metabasites, depending on their bulk composition, may contain (1) the blueschist assemblage, (2) the greenschist assemblage, or (3) the reaction assemblage epidote, actinolite, and sodic amphibole where the composition of all three phases is fixed. If pressure decreases continuously, the $Fe^{3+}$/Al ratio of epidote and sodic amphibole varies systematically. When pressure is lowered to $P_1$, a discontinuous reaction occurs and compositions of the participating phases are fixed. At pressures lower than $P_1$, basaltic rocks with high $Fe^{3+}$/Al ratios contain the greenschist assemblage epidote, tremolite, and hematite whereas ironstones may have epidote, tremolite, and hematite or tremolite, magnesioriebeckite, and hematite depending on their bulk composition. Compositions of sodic amphibole from the basaltic 3-phase reaction assemblage epidote, actinolite, and sodic amphibole may differ significantly from those of ironstones. The composition of the former is shown as a solid line in Figure 30 for the basaltic sodic amphiboles and the latter compositions (ironstone amphiboles) as a dashed line.

The $Fe^{3+}$-Al partitioning between epidote and sodic amphibole for the reaction assemblage is complex (Maruyama et al. 1986). The effect of $Fe^{3+}$-Al substitutions of glaucophane and epidote solid solutions on the Al-end member reaction (1) can be expressed as

$$\ln K_{P_2} - \ln K_{P_1} = -\Delta V^\circ (P_2 - P_1)/RT$$

where $K_{P_2}$ and $K_{P_1}$ stand for the equilibrium constants for the reaction at $P_1$ and $P_2$ respectively, $\Delta V^\circ$ denotes the standard volume change for the reaction, and R and $T$ represent gas constant

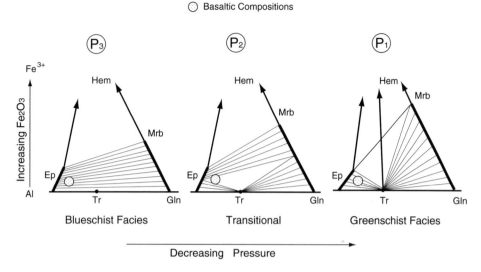

**Figure 31.** Schematic phase relations between hematite-epidote-amphibole in three isobaric sections at $T =$ 300°C for the greenschist-blueschist transition (modified from Fig. 3.12 of Liou et al. 1987). For discussion as a geobarometer see text.

and $T$ (in Kelvin) respectively. $\Delta V°$ is not constant in the dehydration reaction (1). As a first approximation, $\Delta V_{H2O}$ is assumed to be constant within a limited range of pressure (0.3–0.4 GPa) at a given temperature. This assumption results in error up to about 0.05 GPa in pressure estimates at 300–400°C, but may lie within the uncertainties of the experimental studies.

The intercrystalline partitioning of $Fe^{3+}$-Al cations between epidote and glaucophane of the buffered assemblage has to be maintained under equilibrium conditions. Hence, an additional constraint can be introduced by considering the exchange reaction such as

Gln + Ep = Mrb + Cz.

The isopleths of this sliding equilibrium were estimated assuming ideal substitution of $Fe^{3+}$-and Al cations in M(2)-sites for glaucophane solid solution and in both M(1) and M(3)-sites for epidote solid solution (for details, see Maruyama et al. 1986). Justification for such an assumption for epidote has been described by Nakajima et al. (1977). The isopleths illustrated in Figure 29 are drawn for constant composition for both epidote and sodic amphiboles in the buffered assemblage glaucophane, epidote, and actinolite (+ albite, chlorite, and quartz). The results indicate a rapid increase in Fe-content with decreasing pressure, especially at lower pressures. These isopleths of constant $X_{Gln}$ and $X_{Fe}$ do not give estimates of pressure consistent with those obtained from natural paragenesis. This apparent inconsistency may be due to the adopted ideal-composition relations as well as to other assumptions in the calculation. The isopleths have very gentle negative slopes at low pressures. Such a relation is very similar to those of the jadeite-albite-quartz sliding equilibrium. The gentle slope for these isopleths confirms their suitability as a geobarometer for the blueschist-greenschist facies transition. Both sodic amphibole and epidote in the buffered assemblage systematically increase in Al content with increasing pressure.

Also shown in Figure 29 are the isopleths related to a discontinuous reaction for a buffered assemblage epidote, actinolite, and pumpellyite (+ albite, chlorite, and quartz) from Nakajima et al. (1977). In contrast to the blueschist-greenschist transition, the isopleths for the pumpellyite-actinolite and greenschist facies transition are very sensitive to temperature change, hence are good for geothermometer. Combination of these two sets of isopleths shown in Figure 29 provided better constraints for *P-T* relations in high *P/T* metamorphic facies series.

Pressure estimates from this geobarometer (Fig. 32; Maruyama et al. 1986) for epidote with $X_{Fe}$ values of 0.13 to 0.19 in Franciscan metabasites yield ~0.7 GPa, whereas those with $X_{Fe}$ = 0.31 to 0.38 in Mikabu of the Sanbagawa belt and New Zealand give pressure ~ 0.4 GPa. Glaucophane and clinozoisite were stable at higher pressures in Franciscan metabasites than were higher $Fe^{3+}$ sodic amphibole and epidote in the Sanbagawa and New Zealand metabasites.

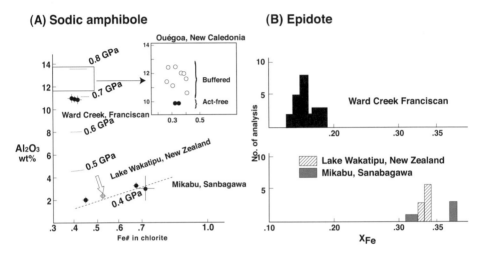

**Figure 32.** Mineral compositions of (A) sodic amphibole, chlorite and (B) epidote from the Sanbagawa, Franciscan, New Zealand and New Caledonia. $Al_2O_3$ wt% of sodic amphibole of the buffered assemblage sodic Amph + Act + Ep + Ab + Chl + Qtz is plotted against Fe# (= Fe/Fe + Mg) of chlorite. Pressure was estimated from pyroxene geobarometry. The composition of epidote from three of the blueschist terranes depends on pressure (modified from Fig. 3.17 of Liou et al. 1987).

## EPIDOTE MINERALS AS CARRIERS OF
## TRACE ELEMENTS DURING SUBDUCTION

Epidote minerals contain several petrogenetically important trace elements as major elements (Frei et al. 2004) and are therefore important for the trace element budget of HP-UHP rocks. Allanite (see Gieré and Sorensen 2004) has also been reported in some UHP schists and orthogneisses hosting coesite-bearing eclogites from the Dabie-Sulu terrane of China (Carswell et al. 2000). The rare epidote mineral, dissakisite-(Ce), a Mg-analogue of allanite-(Ce) with up to 7.8 wt% MgO, was reported from a garnet-corundum rock (Enami and Zang 1988; Zhang et al. 2004) and garnet lherzolite (Yang and Enami 2003) of the Donghai area, Sulu terrane. It is the only REE-enriched phase in the examined samples, thus it is probably an important reservoir and carrier of REE in subducted and exhumed ultramafic rocks. The dissakisite-(Ce) in garnet lherzolite contains up to 5.4 wt% $Cr_2O_3$, and suggests the existence of a possible end-member "$CeCa(Mg,Fe)CrAlSi_3O_{12}(OH)$", in which both the highly incompatible REE as well as the highly compatible element Cr are incorporated.

As clearly indicated by occurrence of niigataite in a HP Omi-Renge metamorphic belt, Japan (Miyajima et al. 2003), epidote minerals in HP-UHP metamorphic rocks have the potential to incorporate Sr as major component. For example, Sr-bearing epidote minerals are common constituents of eclogites and associated orthogneisses throughout the Sulu UHP terrane (Table 3: Nagasaki and Enami 1998). Prograde zoisite, epidote, and allanite have up to 3.2, 2.7 and 4.1 wt% SrO, respectively. They commonly show chemical zoning in which SrO content decreases from core to margin (Fig. 33). Retrograde epidote is poor in SrO (< 0.1 wt%). Apatite is always more depleted in SrO (0.10-0.59 wt% on average) than coexisting zoisite and epidote, and Sr-Ca partition coefficients for zoisite/epidote and apatite [(Sr/Ca)zo/ep-ap] range from 5 to 20. An evaluation of the SrO content in zoisite and epidote and their modal abundance indicates that > 70% of the whole-rock SrO is contained in epidote minerals. The characteristic Sr zonal structure of epidote minerals and high concentration of Sr into epidote minerals strongly suggest that zoisite and epidote are the main Sr reservoirs at HP-UHP conditions where calcic plagioclase is unstable and epidote minerals are main Ca-Al-silicates. Ca-poor, possibly Sr-rich zoisite and epidote have been commonly reported from the Dabie-Sulu terrane (e.g., Zhang and Liou 1994; Okay 1995). Thus, occurrences of Sr-bearing zoisite and epidote in UHP provinces may be widespread, and the presence or absence of these phases may strongly control the geochemical cycle of Sr in deep crustal and mantle environments (Hickmott et al. 1992).

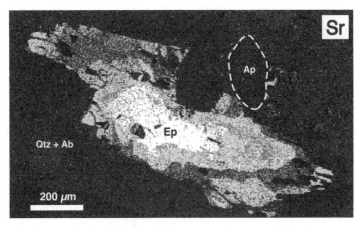

**Figure 33.** SrKα X-ray mapping image of epidote and coexisting minerals in an orthogneiss from the Sulu UHP terrane (after Fig. 3 of Enami 1999). Brighter shades indicate higher element concentration. SrO content ranges from 1.5-2 wt% in the core to less than 0.5 wt% in the margin.

## CONCLUSIONS

Zoisite and epidote exhibit wide range of *P-T* stability and occurrences. Coexisting epidote and sodic amphibole, which is higher *T* equivalent of lawsonite + glaucophane, characterize mineral paragenesis of high-grade blueschist facies. Similarly, kyanite + zoisite + jadeite + $SiO_2$ assemblage is a common HP/UHP product of crustal plagioclase + $H_2O$ in deeply subducted oceanic and continental materials. The maximum pressure limit of natural zoisite and epidote is not known; however, synthetic zoisite is stable to almost 7 GPa and natural epidote minerals coexist with coesite and/or microdiamond in UHP metamorphism. Thus epidote minerals are common Ca-Al-silicate in a variety of HP-UHP rocks recrystallized at the blueschist, epidote-amphibolite and eclogite facies. Only epidote minerals with inclusions of coesite or coesite pseudomorphs should be considered to be stable at UHP conditions. Previous descriptions list zoisite or clinozoisite in the peak UHP assemblage based on petrographic observation and experimental studies; many reported analyses of epidote minerals do not include trace and REE element concentrations.

Epidote minerals in HP-UHP rocks have wide compositional ranges controlled by element substitutions in octahedral M(3)-site. Element proportion in the M(3)-site strongly depends upon $Fe^{3+}$/Al value of host rocks, which roughly correlates with the extent of oxidation during metamorphism. Epidote in low-$f_{O_2}$-rocks, such as graphite-bearing pelitic schists and pyrrhotite-bearing mafic schists, has $X_{Fe} < 0.2$, whereas most epidotes coexisting with hematite have $X_{Fe} > 0.3$. Piemontite is restricted to siliceous schist equilibrated under unusually high-$f_{O_2}$ conditions that are inferred from the common occurrence of braunite and other highly oxidized phases. $X_{Fe}$ of epidote minerals generally decreases with increasing metamorphic grade. In high-$P$ pumpellyite-actinolite facies and low-grade blueschist facies, the decreasing $X_{Fe}$ is probably advanced by continuous decomposition of pumpellyite and lawsonite, respectively. The principal factor responsible for decreasing $X_{Fe}$ under high-grade greenschist facies and higher metamorphic grade conditions is not well known. Goto et al. (2002), however, emphasizes that calcite occurs in most Sanbagawa pelitic schists, and decarbonation reactions are the main factor responsible for increasing clinozoisite component in epidote with increasing metamorphic grade. The two-phase loop between zoisite and clinozoisite has funnel-like form, shifts towards $Fe^{3+}$-rich composition with increasing metamorphic temperature and pressure; zoisite represents the higher $T$ and $P$ phase in relation to epidote. Thus the frequency of appearance of zoisite increases with increasing metamorphic grade and zoisite commonly occurs in UHP metamorphic rocks.

Element substitutions into A(2)-site also control chemical composition of epidote minerals. Allanite is stable in HP-UHP metamorphic rocks and fulfills the role of an important REE reservoir for subducting oceanic and continental materials. Similarly, Sr-bearing epidote minerals are also common in HP-UHP metamorphic rocks, and are the main Sr reservoir. Therefore, epidote minerals may play the most important role in the cycling of some important trace elements in deep crustal and upper mantle environments. Moreover, as epidote minerals are stable to mantle depths, progressive changes of epidote-bearing assemblages at various depths in subducting oceanic and continental materials may affect the extent of mantle metasomatism, magma genesis and fluid transport above the subduction zone. Systematic microanalyses of REE and other trace element abundances of epidote minerals in HP and UHP rocks are essential to substantiate this conclusion.

Only the compositions of epidote from low-variance assemblages vary systematically with intensive variables and are controlled by continuous reactions, hence can be used as $P$ or $T$ indicators. Examples of such application have been suggested for blueschist and subgreenschist facies metabasite assemblages. Similar approaches should be extended to UHP conditions using both thermochemical calculations of pseudosections of the basaltic system and experimentally delineated phase equilibria as very limited geobarometers are available for UHP rocks.

## ACKNOWLEDGMENTS

During the preparation of this review, many colleagues have contributed their ideas, data, figures and photos. We thank Ruyuan Zhang for photomicrographs of epidote with inclusions of coesite pseudomorphs, Diane Moore and Chang-Whan Oh for Franciscan epidote compositions and diagrams, and Liz Eide and Yupeng Yao for paragenetic diagrams for the Dabie-Sulu UHP belt. We also thank G. Franz, A. Liebscher, J. Selverstone, and R. Grapes for their constructive and critical reviews of this manuscript, and Shio Watanabe for drafting several diagrams. Preparation of this manuscript is supported in part by JSPS-14540448 and NSF EAR 0003355.

# REFERENCES

Aoya M (2001) *P-T-*D path of eclogite from the Sambagawa belt deduced from combination of petrological and microstructural analyses. J Petrol 42:1225-1248

Baker J, Matthews A, Mattey D, Rowley D, Xue F (1997) Fluid-rock interactions during ultra-high pressure metamorphism, Dabie Shan, China. Geochim Cosmochim Acta 61:1685-1696

Ballevre M, Pitra P, Bohn M (2003) Lawsonite growth in the epidote blueschists from the Ile de Groix (Armorican Massif, France): a potential geobarometer. J Metamorphic Geol 21:723-736

Banno S, Sakai C (1989) Geology and metamorphic evolution of the Sanbagawa metamorphic belt, Japan. *In*: The Evolution of Metamorphic Belts. JS Daly, RA Cliff, BWD Yardley (eds) Blackwell Scientific Publications, Oxford, p 519-532

Banno S, Shibakusa H, Enami M, Wang Q, Ernst WG (2000) Chemical fine structure of Franciscan jadeitic pyroxene from Ward Creek, Cazadero area, California. Am Mineral 85:1795-1798

Banno S, Yokoyama K, Enami M, Iwata O, Nakamura K, Kasashima S (1976a) Petrology of the peridotite-metagabbro complex in the vicinity of Mt. Higashi-akaishi, Central Shikoku. Part I. Megascopic textures of the Iratsu and Tonaru epidote amphibolite masses. Sci Rept Kanazawa Univ 21:139-159

Banno S, Yokoyama K, Iwata O, Terashima S (1976b) Genesis of epidote amphibolite masses in the Sanbagawa metamorphic belt of central Shikoku. J Geol Soc Japan 82:199-210 (in Japanese with English abstract)

Banno S, Yoshizawa H (1993) Sector-zoning of epidote in the Sanbagawa schists and the question of an epidote miscibility gap. Mineral Mag 57:739-743

Barrientos X, Selverstone J (1993) Infiltration vs. thermal overprinting of epidote blueschists, Ile de Groix, France. Geology 21:69-72

Bird D, Spieler AR (2004) Epidote in geothermal systems. Rev Mineral Geochem 56:235-300

Black PM (1977) Regional high-pressure metamorphism in New Caledonia: phase equilibria in the Ouégoa district. Tectonophysics 43:89-107

Black PM, Brothers RN, Yokoyama K (1988) Mineral paragenesis in eclogite-facies meta-acidites in northern New Caledonia. *In*: Eclogites and Eclogite-facies Rocks. DC Smith, (ed) Elsevier, Amsterdam, p 271-289

Boundy TM, Austrheim H, Donohue CL, Essene EJ, Mezger K (2002) Discovery of eclogite facies carbonate rocks from the Lindås Nappe, Caledonides, Western Norway. J Metamorphic Geol 20:649-667

Brothers RN (1974) High-pressure schists in northern New Caledonia. Contrib Mineral Petrol 46:109-127

Brown EH (1974) Comparison of the mineralogy and phase relations of blueschists from the North Cascades, Washington and greenschists from Otago, New Zeland. Bull Geol Soc Amer 85:333-344

Brown EH (1977) Phase equilibria among pumpellyite, lawsonite, epidote and associated minerals in low grade metamorphic rocks. Contrib Mineral Petrol 64:123-136

Brown EH (1986) Geology of the Shukusan Suite, north Cascades, Washington, U.S.A. *In*: Blueschists and Eclogites. BE Evans, EH Brown, (eds) Geol Soc Am Memoir 164, p 143-154

Brown EH, Ghent ED (1983) Mineralogy and phase relations in the blueschist facies of the Black Butte and Ball Rock areas, northern California Coast Ranges. Am Mineral 68:365-372

Brown EH, Wilson DL, Armstrong RL, Harakal JE (1982) Petrologic, structural, and age-relations of serpentinite, amphibolite, and blueschist in the Shukusan Suite of the Iron Mountains-Gee Point area, north Cascades. Geol Soc Am Bull 92:1087-1098

Bruno M, Compagnoni R, Rubbo M (2001) The ultra-high pressure coronitic and pseudomorphous reactions in a metagranodiorite from the Brossasco-Isasca Unit, Dora-Maira Massif, Western Italian Alps: A petrographic study and equilibrium thermodynamic modeling. J Metamorphic Geol 19:33-43

Brunsmann A, Franz G, Erzinger J, Landwehr D (2000) Zoisite- and clinozoisite-segregations in metabasites (Tauern Window, Austria) as evidence for high-pressure fluid-rock interaction. J Metamorphic Geol 18:1-21

Carswell DA, Wilson RN, Zhai M (2000) Metamorphic evolution, mineral chemistry and thermobarometry of schists and orthogneisses hosting ultra-high pressure eclogites in the Dabieshan of central China. Lithos 52:121-155

Castelli D, Rolfo F, Compagnoni R, Su X (1998) Metamorphic veins with kyanite, zoisite and quartz in the Zhu-Jia-Chong eclogite, Dabie Shan, China. Isl Arc 7:159-173

Clarke GL, Aitchison JC, Cluzel D (1997) Eclogites and blueschists of the Pam Peninsula, NE New Caledonia: a reappraisal. J Petrol 38:843-876

Coleman RG, Lanphere MA (1971) Distribution and age of high-grade blueschists, associated eclogites and amphibolites from Oregon and California. Geol Soc Am Bull 82:2397-2412

Coleman RG, Lee DE (1962) Metamorphic aragonite in the glaucophane schists of Cazedero, California. Am J Sci 260:577-595

Coleman RG, Lee DE (1963) Glaucophane-bearing metamorphic rock types of the Cazadero area, California. J Petrol 4:260-301

Compagnoni R, Rolfo F (1999) Characteristics of UHP pelites, gneisses, and other unusual rocks. Inter Geol Rev 41:552-570

Cong B, Zhai M, Carswell DA, Wilson RN, Wang Q, Zhao Z, Windley BF (1995) Petrogenesis of ultrahigh-pressure rocks and their country rocks at Shuanghe in Dabieshan, central China. Eur J Mineral 7:119-138

Dal Piaz GV, Lombardo B (1986) Early Alpine eclogite metamorphism in the Pennidic Monte Rosa-Gran Paradiso basement nappes of the northwestern Alps. *In*: Blueschists and Eclogites. BE Evans, EH Brown, (eds) Geol Soc Am Memoir 164, p 249-265

Deer WA, Howie RA, Zussman J (1986) Disilicates and Ring Silicates. Longman, London, 1B, p 629

Donohue CL, Essene EJ (2000) An oxygen barometer with the assemblage garnet-epidote. Earth Planet Sci Lett 181:459-472

Droop GTR (1985) Alpine metamorphism in the south-east Tauern window, Austria; 1. *P-T* variations in space and time. J Metamorphic Geol 3:371-402

Droop GTR, Lombardo B, Pognante U (1990) Formation and distribution of eclogite facies rocks in the Alps. *In*: Eclogite Facies Rocks. DA Carswell (ed) Blackie, Glasgow, p 225-259

Dungan MA, Vance JA, Blanchard DP (1983) Geochemistry of the Shuksan greenschists and blueschists, North Cascades, Washington; variably fractionated and altered metabasalts of oceanic affinity. Contrib Mineral Petrol 82:131-146

Eide EA, Liou JG (2000) High-pressure blueschists and eclogites in Hong'an: A framework for addressing the evolution of high- and ultrahigh-pressure rocks in central China. Lithos 52:1-22

Enami M (1977) Sector zoning of zoisite from a metagabbro at Fujiwara, Sanbagawa metamorphic terrain in central Shikoku. J Geol Soc Japan 83:693-697

Enami M (1978) Petrological study of epidote amphibolites and basic schists in the Besshi area, Sanbagawa metamorphic terrain in central Shikoku. Master's Thesis, Kanazawa University, Kanazawa, p 152

Enami M (1980) Notes on petrography and rock-forming mineralogy (8) Margarite-bearing metagabbro from the Iratsu mass in the Sanbagawa belt, central Shikoku. J Japan Assoc Mineral Petrol Econ Geol 75: 245-253

Enami M (1983) Petrology of pelitic schists in the oligoclase-biotite zone of the Sanbagawa metamorphic terrain, Japan: phase equilibria in the highest grade zone of a high-pressure intermediate type of metamorphic belt. J Metamorphic Geol 1:141-161

Enami M (1986) Ardennite in a quartz schist from the Asemi-gawa area in the Sanbagawa metamorphic terrain, central Shikoku, Japan. Mineral J 13:151-160

Enami M (1999) CaAl-silicates: an important Sr container in subducted slab. J Geogr 108:177-187 (in Japanese with English abstract)

Enami M, Banno S (1980) Zoisite-clinozoisite relations in low- to medium-grade high-pressure metamorphic rocks and their implications. Mineral Mag 43:1005-1013

Enami M, Banno Y (2001) Partitioning of Sr between coexisting minerals of the hollandite- and piemontite-groups in a quartzose schist from the Sanbagawa metamorphic belt, Japan. Am Mineral 86:204-215

Enami M, Mizukami T, Yokoyama K (2004) Metamorphic evolution of garnet-bearing ultramafic rocks from the Gongen area, Sanbagawa belt, Japan. J Metamorphic Geol 22:1-15

Enami M, Wallis SR, Banno Y (1994) Paragenesis of sodic pyroxene-bearing quartz schists: implications for the P-*T* history of the Sanbagawa belt. Contrib Mineral Petrol 116:182-198

Enami M, Zang Q (1988) Magnesian staurolite in garnet-corundum rocks and eclogite from the Donghai district, Jiangsu province, east China. Am Mineral 73:48-56

Enami M, Zang Q, Yin Y (1993) High-pressure eclogites in northern Jiangsu-southern Shandong province, eastern China. J Metamorphic Geol 11:589-603

Ernst WG (1965) Mineral paragenesis in Franciscan metamorphic rocks, Panoche Pass, California. Geol Soc Am Bull 76:879-914

Ernst WG (1973) Interpretative synthesis of metamorphism in the Alps. Geol Soc Am Bull 84:2053-2078

Ernst WG, Seki Y, Onuki H, Gilbert MC (1970) Comparative study of low-grade metamorphism in the California coast ranges and the outer metamorphic belt of Japan, Geol Soc Amer Mem 124, p 276

Evans BW (1990) Phase relations of epidote-blueschists. Lithos 25:3-23

Franz G, Selverstone J (1992) An empirical phase diagram for the clinozoisite-zoisite transformation in the system $Ca_2Al_3Si_3O_{12}(OH)$-$Ca_2Al_2Fe^{3+}Si_3O_{12}(OH)$. Am Mineral 77:631-642

Franz G, Smelik EA (1995) Zoisite-clinozoisite bearing pegmatites and their importance for decompressional melting in eclogites. Eur J Mineral 7:1421-1436

Franz G, Spear FS (1983) High pressure metamorphism of siliceous dolomites from the central Tauern Window, Austria. Amer J Sci 283A:396-413

Frei D, Liebscher A, Franz G, Dulski P (2004) Trace element geochemistry of epidote minerals. Rev Mineral Geochem 56:553-605

Getty SR, Selverstone J (1994) Stable isotopic and trace element evidence for restricted fluid migration in 2 GPa eclogites. J Metamorphic Geol 12:747-760

Ghent ED, Black PM, Brothers RN, Stout MZ (1987) Eclogites and associated albite-epidote-garnet paragneisses between Yambe and Cape Colnett, New Caledonia. J Petrol 28:627-643

Gibbons W, Gyopari M (1986) A greenschist protolith for blueschist in Anglesey, U.K. *In*: Blueschists and Eclogites. BE Evans, EH Brown, (eds) Geol Soc Amer Memoir 164, p 217-228

Gieré R, Sorensen SS (2004) Allanite and other REE-rich epidote-group minerals. Rev Mineral Geochem 56: 431-494

Gomez-Pugnairea MT, Visonab D, Franz G (1985) Kyanite, margarite and paragonite in pseudomorphs in amphibolitized eclogites from the Betic Cordilleras, Spain. Chem Geol 50:129-141

Goto A, Banno S, Higashino T, Sakai C (2002) Occurrence of calcite in Sanbagawa pelitic schists: Implications for the formation of garnet, rutile, oligoclase, biotite and hornblende. J Metamorphic Geol 20:255-262

Grapes RH, Hoskin PWO (2004) Epidote group minerals in low–medium pressure metamorphic terranes. Rev Mineral Geochem 56:301-345

Heinrich CA (1982) Kyanite-eclogite to amphibolite facies evolution of hydrous mafic and pelitic rocks, Adula Nappe, Central Alps. Contrib Mineral Petrol 81:30-38

Hickmott DD, Sorensen SS, Rogers PSZ (1992) Metasomatism in a subduction complex: constraints from microanalysis of trace elements in minerals from garnet amphibolite from the Catalina Schist. Geology 20:347-350

Higashino T (1990) The higher-grade metamorphic zonation of the Sambagawa metamorphic belt in central Shikoku, Japan. J Metamorphic Geol 8:413-423

Higashino T, Sakai C, Kurata H, Enami M, Hosotani H, Enami M, Banno S (1984) Electron microprobe analyses of rock-forming minerals from the Sanbagawa metamorphic rocks, Shikoku Part II. Sazare, Kotu and Bessi areas. Sci Rept Kanazawa Univ 29:37-64

Higashino T, Sakai C, Otsuki M, Itaya T, Banno S (1981) Electron microprobe analyses of rock-forming minerals from the Sanbagawa metamorphic rocks, Shikoku Part I. Asemi River area. Sci Rept Kanazawa Univ 26:73-122

Hirajima T, Wallis SR, Zhai M, Ye K (1993) Eclogitized metagranitoid from the Su-Lu ultra-high pressure (UHP) province, eastern China. Proc Japan Acad, Ser B 68:249-254

Hirajima T, Zhang R, Li J, Cong B (1992) Petrology of the nyböite-bearing eclogite in the Donghai area, Jiangsu province, eastern China. Mineral Mag 56:37-46

Holland TJB, Powell R (1990) An enlarged and updated internally consistent thermodynamic dataset with uncertainties and correlations: The system $K_2O$-$Na_2O$-$CaO$-$MgO$-$MnO$-$FeO$-$Fe_2O_3$-$Al_2O_3$-$TiO_2$-$SiO_2$-$C$-$H_2$-$O_2$. J Metamorphic Geol 8:89-124

Holland TJB, Powell R (1998) An internally consistent thermodynamic data set for phases of petrological interest. J Metamorphic Geol 16:309-343

Hoscheck G (2001) Thermobarometry of metasediments and metabasites from the Eclogite zone of the Hohe Tauern, Eastern Alps, Austria. Lithos 59:127-150

Itaya T (1981) Carbonaceous material in pelitic schists of the Sanbagawa metamorphic belt in central Shikoku, Japan. Lithos 14:215-244

Itaya T, Brothers RN, Black PM (1985) Sulfides, oxides and sphene in high-pressure schists from New Caledonia. Contrib Mineral Petrol 91:151-162

Jayko AS, Blake MCJ, Brothers RN (1986) Blueschist metamorphism of the eastern Franciscan belt, northern California. *In*: Blueschists and Eclogites. BE Evans, EH Brown, (eds) Geol Soc Am Memoir 164, p 107-123

Korsakov AV, Shatsky VS, Sobolev NV, Zayachokovsky AA (2002) Garnet-biotite-clinozoisite gneiss: A new type of diamondiferous metamorphic rock from the Kokchetav Massif. Eur J Mineral 14:915-928

Kretz R (1983) Symbols for rock-forming minerals. Am Mineral 68:277-279

Krogh EJ (1982) Metamorphic evolution of Norwegian country-rock eclogites, as deduced from mineral inclusions and compositional zoning in garnets. Lithos 15:305-321

Krogh EJ, Oh CW, Liou JG (1994) Polyphase and anticlockwise *P-T* evolution for Franciscan eclogites and blueschists from Jenner, California, USA. J Metamorphic Geol 12:121-134

Kunugiza K, Takasu A, Banno S (1986) The origin and metamorphic history of the ultramafic and metagabbro bodies in the Sanbagawa metamorphic belt. *In*: Blueschists and Eclogites. BE Evans, EH Brown, (eds) Geol Soc Am Memoir 164, p 375-385

Lee DE, Thomas HH, Marvin RF, Coleman RG (1964) Isotopic ages of glaucophane schists from the area of Cazadero, California. US Geol Surv Prof Paper 475-D:105-107

Liou JG (1973) Synthesis and stability relations of epidote $Ca_2Al_2FeSi_3O_{12}(OH)$. J Petrol 14:381-413

Liou JG (1993) Stabilities of natural epidotes. Abhand Geol Bund 49:7-16

Liou JG, Maruyama S (1987) Paragenesis and compositions of amphiboles from Franciscan jadeite-glaucophane type facies series metabasites at Cazadero, California. J Metamorphic Geol 5:371-395

Liou JG, Maruyama S, Cho M (1987) Very low-grade metamorphism of volcanic and volcaniclastic rocks - mineral assemblages and mineral facies. *In*: Very Low-Grade Metamorphism. M Frey, (ed) Blackie Publishing Co., p 59-113

Liou JG, Zhang RY (2002) Ultrahigh-pressure metamorphic rocks. Ency Phys Sci Tech (Third Edition). 17, Academia Press, p 227-244

Maruyama S, Cho M, Liou JG (1986) Experimental investigations of blueschist-greenschist transition equilibria: pressure dependence of $Al_2O_3$ contents in sodic amphiboles - a new geobarometer. *In*: Blueschists and Eclogites. BW Evans, EH Brown, (eds) Geol Soc Am Memoir 164, p 1-16

Maruyama S, Liou JG (1987) Clinopyroxene-a mineral telescoped through the processes of blueschist facies metamorphism. J Metamorphic Geol 5:529-552

Maruyama S, Liou JG (1988) Petrology of Franciscan metabasites along the jadeite-glaucophane type facies series, Cazadero, California. J Petrol 29:1-37

Maruyama S, Liou JG, Terabayashi M (1996) Blueschists and eclogites of the world and their exhumation. Inter Geol Rev 38:485-594

Massonne HJ, O'Brien P (2003) Reviews of representative UHPM terranes: The Bohemian Massif and the NW Himalaya. *In*: Ultra-High Pressure Metamorphism. Notes in Mineralogy 5. DA Carswell, R Compagnoni, (eds) Eur Mineral Union, p 145-188

Masuda A, Nakamura N, Tanaka T (1973) Fine structures of mutually normalized rare-earth patterns of chondrits. Geochim Cosmochim Acta 37: 239-248

Mattinson CG, Zhang RY, Tsujimori T, Liou JG (in review) Epidote-rich talc-kyanite-phengite eclogites, Sulu terrane, eastern China. Am Mineral

Miyajima H, Matsubara S, Miyawaki R, Hirokawa K (2003) Niigataite, $CaSrAl_3(Si_2O_7)(SiO_4)O(OH)$: Sr-analogue of clinozoisite, a new member of the epidote group from the Itoigawa-Ohmi district, Niigata Prefecture, central Japan. J Mineral Petrol Sci 98:118-129

Miyashiro A (1994) Metamorphic Petrology. UCL Press, London, p 404

Miyashiro A, Banno S (1958) Nature of glaucophanitic metamorphism. Am J Sci 256:97-110

Miyashiro A, Seki Y (1958) Enlargement of the composition field of epidote and piemontite with rising temperature. Am J Sci 256:423-430

Moore DE (1984) Metamorphic history of a high-grade blueschist exotic block from the Franciscan complex, California. J Petrol 25:126-150

Moore DE, Blake MCJ (1989) New evidence for polyphase metamorphism of glaucophane schist and eclogite exotic blocks in the Franciscan Complex, California and Oregon. J Metamorphic Geol 7:211-228

Morrison J (2004) Stable and radiogenic isotope systematics in epidote group minerals. Rev Mineral Geochem 56:607-628

Nagasaki A, Enami M (1998) Sr-bearing zoisite and epidote in ultra high-pressure (UHP) metamorphic rocks from the Su-Lu province, eastern China: an important Sr-reservoir under UHP conditions. Am Mineral 83:240-247

Nakajima T (1982) Phase relations of pumpellyite-actinolite facies metabasites in the Sanbagawa metamorphic belt in central Shikoku, Japan. Lithos 15:267-280

Nakajima T, Banno S, Suzuki T (1977) Reactions leading to the disappearance of pumpellyite in low-grade metamorphic rocks of the Sanbagawa metamorphic belt in central Shikoku, Japan. J Petrol 18:263-284

Newton RC, Kennedy GC (1963) Some equilibrium reactions in the join $CaAl_2Si_2O_8$-$H_2O$. J Geophys Res 68: 2967 - 2983

Oberhänsli R (1986) Blue amphiboles in metamorphosed Mesozoic mafic rocks from the central Alps. *In*: Blueschists and Eclogites. BE Evans, EH Brown, (eds) Geol Soc Amer Mem 164, p 239-247

O'Brien PJ (1993) Partially retrograded eclogites of the Münchberg Massif, Germany; records of a multi-stage Variscan uplift history in the Bohemian Massif. J Metamorphic Geol 11:241-260

Oh CW, Liou JG (1990) Metamorphic evolution of two different eclogites in the Franciscan Complex, California, USA. Lithos 25:41-53

Oh CW, Liou JG, Maruyama S (1991) Low-temperature eclogites and eclogitic schists in Mn-rich metabasites in Ward Creek, California; Mn and Fe effects on the transition between blueschist and eclogite. J Petrol 32:275-301

Okamoto K, Maruyama S (1999) The high-pressure synthesis of lawsonite in the MORB+$H_2O$ system. Am Mineral 84:362-373

Okay AI (1995) Paragonite eclogites from Dabie Shan, China: re-equilibration during exhumation? J Metamorphic Geol 13:449-460

Otsuki M (1980) Petrological study of the basic Sanbagawa metamorphic rocks in central Shikoku, Japan. Doctoral Dissertation, University of Tokyo, Tokyo, p 286

Otsuki M, Banno S (1990) Prograde and retrograde metamorphism of hematite-bearing basic schists in the Sanbagawa belt in central Shikoku. J Metamorphic Geol 8:425-439

Philippot P, Rumble ID (2000) Fluid-rock interactions during high-pressure and ultrahigh-pressure metamorphism. Inter Geol Rev 42:312-327

Poli S, Schmidt MW (2004) Experimental subsolidus studies on epidote minerals. Rev Mineral Geochem 56: 171-195

Potel S, Schmidt ST, de Capitani C (2002) Composition of pumpellyite, epidote and chlorite from New Caledonia - How important are metamorphic grade and whole-rock composition? Schweiz Mineral Petrogr Mitt 82:229-252

Raith M (1976) The Al-Fe(III) epidotes miscibility gap in a metamorphic profile through the Pennnine series of Tauern, Austria. Contrib Mineral Petrol 57:99-117

Reinecke T (1991) Very-high-pressure metamorphism and uplift of coesite-bearing metasediments from the Zermatt-Saas zone, Western Alps. Eur J Mineral 3:7-17

Reinecke T (1998) Prograde high- to ultrahigh-pressure metamorphism and exhumation of oceanic sediments at Lago di Cignana, Zermatt-Saas Zone, Western Alps. Lithos 42:147-189

Rolfo F, Compagnoni R, Xu S, Jiang L (2000) First report of felsic whiteschist in the ultrahigh-pressure metamorphic belt of Dabie Shan, China. Eur J Mineral 12:883-898

Rumble D, Giorgis D, Ireland T, Zhang Z, Xu H, Yui TF, Yang J, Xu Z, Liou JG (2002) Low $\delta^{18}O$ zircons, U-Pb dating, and the age of the Qinglongshan oxygen and hydrogen isotope anomaly near Donghai in Jiangsu Province, China. Geochim Cosmochim Acta 66:2299-2306

Rumble D, Wang Q, Zhang R (2000) Stable isotope geochemistry of marbles from the coesite UHP terrains of Dabieshan and Sulu, China. Lithos 52:79-95

Rumble D, Yui TF (1998) The Qinglongshan oxygen and hydrogen isotope anomaly near Donghai in Jiangsu Province, China. Geochim Cosmochim Acta 62:3307-3321

Sakai C, Higashino T, Enami M (1984) REE-bearing epidote from Sanbagawa pelitic schists, central Shikoku, Japan. Geochem J 18:45-53

Schmidt MW, Poli S (1994) The stability of lawsonite and zoisite at high pressures: Experiments in CASH to 92 kbar and implications for the presence of hydrous phases in subducted lithosphere. Earth Planet Sci Lett 124:105-118

Schmidt MW, Poli S (1998) What causes the position of the volcanic front? Experimentally based water budget for dehydrating slabs and consequences for arc magma generation. Earth Planet Sci Lett 163:361-379

Selverstone J, Franz G, Thomas S, Getty S (1992) Fluid variability in 2 GPa eclogites as an indicator of fluid behavior during subduction. Contrib Mineral Petrol 112:341-357

Selverstone J, Spear FS (1985) Metamorphic P-T paths from pelitic schists and greenstones in the southwest Tauern Window, Eastern Alps. J Metamorphic Geol 3:439-465

Shibakusa H, Maekawa H (1997) Lawsonite-bearing eclogitic metabasites in the Cazadero area, northern California. Mineral Petrol 61:163-180

Sobolev NV, Shatsky VS (1990) Diamond inclusions in garnets from metamorphic rocks; a new environment for diamond formation. Nature 343:742-746

Sorensen SS (1986) Petrologic and geochemical comparison of the blueschist and greenschist units of the Catalina schist terrane, southern California. *In*: Blueschists and Eclogites. BE Evans, EH Brown, (eds) Geol Soc Am Memoir 164, p 59-75

Sorensen SS, Grossman JN (1989) Enrichment of trace elements in garnet amphibolites from a paleo-subduction zone: Catalina Schist, southern California. Geochim Cosmochim Acta 53:3155-3177

Spear FS (1993) Metamorphic Phase Equilibria and Pressure-Temperature-Time Paths. Mineral Soc Amer Monograph, Washington, D.C., p 799

Spear FS, Franz G (1986) P-T evolution of metasediments from the eclogite zone, South-Central Tauern Window, Austria. Lithos 19:219-234

Strens RGJ (1965) Stability and relations of the Al-Fe epidotes. Mineral Mag 35:464-475

Tagiri M (1985) A comparison of graphitizing-degree and metamorphic zones of the Sanbagawa metamorphic belt in central Shikoku. J Japan Assoc Mineral Petrol Econ Geol 80:503-506

Takasu A (1989) P-T histories of peridotite and amphibolite tectonic blocks in the Sambagawa metamorphic belt, Japan. *In*: The Evolution of Metamorphic Belts. JS Daly, RA Cliff, BWD Yardley, (eds) Blackwell Scientific Publications, Oxford, p 533-538

Thurston SP (1985) Structure, petrology, and metamorphic history of the Nome Group blueschist terrane, Salmon Lake area, Seward Peninsula, Alaska. Geol Soc Am Bull 96:600-617

Vrána S, Gryda J (2003) Ultrahigh-pressure grossular-rich garnetite from the Moldanubian Zone, Czech Republic. Eur J Mineral 15:43-54

Wallis S, Aoya M (2000) A re-evaluation of eclogite facies metamorphism in SW Japan: Proposal for an eclogite nappe. J Metamorphic Geol 18:653-664

Wallis S, Takasu A, Enami M, Tsujimori T (2000) Eclogite and related metamorphism in the Sanbagawa belt, Southwest Japan. Bull Res Inst Nat Sci, Okayama Univ Sci 26:3-17

Yang JJ, Enami M (2003) Chromian dissakisite-(Ce) in a garnet lherzolite from the Chinese Su-Lu UHP metamorphic terrane: Implications for Cr incorporation in epidote minerals and recycling of REE into the Earth's mantle. Am Mineral 88: 604-610

Yao Y, Cong B, Wang Q, Ye K, Liu J (2000) A transitional eclogite- to high pressure granulite-facies overprint on coesite-eclogite at Taohang in the Sulu ultrahigh-pressure terrane, eastern China. Lithos 52: 109-120

Yoder HS (1955) Role of water in metamorphism. Geol Soc Am Spec Pap 62: 505-523

Yokoyama K (1976) Finding of plagioclase-bearing granulite from the Iratsu epidote amphibolite mass in central Shikoku. J Geol Soc Japan 82: 549-551

Yokoyama K, Brothers RN, Black PM (1986) Regional eclogite facies in the high-pressure metamorphic belt of New Caledonia. *In*: Blueschists and Eclogites. BE Evans, EH Brown, (eds) Geol Soc Am Memoir 164, p 407-423

Yoshizawa H (1984) Notes on petrography and rock-forming mineralogy; (16), Sector-zoned epidote from Sanbagawa Schist in central Shikoku, Japan. J Japan Assoc Mineral Petrol Econ Geol 79: 101-110

Yui TF, Rumble D, Lo CH (1995) Unusually low $\delta^{18}O$ ultra-high-pressure metamorphic rocks from the Sulu-terraine, eastern China. Geochim Cosmochim Acta 59: 2859-2864

Yui TF, Rumble D, Chen CH, Lo CH (1997) Stable isotope characteristics of eclogites from the ultra-high-pressure metamorphic terrain, east-central China. Chem Geol 137:135-147

Zhang RY (1992) Petrogenesis of high pressure metamorphic rocks in the Su-Lu and Dianxi regions, China. Ph. D. Thesis, Kyoto University, p 176

Zhang RY, Hirajima T, Banno S, Cong B, Liou JG (1995a) Petrology of ultrahigh-pressure rocks from the southern Su-Lu region, eastern China. J Metamorphic Geol 13:659-675

Zhang RY, Liou JG (1994) Coesite-bearing eclogite in Henan Province, central China: detailed petrography, glaucophane stability and PT-path. Eur J Mineral 6:217 - 233

Zhang RY, Liou JG (1997) Partial transformation of gabbro to coesite-bearing eclogite from Yangko, the Sulu terrane, eastern China. J Metamorphic Geol 15:183-202

Zhang RY, Liou JG, Cong B (1995b) Talc-, magnesite- and Ti-clinohumite-bearing ultrahigh-pressure meta-mafic and ultramafic complex in the Dabie Mountains, China. J Petrol 36:1011-1037

Zhang RY, Liou JG, Ernst WG (1995c) Ultrahigh-pressure metamorphism and decompressional *P-T* paths of eclogites and country rocks from Weihai, eastern China. Isl Arc 4:293-309

Zhang RY, Liou JG, Ernst WG, Coleman RG, Sobolev NV, Shatsky VS (1997) Metamorphic evolution of diamond-bearing and associated rocks from the Kokchetav Massif, northern Kazakhstan. J Metamorphic Geol 15:479-496

Zhang RY, Liou JG, Katayama I (2002) Petrologic characteristics and metamorphic evolution of diamond-bearing gneiss from Kumdy-kol. *In*: The Diamond-bearing Kokchetav Massif, Kazakhstan: Petrochemistry and Tectonic Evolution of an Unique Ultrahigh-Pressure Metamorphic Terrane. CD Parkinson, I Katayama, JG Liou, S Maruyama, (eds) Universal Academy Press, Inc., Tokyo, p 213-234

Zhang RY, Liou JG, Zhang YF, Fu B (2003) Transition of UHP eclogites to gneissic rocks of low-grade amphibolite facies during exhumation: Evidence from the Dabie terrane, central China. Lithos 70:269-291

Zhang RY, Liou JG, Zheng JP (2004) Ultrahigh-pressure corundum-rich garnetite in garnet peridotite, Sulu terrane, China. Contrib Mineral Petrol 146:21-31

Zheng YF, Cong B, Li S (1996) Extreme $^{18}O$ depletion in eclogite from the Su-lu terrane in east China. Eur J Mineral 8:317-323

Zheng YF, Fu B, Gong B, Li L (2003) Stable isotope geochemistry of ultrahigh pressure metamorphic rocks from the Dabie–Sulu orogen in China: implications for geodynamics and fluid regime. Earth Sci Rev 62: 105-161

Zheng YF, Li Y, Gong B, Fu B, Xiao Y (1999) Hydrogen and oxygen isotope evidence for fluid-rock interactions in the stages of pre- and post-UHP metamorphism in the Dabie Mountains. Lithos 46:677-693

Reviews in Mineralogy & Geochemistry
Vol. 56, pp. 399-430, 2004
Copyright © Mineralogical Society of America

# 9

# Magmatic Epidote

## Max W. Schmidt

*Institute for Mineralogy and Petrology*
*ETH*
*8092 Zürich, Switzerland*

## Stefano Poli

*Dipartimento Scienze della Terra*
*Via Botticelli 23*
*Universitàdegli Studi di Milano*
*20133 Milano, Italy*

## INTRODUCTION

Epidote was first recognized as a magmatic mineral in the alpine Bergell tonalite by Cornelius (1915). Field observations and microscopic textures let Cornelius to conclude "... the only possibility is, that epidote is a primary mineral in our tonalite, crystallizing early from the magma, i.e., before (in part also contemporaneous with) biotite" *(translated from German, Cornelius (1915), p. 170)*. This knowledge disappeared and for the following 70 years, epidote and zoisite were categorized as metamorphic minerals. The petrologic significance of magmatic epidote was then rediscovered when Zen and Hammarstrom (1984) identified epidote as an important magmatic constituent of intermediate calc-alkaline intrusives in plutons of the North American Cordillera. Zen and Hammarstrom (1984) also suggested that epidote indicates a minimum intrusive pressure of about 0.5 to 0.6 GPa. Subsequently, magmatic epidote was described from many granodioritic to tonalitic plutons, but also from monzogranite (e.g., Leterrier 1972), dikes of dacitic composition (Evans and Vance 1987), and orbicular diorite (Owen 1991, 1992). Furthermore, epidote was not only recognized in crystallizing plutons or dikes but also in high pressure migmatites and pegmatites derived from eclogites (Nicollet et al. 1979; Franz and Smelik 1995).

The role of epidote during magmatic crystallization is relatively well understood, and crystallization temperatures and sequences involving epidote in intermediate magmas (granodiorite-tonalite-trondhjemite, TTG) are experimentally determined and confirmed from natural intrusives. In contrast, little attention is directed towards the inverse process, i.e., melting of epidote bearing lithologies. Epidote is omnipresent in eclogite of intermediate temperature (Enami et al. 2004) and denominates three subfacies (i.e., epidote-blueschist, epidote-amphibolite, and epidote-eclogite facies). Indeed the epidote-amphibolite facies intersects the wet granite solidus near 0.5 GPa at 680°C, defining the pressure above which epidote may be present during melting processes. Experiments on natural compositions have confirmed that epidote is stable above the wet granite solidus in the pressure range 0.5 to 3.0 GPa (Poli and Schmidt 1995, 2004), and thus is involved in partial melting processes. Unfortunately, it is difficult to recognize the participation of epidote during partial melting in nature, as epidote is one of the first phases to "melt out." On the other hand, it is exactly the relatively narrow temperature interval of epidote + melt, which makes epidote a significant provider for $H_2O$ during fluid-absent melting (Vielzeuf and Schmidt 2001).

1529-6466/04/0056-0009$05.00

In this chapter we use the term "epidote" or "epidote minerals" in a general sense for all minerals of the epidote group including zoisite, and "epidote$_{ss}$" for the monoclinic solid solution between $Ca_2Al_3Si_3O_{12}(OH)$ and $Ca_2Al_2Fe^{3+}Si_3O_{12}$ (OH) ("ps"). Solid solutions with significantly more than one Fe per formula unit have not been reported as magmatic epidote. "Zoisite" is used only to specifically designate the orthorhombic polymorph. The review is limited to epidote with relatively low REE contents. The stability and role of allanite, a common early accessory mineral in granitoid intrusions, is discussed by Gieré and Sorensen (2004). We first review natural occurrences of magmatic epidote starting with criteria to identify a magmatic origin of epidote. Our compilation of magmatic epidote occurrences focuses on the oddities, i.e., the <5% of magmatic epidote which are not part of the widespread "epidote in TTG" (i.e., tonalite-trondhjemite-granodiorite) plutons. We then review experimentally determined phase relations of epidote minerals in coexistence with melt. This includes melting and crystallization reactions as well as the bulk composition effect on the magmatic occurrence. The factors influencing the variation in "minimum pressure" indicated by magmatic epidote in intrusions receive particular intention. Finally we investigate the role of epidote during fluid-absent melting processes.

## MAGMATIC EPIDOTE IN INTRUSIVES

Epidote crystallizes from intermediate magmas above a certain pressure (Zen and Hammarstrom 1984) that, mainly in function of bulk composition and oxygen fugacity, may vary from 0.3 to 0.7 GPa. The dependence of epidote crystallization on magma composition is somewhat masked by the very uniform chemistry of tonalite-granodiorite intrusions. These account for >90% of the magmatic epidote occurrences and result in the inappropriate impression that their 0.5 GPa minimum pressure for magmatic epidote is generally applicable, which, as will be discussed below, is not the case. The fairly late rediscovery of the possible magmatic character of epidote is mainly due to the unfounded assumption that epidote is a low temperature mineral and thus metamorphic or hydrothermal, but also it is indeed not always easy to distinguish epidote as magmatic on a textural basis. Textural criteria play a central role, as compositional criteria are not very helpful to distinguish magmatic from metamorphic epidote: the metamorphic compositional array ($ps_0$ to $ps_{100}$) encompasses the magmatic compositional array (typically $ps_{30}$ to $ps_{70}$), and the only deviations from the pseudobinary epidote$_{ss}$ chemistry, such as high concentrations of Ce and other REE, are characteristic for a magmatic origin, though they modify the pressure-temperature range of crystallization. If available, stable isotope data can help to distinguish magmatic from subsolidus epidote (see discussion in Morrison 2004).

### Textural evidence for magmatic epidote

Several features are considered to be characteristic of a magmatic origin of epidote (Zen and Hammarstrom 1984; Moench 1986; Tulloch 1986; Zen and Hammarstrom 1986; Zen 1988):

- A reliable indication is strong zonation with allanite-rich cores. However a lack of such a zonation does not exclude a magmatic origin.

- An ophitic texture is considered to be typically magmatic. In the crystallization sequence epidote seems to appear in tonalite after hornblende but before or contemporaneous with biotite.

- Magmatic epidote is sometimes embayed where in contact with the quartzofeldspatic matrix, pointing towards not being stable at the final crystallization of the magma.

- Magmatic epidote may be embedded as single euhedral crystals in a quartz-feldspar matrix which shows graphic intergrowths.

- The lack of biotite alteration to chlorite and a fresh appearance of plagioclase mostly exclude a later retrograde greenschist facies or hydrothermal overprint and therefore make it unlikely that epidote formed through a subsolidus reaction.

A few examples of such textural features are shown in Figure 1. Of course both primary magmatic and secondary metamorphic epidote might occur together and, in particular cases, an unequivocal identification can be difficult. However, for an estimate of intrusion conditions, the presence of magmatic epidote represents a powerful tool. In tonalite and granodiorite epidote is macroscopically easily identifiable - and its presence excludes a shallow intrusion level for the tonalitic or granodioritic magma.

## Natural magmatic epidote – the crowd

Since magmatic epidote has been revived, a large number of intermediate to deep seated plutons were described to contain magmatic epidote. Intrusion pressures for these plutons were derived by means of the Al-in-hornblende barometer (Hammarstrom and Zen 1986; Schmidt 1992) and by contact aureole pressures. In the context of this chapter, we focus on pressures from the magmatic systems and thus rely on the Al-in-hornblende barometer (see below), with the advantage that these are directly comparable for all TTG compositions. Such intrusion pressures yield in most cases 0.5 to 0.8 GPa, in accordance with the experimental studies discussed in the subsequent sections. Reactions and observed textures of the "common" magmatic epidote are discussed in the section "Epidote in $H_2O$ saturated magmas" below. Typical volumetrically dominant granodiorite-tonalite suites of large calc-alkaline intrusions containing epidote are described from the following regions (non-exhaustive list):

*Europe:*

- Bergell, Central Alps (Cornelius 1915; Davidson et al. 1996; Fig. 1a)
- Porhorje Mountains, Slovenia (Altherr et al. 1995)
- Retezat and Mala Fatra granitoids, Carpathians, Slovakia-Romania (Broska et al. 1997)
- Syrostan and Verkhisetsk batholiths, Ural, Russia (Bea et al. 1997; Montero et al. 2000; Popov et al. 2001)
- Velfjord, Norway (Barnes and Prestvik 2000)
- Jerissos and Verdikoussa plutons, Greece (Pepiper et al. 1993; Frei 1996)

*Africa:*

- Bas Draa, Anti-Atlas, Marocco (Mortaji et al. 2000)

*New Zealand:*

- Ridge, Table Hill, and Campsite plutons, New Zealand (Allibone and Tulloch 1997; Tulloch and Challis 2000)

*North America:*

- Alaska (>10 batholiths, Zen and Hammarstrom 1984; Cook et al. 1991; Zen 1988)
- British-Columbia (at least 5 batholiths, Hammarstrom and Zen 1986; Hollister et al. 1987; Ghent et al. 1991, Zen 1988)
- Washington, Oregon, Idaho (at least 5 batholiths, Zen and Hammarstrom 1984; Zen 1985, 1988)
- California (at least 5 batholiths, Zen and Hammarstrom 1984; Anderson et al. 1988; Keane and Morrison 1997)
- Santa Catalina batholith, Arizona (Anderson et al. 1988)
- Minto block, Canada (Bedard 2003)
- Kaipokak domain and Cape-Breton highlands, Novia Scotia and Labrador, respectively (Farrow and Barr 1992; Barr et al. 2001)
- Appalachians (several plutons, Vynhal et al. 1991)

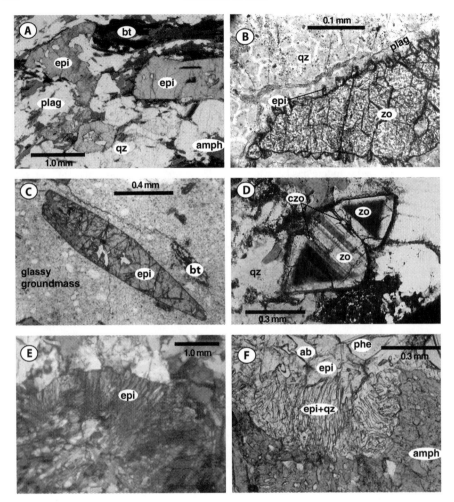

**Figure 1**. Photographs of natural magmatic epidote. (a) Tonalite of the Bergell intrusion, CH/I, where magmatic epidote was first identified by Cornelius (1915). Typical texture of magmatic epidote in the TTG series. In this case, epidote is idiomorphic where growing into amphibole and sometimes embayed where in contact with the quartzo-feldspatic matrix. (b) Magmatic zoisite with overgrowth of subsolidus epidote (small shortprismatic crystalls almost perpendicular to the zoisite grain) and with an oligoclase corona. The photograph represents a migmatite formed in eclogites from the Massif Central, France (Nicollet et al. 1979). (c) Epidote phenocryst in fine-grained matrix from a dacite dike in Boulder County, Colorado. The matrix contains quartz and plagioclase phenocrysts and partly chloritized biotite (Evans and Vance 1987). (d) Zoned zoisite from the Weissenstein pegmatite, Münchberg Massiv, Germany. This pegmatite is derived from high pressure melting of eclogite at ≈2.0 GPa (Franz and Smelik 1995). (e) Corona of fibroradial to vermicular magmatic epidote within an orbicule, from an orbicular diorite from the Greenville Front zone in Labrador (Owen 1991, 1992). (f) Vermicular epidote-quartz intergrowth from the Sanagawa-gabbro. The intergrowth occurs interstitially and consists, beside of epidote and quartz, also of albite and phengite indicating an envolved bulk composition of the last crystallizing melt. Thin section (a) and photographs (b) to (f) courtesy of C. Nicollet, B. Evans, G. Franz, J.V. Owens, and P. Ulmer, respectively.

*Abbreviations for all figures: amph: amphibole_{ss}, an: anorthite, bt: biotite_{ss}, cor: corundum, cpx: clinopyroxene_{ss}, cs: coesite, epi: epidote_{ss}, gar: garnet_{ss}, gr: grossular, hbl: hornblende, ky: kyanite, mgt: magnetite, ms: muscovite, or: orthoclase, omph: omphacite, opx: orthopyroxene_{ss}, ph: phlogopite, phe: phengite_{ss}, plag: plagioclase_{ss}, qz: quartz, tr: tremolite, v: vapor = aqueous fluid, wo: wollastonite, zo: zoisite.*

*Middle America:*

- Cuzahuico granite, Oaxacan complex, Mexico (Elias-Herrera and Ortega Gutierrez 2002)

*South America:*

- Argentina-Chile (at least 5 plutons, Saavedra et al. 1987; Cerredo and De Luchi 1998; Sial et al. 1999a; Dahlquist 2001a,b)
- Serra Negra do Norte Pluton, Borborema Province, and Rio Grande do Norte State, NE Brazil (Galindo et al. 1995; Sial et al. 1999a,b; Campos et al. 2000)

For almost all of the intrusions in these batholiths and plutons, the features described above apply and intrusions with such magmatic epidote shall not be discussed any further. The above list contains mostly intrusions from orogenic settings, pointing to the common occurrence of deep seated intrusions in such settings. In contrast, epidote-bearing intrusions in Archean shields are rare.

## Natural magmatic epidote – the odd one's

This section focuses on the out-of-the-normal natural occurrences of magmatic epidote that we divide into those, where pressures are such that magmatic epidote in TTG's should not occur, unusual bulk compositions with magmatic epidote (e.g., dacite, diorite) sometimes occurring in dikes, and epidote occurrences during migmatisation processes.

***Magmatic epidote at <0.5 GPa.*** Using the calibration of Johnson and Rutherford (1989) for the Al-in-hornblende geobarometer, Vyhnal et al. (1991) concluded a minimum pressure of around 0.28 GPa for epidote formation in epidote-bearing Appalachian monzogranites. However, the Johnson and Rutherford calibration was performed with a $CO_2$-$H_2O$-fluid that raised solidus temperatures, and should only be applied were evidence is pointing towards an elevated solidus temperature (see "Al-in-hornblende barometer" below). Recalculating the intrusion pressures with the calibration for water-saturated systems by Schmidt (1992), the data of Vynhal et al. (1991) suggest that in monzogranite magmatic epidote appears between 0.32 and 0.40 GPa, still significantly lower than in average tonalite and granodiorite. In the case of the Appalachian monzogranite, this might be attributed to the monzogranitic bulk composition but possibly also to an elevated oxygen fugacity as epidotes are very Fe rich (up to $ps_{100}$). Also the Querigut intrusion in the French Pyrenees has epidote in monzogranite but neither in granodiorite nor in tonalite (Leterrier 1972; Roberts et al. 2000). Tulloch (1986) and Tulloch and Challis (2000) discuss many New Zealand epidote-bearing plutons, part of them also monzonitic, which were emplaced between 0.31 and 0.5 GPa. Tulloch (1986) underlines the corroded appearance of epidote where it is not enclosed in biotite in many of these plutons, and points out that substantial upraise of a partially crystallized magma that already contained epidote might lead to the relatively low pressure of solidification, which might then be mistaken as apparent low pressure of epidote crystallization. In fact, as will be discussed below, a discrepancy between epidote crystallization pressure and final intrusion pressure is limited in tonalite, as epidote crystallizes relatively late, but is more probable in granodiorite, and definitively possible towards trondhjemitic compositions where epidote becomes a liquidus phase.

***Unusual bulk compositions.*** Magmatic epidote is also reported from bulk compositions outside the monzogranite to TTG range. Dacite and rhyodacite dikes with up to 71 wt% $SiO_2$ from the Front range of Colorado have idiomorphic epidote phenocrysts ($ps_{63}$), magmatic garnet and muscovite (Fig. 1c) in a very fine grained matrix (Evans and Vance 1987; Dawes and Evans 1991). The unusual phase assemblage, that also includes biotite and plagioclase, together with xenoliths in the magma, were employed to derive a magma chamber pressure of 0.72 to 1.2 GPa (Dawes and Evans 1991).

Magmatic epidote was also found in alkaline granite from the Bhela-Rajna complex, Orissa, India (Pattnaik 1996). Some of the intrusions in this complex contain riebeckite and riebeckite + epidote coexist in one rock type. It is likely that these magmas were relatively oxidized, leading to a potential expansion of the epidote-field in *P-T* space. On the $SiO_2$-poor end of the compositional spectrum, magmatic epidote is contained in an orbicular diorite and interstitially in a coarse grained gabbro. The shell structure of the orbicules in the orbicular diorite from the Grenville Front zone, Labrador (Owen 1991) is built by alternately enriched and depleted biotite, epidote, and magnetite (Fig. 1e). Idiomorphic epidote crystals are enclosed by plagioclase or hornblende, or occur in a vermicular texture. Owen (1991) suggested that the orbicules originated from supercooled water saturated globules within the dioritic magma, and that supercooling suppressed the crystallization of hornblende. The coarse grained garnet and rutile containing amphibole-plagioclase-clinopyroxene Saranga-gabbro, which is part of the lower crust of the Kohistan arc, Northern Pakistan, exhibits interstitial areas with vermicular intergrowth of quartz, albite, and epidote together with muscovite (P. Ulmer, pers. comm.; Fig. 1f). This interstitial intergrowth is interpreted as the crystallization product of the very last melt fraction of intermediate composition formed at high pressure probably under water rich conditions. However a detailed study is not available.

***Epidote in subvolcanic dikes.*** This category is represented by two occurrences. The first one consists of ~20 dacitic to rhyodacitic dikes from Colorado, as already mentioned above (Evans and Vance 1987). The truly remarkable fact is that these dikes were emplaced and quenched at no more than 0.2 GPa and 250°C (conditions derived from the country rock), and thus idiomorphic epidote ($ps_{59}$ to $ps_{72}$) was conserved in a holocrystalline, aphanytic groundmass (Dawes and Evans 1991). The dikes contain 65 to 75 vol% groundmass with phenocrysts of plagioclase, quartz, biotite, and epidote (1.5 to 1.8 vol%). This modal distribution provides evidence that at around 1.0 GPa, epidote crystallizes near the liquidus in dacitic compositions. The crystallization conditions of the magma before injection into the dikes (≥800°C, 0.72 to 1.2 GPa) were determined from 2-feldspar, garnet-plagioclase-biotite-muscovite, and garnet-hornblende equilibria, as well as from the Al-in-hornblende geobarometer (in this case, i.e., at elevated temperatures compared to the granitic solidus, the calibration of Johnson and Rutherford (1989) is appropriate). The second occurence is consituted by idiomorphic epidote ($ps_{71}$ to $ps_{81}$) in a late lamproitic dike that intruded a tonalite of the southern Adamello, N-Italy (J. Blundy, pers. comm.). The intrusion pressure of the epidote-free tonalite host is around $0.3 \pm 0.1$ GPa (by Al-in-hornblende barometry, P.Ulmer, pers. comm.) which suggests that phenocrysts of epidote have been transported from significantly larger depths.

***Epidote in high pressure migmatites and related dikes.*** Epidote-bearing partial melts from eclogites or high pressure granulites are observed in a few high pressure migmatitic terranes. These include migmatites derived from amphibole-bearing eclogites from the Rouergue complex in the French Massif Central (Nicollet et al. 1979; Fig. 1b), from the Niedewitz amphibolite massif in Poland (Puziewicz and Koepke 2001), the Eseka migmatites from Cameron (Nedelec et al. 1993), and migmatized xenoliths contained in the Cuzahuico granite in the Oaxaca terrane, Mexico (Elias-Herrera and Ortega-Gutierrez 2002). In all of these cases, epidote and/or zoisite occur in leucocratic bands, schlieren, or veins that are mostly constituted by quartz and oligoclase with some minor biotite present. These leucosomes are often trondhjemitic in composition and represent typical minimum melts from metabasalt at high pressure. Pressure estimates are 1.3 to 2.0 GPa for the granulites from the Rouergue complex, 1.3 GPa for the Niedewitz amphibolites, and 0.9 GPa for the Eseka migmatites. Furthermore, Franz and Smelik (1995) described pegmatites that are segregated leucosomes of eclogite melting in the Münchberg Massif, Germany. Franz and Smelik (1995) showed that these melts were formed during decompression from 2.5 to 1.0 GPa and have zoisite coexisting with clinozoisite (Fig. 1d) in a leuco-tonalitic bulk composition.

These high pressure melts have an enormous significance for the initial formation of melts during anatexis of mafic bulk compositions. They testify that epidote is involved in high pressure melting, and experimental stability relations in metabasalt (discussed in the last two sections of this chapter) support that epidote might indeed be critical for dehydration melting in the pressure range 1.0 to 3.0 GPa. The natural and experimental evidence on the involvement of epidote clearly demonstrates that metabasalt melting models uniquely based on amphibole are missing the critical phase responsible for the first occurrence of melt at elevated pressures.

## MELTING REACTIONS AND ZOISITE STABILITY AS DEDUCED FROM EXPERIMENTAL STUDIES IN Fe-FREE MODEL SYSTEMS

Experiments in simple systems outline the potential pressure-temperature and $f_{O_2}$ limits of magmatic zoisite and/or epidote. The maximum temperature stability of at least 1182°C occurs in the $CaO-Al_2O_3-SiO_2-H_2O$ system and a minimum melting temperature of 680°C, i.e., close to the granitic minimum in natural complex systems, is observed in the $K_2O-CaO-Al_2O_3-SiO_2-H_2O$ system. Many melting reactions can be unambiguously defined in four or five component systems (e.g., in $CaO-MgO-Al_2O_3-SiO_2-H_2O$), which then serve as model reactions for natural systems. In the latter, melting reactions are often obscured by extensive solid solutions in, e.g., amphibole, clinopyroxene, garnet, and by little mass transformation achieved with epidote involving reactions due to relatively low modal abundances of epidote in most rock types (typically <10 vol%).

### $CaO-Al_2O_3-SiO_2-H_2O$ (CASH)

The phase diagram outlining the stability of zoisite + melt (Fig. 2) in the most simple system in which epidote-group minerals appear, is well defined in the low-pressure region to 2.5 GPa (Boettcher 1970) and along the wet solidus to 7.0 GPa (Poli and Schmidt 1998). The subsolidus reaction limiting the low pressure stability of zoisite + quartz (Fig. 2a, Nitsch and Winkler 1965; Newton 1966) is

$$\text{zoisite} + \text{quartz} = \text{anorthite} + \text{grossular} + H_2O \tag{1}$$

According to Boettcher (1970), this reaction intersects the appropriate eutectic $H_2O$-saturated solidus reaction near 0.9 GPa, 775°C (Fig. 2); the eutectic melting reaction at lower pressure being

$$\text{anorthite} + \text{grossular} + \text{quartz} + H_2O = \text{melt} \tag{2}$$

and at higher pressure, i.e., at pressures above Reaction (1),

$$\text{zoisite} + \text{anorthite} + \text{quartz} + H_2O = \text{melt} \tag{3}$$

The system CASH, serving also as a model for metamorphism and melting in some calc-alkaline assemblages, apparently generates eutectic melt compositions (at $SiO_2$- and $H_2O$-saturated conditions) projecting slightly to the $SiO_2$-side of the join anorthite - zoisite (Boettcher 1970, Reaction 13c) at crustal conditions. At pressures above 1.1 GPa, anorthite + $H_2O$ are unstable and

$$\text{zoisite} + \text{kyanite} + \text{quartz/coesite} + H_2O = \text{melt} \tag{4}$$

is the eutectic melting reaction. This reaction remains valid until the maximum pressure stability of zoisite is reached at 7 GPa (Poli and Schmidt 1998) where melting occurs at 1070°C. The highest temperature for zoisite was determined to 1180°C at 2.6 GPa by Boettcher (1970), the final breakdown reaction results in anorthite + grossular + corundum + melt (Fig. 2a). Between this reaction and the intersection of the zoisite pressure stability with the wet solidus

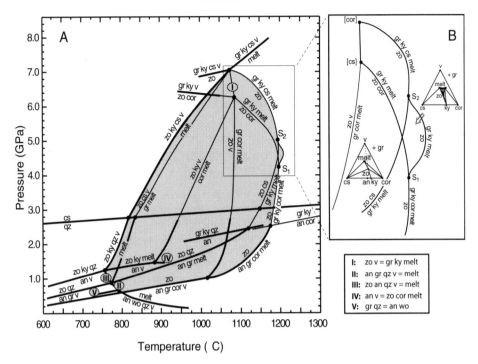

**Figure 2.** (a) Zoisite + melt stability field (grey) in CASH; thick lines are experimentally determined, others result from Schreinemakers' analysis and are uncertain in temperature. (b) Detail of the singular points in (a) where both corundum-saturated and coesite-saturated melts become possible. Abbreviations see Figure 1.

in CASH at 7 GPa, the limit of magmatic zoisite was drawn fairly conservative in Figure 2, as this high pressure - high temperature portion remains experimentally undetermined. Reactions delimiting the magmatic zoisite region (thin lines) under these conditions are derived from Schreinemakers analysis.

The occurrence of the $H_2O$-conserving fluid-absent melting reactions

$$\text{zoisite} = \text{grossular} + \text{kyanite} + \text{corundum} + \text{melt} \tag{5}$$

at pressures above the equilibrium anorthite + corundum = grossular + kyanite (see Fig. 2), and

$$\text{zoisite} = \text{grossular} + \text{kyanite} + \text{coesite} + \text{melt} \tag{6}$$

at ca. 7 GPa implies that the terminal zoisite breakdown with temperature produces $Al_2O_3$-saturated melts at low pressures but also $SiO_2$-saturated melts at higher pressures. This can be easily explained assuming that singularities occur along the terminal zoisite melting reactions (Fig. 2b). The location of such singular points is unknown and the width of the two phase zoisite + melt field defined by Schreinemakers' rules (chemography on the right in Fig. 2b) is purely speculative. Nevertheless it is appealing to formulate a degenerate reaction as the extreme temperature limit for zoisite:

$$\text{zoisite} = \text{grossular} + \text{kyanite} + \text{melt} \tag{7}$$

## K₂O-CaO-Al₂O₃-SiO₂-H₂O (KCASH)

In the KCASH-system the relatively high temperature for minimum melting in CASH is reduced by about 100°C and consequently, the temperature interval of zoisite + melt is shifted to lower temperature (Fig. 3). In KCASH, the eutectic melting reaction at pressures below the zoisite + muscovite + quartz stability is (Johannes 1980; Schliestedt and Johannes 1984)

$$\text{anorthite + muscovite + orthoclase + quartz + H}_2\text{O = melt} \tag{8}$$

and above 0.8 GPa at 680°C it is

$$\text{zoisite + muscovite + orthoclase + quartz + H}_2\text{O = melt} \tag{9}$$

The latter reaction is studied experimentally to 2 GPa, but phase relations at higher pressure are unknown in KCASH. Reaction (9) represents an eutectic minimum melting reaction close to that in natural systems. Additional components such as Na₂O causes involvement of plagioclase on the left hand side of Reaction (9), and, together with Fe- incorporation in zoisite and muscovite, slightly lowers the melting temperature.

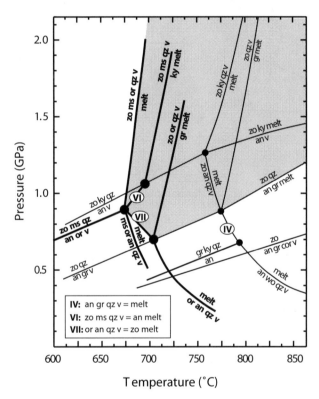

**Figure 3.** Zoisite-melting reactions in CASH and KCASH for pressures to 2 GPa; bold lines from Schliestedt and Johannes (1984) with some reactions added through Schreinemakers' analysis. The grey area denotes the field of possible coexistence of zoisite + quartz + melt. Abbreviations see Figure 1.

## CaO-MgO-Al₂O₃-SiO₂-H₂O (CMASH)

Addition of MgO to CASH provides a model system appropriate to the melting of mafic bulk compositions (e.g., Ellis and Thompson 1986; Thompson and Ellis 1994). Eutectic melting in this system (Fig. 4) involves zoisite only at pressures above the breakdown of anorthite + H₂O to zoisite + kyanite + quartz. From the intersection of this latter reaction with

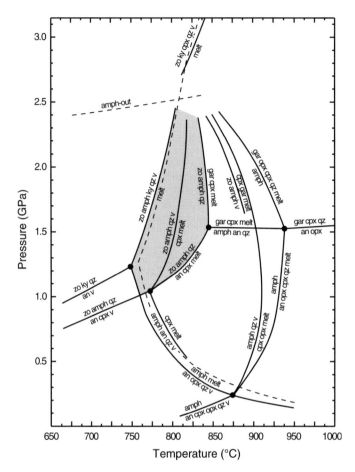

**Figure 4.** Amphibole + zoisite + quartz reactions (grey area) and amphibole stability in CMASH (after Thompson and Ellis 1994 and Quirion and Jenkins 1998). Stippled line: CASH solidus for comparison. Note that a large number of reactions focus into an area near 2.5 GPa, 800°C and at present, the experimental data are insufficient to select between a large number of possible topologies (e.g., Thompson and Ellis 1994). Nevertheless, zoisite exceeds amphibole in pressure stability as depicted by the eutectic solidus reaction at >2.8 GPa. Abbreviations see Figure 1.

the solidus to the pressure stability limit of amphibole, the eutectic melting reaction is then

$$\text{zoisite + amphibole + kyanite + quartz + H}_2\text{O = melt} \tag{10}$$

In bulk compositions representative of mafic rocks, the limiting assemblage for zoisite stability is zoisite + amphibole + quartz. Fluid-saturated melting in such compositions takes place at 770 to 820°C when the reaction

$$\text{zoisite + amphibole + quartz = anorthite + clinopyroxene + H}_2\text{O} \tag{11}$$

(Quirion and Jenkins 1998) encounters the wet solidus at 1.0 GPa, 780°C (Fig. 4) and transforms into the peritectic melting reaction

$$\text{zoisite + amphibole + quartz + H}_2\text{O = clinopyroxene + melt} \tag{12}$$

valid at moderate pressures of 1.0 to 1.5 GPa. The upper temperature stability and the fluid-absent solidus in systems which have amphibole + quartz in excess relative to zoisite and zoisite-excess relative to fluid is then defined by

$$\text{zoisite + amphibole + quartz = anorthite + clinopyroxene + melt} \qquad (13)$$

which transforms into

$$\text{zoisite + amphibole + quartz = garnet + clinopyroxene + melt} \qquad (14)$$

above 1.5 GPa. In quartz + kyanite saturated mafic model compositions, fluid saturated eutectic melting through Reaction (10) takes place about 40°C lower, but the minimum pressure for zoisite + melt in such systems is raised by 0.3 GPa to about 1.2 GPa (see Fig. 4).

The upper pressure limit of amphibole + zoisite + quartz is delimited through amphibole breakdown. However, reactions in the region around 2.5 GPa, 800°C are extremely complicate and at present it is not possible to decide which of the many possible topologies is the stable one as within a narrow region of 0.3 GPa and 100°C at least 10 phases occur in this five component system. A number of possible solutions were presented by Thompson and Ellis (1994) who also point out, that further complexity arises from talc that becomes stable instead of orthopyroxene + quartz in the region of interest. Nevertheless, from the experiments in CMASH (Thompson and Ellis 1994) it is certain that the zoisite pressure stability exceeds amphibole stability and that the melting reaction above 2.5 GPa will contain zoisite but not amphibole (or talc) as a hydrous phase. As discussed in Poli and Schmidt (2004), this is confirmed by experiments in more complex systems (KCMASH, Hermann 2002), natural systems (Poli and Schmidt 1997; Schmidt and Poli 1998), and from natural eclogites (Enami et al. 2004).

### $K_2O$-CaO-MgO-$Al_2O_3$-$SiO_2$-$H_2O$ (KCMASH)

Reactions in KCMASH were studied by Hoschek (1990) at 1.0 to 2.1 GPa and by Hermann (2002) at 2.0 to 4.5 GPa. With respect to KCASH, minimum melting temperatures remain almost identical (Fig. 5) at pressures up to 2.5 GPa but melting reactions now involve tremolite$_{ss}$ and phlogopite$_{ss}$. In this system, the orthoclase absent melting reaction between 1.4 and 2.1 GPa (Fig. 5) produces zoisite (Hoschek 1990), whereas in all of the previous systems, zoisite is consumed by melting reactions. Hermann (2002) found zoisite + amphibole + quartz to coexist with melt just above the solidus to 2.5 GPa in a KCMASH system containing traces of Fe and Ti and in part being enriched in REE. In this particular bulk composition, zoisite (containing minor REE-contents) is delimited to about 800°C, which is 20 to 50°C above the amphibole melting temperature (at pressures of 2.0 to 2.5 GPa). However, the interpretation of phase relations in this latter study is complicated by the possible occurrence of non-quenchable K-rich melts and by the persistence of allanite + melt to pressures of 3 GPa and temperatures to 850°C.

## EPIDOTE IN FLUID-SATURATED MAGMAS

In this contribution we deal only with $H_2O$ fluids, as the amount of $CO_2$ soluble in silicic magmas at pressures typical for crustal settings is negligible. As a consequence a "fluid" is always intended here as an aqueous fluid with some silicates dissolved in it.

### The wet solidus of tonalite and granodiorite

The wet (saturated in an aqueous fluid) solidus of tonalite was determined experimentally by Piwinskii (1968) at 0.1 to 0.3 GPa, by Lambert and Wyllie (1974) at 1.0 to 3.0 GPa, and slightly modified by Schmidt (1993). The wet liquidus of tonalite (Fig. 6) was determined by Eggler (1972) at 0.05 to 0.63 GPa, and by Allen and Boettcher (1983) at 1.0 to 2.5 GPa.

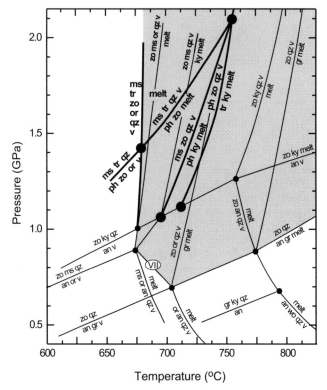

**Figure 5**. Zoisite-melting reactions from Hoschek (1990) in KCMASH for pressures to 2 GPa; thin lines as in Figure 2, grey area denotes possible coexistence of zo + qz + melt. Abbreviations see Figure 1, for reaction VII see Fig. 3.

Most of the reported experiments on tonalitic compositions are generally consistent, however, because similar but not identical bulk compositions were investigated, small differences are to be expected. In addition, many of the earlier experiments have problems with Fe-loss to the capsule material and uncontrolled $f_{O_2}$ (see discussions in Lambert and Wyllie 1974 and Allen and Boettcher 1983), and this resulted in large uncertainties on the phase relations of Fe-Mg phases.

Experiments on the granodiorite solidus are limited to 1.0 GPa. Piwinskii (1968, 1973) determined the solidus of natural amphibole bearing and amphibole free granodiorites at 0.1 to 1.0 GPa to be about 20°C lower than for a tonalite. Naney (1983) found a solidus at 0.2 and 0.8 GPa again 20°C lower for a synthetic granodiorite.

### Epidote-out in tonalite with increasing pressure

The wet solidus for tonalite intersects the subsolidus epidote dehydration reaction near 0.5 GPa, 660°C. Mass balance calculations (Schmidt 1993), using mineral compositions obtained in the close vicinity of the fluid-saturated solidus, suggest that the subsolidus epidote dehydration reaction appropriate to tonalite bulk compositions is

epidote + biotite + quartz =
$$\text{hornblende + K-feldspar + plagioclase + magnetite} + H_2O \qquad (15)$$

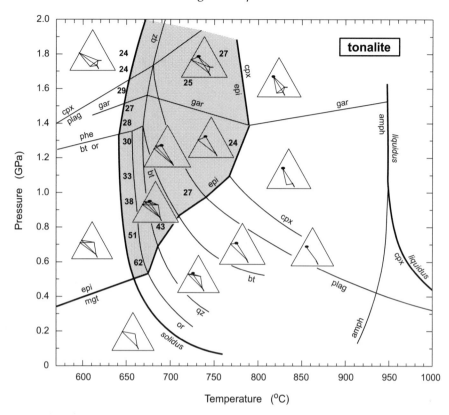

**Figure 6**. Pressure-temperature diagram for the magmatic domain of epidote in tonalite (grey field) to 1.9 GPa. Triangles are ACF-deluxe projections after Thompson (1982, see Fig. 7) and give tielines for coexisting phases in the different stability fields (projected from quartz). Numbers are experimental epidote compositions in terms of Fe pfu; after Schmidt and Thompson (1996). Abbreviations see Figure 1.

the left hand side corresponding to the higher pressure and lower temperature side (Fig. 6). Figure 6 gives *P-T* conditions of the delimiting reactions in the tonalite-$H_2O$ system (after Schmidt and Thompson 1996) and draws assemblages for each field in the ACF-deluxe projection of Thompson (1982, Fig. 7). As concluded from experiments (Schmidt 1993) and as indicated by field evidence (Hammarstrom and Zen 1992; see also Ishihara 1981), magnetite is significantly more abundant in epidote-free than in epidote-bearing granitoids. This was also reported by Drinkwater et al. (1991) for the Coast Plutonic-Metamorphic Complex near Juneau, Alaska, and indicates that at low pressure, magnetite is the principal phase containing $Fe^{3+}$ at temperatures above epidote stability. Between 0.5 to 0.9 GPa, melting reactions involving similar phases as in Reaction (15) occur at temperatures above the fluid-saturated tonalite solidus and Schmidt and Thompson (1996) suggested that, in this pressure range, the reaction delimiting epidote stability in tonalite melts with excess $H_2O$ is

epidote + biotite + (quartz$_{melt}$) + $H_2O$ =
$\qquad$ hornblende + (K-feldspar$_{melt}$) + plagioclase + magnetite + melt $\qquad$ (16)

Quartzofeldspathic components are written in parentheses because at fluid-saturated conditions these minerals sequentially melt at temperatures not much higher than the fluid-

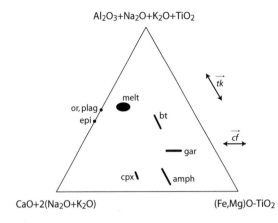

**Figure 7**. ACF-deluxe diagram after Thomson (1982, see Fig. 6). *tk*: *tschermak*-exchange vector $(Mg_{-1}Si_{-1}Al^{VI}Al^{IV})$, *cf.* $Ca_{-1}Fe$, other abbreviations see Figure 1.

saturated solidus. More difficult to ascertain is the nature of the multisystem epidote-melting reactions between biotite-out (ca. 0.9 GPa, 720°C), plagioclase-out (ca. 1.0 GPa, 740°C), and clinopyroxene-in (ca. 1.1 GPa, 770°C). In fact discontinuous Reaction (16) removes biotite from the tonalite bulk composition such that epidote would be stabilized in more mafic (quartz-diorite to gabbro) compositions. The clinopyroxene-in reaction in tonalite-$H_2O$ melts intersects the epidote-melting reaction near 1.1 GPa, 770°C (Fig. 6). Melting of epidote at pressures above the clinopyroxene- in reaction is directly related to the appearance of clinopyroxene. With increasing temperature, modal increase in clinopyroxene is directly proportional to epidote decrease. The appropriate multisystem discontinuous reaction can be considered as

$$\text{epidote} + \text{hornblende} + H_2O = \text{clinopyroxene} + \text{melt} \qquad (17)$$

This reaction is analogous to the peritectic melting Reaction (12) in the model system CMASH (Thompson and Ellis 1994; Fig. 5), only that in wet tonalite quartz disappears at about 30°C above the solidus. At pressures along the epidote melting curve for $H_2O$-saturated tonalite, amphibole has changed from the high temperature side (Rxn. 16) to the low temperature side (Rxn. 17) when clinopyroxene appears.

In water-saturated tonalite, garnet is formed at pressures ranging from 1.4 to 1.6 GPa through a series of different reactions from 600 to 950°C. The maximum temperature stability of 790°C of magmatic epidote in tonalite is reached near 1.4 GPa, i.e., at the intersection with the mostly pressure dependent garnet-in reaction. This maximum temperature results from the change in Clapeyron slopes (dP/dT) of the epidote-melting reactions from positive at low pressure, to negative at higher pressure when garnet + clinopyroxene become a product of epidote-melting (Fig. 6) through the reaction

$$\text{epidote} + \text{amphibole} + H_2O = \text{garnet} + \text{clinopyroxene} + \text{melt} \qquad (18)$$

that is analogous to Reaction (13) in the simple CMASH system. The magmatic domain of epidote in tonalite ends, when the epidote delimiting reaction at high pressure intersects the solidus at 3.2 GPa. This is about 0.7 GPa higher than the amphibole breakdown reaction (Poli and Schmidt 1995). The implications of the relative pressure stabilities of epidote and amphibole are discussed in the section on melting of zoisite bearing eclogites.

***Interpreting textural features of epidote-bearing tonalites.*** The reactions derived from experimental results can be applied to the following petrographic observation. The epidote-melting Reaction (16) in the range 0.5 to 0.9 GPa for tonalite-$H_2O$ ($f_{O_2}$ = NNO) may also be considered, in reverse, as the crystallization of epidote + biotite contemporaneous with peritectic resorption of some hornblende by the melt. Cornelius (1915) and Hammarstrom and Zen (1986) observed such crystallization textures in natural tonalite. At pressures below 0.9 GPa, biotite begins to crystallize before epidote (Fig. 6). Hornblende is consumed by reaction with the melt (Rxn. 16), and its modal decrease should be mirrored by proportional increase in modes of epidote and biotite. Consequently, the local resorption of hornblende through the appearance of epidote and growth of biotite does not imply a general destabilization of hornblende in the tonalite melt (Fig. 6), but a decrease in modal abundance.

### Epidote-out in granodiorite and granite

Experiments by Naney (1983) first demonstrated that epidote is stable above the solidus in granite and granodiorite (Fig. 8). At 0.8 GPa, synthetic granodiorite was found to crystallize epidote up to 700°C under fluid-saturated conditions (>12 wt% $H_2O$). Synthetic granite has epidote present to about 610°C, i.e., 20°C above its solidus. At $H_2O$-undersaturated conditions, the epidote-out reaction was lowered by approximately 20°C. In experiments at 0.2 GPa, epidote did not occur (Fig. 8). Schmidt and Thompson (1996) also performed some experiments on a granodiorite composition. In the range from 0.7 to 1.0 GPa the assemblage hornblende + plagioclase + biotite + melt is the same in granodiorite and tonalite, and the locations of the epidote melting reactions are also almost identical. In this pressure range, a reaction similar to that in tonalite (Rxn. 16) is probable for the granodiorite. At 1.0 to 1.5 GPa, the stability limit

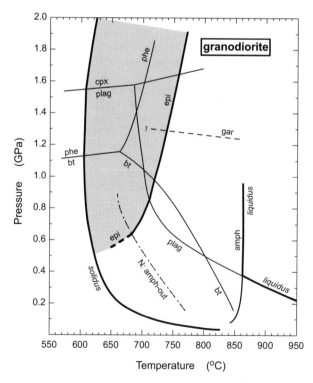

**Figure 8**. Pressure-temperature diagram for the magmatic domain of epidote in granodiorite (grey field) to 1.8 GPa. Compiled from experiments of Naney (1983) and Schmidt and Thompson (1996). Note that in the granodiorite of Naney (N), amphibole reacts out before the magma reaches the solidus. Abbreviations see Figure 1.

of epidote is about 30°C lower in granodiorite than in tonalite. However, the experimental data are not sufficient to constrain the epidote melting reactions in granodiorite at high pressures under $H_2O$-saturated conditions. Differently to tonalite, clinopyroxene does not appear to 1.8 GPa, but garnet appears at and above 1.3 GPa. It is above 1.8 GPa, that granodiorite again contains the same assemblage as tonalite, i.e., hornblende + garnet + clinopyroxene ± epidote + melt. Thus it is expected that around 1.8 GPa the epidote-melting reaction in the granodiorite will also change to a negative slope in *P-T* space. At low temperature, the experiments of Naney (1983) on a synthetic granodiorite composition indicate that hornblende is not stable below approximately 680°C (stippled line marked "N" in Fig. 8). This would mean that at low temperature there are fundamental differences in the fluid-saturated melting reactions in granodiorite and tonalite (at least for Naney's synthetic granodiorite). In tonalite, amphibole forms part of the subsolidus assemblage, whereas in Naney's granodiorite, amphibole is only generated above the solidus (both at 0.2 and 0.8 GPa) by reactions involving biotite + feldspar + quartz. The experiments by Piwinskii (1968) and Schmidt and Thompson (1996), however, showed amphibole to be stable down to subsolidus temperatures in several granodiorite compositions, pointing towards a strong compositional dependence of amphibole-occurrence when moving towards more granitic compositions. Because of the paucity of experiments near the fluid-saturated granodiorite solidus between 0.2 and 0.8 GPa, it is not yet possible to identify the precise epidote melting reaction in granodiorite in this pressure range.

### Epidote-out in trondhjemite

Johnston and Wyllie (1988) investigated experimentally an Archean trondhjemite at 1.5 GPa. Under water saturated conditions, epidote + biotite were observed to be the liquidus phases (approximately 740°C) and amphibole-only crystallized in a narrow field 100 to 130°C below the liquidus. At moderately water undersaturated conditions (>9 wt% $H_2O$), epidote remained present up to 775°C. Van der Laan and Wyllie (1992) studied the same trondhjemite at 1.0 GPa, the experiments yielded epidote to 700°C at fluid-saturated conditions (>10 wt% $H_2O$, solidus at 675°C). Both studies presented phase relationships in isobaric sections with varying temperature and water content. No particular emphasis was placed on epidote stability in these calc-alkaline magmatic rock compositions, and oxygen fugacity was left to what the experimental apparatus might or might not impose. The results of these two studies can be assembled into a coherent *P-T* diagram (Fig. 9) outlining the stability of magmatic epidote in trondhjemitic composition from about 0.9 to 1.6 GPa. Interestingly, the results of Naney (1983) at 0.8 GPa on a synthetic granodioritic composition are also coherent with the above results of Wyllie and coworkers. The crystallization temperatures of epidote in trondhjemite and granodiorite magmas are almost identical and in the granodiorite used by Naney (1983), amphibole dissolves through a peritectic reaction before the magma reaches its solidus temperature. In the absence of amphibole, the most likely reaction for the melting out of epidote in granodiorite and trondhjemite is the amphibole-absent continuous reaction

$$\text{biotite + epidote} + (\text{quartz}_{melt}) + Al_2Mg_{-1}Si_{-1} \text{ [bt]} =$$
$$\text{plagioclase} + (\text{K-feldspar}_{melt}) + H_2O \qquad (20)$$

which forms plagioclase from epidote by consuming $Al_2Mg_{-1}Si_{-1}$ (*tschermak*-exchange) in biotite. This reaction also acts in the subsolidus region, where quartz + feldspars + epidote buffer *tschermak*-exchange in biotite. In any particular multisystem (e.g., KCMASH as in Rxn. 16), this amphibole-absent continuous Reaction (20) will occur at higher pressure and lower temperature than the discontinuous Reaction (16) in tonalite. Field evidence indicates that below 0.7 GPa epidote is less stable in trondhjemite than in granodiorite than in tonalite because in many intrusions epidote bearing tonalite is found adjacent to epidote free granodiorite (e.g., Bergell, Moticska 1970) or epidote bearing tonalite and granodiorite adjacent to epidote free trondhjemites (Archean Minto Block, Canada, Bedard 2003).

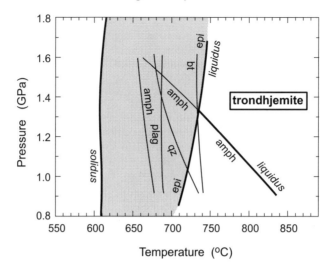

**Figure 9**. Pressure-temperature diagram for the magmatic domain of epidote in trondhjemite between 0.9 and 1.6 GPa (grey field). Compiled from experiments of Johnston and Wyllie (1988) and Van der Laan and Wyllie (1992). Note that epidote is the liquidus phase above 1.3 GPa. The experiments on granodiorite at 0.8 GPa of (Naney 1983, Fig.8) would also be compatible with the experiments at 1.0 GPa presented here. Abbreviations see Figure 1.

### Epidote-out in dioritic and gabbroic compositions

There is no apparent compositional reason, why epidote should be less stable in diorite or basalt than in tonalite. Indeed, subsolidus experiments (Poli 1993) show a slightly lower pressure in MORB than in tonalite necessary for epidote stability in the subsolidus. Nevertheless, to our knowledge, there is only one diorite (Owen 1991) described, that contains magmatic epidote, in addition, this diorite is orbicular and very heterogeneous. There is also the Saranga-gabbro with magmatic epidote (from the Kohistan arc, see above; Ulmer pers. comm.), but epidote only crystallized in a very late stage from a quartz + albite + muscovite saturated interstitial melt which had an intermediate or even granodioritic bulk composition. Apart from these two exceptions, to our knowledge, epidote is not reported from other dioritic or gabbroic intrusions. However, this is likely to be a mere consequence of the high temperature of solidification of such intrusions (typically 800°C), which exceeds the stability field of epidote, and inadequate *P-T* trajectories during the crystallization history. Only a few experiments provide constraints on the magmatic stability field for epidote in mafic rocks. Some experiments on metabasalt yielded zoisite + melt + amphibole + garnet + clinopyroxene at 2.0 to 2.5 GPa, 700 to 750°C (Pawley and Holloway 1993; Poli and Schmidt 1995). These authors did not report supersolidus experiments at lower pressure. In a gabbroic composition, zoisite was found to coexist with melt in five experiments in the range 1.5 to 2.0 GPa, 685 to 775°C (Lambert and Wyllie 1972).

### Epidote-out in granites and granitic dikes

With decreasing CaO-content, normative anorthite component, and anorthite activity in plagioclase and melt, the epidote stability field retracts to higher pressure. The experiments of Naney (1983) show an extremely narrow (20°C) epidote + melt field in granite at 0.8 GPa, which evidently represents the minimum pressure for magmatic epidote in granites. Epidote does occur in some high pressure granitic dikes but, to our knowledge, has not been reported in massive granites.

### The temperature stability of epidote: the role of anorthite component in magmas

The epidote-melting reactions are located at similar *P-T* conditions in $H_2O$-saturated experiments on tonalite, granodiorite, and trondhjemite (Fig. 10). Nevertheless, a systematic variation of epidote-stability with bulk normative anorthite for the investigated tonalite ($an^{CIPW}$ = 27 wt%), granodiorites ($an^{CIPW}$ = 22 to 18 wt%), trondhjemite ($an^{CIPW}$ = 14 wt%) and granite ($an^{CIPW}$ = 7 wt%) is evident (Fig. 10). On the very anorthite-normative end is an anorthosite ($an^{CIPW}$ = 41 wt%), which was investigated by Selbekk and Skjerlie (2002). This composition has plagioclase with $an_{70}$ to $an_{84}$ and crystallized zoisite with 0.13 Fe pfu. The high normative anorthite contents, anorthite-rich plagioclase compositions, and Fe-poor zoisite cause an extension of the magmatic epidote$_{ss}$ stability field to extreme temperatures (but still far below temperatures in the CASH system), which cannot be expected in most crustal rock compositions and in particular not in the TTG series.

The correlation of epidote stability with normative anorthite content may be understood if we accept that normative anorthite contents grossly correlates with anorthite activity in plagioclase and melt. Considering subsolidus and melting reactions such as Reactions (12) and (13), decreasing anorthite activities in plagioclase (and melt) shift these reactions to lower temperature and thus cause the magmatic epidote field to shrink with decreasing

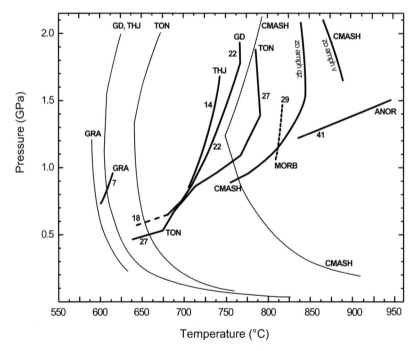

**Figure 10.** Solidi (thin lines) and reactions delimiting magmatic epidote stability (bold lines) at $H_2O$-saturated conditions, except for MORB. There is a clear correlation between the temperature of the epidote-out melting reaction and the normative anorthite content of the bulk composition (numbers are CIPW-normative anorthite contents in wt%). Note that from 1.0 to 1.8 GPa a different granodiorite was used than below 1 GPa. The solidus for trondhjemite is within error identical to that of granodiorite. ANOR: anorthosite, GD: granodiorite, GRA: granite, MORB: mid-ocean ridge basalt, THJ: trondhjemite, TON: tonalite, CMASH: synthetic system $CaO-MgO-Al_2O_3-SiO_2-H_2O$. Epidote$_{ss}$ is stable towards the low temperature side of each bold line. For details see Figures 4, 6, 8, and 9.

anorthite activity. However, this qualitative approach has to be cautioned: in a natural granite-trondhjemite-granodiorite-tonalite-diorite-gabbro series, not only normative anorthite content decreases with $SiO_2$-content but also a number of other compositional parameters vary systematically (which for example lead to a decrease of hornblende content). Thus, synthetic systems must not necessarily comply to the normative anorthite content scheme. The synthetic greywacke used by Singh and Johannes (1996b) with $an^{CIPW} = 23$ wt% has epidote slightly more stable than the tonalite with $an^{CIPW} = 27$ wt%. Whether this is because of variation of compositional parameters other than normative anorthite, or to a slight expansion of the epidote field due to $H_2O$-undersaturated conditions, or due to an unbuffered and possibly higher oxygen fugacity than NNO in the experiments by Singh and Johannes (1996b), cannot be decided without additional experiments.

Below 0.8 GPa, the nature of the epidote-out reaction will depend on the stability of amphibole in the various rock compositions. In trondhjemite (Johnston and Wyllie 1988; van der Laan and Wyllie 1992) and in granodiorite of Naney (1983) ($an^{CIPW} = 18$ wt%), amphibole is not stable at the $H_2O$-saturated solidus. In contrast, in Piwinskii's (1968) and Schmidt and Thompson's (1996) granodiorite ($an^{CIPW} = 16$ wt% and 22 wt%, respectively) amphibole appears at the fluid-saturated solidus. In the presence of amphibole, the epidote-out reactions in granodiorite will be similar to that discussed above for tonalite.

### How likely is epidote fractionation?

Modal abundances of epidote in tonalite are generally low near the epidote-out reactions (1 to 5 vol%, Schmidt and Thompson 1996), but increase to 11 to 14 vol% near the solidus (Fig. 12 in Schmidt 1993) The amount of melt is in the order of 60 to 70 vol% when epidote starts to crystallize. Thus, fractionation of epidote in tonalite would be possible at water saturated conditions, however, epidote contents are low and a significant effect of epidote fractionation on bulk composition appears to be unlikely. By contrast, fractionation of LREE enriched epidote or allanite may modify the trace element patterns in specific cases. The possibility of REE-fractionation through fractionation of early crystallizing epidote was discussed by Johnston and Wyllie (1988). In trondhjemite, the amphibole crystallization field is retracted with respect to tonalite and granodiorite and epidote becomes the liquidus phase at 1.3 GPa, thus improving the chances of epidote to be fractionated.

## THE ROLE OF OXYGEN FUGACITY AND $Fe^{3+}$ IN CONTROLLING EPIDOTE IN MAGMATIC SYSTEMS

### Oxygen fugacity in granitoid bodies

Intermediate (calc-alkaline) granitoid magmas mostly crystallize quartz, magnetite, and titanite. Quartz + magnetite define the lower limit of $f_{O_2}$ close to the quartz-fayalite-magnetite (QFM) buffer. An upper limit of oxidation is provided by the presence of ilmenite in many intermediate granitoids (Ishihara 1977, 1981; Murata and Itaya 1987; Hammarstrom and Zen 1992), which defines an $f_{O_2}$ below the magnetite-ilmenite-rutile buffer (MIR). Wones (1989) showed that the assemblage quartz + magnetite + titanite requires $f_{O_2}$ slightly above QFM. Thus, for most granitoid intrusions, the intrinsic oxidation conditions are experimentally reproduced by the NNO-buffer and the results from these experiments are therefore directly applicable to most natural tonalite intrusions. However, more oxidizing conditions might occur in some plutons (documented by Fe-Ti-oxides, or in extreme cases by the presence of hematite, e.g., Ishihara 1981). While $f_{O_2}$ conditions are not likely to fluctuate extremely during magma crystallization, the $f_{O_2}$ may step from buffer to buffer as particular $Fe^{3+}$-bearing minerals are replaced by others, e.g., magnetite by epidote (Rxn. 15 and references above), or in some extreme cases by acmite or riebeckite (magmatic riebeckite + epidote is reported from

Pattnaik 1996). Alternatively, $f_{O_2}$ could change when the fluid composition changes drastically (physically by phase separation, or chemically through C-O-H-S reactions) during late stage crystallization. This latter effect is most likely to happen at the borders of intrusions where interaction with the surrounding country rock is facilitated.

## Oxygen fugacity, Fe oxidation state and the stability field of magmatic epidote

Experiments on the $CaO-FeO-Al_2O_3-SiO_2-H_2O-O_2$ system at subsolidus temperatures have been performed by Holdaway (1972) and by Liou (1973) at 0.2 to 0.5 GPa and are discussed in Poli and Schmidt (2004). For our purpose here, it should be noted that in both studies the thermal stability of epidote increases with increasing $f_{O_2}$. Furthermore, not only the *P-T* location but also the stoichiometry of epidote subsolidus dehydration reactions (e.g., Rxns. 1,11,15,20), as well as epidote composition, strongly depend on $f_{O_2}$. As minimum pressure of magmatic epidote is defined by the intersection of the latter reactions with the solidus, a shift of these reactions in *P-T*-space is of direct significance for the interpretation of natural phase relations. The breakdown reaction of epidote as determined by Holdaway (1972) and Liou (1973) for $f_{O_2}$ buffered to NNO would intersect the granitic solidus at around 0.5 GPa (Fig. 11), which would then be the lower pressure limit of magmatic epidote; the breakdown reaction of epidote for $f_{O_2}$ buffered to HM is shifted to higher temperature by about 100°C and would thus intersect the wet granite solidus at around 0.3 GPa. This pressure difference indicates the amplitude of the effect of oxidizing a magma on the minimum pressure necessary for magmatic epidote. Schmidt and Thompson (1996) found for natural tonalite at a temperature above the water-saturated solidus that the stability and composition of epidote strongly depends on $f_{O_2}$. They determined the epidote-out reaction to intersect with the wet tonalite solidus at 0.3 instead of 0.5 GPa (Fig. 11) when $f_{O_2}$ changes from NNO to HM. Both epidote melting reactions in the tonalite at NNO and at HM, with excess $H_2O$, lie close to extrapolated epidote subsolidus dehydration reactions in the synthetic CFASH-system at equivalent $f_{O_2}$ (as determined by Holdaway 1972 and Liou 1973). The experimental $f_{O_2}$-range corresponds to the range expected for somewhat oxidized calc-alkaline magmas (Ishihara 1981). The inverse effect in anomalously reduced magmas could explain a possible absence of epidote in intrusions that otherwise should contain magmatic epidote.

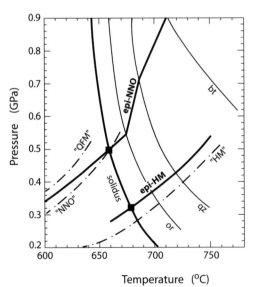

**Figure 11.** Pressure temperature diagram depicting the epidote-out reaction near the lower intersection with the solidus for varying oxygen fugacities (QFM = quartz-fayalite-magnetite, NNO = nickel-bunsenite, HM = hematite-magnetite). Solid lines are from Schmidt and Thompson (1996), stippled lines represent subsolidus epidote-out reactions from Liou (1973) and Holdaway (1972) for a synthetic $CaO-FeO-Al_2O_3-SiO_2-H_2O-O_2$ system.

More oxidizing conditions favor a higher temperature of epidote and thus an intersection of the epidote stability with the wet solidus at lower pressure. In the presence of $Fe^{3+}$, Reaction (1) is displaced to lower pressure because epidote has a higher $Fe^{3+}/Al$ ratio compared to garnet. This effect was anticipated by Strens (1965) and demonstrated experimentally by Holdaway (1972), Liou (1973), and Brunsmann et al. (2002). The same effect of an increasing $Fe^{3+}$-fractionation is expected for epidote-melt equilibria where $Fe^{3+}$ strongly partitions into epidote and thus enlarges its stability. Thus while epidote in many intermediate calc-alkaline plutons is a pressure indicator, its exact pressure significance cannot be assessed without determining the $f_{O_2}$ of the crystallizing magma (which in most intrusions is possible through oxide mineral equilibria).

Singh and Johannes (1996a,b) investigated fluid-absent melting of biotite-plagioclase-quartz starting mixtures. One mixture had synthetic Fe-free phlogopite, while the others had Fe present; all mixtures employed the same plagioclase ($An_{45}$) such that normative anorthite contents are identical ($an^{CIPW} = 23$ wt%). This study isolates the effect of Fe on the magmatic occurrence of epidote: in the Fe-free system, epidote occurs between 1.0 and 1.2 GPa at 750°C, whereas the minimum pressure for magmatic epidote is lowered to between 0.8 and 0.9 GPa at 750°C in the Fe-bearing starting materials.

## EPIDOTE PHASE RELATIONSHIPS AS A TOOL FOR EXTRACTING THE INTRUSION DEPTH

Crawford and Hollister (1982) predicted magmatic epidote to occur at a minimum pressure of approximately 0.6 GPa on the basis of the intersection of Liou's (1973) low pressure subsolidus curve for epidote stability (Fig. 11) with the melting curve for $H_2O$-saturated granite. Zen and Hammarstrom (1984) established that epidote appears in moderate to high pressure intrusions and estimated a minimum pressure for the crystallization of magmatic epidote between 0.6 and 0.8 GPa on the basis of Naney's (1983) experiments. The low pressure limit of 0.6 to 0.8 GPa for magmatic epidote was questioned by Moench (1986), who described several epidote bearing intrusions where pressure estimates from the contact aureoles yielded around 0.4 GPa.

*The Al-in-hornblende geobarometcr.* This geobarometer is based on the total Al-content in magmatic hornblende buffered by the 8-phase assemblage plagioclase + K-feldspar + quartz + biotite + epidote or magnetite + rutile + ilmenite + melt. If all of these phases are present, the 9 component system is fully buffered, and all exchange vectors in hornblende are buffered (in particular *tschermak*, plagioclase, and *edenite* exchange), the Al-content should then only depend on pressure (as the wet solidus is almost constant in temperature). This barometer has been invented and calibrated on the basis of pressures from the contact aureole of intrusions by Hammarstrom and Zen (1986), improved by including additional intrusions by Hollister et al. (1987), and experimentally calibrated for a mixed-volatile, "high"-temperature (i.e., 100-150°C above the wet solidus) situation by Johnson and Rutherford (1989) and for a fluid-saturated situation by Schmidt (1992).

### Discrepancies between estimated intrusion depths and minimum pressure conditions recorded by epidote-bearing assemblages

The minimum pressure for magmatic epidote of 0.5 GPa was criticized by several authors investigating epidote bearing granitoids where both Al-in-hornblende barometry and pressures derived from the contact aureole indicate intrusion levels significantly less than 15 km. This apparent contradiction to the experimental results may result from several differences between the natural and experimental systems:

- The crystallization depth of 70 to max. 80% of the crystals within a magma does not necessarily correspond to the emplacement depth. Examples of liquid state deformation of an intrusion and possible uplift of a crystal-melt mush are observed in the Bergell intrusion (Davidson et al. 1996), in New Zealand (Tulloch 1986), and in the Archean Minto block, Canada (Bedard 2003; see also discussion on the Great tonalite sill at Mount Juneau, Alaska by Drinkwater et al. 1991). In the Bergell tonalite, amphibole, biotite, plagioclase, and epidote, but not intergranular quartz and K-feldspar are deformed, pointing to late stage crystallization of quartz and K-feldspar after upraise and deformation of a crystal-melt mush. In this case, the Al-in-hornblende barometer yields slightly higher pressure than suggested by the contact aureole (e.g., the western end of the Bergell intrusion, where regional metamorphism indicates 0.6 to 0.7 GPa and Al-in-hornblende intrusion pressures 0.75 to 0.8 GPa), although differences of <0.15 GPa are at the resolution limit of geobarometrical methods. Nevertheless, such a difference might be taken as an indication that the Al-in-hornblende barometer may sometimes document pressures of crystallization of a crystal melt mush rather than final solidification and that occasionally, such crystal-melt mushes could have moved to somewhat higher levels.

- If cooling is rapid enough, epidote may not react with the remnant melt. The kinetics of epidote dissolution was investigated by Brandon et al. (1996) and was found to be relatively rapid. At the relevant temperature of 700 to 800°C grain sizes of 0.2 to 3 mm would need 2 to 2000 years for dissolution.

- In many plutons, epidote shows resorption textures. Armoring of epidote within later crystallized biotite or plagioclase (as described by Tulloch 1986) or other minerals may prevent equilibration of a higher pressure epidote during final full crystallization of a magma.

- An increased $f_{O_2}$ increases the stability field of epidote. This might be testified by Fe-Ti-oxides or by an increased Fe-content in epidote. Magmatic epidote from the Appalachian monzogranites have $ps_{73}$ to $ps_{100}$ (Vynhal et al. 1991), i.e., compositions that are at the Fe-rich end of the solid solution, and which are likely to have formed at elevated $f_{O_2}$.

- Bulk composition outside the TTG series may result in a shift of the intersection of epidote-in and wet solidus to lower pressure (see next paragraph).

It is conspicuous that several plutons that lead the investigators to challenge the minimum epidote-pressure of 0.5 GPa were monzogranites (Appalachian granitoids, Vynhal et al. 1991; Querigut-complex, French Pyrenees, Leterrier 1972; Roberts et al. 2000; partly also in the Median batholith in New Zealand, Tulloch 1986; Tulloch and Challis 2000). The particular role of monzogranites is evident in the Querigut massif, which consists of about 10 individual intrusions. In Querigut, only the monzogranites, but not the granodiorites and tonalites bear epidote. Epidote-bearing monzogranites have normative anorthite contents of 13 to 17 wt% and in most cases, oxidation states appear to be within the average range. On the basis of the bulk composition of monzogranites, there is no evident reason, why monzogranites should have a distinctly lower pressure stability of magmatic epidote, and it is necessary to investigate this systematically.

## Epidote in the crystallization sequence: a sensitive tool for estimating pressure conditions

The first appearance of epidote during the crystallization history of a cooling magma could provide a more detailed geobarometer. At fluid-saturation and pressures to about 1.1 GPa (Fig. 6) the sequence of crystallization in tonalite is

above 1.0 GPa   hornblende → **epidote** → plagioclase → biotite → quartz → K-feldspar;

1.0 to 0.8 GPa   hornblende → plagioclase → **epidote** → biotite → quartz → K-feldspar;

0.8 to 0.6 GPa   hornblende → plagioclase → biotite → **epidote** → quartz → K-feldspar;

0.6 to 0.5 GPa   hornblende → plagioclase → biotite → quartz → **epidote** → K-feldspar.

The position of epidote moves from left to right in this sequence as pressure decreases. Some of these various crystallization sequences for epidote have been reported from natural rocks in the literature (for example the references cited by Tulloch, by Moench, and by Zen and Hammarstrom in their exchange of correspondence in 1986). The above pressures result from $H_2O$-saturated experiments with $f_{O_2}$ buffered to NNO. The actual temperature of epidote crystallization would be increased in oxidized magmas, such that, at a given pressure, epidote might move to the left in the crystallization sequence.

## PHASE RELATIONSHIPS IN $H_2O$-UNDERSATURATED SYSTEMS

The maximum temperature limit of magmatic epidote in tonalite has been experimentally determined at fluid-saturated conditions to about 750°C. $H_2O$-undersaturated crystallization of epidote has been investigated at 0.8 GPa in a granite and granodiorite (Naney 1983), and at 1.0 GPa (van der Laan and Wyllie 1992) and 1.5 GPa (Johnston and Wyllie 1988) in a trondhjemite. These studies have defined *T*- $H_2O$ sections (Fig. 12) and show that the $H_2O$-saturated melting temperature changes only slightly at $H_2O$-undersaturated conditions, i.e., decreases by about 20°C in Naney's (1983) granodiorite at 0.8 GPa, and increases from 740 to 775°C in the trondhjemite at 1.5 GPa. Experiments above 0.8 GPa on a hornblende-bearing dacite (very similar in composition to the granodiorites studied by Naney 1983 and Schmidt and Thompson 1996) at 3 and 5 wt% water content (Green 1992) did not produce epidote down to 800°C. This indicates that in granodioritic compositions (at 1.5 GPa) the epidote stability field does not substantially enlarge with decreasing water content. Experiments on an epidote-bearing tonalite with no water added (Skjerlie and Johnston 1993) did not result in stable epidote down to 875°C at 1.0 GPa. Two experimental brackets at $H_2O$-undersaturated conditions exist: Singh and Johannes (1996b) in a synthetic bulk composition (biotite-plagioclase-quartz, 0.8 wt% bulk $H_2O$) defined zoisite and epidote to melt out between 750 and 800°C at 1.0 to 1.5 GPa, in both an Fe free and an Fe bearing system, respectively. Lopez and Castro (2001) studied an amphibolite *sensu strictu* of MORB composition and formed new epidote at 800°C, 1.1 to 1.2 GPa but epidote had melted out at 850°C. Thus, epidote was found to be unstable in experiments on natural or close-to-natural compositions at temperatures above 800°C (Lambert and Wyllie 1972; Winther and Newton 1991; Pawley and Holloway 1993; Skjerlie and Johnston 1993; Sen and Dunn 1994; Poli and Schmidt 1995; Schmidt and Thompson 1996; Lopez and Castro 2001) and there is no evidence for a significantly enlarged stability field of epidote + melt at $H_2O$-undersaturated conditions. Further comments on epidote stability in the $H_2O$-undersaturated region cannot be made without additional experiments. Unfortunately, it is a difficult experimental task to perform fluid-absent experiments at ≈ 800°C. Skjerlie and Patino-Douce (2002) suggested zoisite to be stable up to 1000°C at 2.6 GPa. However, this interpretation of their experiments is at odds with any other study on basaltic compositions in this *P-T* region (Lambert and Wyllie 1972; Winther and Newton 1991; Pawley and Holloway 1993; Sen and Dunn 1994; Poli and Schmidt 1995; Schmidt and Thompson 1996). Skjerlie and Patino-Douce (2002) (i) had zoisite in their starting material and did not reverse their experiments, (ii) the zoisite present in their experimental products had a composition identical to the one in the starting material, (iii) they used a very low $H_2O$ content of 0.2 wt% leading to fluid-absent conditions, and (iv) employed a very coarse grained starting material (>50 μm). Thus, Skjerlie and Patino-Douce (2002) had an extremely unfavorable setup for reaching equilibrium and their interpretation of zoisite-remnants in experimental charges is incompatible with other studies. Further experimental

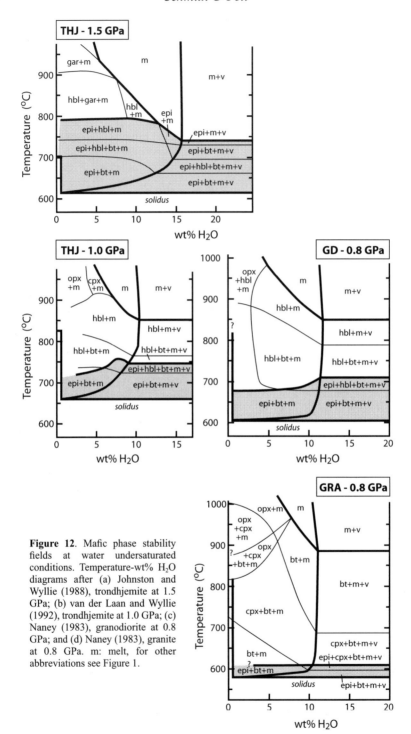

**Figure 12**. Mafic phase stability fields at water undersaturated conditions. Temperature-wt% H$_2$O diagrams after (a) Johnston and Wyllie (1988), trondhjemite at 1.5 GPa; (b) van der Laan and Wyllie (1992), trondhjemite at 1.0 GPa; (c) Naney (1983), granodiorite at 0.8 GPa; and (d) Naney (1983), granite at 0.8 GPa. m: melt, for other abbreviations see Figure 1.

work, demonstrating equilibrium, is required before accepting an extension of epidote stability at $H_2O$-undersaturated conditions.

It was suggested that epidote inclusions in biotite (e.g., in high level oxidized granitoids; Tulloch 1986) and epidotes, which crystallized when the melt proportion was 50 to 80 vol% (as in the rhyodacite dykes described by Evans and Vance 1987), are xenocrysts carried from depth. Between 1.0 to 1.5 GPa epidote crystallizes before biotite in tonalite-$H_2O$ ($f_{O_2}$ = NNO, Fig. 6) and in granodiorite-$H_2O$ ($f_{O_2}$ = NNO, Fig. 8). However, in the early stages of magmatic crystallization it is not likely that fluid-saturation is reached, so the question arises as to the (yet unknown) crystallization sequences at $H_2O$-undersaturated conditions.

### Epidote in fluid-absent melting processes

Nicholls and Ringwood (1973) and Vielzeuf and Schmidt (2001) suggested involvement of epidote in fluid-absent melting reactions between 0.8 and 3.0 GPa. Vielzeuf and Schmidt (2001) compiled available experiments on melting of basaltic compositions into a melting phase diagram, and showed that epidote is indeed the hydrous phase responsible for the first occurrence of fluid-absent melting in basaltic to andesitic bulk compositions.

***Hydrous phases at the wet solidus.*** In the early nineties, several experimental studies investigated the fluid-absent melting of metabasalts (Beard and Lofgren 1991; Rapp et al. 1991; Rushmer 1991; Winther and Newton 1991; Wolf and Wyllie 1994). In these studies, the appearance of first melts in mafic compositions is attributed to the decomposition of amphibole as the only hydrous phase. However, above ca 1.0 GPa this interpretation is no longer valid as epidote is present at the solidus (compare to Poli and Schmidt 2004; Enami et al. 2004). With increasing pressure, the succession of stable mineral assemblages in MORB at the fluid-present solidus is (Vielzeuf and Schmidt 2001):

- to approximately 0.8 GPa, metabasalts are amphibolites sensu stricto composed of amphibole + plagioclase + quartz;

- at 0.8 to 1.0 GPa epidote and garnet crystallize (Apted and Liou 1983; Poli 1993) leading to an assemblage of amphibole + plagioclase + epidote + quartz ± garnet, defining the epidote-amphibolite facies;

- omphacite forms near 1.5 GPa and amphibolite transforms into amphibole-eclogite with an assemblage of amphibole + epidote + clinopyroxene + garnet + quartz (Lambert and Wyllie 1972; Poli 1993);

- around 2.5 GPa, amphibole decomposes, leaving a zoisite-eclogite with a characteristic assemblage of zoisite + clinopyroxene + garnet + quartz/coesite (Pawley and Holloway 1993; Poli 1993);

- finally, near 3.0 GPa, zoisite decomposes (Poli and Schmidt 1995) and a K-poor metabasalt transforms into an eclogite sensu stricto with the anhydrous mineral assemblage omphacite + garnet + coesite.

The pressures in the succession above are valid for MOR basalts, but are very similar in intermediate tonalitic compositions (Schmidt 1993; Poli and Schmidt 1995). The upper pressure limit of epidote in tonalite as defined by Schmidt (1993) is in error, because the critical experiment of Schmidt (1993) at 2.6 GPa, 650°C, that did not yield epidote, was inconsistent with subsequent experiments (Poli and Schmidt 1995) and was thus repeated. The second experiment at the same conditions did then yield 12 wt% epidote. The reason for the inconsistency of the first experiment is unclear (and was not searched for).

The sequence of assemblages from the experiments is in agreement with natural observations. Lardeaux and Spalla (1991) describe coronae of clinopyroxene and zoisite (± garnet) at the boundaries between plagioclase and amphibole in amphibolites metamorphosed under Alpine eclogite facies conditions (Sesia Zone, Western Alps). The upper pressure stability of amphibole relative to zoisite in metabasalts is documented in coesite bearing rocks: coesite coexists with zoisite but never with barroisitic amphibole (e.g., Dabie-Shan, Zhang et al. 1995). This confirms that zoisite is stable to higher pressure than amphibole at the wet solidus. Between 1.0 and 2.5 GPa, epidote is the first hydrous phase to decompose and thus to melt on a prograde *P-T* path (Fig. 13).

The amount of epidote in a metabasalt at subsolidus conditions can be estimated between 5 and 15 wt% (Poli 1993) and thus, only a small amount of melt will be formed by the

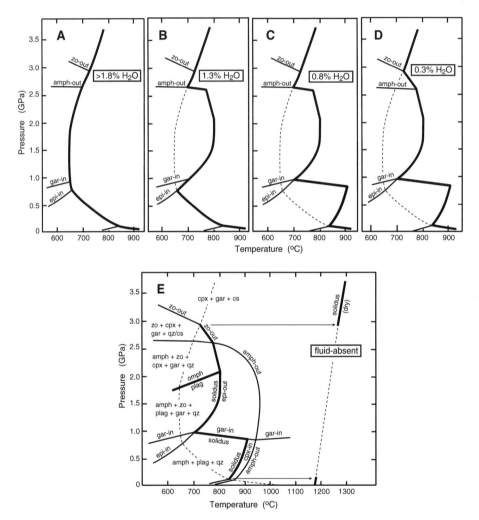

**Figure 13**. Melting in an epidote bearing MOR basalt. (a-d): Solidi for MOR basalt with fixed bulk water contents: (a) >1.8 wt% H$_2$O, fluid-saturated at all pressures (b) 1.3 wt% H$_2$O, assemblages with epidote + amphibole are not fluid saturated; (c) 0.8 wt% H$_2$O, assemblages with amphibole are generally fluid-undersaturated; (d) 0.3 wt% H$_2$O, any assemblage containing hydrous phases is water undersaturated. (e) System open to volatiles: MOR basalt fully water saturated below the wet solidus but following a prograde *P-T* path along which a fluid-absent situation is strictly maintained. Abbreviations see Figure 1.

incongruent epidote melting reaction. Evidence for high pressure melting in eclogite areas has been documented by Nicollet et al. (1979), Franz and Smelik (1995), Puziewitcz and Koepke (2001), and Elias-Herrera and Ortega-Gutierrez (2002). At 2.5 to 3.0 GPa, i.e., beyond the amphibole-out curve, epidote is the major hydrous phase present in metabasalt.

*A generalized phase diagram for partial melting of metabasalt.* The phase relationships outlined above result in a generalized phase diagram for the melting of $SiO_2$-saturated metabasalt (Fig. 13e; Vielzeuf and Schmidt 2001). The upper sequence of diagrams (Fig. 13a-d) delineates the solidus for fixed bulk $H_2O$-contents. The rational of these diagrams is: the $H_2O$-content stored in amphibole and epidote is subtracted from the bulk $H_2O$-content, the difference resulting in a fluid-present or fluid-absent melting situation. If fluid is present, the wet solidus is responsible for the first occurrence of melt, if no free fluid phase results, fluid-absent melting reactions of epidote and amphibole are designated as part of the solidus. From this series of diagrams it is evident that between 1.0 and 3.0 GPa, the fluid-absent melting of epidote marks an intermediate step. The shape of the solidus (Fig. 13c) was confirmed by the experimental study of Lopez and Castro (2001). Their experiments with amphibolite containing 1 wt% $H_2O$ resulted in a strong depression of the solidus temperature when overstepping the garnet-in reaction with pressure, and in epidote-melting above 1.0 GPa.

Nevertheless, such a fixed bulk $H_2O$ is fairly hypothetical. A more realistic endmember situation is, that any significant amount of fluid-phase produced during prograde metamorphism will leave the rock volume in question, thus maintaining at subsolidus conditions a fluid-absent situation. The melting curve for such a process is designated in Figure 13e. The maximum temperature for the stability of epidote (at 1.5 to 2.0 GPa) was modified from Vielzeuf and Schmidt (2001) and now placed at 800°C in agreement with the study of Lopez and Castro (2001). This latter upper temperature limit of epidote was chosen (Fig. 13) irrespective of the bulk fluid-content. Although this temperature must vary at least slightly with $H_2O$-activity in the melt, this is consistent with the absence of epidote in the fluid absent melting experiments conducted above 800°C. We conclude, that during high pressure melting of basaltic to intermediate bulk compositions, epidote will be involved in the melting process. Whereas in tonalite, the fluid-absent melting of potassic hydroxylated phases (i.e., micas) results in a complex interplay with the fluid-absent melting of calcic hydroxylated phases (i.e., epidote and amphibole), epidote will be dominant for forming the first high-pressure melts in potassium-poor rocks. Nevertheless, the amount of melt formed by reactions involving epidote is small (generally <10%), a more significant melt volume is formed by the decomposition of amphibole.

## OPEN PROBLEMS

Whereas the "Deer-Howie-Zussmann" (Deer et al. 1986, section "clinozoisite-epidote", p. 121-123) still casts a dubious eye on the possibility of magmatic epidote, field and experimental petrologists are now accustomed to the coexistence of epidote with silicate melts. Although epidote minerals are omnipresent near the solidus, a quantitative appreciation of their role during melting is still lacking. Experimental studies did not yet define the

- upper temperature stability of zoisite from 1 to 7 GPa;

- zoisite melting relations in simple systems more complex than CASH above 2 GPa (apart from studies by Thompson and Ellis 1994 and Hermann 2002);

- influence of oxygen fugacity in relevant simple systems;

- fluid-absent melting relations of zoisite at temperatures between the wet solidus and ca. 800°C.

The latter appears to be the most interesting and rewarding task, however, will have to overcome the kinetic problems in fluid-absent systems below 800°C and any result will only be credible when truly reversed. One way to go might be seeded and unseeded starting materials, possibly involving seeded or/and unseeded gels. As such experiments are difficult, much information could be gained by field studies of migmatitic terranes that contain epidote. It appears to us, that the few reported involvements of epidote in natural melting processes would not constitute a close-to-complete list. Defining conditions and reactions of epidote melting in prograde amphibolite to granulite facies terrains would complete one of the major deficiencies in natural melting processes.

A further task that remains wide open is to understand epidote compositions and the relation of epidote and plagioclase compositions as a function of bulk composition, intrusion pressure and oxygen fugacity. For this purpose, plagioclase compositions in equilibrium with epidote need to be known (which is obviously not always straightforward in plutons), and thus a detailed comprehensive study would be necessary.

## EPILOGUE

This chapter has a sad history. Initially it was planned to be written together with Kjell Skjerlie from University of Tromsø, Norway. Tragically, our colleague Kjell Petter Skjerlie died last summer in an accident during a mountain hike. He was hit by a falling rock and died immediately. Kjell would just have become 42 years, he is survived by his wife and their three children.

## ACKNOWLEDGMENTS

We thank G. Franz, A.Wittenberg and T. Fehr for reviews of this chapter.

## REFERENCES

Allen JC, Boettcher AL (1983) The stability of amphibole in andesite and basalt at high pressures. Am Mineral 68:307-314
Allibone AH, Tulloch AJ (1997) Metasedimentary, granitoid, and gabbroic rocks from central Stewart Island, New Zealand. New Zeal J Geol Geop 40(1):53-68
Altherr R, Lugovic B, Meyer HP et al. (1995) Early-miocene postcollisional calc-alkaline magmatism along the easternmost segment of the periadriatic fault system (Slovenia and Croatia). Miner Petrol 54:25-247
Anderson JL, Barth AP, Young ED (1988) Mid-crustal cretaceous roots of Cordilleran metamorphic core complexes. Geology 16:366-369
Apted MJ, Liou JG (1983) Phase relations among greenschist, epidote-amphibolite, and amphibolite in a basaltic system. Am J Science 283-A:328-354
Barnes CG, Prestvik T (2000) Conditions of pluton emplacement and anatexis in the Caledonian Bindal Batholith, north-central Norway. Norsk Geol Tidsskr 80:259-274
Barr SM, White CE, Culshaw NG (2001) Geology and tectonic setting of Paleoproterozoic granitoid suites in the Island Harbour Bay area, Makkovik Province, Labrador. Can J Earth Sci 38:441-463
Bea F, Fershtater G, Montero P (1997) Generation and evolution of subduction-related batholiths from the central Urals: constraints on the P-T history of the Uralian orogen. Tectonophys 276:103-116
Beard JS, Lofgren GE (1991) Dehydration melting and water-saturated melting of basaltic and andesitic greenstones and amphibites at 1, 3, 6.9 kb. J Petrol 32:365-401
Bedard JH (2003) Evidence for regional-scale, pluton-driven, high-grade metamorphism in the Archaen Minto Block, northern Superior Province, Canada. J Geol 111:183-205
Boettcher AL (1970) The system CaO-Al$_2$O$_3$-SiO$_2$-H$_2$O at high pressures and temperatures. J Petrol 11:337-379
Brandon AD, Creaser RA, Chacko T (1996) Constraints on rates of granitic magma transport from epidote dissolution kinetics. Science 271:1845-1848

Broska I, Petrik I, Benko P (1997) Petrology of the Mala Fatra granitoid-rocks (Western Carpathians, Slovakia). Geol Carpath 48:27-37

Brunsmann A, Franz G, Heinrich W (2002) Experimental investigation of zoisite-clinozoisite phase equilibria in the system CaO-Fe$_2$O$_3$-Al$_2$O$_3$-SiO$_2$-H$_2$O. Contrib Mineral Petrol 143:115-130

Campos TFC, Nieva AMR, Nardi LSV (2000) Geochemistry of granites and their minerals from Serra Negra do Norte Pluton, northeastern Brazil. Chem Erde-Geochem 60:279-303

Cerredo ME, De Luchi MGL (1998) Mamil Choique Granitoids, southwestern North Patagonian Massif, Argentina: magmatism and metamorphism associated with a polyphasic evolution. J Am Earth Sci 11: 499-515

Cook RB, Crawford ML, Omar GI et al (1991) Magmatism and deformation, southern Revillagigedo Island, southeastern Alaska. Geol Soc Am Bull 103:829-841

Cornelius HP (1915) Geologische Beobachtungen im Gebiet des Forno-Gletschers (Engadin). Centralblatt für Mineral Geol Paläontol 1913, 8:246-252

Crawford ML, Hollister LS (1982) Contrast of metamorphic and structural histories across the Work Channel lineament, Coast Plutonic Complex, British Columbia. J Geophys Res 87:3849-3860

Dahlquist JA (2001a) Low-pressure emplacement of epidote-bearing metaluminous granitoids in the Sierra de Chepes (Famatinian Orogen, Argentina) and relationships with the magma source. Rev Geol Chile 28: 147-161

Dahlquist JA (2001b) REE fractionation by accessory minerals in epidote-bearing metaluminous granitoids from the Sierras Pampeanas, Argentina. Mineral Mag 65:463-475

Davidson C, Rosenberg C, Schmid SM (1996) Synmagmatic folding of the base of the Bergell pluton, Central Alps. Tectonophys 265:213-238

Dawes RL, Evans BW (1991) Mineralogy and geothermobarometry of magmatic epidote-bearing dikes, front range, Colorado. Geol Soc Am Bull 103:1017-1031

Deer WA, Howie RA, Zussman J (1986) Clinozoisite-Epidote. *In:* Disilicates and Ring Silicates, Rock Forming Minerals 1B. Deer WA, Howie RA, Zussman J (eds) Longman Scientific & Technical, Burnt Mill, Harlow. 44-134

Drinkwater JL, Ford AB, Brew DA (1991) Magnetic susceptibilities and iron content of plutonic rocks across the Coast Plutonic -Metamorphic Complex near Juneau, Alaska. *In:* Geologic studies in Alaska by the U.S. Geological Survey. Bradley DC, Dusel-Bacon C. (eds) U.S. Geol Survey Bull 2041:125-139

Eggler DH (1972) Water-saturated and undersaturated melting relations in a Paricutin andesite and an estimate of water content in the natural magma. Contrib Mineral Petrol 34:261-271

Elias-Herrera M, Ortega-Gutierrez F (2002) Caltepec fault zone: An Early Permian dextral transpressional boundary between the Proterozoic Oaxacan and Paleozoic Acatlan complexes, southern Mexico, and regional tectonic implications. Tectonics 21:10.1029/2000TC001278

Ellis DJ, Thompson AB (1986) Subsolidus and partial melting reactions in the quartz excess CaO+MgO+Al$_2$O$_3$+SiO$_2$+H$_2$O system under water-excess and water-deficient conditions to 10 kbar: some implications for the origin of peraluminous melts from mafic rocks. J Petrol 27:91-121

Enami M, Liou JG, Mattinson CG (2004) Epidote minerals in high P/T metamorphic terranes: Subduction zone and high- to ultrahigh-pressure metamorphism. Rev Mineral Geochem 56:347-398

Evans BW, Vance JA (1987) Epidote phenocrysts in dacitic dikes, Boulder County, Colorado. Contrib Mineral Petrol 96:178-185

Farrow CEG, Barr SM (1992) Petrology of high-Al-hornblende-plutons and magmatic-epidote bearing plutons in the southeastern Cape-Breton highlands, Nova-Scotia. Can Mineral 30:377-392

Franz G, Smelik EA (1995) Zoisite-clinozoisite bearing pegmatites and their importance for decompressional melting in eclogites. Eur J Mineral 7:1421-1436

Frei R (1996) The extent of inter-mineral isotope equilibrium: A systematic bulk U-Pb and Pb step leaching (PbSL) isotope study of individual minerals from the Tertiary granite of Jerissos (northern Greece). Eur J Mineral 8:1175-1189

Galindo AC, Dallagnol R, McReath I et al. (1995) Evolution of Brasiliano-age granitoid types in a shear-zone environment, Umarizal-Caraubas region, Rio-Grande-Do-Norte, northeast Brazil. J S Am Earth Sci 8: 79-95

Gieré R, Sorensen SS (2004) Allanite and other REE-rich epidote-group minerals. Rev Mineral Geochem 56: 431-494

Green TH (1992) Experimental phase equilibrium studies of garnet-bearing I-type volcanics and high level intrusives from Northland, New Zealand. Transact Royal Soc Edinburgh: Earth Sciences 83:429-438

Ghent ED, Nicholls J, Simony PS, et al. (1991) Hornblendegeobarometry of the Nelson-batholith, southeastern British-Columbia - tectonic implications. Can J Earth Sci 28:1982-1991

Hammarstrom JM, Zen E (1986) Aluminium in hornblende: an empirical igneous geobarometer. Am Mineral 71:1297-1313

Hammarstrom JM, Zen E (1992) Petrological characteristics of magmatic epidotebearing granites of the western cordillera of America. Abstract. Transact Royal Soc Edinburgh: Earth Sciences 83:490-491

Hermann J (2002) Experimental constraints on phase relations in subducted continental crust. Contrib Mineral Petrol 143:219-235

Holdaway MJ (1972) Thermal stability of Al-Fe-epidote as a function of $f_{O_2}$ and Fe content. Contrib Mineral Petrol 37:307-340

Hollister LS, Grissom GC, Peters EK, Stowell HH, Sisson VB (1987) Confirmation of the empirical correlation of Al in hornblende with pressure of solidification of calc-alkaline plutons. Am Mineral 72:231-239

Hoschek (1990) Melting and subsolidus reactions in the system $K_2O$-CaO-MgO-$Al_2O_3$-$SiO_2$-$H_2O$: experiments and petrologic application. Contrib Mineral Petrol 105: 393-402

Ishihara S (1977) The magnetite-series and ilmenite-series granitic rocks. Mining Geol 27: 293-305

Ishihara S (1981) The granitoid series and mineralization. Economic Geol 75th anniversary volume: 458-484

Johannes W (1980) Melting and subsolidus reactions in the system $K_2O$-CaO-$Al_2O_3$-$SiO_2$-$H_2O$. Contrib Mineral Petrol 74:29-34

Johnson MC, Rutherford MJ (1989) Experimental calibration of the aluminium-inhornblende geobarometer with application to Long Valley caldera (California) volcanic rocks. Geology 17:837-841

Johnston AD, Wyllie PJ (1988) Constraints on the origin of Archean trondhjemites based on phase relationship of Nuk gneiss with H2O at 15 kbar. Contrib Mineral Petrol 100:35-46

Keane SD, Morrison J (1997) Distinguishing magmatic from subsolidus epidote: Laser probe oxygen isotope compositions. Contrib Mineral Petrol 126: 265-274

Lambert IB, Wyllie PJ (1972) Melting of gabbro (quartz eclogite) with excess water to 35 kilobars, with geological applications. J Geol 80:693-708

Lambert IB, Wyllie PJ (1974) Melting of tonalite and crystallization of andesite liquid with excess water to 30 kilobars. J Geol 82:88-97

Lardeaux JM, Spalla MI (1991) From granulites to eclogites in the Seisa zone (Italian Western Alps): a record of the opening and closure of the Piedmont ocean. J metam Geol 9:35-59

Leterrier J (1972) Etude petrographique et geochimique du massif granitique de Querigut (Ariege). These docteur es-sciences naturelles. Universite de Nancy, France, 292 p.

Liou JG (1973) Synthesis and stability relations of epidote, $Ca_2Al_2FeS_3O_{12}(OH)$. J Petrol 14:381-413

Lopez S, Castro A (2001) Determination of the fluid-absent solidus and supersolidus phase relationships of MORB-derived amphibolites in the range 4-14 kbar. Am Mineral 86: 1396-1403

Moench RH (1986) Comment on "Implications of magmatic epidote-bearing plutons on crustal evolution in the accreted terranes of northwestern North America" and "Magmatic epidote and its petrologic significance". Geology 14: 187-188

Montero P, Bea F, Gerdes A, et al. (2000) Single -zircon evaporation ages and Rb-Sr dating of four major Variscan batholiths of the Urals - A perspective on the timing of deformation and granite generation. Tectonophys 317:93-108

Mortaji A, Ikenne M, Gasquet D et al. (2000) Palaeoproterozoic granitoids from the Bas Draa and Tagragra d'Akka Inliers (western Anti-Atlas, Morocco): part of the jigsaw puzzle concerning the West African Craton. J Afr Earth Sci 31:523-538

Morrison J (2004) Stable and radiogenic isotope systematics in epidote group minerals. Rev Mineral Geochem 56:607-628

Moticska P (1970) Petrographie und Strukturanalyse des westlichen Bergeller Massivs und seines Rahmens. Schweiz Mineral Pet Mitt 50:355-443

Murata M, Itaya T (1987) Sulfide and oxide minerals from S-type and I-type granitic rocks. Geochim Cosmochim Acta 51:497-507

Naney MT (1983) Phase equilibria of rock-forming ferromagnesian silicates in granitic systems. Am J Sci 283: 993-1033

Nedelec A, Minyem D, Barbey P (1993) High-P-high-T anatexis of archean tonalitic gray gneisses - the Eseka migmatites, Cameroon. Precambrian Res 62:191-205

Newton RC (1966) Some calc -silicate equilibrium relations. Am J Sci 264:204-222

Nicholls IA, Ringwood AE (1973) Effect of water on olivine stability in tholeiites and the production of silica-saturated magmas in the island-arc environment. J Geol 81:285-300

Nicollet C, Leyreloup A, Dupuy C (1979) Petrogenesis of high pressure trondhjemitic layers in eclogites and amphibolites from Southern Massif Central, France. In: Trondhjemites, dacites, and related rocks. Developements in Petrology. Vol.6 Barker F (ed) Elsevier, Amsterdam Oxford New York, p.435-463

Nitsch KH, Winkler HGF (1965) Bildungsbedingungen von Epidot und Orthozoisit. Beiträge Mineral Pet 11: 470-486

Owen JV (1991) Significance of epitote in orbicular diorite from the Grenville front zone, eastern Labrador. Mineral Mag 55:173-181

Owen JV (1992) Geochemistry of orbicular diorite from the Grenville-front zone, eastern Labrador. Mineral Mag 56(385):451-458

Pattnaik SK (1996) Petrology of the Bhela -Rajna alkaline complex, Nuapara District, Orissa. J Geol Soc India 48:27-40

Pawley AR, Holloway JR (1993) Water sources for subduction zone volcanism: new experimental constraints. Science 260:664-667

Pepiper G, Doutsos T, Mporonkay C (1993) Structure, geochemistry and mineralogy of hercynian granitoid rocks of the Verdikoussa area, northern Thessaly, Greece and their regional significance. Neues Jb Miner Abh 165:267-296

Piwinskii AJ (1968) Experimental studies of igneous rock series: central Sierra Nevada batholith, California. J Geol 76:548-570

Piwinskii AJ (1973) Experimental studies of igneous rock series, central Sierra Nevada batholith, California: Part II. Neues Jb Miner 5:193-215

Poli S (1993) The amphibolite-eclogite transformation: An experimental study on basalt. Am J Sci 293:1061-1107

Poli S, Schmidt MW (1995) $H_2O$ transport and release in subduction zones: experimental constraints on balsaltic and andesitic systems. J Geophy Res 100:22299-22314

Poli S, Schmidt MW (1997) The high-pressure stability of hydrous phases in orogenic belts: an experimental approach on eclogite-forming processes. Tectonophys 273:169-184

Poli S, Schmidt MW (1998) The high-pressure stability of zoisite and phase relationships of zoisite-bearing assemblages. Contrib Mineral Petrol 130:162-175

Poli S, Schmidt MW (2004) Experimental subsolidus studies on epidote minerals. Rev Mineral Geochem 56:171-195

Popov VS, Nikiforova NF, Bogatov VI et al. (2001) Multiple gabbro-granite intrusive series of the Syrostan pluton, southern Urals: Geochemistry and petrology. Geochem Int 39:732-747

Puziewicz J, Koepke J (2001) Partial melting of garnet-hornblende granofels and the crystallisation of igneous epidote in the Niedzwiedz Amphibolite Massif (Fore-Sudetic Block, SW Poland) Neues Jb Miner Monat 12:529-547

Quirion DM, Jenkins DM (1998) Dehydration and partial melting of tremolitic amphibole coexisting with zoisite, quartz, anorthite, diopside, and water in the system $H_2O$-CaO-MgO-$Al_2O_3$-$SiO_2$. Contrib Mineral Petrol 130:379-389

Rapp RP, Watson EB, Miller CF (1991) Partial melting of amphibolite/eclogite and the origin of archean trondhjemite and tonalite. Precambrian Res 51:1-25

Roberts MP, Pin C, Clemens JD, Paquette JL (2000) Petrogenesis of mafic to felsic plutonic rock associations: the calc -alkaline Quérigut complex, French Pyrenees. J Petrol 41:809-844

Rushmer T (1991) Partial melting of two amphibolites: contrasting experimental results under fluid-absent conditions. Contrib Mineral Petrol 107:41-59

Saavedra J, Toselli AJ, Rossi de Toselli JN, Rapela CW (1987) Role of tectonism and fractional crystallisation in the origin of lower Paleozoic epidote-bearing granitoids, northwestern Argentina. Geology 15:709-713

Schliestedt M, Johannes W (1984) Melting and subsolidus reactions in the system $K_2O$-CaOAl$_2O_3$-$SiO_2$-$H_2O$: corrections and additional experimental data. Contrib Mineral Petrol 88:403-405

Schmidt MW (1992) Amphibole composition in tonalite as a function of pressure: an experimental calibration of the Al-in-hornblende-barometer. Contrib Mineral Petrol 110:304-310

Schmidt MW (1993) Phase relations and compositions in tonalite as a function of pressure: An experimental study at 650°C. Am J Science 293:1011-1060

Schmidt MW, Thompson AB (1996) Epidote in calc-alkaline magmas: An experimental study of stability, phase relationships, and the role of epidote in magmatic evolution. Am Mineral 81:462-474

Selbekk RS, Skjerlie KP (2002) Petrogenesis of the anorthosite dyke swarm of Tromsø, North Norway: Experimental evidence for hydrous anatexis of an alkaline mafic complex. J Petrol 43:943-962

Sen C, Dunn T (1994) Dehydration melting of a basaltic composition amphibolite at 1.5 and 2.0 GPa: implications for the origin of adakites. Contrib Mineral Petrol 117:394-409

Sial AN, Toselli AJ, Saavedra J et al. (1999a) Emplacement, petrological and magnetic susceptibility characteristics of diverse magmatic epidote-bearing granitoid rocks in Brazil, Argentina and Chile. Lithos 46:367-392

Sial AN, Dall'Agnol R, Ferreira VP et al. (1999b) Precambrian granitic magmatism in Brazil. Episodes 22:191-198

Singh J, Johannes W (1996a) Dehydration melting of tonalites. Part I. Beginning of melting. Contrib Mineral Petrol 125:16-25

Singh J, Johannes W (1996b) Dehydration melting of tonalites. Part II. Composition of melts and solids. Contrib Mineral Petrol 125:26-44

Skjerlie KP, Douce AEP (2002) The fluid-absent partial melting of a zoisite-bearing quartz eclogite from 1.0 to 3.2 GPa; Implications for melting in thickened continental crust and for subduction-zone processes. J Petrol 43:291-314

Skjerlie KP, Johnston AD (1993) Fluid-absent melting behaviour of an F-rich tonalitic gneiss at mid-crustal pressures: Implications for the generation of anorogenic granites. J Petrol 34:785-815

Strens RGJ (1965) Stability and relations of the Al-Fe epidotes. Mineral Mag 35:464-475

Thompson AB, Ellis DJ (1994) $CaO+MgO+Al_2O_3+SiO_2+H_2O$ to 35 kbar: amphibole, talc and zoisite dehydration and melting reactions in the silica-excess part of the system and their possible significance in subduction zones, amphibolite melting and magma fractionation. Am J Sci 294:1229-1289

Thompson JB (1982) Composition space: An algebraic and geometric approach. Mineral Soc Am Reviews 10:1-32

Tulloch AJ (1986) Comment on "Implications of magmatic epidote-bearing plutons on crustal evolution in the accreted terranes of northwestern North America" and "Magmatic epidote and its petrologic significance". Geology 14:187-188

Tulloch AJ, Challis GA (2000) Emplacement depths of Paleozoic -Mesozoic plutons from western New Zealand estimated by hornblende-Al geobarometry. New Zeal J Geol Geophys 43:555-567

van der Laan SR, Wyllie PJ (1992) Constraints on archean trondhjemite genesis from hydrous crystallization experiments on Nuk Gneiss at 10-17 kbar. J Geol 100:57-68

Vielzeuf D, Schmidt MW (2001) Melting relations in hydrous systems revisited: applications to metapelites, metagreywackes and metabasalts. Contrib Mineral Petrol 141:251-267

Vyhnal CR, McSween HY, Speer JA (1991) Hornblende chemistry in southern Appalachian granitoids: implications for aluminium hornblende thermobarometry and magmatic epidote stability. Am Mineral 76:176-188

Winther KT, Newton RC (1991) Experimental melting of hydrous low-K tholeiite: evidence on the origin of Archaen cratons. Bull Geol Soc Denmark 39:213-228

Wolf MB, Wyllie PJ (1994) Dehydration-melting of amphibolite at 10 kbar: effects of temperature and time. Contrib Mineral Petrol 115:369-383

Wones DR (1989) Significance of the assemblage titanite + magnetite + quartz in granitic rocks. Am Mineral 74:744-749

Zhang RY, Liou JG, Coney BL (1995) Talc-, magnesite- and Ti-clinohumite-bearing ultrahigh pressure meta-mafic and ultramafic complex in the Dabie Mountains, China. J Petrol 36:1011-1037

Zen E (1988) Tectonic significance of high-pressure plutonic rocks in the western cordillera of North America. *In*: Metamorphism and Crustal Evolution of the Western United States. Rubey Vol.VII. W.G.Ernst (ed) Prentice-Hall, Englewood Cliffs, New Jersey, p 41-67

Zen E (1985) Implications of magmatic epidote-bearing plutons on crustal evolution in the accreted terranes of northwestern North-America. Geology 13:266-269

Zen E-an, Hammarstrom JM (1984) Magmatic epidote and its petrologic significance. Geology 12:515-518

Zen E, Hammarstrom JM (1986) Reply on the comments on "Implications of magmatic epidote-bearing plutons on crustal evolution in the accreted terranes of northwestern North America" and "Magmatic epidote and its petrologic significance" by AJ Tulloch and by RH Moench. Geology 14:187-188

Reviews in Mineralogy & Geochemistry
Vol. 56, pp. 431-493, 2004
Copyright © Mineralogical Society of America

# Allanite and Other
# REE-Rich Epidote-Group Minerals

### Reto Gieré

*Institut für Mineralogie, Petrologie und Geochemie*
*Universität Freiburg*
*Albertstrasse 23b*
*D-79104 Freiburg, Germany*
*giere@uni-freiburg.de*

### Sorena S. Sorensen

*Department of Mineral Sciences*
*National Museum of Natural History*
*Smithsonian Institution*
*Washington, D.C. 20560, U.S.A.*
*sorena@volcano.si.edu*

## INTRODUCTION

Epidote-group minerals rich in rare earth elements (REE), in particular allanite, are common accessory phases in igneous, metamorphic, metasomatic, and sedimentary rocks. Small amounts of REE are present in most epidote-group minerals, but in allanite—and the related minerals dissakisite, ferriallanite, dollaseite, khristovite and androsite—the REE are essential structural constituents. An important characteristic of REE-rich epidote-group minerals is that their octahedrally coordinated M sites contain major amounts of divalent cations. This paper summarizes literature data for these minerals and discusses their chemistry, occurrence, phase relations, and petrologic and geologic significance. The chapter emphasizes allanite, because it is the most common and best-studied of the REE-rich epidote-group minerals.

## MINERAL CHEMISTRY AND NOMENCLATURE

Epidote-group minerals contain isolated silicon tetrahedra and corner-sharing groups of two tetrahedra, and are thus assigned to the disilicate or sorosilicate structural family (for a detailed description of the structure, see Franz and Liebscher 2004). The epidote-group structural formula is $A_2M_3(SiO_4)(Si_2O_7)(O,F)(OH)$, or in a simplified form $A_2M_3Si_3O_{11}(O,F)(OH)$, in which $A$ = Ca, Sr, $Pb^{2+}$, $Mn^{2+}$, Th, $REE^{3+}$, and U, and $M$ = Al, $Fe^{3+}$, $Fe^{2+}$, $Mn^{3+}$, $Mn^{2+}$, Mg, $Cr^{3+}$, and $V^{3+}$ (Deer et al. 1986). There are two structurally different A sites, A(1) and A(2), with different coordination numbers, and there are three different M sites, M(1), M(2), and M(3), which are all octahedrally coordinated (Ueda 1955; Dollase 1971).

In epidote-group minerals, trivalent REE are accommodated in the A sites, which in endmember epidote both contain Ca. The incorporation of $REE^{3+}$ is commonly charge balanced by a divalent cation ($Fe^{2+}$, $Mn^{2+}$, Mg) substituted for a trivalent one in the M sites (Table 1). In an attempt to clarify the nomenclature of the REE-dominant members of the

1529-6466/04/0056-0010$10.00

**Table 1.** Idealized formulae for clinozoisite, epidote, and piemontite, and for the REE-rich epidote-group minerals

|  | A(1) | A(2) | M(1) | M(2) | M(3) | T | O(4) | |
|---|---|---|---|---|---|---|---|---|
| **Clinozoisite** | $Ca^{2+}$ | $Ca^{2+}$ | $Al^{3+}$ | $Al^{3+}$ | $Al^{3+}$ | $Si_3O_{11}$ | O | OH |
| **Epidote** | $Ca^{2+}$ | $Ca^{2+}$ | $Al^{3+}$ | $Al^{3+}$ | $Fe^{3+}$ | $Si_3O_{11}$ | O | OH |
| **Piemontite** | $Ca^{2+}$ | $Ca^{2+}$ | $Al^{3+}$ | $Al^{3+}$ | $Mn^{3+}$ | $Si_3O_{11}$ | O | OH |
| **Allanite** | $Ca^{2+}$ | $REE^{3+}$ | $Al^{3+}$ | $Al^{3+}$ | $Fe^{2+}$ | $Si_3O_{11}$ | O | OH |
| **Dissakisite** | $Ca^{2+}$ | $REE^{3+}$ | $Al^{3+}$ | $Al^{3+}$ | $Mg^{2+}$ | $Si_3O_{11}$ | O | OH |
| **Ferriallanite** | $Ca^{2+}$ | $REE^{3+}$ | $Fe^{3+}$ | $Al^{3+}$ | $Fe^{2+}$ | $Si_3O_{11}$ | O | OH |
| **Oxyallanite** | $Ca^{2+}$ | $REE^{3+}$ | $Al^{3+}$ | $Al^{3+}$ | $Fe^{3+}$ | $Si_3O_{11}$ | O | O |
| **Dollaseite** | $Ca^{2+}$ | $REE^{3+}$ | $Mg^{2+}$ | $Al^{3+}$ | $Mg^{2+}$ | $Si_3O_{11}$ | F | OH |
| **Khristovite** | $Ca^{2+}$ | $REE^{3+}$ | $Mg^{2+}$ | $Al^{3+}$ | $Mn^{2+}$ | $Si_3O_{11}$ | F | OH |
| **Androsite** | $Mn^{2+}$ | $REE^{3+}$ | $Mn^{3+}$ | $Al^{3+}$ | $Mn^{2+}$ | $Si_3O_{11}$ | O | OH |

epidote group, Ercit (2002) proposed the (unnecessary) term "allanite subgroup" for all REE-dominant varieties. Because Y is commonly found in allanite, in the rest of this chapter the abbreviation "REE" will include it.

The chemical data presented here were extracted from the literature (see Reference List) to produce a database for allanite and the other REE-rich epidote-group minerals. The database contains more than 1700 chemical analyses, comprising both older data obtained primarily via wet-chemical techniques and newer data obtained mostly by electron probe microanalysis. If $Fe^{2+}$ and $Fe^{3+}$ have been directly analyzed, we have used those values for the calculation of the molar values of each component in allanite. For all other cases, we have estimated $Fe^{2+}/Fe^{3+}$ by normalizing the analysis to 12.5 oxygens and 8 cations. The database includes a large number of unpublished electron microprobe analyses provided by Dr. James Beard, Virginia Museum of Natural History (personal communication; 2004) as well as some of our own unpublished data. Literature sources containing data used in the diagrams of this chapter, but not referred to in the text, are provided in the Reference List and specially marked by an asterisk.

## Allanite

Allanite, which is named after the Scottish mineralogist Thomas Allan (1777-1833), can be represented by the idealized formula $CaREEAl_2Fe^{2+}Si_3O_{11}O(OH)$. It is related to epidote by the coupled substitution

$$REE^{3+} + Fe^{2+} \leftrightarrow Ca^{2+} + Fe^{3+} \tag{1}$$

(Khvostova 1963; Ploshko and Bogdanova 1963) and to clinozoisite by

$$REE^{3+} + Fe^{2+} \leftrightarrow Ca^{2+} + Al^{3+} \tag{2}$$

Allanite and ferriallanite (Kartashov et al. 2002) are the only members of the epidote group in which $Fe^{2+}$ is an essential constitutent (Table 1).

The structure of allanite is almost identical to that of other epidote-group minerals (Ueda 1955; Pudovkina and Pyatenko 1963; Dollase 1971; see also Franz and Liebscher 2004). It consists of two sets of chains of edge-sharing octahedra (M sites) arranged parallel to [010]. One of these is a composite chain that consists of M(1) octahedra with greatly distorted M(3) sites attached on alternate sides along its length (Fig. 1a). The other chain contains the M(2) octahedra, which are occupied only by Al (Dollase 1971, 1973). The chains are cross-linked by two types of tetrahedral (T) sites, i.e., by isolated $SiO_4$ tetrahedra and by corner-sharing

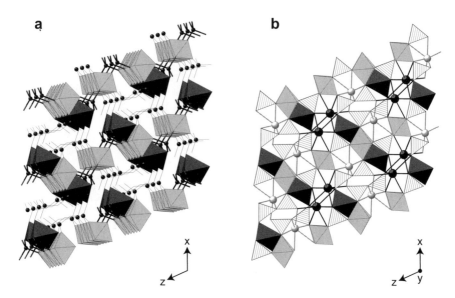

**Figure 1.** Crystal structure of allanite, drawn with CrystalMaker® using data from Dollase (1971). The structure consists of chains of edge-sharing octahedra, which are parallel to [010]: M(1) and M(2) sites are the light gray octahedra; M(3) octahedra are shown as dark octahedra. These chains are linked by single SiO₄ tetrahedra and double-tetrahedral groups (Si₂O₇). **a)** Simplified diagram showing M and T sites only, viewed slightly off the b axis: Si(3) forms an isolated SiO₄ group (shown as sphere with stick bonds); Si(1) and Si(2) share an oxygen, forming an Si₂O₇ group (shown as spheres with wire bonds). **b)** View parallel to the b axis: the SiO₄ and the Si₂O₇ groups are both shown as striated tetrahedra. The large cavities in the framework contain Ca²⁺ (small light gray spheres) and REE³⁺ (black spheres) in 9-fold and 11-fold coordination by oxygen, respectively.

pairs of tetrahedral groups ($Si_2O_7$). The A(1) and A(2) sites reside between the chains and the cross-links (Fig. 1b). In allanite, the A(1) site is occupied by Ca and is 9-fold coordinated, as it is in other epidote-group minerals, but the coordination polyhedron is less regular than in epidote. The A(2) site is 11-fold coordinated in allanite and contains the $REE^{3+}$ (Dollase 1971). Nearly all $Fe^{2+}$ occupies the distorted M(3) positions, but some $Fe^{2+}$ may be present in the M(1) site (Dollase 1973). Even though the $REE^{3+}$ typically enter the A(2) site, extended X-ray absorption fine structure spectra of $Er^{3+}$ and $Lu^{3+}$ environments in synthetic epidotes were interpreted as evidence of smaller REE accommodated in other sites, i.e., $Er^{3+}$ in the A(1) site and $Lu^{3+}$ in the M(3) octahedron (Cressey and Steel 1988). This conclusion, however, has not been corroborated by other data. The unit cell of allanite (Table 2) is larger than those of the clinozoisite-epidote series, and the volume increases with the REE and Fe contents (Fig. 2; Kumskova and Khostova 1964; Bonazzi and Menchetti 1995). This volume increase is correlated with the increase in the *b* value of the unit cell that results from incorporation of REE (Fig. 2) and Fe. Bonazzi and Menchetti (1995) showed that the presence of $REE^{3+}$ decreases the unit-cell parameters *c* and β (Fig. 2). These authors derived equations that can be used to estimate the number of REE and Fe atoms per formula unit in allanite:

$$REE = 1.19 \times (115.41 - \beta) + 2.74 \times (10.160 - c) \tag{3}$$

$$Fe_{total} = 13.22 \times (b - 5.571) - 1.32 \times REE \tag{4}$$

The most common species in the allanite group is allanite-(Ce), which was first described by Thomson (1810). As is commonly observed in REE-rich minerals, however, other REE

**Table 2.** Lattice parameters, crystal system and space group for allanite and other REE-rich epidote-group minerals

| | Epidote | Allanite-(Ce) * | Dissakisite-(Ce) | Ferriallanite-(Ce) | Oxyallanite ** | Dollaseite-(Ce) | Khristovite-(Ce) | Androsite-(La) |
|---|---|---|---|---|---|---|---|---|
| **Unit Cell Parameters** | | | | | | | | |
| a (Å) | 8.914 | 8.927 | 8.905 | 8.962 | 8.893 | 8.934 | 8.903 | 8.896 |
| b (Å) | 5.640 | 5.761 | 5.684 | 5.836 | 5.683 | 5.721 | 5.748 | 5.706 |
| c (Å) | 10.162 | 10.150 | 10.113 | 10.182 | 10.369 | 10.176 | 10.107 | 10.083 |
| $\beta$ (°) | 115.4 | 114.77 | 114.62 | 115.02 | 115.75 | 114.31 | 113.41 | 113.88 |
| V (Å$^3$) | 461.5 | 474.0 | 465.3 | 482.6 | 472.0 | 474.0 | 474.6 | 468.0 |
| Z | 2 | 2 | 2 | 2 | 2 | 2 | 2 | 2 |
| **Crystal System** | monoclinic | monoclinic | monoclinic | monoclinic | monoclinic | monoclinic | monoclinic | monoclinic |
| **Space Group** | P2$_1$/m | P2$_1$/m | P2$_1$/m | P2$_1$/m | P2$_1$/m | P2$_1$/m | P2$_1$/m | P2$_1$/m |
| **Reference** | Dollase (1971) | Dollase (1971, 1973) | Rouse and Peacor (1993) | Kartashov et al. (2002) | Dollase (1973) | Peacor and Dunn (1988) | Pautov et al. (1993) | Bonazzi et al. (1996) |

* Allanite from Pacoima Canyon

** Allanite from Pacoima Canyon, CA, heat-treated in air for 118 h at 680°C. $Fe^{2+}/(Fe^{2+}+Fe^{3+}) = 0.12$

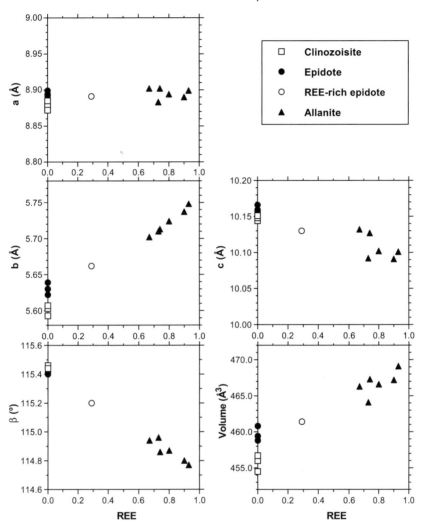

**Figure 2.** Change in values of the unit-cell parameters *a, b, c,* and β and of the unit-cell volume of allanite as a function of REE content (in atoms per formula unit). For comparison, the respective values for clinozoisite and epidote are shown. All data from Bonazzi and Menchetti (1995).

may predominate in certain specimens (e.g., Bayliss and Levinson 1988). La-dominant and Y-dominant allanite were reported in the literature before the introduction by Levinson (1966) of the currently valid nomenclature for rare earth minerals (Hutton 1951a,b; Zhirov et al. 1961; Neumann and Nilssen 1962; Semenov and Barinskii 1958, Hugo 1961). These species are now known as allanite-(La) and allanite-(Y), respectively. Additional occurrences of allanite-(La) and allanite-(Y) have been reported more recently (e.g., Pan and Fleet 1991; Peterson and MacFarlane 1993).

Allanite typically contains small amounts of Th and U and thus may be used as a geochronometer (see below). The presence of isotopes that emit α-particles, however, may lead to the partial or complete transformation of allanite from a crystalline to a metamict state, in which the mineral is more susceptible to alteration (see below). The name orthite was introduced to emphasize a difference in habit from the characteristic tabular form of allanite: it has been used to describe crystals with prismatic habits, yet considerable degrees of alteration and hydration (Deer et al. 1986). This nomenclature, however, is not consistent with the occurrence of isotropic and birefringent varieties of both unaltered and altered hydrated allanite (Deer et al. 1986). Hutton (1951b) suggested that the name orthite be discontinued, and indeed it is no longer in use. The term "allanite", however, has been used quite loosely and inconsistently mainly because: 1) there is still no IMA-approved nomenclature in place for the epidote-group minerals; and 2) REE-rich epidote typically also contains substantial amounts of Th. The latter means that an analysis with REE+Th > 0.5 atoms per formula unit may be called allanite by some authors, whereas others would restrict the name for analyses with REE > 0.5 atoms per formula unit. In this review, we use the former definition. The term "REE-rich epidote" is applied to minerals in which REE are present in weight percent quantities.

Individual physical properties of allanite range widely (Table 3; see also Deer et al. 1986) because the mineral exhibits considerable compositional variability (Ploshko and Bogdanova 1963; Hickling et al. 1970) and various degrees of metamictization (e.g., Zenzén 1916; Tempel 1938; Lima de Faria 1964; Pavelescu and Pavelescu 1972). The refractive indices, birefringence, and density, for example, increase with REE and Fe contents (Tempel 1938; Nesse 2004). The density of metamict allanite is considerably less than that of the crystalline equivalents (see below). Extensive miscibility between epidote-clinozoisite and allanite was pointed out by V.M. Goldschmidt nearly 100 years ago (see comment in Goldschmidt and Thomassen 1924) and later suggested by large suites of analyses (Khvostova 1963; Deer et al. 1986; Pan and Fleet 1990; Grew et al. 1991; Carcangiu et al. 1997). This conclusion is also documented by Figures 3 and 4, which display analyses of epidote-group minerals with widely variable REE contents. The plotted data further demonstrate that epidote-group minerals are in many cases characterized by M-site $Me^{2+}$ contents that exceed the theoretical value of 1 $Me^{2+}$ per formula unit for allanite. It should also be stressed that the plotted data represent analyses of minerals that formed at different pressure and temperature conditions and in a wide variety of geologic environments.

Allanite is a common accessory phase in granite, granodiorite, monzonite, syenite, and granitic pegmatite (e.g., Dollase 1971, 1973; Deer et al. 1986; Buda and Nagy 1995; Broska et al. 2000), and is found in diorite and gabbro as well (Kosterin et al. 1961). Allanite has also been found as phenocrysts in acid volcanic rocks (Izett and Wilcox 1968; Duggan 1976; Brooks et al. 1981; Mahood and Hildreth 1983; Chesner and Ettlinger 1989). The mineral occurs in different types of schist, gneiss, and amphibolite, as well as in metavolcanic rocks and metacarbonates of various metamorphic grades, including those metamorphosed at high-pressure and ultrahigh-pressure conditions (Sakai et al. 1984; Deer et al. 1986; Sorensen 1991; Banno 1993; Cappelli et al. 1993; Smulikowski and Kozlowski 1994; Bingen et al. 1996; Braun 1997; Pan 1997; Gieré et al. 1998; Liu et al. 1999; Ferry 2000; Boundy et al. 2002; Hermann 2002; Spandler et al. 2003; Wing et al. 2003). Detrital allanite is a minor constituent of heavy mineral sands, such as those in Idaho, eastern Greenland, and in Arctic submarine deposits (Frye 1981; Dayvault et al. 1986), and has been observed as detrital grains in clastic sedimentary rocks, such as the Torridonian arkoses (Exley 1980). The mineral also occurs in substantial quantities in the stratiform copper deposit of Talate n'Ouamane, Morocco, where it is associated with various Cu sulfides (Demange and Elsass 1973). Metasomatic allanite has been reported from many geological environments (Gieré 1996; Sorensen 1991), including: limestone skarn (Rudashevskiy 1969; Pavelescu and Pavelescu 1972; Gieré 1986; Smith et al. 2002); altered granite (Ward et al. 1992); altered lujavrite (Coulson 1997); regional metamorphosed calc-silicate rocks (Sargent 1964); carbonate veins (Olson et al. 1954; Peterson and MacFarlane 1993); quartz veins (Banks et al. 1994); and in various other hydrothermal settings, including Alpine clefts (e.g., Exley 1980; Weiss 2002; Weiss and Parodi

2002). As already pointed out by Goldschmidt and Thomassen (1924), allanite is, together with monazite, the main host, and thus repository, of light REE (LREE) in the continental crust (see also Gromet and Silver 1983; Bea 1996).

In some granite pegmatites, alkali granite environments and skarns, allanite occurs in abundances that approach economic potential (Mariano 1989; Möller 1989). Allanite has been mined at the Mary Kathleen uranium deposit in Queensland, where it occurs in a skarn consisting of small uraninite grains in a decussate mass of allanite and apatite (80–90% of the ore; Hawkins 1975; Maas et al. 1987). During the period of 1956 to 1982, approximately 10,000 tons of U and 200,000 tons of REE-oxides were extracted from the Mary Kathleen ore (Scott and Scott 1985). Considerable potential for mining of allanite as a source of LREE exists also in other areas (Hugo 1961; Ehlmann et al. 1964; O'Driscoll 1988; Heinrich and Wells 1980; Halleran and Russell 1996).

### Dissakisite

Dissakisite is the Mg-analogue of allanite, with the ideal formula $CaREEAl_2MgSi_3O_{11}O(OH)$. This endmember was described as a new species in the epidote group by Grew et al. (1991), but allanite with $Mg/(Mg+Fe^{2+}) > 0.5$ has long been known (Meyer 1911; Geijer 1927; Hanson and Pearce 1941, Kimura and Nagashima 1951; Khvostova and Bykova 1961; Treolar and Charnley 1987; Enami and Zang 1988). The name is from the Greek for "twice over," because this Mg-analogue of allanite was, in essence, described twice. Even though additional occurrences of dissakisite have been reported recently (Zakrzewski et al. 1992; de Parseval et al. 1997; Yang and Enami 2003), the mineral is relatively rare. It occurs as an accessory constituent of Si-undersaturated, Mg-and Ca-rich metamorphic rocks, including metacarbonate rock, skarn, garnet lherzolite, and spinel peridotite. At the Trimouns talc-chlorite deposit in the French Pyrenees, dissakisite-(Ce) is associated with bastnäsite; it is the most abundant of all REE minerals which formed from fluids within geodes in dolomite rocks (de Parseval et al. 1997). Dissakisite is related to epidote via the substitution

$$REE^{3+} + Mg^{2+} \leftrightarrow Ca^{2+} + Fe^{3+} \tag{5}$$

and to clinozoisite by

$$REE^{3+} + Mg^{2+} \leftrightarrow Ca^{2+} + Al^{3+} \tag{6}$$

Substitution (6) was suggested by Enami and Zang (1988), who described allanite with up to 6.2 wt% MgO (dissakisite) in garnet-corundum rocks, which are characterized by an extremely high $Al_2O_3/SiO_2$ value as well as by high MgO and CaO contents. Although there is no *a priori* reason to rule out complete miscibility between epidote-clinozoisite and dissakisite, there are few data that demonstrate this relationship (Fig. 4). Grew et al. (1991) pointed out that this lack of data might reflect a more limited extent of solid solution in the Mg-rich, than in the Fe-rich system.

The unit cell of dissakisite-(Ce) is slightly smaller than that of allanite-(Ce) (Table 2; Rouse and Peacor 1993). The color of dissakisite is much paler, its pleochroism weaker, and its density less, than those of allanite (Table 3).

### Ferriallanite

This mineral was discovered in an alkaline granite pegmatite from Mount Ulyn Khuren, which is part of the Khaldzan Buragtag peralkaline granite massif in the Mongolian Altai (Kartashov et al. 2002). Ferriallanite has the idealized formula $CaREEFe^{3+}AlFe^{2+}Si_3O_{11}O(OH)$ and is the $Fe^{3+}$-analogue of allanite (Table 1). The mineral from the type locality is Ce-dominant and thus described as ferriallanite-(Ce). It exhibits both the highest density and birefringence of all REE-rich epidote-group minerals (Table 3).

The structure of ferriallanite shows all the characteristics of monoclinic epidote, and is almost identical to that of allanite (Fig. 1). The reader interested in the atomic parameters

**Table 3.** Physical properties of allanite and other REE-rich epidote-group minerals

| | Allanite-(Ce) * | Dissakisite-(Ce) | Ferriallanite-(Ce) | Oxyallanite | Dollaseite-(Ce) | Khristovite-(Ce) | Androsite-(La) |
|---|---|---|---|---|---|---|---|
| **Indices of Refraction** | | | | | | | |
| α | 1.690 - 1.813 | 1.735 | 1.825 | | 1.715 | 1.773 | |
| β | 1.700 - 1.857 | 1.741 | 1.855 | | 1.718 | 1.790 | |
| γ | 1.706 - 1.891 | 1.758 | 1.880 | | 1.733 | 1.803 | |
| n | 1.53 - 1.70, when metamict | | | | | | 1.877 ** |
| **Birefringence δ** | 0.013 - 0.036 | 0.023 | 0.055 | | 0.018 | 0.030 | |
| **Optic Axial Angle** | | | | | | | |
| $2V_Z$ (calculated) | 40 - 123° | 64.2° | 83° | | | | |
| $2V_Z$ (measured) | | 62° | | | | 83° | |
| **Optic Sign** | (+) or (−) | (+) | (−) | | | (−) | |
| **Dispersion** | r > v; distinct | r < v; medium | r < v; strong | | | r < v; medium | |
| **Color** | brown to black | yellow-brown | black, opaque or translucent | | brown | brown to dark brown | brown-red, transparent |
| **Streak** | gray | white | brown | | | light brown | |
| **Luster** | vitreous, resinous to submetallic | vitreous | resinous | | | vitreous | vitreous |

| Pleochroism | Z ≥ Y > X | weak; X < Y = Z | Z > Y > X | | strong; Y > Z >> X | very strong |
|---|---|---|---|---|---|---|
| X | pale olive-green to reddish brown | pale brown | greenish gray | | very light yellow | pale orange-brown |
| Y | dark brown to brownish yellow | light yellow-brown | brown | | dark reddish brown | |
| Z | greenish brown to dark reddish brown | light yellow-brown | dark red-brown | | brown | deep brown-red |
| **Density (g/cm³)** | | | | | | |
| ρ (measured) | 3.5 - 4.2; when metamict, as low as 2.7 | 3.75 | 4.22 | 3.9 | 4.05 | >4.03 |
| ρ (calculated) | | 3.97 - 4.02 | | | 4.11 | 4.21 |
| **Mohs Hardness** | 5-6½ | | 6 | | 5 | |
| Cleavage | {001} imperfect; {100} and {110} poor | not observed | not observed | not observed | not observed | |
| **Type Locality** | Qeqertsuatsiaq, Aluk, South Greenland | Balchen Mountain, E Antarctica | Mount Ulyn Khuren, Altai, Mongolia | Experimental run product | Inyl'chek Massif, Tien-shan, Kirgiziya | Petalon Mountain, Andros Island, Greece |
| **References** | Deer et al. (1986), Anthony et al. (1995), Zenzén (1916) | Grew et al. (1991) | Kartashov et al. (2002) | Dollase (1973) | Sokolova et al. (1991); Pautov et al. (1993) | Bonazzi et al. (1996) |

\* large ranges due to compositional variation

\** mean refractive index (calculated)

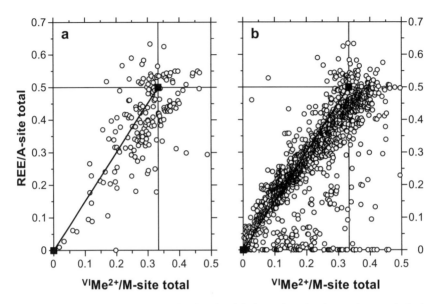

**Figure 3.** Fraction of REE in the A sites vs. fraction of $Me^{2+}$ in the M sites, showing continuous substitutions along a join between the projected compositions (marked by full squares) of ideal clinozoisite or epidote, and ideal allanite or dissakisite. **a)** analyses, where $Fe^{2+}$ and $Fe^{3+}$ have been analyzed (n = 215), **b)** samples, for which $Fe^{2+}/Fe^{3+}$ was calculated (n = 1521). Diagram is based on Figure 37 of Khvostova (1962).

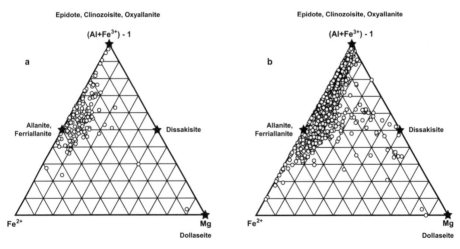

**Figure 4.** Extent of solid solution in the $Fe^{2+}$-$Mg^{2+}$-$(Al+Fe^{3+})$ subsystem. Asterisks show projections of endmember compositions into this space. **a)** analyses, where $Fe^{2+}$ and $Fe^{3+}$ have been directly analyzed (n = 215); **b)** analyses, for which $Fe^{2+}$ and $Fe^{3+}$ have been calculated (n = 1521). Early studies (which generally relied on gravimetric analysis for determination of $Fe^{2+}$ and $Fe^{3+}$) did not focus on the Mg-rich part of the system, whereas modern ones (which rely on the electron microprobe, and thus calculate $Fe^{2+}/Fe^{3+}$) have shown a more substantial solution toward the Mg-endmembers. Figure ignores the possible presence of $Fe^{2+}$ in the A sites.

and interatomic distances for ferriallanite-(Ce) should consult the paper of Kartashov et al. (2003), which lists values that have been corrected relative to those published in the original description (Kartashov et al. 2002). Ferriallanite-(Ce) has the largest unit cell of all the REE-rich epidote minerals (Table 2). Mössbauer spectroscopy indicates that $Fe^{2+}$ is present in only the strongly distorted M(3) octahedral site (Kartashov et al. 2002; Holtstam et al. 2003). Although $Fe^{3+}$ partitions between the other two octahedral sites, it is the dominant cation in M(1). This feature appears to be unique in the epidote-group minerals and is in contrast to epidote, where $Fe^{3+}$ prefers the M(3) site (Dollase 1973; Nozik et al. 1978; Kvick et al. 1988; see also Franz and Liebscher 2004).

The chemical relationships among ferriallanite, allanite, epidote, and clinozoisite are illustrated in an REE vs. Al diagram, first proposed by Petrík et al. (1995). This diagram (Fig. 5) can be used to estimate the proportions of $Fe^{3+}$ and $Fe^{2+}$ in the solid solutions defined by these four endmembers. The diagram graphically represents substitution (1), relating epidote and allanite, and the exchange vector $Fe^{3+}Al_{-1}$, which connects clinozoisite to epidote and allanite to ferriallanite. The data in Figure 6 show that there are almost no analyses on the Al-rich side of the tie-line between clinozoisite and allanite; data in this area of the diagram would require REE substitution to be balanced by some other mechanism than $Me^{2+}$ for $Me^{3+}$. In contrast, quite a few analyses plot on the Fe-rich side of the epidote-ferriallanite join, and some analyses are characterized by REE+Th > 1 (compare with Fig. 3). The diagrams further document that ferriallanite is an important component of many epidote-group minerals with $Fe^{3+}/Fe_{tot} < 0.5$.

At the Mongolian type locality, ferriallanite is of metasomatic origin and is associated with zircon, quartz, aegirine, magnetite, fayalite, fluorite, and the REE minerals kainosite-(Y), β-fergusonite-(Y), hingganite-(Ce), and allanite-(Ce). To date, there is only one other ferriallanite-(Ce) locality, the Bastnäs Fe-Cu-REE deposit in the Skinnskatteberg district, Västmanland, Sweden (Holtstam et al. 2003). There, ferriallanite-(Ce) is closely associated with cerite-(Ce), bastnäsite-(Ce), bastnäsite-(La), törnebohmite-(Ce), quartz, fluocerite-(Ce), and various sulfide minerals. Next to cerite-(Ce), ferriallanite-(Ce) is the most common REE mineral at this deposit, where it was discovered long before its true nature was established: it was first reported in 1781, described as "black hornblende" in 1804, and then recognized as a

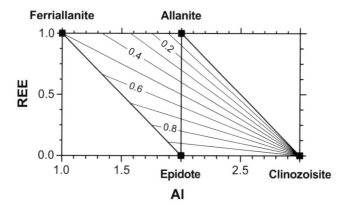

**Figure 5.** REE vs. Al (cations per formula unit) diagram showing the chemical relationships in the system allanite-ferriallanite-epidote-clinozoisite. Lines radiating from the clinozoisite endmember represent lines of constant Fe-oxidation state and are labeled for $Fe^{3+}/Fe_{total}$. This diagram, introduced by Petrík et al. (1995), can be used to estimate the proportions of $Fe^{2+}$ and $Fe^{3+}$ in compositions within this chemical system.

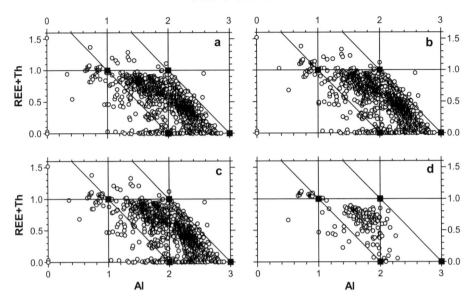

**Figure 6.** Allanite compositions from the database plotted in a diagram similar to the one shown in Figure 5 to estimate $Fe^{3+}/Fe_{total}$ (Petrík et al. 1995). Full squares show projections of endmember compositions into this space (see Fig. 5). **a)** all ferric, with no estimate of $Fe^{2+}/Fe^{3+}$; **b)** all ferrous, with no estimate of $Fe^{2+}/Fe^{3+}$; **c)** all ferrous, with $Fe^{2+}/Fe^{3+}$ estimated; **d)** data points represent analyses, where $Fe^{2+}$ and $Fe^{3+}$ have been analyzed (n = 215). Although the diagram was used for formula parameters calculated on the basis of 12.5 oxygens, and all iron as $Fe^{2+}$ or as $Fe^{3+}$, the positions of points in a-c (n = 1521) are relatively insensitive to the formula calculation procedure. This is likely because both parameters total to relatively large fractions of the formula.

new Ce-bearing mineral named "cerine" in 1815 (see discussion and references in Holtstam et al. 2003). The samples from Bastnäs contain Mg and are richer in REE and F than the Mg-free specimens from Mongolia.

## Oxyallanite

Oxidation of $Fe^{2+}$ to $Fe^{3+}$, the release of $H_2$, and the concomitant replacement of $OH^-$ by $O^{2-}$ would produce an oxy-equivalent of allanite. The reaction

$$Fe^{2+} + OH^- \leftrightarrow Fe^{3+} + O^{2-} + 1/2 \, H_2 \tag{7}$$

is formally equivalent to the oxy-reaction observed in other hydrous $Fe^{2+}$-bearing silicate minerals, e.g., mica and amphibole (Hogg and Meads 1975; Ferrow 1987; Popp et al. 1995a, b). In contrast to micas and amphiboles, the $OH^-$ group, which loses a H atom, is not directly bonded to the $Fe^{2+}$, which is being oxidized; rather, the $OH^-$ group is bonded to M(2) only, a site that is almost entirely filled by Al (Dollase 1973; Nozik et al. 1978; Kvick et al. 1988; Giuli et al. 1999; Franz and Liebscher 2004). Anhydrous oxyallanite $CaREEAl_2Fe^{3+}Si_3O_{11}O(O)$ is a theoretical endmember (Table 1), for which there is good experimental evidence: Dollase (1973) heated allanite from a pegmatite in Pacoima Canyon, California in air at different temperatures and run durations, and analyzed $Fe^{3+}$ and $Fe^{2+}$ of the run products with Mössbauer spectroscopy. Heating at $T < 400°C$ did not noticeably change the value of $Fe^{2+}/(Fe^{2+}+Fe^{3+})$, even if samples were heated for several days. At $T > 400°C$, the $Fe^{2+}$ contents decreased with increasing temperature, such that the sample was almost completely oxidized at 700°C (Fig. 7a). The degree of oxidation was apparently independent of duration of heating. Dollase (1973) further showed that heating in a reducing atmosphere could reverse the oxidation. The changes in

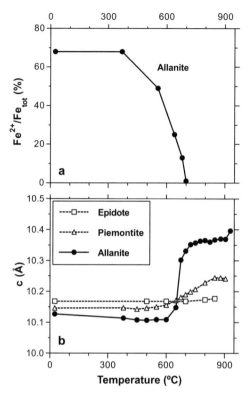

**Figure 7.** Effects of heating allanite, epidote and piemontite in air. **a)** $Fe^{2+}/(Fe^{2+}+Fe^{3+})$ vs. temperature for allanite from Pacoima Canyon (data from Dollase 1973); **b)** Unit-cell parameter $c$ vs. temperature for allanite, epidote, and REE-rich piemontite (data from Bonazzi and Menchetti 1994).

unit-cell dimensions associated with the transformation of allanite into oxyallanite are small (Table 2). Kumskova and Khvostova (1964) and Khvostova (1962), who published weight-loss curves obtained from allanite dehydration experiments, reported similar observations.

Bonazzi and Menchetti (1994) heated single crystals of allanite from granite of the Rhodope Massif, Bulgaria in air for 48 hours at temperatures in the range of 380–900°C. These authors found that the oxidation-dehydrogenation of allanite begins at ~600°C and is essentially complete at about 700–725°C (Fig. 7b). Above 725°C, unit-cell parameters do not change significantly until ~905°C. At the latter temperature, there is a marked change in the unit-cell parameters, which Bonazzi and Menchetti (1994) ascribed to oxidation of $Ce^{3+}$ to $Ce^{4+}$. This oxidation ends with the breakdown of the allanite structure, via the precipitation of $CeO_2$. Earlier studies also reported that cerianite is a breakdown product of allanite at high temperatures (see below).

Bonazzi and Menchetti (1994) also heated single crystals of REE-rich piemontite from Monte Brugiana in the Alpi Apuane, Italy. The piemontite behaved similarly to allanite upon heating, although the oxidation-dehydrogenation process

$$Mn^{2+} + OH^- \leftrightarrow Mn^{3+} + O^{2-} + 1/2\ H_2 \tag{8}$$

took place more gradually, i.e., over a larger $T$ interval, and complete oxidation of $Mn^{2+}$ did not occur until 850°C (Fig. 7b). In contrast, epidote, appeared to be unaffected by heat treatment.

The presence of an oxyallanite component in natural members of the epidote group is difficult to verify, because of analytical uncertainties and the general lack of direct determination of $H_2O$ contents. Based upon chemical analyses of natural samples, Grew et al. (1991) proposed that substitution (7) is important in numerous natural allanite, dissakisite and dollaseite compositions, for which analytical data for both $Fe^{3+}$ and $Fe^{2+}$ were available. As will be discussed, the oxidation-dehydrogenation process may be of considerable petrological importance.

### Dollaseite

Dollaseite-(Ce) was first reported by Geijer (1927). He named it magnesium orthite, because it appeared to be the Mg-analogue of orthite. However, Peacor and Dunn (1988) showed that the specimen studied by Gejier (1927) was not the Mg-analogue of what is now termed allanite (i.e., dissakisite). These authors named it to honor Wayne Dollase, who refined the structures of epidote, allanite, and hancockite (Dollase 1971). Dollaseite-(Ce) contains Mg on both the M(3) and M(1) sites, and compensates for the resulting charge imbalance by the substitution of F for O(4), i.e., the O atom that is not coordinated to Si (Peacor and Dunn 1988). The substitution

$$Mg^{2+} + F^- \leftrightarrow Al^{3+} + O^{2-} \qquad (9)$$

relates dollaseite, with the idealized formula $CaREEAlMg_2Si_3O_{11}F(OH)$, to dissakisite. The relationship between dollaseite and clinozoisite can be described by an overall substitution

$$REE^{3+} + 2\,Mg^{2+} + F^- \leftrightarrow Ca^{2+} + 2\,Al^{3+} + O^{2-} \qquad (10)$$

(Burt 1989), which is a combination of (6) and (9).

The structure refinement of Peacor and Dunn (1988) showed that the dollaseite-(Ce) structure displays only minor shifts from the atomic positions of epidote or allanite. These refinement data are compatible with nearly complete ordering of Mg in M(1), Al in M(2), and Mg and $Fe^{2+}$ in M(3) (Table 1). The unit cell is approximately the same size as that of allanite (Table 2).

Dollaseite-(Ce) is found in a skarn at the Östanmossa Mine in the Norberg district of Sweden, where it is associated with tremolite, norbergite, and calcite (Geijer 1927).

### Khristovite

The second REE-rich species of the epidote group with two divalent cations in octahedral sites was discovered by Sokolova et al. (1991), and was subsequently named khristovite, in honor of Evgenia Valdimirovicha Khristova, a Russian geologist and specialist in Tien-shan geology (Pautov et al. 1993). As in the case of dollaseite-(Ce), a charge imbalance results from two divalent cations on the M sites, which is compensated for by replacing O(4) with F. In contrast to dollaseite-Ce, however, khristovite contains $Mn^{2+}$ rather than Mg on the M(3) site (Table 1, Fig. 8), and thus has the idealized formula $CaREEMgAlMn^{2+}Si_3O_{11}F(OH)$. According to the structural refinement of Sokolova et al. (1991), the A(1) site in khristovite-(Ce) is not occupied exclusively and completely by Ca. Instead, 20% of the site is vacant, and 20% occupied by La. These data indicate that REE are not restricted to the A(2) sites in all epidote-group minerals (see also Cressey and Steel 1988).

Khristovite-(Ce) is found in a Mn-rich rock from the Tien-shan in Kirgizia, where it is associated with other Mn minerals, including rhodonite, tephroite, and rhodochrosite (Sokolova et al. 1991; Pautov et al. 1993). The unit cells of dollaseite-(Ce) and khristovite-(Ce) display almost identical volumes (Table 2), but the density of khristovite is greater (Table 3). Khristovite-(Ce) also exhibits higher indices of refraction than dollaseite-(Ce).

### Androsite

Androsite is extremely rich in Mn and poor in Fe, and has the idealized formula $Mn^{2+}REEMn^{3+}AlMn^{2+}Si_3O_{11}O(OH)$ (Bonazzi et al. 1996). Its name is derived from the type

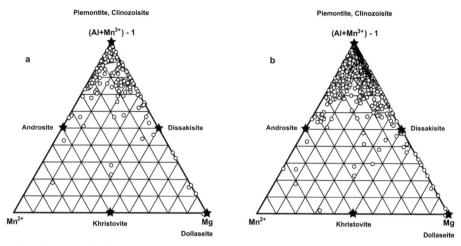

**Figure 8.** Extent of solid solution in the $Mn^{2+}$-$Mg^{2+}$-$(Al+Mn^{3+})$ subsystem. Asterisks show projections of endmember compositions into this space. **a)** analyses, where $Fe^{2+}$ and $Fe^{3+}$ have been directly analyzed (n = 215); **b)** analyses, for which $Fe^{2+}$ and $Fe^{3+}$ have not been analyzed (n = 1521). For the calculation of $Mn^{2+}/Mn^{3+}$, all Fe was first assumed to be $Fe^{3+}$; then, the analyses were normalized to 12.5 oxygens and 8 cations by adjusting $Mn^{2+}/Mn^{3+}$. Figure ignores the possible presence of $Mn^{2+}$ in the A sites.

locality, the Cycladic Island of Andros in Greece. As seen in the structural formula, androsite contains Mn in three different sites, and in both the divalent and trivalent states (Table 1). Androsite is related to allanite by

$$Mn^{2+} \leftrightarrow Ca^{2+} \text{ on } A(1) \tag{11}$$

$$Mn^{3+} \leftrightarrow Al^{3+} \text{ on } M(1) \tag{12}$$

$$Mn^{2+} \leftrightarrow Fe^{2+} \text{ on } M(3) \tag{13}$$

The structure of androsite is virtually identical to that of allanite. In contrast to the epidote-piemontite solid solution, in which $Mn^{3+}$ prefers the more distorted M(3) site (Dollase 1973; Ferraris et al. 1989), $Mn^{3+}$ is accommodated in the M(1) site of androsite. The unit cell of androsite-(La) is smaller than those of most other REE-dominant epidote-group minerals (Table 2). Its calculated density is almost identical to that of ferriallanite (Table 3).

The described specimen from the type locality in Greece, a Mn-ore deposit, contains La as the predominant REE, and thus, is androsite-(La). The Mn-rich, Fe-poor composition is interpreted to reflect the unusual bulk composition of the host rock, in which androsite coexists with braunite, rhodonite, rhodochrosite, spessartine-rich garnet, and quartz.

Until the discovery of khristovite-(Ce), Mn was thought to be a common, but typically relatively minor component of REE-rich epidote-group minerals. Substantial MnO contents (5 to 7 wt%), however, have been reported for allanite (Hutton 1951a; Ovchinnikov and Tzimbalenko 1948; Kimura and Nagashima 1951; Hagesawa 1957, 1958; Pavelescu and Pavelescu 1972). For many of these specimens, described as manganoan allanite and manganorthite, the host rock is a granitic pegmatite. Deer et al. (1986) suggested that $Mn^{2+}$ probably replaced Ca in these minerals, but speculated that, in some cases, Mn might also enter the octahedral sites. This conclusion is consistent with rare occurrences of REE-rich piemontite in Mn-deposits and other Mn-rich rock types (Williams 1893; Kramm 1979; Schreyer et al. 1986; see also Bonazzi and Menchetti 1994, 1995, 2004). These examples provide evidence for extensive solid solution between piemontite and allanite (Bonazzi et al. 1992; Bonazzi and

Menchetti 1994; Bermanec et al. 1994). The substitution is

$$REE^{3+} + Fe^{2+} \leftrightarrow Ca^{2+} + Mn^{3+} \tag{14}$$

and represents the Mn-equivalent of (1). Bonazzi et al. (1996) also reported an intermediate composition between piemontite and androsite-(La). The compositional relationships for REE-rich epidote in the Mn-Mg-Al system are shown in Figure 8, which ignores the possibility of $Mn^{2+}$ in A sites. The substitution of $Mn^{2+}$ for Ca in the A site (Eqn. 11) is not only important in androsite, but has also been proposed for other members of the epidote group (Bonazzi et al. 1992).

## MINOR AND TRACE ELEMENTS IN ALLANITE

In addition to REE, allanite incorporates many other elements, which, depending on the geological environment, can become essential structural constituents present in weight percent quantities (Deer et al. 1986). The wide range of chemical components that can be incorporated into allanite documents that the structure is able to accept ions with widely different radii and charges. These ions are accommodated primarily in the A and M sites, but some have been interpreted to replace Si. Although some elemental abundances and element distributions within grains may reflect secondary processes, such as metamictization and weathering, most minor elements appear to have been incorporated during mineral growth, and thus testify to diverse environments of mineral formation.

The advent of new instrumental techniques designed specifically for the analysis of trace elements has led to a wealth of extensive data sets over the past few years. Most significant among these new instrumental techniques is *in situ* analysis by laser-ablation inductively-coupled plasma mass spectrometry (LA-ICP-MS), a trace element technique that has also been used for epidote-group minerals (e.g., Bea 1996; Hermann 2002; Holtstam et al. 2003; Spandler et al. 2003).

### Substitutions in the A sites

Most minor substitutions in the A sites of allanite and other REE-rich epidote-group minerals are on the A(2) site, but rarely the A(1) site is also involved (Sokolova et al. 1991).

***Thorium.*** This element is a commonly observed A-site constituent present in trace to minor amounts (e.g., Smith et al. 1957; Rao and Babu 1978; Rao et al. 1979; Gieré 1986; Janeczek and Eby 1993; Banno 1993; Barth et al. 1989, 1994; Liu et al. 1999; Wood and Ricketts 2000; Yang and Enami 2003; Oberli et al. 2004). Reported $ThO_2$ contents in allanite are often in the range of 2–3 wt% (Hagesawa 1959, 1960; Khvostova 1962; Ploshko and Bogdanova 1963; Kalinin et al. 1968; Oberli et al. 1981; Peterson and MacFarlane 1993; Buda and Nagy 1995; Bea 1996), and the maximum concentration reported so far is 4.9 wt%, reached at two different localities (Hagesawa 1960; Exley 1980). Two analyses presented by Oberli et al. (1981) show $ThO_2$ contents of 22.2 wt% and 37.3 wt%, and large amounts of $UO_2$ (5.5 wt%, 23.5 wt%) and $P_2O_5$ (2.28 wt%, 2.0 wt%), but the calculated formulae are not consistent with those of epidote-group minerals. The substitution mechanisms responsible for Th incorporation are speculative; proposals for the Mg-free system include (Gieré et al. 1999)

$$Th^{4+} + 2\,Fe^{2+} \leftrightarrow Ca^{2+} + 2\,Fe^{3+} \tag{15}$$

$$Th^{4+} + Fe^{2+} \leftrightarrow REE^{3+} + Fe^{3+} \tag{16}$$

$$Th^{4+} + Fe^{2+} \leftrightarrow REE^{3+} + Al^{3+} \tag{17}$$

All are accompanied by a change in $Fe^{2+}/Fe^{3+}$. Accommodation of Th might, however, also be achieved without involving Fe (or Mg or $Mn^{2+}$), for example,

$$Th^{4+} + Ca^{2+} \leftrightarrow 2\ REE^{3+} \tag{18}$$

(Gromet and Silver 1983; Chesner and Ettlinger 1989; Gieré et al. 1999; Wood and Ricketts 2000), which is an important substitution in other minerals as well (e.g., monazite group), and

$$3\ Th^{4+} + \square \leftrightarrow 4\ REE^{3+} \tag{19}$$

in which $\square$ = vacancy. With the exception of (15), these substitutions require REE-bearing compositions to effect charge balance.

**Uranium.** A parallel set of substitutions to those for Th could be formulated for $U^{4+}$, but U is typically present in much smaller concentrations than Th (Hickling et al. 1970; Smith et al. 1957; Sawka 1988; Sawka and Chappell 1988; Barth et al. 1989, 1994; Zakrzewski et al. 1992; Oberli et al. 2004). Allanite only rarely contains more U than Th; an example of this unusual feature is seen in the metasomatically formed crystals from the Mary Kathleen U-REE skarn in Queensland (Maas et al. 1987). The predominance of Th relative to U in REE-rich epidote-group minerals is clearly displayed in Figure 9, which contains all analyses from our dataset that list values for both U and Th, and which shows that U is typically present in the < 1000 ppm range. However, Janeczek and Eby (1993) measured up to 0.21 wt% $UO_2$ for allanite from pegmatites. Banno (1993) described allanite with even higher $UO_2$ contents (up to 0.36 wt%). The latter occurs in blueschists from the Sanbagawa belt of Japan, and is unusual as it appears to be one of only very few reported allanites with U/Th > 1 (see Figure 9). The maximum $UO_2$ content observed is 0.82 wt% in a crystal containing 1.09 wt% $ThO_2$ from the Cacciola granite in the central Gotthard massif, Switzerland (Oberli et al. 1981).

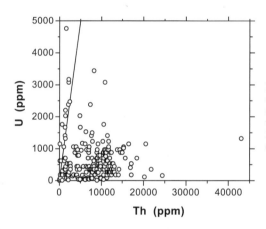

**Figure 9.** U vs. Th diagram showing analyses that contain both U and Th in detectable concentrations (n = 220). Of these analyses, only 16 plot above the line of slope 1 and thus contain U in excess of Th.

**Strontium.** This element has been observed as a minor or trace component of allanite at various localities (e.g., Ploshko and Bogdanova 1963; Bocquet 1975; Rao and Babu 1978). Papunen and Lindsjö (1972) found allanite with 0.87 wt% SrO from a skarn at Korsnäs, Finland. Exley (1980) described an allanite containing 1.5 wt% SrO from altered basalt on the island of Skye, but noted that the Sr content is mostly less than 1000 ppm. Sorensen (1991) reported 1.1–1.8 wt% SrO in zoned allanite from tectonic blocks of metasomatized garnet amphibolite from the Catalina Schist, southern California. Zdorik et al. (1965) observed 3.72 wt% for allanite from carbonatite. Allanite containing 4.28 wt% SrO has been described from a fenite associated with the Mount Bisson REE deposits, British Columbia (Halleran and Russell, 1996). REE-rich epidote that contains 2.2–4.0 wt% SrO is present in lawsonite blueschist from New Caledonia (Spandler et al. 2003), and allanite grains from a leucosyenite in China contain an average of 1.65±0.34 wt% SrO, with values that range to 4.5 wt% (Jiang et al. 2003).

**Lead.** Bermanec et al. (1994) reported Pb-rich piemontite with total $REE_2O_3$ contents of 6.09 to 8.56 wt%, up to 10.4 wt% PbO, and ZnO from 0.6 to 0.75 wt%, as an accessory mineral associated with ardennite, gahnite, and franklinite (?) in mica schist from Nezilovo, Macedonia. This piemontite is characterized by a significant allanite component, a component of the Pb-epidote mineral hancockite, and minor Zn-substitution.

**Alkali elements.** Bermanec et al. (1994) determined up to 0.02 wt% $K_2O$ and 0.12 to 0.24 wt% $Na_2O$ for the aforementioned Pb- and REE-rich piemontite. Similarly, all analyses given by Yang and Enami (2003) reported between 0.14 and 0.21 wt% of $Na_2O$ in dissakisite-(Ce) from China. These values are significantly larger than the 0.02 wt% detection limit of the electron microprobe used by these researchers. Furthermore, all compositions listed by Bea (1996) for allanites from granitoid rocks appear to contain small amounts of $Na_2O$ (0.06 to 0.56 wt%; determined by electron microprobe). Alkali element substitution seems to be rather uncommon and limited (cf. Deer et al. 1986), because of the cation sizes and charge densities. However, Semenov et al. (1978) observed a $Na_2O$ content of 1.36 wt% for allanite from a carbonatite complex in Tamil Nadu. Coulson (1997) reported an analysis of allanite with 3.28 wt% $Na_2O$ (= 0.61 cations, based upon a 12.5 oxygen formula); this allanite was produced by alteration of eudialyte in a lujavrite (described by the author as eudialyte microsyenite) from the Igaliko complex in the Gardar province of South Greenland. Although the overall alteration process is described as Na-loss from the system, the original rock and the primary minerals are extremely sodic, which may account for this exceptional value for Na in allanite. Possible mechanisms for Na incorporation are (Yang and Enami 2003)

$$Na^+ + Ti^{4+} \leftrightarrow Ca^{2+} + Fe^{3+} \tag{20}$$

$$Na^+ + Fe^{3+} \leftrightarrow Ca^{2+} + Fe^{2+} \tag{21}$$

which could also be formulated analogously for Al instead of $Fe^{3+}$. Another possible substitution is

$$Na^+ + REE^{3+} \leftrightarrow 2\ Ca^{2+} \tag{22}$$

which is common in minerals, such as the perovskite and apatite groups (e.g., Burt 1989; Rønsbo 1989; Mitchell 1996; Campbell et al. 1997).

**Manganese and iron.** Divalent Mn and Fe may also substitute for Ca. However, with the exception of androsite-(La) and some altered allanites, the substitutions are not significant (Bonazzi et al. 1992; Bonazzi and Menchetti 1995; Catlos et al. 2000; Poitrasson 2002).

**Vacancies.** Although it is possible that the A sites of some allanites may be partially vacant, the presence of vacancies is difficult to document due to the chemical complexity of the mineral, and especially the viscissitudes of the calculation of $Fe^{2+}/Fe^{3+}$ from electron microprobe analyses. As discussed above and in this section, allanite typically contains several elements with multiple oxidation states as well as water. Furthermore, as pointed out by Catlos et al. (2000), allanite can contain minor amounts of many elements, not all of which may be detected without detailed wavelength scans by electron microprobe, prior to analysis. Therefore, the calculation of vacancies based on electron microprobe data can be associated with large uncertainties (see, for example, Ercit 2002). However, a small body of evidence exists to support the idea that vacancies may occur in the A sites of allanite and related minerals. In conjunction with microprobe data, the structural refinement of khristovite-(Ce) yielded considerable amounts of vacancies in the A sites (Sokolova et al. 1991), particularly in the A(1) site (20% vacant). Peterson and MacFarlane (1993) also inferred existence of vacancies in allanite from granite pegmatites and calcite veins in the Grenville Province,

Ontario, Canada. These authors observed that most of the studied allanites exhibited A-site vacancies, but to varying degrees. Moreover, they concluded that in addition to substitutions (1) and (2), an omission-type substitution:

$$3 \, Ca^{2+} \leftrightarrow 2 \, REE^{3+} + \square \tag{23}$$

was required to explain their data. However, the average formula (based on 12.5 oxygens and 8 cations) for the allanites studied by Peterson and MacFarlane (1993) is $A_{1.88}M_{3.06}T_{3.06}O_{11}O(OH)$, and thus shows excess charge on the M and T sites. This too is a common feature of "A-site-deficient" allanite (Sorensen 1991; Catlos et al. 2000). Although the M and T excess charges could be dismissed as being within the error of a microprobe determination, hundreds of analyses show them; they may hold a clue for explaining the A-site deficiencies.

## Substitutions in the M sites

*Titanium.* This element is a minor constituent of REE-rich epidote-group minerals (Ploshko and Bogdanova 1963; Exley 1980; Hagesawa 1960, Deer et al. 1986; Gieré 1986; Rudashevskiy 1969; Grew et al. 1991; Banno 1993; Liu et al. 1999; Jiang et al. 2003; Yang and Enami 2003). $TiO_2$ contents of 1 to 2 wt% are often observed for allanite from volcanic and plutonic rocks (e.g., Ploshko and Bogdanova 1963; Ghent 1972; Duggan 1976; Brooks et al. 1981; Chesner and Ettlinger 1989; Catlos et al. 2000; Wood and Ricketts 2000; Jiang et al. 2003). Similarly, khristovite-(Ce) contains 1.5 to 1.8 wt% Ti (Pautov et al. 1993). Poitrasson (2002) reported $TiO_2$ contents that range from 1.44 to 4.67 wt% for variably altered allanite occurring in Paleozoic granitoid rocks on Corsica. Besides the 7.46 wt% $TiO_2$ observed by Khvostova (1962), the maximum $TiO_2$ content reported so far is 5.2 wt% for allanite from a fayalite-hedenbergite syenite (Krivdik et al. 1989). Titanium can be incorporated into epidote-group crystal structures via

$$Ti^{4+} + Fe^{2+} \leftrightarrow 2 \, Fe^{3+} \tag{24}$$

and

$$Ti^{4+} + Fe^{2+} \leftrightarrow 2 \, Al^{3+} \tag{25}$$

These substitutions may also involve $Mg^{2+}$ and $Mn^{2+}$, as well as less abundant divalent cations. As is the case for incorporation of Ti in amphibole and mica structures, octahedral-site substitutions such as (24) will be strongly influenced by $f_{O_2}$. Alternatively, Ti can enter the allanite structure along with Na according to substitution (20).

*Vanadium.* As in REE-free epidote-group minerals (Shepel and Karpenko 1969; Franz and Liebscher 2004), V can be incorporated into REE-rich epidotes (Ovchinnikov and Tzimbalenko 1948; Papunen and Lindsjö 1972; Bocquet 1975; Rao and Babu 1978; Rao et al. 1979). In khristovite-(Ce), $V_2O_3$ ranges from 1.0 to 1.7 wt% (Pautov et al. 1993), and Pan and Fleet (1991) described allanite and REE-rich epidote with up to 9.07 wt% $V_2O_3$ from Hemlo in Ontario, Canada. Kato et al. (1994) reported core and rim analyses of a vanadian allanite from the Mn-Fe ore of the Odaki ore body, Kyurazawa Mine, Ashio, Tochigi Prefecture, Japan, with 9.58 and 8.83 wt% $V_2O_3$, respectively. Vanadium probably enters the M site of the allanite structure as $V^{3+}$, as in the case of mukhinite (Shepel and Karpenko 1969); an extremely oxidizing environment would probably be needed to stabilize $V^{5+}$ in allanite.

*Chromium.* Typically, Cr is a minor constituent of the epidote-group minerals, but in some cases, it may be present in relatively large amounts (e.g., Papunen and Lindsjö 1972; Grapes 1981; Sorensen 1991; Grapes and Hoskin 2004; Franz and Liebscher 2004). For example, khristovite-(Ce) contains 1.3 to 2.1 wt% $Cr_2O_3$ (Pautov et al. 1993). Allanite with even greater amounts of Cr has been reported from the Outokumpu mining district in Finnish Karelia, where it occurs in mica- and spinel-rich layers of mica schist (Treolar and Charnley 1987). These authors concluded that the mica schist had become enriched in Cr that had ultimately been derived from underlying ultramafic and volcanic rocks. The Cr-rich allanite from Outokumpu contains 3.63–5.40 wt% $Cr_2O_3$ (= 0.51–0.77 atoms per formula unit), exhibits $Mg/(Mg+Fe^{2+}) > 0.5$, and is thus a Cr-rich dissakisite (for the compositional limits of dissakisite, see Grew et al. 1991). The apparently coexisting spinel and biotite are Cr-rich. From the same area, Treolar (1987) reported epidote with > 15 wt% $Cr_2O_3$, i.e., close to the endmember tawmawite, $Ca_2CrAl_2Si_3O_{11}O(OH)$. Thus, there appears to be at least limited solid solution between tawmawite and dissakisite, via the substitution

$$REE^{3+} + Mg^{2+} \leftrightarrow Ca^{2+} + Cr^{3+} \qquad (26)$$

Chromian dissakisite-(Ce) that contains 4.6 to 5.4 wt% $Cr_2O_3$ was described by Yang and Enami (2003). It is an inclusion in a clinopyroxene grain in garnet lherzolite from Zhimafang, eastern China. Yang and Enami (2003) concluded from microstructural relationships that chromian dissakisite-(Ce) had reacted with olivine to form clinopyroxene. In addition to Outokumpu and Zhimafang, two other localities have yielded Cr-rich allanite compositions. At the Orford Ni mine in Québec, Canada, allanite with 7.11 wt% $Cr_2O_3$ is associated with Cr-rich grossular (Tarassoff and Gault 1994), but this allanite analysis does not list MgO. Blueschists interbedded with serpentinite conglomerates in the Ise area of the Sanbagawa belt, Japan, contain aggregates of chromian phengite, sodic pyroxene, and Cr-rich spinel. Allanite from one such aggregate contains 7.49 wt% $Cr_2O_3$ and 0.03 to 0.09 wt% MgO (Banno 1993).

### Substitutions in T sites

Many analyses of epidote-group minerals yield formulae with <3 Si atoms per formula unit (e.g., Deer et al. 1986). In our database of allanite and related minerals, about 480 analyses (i.e., ~1/3) show Si < 2.95, indicating that substitution of other components for Si is possible. However, extensive substitution for Si in the T sites is not observed in REE-poor epidote minerals. Neutron diffraction of epidote (Nozik et al. 1978; Kvick et al. 1988) and strontian piemontite (Ferraris et al. 1989) has not yielded evidence for substantial Al in tetrahedral coordination. On the other hand, Poitrasson (2002) presented electron microprobe data for altered allanite, which are characterized by overfilling of the M sites and by a Si deficiency of $\approx 0.15$ atoms per formula unit. Based on these results, he suggested that some Al should be present in the T sites. A few studies indicate that, in some rare cases, Si may be partially replaced by $P^{5+}$, $Ge^{4+}$, or possibly $Be^{2+}$.

*Phosphorus.* This element is present in some REE-rich epidote group minerals, but typically at low concentrations (Hagesawa 1960; Mahood and Hildreth 1983; Grew et al. 1991; Petrík et al. 1995; Hermann 2002). Bea (1996) noted that allanite in granitic rocks nearly invariably contains some P (up to 0.2 wt% $P_2O_5$). Bocquet (1975) described a specimen with 0.36 wt% $P_2O_5$ from a hydrothermal allanite-albite-hematite vein, and similar concentrations have been reported for altered allanite occurring in leucosyenite in northern China (Jiang et al. 2003). In a pegmatite, Iimori et al. (1931) found an unusual specimen that contains 6.48 wt% $P_2O_5$, which corresponds to 0.54 P atoms per formula unit. Based on physical properties (including optical) and molar ratios of all chemical elements present, these authors concluded that the specimen resembles allanite, but provided no crystallographic data to corroborate their interpretation. The name proposed for this unusual composition was "nagatelite", after the Japanese locality where it was discovered.

*Germanium.* Typical Ge concentrations in epidote-group minerals range from 1 to 20 ppm (Bernstein 1985; Hermann 2002). However, what appears to be a Ge-rich allanite crystal has been described from a sphalerite ore deposit in Haute Garonne, French Pyrenees (Johan et al. 1983). This specimen contains 10.63 wt% $GeO_2$ in addition to 2.52 wt% $Ga_2O_3$, suggesting considerable replacement of $Si^{4+}$ by the larger $Ge^{4+}$ (26 pm vs. 39 pm; Shannon 1976) in the T sites, and of Al and Fe by $Ga^{3+}$ in the M sites. Because the original analysis does not give quantitative data for the REE, however, the exact amount of these cations in each of the sites cannot be calculated. It is of note that this Ge-rich specimen also contains Zn (see below), reflecting the well-known geochemical association of Ge and Zn in sphalerite deposits (Bernstein 1985).

**Beryllium.** In some pegmatites, allanite incorporates trace to minor amounts of Be (Deer et al. 1986; Hagesawa 1960). Kimura and Nagashima (1951) reported a BeO content of 1.35 wt% for an allanite from Japan. Another relatively fresh specimen from a pegmatite at Iisaka, Fukushima Prefecture, Japan, has been reported to contain 2.49 wt% BeO (Iimori 1939). The Be content of the latter crystal, which also contains 2.05 wt% MnO, is significantly greater than values from other localities (see additional references in, e.g., Iimori 1939; Quensel 1945; Hagesawa 1960; Grew 2002). The effective ionic radius of $Be^{2+}$ in tetrahedral coordination is 27 pm, which is almost identical to that of $Si^{4+}$ (26 pm; Shannon 1976). This close similarity between $Si^{4+}$ and $Be^{2+}$ suggests the latter could possibly be incorporated into the T sites of silicate minerals, including allanite. Iimori (1939) noted, however, that the ionic radius of $Be^{2+}$ is similar to that of $Al^{3+}$ and concluded that Be probably substitutes for Al in the octahedral sites. Iimori (1939) did not provide crystallographic data to corroborate his conclusion that the studied specimen was allanite. Another unusual specimen from a pegmatite at Skuleboda, western Sweden, contains 3.83 wt% BeO, but the analyzed separate is probably not pure. It contains nearly 9 wt% $CO_2$ and approximately 8 wt% $H_2O$; it is also isotropic (n = 1.663), and strongly altered (Quensel 1945).

The presence of Be in substantial quantities in allanite has never been confirmed by modern *in situ* analytical techniques. The existence of Be-rich allanite therefore remains unproven (for discussion, see Grew 2002). The few allanite crystals that have been analyzed by LA-ICP-MS did not contain detectable amounts of Be (e.g., Hermann 2002). Moreover, Lee and Bastron (1962, 1967) have documented that allanite did not incorporate Be (< 2 ppm; one out of 21 analyses contained 3 ppm), even though the studied crystals occur in a quartz-monzonite near to a Be mineralization.

## Other substitutions

**Halogens.** In many cases, allanite displays halogen substitution for $O^{2-}$. Fluorine is typically more abundant than Cl (Rao et al. 1979; Mahood and Hildreth 1983). Chlorine contents of 0.95 wt% were reported for allanite from a skarn in the Hemlo area, Ontario (Pan and Fleet 1990), where one analysis shows Cl > F. Allanite from a granite and its endoskarn in the Tertiary igneous complex on Skye contains substantial amounts of F, with average values that range from 0.21 to 0.44 wt% (Smith et al. 2002). A similar range of F contents (0.11–0.49 wt%) was also described for ferriallanite-(Ce) from the Bastnäs deposit in Sweden (Holtstam et al. 2003). There, F increases with the Mg content. This relationship was not attributed to substitution (9), but rather to the "Fe-F avoidance" phenomenon, which has been observed in several minerals (e.g., Mason 1992). Fluorine contents up to 0.61 wt% are also typical for allanite in peraluminous granitoid intrusions along the Velence-Balaton Line in Hungary (Buda and Nagy 1995). Ivanov et al. (1981) found 0.9 wt% F in allanite from a tungsten deposit on the Chukchi Peninsula. The greatest F content reported for unaltered specimens is 1.10 wt%. It was measured in allanite that occurs at the contact between a microcline pegmatite and calc-granulites (Rao et al. 1979), corresponds to 0.33 F atoms per formula unit, and indicates the sample may be intermediate between allanite and dollaseite. Even larger F contents have been described for altered allanite: Jiang et al. (2003) observed average F contents of 1.7±1.1 wt% in allanite crystals that are intergrown with melanite garnet in leucosyenite. These allanite grains also contain 0.4±0.5 wt% $P_2O_5$.

**Scandium.** Dissakisite (3.8 wt% MgO) enriched in Sc was reported by Meyer (1911) from a granite pegmatite at Impilaks, Finland. This author observed $Sc_2O_3$ contents of 0.8 wt% and 1.0 wt% in relatively fresh and altered material, respectively.

***Zinc.*** Considerable amounts of Zn may be present in REE-rich epidote-group minerals. Bermanec et al. (1994) reported ZnO contents of 0.6 to 0.75 wt%, values similar to the 0.72 wt% determined by Ovchinnikov and Tzimbalenko (1948). However, these values may not be accurate, because Johan et al. (1983) observed only 0.21 wt% ZnO in allanite from a Zn deposit. Ivanov et al. (1981) found 0.24 wt% ZnO in allanite from a tungsten deposit on the Chukchi Peninsula.

***Gallium.*** One allanite sample from a sphalerite ore deposit in Haute Garonne, French Pyrenees has been reported to contain Ga (Johan et al. 1983; presence of REE verified, but REE not analyzed for). The specimen contains 2.52 wt% $Ga_2O_3$, in addition to 10.63 wt% $GeO_2$. Judging from the effective ionic radii of $Ga^{3+}$ ($^{[VI]}r = 62$ pm, $^{[IV]}r = 47$ pm; Shannon 1976), it most likely replaces Al or Fe in the octahedrally coordinated M sites.

***Zirconium.*** A trace element that is only rarely reported for REE-rich epidote minerals is Zr (Hagesawa 1960; Rao and Babu 1978; Rao et al. 1979; Coulson 1997). Hermann (2002) observed Zr at the 2 ppm level, and also lists a wide range of other trace element contents, determined by LA-ICP-MS for two allanite samples from subducted eclogites from the Dora Maira massif in the Western Alps. In blueschist and eclogite samples from New Caledonia, Spandler et al. (2003) found Zr contents that range from 1 to 16 ppm (LA-ICP-MS data). There are only a few localities where considerably larger concentrations have been reported. Gieré (1986) described metasomatically formed allanite-(Ce) from a skarn in the Bergell contact aureole, Italy, which contains an average of 0.51 wt% $ZrO_2$, and a maximum value of 0.58 wt%. This value is similar to that observed by Iimori (1939) for a Be-rich allanite from Fukushima Prefecture, Japan. Zirconium-rich allanite (0.76 wt%) has also been reported by Khvostova (1962), but the largest $ZrO_2$ content observed so far is 2 wt% (Bea 1996). This author states that allanite in granitic rocks typically contains relatively large amounts of $ZrO_2$. Because its ionic radius is similar to that of $Ti^{4+}$, $Zr^{4+}$ most likely enters the M sites in epidote-group minerals, probably by exchange mechanisms that are analogous to substitutions (20), (24), and (25).

***Tin.*** This element is a fairly common constituent of allanite, but $SnO_2$ concentrations are generally relatively small (<0.85 wt%; Kimura and Nagashima 1951; Ueda 1955; Hagesawa 1960; Khvostova 1962). The mechanism for Sn incorporation has been described as

$$Sn^{4+} + Fe^{2+} \leftrightarrow 2\ (Fe^{3+}, Al^{3+}) \qquad (27)$$

for epidote containing up to 2.84 wt% $SnO_2$ that occurs in a skarn (van Marcke de Lummen 1986).

***Barium.*** Semenov et al. (1978) reported that allanite contains considerable amounts of Ba in syenite associated with carbonatite complexes in Tamil Nadu. They found concentrations up to 0.69 wt% BaO and noted that these allanites are also enriched in MnO and $TiO_2$. Sawka (1988) found up to 0.28 wt% BaO in allanite from the McMurray Meadows granodiorite, Sierra Nevada, California. Sawka's data suggest that BaO should be routinely analyzed in pluton-hosted allanite with the electron microprobe by using wavelength-dispersive techniques. Hermann (2002) found Ba in two allanites from eclogites, albeit in much smaller amounts (214 and 11 ppm Ba, LA-ICP-MS data). Similarly, Spandler et al. (2003) measured ~250 ppm Ba in allanite from a lawsonite blueschist (LA-ICP-MS data).

## RARE EARTH CHARACTERISTICS

The compiled database with more than 1700 chemical analyses was used to explore the REE characteristics of allanite and related members of the epidote group. Most of the analyses were made by electron microprobe, and therefore, show scant data for REE heavier than Sm. Exley (1980), Gieré (1986), and Sorensen (1991) pointed out that interference corrections become sizeable for REE heavier than Sm, and microprobe detection limits therefore rise for these elements. Accordingly, many of the chondrite-normalized REE patterns discussed in this section display data that were obtained in the last 20 years, mostly with analytical techniques that have low detection limits for REE (e.g., by LA-ICP-MS or instrumental neutron activation analysis [INAA]). These diagrams were specifically selected to discuss the characteristics of allanite across the entire lanthanide series. For the LREE, the available electron microprobe data confirm the trends discussed below.

The total REE content of allanite can exceed 1 atom per formula unit (see Fig. 3). In these cases, however, the total number of $Me^{2+}$ cations in M sites is not always greater than 0.333, consistent with REE charge balance by substitutions (1), (2), (5), (6), or (14). Together, these two observations suggest that there must be additional mechanisms of REE incorporation into "excess REE" members of the epidote group. An example of this problem is presented by the ferriallanite-(Ce) analyses of Holtstam et al. (2003). To accommodate excess REE, the authors invoked the substitution

$$Ca + Si \leftrightarrow REE + {}^{IV}Al \qquad (28)$$

which would not affect the $Me^{2+}$ content of the M sites. As mentioned above, accommodation of Al by the T sites is not very common, and therefore, the excess REE can probably in most cases be accommodated by substitutions, such as, (18), (19), (22), or (23).

The bulk of the REE data show that chondrite-normalized values of La range from about 10,000 to 200,000, whereas chondrite-normalized Sm values range from 4,000 (the lower values would represent an REE-rich epidote) to 60,000. All chondrite-normalized diagrams in this chapter were calculated with the chondrite data of Wakita et al. (1971). The data show that not only is allanite rich in LREE, but that it appears to be richest in La, Ce, Pr, and Nd, which have the largest ionic radii. The preference of allanite for LREE was described by Goldschmidt and Thomassen (1924) in their pioneering work on element distribution in minerals. Semenov (1958) attributed this preference to a crystal-chemical effect, arguing that the large cation site available in allanite preferentially accommodates the larger REE. Classifying allanite as a "complex mineral with a Ce tendency," this author noted that, in contrast to "selective minerals" (e.g., monazite, bastnäsite, and xenotime), "the REE composition maxima are subject to changes" in allanite. Semenov (1958) concluded that allanite's REE composition reflects the proportions of the lanthanides in the petrogenetic environment. Several investigators have, in fact, observed that the REE content of allanite depends on the bulk chemical composition of the host rocks. Lee and Bastron (1967), for example, described that the degree of REE fractionation varies with the CaO content of granitic rocks in the Mt. Wheeler area, Nevada. Similarly, Murata et al. (1957) and Fleischer (1965) have shown that the average relative enrichment of LREE in allanite varies with rock type, increasing from granitic pegmatite to granodiorite to carbonatite (Fig. 10). Kosterin et al. (1961) pointed out that allanite from granodiorite and diorite has the largest La/Nd and Ce/Nd ratios, those from syenites the lowest; allanite from granite and alaskite displays intermediate ratios.

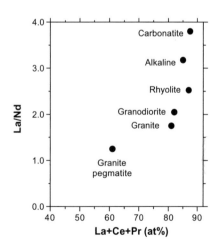

**Figure 10.** La/Nd vs. (La+Ce+Pr) showing *average* compositions (in at%) of allanites in different rock types. Modified after Fleischer (1965).

The REE characteristics of most allanites from igneous rocks are basically similar. As demonstrated by chondrite-normalized REE patterns, allanite strongly fractionates the LREE from the heavy REE (HREE), and typically exhibits a pronounced negative Eu anomaly. In rhyolitic volcanic rocks, allanite phenocrysts are characterized by chondrite-normalized La/Yb values between ~250 and 2340, in strong contrast to the La/Lu values of the glass matrix ( Figs. 11-13; Table 4), which are two orders of magnitude lower. The great affinity of allanite for the LREE is also emphasized by the phenocryst/glass partition coefficients for individual elements (Mahood and Hildreth 1983; Brooks et al. 1981; Chesner and Ettlinger 1989). These coefficients are defined as $D(i) = X_i^{\text{Allanite}}/X_i^{\text{Glass}}$, where $X_i$ is the weight fraction of a specific element $i$ in each phase. The $D(i)$ values decrease by about two orders of magnitude across the lanthanide series, irrespective of the rhyolite type (Fig. 14); a rapid decrease in $D(i)$ with atomic number is observed even within the LREE group. There is a striking similarity in the $D(i)$ *trends* of the Sandy Braes obsidian (Northern Ireland) and the Bishop tuff (California). However, the *values* of $D(i)$ for any given REE are markedly higher (except

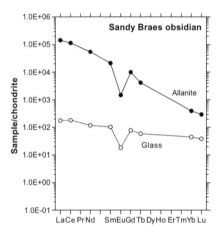

**Figure 11.** REE patterns for allanite phenocrysts and glass matrix in the Sandy Braes obsidian, Northern Ireland. Data from Brooks et al. (1981).

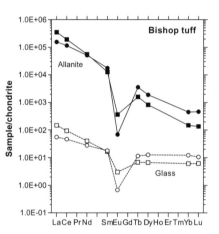

**Figure 12.** REE patterns for allanite phenocrysts and coexisting glass (corresponding open symbols) in the Bishop tuff, California. Data from Mahood and Hildreth (1983).

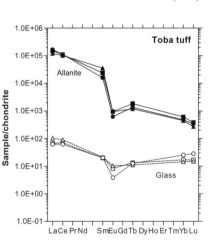

**Figure 13.** REE patterns for allanite phenocrysts and coexisting glass (corresponding open symbols) in the Toba tuff, Sumatra. Dot/open dot symbol = Youngest Toba tuff, other symbols = Middle Toba tuff. Data from Chesner and Ettlinger (1989).

**Table 4.** Chemical characteristics of allanite phenocrysts and their rhyolitic host rocks

| Locality | SiO₂ in host rock (wt%) | Al₂O₃ in host rock (wt%) | CaO in host rock (wt%) | ASI # | Total REE in allanite (ppm) * | Total REE in glass (ppm) * | La/Yb in allanite | La/Yb in glass | Th in allanite (ppm) | Th in glass (ppm) | U in allanite (ppm) | U in glass (ppm) | Ref. |
|---|---|---|---|---|---|---|---|---|---|---|---|---|---|
| Sandy Braes obsidian, Northern Ireland | 72.8 | 12.4 | 0.91 | 0.995 | 195490 | 355 | 366 | 4 | 7600 | 45.2 | <62 | 9.3 | (1) |
| Bishop tuff, California | 75.7 | 13.0 | 0.83 | 0.986 | 332235 | 168 | 2342 | 24 | 5544 | 13.2 | 47.6 | 3.4 | (2) |
| Bishop tuff, California | 77.4 | 12.3 | 0.45 | 0.990 | 194948 | 90 | 344 | 4 | 11914 | 20.4 | 119.0 | 7.0 | (2) |
| Youngest Toba tuff, Sumatra | | | | | 160594 ** | 89 | 348 ** | 2 | | | | | (3) |
| Middle Toba tuff, Sumatra | | | | | 140390 ** | 126 | 270 ** | 7 | | | | | (3) |
| Middle Toba tuff, Sumatra | | | | | 156962 ** | 97 | 249 ** | 4 | | | | | (3) |

References: (1) Brooks et al. 1981; (2) Mahood and Hildreth 1983; (3) Chesner and Ettlinger 1989

\#  Aluminum saturation index: ASI = Al₂O₃/(CaO+Na₂O+K₂O), in molar units

\*  minimum value (sum of all REE reported in original paper)

\*\*  calculated from partition coefficients and coexisting glass compositions (Table 4 and Table 3B in original paper, respectively)

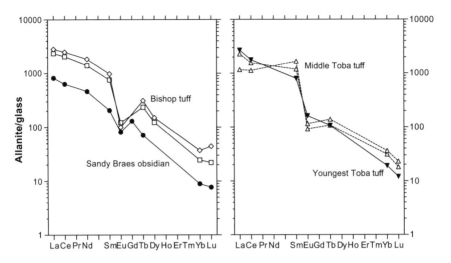

**Figure 14.** Variation of the allanite/glass partition coefficients for rhyolitic rocks across the lanthanide series. Data represent measured concentrations in allanite phenocrysts and coexisting glass (from Brooks et al. 1981; Mahood and Hildreth 1983; Chesner and Ettlinger 1989).

for Eu) for the Bishop tuff, which exhibits $D(i)$ values of similar magnitude to those observed for the Toba tuff (Fig. 14). Allanite was therefore more effective at fractionating REE from the melts of the Bishop and Toba tuffs. Because the total REE content of the tuff glasses is considerably lower than that of the Sandy Braes glass (Table 4), the total REE contents and chondrite-normalized patterns of the allanite phenocrysts are similar in all three rhyolites (Figs. 11-13). Differences in $D(i)$ values between the different rock types are also observed for other elements, for example Th and U. In the Bishop tuff, these elements have again been more effectively partitioned into allanite compared with the Sandy Braes obsidian, whose glass is richer in both elements (Fig. 15). In both rhyolites, however, allanite preferentially accommodated Th, documenting that it strongly fractionates Th from U during crystallization.

The vast majority of published REE data are available for allanite from plutonic rocks, primarily different types of granite and granitic pegmatite. Most analyses of such allanites,

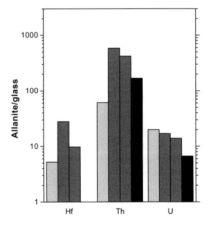

**Figure 15.** Allanite/glass partition coefficients for Hf, Th, and U. Data represent measured concentrations in allanite phenocrysts and coexisting glass. Symbols: light grey = experimental data for granodioritic melt (Hermann 2002); dark grey = data for rhyolitic Bishop tuff, California (Mahood and Hildreth 1983); black = data for Sandy Braes obsidian, Northern Ireland (Brooks et al. 1981).

however, only list the LREE (typically La-Sm). The few fairly recent datasets containing the entire lanthanide series (or a large part thereof) show REE characteristics that are broadly similar to those of allanite from rhyolite. These data have been used to construct chondrite-normalized REE patterns (Fourcade and Allègre 1981; Buda and Nagy 1995; Bea 1996;

Poitrasson 2002). Where available, the REE patterns of the host rocks have also been plotted. The diagrams, shown in Figures 16-21, reveal pronounced negative Eu anomalies for examples of allanite from plagiogranite, tonalite, granodiorite, and monzogranite. It seems likely that negative Eu anomalies also exist in allanite from granite (Fig. 22). The enrichment of LREE relative to HREE is variable, with chondrite-normalized La/Yb values ranging from ~50–300. Only the allanite hosted by the Urbalacone calc-alkaline granodiorite, Corsica (Poitrasson 2002) displays extremely fractionated patterns (La/Yb ≈ 900; Fig. 19).

The basic similarities observed for allanite are not universal within the population of REE-rich epidote minerals, as illustrated by Sorensen (1991) and by the REE-patterns of various endmembers of this group (Fig. 23). However, it is not clear what dictates the

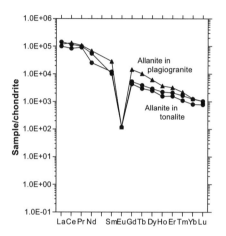

**Figure 16.** REE patterns for allanite in plagiogranite (Khabarny, Southern Urals) and tonalite (Sirostan). LA-ICP-MS data from Bea (1996).

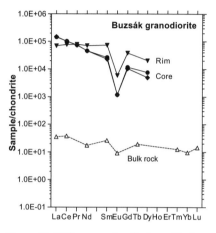

**Figure 17.** REE patterns for allanite and its host rock, Buzsák granodiorite, Hungary. Electron microprobe data from Buda and Nagy (1995).

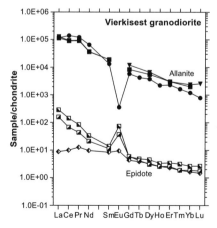

**Figure 18.** REE patterns for allanite and magmatic epidote in the Vierkisest granodiorite, Central Urals. LA-ICP-MS and electron microprobe data from Bea (1996).

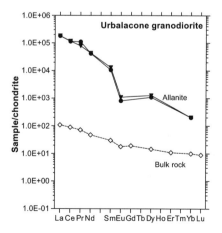

**Figure 19.** REE patterns for allanite and its host rock, Urbalacone granodiorite, Corsica. LA-ICP-MS and electron microprobe (allanite) and ICP-AES (bulk rock) data from Poitrasson (2002).

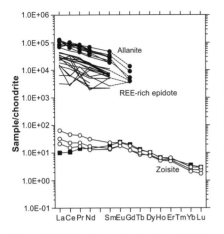

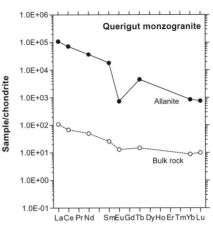

**Figure 20.** REE patterns for allanite and REE-rich epidote in the granodiorite complex of Central Sardinia (electron microprobe data from Carcangiu et al. 1997). Zoisite patterns for mineral separates from two amphibolite samples in the Tauern Window, Austria (X-ray fluorescence data from Brunsmann et al. 2000).

**Figure 21.** REE patterns for allanite and its host rock, Querigut monzogranite, French Pyrenees. INAA data from Fourcade and Allègre (1981).

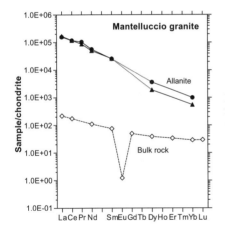

**Figure 22.** REE patterns for allanite and its host rock, Mantelluccio granite, Corsica. LA-ICP-MS and electron microprobe data from Poitrasson (2002).

**Figure 23 (*on facing page*).** Chondrite-normalized REE patterns for the near-endmember or type specimens of REE-rich epidote-group minerals. *Allanite-(La):* dots = electron microprobe data from Pan and Fleet (1991); diamonds = electron microprobe and SIMS data from Bonazzi et al. (1992). Note that the Bonazzi et al. (1992) data represent REE-rich piemontite ($\Sigma$REE < 0.5). *Dissakisite-(Ce):* dots = electron microprobe data from Grew et al. (1991); diamond = electron microprobe data from de Parseval et al. (1997); x = detection limit for Grew et al. (1991) data. *Ferriallanite-(Ce) :* dots = LA-ICP-MS data for sample #882234, from Holtstam et al. (2003); other symbols = electron microprobe data from Kartashov et al. (2002), diamond = data for crystal used for structure refinement. *Androsite-(La):* electron microprobe data from Bonazzi et al. (1996). *Khristovite-(Ce):* dots = electron microprobe data from Sokolova et al. (1991); diamond = electron microprobe data from Pautov et al. (1993). *Dollaseite-(Ce):* electron microprobe data from Peacor and Dunn (1988).

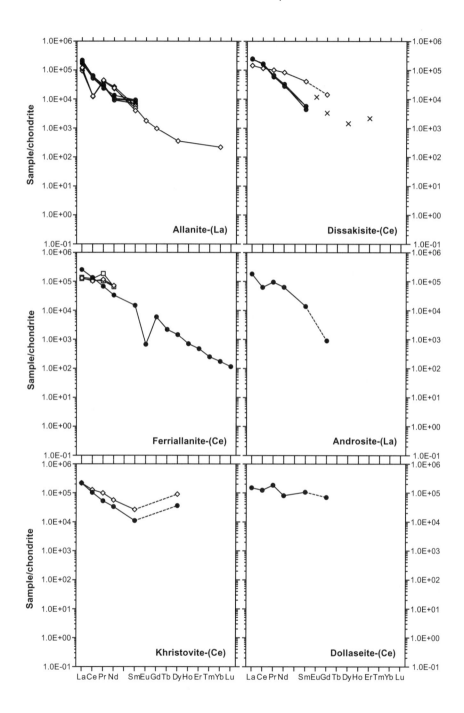

**Figure 23.** *caption on facing page*

extremely fractionated patterns of allanite-(La), ferriallanite-(Ce), and dissakisite-(Ce) versus the relatively flat one of dollaseite-(Ce). Zhirov et al. (1961), who studied pegmatites in north Karelia, offered a possible explanation for this disparity. They observed that if allanite is associated with monazite and xenotime, it is LREE-enriched, but that it is Y-dominant in the absence of phosphate minerals. These data suggest that the REE pattern of allanite is strongly influenced by the crystallization sequences of other REE minerals.

The effect of competitive crystallization of other REE-bearing minerals is not restricted to igneous rocks, but is also observed in metamorphic and metasomatic environments. An example of metamorphic allanite is shown in Figure 24: the two allanites, which occur in eclogites from Dora Maira, Western Alps, exhibit the most strongly fractionated REE patterns reported for allanite (chondrite-normalized La/Yb = 54,686 and 22,498), and they lack an Eu anomaly. Furthermore, the crossover of the allanite patterns with those of the respective bulk rocks implies that HREE-rich phases must be present. Such phases are indeed part of the assemblage and were identified as garnet and zircon, which both show typical HREE-enriched patterns (Hermann 2002). In their study of REE redistribution during high-pressure, low-temperature metamorphism of Fe-gabbros from Liguria (Italy), Tribuzio et al. (1996) also observed that allanite is strongly enriched in LREE and depleted in HREE, but that the chondrite-normalized REE patterns exhibit a peak at Nd (Fig. 25). As indicated in this diagram, the concentrations of Er and Yb are detectable in the eclogite host rock, but not in allanite. Tribuzio et al. (1996) showed that the HREE in the studied Ligurian eclogite primarily reside in garnet, as is the case for Dora Maira eclogite, and also that a significant portion of the bulk-rock REE have been incorporated in apatite. Spandler et al. (2003) published REE patterns of allanites from high-pressure metamorphic rocks in New Caledonia. These authors showed for allanites from lawsonite blueschist that the chondrite-normalized REE values are approximately 60,000 from La to Nd, and then decrease smoothly, without an Eu anomaly, to ~50 for Lu (Fig. 26). The other REE patterns reported by Spandler et al. (2003) are for REE-rich epidote and zoisite, rather than allanite (Fig. 27). The REE-rich epidotes, which occur in an epidote blueschist, display fairly flat, smooth patterns, with chondrite-normalized values ranging from ~1200 (La) to ~20 (Lu); those from a quartz-phengite schist are richer in LREE, with chondrite-normalized values ranging from ~10,000 (La) to ~2 (Lu). None of the metamorphic allanites, for which datasets spanning the entire lanthanide series are available, exhibits an Eu anomaly. These allanites, all from high-pressure rocks, are thus clearly distinct from the igneous allanites.

Another mechanism that may lead to changes in the REE patterns is fractional crystallization. Evidence for this mechanism exists in igneous allanite, which is often strongly zoned, as documented by numerous published backscattered electron (BSE) images (e.g., Pantó 1975; Buda and Nagy 1995; Petrík et al. 1995; Catlos et al. 2000; Poitrasson 2002; Oberli et al. 2004). Zoning is also responsible for the variable chondrite-normalized REE patterns shown for allanite from the Buszák granodiorite in Hungary (Fig. 17; Buda and Nagy 1995).

An example of distinct magmatic zoning in allanite has recently been described by Oberli et al. (2004), who studied allanite occurring as a relatively abundant (~0.9 vol%) phase in the tonalite of the Tertiary Bergell pluton in the eastern Central Alps. All crystals are optically zoned. They exhibit dark brown allanite cores, surrounded by a series of lightly colored zones, and an outermost rim of colorless REE-free epidote. A BSE image of part of a typical allanite grain (Fig. 28) shows that both the core and surrounding area consist of at least two individual sub-zones (labeled A and B, and C and D, respectively). The rim zones C and D are separated by a distinct straight boundary (dashed line in Fig. 28, top left), which traces an euhedral shape that is in marked contrast to the boundary between zones A and B in the core. The BSE image reveals that zone B can be further subdivided into at least two additional zones, B1 and B2 (Fig. 28, top right). The complex dentate, serrate and embayed boundaries observed in the core

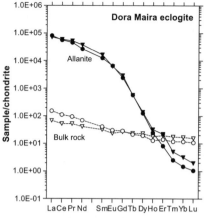

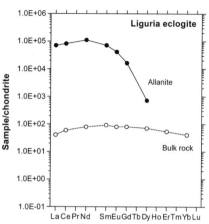

**Figure 24.** REE patterns for allanite and its phengite eclogite host rocks from Dora Maira, Western Alps. LA-ICP-MS (allanite) and solution ICP-MS (bulk rock) data from Hermann (2002).

**Figure 25.** REE patterns for allanite and its eclogite host rock from Liguria, northwestern Italy. Ion microprobe (allanite; mean value) and solution ICP-MS (bulk rock) data from Tribuzio et al. (1996).

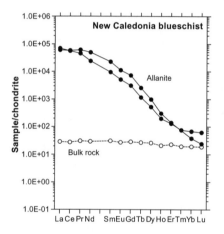

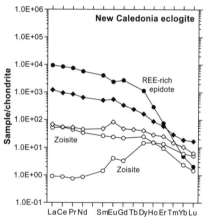

**Figure 26.** REE patterns for allanite and its lawsonite blueschist host rock from New Caledonia. LA-ICPS-MS data from Spandler et al. (2003).

**Figure 27.** REE patterns for zoisite and REE-rich epidote in eclogite from New Caledonia. LA-ICPS-MS data from Spandler et al. (2003). Symbols: diamonds = garnet-bearing epidote blueschist; dots = quartz-phengite schist.

area possibly reflect various growth and corrosion stages during allanite crystallization from a melt. Similar observations can also be made in the element distribution maps of a neighboring area of the same crystal (Fig. 28, bottom; for color version of the element distribution maps, see back cover). To quantify the pronounced zoning, Oberli et al. (2004) have performed a series of electron microprobe analyses along various traverses. As shown in Figures 29-31, the core is characterized by high contents of Th and REE, and is also enriched in Ti and Mg relative to the rim. Figure 29 further shows estimated $Fe^{3+}/Fe_{tot}$ values of around 0.4 for the core, in contrast to the essentially $Fe^{2+}$-free epidote rim. Density separates that proxy core to rim aliquots of the same sample used by Oberli et al. (2004) have been analyzed by $^{57}Fe$

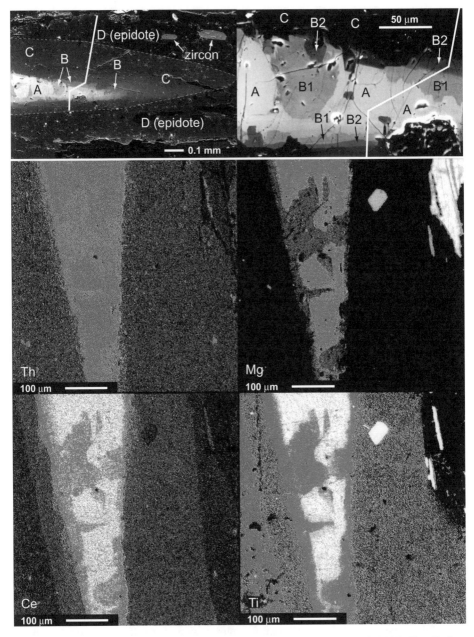

**Figure 28.** BSE images (top) of a part of an allanite crystal studied by Oberli et al. (2004). The allanite occurs in tonalite of the Bergell pluton, Central Alps. Solid line shows trace of electron microprobe traverse. The bright core (zones A and B) displays complex zoning (top left). A detail of the zoning in the core is shown in the top right BSE image. Dashed line marks the euhedral boundary between the allanite rim zone (C) and epidote (D). The X-ray maps (bottom) show the distribution of Th, Mg, Ce, and Ti in an area that is adjacent to the view seen in the BSE images (Gieré, unpublished data). For a color version of the element distribution maps, see back cover.

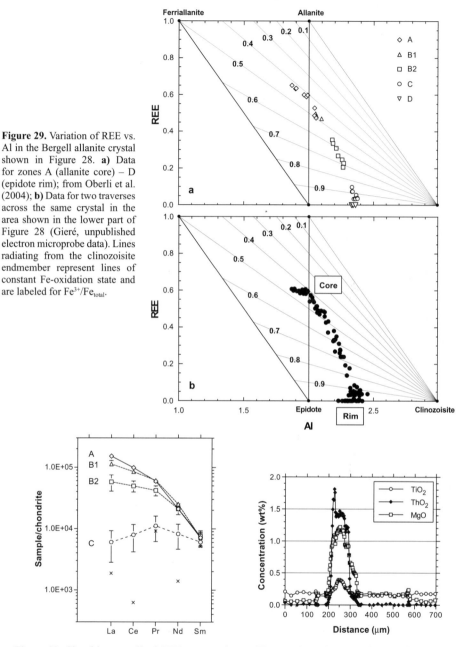

**Figure 29.** Variation of REE vs. Al in the Bergell allanite crystal shown in Figure 28. **a)** Data for zones A (allanite core) – D (epidote rim); from Oberli et al. (2004); **b)** Data for two traverses across the same crystal in the area shown in the lower part of Figure 28 (Gieré, unpublished electron microprobe data). Lines radiating from the clinozoisite endmember represent lines of constant Fe-oxidation state and are labeled for $Fe^{3+}/Fe_{total}$.

**Figure 30.** Chondrite-normalized REE patterns for the individual zones in allanite from the Bergell tonalite, changing from highly LREE enriched in the core (A) to nearly flat in the rim (C). Allanite data are averages for each zone shown in Figure 28, and are from Oberli et al. (2004). Electron microprobe detection limit shown by × symbol.

**Figure 31.** Concentration profile for $TiO_2$, $ThO_2$, and MgO across the zoned allanite crystal shown in Figure 28 (Gieré, unpublished electron microprobe data).

Mössbauer spectroscopy (Virgo, unpublished data), and the results confirmed the trend in $Fe^{3+}/Fe_{tot}$ inferred from microprobe data (Gieré et al. 1999). The total $REE_2O_3$ content decreases dramatically from core (17.6 wt%) to rim (not detectable). This decrease in REE content is accompanied by strong fractionation effects (Fig. 30): Zone A exhibits a chondrite-normalized pattern that is typical for many igneous allanites (compare, for example, with Fig. 16, which shows the pattern of allanite from another tonalite). In comparison to zone A, the LREE enrichment is less pronounced in zone B, which is particularly evident for zone B2. The younger zone C is poor in REE and exhibits a nearly flat chondrite-normalized pattern. In zone D, the youngest part of the crystal, the REE contents are generally below electron microprobe detection limits (see also Fig. 30) and the composition corresponds to epidote (Fig. 29). The regular, correlated changes in major and trace element compositions are distinct from those resulting from hydrothermal overprint (e.g., Poitrasson 2002) and thus, suggest that the observed zonation in allanite is an expression of a magmatic crystallization sequence.

Allanite also exhibits pronounced zoning in many metamorphic and hydrothermal environments (e.g., Sorensen 1991; Catlos et al. 2000; Boundy et al. 2002; Poitrasson 2002; Spandler et al. 2003). Here, zoning is due primarily to release or consumption of REE during metamorphic reactions involving other REE-bearing minerals (see below) or to multiple interactions with hydrothermal fluids of variable REE contents (e.g., Smith et al. 2002; Exley 1980). In several cases, zoning ranges from allanite to zoisite or epidote.

The available REE data, obtained recently for epidote and zoisite by sensitive new techniques and thus spanning the entire lanthanide series, demonstrate that these minerals have chondrite-normalized REE patterns that are distinct from those of allanite and other REE-rich epidote-group minerals. Figure 18 shows chondrite-normalized REE patterns of magmatic epidotes from a granodiorite, which also contains allanite (Bea 1996); two epidotes have chondrite-normalized La/Yb values that are similar to those of allanite, but they exhibit a distinctly positive Eu anomaly. The third epidote pattern is nearly flat, demonstrating that significant variation in the REE fractionation of epidote is possible within a single rock. Two epidote patterns reported by Tribuzio et al. (1996) from an eclogite show yet another distinct shape: a maximum in the middle REE (MREE) with chondrite-normalized values of ~100, and La/Yb values between ~2 and ~8. These patterns are also in marked contrast to those of allanite from the same rock. Similar variability in REE patterns is found for zoisite as well: Brunsmann et al. (2000) described chondrite-normalized REE patterns with a slight LREE enrichment (La/Yb ≈ 10–20) and others with a maximum for the MREE (Fig. 20; see also Frei et al. 2003). Enrichment in either MREE or HREE is displayed by zoisites from eclogites in New Caledonia (Fig. 27; Spandler et al. 2003).

## GEOCHRONOLOGY

Allanite has been shown to be a useful phase for geochronological purposes (Mezger et al. 1989; von Blanckenburg 1992; Barth et al. 1989, 1994; Davis et al. 1994; see also Morrison 2004). The use of allanite as a geochronometer remains relatively restricted in comparison to zircon and monazite, primarily due to its tendency to incorporate common Pb during crystallization and alteration (e.g., Poitrasson 2002; Romer and Siegesmund 2003). Most U-Th-Pb dating studies of allanite have been performed by applying isotope dilution methods to multi-grain samples. To overcome some of the disadvantages of these methods, the use of secondary ion mass spectrometry (SIMS) for Th-Pb dating *in situ* (Catlos et al. 2000; Wing et al. 2003), and in particular the dating of individual growth zones of allanite (Oberli et al. 1999, 2004) has yielded promising results. In addition, U-series dating can be applied to young samples (Vazquez and Reid 2001, 2003).

Catlos et al. (2000) employed a planar solution to matrix corrections for SIMS analyses of allanite that improved the precision of this method to yield about ±10% age accuracy. Their calibration procedure used $^{208}$Pb*/Th$^+$ versus ThO$_2$$^+$/Th$^+$ versus FeO$^+$/SiO$^+$. Although less accurate than U-Pb dating of monazite by ion microprobe, this procedure allows the examination of many types of geological problems for which monazite ages cannot be obtained—as discussed below, detrital monazite reacts in certain types of metamorphic rocks to form allanite, which can then be used in some metamorphic terrains to determine the age of this isograd. Catlos et al. (2000) used SIMS-based geochronology to determine ages of metamorphic allanite grains from garnet-zone pelitic rocks from the footwall of the Himalayan Main Central Thrust in Nepal. There, allanite inclusions in garnet cores are significantly older than monazite found as inclusions in garnet rims and the rock matrix. Monazite appears to record Tertiary tectonic events, whereas allanite formed in the Paleozoic era, perhaps during the Pan-African orogeny. Catlos et al. (2000) also examined an allanite grain from a famous allanite mineral locality, the Pacoima Canyon pegmatite in California. This pegmatite is renowned for cm-long, gemmy, purple crystals of zircon that appear to coexist with equally large, black, apparently non-metamict allanite grains. A well-determined, U-Pb age of 1191 ± 4 Ma (e.g., Barth et al. 1994) for the zircon is significantly older than the Catlos et al. (2000) Th-Pb SIMS age of allanite (1006 ± 37 Ma). The latter authors attributed the difference to Pb loss from allanite during slow cooling.

In contrast to the predominant practice of dating accessory minerals exclusively by the U-Pb method, Oberli et al. (1999, 2004) presented an approach that makes full use of both U-Pb and Th-Pb isotope systematics. In order to investigate the potential of combined Th-U-Pb isotope and $^{230}$Th/$^{238}$U disequilibrium systematics for tracing magmatic crystallization and melt evolution, these authors applied conventional high-resolution single-crystal thermal ionization mass spectrometry (TIMS) techniques to a suite of accessory minerals, including allanite. Because allanite is typically characterized by high Th/U, the observed (apparent) $^{206}$Pb/$^{238}$U ages can be considerably enhanced by excess $^{206}$Pb derived from $^{230}$Th, an intermediate daughter nuclide of the $^{238}$U decay series incorporated in excess of its secular equilibrium ratio. In relatively young rocks, high-resolution $^{206}$Pb/$^{238}$U data require correction for radioactive disequilibrium. Such corrections are commonly based on the assumption that Th/U of the melt, from which the minerals grew, can be approximated by Th/U of the host rock. In addition, the crystallizing melt is assumed to have remained close to radioactive equilibrium. These assumptions, however, do not necessarily hold if there has been fractional crystallization of Th- or U-enriched phases (Oberli et al. 2004). If fractionation occurs at time scales similar to or shorter than the half-life of $^{230}$Th (75,380 yr), it will also cause $^{230}$Th/$^{238}$U in the residual melt to deviate from secular equilibrium.

In view of these difficulties, $^{235}$U-$^{207}$Pb dating would be a viable alternative to the use of the $^{238}$U-$^{206}$Pb system, but the presence of even a moderate common Pb component in young allanite results in imprecise $^{235}$U-$^{207}$Pb ages. On the other hand, Th-Pb dating is not affected by these problems, and thus is the method of choice for precise and accurate dating of allanite or other Th-rich minerals, because there are no long-lived intermediate daughter nuclides in the $^{232}$Th-$^{208}$Pb decay system, and the effect of common Pb is mitigated by relatively high Th concentrations (e.g., Barth et al. 1989, 1994).

Oberli et al. (2004) studied zircon, titanite and fragments of chemically characterized growth zones of allanite in a tonalite sample from the feeder zone of the Tertiary Bergell pluton, eastern Central Alps. The isotopic data obtained for these crystals and crystal fragments document crystallization and melt evolution during at least 5 m.y. Zircon ages range from 33.0 to 32.0 Ma. Crystallization of zircon was followed by the formation of zoned allanite between 32.0 and 28.0 Ma, and the crystallization of magmatic epidote possibly as late as 26 Ma. Trace and major element patterns in the zoned allanite closely mirror melt evolution;

they are characterized by a progressive increase of U concentration and sharp decrease of Th and LREE during grain formation (Figs. 28-32). These zoning patterns reflect the early crystallization of phases low in U and document the dominating control of Th by allanite precipitation. Preservation of substantial amounts of excess $^{206}$Pb derived from initial excess $^{230}$Th in all analyzed allanite grains indicates that their isotopic systems have not been reset by loss of radiogenic Pb during prolonged residence at magmatic conditions and regional-metamorphic cooling, and that the measured sequence of $^{208}$Pb/$^{232}$Th dates translates into a real age sequence. Based on these results, Oberli et al. (2004) suggested closure temperatures ≥700°C for magmatic allanite, consistent with the estimates of von Blanckenburg (1992).

The observed $^{230}$Th/$^{238}$U disequilibrium relationships reveal a smooth, initially steep decrease of $(Th/U)_{magma}$ from values of 2.9 at 32.0 Ma to less than 0.1 at 28.0 Ma in equilibrium with sequential allanite zoning (Fig. 33). Comparison of calculated $(Th/U)_{magma}$ with $(Th/U)_{bulk rock}$, measured at 0.79, requires fractional crystallization of allanite at an early stage. Removal of allanite, and thus Th, from the melt took place until ~31.5 Ma, which provides an upper time limit for emplacement of the studied magma batch. Because zircon and much of allanite crystallization predate emplacement, ages determined on refractory minerals from deep-seated plutons should not be equated with their emplacement ages.

A different approach at unraveling the history of magmatic evolution with the aid of zoned allanite has been taken by Vazquez and Reid (2001, 2003). These authors used electron and ion microprobe techniques for *in situ* analysis of compositional and isotopic zoning in volcanic allanite occurring in the 75,000 year-old Youngest Toba tuff, Indonesia. Core-to-rim zoning in most crystals shows decreases in both their LREE and MgO contents, and increases of MREE, FeO and ThO$_2$. By coupling the observed chemical zoning with *in situ* $^{238}$U-$^{230}$Th age determinations and other features, such as resorbed boundaries, Vazquez and Reid (2001, 2003) concluded that single allanite crystals from the Toba tuff record a complex differentiation history of fractionation and episodic mixing during protracted residence (up to 150,000 yr) in a large rhyolitic magma chamber.

Similar chemical zoning was also described for allanite in a granodiorite from Southern California (Gromet and Silver 1983). The increase in Th concentration from core to rim observed at this locality as well as in allanite from the Toba tuffs (Vazquez and Reid 2001, 2003) is in marked contrast to the pronounced decrease observed in allanite from the Bergell Tonalite (see Figs. 31, 32).

Romer and Siegesmund (2003) studied texturally and chemically heterogeneous allanite from the Riesenferner Pluton, Austria, and found that samples from two rocks scatter in the concordia diagram, defining discordias from 31.8 ± 0.4 Ma and 32.2 ± 0.4 Ma to ~540 Ma. These authors were able to show that the apparent inheritance is not due to the presence of inclusions of older allanite or other minerals (e.g., zircon, monazite, xenotime), but results from incorporation of radiogenic Pb originating from a precursor phase. The inheritance is thus chemical rather than physical. Because crystallization of allanite requires the availability of REE and Th, it will preferentially occur where REE- and Th-rich precursor minerals have been dissolved in the melt. Romer and Siegesmund (2003) concluded that monazite, originating from assimilated Paleozoic rocks, was the precursor and gave rise to the localized enrichment of REE and Th. They further suggested that similar chemical inheritance might also be observed in metamorphic rocks, where allanite can form at the expense of monazite (see below).

## RADIATION DAMAGE AND THERMAL ANNEALING

Natural radioactivity is a process in which an atomic nucleus spontaneously disintegrates (or decays). Radioactive decay is accompanied by emission of α-, β-, or γ-radiation. A stream

**Figure 32.** Plot of **a)** Th, **b)** U and **c)** Pb concentrations, and of **d)** Th/U vs. $^{208}Pb/^{232}Th$ age for density-separated allanite fragments from the Bergell tonalite (data symbols and names as shown in Figs. 28-30). The results define an extended crystallization interval of 4 m.y. The marked decrease of Th and increase of U and Pb concentrations with decreasing age are interpreted to mirror incompatible element concentration in an evolving melt. **e)** Difference between $^{206}Pb/^{238}U$ and $^{208}Pb/^{232}Th$ age. Positive age differences of 2-6 Ma between the U-Pb and Th-Pb ages for all analyses are proof for preservation of excess $^{230}Th$-derived radiogenic $^{206}Pb$, which excludes major resetting of the U-Th-Pb isotopic system in allanite after crystallization. **f)** Allanite/melt fractionation coefficient for Th/U. Diagrams modified after Oberli et al. (2004).

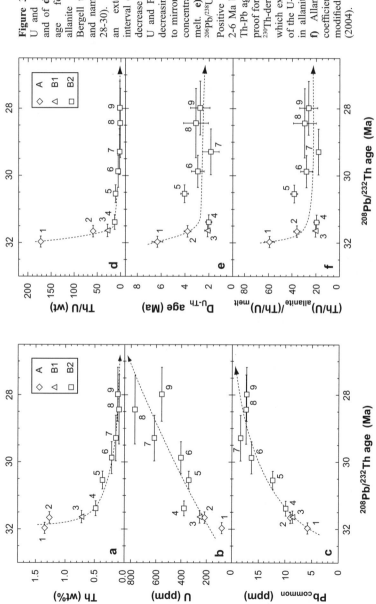

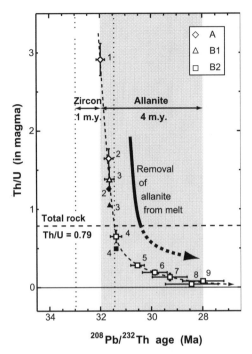

**Figure 33.** Evolution of Th/U in the magma versus time calculated from the allanite data shown in Figure 32d-f. Open and solid symbols (shown only when not overlapped) denote data points uncorrected and corrected for disequilibrium in $(^{230}Th/^{238}U)_{magma}$, respectively. Diagram modified after Oberli et al. (2004). Data symbols and names as shown in Figures 28-30.

of α-particles ($^4$He nuclei) is α-radiation; β-radiation consists of β-particles (electrons or positrons); and γ-radiation is made up of high-energy electromagnetic waves. Both α- and β-decay cause significant radiation damage in radionuclide-containing minerals, primarily via collisions between the nuclear particles and the neighboring atoms in the host. Most of the damage is produced by recoil nuclei, which collide with and displace neighboring atoms in the crystal structure, thus leading to amorphization.

Radiation damage in minerals results essentially from the α-decay of U and Th. When α-particles are ejected from a parent nucleus, the resulting product nucleus experiences recoil. Because an α-particle has a small mass (4 g/mol) and a high energy (~4.5–5.5 MeV), most of the energy is deposited by ionization. The radiation damage occurs primarily near the end of the path (10–20 μm) and causes approximately 100–200 atomic displacements, mostly Frenkel defects. The recoil nuclei, on the other hand, are much heavier (206–234 g/mol) and have a lower energy (~70–100 keV) than the α-particles. Most of their energy is lost via collisions, which cause approximately 1000–2000 atomic displacements along a track of 10–20 nm length (recoil track). These processes cause physical and chemical changes in materials (for recent reviews, see Ewing et al. 1995; Weber et al. 1998).

In many crystalline substances, collisions resulting from α-decay may transform a periodic, crystalline substance into an aperiodic, amorphous (metamict) material. High-resolution transmission electron microscope (TEM) imaging of crystalline, slightly damaged, and metamict materials, shows that the crystalline-to-metamict transition takes place in various stages (e.g., Ewing et al. 1987). The principal long-term effects of this structural transformation are a volume increase and density decrease. The expansion, which is strongly dependent on the structure type of the material, can lead to the development of two types of microfractures: 1) internal fractures, generated by differential volume expansion between zones of different actinide content; and 2) external fractures, or the development of cracks in neighboring minerals. Microfracturing increases the surface area of a metamict grain and creates pathways that permit fluids to more easily interact with the mineral. The amorphous form of a mineral may be metastable and therefore is often more susceptible to leaching and/or dissolution (e.g., Mitchell 1973; Geisler and Schleicher 2000; Geisler et al. 2001; Lumpkin 2001; Lumpkin et al. 2004).

**Metamict allanite**

Studies of optical properties, water content, density, and crystallinity of allanite specimens with various levels of radiation damage have shown that increasing amorphization leads to optical and other physical isotropy, a decrease in mean refractive index and density, and a progressive hydration.

Crystalline allanite shows pleochroism, and has three principal refractive indices ($\alpha$, $\beta$, $\gamma$), which vary between 1.690 and 1.891 (Table 3). In contrast, fully amorphous allanite is isotropic and thus has only one refractive index (n). The mean refractive index of allanite decreases with decreasing density ($\rho$), as was first noted by Zenzén (1916). By examining a series of allanite samples with specific gravities between 4.15 and 2.68 and mean refractive indices of 1.78 to 1.53, Zenzén (1916) discovered that isotropic allanite displays $\rho < 3.50$ g/cm$^3$ and n < 1.70. This effect was also observed by Khvostova (1962), who, like Zenzén (1916), attributed it to increasing metamictization. Frondel (1964), on the other hand, showed that heating metamict or partially metamict allanite increased the indices of refraction, due to the restoration of the allanite structure (i.e., annealing, see below). Tempel (1938) showed that the mean refractive index of allanite decreases with increasing water contents.

Radiation damage in minerals is commonly accompanied by hydration (e.g., Mitchell 1973). Partially or fully metamict allanite specimens may contain substantial amounts of water, and therefore exhibit low microprobe-determined oxide totals (e.g., Khvostova 1962; Deer et al. 1986). Low oxide totals in metamict parts of allanite have, in fact, often been used to infer the presence of substantial amounts of $H_2O$ in the mineral (e.g., Campbell and Ethier 1984; Buda and Nagy 1995). Large $H_2O$ contents are associated with low densities (Tempel 1938). The density of metamict allanite may be as low as 2.68 g/cm$^3$ (Zenzén 1916), and such allanites are often strongly altered (see below).

In their pioneering studies of metamictization of radioactive minerals, including allanite, Ueda and Korekawa (1954) and Ueda (1957) examined a series of allanites with density varying between 3.65 and 4.08 g/cm$^3$. They observed that, as the density of allanite decreases, its X-ray diffraction (XRD) peaks decrease in both sharpness and intensity, shift towards lower 2$\theta$ angles (i.e., show larger d-spacings), and finally fade away. The least dense allanite specimens yield XRD patterns with a single smooth hump that is produced by diffuse scattering of X-rays by aperiodic material. Ueda (1957) further discussed how the unit-cell dimensions of allanite, as calculated from the XRD patterns, increase with the degree of metamictization. The total volume expansion associated with metamictization of minerals can, in certain cases, form radial cracks in surrounding minerals or host phases, particularly the more brittle ones (e.g., Lumpkin 2001). This typical metamictization feature has been reported for: (1) quartz and feldspar around allanite in veins within granulite (Fig. 1 in Hugo 1961), and in pegmatites from Texas (Ehlmann et al. 1964, their Fig. 3) and North Carolina (Fig. 4 in Mitchell 1973); (2) fluorite that hosts allanite inclusions in a hydrothermal vein at the Buffalo fluorite mine, South Africa (Watson and Snyman 1975); and (3) clinopyroxene around dissakisite-(Ce) with a ThO$_2$ content of 1.7–2.1 wt% (Yang and Enami 2003, their Fig. 2).

Many investigations of radiation damage in minerals have been carried out with a TEM. This instrument can be used to observe: 1) $\alpha$-recoil damage in high-resolution bright-field images showing discontinuous lattice fringes and localized amorphous material; and 2) the presence of amorphous domains in selected area electron diffraction (SAED) images on the basis of diffuse haloes, which result from diffuse electron scattering. Moreover, the chemical composition of the same area that has been imaged by TEM can be determined by energy-dispersive X-ray analytical techniques. Janeczek and Eby (1993) studied three samples of allanite that exhibited different degrees of metamictization as a result of different radionuclide content and age. But even within individual crushed allanite grains, variable amounts of $\alpha$-recoil damage were observed by TEM, documenting that the radionuclides are not homogeneously distributed. Two of the samples examined by Janeczek and Eby (1993) exhibit the full range of amorphization, from minor radiation damage to complete metamictization, as documented by SAED images. Two parameters that are important for the characterization of radiation damage in minerals, however, have not yet been determined for allanite: the first is the *initial amorphization dose*, i.e., the dose at which a decrease in the total Bragg diffraction intensity is first observed; the second parameter is the *critical amorphization dose*, i.e., the dose where the total Bragg intensity goes to zero. These two parameters could be determined fairly easily for allanite crystals of known age and with pronounced radionuclide zoning (particularly Th).

## Annealing of metamict allanite

Thermal annealing of metamict minerals is typically carried out to restore their original crystal structures for phase identification and to determine thermal stability. Ueda (1957) studied the effect of heating on partially metamict allanites. After heat treatment at 400°C and 800°C in vacuum, the XRD peaks were sharper and more intense, and shifted towards higher 2θ angles, consistent with increased crystallinity and unit-cell contraction. Thermal annealing was also investigated by Lima de Faria (1964) for several metamict allanite specimens. During heating at 700°C, this author observed recrystallization of metamict allanite both in air and in a $N_2$ atmosphere, but the experiments in air produced a slightly different allanite structure. This effect was attributed to oxidation, and indeed probably produced oxyallanite (see Fig. 7). After heating metamict allanite at 1000°C, Lima de Faria (1964) noted new phases, including: magnetite; a phase similar to lessingite[†], $(REE,Ca)_5(SiO_4)_3(OH,F)$; and, under oxidizing conditions, cerianite ($CeO_2$), which was the main phase after heat treatment in air. Some allanite crystals started melting at 1000°C and had melted completely at 1300°C. Kumskova and Khvostova (1964) observed restoration of the allanite structure at 800°C, but only for slightly metamict specimens. Heating of more extensively damaged samples at the same temperature only partially restored the allanite structure, but additionally generated $CeO_2$ (see also Khvostova 1962; Ehlmann et al. 1964). Similar results were obtained by Vance and Routcliffe (1976), who heated severely damaged allanite specimens in air. These authors also observed $CeO_2$ formation at temperatures < 1000°C. This result led Vance and Routcliffe (1976) to conclude that the samples examined by Lima de Faria (1964) were not sufficiently damaged for $CeO_2$ precipitation to occur below 1000°C. At 1200°C, Vance and Routcliffe (1976) observed complete decomposition of allanite, yielding crystalline polyphase mixtures of lessingite + hematite + cerianite in air, and lessingite + magnetite + cerianite in vacuum. Neither Lima de Faria (1964) nor Vance and Routcliffe (1976) discussed the fate of Al upon decomposition of allanite. Mitchell (1966), on the other hand, discovered the presence of anorthite as an additional phase in the breakdown assemblage consisting of magnetite, cerianite, and an apatite-structured silicate. These crystalline polyphase assemblages were formed upon heating of metamict allanite specimens in air at temperatures above 800°C. Similarly, Janeczek and Eby (1993) observed that allanite decomposed above 850°C to an assemblage that included anorthite, hematite and a britholite-like phase (related to lessingite), which was predominant in the mixture of breakdown phases. Janeczek and Eby (1993), however, did not observe formation of $CeO_2$ during their annealing experiments, which were performed in an Ar atmosphere, thus preventing oxidation of $Ce^{3+}$.

---

[†] The name lessingite has a questionable status, and its IMA-approved synonym is britholite, $(REE,Ca)_5(SiO_4,PO_4)_3(OH,F)$ (de Fourestier 1999). However, in this chapter we are following the terminology used in the original publications referred to.

Thermal annealing experiments can additionally provide information on both the mechanisms and kinetics of recrystallization. For allanite, these aspects of recrystallization were investigated by Janeczek and Eby (1993), Paulmann and Bismayer (2001a), and Paulmann et al. (2000) in progressive and isothermal annealing experiments. *Progressive annealing* experiments showed that the onset of recrystallization took place at 200 to 300°C. At higher temperatures, changes in XRD peak intensities testify to a two-stage annealing mechanism for some partially metamict allanites: the first stage, with pronounced changes in crystallinity occurs between 500 and 600°C; the second stage occurs between 700 and 800°C, and is followed by the breakdown of the allanite structure at about 850°C (Fig. 34). These changes are reflected by associated variations of density (Fig. 34b) and unit-cell parameters (Figs. 34c-e). The response to thermal annealing of the *a*- and *b*-parameters is anisotropic: the unit-cell contraction was more pronounced along *b* than *a* between 500 and 700°C, but the opposite was found above 700°C (Fig. 34c,d). The annealing path of allanites with different degrees of metamictization is similar if they are not fully amorphous. Janeczek and Eby (1993) found that the rate of recrystallization depends on the amount of α-recoil damage: heavily damaged samples anneal faster. These authors also concluded that the activation energy of allanite at 600°C is sufficient to induce heterogeneous recrystallization of amorphous material adjacent to crystalline areas, but not high enough to initiate homogeneous nucleation within larger volumes of amorphous material. *Isothermal annealing* at fixed temperatures between 400 and 700°C indicated a two-stage mechanism. This mechanism consists of a short initial period of rapid recrystallization, followed by a longer time period during which only slight changes in crystallinity are observed. The annealing behavior of allanite is thus similar to that of other partially metamict minerals, such as zircon (Weber 1990) and titanite (Lumpkin et al. 1991). Anisotropic contraction of the *a*- and *b*-parameters of the unit cell is also observed during isothermal annealing, whereby the decrease of *a* is faster than that of *b* at 400 and 500°C (Paulmann and Bismayer 2001b).

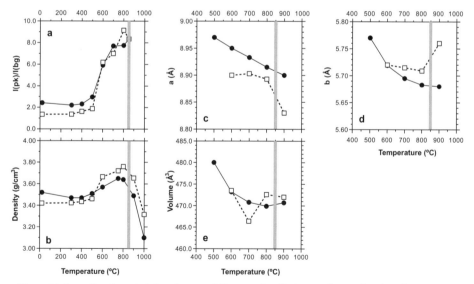

**Figure 34.** Annealing characteristics of two partially metamict allanite samples as a function of temperature (*T*). **a)** XRD-intensity ratio of the (311) peak to background vs. *T*; **b)** Density vs. *T*; **c)** Unit-cell parameter *a* vs. *T*; **d)** Unit-cell parameter *b* vs. *T*; and **e)** Unit-cell volume vs. *T*. The plots were generated using data from Janeczek and Eby (1993) and show the behavior of their most metamict (open squares) and their least metamict (solid dots) samples. Vertical gray bar shows temperature where allanite starts to decompose.

## Annealing of fission tracks

The annealing characteristics of fission tracks in allanite have been studied by Saini et al. (1975). These authors heated allanite specimens from an Indian pegmatite at 800°C to erase fossil fission tracks, irradiated them with thermal neutrons, and then etched the specimens. The annealing experiments were carried out at 640°C to 730°C, and it was found that all fission tracks faded at 720°C if the specimens were heated for one hour. Extrapolation of Arrhenius plots to geological time periods indicates that allanite will lose all fission tracks in 1 m.y. at 380°C, or in 1000 m.y. at 320°C. The authors concluded that allanite is suitable for dating of thermal geological events by fission track dating techniques.

## PHASE RELATIONS

### Allanite in igneous systems

Much of the interest in igneous epidote has centered on whether there is a threshold pressure ($P$) required for epidote to crystallize in silicic systems (e.g., Zen and Hammarstrom 1984, 1986; Moench 1986; Tulloch 1986; Vyhnal et al. 1991). The cores of epidotes in calc-alkaline granitoids (including pegmatites), dacitic dikes, and volcanic tuffs are commonly REE-rich solid solutions of ferriallanite, allanite, clinozoisite, and epidote (e.g., Gromet and Silver 1983; Chesner and Ettlinger 1989; Dawes and Evans 1991; Buda and Nagy 1995; Petrík et al. 1995; Bea 1996; Oberli et al. 2004). As discussed above, the total REE content in allanite commonly decreases from core to rim, and the rim is often a REE-poor epidote. Such relationships between allanite and epidote have been long recognized: Fersman (1931), for example, concluded that allanite crystallization in granitic systems takes place above ≈600°C (his "stage épimagmatique"), is followed by the formation of epidote rims on allanite, and finally by crystallization of separate epidote grains. It has further been shown (Petrík et al. 1995; Oberli et al. 2004) that the decrease in REE from core to rim in allanite is accompanied by an increase in $Fe^{3+}/Fe_{tot}$. Moreover, an analogous relationship exists between Th and $Fe^{3+}/Fe_{tot}$ as well as between Ti and $Fe^{3+}/Fe_{tot}$ (see, e.g., Figs. 28, 31). The relationship between oxidation state of Fe and REE content have in two cases also been confirmed by Mössbauer spectroscopy (Petrík et al. 1995; Gieré et al. 1999).

One possibility that explains these chemical relationships between REE, Th, Ti, and $Fe^{3+}/Fe_{tot}$ is that ferriallananite-allanite-epidote-clinozoisite solid solutions only crystallize in silicic melts as a result of elevated REE, Th and Ti abundances (Gromet and Silver 1983; Sawka et al. 1984). Chesner and Ettlinger (1989), however, concluded in their study of allanite phenocrysts in the Toba tuffs that high concentrations of REE in the melt are not necessary for allanite crystallization to occur. Instead, these authors suggested that temperature is a critical factor for allanite crystallization, because Toba tuffs that erupted below 800°C contain allanite regardless of bulk-rock composition, whereas those that erupted above 800°C contain none. Chesner and Ettlinger (1989) concluded that magmatic allanite can exist only within a restricted temperature range in calc-alkaline volcanic rocks. Moreover, if the oxidation state of Fe in natural allanite is controlled by the oxy-reaction (Eqn. 7), then the stability of allanite in magmatic environments may more closely reflect the relationship between $T$ and $f_{H_2}$, rather than $P$ and the REE content of the melt (Gieré et al. 1999). Although Fe is present primarily as $Fe^{3+}$ in clinozoisite-epidote solid solutions in metamorphic rocks, the situation may be different in magmatic, REE-free epidotes, as evidenced by the $f_{O_2}$-dependent supersolidus stability curve for epidote over a range of oxygen fugacities (Schmidt and Thompson 1996).

Chesner and Ettlinger (1989) have shown petrographic evidence suggesting that allanite is a liquidus phase in the Toba tuffs. They also stated that allanite is a fractionating phase since it is present in the tuffs with a content of $SiO_2$ that ranges from 68 to 76 wt%. Allanite fractionation caused the LREE to behave compatibly in the magma; the HREE, on the other hand, are incompatible in the Toba magmas. Similar observations have also been reported for the Bishop tuff (Michael 1983; Cameron 1984), and document the importance of allanite crystallization in controlling the behavior of REE in granitic magmas. As discussed by Oberli et al. (2004), allanite also has a considerable impact on the evolution of Th and U concentrations in the melt, from which it crystallizes (see Fig. 33). Their data were obtained from allanite that exhibited a sharp decrease in Th content, and an increase in U content from core to rim (see Figs. 31, 32). Such a core-to-rim decrease in Th, however, is not observed in all magmatic allanites, as demonstrated by zoned phenocrysts in the Toba tuffs (Vazquez and Reid 2001, 2003) and in a granodiorite from the eastern Peninsular Ranges batholith, southern California (Gromet and Silver 1983), which both show an opposite trend. Gromet and Silver (1983) explained this increase in Th from core to rim by allanite growth via substitution of other components to replace the less available (depleted) REE, e.g., substitution (18).

In his study of the residence of REE, Th and U, Bea (1996) observed that allanite is, after monazite, the most important LREE carrier in granitoid rocks. He found that primary allanite occurs in all granite types, except the most peraluminous, P-rich varieties, and that it is particularly abundant in rocks that contain magmatic epidote. The fraction of LREE that reside in allanite depends on the bulk-rock $Al_2O_3$ concentration relative to the contents of $Na_2O$, $K_2O$, and CaO (see also Broska et al. 2000): in metaluminous granite, the LREE fraction that resides in allanite is 50–60 wt%, whereas this value is 20–45 wt% in peralkaline granite, in which LREE-fluorocarbonates or aeschynite may be dominant. The respective fractions of Th are 15–42 wt% (metaluminous) and 10–15 wt% (peralkaline). The maximum fraction of U residing in allanite (23 wt%) was observed for a metaluminous granodiorite.

Petrík et al. (1995) and Broska et al. (2000) too noted that allanite is a typical accessory mineral in metaluminous granitoids, where it is often associated with magnetite. However, Broska et al. (2000) reported that allanite can also occur in peraluminous granitoids. They found the mineral in a peraluminous biotite granodiorite and a tonalite from the western Carpathians. In these rocks, allanite occurs together with apatite in polymineralic inclusions within monazite, which is the characteristic LREE phase in peraluminous granites. Further present in the inclusions are albite, potassium feldspar, white mica, and biotite, which probably represent crystallized melt trapped by monazite. The petrographic observations point to early crystallization of allanite, which was subsequently enclosed by monazite. The included assemblage of allanite+apatite indicates that crystallization of monazite began specifically at those locations with high activities of REE and P. Broska et al. (2000) suggested that the replacement of allanite by monazite was due to a decrease in Ca concentration, which resulted from plagioclase crystallization. Based on the monazite saturation equation of Montel (1993), Broska et al. (2002) calculated that the crystallization temperature of allanite must have been higher than ~850°C and 790°C for the two samples they examined. Their study suggests that Ca-rich granitoid melts may precipitate allanite, but that early crystallization of plagioclase lowers the Ca concentration in the melt and induces formation of monazite, which becomes the dominant LREE mineral in these rocks.

## Allanite in metamorphic systems

Allanite is found in pelitic (e.g., Sakai et al. 1984; Wing et al. 2003) and mafic rocks (e.g., Banno 1993; Sorensen and Grossman 1989; Hermann 2002), as well as in granitic gneisses (e.g., Liu et al. 1999) and in carbonate rocks (Boundy et al. 2002) in Buchan, Barrovian, contact, collisional, ultrahigh-pressure collisional and subduction-zone metamorphic settings. Much of it probably forms by greenschist-facies metamorphic reactions that consume detrital or igneous monazite. Smith and Barreiro (1990) first suggested that metamorphic allanite breaks down to form monazite via prograde reactions at temperatures >525°C. Monazite as a breakdown product of allanite has also been described in amphibolite-facies metasedimentary rocks from the Monte Giove area in Italy (Gieré et al. 1998). These observations have led to an extended discussion of allanite-forming and allanite-consuming reactions in various metamorphic systems.

In both Buchan and Barrovian terrains of northern New England, Wing et al. (2003) determined that detrital monazite breaks down and euhedral metamorphic allanite forms in pelitic rocks by means of the reaction

$$3 \text{ REEPO}_4 + 3 \text{ KAl}_2\text{AlSi}_3\text{O}_{10}(\text{OH})_2 + 8 \text{ Ca(Fe,Mg)(CO}_3)_2 + 4 \text{ (Fe,Mg)CO}_3 + 9 \text{ SiO}_2 + 2 \text{ H}_2\text{O} \leftrightarrow$$

(*Monazite*)    (*Muscovite*)    (*Ankerite*)    (*Siderite*)    (*Quartz*)

$$3 \text{ CaREEAl}_2\text{FeSi}_3\text{O}_{11}\text{O(OH)} + 3 \text{ K(Fe,Mg)}_3\text{AlSi}_3\text{O}_{10}(\text{OH})_2 + \text{Ca}_5(\text{PO}_4)_3(\text{OH}) + 20 \text{ CO}_2 \quad (29)$$

(*Allanite*)    (*Biotite*)    (*Apatite*)

In the contact aureole of the Onawa pluton in south-central Maine, in which pelitic rocks lack biotite, ankerite, and siderite, and contain chlorite, plagioclase, and calcite, Wing et al. (2003) proposed a different allanite-forming reaction:

$$15 \text{ REEPO}_4 + 3 \text{ Fe}_5\text{Al}_2\text{Si}_3\text{O}_{10}(\text{OH})_8 + 12 \text{ CaAl}_2\text{Si}_2\text{O}_8 + 28 \text{ CaCO}_3 + 12 \text{ SiO}_2 \leftrightarrow$$

(*Monazite*)    (*Chlorite*)    (*Plagioclase*)    (*Calcite*)    (*Quartz*)

$$15 \text{ CaREEAl}_2\text{FeSi}_3\text{O}_{11}\text{O(OH)} + 5 \text{ Ca}_5(\text{PO}_4)_3(\text{OH}) + 2 \text{ H}_2\text{O} + 28 \text{ CO}_2 \quad (30)$$

(*Allanite*)    (*Apatite*)

Wing et al. (2003) also showed that allanite and biotite isograds coincide in the studied area in New England. Here, Ferry (1984, 1988) concluded that biotite-producing reactions had been driven by aqueous fluid infiltration. Reaction (29) requires the addition of $H_2O$, and both (29) and (30) are decarbonation reactions. Wing et al. (2003) therefore suggested that the reactions that produce the allanite isograd in the pelites were driven by aqueous fluid infiltration. These allanite-forming reactions are similar to those responsible for monazite breakdown during amphibolite-facies metamorphism of granitic rocks in the eastern Alps and in orthogneisses from the west Carpathians (Finger et al. 1998; Broska and Siman 1998). The latter authors proposed the reaction

$$3 \text{ REEPO}_4 + \text{KFe}_3\text{AlSi}_3\text{O}_{10}(\text{OH,F})_2 + 4 \text{ CaAl}_2\text{Si}_2\text{O}_8 + 3 \text{ SiO}_2 + 4 \text{ Ca}^{2+} + 2 \text{ H}^+ \leftrightarrow$$

(*Monazite*)    (*Annite*)    (*Plagioclase*)    (*Quartz*)    (*Fluid*)

$$\text{Ca}_5(\text{PO}_4)_3(\text{OH,F}) + 3 \text{ CaREEAl}_2\text{FeSi}_3\text{O}_{11}\text{O(OH,F)} + \text{KAl}_2\text{AlSi}_3\text{O}_{10}(\text{OH})_2 \quad (31)$$

(*Apatite*)    (*Allanite*)    (*Muscovite*)

They suggested that the reaction occurred in "a gel form," with apatite nucleating on monazite rims via subsolidus retrogression of granite during Alpine metamorphism. Finger et al. (1998) reported that igneous monazite grains in granite from the Tauern Window, central eastern Alps, had been partly replaced by apatite-allanite-epidote coronae during amphibolite-facies Alpine metamorphism. The replacement textures are annular: a core of magmatically zoned monazite is successively surrounded by rings of apatite + thorite, allanite, and epidote. These authors concluded that a metamorphic fluid was required to create the coronae, the kinetics of the reaction were controlled primarily by diffusion, and that a balanced reaction to explain the entire association was difficult to write. However, they noted that plagioclase, biotite, and muscovite could all have been involved in such a reaction.

Aluminosilicate minerals and metamorphic monazite appear at roughly the same pressure-temperature conditions in pelitic rocks. Wing et al. (2003) reported that in all three types of terrains they studied (Buchan, Barrovian, and contact aureole settings), monazite is formed via breakdown of metamorphic allanite at conditions that are recorded by the first appearance of either andalusite, kyanite, or staurolite, along with apatite, muscovite, biotite, plagioclase, and quartz. These authors constructed an allanite-to-monazite reaction that conserves REE in allanite and monazite:

$$3 \; CaREEAl_2FeSi_3O_{11}O(OH) + Ca_5(PO_4)_3(OH) + KAl_2AlSi_3O_{10}(OH)_2 + 4 \; Al_2SiO_5 + 3 \; SiO_2 \leftrightarrow$$

$$\quad (Allanite) \qquad\qquad (Apatite) \qquad\qquad (Muscovite) \qquad (Andalusite/ \quad (Quartz)$$
$$\qquad\qquad\qquad\qquad\qquad\qquad\qquad\qquad\qquad\qquad\qquad Kyanite)$$

$$3 \; REEPO_4 + KFe_3AlSi_3O_{10}(OH)_2 + 8 \; CaAl_2Si_2O_8 + 2 \; H_2O \qquad\qquad (32)$$
$$(Monazite) \qquad (Biotite) \qquad\qquad (Plagioclase)$$

Wing et al. (2003) noted that either cordierite or staurolite could play the same role as andalusite or kyanite in a monazite-forming reaction. In addition, the appearance of new monazite in rocks that both contain and lack aluminosilicate minerals marks similar pressure-temperature conditions for both the monazite and aluminosilicate isograds. Thus, the sequence

detrital or igneous monazite → metamorphic allanite → metamorphic monazite

should be expected during prograde greenschist- to amphibolite-facies metamorphism of pelitic rocks.

In contrast to these observations in rocks of low- and medium-pressure facies series, metasedimentary rocks of the high-pressure/low-temperature Sanbagawa subduction complex and granitic gneiss from the Dabie Shan ultrahigh-pressure metamorphic terrane are reported to preserve igneous or detrital cores of allanite through blueschist- and epidote-amphibolite-facies events, respectively. Sakai et al. (1984) discussed the contrasting optical properties and compositions of allanite core-epidote rim grains from Sanbagawa mica schists. These authors concluded that REE had become locally mobile as allanite broke down in metamorphic fluids, and that these REE had been deposited as REE-rich epidote rims upon allanite grains elsewhere in the fluid-flow system. The authors noted that this was a self-limiting process, because once REE-rich metasomatic epidote armored allanite, REE could not be released any more by the allanite grain. Liu et al. (1999) described dark grains of allanite surrounded by epidote within four samples of granitic gneiss from Dabie Shan. On the basis of core and rim analyses of these grains, they concluded that the allanite had little zoning, and therefore it was of igneous origin. Because the allanite was igneous, they also concluded that the studied Dabie Shan gneiss could not have undergone metamorphism at ultrahigh-pressure conditions. However, two of the four analyses of epidote rims contain weight percent quantities of total REE and thus, suggest a Sanbagawa-like, fluid-buffered origin for the REE-rich rims. Without BSE images, it is not possible to assess whether the allanite grain cores are of igneous origin. It is difficult to argue about the origin of composite grains of allanite, REE-rich-epidote, and epidote in these terranes, if the textures are not adequately documented and in the absence of Th-Pb SIMS geochronology.

The fate of metamorphic allanite or monazite at the amphibolite- to granulite-facies transition has been studied by Bingen et al. (1996), albeit more for the purpose of understanding the redistribution of LREE, medium-heavy REE (M-HREE), Th, and U in accessory mineral assemblages than as a possible indicator of metamorphic conditions. These authors examined accessory minerals in orthogneiss through the transition from the amphibolite to the granulite facies in southwestern Norway. They proposed three reactions that required the contribution of components from both rock-forming and accessory minerals. At the clinopyroxene-in isograd, the proposed schematic reaction is

$$3 \; (M\text{-}HREE)_2O_3 + 3 \; LREE_2O_3 + 2 \; Ca_5(PO_4)_3(F,OH) + 6 \; SiO_2 \leftrightarrow$$
$$(in \; hornblende, \; titanite) \quad (in \; allanite) \qquad\qquad (Apatite) \qquad\qquad (Quartz)$$

$$6 \; LREEPO_4 + 2 \; Ca_2(M\text{-}HREE)_3(SiO_4)_3(F,OH) + 6 \; CaO \qquad\qquad (33)$$
$$(Monazite) \qquad\qquad (in \; apatite) \qquad\qquad\qquad (in \; plagioclase)$$

This type of reaction would initiate the breakdown of either igneous or metamorphic allanite, to produce some metamorphic monazite, along with lessingite-rich apatite. At the orthopyroxene-in isograd, apatite becomes further enriched in LREE and Th, either due to the partial breakdown of monazite according to:

$$3\ LREEPO_4\ +\ 3\ SiO_2\ +\ 7\ CaO\ +\ (F_2, H_2O)\ \leftrightarrow$$
(*Monazite*)     (*Quartz*)    (*in plagioclase*)   (*in fluid*)

$$Ca_5(PO_4)_3(F,OH) + Ca_2LREE_3(SiO_4)_3(F,OH) \qquad (34)$$
(*Apatite*)        (*in apatite*)

or the final disappearance of allanite:

$$3\ LREE_2O_3\ +\ 6\ SiO_2\ +\ 4\ CaO\ +\ (F_2, H_2O)\ \leftrightarrow\ 2\ Ca_2LREE_3(SiO_4)_3(F,OH) \qquad (35)$$
(*in allanite*)  (*Quartz*)  (*in plagioclase*)  (*in fluid*)       (*in apatite*)

Bingen et al. (1996) proposed that the breakdown of allanite in orthogneiss would most likely yield an increased component of lessingite in apatite, rather than the wholesale transformation of apatite + allanite into monazite.

Allanite has been reported in metacarbonate rocks from the Lindås nappe in the Caledonides of western Norway (Boundy et al. 2002). These rocks are found in the Bergen Arc system, and are associated there with the transition between eclogite and granulite facies. The allanite-bearing marble layers consist of calcite, calcian strontianite, clinopyroxene, epidote/allanite, titanite, garnet, barite, and celestine. They occur along eclogite-facies shear zones, and some are interlayered with eclogite on the scale of centimeters. In the marble layers, the epidote-allanite crystals are up to 0.5 cm in size and display oscillatory zoning, but lack systematic core-to-rim trends of REE-zoning. Titanite grains appear to be unzoned in REE. The temperature of marble formation is estimated at ~600°C, based on C-isotopes and a calcsilicate rock that contains garnet + omphacite (580-600°C). In these rocks allanite is interpreted to have crystallized from an $H_2O$-rich fluid that moved along the shear zones under eclogite-facies conditions in the lowermost continental crust. Thus, the REE contents of the allanite are tentatively interpreted to reflect fluid compositions. (See below for other examples of fluid-sourced REE in allanite from both metamorphic and metasomatic systems.)

Allanite is the principal residence site for LREE and Th in metabasites (i.e, blueschist, garnet amphibolite, and eclogite) and metasedimentary rocks from high-pressure/low-temperature terrains (e.g., Sorensen and Grossman 1989, 1993; Sorensen 1991; Tribuzio et al. 1996; Hermann 2002; Spandler et al. 2003). Such allanite is typically complexly zoned, but not in ways characteristic of igneous crystallization (for example, compare the BSE images of Dawes and Evans 1991 to those of Sorensen 1991). REE patterns of zoned allanite grains in garnet amphibolite from the Catalina Schist and the Gee Point locality of the Shuksan Schist (Sorensen and Grossman 1989, 1993; Sorensen 1991) display features that indicate differences in both LREE abundance and fractionation from zone to zone that cannot be readily modeled by igneous crystallization. Although some of the allanite samples studied from the Catalina Schist occur in rocks that appear to be migmatitic, several rocks from this locality and none of the Shuksan examples show evidence for anatexis. For these reasons, Sorensen (1991) interpreted the allanite zoning to have been acquired during subsolidus metasomatic reactions. Both pelitic and mafic rock samples from New Caledonia show zoning features similar to those seen at Catalina and Gee Point (compare the SEM images of Sorensen 1991 to Spandler et al. 2003).

Mass-balance arguments have been used to determine the residence sites of REE in mafic and pelitic rocks from subduction complexes. This method can be problematic for allanite. For example, such small amounts of allanite are required to house all of the LREE that it is difficult to verify that a mode agrees with a mass fraction estimate. Furthermore, allanite in these rocks is generally strongly zoned in LREE, which means it can be difficult to calculate an average bulk composition for the mineral (Sorensen 1991). Nevertheless, Sorensen and Grossman (1989), Tribuzio et al. (1996), Hermann (2002), and Spandler et al. (2003) all used a mass balance approach to address the issue of the residence of LREE and Th in subducted rocks, and they concluded that allanite of metamorphic origin performed this role (see Figs. 35, 36). These figures emphasize the importance of allanite as an LREE host and, at the same time, demonstrate that the HREE mainly reside in other minerals (e.g., zircon, garnet).

In addition to examining eclogites from the Dora Maira massif in the Western Alps, Hermann (2002) conducted piston-cylinder synthesis experiments on a model crustal composition (granodioritic) doped with trace elements to address the allanite stability. He found that accessory allanite forms at the expense of zoisite above 700°C and 2 GPa. Allanite is stabilized by the presence of LREE and exists up to 1050°C and at least 4.5 GPa. The mineral is thus expected as a residual phase in subducted crust in the region of liquid extraction. Because allanite contained essentially all of the REE in his experimental composition, and its disappearance is caused by dissolution in the coexisting hydrous granodioritic melt (see

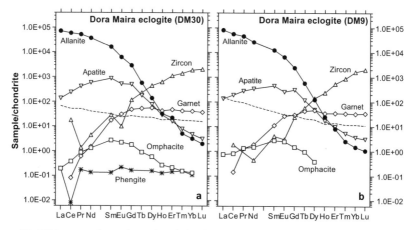

**Figure 35.** REE patterns for various minerals in phengite eclogite from Dora Maira, Western Alps. **a)** sample DM30; **b)** sample DM9. REE were not detected in phengite from sample DM9. Dashed line represents bulk-rock composition of the eclogite samples. LA-ICP-MS (minerals) and solution ICP-MS (bulk rock) data from Hermann (2002).

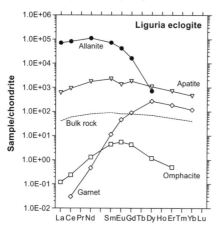

**Figure 36.** REE patterns for various minerals in eclogite from Liguria, northwestern Italy. Dashed line represents bulk-rock composition. Ion microprobe (allanite) and solution ICP-MS (bulk rock) data from Tribuzio et al. (1996).

also Broska et al. 1999), Hermann (2002) concluded that LREE cannot be significantly transported by hydrous fluids in subduction zones. This conclusion, however, is not consistent with the measured REE contents of some hydrothermal fluids (Banks et al. 1994) or the evidence of REE mobility in crustal fluids (e.g., Gieré 1996). Hermann (2002) further concluded that the REE characteristics of arc lavas must be derived from portions of the mantle wedge that had been contaminated by granitic melts. Unfortunately, some of his model experiments for extracting partitioning data have unusual melt and allanite compositions (compare Fig. 37 with Figs. 11-13, 17-20), so it is not clear how applicable the D(i) values obtained from these materials are to the questions he poses.

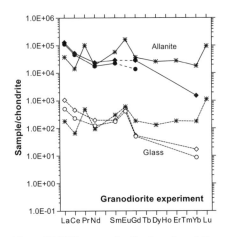

**Figure 37.** REE patterns for allanite and coexisting granodioritic glass. Experimental data from Hermann (2002). Symbols: dots and diamonds for experiments at 1000°C, 4.5 GPa and 1000°C, 3.5 GPa, respectively; asterisks for experiment 900°C, 2 GPa (these data are in contrast to all other granodiorite data, see Figs. 17-20).

**Allanite in metasomatic systems**

Metasomatically formed allanite has been reported from many geological environments (Söhnge 1945; Bocquet 1975; Watson and Snyman 1975; Exley 1980; Campbell and Ethier 1984; Moore and McStay 1990; Pan and Fleet 1990; Sorensen 1991; Ward et al. 1992; Zakrzewski et al. 1992; Gieré 1986, 1996; Smith et al. 2002; Weiss 2002; Weiss and Parodi 2002). Pantó (1975) described secondary allanite formed by the decomposition of feldspar. Similarly, Ward et al. (1992) observed that pervasive hydrothermal alteration of the Dartmoor granite (SW England) had led to mobilization of the REE from feldspar and biotite (with associated accessory monazite, xenotime, apatite and zircon), but noted that the REE were accommodated *in situ* by secondary minerals, including allanite and epidote. Allanite has also been described as a breakdown product of eudialyte in microsyenite of the Gardar province, South Greenland (Coulson 1997). There, eudialyte reacted with hydrothermal fluids to produce various assemblages via, for example, the reaction

$$\text{eudialite} + \text{fluid} \rightarrow \text{allanite} + \text{nepheline} \tag{36}$$

The breakdown products form pseudomorphs after eudialite.

Banks et al. (1994) directly determined the distribution coefficients of LREE between allanite in quartz veins and REE-rich hydrothermal solutions (Table 5). These fluids originated from the Capitan pluton, New Mexico, and were subsequently trapped in fluid inclusions. The distribution coefficients can be used to estimate the concentration of REE in fluids in similar geological settings. In the Bergell contact aureole, allanite occurs in an exoskarn formed during the intrusion of the Bergell pluton at conditions similar to those in the Capitan pluton (Gieré 1986). Using the distribution coefficients of Banks et al. (1994), Gieré (1996) estimated total LREE (La-Sm) contents of 65–115 ppm in the fluid, with Ce as most abundant REE (30-55 ppm) for the Bergell aureole. The respective total concentrations of REE at the Capitan Pluton are 185 ppm. Both these LREE values, the *analyzed* REE content of fluids extracted from the Capitan fluid inclusions and the *calculated* REE content of the Bergell fluids, are very large compared to other hydrothermal fluids. To our knowledge, higher REE concentrations in fluids have only been reported from other fluid inclusions at the Capitan pluton (ΣREE = 1290 ppm; Banks et al. 1994). The REE patterns of allanite and inferred metasomatic fluids

**Table 5.** REE concentrations in hydrothermal fluid and calculated allanite/fluid partition coefficients for some LREE. Data derived from fluid inclusions and allanite in quartz veins from the Capitan pluton, New Mexico (Banks et al. 1994).

|  | REE concentration (ppm) | | Allanite/fluid partition coefficients | |
|---|---|---|---|---|
|  | Fluid | Std Dev | Upper value | Lower value |
| **La** | 72.1 | 5.3 | 1559 | 1154 |
| **Ce** | 84.0 | 6.1 | 1633 | 1116 |
| **Pr** | 6.3 | 0.4 | 844 | 743 |
| **Nd** | 19.3 | 1.8 | 957 | 647 |
| **Sm** | 3.4 | 0.4 | 682 | 540 |

in the Bergell skarn (Fig. 38) emphasize the strong partition of REE into allanite at the conditions of contact metasomatism. Perhaps allanite could be used as a monitor of REE contents in metasomatic fluids.

Fluid $f_{O_2}$ may exert controls upon the development of metasomatic allanite. Smith et al. (2002) described the formation and alteration of allanite in skarn within the Beinn an Dubhaich granite aureole on the Island of Skye. There, allanite-(Ce) is present both as an igneous phase in the granite and as a metasomatic phase in the associated endoskarn. The metasomatic allanite exhibits distinct zoning. It occurs either as large single crystals, or as skeletal intergrowths with amphibole pseudomorphs of pyroxene. The skarn developed from highly saline, magmatic brines, which interacted with carbonate country rocks and led to the formation of hedenbergite (endoskarn) and diopside (exoskarn) at temperatures between 600 and 700°C. Allanite and amphibole

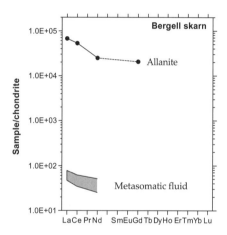

**Figure 38.** Chondrite-normalized REE patterns for allanite and coexisting metasomatic fluid in the Bergell contact aureole. Allanite data (electron microprobe) from Gieré (1986); data for fluid calculated from the REE partition coefficients between allanite and fluid trapped in fluid inclusions (Banks et al. 1994).

(hastingsite) formed subsequently, when hedenbergite reacted with anorthite and REE-bearing, F-rich aqueous fluids during cooling from 600 to 540°C. The data of Smith et al. (2002) indicate that metasomatic allanite contains a significant component of ferriallanite and epidote, which indicates that oxidation accompanied the allanite-forming process. In contrast to its metasomatic counterpart, allanite in the granite is present either as unaltered, euhedral crystals or altered, sub- to euhedral grains associated with a fine-grained REE-rich phase. The altered allanite is poorer in Fe, REE, and Th, which suggests it formed via an interaction between early-crystallized allanite and either residual-magmatic or early-hydrothermal fluids. Smith et al. (2002) concluded, however, that the fluids from which the endoskarn allanite precipitated acquired their elevated REE contents mainly through partitioning from the melt, and only to a lesser extent through leaching from primary igneous allanite.

## Alteration of allanite

Both hydrothermal and magmatic allanite commonly react with fluids and are transformed into a variety of alteration products. These alteration processes take place under hydrothermal and low-temperature to surficial conditions. Moreover, it has long been known that metamict allanite is in general strongly altered. This observation was initially made by Zenzén (1916) for specimens with $\rho < 3.1$ g/cm³, which were all isotropic, i.e., metamict.

From his study of allanite in granitic pegmatites from the Precambrian basement of eastern Egypt, Gindy (1961) concluded that the original composition of allanite strongly influences its susceptibility to hydrothermal alteration. He noted that generally more radioactive crystals are more strongly altered than less radioactive ones, which suggests that the degree of metamictization plays a role in the alteration of allanite. Gindy (1961) further observed that the more altered parts of allanite are less radioactive than the unaltered ones. He interpreted this feature to be evidence for leaching of radioactive elements during alteration. Leaching of Th and LREE during the alteration of allanite was also observed by Morin (1977) for allanite in granitic rocks from Ontario, Canada.

Such observations can be quantified by using SEM, electron microprobe, and LA-ICP-MS. For example, Poitrasson (2002) examined the alteration of magmatically zoned allanite from both a granodiorite and a fayalite-bearing granite from Corsica. In BSE images, he observed that altered areas within individual allanite crystals typically appeared patchy, exhibited irregular limits, and had lower mean atomic numbers than unaltered portions. Alteration appeared to proceed from fractures penetrating the crystals, showing that fractures are preferential fluid paths. Moreover, some of the strongly altered areas displayed large numbers of small, irregular fractures, which are similar to those seen in heavily radiation-damaged areas of other minerals (e.g., Gieré et al. 2000; Lumpkin 2001). With a combination of electron microprobe and LA-ICP-MS data, Poitrasson (2002) showed that altered areas are generally characterized by lesser LREE contents, larger concentrations of common Pb and HREE (Fig. 39), and low analytical totals. The latter points to the incorporation of water into the altered areas of allanite. The behavior of Th and Ti was opposite in the samples from the two different host rocks. This study has documented that substantial leaching of A-site cations can take place during the alteration of allanite prior to its decomposition and transformation into other phases.

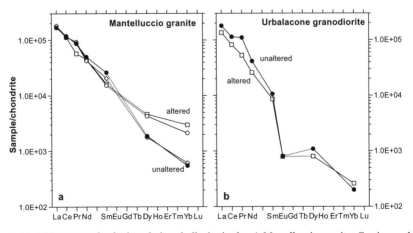

**Figure 39.** REE patterns for fresh and altered allanite in the **a)** Mantelluccio granite, Corsica, and **b)** Urbalacone granodiorite, Corsica. LA-ICP-MS and electron microprobe data from Poitrasson (2002).

The incorporation of water into allanite has long been recognized as an important early step of the weathering process (Meyer 1911). An indirect observation of this feature is seen in microprobe analyses of altered allanites. Low analytical totals, as well as reduced occupancy of the A, but not the M sites, have been reported for altered areas within allanite crystals in a quartz monzonite from Antarctica (Ghent 1972). Indeed, low analytical totals and reduced REE contents appear to be typical of altered allanite in various geological environments, as indicated also by other studies (e.g., Khvostova 1962; Pan and Fleet 1990; Jiang et al. 2003; Smith et al. 2002). Petrík et al. (1995) attributed the patchy appearance of allanite-(Ce) in various granitoids to interaction with late- to post-magmatic fluids. Their data indicated that this interaction led to an oxidation of allanite and a partial escape of REE.

Allanite commonly decomposes or transforms into other phases during its alteration (e.g., Gieré 1996). It is typically replaced by REE fluorocarbonate minerals, most commonly by bastnäsite ($REECO_3F$), and more rarely by synchisite ($CaREE(CO_3)_2F$; e.g., Sverdrup et al. 1959; Adams and Young 1961; Mineyev et al. 1962, 1973; Ehlmann et al. 1964; Perhac and Heinrich 1964; Mitchell 1966; Sakurai et al. 1969; Černý and Černá 1972; Mitchell and Redline 1980; Littlejohn 1981a, 1981b; Rimsaite 1984; Caruso and Simmons 1985; Lira and Ripley 1990; Pan et al. 1994; Buda and Nagy 1995). The bastnäsitization of allanite can be described schematically by the reaction

$$\text{allanite} + \text{fluid} \leftrightarrow \text{bastnäsite} + \text{clay minerals} + \text{thorite} \pm \text{fluorite} \pm \text{magnetite} \qquad (37)$$

in which the term "clay minerals" represents kaolinite, montmorillonite, or illite-series micas. The formation of thorite as a reaction product is due to the elevated Th contents commonly observed in allanite.

Less common is the replacement of allanite by phosphate phases during hydrothermal or epigene alteration. Mitchell and Redline (1980) described weathering of allanite into a mixture of bastnäsite, monazite, $CeO_2$, clay minerals, and Fe-oxyhydroxides in a Virginia granite pegmatite. Similarly, intense weathering of allanite in another Virginia pegmatite resulted in an assemblage consisting primarily of monazite and $CeO_2$ (Meintzer and Mitchell 1988). The overall decomposition of allanite under surficial weathering (i.e., epigene) conditions can be described schematically as

$$\text{allanite} + \text{fluid} \leftrightarrow \text{cerianite} + \text{monazite} + \text{clay minerals} + \text{goethite} \qquad (38)$$

where Th is trapped in both monazite and $CeO_2$ (Meintzer and Mitchell 1988). The formation of phosphates through allanite decomposition has also been described in a hydrothermal environment (Wood and Ricketts 2000). These authors studied the response of allanite-(Ce) to attack by low-salinity (<1.4 wt% NaCl equivalent), fluoride- and phosphate-bearing fluids in the Casto granite pluton, Idaho. These hydrothermal fluids interacted with igneous allanite-(Ce) at fairly low temperatures (100 to 200°C) to produce moderately altered crystals characterized by enrichment in Th, and depletion of La and Ce along fractures and rims. Substantial amounts of P were also observed along both fractures and rims. The most highly altered allanite grains exhibit substantial corrosion effects and are replaced by three phases: Y-bearing fluorite; a phosphate mineral rich in REE and Th, most likely monazite; and minor amounts of a Th-rich phase, probably thorianite. Neglecting the small amounts of the Th phase, Wood and Ricketts (2000) expressed the observed allanite replacement by the schematic overall alteration reaction

$$\text{allanite} + 2\ F^- + 5\ H^+ + 5\ H_2O + H_2PO_4^-$$
$$\leftrightarrow \text{fluorite} + \text{monazite} + Fe^{2+} + 3\ H_4SiO_4^0 + 2\ Al(OH)_3^0, \qquad (39)$$

and suggested that Fe and Al may eventually be fixed in Fe-rich phases and clays, chlorite or white mica, respectively. These authors have not observed cerianite as replacement product of allanite, probably because the redox potential under the hydrothermal conditions in the

Casto pluton was lower than in the epigene environment of the Virginia pegmatites. REE fluorocarbonates are also absent in the breakdown products of the Casto allanites, probably as a result of high phosphate activity or low carbonate and fluoride activities in the hydrothermal fluid.

Several other alteration products of allanite have also been reported (see compilation of Meintzer and Mitchell 1988), including unspecified phases or amorphous material (Morin 1977); epidote (Carcangiu et al. 1997); lanthanite, $REE_2(CO_3)_3 \cdot 8H_2O$ (Saebø 1961); and britholite (Smith et al. 2002).

Many studies of allanite alteration document that REE and Th are removed almost completely during allanite replacement. However, the presence of breakdown products such as monazite, thorianite, thorite, cerianite, and REE fluorocarbonates indicates that REE and Th were captured again locally before they were transported any significant distance by hydrothermal or surficial fluids (e.g., Rimsaite 1982). Moreover, alteration of allanite involves in many cases fluids that are enriched in fluoride, which is known to form strong complexes with both REE (Wood 1990) and Th (Langmuir and Herman 1980). The activities of phosphate, carbonate and fluoride in the fluid, together with the activity of $Ca^{2+}$, are essential factors in determining whether allanite will be transformed into bastnäsite or monazite.

## OUTLOOK

Many directions for future research are suggested by the current state of knowledge of allanite and related minerals, and by recent technological improvements of microanalysis of trace elements, in particular the REE. Existing studies have documented a large number of the most critical variables for understanding the substitutions of various elements into the allanite structure. However, the residence sites and compositional limits of three petrogenetically significant elements, namely P, Be, and V, are at present poorly understood. The V-richer allanite minerals could add to the conclusion that is suggested by Mn and Fe behavior, namely: $f_{O_2}$ and $T$, rather than variations in $P$ and element concentrations appear to be the most critical variables for determining transition metal and REE ratios in the epidote-group minerals (see Bonazzi and Menchetti 2004).

In particular, the characterization of allanite and REE-rich epidote minerals for their full complement of trace elements will likely be greatly enhanced as SIMS and LA-ICP-MS technology becomes more widely available. Many compositional parameters discussed in this chapter are poorly understood simply because they cannot readily be analyzed by the electron microprobe. High-contrast BSE imaging is also needed to conduct a first-rate study of these minerals, because it is so common for allanite and REE-rich epidote minerals to be zoned and altered.

The ratio $Fe^{3+}/Fe_{tot}$ in allanite can be changed via the oxidation-dehydrogenation reaction (Eqn. 7), as demonstrated by Dollase (1973) and Bonazzi and Menchetti (1994). Both of these experimental studies, however, were carried out under metastable conditions. The $Fe^{3+}/Fe^{2+}$ and H contents, therefore, need to be accurately measured for epidote-group minerals equilibrated within their stability fields, so that a thermodynamic model applicable to natural magmatic samples can be formulated. Such experimental data would be invaluable for a better understanding of igneous rocks, in particular rocks that contain coexisting amphibole, mica, and ferriallanite-allanite-epidote-clinozoisite solid solutions. In examples such as the Bergell tonalite, where detailed chronological data for allanite are available (Oberli et al. 2004), one can hope to additionally gain insight into how some petrologically important parameters of the tonalitic melt (e.g., $T$, $P$, $f_{H_2O}$) changed with time. Allanite, therefore, could become a powerful geothermobarometer as well as a hygrometer and chronometer for calc-alkaline systems.

In addition to such experiments, we suggest that the following topics be studied in the future.

1. Radiation damage: We anticipate that both the initial and the critical amorphization doses of allanite could be easily determined, because allanite is commonly zoned with respect to Th, and different zones thus would have received different α doses. Moreover, such studies would help in understanding the relationship between metamictization, hydration and release of REE, Th and Pb to fluids during alteration. Examples of such studies are given for other minerals by Lumpkin (2001) and Lumpkin et al. (2004).

2. Incorporation of Eu: the presence or absence of distinct Eu anomalies, both positive and negative, for different REE-rich epidote-group minerals in different environments has so far not been much discussed. This is in part because low values of Eu have been difficult to measure *in situ*. The use of either SIMS or LA-ICP-MS for microanalytical studies of Eu-incorporation into REE-rich epidote and allanite could reveal the behavior of this element in different petrogenetic environments.

3. Experiments are needed to determine the partition coefficients between allanite and various fluids at different pressures and temperatures. We anticipate that the composition of allanite could be used as a tool for monitoring REE compositions of fluids, once such data are available.

4. Thorium zoning in igneous allanite: it is not clear why some allanites display core-to-rim decreases in Th content (e.g., Oberli et al. 2004), whereas others show the opposite zoning (e.g., Vazquez and Reid 2001, 2003; Gromet and Silver 1983). Since allanite exerts such an important control on the behavior of Th in some igneous rocks, it would be desirable to study this feature in more detail and in various types of igneous rocks (e.g., with/without monazite).

## ACKNOWLEDGMENTS

The authors would like to thank the reviewers of this manuscript, Jean Morrison (University of Southern California) and Jörg Hermann (Australian National University), for their helpful comments and constructive critcism. We are particularly grateful to Wayne Dollase (University of California, Los Angeles) and Felix Oberli (ETH Zürich), who both provided very thorough (unofficial) reviews of the manuscript. The present chapter also benefited from valuable suggestions by the volume editors Axel Liebscher and Gerhard Franz, and by Gregory Lumpkin (University of Cambridge). We further would like to thank Dr. James Beard (Virginia Museum of Natural History, Martinsburg, Virginia) for kindly permitting us to use his extensive, unpublished dataset for allanite from the Appalachians in the figures, and Dr. Jeffrey N. Grossman, USGS, for his assistance in data management. R.G. is grateful to the Purdue University Libraries for crucial help during various stages of this study, and to Dr. Alex Gluhovski for his translations of some Russian papers.

## REFERENCES

Adams JW, Young EJ (1961) Accessory bastnäsite in the Pikes Peak granite, Colorado. U.S. Geol Survey Prof Paper 424-C:292-294

Anthony JW, Bideaux RA, Bladh KW, Nichols MC (1995) Handbook of Mineralogy. Vol. III, Silica, Silicates, Part 1. Mineral Data Publishing, Tuscon, Arizona

Banks DA, Yardley BWD, Campbell AR, Jarvis KE (1994) REE composition of an aqueous magmatic fluid: a fluid inclusion study from the Capitan Pluton, New Mexico. Chem Geol 113:259-272

Banno Y (1993) Chromian sodic pyroxene, phengite and allanite from the Sanbagawa blueschists in the eastern Kii Peninsula, central Japan. Mineral J 16/6:306-317

Barth S, Oberli F, Meier M (1989) U-Th-Pb systematics in morphologically characterized zircon and allanite: a high-resolution isotopic study of the Alpine Rensen pluton (northern Italy). Earth Planet Sci Lett 95: 235-254

Barth S, Oberli F, Meier M (1994) Th-Pb versus U-Pb isotope systematics in allanite from co-genetic rhyolite and granodiorite: implications for geochronology. Earth Planet Sci Lett 124:149-159

Bayliss P, Levinson AA (1988) A system of nomenclature for rare earth mineral species: revision and extension. Am Mineral 73:422-423

Bea F (1996) Residence of REE, Y, Th and U in granites and crustal protoliths; implications for the chemistry of crustal melts. J Petrology 37:521-552

Bermanec V, Armbruster T, Oberhänsli R, Zebec V (1994) Crystal chemistry of Pb- and REE-rich piemontite from Nezilovo, Macedonia. Schweizerische Mineralogische und Petrographische Mitteilungen 74:321-328

Bernstein LR (1985) Germanium geochemistry and mineralogy. Geochim Cosmochim Acta 49:2409-2422

Bingen B, Demaiffe D, Hertogen J (1996) Redistribution of rare earth elements, thorium, and uranium over accessory minerals in the course of amphibolite to granulite facies metamorphism: the role of apatite and monazite in orthogneisses from southwestern Norway. Geochim Cosmochim Acta 60:1341-1354

*Black P (1970) A note on the occurrence of allanite in hornfelses at Paritu, Coromandel County. New Zealand J Geology Geophys 13/2:343-345

Bocquet J (1975) Sur une allanite filonienne, à Bramans en Maurienne (Alpes occidentals, Savoie). Bulletin de la Société française de Minéralogie et de cristallographie 98:171-174

Bonazzi P, Garbarino C, Menchetti S (1992) Crystal chemistry of piemontites: REE-bearing piemontite from Monte Brugiana, Alpi Apuane, Italy. Eur J Mineral 4:23-33

Bonazzi P, Menchetti S (1994) Structural variations induced by heat treatment in allanite and REE-bearing piemontite. Am Mineral 79:1176-1184

Bonazzi P, Menchetti S (1995) Monoclinic members of the epidote group: Effects of the Al = $Fe^{3+}$ = $Fe^{2+}$ substitution and of the entry of REE$^3$. Mineral Petrol 53:133-153

Bonazzi P, Menchetti S (2004) Manganese in monoclinic members of the epidote group: piemontite and related minerals. Rev Mineral Geochem 56:495-552

Bonazzi P, Menchetti S, Reinecke T (1996) Solid solution between piemontite and androsite-(La), a new mineral of the epidote group from Andros Island, Greece. Am Mineral 81:735-742

Boundy TM, Donohue CL, Essene EJ, Mezger K, Austrheim H (2002) Discovery of eclogite-facies carbonate rocks from the Lindås Nappe, Caledonides, Western Norway. J Metamorph Geol 20:649-667

Brandon AD, Creaser RA, Chacko T (1996) Constraints on rates of granitic magma transport from epidote dissolution kinetics. Science 271:1845-1848

Braun M (1997) REE-Ungleichgewichte in Gesteinen der Lebendun Serie am Monte Giove (Val Formazza, Novara, Italien). Ph.D. thesis, University of Basel, Basel, Switzerland

Brooks CK, Henderson P, Rønsbo JG (1981) Rare-earth partition between allanite and glass in the obsidian of Sandy Braes, Northern Ireland. Mineral Mag 44:157-160

Broska I, Chekmir AS, Határ J (1999) Allanite solubility and the role of accessory mineral paragenesis in the Carpathian granite petrology. Geologica Carpathica 50:90-91

Broska I, Petrík I, Williams CT (2000) Coexisting monazite and allanite in peraluminous granitoids of the Tribec Mountains, Western Carpathians. Am Mineral 85:22-32

Broska I, Siman P (1998) The breakdown of monazite in the West-Capathian Veporic orthogneisses and Tatric granites. Geologica Carpathica 49:161-167

Brunsmann A, Franz G, Erzinger J, Landwehr D (2000) Zoisite- and clinozoisite-segregations in metabasites (Tauern Window, Austria) as evidence for high-pressure fluid-rock interaction. J Metamorph Geol 18: 1-21

Buda G, Nagy G (1995) Some REE-bearing accessory minerals in two rock types of Variscan granitoids, Hungary. Geologica Carpathica 46:67-78

Burt DM (1989) Compositional and phase relations among rare earth element minerals. Rev Mineral 21:259-307

Cameron KL (1984) The Bishop Tuff revisited: New rare earth element data consistent with crystal fractionation. Science 224:1338-1340

Campbell FA, Ethier VG (1984) Composition of allanite in the footwall of the Sullivan orebody, British Columbia. Can Mineral 22:507-511

Campbell L, Henderson P, Wall F, Nielsen TFD (1997) Rare earth chemistry of perovskite group minerals from the Gardiner Complex, East Greenland. Mineral Mag 61:197-212

Cappelli B, Franceschelli M., Memmi I (1993) LREE zoning in allanite-(Ce) from low-temperature metamorphics, Alpi Apuane, Italy. Terra: rivista di scienze ambientali e territoriali 5:414-415

Carcangiu G, Palomba M, Tamanini M (1997) REE-bearing minerals in the albitites of central Sardinia, Italy. Mineral Mag 61:271-283

Caruso L, Simmons G (1985) Uranium and microcracks in a 1000-meter core, Redstone, New Hampshire. Contrib Mineral Petrol 90:1-17

Catlos EJ, Sorensen SS, Harrison TM (2000) Th-Pb ion-microprobe dating of allanite. Am Mineral 85:633-648

Cech F, Vrána S, Povondra P (1972) A non-metamict allanite from Zambia. Neues Jahrbuch für Mineralogie, Abhandlungen 116(2):208-223

Černý P, Černá I (1972) Bastnaesite after allanite from rock lake, Ontario. Can Mineral 11:541-543

Chesner CA, Ettlinger AD (1989) Composition of volcanic allanite from the Toba Tuffs, Sumatra, Indonesia. Am Mineral 74:750-758

Coulson IM (1997) Post-magmatic alteration in eudialyte from the North Qoroq center, South Greenland. Mineral Mag 61:99-109

Cressey G, Steel AT (1988) An EXAFS Study of Gd, Er and Lu Site Location in the Epidote Structure. Phys Chem Min 15:304-312

*Dahlquist JA (2001) Low-pressure emplacement of epidote-bearing metaluminous granitoids in the Sierra de Chepes (Famatinian Orogen, Argentina) and relationships with the magma source. Revista Geologica de Chile 28 2:147-161

Davis DW, Schandl ES, Wasteneys HA (1994) U-Pb dating of minerals in alteration halos of Superior Province massive sulfide deposits: syngenesis versus metamorphism. Contrib Mineral Petrol 115:427-437

Dawes RL, Evans BW (1991) Mineralogy and geothermobarometry of magmatic epidote-bearing dikes, Front Range, Colorado. Geol Soc Am Bull 103:1017-1031

Dayvault RD, Krabacher JE, Hardy LC, Colby RJ (1986) Radiological characterization of the Lowman, Idaho, uranium mill tailings remedial action site, GJ-53, UNC Technical Services, Inc, prepared for the U.S. Department of Energy UMTRA Site Characterization Project, Grand Junction, Colorado, 29 p

Deer WA, Howie RA, Zussman J (1986) Rock-forming minerals, vol. 1B: Disilicates and ringsilicates (2nd edition). Longman, Harlow, United Kingdom

Demange M, Elsass P (1973) Présence d'allanite dans le gisement stratiforme cuprifère de Talate n'Ouaman (Maroc). Comptes rendus hébdomadaires des Séances de l'Académie des Sciences, Sciences naturelles 277D:1969-1972

de Fourestier J (1999) Glossary of mineral synonyms. Canadian Mineralogist Special Publication 2. Mineralogical Association of Canada, Ottawa

de Parseval P, Fontan F, Aigouy T (1997) Composition chimique des minéraux de terres rares de Trimouns (Ariège, France). Comptes rendusde l'Académie des Sciences, Paris. Série IIa: Sciences de la Terre et des Plantes 324:625-630

*Ding K, Zhang P, Li Z (1994) First discovery of Nd-rich allanite in xinjiang and its characteristics (in Chinese, English abstract and analysis). Scientia Geologica Sinica 29 1:95-104

Dollase WA (1971) Refinement of the crystal structures of epidote, allanite and hancockite. Am Mineral 56:447-464

Dollase WA (1973) Mössbauer spectra and iron distribution in the epidote-group minerals. Z Kristallogr 138:41-63

Duggan MB (1976) Primary allanite in vitrophyric rhyolites from the Tweed Shield Volcano, north-eastern New South Wales. Mineral Mag 40:652-653

Ehlmann AJ, Walper JL, Williams J (1964) A new, Baringer Hill-type, rare-earth pegmatite from the Central Mineral Region, Texas. Econ Geol 59:1348-1360

Enami M, Zang Q (1988) Magnesian staurolite in garnet-corundum rocks and eclogite from the Donghai district, Jiangsu province, east China. Am Mineral 73:48-56

Ercit TS (2002) The mess that is "allanite". Can Mineral 40:1411-1419

Ewing RC, Weber WJ, Clinard FW (1995) Radiation effects in nuclear waste forms for high-level radioactive waste. Prog Nucl Energy 29(2):63-127

Ewing RC, Chakoumakos BC, Lumpkin GR, Murakami T (1987) The metamict state. Mater Res Bull 12:58-66

Exley RA (1980) Microprobe studies of REE-rich accessory minerals: implications for Skye granite petrogenesis and REE mobility in hydrothermal systems. Earth Planet Sci Lett 48:97-110

Ferraris G, Ivaldi G, Fuess H, Gregson D (1989) Manganese/iron distribution in a strontian piemontite by neutron diffraction. Z Kristallogr 187:145-151

Ferry JM (1984) A biotite isograd in south-central Maine, USA: mineral reactions, fluid transfer and heat transfer. J Petrol 25:871-893

Ferry JM (1988) Infiltration-driven metamorphism in northern New England, USA. J Petrol 29:1121-1159

Ferry JM (2000) Patterns of mineral occurrence in metamorphic rocks. Am Mineral 85:1573-1588

Ferrow E (1987) Mössbauer and X-ray studies on the oxidation of annite and ferriannite. Phys Chem Min 14: 270-275

Fersman AE (1931) Les pegmatites. – leur importance scientifique et pratique. Tomes I-III: Les pegmatites granitiques. Académie des Sciences de l'U.R.S.S, Leningrad, 675 p

Finger F, Broska I, Roberts MP, Schermaier A (1998) Replacement of primary monazite by apatitle-allanite-epidote coronas in an amphibolite facies granite gneiss from the eastern Alps. Am Mineral 83:248-258

Fleischer M (1965) Some aspects of the geochemistry of yttrium and the lanthanides. Geochim Cosmochim Acta 29:755-772

Fourcade S, Allègre CJ (1981) Trace Elements Behavior in Granite Genesis: A Case Study. The Calc-Alkaline Plutonic Association from the Querigut Complex (Pyrénées, France). Contrib Mineral Petrol 76:177-195

Franz G, Liebscher A (2004) Physical and chemical properties of the epidote minerals–an introduction. Rev Mineral Geochem 56:1-82

Frei D, Liebscher A, Wittenberg A, Shaw CSJ (2003) Crystal chemical controls on rare earth element partitioning between epidote-group minerals and melts: an experimental and theoretical study. Contrib Mineral Petrol 146:192-204

Frondel JW (1964) Variation of some rare earths in allanite. Am Mineral 49:1159-1177

Frye K (1981) Encyclopedia of Earth Sciences, Volume IVB: The Encyclopedia of Mineralogy. Hutchinson Ross, Stroudsburg, Pennsylvania

Geijer P (1927) Some mineral associations from the Norberg District. Sveriges Geologiska Undersokning. Avhandlingar och uppsatser series C 343 (Arsbok 20, 1926, No.4):1-32

Geisler T, Schleicher, H (2000) Improved U-Th-total Pb dating of zircons by electron microprobe using a simple new background modeling procedure and Ca as a chemical indicator of fluid-induced U-Th-Pb discordance in zircon. Chem Geol 163:269-285

Geisler T, Ulonska M, Schleicher H, Pidgeon RT, van Bronswijk W (2001) Leaching and differential recrystallization of metamict zircon under experimental hydrothermal conditions. Contrib Mineral Petrol 141:53-65

Ghent ED (1972) Electron Microprobe Study of Allanite from the Mt. Falconer Quartz Monzonite Pluton, Lower Taylor Valley, South Victoria Land, Antarctica. Can Mineral 11:526-530

Gieré R (1986) Zirconolite, allanite and hoegbomite in a marble skarn from the Bergell contact aureole: implications for mobility of Ti, Zr and REE. Contrib Mineral Petrol 93:459-470

Gieré R (1996) Formation of Rare Earth minerals in hydrothermal systems. *In:* Rare Earth Minerals: Chemistry, Origin and Ore Deposits. Jones AP, Williams CT, Wall F (eds) Mineralogical Society Series, Chapman & Hall, p 105-150

Gieré R, Williams CT, Braun M, Graeser S (1998) Complex Zonation Patterns in Monazite-(Nd) and Monazite-(Ce). International Mineralogical Association, 17th General Meeting, Toronto, Abstract Volume, p. A84

Gieré R, Virgo D, Popp R.K (1999) Oxidation state of iron and incorporation of REE in igneous allanite. Journal of Conference Abstracts 4: p. 721

Gieré R, Swope RJ, Buck EC, Guggenheim R, Mathys D, Reusser E (2000) Growth and alteration of uranium-rich microlite. *In:* Scientific Basis for Nuclear Waste Management XXIII. Smith R.W, Shoesmith D.W (eds) Materials Research Society, Symposium Proceedings 608:519-524

Gindy AR (1961) Allanite from Wadi el Gemal area, eastern desert of Egypt, and its radioactivity. Am Mineral 46:985-993

Giuli G, Bonazzi P, Menchetti S (1999) Al-Fe disorder in synthetic epidotes: a single-crystal X-ray diffraction study. Am Mineral 84:933-936

Goldschmidt VM, Thomassen L (1924) Geochemische Verteilungsgesetze der Elemente. III. Röntgenspektrographische Untersuchungen über die Verteilung der Seltenen Erdmetalle in Mineralen. Videnskapsselskapets Skrifter I. Mathematisk-Naturvidenskabelig Klasse 5:1-58

Grapes RH (1981) Chromian epidote and zoisite in kyanite amphibolite, Southern Alps, New Zealand. Am Mineral 66:974-975

Grapes RH, Hoskin PWO (2004) Epidote group minerals in low–medium pressure metamorphic terranes. Rev Mineral Geochem 56:301-345

Grew ES (2002) Mineralogy, petrology and geochemistry of beryllium: an introduction and list of beryllium minerals. Rev Mineral Geochem 50:1-76

Grew ES, Essene EJ, Peacor DR, Su S-C, Asami M (1991) Dissakisite-(Ce), a new member of the epidote group and the Mg analogue of allanite-(Ce), from Antarctica. Am Mineral 76:1990-1997

Gromet LP, Silver LT (1983) Rare earth element distributions among minerals in a grandiorite and their petrogenetic implications. Geochim Cosmochim Acta 47:925-939

Hagesawa S (1957) Chemical studies of allanites and their associated minerals from the pegmatites in the northern part of the Abukuma Massif. The Science Reports of Tohoku University, third series (Mineralogy, Petrology, Economic Geology) 5:345-371

Hagesawa S (1958) Chemical studies of allanites from the new localities in Fukushima and Kagawa Prefectures. Scientific Reports of Tohoku University, third series (Mineralogy, Petrology, Economic Geology) 6:39-56

Hagesawa S (1959) Allanites from the pegmatites of several localities in southwestern Japan. The Science Reports of Tohoku University, third series (Mineralogy, Petrology, Economic Geology) 6:209-226

Hagesawa S (1960) Chemical composition of allanite. The Science Reports of Tohoku University, third series (Mineralogy, Petrology, Economic Geology) 6(3):331-387

Halleran AAD, Russell JK (1996) REE-bearing alkaline pegmatites and associated light REE-enriched fenites at Mount Bisson, British Columbia. Econ Geol 91:451-459

Hanson RA, Pearce DW (1941) Colorado cerite. Am Mineral 26:110-120

*Hata S (1939) Studies on allanite from the Abukuma granite region. Scientific Papers of the Institute of physical and chemical Research, Tokyo 36:301-311

Hawkins BW (1975) Mary Kathleen uranium deposit. *In:* Economic geology of Australia and Papua New Guinea. Australasian Institute of Mining and Metallurgy, Monograph Series 5-8: 398-402. Parkville, Vic

Heinrich EW, Wells RG (1980) The diversity of rare-earth mineral deposits and their geological domains. The Rare earths in modern science and technology 2:511-516

Hermann J (2002) Allanite: thorium and light rare earth element carrier in subducted crust. Chem Geol 192: 289-306

Hickling NL, Phair G, Moore R, Rose Jr HJ (1970) Boulder Creek batholith, Colorado. Part I: allanite and its bearing upon age patterns. Geol Soc Am Bull 81:1973-1994

Hogg CS, Meads RE (1975) A Mössbauer study of the thermal decomposition of biotites. Mineral Mag 40: 79-88

Holtstam D, Andersson UB, Mansfeld J (2003) Ferriallanite-(Ce) from the Basntnäs deposit, Västmanland, Sweden. Can Mineral 41:1233-1240

Hugo PJ (1961) The allanite deposits on Vrede, Gordonia district, Cape Province. Republiek van Suid-Afrika, Departement van Mynwese, Geologiese Opname Bulletin 37:1-65

Hutton CO (1951a) Allanite from Yosemite National Park, Tuolumne Co, California. Am Mineral 36:233-248

Hutton CO (1951b) Allanite from Wilmot Pass, Fjordland, New Zealand. Amer J Sci 249:208-214

Iimori T (1939) A Beryllium-bearing variety of allanite. Scientific Papers of the Institute of physical and chemical Research, Tokyo 36:53-55

Iimori T, Yoshimora J, Hata S (1931) A new radioactive mineral found in Japan. Scientific Papers of the Institute of physical and chemical Research, Tokyo 15:83-88

Ivanov OP, Vorob'ev YuK, Efremenko LYa, Knyazeva DN (1981) Acicular allanite from the Itulin deposit veins. Zapiski Vsesoiuznogo mineralogicheskogo obshchestva 110:361-366 (in Russian)

Izett GA, Wilcox RE (1968) Perrierite, chevkinite, and allanite in Upper Cenozoic ash beds in the Western United States. Am Mineral 53:1558-1567

Janeczek J, Eby RK (1993) Annealing of radiation damage in allanite and gadolinite. Phys Chem Min 19: 343-356

Jiang N, Sun S, Chu X, Mizuta T, Ishiyama D (2003) Mobilization and enrichment of high-field strength elements during late- and post-magmatic processes in the Shuiquangou syenitic complex, Northern China. Chem Geol 200:117-128

Johan Z, Oudin E, Picot P (1983) Analogues germanifères et gallifères des silicates et oxydes dans les gisements de zinc des Pyrénées centrales, France; argutite et carboirite, deux nouvelles espèces minerals. Tschermaks Mineralogische und Petrographische Mitteilungen 31:97-119

Kalinin YeP, Yushkin, N.P, Goldin, B.A (1968) The influence of chemical composition on hardness of crystals of allanite. Zapiski vsesoyuznogo mineralogicheskogo Obshchestva 97:647-652 (in Russian)

Kartashov PM, Ferraris G, Ivaldi G, Sokolova E, McCammon CA (2002) Ferriallanite-(Ce), CaCeFe$^{3+}$AlFe$^{2+}$($SiO_4$)($Si_2O_7$)O(OH), a new member of the epidote group: Description, X-Ray and Mössbauer study. Can Mineral 40:1641-1648

Kartashov PM, Ferraris G, Ivaldi G, Sokolova E, McCammon CA (2003) Ferriallanite-(Ce), CaCeFe$^{3+}$AlFe$^{2+}$($SiO_4$)($Si_2O_7$)O(OH), a new member of the epidote group: Description, X-Ray and Mössbauer study: Errata. Can Mineral 41:829-830

Kato A, Shimizu M, Okada Y, Komuro Y, Takeda K (1994) Vanadium-bearing spessartine and allanite in the manganese-iron ore from the Odaki Orebody of the Kyurazawa Mine, Ashio town, Tochigi Prefecture, Japan. Bulletin National Science Museum, Tokyo, Series C 20 1:1-12

Khvostova VA (1962) Mineralogy of Orthite. Institut mineralogii, geokhimii i kristallokhimii redkikh elementov Akademii nauk SSSR, Trudy 11:119 p (in Russian)

Khvostova VA (1963) On the isomorphism of epidote and orthite. Doklady Academy of Sciences U.S.S.R, Earth Sciences Section 141:1307-1309

Khvostova VA, Bykova AV (1961) Accessory orthite of southern Yakutia. Institut mineralogii, geokhimii I kristallokhimii redkikh elementov. Akademii nauk SSSR, Trudy 7:130-137 (in Russian)

Kimura K, Nagashima K (1951) Chemical investigations of Japanese minerals containing rarer elements. XLII Journal of the Chemical Society of Japan, Pure Chemistry Sections 72:52-54 (in Japanese)

Kosterin AV, Kizyura VE, Zuev VN (1961) Ratios of rare earth elements in allanites from some igneous rocks of northern Kirgiziya. Geochemistry 5:481-484

Kramm U (1979) Kanonaite-rich viridines from the Venn-Stavelot Massif, Belgian Ardennes. Contrib Mineral Petrol 69:387-395

Krivdik SG, Tkachuk VI, Maximchuk IG, Michnik TL (1989) Accessory orthite from alkaline rocks and carbonatites of the Ukrainian Shield. Mineralogicheskii zhurnal 11(1):34-42 (in Russian)

Kumskova NM, Khvostova VA (1964) X-ray study of the epidote-allanite group of minerals. Geochem Int 4: 676-686

Kvick KA, Pluth JJ, Richardson JW Sr, Smith JV (1988) The ferric iron distribution and hydrogen bonding in epidote: a neutron diffraction study at 15 K. Acta Crystallogr Sec B Struc Sci 44:351-355

Langmuir D, Herman JS (1980) The mobility of thorium in natural waters at low temperatures. Geochim Cosmochim Acta 44:1753-1766

Lee DE, Bastron H (1962) Allanite from the Mt. Wheeler area, White Pine County, Nevada. Am Mineral 47: 1327-1331

Lee DE, Bastron H (1967) Fractionation of rare-earth elements in allanite and monazite as related to geology of the Mt. Wheeler mine area, Nevada. Geochim Cosmochim Acta 31:339-356

Levinson AA (1966) A system of nomenclature for rare-earth minerals. Am Mineral 51:152-158

Lima de Faria J (1964) Identification of metamict minerals by X-ray powder photographs. Junta de Investigaçoes do Ultramar; Estudos, Ensaios e Documentos, No. 112, Lisbon, Portugal, 74 p

Lira R, Ripley EM (1990) Fluid inclusion studies of the Rodeo de Los Molles REE and Th deposit, Las Chacras Batholith, Central Argentina. Geochim Cosmochim Acta 54:663-671

Littlejohn AL (1981a) Alteration products of accessory allanite in radioactive granites from the Canadian Shield. Curr Res Geol Sur Canada 81-1B:95-104

Littlejohn AL (1981b) Alteration products of accessory allanite in radioactive granites from the Canadian Shield: Reply. Curr Res Geol Sur Canada 81-1C:93-94

Liu X, Dong S, Xue H, Zhou J (1999) Significance of allanite-(Ce) in granitic gneisses from the ultrahigh-pressure metamorphic terrane, Dabie Shan, central China. Mineral Mag 63/4:579-586

Lumpkin GR (2001) Alpha-decay damage and aqueous durability of actinite host phases in natural systems. J Nucl Mat289:136-166

Lumpkin GR, Eby RK, Ewing RC (1991) Alpha-recoil damage in titanite ($CaTiSiO_5$) direct observation and annealing study using high-resolution transmission electron microscopy. J Mater Res 6:560-564

Lumpkin GR, Smith KL, Gieré R, Williams CT (2004) The geochemical behaviour of host phases for actinides and fission products in crystalline ceramic nuclear waste forms. *In:* Energy, Waste, and the Environment – a geochemical approach. Gieré R, Stille P (eds) Geological Society Special Publications, London (in press)

Maas R, McCulloch MT, Campbell IH (1987) Sm-Nd isotope systematics in uranium-rare earth element mineralization at the Mary Kathleen Uranium Mine, Queensland. Econ Geol 82:1805-1826

Mahood G, Hildreth W (1983) Large partition coefficients for trace elements in high-silica rhyolites. Geochim Cosmochim Acta 47:11-30

*Marble JP (1940) Allanite from the Barringer Hill, Llano County, Texas. Am Mineral 25:168-173

*Marble JP (1943) Possible age of allanite from Whiteface mountain, Essex Co., New York. Amer J Sci 241: 32-42

Mariano AN (1989) Economic geology of rare earth elements. Rev Mineral 21:309-337

Mason RA (1992) Models of order and iron-fluorine avoidance in biotite. Can Mineral 30:343-354

Meintzer RE, Mitchell RS (1988) The epigene alteration of allanite. Can Mineral 26:945-955

Mezger K, Hanson GN, Bohlen SR (1989) U-Pb systematics of garnet: dating the growth of garnet in the late Archean Pikwitonei granulite domain at Cauchon and Natawahunan Lakes, Manitoba, Canada. Contrib Mineral Petrol 101:136-148

Meyer RJ (1911) Über einen skandiumreichen Orthit aus Finnland und den Vorgang seiner Verwitterung. Sitzungsberichte der königlichen preussischen Akademie der Wissenschaften, Berlin 105:379-384 (in German)

Michael PJ (1983) Chemical differentiation of the Bishop Tuff and other high-silica magmas through crystallization processes. Geology 11:31-34

*Michael PJ (1984) Chemical differentiation of the Cordillera Paine granite (southern Chile) by in situ fractional crystallization. Contrib Mineral Petrol 87:179-195

Mineyev DA, Makarochkin BA, Zhabin AG (1962) On the behavior of lanthanides during alteration of rare earth minerals. Geochemistry 7:684-693

Mineyev DA, Rozanov KI, Smirnova NV, Matrosova TI (1973) Bastnaesitization products of accessory orthite. Doklady Akademia Nauk SSSR, Earth Science Sections 210:149-152

Mitchell RS (1966) Virginia metamict minerals: allanite. Southeastern Geology 7:183-195

Mitchell RS (1973) Metamict minerals: a review. Mineral Record 4:177-182

Mitchell RH (1996) Perovskite: a revised classification scheme for an important rare earth element host in alkaline rocks. *In:* Rare Earth Minerals: Chemistry, Origin and Ore Deposits. Jones AP, Williams CT, Wall F (eds) Mineralogical Society Series, Chapman & Hall, London, p 41-76

Mitchell RS, Redline GE (1980) Minerals of a weathered allanite pegmatite, Amherst County, Virginia. Rocks & Minerals 55:245-249

Moench RH (1986) Comment on "Implications of magmatic epidote-bearing plutons on crustal evolution in the accreted terranes of northwestern North America" and "Magmatic epidote and its petrologic significance." Geology 14:187-188

Möller P (1989) Rare earth mineral deposits and their industrial importance. *In:* Lanthanides, Tantalum and Niobium. Möller P, Černý P, Saupé F (eds) Special Pub Soc Geology Appl Mineral Dep 7:171-188

Montel J-M (1993) A model for monazite/melt equilibrium and application to the generation of granitic magmas. Chem Geol 110:127-146

Moore JM, McStay JH (1990) The formation of allanite-(Ce) in calcic granofelses, Namaqualand, South Africa. Can Mineral 28:77-86

Morin JA (1977) Allanite in granitic rocks of the Kenora-Vermilion Bay area, Northwestern Ontario. Can Mineral 15:297-302

Morrison J (2004) Stable and radiogenic isotope systematics in epidote group minerals. Rev Mineral Geochem 56:607-628

Murata KJ, Rose HJ Jr, Carron MK, Glass JJ (1957) Systematic variation of REE in cerium-earth minerals. Geochim Cosmochim Acta 11:141-161

*Nagasaki A, Enami M (1998) Sr-bearing zoisite and epidote in ultra-high pressure (UHP) metamorphic rocks from the Su-Lu province, eastern China: An important Sr reservoir under UHP conditions. Am Mineral 83:240-247

Nesse WD (2004) Introduction to optical mineralogy. Oxford University Press, 3$^{rd}$ edition. New York, London

Neumann H, Nilssen B (1962) Lombaardite, a rare earth silicate, identical with, or very closely related to allanite. Norsk Geologisk Tidsskrift 42:277-286

Nozik YK, Kanepit VN, Fykin LY, Makarov YS (1978) A neutron diffraction study of the structure of epidote. Geochem Int 15:66-69

Oberli F, Sommerauer J, Steiger RH (1981) U-(Th)-Pb systematics and mineralogy of single crystals and concentrates of accessory minerals from the Cacciola granite, central Gotthard massif, Switzerland. Schweizerische Mineralogische und Petrographische Mitteilungen 61:323-348

Oberli F, Meier M, Berger A, Rosenberg C, Gieré R (1999) U-Th-Pb isotope systematics in zoned allanite: a test for geochronological significance. Journal of Conference Abstracts 4: p. 722

Oberli F, Meier M, Berger A, Rosenberg C, Gieré R (2004) U-Th-Pb and $^{230}$Th/$^{238}$U disequilibrium isotope systematics: precise accessory mineral chronology and melt evolution tracing in the Alpine Bergell intrusion. Geochim Cosmochim Acta (in press)

O'Driscoll M (1988) Rare earths – enter the dragon. Industrial Minerals 254:21-55

Olson JC, Shawe DR, Pray LC, Sharp WN (1954) Rare earth mineral deposits of the Mountain Pass District, San Bernardino County, California. US Geol Sur Prof Paper 261:1-75

Ovchinnikov LN, Tzimbalenko MN (1948) Mangan-orthite from Vishnevy Mountains. Doklady Acad. Sci, USSR 63:191-194 (in Russian)

Pan Y (1997) Zircon- and monazite-forming metamorphic reactions at Manitouwadge, Ontario. Can Mineral 35:105-118

Pan Y, Fleet ME (1990) Halogen-bearing allanite from the White River gold occurrence, Hemlo area, Ontario. Can Mineral 28:67-75

Pan Y, Fleet ME (1991) Vanadian allanite-(La) and vanadian allanite-(Ce) from the Hemlo gold deposit, Ontario, Canada. Mineral Mag 55:497-507

Pan Y, Fleet ME, Barnett RL (1994) Rare earth mineralogy and geochemistry of the Mattagami Lake volcanogenic massive sulfide deposit, Québec. Can Mineral 32:133-147

Pantó G (1975) Trace minerals of the granitic rocks of the Valence and Mecsek Mountains. Acta Geologica Academiae Scientiarum Hungaricae 19:59-93

Papunen H, Lindsjö O (1972) Apatite, monazite and allanite; three rare earth minerals from Korsnäs, Finland. Bull Geol Soc Finland 44:123-129

Paulmann C, Bismayer U (2001a) Thermal recrystallization of metamict allanite: a synchrotron radiation study. Eur J Mineral Beihefte 13:137

Paulmann C, Bismayer U (2001b) Anisotropic recrystallization effects in metamict allanite on isothermal annealing. *In:* Gehrke R, Krell U, Schneider JR (eds) HASYLAB Annual Report 2001 I: 435-436 (*www-hasylab.desy.de/science/annual_reports/2001_report/index.html*)

Paulmann C, Schmidt H, Kurtz R, Bismayer U (2000) Thermal recrystallization of metamict allanite on progressive and isothermal annealing. *In:* Dix, W, Kracht, T, Krell, U, Materlik, G, Schneider, J.R (eds) HASYLAB Annual Report 2000 I: 625-626 (*www-hasylab.desy.de/science/annual_reports/2000_report/index.html*)

Pautov LA, Khorov PV, Ignatenko KI, Sokolova EV, Nadezhina TN (1993) Khristovite-(Ce) – (Ca,RE E)REE(Mg,Fe)AlMnSi₃O₁₁(OH)(F,O) A new mineral in the epidote group. Zapiski Vserossiskogo mineralogicheskogo obshchestva, 122(3), 103-111 (in Russian). English abstract available in: Jambor JL, Puziewicz J, Roberts AC (1995) New mineral names. Am Mineral 80:404-409

Pavelescu L, Pavelescu M (1972) Study of some allanites and monazites from the South Carpathians (Romania). Tschermaks Mineralogische und Petrographische Mitteilungen 17:208-214

Peacor DR, Dunn PJ (1988) Dollaseite-(Ce) (magnesium orthite redefined) Structure refinement and implications for F + M²⁺ substitutions in epidote-group minerals. Am Mineral 73:838-842

Perhac RM, Heinrich EWM (1964) Fluorite-bastnäsite deposits of the Gallinas Mountains, New Mexico and bastnäsite paragenesis. Econ Geol 59:226-239

Peterson RC, MacFarlane DB (1993) The rare-earth-element chemistry of allanite from the Grenville Province. Can Mineral 31:159-166

Petrík I, Broska I, Lipka J, Siman P (1995) Granitoid Allanite-(Ce) Substitution Relations, Redox Conditions and REE Distributions (on an Example of I-Type Granitoids, Western Carpathians, Slovakia). Geologica Carpathica 46:79-94

Ploshko VV, Bogdanova VI (1963) Isomorphous substitutions in minerals of the epidote group from the northern Caucasus. Geochemistry 1:61-71

Poitrasson F (2002) *In situ* investigations of allanite hydrothermal alteration: examples from calc-alkaline and anorogenic granites of Corsica (southeast France). Contrib Mineral Petrol 142:485-500

Popp RK, Virgo D, Hoering TC, Yoder HS Jr, Phillips MW (1995a) An experimental study of phase equilibria and iron oxy-component in kaersutitic amphibole: Implications for $f_{H2}$ and $a_{H2O}$ in the upper mantle. Am Mineral 80:543-548

Popp RK, Virgo D, Phillips MW (1995b) H-deficiency in kaersutitic amphibole: Experimental verification. Am Mineral 80:1347-1350

Pudovkina ZV, Pyatenko IA (1963) Crystal structure of non-metamict orthite. Doklady Akad. Nauk SSSR 153: 695-698 (in Russian)

Quensel P (1945) Berylliumorthit (muromontite) från Skuleboda fältspatbrott. Arkiv för Kemi, Mineralogi och Geologi 18A (22):1-17 (in Swedish)

Rao AT, Babu VRRM (1978) Allanite in charnockites from Air Port Hill, Visakhapatnam, Andra Pradesh, India. Am Mineral 63:330-331

Rao AT, Rao A, Rao PP (1979) Fluorian allanite from calc-granulite and pegmatite contacts at Garividi, Andhra Pradesh, India. Mineral Mag 43:312

Rimsaite J (1982) The leaching of radionuclides and other ions during alteration and replacement of accessory minerals in radioactive rocks. Geol Sur Canada Paper 81-1B:253-266

Rimsaite J (1984) Selected mineral associations in radioactive and REE occurrences in the Baie-Johan-Beetz area, Québec: a progress report. Geol Sur Canada Paper 84-1A:129-145

Romer RL, Siegesmund S (2003) Why allanite may swindle about its true age. Contrib Mineral Petrol 146: 297-307

Rønsbo JG (1989) Coupled substitutions involving REEs and Na and Si in apatites in alkaline rocks from the Ilimaussaq intrusion, South Greenland, the petrological implications. Am Mineral 74:896-901

Rouse RC, Peacor DR (1993) The crystal structure of dissakisite-(Ce), the Mg analogue of allanite-(Ce). Can Mineral 31:153-157

Rudashevskiy NS (1969) Epidote-allanite from metasomatites of southern Siberia. Zapiski Vserossiskogo mineralogicheskogo obshchestva 98(6):739-749 (in Russian)

Saebø PC (1961) Contributions to the mineralogy of Norway. No. 11. On lanthanite in Norway. Norsk Geologisk Tidsskrift 41:311-317

Saini HS, Lal N, Nagpaul KK (1975) Annealing studies of fission tracks in allanite. Contrib Mineral Petrol 52:143-145

Sakai C, Higashino T, Enami M (1984) REE-bearing epidote from Sanbagawa pelitic schists, central Shikoku, Japan. Geochem J 18:45-53

Sakurai K, Wakita H, Kato A, Nagashima K (1969) Chemical studies of minerals containing rarer elements from the Far East. LXIII. Bastnäsite from Karasugawa, Fukushima Prefecture, Japan. Bull Chem Soc Japan 42:2725-2728

Sargent KA (1964) Allanite in metamorphic rocks, Horn Area, Bighorn Mountains, Wyoming. Geol Soc Am Spec Paper 76:143

Sawka WN (1988) REE and trace element variations in accessory minerals and hornblende from the strongly zoned McMurry Meadows Pluton, California. Trans R Soc Edinburgh: Earth Sci 79:157-168

Sawka WN, Chappell BW (1988) Fractionation of uranium, thorium and rare earth elements in a vertically zoned granodiorite: Implictions for heat production distributions in the Sierra Nevada batholith, California, U.S.A. Geochim Cosmochim Acta 52:1131-1143

Sawka WN, Chappell BW, Norrish K (1984) Light-rare-earth-element zoning in sphene and allanite during granitoid fractionation. Geology 12:131-134

Schmidt MW, Thompson AB (1996) Epidote in calc-alkaline magmas: An experimental study of stability, phase relationships, and the role of epidote in magmatic evolution. Am Mineral 81:462-474

Schreyer W, Fransolet AM, Abraham K (1986) A miscibility gap in trioctahedral Mn-Mg-Fe chlorites: Evidences from the Lienne Valley manganese deposit, Ardennes, Belgium. Contrib Mineral Petrol 94: 333-342

Scott AK, Scott AG (1985) Geology and genesis of uranium-rare earth deposits at Mary Kathleen, North Queensland. Australasian Inst Mining Metall Proc 290:79-89

Semenov EI (1958) Relationship between composition of rare earths and composition and structures of minerals. Geochemistry 5:574-586

Semenov EI, Barinski RL (1958) The composition characteristics of the rare-earths in minerals. Geokhimiia, Akademiia nauk SSSR 4:314-333 (in Russian)

Semenov EJ, Upendran R, Subramanian V (1978) Rare earth minerals of carbonatites of Tamil Nadu. J Geol Soc India 19:550-557

Shannon RD (1976) Revised effective ionic radii and systematic studies of interatomic distances in halides and chalcogenides. Acta Crystallogr Sect A 32:751-767

Shepel AB, Karpenko MV (1969) Mukhinite, a new vanadian of epidote. Doklady Akad Nauk SSSR 185/6: 1342-1345 (in Russian)

Smith HA, Barreiro B (1990) Monazite U-Pb dating of staurolite grade metamorphism in pelitic schists. Contrib Mineral Petrol 105:602-615

Smith MP, Henderson P, Jeffries T (2002) The formation and alteration of allanite in skarn from the Beinn an Dubhaich granite aureole, Skye. Eur J Mineral 14:471-486

Smith WL, Franck ML, Sherwood AM (1957) Uranium and thorium in the accessory allanite of igneous rocks. Am Mineral 42:367-378

Smulikowski W, Kozlowski A (1994) Distribution of cerium, lanthanum and yttrium in allanites and associated epidotes of metavolcanic rocks of Hornsund area, Vestspitsbergen. Neues Jahrbuch fur Mineralogie–Abhandlungen 166:295-324

Söhnge PG (1945) The structure, ore genesis and mineral sequence of the cassiterite deposits in the Zaaiplaats tin mine, Potgietersrust district, Transvaal. Trans Geol Soc South Africa 47:157-181

Sokolova EV, Nadezhina TN, Pautov LA (1991) Crystal structure of a new natural silicate of manganese from the epidote group. Soviet Physics - Crystallography 36:172-174

Sorensen SS (1991) Petrogenetic significance of zoned allanite in garnet amphibolites from a paleo-subduction zone: Catalina Schist, southern California. Am Mineral 76:589-601

Sorensen SS, Grossman JN (1989) Enrichment of trace elements in garnet amphibolites from a paleosubduction zone: Catalina Schist, southern California. Geochim Cosmochim Acta 53:3155-3177

Sorensen SS, Grossman JN (1993) Accessory minerals and subduction zone metasomatism: a geochemical comparison of two mélanges (Washington and California, U.S.A.). Chem Geol 110:269-297

Spandler C, Hermann J, Arculus R, Mavrogenes J (2003) Redistribution of trace elements during prograde metamorphism from lawsonite blueschist to eclogite facies; implications for deep subduction-zone processes. Contrib Mineral Petrol 146:205-222

Sverdrup TL, Bryn KØ, Saebø PC (1959) Contrib the mineralogy of Norway. No. 2. Bastnäsite, a new mineral for Norway. Norsk Geologisk Tidsskrift 39:237-247

Tarassoff P and Gault RA (1994) The Orford Nickel Mine, Quebec, Canada. Mineral Record 25:327-345

Tempel H-G (1938) Der Einfluss der seltenen Erden und einiger anderer Komponenten auf die physikalisch-optischen Eigenschaften innerhalb der Epidot-Gruppe. Chemie der Erde 11:525-551 (in German)

Thomson T (1810) Experiments on allanite, a new mineral from Greenland. Trans R Soc Edinburgh 8(6): 371-386

Treolar PJ (1987) Chromian muscovites and epidotes from Outokumpu, Finland. Mineral Mag 51:593-599

Treolar PJ, Charnley NR (1987) Chromian allanite from Outokumpu, Finland. Can Mineral 25:413-418

Tribuzio R, Messiga B, Vannucci R, Bottazzi P (1996) Rare earth element redistribution during high-pressure-low-temperature metamorphism in ophiolitic Fe-gabbros (Liguria, northwestern Italy: implications for light REE mobility in subduction zones. Geology 24:711-714

Tulloch AJ (1986) Comment on "Implications of magmatic epidote-bearing plutons on crustal evolution in the accreted terranes of northwestern North America" and "Magmatic epidote and its petrologic significance". Geology 14:186-187

Ueda T (1955) The crystal structure of allanite, $OH(Ca,Ce)_2(Fe^{III}Fe^{II})Al_2OSi_2O_7SiO_4$. Memoirs of the College of Science, University of Kyoto Series B22/2:145-163

Ueda T (1957) Studies on the metamictization of radioactive minerals. Memoirs of the College of Science, University of Kyoto, Series B24/2:81-120

Ueda T, Korekawa M (1954) On the metamictization. Memoirs of the College of Science, University of Kyoto Series B21/2:151-162

van Marcke de Lummen G (1986) Tin-bearing epidote from skarn in the Land's End aureole, Cornwall, England. Can Mineral 24:411-415

Vance ER, Routcliffe P (1976) Heat treatment of some metamict allanites. Mineral Mag 40:521-523

Vazquez JA, Reid MR (2001) Timescales of magmatic evolution by coupling core-to-rim $^{238}U$-$^{230}Th$ ages and chemical compositions of mineral zoning in allanite from the youngest Toba Tuff. EOS Trans, Am Geophys Union, 82/47, Fall Meeting Supplement, Abstract V32D-1019

Vazquez JA, Reid MR (2003) The protracted history of magmatic evolution recorded by zoning in allanites. EOS Trans, Am Geophys Union, 84/46, Fall Meeting Supplement, Abstract V11F-08

von Blanckenburg F (1992) Combined high-precision chronometry and geochemical tracing using accessory minerals applied to the Central-Alpine Bergell intrusion (Central Europe). Chem Geol 100:19-40

Vyhnal CR, McSween HY Jr, Speer JA (1991) Hornblende chemistry in southern Appalachian granitoids: implications for aluminum hornblende thermobarometry and magmatic epidote stability. Am Mineral 76:176-188

Wakita H, Rey P, Schmitt RA (1971) Abundances of 14 rare earth elements and 12 other elements in Apollo 12 samples: five igneous and one breccia rocks and four soils. 2nd Lunar Science Conference, Supplement 2 (2):1319-1329. Geochim Cosmochim Acta

Ward CD, McArthur JM, Walsh JN (1992) Rare earth element behaviour during evolution and alteration of the Dartmoor granite, SW England. J Petrology 33:785-815

Watson MD, Snyman CP (1975) The geology and the mineralogy of the fluorite deposits at the Buffalo fluorspar mine on Buffelsfontein, 347KR, Naboomspruit district. Trans Geol Soc South Africa 78:137-151

Weber WJ (1990) Radiation-induced defects and amorphization in zircon. J Mater Res 5:2687-2697

Weber WJ, Ewing RC, Catlow CRA, Diaz de la Rubia T, Hobbs LW, Kinoshita C, Matzke H, Motta AT, Nastasi M, Salje EKH, Vance ER, Zinkle SJ (1998) Radiation effects in crystalline ceramics for the immobilization of high-level nuclear waste and plutonium. J Mater Res 13(6):1434-1484

Weiss S (2002) Allanit aus alpinen Klüften (I). Lapis 9:29-43 (in German)

Weiss S, Parodi GC (2002) Allanit aus alpinen Klüften (II). Lapis, 10, 24-28 (in German)

Williams GH (1893) Piedmontite and scheelite from the ancient rhyolite of South Mountain, Pennsylvania. Amer J Sci 46:50-57

Wing B, Ferry JM, Harrison TM (2003) Prograde destruction and formation of monazite and allanite during contact and regional metamorphism of pelites: petrology and geochronology. Contrib Mineral Petrol 145: 228-250

Wood SA (1990) The aqueous geochemistry of rare earth elements and yttrium. 2. Theoretical predictions of speciation in hydrothermal solutions to 350°C at saturated water pressure. Chem Geol 88:99-125

Wood SA, Ricketts A (2000) Allanite-(Ce) from the Eocene Casto Granite, Idaho: Response to Hydrothermal Alteration. Can Mineral 38:81-100

Yang JJ and Enami M (2003) Chromian dissakisite-(Ce) in a garnet lherzolite from the Chinese Su-Lu UHP metamorphic terrane: Implications for Cr incorporation in epidote minerals and recycling of REE into the Earth's mantle. Am Mineral 88:604-610

Zakrzewski MA, Lustenhouwer WJ, Nugteren HJ, Williams CT (1992) Rare-earth mineral yttrian zirconolite and allanite-(Ce) and associated minerals form Koberg mine, Bergslagen, Sweden, Mineral Mag 56: 27-35

Zdorik TB, Kupriyanova II, Kumskova NM (1965) Crystalline orthite from some metasomatic formations in Siberia. Mineraly SSSR 15:208-214, Izd. Nauka, Moscow (in Russian)

Zen E-an, Hammarstrom JM (1984) Magmatic epidote and its petrologic significance. Geology 12:515-518

Zen E-an, Hammarstrom JM (1986) Reply on the comments on "Implications of magmatic epidote-bearing plutons on crustal evolution in accreted terranes of northwestern North America" and "Magmatic epidote and its petrologic significance" by A.J. Tulloch and by R.H. Moench. Geology 14:188-189

Zenzén N (1916) Determinations of the power of refraction of allanites. Acta Universitatis Upsaliensis, Bulletin of the Geological Institute 15:61-76

Zhirov KK, Bandurkin GA, Lavrentiev YuG (1961) To the geochemistry of rare-earth elements in pegmatites of north Karelia. Geokhimiya 11:995-1004 (in Russian)

Reviews in Mineralogy & Geochemistry
Vol. 56, pp. 495-552, 2004
Copyright © Mineralogical Society of America

# Manganese in Monoclinic Members of the Epidote Group: Piemontite and Related Minerals

**Paola Bonazzi and Silvio Menchetti**

*Dipartimento di Scienze della Terra*
*Università di Firenze*
*50121 Firenze, Italia*

## INTRODUCTION

According to Mayo (1932), who provided a brief historical review of the names used for piemontite, the first researcher who described this mineral has been Cronstedt in 1758 who named it "röd Magnesia." In 1790 Chevalier Napione analyzed the sample described by Cronstedt and termed it "Manganèse rouge." On the basis of his chemical data Haüy designated the substance as "Manganèse oxidé violet silicifère" in 1801. Later, in his Traité de Mineralogie, Haüy (1822) adopted the name proposed by Cordier (1803) who first recognized the mineral as an "Épidote manganésifere." The name piedmontite was proposed in 1853 by Kenngott the basis of the type locality and more recently transformed into piemontite.

According to the standard guidelines for mineral nomenclature, the name piemontite should be reserved to members of the ternary solid solution $Ca_2Al_2(Mn,Fe,Al)(Si_2O_7)(SiO_4)O(OH)$ that basically contains $Mn^{3+}$ dominant at one site. Nonetheless, the use of this name for any monoclinic manganiferous epidote-group members showing the characteristic strong red-yellow-violet pleochroism is very common and probably convenient with special regard to petrographic purposes. Indeed, the color of manganian (i.e., $Mn^{3+}$ bearing) epidote or clinozoisite ranges to red to pinkish, while manganoan (i.e., $Mn^{2+}$ bearing) members do not exhibit the characteristic reddish hue.

The discredited name "withamite" was used to describe poorly manganiferous piemontite (Hutton 1938; Yoshimura and Momoi 1964) but corresponds, on the basis of the current nomenclature, to a manganian clinozoisite. The name "thulite," sometimes erroneously used for pinkish clinozoisite, should be reserved to $Mn^{3+}$ bearing orthorhombic members.

In this chapter we focus on piemontite *sensu stricto* with the ideal formula $Ca_2Al_2Mn^{3+}(Si_2O_7)(SiO_4)O(OH)$, but also include for the reasons explained above those members of the clinozoisite-epidote-piemontite solid solution series which have been described as piemontite for their peculiar optical properties in many petrological papers. Furthermore, other $Mn^{3+}$-rich members of the epidote group that differ from piemontite for the occupancy of one or more sites will be taken into account. In particular, strontiopiemontite with the ideal formula $CaSrAl_2Mn^{3+}(Si_2O_7)(SiO_4)O(OH)$ is related to piemontite by the homovalent exchange $^{A1}Ca \leftrightarrow {}^{A1}Sr$. A further substitution $^{M1}Al \leftrightarrow {}^{M1}Mn^{3+}$ relates strontiopiemontite to tweddillite with the ideal formula $CaSrAlMn^{3+}{}_2(Si_2O_7)(SiO_4)O(OH)$. On the other hand, androsite-(La), ideal formula $Mn^{2+}REE^{3+} Mn^{3+}Al Mn^{2+}(Si_2O_7)(SiO_4)O(OH)$, is related to piemontite by the heterovalent exchange $^{A2}Ca \leftrightarrow {}^{A2}REE$ coupled to $^{M3}Mn^{3+} \leftrightarrow {}^{M3}Mn^{2+}$ besides the substitutions $^{A1}Ca \leftrightarrow {}^{A1}Mn^{2+}$ and $^{M1}Al \leftrightarrow {}^{M1}Mn^{3+}$. Furthermore, the exchange mechanism

1529-6466/04/0056-0011$10.00

$^{M3}(Mn^{2+},Fe^{2+}) + {}^{O10}OH^- \leftrightarrow {}^{M3}(Mn^{3+},Fe^{3+}) + {}^{O10}O^{2-}$ describes the oxidation of octahedral divalent cations and the corresponding deprotonation when REE-bearing piemontites are heated in air.

## STRUCTURE AND CRYSTAL CHEMISTRY

The crystal structure of piemontite is topologically identical to that of the other monoclinic members of the epidote group (space group $P2_1/m$). It consists of $Si_2O_7$ and $SiO_4$ units linked to two independent edge-sharing octahedral chains, both extending parallel to the *b*-axis (Fig. 1). One chain consists of M2 octahedra while the other chain is formed by M1 octahedra with M3 octahedra attached on alternate sides along its length. This arrangement gives rise to two additional types of cavities, the smaller nine-fold coordinated A1 and the larger ten-fold coordinated A2 sites. For an ideal piemontite endmember the site occupancy can be schematically described as A1 and A2 = Ca, M1 and M2 = Al, and M3 = $Mn^{3+}$.

**Figure 1.** Piemontite structure projected onto (010); dots = Ca.

### M sites

Since the first structural and spectroscopic studies on members of the epidote group (Ito et al. 1954; Strens 1966; Burns and Strens 1967) it is generally assumed that $Fe^{3+}$ and $Mn^{3+}$ mainly substitute Al in M3, which is significantly larger and more distorted (axially compressed) than the other octahedral sites, even in the unsubstituted clinozoisite endmember. In particular, the tetragonal distortion of M3 makes this site very suitable to house the $3d^4$-configurated $Mn^{3+}$ ion, in agreement with the requirements of the Jahn-Teller effect. However, for appropriate compositions, a significant $M^{3+}$ ($M^{3+} = Mn^{3+} + Fe^{3+}$) substitution in at least one of the other two octahedral sites was assumed. Based on the structural model of Ito et al. (1954) who had erroneously determined decreasing octahedral distortion in the order M3 > M2 > M1, Burns and Strens (1967) in their spectroscopic study assigned the additional $Mn^{3+}$ to the M2 site. The first structural approach to the question of cation ordering in piemontite stems from Dollase (1969) who refined the structure of a piemontite with $M^{3+} = 1.03$ per formula unit (pfu) and analyzed the geometrical features distinctive of piemontite. It is evident from Dollase's study that the transition metals occupy preferentially the M3 site ($M^{3+} = 0.83$ pfu) and to a lesser extent the M1 site ($M^{3+} = 0.20$ pfu). This result does not contradict the spectroscopic interpretation of Burns and Strens (1967) as the M1 octahedron was found to be more distorted than M2, which is always the smallest and most regular octahedron of the structure throughout the compositional field of the epidote minerals (see Franz and Liebscher 2004). Because of the comparable X-ray scattering power of $Mn^{3+}$ and $Fe^{3+}$, however, any possible ordering of $Fe^{3+}$ and $Mn^{3+}$ between M3 and M1 remained undetermined. To determine the site assignment of $Fe^{3+}$ and $Fe^{2+}$ and to explore possible variations in the degree of ordering, Dollase (1973) investigated several synthetic and natural samples of the epidote group including three piemontite crystals by means of $^{57}Fe$ Mössbauer spectroscopy. $Fe^{3+}$ was found to occupy both M3 and M1 site with a marked preference for M3 site. In particular, the three piemontite samples analyzed showed $Fe^{3+}$ (M3) = 0.287, 0.097, and 0.106 pfu and $Fe^{3+}$ (M1) = 0.043, 0.013, and 0.014 pfu, respectively. If it is arbitrarily assumed that $Mn^{3+}$ fractionates between M3 and M1 exactly like $Fe^{3+}$, the distribution coefficient $K_D$ ($M^{3+}$ - Al) for the intracrystalline exchange reaction

$$^{M3}\left(Fe^{3+} + Mn^{3+}\right) + {}^{M1}Al = {}^{M1}\left(Fe^{3+} + Mn^{3+}\right) + {}^{M3}Al \tag{1}$$

is

$$K_D\left(M^{3+} - Al\right) = \frac{{}^{M1}\left(Fe^{3+} + Mn^{3+}\right){}^{M3}Al}{{}^{M3}\left(Fe^{3+} + Mn^{3+}\right){}^{M1}Al} \tag{2}$$

For these piemontite samples the distribution coefficients $K_D$ (= 0.015, 0.024, and 0.021, respectively) are similar to those estimated for synthetic epidote on the basis of Mössbauer spectroscopy (Dollase 1973). However, they are significantly lower than those determined by a single crystal diffraction study (Giuli et al. 1999) in synthetic epidote. For a sample of piemontite from the type locality (St. Marcel, Piemonte, Italy) the spectroscopic results (Dollase 1973) were tentatively combined with the structural data previously obtained (Dollase 1969). On this basis, the following site occupancies were derived: M3 = 0.54 $Mn^{3+}$ + 0.29 $Fe^{3+}$ + 0.17 Al and M1 = 0.16 $Mn^{3+}$ + 0.04 $Fe^{3+}$ + 0.80 Al. For the intracrystalline partitioning of $Fe^{3+}$ and $Mn^{3+}$ between M1 and M3

$$^{M3}\left(Fe^{3+}\right) + {}^{M1}\left(Mn^{3+}\right) = {}^{M1}\left(Fe^{3+}\right) + {}^{M3}\left(Mn^{3+}\right) \tag{3}$$

these values correspond to

$$K_D\left(Mn - Fe\right) = \frac{{}^{M1}\left(Fe^{3+}\right) + {}^{M3}\left(Mn^{3+}\right)}{{}^{M3}\left(Fe^{3+}\right) + {}^{M1}\left(Mn^{3+}\right)} = 0.47 \tag{4}$$

and indicate a stronger preference of $Fe^{3+}$ for the more distorted M3 site compared to $Mn^{3+}$. According to Dollase (1973) such a conclusion "must be considered suspect" being in contrast with the known preference of $Mn^{3+}$ for tetragonally distorted sites. In fact, the axial compression of the M3 octahedron, involving the electron hole of $3d^4$ in $d_{z^2}$ orbital, yields a rather small crystal field stabilization energy (CFSE) compared with those typical of $Mn^{3+}$ in other crystal structures (e.g., $Mn^{3+}$ substituting Al in the axially elongated octahedra of the andalusite structure type; Langer et al. 2002). Ferraris et al. (1989) determined the $Mn^{3+}$-$Fe^{3+}$ distribution in a strontian piemontite by single-crystal neutron-diffraction. Their site occupancy refinement led to M3 = 0.61 $Mn^{3+}$ + 0.33 $Fe^{3+}$ + 0.06 Al and M1 = 0.17 $Mn^{3+}$ + 0.02 $Fe^{3+}$ + 0.81 Al. The corresponding $K_D$ (Mn-Fe) of 0.23 seems to confirm the above-mentioned odd behavior of preferential incorporation of $Fe^{3+}$ in M3. On the other hand Ferraris et al. (1989) pointed out that a look at the absolute quantities of $Mn^{3+}$ in the respective sites would eliminate the apparent oddity. Catti et al. (1989) refined the structure of additional five strontian piemontite crystals and confirmed the previous findings that $M^{3+}$ (= $Mn^{3+}$ + $Fe^{3+}$) has a stronger preference for M3 than for M1. Furthermore, Catti et al. (1989) observed a departure from linearity in the function $^{M1}M^{3+} = f(^{M3}M^{3+})$ indicating a saturation effect with increasing total $M^{3+}$ content. No correlation was found between the site occupancy at M3 and M1 and the total Mn content determined by microprobe analysis. This may suggest that the crystal chemical behavior of $Mn^{3+}$ and $Fe^{3+}$ in piemontite is not so different despite their different electronic configurations ($3d^4$ and $3d^5$, respectively).

Recently, Langer et al. (2002) discussed the structural and geometrical changes in the structure of piemontite induced by increased $Mn^{3+}$ or $M^{3+}$ incorporation in synthetic and natural crystals, respectively. There is no significant difference in the distribution behavior of $Mn^{3+}$ or other $M^{3+}$ ions between M1 and M3 if the intracrystalline distribution is plotted as a function of the total $Mn^{3+}$ or $M^{3+}$ content. However, when $^{M1}M^{3+}$ is plotted as a function of $^{M3}M^{3+}$ it becomes evident that synthetic piemontite in which $M^{3+} = Mn^{3+}$ exhibits a quite different order-disorder behavior from that of natural piemontite in which $M^{3+} = Mn^{3+} + Fe^{3+}$. In Figure 2 the available data for the partitioning of $M^{3+}$ between M1 and M3 are plotted together with curves of constant $K_D$ (= 0.005, 0.010, 0.020, 0.050, and 0.10, respectively). Most of the natural piemontites have a $K_D$ between 0.005 and 0.020, higher $K_D$ values close to 0.050 being found only for the piemontite crystal refined by Dollase (1969) and for a manganian epidote from Lom, Norway. By contrast, all synthetic piemontites exhibit $K_D$ values between 0.064 and 0.080 (Fig. 2). Two reasons might explain the higher disorder observed in pure synthetic piemontites. The first one is that the entry of $Mn^{3+}$ in M1 stabilizes the structure more than $Fe^{3+}$ does, possibly due to the axial compression, which also affects the M1 octahedron, although to a lesser extent than M3. A second cause of the higher disorder observed is the crystallization temperature used for the synthetic crystals (700°C), rather high compared to the natural one. Nevertheless, Langer et al. (2002) claimed that "an effect of a temperature difference of about 200°C between natural and synthetic crystals will not greatly change the general distribution pattern". By analogy with the results obtained for the intracrystalline $Fe^{3+}$ distribution in epidote (e.g., Bird and Helgeson 1980; Giuli et al. 1999; see Franz and Liebscher 2004), the

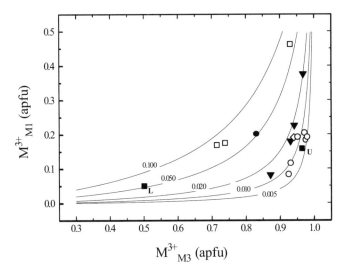

**Figure 2.** Relationship between $M^{3+}$ (= $Mn^{3+} + Fe^{3+}$) content in the octahedral sites M1 and M3 in piemontite. Lines refer to constant fractionation values ($K_D$ = 0.005, 0.010, 0.020, 0.050 and 0.10, respectively; see text). Symbols refer to data from literature as follows: solid circle = type locality (Dollase 1969); solid downward triangles = type locality (Catti et al. 1989); open diamond = type locality (Ferraris et al. 1989); open circles = different localities (Bonazzi 1990); solid squares = different localities (Langer et al. 2002); L = Lom, Norway; open squares = synthetic piemontite (Langer et al. 2002).

higher disorder in synthetic piemontite may be caused by the high crystallization temperature and/or a possible metastable rapid growth.

The mean M3-O and M1-O bond distances increase as a function of their $M^{3+}$ content (Fig. 3a,b). However, because most data for M3 cluster at $M^{3+} > 0.90$ pfu, the regression line is not reliable to predict the <M3-O> behavior. Thus, the extrapolation towards $M^{3+} = 0$ does not match the <M3-O> value of clinozoisite (1.977 Å; Dollase 1968). On the contrary, the data for <M1-O> cover a wide range of $M^{3+}$ values and the linear model obtained

$$\text{<M1-O> (Å)} = 1.910(1) + 0.101(3)\ M^{3+}\ \text{(pfu)}\ (r = 0.989)$$

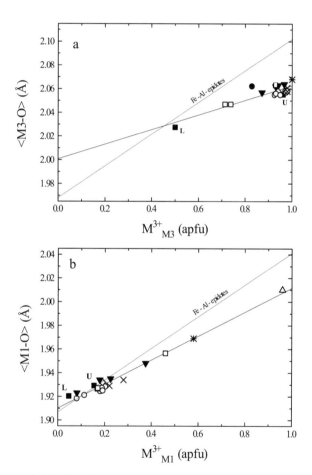

**Figure 3.** a) Mean octahedral M3-O distance plotted vs. the $M^{3+}$ (= $Mn^{3+} + Fe^{3+}$) content in the M3 site: Data fit the equation (solid regression line) <M3-O> = 2.001(7) + 0.062(8) $M^{3+}_{M3}$ ($r$ = 0.874). b) Mean octahedral M1-O distance (Å) plotted *vs.* the $M^{3+}$ (= $Mn^{3+} + Fe^{3+}$) content in the M1 site (pfu): Data fit the equation (solid regression line) <M1-O> = 1.910(1) + 0.101(3) $M^{3+}_{M1}$ ($r$ = 0.989). Symbols: solid circle = piemontite, type locality (Dollase 1969); solid downward triangles = piemontite, type locality (Catti et al. 1989); open diamond = piemontite, type locality (Ferraris et al. 1989); open circles = piemontite, different localities (Bonazzi 1990); crosses = strontiopiemontites, type locality (Bonazzi et al. 1990); star = tweddillite, type locality (Armbruster et al. 2002); solid squares = piemontite, different localities (Langer et al. 2002); open squares = synthetic piemontite (Langer et al. 2002); open upward triangle = androsite-(La), type locality (Bonazzi et al. 1996). Dashed lines refer to natural Fe-Al epidotes (Bonazzi and Menchetti 1995).

describes well the increase of the mean bond distance as a function of the $M^{3+}$ content. For $M^{3+}(M1) = 0$ the regression line obtained for piemontite, including data for strontiopiemontite, tweddillite, and androsite-(La), matches closely both the value found in clinozoisite (1.906 Å; Dollase 1968) and the value extrapolated from the regression line obtained for the clinozoisite-epidote solid-solution series (1.907 Å; Bonazzi and Menchetti 1995). With the increase of $M^{3+}$ content the regression lines for piemontite and clinozoisite-epidote solid solution diverge. For $M^{3+}$ (M1) = 1, the distance <M1-O> is 2.011 Å in piemontite and 2.041 Å in clinozoisite-epidote (Fig. 3b). Although this different behavior is reasonably due to the different structural influence of $Mn^{3+}$ and $Fe^{3+}$, it is difficult to quantify the proportion of $Mn^{3+}$ in M1 only on the basis of the shift of data points from the regression lines. This shift is indeed also affected by differences in the instruments and/or refining methods such as the use of scattering curves for neutral rather than ionized atomic species.

Some ideas of the different structural effects of $Mn^{3+}$ and $Fe^{3+}$ can be obtained from the individual M3-O and M1-O bond distances (Langer et al. 2002). Because the M3 octahedron is compressed along the O4-M3-O8 axis (Fig. 4) a distinct Jahn-Teller effect of $Mn^{3+}$ occurs in piemontite with the electron hole of the $d^4$ cation occurring in the $d_{z^2}$ orbital along this direction. Therefore, any increase of $Mn^{3+}$ in this site will cause an increase of the octahedral volume but at the same time will enhance the relative compression along the O4-M3-O8 octahedral axis. This effect is more pronounced for pure $Mn^{3+}$ substitution in synthetic piemontite than in natural $Fe^{3+}$ bearing piemontite, since the spherical $d^5$-configurated $Fe^{3+}$ obviously does not cause any compression. Accordingly, in the minerals of the clinozoisite-epidote series ($0.24 < Fe^{3+} < 1.14$ pfu) studied by Bonazzi and Menchetti (1995), the individual M3-O bond distances continuously increase from clinozoisite to Fe-rich epidote as a function of $Fe^{3+}$ substitution (see also Franz and Liebscher 2004), except for a red-colored manganian epidote which exhibits a relative shortening of the M3-O4 and M3-O8 distances due to the entry of 0.20 $Mn^{3+}$ pfu.

The effect of $Mn^{3+}$ on the distortion of the M3 octahedron is also evident from the comparative study on Mn free allanite and REE bearing piemontite (Bonazzi and Menchetti 1994). In both minerals, the presence of $REE^{3+}$ substituting for Ca requires the entry of divalent cations ($Fe^{2+}$, $Mn^{2+}$, Mg) in the octahedral sites. Upon heating in air, oxidation of $Fe^{2+}$ (or $Mn^{2+}$) to $Fe^{3+}$ (or $Mn^{3+}$) mainly results in a shortening of the mean <M3-O> distance and, to a lesser extent, of the mean <M1-O> distance. With the increase of trivalent cations at M3, the octahedral axial distances decrease accordingly. In Mn-free allanite oxidation implies an increase of $Fe^{3+}$ and, therefore, both the equatorial O1-O2 (×2) and the apical O4-O8 distances decrease with the same slope suggesting no additional tetragonal distortion. By contrast, the

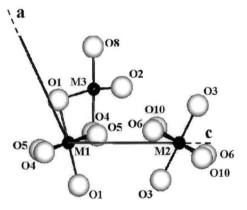

**Figure 4.** The M1, M2 and M3 independent octahedra in the piemontite structure projected along the [010] axis, atoms are labeled according to Dollase (1969).

increase of $Mn^{3+}$ in the case of REE bearing piemontite results in a strong contraction of O4-O8, in accordance with the Jahn-Teller effect, while the equatorial O1-O2 distance does not change significantly, indicating a pronounced additional tetragonal distortion (for a detailed discussion of REE-bearing monoclinic epidote minerals see Gieré and Sorensen 2004).

Langer et al. (2002) also noted a significant distortion of the M1 octahedron with the octahedral axis of compression being O4-M1-O4 (Fig. 4). As in the case of M3, the relative octahedral compression of the M1 octahedron increases with increasing $Mn^{3+}$ content in the site. Therefore it is not surprising that Mössbauer spectroscopy (Dollase 1973) and neutron diffraction (Ferraris et al. 1989) data indicate that not all $Mn^{3+}$ is ordered in M3 at the expense of $Fe^{3+}$. It is worth noting that in tweddillite (Armbruster et al. 2002) where $Mn^{3+} + Fe^{3+} > 1.5$ (with $Mn^{3+} > Fe^{3+}$) the content of trivalent transition metals in M1 is 0.58 pfu and the <M1-O> distance reaches 1.969 Å (Fig. 3b). This value seems to be the highest one for M1 at least in epidote group members having M3 occupied only by trivalent cations. As M3 and M1 share a common O1-O4 edge (Fig. 4), the partitioning of $M^{3+}$ between M3 and M1 also depends on geometrical constraints. Thus, M1 can house even higher amounts of $M^{3+}$ cations when the entry of divalent cations in M3 further increases the M3 volume. If the volume of M3 reaches a value of about 11 $Å^3$ the local strain is lowered by a significant incorporation of $M^{3+}$ in M1 (Bonazzi and Menchetti 1995). In REE bearing piemontite (Bonazzi et al. 1992) the $^{M1}(Mn+Fe)/^{M3}(Mn+Fe)$ ratio therefore correlates positively with the REE content. In androsite–(La) the exceptionally high M3 volume of 12.60 $Å^3$ that is due to the entry of divalent cations, allows a corresponding expansion of the edge-sharing M1 octahedron and results in occupancy of $M^{3+}$ in M1 close to 1 pfu (Bonazzi et al. 1996).

As already mentioned the substitution of $REE^{3+}$ for Ca in REE bearing piemontite requires the entry of divalent cations ($Mn^{2+}$, $Fe^{2+}$, Mg) in the octahedral sites. In Pb and REE rich piemontite from Nezilovo, Macedonia, the observed <M3-O> distance (2.082 Å) is significantly longer than expected for site occupancy by trivalent cations only (Bermanec et al. 1994). The authors therefore assumed that $M^{2+}$ mainly orders in the M3 site. Bonazzi et al. (1992) have drawn a similar conclusion for REE bearing piemontite from the Apuan Alps. In these samples, however, they also found a slight discrepancy between observed and expected <M1-O> values and tentatively attributed it to the entry of the small quantities of Mg in M1 that were detected by electron microprobe analyses (Bonazzi et al. 1992).

**A sites**

The A1 and A2 sites in piemontite (Fig. 5) have been described as seven- and eight-fold coordinated, respectively (Dollase 1969), thus including distances ranging from 2.2 to 2.9 Å. As the charge balance appears more satisfactory if two more neighboring oxygen atoms are included in the first coordination shell, A1 and A2 are better described as nine- and ten-fold coordinated, respectively. The A1 site is generally occupied exclusively by Ca although few samples exhibit a significant substitution by $Mn^{2+}$. Bonazzi et al. (1996) studied the structural effect of this substitution and considered the changes in the arrangement of the O atoms linked to A1. Increasing $Mn^{2+}$ in A1, the distance of the seventh neighbor (O6) increases so that A1 should be more appropriately described as six-fold coordinated ($r_{Mn}2+$ in six-fold coordination = 0.83 Å; Shannon 1976). The distance difference between the sixth and seventh neighbor ($\delta_{6-7}$) increases linearly from 0.316 Å in piemontite (Dollase 1968) to 0.419 Å in REE bearing piemontite from the Varenche mine (Bonazzi et al. 1996). In androsite-(La) in which $Mn^{2+}$ is the prevailing cation in A1, $\delta_{7-6}$ reaches even 0.490 Å (Fig. 6a). The presence of $Mn^{2+}$ also affects the mean <A1-O> distance for six-fold coordination, which linearly decreases with increasing $Mn^{2+}$ content (Fig. 6b). In the sample from Nezilovo (Bermanec et al. 1994) an $Mn^{2+}$ content close to 0.30 pfu in A1 can be assumed on the basis of the average chemical composition (Ca + REE + Pb + Na = 1.70 pfu; Al + Fe + Mn + Mg + Zn = 3.27 pfu). Calculating $\delta_{6-7}$ and <A1-O> in six-fold coordination for the Nezilovo sample the value of 0.30

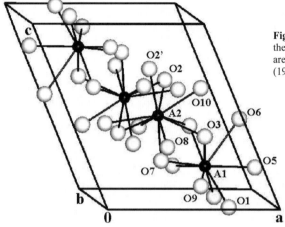

**Figure 5**. A1 and A2 polyhedra in the unit cell of piemontite; atoms are labeled according to Dollase (1969).

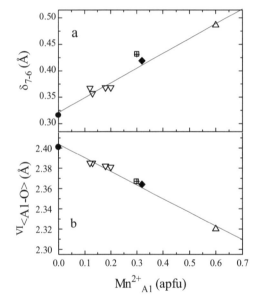

**Figure 6.** Parameters of the A1 polyhedron as a function of the $Mn^{2+}$ content in A1 along the piemontite-androsite-(La) solid solution. a) The difference between the seventh (A1-O6) and the sixth longest (A1-O5) distances is designed $\delta_{7-6}$; the equation of the regression line is $\delta_{7-6} = 0.321(5) + 0.28(2)\ Mn^{2+}{}_{A1}$ ($r = 0.991$). b) The average of the six shortest A1-O distances is designed $^{VI}$<A1-O>; the equation of the regression line is $^{VI}$<A1-O> $= 2.404(2) - 0.135(7)\ Mn^{2+}{}_{A1}$ ($r = -0.993$). Symbols: solid circle = piemontite, type locality (Dollase 1969); open downward triangles = REE bearing piemontites, Apuan Alps (Bonazzi et al. 1992); solid diamond = REE bearing piemontite, Varenche mine (Bonazzi et al. 1996); open upward triangle = androsite-(La), type locality (Bonazzi et al. 1996); crossed open square (not used in the fitting) = piemontite, Nezilovo (Bermanec et al. 1994).

$Mn^{2+}$ pfu in A1 perfectly fits the linear equations modeling the oxygen environment around A1 (see crossed squares in Fig. 6).

Apart from the entry of $Mn^{2+}$, the geometrical changes concerning the A1 polyhedron are minor and related to the chemical substitutions occurring in the adjacent, face-sharing A2 polyhedron (Fig. 5). When heterovalent substitutions take place (i.e., in REE bearing piemontite), variations of individual A1-O distances occur to compensate the local charge imbalance associated with the entry of $REE^{3+}$ in A2 and divalent cations in M3. The homovalent Ca ↔ Sr substitution also causes adjustments within the A1 polyhedron although Sr is completely ordered in the A2 site. Increasing Sr in A2 shortens the A1-O7 distance to compensate the bond strength imbalance on O7 due to the marked lengthening of the A2-O7 distance. A2-O7 is the shortest A2-O distance and correlates linearly ($r = 0.993$) with the Sr content in A2 (Fig. 7). The regression obtained works equally along the piemontite-strontiopiemontite and the piemontite-tweddillite join. The other short distances (i.e., A2-O10 and A2-O2; see Fig. 5) correlate as well with the Sr content although with a flatter slope of the regression lines. On the other hand, the values for the longer distances (i.e., A2-O2′, A2-O3, A2-O8; see Fig. 5) are randomly distributed with respect to the Sr content. A similar although more pronounced effect will reasonably occur with the entry of $Pb^{2+}$ that has an ionic radius greater than Sr ($r_{Sr}$ in ten-fold coordination = 1.36 Å, $r_{Pb}$ in ten-fold coordination = 1.40 Å; Shannon 1976). Indeed, the value for A2-O7 (2.273 Å) observed in the Pb- and REE-rich piemontite crystal from Nezilovo (Bermanec et al. 1994) is significantly greater than that predicted for A2-O7 with A2 = Ca (2.249 Å). Unfortunately, the effect of the Ca $\leftrightarrow$ Pb substitution is difficult to quantify in this sample due to the simultaneous presence of $REE^{3+}$ in A2.

In piemontite $REE^{3+}$ as well as $Th^{4+}$ and $U^{4+}$ appear to be always completely ordered in A2, in spite of the fact that the ionic radii of $LREE^{3+}$ ($r_{La-Ce}$ in ten-fold coordination = 1.27 to 1.25 Å; Shannon 1976) exceed only slightly the ionic radius of Ca ($r_{Ca}$ in ten-fold coordination = 1.23 Å; Shannon 1976). This is mainly due to the bond strength imbalance occurring on O2 when divalent cations occupy the M3 site. Similar to allanite (Bonazzi and Menchetti 1995), the <A2-O> distance is not substantially modified by the Ca $\leftrightarrow$ $REE^{3+}$ substitution but variations are observed in the individual A2-O distances. In particular, the A2-O2 distance correlates negatively with the REE content (Fig. 8). This relationship is related to a charge compensation mechanism rather than to geometrical constraints as shown by the lengthening of the A2-O2 distance with the increase of oxidation in a heated REE bearing piemontite (see Fig. 8). Due to the opposite effect caused by the entry of larger divalent cations, the crystals from Varenche ($^{A2}Sr$ = 0.19 pfu) and Nezilovo ($^{A2}Pb$ = 0.19 pfu) do not fit the model and have been neglected in the calculation of the regression line (Fig. 8) Sr in the other samples does not exceed 0.04 pfu.

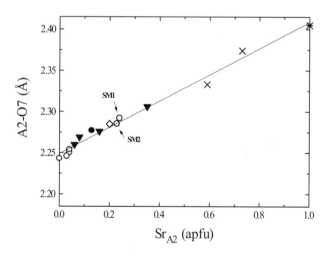

**Figure 7.** The A2-O7 bond distance vs. the Sr content in the A2 site in members of the piemontite-strontiopiemontite and piemontite-tweddillite series. The equation of the regression line is A2-O7 = 2.249(2) + 0.160(5) $Sr_{A2}$ ($r$ = 0.993). Symbols: solid circle = piemontite, type locality (Dollase 1969); solid downward triangles = piemontite, type locality (Catti et al. 1989); open diamond = piemontite, type locality (Ferraris et al. 1989); open circles = piemontite, different localities (Bonazzi 1990); crosses = strontiopiemontite, type locality (Bonazzi et al. 1990); star = tweddillite, type locality (Armbruster et al. 2002).

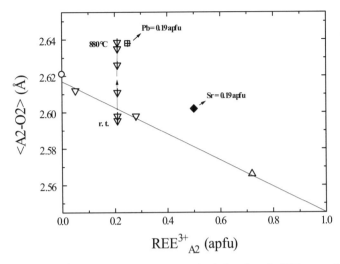

**Figure 8.** The <A2-O2> bond distance (average of four values) plotted vs. the REE content in the A2 site. The equation of the regression line is <A2-O2> = 2.617(3) − 0.072(8) REE$^{3+}_{A2}$ ($r = -0.983$). Symbols: open circle = piemontite, Mt. Corchia, Apuan Alps (Bonazzi 1990); open downward triangles = REE bearing piemontite, Apuan Alps (Bonazzi et al. 1992); solid diamond (not used in the fitting) = REE bearing piemontite, Varenche mine (Bonazzi et al. 1996); open upward triangle = androsite-(La), type locality (Bonazzi et al. 1996); crossed open square (not used in the fitting) = piemontite, Nezilovo (Bermanec et al. 1994). Lined open downward triangles (not used in the fitting) refer to natural (r.t. = room temperature) and heated REE bearing piemontite; heating temperatures 380 to 880°C (Bonazzi and Menchetti 1994).

## T sites

Looking at the tetrahedral bond distances in piemontite as reported in the literature it is evident that the mean distances can be considered as a constant value for every tetrahedron, while individual bond distances and angles fairly change as a consequence of substitutions in the other sites of the structure. Therefore, in close agreement with the results of a neutron diffraction refinement (Ferraris et al. 1989), no significant substitution of Si by other cations affects these sites.

The most significant variation concerning the tetrahedra is the change of the mutual orientation of the Si1 and Si2 tetrahedra, linked together to form a cis-oriented Si$_2$O$_7$ group. As reported by several authors (Dollase 1971; Gabe et al. 1973; Carbonin and Molin 1980; Bonazzi and Menchetti 1995; see also Franz and Liebscher 2004), the bending angle Si1-O9-Si2 in the epidote group minerals decreases with the increase of substitution of larger cations for Al in the M3 octahedron. Bermanec et al. (1994) also found a strong negative linear relationship between the bending angle and the octahedral M3-O8 distance. In piemontite the decrement of the Si1-O9-Si2 value is even greater when M$^{3+}$ = Mn$^{3+}$ (Langer et al. 2002).

## Hydrogen bonding

Like in the other epidote group minerals, the proton in piemontite is bonded to O10 that links two M2 and one A2 cations. Hydrogen bonding between O10 (donor) and O4 (acceptor) constitutes an additional link between the M1 and the M2 octahedral chains. By analogy with dollaseite-(Ce) (Peacor and Dunn 1988), incorporation of F$^-$, if any, occurs at the O4 site.

Using the atomic coordinates from the neutron diffraction refinement of Ferraris et al. (1989), the hydrogen bonding environment in piemontite is H-O10 = 0.966 Å, H-O4 = 2.003 Å, O10-O4 (donor–acceptor) distance = 2.955 Å, and O10-H-O4 bending angle = 168°. The shortest hydrogen-cation distance is H-M2 = 2.475 Å. By heating REE free piemontite in air, in which oxidation of octahedral cations (already in their high valence state) is excluded, no deprotonation occurs below the breakdown temperature of the structure at 880°C (Catti et al. 1988). In contrast, the heating in air of REE bearing piemontite results in a loss of H to compensate the oxidation of $M^{2+}$ to $M^{3+}$. This is evident from the lengthening of the donor acceptor distance that increases almost gradually from 2.959 Å (550°C) to 3.053 Å (880°C) (Bonazzi and Menchetti 1994). With ongoing deprotonation also the amount of $M^{3+}$ in M2 becomes significant (up to 0.18 $M^{3+}$ pfu after the 880°C heat treatment) as the presence of H inhibits the incorporation of $M^{3+}$ in M2 due to the very short M2-H distance.

**High-temperature behavior**

The thermal behavior of Sr bearing piemontite was studied by Catti et al. (1988) by *in-situ* high temperature X-ray diffraction. As expected, the octahedra exhibit positive thermal expansion coefficients and the expansion of M3 is the most pronounced and very anisotropic. With increasing temperature the two shortest distances shrink while the longest ones expand causing a further flattening of the octahedron along O4-M3-O8, in close agreement with the fact that the direction of the least thermal expansion of the lattice is close to the O4-M3-O8 alignment. The thermal expansion of the A1 and A2 polyhedra is less than expected and affects mainly the shortest distances. As a consequence, the regularity of these polyhedra increases as a function of temperature. Both T1 and T2 tetrahedra exhibit negative thermal expansion coefficients. To compensate for the imbalance of bond strength due to the lengthening of weaker bonds some of the Si-O distances (e.g., Si1-O1, Si1-O7, Si2-O3) decrease with the increase of temperature.

**Unit-cell parameters**

The unit-cell volume of piemontite is known to vary as a function of the substitutional degree in both the octahedral (Anastasiou and Langer 1977; Langer et al. 2002) and the A sites (Bonazzi et al. 1990; 1996; Akasaka et al. 2000). Like in Fe-Al epidote (see Franz and Liebscher 2004), *b* increases markedly with the $M^{3+}$ content whereas *a* and *c* increase only slightly. The monoclinic β angle exhibits minor variation also depending on the A site population. Nevertheless, these correlations are not as close as in Fe-Al-epidote, mainly as a consequence of the peculiar crystal chemical behavior of the Jahn-Teller active $Mn^{3+}$ cation. As discussed above an increase of $Mn^{3+}$ causes a further compression of the M3 octahedron along the O4-M3-O8 alignment which is approximately parallel to the *a* axis (Fig. 4). For this reason, the Mn/Fe ratio in piemontite affects both the *a/b* and *a/c* axial ratios resulting in a rather wide spread of values for the unit-cell parameters as a function of the octahedral substitutional degree. The relationships are further complicated by the substitutions at the A sites.

In Figure 9, the individual parameters (*a*, *b*, *c*, and β, respectively) of piemontite and other manganiferous members of the epidote group are plotted against the unit-cell volume. For comparison the corresponding variations in members of the clinozoisite-epidote series are reported. From an overall comparison of the plots it appears that the $Mn^{3+} \leftrightarrow Fe^{3+}$ substitution affects *a* and *b* more sensibly than *c* and β. As expected, Fe free synthetic piemontite (open and dotted squares) exhibits values of the *a* parameter lower than those of Fe-bearing natural piemontite (Fig. 9a). In particular, the "piemontite" from Lom, Norway, studied by Langer et al. (2002) closely approaches the regression line of the Fe-Al-epidotes due to its extremely low Mn/Fe ratio (= 0.33). The *b* parameter shows an opposite behavior (Fig. 9b), with a more pronounced increment as a function of the $M^{3+}$ content in pure synthetic piemontite where $M^{3+} = Mn^{3+}$. Natural piemontite with variable Mn/Fe ratios displays *b* values between synthetic piemontite and Fe-Al epidote. An exception is Sr rich piemontite in which the strong expansion of the unit cell due to the $Sr \leftrightarrow Ca$ substitution mainly occurs along the *c*sinβ axis. The variation of *c* as a function of the unit cell volume (Fig. 9c) differentiates only slightly

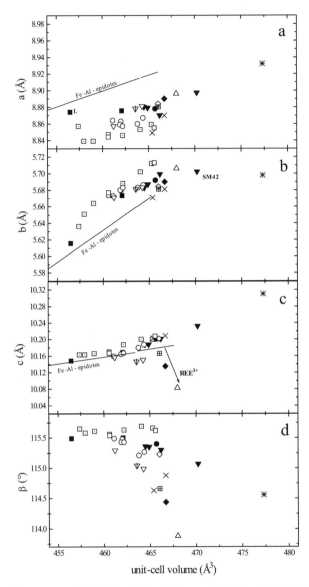

**Figure 9.** Unit cell parameters, *a* (a), *b* (b), *c* (c), and the monoclinic β angle (d) plotted *vs.* the unit cell volume for monoclinic manganiferous epidote-group members. Solid lines refer to natural Fe-Al epidotes (Bonazzi and Menchetti 1995). Symbols: solid circle = piemontite, type locality (Dollase 1969); dotted open squares = synthetic piemontite (Anastasiou and Langer 1977); solid downward triangles = piemontite, type locality (Catti et al. 1989); open diamond = piemontite, type locality (Ferraris et al. 1989); open circles = piemontite, different localities (Bonazzi 1990); crosses = strontiopiemontite, type locality (Bonazzi et al. 1990); open downward triangles = REE bearing piemontites, Apuan Alps (Bonazzi et al. 1992); lined open downward triangle = REE bearing piemontite from Apuan Alps (Bonazzi and Menchetti 1994); crossed open square = piemontite from Nezilovo (Bermanec et al. 1994); solid diamond = REE-bearing piemontite, Varenche mine (Bonazzi et al. 1996); open upward triangle = androsite-(La), type locality (Bonazzi et al. 1996); star = tweddillite, type locality (Armbruster et al. 2002); solid squares = piemontite, different localities (Langer et al. 2002); open squares = synthetic piemontite (Langer et al. 2002).

between the Al-Fe and the Al-Mn members; this parameter, however, is an accurate measure of the androsite-(La) component in members of the piemontite-androsite-(La) series (Bonazzi et al. 1996) as the incorporation of $REE^{3+}$ in A2 and $Mn^{2+}$ in A1 causes a strong contraction along $c$ (solid arrow in Fig. 9c). Furthermore, along the piemontite-androsite-(La) join the $\beta$ angle decreases linearly down to 113.88° in androsite-(La) as a function of the REE content according to the equation

$$\beta(°) = 115.49 - 2.09 \, ^{A2}REE^{3+} \, (pfu) \, (r = -0.979)$$

As apparent from Figure 9d, $\beta$ is also strongly affected by the Sr $\leftrightarrow$ Ca substitution, which causes a decrement of this value (down to 114.56° in tweddillite; Armbruster et al. 2002). Thus very low values of the monoclinic angle are distinctive of high contents of either $REE^{3+}$ or Sr. In the latter case, however, $c$ will be much greater. In the absence of substitutions in the A sites, variations of $\beta$ are almost negligible (Fig. 9d), although synthetic Mn-Al piemontite exhibits slightly higher $\beta$ values (115.54 – 115.69°) than Al-Fe epidote (115.46 - 115.35°).

## OPTICAL AND SPECTROSCOPIC PROPERTIES

Looking at the data widely reported in the literature, it appears that densities (3.30 to 3.61 g/cm³) and refractive indices ($\alpha$ = 1.730 to 1.794, $\beta$ = 1.740 to 1.807, and $\gamma$ = 1.762 to 1.829; Deer et al. 1986) of piemontite are generally higher than those of the members of the clinozoisite-epidote join. This is a consequence of the generally greater substitution of $M^{3+}$ for Al in piemontite. Indeed, the Gladstone-Dale constant of $Fe_2O_3$ is slightly higher than that of $Mn_2O_3$, while both values are higher than that of $Al_2O_3$ (Mandarino 1981). Accordingly, epidote minerals with only small amounts of $Mn^{3+}$ exhibit refractive indices at the lower limits of the above range (e.g., 'withamite' from Glen Coe; Hutton 1938) or even lower indices similar to those of clinozoisite (e.g., $\alpha$ = 1.714, $\beta$ = 1.724, and $\gamma$ = 1.730; Yoshimura and Momoi 1964). Some values reported in the literature lie outside the ranges given above. For instance, $\gamma$ = 1.843 was determined for a piemontite from Goldongri, India (Nayak and Neuvonen 1966) and $\alpha$ = 1.772, $\beta$ = 1.813, and $\gamma$ =1.860 were determined by Strens (1966) on a piemontite from Chikla, India.

Several attempts to correlate the variation of refractive indices (as well as other optical features) of natural samples with their composition were carried out (Short 1933; Guild 1935; Strens 1966; Tanaka et al. 1972). The results obtained were sometimes inconsistent even internally. Strens (1966) considered the variation of the optical properties in relation to site symmetry and occupancy. According to his data, indices and composition correlate linearly in the range 0.8 < (Mn + Fe) < 1.5 pfu although some samples show a wide scatter which cannot be attributed to the Mn/(Mn + Fe) ratio.

To separate the influence of the $Fe^{3+}$ and $Mn^{3+}$ content on the physical properties of natural piemontite, Anastasiou and Langer (1977) measured the refractive indices of a series of synthetic Al-$Mn^{3+}$ piemontites with $Mn^{3+}$ up to 1.75 pfu. They found the refractive indices of the synthetic samples to be generally lower than those of the natural ones and explained this result partly with the lower refractive power of $Mn^{3+}$ compared to $Fe^{3+}$ and partly with the influence of possible replacement of Ca in natural piemontite. Later, Langer et al. (2002) performed a new study on synthetic Al-$Mn^{3+}$ piemontite with $Mn^{3+}$ up to 1.50 pfu. This study shows that all three refractive indices increase as a function of Mn content although with slightly different slopes, the increase of $n$ being in the order $n_\gamma > n_\beta > n_\alpha$. Moreover, no discontinuities at $Mn^{3+}$ ca. 1 pfu as previously observed by Anastasiou and Langer (1977) were found.

As noted by Wieser (1973) "piemontite presents a rare example of a mineral which was distinguished basing on chemical composition (Mn content) but in practice is the

pleochroism which is now decisive feature in distinction between piemontite and epidote".
The pleochroic scheme of Al-Mn-Fe epidote is basically X = lemon yellow or orange yellow,
Y = amethyst, violet or pink, and Z = bright red. Burns and Strens (1967; see also Burns
1993) measured polarized absorption spectra of Al-Fe and Al-Fe-Mn epidotes and provided
interesting structural interpretations thus demonstrating the origin of the pleochroic scheme
and absorption colors. They examined the polarized absorption spectra of a piemontite that
contains 0.625 $Mn^{3+}$ pfu (Fig. 10); as shown in the corresponding energy level diagram (Fig.
11), the number and energy position of the observed bands are well in accord with the presence
of $Mn^{3+}$ in the tetragonally compressed ($D_{4h}$ point symmetry) M3 octahedron. Although the
splitting between $d_{xz}$ and $d_{yz}$ is not resolved in the spectra, an orthorhombic pseudosymmetry
($C_{2v}$), which basically describes the M3 site in piemontite, is best in accord with the number
of the observed bands and their polarization (Langer et al. 2002). When the intensities of
the absorption bands are taken into account, the different vividness of color is also easily
explained. Thus, piemontite displays vivid colors correlating with high values of the molar
extinction coefficients ($\varepsilon$) and originating from the spin-allowed transitions within $Mn^{3+}$ ions

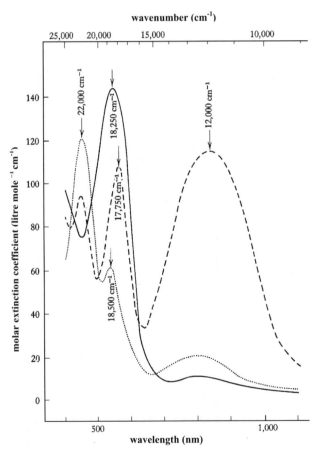

**Figure 10.** Polarized absorption spectra of piemontite from the type locality containing 0.625 $Mn^{3+}$ pfu;
.... = α (X) spectrum; ---- = β (Y) spectrum; —— = γ (Z) spectrum. Reproduced with permission of
Cambridge University Press, from Burns (1993).

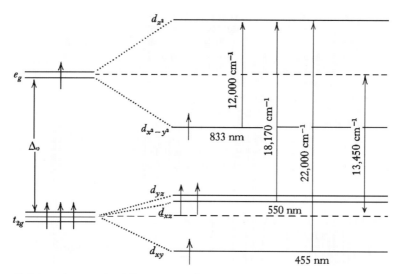

**Figure 11.** Schematic energy level diagram for $Mn^{3+}$ in axially compressed octahedral coordination (M3 site). Observed transitions refer to the polarized absorption spectra given in Figure 10. Reproduced with permission of Cambridge University Press, from Burns (1993).

located in the non-centrosymmetric and very distorted M3 octahedron. On the other hand the Al-Fe epidotes display pastel shades that are characterized by relatively low molar extinction coefficients typical of spin-forbidden transitions within $Fe^{3+}$ ions.

Much spectroscopic work (Wood and Strens 1972; Langer et al. 1976; Langer and Abu-Eid 1977; Smith et al. 1982; see also Liebscher 2004) has been done on this subject. In particular, Wood and Strens (1972) developed a method for calculating the *d*-orbital energy levels of transition-metal ions within coordination polyhedra and applied the derived equations to $Mn^{3+}$ in the M3 site of piemontite. Smith et al. (1982) measured the polarized absorption spectrum of natural piemontite at 295 K and 100 K. Lowering the temperature resulted in a sharpening of broad bands in the 10,000 to 25,000 $cm^{-1}$ regions, supporting their assignment to single $Mn^{3+}$ in the non-centrosymmetric M3 site. Langer et al. (2002) reported single crystal polarized electronic spectra of natural ($0.57 < M^{3+} < 1.17$ pfu) and synthetic ($0.83 < M^{3+} < 1.47$ pfu) piemontites, respectively. Spectra were measured in the range 35,000 to 5,000 $cm^{-1}$ at room temperature with E parallel to all three axes of the optical indicatrix. Spectra are dominated by a slightly polarized absorption edge in the UV region and also by three intense and strongly polarized absorption bands ($v_I$, $v_{II}$, $v_{III}$) that are typically spin-allowed *dd* transitions of $Mn^{3+}$ in the distorted (compressed) M3 octahedron of the piemontite structure. The energy and linear absorption coefficients of these three bands shift as a function of the $Mn^{3+}$ content in M3, so that information about the $Mn^{3+}$–$Fe^{3+}$ ordering between M3 and M1 in natural piemontite can be obtained. Although in the most $Mn^{3+}$-rich synthetic piemontite appreciable amounts of $Mn^{3+}$ enter the M1 site, no clear conclusion could be drawn with respect to the spin-allowed *dd* bands of $Mn^{3+}$ in the compressed M1 octahedron. The reasons are the strong overlapping of the band in the complex spectrum and the low intensities of the $Mn^{3+}$(M1) bands due to the centrosymmetry of the M1 site (Langer et al. 2002).

Infrared (IR) spectroscopic studies of piemontite were preformed by Tanaka et al. (1972), Langer et al. (1976), and Perseil (1987, 1990). By means of FTIR spectroscopy in the OH region, Della Ventura et al. (1996) studied several Sr bearing piemontites (SrO from 0.78 to 5.93 wt%) and some Al-Fe epidotes with low contents of $Mn^{3+}$ and little or no substitution of Sr for Ca. The spectra of the Sr bearing piemontites (Fig. 12) exhibit a well resolvable doublet at almost constant wavenumber (3417 to 3454 cm$^{-1}$) and differ from those of Al-Fe epidotes which show a single OH band, the position of which is a linear function of the Fe$^{3+}$ content (Langer and Raith 1974). Della Ventura et al. (1996) explained this IR feature as a consequence of two simultaneous effects: *i*) the entry of Mn$^{3+}$ in the non OH-coordinated M3 octahedron and its consequent strong anisotropic deformation and *ii*) the entry of Sr substituting for Ca in the adjacent A2 polyhedron (for a more detailed review of spectroscopic studies on epidote minerals see Liebscher 2004).

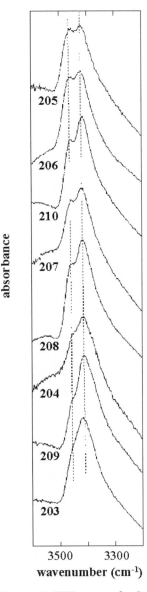

**Figure 12.** FTIR spectra for Sr bearing piemontites. Labeling refers to chemical analyses as reported in Table A of the Appendix. Modified from Della Ventura et al. (1996).

## COMPOSITIONAL VARIABILTY OF PIEMONTITE

This section deals with the compositional variability of piemontite and its related minerals as derived from a number of natural occurrences of piemontite. Chemical variations of this mineral are shown mainly by graphic representations to facilitate the readability of numerical data. 225 chemical analyses were selected from the literature and evaluated by the following criteria.

Chemical formulae were calculated on the basis of 8 cations. Fe was always considered as Fe$^{3+}$ and the content of Mn$^{2+}$ was estimated with respect to Mn$^{3+}$ on the basis of a total number of positive charges equal to 25 (in the absence of F$^{-}$). In some cases, the sum of positive charges was < 25 even if all Mn (and Fe) was considered in the trivalent state. In these cases, the amount of Mn$^{2+}$ was fixed to 0 pfu. Analyses showing a sum of positive charges lower than 24.90 were disregarded. If the estimated Mn$^{2+}$ content exceeds Mn$_{tot}$ (only few analyses) Mn$^{3+}$ was fixed to 0 pfu. Moreover, analyses were only taken into account if they satisfied the criterion 2.90 < Si$^{4+}$ < 3.10 pfu as the T sites in epidote minerals are generally considered to be fully occupied by Si (see above). As a result, all the graphic representations are based on 198 accepted chemical analyses; complete data are given in Tables A and B of the Appendix. In the case of REE bearing piemontite (19 analyses), Mg was added to the octahedral cations (Cat$_{oct}$), as well as Ti, Cu, and Zn if present. The content of Mn$^{2+}$ entering the M sites (Mn$^{2+}_{oct}$) was estimated on the basis of 3 − Cat$_{oct}$. Therefore, Mn$_{oct}$ = Mn$^{3+}$ for all REE free piemontites and Mn$_{oct}$ = Mn$^{3+}$ + Mn$^{2+}_{oct}$ for the REE bearing piemontites. The octahedral cation population in terms of Al, Mn$_{oct}$, and Fe$^{3+}$ is illustrated in Figure 13, whereas Figure 14 reports the frequency distribution plots of Al, Fe$^{3+}$, Mn$_{oct}$, and Ca, respectively.

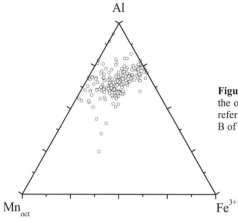

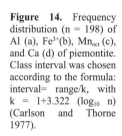

**Figure 13.** Ternary (Al-Fe$^{3+}$-Mn$_{oct}$) diagram showing the octahedral cation population of piemontite. Data refer to the analyses (n = 198) given in Tables A and B of the Appendix.

**Figure 14.** Frequency distribution (n = 198) of Al (a), Fe$^{3+}$(b), Mn$_{oct}$ (c), and Ca (d) of piemontite. Class interval was chosen according to the formula: interval= range/k, with k = 1+3.322 (log$_{10}$ n) (Carlson and Thorne 1977).

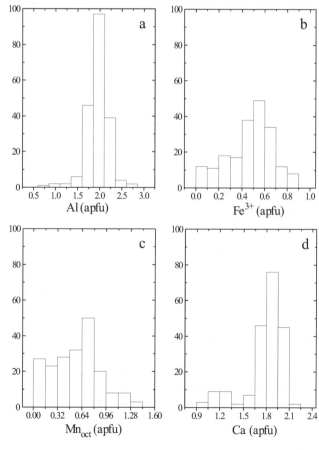

$Al_2O_3$ ranges from 6.25 (Enami and Banno 2001) to 31.50 wt% (Reinecke 1986b) corresponding to 0.74 and 2.80 Al pfu, respectively. The histogram for Al (Fig. 14a) shows a unimodal frequency distribution with a maximum at about 2.0 Al pfu. The minimum value observed (0.74 pfu) in the sample described by Enami and Banno (2001) is too low to fill the M2 site. It is coupled with a high content of the other trivalent cations, $Mn^{3+}$ in particular. This composition (with some differences in the Sr, Pb, and Ba content) is very similar to that of tweddillite, $CaSr(Mn^{3+},Fe^{3+})_2Al(SiO_4)(Si_2O_7)O(OH)$, the new mineral recently described by Armbruster et al. (2002). High $Al_2O_3$ (up to 29.77 wt%, corresponding to 2.73 Al pfu) and low $Mn_2O_3$ contents are typical of the so called *withamite*. $Fe_2O_3$ ranges from 0.10 (= 0.01 Fe pfu) to 14.97 wt% (= 0.89 Fe pfu). The histogram for $Fe^{3+}$ (Fig. 14b) shows an almost unimodal frequency distribution with a maximum in the range 0.50 to 0.60 pfu. However, low values (< 0.20 pfu) are not uncommon. $Fe^{3+}$ (pfu) shows a fairly good negative correlation with $Mn_{oct}$ (Fig. 15).

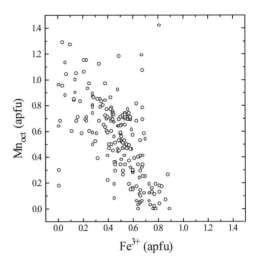

**Figure 15.** $Mn_{oct}$ plotted *vs.* $Fe^{3+}$; data as in Figure 13. For explanation see text.

$Mn_2O_3$ ranges from very low values in manganian clinozoisite to about 20 wt%. Actually, an analysis of piemontite from Langban, Sweden, made by Jakob and quoted by Malmqvist (1929) gives $Mn_2O_3$ = 22.00 wt% which is to our knowledge the maximum content reported for a natural piemontite. This analysis, however, should be disregarded because the sum of positive charges (24.67) is far from the expected value of 25, due to a suspiciously high content of univalent cations (Na = 0.42, K = 0.06 pfu). Strens (1966) and Anastasiou and Langer (1977) have already doubted the accuracy of this analysis. $Mn_{oct}$ shows a negative correlation with Al (Fig. 16). Due to the presence of several analyses of manganian clinozoisite, the histogram of $Mn_{oct}$ (Fig. 14c) exhibits an irregular frequency distribution in its left side. The highest frequency values are between 0.65 and 0.80 with a spread of values for $Mn_{oct}$ extending towards 1.40 pfu. The maximum value, 1.42 pfu, is found in the sample from Sanbagawa belt (Enami and Banno 2001) fulfilling the definition of tweddillite (Armbruster et al. 2002). The expansion of the compositional field to low or very low values of Mn is based on petrological rather than crystal chemical definition of piemontite (see above). In some instances, where strong and complex zoning in Al, $Mn^{3+}$, and $Fe^{3+}$ are observed, Mn-poor compositions may correspond to fluctuations in the $f_{O_2}$ conditions.

Small amounts of CuO (up to 1.85 wt% = 0.13 Cu pfu) and ZnO (up to 0.75 wt% = Zn 0.05 pfu) were found in piemontite from the Varenche mine (Bonazzi et al. 1996) and Nezilovo (Bermanec et al. 1994), respectively.

MgO may be present in piemontite reaching a maximum of 3.7 wt% (= 0.44 Mg pfu) in a sample from a garnet bearing muscovite-piemontite-quartz schist, Nagatoro district, Saitama Prefecture, Japan (Tanaka et al. 1972). The charge balance of the chemical formula requires in this sample, as in any REE free piemontite, Mg in the A sites together with Ca and $Mn^{2+}$. A high MgO content (0.35 pfu) was also found in a piemontite from schist at Chikla, India (Bilgrami 1956). However, the latter analysis was disregarded, due to the low value of Si (2.90 pfu) and the low total positive charge (24.16). MgO contents above the detection limit of the electron

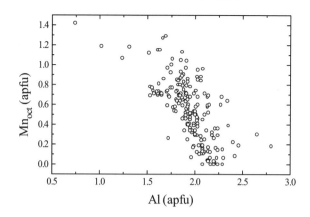

**Figure 16.** $Mn_{oct}$ plotted *vs.* Al; data as in Figure 13. For explanation see text.

microprobe are present in REE bearing piemontite corresponding to Mg contents (pfu) of 0.06 to 0.09 in the Pb and REE rich piemontite from Nezilovo (Bermanec et al. 1994), 0.02 to 0.08 in REE bearing piemontite from the Apuan Alps, Italy (Bonazzi et al. 1992), 0.019 in the REE bearing piemontite from Venn-Stavelot Massif, Ardennes, Belgium (Kramm 1979), and 0.17 for the REE bearing piemontite from the Varenche mine (Bonazzi et al. 1996). Except for the REE bearing piemontites from the Venn-Stavelot Massif (Kramm 1979; Schreyer et al. 1982) a positive correlation is found between Mg and $REE^{3+}$, pointing towards a solid solution with dissakisite-(Ce).

Other interesting chemical variations occur in the A sites. CaO ranges from 9.40 wt% (= 0.97 Ca pfu) in piemontite from Varenche mine (Bonazzi et al. 1996) to 24.75 wt% (= 2.10 Ca pfu, an evidently over-estimated value) in piemontite from Tucson Mountains, Arizona, USA (Guild 1935). The histogram of Ca shows a bimodal frequency distribution (Fig. 14d), with a maximum at Ca = 1.95–2.00 and a second distribution peak in the range 0.90–1.50, corresponding to Sr and REE rich piemontites.

SrO from 0.20 (= 0.01 Sr pfu) to about 18 wt% (= 0.92 Sr pfu) is reported in more than 60 analyses, most of them piemontite and strontiopiemontite from Andros, Greece (Reinecke 1986a, b), St.Marcel, Italy (Mottana and Griffin 1982; Mottana 1986; Della Ventura et al. 1996), Val Graveglia, Italy (Bonazzi et al. 1990), Falotta and Parsettens manganese deposits, Switzerland (Perseil 1990), Tokoro, Hokkaido, Japan (Akasaka et al. 1988; Togari and Akasaka 1987), Shiromaru manganese mine, Tokyo, Japan (Kato and Matsubara 1986), Sanbagawa belt, Japan (Enami and Banno 2001; Minagawa 1992), and Nezilovo, Macedonia (Bermanec et al. 1994). According to Mottana and Griffin (1982) Sr in piemontite from St. Marcel is essentially controlled by local bulk composition, rather than pressure or temperature. The same was observed in rocks from Andros, Greece (Reinecke 1986a) and from the Iberian Massif, Spain (Jiménez-Millán and Velilla 1993). However, Mottana (1986) observed that early-formed piemontite is richer in Sr and poorer in $Mn^{3+}$ than late piemontite. A weak chemical zoning with increasing Sr from core to rim was observed by Enami and Banno (2001) in piemontite from the Sanbagawa metamorphic belt, Japan. Reinecke (1986b) also observed a distinct and complex zoning with respect to the Sr ↔ Ca substitution in piemontite occurring at Vitali (Andros, Greece).

High contents of PbO were found in piemontite from Nezilovo described by Bermanec et al. (1994) with up to 10.35 wt% PbO (= 0.28 Pb pfu) and in a sample described by Enami and Banno (2001) with up to 7.55 wt% PbO (= 0.20 Pb pfu). The latter also exhibits an unusually high BaO content (up to 6.71 wt% BaO, = 0.25 Ba pfu). According to Enami and Banno (2001) such high contents of Pb and Ba in members of the piemontite-strontiopiemontite-tweddillite solid solution points to two theoretical new endmembers for epidote group minerals, "Ba-piemontite" and "Pb-piemontite". BaO up to 0.26 wt% were also observed in piemontite from the type locality, St. Marcel, Italy (Mottana and Griffin 1982).

Na$_2$O ranges from zero (most of the electron microprobe analyses) to a maximum of 2.41 wt% (= 0.37 Na pfu; Bilgrami 1956) and 2.59 wt% (= 0.42 Na pfu; Malmqvist 1929). Likewise, the presence of K$_2$O in piemontite is reported only in a few, often very old, analyses. In principle, a possible heterovalent exchange $^A$Ca $\leftrightarrow$ $^A$Na (or K) could be charge balanced, by analogy with dollaseite-(Ce), with a heterovalent anion substitution at the O4 site $^{O4}$O$^{2-}$ $\leftrightarrow$ $^{O4}$(OH$^-$,F$^-$). However, no recent structure refinement of piemontite supports this hypothesis. Therefore it appears more likely that relatively high amounts of Na and K result from impurities (i.e., intergrown feldspar).

It is well know that variable amounts of REE are common in the Fe rich members (epidote-allanite-ferriallanite series; see Gieré and Sorensen 2004). However, in recent years several analyses revealed the presence of REE also in piemontite. Actually, the first analysis was given by Hillebrand (quoted in Williams 1893 and Clarke 1910) who detected Ce$_2$O$_3$ = 0.89 wt% and other REE = 1.52 wt% (total REE$_2$O$_3$ = 2.41 wt%) in a piemontite from rhyolite of South Mountain, Pennsylvania, USA. As Williams (1893) stated, "this analysis is of especial interest in showing that the South Mountain piemontite is a connecting link between three recognized members of the epidote group" (piemontite, allanite, epidote s.s.). Kramm (1979) and Schreyer et al. (1986) described REE bearing piemontite from Venn-Stavelot Massif. A REE$_2$O$_3$ content close to 8 wt% (with La$_2$O$_3$ = 4.39 wt%) was found in piemontite from the Apuan Alps, Italy (Bonazzi et al. 1992). Several analyses by Bermanec et al. (1994) on piemontite from Nezilovo show REE$_2$O$_3$ varying between 6.09 and 8.56 wt%. The La content prevails on the other REE and exhibits a strong positive correlation with Pb (Bermanec et al. 1994). An REE$_2$O$_3$ content of about 14 wt% (with Ce$_2$O$_3$ largely prevailing on La$_2$O$_3$) was found in piemontite from the Varenche mine (Bonazzi et al. 1996). These occurrences clearly confirm the existence of a solid solution of piemontite towards manganian-allanite and androsite-(La). Data on the chemical composition of allanite and REE bearing epidote (Frondel 1964; Exley 1980; Sakai et al. 1984; Fleischer 1985; Deer et al. 1986) show that Ce is usually the dominant REE and often constitutes over half of the total quantity of lanthanide elements. Dollaseite-(Ce) and dissakisite-(Ce) present a similar trend. On the contrary, the Mn-rich epidotes (e.g., manganian allanite, piemontite and androsite-(La)) exhibit a rather different relative abundance of REE usually with La > Ce. Figure 17 shows the chondrite-normalized REE patterns of piemontite from the Apuan Alps, Nezilovo, and Varenche together with that of androsite-(La). A negative Ce anomaly is evident in all the patterns, with the exception of piemontite from Varenche. All patterns, with this exception, show a strong enrichment in LREE, while the Varenche piemontite exhibits a flatter pattern suggesting a minor depletion of HREE. For the other REE bearing piemontites complete chemical data were not provided, but some considerations can be made. The analysis of piemontite from Pennsylvania provides a Ce$_2$O$_3$ content of 0.89 wt%, while the sum of the other REE is given as 1.52 wt%. Because Ce$_2$O$_3$ is half of the other REE, it is likely that the La$_2$O$_3$/Ce$_2$O$_3$ ratio is about 1, which means that a Ce anomaly is present. In an analysis of piemontite from Venn-Stavelot Massif (Schreyer et al. 1986) only the Ce$_2$O$_3$ content (4.09 wt%) is reported. However, the total wt% of the analysis is very low (93.18), pointing to the possible presence of up to 5 wt% of other REE$_2$O$_3$. In the latter case, the presence of a Ce anomaly is only speculative. Finally, the piemontite from Venn-Stavelot Massif described by Kramm (1979) has a total REE$_2$O$_3$ content of 5.0 wt%, and the author notes that REE oxides are "dominantly La$_2$O$_3$, Pr$_2$O$_3$, and Nd$_2$O$_3$". Therefore, in this case a Ce anomaly is likely. It may also be noted that the La/Ce ratio is often greater than 1 in Mn-rich allanites (Deer et al. 1986).

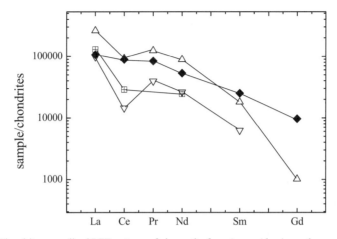

**Figure 17.** Chondrite-normalized REE patterns of piemontite from Apuan Alps (open downward triangles; Bonazzi et al. 1992), Nezilovo (crossed open squares; Bermanec et al. 1994), Varenche mine (solid diamonds; Bonazzi et al. 1996) and androsite-(La) (open upward triangles; Bonazzi et al. 1996). Values used in normalizing REE were taken from Anders and Ebihara (1982).

These data suggest a general trend in epidote group minerals that a Ce anomaly correlates with a high $Mn^{3+}$ content. This correlation is not surprising because the high oxidation conditions, which favor piemontite formation, also favor a high $Ce^{4+}/Ce^{3+}$ ratio, thereby preventing Ce incorporation in piemontite.

## OCCURRENCE AND PARAGENESIS OF PIEMONTITE

Piemontite as a major rock-forming mineral is comparably rare. Nevertheless, as an accessory or minor phase, it occurs in a wide variety of rocks of different bulk compositions that were formed under different physicochemical conditions. Piemontite is a typical metamorphic mineral, while magmatic occurrences are very rare or questionable as truly magmatic. Hydrothermal occurrences are described, but not very frequent.

Guild (1935) determined the Mn concentration in a number of piemontite bearing and piemontite free rock samples and was the first to note that the occurrence of piemontite "must be due to some undetermined peculiarity of the geophysical conditions" rather than to an unusually high Mn content of the host rock. Numerous subsequent occurrences as well as experimental studies have then clarified that these "peculiar conditions" are mainly related to an unusually high oxygen fugacity although high Mn contents enlarge the stability field of piemontite. Thus, piemontite is typically found in highly oxidized manganiferous sedimentary rocks that were affected by low to high-pressure metamorphism in a wide temperature range; the typical protolith for piemontite rocks is Mn-rich chert, former ocean floor sediment. In his review of the alpine occurrences of blueschist facies manganiferous cherts Mottana (1986) pointed out that piemontite occurs in the "oxidized" assemblages associated with Mn rich varieties of sodic clinopyroxene and mica, braunite, and ardennite. The chemistry of piemontite in the metacherts depends on the availability of certain cations rather than on the metamorphic $P$-$T$ conditions and is critically controlled by $f_{O_2}$. In particular, piemontite is richest in $Mn^{3+}$ when it coexists with the other $Mn^{3+}$ bearing phases braunite and ardennite but is significantly depleted in $Mn^{3+}$ when it coexists with spessartine. In the latter case it may contain $Mn^{2+}$ which substitutes for Ca. Furthermore, piemontite from high pressure manganiferous assemblages coexists with vein piemontite, which was formed later at low pressure and usually shows lower Sr contents (Mottana and Griffin 1982). Possible mechanisms which may determine a so high $f_{O_2}$ in the rock to stabilize piemontite range from external buffering by hydrothermal solutions to internally controlled factors like premetamorphic minerals capable of buffering $f_{O_2}$ to a high level (Keskinen and Liou 1987). Although the composition and parageneses of piemontite bearing assemblages indicate that piemontite predominantly forms as a product of low to moderate temperature metamorphism, very high $f_{O_2}$ and low partial pressure of $CO_2$ in the fluid phase together with high Mn contents in the host rock may enhance the stability field of piemontite to higher temperature (Keskinen and Liou 1987, Abs-Wurmbach and Peters 1999).

In this section we review the different occurrences of piemontite reported in the literature with emphasis given to the parageneses and, if available, the physicochemical conditions that favor piemontite formation. Due to for their relative scarcity and often incomplete *P-T* determination we organize piemontites in terms of their regional distribution rather than petrologic characteristics and subdivide them roughly into three sections only: i) metamorphic; ii) contact metamorphic and metasomatic; and iii) magmatic, pegmatitic, and hydrothermal.

## Metamorphic occurrences

***Europe and Russian Federation.*** At the type locality (Praborna mine, St. Marcel Valley, Piemonte, Italy) piemontite occurs in the upper parts of a Mn ore deposit composed of micaschist that include bands of Mn silicates (spessartine, piemontite, Mn phengite) with lenses of Mn oxides intercalated with bands rich in Fe silicates (epidote, clinopyroxene) and hematite (Martin and Kienast 1987). The Praborna deposit records different stages of the Alpine metamorphic cycle. A cretaceous high pressure event at 450–500°C and 0.8–1.0 GPa was followed by a meso-Alpine lower amphibolite to greenschist facies stage that overprinted much of the metapelitic rocks related to the Mn ore bodies and its metabasic cover rocks (Martin and Kienast 1987). Brown et al. (1978) determined the *P-T* conditions for this meso-Alpine overprint to about 300°C and 0.8 GPa and pointed to the unusual suite of minerals accompanying piemontite at St. Marcel including manganoan omphacite ('violan'), albite, quartz, braunite, microcline, hollandite, and strontian calcite. Piemontite at Praborna generally contains significant amounts of Sr substituting for Ca (Mottana and Griffin 1982). A high Sr content in the piemontite bearing horizons at St.Marcel is also confirmed by Sr rich Mn oxides of the cryptomelane-hollandite-coronadite series (Perseil 1988). Piemontite occurrences are also reported from several localities within the metasedimentary cover series (the so called "schistes lustrès") of the ophiolitic sequence of the Zermatt-Saas zone (Switzerland). In the Valsesia-Valtournanche area (Italy) piemontite occurs in manganiferous quartzite and schist and displays a wide compositional variation that reflects the pre-metamorphic heterogeneous Mn distribution and oxidation state within the protoliths rather than different metamorphic conditions (Dal Piaz et al. 1979). A detailed petrologic study of the manganiferous quartzite at Cignana Lake in Valtournanche is provided by Reinecke (1991). At this locality piemontite records two different stages of metamorphism. Piemontite from the high to ultrahigh-pressure assemblage has high Mg content substituting for Ca and coexists with pyrope rich garnet, magnesian braunite, unusually Mg–rich ardennite, kyanite, phengite, talc, paragonite, zoned tourmaline with dravite rich cores, hematite, and rutile. Coesite or quartz pseuodomorphs after coesite are found as inclusions in dravite and pyrope rich garnet. Contrary, piemontite from the low-pressure assemblage is Mg poor and coexists with braunite, tourmaline, garnet, and ardennite that are also characterized by a lower content of Mg. Other Alpine occurrences of piemontite in the "schistes lustrés" are reported from Haute Maurienne (France) where piemontite is restricted to oxidized assemblages in paragenesis with quartz, garnet, braunite, talc, phlogopite, phengite, chlorite, and hematite (Chopin 1978), from Steinen Valley (Switzerland) where it coexists with quartz, spessartine, braunite, and cryptomelan in calcareous schist and siliceous dolomites that were progressively metamorphosed from lower greenschist to higher amphibolite facies conditions (Frank 1983), and from quartzitic marble cropping out in the area of Sant'Andrea di Cotone (Eastern Corsica, France; Caron et al.

1981). The latter occurrence represents metamorphic conditions of about 0.8 GPa and 300°C as derived from the phase assemblage with clinopyroxene, garnet, blue amphibole, phengite, chlorite, stilpnomelane, lawsonite, pumpellyite, and deerite (Caron et al. 1981). Piemontite with a composition intermediate between common piemontite and androsite-(La) is described from a fine grained, reddish-brown gneiss from the dump of the abandoned Mn mine at Varenche, St. Barthélemy Valley (Western Alps, Italy; Bonazzi et al. 1996). The metasedimentary rocks at the Varenche mine probably form part of the Tsatè nappe (Dal Piaz 1988; Marthaler and Stampfli 1989) and were metamorphosed at greenschist facies conditions but locally, like in the gneiss containing the described REE bearing piemontite, show relics of a low grade, high pressure metamorphism. At Piz Cam (Bergell Alps, Switzerland) bladed crystals of piemontite up to 1 cm long occur together with spessartine, rhodonite, rhodochrosite, and kutnahorite in quartzite layers, which are intercalated in a massive sequence of chlorite-muscovite-clinozoisite-albite schist and marble metamorphosed under greenschist facies conditions (Wenk and Maurizio 1978).

Several occurrences of piemontite are reported from the Apennine, Italy. In Val Graveglia, Northern Apennine, piemontite occurs in manganese ore deposits within the metacherts on top of an ophiolite sequence. Here, a prehnite-pumpellyite facies metamorphism produced complex unusual Mn rich mineral assemblages including a wide variety of rare mineral species (Cortesogno and Venturelli 1978; Cortesogno et al. 1979). At the Cerchiara mine piemontite occurs with sodic-clinopyroxenes (aegirine, aegirine-augite, and namansilite), alkali and sodic-calcic amphibole, andraditic garnet, pectolite, ganophyllite, calcite, manganoan calcite, calcian rhodochrosite, and rhodochrosite (Lucchetti et al. 1988). At the Molinello and Cassagna mines strontiopiemontite was found in veins containing calcite, rhodonite, rhodochrosite, and ganophyllite (Bonazzi et al. 1990). Within Mn rich beds of the "Brecce di Seravezza" formation of the Apuan Alps piemontite is associated either with braunite and rare hausmannite or with braunite, hollandite, and minor amounts of rhodochrosite and kutnahorite. Both assemblages formed during alpine metamorphism at 0.4–0.6 GPa, 350–380°C and $f_{O_2}$ values higher than those of the hematite-magnetite buffer (Franceschelli et al. 1996). Bonazzi et al. (1992) described REE bearing piemontite from braunite mineralizations in calcschist horizons within marble of the Apuan Alps at the slope of Mt. Brugiana - La Rocchetta. The piemontite occurs as minute prismatic crystals in bent veinlets associated with quartz and minor amounts of braunite, Mn-oxides, titanite, apatite, mica, and manganoan calcite.

In glaucophane schist and associated rocks from the Ile de Groix (Brittany, France) piemontite coexists with ottrelite and is only found in rocks that lack garnet and possibly reequilibrated at lowest greenschist facies conditions. In these rocks the Mn content of piemontite directly reflects that of its precursor mineral (Makanjuola and Howie 1972).

In Spain piemontite is described from the Ossa-Morena central belt (Iberian Massif) and provides an example of the interplay between $f_{O_2}$, rock composition, mineral assemblage, and piemontite composition. Here, successive greenschist facies metamorphic events progressively reduced initial, highly oxidized braunite nodules and produced minute euhedral crystals of piemontite in paragenesis with Ti rich braunite, hematite, and spessartine (Velilla and Jiménez-Millán 2003). Depending on the physicochemical conditions piemontite from Ossa-Morena displays a wide chemical variation ranging from manganian epidote to piemontite (Jiménez-Millán and Velilla 1993). The $Mn^{2+}$ content of piemontite correlates inversely with $f_{O_2}$. The control of the mineral assemblage on piemontite composition is evident from braunite vs. hematite bearing assemblages. The strong fractionation of $Mn^{3+}$ into braunite prevents high $Mn^{3+}$ contents in coexisting piemontite. On the other hand, a high modal amount of hematite leads to $Mn^{3+}/Fe^{3+}$ ratios in coexisting piemontite up to three times higher than in piemontite coexisting with braunite. Finally, a comparison between the bulk composition of piemontite bearing and piemontite free lithologies from Ossa-Morena suggests that high Ca contents enhance the stability of piemontite (Jiménez-Millán and Velilla 1993).

In the highly oxidized Mn and Al rich layers of the low-grade metamorphic metasediments of Salmchâteau, Venn-Stavelot Massif, Ardennes (Belgium), a Mg and REE rich piemontite occurs as a rare phase forming a peculiar assemblage with viridine (solid solution between andalusite and kanonaite), muscovite, paragonite, Mg chlorite, braunite, hematite, quartz, rutile, and apatite (Kramm 1979). A similar REE rich but Mg free piemontite occurs together with spessartine, quartz, and kutnahorite in the low-temperature chlorite veinlets that crosscut the low-grade manganiferous deposit of the Lienne Valley in the southwestern part of the Venn-Stavelot Massif (Schreyer et al. 1986).

Two extensively studied piemontite occurrences are from the Islands of Andros and Evvia (Greece) that both belong to the Attic-Cycladic blueschist belt. These occurrences provide one of the few opportunities to study the low to medium grade evolution of piemontite bearing silicate assemblages in response to changes in temperature and pressure (Reinecke 1986a) and will be discussed in detail below. Besides the more common piemontite studied by Reinecke (1986a) extremely $Mn^{2+}$ and $Mn^{3+}$ rich but $Fe^{3+}$ poor REE bearing piemontite with composition extending to androsite-(La) occurs as minor or accessory constituent in Ca poor assemblages (e.g., spessartine quartzite, rhodonite rhodochrosite braunite fels) at Petalon Mountain, central Andros (Bonazzi et al. 1996).

In Mn rich metasediments from the Lower Paleozoic formation of the Inner West Carpathians (Slovakia) zoned piemontite occurs together with Fe-oxides in the rim of spessartine nodules that were formed by greenschist facies metamorphism of sedimentary Mn concretions (Spisiak et al. 1989).

An unusually REE and Pb rich piemontite was found at Nezilovo, Macedonia, in the upper part of Babuna River, within a greenschist facies mica schist containing red mica, quartz, ardennite, gahnite, and franklinite (Bermanec et al. 1994).

Piemontite in paragenesis with manganoan muscovite and quartz is reported from the manganiferous mica schists at Dealul Negru (Lotru Mountains, Romania) that also contain manganoan biotite, sericite, chlorite, orthoclase, microcline, albite, spessartite-calderite, ferrian braunite, bixbyite, hematite, rutile, apatite, and zircon (Balan and David 1978).

An extraordinary piemontite occurrence is reported from the Vestpolltind iron manganese deposit (Lofoten-Vesterålen province, Norway). In the jasper banded hematite ore of this deposit piemontite is found in layers closely associated with tremolite-actinolite, diopside, quartz, and minor amounts of manganophyllite, garnet, and sillimanite (Krogh 1977). The observed mineral assemblage, suggesting textural equilibrium, was formed under granulite facies conditions of 0.9–1.1 GPa and 770°C with a local temperature peak of 940 ± 50°C (Krogh 1977). It thus represents, at least to the authors' knowledge, the highest-grade occurrence of piemontite so far reported.

Several piemontite occurrences are reported from the Ural Mountains. In the lowest grade metamorphic jasper deposits of the southern Urals piemontite occurs as an accessory mineral together with minor amounts of chlorite, actinolite, magnetite, pumpellyite, stilpnomelane, calcite, and albite (Krizhevskikh 1995). At the Uchaly deposit (Southern Urals) piemontite coexists with high aluminum and high fluorine titanite in Mn-rich quartzites (the so called "gondites"; see Roy and Purkait 1968 for nomenclature) associated to metavolcanics that probably experienced a high-pressure metamorphism that was followed by a prehnite-pumpellyite facies metamorphism (Pletnev et al. 1999). In metamorphic rocks of the Maldynyrd Range (Subpolar Urals) piemontite and Mn rich varieties of allanite occur in nodule shaped segregations (Yudovich et al. 2000). The host rocks of these nodules are different types of metarhyolite characterized by various degrees of alteration (Yudovich et al. 2001). In the aporhyolite schist complex of the Al'kesvozh Sequence piemontite forms small prismatic crystals and is dispersed together with spessartine in nodules that are made up of alternating layers of fine grained sericite and coarse grained muscovite and quartz. In the sericite quartz schists of the Lake Grubependity cirque nodules mainly consist of piemontite, braunite, and quartz, with minor amounts of hematite, ilmenite, zircon, ardennite, and xenotime. Lenses containing the rare minerals skorodite and arsenosiderite are also enclosed in these nodules. The peculiar mineral chemistry of these nodules, mirrored by anomalously high concentrations of Sr, Ba, As, and Zr in piemontite, is probably due to an internal mobilization of schist derived elements as well as an infiltration of external material (Kozyreva et al. 2001). In the valley of the Alda-Ishkin River and the Khar-Taiga Mountains piemontite occurs in albite-sericite-chlorite-quartz schists (Voznesenskii 1961).

### *Africa.*

*Tanzania.* At Mautia Hill at the western margin of the Monzambique Orogenic belt small subhedral piemontite occurs in talc phlogopite and in talc edenitic hornblende rich layers with subordinate amounts of viridine and yoderite (Basu and Mruma 1985).

Several occurrences of amphibolite facies piemontite in a variety of parageneses have been reported from the Mn rich metasedimentary rocks of the Konse Series in central Tanzania (Meinhold and Frish 1970). In particular, piemontite is a major constituent of amphibolites and quartz schists whereas it is a minor constituent in marble, calc-silicate rocks, quartz-chlorite-mica schist, and quartz-viridine-mica schist.

### *Asia.*

*China.* In the braunite ore deposits widely distributed along the western margin of the Yangtze Platform, piemontite commonly occurs together with rhodonite, spessartine and other metamorphic silicate minerals in layers, which rhythmically alternate to primary Mn oxide and carbonate layers (Liu and Xue 1999). In the felsic metavolcanic layers from Mulan Mountain (Hubei Province) fine-grained piemontite is rimmed by epidote and records metamorphic conditions transitional between blueschist and greenschist facies (Zhou et al. 1993). On Hainan Island, Guangdong province, a high $f_{O_2}$ under alkaline conditions favored the formation and stability of piemontite (Peng and Zheng 1984). One of the few high to ultrahigh pressure occurrences of piemontite is reported from gneisses that are associated with the eclogites at Sulu-Dabie Mountains (southern Henan). Here, piemontite besides quartz, apatite, phengite, albite, epidote, titanite, and omphacite occurs as inclusion in zircon and is interpreted as a relic of the regional ultrahigh pressure metamorphism (Liu et al. 1998; for a detailed review of epidote minerals in high to ultrahigh pressure rocks see Enami et al. 2004).

*India.* Piemontite that formed during a lower amphibolite facies metamorphism occurs in manganiferous metasediments of the Manbazar area (Purulia District, West Bengal) coexisting with quartz, spessartine, manganophyllite, braunite, hematite, and oligoclase (Acharyya et al. 1990). A later contact metamorphism overprinted this assemblage and formed manganian andalusite at the expense of piemontite spessartine, and braunite. Amphibolite facies piemontite is also described from the manganiferous silicate rocks of Gowari Wadhona in the manganese belt of the Sausar Group (Roy and Purkait 1968). In these rocks, first named "gondites" by Fermor (1909), amphibolite facies metamorphism on the original manganiferous sediments produced piemontite-bearing assemblages with unusual manganese rich spessartine, rhodonite, clinopyroxenes, amphibole, manganophillite, alurgite, braunite, and hollandite besides quartz and minor amounts of apatite, plagioclase, calcite, dolomite, and microcline. In the calc-silicate rocks from the Sausar Group piemontite coexists with garnet in braunite bearing and braunite free assemblages

(Fukuoka et al. 1990). Piemontite and garnet are both Mn enriched in the braunite bearing compared to the braunite free assemblages in contradiction to the findings of Jiménez-Millán and Velilla (1993) from Ossa-Morena, Spain (see above).

*Pakistan.* The only piemontite occurrence from Pakistan so far reported is close to the contact between the Lower Swat-Buner schistose group and the epidote amphibolites of the Kohistan basic complex (North-Western Pakistan). Here, piemontite occurs in schists together with quartz, albite, margarite, tourmaline, muscovite, rare spessartine, Mn rich chlorite, rutile, and magnetite, as euhedral, elongated crystals that form laminae within the plane of schistosity (Jan and Symes 1977).

*Thailand.* Piemontite occurs in quartz schists from the Nam Suture Zone in Northen Thailand that were derived from hemipelagic to pelagic Mn rich sediments and metamorphosed under the blueschist facies conditions (Singharajwarapan and Berry 2000).

*Japan.* Within the different metamorphic areas of Japan piemontite occurs in a variety of lithologies and records diverse physicochemical conditions. In veins from the manganiferous iron ore deposits of the Tokoro Belt, Hokkaido, Sr rich piemontite coexists with Mn bearing pumpellyite and/or okhotskite, hematite, and bixbyite (Togari et al. 1988, Akasaka et al. 1988). These veins occur in the ore bodies themselves and in the associated radiolarian cherts. Except for the Fukuyama mine, piemontite from veins in the ore bodies tends to contain more $Mn^{3+}$ and $Fe^{3+}$ than that from veins in the cherts, whereas its Sr content is higher in the latter, therefore suggesting a chemical control by the host rock (Akasaka et al. 1987). Based on the metamorphic conditions of about 250°C and 0.4–0.5 GPa at very high $f_{O_2}$, Akasaka et al. (1988) interpreted the assemblage piemontite + Mn pumpellyite as the low *P*, low *T* equivalent of the piemontite sursassite assemblages discussed by Reinecke (1986a). Metamorphic conditions of the prehnite-pumpellyite facies were also estimated for fine disseminated piemontite that coexists with manganoan grossular in the Dainichi manganese ore deposit (Kanagawa Prefecture; Hirata et al. 1995). Most piemontite occurrences of Japan, however, record greenschist, blueschist to epidote amphibolite facies conditions and belong to the different high *P/T* metamorphic belts of Japan. In the Yamagami metamorphic rocks of the northeastern Abukuma Plateau piemontite coexists with phengitic muscovite, albite, chlorite, tourmaline, rutile, and manganian hematite (Akasaka et al. 1993) and is often microboudinaged defining a stretching lineation (Masuda et al. 1995). Comparable microboudinaged piemontite is also found in quartzites from Nuporomaporo (northern part of Kamuikotan belt) with muscovite and apatite, from Asemi (Sanbagawa belt) with albite, chlorite, muscovite, and apatite, and from Matsunosako (Nagasaki belt) with muscovite, albite, and tourmaline (Masuda et al. 1990).

A detailed description of blueschist to epidote amphibolite facies piemontite from two localities of the Sanbagawa belt stems from Izadyar (2000) and Izadyar et al. (2000, 2003) who studied the relationship between chemical variation in piemontite, host rock bulk chemistry, parageneses, and metamorphic conditions at Asemi-gawa and Besshi. Their results will be presented and discussed below in comparison to piemontite from Evvia and Andros Islands (Greece). In the Shirataki and Oboke areas of the Sanbagawa belt piemontite occurs in siliceous schists that show increasing metamorphic conditions from greenschist to blueschist up to epidote amphibolite facies (Ernst and Seki 1967). These schists are characterized by the ubiquitous coexistence of piemontite, quartz, and stilpnomelane and show an increasing piemontite/stilpnomelane ratio with increasing metamorphic grade. A very peculiar piemontite occurrence of the Sanbagawa belt is described from epidote amphibolite facies rocks of central Shikoku (Enami and Banno 2001). Here, common piemontite as well as barian plumboan strontiopiemontite occur in Mn rich lenses within a layer of quartz schist that was metamorphosed at 480–580°C and 0.7–1.0 GPa (Enami et al. 1994). The Mn rich lenses are mainly composed of piemontite, quartz, hematite, dolomite, and calcite with minor amounts of muscovite, chlorite, talc, and apatite. Small nodules within these lenses consist of an abswurmbachite rich inner zone with minute inclusions of noélbensonite and a hollandite rich outer zone with subordinate amounts of calcite, muscovite, quartz, and piemontite. The barian plumboan strontiopiemontite occurs in clusters of piemontite in occasional mutual grain contact with the latter and gives no textural evidence for chemical disequilibrium between both phases (Enami and Banno 2001). In schists of the Kotu-Bizan district (Eastern Shikoku) piemontite is associated with the typical blueschist facies paragenesis glaucophane, rare lawsonite, and sodic pyroxene (Ernst 1964). Lower amphibolite facies (>0.3 GPa and 530–560°C) piemontite with Mn white mica, viridine, spessartine, Mn tourmaline, and Ti-Mn hematite is reported from a highly oxidized manganiferous layer within the quartz mica schists of the Hidaka Mountains, Hokkaido (Grapes and Hashimoto 1978). It is typically restricted to areas where most viridine has been altered to sericite and occurs disseminated in the matrix as well as in monomineralic veins. These veins do not extend outside the manganiferous layer and crosscut its foliation. Comparable *P-T* conditions (~0.5 GPa and 550°C) were derived for the highly oxidized siliceous schist of the Mineoka tectonic zone (Boso Peninsula) where piemontite coexists with cuprian phlogopite, aegirine-augite, riebeckitic tirodite, quartz, albite, K-feldspar, apatite and hematite (Ogo and Hiroi 1991; Hiroi et al. 1992).

Piemontite associated with braunite (and hematite) ores occurs coexisting with ardennite and/or sursassite in several manganese deposits of the Sanbagawa belt (Enami 1986; Minakawa and Momoi 1987) and in the Tone mine, Nagasaki Prefecture (Sasaki et al. 2002), and as subordinate phase together with aegirine, rhodonite, spessartine, and jacobsite in veinlets within the ore at the Kamisugai mine, Ehime Prefecture (Kato et al. 1982).

### *America.*

*U.S.A.* In the Shadow Lake tuffs (eastern Sierra Nevada, California) piemontite bearing assemblages, first described by Short (1933), are the product of an upper greenschist to lower amphibolite facies overprint of dacitic to rhyodacitic pyroclastic deposits (Keskinen 1981). Besides its occurrence in shaly and schistose layers, piemontite is found as well in late hydrothermal veins at this locality. The piemontite bearing assemblage consists mainly of quartz, albite, phengitic muscovite, phlogopite, and tremolite and is characterized by sporadic piemontite, absence of other epidote group minerals, and a relatively high concentration of finely grained disseminated hematite. Although this mineral assemblage is normally indicative of the greenschist facies, Keskinen (1981) speculated that it might have persisted to temperatures higher than those usually attributed to greenschist facies rocks due to the high $f_{O_2}$ conditions. The Shadow Lake occurrence also provides further evidence for the overwhelming importance of the $f_{O_2}$ in stabilizing piemontite compared to the bulk composition, as the metavolcanic rocks, although characterized by a variety of Mn rich minerals, do not exhibit unusually high Mn contents.

Studying the amphibolite facies occurrences from San Gorgonio Pass (southern California) and Las Tablas and Picuris Range (northern New Mexico), respectively, Smith and Albee (1967) and then Stensrud (1973) showed that the occurrence of piemontite under amphibolite facies conditions is not as unusual as stated by different authors (e.g., Makanjuola and Howie 1972). The microcline-plagioclase-quartz gneiss cropping out near San Gorgonio Pass locally contains piemontite bearing layers that were formed under highly oxidizing conditions and are characterized by hematite, ferrian spessartine, ferrian muscovite, and ferromagnesian silicates (phlogopite, pyroxene, and amphibole) that exhibit high Mg/Fe ratios and unusually high $Fe^{3+}$ contents (Smith and Albee 1967). The piemontite bearing schists from Las Tablas and Picuris Range rarely contain garnet and consists of piemontite, hematite, ferrian muscovite, phlogopite, and microcline. Assuming amphibolite facies temperatures they indicate an $f_{O_2}$ of probably higher than 10–15 bars (Stensrud 1973).

Amphibolite facies piemontite with spessartine, manganian andalusite, braunite, abundant tourmaline, and rare manganiferous zincian staurolite is also reported from a manganiferous layer that crops out along the contact between the Vadito and Ortega Groups (northern New Mexico) was metamorphosed at 500–540°C and 0.38–0.46 GPa (Grambling and Williams 1985).

*Argentina.* Descriptions of piemontite from South America are rare. Piemontite has been mentioned by Gelos et al. (1988) as an important mineral in sediments of the continental platform in Argentina, and is known from skarn deposits of Alta Gracia, Sierras de Gordoba and from hydrothermal alteration zones in the Jujuy province (Viramonte and Sureda, pers. comm.)

## *Oceania.*

*New Zealand.* In a quartz-albite-ardennite-spessartine-phengite-hematite-chlorite-rutile-tourmaline schist of the Haast Schist Group near Arrow Junction, western Otago, piemontite with about 0.7 wt% SrO and 0.06 wt% PbO occurs in crack-seal quartz veins (Coombs et al. 1993). These veins formed shortly after peak metamorphism in the chlorite zone of the greenschist facies at about 0.45 GPa and 390°C. The protolith of the schist has been a highly oxidized (Fe, Mn)–oxide and –hydroxide bearing siliceous pelagic sediment (Coombs et al. 1993). The manganiferous members of the epidote group in the rocks of the Haast Schist Group display a continuous compositional range in terms of total Mn content from $Mn^{3+}$-rich piemontite through manganian epidote to manganoan epidote in which $Mn^{2+}$ substitutes for Ca (Kawachi et al. 1983). As usually, $Mn^{3+}$ richer piemontite tends to occur in rocks with a higher oxidation ratio, confirming that the critical factor to form piemontite is a high oxidation state rather than a very high Mn content. Geochemical investigations show that the formation of piemontite in the schists from Arrow Junction is also closely related to the excess oxygen inherited from the original Mn oxides (Coombs et al. 1985). Contrary to the schists, piemontite from the metacherts of the Haast River Area coexists with quartz, albite, muscovite, phlogopite, actinolitic-tremolitic amphibole, spessartine, tourmaline, hematite, apatite, and titanite and corresponds to a higher metamorphic grade than in the other occurrences of the Haast Schist Group (Hutton 1940; Kawachi et al. 1983; Cooper 1971, Coombs et al. 1993).

*Australia.* In the Wilyama Orogenic Domain (South Australia) piemontite occurs in a manganiferous unit within the albite rich metavolcanic sedimentary sequence of the Olary Block and ranges in composition from manganian epidote ($Mn^{3+} = 0.14$ pfu) to extreme Mn rich specimen ($Mn^{3+}$ up to 1.05 pfu) (Ashley 1984). Geothermometry leads to metamorphic temperatures of 400 to 500°C. The assemblage includes spessartine rich garnet, albite, phengite, manganoan tremolite, phlogopite, quartz, and Mn bearing hematite, and indicates a $f_{O_2}$ above the hematite-magnetite buffer.

## Contact metamorphic and metasomatic occurrences

To the authors' knowledge, no piemontite occurrence that could be unequivocally attributed to a classical contact metamorphism has so far been described. In the Odenwald (near Darmstadt/Germany) piemontite occurs in a viridine hornfels but was probably formed during a retrograde alteration (Abraham and Schreyer 1975). In the medium grade schists of the contact aureole of the Brezovica peridotite (Serbia) piemontite occurs in a talc phengite assemblage (Abraham and Schreyer 1976). But contrary to classical contact metamorphism this occurrence indicates very high $pH_2O$ and is attributed to a subduction zone emplacement of the ultramafic mass at 25 to 35 km depth (Schreyer and Abraham 1977).

Metasomatic piemontite is described from rare Mn bearing skarn deposits associated with gabbro to dioritic plutons of northwestern Iran (Tikmeh Dash pluton and Bostan Abad area). Here, piemontite is primarily found in epidote exoskarns with bixbyite and hematite as the main ore minerals (Somarin and Moayyed 2002). In the Upper Proterozoic Nyarovoisk suite near the Kharbei molybdenite ore deposit (Polar Urals, Russia) piemontite occurs in metasomatized zones of quartz feldspar sericite chlorite schists (Litoshko and Nikitina 1984). In the metasomatic gold pyrite deposits of the Baimak area (Southern Urals, Russia) piemontite occurs with pumpellyite and clinozoisite-epidote (Ismagilov 1976). In (meta)rodingites that are associated with the Bou Azzer ophiolite (Anti-Atlas Mountains, Morocco) piemontite coexists with grossular, salite, and prehnite and formed during an Mg-Ca-Mn metasomatism. This metasomatic event can be related to the serpentinization process that occurred in the associated ophiolite at 200–350°C and 0.2 GPa (Leblanc and Lbouabi 1988).

## Magmatic, pegmatitic, and hydrothermal occurrences

Piemontite in intermediate and acid volcanic rocks has been only rarely reported (see Deer et al. 1986; e.g., Lausen 1927; Guild 1935, Lyashkevich 1958). Nonetheless, it was most probably formed during postmagmatic hydrothermal activity rather than as a primary magmatic phase. In the volcanic sequence at Sulphur Spring Valley (Arizona, USA) piemontite occurs in narrow veinlets and fractures in andesite with considerable amounts of quartz and some kaolinite and was probably deposited by postmagmatic hydrothermal solutions (Lausen 1927). The manganian clinozoisite ('withamite') described by Hutton (1938) from veinlets and vesicles within andesite at Glen Coe, Scotland, also points to a hydrothermal origin. In the Tucson Mountains (Arizona, USA) described by Guild (1935) piemontite is restricted to cavities and fissures that are primarily filled by calcite and quartz and clearly indicates a hydrothermal origin. Likewise, piemontite from the nepheline syenite of Alain Range described by Lyashkevich (1958) generally occurs as replacement of biotite and thus implies postmagmatic alteration. In the fine-grained groundmass of the "Imperial Porphyry" that belongs to the Dokhan volcanic sequence, Eastern Desert, Egypt, piemontite is associated with basaltic hornblende (Basta et al. 1980, 1981).

Piemontite from pegmatites is only rarely reported as a late phase in the crystallization sequence (Deer et al. 1986). It was first described by Bilgrami (1956) from the Sitasaongi mine in the Bhandara district (India) where it coexists with other accessory Mn minerals like bixbyite, braunite, hollandite, manganite, pyrolusite, and cryptomelane in a calc-alkaline pegmatite vein crosscutting the manganese ore horizons (Mitra 1964). Similar occurrences from India are reported from the manganese ore deposits at Kajlidori mine in the Jhabua district (Nayak 1969) and Goldongri mine in the Panchmahal district (Nayak and Neuvonen 1966). In the latter, piemontite occurs at the contact between pegmatite and ore body. In Notodden and Tinnsjo (Norway) manganian clinozoisite (described as withamite) occurs as a late stage crystallization or replacement product in pegmatites and quartz veins that crosscut higher-grade metamorphic rocks (Morton and Carter 1963). Piemontite bearing "pegmatoidal" feldspar segregations are described from gneisses of the Sudeten Mountains (Poland) that are associated with amphibolites (Wieser 1973). Despite their association with amphibolite facies rocks these segregations are probably due to a lower grade hydrothermal activity as testified by the presence of microcline and low temperature albite and therefore let Wieser (1973) to question the stability of piemontite at amphibolite facies conditions.

At Rémigny, Quebec (Canada), fractured, prehnitized, and epidotized dioritic rocks are cut by a series of small veins that were probably precipitated during the terminal stage of the Lac-Rémigny intrusive complex from low temperature (< 300°C) hydrothermal solutions. The veins consist of albite, armenite, Mn zoisite ('thulite'), and piemontite with minor amounts of natrolite, kaolinite, calcite, quartz, and hematite as accessory phases (Pouliot et al. 1984). The primary epidote group mineral in the veins is manganiferous zoisite that locally contains a core of iron rich zoisite. Piemontite formed latest in the sequence of epidote group minerals and typically replaces manganiferous zoisite in irregular patches and along the grain boundaries (Pouliot et al. 1984).

At the Kalahari manganese field of the Transvaal Supergroup (South Africa) fluid rock interaction during high-grade diagenesis and lower greenschist facies metamorphism was followed by up to three hydrothermal events and superficial weathering and led to a complex mineral assemblage. Earliest minerals include sedimentary and diagenetic kutnahorite, Mn rich calcite, braunite, and hematite. Piemontite and Sr-rich piemontite occur in calc-silicate bodies and formed during the oldest of the three hydrothermal events at about T = 450°C together with albite, orthoclase, banalsite, andradite, henritermierite, tephroite, minerals of the pectolite-serandite series, pyroxenes (acmite, Mn bearing diopside), rhodonite, and ruizite (Gutzmer and Beukes 1996). More recently tweddillite was described in the hydrothermally altered calc-silicate rocks of this locality associated with serandite-pectolite and braunite (Armbruster et al. 2002). In the sedimentary to hydrothermal lead manganese ore deposit of Ushkatyn III (Central Kazakhstan) piemontite occurs in an unusual mineral assemblage including braunite, coronadite, kentrolite, garnet, tephroite, barite, and calcite (Dzhaksybaev et al. 1991).

Hydrothermal piemontite partially to completely replaces biotite in the quartz sericite biotite schist and quartzite of the metamorphic series at Sierra Pelona (California, USA; Simonson 1935). Here, the hydrothermal activity is related to the intrusion of a quartz dioritic dike. Extensive development of hydrothermal Mn poor piemontite, epidote, and chlorite characterizes a quartzitic explosion breccia in the Archean migmatites of Labrador, Canada (Taylor and Baer 1973).

## PIEMONTITE COMPOSITION
## AS A FUNCTION OF *P*, *T* AND HOST ROCK COMPOSITION

Compared to the Al-Fe epidotes (see e.g., Grapes and Hoskin 2004), systematic studies of the interplay between mineral composition, host rock composition, and metamorphic conditions are rather rare for piemontite. This is mostly due to the spot character of the occurrence of piemontite bearing lithologies that normally does not allow tracing and studying the evolution of these assemblages along metamorphic arrays in individual metamorphic terranes. The following discussion therefore relies only on the two well-studied examples from the Sanbagawa belt (Japan) and the Attic-Cycladic belt (Greece).

### Sanbagawa belt

Izadyar (2000) and Izadyar et al. (2000, 2003) studied the relationship between chemical variations in piemontite, host rock bulk chemistry, parageneses, and metamorphic conditions at the Asemi-gawa and Besshi areas of the Sanbagawa belt. Samples from Asemi-gawa come from the garnet zone (~0.8–0.9 GPa and 400°C) whereas samples from Besshi come from the slightly higher grade-albite biotite zone (~1.0 GPa and 500°C). At both localities piemontite coexists with quartz, albite, phengite, chlorite, braunite, apatite, hematite, ± garnet, ± talc, and ± crossitic or barrositic amphibole. Based on the presence or absence of talc, the piemontite schists can be subdivided into talc bearing and talc free assemblages at Asemi-gawa as well as at Besshi. The two assemblages are also distinguished by different compositions of the coexisting minerals in both areas. As the bulk chemistry and $f_{O_2}$ of each of the two assemblages are similar at both localities, differences in the composition of piemontite from talc bearing and talc free assemblages, respectively, between Asemi-gawa and Besshi may be attributed to the different metamorphic conditions. Within both assemblages piemontite from Asemi-gawa tends to have higher Mn contents than that from Besshi (Fig.18) and suggests a decreasing Mn content in piemontite with increasing grade (Izadyar 2000). Although the zonation patterns of individual piemontite crystals are generally not very clear and distinct at both localities, two zonation trends may be observed at Asemi-gawa: In the talc bearing assemblage piemontite displays a large inner zone (core) and a narrow outer zone (rim). This rim is enriched in $Fe^{3+}$ and slightly depleted in $Mn^{3+}$ compared to the core (Fig. 18). Piemontite from the talc free assemblage displays a somewhat more complex zonation with a core, an inner rim (mantle), and an outer rim. From core to mantle $Mn^{3+}$ significantly decreases whereas from mantle to outer rim it increases again (Fig.18). With respect to the relationship between Mn content in piemontite and metamorphic grade outlined above, Izadyar (2000) interpreted these zonation trends as reflecting the prograde (core to rim/core to mantle) or the retrograde (mantle to outer rim) path, respectively.

## Attic-Cycladic belt

Reinecke (1986a,b) provides detailed descriptions of the low to medium grade evolution of piemontite bearing silicate parageneses from highly oxidized Mn rich rocks in response to variable metamorphic grade and physicochemical conditions from Evvia and Andros Islands (Greece). Both areas belong to the Attic-Cycladic blueschist belt and indicate an early (Eocene)

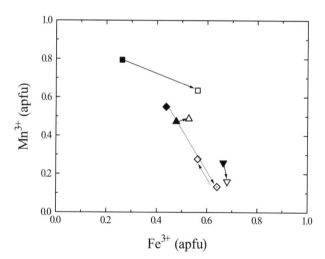

**Figure 18.** $Mn^{3+}$ *vs.* $Fe^{3+}$ contents in piemontite from Sanbagawa quartz schists (central Shikoku, Japan; Izadyar (2000). Solid arrows indicate zonation trend from inner (= core = solid symbols) to outer zone (= rim = empty symbols). Dashed arrows indicate zonation trend from core to intermediate zone (= mantle = dotted symbol) and from mantle to rim, respectively (see text). Symbols refer to different assemblages and localities: squares = talc bearing assemblage, Asemi-gawa; diamonds = talc free assemblage, Asemi-gawa; upward triangles = talc bearing assemblage, Besshi; downward triangles = talc free assemblage, Besshi.

high *P*/low *T* blueschist facies metamorphism that was followed and variably overprinted by a later (Oligocene/Miocene) greenschist facies metamorphism. The corresponding metamorphic conditions were ~400°C and >0.8–0.9 GPa and 400–500°C and >1.0 GPa for the blueschist and <400°C and <0.5–0.8 GPa and 400–500°C and 0.5–0.6 GPa for the greenschist stage on Evvia and Andros, respectively. Thus, Andros reports slightly higher metamorphic conditions. Piemontite occurs in quartzite (Evvia and Andros) and in metapelitic and metacalcareous schists (Andros).

The characteristic assemblages in piemontite bearing quartzite are piemontite, quartz, sursassite, braunite and/or hematite, and Mg chlorite on Evvia and piemontite, quartz, spessartine, braunite and/or hematite, chlorite, and rutile with minor amounts of phengite, crossite, ardennite, albite, tourmaline, apatite, and rare titanite and phlogopite on Andros. The main mineralogical difference between both areas is the ubiquitous occurrence of sursassite on Evvia and of spessartine on Andros. On Evvia spessartine is very rare and replaces sursassite whereas on Andros sursassite only occurs as a relic phase in some of the samples. Compositional zoning is generally absent. In both areas, piemontite in braunite bearing, hematite free assemblages show higher $Mn^{3+}/Fe^{3+}$ ratios (from 2.1–8.3) than in hematite bearing assemblages (from 0.7–2.2). This is the opposite trend to that observed by Jiménez-Millán and Velilla (1993) in rocks from Ossa-Morena but is in accordance with the finding of Fukuoka et al. (1990) in the calc-silicate rocks from the Sausar Group, India (see above). Significant amounts of Sr (up to 0.17 pfu) and $Mn^{2+}$ (up to 0.28 pfu) can substitute for Ca in piemontite from both localities. Nevertheless, $Mn^{2+}$ is systematically higher in the spessartine bearing assemblages of Andros. The higher $Mn^{2+}$ content on Andros is probably governed by a Ca–$Mn^{2+}$ exchange reaction between piemontite and spessartine as it corresponds to Ca enriched rims in spessartine.

Contrary to piemontite in quartzite, piemontite in metapelitic and metacalcareous schist from Andros may show complex compositional zoning. Based on textural and chemical criteria piemontite generations I and II can be distinguished. Piemontite I, formed during the blueschist facies metamorphism, coexists with braunite, manganian phengite (alurgite), $Mn^{3+}$- $Mn^{2+}$ bearing Na pyroxene ('violan'), carbonate, quartz, hollandite, and hematite, and often shows a continuous zonation trend. On the contrary, piemontite II grew during the greenschist facies overprint as narrow, chemically discontinuous rims on piemontite I or as small separate grains and may coexist with $Mn^{3+}$ bearing zoisite ('thulite') in braunite and hematite free assemblages. Like in piemontite from the quartzite on Evvia and Andros, the $Fe^{3+}$ and $Mn^{3+}$ contents of piemontite I largely depend on the presence or absence of hematite and/or braunite and reflect slightly different bulk compositions. Piemontite I from assemblages with hematite but without braunite has ~ 0.63 $Fe^{3+}$ pfu and ~ 0.26 $Mn^{3+}$ pfu, that from braunite bearing assemblages without hematite ~ 0.05 $Fe^{3+}$ pfu and ~ 0.78 $Mn^{3+}$ pfu, and piemontite I from assemblages with hematite and braunite has ~ 0.30 $Fe^{3+}$ pfu and ~ 0.61 $Mn^{3+}$ pfu. Independent of assemblage and absolute element concentrations, the continuous zonation trend of piemontite I, when observed, generally shows an enrichment of $Fe^{3+}$ and Al over $Mn^{3+}$ towards the rims. Reinecke (1986b) interpreted this trend as reflecting continuous equilibration between piemontite I and coexisting minerals on the prograde path of the blueschist facies event. Such a decrease of $Mn^{3+}$ with increasing metamorphic grade is in good accordance with the results of Izadyar (2000) from the Sanbagawa belt (see above). Piemontite II, formed during the greenschist facies overprint, normally occurs as discontinuous rims on piemontite I but also as small discrete grains. In 'thulite' free, hematite and/or braunite bearing assemblages piemontite II is generally Al poorer but $Mn^{3+}$ and $Fe^{3+}$ richer compared to the outer parts of piemontite I on which it grows; furthermore, its $Fe^{3+}/Mn^{3+}$ ratio does not show any correlation with the $Fe^{3+}/Mn^{3+}$ ratio of piemontite I and may vary considerably. Additionally, piemontite II often shows high Sr contents (~ 0.21 pfu) compared to the corresponding piemontite I (~ 0.08 pfu), which can be explained by the transformation of strontian aragonite to strontian poor or free calcite during decompression. Contrary, in 'thulite' bearing, hematite and braunite free assemblages piemontite II is significantly enriched in Al ($\sim$ 2.57 pfu) and depleted in $Mn^{3+}$ ($\sim$ 0.27 pfu) compared to the corresponding piemontite I ($\sim$ 2.28 Al and 0.73 $Mn^{3+}$ pfu). The difference in Sr content between piemontite II and I is not so distinct as in the 'thulite' free assemblages but follows the same trend. In both assemblages piemontite II formed during the decompression from blueschist to greenschist facies conditions due to the breakdown of manganian omphacite. This breakdown reaction may also involve braunite and/ or hematite as educt and tremolite, chlorite, albite, calcite, and quartz as product phases. Some of the piemontite bearing metapelitic and metacalcareous rocks from Andros show additional spessartine. In these cases coexisting piemontite may contain considerable amounts of $Mn^{2+}$ (0.05 to 0.25 pfu) substituting for Ca. Textural relations indicate that spessartine formed at a late stage of the greenschist facies event at the expense of piemontite and probably braunite and other minor phases. Possible spessartine forming and piemontite consuming reactions include

$$\text{piemontite (I,II)} + \text{braunite} + CO_2 =$$
$$\text{spessartine} + \text{calcite} + \text{quartz} \pm \text{hematite} + H_2O + O_2$$

in calcite bearing assemblages and

$$\text{piemontite (I,II)} + \text{braunite} + \text{rutile} + \text{quartz} =$$
$$\text{spessartine} + \text{titanite} + \text{hematite} + H_2O + O_2$$

in calcite free assemblages. As the textures of coexisting piemontite and braunite do not suggest any instability between these two phases and spessartine occurs only in some samples, the breakdown of piemontite and formation of spessartine was most probably triggered by the infiltration of localized, non-pervasive $CO_2$-rich fluids rather than by changes in $P/T$ conditions.

## MONOCLINIC – ORTHORHOMBIC RELATIONS
## IN MANGANIAN EPIDOTE GROUP MINERALS

Although this chapter is primarily concerned with the monoclinic Mn members of the epidote group, we will shortly address the problem of the monoclinic to orthorhombic transition in Mn bearing systems. Natural coexisting orthorhombic ('thulite') and monoclinic Mn bearing epidote minerals are only rarely reported. This might be partly due to the fact that especially in older studies pink to purplish colored epidote minerals were generally referred to as 'thulite' without further specification of symmetry. Abrecht (1981) showed by means of X-ray powder diffraction that 'thulites' from six out of nine localities reported in the literature are actually monoclinic. It is therefore reasonable to assume that a careful inspection of the different 'thulite' occurrences might show that coexisting orthorhombic and monoclinic Mn bearing epidote minerals are much more widespread than recently known. Assemblages with coexisting Mn zoisite and piemontite are reported from e.g. different localities in Norway like Lom and Lexviken and from Andros Island (Greece). Reinecke (1986b) studied in detail the textural and chemical relationships between coexisting Mn zoisite and piemontite from Andros Island, so that we will mostly refer to his data in the following text. Mn zoisite on Andros Island most probably grew synchronously with piemontite II during the greenschist facies event (see above). It is generally $Fe^{3+}$ poor (< 0.04 $Fe^{3+}$ pfu) and contains less $Mn^{3+}$ (0.09 to 0.22 pfu) than the coexisting piemontite (0.18 to 0.31 pfu) although the absolute $Mn^{3+}$ contents vary from sample to sample. The data therefore prove a compositional gap or transition loop between Mn zoisite and piemontite comparable to the Al – $Fe^{3+}$ system. The width of this gap however is variable in the different samples from Andros Island and range from $\Delta$ ($Mn^{3+} + Fe^{3+}$) = 0.07 to 0.15 pfu. As the metamorphic history and phase assemblages of the different samples are almost identical, parameters other than $P$ and $T$ must account for this different width of the transition loop. Reinecke (1986b) proposed the continuous Mn zoisite forming reaction

$$\text{piemontite} + CO_2 = \text{Mn zoisite} + \text{calcite} + \text{quartz} + Mn^{2+}Ca_{-1} \text{ [calcite]} + H_2O + O_2$$

to explain his results. According to this reaction, piemontite and coexisting Mn zoisite should become enriched in Al with increasing $f_{CO_2}$ and decreasing $f_{O_2}$ whereas calcite should become enriched in $Mn^{2+}$. In good agreement calcite that coexists with Mn zoisite and piemontite displays increasing $Mn^{2+}$ content with decreasing $Mn^{3+}$ contents in coexisting Mn zoisite and piemontite.

Systematic experimental studies on the crystal chemical relationships between orthorhombic and monoclinic Mn bearing epidote minerals are lacking. Langer et al. (2002) synthezised coexisting Mn zoisite and piemontite at 800°C, 1.5 GPa and $f_{O_2}$ buffered to $Mn_2O_3/MnO_2$ in the Fe free system in a run with a bulk composition of $X_{Mn} = 0.6$ ($X_{Mn} = Mn/(Mn+Al)$). Unfortunately, they only provided the composition of the Mn zoisite. With ~ 0.51 $Mn^{3+}$ pfu this zoisite is significantly richer in $Mn^{3+}$ than the natural samples from Andros Island. If this difference in $Mn^{3+}$ reflects the higher $P$ and $T$ conditions of the experimental study or different $f_{O_2}$ is not clear. Nevertheless, this discrepancy clearly shows that further studies are necessary to establish the relationships between orthorhombic and monoclinic Mn bearing epidote minerals.

## EXPERIMENTAL STUDIES

The first attempts to synthesize piemontite are from Strens (1964) who used a glass as starting material that was seeded with epidote crystals. He performed synthesis runs at 550 to 650°C and 0.21 to 0.4 GPa with the $f_{O_2}$ internally buffered to $Mn_2O_3/MnO_2$. He obtained piemontite crystals that contain up to 1.2 ($Mn^{3+}$+ $Fe^{3+}$) pfu and found piemontite to be the stable phase throughout the $P$-$T$ region investigated. For this reason, he argued that the progressive disappearance of piemontite towards higher-temperature metamorphic conditions normally found in natural rocks most probably reflects a decrease of $f_{O_2}$ with the increase of temperature, rather than temperature conditions outside the stability field of piemontite.

To study the relationships between chemical composition, stability, and physical properties of piemontite Anastasiou and Langer (1976, 1977) performed experiments in the pure $CaO$-$Al_2O_3$-$Mn_2O_3$-$SiO_2$-$H_2O$ system. Synthesis runs were carried out at 800°C and 1.5 GPa with $f_{O_2}$ buffered to $Mn_2O_3/MnO_2$ in order to prevent an incorporation of $Mn^{2+}$ to replace for Ca. Starting material consisted of calcium-silicate glass, $Mn_2O_3$, and $\gamma$-$Al_2O_3$ powdered and mixed in stoichiometric amounts according to the piemontite formula $Ca_2(Al_{3-x}Mn^{3+}_x)Si_3O_{11}O(OH)$. Bulk compositions ranged from x = 0.25 to x = 3.0. $Mn^{3+}$ bearing zoisite formed at x = 0.25 whereas piemontite was the main phase in runs with starting compositions x = 0.50 to 1.75. In runs with x ≥ 2.00 piemontite was found to coexist with braunite and minor amounts of pyrolusite, wollastonite, quartz, and a phase with the formula $Ca_3 Mn^{3+}_2(Si_2O_7)_2$ that probably formed during quenching (Anastasiou and Langer 1976). On the basis of discontinuities observed in the lattice parameters of piemontite from runs with x = 1.75 and 2.00 Anastasiou and Langer (1977) supposed an upper limit for the piemontite solid solution series of about 1.9 $Mn^{3+}$ pfu at the $P$-$T$-$f_{O_2}$ conditions of their experiments. According to them the high temperature and pressure as well as the very high $f_{O_2}$ were probably responsible for such a high $Mn^{3+}$ content in synthetic piemontite, noticeably higher than that of natural piemontite. Unfortunately, the Mn content in run products was not determined directly, but inferred indirectly from the variations of physical and optical properties as well as the relative amounts of run products. Thus, with respect to the inferred chemical compositions, their results are at least questionable.

More recently, Langer et al. (2002) carried out a new set of synthesis experiments under the same physical conditions of 1.5 GPa, 800°C, $f_{O_2}$ = $Mn_2O_3/MnO_2$ buffer). The synthesized crystals reached a size that allowed a complete characterization of the run products by microchemical, structural, and spectroscopic methods. For bulk compositions with x $_{Mn}^{3+}$ from 0.60 to 2.00, the $Mn^{3+}$ content in piemontite ranges from 0.83 to 1.47 pfu, thus lowering the upper substitutional limit of the piemontite solid solution series to ~ 1.5 $Mn^{3+}$ pfu which is more realistic when compared to natural samples.

Keskinen and Liou (1979) studied the upper thermal stability of piemontite with the composition $Ca_2Al_2Mn^{3+}Si_3O_{11}O(OH)$. The breakdown reaction of piemontite can be represented by the redox dehydration reaction

$$Ca_2Al_2 Mn^{3+}Si_3O_{11}O(OH) = Ca_2Mn^{2+}Al_2Si_3O_{12} + 0.5\ H_2O + 0.25\ O_2$$

It is evident that this reaction is strongly affected by $f_{O_2}$ and that a high $f_{O_2}$ should enhance the stability field of piemontite. To keep the experimental conditions as close as possible to those existing in natural metamorphic environments and to study the influence of different $f_{O_2}$, Keskinen and Liou (1979) carried out synthesis as well as reversal experiments at 0.1 to 0.82 GPa, 212 to 764°C and $f_{O_2}$ defined by the hematite-magnetite (HM), cuprite-tenorite (CT), and copper-cuprite (CC) buffers, respectively. The starting material for the synthesis runs was an oxide-carbonate mixture in the stoichiometric proportions of piemontite composition $Ca_2Al_2Mn^{3+}Si_3O_{11}O(OH)$, tempered at 900°C to break down the carbonates. The starting material for the reversal runs was a mixture of synthetic piemontite and its high temperature equivalents in sub equal proportions. At an $f_{O_2}$ defined by the HM buffer, garnet rather than piemontite was the stable phase under any experimental conditions indicating an upper limit of piemontite thermal stability at 250°C under these conditions. This is in accordance with evidence from natural assemblages in which piemontite rarely, if ever, occurs with magnetite. On the contrary, in the reversal experiments under higher $f_{O_2}$ defined by the CC or CT buffers piemontite was found to be stable up to 402 ± 10°C and 0.1 GPa and 404 ± 10°C and 0.2 GPa under CC buffer conditions and up to 591 ± 10°C and 0.1 GPa and 617 ± 10°C and 0.2 GPa under CT buffer conditions. The data indicate that pressure has only a minor effect on the thermal stability of piemontite compared to $f_{O_2}$. This is in accordance with the natural parageneses that generally indicate that the $f_{O_2}$ is the most important factor to control piemontite stability.

From a comparison with the stability field of Mn free epidotes (Liou 1973) Keskinen and Liou (1979) predicted that incorporation of Fe in piemontite would expand its stability field to lower $f_{O_2}$ and/or higher temperature. Later, Keskinen and Liou (1987) performed additional experiments starting from a bulk composition that corresponds to the intermediate member $Ca_2Al_2Mn^{3+}{}_{0.5}Fe^{3+}{}_{0.5}Si_3O_{11}O(OH)$ of the epidote-piemontite solid solution series. As expected by the authors the incorporation of Fe expands the stability field of piemontite to higher temperatures. At 0.2 GPa the upper thermal stability limit of the investigated piemontite (477 ± 10°C for the $f_{O_2}$ of the CC buffer and 645 ± 10°C for the $f_{O_2}$ of the CT buffer) is slightly higher compared to the upper thermal stability limit of pure piemontite determined in their previous study (404 ± 10°C and 617 ± 10°C, respectively). At the conditions of the hematite-magnetite buffer the data indicate a breakdown of piemontite at 365 ± 10°C compared to < 250°C in the Fe free system (Fig.19). In keeping with field observations, the growth of piemontite at the expense of garnet + anorthite is favored by decreasing temperature and/or increasing $f_{O_2}$. The addition of iron to the system makes the breakdown reaction for the intermediate piemontite more complex than that of pure piemontite. Keskinen and Liou (1987) derived a balanced breakdown reaction of the general form

$$1.6\ Ca_2Al_2Mn^{3+}{}_{0.5}Fe^{3+}{}_{0.5}Si_3O_{11}O(OH) + 0.2\ SiO_2 =$$
$$Ca_{2.2}Mn^{2+}{}_{0.8}Al_{1.2}Fe^{3+}{}_{0.8}Si_3O_{12} + CaAl_2Si_2O_8 + 0.8\ H_2O + 0.2\ O_2$$

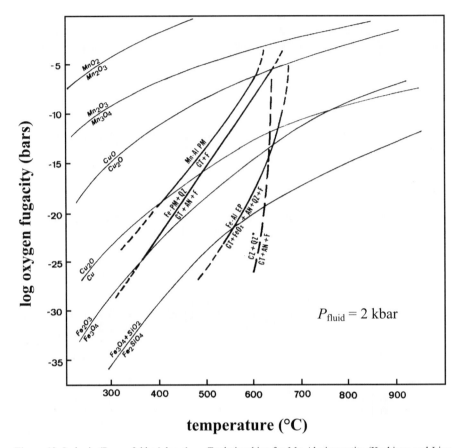

**Figure 19.** Isobaric ($P_{fluid}$ = 2 kbar) log $f_{O_2}$ – $T$ relationships for Mn-Al piemontite (Keskinen and Liou 1979), Fe bearing piemontite (Keskinen and Liou 1987), and Al-Fe epidote (solid line: Liou 1973; dashed line: Holdaway 1966). Modified from Keskinen and Liou (1987).

Their data indicate that this reaction is discontinuous. Unfortunately, due to the very small crystal sizes Keskinen and Liou (1987) were unable to determine the compositions of the participating phases but inferred them from their modal abundances and X-ray data. As piemontite and garnet can show extensive solid solution in terms of their Mn/Fe and (Mn+Fe)/Al ratios, the breakdown of Mn-Fe piemontite might also represent a continuous equilibrium between coexisting piemontite and garnet. As the Mn/Fe and (Mn+Fe)/Al ratios in piemontite and garnet are sensitive to temperature and/or $f_{O_2}$ such a potential continuous breakdown reaction would take place over a narrow $f_{O_2}$-$T$ region wherein the proportion and composition of the reacting phases would show a continuous change. Such a continuous change of piemontite composition would be in good accordance with the continuous or sliding reactions typical for epidote minerals of the Al-Fe solid solution series. Independent of the actual nature of the breakdown reaction the data clearly show that the piemontite breakdown reactions generally have a positive slope in $f_{O_2}$-$T$ space and that the stability field of piemontite is enlarged in the Fe bearing system to higher temperature and lower $f_{O_2}$ (Fig. 19). However, the stability curves of pure piemontite and Mn free epidote converge with the increase of $f_{O_2}$, indicating that at extremely high $f_{O_2}$ the two curves may cross, so that pure $Mn^{3+}$ piemontite might have a higher temperature stability than Mn poor epidote minerals (Fig. 19). Additional studies, both experimental and on natural systems, are necessary to determine the exact breakdown reactions of piemontite as a function of P, T, $f_{O_2}$, and bulk composition.

Piemontite also formed during hydrothermal experiments carried out by Akasaka et al. (2003) in order to synthesize minerals of the pumpellyite-okhotskite series and to investigate their stability relationships. Using a $Ca_4MgMn^{3+}_3Al_2Si_6O_{24.5}$ oxide mixture + excess $H_2O$ as starting material piemontite was synthesized together with Mn pumpellyite in runs at 0.2 GPa and 400°C, 0.3 GPa and 300–400°C and $f_{O_2}$ buffered to $Mn_2O_3/MnO_2$ and $Cu/Cu_2O$. Runs at 0.2 GPa and 300°C ($f_{O_2}$ = $Mn_2O_3/MnO_2$ and $Cu/Cu_2O$) and at 0.3 and 0.4 GPa and 300°C ($f_{O_2}$ = $Cu/Cu_2O$) did not yield any piemontite. In the runs at 0.3 GPa and 300 and 400°C and $f_{O_2}$ buffered to $Mn_2O_3/MnO_2$ and $Cu/Cu_2O$ the modal amount of piemontite increases from 300–400°C whereas the amount of Mn pumpellyite decreases. The data suggest a continuous reaction between Mn pumpellyite and piemontite with a negative slope in P-T space and an increased stability field of piemontite with increasing $f_{O_2}$. Piemontite in the different runs has a composition ranging from $Ca_2Mn^{3+}Al_2Si_3O_{11}O(OH)$ to $Ca_2Mn^{3+}_{1.5}Al_{1.5}Si_3O_{11}O(OH)$, very similar to that found in natural piemontite coexisting with Mn-pumpellyite (Akasaka et al. 1988). To further study the stability relations between piemontite and Mn pumpellyite Akasaka et al. (2003) conducted additional experiments at 0.3 GPa and 250, 300, 400, and 500°C and $f_{O_2}$ buffered to $Mn_2O_3/MnO_2$, $Cu_2O/CuO$, and $Cu/Cu_2O$ using synthetic pure piemontite + excess $H_2O$ as starting material. At 500°C and $f_{O_2}$= $Mn_2O_3/MnO_2$ and $Cu_2O/CuO$ no reaction was observed and piemontite was the only phase stable whereas at T < 500°C piemontite always reacted to form Mn pumpellyite + other phases for all $f_{O_2}$-conditions. The lower T stability at 0.3 GPa of pure Mn piemontite is therefore located between 400 and 500°C. On the basis of the results obtained by Akasaka et al. (2003), the natural assemblages including piemontite and Mn pumpellyite or okhotskite, such as those occurring in the Tokoro manganiferous iron ores (e.g., Togari and Akasaka 1987; Akasaka et al. 1988), can be interpreted as the result of decreasing temperature from above the upper limit of the Mn pumpellyite stability field (crystallization of piemontite alone) through the field of coexistence of piemontite and Mn pumpellyite down to below the lower limit of piemontite stability field (okhotskite).

Sr bearing piemontite was investigated by Akasaka et al. (2000) who performed hydrothermal synthesis experiments at 0.2–0.3 GPa, 500–600°C and $f_{O_2}$ at the $Mn_2O_3/MnO_2$ and $CuO/Cu_2O$ buffer conditions. Following the method of Keskinen and Liou (1979) the starting material was prepared from $CaCO_3$, $SrCO_3$, $Al_2O_3$, $MnO_2$, and $SiO_2$ mixed in appropriate amounts to yield $Ca_{2-x}Sr_xAl_2Mn^{3+}Si_3O_{11}O(OH)$ with $x$ = 0.0, 0.4, 0.8, 1.0, 1.2, and 2.0. In the Sr free system, run products were almost single-phase piemontite. Starting materials with $0 < x \leq 1.2$ yielded Sr bearing piemontite with variable amounts of lawsonite, bixbyte, garnet, and other minor phases depending on the different physical run conditions. The highest Sr content found in piemontite was close to 1.0 pfu even in the runs with $x = 1.2$ suggesting that the upper limit of the Ca-Sr substitution was reached. From the Ca free oxide mixture ($x = 2$) piemontite did not form under any synthesis conditions. By comparing their experimental results with those obtained by Keskinen and Liou (1979) on pure piemontite Akasaka et al. (2000) concluded that the stability of piemontite is not influenced significantly by Sr incorporation.

## CONCLUDING REMARKS

Although piemontite rarely occurs as major rock forming mineral, it is found as a minor phase in a wide variety of rocks formed under different *T*, *P* and *X* conditions. In accordance with the natural parageneses that generally indicate that $f_{O_2}$ is the most important factor to control the piemontite stability, experimental data indicate that pressure has only a minor effect on the thermal stability of piemontite. Both experimental and field evidence clearly indicates that piemontite, although predominantly occurring as a product of low to moderate temperature metamorphism, may persist to temperatures higher than those usually attributed to greenschist facies rocks due to the high $f_{O_2}$ conditions. At $f_{O_2}$ defined by the hematite-magnetite buffer, garnet rather than piemontite is the stable phase at temperatures >250°C. Accordingly, piemontite rarely, if ever, occurs with magnetite in natural assemblages. Mechanisms, which may induce a so high $f_{O_2}$ to stabilize piemontite in the rock, range from external buffering by hydrothermal solutions to internally controlled factors as premetamorphic minerals capable of buffering $f_{O_2}$ to a high level (Keskinen and Liou 1987). Low partial pressure of $CO_2$ in the fluid phase together with high Mn contents in the host rock may extend the stability field of piemontite to higher temperature. In accordance with the wide variability of forming conditions, a wide range of chemical compositions of piemontite has been observed.

The presence of $Fe^{3+}$ substituting $Mn^{3+}$ at the octahedral sites is very common, mainly depending on temperature, bulk-rock $Fe^{3+}$ and $Mn^{3+}$ contents, and $f_{O_2}$. In particular, incorporation of iron in piemontite expands its stability field to lower $f_{O_2}$ and/or higher temperature. Additionally, the presence of Fe in the system makes the breakdown reaction of the intermediate piemontite more complex than that of pure piemontite. Nevertheless, both additional experimental studies and further observation of natural systems are necessary to determine the exact breakdown reactions of piemontite as a function of $P$, $T$, $f_{O_2}$, and bulk composition (Keskinen and Liou, 1987).

In recent years, the presence of trivalent REE substituting for Ca in piemontite has been detected more commonly than expected, thus indicating solid solutions of piemontite towards allanite ($Fe^{2+}$), dissakisite (Mg), and manganoan allanites ($Fe^{2+}$, $Mn^{2+}$) or androsite-(La) ($Mn^{2+}$), the latter case implying also a partial replacement of Ca by $Mn^{2+}$ in the A-sites. The presence of $Mn^{2+}$ replacing Ca is usual in piemontite coexisting with spessartine and significantly depleted in $Mn^{3+}$ (Mottana and Griffin 1982).

In REE free piemontite the presence of detectable amounts of Mg replacing Ca in the A sites seems to be distinctive of high to ultrahigh pressure assemblage (Reinecke 1991).

Appreciable contents of Sr or, to a lesser extent, Pb, indicate solid solution towards strontiopiemontite and tweddillite or hancockite, respectively. So far, no experimental studies of the effect of substitutions of Ca by other cations at the A sites on the stability field of piemontite have been published, except that of Akasaka et al. (2000), who concluded that the stability of piemontite is not influenced significantly by Sr incorporation. The positive correlation between Sr and ($Mn^{3+}$ + $Fe^{3+}$) often observed in piemontite would indicate that significant Sr incorporation in piemontite is favored by the expansion of the unit cell volume due to incorporation of ($Mn^{3+}$ + $Fe^{3+}$) at the octahedral sites (Enami and Banno 2001). Therefore, the content of Sr might be only indirectly related to metamorphic conditions.

Structural adjustments as a response of both homovalent and heterovalent substitutions in the epidote group minerals have been studied in great details. Major difficulties arise in determining cation distribution in piemontites and other $Mn^{3+}$ bearing members, because of the simultaneous presence of $Fe^{3+}$, which exhibits a comparable X-ray scattering power. Few attempts have been made to determine any possible ordering of $Mn^{3+}$ and $Fe^{3+}$ between the M3 and M1 sites in natural piemontites by combining both crystallographic and spectroscopic methods (Dollase 1973; Langer et. al. 2002) or by neutron diffraction techniques (Ferraris et al. 1989). Alternatively, this issue has been approached by studying crystal chemistry and physics of Fe-free synthetic piemontites and clinozoisites (Langer et al. 2002). Nonetheless, the comparable scarcity of data and the high number of variables that can play a role in the degree of ordering (i.e., Fe/Mn ratio, crystallization temperature, additional substitutions in other sites of the structure) show that considerable crystal-chemical research is still necessary to perform a quantitative model for $Mn^{3+}$-$Fe^{3+}$ intracrystalline distribution in piemontite. Last but not least, the chemically controlled, monoclinic to orthorhombic phase transition in the $Mn^{3+}$ bearing systems requires further characterization of natural and synthetic systems wherein the two phases coexist.

## ACKNOWLEDGMENTS

The authors are deeply indebted to the editors for the substantial contribution to the paragraphs "Piemontite composition as a function of *P*, *T* and host rock composition" and "Monoclinic-orthorhombic relations in manganian epidote group minerals", as well as the general editing of the manuscript. The authors also thank Prof. Dr. Irmgard Abs-Wurmbach, Technische Universität Berlin, for the helpful reviewing of the manuscript.

## REFERENCES

Abraham K, Schreyer W (1975) Minerals of the viridine hornfels from Darmstadt, Germany. Contrib Mineral Petrol 49:1-20
Abraham K, Schreyer W (1976) A talc-phengite assemblage in piemontite schist from brezovica, Serbia, Yugoslavia. J Petrol 17:421-439
Abrecht J (1981) Pink zoisite from the Aar Massif, Switzerland. Mineral Mag 44:45-49
Abs-Wurmbach I, Peters Tj (1999) The Mn-Al-Si-O system: an experimental study of phase relations applied to parageneses in manganese-rich ores and rocks. Eur J Mineral 11:45-68
Acharyya KS, Mukherjee S, Basu A (1990) Manganian andalusite from Manbazar, Purulia District, West Bengal, India. Mineral Mag 54:75-80
Akasaka M, Sakakibara M, Togari K (1987) Sr-piemontite from manganiferous hematite ore deposit, Tokoro belt. Mineral Soc of Japan Trans:96 (in Japanese, English abstr.)
Akasaka M, Sakakibara M, Togari K (1988) Piemontite from the manganiferous hematite ore deposits in the Tokoro Belt, Hokkaido, Japan Mineral Petrol 38:105-116
Akasaka M, Suzuki Y, Watanabe H (2003) Hydrothermal synthesis of pumpellyite-okhotskite series minerals. Mineral Petrol 77:25-37
Akasaka M, Watanabe J, Togari K, Kawamura M (1993) Mineralogy of piemontite-bearing schist in the Yamagami metamorphic rocks of northeastern Abukuma Plateau. Ganko 88:141-156 (in Japanese, English abstr.)
Akasaka M, Zheng Y, Suzuki Y (2000) Maximum strontium content of piemontite formed by hydrothermal synthesis. J Mineral Petrol Sci 95:84-94
Anastasiou P, Langer K (1976) Synthese und Stabilität von Piemontit, $Ca_2Al_{3-p}Mn^{3+}_p(Si_2O_7/SiO_4/O/OH)$. Fortschr Mineral 54:3-4
Anastasiou P, Langer K (1977) Synthesis and physical properties of piemontite $Ca_2Al_{3-p}Mn^{3+}_p(Si_2O_7/SiO_4/O/OH)$. Contrib Mineral Petrol 60:225-245
Anders E, Ebihara M (1982) Solar-system abundances of the elements. Geochim Cosmoch Acta 46: 2363-2380
Armbruster T, Gnos E, Dixon R, Gutzmer J, Hejny C, Döbelin N, Medenbach O (2002) Manganvesuvianite and tweddillite, two new $Mn^{3+}$—silicate minerals from the Kalahari manganese fields, South Africa. Mineral Mag 66:137–150
Ashley PM (1984) Piemontite-bearing rocks from the Olary District, South Australia. Austral J Earth Sci 31: 203-216
Balan M, David M (1978) Mineralogical data on piemontite paragenesis at Dealul Negru (Lotru Mts). Studii si Cercetari de Geologie, Geofizica, Geografie: Geologie 23:229-238 (in Romanian, English abstr.)
Banno S (1964) [Petrologic studies on Sanbagawa crystalline schists in the Bessi-Ino district, Central Sikoku, Japan. J Fac Sci Univ Tokyo, sect II, 15: 203-319]. *Data reported in:* Rock-Forming Minerals. Vol. 1B Disilicates and Ring Silicates.Deer WA, Howie RA, Zussman J (eds) Longman, Harlow, UK 1986
Basta EZ, Kamel OA, Awadallah MF (1981) Petrography of Gabal Dokhan Volcanics, Eastern Desert, Egypt. Egyptian J Geol 22:145-171
Basta EZ, Kotb H, Awadallah MF (1980) Petrochemical and geochemical characteristics of the Dokhan formation at the type locality, Jabal Dokhan, Eastern Desert, Egypt. I.A.G. Bull 3:121-140

Basu N, Mruma AH (1985) Mineral chemistry and stability relations of talc-piemontite-viridine bearing quartzite of Mautia Hill, Mpwapwa district, Tanzania. Indian J Earth Sci 12:223-230

Battaglia S, Nannoni R, Orlandi P (1977) La piemontite del Monte Corchia (Alpi Apuane), Atti Soc Tosc Sci Nat A84:174-178

Bermanec V, Armbruster T, Oberhansli R, Zebec V (1994) Crystal chemistry of Pb- and REE-rich piemontite from Nezilovo, Macedonia. Schweiz Mineral Petrog Mitt 74:321-328

Bilgrami SA (1956) Manganese silicate minerals from Chikla, Bhandara district, India. Mineral Mag 31:236-244

Bird DK, Helgeson HC (1980) Chemical interaction of acqueous solutions with epidote-feldspar mineral assemblages in geological system, I: thermodynamic analysis of phase relations in the system CaO-FeO-$Fe_2O_3$-$Al_2O_3$-$SiO_2$-$H_2O$-$CO_2$. Amer J Sciences 280:907-941

Bonazzi P (1990) Nuovi dati chimici e cristallografico-strutturali ricavati da campioni di clinozoisite-epidoto e piemontite: considerazioni inerenti alla cristallochimica del gruppo. Tesi di Dottorato, Dipartimento di Scienze della Terra, Università di Firenze 321 pp

Bonazzi P, Garbarino C, Menchetti S (1992) Crystal chemistry of piemontites: REE-bearing piemontite from Monte Brugiana, Alpi Apuane, Italy. Eur J Mineral 4:23-33

Bonazzi P, Menchetti S (1994) Structural variations induced by heat treatment in allanite and REE-bearing piemontite. Amer Mineral 79:1176-1184

Bonazzi P, Menchetti S (1995) Monoclinic members of the epidote group: effects of the Al $\leftrightarrow$ $Fe^{3+}$ $\leftrightarrow$ $Fe^{2+}$ substitution and of the enty of $REE^{3+}$. Mineral Petrol 53:133-153

Bonazzi P, Menchetti S, Palenzona A (1990) Strontiopiemontite, a new member of the epidote group, from Val Graveglia, Liguria, Italy. Eur J Mineral 2:519-523

Bonazzi P, Menchetti S, Reinecke T (1996) Solid solution between piemontite and androsite-(La), a new mineral of the epidote group from Andros Island, Greece. Amer Mineral 81:735-742

Brown P, Essene EJ, Peacor DR (1978) The mineralogy and petrology of manganese-rich rocks from St. Marcel, Piedmont, Italy. Contrib Mineral Petrol 67:227-232

Burns RG (1993) Mineralogical Applications of Crystal Field Theory. II edition. University Press, Cambridge, UK

Burns RG, Strens RGJ (1967) Structural interpretation of polarized absorption spectra of the Al-Fe-Mn-Cr epidotes. Mineral Mag 36:204-226

Carbonin S, Molin G (1980) Crystal-chemical considerations on eitgh metamorphic epidotes. N Jb Miner Mh 205-215

Carlson WL, Thorne B (1997) Applied Statistical Methods. Prentice Hall, Upple Saddle River. New Jersey

Caron JM, Kienast JR, Triboulet C (1981) High-pressure-low-teperature metamorphism and polyphase Alpine deformation at Sant'Andrea di Cotone (Eastern Corsica, France). Tectonophysics 78:419-451

Catti M, Ferraris G, Ivaldi G (1988) Thermal behaviour of the crystal structure of strontian piemontite. Amer Mineral 73:1370-1376

Catti M, Ferraris G, Ivaldi G (1989) On the crystal chemistry of strontian piemontite with some remarks on the nomenclature of the epidote group. N Jb Miner Mh 357-366

Chopin C (1978) Les paragenèses réduites ou oxydées de concentrations manganésifères des « schistes lustrés » de Haute-Maurienne (Alpes françaises). Bull Minéral 101:514-531

Clarke FW (1910) Analyses of rocks and minerals from the Laboratory of the U.S. Geolog Sur Bulletin 419: 272

Coombs DS, Dowse M, Grapes RH, Kawachi Y, Roser B (1985) Geochemistry and origin of piemontite - bearing and associated manganiferous schists from Arrow Junction, western Otago, New Zealand. Chem Geol 48:57-78

Coombs DS, Kawachi Y, Reay A (1993) An occurrence of ardennite in quartz veins in piemontite schist, western Otago, New Zealand. Mineral Petrol 48:295-308

Cooper AF (1971) Piemontite schists from Haast River, New Zealand. Mineral Mag 38:64-71

Cordier L (1803) Analyse du minéral connu sus le nom de Mine de Manganese violet du Piédmont, faite au Labratoire de l'Ecole de Mines. Jour Min 13:135

Cortesogno L, Lucchetti G, Penco AM (1979) Manganese mineralizations in the "Diaspri di M.Alpe" formation of the Ligurian ophiolites: mineralogy and formation. Rend Soc Ital Mineral Petrol 35:151-197

Cortesogno L, Venturelli G (1978) Metamorphic evolution of the Ophiolite sequences and associated sediments in the Northern Apennines-Voltri Group, Italy. IUGS Sci Rep 38:253-360

Dal Piaz GV (1988) Revised setting of the Piedmont zone in the northern Aosta Valley, Western Alps. Ofioliti 13:157-162

Dal Piaz GV, Di Battistini G, Kienast JR, Venturelli G (1979) Manganiferous quartzitic schists of the Piemonte ophiolite nappe in the Valsesia-Valtournanche area (Italian western Alps). Memorie Sci Geol 32:1-24

Deer WA, Howie RA, Zussman J (eds) (1986) Rock-forming minerals. Vol. 1b: Disilicates and ring silicates. 2nd Ed. Longman, Harlow, UK

Della Ventura G, Mottana A, Parodi GC, Griffin WL (1996) FTIR spectroscopy in the OH-stretching region of monoclinic epidotes from Praborna (St. Marcel, Aosta valley, Italy). Eur J Mineral 8:655-665

Dollase WA (1968) Refinement and comparison of the structures of zoisite and clinozoisite. Amer Mineral 53: 1882-1898

Dollase WA (1969) Crystal structure and cation ordering of piemontite. Amer Mineral 54:710-717

Dollase WA (1971) Refinement of the crystal structures of epidote, allanite and hancockite. Amer Mineral 56: 447-464

Dollase W A (1973) Mössbauer spectra and iron distribution in the epidote-group minerals. Z Kristallogr 138: 41-63

Dzhaksybaev ES, Kaun AV, Kovrov LE (1991) [Lead-manganese ores from the Ushkatyn III deposit. Geologiya Rudnykh Mestorozhdenii 33:107-111]. (in Russian; English abstr.)

Enami M (1986) Ardennite in a quartz schist from the Asemi-gawa area in the Sanbagawa metamorphic terrain, central Shikoku, Japan. Mineral J 13:151-160

Enami M, Banno Y (2001) Partitioning of Sr between coexisting minerals of the hollandite- and piemontite -groups in a quartz-rich schist from the Sanbagawa metamorphic belt, Japan. Amer Mineral 86:205-214

Enami M, Liou JG, Mattinson CG (2004) Epidote minerals in high P/T metamorphic terranes: Subduction zone and high- to ultrahigh-pressure metamorphism. Rev Mineral Geochem 56:347-398

Enami M, Wallis S, Banno Y (1994) Paragenesis of sodic pyroxene-bearing quartz schists: implications for the P-T hystory of the Sambagawa belt. Contrib Mineral Petrol 116:182-198

Ernst WG (1964) Petrochemical study of coexisting minerals from low-grade schists, Eastern Japan. Geochim Cosmochim Acta 28:1631-1668

Ernst WG, Seki Y (1967) Petrologic comparison of Franciscan and Sanbagawa metamorphic terranes. Tectonophysics 4:463-478

Ernst WG, Seki Y, Onuki H, Gilbert MC (1970) Comparative study of low-grade metamorphism in the California Coast Ranges and the outer Metamorphic Belt of Japan. Geol Soc of America Memoir 124: 1-276

Exley RA (1980) Microprobe studies of REE-rich accessory minerals: implications for Skye granite petrogenesis ans REE mobility in hydrothermal systems. Earth Planet Sci Lett 48:97-110

Fermor LL (1909) The manganese ore deposits of India. Mem Geol Survey India 37:78-157

Ferraris G, Ivaldi G, Fuess H, Gregson D (1989) Manganese/iron distribution in a strontian piemontite by neutron diffraction. Z Kristallogr 187:145-151

Fleischer M (1985) A summary of the variations in relative abundances of the lanthanide and yttrium in allanites and epidotes. Bull Geol Soc Finl 57:151-155

Franceschelli M, Puxeddu M, Carcangiu G, Gattiglio M, Pannuti F (1996) Breccia-hosted manganese-rich minerals of Alpi Apuane, Italy: A marine, redox-generated deposit. Lithos 37:309-333

Frank E (1983) Alpine metamorphism of calcareous rocks along a cross-section in the Central Alps: occurrence and breakdown of muscovite, margarite and paragonite. Schweiz Mineral Petrogr Mitt 63:37-93

Franz G, Liebscher A (2004) Physical and chemical properties of the epidote minerals–an introduction. Rev Mineral Geochem 56:1-82

Frondel W (1964) Variations of some rare earths in allanite. Amer Mineral 49:1159-1177

Fukuoka M, Mondal A, Guha D, Chattopadhyay G (1990) Petrochemistry of piemontite -bearing assemblages in calc-silicate rocks from Sausar Group, India. J Geol Soc India 36:403-412

Gabe EJ, Portheine JC, Whitlow SH (1973) A reinvestigation of the epidote structure: Confirmation of the iron location. Amer Mineral 58:218-223

Gaudefroy C, Laurent Y, Permingeat F (1965) La piémontite du gîte de manganèse de Tachgagalt (Anti-Atlas) et sa signification génétique. Notes serv géol Maroc 24:99-102

Gelos EM, Spagnuolo JO, Lizasoin GO (1988) Mineralogia y caracterizacion granulometrica de sedimentos actuales de la plataforma Argentina. Rev Asoc Geol Argentina 43:3-79

Gieré R, Sorensen SS (2004) Allanite and other REE-rich epidote-group minerals. Rev Mineral Geochem 56: 431-494

Giuli G, Bonazzi P, Menchetti S (1999) Al-Fe disorder in synthetic epidotes: a single-crystal X-ray diffraction study. Amer Mineral 84:933-936

Grambling JA, Williams ML (1985) The effects of $Fe^{3+}$ and $Mn^{3+}$ on aluminium silicate phase relations in North.central New Mexico, U.S.A. J Petrol 26:324-354

Grapes RH, Hashimoto S (1978) Manganesiferous schists and their origin, Hidaka Mountains, Hokkaido, Japan. Contrib Mineral Petrol 68:23-35

Grapes RH, Hoskin PWO (2004) Epidote group minerals in low–medium pressure metamorphic terranes. Rev Mineral Geochem 56:301-345

Guild FN (1935) Piedmontite in Arizona. Amer Mineral 20:679-692

Gutzmer J, Beukes NJ (1996) Mineral paragenesis of the Kalahari manganese field, South Africa. Ore Geol Rev 11:405-428

Haüy R J (1822) Traité de Mineralogie. 2:575

Hirata D, Yamashita H, Imanaga I, Takahashi H, Kato A (1995) Manganoan grossular-piemontite association in a low grade metamorphic manganese ore from the Dainichi mine, Hadano City, Kanagawa Prefecture, Japan. Mineral J 17:211-218

Hiroi Y, Harada-Kondo H, Ogo Y (1992) Cuprian manganoan phlogopite in highly oxidized Mineoka siliceous schists from Kamogawa, Boso Peninsula, central Japan. Amer Mineral 77:1099-1106

Holdaway MJ (1966) Hydrothermal stability of clinozoisite plus quartz. Amer J Sci 264:643-667

Hutton CO (1938) On the nature of withamite from Glen Coe, Scotland. Mineral Mag 25:119-124

Hutton CO (1940) [Metamorphism in the Lake Wakatipu region, western Otago, New Zealand. Dept Sci Ind Res, New Zealand, Geol Mem 5] (abstr).

Ishibashi K (1969) [On the paragenesis of piemontite and hematite in the Sambagawa cristalline schist. Sci Fac Sci Rept Kyushu Univ, Geol 9:147-158.] (English abstract) Zentralblatt f Mineralogie 1971. I: 145

Ismagilov MI (1976) Pumpyellite and minerals of the epidote group in ores and wall-rock metasomatites of Baimak gold-pyrite deposits (Southern Urals). *In:* Vopr. Mineral. Geokhim Rud Gorn Porod Yuzhn. Pshenichnyi GN (eds) Akad Nauk SSSR, Bashkir Fil, Inst Geol, Ufa, USSR, p 17-23 (in Russian; English abstr.)

Ito T, Morimoto N, Sadanaga R (1954) On the structure of epidote. Acta Cryst 7:53-59

Izadyar J (2000) Chemical composition of piemontites and reaction relations of piemontite and spessartine in piemontite-quartz schists of central Shikoku, Sanbagawa metamorphic belt, Japan. Schweiz Mineral Petrogr Mitt 80:199-211

Izadyar J, Hirajima T, Nakamura D (2000) Talc-phengite-albite assemblage in piemontite -quartz schist of the Sanbagawa metamorphic belt, central Shikoku, Japan. Island Arc 9:145-158

Izadyar J, Tomita K, Shinjoe H (2003) Geochemistry and origin of piemontite-quartz schist in the Sanbagawa Metamorphic Belt, central Shikoku, Japan. J Asian Earth Sci 21:711-713

Jan MQ, Symes RF (1977) Piemontite schists from Upper Swat, north-west Pakistan. Mineral Mag 41:537-540

Jiménez-Millán J, Velilla N (1993) Compositional variation of piemontite s from different manganese-rich rock-types of the Iberian Massif (SW Spain). Eur J Mineral 5:961-970

Kato A, Matsubara S (1986) Strontian piemontite. 1986 Joint Annual Meeting of The Japanese Association of Mineralogy, Petrology and Economic Geology, Mineralogical Society of Japan, and Society of Resource Geology. 69. (In Japanese).

Kato A, Matsubara S, Tiba T (1982) A pale colored aegirine from the Kamisugai Mine, Ehime Prefecture, Japan. Bull Nat Sci Museum, Series C: Geology Paleontol 8:37-42

Kawachi Y, Grapes RH, Coombs DS, Dowse M (1983) Mineralogy and petrology of a piemontite -bearing schist, western Otago, New Zealand. J Metamorphic Geol 1:353-372

Keskinen M (1981) Petrochemical investigation of the Shadow Lake Piemontite Zone, eastern Sierra Nevada, California. Amer J Science 281:896-921

Keskinen M, Liou JG (1979) Synthesis and stability relations of Mn-Al piemontite, $Ca_2MnAl_2Si_3O_{12}(OH)$. Amer Mineral 64:317-328

Keskinen M, Liou JG (1987) Stability relations of manganese-iron-aluminum piemontite. J Metamorphic Geol 5: 495-507

Kozyreva IV, Shvetsova IV, Ketris MP (2001) Find of Mn-scorodite in schist of the circum-Polar Urals. Dokl Akademii Nauk 376:224-228

Kramm U (1979) Kanonaite-rich viridines from the Venn-Stavelot Massif, Belgian Ardennes. Contrib Mineral Petrol 69:387-395

Krizhevskikh Yu.G (1995) Jasper belt in the Urals. Izvestiya Vysshikh Uchebnykh Zavedenii, Gornyi Zhurnal 8:88-97 (in Russian; English abstr.)

Krogh EJ (1977) Origin and metamorphism of iron formations and associated rocks, Lofoten-Vesteralen, N Norway. I The Vestpolltind iron-manganese deposit. Lithos 10:243-255

Langer K, Abu-Eid RM (1977) Measurement of the polarized absorption spectra of synthetic transition metal-bearing silicate microcrystals in the spectral range 44000-4000 cm$^{-1}$. Phys Chem Min 1:273-299

Langer K, Abu-Eid RM, Anastasiou P (1976) Absorptionsspektren synthetischer Piemontite in den Bereichen 43000-11000 cm$^{-1}$ (232.6–909.1 nm) und 4000–250 cm$^{-1}$ (2.5–40 μm) Z Kristallogr 144:434-436

Langer K, Raith M (1974) Infrared spectra of Al-Fe(III)-epidotes and zoisites, $Ca_2(Al_{1-p}Fe^{3+}_p)Al_2O(OH)$ $[Si_2O_7][SiO_4]$. Amer Mineral 59:1249-1258

Langer K, Tillmanns M, Kersten M, Almen H, Arni RK (2002) The crystal chemistry of $Mn^{3+}$ in the clino- and orthozoisite structure types, $Ca_2M^{3+}_3[OH|O|SiO_4|Si_2O_7]$: A structural and spectroscopic study of some natural piemontites and "thulites" and their synthetic equivalents. Z Kristallogr 217:563-580

Lausen C (1927) Piedmontite from the Sulphur Spring Valley, Arizona. Amer Mineral 12:283-287

Leblanc M, Lbouabi M (1988) Native silver mineralization along a rodingite tectonic contact between serpentinite and quartz diorite (Bou Azzer, Morocco). Econ Geol 83:1379-1391

Liebscher A (2004) Spectroscopy of epidote minerals. Rev Mineral Geochem. 56:125-170

Liou JG (1973) Synthesis and stability relations of epidote, $Ca_2Al_2FeSi_3O_{12}(OH)$. J Petrol 14:381-413

Litoshko DN, Nikitina VD (1984) Manganese-rich piemontite from metasomatites of the Polar Urals. Trudy Instituta Geologii, Rossiiskaya Akademiya Nauk, Ural'skoe Otdelenie, Komi Nauchnyi Tsentr 46:110-112

Liu H, Xue Y (1999) Sedimentology of Triassic Dounan-type manganese deposits, western margin, Yangtze Platform, China. Ore Geol Rev 15:165-176

Liu J, Wu Y, Guo L (1998) Eclogites and their country gneisses: studies on inclusions in the accessory minerals of country gneisses. Chinese Sci Bull 43:65-68

Lucchetti G, Cortesogno L, Palenzona A (1988) Low-temperature metamorphic mineral assemblages in manganese-iron ores from Cerchiara Mine (northern Apennine, Italy). N Jb Miner Mh 367-383

Lyashkevich ZM (1958). On piemontite from the Alai Range. Min Mag Lvov Geol Soc 12:323-331 (Mineral Abstracts, 15, 144)

Makanjuola AA, Howie RA (1972) Mineralogy of the glaucophane schists and associated rocks from Ile de Groix, Brittany, France. Contrib Mineral Petrol 35:83-118

Malmqvist D (1929) [Studien innerhalb der Epidotgruppe mit besonderer Rücksicht auf die manganhältigen Glieder. Bull Geol Inst 22:223-280]. *In:* Rock-Forming Minerals. Vol. 1B Disilicates and Ring Silicates. Deer WA, Howie RA, Zussman J (eds) Longman, Harlow, UK

Mandarino JA (1981) The Glastone-Dale relationship: Part IV. The compatibility concept and its application. Canad Mineral 19:441-450

Marmo V, Neuvonen KJ, Ojanpera P (1959) The piedmontites of Piedmont (Italy), Kajlidongri (India), and Marampa Sierra Leone). Bull Comm géol Finlande 184:11-20

Marthaler M, Stampfli GM (1989) Les schistes lustrés à ophiolites de la nappe du Tsaté: un ancien prisme d'accrétion issu de la marge active apulienne? Schweiz Mineral Petrogr Mitt 69:211-216

Martin S and Kienast JR (1987) The HP-LT manganiferous quartzites of Praborna, Piemonte ophiolite nappe, Italian western Alps. Schweiz Mineral Petrogr Mitt 67: 339-360

Masuda T, Shibutani T, Kuriyama M, Igarashi T (1990) Development of microboudinage: an estimate of changing differential stress with increasing strain. Tectonophysics 178:379-387

Masuda T, Shibutani T, Yamaguchi H (1995) Comparative rheological behaviour of albite and quartz in siliceous schists revealed by the microboudinage of piemontite. J Struct Geol 17:1523-1533

Mayo EB (1932) Two new occurrences of piedmontite in California. Amer Mineral 17:238-248

Meinhold KD, FrischT (1970) Manganese-silicate-bearing metamorphic rocks from central Tanzania. Schweiz Mineral Petrogr Mitt 50:493-507

Minakawa T, Momoi H (1987) Occurrences of ardennite and sursassite from the metamorphic manganese deposits in the Sanbagawa belt, Shikoku, Japan. Kobutsugaku Zasshi 18:87-98 (in Japanese, English abstr.)

Minagawa T (1992) Study on characteristic mineral assemblages and formation process of metamorphosed manganese ore deposits in the Sanbagawa belt. Mem.Fac.Sci. Ehime Univ. 1:1-74 (in Japanese, English abstr.)

Mitra FN (1964) Manganese minerals in a calc-alkaline pegmatite vein cutting across the manganese ore band, Sitasaongi Mine, Bhandara district, Maharastra. Indian Mineral 5:1-16

Morton RD, Carter NL (1963) Contributions to the mineralogy of Norway. On the occurrence of Mn-poor piemontite and withamite in Norway. Norsk Geol Tidsskr 43: 445-455 (English abstr.)

Mottana A (1986) Blueschist-facies metamorphism of manganiferous cherts: A review of the alpine occurrence. *In:* Blueschists and Eclogites. Evans, BW, Brown EH (eds) Mem Geol Soc America 164:267-300

Mottana A, Griffin WL (1982) The crystallochemistry of piemontite from the type-locality (St. Marcel, val d'Aosta, Italy) Reports 13 IMA Meeting:635-640

Nayak VK (1969) Piemontite from the manganese ore deposit of Kajlidongri mine, Jhabua district, Madhya Pradesh. Indian Mineral 10:174-180

Nayak VK Neuvonen KJ (1966) Some manganese minerals from India. Bull Comm Géol Finlande 33:27-37

Ödman OH (1950) [Manganese mineralization in the Ultevis district, Jokkmokk, north Sweden: Part 2. Mineralogical notes. Arsbok Sveriges Geol Undersok 44: n°2]. (English abstract) Mineralogical Abstracts 11:472-473

Ogo Y, Hiroi Y (1991) Origin of various mineral assemblages of the Mineoka Metamorphic rocks from Kamogawa, Boso Peninsula, central Japan. With special reference to the effect of high oxygen fugacity. Ganko 86:226-240 (in Japanese, English abstr.)

Peacor DR, Dunn PJ (1988) Dollaseite-(Ce) (magnesium orthite redefined): structure refinement and implications for $F^-$ + $M^{2+}$ substitutions in epidote-group minerals. Amer Mineral 73:838-842

Peng M, Zheng C (1984) Stability field of piemontite from Hainan Island, Guangdong Province (China). Dizhi Xuebao, 58:63-72 (in Chinese, English abstr.)

Perseil EA (1987) Particularités des piémontites de Saint-Marcel-Praborna (Italie); spectres I.R. Actes du 112ème Congrès National des Sociétés Savantes. Edition du CTHS, Paris

Perseil EA (1988) Presence of strontium in manganese oxides of the St. Marcel-Praborna-Val d'Aosta ore deposit, Italy. Mineralium Deposita 23:306-308

Perseil EA (1990) Sur la présence du strontium dans les minéralisations manganésifères de Falotta et de Parsettens (Grisons-Suisse) – Evolution des paragenèses. Schweiz Mineral Petrog Mitt 70:315-320

Perseil EA (1991) La presence de Sb-rutile dans les concentrations manganésifères de St. Marcel-Praborna (V. Aoste – Italie). Schweiz Mineral Petrog Mitt 71:341-347

Pletnev PA, Kulikova IM, Spiridonov EM (1999) High-aluminum and high-fluorine titanite in gondites and metavolcanites of prehnite-pumpellyite facies from Uchaly deposit, the Southern Urals. Zapiski Vserossiiskogo Mineralogicheskogo Obshchestva 128:69-71

Pouliot G, Trudel P, Valiquette G, Samson P (1984) Armenite - thulite - albite veins at Remigny, Quebec: the second occurrence of armenite. Canad Mineral 22:453-464

Reinecke T (1986a) Phase relationships of sursassite and other manganese-silicates in highly oxidized low-grade, high-pressure metamorphic rocks from Evvia and Andros Islands, Greece. Contrib Mineral Petrol 94:110-126

Reinecke T (1986b) Crystal chemistry and reaction relations of piemontite s and thulites from highly oxidized low grade metamorphic rocks at Vitali, Andros Island, Greece. Contrib Mineral Petrol 93:56-76

Reinecke T (1991) Very-high-pressure metamorphism and uplift of coesite-bearing metasediments from the Zermatt-Saas zone, Western Alps. Eur J Mineral 3:7-17

Reinecke T, Tillmans E, Bernhardt HJ (1991) Abswurmbachite, $Cu^{2+}Mn^{3+}{}_6$ [$O_8/SiO_4$], a new mineral of the braunite group: natural occurrence, synthesis, and crystal structure. N Jb Miner Abh 163:117-143

Roy SR, Purkait PK (1968) Mineralogy and genesis of the metamorphosed manganese silicate rocks (gondite) of Gowari Wadhona, Madhya Pradesh, India. Contrib Mineral Petrol 20:86-114

Sakai C, Higashino T, Enami M (1984) REE-bearing epidote from Sanbagawa pelitic schist, central Shikoku, Japan. Geochem Journ 18:45-53

Sasaki N, Yano M, Matsuyama F (2002) Ardennite from the Tone mine, Nagasaki Prefecture, Japan. Chigaku Kenkyu 51:67-72 (in Japanese, English abstr.)

Shannon RD (1976) Revised effective ionic radii and systematic studies on interatomic distances in halides and chalcogenides. Acta Cryst A32:751-757

Short AM (1933) A chemical study of piedmontite from Shadow Lake, Madera County, California. Amer Mineral 18:493-500

Schreyer W, Abraham K (1977) Howieite and other high-pressure indicators from the contact aureole of the Brezovica, Yugoslavia, peridotite. N Jb Miner Abh 130:114-133

Schreyer W, Fransolet AM, Abraham K (1986) A miscibility gap in trioctahedral manganese-magnesium-iron chlorites: Evidence from the Lienne Valley manganese deposit, Ardennes, Belgium. Contrib Mineral Petrol 94:333-342

Simonson RR (1935) Piedmontite from Los Angeles County, California. Amer Mineral 20:737-738

Singharajwarapan S, Berry R (2000) Tectonic implications of the Nam Suture Zone and its relationship to the Sukhothai Fold belt, Nothern Thailand. J Asian Earth Sci 18:663-673

Smith D, Albee AL (1967) Petrology of a piemontite-bearing gneiss, San Gorgonio Pass, California. Contrib Mineral Petrol 16:189-203

Smith G, Halenius U, Langer K (1982) Low temperature spectral studies of $Mn^{3+}$-bearing andalusite and epidote type minerals in the range 30000-5000 $cm^{-1}$. Phys Chem Minerals 8:136-142

Somarin AK, Moayyed M (2002) Granite- and gabbrodiorite-associated skarn deposits of NW Iran. Ore Geology Rev 20:127-138

Spisiak J, Hovorka D, Rybka R Turan J (1989) Spessartine and piemontite in Lower Paleozoic metasediments of the Inner West Carpathians. Casopis pro Mineralogii a Geologii 34:17-30

Stensrud HL (1973) Does piemontite represent only greenschist facies metamorphism? Contrib Mineral Petrol 40:79-82

Strens RGJ (1964) Synthesis and properties of piemontite. Nature 201:175-176

Strens RGJ (1966) Properties of the Al-Fe-Mn epidotes. Mineral Mag 35:928-944

Tanaka K, Imai N, Nakamura T (1972) A high magnesian piemontite from the Nagatoro disctrict, Saitama Prefecture, Japan. J Jap Assoc Min Petr Econ Geol 67:117-127

Taylor FC, Baer AJ (1973) Piemontite-bearing explosion breccia in Archean rocks, Labrador, Newfoundland. Geol Surv Canad J Earth Sc 10:1397-1402

Togari K, Akasaka M (1987) Okhotskite, a new mineral, an $Mn^{3+}$-dominant member of the pumpellyite group, from the Kokuriki mine, Hokkaido, Japan. Mineral Mag 51:611-614

Togari K, Akasaka M, Sakakibara M, Watanabe T (1988) Mineralogy of manganesiferous iron ore deposits and chert from the Tokoro belt, Hokkaido. Mininig Geology Special Issue 12:115-126

Velilla N, Jiménez-Millán J (2003) Origin and metamorphic evolution of rocks with braunite and pyrophanite from the Iberian Massif (SW Spain). Mineral Petrol 78:73-91

Voznesenskii SD (1961) Piemontite schists from the left bank of the Khemchik River, Western Tuva. Zapiski Vsesoyuz. Mineral. Obshchestva 90:345-348 (English abstr.)

Wenk HR, Maurizio R (1978) Kutnahorite, a rare manganese mineral from Piz Cam (Bergell Alps). Schweiz Mineral Petrog Mitt 58:97-100

Wieser T (1973) Piemontite and associated minerals from Sudeten Mountains (Poland) and Euboea Island (Greece). Mineralogia Polonica 4:3-19.

Williams GH (1893) Piedmontite and scheelite from the ancient rhyolite of South Mountain, Pennsylvania. Amer J Science 46:50-57

Wood BJ, Strens RGJ (1972) Calculation of crystal field splitting in distorted coordination polyhedra: spectra and thermodynamic properties of minerals. Mineral Mag 38:909-917

Yoshimura T, Momoi H (1964) [Withamite from the Yamanaka mine, Hyogo Prefecture. Sci Rept Kyushu Univ, Geol 6:201-206]. (English abstract) Mineralogical Abstracts 17:299

Yudovich YaE, Kozyreva IV, Ketris MP, Shvetsova IV (2001) Geochemistry of rare earth elements in the zone of interformational contact in the Maldynyrd Range (Subpolar Urals). Geochem Internat 39:1-12

Yudovich YaE, Kozyreva IV, Shvetsova IV, Efanova LI, Filippov VN (2000) Manganiferous REE-bearing nodules in metamorphic schists from Polar Urals. Dokl Earth Sci 371:233-235

Zhou G, Liu YJ, Eide EA, Liou JG, Ernst WJ (1993) High-pressure/low-temperature metamorphism in northern Hubei Province, central China. J Metamorphic Geol 11:561-574

# APPENDIX

## Table A.  Chemical analyses (oxides wt.%) of REE free piemontites from literature.

| Ref. # | 2 | 3 | 4 | 5 | 6 | 7 | 8 | 9 | 10 | 11 |
|---|---|---|---|---|---|---|---|---|---|---|
| $SiO_2$ | 36.60 | 36.63 | 35.57 | 38.64 | 37.54 | 37.16 | 36.82 | 36.08 | 37.28 | 37.85 |
| $TiO_2$ | 0.60 | 0.21 | — | — | 0.54 | 0.04 | 0.07 | — | 0.23 | 0.13 |
| $Al_2O_3$ | 15.24 | 17.21 | 18.27 | 15.03 | 19.80 | 19.96 | 19.17 | 19.19 | 20.53 | 22.19 |
| $Fe_2O_3$ | 7.04 | 6.85 | 7.06 | 8.38 | 10.46 | 6.47 | 8.03 | 3.87 | 10.81 | 12.58 |
| $Mn_2O_3$ | 14.93 | 17.78 | 12.43 | 15.00 | 7.32 | 11.11 | 10.80 | 14.91 | — | — |
| FeO | — | — | — | — | — | — | — | — | — | — |
| MnO | — | — | 2.94 | — | 2.00 | 0.45 | 0.51 | — | 5.89 | 3.22 |
| MgO | 2.92 | 0.85 | 0.96 | — | 0.08 | 0.17 | 0.04 | 0.60 | 0.94 | 0.54 |
| CaO | 17.47 | 18.98 | 19.53 | 22.19 | 20.47 | 22.6 | 22.29 | 22.10 | 22.73 | 21.88 |
| SrO | — | — | — | — | — | — | — | — | — | — |
| $Na_2O$ | 2.41 | — | 1.14 | — | 0.13 | 0.05 | 0.10 | 0.14 | 0.09 | 0.09 |
| $K_2O$ | 0.95 | — | 0.87 | — | 0.01 | — | — | — | 0.11 | 0.10 |
| $H_2O$ | 1.95 | 1.75 | 0.85 | 1.78 | 1.46 | 1.75 | 1.85 | 2.54 | 1.80 | 1.80 |
| CuO | 0.31 | — | — | — | — | 0.01 | 0.04 | — | — | — |
| PbO | — | — | — | — | — | — | 0.01 | — | — | — |
| BaO | — | — | — | — | — | — | — | — | — | — |
| total | 100.42 | 100.26 | 99.62 | 101.02 | 99.81 | 99.77 | 99.74 | 99.43 | 100.45 | 100.42 |

| Ref. # | 12 | 13 | 14 | 15 | 16 | 17 | 18 | 19 | 20 | 21 |
|---|---|---|---|---|---|---|---|---|---|---|
| $SiO_2$ | 37.30 | 36.90 | 38.15 | 36.75 | 35.29 | 36.40 | 35.90 | 36.50 | 37.00 | 37.00 |
| $TiO_2$ | 0.04 | 0.05 | 0.12 | 0.02 | 0.05 | — | — | — | — | — |
| $Al_2O_3$ | 23.70 | 21.80 | 23.56 | 19.59 | 17.51 | 20.00 | 19.30 | 21.10 | 19.80 | 20.30 |
| $Fe_2O_3$ | 12.20 | 12.80 | 8.70 | 8.17 | 6.06 | 4.10 | 7.10 | 5.40 | 6.20 | 7.10 |
| $Mn_2O_3$ | 0.95 | 4.30 | — | 12.55 | 15.42 | 14.50 | 11.20 | 10.50 | 11.80 | 10.60 |
| FeO | — | — | — | — | — | — | — | — | — | — |
| MnO | — | — | 7.24 | — | — | 2.10 | 3.00 | 3.60 | 3.30 | 4.10 |
| MgO | 0.24 | 0.24 | 0.14 | 0.07 | 0.72 | 0.22 | 0.20 | 0.05 | 0.13 | 0.01 |
| CaO | 23.20 | 21.70 | 20.33 | 19.90 | 21.77 | 21.40 | 20.10 | 20.00 | 20.10 | 20.00 |
| SrO | — | — | — | — | — | — | — | — | — | — |
| $Na_2O$ | 0.02 | 0.01 | 0.20 | — | — | — | — | — | — | — |
| $K_2O$ | — | — | 0.13 | 0.13 | 0.52 | — | — | — | — | — |
| $H_2O$ | — | — | 1.68 | — | 2.66 | — | — | — | — | — |
| CuO | — | — | — | — | — | — | — | — | — | — |
| PbO | — | — | — | — | — | — | — | — | — | — |
| BaO | — | — | — | — | — | — | — | — | — | — |
| total | 97.65 | 97.80 | 100.25 | 97.18 | 100.00 | 98.72 | 96.80 | 97.15 | 98.33 | 99.11 |

**Table A (continued). Chemical analyses (oxides wt.%)
of REE free piemontites from literature.**

| Ref. # | 22 | 23 | 24 | 25 | 26 | 27 | 28 | 29 | 30 | 31 |
|---|---|---|---|---|---|---|---|---|---|---|
| $SiO_2$ | 35.90 | 37.50 | 36.60 | 37.20 | 37.50 | 38.20 | 36.60 | 37.00 | 37.00 | 36.80 |
| $TiO_2$ | — | — | 0.03 | 0.03 | — | — | — | — | 0.05 | — |
| $Al_2O_3$ | 19.20 | 22.20 | 19.90 | 23.30 | 23.10 | 24.20 | 20.60 | 20.20 | 19.70 | 19.80 |
| $Fe_2O_3$ | 8.70 | 11.20 | 9.16 | 7.54 | 8.69 | 13.00 | 8.05 | 8.48 | 11.30 | 8.46 |
| $Mn_2O_3$ | 9.70 | 3.50 | 11.70 | 6.50 | 5.24 | 0.91 | 11.20 | 11.30 | 11.50 | 11.90 |
| FeO | — | — | — | — | — | — | — | — | — | — |
| MnO | 3.40 | 0.54 | — | — | — | — | — | — | — | — |
| MgO | 0.08 | 0.03 | — | 0.06 | 0.04 | — | 0.07 | — | — | — |
| CaO | 19.80 | 22.40 | 18.80 | 22.50 | 23.00 | 22.80 | 20.70 | 20.50 | 17.90 | 19.30 |
| SrO | — | — | — | — | — | — | — | — | — | — |
| $Na_2O$ | 0.03 | 0.28 | — | 0.04 | — | 0.03 | — | — | — | — |
| $K_2O$ | 0.09 | 0.04 | — | — | — | — | — | — | — | — |
| $H_2O$ | — | — | — | — | — | — | — | — | — | — |
| CuO | — | — | — | — | — | — | — | — | — | — |
| PbO | — | — | — | — | — | — | — | — | — | — |
| BaO | — | — | — | — | — | — | — | — | — | — |
| total | 96.90 | 97.69 | 96.19 | 97.17 | 97.57 | 99.14 | 97.22 | 97.48 | 97.45 | 96.26 |

| Ref. # | 32 | 33 | 34 | 35 | 36 | 37 | 38 | 39 | 40 | 41 |
|---|---|---|---|---|---|---|---|---|---|---|
| $SiO_2$ | 36.60 | 36.80 | 36.90 | 36.80 | 36.70 | 36.70 | 36.58 | 36.97 | 37.43 | 37.29 |
| $TiO_2$ | — | 0.03 | — | 0.03 | 0.03 | — | — | — | — | — |
| $Al_2O_3$ | 21.20 | 21.30 | 20.70 | 20.90 | 20.90 | 20.70 | 21.75 | 21.25 | 23.41 | 22.41 |
| $Fe_2O_3$ | 10.40 | 9.70 | 9.60 | 8.28 | 7.68 | 7.77 | — | — | — | — |
| $Mn_2O_3$ | 7.30 | 7.80 | 8.16 | 10.70 | 11.10 | 11.40 | — | — | — | — |
| FeO | — | — | — | — | — | — | 3.40 | 5.81 | 8.94 | 9.33 |
| MnO | — | — | — | — | — | — | 10.20 | 10.15 | 4.43 | 4.65 |
| MgO | 0.10 | 0.07 | 0.06 | 0.27 | 0.23 | 0.42 | 0.06 | 0.09 | 0.05 | 0.12 |
| CaO | 20.10 | 20.60 | 20.40 | 19.40 | 19.70 | 19.00 | 22.02 | 21.70 | 21.59 | 21.39 |
| SrO | — | — | — | — | — | — | — | — | — | — |
| $Na_2O$ | 0.04 | — | 0.04 | — | 0.07 | 0.06 | 0.98 | — | — | — |
| $K_2O$ | — | — | — | — | — | — | — | — | — | — |
| $H_2O$ | — | — | — | — | — | — | — | — | — | — |
| CuO | — | — | — | — | — | — | — | — | — | — |
| PbO | — | — | — | — | — | — | — | — | — | — |
| BaO | — | — | — | — | — | — | — | — | — | — |
| total | 95.74 | 96.30 | 95.86 | 96.38 | 96.41 | 96.05 | 94.99 | 95.97 | 95.85 | 95.19 |

## Table A (continued). Chemical analyses (oxides wt.%)
## of REE free piemontites from literature.

| Ref. # | 42 | 44 | 45 | 46 | 47 | 48 | 49 | 50 | 51 | 52 |
|---|---|---|---|---|---|---|---|---|---|---|
| $SiO_2$ | 37.43 | 37.40 | 35.9 | 36.57 | 37.56 | 36.66 | 36.69 | 35.25 | 35.47 | 36.82 |
| $TiO_2$ | 0.10 | 0.03 | 0.24 | 0.10 | 0.02 | 0.06 | 0.06 | — | — | — |
| $Al_2O_3$ | 21.27 | 22.85 | 19.97 | 21.17 | 21.04 | 19.85 | 19.33 | 17.95 | 16.71 | 19.81 |
| $Fe_2O_3$ | 3.80 | 13.60 | 8.53 | 8.17 | 10.81 | 8.64 | 2.47 | 7.72 | 10.08 | 7.86 |
| $Mn_2O_3$ | 11.80 | 1.11 | 9.94 | 8.40 | 5.70 | 8.76 | 18.94 | 14.61 | 12.93 | 10.64 |
| FeO | — | — | — | — | — | — | — | — | — | — |
| MnO | — | 1.22 | 2.42 | 3.16 | 1.82 | 3.55 | — | — | — | — |
| MgO | — | 0.02 | 0.05 | 0.03 | 0.01 | 0.03 | — | — | — | — |
| CaO | 24.75 | 22.69 | 21.23 | 20.68 | 21.21 | 19.83 | 21.44 | 22.26 | 22.79 | 22.64 |
| SrO | — | — | — | — | — | — | — | — | — | — |
| $Na_2O$ | — | 0.05 | 0.02 | 0.05 | 0.03 | 0.01 | — | — | — | — |
| $K_2O$ | — | 0.02 | — | — | 0.01 | — | — | 0.04 | — | — |
| $H_2O$ | 0.92 | — | — | — | — | — | — | — | — | — |
| CuO | — | — | — | — | — | — | — | — | — | — |
| PbO | — | — | — | — | — | — | — | — | — | — |
| BaO | — | — | — | — | — | — | — | — | — | — |
| total | 100.07 | 98.99 | 98.3 | 98.33 | 98.21 | 97.39 | 98.93 | 97.83 | 97.98 | 97.77 |

| Ref. # | 53 | 54 | 55 | 56 | 57 | 58 | 59 | 60 | 61 | 62 |
|---|---|---|---|---|---|---|---|---|---|---|
| $SiO_2$ | 36.44 | 36.95 | 37.93 | 36.93 | 37.01 | 37.16 | 37.72 | 37.54 | 38.5 | 38.80 |
| $TiO_2$ | 0.06 | — | — | — | — | — | — | — | 0.19 | 0.14 |
| $Al_2O_3$ | 18.42 | 23.65 | 24.77 | 20.65 | 20.71 | 22.21 | 24.23 | 22.23 | 26.2 | 23.80 |
| $Fe_2O_3$ | 11.40 | 7.37 | 10.15 | 6.64 | 6.52 | 2.42 | 0.32 | 3.04 | 7.70 | 12.8 |
| $Mn_2O_3$ | 9.63 | 6.29 | 3.35 | 9.36 | 8.73 | 13.91 | 12.74 | 13.51 | 1.86 | 0.21 |
| FeO | — | — | — | — | — | — | — | — | — | — |
| MnO | — | — | — | 3.31 | 3.45 | — | — | — | — | — |
| MgO | 0.05 | — | — | — | — | 0.10 | — | — | 0.10 | 0.10 |
| CaO | 21.66 | 22.58 | 22.94 | 19.47 | 18.33 | 20.99 | 22.39 | 20.60 | 23.60 | 23.40 |
| SrO | — | — | — | 0.20 | 0.52 | — | — | — | — | — |
| $Na_2O$ | 0.06 | — | — | — | — | — | — | — | — | — |
| $K_2O$ | — | — | — | — | — | — | — | — | — | — |
| $H_2O$ | — | — | — | — | — | — | — | — | — | — |
| CuO | — | — | — | — | — | — | — | — | — | — |
| PbO | — | — | — | — | — | — | — | — | — | — |
| BaO | — | — | — | — | — | — | — | — | — | — |
| total | 97.72 | 96.84 | 99.14 | 96.56 | 95.27 | 96.79 | 97.40 | 96.92 | 98.15 | 99.25 |

## Table A (continued).  Chemical analyses (oxides wt.%)
## of REE free piemontites from literature.

| Ref. # | 63 | 64 | 65 | 66 | 67 | 68 | 69 | 70 | 71 | 72 |
|---|---|---|---|---|---|---|---|---|---|---|
| $SiO_2$ | 38.00 | 37.70 | 37.00 | 36.50 | 36.40 | 36.80 | 36.10 | 36.50 | 36.50 | 37.74 |
| $TiO_2$ | 0.10 | — | — | — | — | — | — | — | 0.03 | 0.07 |
| $Al_2O_3$ | 24.00 | 23.00 | 19.50 | 19.60 | 19.80 | 19.40 | 19.60 | 19.00 | 19.60 | 26.73 |
| $Fe_2O_3$ | 7.70 | 3.70 | 8.86 | 7.15 | 4.94 | 6.07 | 8.82 | 9.34 | 8.50 | 7.66 |
| $Mn_2O_3$ | — | — | 11.20 | 12.90 | 14.1 | 13.60 | 11.90 | 11.50 | 11.70 | — |
| FeO | — | — | — | — | — | — | — | — | — | — |
| MnO | 5.90 | 11.60 | — | — | — | — | — | — | — | 3.52 |
| MgO | 0.10 | 0.20 | 0.08 | 0.07 | 0.06 | 0.08 | 0.09 | 0.10 | 0.10 | 0.09 |
| CaO | 22.30 | 22.10 | 20.5 | 19.90 | 24.40 | 20.10 | 19.50 | 19.50 | 19.80 | 22.61 |
| SrO | — | — | — | — | — | — | — | — | — | — |
| $Na_2O$ | — | — | — | — | — | 0.08 | — | 0.06 | — | 0.01 |
| $K_2O$ | — | — | — | — | — | — | — | — | — | — |
| $H_2O$ | — | — | — | — | — | — | — | — | — | — |
| CuO | — | — | — | — | — | — | — | — | — | — |
| PbO | — | — | — | — | — | — | — | — | — | — |
| BaO | — | — | — | — | — | — | — | — | — | — |
| total | 98.10 | 98.30 | 97.14 | 96.12 | 99.70 | 96.13 | 96.01 | 96.00 | 96.23 | 98.43 |

| Ref. # | 73 | 74 | 75 | 76 | 77 | 78 | 79 | 80 | 81 | 82 |
|---|---|---|---|---|---|---|---|---|---|---|
| $SiO_2$ | 37.43 | 37.06 | 37.38 | 38.10 | 38.29 | 38.31 | 37.57 | 37.58 | 38.05 | 35.26 |
| $TiO_2$ | 0.33 | 0.24 | — | — | — | — | — | — | — | 0.12 |
| $Al_2O_3$ | 22.18 | 21.86 | 23.83 | 23.89 | 21.95 | 25.99 | 22.15 | 22.54 | 23.53 | 23.50 |
| $Fe_2O_3$ | 2.57 | 2.21 | 11.08 | 12.97 | 12.9 | 6.72 | 8.19 | 13.36 | 12.49 | 4.65 |
| $Mn_2O_3$ | 13.65 | 15.44 | 3.22 | 2.52 | 2.96 | 3.96 | 7.53 | 2.25 | 1.17 | 12.13 |
| FeO | — | — | — | — | — | — | — | — | — | — |
| MnO | — | — | — | — | — | — | — | — | — | — |
| MgO | 0.11 | 0.05 | 0.09 | 0.20 | 0.09 | 0.20 | 0.08 | 0.08 | 0.05 | 0.21 |
| CaO | 21.86 | 21.73 | 22.38 | 21.91 | 21.89 | 23.27 | 23.14 | 22.10 | 23.19 | 22.73 |
| SrO | — | — | — | — | — | — | — | — | — | — |
| $Na_2O$ | 0.03 | 0.02 | — | — | 0.01 | 0.03 | — | 0.04 | 0.01 | — |
| $K_2O$ | — | — | 0.05 | 0.03 | 0.04 | — | — | 0.05 | 0.04 | — |
| $H_2O$ | — | — | — | — | — | — | — | — | — | 1.37 |
| CuO | — | — | — | — | — | — | — | — | — | — |
| PbO | — | — | — | — | — | — | — | — | — | — |
| BaO | — | — | — | — | — | — | — | — | — | — |
| total | 98.16 | 98.61 | 98.03 | 99.62 | 98.13 | 98.48 | 98.66 | 98.00 | 98.53 | 99.97 |

**Table A (continued). Chemical analyses (oxides wt.%)**
**of REE free piemontites from literature.**

| Ref. # | 83 | 84 | 85 | 86 | 87 | 88 | 89 | 90 | 91 | 92 |
|---|---|---|---|---|---|---|---|---|---|---|
| $SiO_2$ | 36.40 | 35.83 | 34.78 | 37.20 | 37.30 | 36.40 | 36.60 | 36.90 | 36.70 | 37.60 |
| $TiO_2$ | 0.01 | — | — | — | — | — | — | — | — | — |
| $Al_2O_3$ | 18.92 | 17.28 | 16.42 | 21.40 | 22.90 | 18.70 | 18.90 | 20.50 | 18.70 | 24.00 |
| $Fe_2O_3$ | 5.56 | 0.59 | 1.59 | 0.60 | 11.40 | 7.20 | 7.90 | 10.90 | 7.20 | 1.80 |
| $Mn_2O_3$ | 13.24 | 20.43 | 19.21 | 15.30 | 1.90 | 11.80 | 10.20 | 5.20 | 11.70 | 10.00 |
| FeO | — | — | — | — | — | — | — | — | — | — |
| MnO | 2.58 | — | 0.15 | 0.90 | — | — | — | — | — | — |
| MgO | 0.19 | 0.11 | — | — | — | — | — | — | — | — |
| CaO | 20.47 | 22.18 | 17.03 | 22.60 | 23.80 | 21.60 | 22.30 | 22.90 | 22.00 | 21.90 |
| SrO | — | 0.78 | 8.53 | — | — | 2.90 | 0.90 | 0.60 | 2.00 | 2.70 |
| $Na_2O$ | 0.02 | — | — | — | — | — | — | — | — | — |
| $K_2O$ | — | — | — | — | — | — | — | — | — | — |
| $H_2O$ | — | — | — | — | 1.90 | 1.80 | 1.80 | 1.80 | 1.80 | 1.90 |
| CuO | — | — | — | — | — | — | — | — | — | — |
| PbO | — | — | — | — | — | — | — | — | — | — |
| BaO | — | — | — | — | — | — | — | — | — | — |
| total | 97.39 | 97.20 | 97.71 | 98.00 | 99.20 | 100.4 | 98.60 | 98.80 | 100.10 | 99.90 |

| Ref. # | 93 | 94 | 95 | 96 | 97 | 98 | 99 | 101 | 102 | 103 |
|---|---|---|---|---|---|---|---|---|---|---|
| $SiO_2$ | 36.50 | 35.70 | 36.90 | 37.70 | 37.10 | 38.80 | 39.80 | 36.99 | 35.31 | 36.61 |
| $TiO_2$ | — | — | — | — | — | — | — | 0.15 | 0.09 | 0.05 |
| $Al_2O_3$ | 21.30 | 18.80 | 20.10 | 25.00 | 21.00 | 28.90 | 31.50 | 22.04 | 21.32 | 20.85 |
| $Fe_2O_3$ | 1.00 | 7.80 | 11.20 | 0.20 | 0.90 | 0.10 | 0.20 | 0.22 | 4.38 | 10.75 |
| $Mn_2O_3$ | 14.30 | 10.80 | 9.00 | 10.50 | 14.60 | 5.20 | 3.30 | 15.69 | 9.51 | 6.91 |
| FeO | — | — | — | — | — | — | — | — | — | — |
| MnO | — | — | — | — | — | — | — | — | — | — |
| MgO | — | — | — | — | — | — | — | — | 0.08 | 0.15 |
| CaO | 20.20 | 17.90 | 21.30 | 22.70 | 22.20 | 24.10 | 24.70 | 23.09 | 16.00 | 22.10 |
| SrO | 4.40 | 6.00 | — | 1.60 | 1.20 | — | — | — | 10.25 | — |
| $Na_2O$ | — | — | — | — | — | — | — | — | 0.04 | 0.07 |
| $K_2O$ | — | — | — | — | — | — | — | 0.04 | — | 0.02 |
| $H_2O$ | 1.90 | 1.80 | 1.90 | 1.90 | 1.80 | 1.90 | 2.00 | — | — | — |
| CuO | — | — | — | — | — | — | — | — | — | — |
| PbO | — | — | — | — | — | — | — | — | — | — |
| BaO | — | — | — | — | — | — | — | — | — | — |
| total | 99.60 | 98.80 | 100.4 | 99.60 | 98.80 | 99.00 | 101.50 | 98.22 | 96.98 | 97.51 |

**Table A (continued). Chemical analyses (oxides wt.%)
of REE free piemontites from literature.**

| Ref. # | 104 | 105 | 106 | 107 | 108 | 109 | 110 | 111 | 112 | 113 |
|---|---|---|---|---|---|---|---|---|---|---|
| $SiO_2$ | 37.86 | 36.85 | 37.35 | 37.25 | 35.52 | 36.91 | 38.25 | 43.51 | 36.93 | 34.27 |
| $TiO_2$ | 0.09 | 0.29 | 0.05 | 0.06 | 0.42 | 0.17 | 0.09 | 0.05 | 0.03 | — |
| $Al_2O_3$ | 19.18 | 20.04 | 18.25 | 17.58 | 15.37 | 15.77 | 23.24 | 18.55 | 25.06 | 17.56 |
| $Fe_2O_3$ | 4.65 | 1.17 | 2.32 | 7.26 | 5.24 | 11.05 | — | — | — | 4.89 |
| $Mn_2O_3$ | 14.48 | 17.1 | 17.51 | 13.10 | 17.50 | 12.48 | — | — | — | 12.93 |
| FeO | — | — | — | — | — | — | 13.47 | 2.26 | 4.87 | — |
| MnO | — | — | — | — | — | — | 0.38 | 11.67 | 5.74 | 2.87 |
| MgO | 0.19 | 0.22 | 0.34 | 0.20 | 0.80 | 0.07 | 0.14 | 0.27 | 0.15 | — |
| CaO | 22.30 | 22.64 | 22.00 | 22.07 | 20.89 | 22.12 | 22.03 | 19.27 | 23.16 | 11.69 |
| SrO | — | — | — | — | — | — | — | — | — | 13.45 |
| $Na_2O$ | 0.07 | — | — | 0.03 | 0.25 | — | — | — | — | — |
| $K_2O$ | 0.03 | — | — | 0.02 | 0.03 | — | 0.01 | — | — | — |
| $H_2O$ | — | — | — | — | — | — | — | — | — | 1.72 |
| CuO | — | — | — | — | — | — | — | — | — | — |
| PbO | — | — | — | — | — | — | — | — | — | — |
| BaO | — | — | — | — | — | — | — | — | — | — |
| total | 98.85 | 98.31 | 97.82 | 97.57 | 96.02 | 98.57 | 97.61 | 95.58 | 95.94 | 99.38 |

| Ref. # | 114 | 115 | 116 | 117 | 118 | 119 | 120 | 121 | 122 | 123 |
|---|---|---|---|---|---|---|---|---|---|---|
| $SiO_2$ | 36.00 | 34.40 | 35.54 | 35.43 | 35.58 | 35.57 | 36.40 | 36.70 | 36.79 | 36.94 |
| $TiO_2$ | — | — | — | — | — | — | — | — | — | — |
| $Al_2O_3$ | 19.49 | 18.00 | 18.85 | 18.94 | 19.95 | 19.83 | 20.00 | 20.20 | 18.63 | 18.29 |
| $Fe_2O_3$ | 6.78 | 9.01 | 8.96 | 8.88 | 12.75 | 12.97 | 7.40 | 8.50 | 8.34 | 9.65 |
| $Mn_2O_3$ | 10.70 | 9.27 | 8.08 | 8.86 | 4.17 | 4.17 | 11.40 | 11.00 | 10.60 | 10.42 |
| FeO | — | — | — | — | — | — | — | — | — | — |
| MnO | — | — | — | — | — | — | — | — | — | — |
| MgO | — | — | — | — | — | — | 0.66 | 0.06 | — | — |
| CaO | 19.54 | 19.31 | 19.14 | 18.32 | 18.13 | 17.77 | 19.50 | 20.80 | 22.85 | 22.15 |
| SrO | 3.62 | 3.30 | 4.01 | 4.87 | 5.64 | 6.19 | — | — | — | — |
| $Na_2O$ | — | — | — | — | — | — | — | — | — | — |
| $K_2O$ | — | — | — | — | — | — | — | — | — | — |
| $H_2O$ | 3.44 | 3.33 | 3.38 | 3.41 | 3.42 | 3.40 | — | — | — | — |
| CuO | — | — | — | — | — | — | — | — | — | — |
| PbO | — | 2.34 | 1.05 | 1.11 | — | — | — | — | — | — |
| BaO | — | — | — | — | — | — | — | — | — | — |
| total | 99.57 | 98.96 | 99.01 | 99.82 | 99.64 | 99.90 | 95.36 | 97.26 | 97.21 | 97.45 |

**Table A (continued).  Chemical analyses (oxides wt.%)
of REE free piemontites from literature.**

| Ref. # | 124 | 125 | 126 | 127 | 128 | 129 | 130 | 131 | 132 | 133 |
|---|---|---|---|---|---|---|---|---|---|---|
| $SiO_2$ | 37.06 | 37.34 | 36.93 | 36.39 | 37.01 | 36.75 | 36.73 | 36.50 | 36.88 | 37.13 |
| $TiO_2$ | — | — | — | — | — | — | — | — | — | — |
| $Al_2O_3$ | 20.89 | 21.00 | 20.65 | 19.64 | 19.94 | 19.21 | 18.65 | 17.77 | 18.73 | 21.87 |
| $Fe_2O_3$ | 6.98 | 6.62 | 6.64 | 8.18 | 4.52 | 4.51 | 8.15 | 9.33 | 10.78 | 11.94 |
| $Mn_2O_3$ | 9.07 | 9.26 | 9.43 | 9.90 | 12.78 | 14.03 | 11.02 | 11.17 | 8.78 | 2.74 |
| FeO | — | — | — | — | — | — | — | — | — | — |
| MnO | 1.16 | 1.09 | 3.25 | — | — | — | 1.01 | 1.70 | 0.57 | 2.17 |
| MgO | — | — | — | — | — | — | — | — | — | — |
| CaO | 21.41 | 21.27 | 19.47 | 21.47 | 22.98 | 22.52 | 21.52 | 21.13 | 22.26 | 21.53 |
| SrO | 0.30 | 0.25 | 0.20 | 0.93 | 0.33 | 0.71 | 0.18 | 0.17 | 0.23 | 0.20 |
| $Na_2O$ | — | — | — | — | — | — | — | — | — | — |
| $K_2O$ | — | — | — | — | — | — | — | — | — | — |
| $H_2O$ | — | — | — | — | — | — | — | — | — | — |
| CuO | — | — | — | — | — | — | — | — | — | — |
| PbO | — | — | — | — | — | — | — | — | — | — |
| BaO | — | — | — | — | — | — | — | — | — | — |
| total | 96.87 | 96.83 | 96.57 | 96.51 | 97.56 | 97.73 | 97.26 | 97.77 | 98.23 | 97.58 |

| Ref. # | 134 | 135 | 136 | 152 | 153 | 154 | 155 | 156 | 157 | 158 |
|---|---|---|---|---|---|---|---|---|---|---|
| $SiO_2$ | 37.16 | 37.15 | 36.54 | 36.60 | 31.10 | 36.70 | 30.00 | 31.10 | 32.10 | 37.40 |
| $TiO_2$ | — | — | — | 0.02 | 0.48 | — | 0.27 | 0.53 | 0.23 | — |
| $Al_2O_3$ | 21.92 | 21.83 | 20.67 | 19.4 | 8.92 | 19.10 | 6.25 | 8.95 | 11.90 | 19.40 |
| $Fe_2O_3$ | 11.86 | 12.66 | 7.91 | 5.50 | 9.10 | 5.85 | 10.70 | 9.19 | 6.97 | 6.60 |
| $Mn_2O_3$ | 2.85 | 2.39 | 8.44 | 12.60 | 17.5 | 12.80 | 18.70 | 17.40 | 16.90 | 11.80 |
| FeO | — | — | — | — | — | — | — | — | — | — |
| MnO | 2.54 | 1.90 | 3.41 | — | — | — | — | — | — | — |
| MgO | — | — | — | 0.04 | 0.02 | 0.06 | 0.01 | 0.01 | 0.02 | 0.09 |
| CaO | 20.18 | 21.07 | 19.59 | 22.20 | 10.2 | 22.3 | 10.10 | 10.30 | 11.00 | 22.80 |
| SrO | 0.64 | 0.37 | 0.77 | 0.97 | 8.59 | 1.00 | 9.98 | 8.68 | 13.10 | 0.48 |
| $Na_2O$ | — | — | — | 0.01 | 0.03 | — | 0.01 | — | 0.02 | 0.01 |
| $K_2O$ | — | — | — | — | — | — | — | — | 0.01 | — |
| $H_2O$ | — | — | — | — | — | — | — | — | — | — |
| CuO | — | — | — | — | — | 0.02 | — | — | — | — |
| PbO | — | — | 0.06 | — | 5.44 | — | 7.55 | 5.61 | 2.79 | — |
| BaO | — | — | — | 0.06 | 6.44 | — | 3.66 | 6.71 | 2.71 | 0.01 |
| total | 97.15 | 97.37 | 97.39 | 97.40 | 97.82 | 97.83 | 97.23 | 98.48 | 97.75 | 98.59 |

**Table A (continued). Chemical analyses (oxides wt.%)
of REE free piemontites from literature.**

| Ref. # | 159 | 160 | 161 | 162 | 163 | 164 | 165 | 166 | 167 | 168 |
|---|---|---|---|---|---|---|---|---|---|---|
| $SiO_2$ | 37.20 | 36.90 | 37.74 | 38.55 | 38.12 | 39.00 | 37.61 | 38.43 | 37.50 | 37.82 |
| $TiO_2$ | 0.08 | — | 0.11 | 0.07 | 0.06 | 0.07 | 0.04 | 0.06 | — | — |
| $Al_2O_3$ | 19.20 | 20.00 | 20.25 | 19.07 | 20.59 | 19.51 | 20.30 | 18.81 | 20.87 | 21.59 |
| $Fe_2O_3$ | 7.02 | 10.40 | 2.90 | 9.78 | 3.82 | 9.34 | 6.11 | 9.10 | 11.12 | 10.61 |
| $Mn_2O_3$ | 11.50 | 6.89 | 14.58 | 11.75 | 13.28 | 12.70 | 11.35 | 12.34 | 6.69 | 6.32 |
| FeO | — | — | — | — | — | — | — | — | — | — |
| MnO | — | — | — | — | — | — | — | — | — | — |
| MgO | 0.01 | 0.07 | — | — | — | — | — | — | — | — |
| CaO | 23.00 | 22.50 | 22.99 | 19.81 | 22.48 | 18.98 | 23.04 | 19.64 | 20.29 | 20.63 |
| SrO | 0.22 | 0.17 | — | — | — | — | — | — | — | — |
| $Na_2O$ | — | — | — | — | — | — | — | — | — | — |
| $K_2O$ | — | — | — | — | — | — | — | — | — | — |
| $H_2O$ | — | — | — | — | — | — | — | — | — | — |
| CuO | — | — | — | — | — | — | — | — | — | — |
| PbO | — | — | — | — | — | — | — | — | — | — |
| BaO | — | 0.04 | — | — | — | — | — | — | — | — |
| total | 98.23 | 96.97 | 98.57 | 99.03 | 98.35 | 99.60 | 98.45 | 98.38 | 96.47 | 96.97 |

| Ref. # | 169 | 170 | 171 | 172 | 173 | 174 | 175 | 176 | 177 | 178 |
|---|---|---|---|---|---|---|---|---|---|---|
| $SiO_2$ | 37.89 | 37.89 | 37.58 | 37.47 | 38.21 | 37.89 | 39.96 | 38.09 | 38.28 | 38.65 |
| $TiO_2$ | — | — | — | — | — | 0.08 | — | — | — | 0.05 |
| $Al_2O_3$ | 21.25 | 21.19 | 21.86 | 20.15 | 20.04 | 22.61 | 22.11 | 21.32 | 22.86 | 22.00 |
| $Fe_2O_3$ | 8.89 | 9.31 | 4.60 | 7.17 | 7.59 | 10.88 | 8.49 | 6.52 | 10.34 | 9.63 |
| $Mn_2O_3$ | 7.03 | 7.42 | 10.42 | 10.93 | 9.44 | 2.04 | 5.38 | 8.90 | 2.52 | 4.68 |
| FeO | — | — | — | — | — | — | — | — | — | — |
| MnO | — | — | — | — | — | — | — | — | — | — |
| MgO | — | — | — | — | — | — | — | — | — | — |
| CaO | 21.30 | 20.85 | 21.98 | 20.95 | 21.27 | 23.05 | 22.25 | 22.47 | 23.01 | 22.83 |
| SrO | — | — | — | — | — | — | — | — | — | — |
| $Na_2O$ | — | — | — | — | — | — | — | — | — | — |
| $K_2O$ | — | — | — | — | — | — | — | — | — | — |
| $H_2O$ | — | — | — | — | — | — | — | — | — | — |
| CuO | — | — | — | — | — | — | — | — | — | — |
| PbO | — | — | — | — | — | — | — | — | — | — |
| BaO | — | — | — | — | — | — | — | — | — | — |
| total | 96.36 | 96.66 | 96.44 | 96.67 | 96.55 | 96.55 | 98.19 | 97.30 | 97.01 | 97.84 |

## Table A (continued). Chemical analyses (oxides wt.%) of REE free piemontites from literature.

| Ref. # | 179 | 180 | 181 | 182 | 183 | 184 | 185 | 186 | 187 | 188 |
|---|---|---|---|---|---|---|---|---|---|---|
| $SiO_2$ | 38.38 | 38.48 | 39.24 | 38.35 | 38.20 | 37.79 | 38.33 | 39.04 | 40.18 | 38.58 |
| $TiO_2$ | 0.04 | — | 0.03 | — | — | — | — | — | — | — |
| $Al_2O_3$ | 20.97 | 22.64 | 22.46 | 22.22 | 23.02 | 21.20 | 22.97 | 22.23 | 21.32 | 22.67 |
| $Fe_2O_3$ | 7.52 | 10.77 | 10.57 | 11.72 | 12.41 | 10.77 | 11.76 | 12.31 | 10.63 | 10.54 |
| $Mn_2O_3$ | 8.58 | 2.23 | 3.87 | 3.24 | 2.16 | 5.01 | 1.32 | 4.41 | 6.45 | 4.07 |
| FeO | — | — | — | — | — | — | — | — | — | — |
| MnO | — | — | — | — | — | — | — | — | — | — |
| MgO | — | — | — | — | — | — | — | — | — | — |
| CaO | 22.18 | 23.14 | 22.89 | 21.89 | 21.69 | 21.47 | 22.19 | 20.97 | 20.57 | 21.85 |
| SrO | — | — | — | — | — | — | — | — | — | — |
| $Na_2O$ | — | — | — | — | — | — | — | — | — | — |
| $K_2O$ | — | — | — | — | — | — | — | — | — | — |
| $H_2O$ | — | — | — | — | — | — | — | — | — | — |
| CuO | — | — | — | — | — | — | — | — | — | — |
| PbO | — | — | — | — | — | — | — | — | — | — |
| BaO | — | — | — | — | — | — | — | — | — | — |
| total | 97.67 | 97.26 | 99.06 | 97.42 | 97.48 | 96.24 | 96.57 | 98.96 | 99.15 | 97.71 |

| Ref. # | 189 | 190 | 191 | 192 | 193 | 194 | 197 | 198 | 199 | 200 |
|---|---|---|---|---|---|---|---|---|---|---|
| $SiO_2$ | 38.57 | 37.10 | 36.20 | 35.51 | 37.26 | 36.76 | 36.40 | 37.30 | 38.75 | 37.57 |
| $TiO_2$ | — | 0.05 | — | — | — | — | 0.11 | 0.51 | 0.41 | 0.24 |
| $Al_2O_3$ | 23.3 | 19.50 | 19.50 | 12.68 | 20.76 | 21.00 | 15.89 | 20.55 | 24.71 | 24.7 |
| $Fe_2O_3$ | 12.07 | 2.20 | 5.70 | 10.88 | 11.28 | 4.56 | 11.06 | 13.50 | 10.77 | 11.31 |
| $Mn_2O_3$ | 1.40 | 16.00 | 15.30 | 16.83 | 6.21 | 12.08 | 11.99 | — | 0.83 | 0.96 |
| FeO | — | — | — | — | — | — | — | 0.50 | — | — |
| MnO | — | — | — | — | — | — | 0.55 | 2.84 | — | — |
| MgO | — | 0.02 | — | — | 0.34 | — | — | 0.16 | 0.43 | 0.09 |
| CaO | 22.34 | 22.10 | 21.00 | 23.12 | 22.21 | 22.69 | 22.25 | 21.87 | 21.86 | 23.32 |
| SrO | — | — | — | — | — | — | — | — | 0.31 | 0.09 |
| $Na_2O$ | — | — | — | — | — | — | 0.41 | 0.04 | 0.26 | — |
| $K_2O$ | — | — | — | — | — | — | 0.04 | — | 0.11 | — |
| $H_2O$ | — | — | — | 1.56 | 1.88 | 1.92 | 0.93 | 2.09 | 1.93 | 1.83 |
| CuO | — | — | — | — | — | — | — | — | — | — |
| PbO | — | — | — | — | — | — | 0.04 | — | — | — |
| BaO | — | — | — | — | — | — | — | — | — | — |
| total | 97.68 | 96.97 | 97.70 | 100.58 | 99.94 | 99.01 | 99.80 | 99.36 | 100.37 | 100.11 |

**Table A (continued). Chemical analyses (oxides wt.%)
of REE free piemontites from literature.**

| Ref. # | 201 | 202 | 203 | 204 | 205 | 206 | 207 | 208 | 209 |
|---|---|---|---|---|---|---|---|---|---|
| $SiO_2$ | 38.42 | 36.90 | 35.65 | 36.29 | 35.46 | 35.84 | 35.87 | 35.83 | 36.08 |
| $TiO_2$ | — | 0.05 | — | — | — | — | — | — | 0.19 |
| $Al_2O_3$ | 20.53 | 20.30 | 18.47 | 19.20 | 16.13 | 16.47 | 18.00 | 17.28 | 20.86 |
| $Fe_2O_3$ | 8.86 | — | 6.04 | 4.59 | 3.64 | 3.26 | 3.36 | 0.59 | 2.17 |
| $Mn_2O_3$ | 5.31 | — | 13.16 | 14.29 | 18.31 | 18.55 | 16.8 | 20.43 | 13.53 |
| FeO | — | 4.36 | — | — | — | — | — | — | — |
| MnO | 3.60 | 12.65 | — | — | — | — | — | — | — |
| MgO | 0.36 | 0.10 | 0.24 | 0.10 | — | — | — | 0.11 | 0.10 |
| CaO | 20.81 | 22.80 | 19.09 | 21.36 | 18.45 | 18.76 | 19.93 | 22.18 | 19.89 |
| SrO | — | — | 5.32 | 1.89 | 5.93 | 5.86 | 4.26 | 0.78 | 4.98 |
| $Na_2O$ | 0.07 | — | — | — | — | — | — | — | — |
| $K_2O$ | 0.07 | 0.06 | — | — | — | — | — | — | — |
| $H_2O$ | 1.80 | — | — | — | — | — | — | — | — |
| CuO | — | — | — | — | — | — | — | — | — |
| PbO | — | — | — | — | — | — | — | — | — |
| BaO | — | — | — | — | 0.13 | — | — | — | — |
| total | 99.89 | 97.22 | 97.97 | 97.72 | 98.05 | 98.74 | 98.22 | 97.20 | 97.80 |

| Ref. # | 210 | 211 | 212 | 213 | 214 | 215 | 216 | 217 | 218 |
|---|---|---|---|---|---|---|---|---|---|
| $SiO_2$ | 36.20 | 36.71 | 36.55 | 37.97 | 38.94 | 39.27 | 37.93 | 37.39 | 38.02 |
| $TiO_2$ | 0.14 | — | 0.01 | 0.36 | — | — | — | — | 0.24 |
| $Al_2O_3$ | 19.19 | 20.68 | 29.77 | 22.27 | 19.98 | 22.40 | 26.38 | 17.70 | 22.72 |
| $Fe_2O_3$ | 0.90 | 9.98 | 7.01 | 12.27 | 14.49 | 10.30 | 5.17 | 14.20 | 8.16 |
| $Mn_2O_3$ | 18.50 | 7.21 | — | — | 2.16 | — | 3.90 | 4.09 | — |
| FeO | — | — | 0.36 | 0.24 | — | — | 0.26 | — | 0.18 |
| MnO | — | — | 1.08 | 1.90 | — | 3.92 | — | — | 5.13 |
| MgO | — | — | 0.06 | 0.69 | 0.74 | 0.14 | 0.36 | 0.63 | 0.57 |
| CaO | 20.29 | 20.15 | 21.54 | 22.60 | 21.33 | 21.87 | 23.15 | 21.91 | 22.95 |
| SrO | 3.52 | 3.71 | — | — | — | — | — | — | — |
| $Na_2O$ | — | — | 0.41 | 0.09 | 0.36 | — | 0.89 | 0.43 | 0.09 |
| $K_2O$ | — | — | 0.10 | 0.02 | 0.32 | — | 0.34 | 0.65 | 0.07 |
| $H_2O$ | — | — | 2.51 | 1.74 | 1.89 | 2.38 | 2.08 | 2.60 | 1.72 |
| CuO | — | — | — | — | — | — | — | — | — |
| PbO | — | — | — | — | — | — | — | — | — |
| BaO | — | — | — | — | — | — | — | — | — |
| total | 98.74 | 98.44 | 99.40 | 100.15 | 100.21 | 100.41 | 100.46 | 99.60 | 99.85 |

### Table A (continued). Chemical analyses (oxides wt.%) of REE free piemontites from literature.

| Ref. # | 219 | 220 | 221 | 222 | 223 | 224 | 225 |
|---|---|---|---|---|---|---|---|
| $SiO_2$ | 36.19 | 36.82 | 36.56 | 36.55 | 33.79 | 33.47 | 33.22 |
| $TiO_2$ | 0.13 | 0.61 | — | 0.31 | — | — | — |
| $Al_2O_3$ | 18.62 | 20.65 | 20.79 | 12.43 | 16.49 | 16.48 | 17.51 |
| $Fe_2O_3$ | 11.3 | 9.68 | 7.20 | 6.43 | 8.06 | 9.39 | 8.44 |
| $Mn_2O_3$ | 8.98 | 9.99 | 12.77 | 22.00 | 10.98 | 10.52 | 10.22 |
| FeO | — | — | — | — | — | — | — |
| MnO | 0.28 | — | — | — | — | — | — |
| MgO | 0.17 | 0.04 | 3.70 | — | — | — | — |
| CaO | 22.34 | 18.96 | 17.36 | 16.10 | 12.18 | 10.94 | 10.85 |
| SrO | — | 1.23 | — | — | 16.79 | 16.99 | 17.51 |
| $Na_2O$ | — | — | — | 2.59 | — | — | — |
| $K_2O$ | — | — | — | 0.59 | — | — | — |
| $H_2O$ | 2.2 | 2.06 | 2.04 | 3.02 | — | — | — |
| CuO | — | — | — | — | — | — | — |
| PbO | — | — | — | — | — | — | — |
| BaO | 0.01 | — | — | — | 0.60 | 0.73 | 0.57 |
| total | 100.27 | 100.04 | 100.42 | 100.02 | 98.89 | 98.52 | 98.32 |

*References*: 2, 3, 4, 5 Bilgrami (1956); 6, 7, 8 (includes SnO 0.01) Marmo et al. (1959); 9 Gaudefroy et al. (1965); 10 (includes $P_2O_5$ 0.04), 11 (includes $P_2O_5$ 0.04), 12, 13 Smith and Albee (1967); 14 Cooper (1971); 15 Jan and Symes (1977); 16 Battaglia et al. (1977); 17, 18, 19, 20, 21, 22, 23 Chopin (1978); 24, 25, 26, 27, 28, 29, 30, 31, 32, 33, 34, 35, 36, 37, 38, 39, 40, 41 Dal Piaz et el. (1979); 42 Guild (1935); 44, 45, 46, 47, 48 Kawachi et al. (1983); 49, 50, 51, 52, 53, 54, 55 Ashley (1984); 56, 57 Velilla and Jiménez—Millán (2003); 58, 59, 60 Abraham and Schreyer (1975); 61, 62 Taylor and Baer (1973); 63, 64 Krogh (1977); 65, 66, 67, 68, 69, 70, 71 Dal Piaz et al. (1979); 72, 73, 74 Grapes and Hashimoto (1978); 75, 76, 77, 78, 79, 80, 81 Keskinen (1981); 82 Short (1933); 83 Smith et al. (1982); 84, 85, 86 Mottana (1986); 87, 88, 89, 90, 91, 92, 93, 94, 95, 96, 97, 98, 99 Reinecke (1986b); 101 Lucchetti et al. (1988); 102, 103, 104, 105, 106, 107, 108, 109 Akasaka et al. (1988); 110, 111 Leblanc and Lbouabi (1988); 112 Basu and Mruma (1985); 113 Bonazzi et al. (1990); 114, 115, 116, 117, 118, 119 Perseil (1991); 120, 121 Reinecke (1991); 122, 123 Hiroi et al. (1992); 124, 125, 126, 127, 128, 129, 130, 131, 132, 133, 134, 135 Jiménez—Millán and Velilla (1993); 136 Coombs et al. (1993); 152, 153, 154, 155, 156, 157, 158, 159, 160 Enami and Banno (2001); 161, 162, 163, 164, 165, 166, 167, 168, 169, 170, 171, 172, 173, 174, 175, 176, 177, 178, 179, 180, 181, 182, 183, 184, 185, 186, 187, 188, 189 Izadyar (2000); 190, 191 Reinecke et al. (1991); 192, 193, 194 Hirata et al. (1995); 197 (includes $P_2O_5$ 0.13) Nayak and Neuvonen (1966); 198 Ernst et al. (1970); 199, 200 Hutton (1938); 201 (includes $P_2O_5$ 0.06) Ernst (1964); 202 Franceschelli et al. (1996); 203, 204, 205, 206, 207, 208, 209, 210, 211 Della Ventura et al. (1996); 212 Yoshimura and Momoi (1964); 213, 218 Ishibashi (1969); 214, 217 Wieser (1973); 215 (includes $P_2O_5$ 0.13) Banno (1964); 216 Lyashkevich (1958); 219 (includes F 0.05) Ödman (1950); 220 Hutton (1940); 221 Tanaka et al. (1972); 222 Malmqvist (1929); 223, 224, 225 Kato and Matsubara (1986).

## Table B. Chemical analyses (oxides wt.%) of
## REE bearing piemontite from literature.

| Ref. # | 1 | 43 | 100 | 137 | 138 | 139 | 140 | 141 | 142 | 143 |
|---|---|---|---|---|---|---|---|---|---|---|
| $SiO_2$ | 37.37 | 35.6 | 34.61 | 31.81 | 32.06 | 31.73 | 32.29 | 30.95 | 30.77 | 31.75 |
| $TiO_2$ | — | 0.10 | — | 0.03 | — | 0.06 | 0.04 | 0.02 | 0.04 | 0.03 |
| $Al_2O_3$ | 22.07 | 19.27 | 20.54 | 14.34 | 14.63 | 14.29 | 14.48 | 13.21 | 13.41 | 14.58 |
| $Fe_2O_3$ | 4.78 | 4.92 | 3.71 | 7.52 | 7.66 | 7.55 | 7.73 | 7.48 | 7.62 | 7.27 |
| $Mn_2O_3$ | 8.15 | 10.08 | 17.33 | 13.93 | 13.25 | 13.68 | 13.91 | 12.88 | 13.27 | 13.87 |
| FeO | — | — | — | — | — | — | — | — | — | — |
| MnO | 2.29 | — | — | — | — | — | — | — | — | — |
| MgO | 0.30 | 1.48 | — | 0.56 | 0.44 | 0.52 | 0.45 | 0.51 | 0.54 | 0.49 |
| CaO | 18.83 | 19.48 | 12.90 | 12.08 | 12.80 | 11.88 | 12.43 | 10.99 | 11.16 | 12.18 |
| SrO | — | — | — | — | — | — | — | — | — | — |
| $Na_2O$ | 0.27 | — | — | 0.19 | 0.18 | 0.17 | 0.13 | 0.12 | 0.13 | 0.24 |
| $K_2O$ | 0.81 | — | — | 0.02 | — | 0.02 | — | 0.01 | — | 0.02 |
| $H_2O$ | 2.48 | — | — | — | — | — | — | — | — | — |
| $La_2O_3$ | — | 5.00 | — | 3.88 | 2.64 | 3.87 | 3.61 | 4.36 | 4.55 | 3.26 |
| $Ce_2O_3$ | 2.41 | — | 4.09 | 3.11 | 2.63 | 1.96 | 1.35 | 1.59 | 1.20 | 2.46 |
| $Pr_2O_3$ | — | — | — | — | — | — | — | — | — | — |
| $Nd_2O_3$ | — | — | — | 1.57 | 1.43 | 1.26 | 1.11 | 1.10 | 1.07 | 1.51 |
| $Gd_2O_3$ | — | — | — | — | — | — | — | — | — | — |
| $Sm_2O_3$ | — | — | — | — | — | — | — | — | — | — |
| $ThO_2$ | — | — | — | — | — | — | — | — | — | — |
| CuO | 0.11 | — | — | — | — | — | — | — | — | — |
| PbO | 0.14 | — | — | 6.08 | 6.27 | 6.88 | 6.77 | 10.35 | 9.86 | 5.78 |
| ZnO | — | — | — | 0.74 | 0.65 | 0.68 | 0.60 | 0.65 | 0.69 | 0.75 |
| BaO | — | — | — | — | — | — | — | — | — | — |
| F | — | — | — | — | — | — | — | — | — | — |
| total | 100.00 | 95.93 | 93.18 | 95.86 | 94.64 | 94.55 | 94.9 | 94.22 | 94.31 | 94.19 |

*References*: 1 Hillebrand in Williams (1893); 43 Kramm (1979); 100 Schreyer et al. (1986); 137, 138, 139, 140, 141, 142, 143 Bermanec et al. (1994)

**Table B (continued). Chemical analyses (oxides wt.%) of
REE bearing piemontite from literature.**

| Ref. # | 144 | 145 | 146 | 147 | 148 | 149 | 150 | 151 | 195 | 196 |
|---|---|---|---|---|---|---|---|---|---|---|
| $SiO_2$ | 32.52 | 31.54 | 31.97 | 30.52 | 34.06 | 35 | 35.36 | 33.85 | 31.33 | 30.41 |
| $TiO_2$ | — | 0.05 | 0.04 | 0.04 | — | — | — | — | — | — |
| $Al_2O_3$ | 14.78 | 14.96 | 14.48 | 13.7 | 17.02 | 18.5 | 18.49 | 16.86 | 15.11 | 8.54 |
| $Fe_2O_3$ | 7.53 | 7.14 | 8.19 | 7.86 | 5.93 | 7.02 | 6.92 | 6.32 | 5.52 | 1.17 |
| $Mn_2O_3$ | 12.65 | 13.98 | 13.95 | 12.98 | 14.36 | 13.54 | 13.62 | 15.18 | — | — |
| FeO | — | — | — | — | — | — | — | — | — | — |
| MnO | — | — | — | — | — | — | — | — | 11.35 | 28.29 |
| MgO | 0.43 | 0.56 | 0.48 | 0.6 | 0.63 | 0.18 | 0.26 | 0.49 | 1.16 | — |
| CaO | 12.67 | 13.02 | 12.83 | 11.32 | 16.29 | 18.64 | 18.48 | 15.97 | 9.40 | 5.81 |
| SrO | — | — | — | — | 0.48 | 0.62 | 0.68 | 0.69 | 3.49 | 0.66 |
| $Na_2O$ | 0.17 | 0.24 | 0.18 | 0.21 | — | — | — | — | — | — |
| $K_2O$ | 0.02 | — | — | — | — | — | — | — | — | — |
| $H_2O$ | — | — | — | — | — | — | — | — | — | — |
| $La_2O_3$ | 2.2 | 2.68 | 3.62 | 5.09 | 4.39 | 0.97 | 1.77 | 3.73 | 2.95 | 7.22 |
| $Ce_2O_3$ | 3.06 | 3.40 | 1.47 | 1.82 | 1.29 | 0.97 | 0.59 | 1.33 | 6.43 | 6.67 |
| $Pr_2O_3$ | — | — | — | — | 0.58 | 0.19 | 0.36 | 0.59 | 0.91 | 1.34 |
| $Nd_2O_3$ | 1.58 | 1.77 | 1.00 | 0.98 | 1.69 | 0.63 | 1.32 | 1.98 | 2.84 | 4.72 |
| $Gd_2O_3$ | — | — | — | — | 0.11 | 0.09 | 0.1 | 0.13 | 0.45 | 0.31 |
| $Sm_2O_3$ | — | — | — | — | — | — | — | — | 0.21 | 0.03 |
| $ThO_2$ | — | — | — | — | 0.15 | — | — | — | 0.94 | 0.03 |
| CuO | — | — | — | — | — | — | — | — | 1.81 | 0.16 |
| PbO | 6.37 | 5.57 | 6.27 | 10.28 | — | — | — | — | — | — |
| ZnO | 0.70 | 0.73 | 0.62 | 0.68 | — | — | — | — | — | — |
| BaO | — | — | — | — | — | — | — | — | — | — |
| F | — | — | — | — | — | — | — | — | 0.11 | 0.10 |
| total | 94.68 | 95.64 | 95.10 | 96.08 | 96.98 | 96.35 | 97.95 | 97.12 | 94.01 | 95.47 |

*References*: 144, 145, 146, 147 Bermanec et al. (1994); 148, 149, 150, 151 Bonazzi et al. (1992); 195, 196
Bonazzi et al. (1996).

Reviews in Mineralogy & Geochemistry
Vol. 56, pp. 553-605, 2004
Copyright © Mineralogical Society of America

# Trace Element Geochemistry
# of Epidote Minerals

### Dirk Frei

*Geological Survey of Denmark and Greenland*
*Øster Voldgade 10*
*DK 1350 København K, Denmark*

### Axel Liebscher

*GeoForschungsZentrum Potsdam*
*Department 4, Chemistry of the Earth*
*Telegrafenberg*
*D-14407 Potsdam, Germany*

### Gerhard Franz

*Technische Universität Berlin*
*Fachgebiet Petrologie, Sekretariat BH 1*
*Ernst-Reuter-Platz 1*
*D-10587 Berlin, Germany*

### Peter Dulski

*GeoForschungsZentrum Potsdam*
*Department 3 Geodynamik*
*Telegrafenberg*
*D-14407 Potsdam, Germany*

## INTRODUCTION

One of the most striking features of epidote minerals is their ability to incorporate significant amounts of geochemically important trace elements such as large ion lithophile elements (LILE), especially Sr and Pb, transition metals, actinides, and rare earth elements (REE). Epidote minerals are common in a broad range of whole rock compositions and they can be the most important reservoir for these elements in a variety of crustal rocks. We summarize the available trace element data of epidote minerals including zoisite from the literature and discuss their geochemical significance. Additionally, we present a set of new data from a wide range of geological environments.

We focus on the orthorhombic polymorph zoisite [$Ca_2Al_3Si_3O_{11}O(OH)$], which shows a very limited variation in major element chemistry, and the monoclinic epidote minerals along the join $Ca_2Al_3Si_3O_{11}O(OH) - Ca_2Fe^{3+}_3Si_3O_{11}O(OH)$, which is typically constrained to the Al-rich part, i.e., the $Fe^{3+}$ content rarely exceeds one cation per formula unit (pfu). The term "trace element" is problematic and ambiguous for the epidote minerals because they form solid solutions with actual end members whose components are usually abundant only as minor or trace elements, such as

1529-6466/04/0056-0012$10.00

| piemontite | $Ca_2Al_2(\mathbf{Mn^{3+}},Fe^{3+})Si_3O_{11}O(OH)$, |
|---|---|
| mukhinite | $Ca_2Al_2\mathbf{V^{3+}}Si_3O_{11}O(OH)$, |
| tawmawite | $Ca_2Al_2\mathbf{Cr^{3+}}Si_3O_{11}O(OH)$, |
| niigataite | $Ca\mathbf{Sr}Al_3Si_3O_{11}O(OH)$, |
| hancockite | $Ca\mathbf{Pb}Al_2(Al, Fe^{3+})Si_3O_{11}O(OH)$, |
| allanite | $Ca\mathbf{REE}Al_2Fe^{2+}Si_3O_{11}O(OH)$ |
| dissakisite | $Ca\mathbf{REE}Al_2MgSi_3O_{11}O(OH)$, |
| dollaseite | $Ca\mathbf{REE}Al_2MgSi_3O_{11}\mathbf{F}(OH)$. |

Therefore, these elements can be present in a continuous spectrum from the ppm to the wt% level.

Only few occurrences have been reported for the Sr, V, Cr, and Pb endmembers or for epidotes where they are major elements (see Grapes and Hoskin 2004, Enami et al. 2004), and possibly there is complete miscibility (e.g. Franz and Liebscher 2004; Grapes and Hoskin 2004). Mn- and REE-rich epidote minerals are covered by Gieré and Sorensen (2004) and Bonazzi and Menchetti (2004). The data suggest complete miscibility between clinozoisite, epdiote and allanite and clinozoisite, epidote and piemontite. Hence, the maximum REE- and Mn-content up to which analyses have been considered here has been chosen arbitrarily at a level of 8 and 3 wt%, respectively.

## CRYSTAL-CHEMICAL FRAMEWORK FOR THE INCORPORATION OF TRACE ELEMENTS INTO EPIDOTE MINERALS

The most important mechanism for the incorporation of trace elements into epidote minerals is the substitution on lattice sites for major cations because of the continuous transition between trace, minor, and major elements. There is no cation position which is especially suited for vacancies such as the A site in amphibole, and therefore we neglect substitutions with vacancies as a first approximation, but want to emphasize that they might possibly be important in certain cases such as the REE epidotes. Similarly, anion substitution is rare and only briefly touched.

The principal controls on trace element incorporation into crystals via substitution for a major cation at a given $P$ and $T$ are the crystal chemistry and structure of the mineral of interest, the mismatch in valence and ionic radius between the trace and major cation, and the availability of suitable mechanisms to charge balance heterovalent substitutions. In the case of ions with variable valence state, the capability to charge balance heterovalent substitutions can be highly dependent on oxygen fugacity ($f_{O_2}$). If a fluid phase is part of the system the composition and structure of the fluid might also exert important controls on the incorporation of trace elements. However, because crystal structures are much more regular and rigid relative to fluid phases including melt, they accept onto their lattice sites only ions of similar radius and charge as the major cations normally occupying these sites. Hence, crystal chemical constraints are of first-order importance in understanding the trace element signature of minerals.

### Crystal structure and chemistry of epidote minerals

The structure is composed of endless chains of edge-sharing octahedra parallel $b$ that are crosslinked by isolated $SiO_4$ tetrahedra and $Si_2O_7$ groups. Orthorhombic zoisite has only one type of chain with two nonequivalent octahedra M1,2 and M3 whereas the monoclinic forms have two types of octahedra chains with three nonequivalent octahedra M1, M2, and M3. In zoisite all octahedral sites are normally occupied by Al. The $Fe^{3+}$ content does not exceed 0.15 $Fe^{3+}$ pfu and is exclusively incorporated into the strongly distorted M3 octahedron in contrast to the more regular M1,2 octahedron. In clinozoisite and epidote, the M1 and M2 sites are normally occupied by Al, whereas the tetragonal distorted M3 site is occupied by Al and $Fe^{3+}$. For compositions exceeding 0.6 $Fe^{3+}$ pfu, Fe is also incorporated into the M1 site and small amounts of Fe might even be present in the most regular M2 site.

The large irregular cavities between the cross-linked octahedral chains form the nonequivalent A1 and A2 polyhedral sites and are very important for the substitution of a large number of trace elements. Because of the irregular character of the polyhedron, the coordination number can vary between 7 and 11. In orthorhombic zoisite both the smaller A1 and the larger A2 site are described as 7-fold coordinated (Liebscher et al. 2002). For Fe-poor clinozoisite, however, the A1 site is described as 7-fold and the A2 site as 8-fold coordinated (Dollase 1968). With increasing Fe content (and Mn; Bonazzi and Menchetti 2004) charge balance is more satisfactory if two more neighboring oxygen atoms are included in the first coordination shell and the A1, A2 sites are better described as 9-fold and 10-fold coordinated, respectively. In allanite A2 is 11-fold coordinated.

### Trace element incorporation into A sites

*Alkali elements.* The alkali elements are incorporated into the A sites most likely via coupled substitutions involving the M and A sites to maintain charge balance. Possible substitution mechanisms are

$$^A Ca^{2+}_{-1}\, ^M(Al,Fe)^{3+}_{-1}\, ^A(Me)^{1+}_{1}\, ^M(Me)^{4+}_{1} \tag{1}$$

$$^A Ca^{2+}_{-1}\, ^M(Al,Fe)^{3+}_{-2}\, ^A(Me)^{1+}_{1}\, ^M(Me)^{2+}_{1}\, ^M(Me)^{5+}_{1} \tag{2}$$

$$^A Ca^{2+}_{-2}\, ^A(Me)^{1+}_{1}\, ^A(REE)^{3+}_{1} \tag{3}$$

where $^A(Me)^{1+} = Na^{1+}$, $K^{1+}$, $Rb^{1+}$, and $Cs^{1+}$; $^M(Me)^{2+} = Mn^{2+}$, $Mg^{2+}$, $Fe^{2+}$, and $Zn^{2+}$; $^M(Me)^{4+} = Ti^{4+}$, $Zr^{4+}$, $Mo^{4+}$, $Sn^{4+}$, and $Hf^{4+}$; and $^M Me^{5+} = Nb^{5+}$ and $Ta^{5+}$. Because of their larger ionic radius relative to $Ca^{2+}$, alkali elements except Li are expected to strongly prefer the A2 site in both orthorhombic and monoclinic epidotes. No information is available for the site occupancy of Li. The combined substitution of $Na^{1+}$ and $REE^{3+}$ for $Ca^{2+}$ is commonly observed in perovskite (e.g., Campell et al. 1997) and apatite (Rønsbo 1989; Cherniak 2000; Klemme and Dalpé 2003). In epidote minerals, however, substitution of significant amounts of Na is uncommon (cf. Deer et al. 1986) because of unfavorable charge density.

*Alkaline earth elements and other divalent cations.* They are accommodated into the A sites by homovalent substitution for $Ca^{2+}$, expressed in a general form as

$$^A Ca^{2+}_{-1}\, ^A(Me)^{2+}_{1} \tag{4}$$

where $^A(Me)^{2+} = V^{2+}$, $Mn^{2+}$, $Fe^{2+}$, $Sr^{2+}$, $Ba^{2+}$, and $Pb^{2+}$. Because $Sr^{2+}$, $Ba^{2+}$, and $Pb^{2+}$ have an ionic radius larger than $Ca^{2+}$ in equal coordination, they show a strong preference for the larger and more distorted A2 site in both the orthorhombic and monoclinic forms.

Substitution of $Mn^{2+}$ (and subordinate $Fe^{2+}$ and $V^{2+}$) for $Ca^{2+}$ is indicated in some cases by stoichiometry. Because of their smaller ionic radius relative to $Ca^{2+}$ in equal coordination, both $Mn^{2+}$ and $Fe^{2+}$ are expected to be incorporated into the smaller A1 site in these cases in both orthorhombic and monoclinic forms (Tsang and Ghose 1971; Bonazzi et al. 1992). In zoisite, $V^{2+}$ might be incorporated into both A sites with a preference for A1 (Tsang and Ghose 1971). However, incorporation of significant amounts of $Mn^{2+}$ and $Fe^{2+}$ into the A1 site is usually restricted to rare compositions such as the REE-rich epidote androsite-(La) and some altered allanites (Gieré and Sorensen 2004).

***Rare earth elements.*** The most prominent mechanism for the incorporation of trivalent REE including Y into the A sites is the concomitant replacement of trivalent for divalent cations into the M sites to maintain charge balance via

$$^A\text{Ca}^{2+}_{-1}\ ^M(\text{Al},\text{Fe})^{3+}_{-1}\ ^A(\text{REE})^{3+}_{1}\ ^M(\text{Me})^{2+}_{1} \tag{5}$$

In most cases incorporation of REE according to (5) is accompanied by a reduction of $Fe^{3+}$ to $Fe^{2+}$ or the introduction of $Mg^{2+}$.

Because the larger A2 site is higher coordinated and more distorted compared to the A1 site, the REE are expected to show a strong preference for the A2 site in monoclinic Al-Fe solid solutions (Dollase 1971; Bonazzi and Menchetti 1995). Strong preference of REE for the A2 site in monoclinic solid solutions is also supported by bond-valence calculations (Pan and Fleet 1996) and the difference in electrostatic energy between the A sites (Smyth and Bish 1988). Cressey and Steel (1988), however, concluded from EXAFS data that in allanite with $\Sigma$REE > 0.9 pfu the heavy REE (HREE = Er to Lu) show a preference for the A1 site with Lu might even be incorporated into the M3 site, whereas the light REE (LREE = La to Nd) and middle REE (MREE = Sm to Ho) are exclusively incorporated into the larger A2 site.

In marked contrast to clinozoisite-epidote-allanite, REE in zoisite are preferentially incorporated into the A1 site and only La and Ce are accommodated in the A2 site in significant amounts (Frei et al. 2003). Because the structure of A1 and A2 in Fe-poor clinozoisite is very similar compared to zoisite, a significant proportion of the middle and heavy REE are expected to substitute into the A1 site in Fe-poor clinozoisite. In both orthorhombic and monoclinic polymorphs all divalent ions incorporated into the M sites have a strong preference for the highly distorted M3 site and only subordinate amounts are incorporated into M1 (Dollase 1971, 1973; Bonazzi et al. 1992).

Besides the combined substitution with alkali elements according to (3) incorporation of REE might be achieved by the substitution of divalent for trivalent cations on both M3 and M1 sites and concomitant replacement of O by F or Cl to compensate the charge imbalance. This leads to the complex coupled substitution

$$^A\text{Ca}^{2+}_{-1}\ ^M(\text{Al},\text{Fe})^{3+}_{-2}\ \text{O}^{2-}_{-1}\ ^A(\text{REE})^{3+}_{1}\ ^M(\text{Me})^{2+}_{2}\ (\text{F},\text{Cl})^{1-}_{1} \tag{6}$$

Like in (5), the introduction of REE is usually achieved by introduction of $Mg^{2+}$ or a valence change of Fe. F or Cl substitute for the oxygen atom that is not coordinated to Si (Peacor and Dunn 1988) and substitution according to (6) is probably the most important mechanism to incorporate significant amounts of F and Cl into the epidote crystal structure.

***Actinides.*** $Th^{4+}$ and $U^{4+}$ might be incorporated into the A sites for $Ca^{2+}$ by concomitant substitution of divalent ions into the M sites to effect charge balance via

$$^A\text{Ca}^{2+}_{-1}\ ^M(\text{Al},\text{Fe})^{3+}_{-2}\ ^A(\text{Th},\text{U})^{4+}_{1}\ ^M(\text{Me})^{2+}_{2} \tag{7}$$

(Gieré et al. 1999). Because the ionic radii of both Th and U are smaller than the ionic radius of Ca in appropriate coordination they are not expected to favor a particular A site. However, because the A2 site is generally underbonded even in ten-fold coordination in the monoclinic Al-Fe solid solutions (Pan and Fleet 1996), Th and U might show preference for the A2 site in these polymorphs due to their high charge.

Other possible incorporation mechanisms might involve substitution of Th and U for REE in A with concomitant substitution of divalent ions into M according to

$$^A(\text{REE})^{3+}_{-1}\ ^M(\text{Al},\text{Fe})^{3+}_{-1}\ ^A(\text{Th},\text{U})^{4+}_{1}\ ^M(\text{Me})^{2+}_{1} \tag{8}$$

or concomitant substitution of $Ca^{2+}$ in A according to

$$^A(REE)^{3+}_{-2} \, ^A(Th,U)^{4+}_1 \, ^ACa^{2+}_1 \tag{9}$$

(Gromet and Silver 1983; Gieré et al. 1999; Wood and Ricketts 2000). Incorporation of actinides according to (7) and (8) is usually achieved by a valence change of Fe. Substitution mechanisms (8) and (9) require REE-bearing compositions to maintain charge balance and therefore might be more significant for the REE-rich epidote minerals (Gieré and Sorensen 2004).

### Trace element incorporation into M sites

***Beryllium.*** The ionic radius of $Be^{[6]}$ is 0.45 Å, close to the ionic radius of $Al^{[6]}$ (0.535 Å) and might thus be substituting for Al in the octahedral sites. Possible substitutions that effect charge balance are coupled with REE into A sites according to

$$^ACa^{2+}_{-1} \, ^MAl^{3+}_{-1} \, ^A(REE)^{3+}_1 \, ^MBe^{2+}_1 \tag{10}$$

a simplification of substitution (5), or concomitant introduction of tetravalent ions into the octahedral sites according to

$$^M(Al,Fe)^{3+}_{-2} \, ^M(Me)^{4+}_1 \, ^MBe^{2+}_1 \tag{11}$$

The small ionic radius of Be suggests preferential incorporation of Be into the smaller and more regular M1 and M2 sites. However, because the ionic radius of $Be^{[4]}$ is similar to that of $Si^{[4]}$ (0.27 Å compared to 0.26 Å, respectively), Be could possibly substitute for Si in tetrahedral sites (see below).

***Transition elements.*** The transition metals and other trivalent elements with small ionic radius, such as Ga, are incorporated into the M sites via

$$^M(Al,Fe)^{3+}_{-1} \, ^M(Me)^{3+}_1 \tag{12}$$

where $^M(Me)^{3+} = Sc^{3+}$, $V^{3+}$, $Cr^{3+}$, $Mn^{3+}$, $Co^{3+}$, $Ni^{3+}$, $Cu^{3+}$, and $Ga^{3+}$. Like $Fe^{3+}$, most of these elements show a strong preference for the highly distorted M3 site in both orthorhombic and monoclinic epidote minerals. The only exception is the incorporation of $Cr^{3+}$ into the monoclinic polymorphs, which shows a strong preference for the M1 site and only minor preference for the M3 and M2 sites (Burns and Strens 1967; Sánchez-Vizcaíno et al. 1995).

***High field strength elements (HFSE).*** The HFSE and other highly charged elements with small ionic radius, such as $Mo^{4+}$, $Sn^{4+}$, $W^{6+}$, and $U^{6+}$ are incorporated into the octahedral sites. The most likely substitution mechanism for the incorporation of tetravalent ions into the M-sites is by concomitant substitution of divalent ions for $Al^{3+}$ and $Fe^{3+}$ to maintain charge balance via

$$^M(Al,Fe)^{3+}_{-2} \, ^M(Me)^{4+}_1 \, ^M(Me)^{2+}_1 \tag{13}$$

Another possible mechanism for the incorporation of tetravalent ions into the M sites is via substitution (1). In addition to the complex coupled substitution (2), the pentavalent HFSE Nb and Ta might be incorporated into the M sites via

$$^M(Al,Fe)^{3+}_{-2} \, ^M(Nb,Ta)^{5+}_1 \, ^MLi^{1+}_1 \tag{14}$$

which involves incorporation of Li in the M sites. Because of their high charge, the HFSE are expected to strongly favor incorporation into the M3 site. However, to date there are no studies available that address the incorporation mechanism of tetravalent and higher valence ions.

### Trace element incorporation into T sites

Structure refinements (Bonazzi and Menchetti 1995) and neutron diffraction (Nozik et al. 1978; Kvick et al. 1988; Ferraris et al. 1989) of epidote minerals do not indicate significant substitution of Si by other cations. The only evidence for substitution of Si comes from stoichiometric considerations, i.e., some analyses yield formulae with <3 Si atoms pfu (Deer et al. 1986), and a few studies report replacement of Si by Al, $Be^{2+}$, $P^{5+}$, or $Ge^{4+}$. However, most of these studies refer to the REE-rich varieties of the epidote group (Gieré and Sorensen 2004).

***Beryllium.*** Be might be incorporated into the T sites by coupled substitution with actinides in A sites according to

$$^{A}Ca_{-1}^{2+}\ ^{T}Si_{-1}^{4+}\ ^{A}(Th,U)_{1}^{4+}\ ^{T}Be_{1}^{2+} \tag{15}$$

However, Be might also be incorporated into the M sites via substitutions (10) and (11).

***Phosphorus.*** The tetrahedral substitution of $P^{5+}$ (ionic radius = 0.17 Å) for $Si^{4+}$ (ionic radius = 0.26) could possibly take place via coupled substitutions involving both A and M sites according to

$$^{A}Ca_{-1}^{2+}\ ^{T}Si_{-1}^{4+}\ ^{A}(Me)_{1}^{1+}\ ^{T}P_{1}^{5+} \tag{16}$$

and

$$^{M}(Al,Fe)_{-1}^{3+}\ ^{T}Si_{-1}^{4+}\ ^{M}(Me)_{1}^{2+}\ ^{T}P_{1}^{5+} \tag{17}$$

***Germanium.*** The larger $Ge^{4+}$ (ionic radius 0.39 Å) might replace $Si^{4+}$ via

$$^{T}Si_{-1}^{4+}\ ^{T}Ge_{1}^{4+} \tag{18}$$

## TRACE ELEMENT VARIATIONS IN EPIDOTE MINERALS

To enlarge and substantiate the database on trace element concentrations in epidote minerals for this review, we have determined the abundances of Rb, Sr, Y, Zr, Nb, Cs, Ba, REE, Hf, Ta, Pb, Th, and U in 33 selected epidote mineral specimens (19 zoisite, 3 clinozoisite, and 11 epidote) and, whenever possible, coexisting mineral phases. They encompass a wide range of chemical compositions and lithologies (Appendix A). Mineral separates were prepared by handpicking to optical purity. The concentrations (Appendix B) have been determined from solutions by inductively coupled plasma-mass spectrometry (ICP-MS; Fisons VG PlasmaQuad 2+) at the GeoForschungsZentrum Potsdam following the methods described in detail by Dulski (2001). The compiled literature data cover igneous, metamorphic, metasomatic, and hydrothermal rocks, and were acquired by either "bulk" or "*in situ*" techniques (sources and techniques in Appendix C).

### Observed trace element variation

Reported are the arithmetical (mean) and geometrical (median) average, the concentration range observed, and the number of determinations (Appendix D), calculated using the literature database (Appendix C). Analysis where V, Cr, Mn, Sr, Pb, and REE are present as major elements have not been used to calculate averages and only the highest observed contents are reported (e.g., Pb in hanckockite; Holtstam and Langhof 1994). For datasets with an insufficient number of analyses no summary is given. The complete compilation is available from the first author upon request.

Inclusion of the geometrical mean proved to be important because analyses of epidote minerals from unusual and rare bulk compositions bias the arithmetical average towards unrealistic high values. This is demonstrated for the actinides Th and U: In Fe-rich monoclinic epidote, the arithmetical mean for Th and U is ~20 and ~100 times higher, respectively, than the geometrical mean. In this case, the arithmetical mean is dominated by some unusual Th- and U-rich epidotes occurring in apatite-allanite-epidote coronas around monazite formed by amphibolite facies breakdown of monazite in metagranite (Finger et al. 1998).

With the exception of EMPA, where the detection limit is usually a few hundred ppm for most elements, all techniques allow the determination of trace element abundances at or below the ppm concentration range. The analytical precision and accuracy are generally better than 20% relative for most of the analyses reported (Appendix C).

Almost half of the reported studies were carried out using an electron microprobe. For V, Cr, Sr, REE, and Pb a significant fraction and for Ti and Mn the majority of the data were acquired by EMPA. Due to the analytical difficulties in determining the REE by EMPA the data (Appendix D) for REE heavier than Sm obtained by EMPA are scant and the data for these elements might be biased towards higher concentrations. Another source for bias is that many studies specifically address to unusual trace element rich compositions (e.g. Grapes 1981; Johan et al. 1983; Grapes and Watanabe 1984; Van Marcke de Lummen 1986; Treloar 1987; Jancev and Bermanec 1998; Miyajima et al. 2003). Also the choice of the materials might produce a bias. Separates produced from relatively large single crystals, often of gem quality, from alpine clefts, veins, vugs, and segregations or very coarse grained rocks such as pegmatites might not be representative for rock forming minerals. Separates of small crystals might still contain optically undetected impurities (usually inclusions of fluids or other mineral phases). In terms of sample volume, these impurities may be judged as negligible, but as demonstrated by e.g. Eggins et al. (1998), Prince et al. (2000), and Thöni (2003) may have pronounced and, especially for highly incompatible elements, even dramatic consequences on measured trace element abundances. Mineral separates cannot account for trace element zonation that may provide essential information on the petrogenesis of the mineral in question and on equilibrium-disequilibrium features, and fluid-rock interactions (Messiga et al. 1995; Kretz et al. 1999; Brunsmann et al. 2000, 2001; Spear and Pyle 2002; Zack et al. 2002; Spandler et al. 2003).

The variations are discussed in groups of elements with similar geochemical behavior:

large ion lithophile elements K, Rb, Cs, Sr, Ba, Pb, including Na;

actinides U, Th;

rare earth elements including Y;

transition elements Sc, Ti, V, Cr, Co, Ni, Cu, Zn, Mo, and W,

high field strength elements Zr, Hf, Nb, and Ta;

Elements that are not covered by these subdivisions are discussed briefly at the end.

### Large ion lithophile elements (K, Rb, Cs, Sr, Ba, Pb, including Na)

*Sodium.* Most reported analyses scatter around the detection limit of EMPA. The highest abundances so far are ~0.5wt% $Na_2O$ for epidote in blueschist from Silesia, Poland (Howie and Walsh 1982); 0.60 wt% $Na_2O$ (0.09 Na pfu) for clinozoisite and 0.17 wt% $Na_2O$ (0.027 Na pfu) for zoisite in eclogite that suffered strong Sr-metasomatism during amphibolitization (Bjørkedalen, West Norway; Brastad 1985).

*Potassium.* The large ionic radius of $K^+$ combined with the need to charge balance, suggests only very limited incorporation of K. Indeed only very few studies report K abundances above the detection limit of the EMPA. Brastad (1985) reports up to 0.25 (0.029 K pfu) and 0.12 wt% $K_2O$ (0.012 K pfu) for epidote and clinozoisite, respectively (see above). The highest observed content so far is 1.2 wt% $K_2O$ (0.137 K pfu) for epidote in hydrothermally altered acidic ignimbrite from the Eastern Rhodopes, Bulgaria (Yanev et al. 1998).

*Rubidium, caesium, barium.* Because of their very large ionic radius compared to Ca they are normally present only at the trace element level. $Ba^{2+}$ has a distinctively lower ionic radius than $Cs^+$ and $Rb^+$ and can be more easily incorporated via simple homovalent substitution

(5). Therefore, Ba contents are an order of magnitude higher than Rb, and two to three orders of magnitude higher than Cs. Rb, Cs, and Ba are equally abundant in clinozoisite-epidote and in zoisite although the orthorhombic structure is much more limited to incorporate $Me^{2+}$. The abundances of Rb, Cs, and Ba reflect the highly lithophile character of these elements, i.e., epidote minerals in felsic crustal rocks are enriched compared to epidote in basic mantle derived rocks. This behavior can be exemplified by the Rb content in clinozoisite-epidote: In metabasic rocks, the observed concentration ranges from ~2 to 8 ppm, whereas the observed concentration in metagranite is about ~20 to ~100 ppm (Fig. 1).

Cs and Rb are positively correlated for clinozoisite-epidote, indicating progressive enrichment of Rb and Cs with increasing degree of differentiation of the protolith (Fig. 1). Ba and Cs are also positively correlated in clinozoisite-epidote (not shown). The few data for zoisite show Cs contents of about 0.1 ppm and Rb between 1 and 10 ppm. These data are

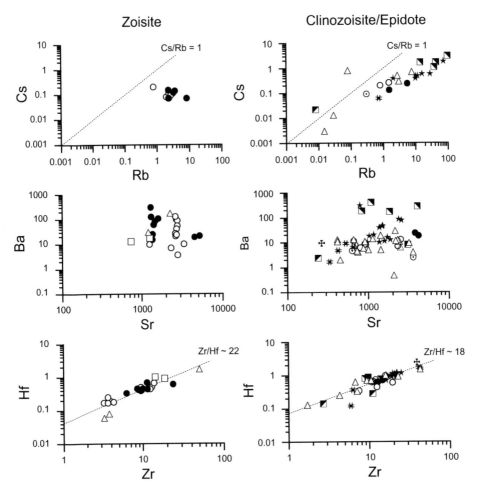

**Figure 1.** Trace element contents in zoisite (left) and in clinozoisite-epidote (right); for explanation see text. The 1:1 line in the diagrams is given as a reference line except in the Hf/Zr diagrams, where the lines are regression lines and indicate a Zr/Hf ratio of 22 and 18, respectively. Data sources in Appendix A, C. See legend on facing page. *Figure continued on facing page.*

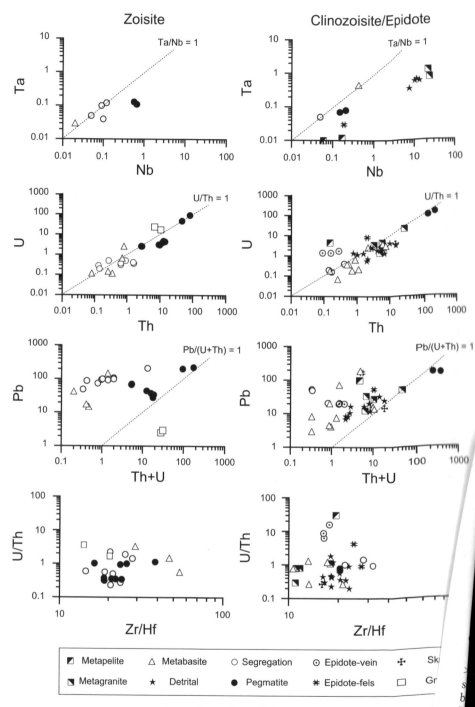

**Figure 1.** *caption on facing page*

from mineral separates and therefore might reflect mica impurities. $Ba^{2+}$ reaches values up to ~200 ppm in zoisite but most data are < 100 ppm. Typical Ba concentrations in epidote are comparable to zoisite but may reach ~ 800 ppm (for Ba as a major element in some epidotes see Franz and Liebscher 2004).

***Strontium***. Sr is a very characteristic and widespread trace, minor and major element in epidote minerals, despite its larger ionic radius compared to Ca. Complete miscibility of Ca and Sr in epidote (see Grapes and Hoskin 2004) is very likely. In most metamorphic and magmatic lithologies, epidote minerals are enriched compared to their host rock, but their absolute concentrations strongly depend on Sr bulk rock, other potential Sr-carriers such as calcic plagioclase, titanite, apatite and calcite, and fluid-rock interaction. This results in an extreme range of contents even within the same rock type. The observed cluster near ~2,000 ppm Sr in detrital epidote grains (Spiegel et al. 2002) reflects the "common" contents.

For igneous rocks, the only data available for epidote are from quartz monzonite, which display homogeneous Sr contents ranging from ~1,000 to ~1,600 ppm (Keane and Morrison 1997). Zoisite in high-$P$ pegmatites in eclogites from the Münchberg Massif, Bavaria, Germany, show homogeneous Sr contents around ~1,400 ppm (Appendix A). In contrast, zoisite in high-$P$ pegmatites associated with eclogites from Saualpe, Carinthia, Austria, show much higher contents ranging from ~2,000 to ~6,000 ppm (Appendix A, and Fig. 18 below). In metagranite and metatonalite, the Sr contents range from ~800 to ~5,000 ppm; however, the database is insufficient to delineate compositional trends as a function of lithology.

In metabasic rocks, the reported Sr contents in clinozoisite-epidote generally range from ~150 to 4,000 ppm, but in parageneses where other potential Sr-bearing minerals are absent, Sr is present at the major element level (e.g., Brastad 1985; Nagasaki and Enami 1998). Neither a distinct variation between rocks from different metamorphic facies (greenschist, amphibolite, blueschist, and eclogite) nor between zoisite and epidote can be observed. Hence, epidote minerals are strongly enriched compared to normal Sr contents in basaltic rocks (N-MORB = 90 ppm; basalts = 280 ppm). Nagasaki and Enami (1998) reported up to 2.66 wt% SrO (= 0.123 Sr pfu) in epidote and 3.04 wt% SrO (= 0.137 Sr pfu) in zoisite in ultra-high-$P$ rocks (peak metamorphic conditions >2.6 GPa and 700–890°C) from the Su-Lu province, eastern China. These high contents are explained by the instability of calcic plagioclase and titanite at ultra-high-$P$ conditions and the strong partitioning of Sr into epidote and zoisite compared to coexisting apatite and K-white mica (see below). In retrograded eclogites, Brastad (1985) describes epidote with up to 8.5 wt% SrO (0.40 Sr pfu), clinozoisite with up to 4.7 wt% SrO (0.22 Sr pfu) and zoisite with up to 7.4 wt% SrO (0.33 Sr pfu). These unusually high values are interpreted due to the infiltration of a Sr-rich fluid derived from Sr-rich neighboring lithologies during amphibolitization of the eclogite.

In metasedimentary rocks Sr variations in epidote minerals are also large, ranging from values of ~240 ppm in pelitic schist (Yang and Rivers 2002), ~800 to ~1,200 ppm in gneiss (Appendix A), to values as high as ~3,000 ppm in mica schist (Spandler et al. 2003). Nagasaki and Enami (1998) report contents of up to 0.64 wt% SrO (0.03 Sr pfu) in epidote and 1.54 wt% SrO (0.07 Sr pfu) in schist associated with eclogite from the ultra-high-$P$ Su-Lu terrane (see above). Unusually high values of up to 8.5 wt% SrO (0.4 Sr pfu) are found in metagreywacke-quartzofeldspathic schist from the Southern Alps, New Zealand (Grapes and Watanabe 1984), apparently due to the breakdown of detrital plagioclase at the pumpellyite-clinozoisite isograd.

There is a weak positive correlation between Sr and Ba for clinozoisite-epidote with Sr > Ba (Fig. 1), but the subsets of the data for metagranite and for zoisite from pegmatites and segregations show that Ba and Sr are geochemically decoupled (Fig. 1). This decoupling can be produced by fractional crystallization, where Sr can be incorporated into plagioclase and Ba into K-feldspar and mica, or during fluid-rock interaction.

**Lead.** For the majority of naturally occurring epidote minerals, Pb contents are confined to the trace element level, mostly one or two orders of magnitude less abundant than Sr, except for the rare cases were Pb is a major element (Franz and Liebscher 2004). In segregations and pegmatites, Pb contents in epidote minerals are highest (Fig. 1), lowest in metabasites and gneiss. They are markedly enriched compared to average Pb abundances in basalt and granite. Pb and U + Th are only weakly correlated (Fig. 1), demonstrating that common lead is a major constituent. The V-bearing variety 'tanzanite' is the only example, where the Pb content is significantly below the 1:1 line.

### Actinides

Abundances of Th and U are typically confined to the trace element level, and usually reflect the content of their host rocks. Hence, epidote minerals in metabasite show ten- to twenty-fold lower U and Th contents compared to epidote minerals in metagranite, metapelite and pegmatites. Segregations are also low in U and Th (Fig. 1). U and Th show a positive linear correlation close to the 1:1 line in both polymorphs (Fig. 1). Hence, compared to chondritic (3.9), N-MORB (2.6), and average crustal ratios of Th/U, epidote minerals are generally enriched in U, but for low concentrations there is a scatter of the Th/U ratios slightly higher and lower than 1. The only pronounced deviation from the 1:1 line is observed for epidote in veins from the Knappenwand, Untersulzbachtal, Austria, and a metapelite. The marked enrichment of U compared to Th in these samples might point to an enhanced mobility of U under highly oxidizing conditions.

REE rich epidotes can contain high amounts of U and Th in certain igneous rocks and in certain textural positions in metamorphic rocks: Keane and Morrison (1997) report up to 0.22 wt% $ThO_2$ (0.004 Th pfu) in quartz-monzonite and Dawes and Evans (1991) report up to 0.10 wt% $ThO_2$ (0.002 Th pfu) in dacitic to rhyodacitic dikes. The highest Th contents of 1.01 wt% $ThO_2$ (0.020 Th pfu) are described by Sakai et al. (1984) in REE-rich epidote formed during metamorphic breakdown of Th-rich detrital allanite in the Sanbagawa pelitic schists, Central Shikoku, Japan (no U analyses given). Finger et al. (1998) reported up to 0.22 wt% $ThO_2$ (0.004 Th pfu) and 0.31 wt% $UO_2$ (0.006 U pfu) in epidote in apatite-allanite-epidote coronas of monazite in granitic gneiss formed during amphibolite facies regional metamorphism (Tauern Window, Austria). These relatively high U and Th contents reflect the much better charge balancing capacity of epidote compared to zoisite, because the U and Th incorporation requires $Me^{2+}$ on an octahedral site via substitutions (7) or (8) or involve REE via substitution (9).

### Rare earth elements and yttrium

The compositional data for epidote minerals point to a complete miscibility between clinozoisite, epidote and allanite (cf. Gieré and Sorensen 2004). Hence, the observed REE contents range from a few hundred ppb to the wt% level. In contrast, the contents in zoisite do not exceed a few hundred ppm (cf. Appendix D). The characteristics of the chondrite-normalized REE patterns (CI chondrite values from McDonough and Sun 1995) in epidote minerals from a variety of important geological environments are described and discussed below. We discarded data sets of microprobe analyses for a presentation in Figures 2–7, because these data are mostly incomplete and generally not accurate enough to derive a REE pattern.

*Igneous rocks.* The few available data from a variety of felsic to intermediate igneous rocks point to a marked enrichment of LREE relative to MREE and HREE. Absolute REE-contents in different igneous rocks are highly variable and range from ~100,000 times chondritic to ~100 times chondritic and reflect the range of REE bearing epidote to allanite. Two patterns of epidote from granodiorite and syenite are similar with $La_N/Sm_N$ and $La_N/Yb_N$ ratios around ~4 and ~70, respectively, and a small positive Eu anomaly (Fig. 2).

*Metabasic rocks.* Zoisites from eclogite and garnet amphibolite display a concave upwards REE pattern (Fig. 3a,b) with an enrichment of MREE relative to LREE ($La_N/Sm_N = ~0.5$) and

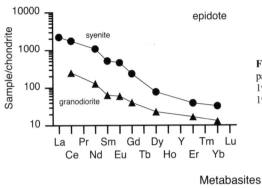

**Figure 2.** Chondrite normalized REE patterns of epidote from syenite (Braun et al. 1993) and granodiorite (Gromet and Silver 1983).

Metabasites

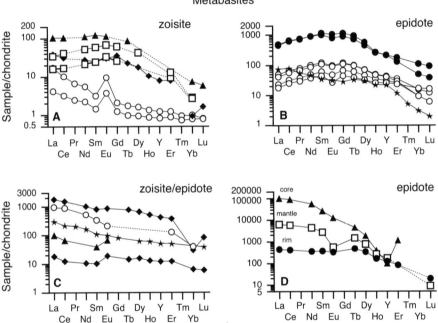

**Figure 3.** Chondrite normalized REE patterns of zoisite (a) and epidote (c-d) from mafic metamorphic rocks. Data sources: (a) open circles: this study, amphibolite samples # 363, 1445; open squares: eclogite (Thöni and Jagoutz 1992); filled triangles: garnet-amphibolite (Sorensen and Grossman 1989); filled diamonds: eclogite (Sassi et al. 2000) (b) all data for eclogite (filled and open circles, same sample) and blueschist (asterisks) from Spandler et al. (2003) first described as zoisite, but refer to epidote (Spandler pers. com. 2004). (c) filled diamonds: eclogite (Sassi et al. 2000); asterisks: blueschist (Howie and Walsh 1982); filled triangles (Sorensen and Grossman 1989); open symbols: zoisite from garnet amphibolite (Lucassen pers. comm.). (d) Zoned allanite-epidote from Sassi et al. (2000).

HREE (with $La_N/Yb_N = \sim7$ to $\sim25$) In two samples from a high temperature amphibolite, zoisite REE patterns are straight LREE enriched ($La_N/Yb_N = 6$ to 17) and show a pronounced positive Eu-anomaly ($Eu_N/Eu_N^* = 2.8$ to 3.8; Fig. 3a). Zoisite may also show a straight pattern (Fig. 3c). The REE data for epidote in eclogite (data from Spandler et al. 2003; Fig. 3b) are similar to the zoisite data from eclogite; one sample from a blueschist displays a straight LREE enriched pattern, similar to other patterns shown in Figure 3c. The extremely low Yb content in the REE pattern reported by Sassi et al. (2000) might be best explained with analytical difficulties during SIMS analysis. Observed REE concentrations are generally in the same order of magnitude in zoisite and in epidote. Sassi et al. (2000) reported a case of a zoned allanite-epidote, where rim analyses (Fig. 3d) have only a few hundreds of ppm LREE.

*Metafelsic and metapelitic rocks.* Only few and mostly incomplete REE data are available. The only complete REE patterns of zoisite (var. "tanzanite") in kyanite-graphite-calcsilicate gneiss (Fig. 4a) show moderate fractionation of LREE from HREE ($La_N/Yb_N$ = 3.1 and 6.5), almost no fractionation of LREE from MREE ($La_N/Sm_N$ = 1.3 and 0.9), and a pronounced negative Eu-anomalie ($Eu_N/Eu_N$* = 0.6 and 0.7).

Although the REE data reported for epidote in metagranite are scant due to analysis with EMPA, the available data points to a marked enrichment of LREE relative to MREE and HREE. La-contents are about three orders of magnitude higher in epidote compared to epidote from schist and to zoisite. The patterns of epidote from metapelite (Fig. 4b) show in one case moderate enrichment of LREE relative to HREE ($La_N/Yb_N$ = 7), only weak enrichment of LREE relative to MREE ($La_N/Sm_N$ = 1.5) and displays a weak positive Eu-anomaly ($Eu_N/Eu_N$* = 1.7), and in the other case an exceptional concave upwards REE pattern with a maximum at Tb, resulting in strong enrichment of MREE relative to LREE ($La_N/Sm_N$ = 0.3) and HREE ($Sm_N/Lu_N$ = 13).

The data base for epidote minerals from calcsilicate rocks and impure marbles, where they are modally important constituents, is very poor. Only two incomplete patterns exist (Fig. 4c) for calcsilicate rocks from high temperature metamorphic terranes in Chile and Argentina (Lucassen, pers. comm.), which show a flat pattern with a negative Eu anomaly.

*Pegmatites.* There are data from two localities, where zoisite and clinozoisite occur in pegmatites in eclogite terranes, the Saualpe/Kärnten, Austria, and the Münchberg massif, Bavaria, Germany. Zoisite from Münchberg (Fig. 5a) displays two types of REE patterns. One has a straight downwards REE pattern with marked enrichment of LREE over HREE

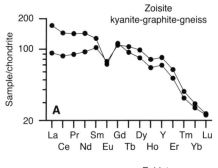

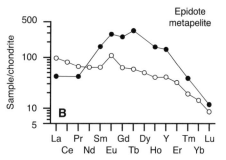

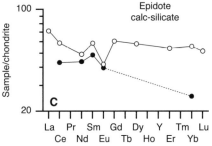

**Figure 4.** Chondrite normalized REE pattern of zoisite (a) from kyanite gneiss (this study, samples Mir 1,2, var. 'tanzanite'); epidote (b) from metapelite (dots: Yang and Rivers 2002; open circles: Spandler et al. 2003) and (c) from calcsilicate rocks (Lucassen pers. comm.; circles: Chile; dots: Argentina).

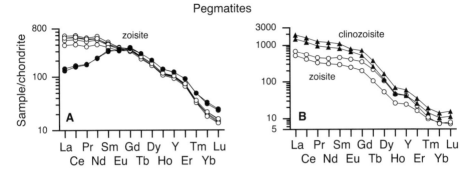

**Figure 5.** Chondrite normalized REE patterns of zoisite and clinozoisite from pegmatites of high-$P$ terranes (a) München massif/Germany (the sample indicated by dots has amphibole present, the other samples are essentially amphibole free) and (b) Saualpe/Austria (clinozoisite is a break-down product of zoisite); all data from this study.

($La_N/Yb_N \approx 30$) and only weak enrichment of LREE over MREE ($La_N/Sm_N \approx 1.5$). The other one displays a concave upwards REE pattern with a maximum at Gd and a much stronger enrichment of MREE over HREE ($Sm_N/Yb_N = 10$) than for LREE ($Sm_N/La_N \approx 2$), generally similar to the patterns from eclogites. At the Saualpe locality (Fig. 5b), zoisite and clinozoisite have almost identical patterns, with a slight enrichment in clinozoisite by a factor of 3 to 4. The patterns are similar to the first type from München with only weak LREE enrichment over MREE ($La_N/Sm_N \approx 1.5$), but a more pronounced LREE enrichment over HREE ($La_N/Yb_N \approx 90$) and a positive hump at Y.

**Segregations.** The REE pattern of zoisite from segregations in high-$P$ amphibolite (massif host rock with abundant garnet; Fig. 6a) is enriched in LREE ($La_N/Yb_N = 9$–23) with a small positive Eu anomaly ($Eu_N/Eu_N^* = 1.1$–1.4). A clinozoisite pattern from the same type of rock, where the clinozoisite was formed from zoisite, is parallel, but slightly enriched. The patterns from segregations in a fine-grained amphibolite (Fig. 6b) show a maximum at Eu and strong MREE enrichment over HREE ($Sm_N/Yb_N \approx 5$) and a slightly weaker MREE enrichment over LREE ($Sm_N/La_N = 2$–3). Other patterns from clinozoisite in high-$P$ segregations (Fig. 6c) are variable, from flat, almost unfractionated to depleted in LREE and enriched in LREE. All data are from samples of the Tauern Window/Austria.

**Metasomatic environments.** The data for epidote minerals from metasomatic environments, such as epidote fels, vugs, and veins, are extremely variable and show all types of patterns. Epidote from vugs in basalt (Fig. 7a) displays straight downward REE patterns with varying enrichment of LREE over MREE ($La_N/Sm_N = 3$–16) and HREE ($La_N/Yb_N = 3$–7). In some samples positive Eu anomalies are indicated. These samples are very similar to the segregations (Fig. 6a). Epidotes from contact metasomatism (skarn and epidote fels at a contact to granite; Fig. 7b) show a straight LREE enriched pattern (similar to the high-$P$ rutile-segregation, Fig. 6c), a flat unfractionated REE pattern, and a straight downward, LREE enriched pattern ($La_N/Sm_N = 1.6$; $La_N/Yb_N = 7$) with a pronounced negative Eu anomaly ($Eu_N/Eu_N^* = 0.6$).

Epidotes from veins at the Knappenwand/Austria are typically enriched in HREE with a limited range of HREE contents, but large variation in LREE contents (Fig. 7c). The resulting patterns vary from only slightly HREE enriched ($La_N/Yb_N \approx 0.5$) to strongly HREE enriched ($La_N/Yb_N = 0.07$–0.15). Most are characterized by a positive Eu anomaly, which is very pronounced ($Eu_N/Eu_N^* = 4$–22) for samples with low LREE contents. Clinozoisite from a hydrothermal vein (Fig. 7d; lower pattern) and epidote from an unspecified metamorphic rock

Segregations

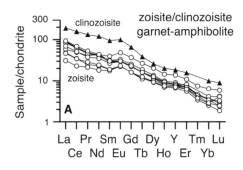

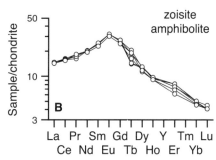

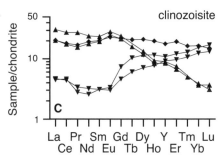

**Figure 6.** Chondrite normalized REE patterns of zoisite and clinozoisite from high-*P* segregations of the Tauern Window/Austria (all data from this study and Brunsmann 2000); (a) zoisite and clinozoisite from garnet-amphibolite; (b) zoisite from a fine grained amphibolite; note the similarity to the one pattern in Figure 5a. (c) clinozoisite; upward triangles from segregation from amphibolite; downward triangles from calc-mica-schist; diamonds from a rutile-epidote-quartz segregation.

(Pan and Fleet 1996) are almost identical. An epidote from Wollaston county (Fig. 7e; Pan and Fleet 1996) and an epidote from a vein filling in amphibolite are similar, as well as two from hematite-quartz and calcite veins (Fig. 7f).

## Transition elements (Sc, Ti, V, Cr, Mn, Co, Ni, Cu, Zn)

*Scandium.* The studies that report Sc contents in epidote minerals point to an enrichment of Sc in epidote relative to the host rock, but the data are far too few to provide conclusive evidence. Yang and Rivers (2002) determined 176 ppm Sc in epidote from a calc-pelitic schist from Labrador, and Sorensen and Grossman (1989) reported 91 ppm Sc for clinozoisite and 29 ppm for zoisite in garnet-amphibolite from the Catalina Schist, southern California. The Sc content of epidote in eclogite from New Caledonia ranges from of 33 to 93 ppm (Spandler et al. 2003), and Brunsmann et al. (2000) report values from 14 to 29 ppm for zoisite in high-*P* segregations in eclogites. Unlike for the REE-rich epidote minerals, where Meyer (1911) reports Sc-enriched allanite with $Sc_2O_3$ contents of up to 1.0 wt%, there is no evidence so far for Sc-rich epidote.

*Titanium.* It is a typical trace and minor constituent and its variation is very similar to that of Mn. Due to the limited ability of the zoisite structure to substitute for Al, Ti contents in zoisite rarely exceed trace element levels and are typically confined to a few hundred ppm. In contrast, in clinozoisite-epidote Ti is normally present as a minor element with values from 0.18 to 0.37 wt% most frequently observed. Similar to Mn, the observed concentration range and median is almost identical for metabasic, metagranitic and igneous rocks ($TiO_2$ content in epidote from metabasite ranges from 0.03 to 2.06 wt% with a median of 0.15 wt%; from metagranite from 0.08 wt% to 1.46 wt% with a median of 0.25 wt%; in igneous rocks from 0.08 to 0.75 wt% with a median of 0.18 wt%). The highest Ti contents found so far are 2.06

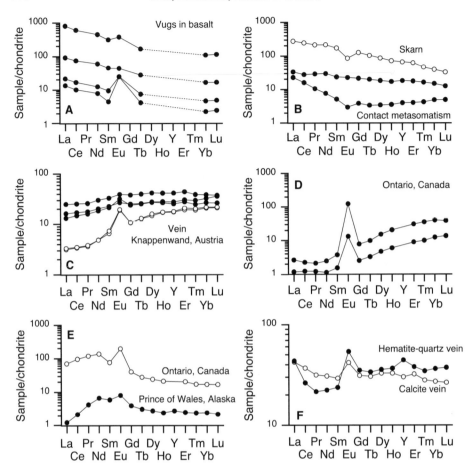

**Figure 7.** Chondrite normalized REE patterns of epidote minerals from (a) vugs in basalt (Nystrøm 1984); (b) skarn from Arendal/Norway; and contact metasomatic rocks Schriesheim/Odenwald (Germany); (c) in veins from Knappenwand, Untersulzbachtal (Austria); the two identical patterns (open circles) are from one sample which is the matrix of freely grown crystals. The other patterns are from euhedral gem quality crystals. (d) clinozoisite from a hydrothermal vein (lower pattern) and epidote from an unspecified metamorphic rock (Pan and Fleet 1996). (e) Epidote from Wollaston county (Pan and Fleet 1996) with a positive Eu anomaly. Lower pattern is for epidote from a vein filling in amphibolite. (f) Epidote from a hematite-quartz vein (Harz/Germany) and from a calcite vein (Sustenpaß/Switzerland).

wt% $TiO_2$ (0.125 Ti pfu) in addition to 0.87 wt% $Mn_2O_3$ (0.053 Mn pfu) in epidote from epidote-glaucophane-blueschist from the West Carpathians, Slovakia (Howie and Walsh 1982). Between zoisite and clinozoisite, Ti is preferentially incorporated into clinozoisite (Franz and Selverstone 1992). Other examples for this enrichment in clinozoisite from pegmatites and amphibolites are shown below (Fig. 18)

*Vanadium.* V can be present as a major element, but is typically only a trace constituent. It is usually enriched compared to the host rock. Observed V contents for epidote in metagranite range from 155 to 624 ppm (Spiegel et al. 2002) and in metabasic rocks from 250 ppm in epidote-glaucophane-blueschist (Howie and Walsh 1982), in eclogites in New Caledonia with relatively uniform V contents ranging from 846 to 985 ppm (Spandler et al. 2003), to a maximum of 1080 ppm in amphibolite (Schreyer and Abraham 1978), with values from 600 to 800 ppm most frequently observed. The higher V contents in metabasic rocks compared to metagranitic rocks thus reflect the higher V contents in the basaltic precursors (18 ppm in granite compared to 270 ppm in basalt). For detrital epidote grains in sandstones of the Swiss Molasse Basin Spiegel et al. (2002) report a variation in V contents from 288 to 577 ppm. Zoisite from high-$P$ segregations in eclogites has extremely uniform V contents of 317 to 320 ppm (Brunsmann et al. 2000). Own unpublished EMP analyses show up to 0.3 wt% $V_2O_3$ in zoisite (var. "tanzanite") from Tanzania in good accordance with data from Hurlbut (1969).

*Chromium*. Like V, Cr can be present as a major element, but only rarely exceeds the trace element level. Similar to V, epidote minerals are usually enriched in Cr compared to their host rocks. The observed Cr abundances for epidote in metabasite (274–550 ppm; Howie and Walsh 1982; Nagasaki and Enami 1998; Sassi et al. 2000) are higher than for epidote in metagranite (12–121 ppm; Spiegel et al. 2002), reflecting the Cr contents of the respective precursor rocks (8 ppm in granite compared to 147 ppm in basalt). A distinctly higher Cr content of 1263 ppm is reported by Yang and Rivers (2002) for epidote in Ca-pelitic schist. The Cr content for detrital epidote grains in sandstones of the Swiss Molasse Basin ranges from 34–337 ppm (Spiegel et al. 2002). Epidote occurrences where $Cr^{3+}$ is a major element are scarce (Bleek 1907; Eskola 1933; Cooper 1980; Grapes 1981; Treloar 1987; Ashley and Martin 1987; Challis et al. 1995; Sánchez-Vizcaíno et al. 1995; Devaraju et al. 1999). They are restricted to parageneses with other Cr-enriched phases (e.g. chromite, Cr-rich K-white mica "fuchsite," Cr-rich margarite). Examples are Cr-rich amphibolite (Grapes 1981), metamorphosed Cr-spinel rich beach placer deposits (Sánchez-Vizcaíno et al. 1995), and metakomatiite that experienced residual concentration of Al, Ti, Cr, and V due to pre-metamorphic hydrothermal alteration (Ashley and Martin 1987).

Only a few studies report data for Cr contents in zoisite: Brunsmann et al. (2000) observed uniform Cr contents of about ~120 ppm for zoisite in high-$P$ segregations in eclogite. Cr contents in eclogite from Dabie Shan that experienced ultra-high-$P$ metamorphism range from 820 to 1370 ppm (Nagasaki and Enami 1998; Sassi et al. 2000). Game (1954) and Grapes (1981) reported zoisite with Cr as minor element (0.23 and 0.40 wt% $Cr_2O_3$; 0.014 and 0.024 Cr pfu) coexisting with chromian epidote (3.72 and 7.12 wt% $Cr_2O_3$; 0.233 and 0.441 Cr pfu) in kyanite-amphibolite from New Zealand. The highest content reported so far is 2.46 wt% $Cr_2O_3$ (0.15 Cr pfu) in zoisite coexisting with fuchsite and margarite in a pseudomoph after kyanite (Cooper 1980). This points to a limited solubility of $Cr^{3+}$ in the zoisite structure, in accordance with the limited incorporation of $Fe^{3+}$ into zoisite.

*Manganese*. Mn is a major cation in piemontite and related minerals and forms extensive solid solution along the $Al-Fe^{3+}-Mn^{3+}$ ternary (for details see Bonazzi and Menchetti 2004; Franz and Liebscher 2004). Hence, Mn is a frequent minor and trace constituent in epidote and zoisite. Because Mn is included in most standard routines for the determination of major elements by electron microprobe, a large number of analyses is available. The majority of the studies, however, refer to Fe-rich epidote, whereas the database for Fe-poor clinozoisite and zoisite is restricted. In zoisite, Mn contents are usually restricted to a few hundred ppm. The highest Mn contents in natural zoisite are 1.6 to 3.7 wt% $Mn_2O_3$ (0.09 to 0.22 $Mn^{3+}$ pfu) found in zoisite from low grade metamorphic rocks metamorphosed at highly oxidizing conditions (Reinecke 1986).

In the majority of monoclinic Al-Fe solid solutions, Mn is normally abundant as a minor element. Within each lithological group Mn contents display large variations and between different lithological units large overlaps are observed ($Mn_2O_3 = 0.02$ to 0.87 wt% in metabasite; 0.11 to 0.43 wt% in metagranite; 0.17 to 3.20 wt% in igneous rocks). Yanev et al. (1998) and Choo (2002) describe Mn-enriched epidote formed during hydrothermal alteration of acidic volcanic tuffs (eastern Rhodopes, Bulgaria: 3.12 wt% $Mn_2O_3$; Bobae sericite deposit, southeastern Korea: 1.95 wt% $Mn_2O_3$). Individual grains show strong oscillatory or patchy zoning, interpreted as fluctuations in the Mn concentration of the hydrothermal fluid responsible for epidote formation.

*Cobalt, nickel, and copper.* The data for Co, Ni, and Cu are very limited. Cobalt abundances are usually restricted to a few ppm. Reported values for epidote range from 1 ppm in epidote from a calc-pelitic schist (Yang and Rivers 2002) to 14 ppm in epidote from a metagranite (Spiegel et al. 2002). Sorensen and Grossman (1989) report 8.1 and 1.6 ppm Co for clinozoisite and zoisite, respectively, in garnet-amphibolite from the Catalina Schist, southern California, USA. Ni contents of epidote in metabasite range from 10 to 31 ppm (Howie and Walsh 1982; Spiegel et al. 2002), and are 19 and 45 ppm in calc-pelitic schist (Yang and Rivers 2002) and metagranite (Spiegel et al. 2002), respectively. Zoisite in garnet-amphibolite (Catalina Schists, southern California, USA; Sorensen and Grossman 1989) contains 22 ppm Ni. Detrital epidote grains of the Swiss Molasse Basin display a range of Ni contents from 7 to 66 ppm (Spiegel et al. 2002). Cu concentrations range from 2 to 73 ppm for epidote in metabasite (Howie and Walsh 1982; Spiegel et al. 2002), from 6.6 to 28.2 ppm in metagranite, and from 3.1 to 19.3 ppm in detrital grains (Spiegel et al. 2002). No studies are available that address Cu abundances in clinozoisite and zoisite.

*Zinc.* Concentrations for epidote range from 4 to 64 ppm in metabasite (Howie and Walsh 1982; Spiegel et al. 2002) and from 29 to 89 ppm in metagranite, reflecting the higher Zn abundance in granitic rocks. Yang and Rivers (2002) report a distinctively lower Zn content of 8 ppm for epidote in calc-pelitic schist. In clinozoisite and zoisite separates from garnet-amphibolite Sorensen and Grossman (1989) determined 95 and 13.4 ppm Zn, respectively. The latter value is in good agreement with the 20 ppm Hickmott et al. (1992) measured in zoisite from the same locality by PIXE. The highest Zn abundances observed so far are 0.49 and 1.14 wt% ZnO (0.027 and 0.066 Zn pfu, respectively) in unusual Ge- and Ga-enriched zoisite from the Trappes zinc deposit, Haute Garonne, Central Pyrenees, France (Johan et al. 1983).

### High field strength elements (Zr, Hf, Nb and Ta)

*Zirconium and hafnium.* Abundances grossly reflect their abundances in the Earth's major reservoirs. Zr contents range from a few ppm to about ~200 ppm, whereas Hf contents range from around ~0.1 ppm to 1 ppm. Hf and Zr show a positive linear correlation for both polymorphs (Fig. 1) with a Zr/Hf ratio near ~20 and thus significantly lower than the ratios for C1-chondrites (37), N-MORB (36) and average continental crust (41), pointing to a marked enrichment of Hf in epidote minerals in all major geological environments. Zr/Hf and U/Th ratios vary independently from each other (Fig. 1), but zoisite has a larger variation in Zr/Hf, whereas epidote has a larger variation in U/Th.

*Niobium and tantalum.* Like Zr and Hf, the observed concentration range for Nb and Ta is confined to the abundances of these elements in the major geochemical reservoirs of the Earth. No clear correlation for Nb and Ta is observed (Fig. 1) and the Nb/Ta ratios scatter between close to unity for zoisite from segregations in high-*P* segregations and in an amphibolite, values around ~5 in metamorphosed pegmatites and near chondritic values of 14 in a mica schist. However, due to the very low concentrations in zoisite the fractionation effect will be small. In some clinozoisite-epidote samples Nb and Ta are positively correlated with Nb/Ta ratios close to the chondritic ratio of 17 (Fig. 1). Hence, epidote minerals are not expected to have a large fractionation effect on Nb and Ta.

### Other elements

*Lithium.* The observed concentration of Li ranges from 0.15 ppm in zoisite in eclogites from Trescolmen, Central Alps, Switzerland (Zack et al. 2002), and 0.2 to 1.1 ppm in epidote from various high-*P* rocks (H. Marschall, pers. comm.) to 15 and 19 ppm in epidote-glaucophane-blueschists from Slovakia (Howie and Walsh 1982). These high concentrations were determined on mineral separates by wet chemical and ICP AES analyses.

***Beryllium.*** Domanik et al. (1993) observed 0.1 and 0.4 ppm Be in epidote, and 0.2 and 0.4 ppm Be in clinozoisite in blueschist and amphibolite from the San Franciscan Complex and Catalina Schists, California, USA. H. Marschall (pers. comm.) determined 0.016 to 0.046 ppm in epidote from various high-*P* rocks.

***Boron.*** In the same samples analyzed for Be, Domanik et al. (1993) reported 8 and 5 ppm B in epidote and 3 and 24 ppm B in clinozoisite, and H. Marschall (pers. comm.) 0.22 to 0.61 ppm in epidote. Lisitin and Khitrov (1962) report 55 ppm to 330 ppm B in epidote from skarn and epidosite from the contact aureole of a co-magmatic series of igneous rocks (diorite, syenite and andesite) of the Ural Mountains, Russia.

***Magnesium.*** It is a major constituent in the REE-epidotes dissakisite $CaREEAl_2MgSi_3O_{11}O(OH)$ and dollaseite $CaREEAlMg_2Si_3O_{11}F(OH)$ and is frequently observed as minor and trace element in REE-bearing zoisite, clinozoisite and epidote. Like Ti and Mn, Mg is included in most standard routines for the determination of major elements in EMPA, and the relatively high detection limits of EMPA will inevitably bias the reported Mg abundances towards higher values. Hence, the lowest reliable MgO contents scatter around the detection limit of EMPA. The highest reported content in epidote is 0.95 wt% MgO (0.12 Mg pfu) for phenocrysts in porphyritic dacitic to rhyodacitic dikes (Front Range, Colorado, USA; Dawes and Evans 1991). The observed variations point to higher Mg contents in clinozoisite-epidote compared to zoisite, in accordance with the limited acceptance of the zoisite structure for divalent cations (Franz and Selverstone 1992; Brunsmann et al. 2001; Liebscher et al. 2002). For all reliable analyses from our data set, Mg pfu is equal or less than the sum of the REE pfu indicating substitution (5) or (6).

***Gallium.*** Abundances are remarkably constant; they vary from ~40 to 60 ppm irrespective of polymorph and of host rock (Hickmott et al. 1992; Brunsmann et al. 2000; Spiegel et al. 2002). Unusually high Ga contents with 0.43 wt% $Ga_2O_3$ (0.022 Ga pfu) and 0.53 wt% $Ga_2O_3$ (0.027 pfu) have been determined in a zoisite sample from the Trappes sphalerite ore deposit (see above) by Johan et al. (1983).

***Tin.*** Analyses have so far only been reported for unusually tin-rich epidote in skarn associated with Sn-deposits (Myer 1965; British Columbia, Canada, Mulligan and Jambor 1968; Cornwall, England, Alderton and Jackson 1978; ibid, Van Marcke de Lummen 1986). The highest Sn content described so far is 2.84 wt% $SnO_2$ (0.09 Sn pfu) for epidote in a skarn from The Crowns, Botallack, Cornwall, England (Van Marcke de Lummen 1986).

***Arsenic, molybdenum, silver, cadmium, antimony and tungsten.*** The only available data have been determined by Spiegel et al. (2002) using Multi Collector - Inductively Coupled Plasma - Mass Spectrometry (MC-ICP-MS). Samples are dominantly detrital grains in sandstone from the Swiss Molasse Basin, but include a few epidote minerals from metabasic and metagranitic pebbles collected from the Bernina river and molasse conglomerates, and three metabasites from the penninic Bündner Schiefer series (Tauern Window, Eastern Alps). Unfortunately, neither the major element chemistry, nor the symmetry of the analysed specimens is reported. The description of the separated grains ("clear, yellow-greenish epidotes" and "milky, often zoned, dark-green-bluish epidotes"; cf. Spiegel et al. 2002) is suggestive for monoclinic epidote. Arsenic abundances scatter around 3 to 5 ppm and only one epidote from a metabasalt displays a higher content of 52 ppm. No systematic differences in concentration levels are observed between epidote from metabasic and metagranitic lithologies. Mo contents above the detection limit are only reported for epidote from metagranite. The observed concentrations scatter around the crustal average of 1100 ppb. Ag contents range from 800 to 2800 ppb with values around 1000 ppb most frequently observed. Although the highest Ag content is reported in a metabasic rock, the average Ag content from metagranite (~2000 ppb) appears to be slightly higher compared to metabasite (~1300 ppb). All observations point to a marked enrichment of epidotes compared to the average crustal Ag abundance of 70 ppb. Cd and Sb abundances are also enriched compared to crustal values. Cd values scatter around 700 ppb while Sb values scatter around 1000 ppb. For both elements there is no evidence for lithological control on abundances. W contents show a relative wide concentration range compared to the aforementioned elements from 1–133 ppm, but values > 50 ppm are most frequently observed. This points to a marked enrichment for W compared to the average crustal abundance of 1 ppm. The observed concentrations suggest significant differences in W contents between epidote from metagranite (51–133 ppm) and metabasite (10–45 ppm).

***Germanium.*** In addition to the high Zn and Ga contents, the zoisite sample from the Trappes sphalerite ore deposit (Johan et al. 1983) contains Ge as a major element with 5.59 wt% $GeO_2$ (0.252 Ge pfu) and 6.42 wt% $GeO_2$ (0.290 Ge pfu), suggesting significant substitution of the larger $Ge^{4+}$ for the smaller $Si^{4+}$ into the T sites.

***Phosphorous.*** Although the presence of P is noted in many EMPA obtained under standard instrument-operating conditions, the bulk of the reported data is below or close to the detection limit and their reliability is highly questionable. Howie and Walsh (1982) found 0.42 wt% $P_2O_5$ in epidote-glaucophane-blueschist from Slovakia. In oscillatory zoned epidote from hydrothermally altered rhyodacitic tuff from the Bobae sericite deposit, Korea, Choo (2002) describes $P_2O_5$ contents ranging from 0.24 to 0.67 wt%. For zoisite, Maaskant (1985) reports 0.30 wt% $P_2O_5$ in zoisite from high grade metamorphic rocks from Cabo Ortegal, Galicia, Spain. The processes that lead to elevated P contents are yet unknown. In contrast to V, Cr, Sr, and Pb, where high abundances are unequivocally linked to high abundances in the host rock, high P content in the host rock leads to the formation of P-bearing phases (mainly apatite, monazite, and xenotime) and P is strongly partitioned into these phases compared to epidote minerals. This behavior is evident from the observations of Finger et al. (1998), who described the formation of apatite-allanite-epidote coronas around monazite (see above). Both epidote and allanite are characterized by high U, Th, and REE contents derived from the breakdown of monazite, but the P contents are below the detection limit (0.03 wt% $P_2O_5$, i. e., ~130 ppm P) of the electron microprobe used (Finger et al. 1998).

For this review we have analyzed a series of epidote and zoisite samples from our collection with the EMP (own unpublished data), taking special care for obtaining reliable results by optimizing the operating conditions. The highest amount found was 0.07 wt% $P_2O_5$ (0.005 P pfu).

### Anions

***Halogens.*** In REE-rich epidotes, F and Cl substitution for $O^{2-}$ is often observed (with F typically more abundant than Cl) and F can even occur as a major constituent in dollaseite $CaREEAlMg_2Si_3O_{11}F(OH)$ (Gieré and Sorensen 2004). In marked contrast, reliable analyses reporting F contents above the detection limit of EMPA are extremely rare for REE-poor epidotes, pointing to overall very low F abundances. Noticeable F contents in monoclinic Al-Fe solid solutions are described for magmatic epidote in quartz-monzonite (0.06 wt% F) at Mount Lowe, Califonia, USA (Keane and Morrison 1997), and in dacitic-rhyodacitic dikes (0.06 to 0.6 wt% F) from the Front Range, Colorado, USA (Evans and Vance 1987; Dawes and Evans 1991). Pan and Fleet (1996) report 0.3 wt% F in REE-bearing clinozoisite occurring as a grain nucleus in F- and REE-free clinozoisite coexisting with F- and Cl-rich allanite (0.76 and 0.95 wt%, respectively) in the zone of anomalous Au abundances at the White River gold occurrence, Hemlo area, Canada. This observation clearly suggests that the incorporation of significant amounts of halogens is preferentially achieved via substitution (7) at high halogen-fugacities and sufficient availability of REE and divalent cations for charge balance. The preference of F for Mg-rich compositions (e.g., dollaseite) might suggest that epidote minerals comply with the Fe-F-avoidance principle observed in other hydrous silicates such as biotite and amphibole (Volfinger et al. 1985). In zoisite, F contents are generally below or near the detection limit of the EMP (e.g. Hoschek 1980). For this review we have analyzed a series of epidote and zoisite samples from our collection with the EMP (own unpublished data), taking special care for obtaining reliable results by optimizing the measuring time. The highest amount was 0.096 wt% F in zoisite and 0.047 wt% in clinozoisite. However, Anisimova et al. (1975) list zoisite with 1.68 wt% F from an eclogite-hydrothermal vein occurrence.

Only two studies report significant Cl contents. Maaskant (1985) described 0.1 wt% Cl in zoisite in high-grade metamorphic rocks from Cabo Ortegal, Galicia, Spain, and Evans and Vance (1987) found 0.11 wt% Cl, in addition to 0.6 wt% F, in epidote phenocrysts in dacitic-rhyodacitic dikes (see above). In our survey we observed that in most cases Cl contents are near 0.01 wt% or below, the highest content was 0.11 wt% in zoisite from pegmatites in an eclogite terrane (Saualpe, Kärnten, Austria).

## PARTITIONING OF TRACE ELEMENTS BETWEEN EPIDOTE MINERALS AND MELTS, FLUIDS AND OTHER MINERALS

For a quantitative description of the role of epidote minerals in petrogenetic processes a thorough understanding of the trace element partitioning between epidote minerals and melts, fluids, and/or other minerals is crucial. In the following, we review the available trace element partitioning data for epidote minerals. We have included the available partitioning data for allanite, because Gieré and Sorensen (2004) address this topic only briefly and the allanite data also yield some information about the partitioning behavior of the structurally very similar monoclinic Al-Fe solid solutions, for which no partitioning data exist. We refer to the *Nernst partition coefficient* (Beattie et al. 1993)

$$D_i^{\alpha/\beta} = \frac{wt\ fraction_i^\alpha}{wt\ fraction_i^\beta}$$

with $\alpha$ = epidote minerals and $\beta$ = melt, fluid, or other mineral phase. Relative element fractionation is described by the exchange coefficient $K_D$ (Beattie et al. 1993)

$$K_{D_{i-j}}^{\alpha/\beta} = \frac{D_i^{\alpha/\beta}}{D_j^{\alpha/\beta}}$$

### Epidote minerals/melt partitioning

The few mineral/melt data mostly refer to allanite/melt partitioning of REE, transition metals (Sc, Cr, Mn, Fe, Co, Cu and Zn), LILE (Sr, Ba and Rb), actinides (Th and U), and HFSE (Zr, Hf, Nb, Ta) derived from coexisting phenocryst-matrix pairs in silica-rich igneous rocks (Brooks et al. 1981; Mahood and Hildreth 1983; Sawka 1988; Ewart and Griffin 1994). Only two studies present experimentally derived allanite/melt partition coefficients for Ba, Th, U, and REE (Hermann 2002a), and zoisite/melt partition coefficients for REE (Frei et al. 2003). Experimental studies addressing the effects of varying $P$, $T$, and crystal and/or bulk compositions on partitioning do not exist.

In allanite (Fig. 8), both Rb and Ba are incompatible with $D_{Rb}^{aln/melt} < D_{Ba}^{aln/melt}$ by about one order of magnitude. Whereas U is moderately compatible, Th is highly compatible and hence it is very effectively fractionated from U by allanite. However, the relative fractionation differs significantly between that determined experimentally ($K_{D\ (Th-U)}^{aln/melt} = 4$) and that derived from naturally occurring allanite/melt pairs ($K_{D\ (Th-U)}^{aln/melt} = \sim30$) due to the about ten times lower $D_U^{aln/melt}$ in the experiments. Hermann (2002a) does not report redox conditions for his experimental runs, and it can be speculated that the low $D_U^{aln/melt}$ reflects oxidation of $U^{4+}$ to $U^{6+}$,

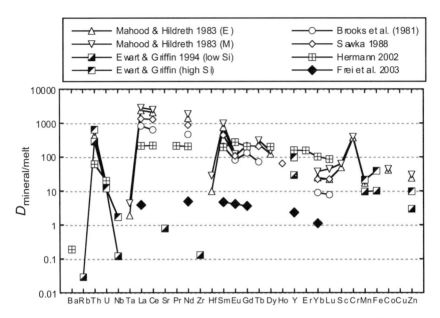

**Figure 8.** Multi-element diagram displaying allanite/melt (open symbols) and zoisite/melt (filled symbols) partition coefficients. Elements on the abscissa have been arranged in order of decreasing compatibility in the Earth's crust from the left to the right according to Hofmann (1988).

which lowers U partition coefficients considerably (Blundy and Wood 2003). High $f_{O_2}$ might prevent the substitution of U and Th into allanite by increasing $Fe^{3+}/Fe^{2+}$ or $Mn^{3+}/Mn^{2+}$ and thus reducing the amount of divalent cations for charge balancing. Partitioning for U and Th between epidote minerals and melts is expected to be strongly dependent on $f_{O_2}$.

For the HFSE $D_{HFSE}^{aln/melt}$ values vary from moderately compatible to moderately incompatible. $D_{HFSE}^{aln/melt}$ for phenocrysts in equilibrium with Si-rich melts are higher than those in equilibrium with Si-poor melts (Mahood and Hildreth 1983; Ewart and Griffin 1994). The higher $D_{HFSE}^{aln/melt}$ in Si-rich samples can be explained by lower equilibration temperature (Mahood and Hildreth 1983; Blundy et al. 1995) and/or a higher degree of polymerization of the coexisting melt (Mysen and Virgo 1980). Contamination of the analyses by rutile, titanite or melt inclusions can be ruled out to explain the high $D_{Nb}^{aln/melt}$ (Ewart and Griffin 1994), because the data were obtained *in situ* by proton-microprobe and have been carefully evaluated for possible contaminations. However for $D_{Hf}^{aln/melt}$ in Si-rich glass Mahood and Hildreth (1983) reported coexisting zircon and hence contamination in the allanite separate analyzed by INAA for Hf by concomitantly analyzed zircon inclusions can not be ruled out.

All REE are highly compatible in allanite and moderately compatible in zoisite. The experimentally determined $D_{LREE}^{aln/melt}$ values (Hermann 2002a) are about one order of magnitude lower than those determined for natural allanite/melt pairs, those for the HREE are higher. $D_{REE}^{zo/melt}$ and $D_{REE}^{aln/melt}$ as a function of ionic radius ('Onuma diagrams') show the marked differences in the relative REE partitioning behavior of zoisite, synthetic allanite, and natural allanite (Fig. 9). The $D_{REE}^{aln/melt}$ partitioning pattern of synthetic allanite (Hermann 2002a) is strikingly different compared to the other samples and shows an almost flat partitioning pattern with only weak ($K_{D\,(La-Yb)}^{aln/melt} = 2.1$) or no REE fractionation ($K_{D\,(La-Sm)}^{aln/melt} = 1.1$). Moreover, the $D_{REE}^{aln/melt}$ values do not show a smooth parabolic dependence on ionic radius and the derived values

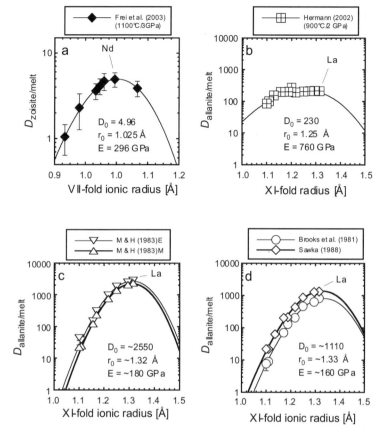

**Figure 9.** Zoisite/melt (a) and allanite/melt (b-d) REE partition coefficients vs. ionic radii ('Onuma diagrams'). Solid curves are non-linear least squares fits to Blundy and Wood's (1994) elastic strain model assuming REE incorporation into both VII-fold coordinated A sites in zoisite and in the large XI-fold coordinated A2 site in allanite. Also reported are best-fit values for $D_0$, $r_0$ and $E$. Best fit values reported for the data displayed in (c) and (d) are average values from Frei et al. (2003); abbreviation M & H (1983) refers to samples E and M, respectively, from Mahood and Hildreth (1983).

for $r_0$ and $E$ are crystallographically and physically unrealistic. The reason for the flat REE partitioning pattern determined by Hermann (2002a; Fig. 9b) and the notable inconsistency with the natural samples remains unresolved. The other data show that with increasing ionic radius, the compatibility of the REE in zoisite (Fig. 9a) increases from Yb ($D_{Yb} = 1.1$) to Nd ($D_{Nd} = 4.9$) by half an order of magnitude and decreases from Nd to La ($D_{La} = 3.9$) resulting in a parabolic partitioning pattern. Hence, zoisite fractionates the LREE and MREE effectively from the HREE ($K_D{}^{zo/melt}_{(La-Yb)} = 3.7$ and $K_D{}^{zo/melt}_{(Sm-Yb)} = 4.5$) and slightly fractionates the MREE from LREE ($K_D{}^{zo/melt}_{(La-Sm)} = 0.8$). The Blundy and Wood (1994) elastic strain model, using a Levenberg-Marquadt-type non-linear least square fit to the data, yields values for $r_0$ and $E$ that are in good agreement with a preferred incorporation of the REE into the A1 site. Only La and Ce are incorporated in significant amounts into the A2 site in zoisite (Frei et al. 2003).

In contrast, the partitioning data derived from allanite phenocrysts (Fig. 9c,d) result in strongly LREE enriched partitioning patterns ($K_D{}^{aln/melt}_{(La-Yb)} \approx 80$), in good agreement with those predicted for the exclusive incorporation of REE into the A2 site of allanite using Blundy and

Wood's (1994) elastic strain model (Frei et al. 2003). The obtained $r_0$ are similar to that for the A2 site in allanite derived from structure refinements (Dollase 1971), and the best-fit values for $E$ are comparable to those predicted from crystallographic data.

Although there are no REE mineral/melt partitioning data available for the clinozoisite-epidote series some general aspects on its partitioning systematic might be drawn from the allanite/melt and zoisite/melt data described above and published trace element pattern in zoisite, clinozoisite, and epidote from (meta)igneous rocks. Their REE pattern of epidote from igneous and metamorphic rocks are generally straight, enriched in LREE with positive La/Sm ratios (see Figs. 2, 3, 5). The patterns for epidote in syenite, granite, most pegmatites and some metabasites are indicative for exclusive partitioning of REE into the A2 site in epidote and reflect the similarities between the epidote and allanite structures (Frei et al. 2003). For clinozoisite, only few REE data are reported and the patterns are diverse, with both LREE (La/Yb >> 1) and HREE enrichment (La/Yb as low as 0.09). However, the REE patterns of clinozoisite and zoisite coexisting in high-$P$ pegmatites (and of clinozoisite formed by the breakdown of zoisite in high-$P$ segregations) indicate that no significant fractionation occurs between these phases (see below). This suggests that clinozoisite shows an REE mineral/melt fractionation pattern comparable to that of zoisite.

### Epidote mineral/hydrous fluid partitioning

The only measured mineral/fluid partitioning data are derived from analyzing fluid inclusion in allanite from a granite ($D_{REE}^{aln/fluid}$; Banks et al. 1994). Brunsmann et al. (2001) obtained $D_{REE}^{zo/fluid}$ from high-$P$ zoisite segregations indirectly by mass-balance calculation. The model, which is the basis for the mass balance, allows however a variation of the absolute $D$ values up to one order of magnitude. The magnitudes and the general patterns of the partition coefficients resemble those for mineral/melt partitioning. The values for $D_{REE}^{aln/fluid}$ (Fig. 10) indicate strongly compatible behavior and a characteristic enrichment of LREE, but a relatively flat partitioning pattern ($K_{D\,(La-Sm)}^{All/fluid} \approx 2$). In zoisite, the REE are moderately incompatible with respect to the fluid, in contrast to the moderately compatible behavior of $D_{REE}^{zo/melt}$, but the pattern show a similar enrichment of MREE compared to the LREE ($K_{D\,(La-Sm)}^{zo/fluid} \approx 0.2$) also found for zoisite/melt partitioning. The absolute values of $D_{REE}$ for allanite and zoisite differ by four orders of magnitude for both silicate melt and aqueous fluid, explained by the restricted capability of the zoisite structure to incorporate divalent cations like $Fe^{2+}$ and $Mg^{2+}$, in contrast to the monoclinic modifications (cf. Brunsmann et al. 2001).

The mineral/fluid $D_{REE}$ for both allanite and zoisite (Fig. 10) do not fit to Blundy and Wood's (1994) elastic strain model and indicate that the fluid chemistry has a strong influence on mineral fluid partitioning. The factor, which plays an important role in controlling mineral/fluid partitioning, is the solubility of REE in fluids, which is known to be controlled by the

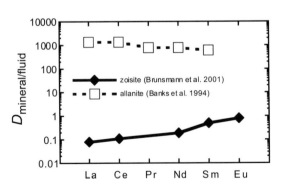

**Figure 10.** REE partition coefficients for allanite/fluid (open symbols) and zoisite/fluid (filled symbols).

availability of ligands (e. g. F⁻, Cl⁻, $CO_3^{2-}$, $PO_4^{3-}$, $SO_4^{2-}$) to form REE complexes (Cantrell and Byrne 1987; Wood 1990a; Haas et al. 1995; Gieré 1996). High concentrations of ligands will stabilize REE complexes in solution and lower mineral/melt partition coefficients. Moreover, stability constants for complexes with ligands of LREE differ from those for HREE (Wood 1990b; Haas et al. 1995) and thus fluid composition can have a significant bearing on relative REE fractionation. The observed HREE enrichments in epidote, clinozoisite and zoisite precipitated in hydrothermal environments and dehydrated high-*P* rocks (Figs. 6, 7) thus might be explained by control of fluid chemistry on mineral/fluid partitioning. Another important factor controlling the REE partitioning behavior between epidote minerals and hydrous fluids is oxygen fugacity of the fluid, which controls $Fe^{2+}/Fe^{3+}$ and $Mn^{2+}/Mn^{3+}$.

Becker et al. (1999) calculated fluid partitioning of $10^4$ to $10^6$ by combining experimentally determined clinopyroxene/fluid partition coefficients with measured $D^{cpx/zo-ep}$ according to the simple relationship $D^{zo-ep/fluid} = D^{cpx/fluid}/D^{cpx/zo-ep}$. These values are several orders of magnitude higher than those derived by mass balance calculation (Brunsmann et al. 2001). The approach by Becker et al. (1999) has the disadvantage that experimental and natural conditions are different. Naturally occurring zoisite/clinopyroxene pairs usually equilibrated at temperatures of ≈600°C, but the experiments were carried out at 900 to 1200°C (Brenan et al. 1995; Stalder et al. 1998). Since the absolute values of mineral/mineral partition coefficients strongly depend on temperature (Eggins et al. 1998) the calculated zoisite/fluid partition coefficients might be in error by orders of magnitudes. Moreover, clinopyroxene from experiments is diopside rich (<10 mol% jadeite component), whereas the clinopyroxene in eclogites has usually >25 mol% jadeite component. Because the major element chemistry of minerals can have a profound effect on relative trace element fractionations (e.g., Van Westrenen et al 1999; Landwehr et al. 2001), this compositional difference might lead to significant errors in calculated relative fractionations.

The extreme differences between the results of Brunsmann et al. (2001) and Becker et al. (1999) can not be explained and it is not clear, which values are closer to the true ones. Consequently, care has to be taken when calculating mineral/fluid partitioning coefficients either from mass balance or from mineral/mineral partitioning coefficients.

**Partitioning between epidote minerals and other minerals**

The relative partitioning of trace elements between epidote minerals and the major rock-forming minerals garnet, omphacite, plagioclase, amphibole (including glaucophane, hornblende, and unspecified amphibole), calcite, a variety of miscellaneous mineral phases (corundum, kyanite, titanite, apatite, rutile, and hematite) and between zoisite and clinozoisite for a variety of different rocks is shown in Figures 11 to 19.

***Epidote - garnet.*** The data for $D^{ep/grt}$ are, apart from the absolute magnitude, very similar in all rock types (Fig. 11). The REE patterns are steep with a marked preference of LREE for epidote minerals and HREE for garnet. U, Th, Sr, and Pb are strongly enriched in epidote. With the exception of $D^{zo/grt}_{Rb}$, which is close to unity, the LILE are preferentially incorporated in epidote minerals. Partition coefficients for HFSE are closer to unity resulting in a marked Nb-Ta and Zr-Hf trough. It is noticeable that Hf relative to Zr shows a preference for epidote compared to garnet in all cases.

***Epidote/zoisite - omphacite.*** Only isolated data exist for zoisite and epidote (Fig. 12), which are enriched in all elements. The highest enrichments are observed for LREE, Pb and Sr, and in one case for Ba. Like for garnet, partition coefficients for Zr and Hf are close to unity and display a distinct negative anomaly, but Hf shows a preference for omphacite.

***Epidote/zoisite - amphibole.*** Although all partition coefficients share some common features (Fig. 13), they are more variable compared to those for garnet and omphacite. Except Cs, Rb, Ba and the HFSE, which scatter around $D = 1$ and go up to 10 all other elements

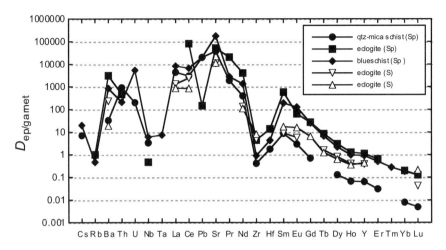

**Figure 11.** Multi-element diagram displaying trace element partition coefficients for epidote/garnet from high-*P* rocks; Sp = Spandler et al. (2003) and S = Sassi et al. (2000).

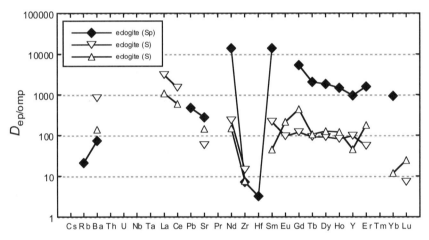

**Figure 12.** Multi-element diagram displaying trace element partition coefficients for epidote/omphacite in eclogites; Sp = Spandler et al. (2003) and S = Sassi et al. (2000).

are enriched in epidote minerals. Apart from the troughs in HFSE the partitioning patterns are characterized by a pronounced negative Pb anomaly. The actinides partition into epidote minerals, but *D*-values vary by three orders of magnitude. The most pronounced deviations are observed for epidote/amphibole pairs in epidote veins (Fig. 13a; samples 245-1, 2), which shows a flat REE pattern and $D_{Cs}^{ep/amp}$ and $D_{Ba}^{ep/amp}$ values > 1, and from an epidote fels (Fig. 13b; sample SM 4) with a concave-upwards REE partitioning pattern and a pronounced negative Y anomaly.

***Zoisite - plagioclase.*** The most remarkable feature of the partitioning pattern is the variation by more than three orders of magnitude for Th, U, REE and HFSE, whereas the values for $D_{LILE}^{zo/plg}$ show a rather limited variation. Zoisite is commonly enriched compared to plagioclase except for Ba, where $D_{Ba}^{zo/plg}$ is always < 1, and Sr and Pb, where most *D*'s are close

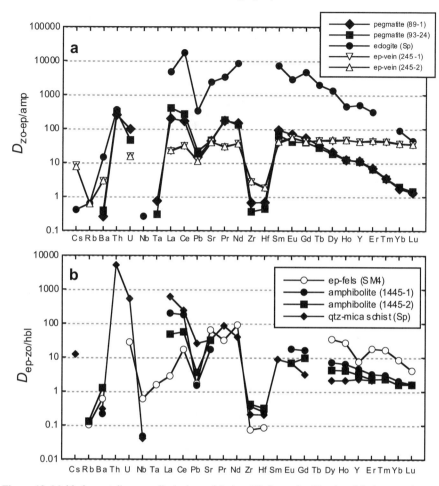

**Figure 13.** Multi-element diagrams displaying zoisite/amphibole and epidote/amphibole trace element partition coefficients in various geological environments; (a) refers to unspecified amphibole modifications whereas (b) refers to hornblende. Sample numbers refer to specimens analyzed for this study given in Appendix A; Sp = Spandler et al. (2003).

to unity. The ability of anorthite (sample 1445, Fig. 14) to incorporate actinides and REE is reflected by the low $D$'s for these elements compared to Na-rich plagioclase (all other samples, Fig. 14); however, the major element composition of plagioclase seems to have only little effect on $D$'s for Ba, Sr and Pb. Two types of $D_{REE}^{zo/plg}$ patterns can be distinguished, a generally flat one with no fractionation within the REE (especially true for the zoisite/anorthite pair) and a concave upward one with MREE enrichment in zoisite. Eu relative to Sm and Gd shows only a very weak or no anomaly.

***Epidote/zoisite - calcite.*** The REE partitioning patterns are steep and indicate an enrichment of LREE in epdiote and in two samples an enrichment of HREE in calcite. The epidote from a vein assemblage (812, Fig. 15) has a strong enrichment of HREE (samples from the same locality behave also different in epidote-amphibole distribution). U and Th are strongly enriched in epidote minerals; with two exceptions, this is also true for Ba, Sr and Pb.

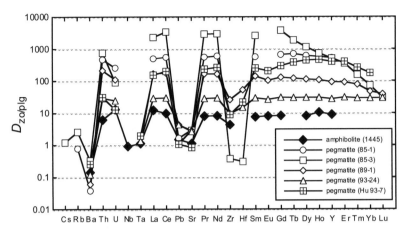

**Figure 14.** Multi-element diagram displaying zoisite/anorthite (filled symbols) and zoisite/albite (open symbols) trace element partition coefficients in various geological environments. Sample numbers refer to specimens analyzed for this study given in Appendix A.

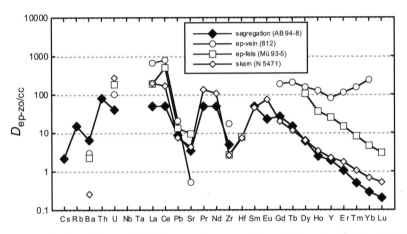

**Figure 15.** Multi-element diagram displaying trace element partition coefficients for zoisite/calcite (filled symbols) and epidote/calcite (open symbols) in various geological environments. Sample numbers refer to specimens analyzed for this study given in Appendix A.

***Epidote minerals - other phases.*** For the oxides corundum, rutile and hematite the distribution patterns (zoisite for corundum, epidote for rutile and hematite) are similar, with the troughs in the HFSE and in Pb (Fig. 16a). The epidote/apatite patterns (Fig. 16b) are the only ones that lack negative Nb-Ta anomalies and display a pronounced positive anomaly for Zr and Hf. For epidote/titanite, the *D* value is extremely low for Nb and Ta, is low for Zr and Hf, and for other elements similar to zoisite/apatite. Two samples for zoisite/kyanite from the same locality ('tanzanite') show strong enrichment for most elements, except for Rb, Zr, and Hf.

***Clinozoisite - zoisite.*** Most data are from pegmatites (Saualpe, Austria, and Münchberg Massif), with some data from hydrous eclogites (Trescolmen/Switzerland; Fig. 17). All patterns show that the trace element content in  the respective clinozoisite-zoisite pairs is very similar, with a slight enrichment of most elements  in clinozoisite, except for Pb and

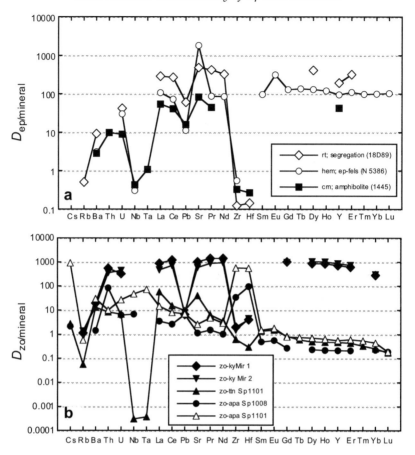

**Figure 16.** Multi-element diagrams displaying trace element partition coefficients between (a) zoisite (filled symbols), epidote (open symbols) and miscellaneous minerals (corundum, kyanite, titanite, apatite, rutile and hematite), and (b) zoisite and miscellaneous minerals (kyanite, titanite, and apatite) in various geological environments. Sample numbers refer to specimens analyzed for this study given in Appendix A.

Sr, which are enriched in zoisite. The REE pattern suggest that their relative fractionation is almost negligible, but there are slight differences in REE enrichment of clinozoisite between individual samples. The almost constant values for $D$ show that the structures of both zoisite and clinozoisite have the same preferences for the REE. As a first order approximation clinozoisite partitioning behavior can therefore be modeled together with zoisite (Zack et al. 2002). The systematic enrichment of zoisite in Sr is shown for two examples (Fig. 18a), one from pegmatites (Saualpe, Austria), and the other one from amphibolite (Longuido Mts., Tanzania). In contrast, the transition element Ti is preferentially distributed in clinozoisite (Fig. 18b).

## CONCLUSIONS AND OUTLOOK

### Trace element budgets and element recycling

*Metamorphic rocks.* The generally high abundances of Sr, Pb, transition metals, actinides, and (light) REE in epidote minerals and the corresponding large $D^{\text{epi/mineral}}$ values indicate that

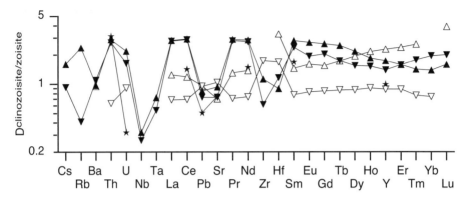

**Figure 17.** Multi-element diagram displaying clinozoisite/zoisite trace element partitioning coefficients in pegmatites (open triangles: Münchberg/Germany; closed triangles: Saualp/Austria) and from eclogite (asterisks; Trescolmen, Switzerland; Zack et al. 2002).

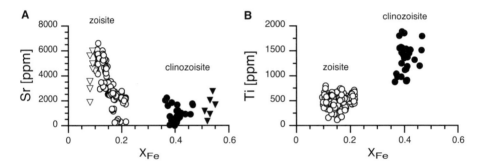

**Figure 18.** Concentrations of Sr (a) and Ti (b) for coexisting zoisite and clinozoisite; circles and dots: data from pegmatites (Saualpe, Austria), triangles: data from amphibolite (Longuido Mts, Tanzania).

epidote minerals are the principal host of these elements in many common mineral paragenesis and may store large parts of the whole rock's trace element content. The LILE with their strong preference for mica and feldspar are not expected in large amounts in epidote minerals, as well as the HFSE, which will be incorporated into Ti-minerals (e. g. rutile and titanite).

For single elements or element groups, e.g. for the concentration of Sr, the element contents of epidote minerals strongly depends on the presence of competitive minerals. If plagioclase and/or calcite are present (Figs. 14, 15), it is expected that in epidote minerals the Sr content is low, but it can increase strongly when these minerals become unstable. Pb will be enriched in epidote minerals, except for plagioclase rich rocks (Fig. 14) and for some hornblende-rich rocks (Fig. 13b). Therefore it can be expected that the whole rock content of Sr and Pb in high-*P* rocks is largely hosted by epidote. The REE can generally be enriched in epidote minerals, but no general rule can be given and REE contents in epidote minerals are extremely variable (Figs. 2-7). In many cases LREE are enriched in epidote minerals in the presence of abundant garnet that incorporates the HREE.

Only few studies address the trace element distribution amongst the minerals of metamorphic rocks quantitatively. In zoisite bearing gabbroic eclogite from Trescolmen, Central Alps (Switzerland), zoisite comprises 8 vol% of the whole rock and contains >80% Sr and Pb, 60 to 80% LREE, and ~30 and 70% Th and U of the whole rock budget (Zack et al. 2002). In garnet amphibolite from Catalina, California (USA), zoisite makes up only about 15 wt% of the whole rock but contributes with ~80% of the Sr content, ~45% of Th, ~60% of U, and ~65 to 85% of light and middle REE to the whole rock trace element budget (Sorensen and Grossman 1989). In eclogite from Junan, Su-Lu (China), the modal abundances of zoisite and epidote are ~28 and <8 vol%, respectively and correlate linearly with the whole rock Sr content (Nagasaki and Enami 1998). Both contribute to over 70% to the whole rock Sr budget.

The data for amphibolite, gneiss and calcsilicate rocks from Chile and Argentina (Fig. 19a,b; Lucassen pers. comm.) provide nice examples for the REE trace element budget in metamorphic rocks. In zoisite-garnet-amphibolite (Lucassen et al. 1999) with plagioclase (~An$_{20}$), titanite present only as an acessory phase, the whole rock pattern is similar to the zoisite and hornblende patterns, but zoisite clearly dominates with La concentrations of 1000 times chondritic compared to hornblende (<100 times). The biotite-gneiss with rutile as the accessory Ti phase shows the same feature; plagioclase has La < 10 times chondritic values, and zoisite dominates the whole rock pattern. In contrast, in two calcsilicate rocks from the Paleozoic basement in Northern Chile and NW Argentina, which equilibrated at conditions transitional between amphibolite and granulite facies (Lucassen et al. 2000), titanite dominates the REE pattern (Fig. 19b).

Spandler et al. (2003) provided trace element contents of minerals and whole rocks from an eclogite-blueschist terrane in New Caledonia, and the calculated distribution, normalized

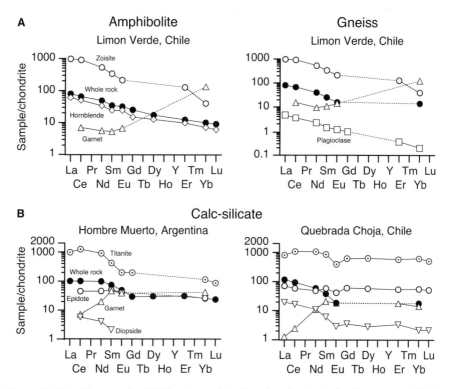

**Figure 19.** Chondrite normalized REE patterns of whole rock and minerals for (a) garnet-amphibolite and biotite-gneiss (Limón Verde, Chile) and (b) calcsilicate rocks from N-Chile and NW-Argentina (data courtesy F. Lucassen).

to the whole rock (Fig. 20) shows that epidote-allanite are the most important minerals for U, Th, Sr, Pb and REE for all rock types. In blueschist, however, lawsonite can also dominate the Sr and Pb contents. Only for Rb, Cs and Ba phengite is dominant, HREE can be dominated by garnet or titanite. The HFSE elements (omitted from the diagram for the sake of clarity) are dominated also by titanite and other phases.

*Melts.* In a melt the concentrations of Sr and REE might be largely controlled by their partitioning behavior between epidote and melt, because epidote minerals occur as magmatic phases (see Schmidt and Poli 2004; Poli and Schmidt 2004) and can be residual phases during partial melting of metabasalts and metasediments (Hermann and Green 2001; Hermann 2002a,b; Skjerlie and Patino Douce 2002). For allanite the only reported $D_{Sr}^{aln/melt}$ is close to unity and points to a marked fractionation of REE from Sr with a $K_{D\,(Ce-Sr)}^{aln/melt}$ of 200 to 2000. Because of the strong compatibility of REE in allanite and the straight LREE enriched partitioning pattern, allanite might strongly fractionate Sr from LREE and to a lesser extent MREE and HREE. Thus, the melt phase will be enriched in Sr relative to LREE. In contrast, the high Sr concentration in zoisite and clinozoisite-epidote and the only moderate compatibility of REE with MREE enriched partitioning pattern might lead to melts that are strongly depleted in Sr relative to LREE and HREE. The *relative* depletion of Sr and also Pb compared to the LREE in allanite (Fig. 20a,d) and the inverse behavior of epidote is also seen in the distribution pattern of eclogites, normalized to whole rock (Fig. 20c,d).

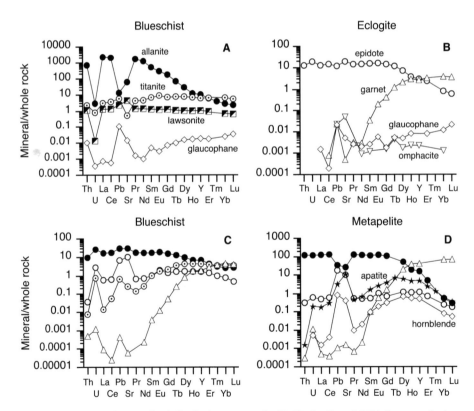

**Figure 20.** Whole rock normalized distribution patterns for U, Th, Sr, Pb and REE from an eclogite-blueschist terrane, New Caledonia (selected data from Spandler et al. 2003). Allanite and epidote dominate the trace element pattern.

The isotope ratio $^{87}Sr/^{86}Sr$ in epidote minerals will be very close to the initial Sr ratio, because the amount of Rb is negligible. This might be interesting for igneous rocks, where the $^{87}Sr/^{86}Sr$ in epidote will directly give an information about the crustal versus mantle signature. In metamorphic rocks, the ratio will give a good approximation for mineral isochrones.

The trace element contents of Pb, U and Th show that the U-Th decay series have to be considered with care. The elements are incorporated independently from each other, however with a general enrichment of U over Th, compared to other crustal rocks.

### Survey of epidote minerals provenance

It is tempting to use the trace element patterns of detrital epidote minerals in sediments as an indicator for the source rock. However, there is so much overlap for individual trace elements in certain rock groups such as felsic igneous rocks, metabasites, metasediments etc. that they can not be used for this purpose. It must also be assumed that in the detrital mineral spectrum of a sediment epidotes from vein assemblages and segregations are important, and especially their trace element patterns are extremely variable. A promising approach, however, seems the combination of radiogenic isotope characteristics with trace elements (Spiegel et al. 2002), which will be even more promising if performed by individual grain analysis. If the epidote characteristics of the source region are well known, the large variation in trace element patterns offers the potential as provenance indicators.

### Open questions

Though a large database about trace elements in epidote minerals is now available, it is far from being sufficient. This is especially true for elements that are usually present only at low concentration levels and hence are difficult to determine. Examples include the light elements Li, Be, and B, or the PGE. For blueschist-amphibolite-eclogite facies rocks the database is quite good, but it is astonishing that for the common occurrences of epidote in greenschist, impure marble, calcsilicate and metasomatic rocks almost no data are published. There is also a large data gap for REE-poor epidote in igneous rocks. Especially missing are complete sets of trace element data for epidote and the accompanying minerals, which allow calculating distribution coefficients and trace element budgets.

A serious impediment for the interpretation of trace element signatures recorded in epidote minerals and for any qualitative and quantitative description of the role of epidote minerals in magmatic and fluid-mediated processes is the almost complete absence of experimental partitioning data. Obviously, there is a need for experiments that establish an understanding of the partitioning of a wide range of geochemically important trace elements between epidote minerals and liquids (fluids and melts) as a function of $P$, $T$, and crystal and liquid composition.

## ACKNOWLEDGMENTS

We thank M. Lewerenz for careful handpicking of mineral separates. F. Lucassen and H. Marschall kindly provided unpublished trace element data. The specimens have been used in a previous study by G. Smelik, donated from mineral collections at the Technische Universität Berlin, Technische Universität München, Smithonian Institution Washington, and Humboldt Universität Berlin, and we thank G. Smelik as well as the curators of these institutions for the material. F. Lucassen is thanked for reviewing this contribution on a rather short notice, and T. Zack for helpful comments. This paper is published with the permission of the Geological Survey of Denmark and Greenland.

## REFERENCES

Alderton DHM, Jackson NJ (1978) Discordant calc-silicate bodies from the St Just aureole, Cornwall. Mineral Mag 42:427-434

Anisimova AS, Nekrasova, LP, Ploshko, VV, Shport WP (1975) Fluorine-zoisite and fluorine-kyanite from the eclogites and hydrothermal veins of the Caucasus. Zap Vses Min Obsch 104:97-99 (in Russian)

Ashley PM, Martin JE (1987) Chromium-bearing minerals from a metamorphosed hydrothermal alteration zone in the Archean of Western Australia. N Jahrb Mineral Abh 157:81-111

Banks DA, Yardley BWD, Campell AR, Jarvis KE (1994) REE composition of an aqueous magmatic fluid: A fluid inclusion study from the Capitan Pluton, New Mexico, USA. Chem Geol 110:299-314

Beattie P, Drake M, Jones J, Leeman W, Longhi J, McKay G, Nielsen R, Palme H, Shaw D, Takahashi E, Watson B (1993) Terminology for trace element partitioning. Geochim Cosmochim Acta 57:1605-1606

Becker H, Jochum KP, Carlson RW (1999) Constraints from high-pressure veins in eclogites on the composition of hydrous fluids in subduction zones. Chem Geol 160:291-308

Bleek AWG (1907) Die Jadeitlagerstätten in Upper Burma. Z prak Geol 341-365

Blundy JD, Falloon TJ, Wood BJ, Dalton JA (1995) Sodium partitioning between clinopyroxene and silicate melts. J Geophys Res 100:15501-15515

Blundy JD, Wood BJ (1994) Prediction of crystal-melt partition coefficients from elastic moduli. Nature 372: 452-454

Blundy JD, Wood BJ (2003) Mineral-melt partitioning of uranium, thorium and their daughters. Rev Mineral Geochem 52:59-123

Bonazzi P, Garbarino C, Menchetti S (1992) Crystal chemistry of piemontites: REE bearing piemontite from Mont Brugiana, Alpi Apuane, Italy. Eur J Mineral 4:23-33

Bonazzi P, Menchetti S (1995) Monoclinic endmembers of the epidote-group: effects of the Al $\leftrightarrow$ Fe$^{3+}$ $\leftrightarrow$ Fe$^{2+}$ substitution and of the entry of REE$^{3+}$. Mineral Petrol 53:133-153

Bonazzi P, Menchetti S (2004) Manganese in monoclinic members of the epidote group: piemontite and related minerals. Rev Mineral Geochem 56:495-552

Brastad K (1985) Sr metasomatism, and partitioning of Sr between the mineral phases of a meta-eclogite from Bjørkedalen, West Norway. Tschermaks Mineral Petrol Mitt 34:87-103

Braun JJ, Pagel M, Herbillon A, Rosin C (1993) Mobilization and redistribution of REEs and thorium in a syenitic lateritic profile: A mass balance study. Geochim Cosmochim Acta 57:4419-4434

Brenan JM, Shaw HF, Ryerson FJ, Phinney DL (1995) Mineral-aqueous fluid partitioning of trace elements at 900°C and 2.0 GPa; constraints on the trace elements chemistry of mantle and deep crustal fluids. Geochim Cosmochim Acta 59:3331-3350

Brooks CK, Henderson P, Rønsbo JG (1981) Rare-earth partition between allanite and glass in the obsidian of Sandy Braes, Northern Ireland. Mineral Mag 44:157-160

Brunsmann A, Franz G, Erzinger J (2001) REE mobilization during small-scale high-pressure fluid-rock interaction and zoisite/fluid partitioning of La to Eu. Geochim Cosmochim Acta 65:559-570

Brunsmann A, Franz G, Erzinger J, Landwehr D (2000) Zoisite- and clinozoisite- segregations in metabasites (Tauern Window, Austria) as evidence for high-pressure fluid-rock interaction. J Metam Geol 18:1-21

Burns RG, Strens GJ (1967) Structural interpretation of polarized absorption spectra of the Al-Fe-Mn-Cr epidotes. Mineral Mag 36:204-226

Cantrell KJ, Byrne RH (1987) Rare earth element complexation by carbonate and oxalate ions. Geochim Cosmochim Acta 51:587-605

Carcangiu G, Palomba M, Tamanini M (1997) REE-bearing minerals in the albitites of Central Sardinia, Italy. Mineral Mag 61:271-283

Challis GA, Grapes R, Palmer K (1995) Chromian muscovite, uvarovite, and zincian chromite; products of regional metasomatism in Northwest Nelson, New Zealand. Can Mineral 33:1263-1284

Cherniak DJ (2000) Rare earth element diffusion in apatite. Geochim Cosmochim Acta 64:3871-3885

Choo CO (2002) Complex compositional zoning in epidote from rhyodacitic tuff, Bobae sericite deposit, southeastern Korea. N Jahrb Mineral Abh 177:181-197

Cooper AF (1980) Retrograde alteration of chromian kyanite in metachert and amphibolite whiteschist from the Southern Alps, New Zealand, with implications for uplift on the Alpine Fault. Contrib Mineral Petrol 75:153-164

Cressey G, Steel AT (1988) An EXAFS study on Gd, Er and Lu site location in the epidote structure. Phys Chem Mineral 15:304-312

Dahlquist JA (2001) REE fractionation by accessory minerals in epidote-bearing metaluminous granitoids from the Sierras Pampeanas, Argentina. Mineral Mag 65:463-475

Dawes RL, Evans BW (1991) Mineralogy and geothermobarometry of magmatic epidote-bearing dikes, Front Range, Colorado. Geol Soc Am Bull 103:1017-1031

Deer RL, Howie RA, Zussman J (1986) Rock-forming minerals, Vol 1B: Disilicates and ringsilicates (2nd edition). Longman, Harlow

Devaraju TC, Raith MM, Spiering B (1999) Mineralogy of the Archaean barite deposit of Ghattihosahalli, Karnataka, India. Can Mineral 37:603-617

Dollase WA (1968) Refinement and comparison of the structure of zoisite and clinozoisite. Am Mineral 53: 1882-1898

Dollase WA (1971) Refinement of the crystal structures of epidote, allanite and hancockite. Am Mineral 56: 447-464

Dollase WA (1973) Mössbauer spectra and iron distribution in the epidote group minerals. Z Krist 138:41-63

Domanik KJ, Hervig RL, Peacock SM (1993) Beryllium and boron in subduction zone minerals: An ion microprobe study. Geochim Cosmochim Acta 57:4997-5010

Dulski P (2001) Reference materials for geochemical studies; new analytical data by ICP-MS and critical discussion of reference values. Geostand Newslett 25:87-125

Dunn P (1985) The lead silicates from Franklin, New Jersey: occurrence and composition. Mineral Mag 49: 721-727

Eggins SM, Rudnick RL, McDonough WF (1998) The composition of peridotites and their minerals: a laser-ablation ICP-MS study. Earth Plant Sci Lett 154:53-71

Enami M, Liou JG, Mattinson CG (2004) Epidote minerals in high P/T metamorphic terranes: Subduction zone and high- to ultrahigh-pressure metamorphism. Rev Mineral Geochem 56:347-398

Eskola P (1933) On the chrome minerals of Outukumpu. Bull Comm Geol Fin 7: 26

Evans BW, Vance JA (1987) Epidote phenocrysts in dacitic dikes, Boulder County, Colorado. Contrib Mineral Petrol 96:178-185

Ewart A, Griffin WL (1994) Application of proton-microprobe data to trace-element partitioning in volcanic rocks. Chem Geol 117:251-284

Fehr KT, Heuss-Aßbichler S (1997) Intracrystalline equilibria and immiscibility gap along the join clinozoisite - epidote: An experimental and $^{57}$Fe Mössbauer study. N Jahrb Mineral Abh 172:43-67

Ferraris G, Ivaldi G, Fuess H, Gregson D (1989) Manganese/iron distribution in a strontian piemontite ny neutron diffraction. Z Krist 187:145-151

Finger F, Broska I, Roberts MP, Schermaier A (1998) Replacement of monazite by apatite-allanite-epidote coronas in an amphibolite facies granite gneiss from the eastern Alps. Am Mineral 83:248-258

Franz G, Liebscher A (2004) Physical and chemical properties of the epidote minerals–an introduction. Rev Mineral Geochem 56:1-82

Franz G, Selverstone J (1992) An empirical phase diagram for the clinozoisite-zoisite transformation in the system $Ca_2Al_3Si_3O_{12}(OH)-Ca_2Al_2Fe^{3+}Si_3O_{12}(OH)$. Am Mineral 77:631-642

Frei D, Liebscher A, Wittenberg A, Shaw, CSJ (2003) Crystal chemical controls on rare earth element partitioning between epidote group minerals and melts: an experimental and theoretical study. Contrib Mineral Petrol 146:192-204

Game PM (1954) Zoisite-amphibolite with corundum from Tanganyika. Mineral Mag 30:458-466

Gieré R (1996) Formation of rare earth minerals in hydrothermal systems. In: Jones AP, Wall F, Williams CT (eds) Rare earth minerals: Chemistry, origin and ore deposits. Mineral Soc Ser 7:105-150

Gieré R, Sorensen SS (2004) Allanite and other REE-rich epidote-group minerals. Rev Mineral Geochem 56: 431-494

Gieré R, Virgo D, Popp RK (1999) Oxidation state of iron and incorporation of REE in allanite. Jour Conf Abstr 4:721

Grapes RH (1981) Chromian epidote and zoisite in kyanite amphibolite, Southern Alps, New Zealand. Am Mineral 66:974-975

Grapes RH, Hoskin PWO (2004) Epidote group minerals in low–medium pressure metamorphic terranes. Rev Mineral Geochem 56:301-345

Grapes RH, Watanabe T (1984) Al-Fe$^{3+}$ and Ca-Sr$^{2+}$ epidotes in metagreywacke-quartzofeldspathic schist, Southern Alps, New Zealand. Am Mineral 69:490-498

Gromet LP, Silver LT (1983) Rare earth element distributions among minerals in a granodiorite and their petrogenetic implications. Geochim Cosmochim Acta 47:925-939

Haas JR, Shock EL, Sassani DC (1995) Rare earth elements in hydrothermal systems: Estimates of standard partial molal thermodynamic properties of aqueous complexes of the rare earth elements at high pressures and temperatures. Geochim Cosmochim Acta 59:4329-4350

Harlow GE (1994) Jadeitites, albitites and related rocks from the Motagua fault zone, Guatemala. J Metam Geol 12:49-68

Hermann J (2002a) Experimental constraints on phase relations in subducted oceanic crust. Contrib Mineral Petrol 143:219-235

Hermann J (2002b) Allanite: thorium and light rare earth element carrier in subducted crust. Chem Geol 192: 289-306

Hermann J, Green DH (2001) Experimental constraints on high pressure melting in subducted crust. Earth Planet Sci Lett 188:149-168

Hickmott DD, Sorensen SS, Rogers PSZ (1992) Metasomatism in a subduction complex: Constraints from microanalysis for trace elements in minerals from garnet amphibolite from the Catalina Schist. Geology 20:347-350

Hofmann AW (1988) Chemical differentiation of the earth: the relationship between mantle, continental crust and oceanic crust. Earth Planet Sci Lett 90:297-314

Holtstam D, Langhof J (1994) Hancockite from Jakobsberg, Filipstad, Sweden; the second world occurrence. Mineral Mag 58:172-174

Hoschek G (1980) The effect of Fe-Mg substitution on phase relations in marly rocks of the Western Hohe Tauern (Austria). Contrib Mineral Petrol 75:123-128.

Howie RA, Walsh JN (1982) The geochemistry and mineralogy of an epidote-glaucophanite from Hačava, Spišskogemerské rudohorie Mountains, West Carpathians, Czechoslovakia. Geol Práce 78:59-64

Hurlbut Jr CS (1969) Gem zoisite from Tanzania. Am Mineral 54:702-709

Jancev S, Bermanec V (1998) Solid solution between epidote and hancockite from Nezilovo, Macedonia. Geol Croat 51:23-26

Johan Z, Oudin E, Picot P (1983) Analogues germanifères et gallifères des silicates et oxydes dans les gisements de zinc des Pyrénées centrales, France; argutite et carboirite, deux nouvelles espèces minérales. Tschermaks Min Petr Mitt 31:97-119

Karev (1974) New finds of vanadium-bearing minerals from metamorphic rocks of Kuznetsk. Geol Geofiz 11: 141-143 (in Russian)

Keane SD, Morrison J (1997) Distinguishing magmatic from subsolidus epidote: laser probe oxygen isotope compositions. Contrib Mineral Petrol 126:265-274

Klemme S, Dalpé C (2003) Trace-element partitioning between apatite and carbonatite melt. Am Mineral 88: 639-646

Kretz R (1983) Symbols for rock forming minerals. Am Mineral 68:277-279

Kvick KA, Pluth JJ, Richardson Sr JW, Smith JV (1988) The ferric iron distribution and hydrogen bonding in epidote: a neutron diffraction study at 15 K. Acta Cryst B44:1753-1766

Landwehr D, Blundy JD, Chamorro-Perez EM, Hill E, Wood BJ (2001) U-series disequilibria generated by partial melting of spinel lherzolite. Earth Planet Sci Lett 188:329-348

Liebscher A, Gottschalk M, Franz G (2002) The substitution $Fe^{3+}$-Al and the isosymmetric displacive phase transition in synthetic zoisite: a powder X-ray and infrared spectroscopy study. Am Mineral 87:909-921

Lisitin AE, Khitrov VG (1962) A microspectrochemical study of the distribution of boron in minerals of some igneous and metamorphic rocks in the middle Urals. Geochem 3:249-258

Liu X, Dong S, Xue H, Zhou J (1999) Significance of allanite-(Ce) in granitic gneisses from the ultrahigh-pressure metamorphic terrane, Dabie Shan, central China. Min Mag 63:579-586

Lucassen F, Franz G, Laber A (1999) Permian high pressure rocks – the basement of the Sierra de Limón Verde in N-Chile. J South Amer Earth Sci 12:183-199

Lucassen F, Wilke HG, Becchio R, Viramonte J, Franz G, Laber A, Wemmer K, Vroon P (2000) The Paleozoic basement of the Central Andes (18°S-26°S) – the metamorphic view. J South Amer Earth Sci 13:697-715

Maaskant P (1985) The iron content and the optic axial angle in zoisites from Galicia, NW Spain. Min Mag 49:97-100

Mahood G, Hildreth W (1983) Large partition coefficients for trace elements in high silica ryholites. Geochim Cosmochim Acta 47:11-30

McDonough WF, Sun SS (1995) The composition of the Earth. Chem Geol 120:223-253

Messiga B, Tribuzio R, Botazzi P, Ottolini L (1995): An ion microprobe study on trace element composition of clinopyroxenes from blueschist and eclogitized Fe-Ti-gabbros, Ligurian Alps, northwestern Italy: some petrologic considerations. Geochim Cosmochim Acta 59:59-75

Meyer RJ (1911) Über einen skandiumreichen Orthit aus Finnland und den Vorgang seiner Verwitterung. Sitzungsberichte der königlichen preussischen Akademie der Wissenschaften zu Berlin. 105:379-384

Miyajima H, Matsubara S, Miyawaki R, Hirokawa K (2003) Niigataite, $CaSrAl_3(Si_2O_7)(SiO_4)O(OH)$: Sr-analogue of clinozoisite, a new member of the epidote group from the Itoigawa-Ohmi district, Niigata Prefecture, central Japan. J Min Petr Sci 98:118-129

Mulligan R, Jambor JL (1968) Tin-bearing silicates from skarn in the Cassiar District, Northern British Columbia. Can Mineral 9:358-370

Myer GH (1965) X-ray determinative curve for epidote. Am J Sci 263:78-86

Myer GH (1966) New data on zoisite and epidote. Am J Sci 264:364-385

Mysen BO, Virgo D (1980) Trace element partitioning and melt structure: an experimental study at 1 atm pressure. Geochim Cosmochim Acta 44:1917-1930

Nagasaki A, Enami M (1998) Sr-bearing zoisite and epidote in ultra-high pressure (UHP) metamorphic rocks from the Su-Lu province, eastern China: an important Sr reservoir under UHP conditions. Am Mineral 83:240-247

Nozik YK, Kanepit VN, Fykin LY, Makarov YS (1978) A neutron diffraction study of the structure of epidote. Geochem Int 15:66-69

Nyström JO (1984) Rare earth element mobility in vesicular lava during low grade metamorphism. Contrib Mineral Petrol 88:328-331

Pan Y, Fleet ME (1996) Intrinsic and external controls on the incorporation of rare-earth elements in calc-silicate minerals. Can Mineral 34:147-159

Peacor RP, Dunn PJ (1988) Dollaseite-(Ce) (magnesium orthite refined): Structure and implications for $F + M^{2+}$ substitutions in epidote group minerals. Am Mineral 73:838-842

Penfield SL, Warren CH (1899) Some new minerals from the zinc mines at Franklin, New Jersey and a note concerning the mineral composition of ganomalite. Am J Sci 8:339-353

Poli S, Schmidt MW (2004) Experimental subsolidus studies on epidote minerals. Rev Mineral Geochem 56: 171-195

Pouliot G, Trudel P, Valiquette G, Samson PN (1984) Armenite-thulite-albite veins at Remigny, Quebec; the second occurrence of armenite. Can Mineral 22:453-464

Prince et al. (2000) Comparison of laser ablation ICP-MS and isotope dilution REE analyses — implications for Sm–Nd garnet geochronology. Chem Geol 168 :255-274

Reinecke T (1986) Crystal chemistry and reaction relations of piemontites and thulites from highly oxidized low grade metamorphic rocks at Vitali, Andros Island, Greece. Contrib Mineral Petrol 93:56-76

Rønsbo JG (1989) Coupled substitutions involving REEs and Na and Si in apatites in alkaline rocks from the Ilimaussaq intrusion, South Greenland, and the petrological implications. Am Mineral 74:896-901

Sakai C, Higashino T, Enami M (1984) REE-bearing epidote from Sanbagawa pelitic schists, central Shikoku, Japan. Geochem J 18:45-53

Sánchez-Vizcaíno VL, Franz G, Gómez Pugnaire MT (1995) The behaviour of Cr during metamorphism of carbonate rocks from the Nevado-Filabride Complex, Betic Cordilleras, Spain. Can Mineral 33:85-104

Sassi R, Harte B, Carswell DA, Yujing H (2000) Trace element distribution in Central Dabie eclogites. Contrib Mineral Petrol 139:298-315

Sawka WN (1988) REE and trace element variations in accessory minerals and hornblende from the strongly zoned McMurray Meadows Pluton, California. Trans R Soc Edinburgh: Earth Sci 79:157-168

Schmidt MW, Poli S (2004) Magmatic epidote. Rev Mineral Geochem 56:399-430

Schreyer W, Abraham K (1978) Prehnite/chlorite and actinolite/epidote bearing mineral assemblages in the metamorphic igneous rocks of La Halle and Challes, Venn-Stavelot-Massif, Belgium. Ann Soc Géol Belgium 101:227-241

Seemann R, Koller F, Höck V (1993) Die Mineralfundstelle Knappenwand – Erweiterte Zusammenfassung. Abh Geol 49:33-37

Shepel AV, Karpenko MV (1969) Mukhinite, a new vanadium species of epidote. Doklady Acad Nauk SSSR 185:1342-1345 (in Russian)

Skjerlie KP, Patino Douce AE (2002) The fluid-absent partial melting of a zoisite-bearing quartz eclogite from 1.0 to 3.2 GPa; implications for melting in thickened continental crust and for subduction-zone processes. J Petrol 43:291-314

Smyth JR, Bish DL (1988) Crystal Structures and Cation Sites of the Rock-forming Minerals. Allen and Unwin, Boston

Sorensen SS, Grossman JN (1989) Enrichment of trace elements in garnet amphibolites from paleo-subduction zone: Catalina Schist, Southern California. Geochim Cosmochim Acta 53:3155-3177

Spandler C, Hermann J, Arculus R, Mavrogenes J (2003) Redistribution of trace elements during prograde metamorphism from lawsonite blueschist to eclogite facies; implications for deep subduction-zone processes. Contrib Mineral Petrol 146:205-222

Spear FS, Pyle JM (2002) Apatite, monazite and xenotime in metamorphic rocks. Rev Mineral Geochem 48: 293-335

Spiegel C, Siebel W, Frisch W, Berner Z (2002) Nd and Sr isotope ratio and trace element geochemistry of epidote group minerals from the Swiss Molasse Basin as provenance indicators: implications for the reconstruction of the exhumation history of the Central Alps. Chem Geol 189:231-250

Stalder R, Foley SF, Brey GP, Horn I (1998) Mineral-aqueous fluid partitioning of trace elements at 900°C-1200°C and 3.0 GPa to 5.7 GPa: new experimental data set for garnet, clinopyroxene, and rutile and implications for mantle metasomatism. Geochim Cosmochim Acta 62:1781-1801

Thomas S, Franz G (1989) Kluftminerale und ihre Bildungsbedingungen in Gesteinen der Eklogitzone/ Südvenedigergebiet (Hohe Tauern, Österreich). Mitt Öster Geol Ges 81:167-188

Thöni M (2003) Sm-Nd isotope systematics in garnet from different lithologies (Eastern Alps): age results, and an evaluation of potential problems for garnet Sm-Nd chronometry. Chem Geol 194:353-379

Thöni M, Jagoutz E (1992) Some new aspects of dating eclogites in orogenic belts: Sm-Nd, Rb-Sr, and Pb-Pb isotopic results from the Austroalpine Saualpe und Koralpe type-locality (Carinthia/Styria, southeastern Austria). Geochim Cosmochim Acta 56:347-368

Treloar PJ (1987) Chromian muscovites and epidotes from Outokumpu, Finland. Min Mag 51:593-599

Tsang T, Ghose S (1971) Electron paramagnetic resonance of $V^{2+}$, $Mn^{2+}$, $Fe^{3+}$ and optical spectra of $V^{3+}$ in blue zoisite, $Ca_2Al_3Si_3O_{12}(OH)$. J Chem Phys 54:856-862

Van Marcke de Lummen G (1986) Tin-bearing epidote from skarn in the Land's End aureole, Cornwall, England. Can Mineral 24:411-415

Van Westrenen W, Blundy JD, Wood BJ (1999) Crystal-chemical controls on trace element partitioning between garnet and anhydrous silicate melt. Am Mineral 84:838-847

Volfinger M, Robert JL, Vielzeuf D, Neiva AMR (1985) Structural control of the chlorine content of OH-bearing silicates (micas and amphiboles). Geochim Cosmochim Acta 49:37-48

Willemse J, Bensch JJ (1966) Inclusions of original carbonate rocks in gabbro and norite of the eastern part of the Bushveld complex. Trans Geol Soc Sout Afr. 67:1-84

Wood SA (1990a) The aqueous geochemistry of rare earth elements and yttrium. 1. Review of available low-temperature data for inorganic complexes and the inorganic REE speciation of natural waters. Chem Geol 82:159-186

Wood SA (1990b) The aqueous geochemistry of rare earth elements and yttrium. 2. Theoretical predictions of speciation in hydrothermal solutions to 350°C at saturation water vapor pressure. Chem Geol 88:99-125

Wood SA, Ricketts A (2000) Allanite-(Ce) from the Eocene Casto Granite, Idaho: Response to hydrothermal alteration. Can Mineral 38:81-100

Yanev Y, Bardintzeff JM, Rakalov K, Jelev G (1998) Mn-bearing and REE-rich epidote (epidote-allanite) from the hydrothermally altered acid volcanics, Eastern Rhodopes (Bulgaria). N Jahrb Mineral Monatsh 5: 221-233

Yang JJ, Enami M (2003) Chromian dissakisite-(Ce) in a garnet lherzolite from the Chinese Su-Lu UHP metamorphic terrane: Implications for Cr incorporation in epidote minerals and the recycling of REE into the Earth's mantle. Am Mineral 88:604-610

Yang P, Rivers T (2002) The origin of Mn and Y annuli in garnet and the thermal dependence of P in garnet and Y in apatite in calc-pelite and pelite, Gagnon Terrane, Western Labrador. Geol Mat Res 4:n1

Zack T, Foley S, Rivers T (2002) Equilibrium and disequilibrium trace element partitioning in hydrous eclogites (Trescolmen, Central Alps). J Petrol 43:1947-1974

**Appendices found on the following pages.**

# APPENDIX A

Minerals that have been analysed for trace element concentrations in this study.
Mineral abbreviations are used according to Kretz (1983).

| Mineral | Sample # | Source[1] | Locality | Lithology | Coex. Minerals | Reference[2] |
|---------|----------|-----------|----------|-----------|----------------|--------------|
| Ep | 1423 | TUB | Hohe Waid, Odenwald, Germany | epidote fels | | |
| Ep | 1427 | TUB | Hohe Waid, Odenwald, Germany | epidote fels | | |
| Ep | 18D89 | TUB | Frosnitztal, Tauern, Austria | segregation | Rt | |
| Ep | 245 | TUB | Knappenwand, Tauern, Austria | epidote vein | Amp | |
| Ep | 812 | TUB | Knappenwand, Tauern, Austria | epidote vein | Cal | |
| Ep | Mü 93-2/1 | TUM | Knappenwand, Tauern, Austria | epidote vein | | S93, FHA97 |
| Ep | Mü 93-2/3 | TUM | Knappenwand, Tauern, Austria | epidote vein | | S93, FHA97 |
| Ep | Mü 93-5 | TUM | Sustenpass, Switzerland | epidote fels | Cal | |
| Ep | N5386 | TUB | Hasseröde, Harz Mts., Germany | epidote fels | Hem | |
| Ep | N5471 | TUB | Arendal, Norway | skarn deposit | Cal | |
| Ep | SM4 | SM | Prince of Wales Isl., Alaska, USA | epidote fels | | FHA97, D71, D73 |
| Czo | HU 93-6 | TUB | Saualpe, Kärnten, Austria | pegmatite | Zo | |
| Czo | HU 93-7 | TUB | Saualpe, Kärnten, Austria | pegmatite | Ab, Zo | |
| Czo | ZDa 1 | TUB | Frosnitztal, Tauern, Austria | segregation | | TF89 |
| Zo | 1445 | TUB | Longido Mts., Matabu, Tanzania | amphibolite | Cm, An, Hbl | |
| Zo | 1445/4 | TUB | Longido Mts., Matabu, Tanzania | amphibolite | Hbl | |

| Zo | Mir | TUB | Tanzania | gneiss | Ky | |
|----|-----|-----|----------|--------|-----|---|
| Zo | 85-1 | TUB | Weissenstein, Münchberg, Germany | pegmatite | Ab | F95 |
| Zo | 85-3 | TUB | Weissenstein, Münchberg, Germany | pegmatite | Ab | F95 |
| Zo | 85-4 | TUB | Weissenstein, Münchberg, Germany | pegmatite | Amp | F95 |
| Zo | 85-5 | TUB | Weissenstein, Münchberg, Germany | pegmatite | Ab | F95 |
| Zo | 89-1 | TUB | Weissenstein, Münchberg, Germany | pegmatite | Amp | F95 |
| Zo | 93-24 | TUB | Weissenstein, Münchberg, Germany | pegmatite | Ab, Amp | F95 |
| Zo | AB 94/3 | TUB | Umbaltal, Tauern, Austria | segregation | | B00, B01 |
| Zo | AB 94/5 | TUB | Umbaltal, Tauern, Austria | segregation | | B00, B01 |
| Zo | AB 94/6 | TUB | Umbaltal, Tauern, Austria | segregation | | B00, B01 |
| Zo | AB 94/8 | TUB | Umbaltal, Tauern, Austria | segregation | Cal, Ms | B00, B01 |
| Zo | AB 94/9 | TUB | Umbaltal, Tauern, Austria | segregation | | B00, B01 |
| Zo | AB 94(16)b | TUB | Umbaltal, Tauern, Austria | segregation | | B00, B01 |
| Zo | AB 94/17 | TUB | Umbaltal, Tauern, Austria | segregation | | B00, B01 |
| Zo | HU 93-6 | TUB | Saualpe, Kärnten, Austria | pegmatite | Czo | |
| Zo | HU 93-7 | TUB | Saualpe, Kärnten, Austria | pegmatite | Ab, Czo | |
| Zo | HU 93-8 | HUB | Faltigels/Sterzing, Südtirol, Italy | unknown (segregation?) | | |

[1] Minerals investigated are from the mineral collections of the Technische Universität Berlin (TUB), Technische Universität München (TUM), Smithonian Institution Washington (SM), and Humboldtuniversität Berlin (HUB).

[2] References refer to studies examining minerals from the same locality, but not necessarily the same composition. References cited: S93 = Seeman et al. 1993; FHA97 = Fehr and Heuss-Assbichler 1997; D71 = Dollase 1971; D73 = Dollase 1973; TF89 = Thomas and Franz 1989; F95 = Franz et al. 1995; B00= Brusmann et al. 2000; B01 = Brunsmann et al. 2001

## APPENDIX B

**(found on following 5 pages)**

Trace element concentrations (in µg/g) of epidote minerals and coexisting minerals that have been analysed for this study (cf. Appendix A).

| Sample # | 1423 | 1427 | 18D89 | 18D89 | 245 | 245 | 245 | 812 | 812 | Mü 93-2/1 | Mü 93-2/3 |
|---|---|---|---|---|---|---|---|---|---|---|---|
| **Mineral** | Ep | Ep | Ep | Rt | Ep 1 | Ep 2 | Amp | Ep | Cal | Ep | Ep |
| **Rb** | 0.73 | < 0.2 | 0.49 | 0.96 | < 0.2 | 0.30 | 0.49 | 0.36 | < 0.1 | < 0.3 | < 0.2 |
| **Sr** | 528.9 | 427.4 | 799.9 | 1.66 | 928.7 | 903.2 | 22.6 | 687.4 | 1308 | 634.7 | 661.7 |
| **Y** | 28.5 | 6.27 | 28.1 | 0.15 | 27.4 | 27.9 | 0.66 | 65.3 | 0.81 | 39.9 | 43.5 |
| **Zr** | 43.1 | 41.7 | 7.47 | 60.5 | 10.7 | 11.0 | 3.86 | 11.9 | 0.69 | 11.3 | 15.6 |
| **Nb** | 0.07 | 0.13 | < 0.1 | 488.6 | < 0.06 | < 0.07 | 0.07 | < 0.06 | 0.02 | < 0.1 | < 0.06 |
| **Cs** | 0.06 | 0.05 | < 0.05 | 0.01 | 0.10 | 0.09 | 0.01 | < 0.03 | < 0.004 | < 0.05 | < 0.03 |
| **Ba** | 9.10 | 4.68 | 6.10 | 0.66 | 8.22 | 8.40 | 2.83 | 7.92 | 2.66 | 4.62 | 11.3 |
| **La** | 7.90 | 5.39 | 4.88 | 0.02 | 0.75 | 0.79 | 0.03 | 5.94 | 0.01 | 3.12 | 3.85 |
| **Ce** | 17.0 | 9.77 | 11.2 | 0.04 | 2.09 | 2.14 | 0.07 | 15.6 | 0.02 | 9.10 | 10.5 |
| **Pr** | 2.84 | 1.03 | 1.68 | 0.00 | 0.36 | 0.37 | 0.01 | 2.55 | < 0.004 | 1.57 | 1.74 |
| **Nd** | 13.5 | 3.49 | 8.86 | 0.03 | 2.22 | 2.21 | 0.06 | 13.8 | < 0.02 | 8.38 | 9.32 |
| **Sm** | 3.48 | 0.77 | 3.00 | < 0.02 | 0.97 | 1.08 | 0.03 | 4.89 | < 0.02 | 3.15 | 3.36 |
| **Eu** | 1.27 | 0.17 | 1.31 | < 0.007 | 1.07 | 1.09 | 0.02 | 2.21 | < 0.007 | 1.79 | 1.52 |
| **Gd** | 4.25 | 0.77 | 4.37 | < 0.02 | 2.13 | 2.14 | 0.05 | 7.59 | 0.04 | 4.76 | 5.02 |
| **Tb** | 0.72 | 0.12 | 0.76 | < 0.004 | 0.46 | 0.47 | 0.01 | 1.44 | 0.01 | 0.92 | 0.93 |
| **Dy** | 4.50 | 0.85 | 5.23 | 0.01 | 3.62 | 3.92 | 0.08 | 10.3 | 0.07 | 6.69 | 6.84 |
| **Ho** | 0.93 | 0.19 | 1.09 | < 0.003 | 0.93 | 0.94 | 0.02 | 2.27 | 0.02 | 1.46 | 1.54 |
| **Er** | 2.82 | 0.65 | 3.12 | 0.01 | 3.07 | 3.32 | 0.07 | 7.06 | 0.06 | 4.45 | 4.89 |
| **Tm** | 0.40 | 0.11 | 0.41 | < 0.003 | 0.47 | 0.50 | 0.01 | 0.96 | 0.01 | 0.61 | 0.74 |
| **Yb** | 2.40 | 0.79 | 2.67 | < 0.01 | 3.35 | 3.47 | 0.09 | 6.19 | 0.03 | 4.31 | 5.24 |
| **Lu** | 0.31 | 0.12 | 0.37 | < 0.004 | 0.52 | 0.54 | 0.02 | 0.92 | < 0.004 | 0.65 | 0.88 |
| **Hf** | 1.57 | 1.68 | 0.34 | 2.31 | 0.62 | 0.65 | 0.34 | 0.73 | < 0.02 | 0.64 | 0.96 |
| **Ta** | < 0.07 | < 0.07 | < 0.09 | 5.48 | < 0.07 | < 0.1 | 0.01 | < 0.07 | < 0.008 | < 0.09 | 0.25 |
| **Pb** | 169.2 | 53.5 | 20.8 | 0.34 | 18.6 | 18.0 | 1.61 | 19.7 | 0.95 | 19.6 | 20.7 |
| **Th** | 2.73 | 1.98 | 0.41 | < 0.007 | < 0.03 | < 0.02 | < 0.003 | 0.28 | < 0.007 | 0.09 | 0.16 |
| **U** | 2.49 | 8.42 | 0.41 | 0.01 | 0.39 | 0.39 | 0.03 | 1.77 | 0.02 | 1.42 | 1.40 |

| Sample # | Mü 93-5 | Mü 93-5 | N 5386 | N 5386 | N 5471 | N 5471 | SM4 | SM4 | 1445 | 1445 | 1445 |
|---|---|---|---|---|---|---|---|---|---|---|---|
| Mineral | Ep | Cal | Ep | Hem | Ep | Cal | Ep | Hbl | Zo | Crn | An |
| Rb | < 0.3 | < 0.06 | < 0.2 | 0.14 | 0.26 | 0.15 | 0.18 | 1.76 | < 0.09 | 0.89 | 0.71 |
| Sr | 656.3 | 69.3 | 3600 | 1.99 | 266.3 | 152.3 | 335.7 | 5.17 | 1213 | 14.7 | 1080 |
| Y | 47.4 | 1.88 | 69.7 | 0.74 | 103.1 | 21.3 | 4.32 | 0.57 | 1.94 | 0.05 | 0.22 |
| Zr | 6.27 | 2.34 | 0.41 | 0.73 | 39.0 | 2.40 | 5.90 | 79.9 | 3.27 | 9.92 | 0.76 |
| Nb | < 0.1 | < 0.01 | 0.16 | 0.51 | < 0.06 | 0.02 | 0.19 | 0.31 | 0.02 | 0.06 | 0.03 |
| Cs | < 0.06 | 0.04 | < 0.03 | 0.04 | < 0.03 | 0.02 | < 0.02 | 0.01 | < 0.02 | 0.02 | < 0.02 |
| Ba | 6.31 | 2.81 | 2.47 | 0.80 | 9.41 | 24.5 | 1.66 | 2.76 | 30.0 | 10.6 | 211.7 |
| La | 10.0 | 0.05 | 10.3 | 0.10 | 65.0 | 0.05 | 0.30 | 0.10 | 4.10 | 0.08 | 0.33 |
| Ce | 22.6 | 0.05 | 16.2 | 0.22 | 149.8 | 0.13 | 1.33 | 0.08 | 6.23 | 0.15 | 0.64 |
| Pr | 3.10 | < 0.004 | 2.12 | 0.02 | 21.0 | 0.02 | 0.42 | 0.01 | 0.65 | 0.01 | 0.08 |
| Nd | 14.1 | < 0.02 | 10.2 | 0.12 | 98.3 | 0.13 | 3.05 | 0.03 | 2.53 | < 0.06 | 0.31 |
| Sm | 4.33 | < 0.02 | 3.52 | 0.04 | 25.5 | 0.09 | 0.87 | < 0.02 | 0.47 | < 0.06 | 0.06 |
| Eu | 2.36 | < 0.007 | 3.03 | 0.01 | 4.82 | 0.03 | 0.45 | < 0.007 | 0.54 | < 0.02 | 0.07 |
| Gd | 6.20 | < 0.02 | 7.00 | 0.05 | 25.0 | 0.31 | 0.78 | < 0.02 | 0.42 | < 0.03 | 0.05 |
| Tb | 1.10 | < 0.004 | 1.22 | 0.01 | 3.72 | 0.10 | 0.11 | < 0.004 | 0.06 | < 0.007 | < 0.007 |
| Dy | 8.09 | 0.08 | 8.79 | 0.07 | 21.3 | 1.28 | 0.68 | 0.02 | 0.40 | < 0.03 | 0.05 |
| Ho | 1.80 | 0.05 | 2.01 | 0.02 | 3.90 | 0.53 | 0.13 | 0.00 | 0.07 | < 0.006 | 0.01 |
| Er | 5.15 | 0.35 | 6.09 | 0.06 | 9.91 | 2.95 | 0.39 | 0.02 | 0.20 | < 0.03 | < 0.03 |
| Tm | 0.69 | 0.08 | 0.85 | 0.01 | 1.14 | 0.66 | 0.06 | 0.00 | 0.02 | < 0.006 | < 0.006 |
| Yb | 4.38 | 0.94 | 5.87 | 0.06 | 6.30 | 6.45 | 0.39 | 0.05 | 0.16 | < 0.03 | < 0.03 |
| Lu | 0.65 | 0.21 | 0.92 | 0.01 | 0.80 | 1.26 | 0.05 | 0.01 | 0.02 | < 0.007 | < 0.007 |
| Hf | 0.36 | 0.05 | < 0.1 | < 0.02 | 2.44 | 0.05 | 0.12 | 1.39 | 0.06 | 0.23 | < 0.04 |
| Ta | < 0.1 | < 0.008 | < 0.07 | 0.01 | < 0.07 | < 0.008 | 0.03 | 0.02 | 0.03 | 0.03 | 0.03 |
| Pb | 47.4 | 3.54 | 7.63 | 0.68 | 14.5 | 6.35 | 2.34 | 1.36 | 17.7 | 1.10 | 11.8 |
| Th | < 0.04 | < 0.007 | < 0.03 | < 0.007 | 15.1 | < 0.007 | < 0.01 | 0.02 | 0.26 | 0.03 | 0.04 |
| U | 1.44 | 0.01 | 3.45 | 0.11 | 3.99 | 0.01 | 1.87 | 0.07 | 0.15 | 0.02 | 0.01 |

| Sample # | 1445 | 1445/4 | 1445/4 | Mir | Mir | Mir | 85-1 | 85-1 | 85-3 | 85-3 | 85-4 | 85-4 | 85-5 |
|---|---|---|---|---|---|---|---|---|---|---|---|---|---|
| Mineral | Hbl | Zo | Hbl | Zo 1 | Zo 2 | Ky | Zo | Ab | Zo | Ab | Zo | Amp | Zo |
| Rb | 1.27 | 0.16 | 1.93 | 0.15 | 0.18 | 0.13 | 1.65 | 2.12 | 8.41 | 3.28 | 3.32 | 12.1 | 3.75 |
| Sr | 71.3 | 2253 | 275.9 | 1261 | 733.5 | 1.23 | 1419 | 478.2 | 1303 | 451.9 | 1610 | 23.2 | 1318 |
| Y | 0.42 | 1.31 | 0.27 | 109.1 | 130.4 | 0.15 | 154.4 | 0.29 | 152.5 | 0.30 | 190.7 | 10.6 | 156.2 |
| Zr | 9.00 | 3.77 | 9.28 | 18.6 | 14.3 | 9.43 | 8.45 | 0.96 | 6.20 | 16.7 | 9.73 | 15.1 | 9.43 |
| Nb | 0.51 | <0.02 | 0.66 | <0.07 | <0.07 | 0.10 | <0.1 | <0.02 | <0.1 | 0.06 | 0.14 | 1.26 | <0.1 |
| Cs | <0.02 | <0.02 | <0.02 | <0.02 | <0.02 | 0.06 | <0.05 | 0.03 | 0.07 | 0.06 | 0.12 | 0.32 | <0.05 |
| Ba | 138.2 | 174.4 | 72.9 | 17.5 | 12.9 | 0.90 | 60.2 | 1577 | 304.4 | 929.5 | 107.9 | 141.3 | 126.4 |
| La | 0.02 | 0.99 | 0.05 | 40.6 | 21.8 | 0.05 | 139.3 | 0.28 | 127.9 | 0.06 | 30.8 | 0.55 | 114.7 |
| Ce | 0.03 | 1.97 | 0.09 | 88.6 | 52.8 | 0.07 | 363.5 | 0.66 | 334.9 | 0.10 | 94.3 | 1.23 | 305.2 |
| Pr | <0.009 | 0.24 | 0.01 | 14.0 | 8.76 | 0.01 | 52.9 | 0.10 | 48.9 | 0.02 | 16.6 | 0.17 | 45.0 |
| Nd | <0.06 | 1.02 | <0.06 | 65.6 | 43.1 | 0.05 | 262.5 | 0.44 | 237.7 | 0.08 | 102.9 | 0.98 | 223.8 |
| Sm | <0.06 | 0.22 | <0.06 | 18.9 | 15.4 | <0.02 | 66.0 | 0.12 | 61.6 | 0.02 | 46.4 | 0.42 | 58.2 |
| Eu | 0.03 | 0.21 | 0.03 | 4.00 | 4.27 | <0.007 | 20.1 | BaO | 19.7 | BaO | 17.3 | 0.16 | 18.6 |
| Gd | 0.03 | 0.25 | 0.03 | 22.7 | 21.8 | 0.02 | 66.5 | 0.10 | 63.0 | 0.02 | 70.6 | 0.94 | 63.7 |
| Tb | <0.007 | 0.04 | <0.007 | 3.37 | 3.83 | <0.004 | 8.69 | 0.01 | 8.18 | 0.00 | 9.77 | 0.20 | 8.56 |
| Dy | 0.05 | 0.23 | 0.02 | 20.2 | 24.2 | 0.02 | 43.4 | 0.07 | 41.0 | 0.04 | 49.9 | 1.59 | 42.9 |
| Ho | 0.01 | 0.05 | 0.01 | 3.59 | 4.34 | 0.00 | 6.21 | 0.01 | 6.01 | 0.01 | 7.55 | 0.41 | 6.36 |
| Er | 0.06 | 0.14 | 0.04 | 8.29 | 10.1 | 0.01 | 11.1 | <0.03 | 11.2 | 0.03 | 14.5 | 1.32 | 11.6 |
| Tm | 0.01 | 0.02 | 0.01 | 0.82 | 0.95 | <0.003 | 0.83 | <0.006 | 0.85 | 0.01 | 1.21 | 0.20 | 0.88 |
| Yb | 0.08 | 0.12 | 0.07 | 4.27 | 4.69 | 0.02 | 3.18 | <0.03 | 3.31 | 0.05 | 5.32 | 1.36 | 3.49 |
| Lu | 0.01 | 0.02 | 0.01 | 0.56 | 0.57 | <0.004 | 0.34 | <0.007 | 0.36 | 0.01 | 0.61 | 0.21 | 0.39 |
| Hf | 0.25 | 0.08 | 0.31 | 0.91 | 1.01 | 0.22 | 0.45 | <0.04 | 0.33 | 1.07 | 0.41 | 0.75 | 0.39 |
| Ta | <0.03 | <0.03 | 0.04 | <0.1 | <0.1 | 0.11 | <0.09 | <0.03 | <0.09 | 0.08 | <0.09 | <0.07 | <0.09 |
| Pb | 11.7 | 42.6 | 12.9 | 2.52 | 2.97 | 0.38 | 34.4 | 20.9 | 30.3 | 27.5 | 70.9 | 3.77 | 37.6 |
| Th | <0.01 | 0.08 | <0.01 | 10.5 | 6.79 | 0.02 | 14.0 | 0.03 | 12.7 | 0.02 | 2.78 | 0.12 | 11.8 |
| U | <0.01 | 0.12 | <0.01 | 17.6 | 24.3 | 0.05 | 4.50 | 0.02 | 4.79 | 0.05 | 2.61 | 0.05 | 4.05 |

| Sample # | 85-5 | 89-1 | 89-1 | 89-1 | 93-24 | 93-24 | 93-24 | AB 94-3 | AB 94/5 | AB 94-6 | AB 94-8 | AB 94-8 | AB 94-8 |
|---|---|---|---|---|---|---|---|---|---|---|---|---|---|
| Mineral | Ab | Zo | Ab | Amp | Zo | Ab | Amp | Zo | Zo | Zo | Zo | Cal | Ms |
| Rb | 1.54 | <0.4 | 12.0 | 5.70 | <0.4 | 1.37 | 6.26 | <0.4 | 3.15 | 2.00 | 2.35 | 0.15 | 4.34 |
| Sr | 395.2 | 1358 | 487.6 | 30.7 | 1372 | 514.6 | 29.9 | 2780 | 2753 | 2703 | 2603 | 759.4 | 1727 |
| Y | 0.38 | 145.7 | 1.59 | 12.9 | 162.6 | 5.91 | 14.8 | 21.6 | 12.5 | 12.9 | 14.2 | 7.37 | 8.81 |
| Zr | 4.55 | 8.80 | 0.34 | 13.3 | 10.0 | 1.05 | 27.6 | 13.0 | 3.93 | 4.28 | 3.23 | 0.65 | 2.26 |
| Nb | <0.02 | <0.05 | 0.05 | 2.07 | <0.05 | 0.05 | 4.20 | 0.10 | <0.05 | <0.1 | <0.05 | 0.02 | 0.08 |
| Cs | 0.04 | <0.02 | 0.27 | 0.08 | <0.02 | 0.11 | 0.10 | 0.15 | 0.11 | 0.08 | 0.07 | 0.03 | 0.74 |
| Ba | 989.4 | 15.2 | 250.8 | 58.6 | 24.8 | 203.8 | 63.9 | 3.67 | 53.9 | 40.4 | 26.8 | 4.15 | 75.5 |
| La | 0.22 | 127.6 | 0.76 | 0.63 | 92.9 | 3.23 | 0.23 | 13.7 | 21.3 | 20.6 | 19.8 | 0.40 | 3.37 |
| Ce | 0.55 | 341.7 | 1.93 | 2.01 | 247.9 | 8.33 | 0.91 | 30.2 | 39.8 | 38.8 | 37.8 | 0.75 | 7.00 |
| Pr | 0.08 | 50.5 | 0.30 | 0.29 | 36.3 | 1.28 | 0.20 | 4.07 | 4.67 | 4.45 | 4.43 | 0.09 | 0.92 |
| Nd | 0.39 | 240.4 | 1.46 | 1.61 | 183.5 | 6.21 | 1.40 | 19.7 | 19.6 | 18.7 | 19.2 | 0.39 | 4.63 |
| Sm | 0.10 | 60.1 | 0.43 | 0.65 | 52.6 | 1.78 | 0.81 | 5.87 | 4.12 | 4.31 | 4.64 | 0.10 | 1.52 |
| Eu | BaO | 18.5 | 0.18 | 0.27 | 17.3 | 0.66 | 0.42 | 2.72 | 1.66 | 1.70 | 1.87 | 0.08 | 0.82 |
| Gd | 0.11 | 59.8 | 0.46 | 1.12 | 62.1 | 2.02 | 1.53 | 7.32 | 4.42 | 4.50 | 5.04 | 0.19 | 2.13 |
| Tb | 0.01 | 7.77 | 0.06 | 0.23 | 8.48 | 0.28 | 0.31 | 0.96 | 0.57 | 0.59 | 0.68 | 0.05 | 0.30 |
| Dy | 0.07 | 39.7 | 0.36 | 1.93 | 43.2 | 1.42 | 2.34 | 4.98 | 2.90 | 3.08 | 3.33 | 0.55 | 1.68 |
| Ho | 0.01 | 5.83 | 0.05 | 0.48 | 6.36 | 0.21 | 0.52 | 0.79 | 0.49 | 0.50 | 0.54 | 0.22 | 0.29 |
| Er | 0.03 | 10.6 | 0.12 | 1.52 | 11.8 | 0.38 | 1.79 | 1.75 | 1.09 | 1.13 | 1.21 | 1.15 | 0.74 |
| Tm | <0.006 | 0.81 | 0.01 | 0.23 | 0.89 | 0.03 | 0.26 | 0.18 | 0.11 | 0.11 | 0.11 | 0.23 | 0.08 |
| Yb | <0.03 | 2.92 | 0.06 | 1.66 | 3.52 | 0.12 | 1.78 | 1.03 | 0.59 | 0.55 | 0.58 | 1.99 | 0.43 |
| Lu | <0.007 | 0.33 | 0.01 | 0.25 | 0.40 | 0.01 | 0.27 | 0.14 | 0.07 | 0.06 | 0.08 | 0.38 | 0.05 |
| Hf | 0.18 | 0.42 | 0.01 | 0.63 | 0.45 | 0.03 | 1.01 | 0.51 | 0.19 | 0.18 | 0.17 | <0.02 | 0.11 |
| Ta | <0.03 | 0.07 | 0.06 | 0.10 | 0.07 | 0.05 | 0.24 | 0.04 | <0.4 | <0.09 | 0.06 | <0.008 | 0.06 |
| Pb | 18.6 | 28.0 | 6.44 | 1.89 | 44.6 | 10.3 | 2.05 | 74.2 | 107.0 | 99.4 | 92.7 | 10.6 | 55.8 |
| Th | 0.03 | 13.1 | 0.06 | 0.05 | 9.18 | 0.32 | 0.03 | 0.27 | 1.59 | 1.53 | 0.90 | 0.01 | 0.09 |
| U | 0.03 | 4.64 | 0.04 | 0.05 | 3.19 | 0.13 | 0.07 | 0.51 | 0.37 | 0.42 | 0.51 | 0.01 | 0.14 |

| Sample # | AB 94-9 | AB 94-(16)b | AB 94/17 | HU 93-6 | HU 93-6 | HU 93-7 | HU 93-7 | Hu 93-7 | HU 93-8 | ZDa1 | ZDA1 |
|---|---|---|---|---|---|---|---|---|---|---|---|
| Mineral | Zo | Zo | Zo | Zo | Czo | Zo | Czo | Ab | Zo | Zo | Czo |
| Rb | 2.38 | <0.4 | 2.70 | 2.31 | 5.47 | 3.67 | 1.51 | 1.00 | <0.4 | 0.80 | 1.48 |
| Sr | 2628 | 3315 | 2693 | 4503 | 4210 | 5204 | 3809 | 1646 | 1254 | 2309 | 2512 |
| Y | 10.5 | 13.2 | 11.7 | 37.5 | 64.9 | 65.6 | 91.4 | 0.41 | 51.9 | 17.2 | 14.5 |
| Zr | 3.63 | 13.8 | 3.63 | 11.3 | 12.7 | 23.6 | 14.5 | 1.13 | 9.46 | 12.4 | 19.3 |
| Nb | <0.1 | 0.33 | 0.12 | 0.66 | 0.21 | 0.57 | 0.15 | 0.05 | 0.09 | 0.05 | 0.03 |
| Cs | 0.11 | <0.02 | 0.09 | 0.15 | 0.24 | 0.14 | 0.13 | 0.05 | 0.14 | 0.20 | 0.26 |
| Ba | 20.9 | 10.1 | 23.7 | 18.9 | 18.4 | 21.3 | 23.4 | 97.3 | 9.83 | 7.00 | 12.9 |
| La | 11.3 | 4.86 | 16.5 | 123.2 | 348.0 | 162.2 | 451.0 | 0.61 | 39.8 | 1.07 | 1.10 |
| Ce | 21.2 | 10.8 | 31.2 | 252.5 | 736.0 | 339.5 | 974.4 | 1.24 | 79.6 | 2.83 | 2.69 |
| Pr | 2.43 | 1.54 | 3.69 | 32.5 | 93.7 | 44.5 | 124.2 | 0.16 | 9.46 | 0.27 | 0.33 |
| Nd | 10.6 | 8.17 | 15.9 | 141.3 | 402.0 | 197.2 | 539.6 | 0.70 | 36.2 | 1.17 | 1.46 |
| Sm | 2.76 | 2.91 | 3.93 | 42.9 | 120.1 | 67.3 | 161.4 | 0.21 | 7.49 | 0.46 | 0.46 |
| Eu | 1.25 | 1.41 | 1.65 | 13.8 | 37.0 | 22.9 | 44.4 | 0.08 | 2.10 | 0.19 | 0.17 |
| Gd | 3.22 | 4.32 | 4.40 | 39.5 | 102.4 | 69.0 | 141.0 | 0.21 | 7.27 | 1.47 | 0.92 |
| Tb | 0.41 | 0.58 | 0.57 | 3.77 | 9.37 | 7.70 | 13.4 | 0.02 | 1.22 | 0.38 | 0.22 |
| Dy | 2.26 | 2.95 | 2.84 | 12.9 | 27.9 | 25.9 | 39.9 | 0.10 | 8.19 | 2.87 | 1.76 |
| Ho | 0.38 | 0.46 | 0.44 | 1.41 | 2.60 | 2.45 | 3.73 | 0.01 | 1.71 | 0.63 | 0.43 |
| Er | 0.89 | 1.08 | 1.00 | 2.57 | 4.06 | 3.47 | 5.48 | 0.03 | 5.46 | 1.95 | 1.55 |
| Tm | 0.08 | 0.12 | 0.09 | 0.24 | 0.34 | 0.26 | 0.46 | 0.00 | 0.76 | 0.32 | 0.27 |
| Yb | 0.40 | 0.62 | 0.45 | 1.19 | 1.65 | 1.13 | 2.21 | 0.02 | 4.61 | 2.50 | 2.02 |
| Lu | 0.05 | 0.08 | 0.10 | 0.17 | 0.27 | 0.19 | 0.38 | <0.004 | 0.60 | 0.42 | 0.33 |
| Hf | 0.17 | 0.67 | 0.25 | 0.69 | 0.62 | 0.61 | 0.71 | 0.02 | 0.50 | 0.44 | 0.60 |
| Ta | <0.09 | <0.4 | 0.12 | 0.11 | 0.08 | 0.13 | 0.07 | 0.04 | 0.10 | 0.05 | <0.03 |
| Pb | 95.4 | 87.7 | 99.6 | 201.1 | 170.3 | 223.4 | 163.5 | 42.1 | 209.8 | 49.1 | 53.2 |
| Th | 0.64 | 0.13 | 0.64 | 46.5 | 133.1 | 81.4 | 218.0 | 0.30 | 9.86 | 0.14 | 0.17 |
| U | 0.32 | 0.29 | 0.38 | 47.7 | 104.3 | 92.4 | 151.2 | 0.25 | 3.10 | 0.20 | 0.16 |

**APPENDIX C**

Sources for trace element analyses of epidote group minerals (arranged in chronological order)

| Authors | Year | Mineral | Environment | Elements analyzed | Methods[a] |
|---|---|---|---|---|---|
| Penfield & Warren | 1899 | Ep | metamorphic | Mn, Sr, Pb | wet chemical |
| Bleek | 1907 | Ep | metamorphic | Cr | wet chemical |
| Eskola | 1933 | Ep | metamorphic | Cr | wet chemical |
| Game | 1954 | Zo | metamorphic | Cr, Mn | wet chemical |
| Lisitin & Khitrov | 1962 | Ep | metamorphic | B | spectroscopic |
| Willemse & Bensch | 1964 | Czo | metamorphic | Ti, Mn, Sr, P | wet chemical |
| Myer | 1965 | Ep | hydrothermal | Sn | spectroscopic |
| Myer | 1966 | Ep; Zo | metamorphic | Be, P, Sc, V, Cr, Mn, Ni, Cu, Sr, Y, Zr, Ag, Sn, Ba, REE (La, Ce, Nd, Yb), Pb, Bi | spectroscopic |
| Mulligan & Jambor | 1968 | Ep | metamorphic | Sn | wet chemical |
| Shepel & Karpenko | 1969 | Ep | metasomatic | Cr, V | wet chemical |
| Karev | 1974 | Ep | metamorphic | V | wet chemical |
| Schreyer & Abraham | 1978 | Ep; Zo | metamorphic | V, Mn, Sr | EMPA |
| Alderton & Jackson | 1978 | Ep | metasomatic | Sn | EMPA |
| Cooper | 1980 | Zo | metamorphic | Cr | EMPA |
| Grapes | 1981 | Ep; Zo | metamorphic | Cr, Mn | EMPA |
| Howie & Walsh | 1982 | Ep | metamorphic | Li, P, V, Cr, Mn, Co, Ni, Cu, Zn, Sr, Y, Nb, Ba, REE (La, Ce, Pr, Nd, Sm, Eu, Gd, Dy, Ho, Er, Yb, Lu), Pb | wet chemical; ICP-AES |
| Gromet & Silver | 1983 | Ep | igneous | REE (Ce, Nd, Sm, Eu, Gd, Dy, Er, Yb) | ID-TIMS |
| Johan et al. | 1983 | Zo | metamorphic | Zn, Ga, Ge | EMPA |
| Grapes & Watanabe | 1984 | Ep | metamorphic | Mn, Sr | EMPA |
| Nyström | 1984 | Ep | hydrothermal | REE (La, Ce, Nd, Sm, Eu, Tb, Yb, Lu, U) | INAA |

| Authors | Year | Mineral | Environment | Elements analyzed | Methods[a] |
|---|---|---|---|---|---|
| Sakai et al. | 1984 | Ep | metamorphic | REE (La, Ce, Pr, Nd, Sm, Gd), Th | EMPA |
| Pouliot et al. | 1984 | Zo | hydrothermal | Cr, Mn | EMPA |
| Brastad | 1985 | Zo | metamorphic | Sr | EMPA |
| Dunn | 1985 | Ep | metamorphic | Sr, Pb | EMPA |
| Maaskant | 1985 | Zo | metamorphic | P, Cl, Sr | EMPA |
| Van Marcke de Lummen | 1986 | Ep | metasomatic | Mn, Sn | EMPA |
| Ashley & Martyn | 1987 | Ep | metasomatic | Ti, Cr | EMPA |
| Treloar | 1987 | Ep | metamorphic | V, Cr, Sr | EMPA |
| Sorensen & Grossmann | 1989 | Czo; Zoi | metamorphic | Na, Sc, Ti, Cr, Co, Ni, Zn, Rb, Sr, Zr, Cs, Ba, REE (La, Ce, Nd, Sm, Eu, Tb, Yb, Lu), Hf, Ta, Th, U | INAA |
| Pan & Fleet | 1990 | Czo | metamorphic | F, REE (La, Ce, Nd) | EMPA |
| Dawes & Evans | 1991 | Ep | igneous | F, Cl, Mn, REE (La, Ce, Nd, Sm), Th | EMPA |
| Hickmott et al. | 1992 | Zo | metamorphic | Ni, Zn, Ga, Rb, Sr, Y, Zr, Nb, REE (Ce, Nd, Sm, Eu, Dy, Er, Yb) | SIMS/PIXE |
| Braun et al. | 1993 | Ep | igneous | REE (La, Ce, Nd, Sm, Eu, Gd, Dy, Er, Yb), Th | ID-TIMS |
| Domanik et al. | 1993 | Ep, Czo | metamorphic | Be, B, Rb, Sr, Y, Ba, Ce | SIMS |
| Harlow | 1994 | Czo, Ep, Zo | metasomatic | Sr | EMPA |
| Holtstam & Langhof | 1994 | Ep | metamorphic | Ti, Mn, Pb | EMPA |
| Bonazzi & Menchetti | 1995 | Ep | various | Ti, Mn, Sr, Y, REE (La, Ce, Pr, Nd, Sm, Gd) | EMPA |
| Sanchez et al. | 1995 | Czo; Ep | metamorphic | Cr, Mn | EMPA |
| Challis et al. | 1995 | Ep | metamorphic | Cr | EMPA |
| Pan & Fleet | 1996 | Czo | hydrothermal, metamorphic | REE (La, Ce, Pr, Nd, Sm, Eu, Gd, Tb, Dy, Ho, Er, Tm, Yb, Lu) | ICP-MS |
| Keane & Morrison | 1997 | Ep | igneous | Mn, Sr, Y, REE (La, Ce, Pr, Nd, Sm, Gd), Th | EMPA |
| Carcangiu et al. | 1997 | Ep | metasomatic | Mn, Y, REE (La, Ce, Pr, Nd, Sm, Gd), Th | EMPA |
| Finger et al. | 1998 | Ep | metamorphic | Y, REE (La, Ce, Pr, Nd, Sm), Th, U | EMPA |

| Authors | Year | Mineral | Environment | Elements analyzed | Methods[a] |
|---|---|---|---|---|---|
| Nagasaki & Enami | 1998 | Zo | metamorphic | Ti, Cr, Mn, Sr, REE (La, Ce, Nd) | EMPA |
| Yanev et al. | 1998 | Ep | hydrothermal | Ti, Y, REE (La, Ce, Pr, Nd, Sm, Dy) | EMPA |
| Jancev & Bermanec | 1998 | Ep | metamorphic | Mn, Pb | EMPA |
| Devaraju et al. | 1999 | Ep | metamorphic | Cr | EMPA |
| Liu et al. | 1999 | Ep | metamorphic | Ti, Mn, Y, REE (La, Ce, Pr, Nd, Sm), Th, | EMPA |
| Brunsmann et al. | 2000 | Zo | metamorphic | Sc, V, Cr, Mn, Co, Ni, Zn, Ga, Sr, Y, Ba, REE (La, Ce, Pr, Nd, Sm, Eu, Gd, Tb, Dy, Ho, Er, Yb, Lu) | XRF/ICP-AES |
| Sassie et al. | 2000 | Ep | metamorphic | Sr, Y, Zr, Nb, REE (La, Ce, Pr, Nd, Sm, Eu, Gd, Tb, Dy, Ho, Er, Yb, Lu) | SIMS |
| Brunsmann et al. | 2001 | Zo | metamorphic | REE (La, Ce, Nd, Sm, Eu, Gd, Tb, Dy, Ho, Er, Yb, Lu) | ICP-AES |
| Dahlquist | 2001 | Ep | igneous | Ti, Mn, Y, Zr, REE (La, Ce, Er, Yb) | EMPA |
| Choo | 2002 | Ep | hydrothermal | P, Ti, Mn | EMPA |
| Spiegel et al. | 2002 | Ep | metamorphic, detrital | Ti, V, Cr, Mn, Co, Ni, Cu, Zn, Ga, As, Rb, Sr, Zr, Nb, Mo, Ag, Cd, Sb, Cs, Ba, Hf, Ta, W, Pb, Th, U | MC-ICP-MS |
| Zack et al. | 2002 | Zo; Czo | metamorphic | Li, Be, B, Sr, Y, Zr, Nb, Ba, REE (Ce, Nd, Sm), Pb, Th, U | LA-ICP-MS |
| Yang & Rivers | 2002 | Ep | metamorphic | P, Sc, Ti, V, Cr, Co, Ni, Zn, Sr, Y, Zr, Nb, Cs, Ba, REE (La, Pr, Sm, Eu, Gd, Tb, Ho, Tm, Lu), Hf, Ta | LA-ICP-MS |
| Miyajima et al. | 2003 | Czo | metamorphic | Ti, Mn, Sr | EMPA |
| Spandler et al. | 2003 | Ep (original | metamorphic | Sc, Ti, V, Mn, Rb, Sr, Y, Zr, Nb, Cs, Ba, REE (La, Ce, Pr, Nd, Sm, Eu, Gd, Tb, Dy, Ho, Er, Tm, Yb, Lu), Hf, Ta, Pb, Th, U | LA-ICP-MS |
| Spandler pers.comm. | 2004 | classif. as Zo) |  |  |  |
| This Study | 2004 | Ep, Czo, Zo | various | Rb, Sr, Y, Zr, Nb, Cs, Ba, REE (La, Ce, Pr, Nd, Sm, Eu, Gd, Tb, Dy, Ho, Er, Tm, Yb, Lu), Hf, Ta, Pb, Th, U | ICP-MS |

Abbreviations are: EMPA = Electron Microprobe Analyser, ID-TIMS = Isotope Dilution - Thermal Ionisation Mass Spectrometry, ICP-AES = Inductively Coupled Plasma - Atomic Emission Spectroscopy, ICP-MS = Inductively Coupled Plasma – Mass Spectrometry, INAA = Instrumental Neutron Activation Analysis, LA-ICP-MS = Laser Ablation - Inductively Coupled Plasma - Mass Spectrometry, MC-ICP-MS = Multi Collector - Inductively Coupled Plasma – Mass Spectrometry, SIMS = Secondary Ion Mass Spectrometry, PIXE = Proton Induced X-Ray Emission, and XRF = X-Ray Fluorescence.

## APPENDIX D

Summary of observed trace element variations in naturally occurring zoisite and clinozoisite and epidote. Reported are mean, median, range, and number of determinations. All concentrations are reported in µg/g except for values in bold italics that denote maximum observed concentrations reported in weight-% oxides. Elements are arranged according to increasing atomic mass.

| Element | ZOISITE Mean | Median | Min. | Max. | n | CLINOZOISITE Mean | Median | Min. | Max. | n | EPIDOTE Mean | Median | Min. | Max. | n |
|---|---|---|---|---|---|---|---|---|---|---|---|---|---|---|---|
| Mg; ***MgO*** | ***0.19*** | ***0.08*** | ***0.04*** | ***0.49*** | 8 (8) | | | ***0.04*** | ***0.13*** | 6 | ***0.25*** | ***0.17*** | ***0.04*** | ***0.95*** | 57 (57) |
| Sc | 42 | 39 | 14 | 93 | 14 | | | | | | | | | | |
| Ti; ***TiO$_2$*** | 521 | 540 | 60 | 1078 | 17 | | | 479 | 839 | 4 | 2040 | 1018 | 60 | ***2.06*** | 126 |
| V; ***V$_2$O$_3$*** | 651 | 846 | 317 | 985 | 11 | | | | | | 507 | 505 | 155 | ***11.29*** | 49 (26) |
| Cr; ***Cr$_2$O$_3$*** | 513 | 134 | 104 | ***2.46*** | 9 (4) | | | 350 | ***5.74*** | 4 (3) | 207 | 126 | 12 | ***15.37*** | 29 (5) |
| Mn; ***MnO*** | 419 | 182 | 68 | 2320 | 12 (8) | | | 464 | 2323 | 4 (4) | 1850 | 1630 | 155 | ***2.88*** | 131 (108) |
| Co | | | | | | | | | | | 3.9 | 3.2 | 1.0 | 13.9 | 26 |
| Ni | | | | | | | | | | | 20.6 | 17.0 | 7.0 | 66.0 | 22 |
| Cu | | | | | | | | | | | 11.3 | 6.7 | 2.0 | 73.4 | 25 |
| Zn; ***ZnO*** | | | | ***1.14*** | 2 (2) | | | | | | 25.8 | 13.0 | 4.0 | 89.0 | 23 |
| Ga | 1060 | 65 | 59 | 3900 | 7 (2) | | | | | | 44 | 45 | 20 | 73 | 23 |
| Ge$_2$O$_3$ | | | | ***6.42*** | 2 (2) | | | | | | | | | | |
| As | | | | | | 2260 | 1970 | 1.0 | 5.5 | 5 | 5.8 | 3.1 | 1.0 | 52.7 | 23 |
| Rb | 2.5 | 2.2 | 0.01 | 10.0 | 24 | | | | | | 12.3 | 3.0 | 0.1 | 98.5 | 31 |
| Sr; ***SrO*** | 2230 | 2300 | 729 | ***7.4*** | 50 (8) | 2260 | 1970 | 420 | ***10.27*** | 11 (4) | 2020 | 1560 | 240 | ***16.33*** | 91 (45) |
| Y; ***Y$_2$O$_3$*** | 83.2 | 40.1 | 1.31 | 426 | 39 | | | 2.00 | 91.4 | 6 | 1740 | 193 | 4.3 | ***1.69*** | 41 (19) |
| Zr | 8.4 | 8.4 | 1.7 | 23.6 | 27 | | | 12.7 | 19.3 | 3 | 21.4 | 15.7 | 0.4 | 229 | 42 |
| Nb | 0.16 | 0.09 | 0.002 | 0.66 | 14 | | | 0.03 | 0.21 | 3 | 6.8 | 2.8 | 0.06 | 35 | 29 |
| Mo | | | | | | | | | | | 0.80 | 0.65 | 0.40 | 1.5 | 4 |
| Ag | | | | | | | | | | | 1.4 | 1.0 | 0.8 | 2.8 | 23 |
| Cd | | | | | | | | | | | 0.66 | 0.62 | 0.36 | 1.45 | 23 |
| Sn; ***SnO$_2$*** | | | | | | | | | | | 1.36 | 1.64 | ***0.16*** | ***2.84*** | 11 (8) |

| Element | Mean | Median | Min. | Max. | n | Mean | Median | Min. | Max. | n | Mean | Median | Min. | Max. | n |
|---|---|---|---|---|---|---|---|---|---|---|---|---|---|---|---|
| Sb | 0.09 | 0.08 | 0.003 | 0.20 | 18 | | | 0.13 | 0.26 | 3 | 1.5 | 1.0 | 0.6 | 5.0 | 14 |
| Cs | 46.1 | 19.9 | 0.5 | 304 | 34 | | | 5.0 | 23.4 | 5 | 0.76 | 0.50 | 0.04 | 3.10 | 20 |
| Ba | 38.0 | 17.7 | 1.0 | 162 | 41 | | | 0.3 | 451 | 5 | 48.6 | 11.3 | 1.7 | 407 | 43 |
| La; $La_2O_3$ | 144 | 40 | 2.0 | 1200 | 47 | | | 0.8 | 975 | 7 | 7410 | 560 | 0.3 | *4.46* | 68 (38) |
| Ce; $Ce_2O_3$ | 15.4 | 4.7 | 0.2 | 73.7 | 39 | | | 0.1 | 124 | 4 | 13360 | 940 | 1.3 | *7.99* | 73 (39) |
| Pr; $Pr_2O_3$ | 108 | 25 | 1.0 | 690 | 47 | | | 0.5 | 540 | 5 | 2400 | 940 | 0.2 | *1.11* | 40 (24) |
| Nd; $Nd_2O_3$ | 23.6 | 6.3 | 0.2 | 170 | 44 | | | 0.2 | 161 | 6 | 4940 | 560 | 1.1 | *3.12* | 68 (39) |
| Sm; $Sm_2O_3$ | 8.29 | 1.87 | 0.19 | 56.9 | 41 | | | 0.17 | 44.4 | 5 | 1570 | 170 | 0.6 | *1.70* | 63 (36) |
| Eu | 29.7 | 7.16 | 0.25 | 226 | 40 | | | 0.51 | 141 | 4 | 34.2 | 2.5 | 0.17 | 708 | 27 (9) |
| Gd; $Gd_2O_3$ | 4.14 | 1.20 | 0.04 | 34.3 | 41 | | | 0.12 | 13.4 | 4 | 1050 | 16.4 | 0.77 | *1.38* | 31 (9) |
| Tb | 19.5 | 6.2 | 0.23 | 140.6 | 40 | | | 1.17 | 39.9 | 4 | 12.23 | 0.97 | 0.11 | 162 | 24 |
| Dy; $Dy_2O_3$ | 2.74 | 0.99 | 0.05 | 14.7 | 40 | | | 0.33 | 3.73 | 4 | 426 | 10.3 | 0.68 | *0.44* | 29 (7) |
| Ho | 5.28 | 1.90 | 0.14 | 28.6 | 39 | | | 1.45 | 5.48 | 4 | 5.01 | 1.54 | 0.13 | 28.5 | 21 |
| Er | 0.68 | 0.32 | 0.02 | 5.60 | 25 | | | 0.25 | 0.46 | 4 | 48.8 | 4.94 | 0.39 | 438 | 23 |
| Tm | 2.30 | 1.13 | 0.12 | 17.80 | 41 | | | 1.65 | 2.21 | 4 | 0.61 | 0.61 | 0.06 | 1.14 | 15 |
| Yb | 0.28 | 0.14 | 0.02 | 2.30 | 41 | | | 0.27 | 0.38 | 4 | 24.9 | 4.3 | 0.36 | 351 | 25 |
| Lu | 0.39 | 0.40 | 0.06 | 1.01 | 26 | | | 0.60 | 0.71 | 3 | 0.65 | 0.50 | 0.05 | 2.80 | 24 |
| Hf | 0.56 | 0.07 | 0.01 | 5.9 | 12 | | | 0.07 | 0.08 | 2 | 0.93 | 0.92 | 0.12 | 2.44 | 35 |
| Ta | | | | | | | | | | | 0.42 | 0.40 | 0.01 | 1.23 | 12 |
| W | | | | | | | | | | | 52.1 | 43.7 | 1.1 | 133 | 23 |
| Pb; $PbO$ | 77 | 72 | 2.5 | 223 | 40 | | | 7.6 | 170 | 3 | 219 | 20 | 2.3 | *32.90* | 37 |
| Th; $ThO_2$ | 7.2 | 1.2 | 0.08 | 81.4 | 22 | | | 0.17 | 218 | 3 | 560 | 24.0 | 0.09 | *1.01* | 51 (24) |
| U | 7.1 | 2.6 | 0.07 | 92.4 | 33 | | | 0.16 | 151 | 4 | 172 | 1.57 | 0.19 | 2700 | 42 (3) |

Reviews in Mineralogy & Geochemistry
Vol. 56, pp. 607-628, 2004
Copyright © Mineralogical Society of America

# Stable and Radiogenic Isotope Systematics in Epidote Group Minerals

## Jean Morrison

*Department of Earth Sciences*
*University of Southern California*
*Los Angeles, California 90089-0740, U.S.A.*
*morrison@usc.edu*

## INTRODUCTION

Epidote minerals (the epidote group together with the orthorhombic polymorph zoisite) occur in a wide range of igneous, metamorphic and sedimentary lithologies. Although often present in only small quantities, epidote group minerals can nonetheless be used to generate important quantitative constraints on processes such as regional metamorphism, deep-seated pluton emplacement and uplift, hydrothermal fluid flow, and sedimentary provenance (e.g., Zen and Hammarstrom 1984; Brandon et al. 1996; Cartwright et al. 1996; Keane and Morrison 1997; Spiegel et al. 2002). Petrologic studies on epidote group minerals formed by these and other processes are reviewed in Bird and Spieler (2004), Grapes and Hoskin (2004), Enami et al. (2004), and Schmidt and Poli (2004). In this chapter, stable and radiogenic isotope studies of epidote group minerals will be reviewed. Although stable and/or radiogenic isotope data exist for epidote group minerals from a number of different geologic settings, such data have produced particularly important insights into 1) fluid flow associated with variable grades of metamorphism, particularly ultra-high-pressure (UHP) metamorphism, 2) intrusion of deep-seated plutonic systems, 3) the nature of hydrothermal alteration in geothermal systems, 4) age relations in hydrothermal systems, and 5) sedimentary provenance.

## STABLE ISOTOPE SYSTEMATICS

Based on the chemical composition of naturally occurring epidote minerals, oxygen, hydrogen, and chlorine stable isotopes all have the potential to be important petrogenetic indicators in systems involving epidote minerals. Oxygen is present in epidote minerals bound in isolated $TO_4$ tetrahedra, $T_2O_7$ groups, and in the hydroxyl site as both $OH^-$ and $O^{-2}$. Hydrogen and chlorine are present in the hydroxyl site as $OH^-$ and $Cl^-$, respectively, although $Cl^-$ is present in only very small quantities (Frei et al. 2004). To date, oxygen and hydrogen isotope systematics have been studied both experimentally and empirically, but little work has been done on chlorine isotopes. This section will focus first on experimental and theoretical data on stable isotope fractionations pertinent to epidote minerals. Then existing empirical data will be considered in the context of three of the major geologic processes described above: 1) fluid flow associated with variable grades of metamorphism particularly ultra-high pressure (UHP) metamorphism, 2) intrusion of deep-seated plutonic systems, and 3) the nature of hydrothermal alteration in various types of geothermal systems.

1529-6466/04/0056-0013$05.00

## Experimental studies

A compilation of published studies involving experimental and calculated determinations of hydrogen and oxygen isotope fractionations in epidote, clinozoisite and/or zoisite is given in Table 1. Well-constrained and accurate determinations of the temperature dependence of equilibrium mineral-mineral and mineral-fluid fractionations are essential in understanding the significance of stable isotope data in rocks. However, among the epidote minerals, significant uncertainty remains regarding equilibrium temperature-fractionation relations in both hydrogen and oxygen isotope systems. These uncertainties limit the interpretations that can be made regarding empirical data on natural samples. In light of the petrologic significance of epidote minerals, additional experimental data to constrain equilibrium temperature-fractionation relations are warranted.

***Hydrogen.*** Pioneering studies by Hugh P. Taylor and his co-workers (e.g., Sheppard and Taylor 1974; Forrester and Taylor 1977; Taylor and Forrester 1979) demonstrated that the hydrogen isotope compositions of hydrous minerals could provide important constraints on large-scale processes of crustal fluid flow. These early studies provided the impetus for

**Table 1.** Compilation of experimental and theoretical studies on oxygen and hydrogen isotope fractionations involving epidote minerals.

| Reference | System | Comments |
|---|---|---|
| ***Hydrogen*** | | |
| Graham and Sheppard (1980) | epidote-NaCl-$H_2O$ epidote-$CaCl_2$-$H_2O$ epidote-seawater | Experiments over $T = 250$–$550°C$ and $P = 0.2$–$0.4$ GPa |
| Graham et al. (1980) | epidote-$H_2O$ zoisite-$H_2O$ clinozoisite-$H_2O$ | Experiments over $T = 200$–$650°C$ and $P = 0.2$–$0.4$ GPa |
| Venneman and O'Neil (1996) | epidote-$H_2$ | Experiments over $T = 150$–$400°C$ and $P = 0.03$–$0.2$ MPa |
| Driesner (1997) | epidote-$H_2O$ | Calculation of pressure effect on hydrogen isotope fractionation |
| Chacko et al. (1999) | epidote-$H_2O$ | Experiments over $T = 300$–$600°C$ and $P = 0.12$–$0.22$ GPa |
| ***Oxygen*** | | |
| Matthews et al. (1983) | zoisite-$H_2O$ | Experiments over $T = 400$–$700°C$ and $P = 1.34$ GPa |
| Zheng (1993a) | quartz-zoisite zoisite-$H_2O$ calcite-zoisite quartz-epidote epidote-$H_2O$ calcite-epidote | Calculated using modified increment method |
| Matthews (1994) | quartz-epidote quartz-zoisite calcite-zoisite albite-zoisite jadeite-zoisite anorthite-zoisite | $A$ values derived by combining existing and new experimental data, Calculation of effect of [Al = $Fe^{3+}$]$^{oct}$ substitution in epidote |

a number of experimental determinations of the hydrogen isotope fractionation between hydrous minerals such as biotite, muscovite, hornblende, epidote and zoisite, and water. Since the late 1970's there have been a number of experimental studies designed to quantify diffusion coefficients and the temperature dependence of the equilibrium fractionation between these common hydrous minerals and water (Suzoki and Epstein 1976; Graham and Sheppard 1980; Graham et al. 1980; Graham 1981; Graham et al. 1984; Venneman and O'Neil 1996). However, a number of different experimental methodologies were employed and the resulting mineral-water fractionations varied so significantly that reliable application of the experimentally determined fractionation was difficult.

Graham et al. (1980) and Venneman and O'Neil (1996) determined equilibrium epidote-$H_2O$ and epidote-$H_2$ isotope fractionations, respectively, and their results varied from each other by as much as ~35‰ below 300°C and by >35‰ above 300°C. Graham et al. (1980) give the following equations for epidote-water exchange

$$1000 \ln \alpha_{\text{epidote-water}} = -35.9 \pm 2.5 \quad (1)$$

valid from 300 to 650°C, and

$$1000 \ln \alpha_{\text{epidote-water}} = 29.2 \, (10^6/T^2) - 138.8 \quad (2)$$

valid below 300°C. Venneman and O'Neil (1996) give the following equation for epidote-$H_2$ exchange

$$1000 \ln \alpha_{\text{epidote-H2}} = 110.756 \times 10^6/T^2 + 149.980 \times 10^3/T - 158.685 \quad (3)$$

Driesner (1997) conducted a theoretical evaluation of the effect of pressure on hydrogen isotope fractionation between hydrous minerals and water and concluded that with decreasing pressure the mineral-water fractionation becomes more negative. Based on his calculations, he applied a correction to the experimental results of Venneman and O'Neil (1996) for the epidote-water exchange. The Venneman and O'Neil (1996) experiments were conducted at pressures ranging from 0.4–2 bars, and Driesner's correction shifted the $1000 \ln \alpha_{\text{epidote-H2}}$ values of Venneman and O'Neil (1996) into good agreement with the data of Graham et al. (1980) at ~300°C. However, above and below 300°C, the "corrected" Venneman and O'Neil (1996) values were still significantly different from those obtained by Graham et al. (1980).

Chacko et al. (1999) devised a new experimental technique for determining the equilibrium hydrogen isotope fractionation between hydrous minerals and water and applied this new approach to the epidote-water system. In their experiments large single crystals of natural, museum quality epidote with $X_{\text{Ps}} = 0.325$ ($X_{\text{Ps}} = Fe^{3+}/(Fe^{3+} + Al^{3+})$) from the Pampa Blanca deposit, Peru, were allowed to exchange isotopically with water at temperatures ranging from 300 to 600°C at pressures of ~0.21 GPa. Isotopic zoning profiles within the epidote crystals were then measured by ion microprobe. With this experimental technique, Chacko et al. (1999) were able to demonstrate that the isotope exchange between epidote and water occurs via diffusion rather than via surface reaction or recrystallization, and to minimize any isotope exchange effects associated with quenching. The resulting equilibrium fractionation between epidote-water at temperatures between 300–600°C is

$$1000 \ln \alpha_{\text{epidote-water}} = 9.3 \times 10^6/T^2 - 61.9 \quad (4)$$

Although some uncertainty still remains regarding the exact causes of the differences among the values determined by Graham et al. (1980), Venneman and O'Neil (1996), and Chacko et al. (1999), the new technique employed by Chacko et al. (1999) does seem to yield reliable equilibrium fractionation factors for hydrogen isotope exchange between epidote and water.

***Oxygen.*** Experimental determinations involving oxygen in epidote minerals are limited. The only direct experimental determination of the oxygen isotope fractionation involving epidote minerals is that for zoisite-water by Matthews et al. (1983). These experiments were conducted using the three-isotope method between 400 to 700°C, at 1.34 GPa. Their results document very slow oxygen isotope exchange between zoisite and water as compared to the oxygen isotope exchange between feldspar and water. By combining their zoisite-water oxygen isotope exchange data with the quartz-water and albite-water fractionation data of Matsuhisa et al. (1979) and the calcite-water fractionation data of O'Neil et al. (1969), Matthews et al. (1983) developed an internally consistent set of mineral-mineral fractionation coefficients (the "$A$" values in the equation $1000 \ln \alpha = A \times 10^6 T^{-2}$) for quartz-zoisite, albite-zoisite, and calcite-zoisite (1.56, 1.06, and 1.06, respectively). Subsequently, Matthews (1994) combined their existing experimental data on zoisite-water with data from mineral-carbonate experiments and derived a new quartz-zoisite fractionation coefficient ($A = 2.0$). In addition, Matthews (1994) estimated the isotopic effect of the $Fe^{3+}$ content in clinozoisite-epidote solid solutions. Assuming that the isotopic effects of the substitution of $Fe^{3+}$ for Al in the octahedral site is similar to the isotopic effects of the $Fe^{3+}$ for Al substitution in garnet, Matthews (1994) derived the equation

$$A_{\text{quartz-epidote}} = 2.00 + 0.75\, X_{Ps} \qquad\qquad 5)$$

where $X_{Ps} = Fe^{3+}/(Fe^{3+} + Al)$. This calibration of the quartz-epidote fractionation has been widely used to estimate metamorphic temperatures from measured quartz and epidote $\delta^{18}O$ values from a wide range of blueschist and eclogite facies rocks. However, empirical data on natural samples (discussed below) appears to indicate that this calibration may yield temperatures that are too low by as much as 150 to 200°C.

### Empirical studies involving hydrogen and oxygen isotopes

Stable isotope studies address a range of different aspects of epidote mineral petrogenesis. Because stable isotope data are particularly good indicators of fluid source characteristics, many of the empirical studies involving epidote minerals focus on determining the source of the fluids that were involved in mineral genesis. Such studies are, however, often limited by complex metamorphic and deformational histories, multiple generations of fluid flow, and uncertainties regarding the timing of fluid flow. Although significant uncertainties persist regarding equilibrium temperature-fractionation relations for oxygen and hydrogen isotopes in epidote minerals, variations in the isotopic composition of natural fluid sources are sufficiently robust such that empirical data can generally be interpreted with confidence. In addition, because stable isotope systematics are also potentially useful as geothermometers, many studies attempt to use stable isotope data, particularly oxygen, to assess the temperatures of mineral growth or equilibration. However, uncertainties in equilibrium temperature-fractionation relations continue to make interpretations of calculated temperatures difficult.

A number of examples were chosen to illustrate the broad range of research that has been conducted using stable isotope compositions of epidote minerals and these are described below. Representative examples of studies of regional metamorphism, UHP metamorphism, the intrusion of deep-seated granitic plutons, and hydrothermal alteration, are considered.

*Regional metamorphism.* Epidote group minerals, in particular epidote and zoisite, are significant and defining metamorphic minerals in the greenschist, epidote-amphibolite, blueschist, and eclogite facies. Metamorphic reactions that involve epidote group minerals and the petrologic constraints that can be derived from existing experimental and empirical data for these reactions are discussed extensively by Bird and Spieler (2004), Grapes and Hoskin (2004), Enami et al. (2004), and Poli and Schmidt (2004). In general, studies of greenschist, epidote-amphibolite, blueschist, and eclogite facies rocks that have utilized the stable isotopic compositions of epidote minerals have focused on assessing 1) metamorphic temperatures, 2) the extent to which pre-metamorphic isotopic compositions have been preserved, and 3) the origins of any fluids associated with metamorphism.

In most studies on regional metamorphic terranes that involve stable isotope data, the data on epidote, clinozoisite and/or zoisite, if any, are often a part of a larger data set that focuses on volumetrically more significant phases such as quartz, calcite, feldspar, biotite and garnet (Matthews and Schliestedt 1984; Thomas et al. 1985; Kohn and Valley 1994, 1998; Cartwright et al. 1997; Putlitz et al. 2000), or, on petrologic and other types of geochemical data (e.g., Brunsmann et al. 2000). For example, Kohn and Valley (1994) conducted an oxygen isotope study of metamorphic fluid flow in amphibolite facies rocks from Townshend Dam, Vermont. They measured $\delta^{18}O$ values in garnet, hornblende, epidote and ilmenite. The focus of the study is, however, on garnet oxygen isotope systematics, and they conclude that fluid flow at this locality was likely either layer parallel or focused along vein systems, rather than pervasive and across strike. Cartwright et al. (1997) conducted a detailed petrologic and stable isotope characterization (including epidote $\delta^{18}O$ values) of the amphibolite facies metamorphism and fluid flow at Breakneck Hill and Bustard Ridge in the Jervois region of the Arunta Inlier (Australia). They demonstrated that $H_2O$-rich fluids derived from nearby crystallizing pegmatites infiltrated marble layers near peak metamorphic conditions. The $\delta^{18}O$ values in epidote from marbles at Breakneck Hill range from 1.5 to 6.2‰, whereas the $\delta^{18}O$ values in epidote from Bustard Ridge range from 10.5 to 10.7‰. The epidotes from Breakneck Hill show textural evidence for late-stage growth and thus Cartwright et al. (1997) interpreted their lower $\delta^{18}O$ values to indicate that they grew in equilibrium with a meteoric fluid that infiltrated into the cooling hydrothermal system.

In a study of zoisite- and clinozoisite-segregations in metabasites in the Tauern Window, Austria, Brunsman et al. (2000) demonstrated that segregations of zoisite + quartz + calcite formed during an early- to pre-Hercynian high pressure metamorphism (at $P \gg 0.6$ GPa and $T = 500$ to 550°C) by hydrofracturing in response to dehydration in the host amphibolites. In contrast, segregations of clinozosite + quartz + omphacite + titanite + chlorite + calcite, formed during the later Eoalpine high pressure metamorphism (at $P = 0.9$ to 1.2 Gpa and $T = 400$ to 500°C) and exhumation. Brunsman et al. (2000) measured ten whole rock $\delta^{18}O$ values that range from 7 to 7.5‰ (n = 8) for the host rocks whereas the zoisite-segregation values are both 9‰ (n = 2). In addition, they measured $\delta^{18}O$ values in five zoisite mineral separates ($\delta^{18}O$ = 5.8 to 7.2‰), $\delta^{18}O$ values in two quartz mineral separates ($\delta^{18}O$ = 10.6 to 10.7‰), and $\delta^{18}O$ and $\delta^{13}C$ values in three calcite mineral separates, all from zoisite-segregations (calcite $\delta^{18}O$ = 10.7 to 10.8, $\delta^{13}C = -7.1$ to -7.2‰) The data are interpreted to indicate that water with $\delta^{18}O$ values of 8.2 to 8.6‰ would have been in equilibrium with zoisite, quartz and host metabasite at 525 to 575°C. This indicates that the fluids were likely derived from dehydration of the host metabasite (Brunsman et al. 2000).

In some metamorphic settings, epidote is the dominant hydrothermal or metamorphic mineral and is the focus of stable isotopic studies. Cartwright et al. (1996) conducted a detailed oxygen isotope study of epidote, quartz, plagioclase, garnet and clinopyroxene to unravel the metamorphic history of a multiply deformed gneiss-calcsilicate terrain at Conical Hill in the Reynolds Range in Australia. Here, the Napperby Gneiss underlies the Lower Calcsilicate Unit, which is characterized by widespread epidotization that decreases in extent away from the contact (Fig. 1). Crosscutting pegmatitic veins grade away from the contact into quartz + garnet veins. Quartz + garnet + epidote veins also occur throughout the Lower Calcsilicate Unit. Values of epidote $\delta^{18}O$ and $\delta D$ were measured in samples from Conical Hill: $\delta^{18}O$ ranges from 3.1 to 3.7‰ (8 samples) and $\delta D$ is -136 and -145‰ (2 samples). The values of $\delta^{18}O$ for vein epidote do not differ from those for epidote that occur in layers in the epidotized Lower Calcsilicate Unit. Calculated $\Delta_{quartz-epidote}$ values range from 3.1 to 3.4‰. Using the calibration of Zheng (1993), Cartwright et al. (1996) inferred temperatures of epidotization to range from 560 to 615°C and interpreted the data to document a complex polyphase fluid flow history. They argued that the generally low $\delta^{18}O$ values for the calcsilicates (average quartz $\delta^{18}O$ = 6.58‰) and the Napperby Gneiss (average whole rock $\delta^{18}O$ = 6.41‰) are the result of an

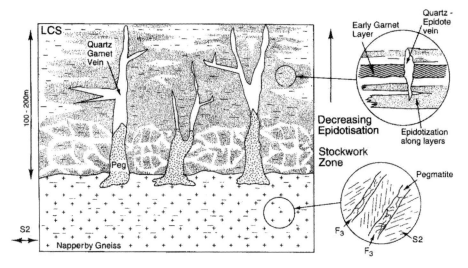

**Figure 1.** Sketch of the outcrop at Conical Hill. Pegmatite segregations in the Napperby Gneiss are locally concentrated in the axial zones of centimeter-scale F3 kink bands (lower inset). Away from the Napperby Gneiss-Lower Calcsilicate (LCS) contact, pegmatite veins grade into garnet + quartz veins that cut the F2 fabric. Epidotization also decreases away from the contact and occurs preferentially along some layers (upper inset). The crosscutting relationships suggest that the epidotization occurred after M2 and was caused by fluid emanating form the crystallizing pegmatite segregations in the Napperby Gneiss. [Used by permission of Elsevier Publishing Company, from Cartwright et al. (1996), Precambrian Research, Vol. 77, Fig. 2, p. 214.].

early contact metamorphism that involved meteoric fluids. Upon cooling from the upper amphibolite facies metamorphism, $H_2O$-rich fluids were exsolved from the melts within the Napperby gneiss and these fluids caused the epidotization and associated veins.

In addition to their use in studies of common regional metamorphism, the stable isotope composition of epidote and/or zoisite has played a significant role in the development of current theories regarding the formation of UHP metamorphic rocks (Yui et al. 1995; Baker et al. 1997; Rumble and Yui 1998; Zheng et al. 1998a; Fu et al. 1999; Zheng et al. 2003). The presence of coesite and/or diamond in these rocks documents metamorphic pressures in considerable excess of 2.5 GPa (e.g., Wang et al. 1989; Wang and Liou 1991; Wang et al. 1992; Liou et al. 1994) and defines UHP metamorphic conditions. It has been the stable isotope compositions of these rocks, including data on epidote minerals, that documented infiltration of meteoric fluids *prior* to the UHP metamorphism thus proving the subduction of upper crustal rocks to depths far exceeding 75 km (for the general role of epidote minerals in UHP rocks see Enami et al. 2004).

In a seminal publication, Yui et al. (1995) reported $\delta^{18}O$ values from one eclogite and one quartzite sample from the UHP Sulu Terrain in China. The quartzite contains quartz, phengite, kyanite, garnet, epidote, and rutile, with corresponding $\delta^{18}O$ values of $-7.3‰$ (quartz), $-8.7‰$ (phengite), $-9.1‰$ (kyanite), $-10.0‰$ (epidote), and $-10.2‰$ (garnet). The eclogite sample is composed predominantly of phengite, omphacite and garnet, with less abundant rutile, kyanite, epidote, quartz and talc. Values of $\delta^{18}O$ in the eclogite are $-9.0‰$ (phengite), $-10.2‰$ (omphacite), and $-10.4‰$ (garnet). Coesite and coesite pseudomorhps have been reported in both rock types from the area (Yui et al. 1995).

The oxygen isotope fractionations among the coexisting minerals in both samples yield widely variable temperatures depending upon which calibration of the temperature dependence of the fractionation is used. For example, phengite-omphacite temperatures calculated using Richter and Hoernes (1988) and Hoffbauer et al. (1994) yield unrealistic temperatures of 42 and $-31°C$, respectively. However, using the calibrations of Javoy (1977), Matthews (1994) and Zheng (1993a,b), the mineral pairs phengite-omphacite, phengite-garnet, quartz-phengite, quartz-kyanite, quartz-epidote and quartz-garnet yield temperatures that range from 521 to 879°C. Among those values are three temperatures calculated from the $\Delta_{quartz-epidote}$ value of 2.7‰ from the quartzite sample: 615°C (Richter and Hoernes 1988), 640°C (Matthews 1994), and 721°C (Zheng 1993a). Cation exchange thermometry yields 700 to 890°C for peak metamorphic temperatures in the region (Yui et al. 1995). The fact that some of the oxygen isotope temperatures are below those indicated by cation exchange thermometry is attributed to partial retrograde oxygen isotope exchange during cooling (Yui et al. 1995). Thus, although the data are subject to uncertainties related to both the reliability of the temperature-fractionation relations and the extent to which retrograde exchange has occurred, the "near oxygen isotope equilibrium" among minerals is interpreted to indicate that oxygen isotope equilibrium was achieved at the peak of UHP metamorphism and that these rocks acquired their unusually low $\delta^{18}O$ values from meteoric water-rock interaction at relatively high temperatures near the surface *prior* to subduction and UHP metamorphism (Yui et al. 1995, and see below).

Subsequent, more detailed studies of the Qinglongshan Province in China further characterized unusually low $\delta^{18}O$ UHP lithologies. Rumble and Yui (1998) reported $\delta^{18}O$ values for epidote in eclogite, quartzite, mafic segregations and altered gneiss, that range from 4.0 to $-11.1$‰ (n = 15), five values of $\Delta_{quartz-epidote}$ (2.65, 2.71, 3.11, 3.21 and 5.39‰), and ten $\Delta_{garnet-epidote}$ values ranging from $-0.35$ to 0.81‰, with an average of $0.1 \pm 0.4$. Of the $\Delta_{quartz-epidote}$ values, the value 5.39‰ is significantly higher than the others. This value is from an altered gneiss and the mineralogical evidence in this sample indicates that it has undergone post metamorphic metasomatism and therefore the $\Delta_{quartz-epidote}$ value of 5.39‰ and the calculated temperature (361°C using Matthews' 1994 calibration) reflect post metamorphic processes rather than peak metamorphic equilibration (Rumble and Yui 1998). Temperatures calculated using Matthews (1994) for the remaining $\Delta_{quartz-epidote}$ values are 632, 622, 562 and 548°C. Peak metamorphic temperatures for these rocks estimated from common exchange thermometers range from 725 to 875°C (Rumble and Yui 1998). Thus, the Matthews (1994) quartz-epidote oxygen isotope temperatures are below those from cation exchange thermometry by as much as $\sim175°C$.

Rumble and Yui (1998) did not find a conclusive explanation for the failure of the many oxygen isotope mineral pairs they evaluated but concluded that the combined effects of 1) incomplete equilibration at peak conditions, 2) problematic calibrations, 3) re-equilibration upon cooling, and 4) retrograde metamorphism were likely responsible.

In a recent paper, which incorporates both existing stable isotope data from the literature and new data, Zheng et al. (2003) review the stable isotope data for the Dabie-Sulu orogen in China. Anomalously low $\delta^{18}O$ values (as low as $-11$‰) for silicate minerals in eclogites and schists and unusually low $\delta D$ values ($\delta D$ as low as $-127$‰ in phengite, ranging from $-66$ to $-49$‰ for zoisite) characterize these UHP lithologies. Oxygen and hydrogen values for hydrous phases, and the range for "normal" metamorphic rocks are shown in Figure 2. This figure illustrates the significant depletion in both $\delta^{18}O$ and $\delta D$ relative to "normal" metamorphic rocks.

The $\delta D$ systematics among biotite and epidote, and phengite and zoisite in eclogites, gneisses and quartz-schists from the Dabi-Sulu orogen are complex and show disequilibrium behavior (Zheng et al. 1998b; Fu et al. 1999). Further, they do not represent one phase of growth or equilibration. Rather, the $\delta D$ systematics are interpreted to reflect 1) early meteoric water interaction, 2) equilibration at UHP conditions, and 3) differential exchange among the hydrous phases and fluids during exhumation. Despite disequilibrium, it is proposed that

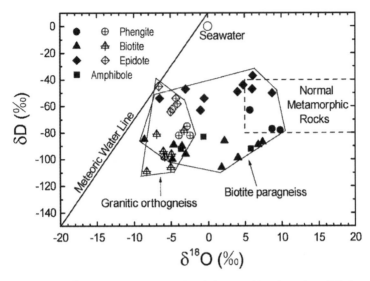

**Figure 2.** Plot of δD vs. δ¹⁸O for hydroxyl-bearing minerals from granitic orthogneiss and biotite paragneiss at Shuanghe in Central Dabie (data from Fu et al. 1999). The δ¹⁸O range of the normal metamorphic rocks is shown for comparison. [Used by permission of Elsevier Publishing Company, from Zheng et al. (2003), Earth-Science Reviews, Vol. 1276, Fig. 6, p. 14.].

the unusually low δD values were initially acquired by meteoric water interaction prior to subduction and UHP metamorphism (Zheng et al. 2003). In Figure 3, values of δ¹⁸O for the Dabi-Sulu eclogites are shown with values from other eclogites (Zheng et al. 2003). This figure illustrates the magnitude of the δ¹⁸O depletion relative to other eclogites and other significant oxygen reservoirs.

The entirety of the stable isotope data are interpreted to indicate that lithologies with anomalously low δ¹⁸O and δD values acquired those values from meteoric-hydrothermal alteration *prior* to subduction and UHP metamorphism. Thus, rocks that underwent hydrothermal alteration relatively near to the earth's surface were then subducted to depths of >200 km, metamorphosed, then rapidly returned to the surface (see Zheng et al. 2003 and references therein).

***Intrusion of deep-seated granitic plutons.*** Until the mid-1980's there were minimal opportunities to constrain the emplacement depths of granitic batholiths due largely to the lack of well-calibrated reactions for granitic mineral assemblages. Zen and Hammarstrom (1984) documented that magmatic epidote occurred in deep-seated plutons and thus constrained emplacement depths to moderately high pressures. This finding renewed interest in the quantifying emplacement conditions for granitic plutons.

The recognition of magmatic epidote in plutons has become common (Zen 1985; Anderson 1990; Carlson et al. 1991; Dawes 1991). For a detailed discussion of the petrologic characteristics of magmatic epidote, see Schmidt and Poli (2004). Zen and Hammarstrom (1984) concluded that the presence of magmatic epidote in plutons of intermediate composition implies a minimum crystallization pressure of 0.6 GPa based on the intersection of the tonalite solidus with the epidote stability curve of Liou (1973). Their conclusions are consistent with the experimental data of Naney (1983) who demonstrated that epidote is stable in a granodioritic melt at a pressure of 0.8 GPa. Experimental and empirical evidence has shown that in plutonic rocks, magmatic epidote normally crystallizes at temperatures just above the

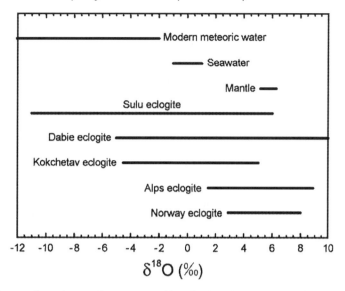

**Figure 3.** A comparison of oxygen isotope compositions for the Dabie-Sulu eclogites with other eclogites in the world (revised after Zheng et al. 1996 by incorporating additional data from Zheng et al. 1998a, 1999 and Masago et al. 2001). The low $\delta^{18}O$ values up to $-11$ to $-4‰$ for the Dabi-Sulu eclogites record pre-metamorphic isotope signature due to meteoric hydrothermal alteration to their protoliths. [Used by permission of Elsevier Publishing Company, from Zheng et al. (2003), Earth-Science Reviews, Vol. 1276, Fig. 8, p. 16.].

solidus, and that it is likely formed from the reaction of hornblende and melt (Naney 1983; Zen and Hammarstrom 1984; Schmidt and Thompson 1996). However, the occurrence of epidote phenocrysts in rapidly quenched liquids suggests that the presence of amphibole may not be necessary for the formation of magmatic epidote (Dawes and Evans 1991; Evans and Vance 1987) and thus epidote crystallizes early. Although controversial (Zen 1985; Moench 1985; Tulloch et al. 1986; Zen and Hammarstrom 1986) available evidence suggests near solidus crystallization of epidote.

Epidote, however, also forms by subsolidus reactions and in fact many magmatic epidote-bearing plutons also contain demonstrably subsolidus epidote suggesting a continuous sequence of epidote formation across the solidus. The textural criteria that Zen and Hammarstrom (1984) offer for magmatic epidote include: euhedral grains, particularly against biotite, vermicular, myrmekitic intergrowths with plagioclase + quartz, unretrogressed igneous parageneses, allanite cores, inclusions of bleb-like (resorbed?) Al-rich hornblende, and involvement of epidote in magmatic or flow banding. Yet many of these textures are also found in rocks containing subsolidus epidote (Keane and Morrison 1997). Thus, unambiguous discrimination between magmatic and subsolidus epidote has proved difficult with the available textural criteria.

Keane and Morrison (1997) studied in detail the oxygen isotope composition of four textural varieties of epidote in the Mt. Lowe intrusion in the San Gabriel Mountains of California using a $CO_2$ laser probe, which enabled measurement of within-grain oxygen isotope variations. The Mt. Lowe crystallized at 0.5 to 0.6 GPa and 640 to 780°C (Barth 1989; Anderson 1996). Keane and Morrison (1997) analyzed rocks consisting of coarse epidote, hornblende, biotite and plagioclase in a slightly finer-grained and partially recrystallized quartzofeldspathic matrix, with minor sphene, apatite, magnetite, zircon, and rare garnet, along with secondary chlorite, calcite, and muscovite. The four major textural varieties of epidote include 1) euhedra, 2) anhedra, 3) intergrowths and, 4) veins. Epidote euhedra meet most of the criteria for magmatic epidote proposed by Zen and Hammarstrom (1984) and have been interpreted as magmatic in previous studies of these rocks (Barth and Ehlig 1988; Barth 1990). Epidote also occurs as equant anhedral crystals normally ~5 mm in diameter. The origin of the anhedra is texturally ambiguous. The presence of allanite cores, although rare, suggests that a magmatic origin is possible, however, the complete lack of euhedral crystal faces could be interpreted to indicate subsolidus growth. The third common epidote texture is fine-grained intergrowths and coronas around and within masses of hornblende and biotite. The textural evidence suggests formation by a solid-state reaction of hornblende altering to epidote although it is not clear petrographically whether biotite is a product or a reactant. Finally, epidote occurs with calcite and chlorite in clearly late-stage subsolidus veins which crosscut all phases in the rock.

In six individual euhedra, four to nine $\delta^{18}O$ measurements were made (Fig. 4). In general, each crystal is relatively homogeneous in $\delta^{18}O$ except for one value in each crystal at the tip. In addition, two epidote anhedra were analyzed (Fig. 5). Values of $\delta^{18}O$ for the anhedra range from 4.50 to 6.08‰, which largely overlap those for the euhedra. In anhedra D-5A (Fig. 5) the values 5.09, 5.27, and 5.32‰ correspond to the allanite core. Anhedra D-5B (Fig. 5) exhibits a striking variation, which is apparent even on the submillimeter scale and the zonation is highly irregular with no relationship to adjacent mineralogy or extent of alteration. Epidote intergrowths have $\delta^{18}O$ values distinctly different from both the euhedra and anhedra: two analyses of epidote intergrowths are 3.90 and 3.56‰. Vein epidote also has $\delta^{18}O$ values markedly different from both the euhedra and anhedra: two analyses of epidote grains from a single vein are nearly identical with values of 3.94 and 3.98‰.

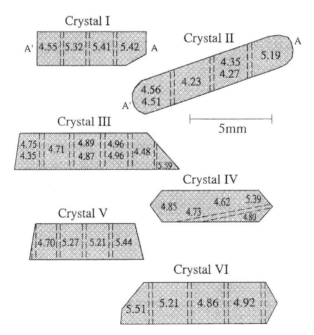

**Figure 4.** Values of $\delta^{18}O$ for six euhedral epidote crystals. Each crystal is relatively homogeneous in $\delta^{18}O$ except for one anomalous value at the tip. Location of electron microprobe traverses (A–A') are shown for Crystals I and II. Dashed lines show the kerf. From Keane and Morrison (1997).

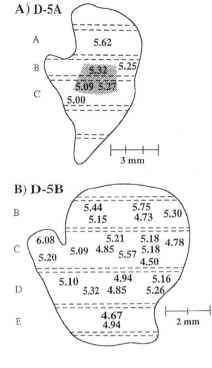

**A) D-5A**

A    5.62

B    5.32  5.25
     5.09  5.27

C    5.00

3 mm

**B) D-5B**

B    5.44    5.75
     5.15    4.73  5.30

C    6.08    5.21    5.18  4.78
     5.20  5.09  4.85  5.57  5.18
                           4.50

D    5.10         4.94    5.16
     5.32  4.85          5.26

E         4.67
          4.94          2 mm

**Figure 5.** (A) $\delta^{18}O$ values for epidote anhedral Crystal D-5A. The shaded region represents the allanite core. (B) $\delta^{18}O$ values for epidote anhedral crystal D-5B. Isotopic zonation is highly irregular and much more erratic than that of the anhedra. Notice, in strip C for example, the submillimeter variation of over 1‰. From Keane and Morrison (1997).

Values of $\Delta_{plagioclase-epidote}$ ($= \delta^{18}O_{plagioclase} - \delta^{18}O_{epidote}$) were calculated for eight touching pairs of epidote and plagioclase in euhedra V and VI (Fig. 4). For all plagioclase-epidote pairs in which epidote $\delta^{18}O$ values are >5‰, a higher temperature plagioclase-epidote fractionation is observed (average $\Delta_{plagioclase-epidote} = 1.47 \pm 0.05‰$). For plagioclase-epidote pairs with epidote $\delta^{18}O$ values <5‰, a range of $\Delta_{plagioclase-epidote}$ values from 1.44 to 2.95‰, which corresponds to a lower temperature, is observed (average $\Delta_{plagioclase-epidote} = 2.18 \pm 0.63‰$). The data were interpreted to indicate that epidote $\delta^{18}O$ values >5‰ coupled with $\Delta_{plagioclase-epidote}$ values <1.6 reflect higher temperature, possibly magmatic oxygen isotope compositions and that epidote $\delta^{18}O$ values <5‰ are the result of subsolidus exchange between epidote and adjacent plagioclase. Thus, intracrystalline variations in euhedral epidote document (1) partial preservation of the original magmatic oxygen isotope composition acquired at the time of crystallization of epidote from the magma and (2) lower temperature or subsolidus oxygen isotope exchange between epidote and adjacent plagioclase, possibly enhanced by the presence of a intergranular fluid.

Keane and Morrison (1997) interpreted the data to indicate that both euhedral and anhedral epidote can be magmatic in origin and that magmatic $\delta^{18}O$ values may have been preserved in both textures despite some resetting during subsolidus cooling. The demonstration of a magmatic origin for both textural varieties of epidote is significant in that the total lack of well-developed crystal faces on epidote may not rule out a magmatic origin. Furthermore, the development of euhedral faces against biotite by epidotes that are subsolidus in origin (the intergrowths) suggests that such a texture is not necessarily a reliable criterion for magmatic epidote. Thus, interpretation of epidote parageneses on the basis of textures alone can be misleading, but detailed texturally controlled oxygen isotope compositions may provide additional useful information in discriminating between magmatic and subsolidus origins.

Morrison et al. (1999), Morrison and Anderson (2000; in prep.) measured mineral $\delta^{18}O$ values (Fig. 6) in two distinct rapidly-cooled or "quenched" lithologies: a biotite rhyolite of the Pleistocene Sutter Buttes volcano of California and Late Cretaceous rhyodacite dikes from Boulder County, Colorado (Evans and Vance 1987). The biotite rhyolite contains phenocrysts of magmatic epidote, quartz, plagioclase and biotite, and is virtually devoid of alteration minerals. Discrete but spatially related epidote, quartz, plagioclase and biotite crystals from six samples yield an average $\Delta_{quartz\text{-}epidote}$ of $3.10 \pm 0.12‰$ ($n = 5$), $\Delta_{plagioclase\text{-}epidote} = 1.75‰$, and $\Delta_{epidote\text{-}biotite} = -0.44‰$. In the rhyodacite, $\Delta_{quartz\text{-}epidote} = 3.12 \pm 0.06‰$ and $\Delta_{epidote\text{-}biotite} = 1.47 \pm 0.34‰$. Based on petrologic data, both of these lithologies likely crystallized at ~750°C. Temperatures inferred from these $\Delta_{quartz\text{-}epidote}$ values using Matthews (1994) are ~550°C.

In contrast, the three Late Cretaceous tonalitic plutons in which Zen and Hammarstrom (1984) originally documented magmatic epidote, plus the Ordovician Ellicott City granodiorite, yield distinctly larger values: $\Delta_{quartz\text{-}epidote} = 4.19 \pm 0.17‰$, $\Delta_{plagioclase\text{-}epidote} = 1.64 \pm 0.19‰$, $\Delta_{epidote\text{-}amphibole} = 0.54 \pm 0.25‰$, and $\Delta_{epidote\text{-}biotite} = 0.85 \pm 0.24‰$ (Fig. 6). All four of these lithologies underwent slow, post-crystallization cooling. Cation exchange thermometry indicates that these plutons crystallized between ~680 and ~770°C. Temperatures inferred from $\Delta_{quartz\text{-}epidote}$ values using Matthews (1994) are ~450°C.

The data are interpreted to indicate that published high temperature fractionation factors involving epidote may be in error, yielding temperatures that are too low by as much as ~200°C. The measured $\Delta_{quartz\text{-}epidote}$ values from the two quenched lithologies range from 2.95 to 3.19‰ with an average of $3.1 \pm 0.1‰$. This value likely provides a more reliable empirical calibration of the quartz-epidote system than the published Matthews (1994) calibration at high

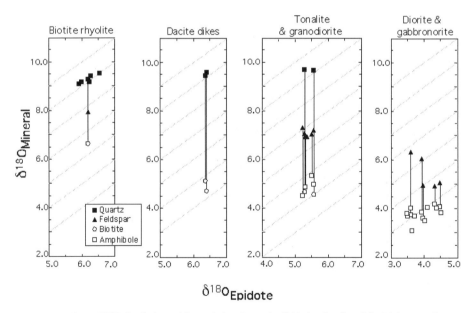

Figure 6. Values of $\delta^{18}O$ for 1) the rapidly-cooled or "quenched" biotite rhyolite of the Pleistocene Sutter Buttes volcano of California (biotite rhyolite), 2) the Late Cretaceous rhyodacite dikes from Boulder County, Colorado (dacite dikes), 3) the Late Cretaceous tonalitic plutons in which Zen and Hammarstrom (1984) originally documented magmatic epidote, plus the Ordovician Ellicott City granodiorite (tonalite and granodiorite), 4) and the diorite and gabbronorites from Milford Sound, New Zealand (diorite and gabbronorites). From Morrison and Anderson (2000; and in prep.)

temperatures. If $\Delta_{quartz-epidote} = 3.1$ corresponds to a temperature of 750°C (based on petrologic estimates of crystallization temperatures for the two lithologies), then the $\Delta_{quartz-epidote}$ values in the Zen and Hammarstrom (1984) samples recalculate to temperatures of ~610°C. This value is below the estimated crystallization temperature range (680 to 770°C) but the difference could readily be explained by diffusional reequilibration during slow cooling (Morrison and Anderson, in prep).

Ferreira et al. (2003) conducted a detailed oxygen isotope study of two epidote-bearing plutons in NE Brazil, the Emas and Sao Rafael plutons. Textural variations and geochemistry indicates that epidote occurs as both magmatic epidote and subsolidus epidote. They analyzed epidote, zircon, titanite, and quartz from eighteen different samples. Epidote $\delta^{18}O$ values range from 4.64 to 5.37‰ (with an average of 5.15 ± 0.25‰) in the Sao Rafael pluton, and from 8.50 to 10.13‰ (with an average of 9.17 ± 0.51‰) in the Emas pluton. Resulting average $\Delta_{quartz-epidote}$ values are large (4.81 ± 0.15‰ and 5.16 ± 0.34‰, respectively) which yield temperatures below 400°C for both plutons using the Matthews (1994) calibration. Ferreira et al. (2003) used the Morrison et al. (1999) $\delta^{18}O$ data from the quenched lithologies (see above) to revise the Matthews (1994) calibration resulting in a $\Delta_{quartz-epidote}$ value of 5.19 at 500°C. Using this revised calibration, they calculate a temperature of 504°C for the Emas pluton and 531°C for the Sao Rafael pluton. These temperatures are still below the inferred crystallization temperature, but the data are interpreted to result from subsolidus diffusive exchange during slow cooling. They also evaluated zircon-epidote systematics. Epidote $\delta^{18}O$ values in both plutons were less than zircon $\delta^{18}O$ values with $\Delta_{zircon-epidote}$ values of ~0.8. They argue that epidote $\delta^{18}O$ values are reversed from what they should be relative to zircon at magmatic temperatures and that this reversal is the result of re-equilibration of epidote with high $\delta^{18}O$ minerals such as quartz and feldspar during cooling.

***Hydrothermal alteration.*** Hydrothermal alteration that results in the growth of epidote, clinozoisite, and zoisite occurs in a wide range of geologic settings including mid-ocean ridges, contact metamorphic aureoles, epizonal plutons, faults, ore systems (e.g., Marshall et al. 1999; Chiaradia, 2003) and shallow geothermal systems such as the Salton Sea hydrothermal system (for a review of epidote in hydrothermal systems see Bird and Spieler 2004). Although epidote, clinozoisite, and zoisite are important minerals in the petrogenesis of many of these systems, stable isotopic data on these phases, particularly oxygen, is often lacking in the seminal publications defining the processes. For example, although epidote is a common product of the hydrothermal alteration of sea floor basalts at temperatures from 250 to 600°C, and stable isotope systematics have been essential in defining the nature and significance of this alteration (e.g., Muehlenbachs and Clayton 1972; Gregory and Taylor 1981), data on epidote in these early publications are uncommon. Stakes and O'Neil (1982), however, published oxygen and hydrogen isotope data on ocean floor basalts collected in dredge hauls. They measured both $\delta^{18}O$ and $\delta D$ values in three samples of epidote-rich greenstone. Values of $\delta^{18}O$ are 2.8, 6.8 and 3.8‰ and $\delta D$ values are −13, −23 and −20‰, respectively. Values of $\Delta_{quartz-epidote}$ for these three samples are 4.6, 4.2 and 4.1‰. They inferred temperatures of at least 350°C from these $\Delta_{quartz-epidote}$ values, and combined these data with other mineral-water data to infer a systematic variation in the temperature of alteration for mineralogically distinct alteration phases in mid-ocean ridge rocks.

Hydrothermal alteration associated with the intrusion of variable composition igneous rocks often results commonly results in alteration assemblages include muscovite (and/ or sericite), albite, chlorite, amphibole and calcite, in addition to epidote minerals. For example, Jénkin et al. (1992) studied the retrograde alteration assemblages in the Connemara metamorphic complex of Ireland. Petrologic and fluid inclusion constraints placed the retrograde alteration at temperatures of ~275°C and pressures <0.15 GPa (Jenkin et al. 1992). They measured both $\delta^{18}O$ and $\delta D$ values in a variety of retrograde phases including epidote.

Epidote $\delta^{18}O$ values from the metagabbro suite ranged from 5.22 to 8.11‰ (n = 4) and $\delta D$ values ranged from −28.9 to −11.4‰ (n = 8). They evaluated the effects of a number of parameters such as $f_{O_2}$, salinity, and Fe content and concluded that the fluid from which epidote precipitated ranged in $\delta^{18}O$ from 1.3 to 7‰ depending on the salinity of the fluid. The $\delta D$ of the retrograde fluid was inferred to have been approximately −20 and −30‰. Jenkin et al. (1992) concluded that the fluid was likely meteoric in origin and that meteoric fluid circulation may have been driven by heat derived from intrusion of the Galway granite.

Recio et al. (1997) conducted a stable isotope and geochemical study of episyenitization (hydrothermal alteration of granite involving removal of quartz and alteration of feldspar) in the Central Iberian Massif of Spain. In addition to loss of quartz and alteration of feldspar, episyenitization often includes chloritization of biotite and amphibole, and growth of epidote. In a detailed study of K-feldspar, plagioclase, quartz, chlorite, epidote, and apatite, Recio et al. (1997) measured $\delta^{18}O$ values for epidote in five samples of altered granite. Values of $\delta^{18}O$ ranged from −2.0 to 2.4‰ (average = 0.0 ± 1.9‰). They interpreted these data and the isotopic compositions of quartz, chlorite and feldspar to indicate that quartz, chlorite and epidote formed late in the hydrothermal alteration process, at a temperature of ~300°C in equilibrium with a fluid with a $\delta^{18}O$ of approximately −1.4‰. Seawater was considered the most likely source. A late-magmatic deuteric fluid was thought to have initiated episyenite formation.

The Idaho batholith serves as an example of a hydrothermally altered plutonic complex in which large-scale meteoric hydrothermal systems were established (Criss and Taylor 1983, 1986). Large portions of the Idaho batholith were infiltrated by meteoric fluids and altered in terms of both their mineralogy and stable isotope geochemistry (Criss and Taylor 1983, 1986; Mora et al. 1999). King and Valley (2001) conducted a detailed oxygen isotope study of zircon, quartz, feldspar, biotite, and epidote from the Idaho batholith and demonstrated that magmatic $\delta^{18}O$ values, defined on the basis of zircon $\delta^{18}O$, have been variably affected by 1) assimilation of high $\delta^{18}O$ country rock, 2) oxygen isotope exchange during closed-system cooling, and 3) subsolidus fluid infiltration. They measured epidote $\delta^{18}O$ values in four samples and the values ranged from 4.34 to 6.76‰. Values of $\Delta_{zircon-epidote}$ ranged from 0.68 to 1.40‰, and show the same "reversal" observed by Ferreira et al. (2003).

Epidote minerals are also common constituents in fault zones that have experienced hydrothermal fluid flow (e.g., Anderson et al. 1983). Morrison and Anderson (1998) reported oxygen isotope compositions of epidote and quartz from chloritic breccias that underlie the detachment fault in the Whipple Mountains metamorphic core complex of California. This metamorphic core complex formed during Cenozoic crustal extension at the lithospheric scale and documents profound lithospheric extension. Extensive mineralogic alteration resulting from hydrothermal fluid flow is characteristic of this regionally extensive detachment fault. Morrison and Anderson (1998) analyzed five discrete quartz-epidote pairs from each of four samples of chlorite breccia that underlie the detachment fault. Quartz and epidote $\delta^{18}O$ data are shown in Figure 7. Mean $\Delta_{quartz-epidote}$ values increase towards the detachment fault from 4.54 ± 0.46‰ at 50 m below the fault to 5.81 ± 0.52‰ at 12 m (Fig. 7a) and the corresponding mean temperatures (estimated using Matthews 1994) decrease from 432 ± 34°C at 50 m below the fault to 350 ± 29°C at 12 m (Fig. 7b). This decrease in temperature is consistent with mineralogic variations that include a decrease in grain size of sericitic alteration and progressive obliteration of original igneous textures. This extreme thermal gradient of 82°C/ 38 m (2160°C/km), which is shown on Figure 7b was interpreted to result from advective heat extraction via circulating surface-derived fluids, and was termed "fault zone refrigeration" by Morrison and Anderson (1998).

In the Salton Sea Geothermal system stable isotope data was essential in defining the systematics of the hydrothermal flow. Bird et al. (1988) measured both $\delta^{18}O$ and $\delta D$ values for

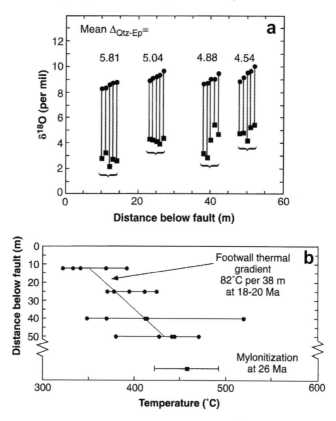

**Figure 7.** (a) Coexisting quartz (circles) and epidote (squares) $\delta^{18}O$ values (in per mil) plotted versus distance below the detachment fault. The five quartz-epidote pairs from each sample represent discrete subvolumes of the sample. For each sample, values are arbitrarily arranged in order of increasing quartz $\delta^{18}O$. The mean $\Delta_{quartz\text{-}epidote}$ values decrease towards the fault from $4.54 \pm 0.46‰$ to $5.81 \pm 0.52‰$. (b) Temperatures estimated from oxygen isotope compositions are shown versus distance below the fault. Ambient pre-detachment faulting temperatures of $\sim458 \pm 35°C$ for these structural levels were estimated from two feldspar thermometry. The mean thermal gradient of $\sim82°C/38$ m is interpreted to result from infiltration of cold, surface-derived fluids into the upper portions of the lower plate during detachment faulting. The thermal gradient is thought to decrease to nominal values at depths $>50$ m below the fault. From Morrison and Anderson (1998).

epidote at variable depths in the State 2-14 drill hole. $\delta D$ values range from $-96$ to $-90‰$ and $\delta^{18}O$ values range from 2.0 to 2.7‰. They also measured of the $\delta^{18}O$ and $\delta D$ of the reservoir fluid and concluded that the epidotes likely grew in equilibrium with borehole meteoric fluids at temperature s of $\sim300°C$, which was the inferred down hole temperature.

## RADIOGENIC ISOTOPE SYSTEMATICS

Radiogenic isotope studies involving epidote minerals include the use of Sm-Nd, Pb-Pb, U-Pb, Rb-Sr, U-Th-Pb systematics to date the growth of or equilibration among mineral assemblages including epidote and/or allanite (e.g., Ludwig and Stuckless 1978; Cliff and Cohen 1980; Mezger et al. 1989; Barth et al. 1989, 1994;Poitrasson et al. 1998; Buick et al.

1999). The more recent use of allanite as a geochronometer is reviewed in Giere and Sorensen (2004). In addition, a recent detailed radiogenic isotope study has been conducted on detrital epidote to constrain sediment provenance (Spiegel et al. 2002).

The use of radiogenic isotope systematics to determine the timing of growth and/or equilibration among mineral phases in hydrothermal alteration assemblages is potentially very significant because such data could provide quantitative constraints on the timing of hydrothermal alteration and fluid flow. In general, age constraints on fluid flow causing hydrothermal alteration are usually derived from petrologic and textural relations and are thus only relative ages. The isochron method in radiogenic isotope systematics requires that the minerals in the system begin with the same "initial" isotopic composition (e.g., $^{87}Sr/^{86}Sr$, $^{143}Nd/^{144}Nd$). Because hydrothermal alteration is often characterized by only partial alteration and replacement of host phases, such systems will seldom be characterized by the same "initial" isotopic compositions. This is a fundamental impediment for the use of the isochron method in determining ages of hydrothermal alteration (e.g., Poitrasson et al. 1998). In order to circumvent this problem, single mineral Pb-Pb geochronology has been used to determine ages of hydrothermal minerals such as garnet and epidote (e.g., Buick et al. 1999). While these studies document the chemical, mineralogic and isotopic complexity of hydrothermally altered lithologies, the magnitude of the complexities illustrates that significantly more research is needed before such systems can be employed with confidence. A few examples from the literature were chosen to illustrate the types of research that have been conducted using radiogenic isotope compositions of epidote minerals and these are described below.

Poitrasson et al. (1998) proposed that a detailed assessment of Sm-Nd systematics in mineral phases from plutonic rocks can be used to infer syn- and post-magmatic hydrothermal evolution. Sm and Nd concentrations and $^{147}Sm/^{144}Nd$ and $^{143}Nd/^{144}Nd$ ratios were determined for allanite, epidote, whole rock, feldspar, zircon, ferrobarroisite, fluorite and fergusonite in a hypersolvus fayalite granite from Mantelluccio, France. The fluorite-bearing granite was emplaced at a depth corresponding to ~0.05 GPa and temperatures of 900 to 700°C. The data and inferred isochrons are shown in Figure 8. Two statistically acceptable isochrons can be derived from the data, one at 331 ± 10 Ma and one at 291 ± 13 Ma (Fig. 8). Poitrasson et al. (1998) proposed that the younger age of 291 ± 13 Ma obtained from allanite, whole rock, ferrobarroisite and inclusion-free zircon, which is consistent with a U-Pb zircon age of 283 ± 1 Ma, is the magmatic emplacement age of the pluton. They then suggest that the older age of 331 ± 10 Ma, obtained from fergusonite, fluorite, zircon, ferrobarroisite, and alkali feldspar, is only an apparent age. Although there are a number of possible interpretations for this five mineral alignment, they suggest that it is most likely the result of a disturbance in the Sm-Nd systematics by high temperature HREE-bearing hydrothermal fluids during the late stages of crystallization. Of the minerals analyzed, epidote has by far the lowest $^{143}Nd/^{144}Nd$ ratio (Fig. 8). Poitrasson et al. (1998) argue that these values in conjunction with the small grain size and anhedral shape of the epidote indicate that it crystallized late and in the presence of hydrothermal fluids. Further, they argue that the low $^{143}Nd/^{144}Nd$ ratio despite a $^{147}Sm/^{144}Nd$ ratio similar to the other phases indicates that epidote isotopic ratios were not acquired via exchange with those minerals. Instead, they argue that the epidote likely acquired its Nd from a source separate from the granite. They concluded that the Nd in the epidote was derived from basement rocks on the order of 1km away, probably in a hydrothermal event associated with Alpine faulting that post-dated crystallization of the granite by as much as ~200 Ma. Thus, this study yielded no direct age determinations of either of the two hydrothermal alteration events and the complex systematics raise some important questions. However, careful analysis of the data may yield important information about the timing of hydrothermal fluid flow.

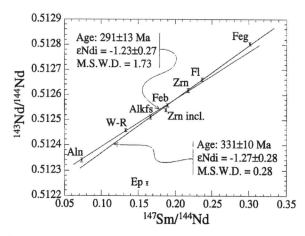

**Figure 8.** $^{147}Sm/^{144}Nd$ vs. $^{143}Nd/^{144}Nd$ diagram showing that alkali-feldspar (Alkfs), ferrobarroisite (Feb), zircon without inclusions (Zrn), fluorite (Fl) and fergusonite (Feg) of the hypersolvus fayalite granite of Mantelluccio (sample CB9202) yield a statistically good isochron at 331 ± 10 Ma. Allanite (Aln), whole rock (W-R), ferrobarroisite (Feb), zircon without inclusions (Zrn) also give a statistically acceptable alignment at 291 ± 13 Ma. The whole-rock is roughly on an intermediate position between the allanite and the other minerals, indicating that the former contribute significantly to the REE budget of the rock. The zircon with inclusions (Zrn incl.) and epidote (Ep) do not plot on any isochrons (open circles). [Used by permission of Elsevier Publishing Company, from Poitrasson et al. (1998), Vol. 146, Fig. 2, p. 192.].

Buick et al. (1999) employed the Pb-Pb stepwise leaching technique (PbSL) of Frei and Kamber (1995) to measure Pb isotope ratios in epidote and garnet from the Conical Hill lithologies studied by Cartwright et al. (1996) that are discussed above (cf. Fig. 1) to constrain the age of high temperature retrogression. They analyzed three samples with coexisting garnet + epidote. The PbSL data on garnet placed garnet growth at 1576 ± 3 Ma. Epidote that occurs both inter-grown with garnet, and partially replacing garnet, yielded two isochrons of 1454 ± 34 Ma and 1469 ± 26 Ma, which are identical within the uncertainties. The surprisingly large age difference (~100 to 120 Ma) between garnet and epidote growth is interpreted to indicate that epidote grew in a later phase of fluid flow within the vein system that may have hosted multiple episodes of fluid flow over the ~100 to 120 m.y. time period. However, as the authors note, this interpretation of the data is subject to many uncertainties due to variables such as possible variations in the isotopic composition of the fluid and the closure temperature for Pb diffusion.

Spiegel et al. (2002) measured the isotopic and trace element compositions of detrital epidote from sandstones (and possible source rocks) from the Swiss Molasse Basin to assess whether such data could be used to determine detrital epidote provenance in terms of crust- versus mantle-derived source regions. They measured $^{87}Sr/^{86}Sr$ and $^{143}Nd/^{144}Nd$ ratios and trace element compositions in epidote from 23 samples. They found that $^{143}Nd/^{144}Nd$ ratios of epidote in the sandstones could be used to discriminate source region (Fig. 9). Ratios of $^{143}Nd/^{144}Nd$ for mantle-derived source rocks range from 0.5130 to 0.5131, and those from crustal-source rocks range from 0.5121 to 0.5125. The data define three types of epidote-bearing sandstones: one composed of epidote from Lower Austroalpine metagranites (a pure crustal-source), one composed of epidote from Pennitic ophiolites (a pure mantle-source), and one composed of epidote derived from both source regions. These data have important implications for the exhumation history of the Central Alps (Spiegel et al. 2002).

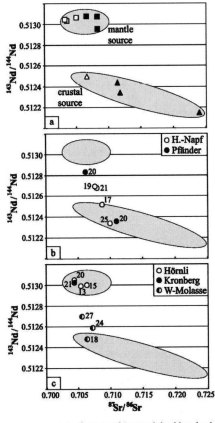

**Figure 9.** (a) Sr and Nd isotopic compositions of epidote from the recent Alpine hinterland and from pebbles of the molasses conglomerates, defining ranges that characterize the mantle or crustal origin of the epidote-bearing source rock. (b) and (c) Sr and Nd isotopic ratios of detrital epidote from molasses sandstones. Numbers next to the data points give deposition ages (Ma). [Used by permission of Elsevier Publishing Company, from Spiegel et al. (2002), Chemical Geology, Vol. 189, Fig. 5, p. 242.]

▲ presently exposed Austroalpine hinterland
■ presently exposed Penninic hinterland
△ metagranitic pebbles from conglomerates
□ metabasic pebbles from conglomerates
⦵ molasse sandstones

# SUMMARY

Stable and radiogenic isotope studies on epidote minerals address a wide range of petrologic problems and have provided significant insights into the nature of these processes. Further insights into the nature of geologic processes that involve epidote group minerals will be possible with additional well-constrained experimental determinations of equilibrium temperature versus mineral and fluid fractionation relations for oxygen isotopes in particular. This data will be essential to advance our ability to interpret empirical data on natural rock systems. In addition, more stable isotope data on multiple coexisting minerals for which textural relations are known will also advance our understanding of stable isotope systematics involving epidote minerals. Finally, additional detailed studies of the radiogenic isotope systematics in hydrothermal alteration assemblages are necessary to understand the complex chemical and isotopic behaviors that characterize these systems.

## ACKNOWLEDGMENTS

Helpful reviews by D. Frei, A. Liebscher, G. Franz and an anonymous reviewer are gratefully acknowledged. The author is also grateful to J. Lawford Anderson who did far more than his fair share at home to enable the completion of this paper.

## REFERENCES

Anderson JL (ed) (1990) The nature & origin of Cordilleran magmatism. Boulder, CO, Geological Society of America - Memoir 174

Anderson JL (1996) Status of thermobarometry in granitic batholiths. The third Hutton symposium on the Origin of granites and related rocks. M. Brown, P. Candela, D. Pecket al. Boulder, CO, Geological Society of America - Special Paper 315:125-138

Anderson JL, Osborne RH, Palmer DF (1983) Cataclastic rocks of the San Gabriel fault-an expression of deformation at deeper crustal levels in the San Andreas Fault Zone. Tectonophys 98:209-251

Baker J, Matthews A, Mattey D, Rowley D, Xue F (1997) Fluid-rock interactions during ultra-high pressure metamorphism, Dabi Shan, China. Geochim Cosmochim Acta 61:1685-1696

Barth AP (1989) Mesozoic rock units in the upper plate of the Vincent thrust fault, San Gabriel Mountains, Southern California. Dept. Earth Sciences. Los Angeles, CA, Univ. of Southern California

Barth AP (1990) Mid-crustal emplacement of Mesozoic plutons, San Gabriel Mountains, California, and implications for the geologic history of the San Gabriel Terrane. The nature & origin of Cordilleran magmatism. J. Anderson. Boulder, CO, Geological Society of America - Memoir 174:33-45

Barth AP, Ehlig PL (1988) Geochemistry and petrogenesis of the marginal zone of the Mount Lowe Intrusion, central San Gabriel Mountains, California. Contrib Mineral Petrol 100:192-204

Barth S, Oberli F, Meier M (1989) U-Th-Pb systematics in morphologically characterized zircon and allanite: a high resolution isotopic study of Alpine Rensen pluton (northern Italy). Earth Planet Sci Lett 95:235-254

Barth S, Oberli F, Meier M (1994) Th-Pb versus U-Pb isotope systematics in allanite from co-genetic rhyolite and granodiorite: implications for geochronology. Earth Planet Sci Lett 124:149-159

Bird D, Spieler AR (2004) Epidote in geothermal systems. Rev Mineral Geochem 56:235-300

Bird DK, Cho M, Janik CJ, Liou JG, Caruso LJ (1988) Compositional, order/disorder, and stable isotope characteristics of Al-Fe epidote, State 2-14 Drill Hole, Salton Sea Geothermal System. J Geophys Res 93(B11):13,135-13,144

Brandon AD, Creaser RA, Chacko T (1996) Constraints on rates of granitic magma transport from epidote dissolution kinetics. Science 271:1845-1848

Brunsmann A, Franz G, Erzinger J, Landwehr D (2000) Zoisite- and clinozoisite segregations in metabasites (Tauern Window, Austria) as evidence of high-pressure fluid-rock interaction. J Metamorph Geol 18:1-21

Buick IS, Frei R, Cartwright I (1999) The timing of high-temperature retrogression in the Reynolds Range, central Australia: constraints from garnet and epidote Pb-Pb dating. Contrib Mineral Petrol 135:244-254

Carlson C, Paterson SR, Geary E (1991) Magmatic (?) epidote in a lower crustal terrain, New Zealand. Geological Society of America, Cordilleran Section, 87th annual meeting Abstracts with Programs, San Francisco, CA

Cartwright I, Buick IS, Mass R (1997) Fluid flow in marbles at Jervois, central Australia: oxygen isotope disequilibrium and zoning produced by decoupling of mineralogical and isotopic resetting. Contrib Mineral Petrol 128:335-351

Cartwright I, Buick IS, Vry JK (1996) Polyphase metamorphic fluid flow in the Lower Calcsilicate Unit, Reynolds Range, central Australia. Precam Res 77:211-229

Chacko T, Riciputi LR, Cole DR, Horita J (1999) A new technique for determining equilibrium hydrogen isotope fractionation factors using the ion microprobe: Application to the epidote-water system. Geochim Cosmochim Acta 63:1-10

Chiaradia M (2003) Formation and evolution processes of the Salanfe W-Au-As-skarns (Aiguilles Rouges Massif, western Swiss Alps). Mineral Deposita 38:154-168

Cliff RA, Cohen A (1980) Uranium-lead systematics in a regionally metamorphosed tonalite from the eastern Alps. Earth Planet Sci Lett 50:211-218

Criss RE, Taylor Jr HP (1983) An $^{18}O/^{16}O$ and D/H study of Tertiary hydrothermal systems in the southern half of the Idaho Batholith. Geol Soc Am Bull 94:640-663

Criss RE, Taylor Jr HP (1986) Meteoric-hydrothermal systems. Stable isotopes in high temperature geological processes. Rev Mineral 16:373-424

Dawes RL (1991) Textural and compositional distinctions between magmatic and subsolidus epidote in granitoid plutons. Geological Society of America, Cordilleran Section, 87th annual meeting Abstracts with Programs - Geol Soc Am

Dawes RL, Evans BW (1991) Mineralogy and geothermobarometry of magmatic epidote-bearing dikes, Front Range, Colorado. Geol Soc Am Bull 103:1017-1031

Driesner T (1997) The effects of pressure on deuterium-hydrogen fractionation in high temperature water. Sci 277:791-794

Enami M, Liou JG, Mattinson CG (2004) Epidote minerals in high P/T metamorphic terranes: Subduction zone and high- to ultrahigh-pressure metamorphism. Rev Mineral Geochem 56:347-398

Evans BW, Vance JA (1987) Epidote phenocrysts in dacitic dikes, Boulder County, Colorado. Contrib Mineral Petrol 96:178-185

Ferreira VP, Valley JW, Sial AN, Spicuzza M (2003) Oxygen isotope compositions and magmatic epidote from two contrasting metaluminous granitoids, NE Brazil. Contrib Mineral Petrol 145:205-216

Forrester RW, Taylor HP (1977) $^{18}O/^{16}O$, D/H, and $^{13}C/^{12}C$, studies of the Tertiary igneous complex of Skye, Scotland. Am J Sci 277:136-177

Frei D, Liebscher A, Franz G, Dulski P (2004) Trace element geochemistry of epidote minerals. Rev Mineral Geochem 56:553-605

Frei R, Kamber BS (1995) Single mineral Pb-Pb dating. Earth Planet Sci Lett 129:261-268

Fu B, Zheng Y, Wang Z, Xiao Y, Gong B, Li S (1999) Oxygen and hydrogen isotope geochemistry of gneisses associated with ultrahigh pressure eclogites at Shuanghe in the Dabie Mountains. Contrib Mineral Petrol 134:52-66

Gieré R, Sorensen SS (2004) Allanite and other REE-rich epidote-group minerals. Rev Mineral Geochem 56: 431-494

Graham CM (1981) Experimental hydrogen isotope studies; III, Diffusion of hydrogen in hydrous minerals, and stable isotope exchange in metamorphic rocks. Contrib Mineral Petrol 76:216-228

Graham CM, Harmon RS, Sheppard SMF (1984) Experimental hydrogen isotope studies; hydrogen isotope exchange between amphibole and water. Am Mineral 69:128-138

Graham CM, Sheppard SMF (1980) Experimental hydrogen isotope studies, II. Fractionations in the systems epidote-NaCl-H$_2$O, epidote-CaCl$_2$-H$_2$O and epidote-seawater, and the hydrogen isotope composition of natural epidotes. Earth Planet Sci Lett 49:237-251

Graham CM, Sheppard SMF, Heaton THE (1980) Experimental hydrogen isotope studies--I. Systematics of hydrogen isotope fractionation in the systems epidote-H$_2$O, zoisite-H$_2$O and AlO(OH)-H$_2$O. Geochim Cosmochim Acta 44:353-364

Grapes RH, Hoskin PWO (2004) Epidote group minerals in low–medium pressure metamorphic terranes. Rev Mineral Geochem 56:301-345

Gregory RT, Taylor Jr HP (1981) An oxygen isotope profile in a section of Cretaceous oceanic crust, Samail ophiolite, Oman; evidence for delta $^{18}O$ buffering of the oceans by deep (>5 km) seawater-hydrothermal circulation at mid-ocean ridges. J Geophys Res B 86:2737-2755

Hoffbauer R, Hoernes S, Fiorentini E (1994) Oxygen isotope thermometry based on a refined increment method and its application to granulite-grade rocks from Sri Lanka. Precam Res 66:199-220

Javoy M (1977) Stable isotopes and geothermometry. J Geol Soc London 133:609-636

Jenkin GRT, Fallick AE, Leake BE (1992) A stable isotope study of retrograde alteration in SW Connemara, Ireland. Contrib Mineral Petrol 110:269-288

Keane SD, Morrison J (1997) Distinguishing magmatic from subsolidus epidote; laser probe oxygen isotope compositions. Contrib Mineral Petrol 126:265-274

King EM, Valley JW (2001) The source, magmatic contamination, and alteration of the Idaho Batholith. Contrib Mineral Petrol 124:72-88

Kohn MJ, Valley JV (1994) Oxygen isotope constraints on metamorphic fluid flow, Townshend Dam, Vermont, USA. Geochim Cosmochim Acta 58:5551-5566

Kohn MJ, Valley JW (1998) Oxygen isotope geochemistry of the amphiboles; isotope effects of cation substitutions in minerals. Geochim Cosmochim Acta 62:1947-1958

Liou JG (1973) Synthesis and Stability Relations of Epidote, Ca$_2$Al$_2$FeSi$_3$O$_{12}$(OH). J Petrol 14:381-413

Liou JG, Zhang R, Ernst WG (1994) An introduction to ultrahigh-pressure metamorphism. The Island Arc 3: 1-24

Liou JG, Zhang RY, Ernst WG, Eide EA, Maruyama S (1994) Metamorphism and tectonics of ultrahigh-P and high-P belts in the Dabie-Sulu region, east central China. AGU 1994 fall meeting Eos, Transactions American Geophysical Union

Ludwig KR, Stuckless JS (1978) Uranium-lead isotope systematics and apparent ages of zircons and other minerals in Precambrian granite rocks, Granite Mountains, Wyoming. Contrib Mineral Petrol 65:243-254

Marshall D, Watkinson D, Farrow C, Molnar F, Fouillac A (1999) Multiple fluid generations in the Sudbury igneous complex: fluid inclusion, Ar, O, H, Rb and Sr evidence. Chem Geol 154:1-19

Masago H, Rumble D, Ernst WG, Maruyama S (2001) A discovery of the very low $\delta^{18}O$ eclogites from the Kokchetav massif and its geological significance. Extended abstract of the UHPM Workshop 2001 at Waseda University, Tokyo: 164-167

Matsuhisa Y, Goldsmith JR, Clayton RN (1979) Oxygen isotopic fractionation in the system quartz-albite-anorthite-water. Geochim Cosmochim Acta 43:1131-1140

Matthews A (1994) Oxygen isotope geothermometers for metamorphic rocks. J Metamorph Geol 12:211-219

Matthews A, Goldsmith JR, Clayton RN (1983) Oxygen isotope fractionation between zoisite and water. Geochim Cosmochim Acta 47:645-654

Matthews A, Schliestedt M (1984) Evolution of the blueschist and greenschist facies rocks of Sifnos, Cyclades, Greece; a stable isotope study of subduction-related metamorphism. Contrib Mineral Petrol 88:150-163

Mezger K, Hanson GN, Bohlen SR (1989) U-Pb systematics of garnet: dating the growth of garnet in the late Archean Pikiwitonei granulite domain at Cauchon and Natawahunan Lakes, Manitoba, Canada. Contrib Mineral Petrol 101:136-148

Moench MT (1985) Comments and reply on "Implications of magmatic epidote-bearing plutons on crustal evolution in the accreted terranes of northwestern North America" and "Magmatic epidote and its petrologic significance." Geology 14:188-189

Mora CI, Riciputi LR, Cole DR (1999) Short-lived oxygen diffusion during hot, deep-seated meteoric alteration of anorthosite. Sci 286:2323-2325

Morrison J, Anderson JL (1998) Footwall refrigeration along a detachment fault: implications for the thermal evolution of core complexes. Sci 279:63-66

Morrison J, Anderson JL (2000) Oxygen isotope systematics of magmatic epidote. 31st International Geological Congress, Abstract Volume (Rio de Janeiro)

Morrison J, Anderson JL (in prep) Oxygen isotope laser probe constraints on the origins of epidote. Contrib Mineral Petrol

Morrison J, Anderson JL, Carlson C, Paterson S (1999) Oxygen isotope systematics in magmatic epidote-bearing volcanic and plutonic rocks. Geological Society of America Abstracts with Programs 31:A416

Muehlenbachs K, Clayton RN (1972) Oxygen Isotope Studies of Fresh and Weathered Submarine Basalts. Can J Earth Sci 9:172-184

Naney MT (1983) Phase equilibria of rock-forming ferromagnesian silicates in granitic systems. Am J Sci 283:993-1033

O'Neil JR, Clayton RN, Mayeda TK (1969) Oxygen isotope fractionation in divalent metal carbonates. J Chem Phys 51:5547-5558

Poitrasson F, Paquette JL, Montel JM, Pin C, Duthou JL (1998) Importance of late-magmatic and hydrothermal fluids on the Sm-Nd isotope mineral systematics of hypersolvus granites. Chem Geol 146:187-203

Poli S, Schmidt MW (2004) Experimental subsolidus studies on epidote minerals. Rev Mineral Geochem 56:171-195

Putlitz B, Matthews A, Valley JW (2000) Oxygen and hydrogen isotope study of high-pressure metagabbros and metabasalts (Cyclades, Greece): implications for the subduction of oceanic crust. Contrib Mineral Petrol 138:114-126

Recio C, Fallick AE, Ugidos JM, Stephens WE (1997) Characaterization of multiple fluid-granite interaction processes in the episyenites of Avila-Bejar, Central Iberian Massif, Spain. Chem Geol 143:127-144

Richter R, Hoernes S (1988) The application of the increment method in comparison with experimentally derived and calculated O-isotope fractionations. Chemie der Erde 48:1-18

Rumble DR, Yui TF (1998) The Qinglongshan oxygen and hydrogen isotope anomaly near Donghai in Jiangsu Province,China. Geochim Cosmochim Acta 62:3307-3321

Schmidt MW, Poli S (2004) Magmatic epidote. Rev Mineral Geochem 56:399-430

Schmidt MW, Thompson AB (1996) Epidote in calcalkaline magmas; an experimental study of stability, phase relationships, and the role of epidote in magmatic evolution. Am Mineral 81:462-474

Sheppard SMF, Taylor HP (1974) Hydrogen and oxygen isotope evidence for the origins of water in the Boulder Batholith and the Butte ore deposits, Montana. Eco Geol 69:926-946

Spiegel C, Siebel W, Frisch W, Berner Z (2002) Nd and Sr isotopic ratios and trace element geochemistry of epidote from the Swiss Molasse Basin as provenance indicators: implications for the reconstruction of the exhumation history of the Central Alps. Chem Geol 189:231-250

Stakes DS, O'Neil JR (1982) Mineralogy and stable isotope geochemistry of hydrothermally altered oceanic rocks. Earth Planet Sci Lett 57:285-304

Suzuoki T, Epstein S (1976) Hydrogen isotope fractionation between OH-bearing minerals and water. Geochim Cosmochim Acta 40:1229-1240